Marriages, Families, and Relationships

MAKING CHOICES IN A DIVERSE SOCIETY

Eleventh Edition

Mary Ann Lamanna
University of Nebraska, Omaha

Agnes Riedmann
California State University, Stanislaus

Ann Strahm, Contributing Author
California State University, Stanislaus

WADSWORTH
CENGAGE Learning

Australia • Brazil • Japan • Korea • Mexico • Singapore • Spain • United Kingdom • United States

WADSWORTH
CENGAGE Learning™

Marriages, Families, and Relationships: Making Choices in a Diverse Society, Eleventh Edition
Mary Ann Lamanna and Agnes Riedmann

Acquisitions Editor: Erin Mitchell

Developmental Editor: Tangelique Williams

Assistant Editor: Linda Stewart

Editorial Assistant: Mallory Ortberg

Media Editor: Melanie Cregger

Marketing Manager: Andrew Keay

Marketing Assistant: Dimitri Hagnéré

Marketing Communications Manager:
  Laura Localio

Content Project Manager: Cheri Palmer

Design Director: Rob Hugel

Art Director: Caryl Gorska

Print Buyer: Mary Beth Hennebury

Rights Acquisitions Specialist: Don Schlotman

Production Service: MPS Limited, a Macmillan
  Company

Text Designer: Diane Beasley

Photo Researcher: PreMedia Global

Text Researcher: Sue Howard

Copy Editor: Heather McElwain

Illustrator: MPS Limited, a Macmillan Company

Cover Designer: Riezebos Holzbaur/Brie Hattey

Cover Image: PHOTOLIBRARY/Imagesource

Compositor: MPS Limited, a Macmillan Company

For product information and technology assistance, contact us at
Cengage Learning Customer & Sales Support, 1-800-354-9706.

For permission to use material from this text or product,
submit all requests online at **www.cengage.com/permissions.**
Further permissions questions can be e-mailed to
permissionrequest@cengage.com

Library of Congress Control Number: 2010929057

Student Edition:
ISBN-13: 978-1-111-30154-5
ISBN-10: 1-111-30154-9

Loose-leaf Edition:
ISBN-13: 978-1-111-72643-0
ISBN-10: 1-111-72643-4

Wadsworth
20 Davis Drive
Belmont, CA 94002-3098
USA

Cengage Learning is a leading provider of customized learning solutions with office locations around the globe, including Singapore, the United Kingdom, Australia, Mexico, Brazil, and Japan. Locate your local office at **www.cengage.com/global.**

Cengage Learning products are represented in Canada by Nelson Education, Ltd.

To learn more about Wadsworth, visit **www.cengage.com/wadsworth**

Purchase any of our products at your local college store or at our preferred online store **www.cengagebrain.com.**

Printed in the United States of America
1 2 3 4 5 6 7 14 13 12 11

dedication

to our families, especially

Larry, Valerie, Sam, Janice, Simon, and Christie

Bill, Beth, Angel, Chris, Natalie, Alex, and Livia

Lucy Field, Ellen and Zeke Martinez, Andrew R. Jones,

and Eileen Engwall

About the Authors

Mary Ann Lamanna is Professor Emerita of Sociology at the University of Nebraska at Omaha. She received her bachelor's degree in political science Phi Beta Kappa from Washington University (St. Louis), her master's degree in sociology (minor in psychology) from the University of North Carolina, Chapel Hill, and her doctorate in sociology from the University of Notre Dame.

Research and teaching interests include family, reproduction, and gender and law. She is the author of *Emile Durkheim on the Family* (Sage Publications, 2002) and co-author of a book on Vietnamese refugees. She has articles in law, sociology, and medical humanities journals. Current research concerns the sociology of literature, specifically "novels of terrorism" and a sociological analysis of Marcel Proust's novel *In Search of Lost Time*. Professor Lamanna has two adult children, Larry and Valerie.

Agnes Riedmann is Professor of Sociology at California State University, Stanislaus. She attended Clarke College in Dubuque, Iowa. She received her bachelor's degree from Creighton University and her doctorate from the University of Nebraska. Her professional areas of interest are theory, family, and the sociology of body image. She is author of *Science That Colonizes: A Critique of Fertility Studies in Africa* (Temple University Press, 1993). Dr. Riedmann spent the academic year 2008–09 as a Fulbright Professor at the Graduate School for Social Research, affiliated with the Polish Academy of Sciences, Warsaw, where she taught courses in family, as well as in social policy and in globalization. She has two children, Beth and Bill; two granddaughters, Natalie and Livia; and a grandson, Alex.

Contributing author for this edition, Ann Strahm, is Assistant Professor of Sociology at California State University, Stanislaus. She received her bachelor's degree and her doctorate from the University of Oregon. Research and teaching interests include family, social inequalities, media, and theory. Her work appears in various journals and other publications, including *Sociological Focus, Journal of Media Sociology,* and *Project Censored,* edited by Peter Phillips.

Brief Contents

Contents

Chapter 4 Our Gendered Identities 78

Chapter 5 Our Sexual Selves 106

Chapter 6 Love and Choosing a Life Partner 136

Chapter 7 Marriage: From Social Institution to Private Relationship 164

Chapter 11   Work and Family 280

Chapter 12   Communication in Relationships, Marriages, and Families 316

Chapter 13   Power and Violence in Families 338

Chapter 14   Family Stress, Crisis, and Resilience 376

Chapter 15 Divorce: Before and After 400

Chapter 16 Remarriages and Stepfamilies 444

Chapter 17 Aging Families 472

How to find the Appendices

The following appendices are available in full color on the CourseMate site for the eleventh edition. To access these appendices, go to **www.cengagebrain.com** and search for this title. Once on the CourseMate site, click on "Appendices" from the left navigation bar. You can access suggested readings in the same manner.

Appendix A: Human Sexual Anatomy

Appendix B: Human Sexual Response

Appendix C: Sexually Transmitted Infections

Appendix D: Sexual Dysfunctions and Therapy

Appendix E: Conception, Pregnancy, and Childbirth

A bonus appendix, Contraception, can also be accessed on the CourseMate website.

Boxes

A Closer Look at Family Diversity

As We Make Choices

Facts about Families

Issues for Thought

My Family

Preface

As we complete our work on the eleventh edition of this text, we look back over ten earlier editions. Together, these represent more than thirty years spent observing—and rethinking—the contemporary American family. Not only have families changed during this time, but so has social science's interpretation of them. It is gratifying to be a part of the enterprise dedicated to studying families and to share this knowledge with students.

Our own perspective on families has developed and changed during this period. Indeed, as marriages and families have evolved over the last three decades, so has this text. Now, for the first time since its original edition, we have altered the book's title. From the beginning, this text has been titled *Marriages and Families*—a title that was the first to purposefully use plurals to recognize the diversity of family forms—a diversity that we noted as early as 1980. This edition is titled, *Marriages, Families, and Relationships*. We added the term *relationships* to recognize the increasing incidence of individuals forming commitments outside of legal marriage. Where appropriate, we have similarly altered chapter titles. At the same time, we continue to recognize and appreciate the fact that the vast majority of Americans are married or will marry. Hence, we consciously persist in giving due attention to the values and issues of married couples. Of course, the concept of marriage itself has changed appreciably. No longer necessarily heterosexual, marriage is now an institution to which same-sex couples in several states presently have legal access.

Meanwhile, the book's subtitle, *Making Choices in a Diverse Society*, continues to speak to the significant changes that have taken place since our first edition. To help accomplish our goal of encouraging students to better appreciate the diversity of today's families, we present the latest research and statistical information on varied family forms, lesbian and gay male families, and families of diverse race and ethnicity, socioeconomic, and immigration status, among other variables.

We continue to take account not only of increasing racial/ethnic diversity, but also of the fluidity of the concepts *race* and *ethnicity* themselves. In this edition, we give greater direct attention to the socially constructed nature of these concepts. We integrate these materials on family diversity throughout the textbook, always with an eye toward avoiding stereotypical, simplistic generalizations and, instead, to explaining data in sociological and sociohistorical contexts.

Interested from the beginning in the various ways that gender plays out in families, we have persistently focused on areas in which gender relations have changed and continue to do so, as well as on areas in which there has been relatively little change. In keeping with our practice of reviewing and reevaluating every single word for a new edition, we have in this revision given concerted attention to discussions that may now be better presented in gender-neutral context and language. However, we hasten to add that assuredly not all topics lend themselves to gender-neutral language. For example, research consistently shows that intimate violence perpetrated by heterosexual men is qualitatively different from that perpetrated by heterosexual women.

In addition to our attention to gender, we have studied demography and history, and we have paid increasing attention to the impact of social structure on family life. We have highlighted the family ecology perspective in keeping with the importance of social context and public policy. We cannot help but be aware of the cultural and political tensions surrounding families today. At the same time, in recent editions and in response to our reviewers, we have given more attention to the contributions of psychology and to a social psychological understanding of family interaction and its consequences.

We continue to affirm the power of families as they influence the courses of individual lives. Meanwhile, we give considerable attention to policies needed to provide support for today's families: working parents, families in financial stress, single-parent families, families of varied racial/ethnic backgrounds, stepfamilies, same-sex couples, and other nontraditional families—as well as the classic nuclear family.

We note that, despite changes, marriage and family values continue to be salient in contemporary American life. Our students come to a marriage and family course because family life is important to them. Our aim now, as it has been from the first edition, is to help students question assumptions and to reconcile conflicting ideas and values as they make choices throughout their lives. We enjoy and benefit from the contact we've had with faculty and students who have used this book. Their enthusiasm and criticism have stimulated many changes in the book's content. To know that a supportive audience is interested in our approach to the study of families has enabled us to continue our work over a long period.

The Book's Themes

Several themes are interwoven throughout this text: People are influenced by the society around them as they make choices, social conditions change in ways that may impede or support family life, there is an interplay between individual families and the larger society, and individuals make family-related choices throughout adulthood.

Making Choices throughout Life

The process of creating and maintaining marriages, families, and relationships requires many personal choices; people continue to make family-related decisions, even "big" ones, throughout their lives.

Personal Choice and Social Life

Tension frequently exists between individuals and their social environment. Many personal troubles result from societal influences, values, or assumptions; inadequate societal support for family goals; and conflict between family values and individual values. By understanding some of these possible sources of tension and conflict, individuals can perceive their personal troubles more clearly and work constructively toward solutions. They may choose to form or join groups to achieve family goals. They may become involved in the political process to develop state or federal social policy supportive of families. The accumulated decisions of individuals and families also shape the social environment.

A Changing Society

In the past, people tended to emphasize the dutiful performance of social roles in marriage and family structure. Today, people are more apt to view committed relationships as those in which they expect to find companionship, intimacy, and emotional support. From its first edition, this book has examined the implications of this shift and placed these implications within social scientific perspective. Individualism, economic pressure, time pressures, social diversity, and an awareness of committed relationships' potential impermanence are features of the social context in which personal decision making takes place today. With each edition, we recognize again that, as fewer social guidelines remain fixed, personal decision making becomes even more challenging.

Then too, new technologies continue to create changes in family members' lives. Discussions about technological developments in communication appear throughout the book—for example, maintaining ties between college students and their parents (Chapter 10), social class differences in Internet access (Chapter 3), sexting and cyber adultery (Chapter 5), Internet matchmaking (Chapter 6), reproductive technology (Chapter 9), parental surveillance of children (Chapter 10), working at home versus being tethered to the office (Chapter 11), and how noncustodial parents keep in touch with children through technology (Chapter 15).

Meanwhile, in this edition, we have added a new section, "Three Societal Trends that Impact Families," to Chapter 1. This section explores the impacts of advancing communication and reproductive technologies, as well as the changing racial/ethnic demographic picture in the United States, and the emergent and troubling situation of economic uncertainty resulting from the recession that began in 2008.

The Themes throughout the Life Course

The book's themes are introduced in Chapter 1, and they reappear throughout the text. We developed these themes by looking at the interplay between findings in the social sciences and the experiences of the people around us. Ideas for topics continue to emerge, not only from current research and reliable journalism, but also from the needs and concerns that we perceive among our own family members and friends. The attitudes, behaviors, and relationships of real people have a complexity that we have tried to portray. Interwoven with these themes is the concept of the life course—the idea that adults may change by means of reevaluating and restructuring throughout their lives. This emphasis on the life course creates a comprehensive picture of marriages, families, and relationships and encourages us to continue to add topics that are new to family texts. Meanwhile, this book makes these points:

- People's personal problems and their interaction with the social environment change as they and their relationships and families grow older.

- People reexamine their relationships and their expectations for relationships as they and their marriages, relationships, and families mature.

- Because family forms are more flexible today, people may change the type or style of their relationships and families throughout their lives.

Marriages and Families—Making Choices

Making decisions about one's family life, either knowledgeably or by default, begins in early adulthood and lasts into old age. People choose whether they will adhere to traditional beliefs, values, and attitudes about gender roles or negotiate more flexible roles and relationships. They may rethink their values about sex and become more knowledgeable and comfortable with their sexual choices.

Women and men may choose to remain single, to form heterosexual or same-sex relationships outside of marriage, or to marry. They have the option today of staying single longer before marrying. Single people make choices about their lives, ranging from decisions about living arrangements to those about whether to engage in sex only in marriage or committed relationships, to engage in sex for recreation, or to abstain from sex altogether. Many unmarried individuals live as cohabiting couples (often with children), an increasingly common family form.

Once individuals form couple relationships, they have to decide how they are going to structure their lives as committed partners. Will the partners be legally married? Will they become domestic partners? Will theirs be a dual-career union? Will they plan periods in which one partner is employed, interspersed with times in which both are wage earners? Will they have children? Will they use new reproductive technology to become parents? Will other family members live with them—siblings or parents, for example, or, later, adult children?

Couples will make these decisions not once, but over and over during their lifetimes. Within a committed relationship, partners also choose how they will deal with conflict. Will they try to ignore conflicts? Will they vent their anger in hostile, alienating, or physically violent ways? Or will they practice supportive ways of communicating, disagreeing, and negotiating—ways that emphasize sharing and can deepen intimacy?

How will the partners distribute power in the marriage? Will they work toward relationships in which each family member is more concerned with helping and supporting others than with gaining a power advantage? How will the partners allocate work responsibilities in the home? What value will they place on their sexual lives together? Throughout their experience, family members continually face decisions about how to balance each one's need for individuality with the need for togetherness.

Parents also have choices. In raising their children, they can choose the authoritative parenting style, for example, in which parents take an active role in responsibly guiding and monitoring their children, while simultaneously striving to develop supportive, mutually cooperative family relationships.

Many partners face decisions about whether to separate or divorce. They weigh the pros and cons, asking themselves which is the better alternative: living together as they are or separating? Even when a couple decides to separate or divorce, there are further decisions to make: Will they cooperate as much as possible or insist on blame and revenge? What living and economic support arrangements will work best for themselves and their children? How will they handle the legal process? The majority of divorced individuals eventually face decisions about recoupling. In the absence of firm cultural models, they choose how they will define stepfamily relationships.

When families encounter crises—and every family will face *some* crises—members must make additional decisions. Will they view each crisis as a challenge to be met, or will they blame one another? What resources can they use to handle the crisis? Then, too, as more and more Americans live longer, families will "age." As a result, more and more Americans will have not only living grandparents but also great grandparents. And increasingly, we will face issues concerning giving—and receiving—family elder care.

An emphasis on knowledgeable decision making does not mean that individuals can completely control their lives. People can influence but never directly determine how those around them behave or feel about them. Partners cannot control one another's changes over time, and they cannot avoid all accidents, illnesses, unemployment, separations, or deaths. Society-wide conditions may create unavoidable crises for individual families. However, families *can* control how they respond to such crises. Their responses will meet their own needs better when they refuse to react automatically and choose instead to act as a consequence of knowledgeable decision making.

Key Features

With its ongoing thorough updating and inclusion of current research and its emphasis on students' being able to make choices in an increasingly diverse society, this book has become a principal resource for gaining insights into today's marriages, relationships, and families. Over the past ten editions, we have had four goals in mind for student readers: first, to help them better understand themselves and their family situations; second, to make students more conscious of the personal decisions that they will make throughout their lives and of the societal influences that affect those decisions; third, to help students better appreciate the variety and diversity among families today; and fourth, to encourage them to recognize the need for structural, social policy support for families. To these ends, this text has become recognized for its accessible writing style, up-to-date research, well-written features, and useful chapter learning aids.

Up-to-Date Research and Statistics

As users have come to expect, we have thoroughly updated the text's research base and statistics, emphasizing cutting-edge research that addresses the diversity of marriages and families, as well as all other topics. In accordance with this approach, users will notice several new tables and figures. Revised tables and figures have been updated with the latest available statistics—data from the U.S. Census Bureau and other governmental agencies, as well as survey and other research data.

Features

The several themes described earlier are reflected in the special features.

Former users will recognize many of these features as having appeared in previous editions. Partly because students today expect boxed material to be brief, several feature boxes have been shortened to more succinctly make their points. Other features, maybe some of your favorites, have been eliminated. (For us, decisions about what to cut are often more difficult than those about what to add.) All the special features in this edition have been rigorously reviewed, edited, and updated. The following sections describe our five feature box categories:

AS WE MAKE CHOICES. We highlight the theme of making choices with a group of boxes throughout the text, for example, "Ten Rules for a Successful Relationship," "Disengaging from Power Struggles," "Selecting a Child Care Facility," and "Tips for Step-Grandparents." These feature boxes emphasize human agency and are designed to help students through crucial decisions.

A CLOSER LOOK AT FAMILY DIVERSITY. In addition to integrating information on cultural and ethnic diversity throughout the text proper, we have a series of features that give focused attention to instances of family diversity—for example, "African Americans and 'Jumping the Broom'," "Diversity and Child Care," "Family Ties and Immigration," and "Parenting LGBT Children," among others.

MY FAMILY. Beginning with the onset of this project, Agnes Riedmann has interviewed individuals in a variety of social categories about their experiences in marriages, families, and relationships. These interviews have comprised many of the My Family boxes. Other feature boxes of this type, such as Chapter 15's "How It Feels When Parents Divorce," are excerpts from previously published material. Some student essays also appear as My Family feature boxes. An example is the one in Chapter 16 titled "My (Step)Family." All the My Family features are designed to balance and complement the chapter's material. We hope that the presentation of these individuals' stories will help students to see their own lives more clearly and will encourage them to discuss and reevaluate their own attitudes and values.

ISSUES FOR THOUGHT. These features are designed to spark students' critical thinking and discussion. As an example, the Issues for Thought box in Chapter 3 explores thought-provoking issues related to "Studying Families and Ethnicity." Chapter 4 includes a new box describing ways that gender differences manifest in a heterosexual couple's wedding planning. As a final example, a new box in Chapter 17, "Filial Responsibility Laws," encourages students to consider what might be the benefits and drawbacks of legally mandating filial responsibility.

FACTS ABOUT FAMILIES. This feature presents demographic and other factual information on focused topics such as family members' time use ("Where Does the Time Go?" in Chapter 11). Other examples include

new boxes on "How Family Researchers Study Religion from Various Theoretical Perspectives" (Chapter 2), on transracial adoption (Chapter 9), on "Foster Parenting" (Chapter 10), and on "Relationship and Family Counseling" (Chapter 12).

 **FOCUS ON CHILDREN.** A sixth important feature is called Focus on Children. When you see this icon, you are being alerted to important material related to children. We include this focus because the sociology of childhood has become an important area of scholarly interest and also due to policy concern about the extent to which contemporary America's children are well nurtured—hence, our desire to encourage students to examine the condition of children today from a sociological perspective. We hope that, as a consequence, students will be better prepared to make informed decisions now and in the future.

Chapter Learning Aids

A series of chapter learning aids help students comprehend and retain the material.

- **Chapter Summaries** are presented in bulleted, point-by-point lists of the key material in the chapter.
- **Key Terms** alert students to the key concepts presented in the chapter. A full glossary is provided at the end of the text.
- **Questions for Review and Reflection** help students review the material. Thought questions encourage students to think critically and to integrate material from other chapters with that presented in the current one. In every chapter, one of these questions is a policy question. This practice is in line with our goal of moving students toward structural analyses regarding marriages, families, and relationships.
- **Footnotes**, although not overused, are presented when we feel that a point needs to be made but might disrupt the flow of the text itself. For example, a new footnote in Chapter 10 (footnote #13) integrates critical thinking about research methods, specifically operational definitions, with findings on the home environments of many low-income single mothers.
- **Suggested Readings** on Sociology CourseMate give students ideas for further reading on topics and issues presented in the chapter.

Key Changes in This Edition

In addition to incorporating the latest available research and statistics—and in addition to carefully reviewing every word in the book—we note that this edition includes many key changes, some of which are outlined here.

In this eleventh edition, we have again revisited and somewhat restructured the chapter outline and order. We have dropped the former Chapter 5, "Loving Ourselves and Others," and combined material from former Chapter 5 with that on mate selection. The result is Chapter 6, now titled, "Love and Choosing a Life Partner." Moreover, in response to reviewers, we have returned this chapter on mate selection to its traditional placement prior to the two chapters (now Chapters 7 and 8) that respectively explore the nature of marriage and relationships other than marital.

As with previous revisions, we have given considerable attention not only to chapter-by-chapter organization, but also to *within*-chapter organization. Our ongoing intents are to streamline the material presented whenever possible and to ensure a good flow of ideas. In this edition, we have also continued to consolidate similar material that had previously been addressed in separate chapters.

Meanwhile, we have substantially revised each and every chapter. Every chapter is updated with the latest research throughout. We mention some (but not all!) specific and important changes here.

Chapter 1, Marriage, Relationships, and Family Commitments: Making Choices in a Changing Society, continues to present the choices and life course themes of the book, as well as points to the significance for the family of larger social forces. In addition to updating this chapter's exploration of conflicting views on the changing family—family "change" versus family "decline"—we have worked to better place this discussion within the perspective of the family as a social institution.

In addition to wrestling with how to define family, this chapter now includes discussion of the distinction between structural and functional family definitions. Accordingly, we have moved discussion of family functions from where it previously existed in Chapter 2 to Chapter 1. As part of our effort to present family change within the context of family as institution, we have added a new section, "Adapting Family Definitions to the Postmodern Family."

Another example of our efforts to enhance this chapter's role of adding perspective for the rest of the book involves the addition of a new section, "Three Societal Trends that Impact Families." This section explores advancing communication and reproductive technologies, the new faces of America's multicultural families (that is, fewer non-Hispanic whites and more people of color), and the dramatic—and, for many, disastrous—economic uncertainty that characterizes many societies today, including ours. We return to this issue of economic uncertainty, and its impact on individuals' and family members' options and decisions, at several points throughout this edition of the book.

Chapter 2, Exploring Relationships and Families, has been completely reworked with updated treatments of both theory and methods. We redrafted this chapter to better portray the integral relationship between family theories and methods for researching families. Hence, we begin with a discussion of science in general, then move to an exploration of family theories, followed by a presentation of research methods. Toward the end of this chapter, we have added a new section, "The Relationship between Theory and Research."

With regard to theoretical perspectives, we have revisited each of our discussions of family theory, especially our discussion of structure-functionalism. We now present concepts and ideas such as *functional alternative*, as well as Merton's classic question, "Functional for whom?" within the context of today's families. In response to some of our reviewers, we have included attachment theory in our presentation of theories prominent in the family literature. The interactionist-constructionist perspective has also been reworked, with greater emphasis given to the idea that societal "realities" and definitions are socially constructed.

In addition, we have reconceptualized our treatment of research methodologies, now presenting methods within the context of scientific norms. Using very current and intriguing studies, we have worked to include what we think are truly interesting examples of the various ways that research can be conducted. Because several of our reviewers asked for greater attention to the impact of religion on family life, we have given greater attention to this topic throughout the text. One example in this chapter is new boxed material, "How Family Researchers Study Religion from Various Theoretical Perspectives." This new feature illustrates ways that researchers from various theoretical perspectives have approached the general topic, religion and family.

Chapter 3, American Families in Social Context, includes significant demographic updates, and the history section has been revised. This chapter has also been revisited to provide a more global analysis, allowing for a contextualization of the cultural and economic changes affecting American families. The section, "Conceptualizing Race and Ethnicity," emphasizes diversity *within* major racial/ethnic groups and notes the increasingly fluid nature of racial/ethnic categories. A section on Arab American families has been added. Significant changes in Latino family patterns are reported, and we have also included more on differences among African American families by class. Coverage of immigrant families takes account of the current politics of immigration, pointing especially to how U.S. immigration policy affects binational families.

Chapter 4, Our Gendered Identities, has been thematically updated to take into account the fact that the more essentialist perspective of "men" versus "women" has given way to the intersection of gender with race, class, sexuality, and globalization. A new box—A Closer Look at Family Diversity: "Gendered Divisions of Labor—Preparing to Wed"—exemplifies the divergent ways that women and men frame gender issues.

Entirely new material on gender differences is presented. A section on the gender similarities hypothesis includes research supporting the hypothesis that there are actually few differences between males and females in traits and abilities. The section "Is Anatomy Destiny?" is included at the end of the gender inequality section. The Issues for Thought feature, "Challenges to Gender Boundaries," has been updated and expanded to discuss the lived experiences of two people who went through gender reassignment.

Chapter 4 also includes discussion of gender inequality in major social institutions. This section includes subsections on "Gender and Health" and "Gender and Education." The section on education addresses the concern about lower rates of college enrollment and graduation for males. A section within the socialization segment looks at the related conflict between advocacy for girls and for boys—is there a "war against boys"? In addition, there is an updated perspective on the women's and men's movements.

Chapter 5, Our Sexual Selves, presents a new feature, As We Make Choices: "Sexting—Five Things to Think about Before Pressing 'Send.'" This box invites students to consider some possible consequences of "sexting," a relatively new phenomenon in which young people are using technology to take and send sexually provocative photographs and text messages over their cell phones. More generally, theoretical perspectives on human sexuality have been expanded to include both micro and macro levels of analysis. Material on infidelity has been updated and expanded. "The Politics of Sex" includes discussion of the politicization of research and of sex education.

Chapter 6, Love and Choosing a Life Partner (formerly Chapter 9), now combines important and updated material on love from the former Chapter 5 with discussions of mate selection. Reconceptualizing this chapter in this manner allows us to place greater emphasis on ways that ides about love and loving influence mate selection decisions, which in turn impact relationship satisfaction and stability. Material on arranged and free-choice marriages, formerly appearing in two separate chapters, has been combined within this chapter. As in the past, this chapter includes discussion of homogamy as well as increasing marital heterogamy.

Chapter 7, Marriage: From Social Institution to Private Relationship, has been thoroughly updated with new statistics and research findings. Former users will recall that this chapter explores the changing picture regarding marriage, noting the social science debate regarding whether this changing picture represents family change or decline. As part of our updated exploration of this question, we have added considerably to our exploration of the selection hypothesis versus the experience hypothesis with regard to the research-based benefits of marriage. To further explicate our discussion, we have added a new figure, Figure 7.4, "Causal Order: Experience Hypothesis, Selection Hypothesis."

Furthermore, this chapter now progresses from the macro to the micro, with a final major section, "Marital Satisfaction and Choices throughout Life," that includes subsections addressing preparation for marriage, the first years of marriage, and the process of creating couple connection.

Chapter 8, Living Alone, Cohabiting, Same-Sex Unions, and Other Intimate Relationships, discusses demographic, economic, technological, and cultural reasons for the increasing proportion of unmarrieds. After describing the various living arrangements of non-marrieds—with an updated and expanded section on "living alone together"—the chapter presents a largely lengthened discussion of cohabitation and family life, including discussion of the cohabiting relationship itself as well as the most recent research on the consequences of raising children in a cohabiting family. This chapter also includes extensive, expanded, and thoroughly updated sections on trends in legal marriage for same-sex couples, as well as new treatment of same-sex couple relationships. The latter includes research that is just beginning on comparisons of legally married same-sex couples with those who are not.

Chapter 9, To Parent or Not to Parent, includes a new box, "Through the Lens of One Transracial Adoptee," in which multiracial and multicultural adoptions are discussed. This piece is a narrative written by an international transracial adoptee on her experiences growing up "different" in a white American household. Data analysis on international and transracial adoptions are included. This chapter also includes new material on childlessness and the well-being of people who choose to not have children. Finally, racial and ethnic differences in fertility rates are updated and discussed.

Chapter 10, Raising Children in a Diverse Society, like all the chapters in this edition, has been thoroughly updated with the most current research. As in recent prior editions, after describing the authoritative parenting style, we note its acceptance by mainstream experts in the parenting field. We then present a critique that questions whether this parenting style is universally appropriate or simply a white, middle-class pattern that may not be so suitable to other social contexts. We also discuss challenges faced by parents who are raising religious- or ethnic-minority children in potentially discriminatory environments.

In this edition, we continue to emphasize the challenges that all parents face in contemporary America. At the same time, we recognize that parents face difficulties unique to their socioeconomic situations and also to the family form in which they find themselves. To enhance our exploration of these issues, we have added sections on single mothers, single fathers, and nonresident fathers. Again, as with all other chapters in this text, we keep in mind the linkage between structural conditions and personal decisions. Hence, there is added discussion of the parenting beliefs and practices in working-class families.

The box (familiar to former users), "Communicating with Children—How to Talk So Kids Will Listen and Listen So Kids Will Talk" has been moved to Chapter 12. Former users will note two new figures in this chapter: Figure 10.1, "Family Groups with Children under Age 18," and Figure 10.2, "Stress Model of Effective Parenting."

Chapter 11, Work and Family, includes an updated discussion of women's leaving the labor force and reentry—a particularly important discussion during this time of economic crisis. There is more on men's labor force patterns, and this frames the discussion of stay-at-home fathers and "househusbands." This chapter also examines the persistence of gender differences in the "second shift" of housework and child care within the context of the wage gap between men and women, and how the impact of the current recession on male employment is affecting these issues. Data on work, family, and leisure are updated, and there are new figures on time spent with children, jobs held by women and men, who works in a married-couple family, and the priority given to work and family by baby boomers, Generation X, and Generation Y. Additionally, there is a new A Closer Look at Family Diversity box, "Extreme Child Care Maneuvers," in which working couples and single parents, who cannot afford child care, cope with busy work and family schedules.

Although we incorporate material from important new books in many chapters of this text, one on the intersection of work and family is worth mentioning here: Gerson's, *The Unfinished Revolution: How a New Generation is Reshaping Family, Work, and Gender in America*. We continue to follow the National Institute of Child Health and Human Development study of child care. In the boxes "Selecting a Child Care Facility" and "Child Care and Children's Outcomes," we present the latest information on child care decisions as they relate to work and family.

Chapter 12, Communication in Relationships, Marriages, and Families places greater emphasis on family cohesion as a function of positive couple communication and emphasizes the components and desirability of supportive couple communication, as well as exploring positive ways to address disagreements and conflict.

This chapter now includes reconsideration and updated discussion of the "female demand/male withdrawal" phenomenon first elucidated by Gottman and colleagues. Here is an example of our intent to use more gender-neutral language wherever possible. In accordance with some recent research findings, we now term this phenomenon simply the "demand/withdrawal" communication pattern.

Discussion of the effects of unresolved family conflict on children has elevated from its former position as boxed material, formerly within Chapter 10, and is integrated here as a chapter section. The As We Make Choices box, "Communicating with Children—How to Talk So Kids Will Listen and Listen So Kids Will Talk," has been moved to this chapter from Chapter 10. In addition, readers will find a new Facts about Families box in this chapter, "Relationship and Family Counseling." This box brings material, having been updated, which was formerly available in an appendix into this chapter. We made this change because we saw this material as increasingly important and better presented with the book itself.

Chapter 13, Power and Violence in Families, now includes a chart on the bases of power for both native-born and immigrant couples. This chapter now consolidates the classic research on family power, while current research on marital and partner power has been expanded to include issues of household work and money management, as well as decision making per se. A discussion of equality and equity concludes the part of the chapter on marital and partner power. Additionally, analysis of power differential between citizens and their immigrant spouses is introduced.

In the section on family violence, the controversial question of gender symmetry in intimate partner violence continues to be considered. We have added a new section on child-to-parent violence. Sibling violence and child sexual abuse are treated in separate subsections. The section on abuse among same-gender, bisexual, and transgender couples has been updated and expanded. Finally, the section on violence among immigrant couples is expanded.

Chapter 14, Family Stress, Crisis, and Resilience, continues to emphasize and expand discussion of the growing body of research on resilience in relation to family stress and crises. Using updated research and newly recognized issues, such as raising children in what seems to be a society-wide "culture of fear," this chapter includes a new box on parenting, titled, "ADHD, Stigma, and Stress." Another new box—"A Closer Look at Family Diversity: Young Caregivers"—examines "early," "late," or "on-time" caregiving within the context of life-cycle expectations. As our readers have come to expect, we end this chapter on a positive, albeit realistic, note with a final section exploring family crisis as disaster or opportunity.

Chapter 15, Divorce: Before and After, has been updated to include both heterosexual and homosexual divorce, the dissolution of civil unions, and the dissolution of long-term relationships. A new box, "Facts about Families: The Rise of the 'Silver Divorce,'" has been included to address the increase in the number of older-age divorces. We present new research on who initiates the divorce and what difference that makes for divorcing individuals. With reference to the most recent research, we also present a somewhat more positive look at divorce outcomes for men, women, and children. A box on "Postdivorce Pathways" explores diversity in outcomes. There is also now a subsection on stable unhappy marriages, as well as positive and negative outcomes for children of divorce and multiple family transitions.

Chapter 16, Remarriages and Stepfamilies, continues to expand our treatment of diversity within stepfamilies. To this end, the chapter now places greater emphasis on stepfamilies as not necessarily *remarried* families. Furthermore, the chapter features a new figure on the various pathways to stepfamily living. Another new figure, "U.S. Children Under Age 18 Living in Stepfamilies," notes the diversity of children's living arrangements within stepfamilies. Yet another new figure (Figure 16.3), illustrates various types of communication channels within stepfamilies. In accordance with our goal of making research activities better understood by and more meaningful to students, we now include a new box that explores "Measuring Everyday Stepfamily Life."

Meanwhile, we have expanded our critical evaluation of the "nuclear-family model monopoly," whereby the cultural assumption is that the first-marriage family is the "real" model for family living, with all other family forms viewed as deficient. A new section on re-wedding ceremonies illustrates the relative devaluing of other than the traditional (first-marriage) nuclear family. Within this context, we give increased attention to boundary and role ambiguities experienced by stepfamily members, as evidenced in language and terms of address, among other ways. The new Table 16.1 further explores the concept *boundary ambiguity*, as measured in one current stepfamily study.

This chapter offers extended and fully updated analysis of children's well-being in remarried and in cohabiting stepfamilies. As with other chapters in this text, we focus on the intersection of the macro with the micro as we give greater attention to the causes and consequences of stepfamily cohesion, followed by suggestions for developing stepfamily cohesion. To this end, we include a box on stepparenting tips as well as a final section, "Creating Supportive Stepfamilies."

Chapter 17, Aging Families, has been updated throughout, not only with the latest research but also with recognition of emerging challenges—and opportunities, such as more time for grandparenthood, and for great-grandparenthood—related to an aging population. The new Figure 17.1, "Older Americans as a Percentage of the Total U.S. Population, 2000 and 2010, with projections for 2025 and 2050" has been added to more directly illustrate the fact that, as the baby boom cohort grows older, populations over age fifty-five, sixty-five, seventy-five, and eighty-five will increase.

In an effort to address the increasingly common phenomenon of stepfamily living in relevant places throughout the book, we have added a new box to this chapter, "As We Make Choices: Tips for Step-Grandparents." This chapter now includes the new box "Issues for Thought: Filial Responsibility Laws," described elsewhere in this Preface.

Appendices. All the appendices, which appear on Sociology CourseMate, have been updated. With regard to

Appendix F in the previous edition, the proliferation and rapid changes in contraceptive methods have led us to refer the reader to more specialized content, accessible at **www.cengagebrain.com**. Materials from former Appendices G, "High-Tech Fertility," and H, "Marriage and Close Relationship Counseling," have been incorporated into Chapters 1 and 12, respectively. Because there is now ample discussion and advice available on the Internet pertaining to the economy, budgeting, and financial planning, Appendix I, "Managing a Budget" has been dropped in this edition.

Supplements

Supplements for the Instructor

Instructor's Resource Manual with Test Bank. This thoroughly revised and updated Instructor's Resource Manual contains detailed lecture outlines; chapter summaries; and lecture, activity, and discussion suggestions; as well as film and video resources. It also includes student learning objectives, chapter review sheets, and Internet and InfoTrac® College Edition exercises. The test bank consists of a variety of questions, including multiple-choice, true/false, completion, short answer, and essay questions for each chapter of the text, with answer explanations and page references to the text.

PowerLecture™ with JoinIn™ and ExamView®. This easy-to-use, one-stop digital library and presentation tool includes the following:

- Preassembled **Microsoft® PowerPoint® lecture slides** with graphics from the text, making it easy for you to assemble, edit, publish, and present custom lectures for your course (also included are all photos from the text along with video clips);

- Polling and quiz questions that can be used with the **JoinIn on TurningPoint®** ("clickers") personal response system;

- Video clips with correlated assessment questions;

- **ExamView testing software** that includes all the test items from the printed test bank in electronic format, enabling you to create customized tests of up to 250 items that can be delivered in print or online.

The Wadsworth Sociology Video Library, Volume 1. This DVD drives home the relevance of course topics through short, provocative clips of current and historical events. Perfect for enriching lectures and engaging students in discussion, many of the segments on this volume have been gathered from BBC Motion Gallery. Ask your Cengage Learning representative for a list of contents.

Lecture Ideas for Courses on the Family, Volumes 1 and 2. This handy booklet, offered free to adopters, contains numerous suggestions for activities, lecture ideas, or classroom discussions, all contributed by instructors around the country. Contact your local Wadsworth rep to find out how you can access this useful teaching resource.

Online Activities for Courses on the Family. Made up of contributions from marriage and family instructors, this online supplement is offered free to adopters of our marriage and family books and features new classroom activities for professors to use. Incorporate these activities in your lectures to get students thinking, or use them as a jumping off point to create your own unique activities!

Supplements for the Student

Study Guide. This study guide includes a chapter summary, learning objectives, key terms with completion exercises, Internet and InfoTrac College Edition activities, and key theoretical perspectives for each chapter. Practice tests also contain multiple-choice, true/false, short answer, and essay questions, complete with answers and page references.

Relationship Skills Exercises. This updated supplement, full of assessments and questionnaires, will help students think more reflectively on important topics related to marriage, such as finances and intimacy. Assignments can be done in class or at home, alone or with a partner. New to this edition: Assessments and questionnaires correspond more closely to the textbook so that students can see the connection between what they read for class and their everyday lives.

The Marriage and Families Activities Workbook. What are your risks of divorce? Do you have healthy dating practices? What is your cultural and ancestral heritage, and how does it affect your family relationships? The answers to these and many more questions are found in this workbook of nearly one hundred interactive self-assessment quizzes designed for students studying marriage and family. These self-awareness instruments, all based on known social science research studies, can be used as in-class activities or homework assignments to help students learn more about themselves and their family experience.

Media-Based Supplements

Sociology CourseMate. Lamanna/Riedmann's *Marriages, Families, and Relationships* includes Sociology CourseMate, which helps you make the grade.

Sociology CourseMate includes:

- an interactive eBook, with highlighting, note-taking, and search capabilities
- interactive learning tools including:
  - quizzes
  - flash cards
  - videos
  - learning objectives

- Internet activities
- and more!

Go to www.cengagebrain.com to access these resources related to your text in Sociology CourseMate.

CengageNOW™. CengageNOW is an online teaching and learning resource that gives you and your students more control in less time and delivers better outcomes—NOW. An online study system, CengageNOW gives students the option of taking a diagnostic pretest for each chapter. The system uses the results of each pretest to create personalized chapter study plans for students. The Personalized Study Plans: help students save study time by identifying areas on which they should concentrate and give them one-click access to corresponding pages of the Cengage Learning eBook, provide interactive exercises and study tools to help students fully understand chapter concepts, and include a posttest for students to take to confirm that they are ready to move on to the next chapter. To login, go to **www.cengagebrain.com**.

CourseReader Sociology is a fully customizable online reader which provides access to hundreds of readings, audio and video selections from multiple disciplines. This easy to use solution allows you to select exactly what content you need for your courses, and features many convenient pedagogical features like highlighting, printing, note taking, and audio downloads. The CourseReader: Sociology is the perfect complement to any class.

WebTutor™ on BB/WebCT®. This web-based learning tool takes the sociology course beyond the classroom. Students gain access to a full array of study tools, including chapter outlines, chapter-specific quizzing material, interactive games and maps, and videos. With WebTutor, instructors can provide virtual office hours, post syllabi, track student progress with the quizzing material, and even customize the content to suit their needs.

Acknowledgments

This book is a result of a joint effort on our part; neither of us could have conceptualized or written it alone. We want to thank some of the many people who helped us. Looking back on the long life of this book, we acknowledge Steve Rutter for his original vision of the project and his faith in us. We also want to thank Sheryl Fullerton and Serina Beauparlant, who saw us through early editions as editors and friends and who had significant importance in shaping the text that you see today.

As has been true of our past editions, the people at Cengage Learning have been professionally competent and a pleasure to work with. We are especially grateful to Erin Mitchell, Acquisitions Editor, who has guided this edition, and to Chris Caldera, former Sociology Editor, who oversaw the initiation of this current revision.

Tangelique Williams, Development Editor, worked with us "hands-on" throughout this edition. Assistant Editor Linda Stewart and Editorial Assistant Mallory Ortberg have been important to the success of this edition. Don Schlotman, Rights Acquisitions Specialist, made sure we were accountable to other authors and publishers when we used their work.

Jill Traut, Project Manager for MPS Content Services, led a production team whose specialized competence and coordinated efforts have made the book a reality. She was excellent to work with, always available and responsive to our questions, flexible, and ever helpful. She managed a complex production process smoothly and effectively to ensure a timely completion of the project and a book whose look and presentation of content are very pleasing to us—and, we hope, to the reader.

The internal production efforts were managed by Cheri Palmer, Content Project Manager. Copyeditors Kjersti Sanders and Heather McElwain did an outstanding job of bringing our draft manuscript into conformity with style guidelines and were amazing in terms of their ability to notice fine details—inconsistencies or omissions in citations, references, and elements of the manuscript. Chris Althof, Photo Researcher (Bill Smith Group) worked with us to find photos that captured the ideas we presented in words.

Diane Beasley developed the overall design of the book, one we are very pleased with. Caryl Gorska, Art Director, oversaw the design of new edition. Heather Mann proofread the book pages, and Edwin Durbin compiled the index. Once it is completed, our textbook needs to find the faculty and students who will use it. Andrew Keay, Marketing Manager, captured the essence of our book in the various marketing materials that present our book to its prospective audience.

Closer to home, Agnes Riedmann wishes to acknowledge her late mother, Ann Langley Czerwinski, PhD, who helped her significantly with past editions. Agnes would also like to acknowledge family, friends, and professional colleagues who have supported her throughout the thirty-five years that she has worked on this book.

Sam Walker has contributed to each edition of this book through his enthusiasm and encouragement for Mary Ann Lamanna's work on the project. Larry and Valerie Lamanna and other family members have enlarged their mother's perspective on the family by bringing her into personal contact with other family worlds—those beyond the everyday experience of family life among the social scientists!

At this point, we would also like to acknowledge one another as coauthors for nearly thirty-five years. Each of us has brought somewhat different strengths to this process. We are not alike—a fact that has continuously made for a better book, in our opinion. At times, we have lengthy email conversations back and forth over the inclusion of one phrase. Many times, we have disagreed over the course of the past thirty years—over how long to make a section, how much emphasis to give a particular topic, whether a certain citation is the best one to use, occasionally over the tone of an anxious or frustrated email. But we have always agreed on the basic vision and character of this textbook. And we continue to grow in our mutual respect for one another as scholars, writers, and authors.

Contributing author for this edition, Ann Strahm, wishes to acknowledge her late mother, Lois Strahm, a working-class single mother whose courage to leave an abusive husband and raise her adopted child on little more than minimum wage remains a model of love, strength, and courage.

Reviewers gave us many helpful suggestions for revising the book. Peter Stein's work over the years as a thorough, informed, and supportive reviewer has been an especially important contribution. Although we have not incorporated all suggestions from reviewers, we have considered them all carefully and used many. The review process makes a substantial, and indeed essential, contribution to each revision of the book.

Eleventh Edition Reviewers

Rachel Hagewen, University of Nebraska, Lincoln; Marija Jurcevic, Triton College; Sheila Mehta-Green, Middlesex Community College; Margaret E. Preble, Thomas Nelson Community College; Teresa Rhodes, Walden University

Tenth Edition Reviewers

Terry Humphrey, Palomar College; Sampson Lee Blair, State University of New York, Buffalo; Lue Turner, University of Kentucky; Stacy Ruth, Jones County Junior College; Shirley Keeton, Fayetteville State University; Robert Bausch, Cameron University; Paula Tripp, Sam Houston State University; Kevin Bush, Miami University; Jane Smith, Concordia University; Peter Stein, William Paterson University.

Of Special Importance

Students and faculty members who tell us of their interest in the book are a special inspiration. To all of the people who gave their time and gave of themselves—interviewees, students, our families and friends—many thanks. We see the fact that this book is going into an eleventh edition as a result of a truly interactive process between ourselves and students who share their experiences and insights in our classrooms; reviewers who consistently give us good advice; editors and production experts whose input is invaluable; and our family, friends, and colleagues whose support is invaluable.

1

Marriage, Relationships, and Family Commitments: Making Choices in a Changing Society

Ariel Skelley/Getty Images

This text is different from others you may read. It isn't intended to prepare you for a particular occupation. Instead, it has three other goals: to help you to (1) appreciate the variety and diversity among families today; (2) understand your past and present family situations and anticipate future possibilities; and (3) be more conscious of the personal decisions you must make throughout your life and of the societal influences that affect those decisions.

Families are central to society and to our everyday lives. Families undertake the pivotal tasks of raising children and providing members with intimacy, affection, and companionship. In recent times, what we think of as "family" has changed dramatically. Indeed, today's is "not your grandmother's" family.

In this chapter, we'll look at the difficulty of defining the family today, partly because the word *family* is a "nice" term—one with which almost everyone wants to be associated (Popenoe 1993, p. 529). We'll explore some definitions of the family. After that, we will discuss three broad social trends that affect our relationships and family life.

Later in this chapter, we'll note that when maintaining committed relationships and families, people need the ability to make knowledgeable decisions. The theme of knowledge plus commitment is an integral part of this book. We begin this chapter with a working definition of *family*—one that we can keep in mind throughout the course.

Defining *Family*

When asked to list their family members, some college students include their pets. Are dogs and cats family members? Some individuals who were conceived by artificial insemination with donor sperm are tracking down their "donor siblings"—half brothers and sisters who were conceived using the same man's sperm. They may define their "donor relatives" as family (Shapiro 2009). People make a variety of assumptions about what families are or should be. Indeed, there are many definitions given for the family today, not only among laypeople but also among family scientists themselves (Weigel 2008). To begin to think more about your own definition, you might examine "Issues for Thought: Which of These Is a Family?"

We, your authors, have chosen to define **family** as follows: A family is any sexually expressive, parent–child, or other kin relationship in which people—usually related by ancestry, marriage, or adoption—(1) form an economic and/or otherwise practical unit and care for any children or other dependents, (2) consider their identity to be significantly attached to the group, and (3) commit to maintaining that group over time.

How did we come to this definition? First, caring for children or other dependents suggests a function that the family is expected to perform. Definitions of many things have both *functional* and *structural* components. Functional definitions point to the purpose(s) for which a thing exists—i.e., what it *does*. For example, a functional definition of an iPhone would emphasize that it allows you to make and receive calls, take pictures, connect to the Internet, and access media. Structural definitions emphasize the form that a thing takes—what it actually *is*. To define an iPhone structurally, we might say that it is an electronic device, small enough to be handheld, with a multimedia screen, and whose components allow for sophisticated satellite communication.

An indirect indicator of the centrality of the family to American life is the degree to which family themes are used as advertising motifs. In the first photo, a family splashing happily in the ocean fronts an ad for "Hottest Hotels," while the text of the second photo from another ad describes the family life of the father and the child pictured.

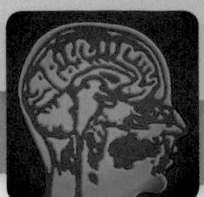

A husband and wife and their offspring.

A single woman and her three young children.

A fifty-two-year-old woman and her adoptive mother.

A man, his daughter, and the daughter's son.

An eighty-four-year-old widow and her dog, Fido.

A man and all of his ancestors back to Adam and Eve.

The 1979 World Champion Pittsburgh Pirates (theme song: "We Are Family").

Three adult sisters living together.

Two lesbians in an intimate relationship and their children from a previous marriage of one woman and a previous

relationship of the other woman with a male friend.

Two children, their divorced parents, the current spouses of their divorced parents, and the children from previous marriages of their stepparents.

A child, his stepfather, and the stepfather's wife subsequent to his divorce from the child's mother.

Two adult male cousins living together.

A seventy-seven-year-old man and his lifelong best friend.

A childless husband and wife who live one thousand miles apart.

A widow and her former husband's grandfather's sister's granddaughter.

A divorced man, his girlfriend, and her child.

Both sets of parents of a deceased married couple.

A married couple, one son and his wife, and the latter couple's children, all living together.

Six adults and their twelve young children, all living together in a communal fashion.

Critical Thinking

Which of these do you consider a family? What is it that makes them "family" or "not family"?

Source: From *Family Theories: An Introduction*, by James K. White and David M. Klein, p. 22. Copyright © 2002 by Sage Publications, Inc. Reprinted by Permission.

Concepts of the family comprise both functional and structural aspects (Weigel 2008). We'll look now at how the family can be recognized by its functions, and then we'll discuss structural definitions of the family.

Family Functions

Social scientists usually list three major functions filled by today's families: raising children responsibly, providing members with economic and other practical support, and offering emotional security.

Family Function 1: Raising Children Responsibly If a society is to persist beyond one generation, it is necessary that adults not only bear children but also feed, clothe, and shelter them during their long years of dependency. Furthermore, a society needs new members who are properly trained in the ways of the culture and who will be dependable members of the group. These goals require children to be responsibly raised. Virtually every society assigns this essential task to families.

A related family function has traditionally been to control its members' sexual activity. Although there are several reasons for the social control of sexual activity, the most important one is to ensure that reproduction takes place under circumstances that help to guarantee the responsible care and socialization of children. The universally approved locus of reproduction remains the married-couple family. "Marriage remains the most common living arrangement for raising children. At any

one time, most American children are being raised by two parents" (Cherlin 2005, p. 37). Still, in the United States and other industrialized countries today, the child-rearing function is often performed by divorced, separated, never-married, and/or cohabiting parents, and sometimes by grandparents.

Family Function 2: Providing Economic and Other Practical Support A second family function involves providing economic support. Throughout much of our history, the family was primarily a practical, economic unit rather than an emotional one (Shorter 1975; Stone 1980). Although the modern family is no longer a self-sufficient economic unit, virtually every family engages in activities aimed at providing for such practical needs as food, clothing, and shelter.

Family economic functions now consist of earning a living outside the home, pooling resources, and making consumption decisions together. In assisting one another economically, family members create some sense of material security. For example, spouses and partners offer each other a kind of unemployment insurance. Family members also care for one another in other practical ways, such as nursing and transportation during an illness.

Family Function 3: Offering Emotional Security Although historically the family was a pragmatic institution involving material maintenance, in today's world the family has grown increasingly important as a source of emotional security (Cherlin 2008; Coontz 2005b).

Many people who own pets think of them as part of the family. Do you think it is appropriate to broaden the definition of *family* to include other than humans? What changes in the family may have encouraged changes in our attitudes about pets and family membership?

This is not to say that families can solve all our longings for affection, companionship, and intimacy. Sometimes, in fact, the family situation itself is a source of stress, as discussed in Chapters 12 and 13. But families and committed relationships are meant to offer important emotional support to adults and children. Family may mean having a place where you can be yourself, even sometimes your worst self, and still belong.

Defining a family by its functions is informative and can be insightful. For example, Laura Dawn, in her book of stories about people who took in survivors of Hurricane Katrina, describes "how strangers became family" (Dawn 2006). But defining a family only by its functions would be too vague and misleading. For instance, neighbors or roommates might help with childcare, provide for economic and other practical needs, or offer emotional support. But we still might not think of them as family. An effective definition of family needs to incorporate its structural elements as well.

Structural Family Definitions

Traditionally, both legal and social sciences have specified that the family consists of people related by blood, marriage, or adoption. In their classic work *The Family:*

From Institution to Companionship, Ernest Burgess and Harvey Locke (1953 [1945]) specified that family members must "constitute a household," or reside together. Some definitions of the family have gone even further to include economic interdependency and sexual–reproductive relations (Murdock 1949).

The U.S. Census Bureau defines a family as "a group of two or more persons related by blood, marriage, or adoption and residing together in a household" (U.S. Census Bureau 2010b, p. 6). It is important to note here that the Census Bureau uses the term **household** for any group of people residing together. Not all households are families by the Census Bureau definition—that is, persons sharing a household must *also* be related by blood, marriage, or adoption to be considered a family.

Family structure, or the form a family takes, varies according to the society in which it is embedded. In preindustrial or *traditional* societies, the family structure involved whole kinship groups. The **extended family** of parents, children, grandparents, and other relatives performed most societal functions, including economic production (e.g., the family farm), protection of family members, vocational training, and maintaining social order. In industrial or *modern* societies, the typical family structure often became the **nuclear family** (husband,

Blend Images/Getty Images

The extended family—grandparents, aunts, and uncles—is often an important source of security. Extended families meet most needs in a traditional society—economic and material needs and child care, for example—and they have strong bonds. In urban societies, specialized institutions such as factories, schools, and public agencies often meet practical needs. But extended families may also help one another materially in urban society, especially during crises.

wife, children). Until about fifty years ago, social attitudes, religious beliefs, and law converged into a fairly common expectation about what form the American family should take: breadwinner husband, homemaker wife, and children living together in an independent household—the *nuclear family model*. Nevertheless, the *extended* family continues to play an important role in many cases, especially among recent immigrants and race/ethnic minorities.[1]

Today, family members are not necessarily bound to one another by legal marriage, blood, or adoption. The term *family* can identify relationships beyond spouses, parents, children, and extended kin. Individuals fashion and experience intimate relationships and families in many forms. As social scientists take into account this structural variability, it is not uncommon to find them referring to the family as *postmodern* (Stacey 1990).

---

[1] The *nuclear* family has lost many functions formerly performed by the traditional extended family (Goode 1963). Economic production now primarily involves working for a nonfamily employer. Police and fire departments, the military, juvenile authorities, and mental health services provide protection and maintain social order. Schools, technical institutes, and universities educate and train the upcoming generation.

Postmodern: There Is No Typical Family

Today, only 7 percent of families fit the 1950s nuclear family ideal of married couple and children, with a husband-breadwinner and wife-homemaker (U.S. Census Bureau 2009e, Tables F1, FG8). Two-earner families are common, and there are reversed-role relationships (working wife, househusband). The past several decades have witnessed a proliferation of relationship and family forms: single-parent families, stepfamilies, cohabitating heterosexual couples, gay and lesbian marriages and families, three-generation families, and communal households, among others. It appears that individuals can construct a myriad of social forms in order to address family functions. The term **postmodern family** came into use in order to acknowledge the fact that families today exhibit a multiplicity of forms and that new or altered family forms continue to emerge and develop.

Figure 1.1 displays the types of households in which Americans live. Just 21.5 percent of households are nuclear families of husband, wife, and children, as compared with 44 percent in 1960 (Casper and Bianchi 2002, p. 8; U.S. Census Bureau 2010b, Tables 59, 63). The most common household type today is that of married couples *without* children, where the children have grown up and left or where the couple has not yet had children or doesn't plan to.

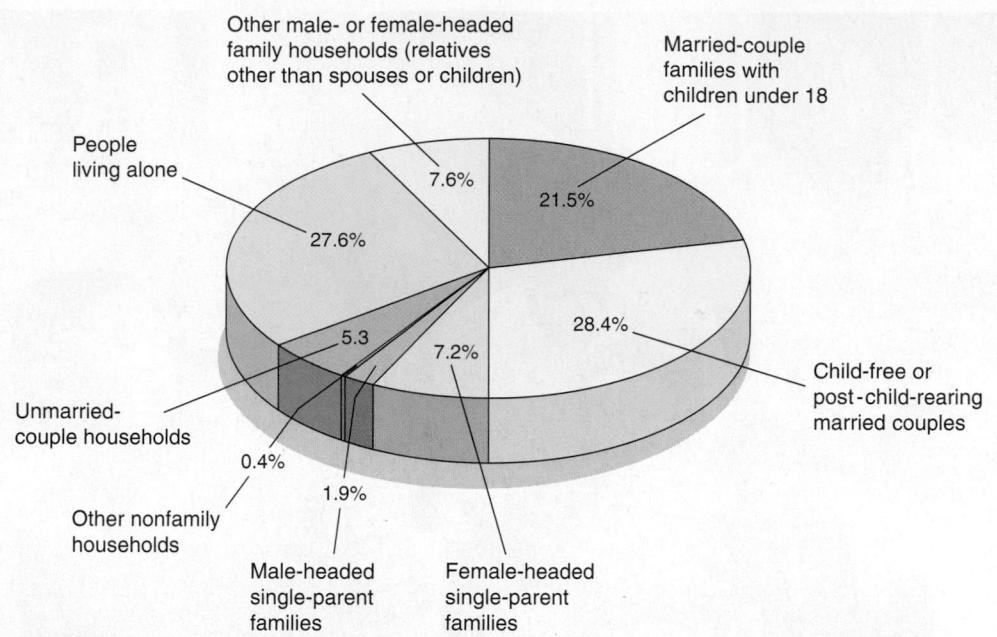

Figure 1.1 The many kinds of American households, 2008.[a] A household is a person or a group of people who occupy a dwelling unit. This figure displays both family and nonfamily households.[b]

Source: U.S. Census Bureau 2010b, Tables 59, 63.

[a] This is the most recent year for which all of the data for this figure are available.

[b] Unmarried-couple households may be composed of two male partners (5.7 percent); two female partners (6.4 percent); or a heterosexual couple (87.9 percent). The Census Bureau classifies unmarried-couple households as "nonfamily households."

More households today (27.6 percent) are maintained by individuals living alone than by married couples with children (U.S. Census Bureau 2008a). There are also female-headed (7.2 percent) and male-headed (1.9 percent) single-parent households, unmarried-couple households (5.3 percent), and family households containing relatives other than spouses or children (7.6 percent). "Facts about Families: American Families Today" presents additional information about other families. Today we see historically unprecedented diversity in family composition, or form.

As one result of this diversity, law, government agencies, and private bureaucracies such as insurance companies must now make decisions about what they once could take for granted—that is, what a family is. If zoning laws, rent policies, employee benefit packages, and insurance policies cover families, decisions need to be made about what relationships or groups of people are to be defined as family. The September 11th Victim Compensation Fund of 2001 struggled with this issue in allocating compensation to victims' survivors. New York State law was amended to allow awards to unmarried gay and heterosexual partners (Gross 2002). President George W. Bush subsequently signed a federal bill extending benefits to domestic partners of firefighters and police officers who lose their lives in the line of duty (Allen 2002).[2]

Adapting Family Definitions to the Postmodern Family

As family forms have grown increasingly variable, social scientists have proposed—and often struggled with—new, more flexible definitions for the family. Sociologist David Popenoe (1993) defines today's family as "a group of people in which people typically live together in a household and function as a cooperative unit, particularly through the sharing of economic resources, in the pursuit of domestic activities" (1993, p. 528). Sociologist Frank Furstenberg writes as follows: "My definition of 'family' includes membership related by blood, legal ties, adoption, and informal ties including *fictive* or socially agreed upon kinship" (2005, p. 810, italics in original).

Legal definitions of family have become more flexible as well. In the past few decades, judges, when

[2] A *domestic partnership* is a legal or policy-defined relationship between two individuals who live together and share a domestic life but are not married. Domestic partnership laws and policies generally apply to both same-sex and heterosexual couples.

What do U.S. families look like today? Statistics can't tell the whole story, but they are an important beginning. As you read this box, it's important to remember that the demographic data presented here are generalizations and do not allow for differences among various sectors of society. Chapter 3 explores social diversity, but for now let's look at these overall statistics.

1. Marriage is important to Americans. About 90 percent of American adults are or have been married, or say that they want to marry (Bergman 2006a; Saad 2006b).

2. Fewer people are married today. Fifty-eight percent of adults were married in 2008, compared to 61 percent in 1990. Twenty-six percent have never married; 10 percent are divorced, and 6 percent widowed (U.S. Census Bureau 2010b, Table 56).

3. People are postponing marriage. In 2009, the median age at first marriage was 25.9 for women and 28.1 for men, as compared with 20.8 for women and 23.2 for men in 1970 (U.S. Census Bureau 2010a, Table MS-2).

4. Cohabitation is an emergent family form as well as a transitional lifestyle choice. The number of cohabiting adults has increased more than ten-fold since 1970 (U.S. Census Bureau 2009e, Table UC-1). Nearly 40 percent of cohabiting couples lived with children under eighteen in 2008— either their own or from a previous relationship or marriage. Unmarried-couple families are only 5 percent of households at any one time, but more

than 50 percent of first marriages are preceded by cohabitation (Fields 2004; Smock and Gupta 2002; U.S. Census Bureau 2009e, Table UC-1).

5. Fertility has declined. After a high of 3.6 in 1957, the total fertility rate— the average number of births that a woman will have during her lifetime— has been at about 2 over the past twenty years (Dye 2008; U.S. Census Bureau 2010b, Table 83). One-fifth of women ages 40–44 are childless—a fraction that is twice as high as thirty years ago (Dye 2008).

6. Parenthood is often postponed. About 19 percent of women reach their forties without bearing a child (Dye 2008).

7. The nonmarital birth rate has risen over the past sixty years. Compared to 4 percent in 1950, more than one-third (38 percent) of all U.S. births today are to unmarried mothers (U.S. Census Bureau 2010b, Tables 80, 85). Between one quarter and one half of nonmarital births today occur to cohabiting couples (Carter 2009; Dye 2008).

8. Same-sex couples—some of them legally married—are increasingly visible. About 565,000 same-sex couple households existed in 2008 (Gates 2009b). It's estimated that about one-fifth of male same-sex partner households and one-third of female same-sex households include children (Rosenfeld and Byung-Soo 2005).[b]

9. The divorce rate is high. The divorce rate doubled from 1965 to 1980. Then it dropped, having fallen more

than 30 percent since 1980 (U.S. Census Bureau 2006c, Tables 72, 117; 2009e, Table 123). Still, it is estimated that between 40 and 50 percent of recent first marriages will end in divorce (Cherlin 2009a; U.S. Census Bureau 2007b, Table 2).

10. The remarriage rate has declined in recent decades but remains significant. Among the divorced, about 52 percent of men and 44 percent of women remarry. In 2004, 12 percent of all adult men and 13 percent of women had been married twice. Three percent of men and of women had married three or more times (Edwards 2007).

11. There are more families with members over age sixty-five today than in the past. The proportion of Americans over age 65 is about 13 percent, and that figure is projected to reach 20 percent by 2050 (U.S. Census Bureau 2010b, Table 8).

Critical Thinking

What do these statistics tell you about the strengths and weaknesses of the contemporary American family and about family change?

a. Figures differ depending on whether *household* or *family* is the unit of analysis. For example, married-couple family households are about 61 percent of all *households*, but about 82 percent of all *families* (U.S. Census Bureau 2010b, Table 59).

b. It is difficult to accurately estimate the number of same-sex-couple households in the United States (O'Connell and Lofquist 2009). See the discussion on this topic in Chapter 8, with attention to footnotes 5 and 6 in Chapter 8.

defining the family in cases that come before them, have used the more intangible qualities of stability and commitment along with the more traditional criteria of common residence and economic interdependency (*Dunphy v. Gregor* 1994). From this point of view, the definition of family "is the totality of the relationship as evidenced by the dedication, caring and self-sacrifice

of the parties" (Judge Vito Titone in *Braschi v. Stahl Associates Company* 1989).

Many employers have redefined *family* with respect to employee benefit packages. Just over half of the Fortune 500 companies, as well as many state and local governments, offer domestic partner benefits. In 2009, legislation was introduced in Congress that may extend domestic partner benefits to all federal civilian employees, and President Obama signed an executive order granting federal employees and their domestic partners some of the rights enjoyed by married couples (Human Rights Campaign 2009; Phillips 2009). Federal practices permit low-income unmarried couples to qualify as families and live in public housing. Several states allow same-sex marriage, and several others provide some spousal rights to same-sex couples. Same-sex marriage is discussed further in Chapter 8.

We, your authors, began this section with our definition of *family*. Our definition integrates important elements of the definitions described previously. We recognize the diversity of postmodern families, while paying heed to the essential functions that families are expected to fill. Our definition combines some structural criteria with a more social–psychological sense of family identity. We include the commitment to maintaining a relationship or group over time as a component of our definition because we believe that such a commitment is necessary in fulfilling basic family functions. It also helps to differentiate the family from casual relationships, such as roommates, or groups that easily come and go.

We have worked to balance an appreciation for flexibility and diversity in family structure and relations—and for freedom of choice—with the increased concern of many social scientists about the family's ongoing functional obligations. We hope that our definition will stimulate your thoughts about what a family is. Ultimately, because society exercises diminished control over what form a family should take, there is no one correct answer to the question, "What is a family?"

Relaxed Institutional Control over Relationship Choices—"Family Decline" or "Family Change"?

Historically, the family has been understood as a **social institution**. Social institutions are patterned and largely predictable ways of thinking and behaving—beliefs, values, attitudes, and norms—that are organized around vital aspects of group life and serve essential social functions. Social institutions are meant to meet people's basic needs and enable the society to survive. Earlier in this chapter, we described three basic family functions. Because social institutions prescribe socially accepted beliefs, values, attitudes, and behaviors, they exert considerable social control over individuals.

Since the 1960s, however, family formation has become less and less predictable. As young people began to postpone marriage and more couples divorced, the proportion of the adult population that was married decreased. Cohabitation and single-mother families increased. In the 1990s, "the number of same-sex cohabiting couples recorded in the U.S. census rose sharply" (Rosenfeld and Byung-Soo 2005, p. 541). More recently, same-sex

marriage has emerged and is now legally available in about 10 percent of states. Combined with increased longevity and lower fertility rates, these changes have meant that a smaller portion of adulthood is spent in traditionally institutionalized marriages and families (Cherlin 2004, 2008).

"Beginning in the late 1950's, Americans began to change their ideas about the individual's obligations to family and society…. [T]his change was away from an ethic of obligation to others and toward an obligation to self" (Whitehead 1997, p. 4). Put another way, we are witnessing an ongoing social trend that involves increasingly relaxed institutional control over relationship choices. Whether this diminished institutional control is harmful or beneficial is a matter of debate among social scientists, policy makers, talk radio hosts, and many of the rest of us.

Critics have described the relaxation of institutional control over relationships and families as "family decline" or "breakdown." Those with a **family decline perspective** claim that a cultural change toward excessive individualism and self-indulgence has led to high divorce rates and could undermine responsible parenting (Whitehead and Popenoe 2006):

> According to a marital decline perspective … because people no longer wish to be hampered with obligations to others, commitment to traditional institutions that require these obligations, such as marriage, has eroded. As a result, people no longer are willing to remain married through the difficult times, for better or for worse. Instead, marital commitment lasts only as long as people are happy and feel that their own needs are being met. (Amato 2004, p. 960)

Moreover, fewer family households contain children. According to the family decline perspective, this situation "has reduced the child centeredness of our nation and contributed to the weakening of the institution of marriage" (Popenoe and Whitehead 2005, p. 23). "Facts about Families: Focus on Children" provides some statistical indicators about the families of contemporary children.

Not everyone concurs that the family is in decline: family change, yes, but not decline. "Marriage has been in a constant state of evolution since the dawn of the Stone Age…. But although the institution of marriage is undergoing a powerful revolution, there is no marriage crisis," declares historian Stephanie Coontz (2005a, p. A17).

Advocates with a **family change perspective** argue that we need to view the family from an historical standpoint. In the nineteenth and early twentieth centuries, American families were often broken up by illness and death, and children were sent to orphanages, foster homes, or already burdened relatives. Single mothers, as well as wives in lower-class, working-class, and immigrant families, did not stay home with children but went out to labor in factories, workshops, or domestic service. The proportion of children living only with their fathers in 1990 wasn't much different from that of a century ago (Kreider and Fields 2005, p. 12).

In many places throughout this text, we focus particularly on children in families. As with our population as a whole, the number of children in the United States is growing. Today there are approximately 84 million children under age eighteen living in the United States. However, the proportion of children under eighteen today—about 24 percent—represents a substantial drop from the 1960s, when more than one-third of Americans were children (U.S. Census Bureau 2010b, Table 8; U.S. Federal Interagency Forum on Child and Family Statistics 2006). Here we look at five statistical indicators regarding U.S. children's living arrangements and overall economic well-being.

1. *At any given time, a majority of children live in two-parent households.* In 2008, 70 percent of children under eighteen lived with two parents—and 68 percent, with two *married* parents. Twenty-six percent of children lived with only one parent (23 percent with mother; 4 percent with father), and another 4 percent did not live with either parent (U.S. Census Bureau 2008a, Table SO901; U.S. Census Bureau 2009c, Table CH-1).

2. *Even in two-parent households, there is considerable variation in children's living arrangements.* A 2001 study on the living arrangements of children in two-parent households found 88 percent of children living with their biological parents (3 percent unmarried), 6 percent with biological mother and stepfather, 1 percent with biological father and stepmother, and smaller numbers with an adoptive parent or stepparent (Kreider and Fields 2005, Table 1).

3. *Individuals experience a variety of living arrangements throughout childhood.* A child may live in an intact two-parent family, a single-parent household, with a cohabitating parent, and in a remarried family in sequence (Raley and Wildsmith 2004). About half of all American children are expected to live in a single-parent household at some point in their lives, most likely in a single-mother household (U.S. Federal Interagency Forum 2005, p. 8, Figure POP6-A).[a]

4. *Children are more likely to live with a grandparent today than in the recent past.* In 1970, 3 percent of children lived in a household containing a grandparent, but by 2008 that rate had more than doubled, to 9 percent. In about a quarter of the cases, grandparents had sole responsibility for raising the child, but many households containing grandparents are extended family households that include other relatives as well (Edwards 2009; Kreider and Fields 2005, Tables 7, 8, 10).

5. *Although most parents are employed, children are more likely than the general population to be living in poverty.* The poverty rate of children has stood at about 18 percent over the past ten years, whereas that of the general adult population is about 12 percent and that of the elderly, about 10 percent. The child poverty rate is lower now than its peak of 22.3 percent in 1983, but higher than in 1970 (U.S. Census Bureau 2009g, Tables 690, 691; 2010b, Table 697). About 13.3 million American children live in poverty, and about 5.8 million of those live in extreme poverty (Children's Defense Fund 2009, p. 14; U.S. Census Bureau 2010b, Table 697).

Critical Thinking

Perhaps the greatest concern Americans have about contemporary family change is its impact on children. What do these family data tell us about the family lives of children today?

a. The census category "single-parent family household" obscures the fact that there may be other adults present in that household, such as grandparents. It is also the case that some households defined by the Census Bureau as single-parent households are actually two-parent households. In about 20 percent of "single-father" households and 10 percent of "single-mother" households there is a cohabitating partner (U.S. Federal Interagency Forum 2005, p. 9). The Census Bureau assumes that such a partner functions as a second parent (Fields 2003, pp. 4–5).

Family change scholars posit that today's family forms need to be seen as historically expected adjustments to changing conditions in the wider society, including the decline in manufacturing jobs that used to provide solid economic support for working-class families, the related need for more education, the entry of women into the labor force, and the increased insecurity of middle- and even upper-class jobs. These economic trends have shaped marital timing and fertility rates, as well as the ability of lower-income individuals to enter into marriage (Edin and Kefalas 2005; McLanahan, Donahue, and Haskins 2005).

Furthermore, "Accompanying...the economic changes was a broad cultural shift among Americans that eroded norms both of marriage before childbearing and of stable lifelong bonds after marriage" (Cherlin 2005, p. 46). These social scientists attribute family changes to economic developments as much as to cultural change, but they do not ignore the difficulties that divorce and nonmarital parenthood present to families and children. However, they view the family as "an adaptable institution" (Amato et al. 2003, p. 21) and argue that it makes more sense to provide support to families as they exist today rather than to attempt to turn back the clock to an idealized past (Cherlin 2009a).

Today's families struggle with new economic and time pressures that affect their ability to realize their family values. Family change scholars "believe that at least part of the increase in divorce, living together, and single parenting has less to do with changing values than with inadequate

One advertisement portrays a happy gay family, making the statement that this is a home like any other. Advertisers have departed from the safe image of the nuclear family to portray nontraditional family forms, as well as family crises such as divorce, as we see in the accompanying photo. The second photo goes so far as to take a rather lighthearted view of divorce. Advertisers say that they are trying to accurately reflect their customers, many of whom "do not fit into the nuclear-family tableau often seen in commercials" (Bosman 2006; see also Lauro 2000).

support for families in the U.S., especially compared to other advanced industrial countries" (Yorburg 2002, p. 33). Many European countries, for example, have paid family leave policies that enable parents to take time off from work to be with young children and that provide more generous economic support for families in crisis.

Some argue further that broad cultural values of individualism and collectivism have not changed all that much. For instance, data from the Longitudinal Study of Generations suggest that an earlier upward trend in individualism may have reversed since the early 1970s and that now the "historical trend is toward greater collectivism" (Bengston, Biblarz, and Roberts 2002, p. 119).

Then, too, most of us wouldn't want to return to an era in which marriage served primarily practical ends while minimizing in importance the happiness of a couple's relationship. Researchers Arland Thornton and Linda Young-DeMarco, who studied attitudes toward the family in the second part of the twentieth century, conclude that "Americans increasingly value freedom and equality in their personal and family lives while at the same time maintaining their commitment to the ideas of marriage, family, and children" (2001, p. 1031).

Scholars, family advocates, and the public continue "this crucial national conversation among Americans struggling to interpret and make sense of the place of marriage and family in today's society" (Nock 2005, p. 13). We return to this debate in Chapter 7. The diversity that we see among families today is the cumulative result of many individuals who have over the years made

personal choices about family living. Next, we explore three societal trends that also influence our options as we make choices about relationships and families.

Three Societal Trends That Impact Families

In addition to relaxed institutional control over the family today, three other society-wide trends have already dramatically changed American family life and will continue to do so. These trends are (1) new communication and reproductive technologies, (2) changes in America's race/ethnic composition, and (3) economic uncertainty.

Advancing Communication and Reproductive Technologies

The pace of technological change has never been faster; new technologies will continue to alter not only family relationships but how we define families as well. Here we highlight two recent technologies that have dramatically impacted families: communication and reproductive technologies.

Communication Technologies What are some ways in which communication technology has impacted families? For one thing, it's only fairly recently that parents and children can be readily reached by cell phone. Today we can video record family events such as a birthday party or a bris (the ritual circumcision of a Jewish son) on our cell phones, then send them to family members around the world ("Family Ties" 2008). Then too, with calls and text messaging, parents can monitor teens who aren't home. Technologies installed in family automobiles allow parents to monitor their children's driving speeds, and Global Positioning Systems (GPS) can tell parents where their children have driven.

Developments such as e-mail, websites, webcams, blogs, Facebook, Skype, and Twitter facilitate communication in ways that we would never have dreamed possible thirty years ago. Sixty-two percent of American households have computers at home, and another 10 percent of Americans access computers elsewhere (U.S. Census Bureau 2010b, Table 1118). Many relationships now begin in cyberspace, minimizing the need for geographical proximity at first meeting. Grandparents and other extended family members stay in contact on Facebook (Grossman 2009). Internet access is changing power relations in some families, as tech-savvy youth become information experts for their families, a skill that can enhance their power relative to other family members (Belch, Krentler, and Willis-Flurry 2005).

Home access to the Internet makes family boundaries more permeable. Types of information that would not have been possible before now come into and leave the home (Turow 2001). Social support for a myriad of personal and family challenges, from infertility to living in stepfamilies to caring for someone with a chronic illness, can be found on the Internet. The Internet also offers access to a wealth of information on physical and mental health, as well as opportunities for online therapy.

However, just as the Internet has been a boon for those seeking social support, it has been the source of conflict and concern for some families who have dealt with greater access to pornography and/or with infidelity initiated on the Internet. Social networking sites such as Facebook have made breaking up or divorce potentially more hurtful as partners publish details on their pages (Luscombe 2009). Moreover, communication technology results in a *digital divide* between those who have access to and use computers and the 29 percent of Americans who don't and therefore cannot access the benefits of computer use ("Digital Divide..." nd.). Impacts of communication technologies on family life are further explored throughout this text.

Reproductive Technologies "Mommy, Mommy, when I grow up, I want to be a mommy just like you. I want to go to the sperm bank just like you and get some sperm and have a baby just like me" (six-year-old, quoted in Ehrensaft 2005, p. 1).

"Mom?...What was the year that you and Dad met our donor?" (Orenstein 2007, p. 35).

Technology has affected pregnancy as modern science continues to develop new techniques to enable couples or individuals to have biological children. The more common infertility interventions involve prescription drugs and microscopic surgical procedures to repair a female's fallopian tubes or a male's sperm ducts (Ehrenfeld 2002). More widely publicized assisted reproductive technology (ART) offers increasingly successful reproductive options (U.S. Centers for Disease Control and Prevention 2008).

In general, ART involves the manipulation of sperm and/or egg in the absence of sexual intercourse, often in a laboratory, and may involve third parties. ART procedures include artificial insemination (male sperm introduced to a female egg without sexual intercourse); donor insemination (artificial insemination with sperm from a donor rather than from the man who will be involved in raising the child); in vitro fertilization (sperm fertilizes egg in a laboratory rather than in the woman's body); surrogacy (one woman gestates and delivers a baby for another individual who intends to raise the child); egg sale or donation (by means of a surgical procedure, a woman relinquishes some of her eggs for use by others); and embryo transfers (a laboratory-fertilized embryo is placed into a woman's womb for gestation and delivery).

Beside allowing otherwise infertile heterosexual couples to have biological children, how do fertility technologies affect family options? For one thing, artificial insemination by donor allows single individuals, as well as lesbian and gay couples, to become biological parents. There are now commercial sperm banks oriented to a lesbian clientele (Mundy 2007). The ability to freeze eggs, sperm, or fertilized embryos enables individuals to become biological parents later in life, after careers are launched, after undergoing medical treatments that will leave them infertile, or even after death. In the last decade, men deployed to Iraq have banked sperm before their departure, anticipating either contact with hazardous materials or death. At least one baby has been conceived by a father who was killed in Iraq prior to his child's conception (Lehmann-Haupt 2009; Oppenheim 2007).

Although ART procedures allow biological parenthood in ways that were unimaginable thirty years ago, these medical advancements do raise family and ethical issues. Embryo transfer and surrogacy create situations in which a child could have as many as three mothers—the genetic mother who donates or sells her egg, the gestational mother who carries and delivers the child, and the social mother who will raise the child—as well as two fathers, a genetic father and a social father (Schwartz 2003). Many states have laws by which sperm donors, with the exception of the husband, have no parental rights, but this barrier between sperm donors and their biological children is gradually being broken. Some sperm donors are sought out by their "children" as they enter adolescence or young adulthood (Harmon 2007b).

Embryo screening—a technology for examining fertilized eggs before implantation to choose or eliminate certain ones—is a boon for prospective parents whose family heritage includes disabling genetic conditions (Harmon 2006). But it also raises the possibility of selection for sex or other traits (Grady 2007; Marchione and Tanner 2006). Those who use sperm banks may choose donor traits they would like to see in their children (Almeling 2007). Especially marketable are eggs and sperm from donors with certain characteristics such as high intelligence, physical attractiveness, athletic ability, or musical talent. Philosophers ponder the implications for parents and children when children are made to order.

Fertilized embryos may be frozen for later implantation in the event that a parent experiences a desire for more children. However, this process raises the issue of what to do with excess frozen embryos. Opponents of abortion argue that to destroy them is murder, but how long can they be saved, and where? (Excess frozen human embryos are sometimes donated to infertile couples.) ART procedures in which several fertilized embryos are implanted into the uterus, in the hope that one will successfully develop into a baby, often involve early abortion of excess developing embryos. When such abortion procedures are not followed, multiple live births—as many as eight—can be the result (Archibold 2009). Multiple births can occur after the use of fertility drugs or after embryo implantation if a woman refuses to undergo abortion of excess developing fetuses on moral grounds.

Reproductive technologies also raise inequality issues. ART is usually not affordable by those with low incomes. Then too, egg donation is an invasive procedure involving some medical risk. Yet selling of eggs offers some women a way to make money. Interestingly, we have seen more selling of eggs in this latest recession (English 2009).

Other issues involve questions about the child's identity if reproductive sperm or eggs are genetically different from those of the social parents. For example, Judaism is traditionally passed down through the mother's genetic line. Therefore, using an egg from a donor who is not Jewish can raise questions about the child's religio-cultural identity (Orenstein 2007). These issues are further explored in Chapter 9.

Like changing technology, another American social trend affecting families is an increasing racial and ethnic diversity.

The New Faces of America's Families: Fewer Non-Hispanic Whites, More People of Color

In 1965, the United States saw the first indications of fertility decline among non-Hispanic white, native-born women. That same year, the United States opened its doors wider to immigrants, the majority of whom are people of color (Mather 2009). Over the subsequent forty-five years, relatively low fertility rates among non-Hispanic whites (compared to higher rates among racial and ethnic minorities) and immigration have combined to "put the United States on a new demographic path" (Mather 2009).

Across the nation, the faces of America's families, particularly America's children, provide evidence of increasing ethnic diversity. Forty-one million African Americans, including those who identify themselves as mixed race, constitute about 14 percent of the U.S. population and are expected to reach 15 percent by 2050 (U.S. Census Bureau 2009c). Although the number of black immigrants is relatively small compared with those arriving from Latin America and Asia, black migration from Africa and the Caribbean has increased in recent years (Kent 2007; Mather 2009). Making up about one-third of the U.S. population today, racial and ethnic minorities are projected to reach 50 percent of the total population by about 2042. Mostly due to rapid growth in Latino families, the population under age eighteen is projected to reach this point by 2023 (Mather 2009).

Recent immigration rates are not climbing as quickly as they did during the 1990s and early 2000s. Still, the United States admitted approximately one million legal immigrants annually in recent decades. Asia and Latin America are the major sending regions. In addition to legal immigrants, an estimated 11 million unauthorized immigrants reside in the United States, the vast majority from Mexico, Central America, and the Caribbean (Mather and Pollard 2009).

Although immigrants were previously concentrated in a few states, now they are much more geographically dispersed. Many U.S. counties have now reached "majority-minority" status, with more than half of their residents identified as a race other than non-Hispanic white (Mather and Pollard 2009). Many refugees (persons outside the country of their nationality who cannot return to their home country due to threats of violence against a social category or group to which they belong) have spread out across the United States to areas that previously had little recent immigration. Nebraska, for example, is home not only to Mexicans and Central Americans who have come to work in meat-packing plants, but also to clusters of Afghani, Cuban, Hmong, Serbian, Somali, Sudanese, Soviet Jewish, and Vietnamese refugees.

Some immigrants are highly educated professionals who could not find suitable employment in their home countries. For the most part, however, immigrants leave a poorer country for a richer one in hopes of bettering their family's economic situation. In the current recession, many immigrants are facing the decision whether to remain in the United States or return to their country of origin (Schuman 2009). Nevertheless, as immigrants establish themselves, they typically begin to send for relatives as ethnic kin and community networks develop here. In fact, the majority of legal immigrants enter the United States through family sponsorship (Martin and Midgley 2006).

As a result, more and more Americans maintain **transnational families** whose members bridge and maintain relationships across national borders. They may experience back-and-forth changes of residence, frequent family visits, business dealings, money transfers to family, placement of children with relatives in the other country, or the search for a marriage partner in the home country. Also, many immigrant families are **binational**, with nuclear family members having different legal statuses. For instance, one partner may be a legal resident, the other not. Children born in the United States are automatically citizens, while one or both parents may be undocumented (illegal) residents. Problematically, the undocumented, or unauthorized immigrant, parents of many native-born, American children in binational families face deportation (Capps et al. 2007). Transnational and binational families are discussed in several places throughout this text, particularly in Chapter 3.

Children born to interracial and inter-ethnic unions further add to America's diversity. Although the growth in race/ethnic intermarriage *rates* for Asians and Hispanics has declined somewhat since the 1990s, their *numbers* continued to rise. Interracial and inter-ethnic

marriage and cohabitation rates involving African Americans have continued to increase significantly (Qian and Lichter 2007). As a result, the proportion of interracial children is significant (Rosenfeld and Byung-Soo 2005, p. 541).

Perhaps nothing better symbolizes American families' changing faces than does our First Family. Race/ethnic inequalities and discrimination assuredly persist (Davis and Bali 2008; Jimenez 2008; Pager, Bonikowski, and Western 2009). However, the 2008 U.S. presidential election of the son of a white mother and a Kenyan father indicates that a majority of voting Americans have grown fairly comfortable with America's changing family faces (Carroll 2007a; Jimenez 2008). President Obama's overall messages are of pride in one's racial heritage and hope for the future (2006, 2007 [1995]). Having appeared on the cover of *Time Magazine* in 2009, First Lady Michelle Obama further symbolizes the accomplishment and potential of diverse American families.

To multiracial children, President Obama's victory was also one for them (Gillman 2008). According to Rice University sociologist Jenifer Bratter, President Obama "embodies the possibility of being welcomed by both

Along with her husband, First Lady Michelle Obama symbolizes pride in one's racial/ethnic heritage and hope for one's future. Shortly after President Obama's election, she visited a Washington, DC, high school where she encouraged students to embrace whatever opportunities avail themselves. She recalled that, sadly, she had lived close to the University of Chicago as a child but never ventured inside. "It was a fancy college, and it didn't have anything to do with me...There are so many kids like that...who are living inches away from power and prestige and fame and fortune, and they don't even know that it exists" (Michelle Obama, in Gibbs and Scherer 2009, p. 24). In this photo, First Lady Obama is having lunch with Head Start students.

sides of the divide that modern interracial families are constantly contesting with" (Bratter, in Gillman 2008). As the U.S. population changes, policy makers need to recognize the complexity and diversity of the growing minority population (Mather 2009, p. 13). We return to issues of racial and ethnic diversity in Chapter 3 and throughout this textbook. Here, we turn to a third society-wide trend that affects families today—economic uncertainty.

Economic Uncertainty

"Recession means worry—all too tangible worry" (Bazelton 2009). Incomes grew little for the middle and working classes even prior to the recession that began in 2008. And although the U.S. economy was good for many Americans during the 1990s, others experienced job insecurity, loss of benefits, longer workdays, and more part-time and temporary work (Teachman, Tedrow, and Crowder 2000). Nonetheless, the recession that began in 2008 has increased unemployment and caused uncertainty and change in virtually all families (Ramo 2009). The following are some implications for relationships and families in today's economy:

- Because many put off marriage until they can earn enough to support a family, more marriages may be delayed or foregone entirely (Gibson-Davis 2009; Roberts 2009; Wang and Morin 2009).

- Fertility decisions will change. Because some individuals are not purchasing contraceptives due to the cost, more unintended pregnancies may be the result (Roan 2009b). On the other hand, bearing and raising children is expensive, so the birth rate could decline at least temporarily (Belkin 2009; Haub 2009; Wang and Morin 2009).

- Although more individuals are selling their eggs or sperm, fertility treatments are down in number (English 2009; Roan 2009a).

- Fewer parents will send their children to costly after-school lessons and other activities.

- Fewer families will send their children to college, and fewer upper-middle-class families will send their children to exclusive universities.

- Young adults' difficulties in finding jobs mean more "boomerang kids" as they return to live in their parents' homes (Trumbull 2009; Wang and Morin 2009).

- Home foreclosures and apartment evictions mean more homeless families, as well as more extended family and intergenerational households as adult children move in with their parents (Palmer 2008; Sard 2008; Spratling 2009).

- Job losses and stock market declines mean more intergenerational households as older parents move into the homes of their adult children (Brandon 2008; Haas 2009).

- Unemployment is expected to raise psychological depression rates, particularly among men (O'Reilly 2009; "Recession Depression ..." 2009).

- Although this was already the case for poverty- and near-poverty-level parents, many more families cut back on paid daycare. Instead, they put together a combination of childcare measures, including an unemployed parent, other relatives, and/or alternating work schedules, to care for their children (Chen 2009).

- More families may relocate as they follow potential job opportunities or choose to be nearer to extended kin (Allen 2009).

- With less money for paid domestic services that previously freed many middle- and upper-class women from household chores, couples may renegotiate housework and childcare duties (Bazelton 2009).

- Because states faced with depleted budgets have cut social services, more families struggle to meet health care and other needs.

- Lost or diminished pensions means that some individuals delay retirement, whereas already retired, geographically distant grandparents make fewer trips to visit grandchildren.

- Children's and grandchildren's inheritances will decline in value ("As Recession Erodes..." 2009).

- Because stress is a causal factor in domestic violence, child and partner abuse has increased, and the consequences for some could be more serious as money for shelters and related services dries up (Lauby and Else 2008; "U.S. Recession Causing..." 2009).

- The divorce rate may drop, at least temporarily. "[F]ewer unhappy couples will risk starting separate households. Furthermore, the housing market meltdown will make it more difficult for them to finance their separations by selling their homes" (Cherlin 2009b; see also Schultz 2008; Wilcox 2009).

- Because financial, social, and psychological resources help families to cope with stress and crises, those with better finances and higher education will better weather recession-related family challenges.

- Families may find new ways to interact together with activities that don't cost money. As one middle-class mother said, "We have more time now. We talk. We may not go anywhere but at least we're all home together"(in Stetler 2009).

- Dating interest is up. Both online and offline matchmakers attribute the jump to the recession: "At a time when money is scarce or uncertain, when people are assessing their priorities, they don't want to go through it alone" (Dr. Pepper Schwartz in Ellin 2009).

This is by no means an exhaustive list. Furthermore, we don't know what the economic future will bring. Can you think of more economy-related implications

"Looks like we invested in a dysfunctional family of funds."

www.CartoonStock.com

A.BACALL

for relationships and families? For most of us, living in a uncertain and problematic economy involves making hard choices. We'll explore the process of making choices next.

The Freedom and Pressures of Choosing

As families have become less rigidly structured, people have made fewer choices "once and for all." Of course, previous decisions do have consequences, and they represent commitments that limit later choices. Nevertheless, many people reexamine their decisions about family—and face new choices—throughout the course of their lives. Thus, choice is an important emphasis of this book.

The best way to make decisions about our personal lives is to make them knowledgeably. It helps to know something about all the alternatives; it also helps to know what kinds of social pressures affect our decisions. As we'll see, people are influenced by the beliefs and values of their society. There are **structural constraints**, economic and social forces, that limit personal choices. In a very real way, we and our personal decisions and attitudes are products of our environment.

But in just as real a way, people can influence society. Individuals create social change by continually offering new insights to their groups. Sometimes social change

occurs because of conversation with others. Sometimes it requires becoming active in organizations that address issues such as racial equality, immigrant rights, gay rights, or stepfamily supports, for example. Sometimes influencing society involves many people living their lives according to their values even when these differ from more generally accepted group or cultural norms.

We can apply this view to the phenomenon of "living together," or cohabitation. Fifty years ago, it was widely believed that cohabiting couples were immoral. But in the 1970s, some college students openly challenged university restrictions on cohabitation, and subsequently many more people than before—students and nonstudents, young and old—chose to live together. As cohabitation rates increased, societal attitudes became more favorable. Over time, cohabitation became "mainstream" (Smock and Gupta 2002). Although some religions and individuals continue to object to living together outside marriage, it is now significantly easier for people to choose this option. We are influenced by the society around us, but we are also free to influence it and we do that every time we make a choice.

Personal Troubles, Societal Influences, and Family Policy

People's private lives are affected by what is happening in the society around them. In his book *The Sociological Imagination* (2000 [1959]), sociologist C. Wright Mills developed the principle that personal troubles are connected to events and patterns in the larger social world. Many times what seem to be personal troubles are shared by others, and these troubles often reflect societal influences. For example, when a family breadwinner is laid off or cannot find work, the cause may not lie in his or her lack of ambition but rather in the economy's inability to provide employment. The difficulty of juggling work and family is not usually just a personal question of individual time management skills but of society-wide influences—the totality of time required for employment, commuting, and family care in a society that provides limited support for working families.

This text assumes that people need to understand themselves and their problems in the context of the larger society. Individuals' choices depend largely on the alternatives that exist in their social environment and on cultural values and attitudes toward those alternatives. Moreover, if people are to shape the kinds of families they want, they must not limit their attention just to their own relationships and families. This is a principal reason why we explore social policy issues in various chapters throughout this text.

Family Policy **Family policy** involves all the procedures, regulations, attitudes, and goals of government, religious institutions, and the workplace that affect families.

The federal government and states have developed programs to encourage and support marriage, to encourage father involvement in fragile families, to discourage teen sexual activity, and to move single mothers from welfare to work.[3] Recently, researchers have explored how policy decisions affect foster parents and grandparent caregivers (McWey, Henderson, and Alexander 2008; Letiecq, Bailey, and Porterfield 2008).

Poverty has been a major focus of policy scholars. American families worry about making ends meet: how we will support ourselves, find comfortable housing, educate our children, get affordable health care, finance our old age. Poverty is a real problem for many U.S. families, and research suggests that deep poverty in early childhood affects outcomes for children (Wagmiller et al. 2006). The United States provides fewer services to families than does any other industrialized nation, while Western Europe offers many examples of a successful partnership between government and families in the interests of family support.

One way to view family policy is to recognize that "laws place some families in the margins of society while privileging others" (Henderson 2008, p. 983). Although this is changing, family policy has privileged heterosexual relationships by defining them as the only acceptable norm while placing same-sex unions in society's margins by defining them as not-marriage. The debates over legal marriage for same-sex families, various legislation, and court rulings all work to create family policy regarding same-sex unions. Issues regarding same-sex couples' separation, divorce, and child custody, as well as determining the legal status for lesbian parents who used ART, are all social policy matters (Hare and Skinner 2008; Oswald and Kuvalanka 2008).

Given the social and political diversity of American society, all parents or political actors are unlikely to agree on the best courses of action. Not only are Americans not in agreement on the role government should play vis-à-vis families, but they are divided on what "family" means in a policy sense. Some argue that only heterosexual, married families should be encouraged, whereas others believe in supporting a variety of families—single-parent, cohabiting heterosexual, or gay and lesbian families, for example (Bogenschneider 2006; Waite 2001). Indeed, the diversity of family lifestyles in the United States makes it extremely difficult to develop family policies that would satisfy all, or even most, of us.

Then, too, more government help to families would be costly. Yet the estimated costs of *not* having family programs might be higher. For instance, disadvantaged children whose adult lives take a bad turn could eventually cost society more in unemployment compensation and incarceration expenses than would preventive investments that help to support these children and their families (Eckholm 2007). Although no social policy can guarantee ideal families, such policies could contribute to a good foundation for family life.

Making knowledgeable family decisions can mean getting involved in national and local political debates and campaigns. One's role as family member, as much as one's role as citizen, has come to require participation in society-wide decisions to create a desirable context for family life and family choices. Concern has arisen about the degree to which Americans do actively participate in attempts to influence neighborhood, community, regional, state, or federal policy. Research indicates that—perhaps with the exception of the 2007–8 presidential campaign—the number of people with whom Americans discuss "important matters" has declined, especially among educated middle-class individuals. The authors speculate that American involvement in community and neighborhood has declined due to longer working hours, the movement of women into the labor force, commuting patterns, more heterogeneous neighborhoods, and the tendency to rely on technological tools for interpersonal contact (McPherson and Smith-Lovin 2006).

But another sociologist argues that Americans have simply changed the form of their community engagement. They are less dependent on the neighborhood and more likely to become involved in professional associations, volunteer in advocacy and service organizations, and participate in self-help groups and religious organizations (Wuthnow 2002).

How Social Factors Influence Personal Choices

Social factors influence people's personal choices in three ways. First, it is usually easier to make the common choice. In the 1950s and early 1960s, when people tended to marry earlier than they do now, it felt awkward to remain unmarried past one's mid-twenties. Now, staying single longer is a more comfortable choice. Similarly, when divorce and nonmarital parenthood were highly stigmatized, it was less common to make these decisions than it is today. As another example, contemporary families usually include fewer children than historical families did, making the choice to raise a large family more difficult than in the past (Zernike 2009).

A second way that social factors can influence personal choices is by expanding people's options. For example, the availability of effective contraceptives makes limiting one's family size easier than in the past, and it enables deferral of marriage with less risk that a sexual relationship will lead to pregnancy. Then too,

[3] Space does not permit a comprehensive review of current and proposed family policies and programs and their effectiveness. However, see, for example, Amato 2005; Capps et al. 2007; Children's Defense Fund 2008, 2009; Dion 2005; Duncan and Chase-Lansdale 2004; Offner 2005; and Ooms 2005.

as we have seen, new forms of reproductive technology provide unprecedented options for becoming a parent.

However, social factors can also limit people's options. For example, American society has never allowed polygamy (more than one spouse) as a legal option. Those who would like to form plural marriages risk prosecution (Janofsky 2001). Until the 1967 *Loving v. Virginia* Supreme Court decision, a number of states prohibited racial intermarriage. As discussed in Chapter 8, the possibility of same-sex marriage is presently being contested in various courts throughout the United States, and outcomes will either expand or limit couples' options. More broadly, economic changes of the last thirty-five years, which make well-paid employment more problematic, have limited some individuals' marital options (Sassler and Goldscheider 2004).

Making Choices

By taking a course in marriage and the family, you may become more aware of your choices, when they are available, and how a decision may be related to subsequent options and choices. All people make choices, even when they are not conscious of it. Let's look more closely at two forms of decision making—choosing by default and choosing knowledgeably—along with the consequences of each.

Choosing by Default

Unconscious decisions are called **choosing by default**. Choices made by default are ones that people make when they are not aware of all the alternatives or when they pursue the proverbial path of least resistance. If you're taking this class but you're unaware that a class in modern dance, which you would have preferred, is meeting at the same time, you have chosen not to take the class in modern dance. But you have done so by default because you didn't find out about all the alternatives before you registered.

Another type of decision by default occurs when people pursue a course of action primarily because it seems the easiest thing to do. Sometimes college students choose even their majors by default. They try to register, only to find that the classes

they had planned to take are closed. So they register for something they hadn't planned on, do well enough, and continue in that program of study.

Many decisions concerning relationships and families are also made by default. For example, partners may focus on career success to the neglect of their relationship simply because this is what society seems to require. For these career-oriented partners, the goal of spending more time together or with family may be on the horizon, but it is never reached because it is not consciously planned for.

Although most of us have made at least some decisions by default, almost everyone can recall having the opposite experience: choosing knowledgeably. Figure 1.2, "The Cycle of Knowledgeable Decision Making," maps this process. You may want to look back at this figure as you go through the course and think about the decisions to be made at various life stages.

Choosing Knowledgeably

Our society offers many options. People can stay single, cohabitate, or marry. They can form communal living groups or family-like ties with others. They can decide

Families are composed of individuals, each seeking self-fulfillment and a unique identity, but individuals can find a place to learn and express togetherness, stability, and loyalty within the family. Families also perform a special archival function: Events, rituals, and histories are created and preserved, and, in turn, become intrinsic parts of each individual. These sisters are sharing memories recorded in family photos.

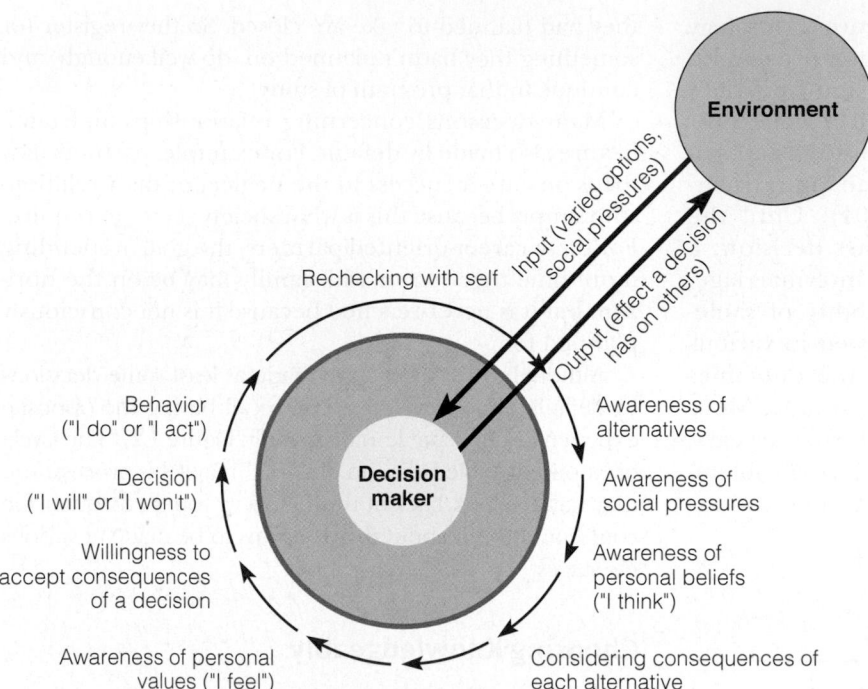

Figure 1.2 The cycle of knowledgeable decision making.

Source: Adapted from *Shifting Gears*, by Nena O'Neill and George O'Neill, p. 157, 1974. Copyright 1974 by Nena O'Neill and George O'Neill. Reprinted by permission of the publisher, M. Evans and Company, New York, NY.

to separate or stay together. They can have children with the aid of reproductive technology. They can parent stepchildren or foster children. One important component of **choosing knowledgeably** is recognizing as many options as possible (Meyer 2007). In part, this text is designed to help you do that.

A second component in making knowledgeable choices is recognizing the social pressures that may influence personal choices. Some of these pressures are economic, whereas others relate to cultural norms that are typically taken for granted. Sometimes people decide that they agree with socially accepted or prescribed behavior. They concur in the teachings of their religion, for example. Other times, people decide that they strongly disagree with socially prescribed beliefs, values, and standards. Whether they agree with such standards or not, once people recognize the force of social pressures, they can choose whether to act in accordance with them.

An important aspect of making knowledgeable choices is considering the consequences of each alternative rather than just gravitating toward the one that initially seems most attractive. For example, a considerable number of young adults live with their parents. Of those between ages 25 and 34, about 15 percent of men and 10 percent of women live with parents (U.S. Census Bureau 2009e, Table AD-1). Someone deciding whether to move back into his or her parents' home may want to list the consequences. In the positive column, moving home might mean being able to save money that would have otherwise

gone for rent, utilities, food, and perhaps childcare. In the negative column, returning to one's parental home might involve giving up some independence, creating a more cramped family space, and risking increased family conflict. Listing positive and negative consequences of alternatives—either mentally or on paper—helps one see the larger picture and thus make a more knowledgeable decision.

Part of this process requires becoming aware of your values and choosing to act consistently with them. Contradictory sets of values exist in American society. For instance, standards regarding nonmarital sex range from abstinence to sex in committed relationships to sex for recreation only. Contradictory values can cause people to feel ambivalent about what they want for themselves.

Clarifying one's values involves cutting through this ambivalence in order to decide which of several standards are more strongly valued. It is important to respect the so-called gut factor—the emotional dimension of decision making. Besides rationally considering alternatives, people have subjective (often almost visceral) feelings about what for them is right or wrong, good or bad. Respecting one's feelings is an important part of making the right decision. Following one's feelings can mean grounding one's decisions in a religious or spiritual tradition or in one's cultural heritage, for these have a great deal of emotional power and often represent deep commitments.

Another important component of decision making is rechecking. Once a choice is made and a person acts on it, the process is not necessarily complete. As Figure 1.2 suggests, people constantly recheck their decisions throughout the entire decision-making cycle, testing these decisions against their experiences and against any changes in the social environment.

Underlying this discussion is the assumption that individuals cannot have everything. Every time people make an important decision or commitment, they rule out alternatives—for the time being and perhaps permanently. People cannot simultaneously have the relative freedom of a child-free union and the gratification that often accompanies parenthood.

It is true, however, that people can focus on some goals and values during one part of their lives, then turn their attention to different ones at other times. Fifty years ago we thought of adults as people who entered adulthood in their early twenties, found work, married, had children, and continued on the same track until the end of the life course. That view has changed. Today we view adulthood as a time with potential for continued personal development, growth, and change.

In a family setting, development and change involve more than one individual. Multiple life courses must be coordinated, and if one member changes, that affects the values and choices of other members of the family. Moreover, life in American families reflects a cultural tension between family solidarity and individual freedom (Amato 2004; Cherlin 2009a).

A Family of Individuals

Americans place a high value on family. It is hardly surprising that a vast majority of Americans report family is extremely important to them (Carroll 2007b, "Marriage" 2008). Why?

Families as a Place to Belong

Families create a place to belong, serving as a repository or archive of family memories and traditions (Cieraad 2006). **Family identity**—ideas and feelings about the uniqueness and value of one's family unit—emerge via traditions and rituals: family dinnertime, birthday and holiday celebrations, vacation trips, and perhaps family hobbies like working together in the garden. Family identities typically include members' cultural heritage. For example, all the children in one family may be given Irish, Hispanic, Asian Indian, or Russian names.

Families provide a setting for the development of an individual's **self-concept**—basic feelings people have about themselves, their abilities, characteristics, and worth. Arising initially in a family setting, self-concept and identity are influenced by significant figures in a young child's life, particularly those in the parent role, together with siblings and other relatives.

How family members and others interact with and respond to us continues to impact self-concept and identity throughout life (Cooley 1902, 1909; Mead 1934; Yeung and Martin 2003). A child who is loved comes to think he or she is a valuable and loving person. A child who is given some tasks and encouraged to do things comes to think of him- or herself as competent.

Early childhood also marks the onset of learning social roles. Children connect certain behaviors to the different roles of mother, father, grandmother, grandfather, sister, brother, and so on. Much of young children's play consists of imitating these roles. *Role-taking*, or playing out the expected behavior associated with a social position, is how children begin to learn behavior appropriate to the roles they may play in adult life. Behavior and attitudes associated with roles become *internalized*, or incorporated into the self.

Meanwhile, expressing our individuality within the context of a family requires us to negotiate innumerable day-to-day issues. How much privacy can each person have at home? What family activities should be scheduled, how often, and when? What outside friendships and activities can a family member sustain?

Midlife changes can be both exhilarating and intimidating, as these college students have probably discovered. Certainly the decision of a middle-aged adult to earn a college degree involves many emotional and practical changes. But by making knowledgeable choices—by weighing alternatives, considering consequences, clarifying values and goals, and continually rechecking—personal decisions and changes can be both positive and dynamic.

© Curtis Willocks/Brooklyn Image Group

Familistic (Communal) Values and Individualistic (Self-Fulfillment) Values

Familistic values such as family togetherness, stability, and loyalty focus on the family as a whole. They are *communal* or *collective values*; that is, they emphasize the needs, goals, and identity of the group. Many of us have an image of the ideal family in which members spend considerable time together, enjoying one another's company. Furthermore, the family is a major source of stability. We believe that the family is the group most deserving of our loyalty (Connor 2007). Those of us who marry vow publicly to stay with our partners as long as we live. We expect our partners, parents, children, and even our more distant relatives to remain loyal to the family unit.

But just as family values permeate American society, so do **individualistic** *(self-fulfillment)* **values**. These values encourage people to think in terms of personal happiness and goals and the development of a distinct individual identity. An individualistic orientation gives more weight to the expression of individual preferences and the maximization of individual talents and options.

The contradictory pull of both familistic and individualistic values creates tension in society (Amato 2004, Cherlin 2009a)—and tension within ourselves that we must resolve. "It is within the family…that the paradox of continuity and change, the problem of balancing individuality and allegiance, is most immediate" (Bengston, Biblarz, and Roberts 2007, p. 323).

American society has never had a remarkably strong tradition of *familism*, the virtual sacrifice of individual family members' needs and goals for the sake of the larger kin group (Sirjamaki 1948; Lugo Steidel and Contreras 2003). Our national cultural heritage prizes individuality, individual rights, and personal freedom. But on the other hand, an overly individualistic orientation puts stress on relationships when there is little emphasis on contributing to other family members' happiness or postponing personal satisfactions in order to attain family goals.

Partners as Individuals and Family Members

The changing shape of the family has meant that family lives have become less predictable than they were in the mid-twentieth century. The course of family living results in large part from the decisions and choices two adults make, moving in their own ways and at their own paces through their lives. Assuming that partners' respective beliefs, values, and behaviors mesh fairly well at first, any change in either can adversely affect the fit.

One consequence of ongoing adult developmental change in two individuals is that the union may be put at risk. If one or both change considerably over time, they may grow apart instead of together. A challenge

In a world of demographic, cultural, and political changes, there is no typical family structure. Today the postmodern family includes cohabiting families, single-parent families, lesbian and gay partners and parents, and remarried families. Whatever their form, families are entrusted with filling three basic functions for society—responsible child rearing, practical and economic support, and emotional support.

for contemporary relationships is to integrate divergent personal change into the relationship while nurturing any children involved.

How can partners make it through such changes and still stay together? Two guidelines may be helpful. The first is for people to take responsibility for their own past choices and decisions rather than blaming previous "mistakes" on their mates. In addition, it helps to recognize that a changing partner may be difficult to live with for a while. A relationship needs to be flexible enough to allow for each partner's individual changes—to allow family members some degree of freedom. At the same time, it's good to remember the benefits of family living and the commitment necessary to sustain it. Individual happiness and family commitment are not inevitably in conflict; research shows that a supportive marriage has a significant positive impact on individual well-being (Waite and Gallagher 2000).

> On the one hand, people value the freedom to leave unhappy unions, correct earlier mistakes, and find greater happiness with new partners. On the other hand, people are concerned about social stability, tradition, and the overall impact of high levels of marital instability on the wellbeing of children. The clash between these two concerns reflects a fundamental contradiction within marriage itself; that is, marriage is designed to promote both institutional and personal goals.... To make marriages with children work effectively, it is necessary for spouses to find the right balance between institutional and individual elements, between obligations to others and obligations to the self. (Amato 2004, p. 962)

Throughout this text we will continue to explore the tension between individualistic and familistic values and discuss creative ways that partners can alter a committed, ongoing relationship in order to meet their changing needs.

Marriages and Families: Four Themes

In this chapter we have defined the term *family* and discussed diversity and decision making in the context of family living. We can now state explicitly the four themes of this text.

1. Personal decisions must be made throughout the life course. Decision making is a trade-off; once we choose an option, we discard alternatives. No one can have everything. Thus, the best way to make choices is knowledgeably.

2. People are influenced by the society around them. Cultural beliefs and values influence our attitudes and decisions. Societal or structural conditions can limit or expand our options.

3. We live in a society characterized by considerable change, including increased ethnic, economic, and family diversity; by tension between familistic and individualistic values; by decreased marital and family permanence; and by increased political and policy concern about the needs of children and families. This dynamic situation can make personal decision making more challenging than in the past—and more important.

4. Personal decision making feeds into society and changes it. We affect our social environment every time we make a choice. Making family decisions can also mean choosing to become politically involved in order to effect family-related social change. Making family choices consciously, according to our values, gives our family lives greater integrity.

We continue our examination of the family in Chapter 2, "Exploring the Family," and in Chapter 3, which discusses the social context in which families make choices.

George and Gaynel Couran were married in 1916. "That was the girl for me. I got the woman I wanted," said George at the couple's eightieth wedding anniversary. Judging by her expression, Gaynel undoubtedly got the man she wanted. The Courans learned to balance individualism and familism over the course of their marriage.

Summary

- This chapter introduced the subject matter for this course and presented the four themes that this text develops. The chapter began by addressing the challenge of defining the term *family*.

- We, your authors, define *family* as any sexually expressive, parent–child, or other kin relationship in which people—usually related by ancestry, marriage, or adoption—(1) form an economic and/or otherwise practical unit and care for any children or other dependents, (2) consider their identity to be significantly attached to the group, and (3) commit to maintaining that group over time.

- Social scientists usually list three major functions filled by today's families: raising children responsibly, providing members with economic and other practical support, and offering emotional security.

- With relaxed institutional control, family diversity has progressed to the point that there is no typical family form today.

- Whether we are in an era of "family decline" or "family change" is a matter of debate.

- In addition to the trend of relaxed institutional control over family formation and family life generally, we examined three other contemporary societal trends that affect families: advancing communication and reproductive technologies, the changing racial and ethnic composition of American families, and economic recession and uncertainty.

- Marriages and families are composed of separate, unique individuals. Our culture values both families and individuals.

- Families provide members a place to belong and help ground identity development. Finding personal freedom within families is an ongoing, negotiated process.

- People make choices, either actively and knowledgeably or by default, that determine the courses of their lives. People must make choices and decisions throughout their life course. Those choices and decisions are limited by social structure and at the same time are causes for change in that structure.

- Change and development continue throughout adult life. Because adults change, relationships, marriages, and families are far from static. Every time one individual in a relationship changes, the relationship itself changes, however subtly.

Questions for Review and Reflection

1. Without looking at ours, write your definition of *family*. Then compare yours to ours. How are the two similar? How are they different? Does your definition have some advantages over ours?

2. Why is the family a major social institution? Does your family fulfill each of the family functions identified in the text? How?

3. What important changes in family patterns do you see today? Do you see positive changes, negative changes, or both? What do they mean for families, in your opinion?

4. What are some examples of a personal or family problem that is at least partly a result of problems in the society?

5. **Policy Question.** What are some changes in law and social policy that you would like to see put in place to enhance family life?

Key Terms

binational family 15
choosing by default 19
choosing knowledgeably 20
extended family 6
familistic (communal, or collective)
 values 21
family 4
family change perspective 10
family decline perspective 10
family identity 21
family policy 17
family structure 6
household 6
individualistic (self-fulfillment) values 22
nuclear family 6
postmodern family 7
self-concept 21
social institution 10
structural constraints 17
transnational family 15

Online Resources

Sociology CourseMate

www.CengageBrain.com

Access an integrated eBook, chapter-specific interactive learning tools, including flashcards, quizzes, videos, and more in your Sociology CourseMate, accessed through **CengageBrain.com.**

www.CengageBrain.com

Want to maximize your online study time? Take this easy-to-use study system's diagnostic pre-test, and it will create a personalized study plan for you. By helping you identify the topics that you need to understand better and then directing you to valuable online resources, it can speed up your chapter review. CengageNOW even provides a post-test so you can confirm that you are ready for an exam.

2

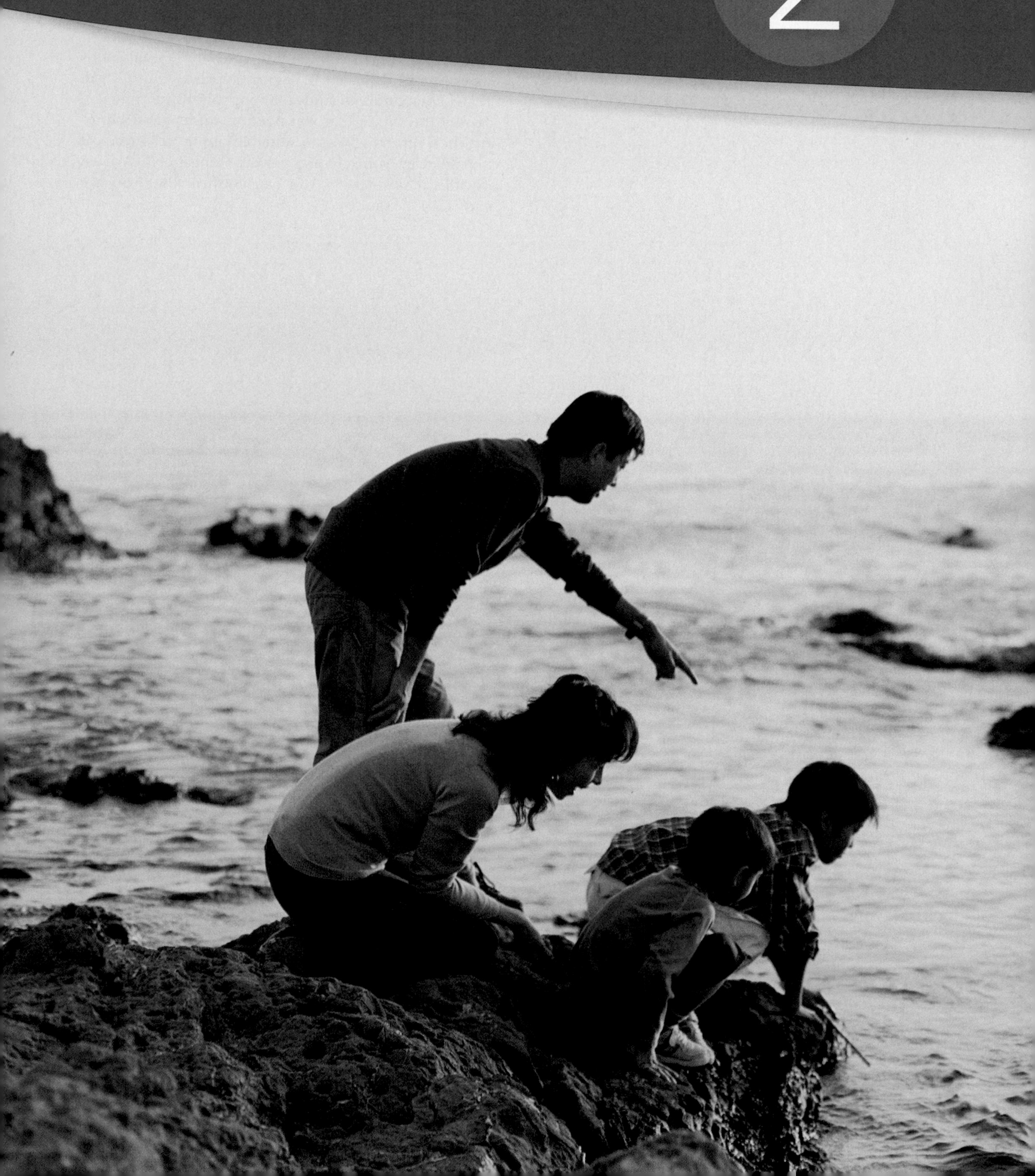

Exploring Relationships and Families

- "What's happening to the family today?"
- "What's a good family?"
- "How do I make that happen?"

In Chapter 1 we said that the best way to make choices is to make them knowledgeably. Throughout this textbook we, your authors, point out many facts that we know to be true about relationships and families as well as things that we know are *not* true although many people may think otherwise.

We base what we write on published information that we trust is accurate.[1] Where does this information come from? Mainly, it results from social scientists' use of theoretical perspectives and research methods designed to explore family life. This chapter invites you into the world of social science so that you can understand and share this way of examining family life.

First we'll discuss how science differs from simply having an opinion or strongly held belief. Next we will examine various theoretical perspectives used by social scientists. After that we'll explore some important things to know about scientific research, then discuss various ways that family scientists gather data. Throughout, we need to keep in mind that studying a phenomenon as close to our hearts as family life can be a knotty challenge.

Science: Transcending Personal Experience

The great variation in family forms and the variety of social settings for family life mean that few of us can rely only on firsthand experience when studying families. Although we "know" about the family because we have lived in one, the beliefs we have about the family based on personal experience may not tell the whole story. We may also be misled by media images and common sense—what "everybody knows." What "everybody knows" can misrepresent the facts.

The Blinders of Personal Experience

Most people grow up in some form of family and know something about what relationships, marriages, and families are. Although personal experience provides us with information, it may also act as blinders. We assume that our own family is normal or typical. If you grew up in a large family, for example, in which a grandparent or an aunt or uncle shared your home, you probably assumed

Families are charged with the pivotal tasks of raising children and providing members with ongoing intimacy, affection, and companionship. Family members consider their identity to be significantly attached to the group. Is this family like the one in which you grew up? If yes, how? If no, how does yours differ?

(for a short time at least) that everyone had a big family. Perceptions like this are usually outgrown at an early age. However, some family styles may be taken for granted or assumed to be universal when they are not.

In looking at family customs around the world, we can easily see the error of assuming that all marriage and family practices are like our own. Not only do common American assumptions about family life fail to hold true in other places, but they frequently don't even describe our own society well. Lesbian or gay male families; black, Latino, and Asian families; Jewish, Protestant, Catholic, Latter-day Saints (Mormon), Islamic, Buddhist, and nonreligious families; upper-class, middle-class, and lower-class families; urban and rural families—all represent some differences in family lifestyle.

Nevertheless, the tendency to use only our experiential knowledge as a yardstick for measuring things is strong. Therefore, science has developed norms for transcending the blinders of personal experience.

Scientific Investigation: Removing Blinders

The central aim of scientific investigation is to find out what is actually going on, as opposed to what we assume is happening. **Science** can be defined as "a logical system that bases knowledge on ... systematic observation" and on *empirical evidence*—facts we verify with our senses (Macionis 2006, p. 15).

The central purpose of the *scientific method* is to overcome researchers' blinders, or biases. (A Chapter 3 box, "Studying Families and Ethnicity," discusses the issue of

---

[1] We provide citations to the sources of our information, then give the complete reference that goes with each citation in the reference section at the back of this book. If you wish, you can find the article or book that we've cited, then read it for yourself and see whether you agree with our interpretation.

racial/ethnic bias in research.) Scientific researchers are ever cognizant of the need to gather data that accurately correspond with reality. "We must be dedicated to finding the truth *as it is* rather than as we think it *should be*" (Macionis 2006, p. 18, italics in original).

Scientific Norms In order to transcend personal biases, scientists follow certain norms (Babbie 2007; Merton 1973 [1942]). Of course, researchers are expected to be honest, never fabricating results. Scientists are expected to publish their research. Publishers are required to evaluate submissions only on merit, never taking into account the researcher's social characteristics, such as race/ethnicity, gender, socioeconomic class, religion, or institutional affiliation. To accomplish this, publishers have reviewers, or "referees," who evaluate submissions "blind" (without knowing the name or anything else about the researcher submitting the article for publication).

Publishing allows research results to be reviewed and critiqued by others. In this way science becomes *cumulative*: findings from various research projects build upon one another. Over time a particular conclusion will be seen to have more evidence behind it than others. It is well established, for example, that marriage carries many benefits for the individual, the couple, and their children (Waite and Gallagher 2000). It is also well established that the arrival of children is associated with at least an initial decline in marital happiness, probably due to having less leisure time as well as the challenges of child raising and concomitant modifications to the couple's relationship (Clayton and Perry-Jenkins 2008).

This last is a conclusion that is not so pleasing to hear, but an important scientific norm involves having *objectivity*: "The ideal of objective inquiry is to let the facts speak for themselves and not be colored by the personal values and biases of the researcher" (Macionis 2006, p. 18). To do this, scientists use rigorous methods that follow a carefully designed research plan. We return to a discussion of scientific methods later in this chapter.

"In reality, of course, total neutrality is impossible for anyone" (Macionis 2006, p. 18). However, following standard research practices and submitting the results to review by other scientists is likely in the long run to correct the biases of individual researchers. At the same time, there are many visions of the family and relationships; what an observer reads into the data depends partly on his or her theoretical perspective.

Theoretical Perspectives on the Family

Theoretical perspectives are ways of viewing reality. As a tool of analysis they are equivalent to lenses through which observers view, organize, then interpret what they see. A theoretical perspective leads family researchers to identify those aspects of families and relationships that interest them and suggests possible explanations for why patterns and behaviors are the way they are.

There are a number of different theoretical perspectives on the family. It is useful to think of each as a point of view. As with a physical object such as a building, when we see a family from different angles, we have a better grasp of what it is than if we look at it only from one, fixed position. Often theoretical perspectives on relationships and families complement one another and may appear together in a single piece of research. In other instances, the perspectives appear contradictory, leading scholars and policy makers into heated debate.

In this section, we describe nine theoretical perspectives related to families:

1. family ecology
2. the family life course development framework
3. the structure–functional perspective
4. the interaction–constructionist perspective
5. exchange theory
6. family systems theory
7. conflict and feminist theory
8. the biosocial perspective
9. attachment theory.

We will see that each perspective illuminates our understanding in its own way. Table 2.1 presents a summary of these theoretical perspectives.

Theoretical perspectives are broad and wide ranging, typically encompassing theoretical subcategories—what social theorist Robert Merton (1968 [1949]) called *theories of the middle range*. Attachment theory, discussed later in this chapter, is an example of a middle-range theory within the discipline of psychology (White and Klein 2008).

The Family Ecology Perspective

The **family ecology perspective** explores how a family is influenced by the surrounding environment. The relationship of work to family life, discussed in Chapter 11, is one example of an ecological focus (Lleras 2008). Sociologists might look at how nonstandard work schedules affect family relationships, for example (Davis et al. 2008). We use the family ecology perspective throughout this book when we stress that, although society does not determine family members' behavior, it does present constraints for families as well as opportunities. Families' lives and choices are affected by economic, educational, religious, and cultural institutions, as well as by historical circumstances such as the development of the Internet, war, recession, or immigration patterns.

Table 2.1 Theoretical Perspectives on the Family

Theoretical Perspective	Theme	Key Concepts	Current Research
Family Ecology	The ecological context of the family affects family life and children's outcomes.	Natural physical–biological environment; Social–cultural environment.	Effect on families of economic inequality in the United States; Racial/ethnic and immigration status variations; Effect on families of the changing global economy; Family policy; Neighborhood effects
Family Life Course Development Framework	Families experience predictable changes over time.	Family life course; Developmental tasks; "On-time" transitions; Role sequencing	Emerging adulthood; Timing of employment, marriage, and parenthood; pathways to family formation
Structure–Functional	The family performs essential functions for society.	Social institution; Family structure; Family functions; Functional alternatives	Cross-cultural and historical comparisons; Analysis of emerging family structures in regard to their comparative functionality; Critique of contemporary family
Interaction–Constructionist	By means of interaction, humans construct sociocultural meanings. The internal dynamics of a group of interacting individuals construct the family.	Interaction; Symbol; Meaning; Role making; Social construction of reality; Deconstruction; Postmodernism	Symbolic meaning assigned to domestic work and other family activities; Deconstruction of reified categories
Exchange Theory	The resources that individuals bring to a relationship or family affect the formation, continuation, nature, and power dynamics of a relationship. Social exchanges are compiled to create networks and social capital.	Resources; Rewards and costs; Family power; Social networks; Social support	Family power; Entry and exit from marriage; Family violence; Network-derived social support
Systems Theory	The family as a whole is more than the sum of its parts.	System; Equilibrium; Boundaries; Family therapy	Family efficacy and crisis management; Family boundaries
Feminist Theory	Gender is central to the analysis of the family; male dominance in society and in the family is oppressive of women.	Male dominance; Power and inequality	Work and family; Family power; Domestic violence; Deconstruction of reified gender categories; Deconstruction of definition of marriage as necessarily heterosexual; Advocacy of women's issues
Biosocial Perspective	Evolution of the human species has put in place certain biological endowments that shape and limit family choices.	Evolutionary heritage; Genes, hormones, and brain processes; Inclusive fitness	Connections between biological markers and family behavior; Evolutionary heritage explanations for gender differences, sexuality, reproduction, and parenting behaviors; Development of research methods that can explore the respective influences of "nature" and "nurture"
Attachment Theory	Early childhood experience with caregiver(s) shape psychological attachment styles.	Secure, insecure/anxious, and avoidant attachment styles	Attachment style and mate choice, jealousy, relationship commitment, separation, or divorce

Every family is embedded in "a set of nested structures, each inside the next, like a set of Russian dolls" (Bronfenbrenner 1979, p. 3). At the foundation is the *natural physical–biological environment*—climate and climate change, soil, plants, animals. The *social–cultural environment* consists of human-made things, or *cultural artifacts*, such as bridges and iPhones, as well as cultural values and products such as language and law or educational and economic systems. All parts of the model are interrelated and influence one another (Bubolz and Sontag 1993; and see Figure 2.1).

Family ecologists stress the interdependence of all the world's families—not only with one another, but also with our fragile physical–biological environment. In this vein, the Family Energy Project at Michigan State University has studied families' energy usage (Bubolz and Sontag 1993). Although it is crucial, the interaction of families with the physical–biological environment is beyond the scope of this text. Our interest centers on families in their sociocultural environments.

The social–cultural ecology of families may be examined historically: "By virtue of when they were born, members of each generation live through unique times shaped by unexpected historical events, changing political climates, and evolving socioeconomic conditions" (Carlson 2009, p. 2). Ways that historical periods affect individuals, relationships, and families are explored in Chapter 3.

This perspective also analyzes the environments of contemporary families at various levels, from the global to the neighborhood. On the global level, for instance, income and job opportunities for American families are affected when tasks are outsourced to other countries. Furthermore, the economic recession that began in 2008 ended many jobs filled by immigrants, who consequently wrestled with decisions about returning to their home countries (Schuman 2009). As another example, the September 11, 2001, terrorist attacks on the United States and the subsequent Afghan and Iraq wars have been part of a global conflict affecting American family life in countless ways.

Family ecologists also stress the importance of neighborhoods to well-being (Bowen et al. 2008). For instance, children in poor neighborhoods are at greater risk for negative social, educational, economic, and health outcomes (Mather and Rivers 2006). In addition to violence, other neighborhood risk factors include high crime rates, low adult educational attainment, and a higher percentage of female-headed households (Knoester and Haynie 2005).

Ecologists have also examined the sociocultural settings of more privileged families (Swartz 2008). Researchers have found close-knit, often suburban neighborhoods that facilitate families' "bringing up kids together" (Bould 2003). Typically, these are racially

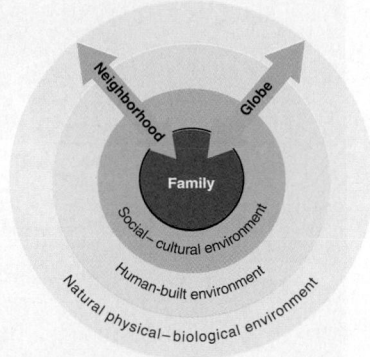

Figure 2.1 The family ecology perspective. The family is embedded in natural physical–biological and social–cultural environments from the global level to the neighborhood level.

Source: Adapted with permission from "Human Ecology Theory," pp. 419–48 in *Sourcebook of Family Theories and Methods*, ed. by Pauline G. Boss et al. Copyright © 1993, Springer-Verlag.

homogeneous neighborhoods, with stay-at-home moms who bonded when children were small and who continue to have a level of trust that permits families to monitor and discipline each other's children. Bould (2003) ponders the trade-off this seems to require in terms of women's role choices and neighborhood diversity.

Examining the kinds of economic and social advantages enjoyed by the middle and upper levels of society may provide insight into the conditions that would enable *all* families to succeed. Moreover, there are sometimes elements in a the social–cultural environment of upper-socioeconomic-level families—excessive achievement pressure or the isolation of children from busy, accomplishment-oriented parents—that are problematic (Luthar 2003). The ecology perspective helps to identify factors that are important to societal and community support for all families. Exploring family life through this perspective leads to interest in family policy (the various laws and other regulations and procedures that impact families), discussed in Chapter 1.

Contributions and Critiques of the Family Ecology Perspective This perspective first emerged in the late nineteenth century, a period marked by concern about family welfare. The family ecology model resurfaced in the 1960s with the War on Poverty, a program directed toward the elimination of the high levels of poverty that then existed. The family ecology perspective makes an important contribution today by challenging the idea that family satisfaction or success depends solely on individual effort (Marks 2001). Furthermore, the

People in this neighborhood join together in activities of benefit to all. This group is organizing a Neighborhood Watch program.

perspective turns our attention to family social policy—what may be done about social issues or problems that affect relationships and families.

A possible disadvantage of the family ecology perspective is that it is so broad and inclusive that virtually nothing is left out. One research agenda can hardly take into account the family's sociocultural environment on all levels, from the global to the neighborhood. More and more, however, social scientists are exploring family ecology in concrete settings. For example, Canadian researchers Phyllis Johnson and Kathrin Stoll (2008) investigated how Sudanese refugee men continued to enact the breadwinner role for their families in Africa while resettling alone in Western Canada. As a second example, U.S. researchers affiliated with the Children's Defense Fund continue to examine the ongoing plight of Louisiana children who suffer the disastrous effects of Hurricane Katrina (Cass 2007).

The Family Life Course Development Framework

Whereas family ecology analyzes relationships, families, and the broader society as interdependent parts of a whole, the **family life course development framework** focuses on the family itself as the unit of analysis (White and Klein 2008). The concept of the *family life course* is central here, based on the idea that the family changes in fairly predictable ways over time.

Typical stages in the *family life course* are marked by (1) the addition or subtraction of family members (through birth, death, and leaving home), (2) the various stages that the children go through, and (3) changes in the family's connections with other social institutions (retirement from work, for example, or a child's entry into school). Each stage has requisite *developmental tasks* that must be mastered prior to transitioning successfully to the next stage. Therefore, this perspective has tended to assume that families perform better when life course stages proceed in orderly fashion.

Traditionally this perspective has assumed that families begin with marriage. The *newly established couple* stage ends when the arrival of the first baby thrusts the couple into the *families of preschoolers* stage, followed later by the *families of primary school children* stage, and still later by the *families with adolescents* stage. *Families in the middle years* help their offspring enter the adult worlds of employment and their own family formation. Later, parents return to a couple focus with (if they are fortunate!) the time and money to pursue leisure activities. Still later, *aging families* must adjust to retirement and perhaps health crises or debilitating chronic illness. The death of a spouse marks the end of the family life course (Aldous 1996).

Role sequencing, the order in which major life course transitions take place, is important to this perspective. The *normative order hypothesis* proposes that the work–marriage–parenthood sequence is best for mental health and happiness (Jackson 2004). Then too, *"on-time" transitions*—those that occur when they are supposed to, rather than "too early" or "late"—are generally considered most likely to result in successful role performance during subsequent life course stages (Booth, Rustenbach, and McHale 2008; Hogan and Astone 1986).[2]

Life course theorists are aware of the contingencies of history on the family life course (Shanahan 2000). In different historical periods and among different generations, "normal" timing of the family life cycle has varied. In the post–World War II era, marriage and parenthood typically occurred at a much younger age than it does today. "There used to be a societal expectation that people in their early twenties would have finished their schooling, set up a household, gotten married, and started their careers… But now that's the exception rather than the norm'" (sociologist Frank Furstenberg Jr. in Lewin 2003; see also Furstenberg et al. 2004).

Researchers have found that now having a baby in one's twenties may be seen as "out of step," or as a risky life course move (Jong-Fast 2003). Using this theoretical framework, sociologists Jeremy Uecker and Charles Stokes (2008) explored the incidence of "early marriage"—that is, marrying before age twenty-three—in the United States today. Uecker and Stokes argue that "[s]cholars and policymakers… should pay adequate attention to understanding and supporting these individuals' marriages" (p. 835). Because marrying in one's early twenties is an exception to the rule today, researchers of the life course perspective have turned their attention to the current, "emerging" transition to adulthood (Bentley 2007).

Emerging adulthood is a stage in individual development that precedes and affects entry into the family life course. The concept conveys a sense of ongoing development, a period "when the scope of independent exploration of life's possibilities is greater for most people than it will be at any other period of the life course" (Arnett 2000, p. 469). Transition to adulthood is now completed more gradually and later than it has been in the past—usually by age thirty (Arnett 2004; Furstenberg 2008). A principal reason for this change: It takes longer today to earn enough to support a family (Gibson-Davis 2009). Emerging adulthood is further explored in Chapters 7 and 8.

In addition to examining the transition to adulthood, researchers using the family life course development framework also extensively study the various transitions, or "pathways," to family formation (Amato et al. 2008; Lichter and Qian 2008). These transitions are far more likely today to include single parenthood and cohabitation.

Contributions and Critiques of the Family Life Course Development Framework The family life course development framework directs attention to various stages that relationships and families encounter throughout life. Hence this perspective encourages us to investigate various family behaviors over time. For instance, research consistently finds that women are more likely to work on maintaining family relationships (see Chapter 11). Building on this research, three Belgian sociologists asked, "Is this true for all life stages?" They found the answer to be yes (Bracke, Christiaens, and Wauterickx 2008). As another example, researchers looked at reasons for calling telephone crisis hotlines across the life course. They found that "issues of loneliness increased with age whereas depression-related calls decreased" (Ingram et al. 2007).

Furthermore, this perspective directs our attention to how particular life course transitions affect relationships and family interaction. For example, researchers have investigated how transitions to parenthood or from cohabitation to marriage affect the time that partners spend on housework (Baxter, Hewitt, and Haynes 2008). Then too, as we have seen, the perspective brings our attention to how transitioning "early," "late," or "on-time" affects relationships and families (Booth, Rustenbach, and McHale 2008). This perspective also prompts researchers to look at interactions among family members who are in different life course stages. An example is a study of ongoing affection between grandparents and young adults (Monserud 2008).

Critics note remnants of the traditional tendency within this perspective toward assumptions of life course standardization, possibly suggesting a white, middle-class bias. Moreover, due to economic, ethnic, and cultural differences, two families in the same life cycle stage may be very different. For these reasons, the family development perspective is somewhat less popular now than it once was.

The Structure–Functional Perspective

The **structure–functional perspective** investigates how a given social structure functions to fill basic societal needs. As discussed in Chapter 1, families are principally accountable for three vital family functions: to raise children responsibly, to provide economic support, and to give family members emotional security. *Social structure* refers to the ways that families are patterned or organized—that is, the form that a family may take.

As discussed in Chapter 1, there is no typical American family structure today. Instead, families evidence a variety of forms including, among others, same-sex families, cohabitation, single-parent families, and transnational families whose members bridge and maintain relationships across national borders. The structure–functional

---

[2] Originating in the work of Charles Horton Cooley (1902, 1909) and George Herbert Mead (1934), interactionism was important to sociology during the 1920s and 1930s, when the field of family studies was establishing itself as a legitimate social science. This discussion of interactionism does not attempt to analyze different traditions such as *symbolic interactionism* (Blumer 1969; Cooley 1909; Mead 1934), *dramaturgical sociology* (Goffman 1959; Lyman and Scott 1975), and a more *structural interactionism* (Gecas 1982; Stryker 2003 [1980]).

According to the family life course development framework, this father is in the *families of primary school children* stage. Like other family life course stages, this stage has particular tasks that need to be performed—tasks for which previous life course stages, if completed successfully, have helped to prepare him.

perspective encourages researchers to ask how well a particular family structure performs a basic family function. For example, there is considerable research into how well single parents or cohabiting couples perform the function of responsible child rearing (Ackerman et al. 2001; Brown 2004; Carlson 2006; Manning and Lamb 2003). Results of this research are explored in Chapters 8, 10, and 15.

The structure–functional perspective may encourage a family researcher to think in terms of *functional alternatives*—alternate structures that might perform a function traditionally assigned to the nuclear family. A study among recent immigrants found that *fictive kin*—relationships "based not on blood or marriage but rather on religious rituals or close friendship ties, that replicates many of the rights and obligations usually associated with family ties"—can serve as a functional alternatives to the nuclear family. Results showed that "functions include assuring the spiritual development of the child and thereby reinforcing cultural continuity, exercising social control, providing material support, and assuring socio-emotional support" (Ebaugh and Curry 2000, pp. 189, 199).

The term *dysfunction* emerged from the structure–functional perspective (Merton 1968 [1949]) as a focus on social patterns or behaviors that fail to fulfill basic family needs. Obviously domestic violence is dysfunctional in that it opposes the family function of providing emotional security. The dysfunctions of parental conflict for childhood development are discussed in Chapter 10's box, "Growing Up in a Household Characterized by Conflict." Although the term "dysfunctional family" is often used by laypeople and in counseling psychology, sociologists seldom use the term, which is considered too vague and imprecisely defined.

The structure–functional perspective might also encourage one to ask, "Functional for whom?" when examining a particular social structure (Merton 1968 [1949]). For instance, traditional, virtually unquestioned male authority and prestige may be functional for fathers—and in some cultures for brothers—but not necessarily for mothers or sisters. Separating may seem to be functional for one or both of the adults involved, but it's not necessarily so for the children (Amato 2000, 2004).

Contributions and Critiques of the Structure–Functional Perspective Virtually all social scientists agree on the one basic premise underlying structure–functionalism: that families are an important social institution performing essential social functions. The structure–functional perspective encourages us to ask how well various family forms do in filling basic family needs. Furthermore, the perspective can be interpreted as encouraging us

"This family is way too functional."

to examine ways in which functional alternatives to the heterosexual nuclear family may perform basic family functions.

However, as it dominated family sociology in the United States during the 1950s, the structure–functional perspective gave us an unrealistic image of smoothly working families characterized only by shared values. Furthermore, the perspective once argued for the functionality of specialized gender roles: the *instrumental* husband-father who supports the family economically and wields authority inside and outside the family, and the *expressive* wife-mother-homemaker whose main function is to enhance emotional relations at home and socialize young children (Parsons and Bales 1955).

Furthermore, the structure–functional perspective has generally been understood to define the heterosexual nuclear family as the only "normal" or "functional" family structure. As a result many social scientists, particularly feminists, rebuke this perspective (Anderson 2005; Stacey 2006). The vast majority of family sociologists today rarely reference structure–functionalism directly.

The Interaction–Constructionist Perspective

As its name implies, the **interaction–constructionist perspective** focuses on *interaction,* the face-to-face encounters and relationships of individuals who act in awareness of one another. Often this perspective explores the daily conversation, gestures, and other behaviors that go on in families. By means of these interchanges, something called "family" appears (Berger and Kellner 1970). Family identity, traditions, and commitment emerge through interaction, with the development of relationships and the generation of *rituals*—recurring practices defined as special and different from the everyday (Byrd 2009; Oswald and Masciadrelli 2008).

Sometimes this perspective explores family *role making* as partners adapt culturally understood roles—for example, uncle, mother-in-law, grandmother, or stepfather—to their own situations and preferences. One study looked at how older Chinese and Korean immigrants remade family roles upon immigrating to the United States (Wong, Yoo, and Stewart 2006). A Korean grandmother described remaking her mother-in-law role:

> Once I immigrated I realized there are cultural differences between the U.S. and Korea especially when it comes to family dynamics. For example, I can't always say what I would like to say to my daughter-in-law. I follow the American ways and have given up trying to tell her what to do… I would like to tell my daughter-in-law to punish the grandchildren when they misbehave. But in America, us elders do not have the right to say this. I just keep these thoughts to myself. (p. S6)

This point of view also examines how family members interact with the outside world in order to manage family identity. An example is a study of interaction strategies used by couples who had chosen to remain childfree. Feeling potentially stigmatized, some claimed that they were biologically unable to have children. Others aggressively asserted the merits of a childfree lifestyle (Park 2002). The couples worked to construct how others would define their not having children.

Reality as Constructed This approach explores ways that people, by interacting with one another, *construct,* or create, meanings, symbols, and definitions of events or situations. A respondent in the study of Chinese and Korean immigrants saw family photographs as symbols of her changing (reconstructed) family role:

> My children got married and started to have a family of their own… We are now no longer the center, but on the peripheries of their families. Even when we take pictures, we don't stand in the center but on the side. It's totally different in China. Even when we took pictures, parents would be pictured in the middle. (Wong, Yoo, and Stewart 2006, p. S6)

As people "put out" or *externalize* meanings, these meanings come to be *reified,* or made to seem real. Once a meaning or definition of a situation is reified, people *internalize* it and take it for granted as "real," rather than viewing it as a human creation (Berger and Luckman 1966). For example, many newlyweds take it for granted that a honeymoon should follow their wedding; they don't think about the fact that the idea of a honeymoon is socially constructed (Bulcroft et al. 1997). Sociologists James Holstein and Jaber Gubrium (2008) combine this perspective with the family life course development framework to investigate how individuals gradually construct their life course.

Unlike structure–functionalism, in which analysis begins with one or more family forms that are understood as given, the interaction–constructionist perspective focuses on the processes through which family forms are constructed and maintained. For instance, we typically think of the "battered woman" as having been abused by a male, thereby maintaining the social construction of domestic life as heterosexual (VanNatta 2005). Our values and beliefs about divorce, childbearing outside marriage, and single-parent families can also be understood as socially constructed (Thornton 2009). Exposing the ways that symbols and definitions are constructed is called *deconstruction*, a process typically identified with postmodern theory.

Postmodern Theory Postmodern theory can be understood as a special focus within the broader interaction–constructionist perspective (Kools 2008). Gaining recognition in the social sciences since the 1980s, postmodern theory largely analyzes *social discourse* or *narrative* (public or private, written or verbal statements or stories). The analytic purpose is to demonstrate that a phenomenon is socially constructed (Gubrium and Holstein 2009). A principal goal involves debunking *essentialism*—the idea that categories really do exist in nature and are not simply reifications. Examples include analyses of the concepts of *gender* and *race*. Formerly taken for granted as essentially "real," these categories are now generally recognized—at least within the social sciences—as social constructions. (Chapter 4's "Issues for Thought: Challenges to Gender Boundaries," further explores the social construction of gender.)

When applied to relationships, postmodern theory posits that beliefs about what constitutes a "real" family are nothing more than socially fabricated narratives, having been constructed through public discourse.

Contributions and Critiques of the Interaction-Constructionist Perspective The interaction–constructionist perspective alerts us to the idea that much in our environment is neither "given" nor "natural," but socially constructed by humans—those in the past and those around us now. In this way the perspective can be liberating. If a social structure, definition, value, or belief is oppressive, it can be challenged: Constructed by human social interaction, phenomena can be changed by such interaction as well. Social movements advocating legalization of same-sex marriage proceed from this beginning point. At the family level, this perspective

This African American family is celebrating Kwanzaa, created in the 1960s by Ron Karenga based on African traditions. An estimated 10 million black Americans now celebrate Kwanzaa as a ritual of family, roots, and community. The experience of adopting or creating family rituals fits the interaction–constructionist perspective on the family.

leads researchers to focus on family members' interaction patterns, along with emergent definitions, symbols, rituals, and the consequences thereof.

Critics argue that the research typically associated with this perspective—that is, qualitative research, which is explained later in this chapter (Holstein and Gubrium 2008)—lacks objectivity. Such critics ask, "Where do we go from here?" (Wasserman 2009). Once the taken-for-granted is deconstructed, then what? "If everything is socially constructed, then we gain nothing by employing the term. It has become a mantra that explains very little" (Stacey 2006, p. 481). Moreover, it is virtually impossible to conduct traditional social science research in the absence of agreed-upon social categories (Cockerham 2007).

Exchange Theory

Exchange theory applies an economic perspective to social relationships. A basic premise is that when engaged in social exchanges, individuals prefer to limit their costs and maximize their rewards. After calculating potential rewards against costs and weighing our alternatives, we choose among options. Those of us with more resources, such as education or good incomes, have a wider range of options from which to choose. This orientation examines how individuals' personal resources, including physical attractiveness and personality characteristics, affect the formation and continuation of relationships.

According to this perspective, an individual's dependence on and emotional involvement in a relationship affects her or his relative power in the relationship. When alternatives to a relationship seem slim, one wields less power in the relationship. According to the *principle of least interest*, the partner with less commitment to the relationship is the one who has more power (Waller 1951). Those with more resources and options can use them to bargain and secure advantages in relationships. People without resources or alternatives to a relationship typically defer to the preferences of the other and are less likely to leave (Brehm et al. 2002; Sprecher, Schmeeckle, and Felmlee 2006). From this point of view, responses to domestic violence and decisions to separate or divorce are affected by partners' relative resources.

The relative resources of participants shapes power and influence in families and impacts household communication patterns, decision making, and division of labor (Sabatelli and Shehan 1993). Relationships based on exchanges that are equal or equitable (fair, if not actually equal) thrive, whereas those in which the exchange balance feels consistently one-sided are more likely to dissolve or be unhappy.

Having flourished during the 1960s and 1970s, this perspective must fight the human tendency to see family relationships in more romantic and emotional terms. Yet dating relationships, marriage and other committed partnerships, divorce, and even parent–child relationships show signs of being influenced by participants' relative resources (Nakonezny and Denton 2008).

Social Networks Exchange theory also focuses on how everyday social exchanges between and among individuals accumulate to create social networks. Elizabeth drives Juan to the airport, Juan babysits for Maria, Maria proofreads an assignment for Elizabeth, and so forth until a network of social exchanges emerges. The Internet offers opportunities for building social networks ranging from the local to the international level, such as those on Facebook.

Among other things, *social network theory*, a middle-range subcategory within the exchange perspective, examines how social networks provide individuals with *social capital*, or resources (friendship, people with whom to exchange favors) that result from their social contacts. Social capital is analogous to financial capital, or money, inasmuch as we can "spend" it to acquire rewards, such as a romantic partner, a job, or emotional support (Benkel, Wijk, and Molander 2009; Wejnert 2008).

Contributions and Critiques of Exchange Theory The exchange perspective provides a framework from which to draw specific hypotheses about weighing alternatives and making decisions regarding relationships. Furthermore, this perspective leads us to recognize that inequality, or an unfavorable balance of rewards and costs, gradually erodes positive feelings in a relationship (Brehm et al. 2002). The perspective also encourages us

to recognize the social capital brought about by membership in social networks. Exchange theory is subject to the criticism that it assumes a human nature that is unrealistically rational and even cynical at heart about the roles of love and responsibility.

Family Systems Theory

Family systems theory views the family as a whole, or *system*, comprised of interrelated parts (the family members) and demarcated by boundaries. Originating in natural science, systems theory was applied to the family first by psychotherapists and was then adopted by family scholars.

A *system* is a combination of elements or components that are interrelated and organized into a whole. Like an organic system (the body, for example), the parts of a family compose a working system that behaves fairly predictably. The ways in which family members respond to one another can show evidence of patterns. For example, whenever Jose sulks, Oscar tries to think of something fun for them to do together.

Furthermore, systems seek *equilibrium*, or stable balance and symmetry. Change in one of the parts sets in motion a process to restore equilibrium. For example, in the body system if one hand becomes disabled, the other must adjust to do the work of both. In family dynamics, this tendency toward equilibrium puts pressure on each member to retain his or her fairly predictable role. A changing family member is subtly encouraged to revert to her or his original behavior within the family system. For change to occur, the family system as a whole must change. Indeed, that is the goal of family therapy based on systems theory. The family may see one member as the problem, but if the psychologist draws the whole family into therapy, the family *system* should begin to change.

Social scientists have moved systems theory beyond its therapeutic origins to employ it in a more general analysis of families. They are especially interested in how family systems process information, deal with challenges, respond to crises, and regulate contact with the outside world. Researchers have elaborated and explored concepts such as *family boundaries* (ideas about who is in and who is outside the family system). This perspective also prompts researchers to investigate such things as *family boundary ambiguity*, wherein it is unclear who is in the family and who isn't. Stepfamilies have been researched from this point of view: Do children of divorced parents belong to two (or more) families? Are former spouses and their relatives part of the family (Boss 1997; Stewart 2005a)?

Contributions and Critiques of Family Systems Theory When working with families in therapy, this perspective has proven very useful. By understanding how their family system operates, individuals can make desired personal and/or family changes. Systems theory often gives family members insight into the effects of their behavior. It may make visible the hidden motivations behind certain family patterns. For example, doctors were puzzled by the fact that death rates were higher among kidney dialysis patients with *supportive* families. Family systems theorists attributed the higher rates to the unspoken desire of the patients to lift the burden of care from the close-knit family they loved (Reiss, Gonzalez, and Kramer 1986).

Envisioning the family as a system can be a creative perspective for research. Rather than seeing only the influence of parents on children, for example, system theorists are sensitized to the fact that this is not a one-way relationship and have explored children's influence on family dynamics (Crouter and Booth 2003).

A criticism of systems theory is that it does not take sufficient note of a family's economic opportunities, racial/ethnic and gender stratification, and other features of the larger society that influence internal family relations. When used by therapists, systems theory has been criticized as tending to diffuse responsibility for conflict by attributing dysfunction to the entire system, rather than to culpable family members within the system. This situation can lead to "blaming the victim," as well as making it difficult to extend social support to victimized family members while establishing legal accountability for others, as in situations involving incest or domestic violence (Stewart 1984).

Conflict and Feminist Theory

We like to think of families as beneficial for all members. For decades sociologists ignored the politics of gender and differentials of power and privilege within relationships and families. Beginning in the 1960s, conflict and feminist theorizing and activism began to change that oversight, as issues of latent conflict and inequality were brought into the open.

A first way of thinking about the **conflict perspective** is that it is the opposite of structure–functional theory. Not all of a family's practices are good; not all family behaviors contribute to family well-being. Family interaction can include domestic violence as well as holiday rituals—sometimes both on the same day.

Conflict theory calls attention to power—more specifically, unequal power. It explains behavior patterns such as the unequal division of household labor in terms of the distribution of power between husbands and wives. Because power within the family derives from power outside it, conflict theorists are keenly interested in the political and economic organization of the larger society.

The conflict perspective traces its intellectual roots to Karl Marx, who analyzed class conflict. Applied to the family by Marx's colleague Friedrich Engels (1942 [1884]), the conflict perspective attributed family and marital problems to class inequality in capitalist society.

In the 1960s, a renewed interest in Marxism sparked the application of the conflict perspective to families in a different way. Although Marx and Engels had focused on economic classes, the emerging feminist movement applied conflict theory to the sex/gender system—that is, to relationships and power differentials between men and women in the larger society and in the family.

Although there are many variations, the central focus of the **feminist theory** is on gender issues. A unifying theme is that male dominance in the culture, society, families, and relationships is oppressive to women. *Patriarchy,* the idea that males dominate females in virtually all cultures and societies, is a central concept (Hesse-Biber 2007).

Unlike the perspectives described earlier, which were developed primarily by scholars, feminist theories emerged from political and social movements over the past fifty years. As such, the mission of feminist theory is to use knowledge to actively confront and end the oppression of

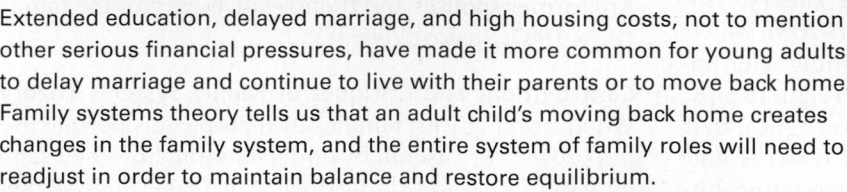

Extended education, delayed marriage, and high housing costs, not to mention other serious financial pressures, have made it more common for young adults to delay marriage and continue to live with their parents or to move back home. Family systems theory tells us that an adult child's moving back home creates changes in the family system, and the entire system of family roles will need to readjust in order to maintain balance and restore equilibrium.

© David Young-Wolff/PhotoEdit

women and related patterns of subordination based on social class, race/ethnicity, age, or sexual orientation.

The feminist perspective has contributed to political action regarding gender and race discrimination in wages; sexual harassment; divorce laws that disadvantage women; rape and other sexual and physical violence against women and children; and reproductive issues, such as abortion rights and the inclusion of contraception in health insurance. Feminist perspectives promote recognition of women's unpaid work; the greater involvement of men in housework and child care; efforts to fund quality day care and paid parental leaves; and transformations in family therapy so that counselors recognize the reality of gender inequality in family life and treat women's concerns with respect (Baker 2008; Diduck and O'Donovan 2006; Friedman and Valenti 2008; Gupta and Ash 2008; Koepke 2007; Mollen 2008). The feminist perspective has combined with the family life course development framework to analyze aging and gender issues (Ross-Sheriff 2008).

Since publication of Naomi Wolf's 1991 classic, *The Beauty Myth: How Images of Beauty Are Used Against Women*, feminist theory has given considerable attention to eating disorders and body image issues (Latham 2008). For example, a study that combines the feminist with the interaction–constructionist perspective investigated the process through which a young woman internalizes an identity as a "fat girl" and thereby "unfit" (Rice 2007). Feminist scholars also consider whether a decision to have cosmetic surgery evidences a woman's agency or the unrecognized influence of a patriarchal construct of feminine beauty (Tiefer 2008).

In recent years, feminist theory has embraced postmodern analyses, deconstructing formerly taken-for-granted concepts such as *gender dichotomy* (the idea that there are naturally two very distinct genders) or the idea that marriage must naturally be heterosexual (Dreger and Herndon 2009). Having co-opted a pejorative term from the popular culture, some feminists refer to this kind of analysis as *queer theory* (Eaklor 2008; Stacey 2006).[3] From the feminist perspective, championing the traditional heterosexual nuclear family at the cost of both heterosexual and lesbian women's equality and well-being is unconscionable (Harding 2007; Stacey 1993).

Contributions and Critiques of Feminist Theoretical Perspectives By calling attention to women's experiences, feminist theory has encouraged us to see things about relationships and family life that had been overlooked before the 1960s. Women's domestic work was largely

invisible in social science until the feminist perspective began to treat household labor as work that has economic value. The feminist perspective brought to light issues of wife abuse, marital rape, child abuse, and other forms of domestic violence.

According to some social scientists, feminist theory is too political, value-laden, or adversarial to be considered a valid academic approach (Landau 2008; Lloyd, Few, and Allen 2007). The concept of patriarchy has been criticized as being unscientifically vague and ahistorical. Posited to exist in virtually all societies, patriarchy loses meaning as an analytic category when it minimizes differences between America in the twenty-first century and ancient Rome, where husbands allegedly had life-and-death power over women. Moreover, inasmuch as some feminist theory embraces postmodernism, it is subject to the same criticisms as postmodernism, which were described above. "Feminist engagements with science have never been straightforward, and are less so all the time" (Valentine 2008, p. 355).

The Biosocial Perspective

The **biosocial perspective** is characterized by "concepts linking psychosocial factors to physiology, genetics, and evolution" (Booth, Carver, and Granger 2000, p. 1018). This perspective argues that human physiology, genetics, and hormones predispose individuals to certain behaviors (Bearman 2008). That is, biology interacts with the social environment to affect much of human behavior and, more specifically, many family-related behaviors (Booth et al. 2006). "[Q]uantitative genetic studies have increasingly… found major interplay between genetic and non-genetic [environmental] factors, such that the outcomes cannot sensibly be attributed just to one or the other, because they depend on both" (Rutter 2002, pp. 1–2).

According to the biosocial (or evolutionary psychology) perspective, much of contemporary human behavior evolved in ways that enable survival and continuation of the human species. Successful behavior patterns are encoded in the genes, and this evolutionary heritage is transmitted to succeeding generations.[4] The survival of one's genetic material into future generations is paramount. Hence human behavior has biologically evolved to be oriented to the survival and reproduction of all close kin who carry those genes (Dawkins 1976; Hamilton 1964).

Evolutionary explanations are offered for many contemporary family patterns. For instance, research suggests that children are more likely to be abused by nonbiologically related parents or caregivers than by biological parents.

---

[3] According to feminist scholar Judith Stacey, "[T]he new gender politics (featuring intersexuality, transgender, and transsexuality) in concert with ongoing, very powerful lesbi-gay initiatives have shifted the center of political and intellectual gravity and passion from feminist to queer theory and scholarship… however controversial this may be" (2006, p. 480).

[4] The biosocial perspective has its roots in Charles Darwin's *The Origin of Species* (1977 [1859]). Darwin proposed that species evolve according to the principle of *survival of the fittest*. Only the strongest, more intelligent, and adaptable members of a species survive to reproduce, a process whereby the entire species is strengthened and prospers over time.

Nonbiological parent figures are less likely to invest money and time in their children's development and future prospects (Case, Lin, and McLanahan 2000). The biosocial perspective explains this by arguing that parents "naturally" protect the carriers of their genetic material (Gelles and Lancaster 1987). Therefore, although he acknowledges that there are many successful stepfamilies and adoptions, sociologist David Popenoe (1994) finds these family forms to be unsupported by our evolutionary heritage. He concludes that "we as a society should be doing more to halt the growth of stepfamilies" (p. 21).

From its early days, some proponents from the biosocial perspective have held that certain human behaviors, because they evolved for the purpose of human survival, were both "natural" and difficult to change. It has been asserted, for example, that traditional gender roles evolved from patterns shared with our mammalian ancestors that were useful in early hunter–gatherer societies. Gender differences—males allegedly more aggressive than females, and mothers more likely than fathers to be primarily responsible for child care—are seen as anchored in hereditary biology (Rossi 1984; Udry 1994, 2000).

However, biosociologists emphasize that biological predisposition does *not* mean that a person's behavior cannot be influenced or changed by social structure (Bearman 2008). "Nature" (genetics, hormones) and "nurture" (culture and social relations), they argue, interact to produce human attitudes and behavior. As an example, research on testosterone levels in married couples found high levels of the husbands' testosterone to be associated with poorer marital quality when their role overload was high, but with better marital quality when role overload was low. In other words, "testosterone enables positive behavior in some instances and negative behavior in others" (Booth, Johnson, and Granger 2005, p. 483; see also Booth et al. 2006).

Contributions and Critiques of the Biosocial Perspective This perspective encourages scientists to investigate research questions regarding relationships and families

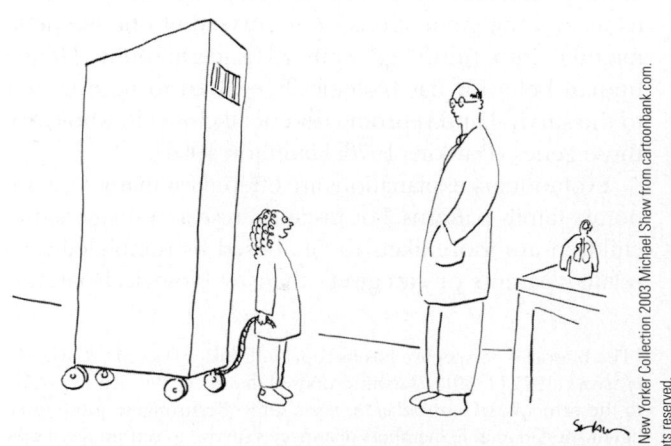

"The title of my science project is 'My Little Brother: Nature or Nurture.'"

that would otherwise be overlooked: Is there a genetic basis for human family and relationship behaviors and attitudes? If so, to what extent can those attitudes and behaviors be changed? To what degree do social forces (nurture) and biological predispositions (nature) interact to result in human behavior and attitudes?

Over the past twenty-five years, the biosocial perspective has emerged as a significant theoretical perspective on the family. Researchers have employed this point of view to examine such phenomena as gender differences, sexual bonding, mate selection, jealousy, parenting behaviors, marital stability, and male aggression against women (Bearman 2008; Booth et al. 2006; Muller and Wrangham 2009; Sagarin 2005; Salmon and Shackelford 2008).

However, this perspective was once used to justify gender inequality as biologically based and hence "natural." More recently, evolutionary perspectives have been the basis for criticism of nonreproductive sexual relationships and the employment of mothers as contrary to nature (Daly and Wilson 2000). It is therefore not surprising that many distrust this perspective or that it has been politically and academically controversial. We explore and appraise the biosocial approach, or evolutionary psychology, when discussing gender (Chapter 4), extramarital sex (Chapter 5), child care (Chapter 11), and children's well-being in stepfamilies (Chapter 16).

Attachment Theory

Counseling psychologists often analyze individuals' relationship choices in terms of attachment style. **Attachment theory** posits that during infancy and childhood a young person develops a general style of attaching to others (Ainsworth 1967; Ainsworth et al. 1978; Bowlby 1982, 1988; Bretherton 1992; Fletcher 2002). Once a youngster's attachment style is established, she or he unconsciously applies that style, or "state of mind," to later adult relationships.

A child's primary caretakers (usually parents and most often the mother) evoke a *style* of attachment in him or her. The three basic attachment styles are *secure, insecure/anxious,* and *avoidant.* Children who can trust that a caretaker will be there to attend to their practical and emotional needs develop a secure attachment style. Children who feel uncared for or abandoned develop either an insecure/anxious or an avoidant attachment style.

In adulthood, a secure attachment style involves trust that the relationship will provide ongoing emotional and social support (Fletcher 2002; Hazan and Shaver 1987, 1994). An insecure/anxious attachment style entails concern that the beloved will disappear, a situation often characterized as "fear of abandonment." Someone with an avoidant attachment style dodges emotional closeness either by avoiding relationships altogether or demonstrating ambivalence, seeming preoccupied, or otherwise establishing distance in intimate situations (Fletcher 2002; Rauer and Volling 2007).

Attachment theory has grown in importance and prominence in family studies over the past several decades. Some researchers combine this perspective with the family life course development framework to look at stability or variability of attachment styles throughout an individual's life (Klohnen and Bera 1998). Attachment theory is also used by counseling psychologists. The assumption is that if a client learns to recognize a problematic attachment style, he or she can change that style (Ravitz, Maunder, and McBride 2008; Weissman, Markowitz, and Klerman 2007).

Contributions and Critiques of Attachment Theory This perspective prompts us to look at how personality impacts relationship choices, from initiating to maintaining them. Attachment theory also encourages us to ask what kind of parenting best encourages a secure attachment style. These are important research questions. Critics argue that an attachment style might depend on the situation in which a person finds him or herself rather than on a consistent personality characteristic (Fleeson and

Noftle 2008). Of course, when therapists employ this point of view, they recognize that even if it is a relatively stable personality characteristic, one's attachment style can be changed over time.

The Relationship Between Theory and Research

Theory and research are closely integrated, ideally at least. Theory should be used to help direct research questions and to suggest useful concepts. Often when designing their research, scientists employ one or more theoretical perspectives from which to generate an *hypothesis* or "educated guess" about the way things are. Scientists then test these hypotheses by gathering data. At other times, to interpret data that has already been gathered, scientists ask themselves what theoretical perspective best explains the facts. Over time our understanding of family phenomena may change as social scientists undertake new research and modify theoretical perspectives. Even when theory is not directly spelled out in a study, it is likely that the research fits into one

© Martin Rodgers/Stock Boston

These folks are waiting for medical attention in a neighborhood clinic. How might scholars from different theoretical orientations see this photograph? *Family ecologists* might remark on the quality of the facilities—or speculate about the family's home and neighborhood—and how these factors affect family health and relations. They might compare this crowded and understaffed clinic for the poor with the better equipped and staffed doctors' offices that provide health care to wealthier Americans. Scholars from the *family life course development framework* would likely note that this woman is in the child-rearing stage of the family life cycle. *Structure–functionalists* would be quick to note the child-raising (and, perhaps, expressive) function(s) that this woman is performing for society. *Interaction–constructionists* might explore the mother's body language: What is she saying nonverbally to the child on her lap? What might he symbolize to her? *Exchange theorists* might speculate about this woman's personal power and resources relative to others in her family. *Family system theorists* might point out that this mother and child are part of a family system: Should one person leave or become seriously and chronically ill, for example, the roles and relationships among all members of the family would change and adapt as a result. *Feminist theorists* might point out that typically it is mothers, not fathers, who are primarily responsible for their children's health—and ask why. The answer from a *biosocial perspective* might be that women have evolved a stronger nurturing capacity that is partly hormonally based. *Attachment theorists* might surmise whether the mother is interacting with her child in a way that promotes a secure, insecure/anxious, or avoidant attachment style. How would you interpret this photo?

Research topics can be studied from different points of view. Here we see how the topic, religion and family life, has been investigated with different theoretical perspectives and by the use of various research methods.

- **The Family Ecology Perspective** Loser et al. (2008) conducted qualitative interviews with highly religious parents and children in sixty-seven families who belonged to the Church of Jesus Christ of Latter-day Saints. The researchers concluded that religion is often personally internalized and should *not* be understood only as a component in a family's sociocultural environment.

- **The Family Life Course Development Framework** Pearce (2002) analyzed longitudinal data from a Detroit survey of white mothers and children to find that early childhood religious exposure later influenced childbearing attitudes during transition to adulthood. Young adults with Catholic mothers or mothers who frequently attended religious services were especially likely to resist the idea of not having children.

- **The Structure–Functional Perspective** Schottenbauer, Spernak, and Hellstron (2007) found that parents' use and modeling of religiously based coping skills, along with family attendance at religious or spiritual programs, was functional in enhancing children's health, social skills, and overall behavior.

- **The Interaction–Constructionist Perspective** Hirsch (2008) used naturalistic observation to understand how young, actively Catholic women in Mexico creatively interpret their religion's proscription against birth control while choosing to use it. As one "grassroots theologian" explained, "[E]ven in the bible it says 'help yourself, so I can help you'; even the priests tell you that" (pp. 98–99).

- **Exchange/Network Theory** Christian Smith (2003) used secondary analysis of the national Survey of Parents and Youth to find that participation in religious congregations increases the likelihood that family members will benefit from sharing a network which includes parents, their children, their children's friends and teachers, and their children's friends' parents.

- **Family Systems Theory** Lambert and Dollahite (2008) conducted qualitative research with fifty-seven religious couples and concluded that these respondents saw God as a third partner in an otherwise dyadic family system—a third system member whose presence enhanced their marital commitment.

- **Feminist Theory** Feminist social historians Carr and Van Leuven (1996) edited a cross-cultural anthology whose works examine the implications of religion for female family members of various religious cultures. Overall, the book argues that women's oppression originates not in religion itself but in the exploitation of religion as a subjugation tool by patriarchal religio-cultural systems.

- **The Biosocial Perspective** Wright (1994) argued that humans have evolved as "the moral animal," a situation that facilitates our species' cooperation toward the goal of survival.

- **Attachment Theory** Reinert (2005) surveyed seventy-five Catholic seminarians, presenting them with "Awareness of God" and "Attachment to Mother" scales. He found that a seminarian's early childhood attachment to his mother to be a key influence in the degree of his attachment to a personal God.

Critical Thinking

Think of a family-related topic and consider how you might study it. What theoretical perspective would you use to help frame your research questions? What research methods and data-gathering techniques would you use?

---

or more of the theoretical perspectives described above. "Facts About Families: How Family Researchers Study Religion from Various Theoretical Perspectives" illustrates ways that researchers have used these theoretical perspectives when studying the broad topic of religion and family. We'll turn our attention now to various methods that researchers use to gather information, or data, on family life.

Doing Family Research

Students take entire courses on research methods, and obviously we can't cover the details of such methods here. However, we do want to explore some major

principles so that you can think critically about published research discussed in this text or in the popular press. As the subtitle of one textbook says, research methods provide "a tool for life" (Beins 2008). We invite you to think this way as well.

Designing a Scientific Study: Some Basic Principles

At the onset of a scientific study, researchers carefully design a detailed research plan. Some research is designed to gather *historical data*. Hanawalt (1986) constructed a rich picture of the medieval family through an examination of death records. By examining old diaries, Hanawalt challenged the assumption that preindustrial family life lacked emotional intimacy (Osment 2001). Historical studies of marriage and divorce in the United States portray a picture of the past that, contrary to common belief, was not necessarily stable or harmonious (Cott 2000; Hartog 2000). Although family history is an important area of research, this textbook does not allow space for us to fully explore it.

Research designs can also be *cross-cultural*, comparing one or more aspects of family life among different societies. A study described in Chapter 6, which asked students in ten different countries whether it's necessary to be in love with the person you marry, is an example of cross-cultural research (Levine et al. 1995).

Scientists consider many questions when designing their research: Will the study be cross-sectional or longitudinal? Deductive, inductive, or a combination of the two? Will the study be mainly quantitative or qualitative? Will the sample be random and the data generalizable? Because the goal of all research is to transcend our personal blinders or biases, as was discussed at the beginning of this chapter, scientists must meticulously define their terms and take care not to overgeneralize. This section looks briefly at these considerations.

Cross-Sectional Versus Longitudinal Data When designing a study, researchers must decide whether to gather cross-sectional or longitudinal data. *Cross-sectional studies* gather data just once, giving us a snapshot-like, one-time view of behaviors or attitudes. *Longitudinal studies* provide long-term information as researchers continue to gather data over an extended period of time.

For example, to understand how psychotherapy may modify attachment insecurity over the life course, researchers designed longitudinal studies that followed respondents' attachment styles over thirty years from childhood into adulthood (Klohnen and Bera 1998). Other researchers monitored nonresident fathers' involvement with their children for three years and found finances and relations with their children's mothers to be significant causes for changes over time (Ryan, Kalil, and Ziol-Guest 2008).[5]

Deductive Versus Inductive Reasoning *Deductive reasoning* in research begins with an hypothesis that has been derived (i.e., deduced) from a theoretical point of view. "Reasoning down" from the abstract to the concrete, a researcher designs a data-gathering strategy in order to test whether the hypothesis can be supported by observed facts. Researchers who use *inductive reasoning* observe detailed facts, then induce, or "reason up," to arrive at generalizations grounded in the observed data. Inductive studies do not begin with a preconceived hypothesis. Instead, researchers begin their observations with open minds about what they'll see and find. Typically, deductive reasoning is associated with quantitative research and inductive reasoning with qualitative research.

Quantitative Versus Qualitative Research In *quantitative research*, the scientist gathers, analyzes, and reports data that can be quantified or understood in numbers. Quantitative research finds numerical incidences in a population—for example, the average size of a family household or what percent of Americans are currently cohabiting. Statistical facts and findings, such as those in Chapter 1's "Facts About Families" boxes, are examples of quantitative data. Quantitative research also uses computer-assisted statistical analysis to test for relationships between phenomena. For instance, quantitative analysis has found a statistically demonstrated correlation between being raised by a single parent and teen pregnancy (Albrecht and Teachman 2003).

When performing *qualitative research*, the scientist gathers, analyzes, and reports data primarily in words or stories. For example, social scientists interviewed eight mothers in three rural trailer parks who described their lives in detail (Notter, MacTavish, and Shamah 2008). In their subsequently published article, the researchers quote the women in their own words:

> I pay attention to how my mother raised me and I try to do it different. I try to teach him [her son] how to take care of himself. He knows how to do chores and how to cook. I had to learn all of that on my own. I try to teach him how to state his opinion. My mother never taught me to do that... (p. 619)

As a second example, sociologist Gina Miranda Samuels (2009) conducted qualitative research with black adults raised by white parents. Samuels's findings are reported in narrative using respondents' own words (p. 89):

> But I remember there was one girl named Ebony and she could not BELIEVE I had been adopted by white people. She was like, "WOW! You were adopted by white people?! Are they nice to you? Do they treat you well?" And that was a shock to me because that was the first time I realized that black people might not get treated well by white people.... (Justine, 28)

> I was in my salon and I didn't even know what a hot comb was. That was my giveaway! And he [the stylist] was like, "Were you raised by white people?" And then, he was like, "OH. I was able to tell that by the way you talked and by the way you carried yourself." (Crystal, 24)

The aim of qualitative research is to gain in-depth understandings of people's experiences, as well as the

---

[5] A difficulty encountered in longitudinal studies, besides cost, is the frequent loss of subjects due to death, relocation, or their loss of interest. Social change occurring over a long period of time may make it difficult to ascertain what, precisely, has influenced family change. Yet cross-sectional data (one-time comparison of different groups) cannot show change in the same individuals over time.

processes they go through when defining, adapting to, and making decisions about their situations. Qualitative research typically employs the interaction–constructionist theoretical perspective, described earlier in this chapter.

When designing studies, researchers must also carefully define their terms: What precisely is being studied and how exactly will it be measured (Bickman and Rog 2009)?

Defining Terms Researchers scrupulously define the concepts that they intend to investigate, then report those definitions in their published studies so that readers know precisely what was investigated and how. For example, researchers once considered all (heterosexual) cohabitators as fitting one general definition. They found cohabitation before marriage to be statistically related to divorce later (Dush, Cohan, and Amato 2003). However, as definitions of cohabitation were further refined to differentiate serial from one-time cohabitators, results began to show that cohabitating only with your future marriage partner was *not* likely to end in divorce (Lichter and Qian 2008; Teachman 2003).

Researchers need for respondents to know exactly what they mean when answering questions. As discussed in Chapter 1, people define *family* in a wide variety of ways. "Researchers need to be aware that what they are trying to measure regarding family may not be what participants are thinking about in their responses" (Weigel 2008, p. 1443; also see Harris 2008). A solution might be to avoid designing questions with the term "family" and instead to use more specific language such as "spouse," "biological sister," "same-sex partner," or "stepbrother."

Samples and Generalization You may have noticed that in the study of women in trailer parks mentioned above, the researchers interviewed only eight mothers. We cannot expect the situations of these few respondents to correspond with all American women living in trailer parks. For one thing, each of the eight mothers was white (Notter, MacTavish, and Shamah 2008). We cannot possibly conclude from this research that all women who live in trailer parks are white. Rather than a nation-wide demographic portrayal, the purpose of interviewing these eight women was to learn about the experiences and processes that mothers can go through when residing in trailer parks.

To gather data that can be *generalized* (applied to a population of people other than those directly questioned), a researcher must draw a sample that accurately reflects, or represents, that population in important characteristics such as age, race, gender, and marital status. Results from a survey in which all respondents are college students, for example, cannot be interpreted as representative of Americans in general.

Gallup polls are examples of research that uses *representative samples* which reflect the national adult population. When a Gallup poll reports that most Americans would be unwilling to forgive an unfaithful spouse, we know that the findings from their sample can be generalized to the whole national population with only a small probability of error (Jones 2008). In order to draw a representative sample of a population, everyone in that population must have an equal chance to be selected. The best way to accomplish this is to have a list of every individual in the population and then randomly choose from the list (see Babbie 2009). A national *random sample* of approximately 1,500 people may validly represent the U.S. population.

Sometimes there are no complete lists of members of a population. For instance, researchers were interested in the ramifications of living with a compulsive hoarder (one who continuously acquires yet fails to discard large numbers of possessions). They located 665 respondents who reported having a family member or friend with hoarding behaviors (Tolin et al. 2008). How did they accomplish this? The researchers had made several national media appearances about hoarding. As a result, over 8,000 people had contacted them for guidance or information. Drawing from this group, the researchers e-mailed potential participants, inviting them to take part in the study and asking them to forward the invitation to others in similar situations.

Ultimately these researchers found that living with the clutter associated with hoarding often causes depression and isolation, partly because one is embarrassed to invite friends home. But although the findings were based on a fairly large sample, they cannot be generalized to all people who live with a compulsive hoarder because the sample was not random: Not everyone who lives with a compulsive hoarder had the same chance of being chosen for the study. It is reasonable to argue that those who contacted the researchers for guidance or information were more distraught than those who didn't. As a result, the findings may show greater difficulty in living with a compulsive hoarder than is generally experienced by all those in this population. Nevertheless, this is valuable research inasmuch as it lends insight into what living with a compulsive hoarder entails, at least for many.

There are many occasions when it is impossible to find a random sample for the topic one wants to investigate. The study of blacks raised by whites, discussed above, provides a second case in point. In this instance, the researcher recruited volunteer respondents by using web-based and print advertisements to African American and multiracial adoption agencies (Samuels 2009). In general, volunteers do not result in a random sample.

In addition to these considerations, designing a study involves decisions about the techniques by which the data will be collected, or gathered.

Data Collection Techniques

We will refer to various **data collection techniques**—interviews and questionnaires, naturalistic observation, focus groups, experiments and laboratory observation, case studies—throughout this text, so we will briefly describe them here. Each technique has strengths and weaknesses. However, the strengths of one technique can compensate for the weaknesses of another. To get around the drawbacks of a given technique, researchers may combine two or more in one study (Clark et al. 2008).

Interviews and Questionnaires The most common data-gathering technique in family research involves personally interviewing respondents or asking them to complete self-report questionnaires about their attitudes and past or present behaviors. When conducting interviews, researchers ask questions in person or by telephone. Gallup polls use telephone surveys. Alternatively, a researcher may distribute paper-and-pencil or web-based questionnaires that respondents complete by themselves. Increasingly viewed as comparable in reliability to paper-and-pencil questionnaires, Internet surveys use e-mail or web-based formats (Coles, Cook, and Blake 2007). Examples of the latter can be found at Surveymonkey.com, which facilitates the design, distribution, and some analysis of online surveys.[6]

Questions can be *structured* (or *closed-ended*). After a statement such as "I like to go places with my partner," the respondent chooses from a set of fixed answers, such as "always," "usually," "sometimes," or "never." Researchers spend much time and energy on the wording of such questions because they want all respondents to interpret them in the same way. Also, word choice can influence responses. For example, respondents tend to be more favorable to the phrase "assistance to the poor" than they are to the term "welfare" (Babbie 2007, p. 251).

A *survey* is a quantitative data-gathering tool that comprises a series of structured, or closed-ended, questions. Once completed, survey responses are tallied and analyzed, usually with computerized coding and statistical programs. Sometimes surveys incorporate previously published *scales*, or sets of closed-ended questions on the same topic. There are many scales to be found in the literature.

The hoarding study described in this chapter used the Hoarding Rating Scale (Tolin et al. 2007).[7]

Sociologists often engage in *secondary analysis* of large data sets—the result of fairly comprehensive surveys administered to a national representative sample. Once completed and tallied, the responses are made public, often via the Internet, so that other researchers (who had nothing to do with designing the questions) can analyze the data. A myriad of data sets are available for secondary analysis. As just one example, the National Survey of Family Growth (NSFG) contains data from a national sample of nearly 11,000 U.S. women between the ages of fifteen and forty-four. To investigate the breadwinner role cross-culturally, researchers analyzed data previously collected in the International Social Survey, conducted annually in over twenty countries (Yodanis and Lauer 2007). The longitudinal national survey, "Marital Instability Over the Life Course," whose findings are described periodically throughout this text, is also available for secondary analysis (Amato and Booth 1997; Hawkins and Booth 2005).[8] Many studies that use secondary analysis on other large data sets are described throughout this textbook.

Surveys are an efficient way to gather data from large numbers of people. Different respondents' standardized answers to structured questions can be readily compared. However, because they allow only predetermined answers to standardized questions, surveys may miss points that respondents would consider important but cannot report. For this reason, some researchers ask unstructured questions.

Unstructured (also called *open-ended*) questions do not offer a limited, preset range of answers. Instead, the purpose is to allow the respondent to talk freely. Interviewers using open-ended questions learn to listen, then probe for more detail. Samuels's study of blacks adopted by whites used unstructured, open-ended questions:

> I began the interviews by asking participants to share their adoption stories, including what they knew about their birth families. Participants described their childhood communities, how they were raised to think about their racial heritages and adoptions, and if their insights or identities changed as they became adults. (2009, p. 84)

Questioning respondents—whether quantitatively or qualitatively, whether interviewing or using a self-report

---

[6] Self-report questionnaires are less time consuming and costly for researchers than are face-to-face or telephone interviews. One problem is that the respondent may quit before finishing the survey or fail to return it. Furthermore, questionnaires cannot be successfully administered to respondents who do not read or who do not understand the language in which the questionnaire was written. Online surveys must be password protected so that only those in the sample can respond, and researchers have to watch for duplicate responses. Questionnaires are inappropriate for young children, of course. Face-to-face interviews and focus groups offer alternatives in studies involving young children (Freeman and Mathison 2009).

[7] Further examples include scales that measure, among other things, self-esteem, body image, romantic attachment, or children's exposure to domestic violence (Edleson et al. 2007; Hazan and Shaver 1987; Souto and Garcia 2002).

[8] As part of their study design, researchers who use secondary analysis determine how they might analyze answers to already conceived questions in order to yield new information and insights. A drawback to secondary analysis is that researchers, because they do not design their own questions, can investigate only topics and details about which survey questions have been designed by others.

questionnaire format—is the most common data-gathering technique used by family researchers. Nevertheless, there are limitations. Valid responses depend on participants' honesty, motivation, and ability to respond. Some individuals—for example, people who suffer from the advanced stages of Alzheimer's disease—are not appropriate subjects for questioning.

Another disadvantage associated with questioning respondents about prior events involves the human tendency to forget the past or to reinterpret what happened in the past—a situation remedied by longitudinal studies. Furthermore, respondents may say what they think they *should* say. If asked whether they spank their children, for example, parents who do might be reluctant to say so. One way to get around this problem is to observe behavior directly.

Naturalistic Observation In **naturalistic observation** (also called "participant observation" or "field research"), the researcher spends extensive time with respondents and carefully records their activities, conversations, gestures, and other aspects of everyday life. This data-gathering technique often accompanies the interaction–constructionist theoretical perspective. The researcher attempts to discern family relationships and communication patterns and to draw implications and conclusions for understanding family behavior in general. The study of women living in trailer parks employed naturalistic observation as well as interviews. Over a period of sixteen months, researchers spent twelve to twenty hours in each woman's home, sometimes taking part in family meals (Notter, MacTavish, and Shamah 2008).

The principal advantage of naturalistic observation is that it allows one to view family behavior as it actually happens in its own natural—as opposed to artificial—setting. The most significant disadvantage is that what is recorded and later analyzed depends on what one or a few observers think is significant. Another drawback is that naturalistic observation requires enormous amounts of time to observe only a few families who cannot be assumed to be representative of family living in general. Moreover, not all research topics lend themselves to naturalistic observation. Explaining her decision to interview rather than directly observe blacks raised by whites, Samuels points out that "[t]here are not 'sides of town' or neighborhoods where multiracials or transracially adopted families and individuals reside, in which a researcher can become immersed, gain access, and conduct naturalistic inquiry" (2009, p. 84).

Focus Groups A focus group is a form of qualitative research in which, in a group setting, a researcher asks a gathering of ten to twenty people about their attitudes or experiences regarding a situation. Researchers have used focus groups to explore how parents feel about their overweight children, for example (Jones et al. 2009). Participants are free to talk with each other as well as to the researcher or group leader. Focus group sessions last between one and two hours and are typically electronically recorded, then transcribed or entered into a computer for later analysis. The study of Chinese and Korean immigrants, discussed earlier in this chapter, is based on data collected in eight San Francisco area focus groups (Wong, Yoo, and Stewart 2006).

Group discussion produces data and insights that would be less forthcoming otherwise. Also, researchers can capture the participants' everyday speech in order to better understand their life situations—and perhaps to include some of their language in subsequent interviews or questionnaires. Focus groups are useful when researchers do not feel they know enough about a topic to design a set of closed-ended questions. Focus groups can also be successful when working with children (Clark 2009; Freeman and Mathison 2009).

However, there are disadvantages to the method. Researchers have less control in a group setting than they do in a one-on-one interview, and hence focus groups can be time consuming, given the amount of usable data recorded. Then too, data can be difficult to analyze because focus group conversation is casual and flows in response to others' comments. Furthermore, the researcher can easily influence responses by inadvertent comments that lead respondents to say things that they may not really mean.

Experiments and Laboratory Observation In an experiment or in laboratory observation, behaviors are carefully monitored or measured under controlled conditions. In an **experiment**, subjects from a pool of similar participants are randomly assigned to groups (*experimental* and *control groups*) that are then subjected to different experiences (*treatments*). For example, families with a child who is undergoing a bone marrow transplant may be asked to participate in an experiment to determine how they may best be helped to cope with the situation. One group of families may be assigned to a support group in which the expression of feelings, even negative ones, is encouraged (the experimental group). A second group may receive no special intervention (the control group). If at the conclusion of the experiment the groups differ according to measures of coping behavior, the outcome is presumed to be a result of the experimental treatment. Because no other differences are presumed to exist among the randomly assigned groups, the results of the experiment provide evidence of the effects of therapeutic intervention.

The experiment just described takes place in a field (real-life) setting, but experiments are often conducted in a laboratory setting because researchers have more control over what will happen. The laboratory setting allows the researcher to plan activities, measure results, determine who is involved, and eliminate outside influences. A true experiment incorporates the features of random assignment and experimental manipulation of the important variable (the treatment).

Laboratory observation, on the other hand, simply means that behavior is observed in a laboratory setting, but it does not involve random assignment or experimental manipulation of a variable. Family members may be asked to discuss a hypothetical problem or to play a game while their behavior is observed and recorded. Later those data can be analyzed to assess the family's interaction style and the nature of their relationships. Laboratory methods are useful in measuring physiological changes associated with anger, fear, sexual response, or behavior that is difficult to report verbally. In the 1970s social psychologist John Gottman began studying newly married couples in a university lab while they talked casually or wrestled with a problem. Video cameras recorded the spouses' gestures, facial expressions, and verbal pitch and tone. Some couples volunteered to let researchers monitor shifts in their heart rates and chemical stress indicators in their blood or urine as a result of their communicating with each other (Gottman 1996). Gottman's findings are discussed in detail in Chapter 12.

Laboratory observation has advantages and disadvantages. An advantage is that social scientists can watch human behavior directly, rather than depending on what respondents *tell* them. A disadvantage is that the behaviors being observed often take place in an artificial situation, and whether an artificial situation is analogous to real life is debatable. A couple asked to solve a hypothetical problem through group discussion may behave differently in a research laboratory than they would at home.

Clinicians' Case Studies We also obtain information about families from *case studies* compiled by clinicians—psychologists, psychiatrists, marriage counselors, and social workers—who counsel people with marital and family problems. As they see individuals, couples, or whole families over a period of time, these counselors become acquainted with communication patterns and other interactions within families. Clinicians offer us knowledge about family behavior and attitudes by describing cases or reporting conclusions based on a series of cases.

The advantages of case studies are the vivid detail and realistic flavor that enable us to experience vicariously the family life of others. Clinicians' insights can provide hypotheses for further research. However, case studies have important weaknesses. There is always a subjective or personal element in the way the clinician views the family. Inevitably, any one person has a limited viewpoint. Clinicians' professional training may lead them to misinterpret aspects of family life. Psychiatrists, for example, used to assume that the career interests of women were abnormal and caused the development of marital and sexual problems (Chesler 2005 [1972]).

Furthermore, people who present themselves for counseling may differ in important ways from those who do not. Most obviously, they may have more problems. For example, throughout the 1950s psychiatrists

A research team plans data collection and analysis for a survey of how families spend their time together.

reported that gays in therapy had many emotional difficulties. Subsequent studies of gay males not in therapy concluded that gays were no more likely to have mental health problems than were heterosexuals (American Psychological Association 2007). Ideally, a number of scientists examine one topic by using several different methods. The scientific conclusions in this text are the results of many studies from various and complementary research tools. Despite the drawbacks and occasional blinders, the total body of information available from sociological, psychological, and counseling literature provides a reasonably accurate portrayal of marriage and family life today. Although imperfect, the methods of scientific inquiry bring us better knowledge of the family than does either personal experience or speculation based on media images.

The Ethics of Research on Families

Exploring the lives of families, the way social science researchers do, carries responsibility. Researchers must do nothing that would negatively impact respondents, a principle summarized as "do no harm." Researchers also must show respect to those being studied, and take into consideration the needs of their respondents. Feminist theorists in particular argue that researchers should be attuned to how their findings might help their respondents as well (McGraw, Zvonkovic, and Walker 2000). To help accomplish these standards, most research plans must be reviewed by a board of experts and community representatives called an *institutional review board (IRB)*. No federally funded research can proceed without an IRB review, and most institutions require one for all research on human subjects (Cohen 2007a).

The IRB scrutinizes each research proposal for adherence to professional ethical standards for the protection

of human subjects. These standards include *informed consent* (the research participants must be apprised of the nature of the research and then give their consent); lack of coercion; protection from harm; confidentiality of data and identities; the possibility of compensation of participants for their time, risk, and expenses; and the possibility of eventually sharing research results with participants and other appropriate audiences. Other than ensuring that the research is scientifically sound enough to merit the participation of human subjects, IRBs do not focus on evaluating the research topic or methodology.

Summary

- Scientific investigation—with its ideals of objectivity, cumulative results, and various methodological techniques for gathering empirical data—is designed to provide an effective and accurate way of gathering knowledge about the family.

- Different theoretical perspectives—family ecology, the family life course development framework, structure–functional, interaction–constructionist, exchange, family systems, feminist, biosocial, and attachment theory—illuminate various features of families and provide a foundation for research.

- Research designs can be historical, cross-cultural, longitudinal, qualitative or quantitative, and inductive or deductive.

- Data collection techniques include various ways of questioning respondents, focus groups, laboratory observation and experiments, naturalistic observation, and clinicians' case studies.

- Researchers need to be guided by professional standards and ethical principles of respect for research participants.

Questions for Review and Reflection

1. Choose one of the theoretical perspectives on the family, and discuss how you might use it to understand something about life in *your* family.

2. Choose a magazine photo and analyze its content from one of the perspectives described in this chapter. Then analyze the photo from another theoretical perspective. How do your insights differ depending on which theoretical perspective is used?

3. Discuss why science is often considered a better way to gain knowledge than is personal experience alone. When might this not be the case?

4. Think of a research topic, then review the data-gathering techniques described in this chapter to decide which of these you might use to investigate your topic.

5. **Policy Question.** What aspect of family life would it be helpful for policy makers to know more about as they make law and design social programs? How might this topic be researched? Is it controversial?

Key Terms

Online Resources

Sociology CourseMate

www.CengageBrain.com

Access an integrated eBook, chapter-specific interactive learning tools, including flashcards, quizzes, videos, and more in your Sociology CourseMate, accessed through CengageBrain.com.

www.CengageBrain.com

Want to maximize your online study time? Take this easy-to-use study system's diagnostic pre-test, and it will create a personalized study plan for you. By helping you identify the topics that you need to understand better and then directing you to valuable online resources, it can speed up your chapter review. CengageNOW even provides a post-test so you can confirm that you are ready for an exam.

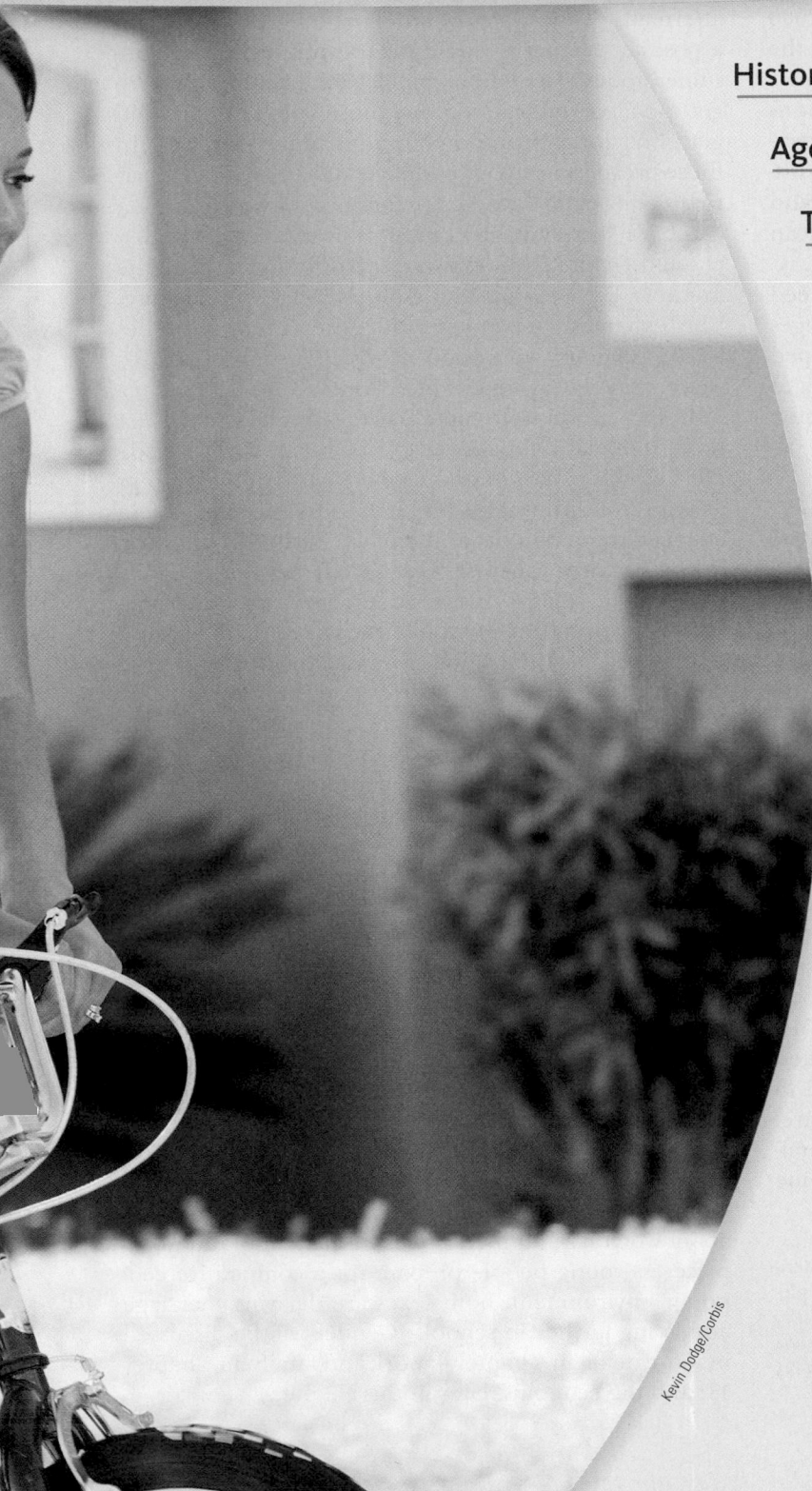

American Families in Social Context

Kevin Dodge/Corbis

Families meet important needs for their members and for society. Family members make commitments to one another and share an identity. Yet Chapter 1 showed us that families are not all alike in form. We also saw in Chapter 1 that social factors influence our personal options and choices. Put another way, individuals and families vary as a result of the social settings in which they exist.

We report U.S. Census Bureau data from the American Community Survey and from Current Population Reports, national surveys that supplement the 2000 decennial census. In this chapter, we will explore the social contexts in which today's families live out their opportunities and decisions. We'll examine variations in family life associated with race/ethnicity, immigration, religion, and the events of our nation's recent history. We'll look at how the economy affects families and at the impact of the changing age structure of American society.

This chapter focuses on U.S. society, but we need to point out that economic and technological changes affect families in other societies in both the developed and developing worlds. We begin here with a look at how historical events in the United States have affected family life.

Historical Events

Historical events and conditions affect options, decisions, and the everyday lives of families. In the early twentieth century,[1] for example, the shift from an agricultural to an industrial economy brought people from farm to city. There was a "great migration" of rural southern African Americans to the urban north.

American family life is experienced differently by people living through the Great Depression, World War II, the optimistic fifties, the tumultuous sixties, the economically constricted seventies and eighties, the time-crunched nineties, war and the threat of terrorism throughout the 2000s, and the continuation of a globalized economy. During the years of the Great Depression, couples delayed marriage and parenthood and had fewer children than they wanted (Elder 1974). During World War II, married couples were separated for long periods. Married women were encouraged to get defense jobs and to place their children in day care. Some husbands and fathers were casualties of war. Families in certain nationality groups—Japanese and some

Italians—were sent to internment camps and had their property seized even though most were U.S. citizens or long-term residents (Mercier n.d.; Taylor 2002c; Tonelli 2004).

The end of World War II was followed by a spurt in the divorce rate, when hastily contracted wartime marriages proved to be mistakes or extended separation led couples to grow apart. World War II was also followed by an uptick in marriages and childbearing. In the 1950s, family life was not overshadowed by national crisis. The aftermath of the war saw an expanding economy and a postwar prosperity based on the production of consumer goods. The GI bill enabled returning soldiers to get a college education, and the less educated could get good jobs in automobile and other factories. In those prosperous times, people could afford to get married young and have larger families. Most white men earned a "family wage" (enough to support a family), and most white children were cared for by stay-at-home mothers. Divorce rates slowed their long-term increase. The expanding economy and government subsidies for housing and education provided a sound basis for white, middle-class family life (Coontz 1992).

It is important to note here, however, that even though the GI bill was available to returning black soldiers as well as to whites, many colleges did not accept African Americans, and one had to be accepted into a college program in order to qualify for the GI bill's college assistance. Likewise, the GI bill did not officially discriminate against African American desire for home-ownership, but the bill was of little use to them because of the many neighborhood covenants against black residents (Binki, Eitelberg, Schexnider, and Smith 1982; Reed and Strum 2008).

The large baby boom cohort, born after World War II (1946–1964), has had a powerful impact on American society, giving us the cultural and sexual revolutions of "the sixties" as they moved from adolescence to young adulthood in the Vietnam War era. Baby boomers are now reshaping both middle age and aging as they move into their senior years (Pew Research Center 2005).

The baby boomers had a relatively secure childhood in both psychological and economic terms. The generations that followed have encountered a more challenging economic and family environment (Bengston, Biblarz, and Roberts 2002). Today, a man is far less likely to earn a family wage. Partly for that reason, more wives seek employment, including mothers of infants and preschool children. Moreover, the feminist movement opened opportunities for women and changed ideas about women's and men's roles in the family and workplace. As young people prepare for a competitive economic environment, both sexes are delaying marriage and going farther in school. (Gender will be discussed in more detail in Chapter 4; work and family, in Chapter 11; and the economy, later in this chapter.)

[1] American families have a rich history prior to the twentieth century. The limited history section of this chapter cannot do justice to American families' experiences in these various historical eras. As supplements, we especially recommend *Domestic Revolutions: A Social History of American Family Life* (Mintz and Kellogg 1988); *Historical Influences on Lives and Aging* (Schaie and Elder 2005); and *Huck's Raft: A History of American Childhood* (Mintz 2004).

In the 1960s and 1970s, marriage rates declined and divorce rates increased dramatically—perhaps in response to a declining job market for working-class men, the increased economic independence of women, and the cultural revolution of the sixties, which encouraged more individualistic perspectives. These trends, as well as the sexual revolution, contributed to a dramatic rise in nonmarital births.

The most recent historical moments focus on adaptation—adapting to a globalized economy, insecurity post-September 11, 2001, and new or continued overseas wars. We are also living with change inasmuch as the people of the United States elected a biracial man to the presidency. Although his election does not suggest that racism and ethnic tensions are no longer an issue, it does point to increased racial tolerance among Americans. President Obama's election played an important role in the increase of black optimism about the future. A CNN poll found that "53 percent . . . [of African Americans] said that life for blacks in the future will be better than it is now. Two years ago, the number was 44 percent" ("Poll" 2010).

Of course, the family has faced the necessity of adapting to demographic, social, economic, and political change throughout its history. Families in the past have also coped with war. "Facts about Families: Military Families" on pages 58–59 focuses attention on those families whose lives are structured by war and military service today.

Age Structure

Historical change involves not only specific events but also the basic facts of human life. One of the most dramatic developments of the twentieth century was the increased longevity of our population. Life expectancy in 1900 was forty-seven years, but an American child born in 2006 is expected to live to seventy-eight (Heron et al. 2009). Racial inequalities remain serious obstacles to the life expectancy of people of color in modern American society. For example, in 2006 the "risk of death for the black population was about 30 percent higher than for the white population" (Heron et al. 2009, p. 4). Asian Americans and Native Americans have seen their life expectancy improve over the past few years; but researchers are still unable to pinpoint life expectancy for Hispanics due to issues related to reporting (Heron et al. 2009).

Aging itself has changed; the years that have been added to our lives have been healthy and active ones (Bergman 2006c). Survival to older ages has meant that men and women over sixty-five are now more likely to be living with spouses than in the past. For those without spouses, maintaining an independent residence has become more feasible economically and in terms of health.

Among the positive consequences of increased longevity are more years invested in education, longer marriages for those who do not divorce, a longer period during which parents and children interact as adults, and a long retirement during which family activities and other interests may be pursued or second careers launched. More of us will have longer relationships with grandparents, and some will know their great-grandparents (Rosenbloom 2006).

At the same time, the increasing numbers of elderly people must be cared for by a smaller group of middle-aged and young adults. Moreover, divorce and remarriage may change family relationships in ways that affect the willingness of adult children to care for their parents (Bergman 2006c). The impact of a growing proportion of elderly will also be felt economically. As the ratio of retired elderly to working-age people grows, so will the problem of funding Social Security and Medicare (Topoleski 2009). With the current economic downturn, multiple family generations often share a household.

At the other end of the age structure, the declining proportion of children is likely to affect social policy support for families raising children. Fewer children may mean less attention and fewer resources devoted to their needs in a society under pressure to provide care for the elderly: "The percentage of American households with children has dropped . . . by 2010, households with children will account for little more than one-quarter of all households—the lowest share in the nation's history . . . the child-dense neighborhood is disappearing in many places. Suburbs . . . are now filling with empty-nesters. And many affluent empty-nesters are abandoning the tree-shaded streets of suburbia for the neon-lit excitement of the city" (Whitehead and Popenoe 2008, p. 20).

In the foregoing discussion of history and of the age structure of our population, there is an underlying theme: the economy. Economic opportunities, resources, and obligations are an important aspect of the American society in which families are embedded. We turn now to a more detailed discussion of the economic foundation of the contemporary family.

The Economy and Social Class

We have been encouraged to think of the United States as a classless society. Yet **life chances**—the opportunities one has for education and work, whether one can afford to marry, the schools that children attend, and a family's health care—all depend on family economic resources. Income and class position also affect access to an important feature of contemporary society: technology.

Forty-two percent of people who make under $30,000 per year have access to broadband, whereas nearly 85 percent of people with incomes of $75,000 or more have access to broadband (U.S. Census Bureau 2010b, Table 1121). These figures are important because they show not only a family's ability to access services (many of which are now available only via the Internet), but also opportunities for the next generation. For example, comfort with a computer is an essential form of knowledge in today's education and workplace settings. Limited or no access to those technologies puts children at a disadvantage and is one of the mechanisms for the generational transmission of inequality.

Class differences in economic resources affect the timing of leaving home, marrying, and assuming caretaking responsibilities. Elderly Americans who were able to purchase property and make investments that funded a long retired life are able to afford home health care rather than relying on their children and grandchildren for care. Other families, who may rely on elderly parents and grandparents for Social Security and subsidized housing, will often lose those resources upon the death of their elderly family member. This is a particularly difficult situation for a familial caregiver who lost work days and then his or her job because an elderly family member needed care.

Money may not buy happiness, but it does afford a myriad of options: sufficient and nutritious food, comfortable residences, better health care, keeping in touch with family and friends through the Internet, education at good universities, vacations, household help, and family counseling.

Economic Change and Inequality

The U.S. economy has not necessarily been good for many Americans. Many have experienced increased job insecurity, loss of benefits, longer workdays, and more part-time and temporary work. Approximately 6.4 percent of families were classified as "working poor" in 2007 (at least one wage earner, but below-poverty-level incomes) (U.S. Bureau of Labor Statistics 2009a).

Programs of assistance for the poor have been cut, and there is increased economic risk and volatility as well as uncertainty about the future of such benefits as pensions and health insurance (Hacker 2006). In 2008, a survey of likely voters found that nearly 65 percent believed the next generation would not be better off (Tarrance Group and Lake Research 2008).

Steve Lohr argues that "dire predictions of job losses from shifting high technology work to low-wage nations with strong education systems [are] greatly exaggerated" because more complex, higher-end employment will replace the lost jobs (2006, p. C-11). Research shows, however, that although the export of white-collar jobs has slowed in recent years, it has done

so only because economic woes are great everywhere. In fact, the so-called "lost jobs" replacements are actually jobs that were previously being outsourced. The current economic conditions are such that the United States has "highly skilled workers who are eager for jobs and often willing to accept lower pay. That's prompting global outsourcing providers to beef up their presence in the U.S., where they can scoop up local talent and offer services for far less than they could have two years ago" (Scott 2009). At the same time, white-collar jobs, especially those in technological and engineering fields and some financial services, continue to be moved overseas to countries such as India and China (A. Bernstein 2004). In other words, families are currently facing very stressful economic times which, as you will see throughout the text, has important consequences for relationships.

Income Regardless of economic change, the overall long-term trend in household income has been upward (see Figure 3.1), though it can drop from 1.7 to 6 percent during periods of recession. However, this picture masks a distribution of income in the United States that is highly unequal. In 2008, the top one-fifth of U.S. households received half the nation's total income, whereas the poorest one-fifth received just 3.4 percent (DeNavas-Walt, Proctor, and Smith 2009, p. 9). Over the past thirty years, this inequality gap has grown. The rich have gotten richer, and the poor have gotten poorer. "The income gap is now as extreme as it was in the 1920s, wiping out decades of rising equality," states Princeton economist Paul Krugman (2006a, p. 46). The gap continues to grow, with the poorest 20 percent of the population earning $20,712 or less, and the 20 percent of the population with the highest incomes earning $100,241 or more (DeNavas-Walt et al. 2009, p. 9). Even these dramatic numbers do not tell the whole story. In 2007, the top 5 percent of income earners (for example, CEOs earning bonuses) took home $177,000 (U.S. Census Bureau 2010b, Table 678).

During the most recent recession, between 2007 and 2008, the top 5 percent of households increased their earnings by 1.4 percent (the top 20 percent increased their earnings by 0.6 percent). At the same time, however, 60 percent of the population saw their incomes drop between 0.4 percent and 1.1 percent (DeNavas-Walt et al. 2009, Table 3).

Income varies by race and ethnicity (see Figure 3.1), but all middle to lower groups show moderate gains at best over the long term. In fact, updated studies, such as those by the Economic Policy Institute, show that over the last two years income dropped 3.7 percent for black workers between the ages of twenty-five and fifty-four (Austin 2009). Most of any economic gain has accrued to those with a college education (Bergman 2006b), but even they have not done well over the last five years,

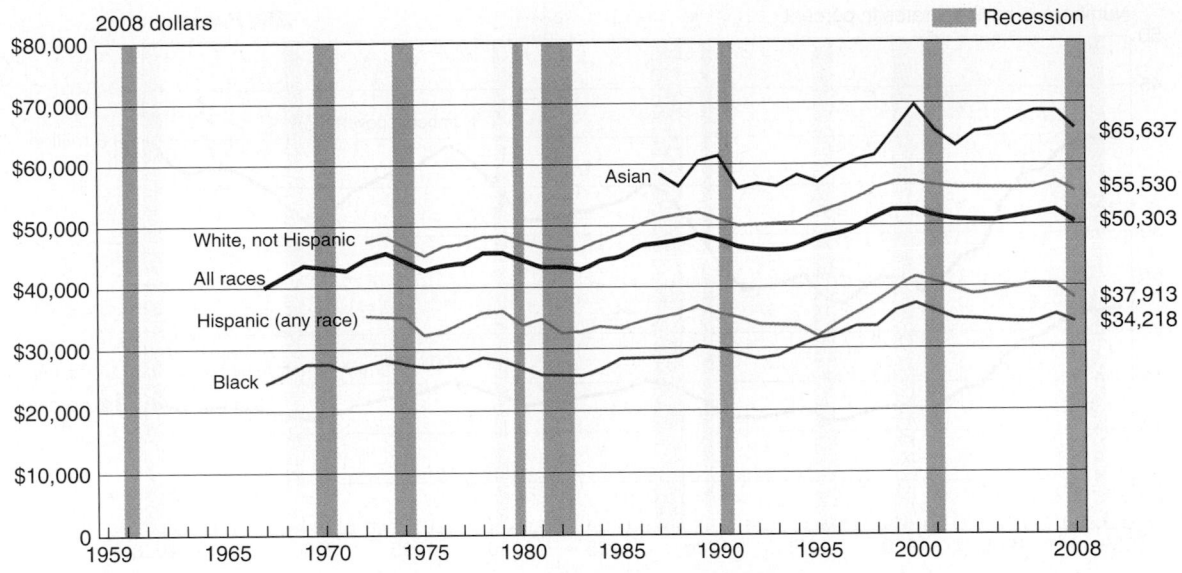

Figure 3.1 Real Median Household Income by Race and Hispanic Origin: 1967 to 2008.

showing gains just a little better than inflation (Leonhardt 2006a, p. 63).

Income varies by gender as well. Women have gained more than men since 1970, while men's wages were largely stagnant (DeNavas-Walt, Proctor, and Lee 2006, Figure 3). Still, access to a male wage remains an advantage. Experts debate the extent to which changes in the family—that is, more female-headed, single-parent households—have contributed to poverty levels (see Chapter 7). New research suggests that single-parent households headed by employed women have lower rates of poverty than do those who continue accessing welfare services (Western et al. 2008). Incomes also vary by family type. Married-couple households had the highest incomes in 2008—$73,010 compared to $49,186 for male-headed households and $33,073 for female-headed households (DeNavas-Walt, Proctor, and Smith 2009, Table 1). Some scholars point to the increasing tendency for well-educated, high-earning men to marry their female counterparts, whereas men and women at the lower end of the economic scale marry each other, creating a "real marriage penalty." Families diverge even more in income because of this multiplier effect (Paul 2006; Schwartz and Mare 2005).

Poverty Poverty rates show somewhat the same pattern as income: long-term improvement but increased disadvantage in the short term. Poverty rates fell dramatically in the 1960s and have varied since then (see Figure 3.2), with the current poverty rate increasing considerably during the current economic downturn. The poverty rate has risen since 2000 to 13.2 percent in 2008. The child poverty rate is 19 percent, higher than child poverty rates in other wealthy industrialized nations. Nearly one in five children in the United States is in poverty (DeNavas-Walt, Proctor, and Smith 2009, p. 13 and Table 4).

Poverty rates vary by racial/ethnic group. Non-Hispanic whites had the lowest poverty rate in 2008 (8.6 percent), followed by Asian Americans (11.8 percent). Hispanics (23.2 percent) and African Americans (24.7 percent) have higher rates of poverty. Although the poverty *rate* of non-Hispanic whites is

Economic inequality is rising in the United States. Not only lower income sectors, but also the middle class have failed to gain ground.

© Jeff Greenberg/ The Image Works

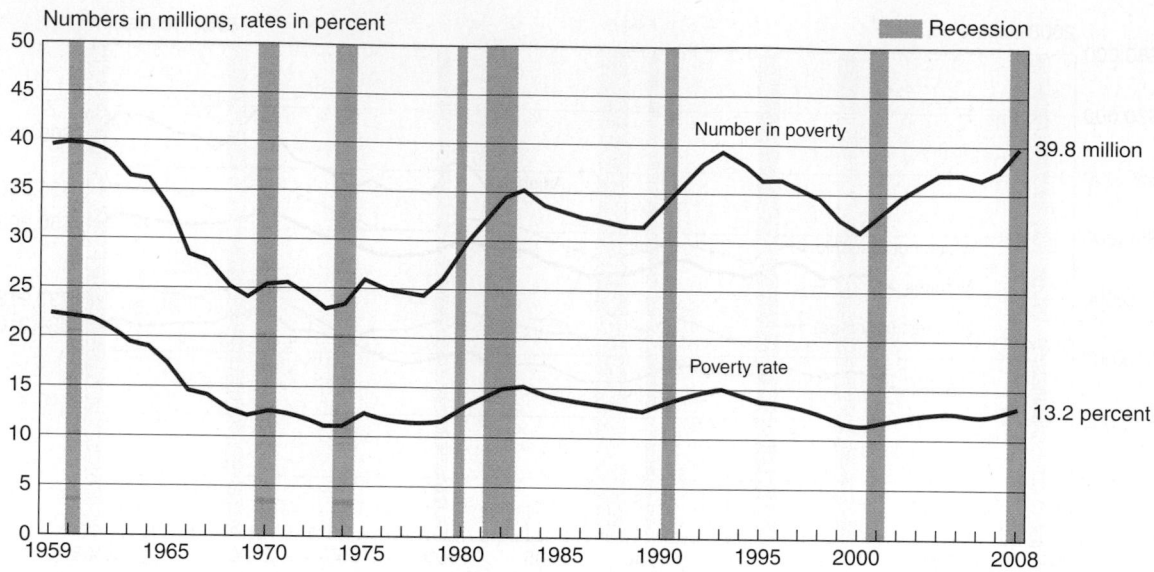

Figure 3.2 Number in Poverty and Poverty Rate: 1959 to 2008.

Note: Income is adjusted for inflation, presented in 2008 dollars

low, they compose 42.7 percent of the *total number of persons in poverty* because they are such a large part of the population (DeNavas-Walt, Proctor, and Smith 2009, p. 14 Table 4).

Blue-, Pink-, and White-Collar Families

Lifestyles vary by **social class** (which is often measured in terms of education, occupation, and income). In studying social class, social scientists have often compared blue-collar and white-collar workers in terms of their values and lifestyles. Working-class people, or blue-collar workers, are employed as mechanics, truckers, machine operators, and factory workers, in jobs typically requiring uniforms or durable work clothes. Some workers, such as police officers, occupy an intermediate position between blue collar and white collar. Pink-collar jobs are primarily lower-paying jobs held mostly by women; waitressing, retail sales, and secretarial positions are examples of pink-collar jobs. White-collar workers include professionals, managers, clerical workers, salespeople, and so forth, who have traditionally worn white shirts to work.

To complicate matters, the nature of some blue-collar jobs has changed dramatically with the advent of computerized manufacturing and the technological transformation of health care support work. At the same time, some professions—medicine, for example—have lost ground in terms of income and autonomy. Job security for both white-collar and blue-collar workers has been eroded by changes in the economy. For example, between 1973 and 2005 real wages fell approximately 14 percent (Collins, Leondar-Wright, and Sklar 1999;

Sawhill and Morton 2009). These effects have an impact on familial relationships.

Blue-, pink-, and white-collar employees, however, may continue to look at life differently, even at similar income levels. Regarding marriage, for example, working-class couples tend to emphasize values associated with parenthood and job stability and may be more traditional in gender role ideology. White-collar couples are more inclined to value companionship, self-expression, and communication. Middle-class parents value self-direction and initiative in children, whereas parents in working-class families stress obedience and conformity (Hochschild 1989; Lareau 2003b; Luster, Rhoades, and Haas 1989; Tom Smith 1999, pp. 12–19).

Social theorist Pierre Bourdieu (1977) refers to **habitus** as "one's experience and perception of the social world" (Gerbrandt 2007, p. 56). The experiences we have are shaped by the social class in which we reside, as well as our race and gender. The perceptions we form via those experiences impact the ways in which we interact with the world, including our families. "A child develops a set of bodily and mental procedures that frames perceptions, appreciations, and actions vis-à-vis familial and intimate external environments" (Gerbrandt 2007, p. 57). In other words, as we see below, the class position and racial characteristics of our family impact our childhood experiences, which will impact the decisions we make and how we experience the world as we mature into adulthood.

Middle-class parenting strategies include involving children in a myriad of stimulating activities and lessons to enhance their development. Although there is no doubt that middle-class parents provide their children with advantages regarding educational success, health care, and housing, sociologist Annette Lareau's study of parenting at different class levels finds certain advantages accruing to children in working-class and poor families. These children see relatives frequently and have much deeper relationships with cousins and older relatives, as well as less time-pressured lives (Lareau 2003a). More highly educated, high-income parents with managerial/professional occupations are less likely than those at lower socioeconomic levels to have dinner with their child every day during a typical week (Dye and Johnson 2007, Tables D7, D8; Roberts 2007). The achievement pressures and social isolation more characteristic of affluent suburban families result in some surprisingly high rates of depression and substance abuse among upper-middle- and upper-class preteens and adolescents (Luthar 2003).

Race and Ethnicity

Social class can be as important as race or ethnicity in shaping people's family lives. The attitudes, behaviors, and experiences of middle-class Americans differ from those who are poor. The study of child socialization often finds class to be more significant than race in terms of parental values and interactions with their children (Lacy and Harris 2008; Lareau 2006).

Yet, racial/ethnic heritage—the family's place within our culturally diverse society—affects preferences, options, and decisions, not to mention opportunities. Moreover, the growth of immigration in recent decades will increase the impact of ethnicity on family life because new immigrants retain more of their ethnic culture than do those who have been in the United States longer.

Conceptualizing Race and Ethnicity

To begin this discussion, we need to consider what is meant by race and ethnicity.

Race Race is a social construction reflecting how Americans think about different social groups. "Race is a real cultural, political, and economic concept, but it's not biological," says biology professor Alan Templeton ("Genetically, Race Doesn't Exist" 2003, p. 4). The term **race** implies a biologically distinct group, but scientific thinking rejects the idea that there are separate races clearly distinguished by biological markers. Features such as skin color that Americans use to place someone in a racial group are superficial, genetically speaking.

In this text, we use the racial/ethnic categories formally adopted by the U.S. government because we draw on statistics collected by the U.S. Census Bureau and other agencies. Racial categorization and how individuals are placed in census categories have varied throughout American history (Lee and Edmonston 2005, pp. 8–9) The 2000 census employed five major racial categories: (1) white, (2) black or African American, (3) Asian, (4) American Indian or Alaska Native, and (5) Hawaiian or Other Pacific Islander (U.S. Office of Management and Budget 1999). In the census, racial identity is based on self-reporting. In 2000, individuals were permitted to indicate more than one race, but only 2.6 percent did so (DeNavas-Walt, Proctor, and Smith 2009, p. 14, note 3; Jones and Smith 2001).[2]

This last point highlights one problem with census racial categories: Many people have mixed ancestry. "White" Americans may have some ancestors who were African American or Native American, whereas most "African Americans" have white as well as black ancestry, and some have Native American or Asian ancestry (Bean et al. 2004; Davis 1991).

Ethnicity You'll notice that "Latino" is not listed as a racial category. That's because Hispanic or Latino is considered an **ethnic identity**, not a *race*. **Ethnicity** has no biological connotations; instead, it refers to cultural distinctions often based in language, religion, foodways, and history.

For census purposes, there are two major categories of ethnicity: Hispanic and non-Hispanic.[3] Hispanics may be of any race (U.S. Census Bureau 2003b).[4] In many statistical analyses, Hispanics are separated out from other whites so that *non-Hispanic white* and *Hispanic* become separate categories.

---

[2] The Census Bureau does *not* include Arab as a separate major racial/ethnic category. In the 2004 American Community Survey, 1,400,000 Americans identified themselves as having Arab ancestry (U.S. Census Bureau 2008f, Table 51).

[3] The Census Bureau does record more finely differentiated cultural identities of both Hispanics and non-Hispanics using terms such as *ancestry* and *national origin*. Examples of these categories include German, Russian, Mexican, and Salvadoran.

[4] The racial composition of the Latino population is not knowable with certainty. Because some Hispanics may not self-identify with white or black, they are likely to choose "other" when asked for their racial classification (Hitlin, Brown, and Elder 2007). Latino countries of origin typically have more nuanced racial vocabularies than does the United States. Moreover, some Latinos view their ethnic identity as a racial one, whereas others view it through national origin (Stokes-Brown 2009).

In the 2000 census, the racial self-identification of Hispanics' ethnic identity was 48 percent white; 2 percent black; 42 percent "some other race"; 6 percent more than one race; and 2 percent specific other races. Demographers infer from these responses that 90 percent of Hispanics would be classified as "white" (Bean et al. 2004; Kent et al. 2001; Lee and Edmonston 2005, p. 19).

Military families today dramatically illustrate how history and current events impact family life.

A Military Family Life

The military can be considered a "total institution" (Goffman 1961). That is, it encompasses all aspects of life—living quarters, associates, schedules, locus of work and social activities, decision-making authority, and, above all, "the sublimation of individual interests to institutional goals [which can extend] to the sacrifice of one's own life" (Lundquist and Smith 2005, p. 1).

Yet one research team argues that the military is a more "family-friendly" setting (at least in peacetime) than the civilian world. To attract an all-volunteer force, a number of benefits and support systems were put in place that include family housing, extensive health insurance, day care, and school-age activity centers for older children. Other advantages include job security, a sense of community and community support— and even discount shopping.

Half of U.S. military service members are married, and almost three-quarters have children. Some couples (12 percent) are dual military. The military, which brings together many men and women of marriageable age, seems to provide an "active marriage market" (McCone and O'Donnell 2006). In fact, new research shows there is a connection between marriage and military service, particularly for African American men. It is suggested that the stability of pay and a good benefits package provided through military service are important, but the "race-neutral" work and living experience is also important as it provides a positive quality of life for military personnel (Teachman 2009). Six percent of military personnel are single parents (Segal and Segal 2004).

Military Families in Wartime

Some effects of war on the family seem obvious—family separation and the risk of death (Alvarez 2006a). Yet as the wars in Afghanistan and Iraq have continued far longer than expected, as well as become more controversial, military personnel and their families have begun to suffer. Health issues have increased for both troops and their families at home, so much so that in 2008 nearly two million soldiers' children needed mental health care (Hefling 2009). Although the troops have strong popular support, sacrifice seems limited to military personnel and their families (Haberman 2006).

Separation of a military parent from spouse or partner and children is not a new feature of wartime. But the Iraq and Afghanistan wars have seen repeated and lengthy deployments. A new study by the Pentagon shows that 60 percent of military personnel's children suffer anxiety, have poor coping mechanisms, and are doing poorly in school. Many of these children attend school with other children who have lost a parent in Iraq or Afghanistan; thus, much of the anxiety comes from the daily encounters with death and the possibility of one's own parents dying (Zoroya 2009). One newly married couple, for example, has spent a total of two weeks together in the fourteen months of their marriage. There are more dual military couples serving in Iraq and Afghanistan than in past military conflicts (Morris 2007).

A large percentage of armed forces stationed overseas are composed of National Guard and Reserve troops, who have not usually anticipated such extensive separation from family. Many of these troops are older and have well-established civilian careers or businesses that are severely disrupted by military service (Skipp and Ephron 2006). On the positive side, technological developments such as e-mail, websites, and webcams facilitate communication between soldiers and their families.

Women make up 15 percent of today's military. The majority are clustered in the Army (78,000), followed closely by the Air Force (67,000) and the Navy (53,000), with only a small number (13,000) in the ranks of the Marines (U.S. Census Bureau 2010b, Table 498). Eleven percent of the troops serving in Afghanistan and Iraq are women (Alvarez 2009, p. 1). Although they are not assigned to combat per se, support service personnel in Iraq and Afghanistan are as much at risk of injury and death as combat infantry. Some counselors have wondered about the difficulty soldier-mothers may have in returning from the battlefield to the "normal" life of parent and child (Alvarez 2006b, 2009; St. George 2006).

There are other problems subsequent to military service in wartime that become more intense with the Iraq and Afghanistan wars. The soldier who went to war may not return as the same person; in fact, 20 to 30 percent of soldiers return from the war with psychological problems. Additionally, soldiers deployed to Iraq and Afghanistan have more severe and long-lasting injuries. Because of the roadside explosives that are used by the insurgents, many more soldiers are returning home with

"Flat Daddies," cardboard cutouts of a parent deployed to Iraq or Afghanistan, serve as symbolic placeholders in the families left behind. Flat Daddies and Flat Mommies may go to school and sports events or be brought to the holiday dinner table to serve as a reminder and emotional focus for the duration of a family member's absence. "Flat Toby" is a real person to his wife and children in the Austin household in Colorado Springs. "'It's nice to see him each day, just to remember that he's still with us. . . . It's one of the best things I've done during this deployment. I really think it's helped us stay connected, to remember that he's still with us'" (Zezima 2006, p. A8; see also MacQuarrie 2006).

Linda Coan O'Kresik/The New York Times/Redux

traumatic brain injuries (Carey 2008; Hefling 2009; Ruane 2006; Zoroya 2006). Also, some families must face the tragedy of losing their loved one. In an era in which cohabitation, divorce, and single parenthood are common, ambiguous or tentative family relationships have led to disputes over remains and burials (Murphy and Marshall 2005), and over insurance and other benefits based on a deceased's military service.

The armed services have made an effort to support families and especially to reach out to the some 600,000 children who have a parent in service. A *Sesame Street* DVD explains deployment and emotions at a child's level, and the military also provides a comic book which assists young people in understanding the difficulties some family members might face after returning home from war (Elfrink 2006; Zoroya 2009). There are also military programs for at-risk youth, free memberships to the YMCA, educational and child care assistance, as well as vouchers for getaways and marriage enrichment programs for couples (Ansay, Perkins, and Nelson 2004; Hefling 2009).

Critical Thinking

Do you know one or more families that have been affected by a family member's service in Iraq or Afghanistan? How do their experiences illustrate the fact that history and events impact individual and family life?

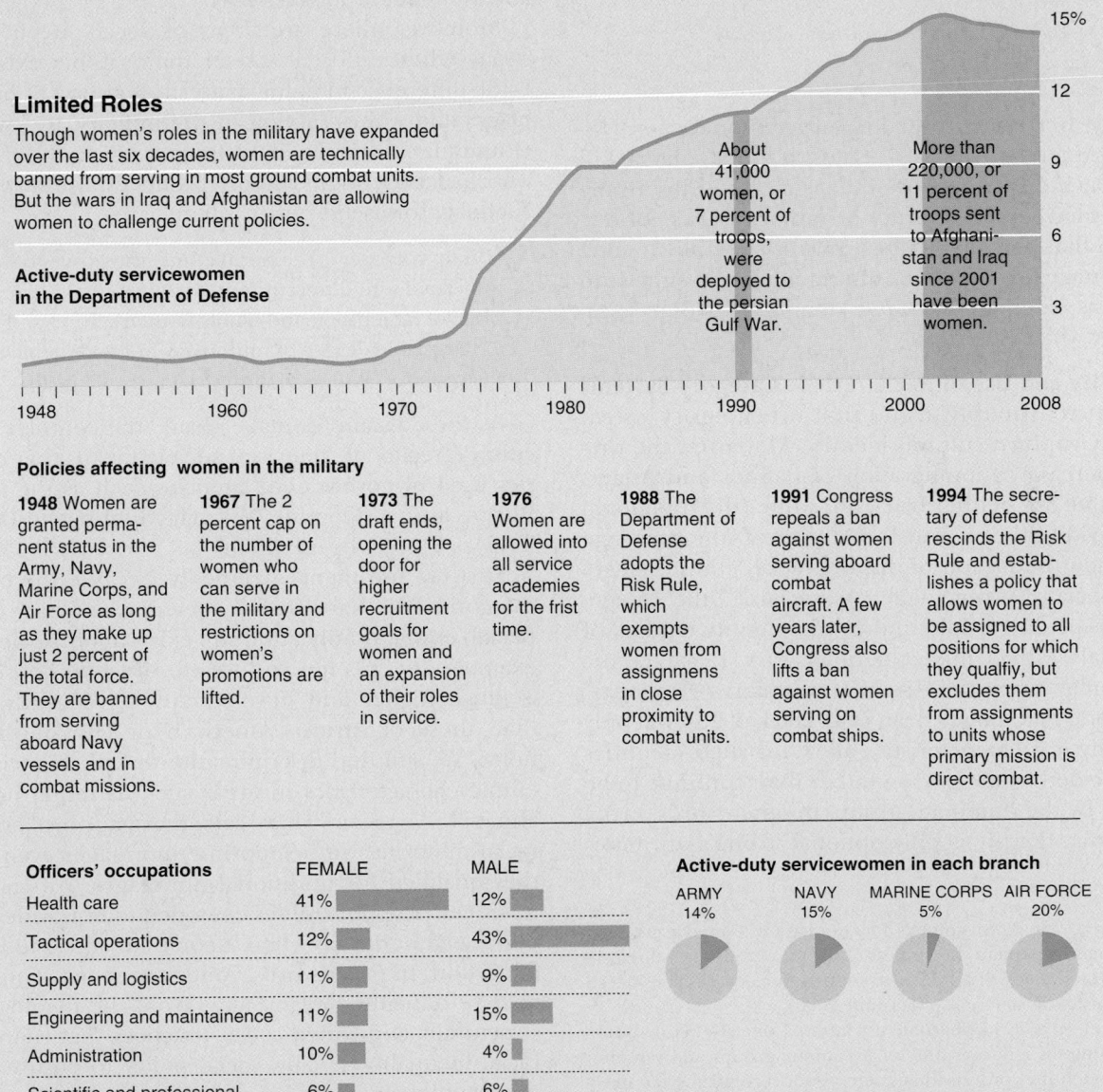

Limited Roles

Though women's roles in the military have expanded over the last six decades, women are technically banned from serving in most ground combat units. But the wars in Iraq and Afghanistan are allowing women to challenge current policies.

**Active-duty servicewomen
in the Department of Defense**

About 41,000 women, or 7 percent of troops, were deployed to the persian Gulf War.

More than 220,000, or 11 percent of troops sent to Afghanistan and Iraq since 2001 have been women.

1948   1960   1970   1980   1990   2000   2008

Policies affecting women in the military

1948 Women are granted permanent status in the Army, Navy, Marine Corps, and Air Force as long as they make up just 2 percent of the total force. They are banned from serving aboard Navy vessels and in combat missions.

1967 The 2 percent cap on the number of women who can serve in the military and restrictions on women's promotions are lifted.

1973 The draft ends, opening the door for higher recruitment goals for women and an expansion of their roles in service.

1976 Women are allowed into all service academies for the frist time.

1988 The Department of Defense adopts the Risk Rule, which exempts women from assignmens in close proximity to combat units.

1991 Congress repeals a ban against women serving aboard combat aircraft. A few years later, Congress also lifts a ban against women serving on combat ships.

1994 The secretary of defense rescinds the Risk Rule and establishes a policy that allows women to be assigned to all positions for which they qualfiy, but excludes them from assignments to units whose primary mission is direct combat.

Officers' occupations	FEMALE	MALE
Health care	41%	12%
Tactical operations	12%	43%
Supply and logistics	11%	9%
Engineering and maintainence	11%	15%
Administration	10%	4%
Scientific and professional	6%	6%

Active-duty servicewomen in each branch

ARMY	NAVY	MARINE CORPS	AIR FORCE
14%	15%	5%	20%

Figure 3.3 Women in the Military: Limited Roles. Though women's roles in the military have expanded over the last six decades, woman are technically banned from serving in most ground combat unibs. But the wars in Iraq and Afghanistan are allowing women to challenge current policies.

"Actually, I prefer the term Arctic-American."

Minority In a final distinction, African Americans, Hispanics, American Indians, Asians, and Hawaiians and Other Pacific Islanders are often grouped into a category termed minority group or minority. This conveys the idea that persons in those groups experience some disadvantage, exclusion, or discrimination in American society as compared to the dominant group: non-Hispanic white Americans.[5]

The Utility and Use of Racial/Ethnic Category Systems
One can reasonably argue that no category system can truly capture cultural identity. Moreover, the dramatic increase in immigration of Latinos and Asians, groups that are neither black nor white (the traditional racial dividing line), and the growth of intermarriage have made the notion of distinct racial/ethnic categories especially problematic. As racial/ethnic categories become more fluid and as the identity choices of individuals with a mixed heritage vary, racial/ethnic identity may come to be seen as voluntary—"optional" rather than automatic (Bean et al. 2004, p. 23), but only if the physical characteristics allow for such identities to be decided by the person rather than continue to be ascribed by the dominant group. In other words, racial and ethnic identity is only optional if one's attributes

are not strongly associated by others with a particular racial and/or ethnic category.

A further point is that there is considerable diversity *within* major racial/ethnic groupings. There are Caribbean and African blacks, for example, as well as those descended from U.S. slave populations. Within each major racial/ethnic category, there are often significant differences in family patterns.

Within-group diversity makes generalizations about racial/ethnic groups somewhat questionable. "Hispanic" or "Latino" categories are "useful for charting broad demographic changes in the United States . . . [but they] conceal variation in the family characteristics of Latino groups [Cubans and Mexicans, for example] whose differences are often greater than the overall differences between Latinos and non-Latinos" (Baca Zinn and Wells 2007, pp. 422, 424).

Moreover, there are areas of social life in which racial/ethnic differences seem minor if they exist at all. Little difference in family patterns is apparent between blacks and whites serving in the military, for example (Lundquist 2004). In his ethnographic study of fathers who had been high school classmates, anthropologist Nicholas Townsend remarks that

> [i]n my interviews, the racial-ethnic category was not associated with different fundamental values about the place of fatherhood and family in men's lives. . . . I found a remarkable degree of uniformity in men's depictions of the central elements of fatherhood. (2002, p. 20)

As these examples make clear, "the complex multicultural reality of American society means that categories used by government agencies such as the Census Bureau are . . . 'illogical'" (Walker, Spohn, and DeLone 2007, p. 9). So why use them?

First, we use them, qualifiedly, because these racial categories do have social meaning in our society. Racial/ethnic stratification still exists in our society. For example, the income and wealth of white households is much higher and poverty rates significantly lower than those of African American or Hispanic households. We still find discrimination based on racial and ethnic characteristics in areas such as rental housing and jobs. We recently have seen mortgage brokers push racial minorities into subprime mortgages even when they qualified for traditional mortgages. We continue to see fewer opportunities extended to minorities from poor neighborhoods when it comes to access to higher education. In other words, American society continues to have difficulty moving away from the ideology and structurally embedded social practices that reproduce racial inequality. Finally, social policy to ensure equal opportunity requires a base of information about group outcomes.

Second, to learn more about minority families than mere speculation can tell us, we need to use the data

---

[5] *Minority* in a sociological context does not have its everyday meaning of fewer than 50 percent. Regardless of size, if a group is distinguishable and in some way disadvantaged within a society, it is considered by sociologists a *minority group* (Ferrante 2000).

The term *minority* has become a contested one, viewed by some as demeaning; as ignoring differences among groups and variation in the self-identities of individuals; and as not recognizing the likely future of the United States as a "majority-minority" nation (Gonzalez 2006a; Wilkinson 2000). We will try to avoid using it other than when speaking of numerical differences or in reporting Census Bureau data so labeled.

collected by the government and other agencies. These data can tell us something about the contexts of family life and the impact that social attitudes about "race" and ethnicity have on life chances. "[R]acial statuses, although not representing biological differences, are of sociological interest in their forms, their changes, and their consequences" (American Sociological Association 2002).

Now it's time to use these racial/ethnic categories to explore the features of family life in various social settings. In doing so, we turn to research rather than rely on assumptions about differences—which may be mistaken. Of course, researchers may themselves be influenced by stereotyped assumptions, but they have become much more conscious of such pitfalls in recent decades, as "Issues for Thought: Studying Families and Ethnicity" indicates.

Racial/Ethnic Diversity in the United States

The United States is an increasingly diverse nation. The most recent national population statistics show that in 2008, the nation was 65.6 percent non-Hispanic white; 12 percent black; 4.3 percent Asian; less than 1 percent American Indian/Alaska Native; and less than 1 percent Native Hawaiian and Other Pacific Islander. The Hispanic population has increased dramatically: Hispanics are now 15.4 percent of the population, surpassing blacks as the largest racial/ethnic group after non-Hispanic whites (U.S. Census Bureau 2010b, Table 6).[6] Hispanics and Asians remain the fastest-growing segment of the population; however, their growth has slowed since 2005. Part of the reason for the slower rates of growth is the reduction in international migration (slowing down to approximately 30 percent of U.S. population growth). As with previous population surveys, the data show that fertility rates among Hispanics remain higher than among their black or white counterparts, with approximately 55 percent of children of immigrants identifying as Hispanic (Mather 2009, p. 3; Mather and Pollard 2009).

The 2010 child population estimate is even more diverse: 55.7 percent are non-Hispanic white; 22 percent Hispanic; 15 percent black; and 4 percent Asian and Pacific Islander. Finally, 4.9 percent of children are reported to be either American Indian, Alaska Native, Native Hawaiian, or of more than one race (U.S. Census Bureau 2010a, Table C3).

Presently, racial/ethnic minorities compose over one-third of the U.S. population and 43 percent of the child population. By 2042 racial/ethnic minorities are expected to make up half of the population (Mather

The child population of the United States is more racially and ethnically diverse than the adult population and will become even more diverse in the future.

Charles Thatcher/Getty Images

and Pollard 2009). In five "majority-minority" states, racial/ethnic minorities already compose over half the population (U.S. Census Bureau 2008d, Table 18). As these trends unfold, they describe a world that younger people find familiar: "'Beginning with Generation X, [for] people in their 20s to early 40s and all the generations that follow, multicultural is normal'" (marketing expert Ann Fishman, cited in El Nasser and Grant 2005, p. 4-A).

African American Families

African Americans have been increasingly divided between a middle class that has benefited from the opportunities opened by the Civil Rights Movement and a substantial sector that remains disadvantaged. The current economic recession, however, has reduced many of the gains made by black families. Between 2007 and 2008, African American households saw their median income decline 2.8 percent to $34,218 (DeNavas-Walt, Proctor, and Smith 2009; Shierholz 2009). Additionally, a higher proportion of black children than those of most other racial/ethnic groups lives in poverty (33.9 percent), although by 2008 more Hispanic children were living in poverty (39.6 percent) (Shierholz 2009). Black women are more than twice as likely as white women to suffer the death of an infant (Heron et al. 2009, Table 31).

Childbearing and child rearing are increasingly divorced from marriage. True of all racial/ethnic groups, this trend is especially pronounced among blacks, with 71 percent of births in 2007 to unmarried mothers (Hamilton, Martin, and Ventura 2009, Table 1). Black divorce rates are higher as well, although the more

---

[6]Percentages may not always add up to 100, due to rounding errors and to the complexities of classifying individuals who indicated more than one race in the census multicategory system.

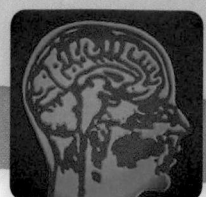

As men and women from diverse racial/ethnic backgrounds came into the field of family studies, they pointed out how limited and biased our theoretical and research perspectives had been. For many years, research on African Americans was focused on poor, single-parent households in the inner city, and research on Latinos focused on Mexican immigrants' alleged "patriarchal" culture and other barriers to economic advancement and assimilation (Baca Zinn and Wells 2007; S. A. Hill 2006; Taylor 2007). This research focus still exists, and may even have intensified in current media and policy attention to the African American marriage "crisis" (U.S. Administration for Children and Families n.d.) and the "caste" barrier between single-parent and married-couple families (Hymowitz 2006).

Following the negative reaction to the earlier, limited portrayal of racial/ethnic family differences, researchers began to report on the strengths of families of color, multiracial families, and multi-ethnic families, pointing to strong extended family support, more egalitarian spousal relationships, and class, regional, and rural/urban diversity. For example, a substantial proportion of African American single-mother households contain other adults who take part in rearing the children (Taylor 2007). Another study of Hispanic and non-Hispanic white families points to the different ways in which extended families function. Although the Hispanic families provide instrumental help, the white families provide financial help. Both families are close in terms of communication (Sarkisian, Gerena, and Gerstel 2006); the image of white families as lacking an extended family context is thus challenged.

Ideas, insights, and concepts developed in the study of families that vary from the majority group are applied to enrich family studies more generally. For example, Annette Lareau (2003a) points to the rich family life of working- and lower-class children, whose parents are less focused on educational and achievement goals and activities so that they have more time to spend with relatives and lead less-stressful lives than seemingly more privileged middle-class children (Levine 2006).

Research using a comparative approach has shown us that the same family phenomenon may have different outcomes in different racial/ethnic settings. For example, premarital cohabitation is associated with future marital disruption among whites but not among African Americans or Latinos, where it may function as a marital substitute and represent more stable unions. Also, communication processes vary by family types, with multiracial and multi-ethnic families developing unique forms of communication that assist in

substantial difference is that black couples are far more likely to have never married than are white couples (U.S. Administration for Children and Families n.d.). As a consequence, only 35 percent of African American children are living with married parents, compared with 75 percent of white (non-Hispanic) and 64 percent of Hispanic children (U.S. Federal Interagency Forum 2009, p. 2; see Chapter 7 for further discussion).

Differences between African Americans and whites in the proportion of two-parent families are not new, but as recently as the 1960s, more than 70 percent of black families were headed by married couples, whereas in 2008 only 31.1 percent were (Billingsley 1968; U.S. Census Bureau 2010a, Table A1). Experts do not agree on the cause of the decline in marriage and two-parent families among African Americans, but economic and employment factors are noted in much of the literature as important components of the issue. In fact, research shows that African Americans value marriage. For example, a Gallup poll taken in 2006 finds 69 percent of blacks agreeing that marriage is "very important" "when a man and woman plan to spend the rest of their lives together as a couple"—*higher than the figure for whites* (Saad 2006c).

Given similar values regarding marriage, research consistently suggests that the primary source of difference in marital patterns is economic. Our economy's shift away from manufacturing has meant the elimination of the relatively well-paying entry-level positions that once sustained black working-class families. Low levels of black male employment and income may preclude marriage or doom it from the start (Burton et al. 2009; Holland 2009; Joshi, Quane, and Cherlin 2009; Taylor 2007; W. J. Wilson 2009).

African American women have traditionally been employed, and they may be less dependent on the earnings of a spouse for economic survival. Married black women tend to have higher employment rates than their white counterparts (Corra et al. 2009; Durr and Hill 2006). But the *economic independence* explanation of low marriage rates is not supported by research; it appears that the better their earnings, the more likely black women are to marry (Berlin 2007; Tucker 2000). Research also indicates that the availability of welfare is not a significant factor in a black woman's decision to marry (Berlin 2007; Teachman 2000).

Another possible explanation for the lower marriage rates of African Americans is the **sex ratio**, the

maintaining solidarity among the family members (Soliz, Thorson, and Rittenour 2009, p. 829).

Additionally, stressors such as poverty and living in disadvantaged neighborhoods impact families of color at greater levels than white families. For example, new comparative research suggests that Mexican American fathers and African American mothers are more prone to depression when they and their families reside in dangerous neighborhoods. Such depression leads to poor parenting—particularly, inconsistent child discipline (White et al. 2009). Other comparative research examining family cohesion finds that Mexican American fathers sense a greater level of family cohesion during times of economic stress, although white mothers perceive less. White mothers, during times of economic stress, engage in inconsistent child discipline, whereas Mexican American mothers maintain consistent discipline strategies regardless of economic hardship (Behnke et al. 2008).

Research on extended family ties illuminates the great amount of instrumental help that Hispanic extended families provide to their members. This means that workplace policies that presume only nuclear family members need the flexibility to provide family care does not take into account the real lives of Hispanic families (Sarkisian, Gerena, and Gerstel 2006).

Today's research on family and ethnicity tends to be more complex and sophisticated than in the past. Concern about family fragility and individual disorganization is balanced by recognition of diversity and of community and family strengths. Multiple influences on racial/ethnic families are acknowledged: (1) mainstream culture; (2) ethnic settings; and (3) the negative impact of disadvantaged neighborhoods or family circumstances that can produce behaviors that are inappropriately viewed as a "minority culture" (S. Hill 2004). Structural influences—that is, economic opportunity—are seen as a powerful influence on family relations and behavior. The role of "agency," or the initiative of families, is recognized: "What happens on a daily basis in family relations and domestic settings also constructs families. . . . Families should be seen as settings in which people are agents and actors, coping with, adapting to, and changing social structures to meet their needs" (Baca Zinn and Wells 2007, p. 426; see also S. Hill 2004).

Critical Thinking

Does your family heritage or your observation of families make you aware of some family patterns that you would see as different from common American assumptions about families? How could these observations be applied to help researchers learn more about families in a variety of family settings?

---

number of eligible men available for women seeking marital partners (Taylor 2007). High rates of incarceration, what some black scholars call "the prisonization of black America" (Clayton and Moore 2003, p. 85),[7] as well as poorer health and higher mortality, have taken many African American men out of circulation.

Many scholarly and policy analyses of the African American family emphasize the "crisis" of marriage among blacks (U.S. Administration for Children and Families n.d.), and we have reported that here. At the same time, African American scholars rightly complain of "sweeping generalizations" and "pejorative characterizations" (Taylor 2002a, p. 19) that often reflect a research focus on lower-income blacks in the inner city (Taylor 2007; Willie and Reddick 2003).

Recent research gives more attention to middle-class blacks (e.g., Pattillo-McCoy 1999) and gives us more nuanced portraits of those families not organized around a married couple. Sociologist Jennifer Hamer (2001) undertook a qualitative study of eighty-eight lower-income black fathers living away from their children. These fathers view spending time with children, providing emotional support and discipline, and serving as role models and guides as among their most important parental functions, although they also tried to do what they could by way of economic support.

Scholars have noted the strengths of black families (Hill 2003 [1972]; S. A. Hill 2004; Taylor 2007), especially strong kinship bonds. Single-parent families or unmarried individuals are often embedded in extended families and experience family-oriented daily lives. As a family system, African American families are child focused. Black families are culturally predisposed to accept children regardless of circumstances: "Children are prized" (Crosbie-Burnett and Lewis 1999, p. 457). In a child-focused family system, the extended family and community are involved in caring for children; their survival and well-being do not depend on the parents alone (Uttal 1999, p. 855).

With regard to couple dynamics, we find that married blacks have more egalitarian gender roles than do

[7] "Racial profiling, mandatory minimum sentences, and especially the disparities in drug laws [which more heavily penalize crimes involving drugs typically used by blacks] have had a dramatic effect on the incarceration rates of young males, especially in urban inner-city neighborhoods" (Clayton and Moore 2003, p. 86). In 2005 African Americans comprised 14 percent of drug users, but represented 33.9 percent of drug arrests and 53 percent of drug convictions (Mauer 2009).

whites, as characterized by role flexibility and power sharing. Recent studies show us that the marital happiness of black wives, traditionally the lowest of all categories surveyed, is increasing (Corra et al. 2009). Other research finds that African American men do more housework and are more supportive of working wives than are other men, but are traditional in other respects. Child socialization is less gender differentiated in African American families (McLoyd et al. 2000; Taylor 2002a).

As we write, the direction of change in the circumstances of African American families is uncertain. Since 1966 there has been a small but steady rise in the percentage of black families headed by married couples ("Married Households" 2003). Yet poverty and racism continue to create stress on many African American families. Segregation persists in much of everyday life, even in the suburban middle-class settings to which many African Americans have moved (El Nasser 2001; Scott 2001). The income of a number of black families has risen, but conditions at the lowest economic levels have not shown improvement, nor have blacks at higher income levels acquired assets that compare with those of non-Hispanic whites (Krivo and Kaufman 2004; Stoll 2004; "Study Says" 2004). Moreover, "[b]lack baby boomers did not close the income gap even though [they came] of age after the civil rights era" (Fears 2004, p. A2). Middle-class blacks have comparatively lower-status jobs and incomes, and some must cope with de facto housing segregation and neighborhoods that often have higher crime rates, poorer schools, and fewer services. Even highly educated, high-income African American families face some problems that represent lingering racism and assumptions based on stereotypes, finding it difficult to obtain nannies or other in-the-home child care, for example (Kantor 2006). "The reality . . . is that even the black and white *middle classes* remain separate and unequal" (Pattillo-McCoy 1999, p. 2).

In analyzing the contemporary situation of African Americans, scholars have begun to note and investigate the increasing geographic dispersion, class separation, and gender differences that characterize the post-civil rights era (S. A. Hill 2004). Blacks are becoming more ethnically heterogeneous, as foreign-born blacks (African and Caribbean immigrants) comprise 8 percent of the black population (compared to 1.3 percent in 1970), and are responsible for approximately 20 percent of the increase in the black population (Kent 2007). Research on "African American" families is beginning to include analysis of those diverse cultural communities—particularly the African principle of Harambee ("let's all pull together"), which is the survival mechanism through which we see the prominence of related concepts such as the extended family and the inclusion of "fictive kin" (unrelated persons given symbolic kinship status) and "Other-Mothering" (Cowdery et al. 2009; S. A. Hill

2004; Taylor 2007). The extent to which African American families are influenced by mainstream middle-class values, African values, and "minority culture" (their disadvantaged position in the American stratification system; S. A. Hill 2004, pp. 15–16) is a likely focus of future research, as is the value of motherhood and productive work roles for women.

Latino (Hispanic) Families

Although the first Spanish settlements in what is now the United States date to the sixteenth century, many Latinos are recent immigrants from Mexico, Central America, the Caribbean, or South America. A majority of U.S. Latinos were born in this country, but 40 percent were foreign born (Mather and Pollard 2009).

Latino families may be binational. **Binational families** are those in which some family members are American citizens or legal residents, while others are **undocumented immigrants** (i.e., not legally in the United States and subject to sudden deportation). Issues affecting families of immigrants from various nations are discussed in "A Closer Look at Diversity: Family Ties and Immigration." Here we examine Latino family circumstances more generally.

Latinos are most likely to be employed in service-level occupations. With the current economic downturn, Hispanic workers have a higher employment rate than any other racial group (U.S. Census Bureau 2009a, Table S2301), yet on average, Latino families are less economically advantaged than non-Hispanic white families. Some 28.3 percent of Latino children are poor, compared to 17.6 percent of all children (U.S. Census Bureau 2010a, Table 696).

Educational levels are relatively low; only 62.3 percent of Latinos have graduated from high school and 13.3 percent from college (U.S. Census Bureau 2010b, Table 224). This may be partially due to a "Hispanic culture of hard work" that draws Latinos into the labor force early to contribute to family welfare. Latino parents, like parents in all racial/ethnic groups, have high educational aspirations for their children (Lopez 2009; Omaha Public Schools [OPS] Dual Language Research Group 2006; Schneider, Martinez, and Owens 2006). But language difficulties, the initial economic disadvantage of Hispanic youth, and low educational levels of parents are often compounded by poor schools and weak relationships with teachers as well as pressures to help out at home (Lopez, Livingston, and Kochhar 2009; Schneider, Martinez, and Owens 2006). Dropping out of high school may also depend on neighborhood context. Although Latino youth drop out more than non-Hispanic youth in the inner city, in suburban areas there is no difference (Lopez 2009).

The second generation is characterized by **segmented assimilation.** Although some second generation

Hispanics are doing quite well in attaining a secure place in the American socioeconomic system, many face a substantial likelihood of downward social mobility. Skill levels overall are low, and those educated for the professions may encounter hostility toward them because of their immigrant heritage, which will affect their economic prospects. Both because of and in spite of the discrimination and hostility faced by immigrants and their progeny, some Hispanics have taken an entrepreneurial route, starting small businesses (Bergman 2006d).[8]

Hispanics tend to marry at young ages—about 24.6 percent of women were married by ages twenty to twenty-four (U.S. Census Bureau 2010c, Table A1). Native-born Hispanic women tend to have the same marriage rates as non-Hispanic women, but Hispanic women who are immigrants tend to have higher rates of marriage (Gonzales 2008). Mexican Americans are more "married" than other disadvantaged groups in the United States, and Cuban and Mexican marriage and marital dissolution rates are similar to those of non-Hispanic whites. Puerto Ricans share a Caribbean tradition of informal marriage—that is, cohabitation that resembles marriage (Rodman 1971; and see Chapter 8, "A Closer Look at Diversity: Cohabiting Means Different Things to Different People—The Meaning of Cohabitation for Puerto Ricans, Compared to Mexican Americans").

Hispanic birth rates are the highest of any racial/ ethnic group, but they vary by within-group ethnicity. Mexican American birth rates are among the highest in the United States, whereas those of Cubans are among the lowest. Interestingly, Central American, Cuban, and Mexican immigrants all have lower **infant mortality rates** than non-Hispanic whites, despite higher levels of poverty and lower levels of education and income (Gonzales 2008; MacDorman and Mathews 2008, Figure 3). One possible explanation, besides extended family support, is that Latinas, especially recent immigrants, are more likely to refrain from smoking, drinking, and drug use (Chung 2006; Pew Hispanic Center 2002).

Half of births to native-born Latinas in 2007 were to unmarried women. Puerto Rican women are especially likely to be unmarried at a child's birth, perhaps a consequence of the Caribbean pattern of acceptance of informal marital ties. Mexican American women

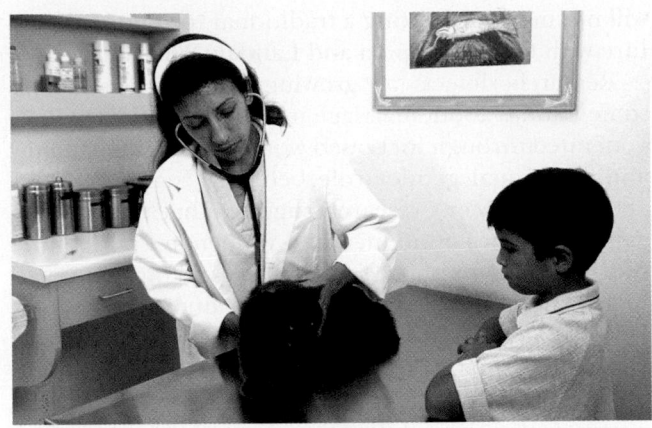

Latinas do not necessarily limit their lives to traditional roles. They enter the labor force and undertake important activities in the community.

typically have their first child in marriage, but some suggest that the nonmarital birth pattern may be becoming characteristic of lower-income Mexican American women (Baca Zinn and Wells 2007; Hamilton, Martin, and Ventura 2006; National Latina Institute for Reproductive Health 2009; Wildsmith and Raley 2006). Other research shows a strong pattern between adolescent pregnancy in Latinas whose closest female relatives were also teenage mothers (East, Reyes, and Horn 2007). The teenage birthrate had declined steadily since the early 1990s, but it has risen approximately 5 percent since 2005 (Hamilton, Martin, and Ventura 2009, Tables 2–5, Figure 2). Hispanics are an exception to this trend. The rate of Latina teen pregnancy dropped by 2 percent in 2007 even though it increased for every other racial/ ethnic group (Hamilton et al. 2009, p. 2).

At the same time, the assumption that Latinos are conservative on sexual issues is supported by survey results indicating high rates of disapproval of abortion and of opposition to the legalization of homosexual relations and same-sex marriage (Baca Zinn and Pok 2002; Baca Zinn and Wells 2007; McLoyd et al. 2000; Perez 2002; Taylor 2002b). Hispanics, especially Mexican immigrants, are seen to have a "pro-nuptial" family culture that some have hoped would leaven the mainstream American culture of a "retreat from marriage" (Brooks 2006). Alas, calculations by Oropesa and Landale (2004) reveal that even as the United States becomes increasingly Hispanic, there would not be that much of an impact on marital rates and marital stability. Moreover, these authors speculate that it is at least as likely that second and third post-immigration generations would become assimilated enough to join the mainstream retreat from marriage. One must also take into account changes in the home countries toward higher divorce rates and more cohabitation—new immigrants

---

[8] Recent scholars of immigration have challenged the classic assumption that over time immigrants assimilate into American society and culture. Portes and Rumbaut (2000) argue that although this is true of some migrants, it is not true of others. Whether or not immigrants secure a place in the mainstream economy depends on the human capital of the migrants (their education and skills), the match to the labor market, and reception of a particular immigrant group at a particular time by the government (e.g., refugee support programs) and by co-ethnics already present in the United States.

will not necessarily bring a traditional "pro-nuptial" culture with them (Oropesa and Landale 2004).

Research detects a "growing ambivalence among some Latinas about marriage in the face of the conflicts generated through increased women's economic power and traditional gender role beliefs" (Tucker 2000, p. 180). More recent research suggests that the younger generations of Latino men and women are putting the pursuit of careers ahead of marriage, and the continued high marriage rates are more likely due to immigrant Latinos rather than native-born men and women (Pew Hispanic Center 2009). Additionally, Latina women are starting to have smaller families, "resisting the social pressures that shaped the Hispanic tradition of big families." They increasingly postpone marriage and limit families in favor of working and getting an education to have better economic prospects (Navarro 2004).

Single-mother families are increasing as nonmarital births are rising among Hispanics, with children in the third generation and beyond being "twice as likely as other Hispanic children to live in a female-headed household" (Fry and Passell 2009, p.14). Possible explanations are that single motherhood in the United States requires less dependency on the extended family and that single mothers do not experience the shame they might in Mexico. Nevertheless, two-parent families remain the most common form of children's living arrangements. In 2007, 64 percent of Hispanic children were living with married-couple parents, and 27 percent lived in a single-mother family (Fry and Passell 2009, Table 2).

Latino families are more likely to be extended and larger than those of non-Hispanic whites (Ramirez and de la Cruz 2003, Figure 6). Both structural factors (economic necessity) and culture seem to shape an extended family co-residence pattern (Sarkisian, Gerena, and Gerstel 2006).

Gender roles are an area of family change that has been shaped by immigration. In the United States, regardless of cultural ideals about the importance of women's maternal and domestic roles, Latina wives must typically enter the labor force to contribute to the family economy. That can lead to increased autonomy and independence and a stronger voice in family decisions. On average, Mexican immigrant women in this country have less power than men in terms of decision making and the division of household labor, but more power than their counterparts in Mexico. Salvadoran women, who often have been in the labor force in their home country, nevertheless cite the greater autonomy they experience here in terms of freedom to come and go without the close monitoring of the husband and the ease of obtaining help against an abusive spouse (Baca Zinn and Pok 2002; Baca Zinn and Wells 2007; Hirsch 2003; Hondagneu-Sotelo and Messner 1994; Oropesa and Landale 2004; Zentgraf 2002).

Given class and cultural diversity among Latinos, continued immigration, family change in the home countries, and increasing intermarriage, the future direction of the Latino family is difficult to predict.

Asian American Families[9]

Although their numbers are relatively small, Asian Americans are one of the fastest-growing racial/ethnic groups, second only to Hispanics (Mather 2009; Mather and Pollard 2009). The majority of the Asian American population is foreign born, but this is likely to change over time as Asian immigrants establish families in this country and as immigration rates continue to slow.

Asian Americans are often termed a "model minority" because of their strong educational attainment (the highest proportion of college graduates), high representation in managerial and professional occupations, and family incomes that are the highest of all racial/ethnic groups (Bergman 2006b; DeNavas-Walt, Proctor, and Smith 2009, Table 1; Reeves and Bennett 2004).

But this image has important negative effects: It encourages institutional quota caps for Asian Americans, and it denies the humanity of Asian Americans—particularly those who do not fit the "model-minority stereotype," such as poor immigrants from Southeast Asian countries. Furthermore, the stereotype ignores the conditions that Asian Americans find themselves in, such as living in urban areas where both the cost of living and incomes are higher. Also, in Asian families often the parents and the children contribute to the family income, thus creating an artificial assumption that Asians earn more (Chou and Feagin 2008). When it comes to children, the "model-minority stereotype is a persistent social issue that has important implications for Asian-American children" (Lott 2004). As Juanita Tamayo Lott, author of a book on Asian Americans, puts it:

> Asian Americans have the dubious distinction of being labeled a "model minority"—based on their stereotype as overachievers and as models to other racial minority groups. . . . First, expectations of all Asian-American children (and adults) are initially higher than for other population groups. . . . Second, many overachieving Asian-American children feel they are never good enough, as the bar for achievement continues to be raised. Third, Asian-American children who do not fit the model-minority stereotype are treated as underachievers, resulting in low self-esteem and self-worth.

A higher percentage of Asian Americans are married than among the general population or non-Hispanic whites.

[9] For the 2000 census, the previous "Asian/Pacific Islander" category was divided into two categories: "Asian" and "Hawaiian Native and Other Pacific Islander."

Asian American children (85 percent) are very likely to be living in married-couple families; only 10 percent live in single-mother families, 2 percent in single-father families, and 2 percent live with neither (U.S. Census Bureau 2009a, Table C-2). Infant mortality rates are low (lower than those of whites), although preterm deaths increased 5 percent between 2000 and 2005 (MacDorman and Mathews 2008, Figure 6). Asian American teen birth rates and nonmarital births are also very low (Hamilton, Martin, and Ventura 2009; Ventura 2009a). Asian Americans are most likely of all groups to be caring for older family members (American Association of Retired Persons 2004).

As with all racial/ethnic groups, there is considerable within-category diversity; in fact, more diversity exists among Asian Americans in terms of language, religion, and customs than in any other broad racial/ethnic category. The contrast among various Asian American ancestry groups is striking. Fifty-five percent of Hmong are under eighteen, whereas only 12 percent of Japanese are. Sixty-seven percent of Asian Indians and Pakistanis are married, but only 49 percent of Cambodians. South Asians tend to marry at young ages, whereas Japanese, Korean, and Chinese women delay marriage. Sixty-four percent of Asian Indians are college graduates, but less than 10 percent of Cambodians, Hmong, and Laotians have college degrees. Income and poverty status also vary (Lichter and Qian 2004, Table 1; Reeves and Bennett 2004, Figures 3, 4, and 9).

Discrimination and hostility toward Asians still exist, and Asian Americans are more likely than whites to be poor (DeNavas-Walt, Proctor, and Smith 2009, Table 4). Even advantaged youth may feel marginalized at school and among peers—that is, may not feel they are fully accepted (Purkayastha 2002). At the same time, Asian Americans have high rates of intermarriage. Native-born Asian Americans are less residentially segregated than most other racial/ethnic groups (Ishii-Kuntz 2000; S. M. Lee 1998), although newer immigrants from poorer, East Asian countries tend toward residential segregation in the form of ethnic enclaves (Xie and Gough 2009).

Asian American families are often more cohesive and less individualistic than are non-Hispanic white families. Scholars explain the survival of extended family commitment among Asian Americans in the United States as the need for family cohesion in the face of economic pressures and discrimination against Asian immigrants (Brenner and Kim 2009). Indo-Americans have strongly **transnational families**, maintaining cohesive relations with the home country through visits, business linkages, remittances, telecommunications, and marriage arrangements (Purkayastha 2002; Şenyürekl and Detzner 2009; Wright 2005).

Ironically, some scholars credit the increased independence of Asian women in the United States to discrimination. To begin with, the image of Asian women as subordinated to men in patriarchal households was not always the reality. In the United States, Asian women entered the labor force because of the low wages of Asian men. The internment of Japanese American citizens and legal residents during World War II undercut men's patriarchal authority over both women and children. Contemporary Japanese married couples evidence greater equality than in the past, although there is still a gendered division of labor (Takagi 2002). Male dominance may continue to be characteristic of more recent immigrants and some subgroups, but not of Chinese, Japanese, and Koreans (Ishii-Kuntz 2000).

Asian American parents worry about whether their children will become too "Americanized," getting divorced and losing the priority of family (Lott 2004). Youth struggle with being "between two worlds," feeling and wanting to be American, yet seen as "'forever foreigners'" because of their appearance. But with maturity, many return to an affirmation of their Asian cultural identity and its strengths (Mustafa and Chu 2006).

Pacific Islander Families

We don't know much yet about the families of Pacific Islanders, now considered separately from Asians. Major groups are Native Hawaiians, Samoans, and Guamians. Hawaiians, of course, are American citizens by birth, and so are American Samoans, Guamians, and those born in the North Mariana Islands. U.S. residents born in the other Pacific islands may have become naturalized citizens or are not citizens.

The Pacific Islander population is relatively young (median age is 29.8 years old) and about 29 percent of the population are children, so it is not surprising that it has a higher proportion of "never married" individuals than the overall American population. Just under half are married (47.4 percent), a figure that closely corresponds to the overall U.S. figure. Pacific Islander children are more likely to reside in family households (31.5 percent) than the U.S. population generally (21 percent), and Pacific Islanders have similar rates of marital stability as compared to non-Hispanic whites (U.S. Census Bureau 2007a, p. 10).

Educational attainment is similar to the United States overall at the high school level, but Pacific Islanders have a smaller proportion of college graduates, and there are fewer professionals and more service workers. Median household income is slightly higher ($55,273 in 2007) and poverty is slightly higher (15.7 percent) than for the general United States population (U.S. Census Bureau 2008e, Tables 1 and 9).

A Closer Look at Diversity

Family Ties and Immigration

There is more racial and ethnic diversity among American families than ever before, and much of this diversity results from immigration. The foreign-born now constitute 12 percent of the U.S. population. Seventeen percent of America's children live in a household headed by a foreign-born parent (Lugaila and Overturf 2004; Martin and Midgley 2006).

The United States admits approximately one million legal immigrants each year. Asia, Latin America, and the Caribbean—not Europe—are now the major sending regions. In addition to legal immigrants, there are approximately 11.9 million undocumented immigrants (not legal residents) residing in the United States, with approximately 8.3 million in the U.S. labor force (Passel and Cohn 2009, p. 2). The vast majority of these are from Mexico, Central America, and the Caribbean, but there are also substantial numbers from such countries as Canada, Poland, and Ireland. Although immigrants were previously concentrated in a few states, now they are much more geographically dispersed (Passel and Cohn 2009, Table 3). Almost one-third of foreign-born Americans have become naturalized citizens ("Illegal Immigrant Population" 2001; Martin and Midgley 2006).

Why do immigrants choose to come here? For the most part, immigrants leave a poorer country for a richer one in hopes of bettering their family's economic situation. Many immigrants have arrived here and spread out across the United States to areas that previously had little immigration. Nebraska, for example, is home to clusters of Vietnamese refugees, Afghani, Cuban, Hmong, Serbian, Somali, Sudanese, Soviet Jewish, and Mexican and Central Americans who have come to work in the meat-packing plants. Immigrants may be single individuals or they may be young or middle-aged married adults who migrate with spouses or children or who plan to bring them here.

Many immigrants experience a variety of challenges, including back-and-forth changes of residence, frequent family visits, international business dealings, money transfers to family abroad, placement of children with relatives in the home country, and seeking marital partners in the home country. As immigrants establish themselves, they begin to send for relatives—in fact, the majority of legal immigrants enter the United States through family sponsorship (Martin and Midgley 2006). Many immigrant families have members with different legal statuses. One spouse may be a legal resident, the other not. Children born here are automatically citizens, but one or both of their parents may be illegal residents of the United States.[a] In fact, almost one-third of all immigrant children come from such mixed-status families (Fortuny et al. 2009). Of concern are the many young adults whose undocumented parents brought them to the United States as children and who are therefore not legal residents but have no connections in their country of origin (Gonzalez 2006b). For example, the Urban Institute notes that 24 percent of children under age five have immigrant parents, as compared with 21 percent of children between ages six and seventeen (Fortuny et al. 2009).

Assimilation and acculturation processes can create tension between immigrant parents and their children (Baca Zinn and Pok 2002). Children of immigrants often feel the push/pull between societal expectations and their parents' more traditional ideals (Pyke 2007).

Marriages of recent immigrants seem less egalitarian than those of couples of similar background whose families have been in the United States longer. Yet migration is likely to change husband–wife roles even in cultures in which family life is experienced as carrying on

American Indian (Native American) Families[10]

A unique feature of Native American families is the relationship of tribal societies to the U.S. government. At present there are over 500 federally recognized tribes (Willeto and Goodluck 2004).

In the latter half of the nineteenth century, American Indians were forcibly removed from their original tribal lands to reservations, and some tribes were dissolved. Assimilation policies led to the creation of boarding schools where young American Indian children were placed for years with little contact with family or tribe. American Indians were encouraged to seek better conditions for their infants by placing them for adoption with white families; many of those adoptions appear in retrospect to have been forced or fraudulent (Fanshel 1972). Given the history of Native American oppression, "it is not surprising . . . that American Indians suffer the highest rates of most social problems in the U.S." (Willeto and Goodluck 2004). Tribes have high rates of teen suicide, school violence, teen pregnancy, and drug and alcohol abuse (Kershaw 2005; Madrigal 2001).

In the 1960s, American Indians successfully advocated for their rights, and a degree of tribal sovereignty was formalized in federal law. The Indian Child Welfare Act of 1978 gave tribes communal responsibility for tribal children. The law favors "placement of tribal

[10] Although Alaska Native tribes are included in this category, for reasons of convenience and because little research has been done on Alaska Natives, we will refer to "Native Americans" or "American Indians." As to the choice between those terms, both are accepted by substantial numbers of respondents surveyed on this point. Others argue that only the individual tribal names should be used because their cultures are "vastly different" (Gaffney 2006).

© Kayte Deioma / Photo Edit

Many immigrants to the United States start small businesses. This immigrant family from Guatemala owns a bakery.

tradition and accepting the decisions of family heads. Male heads of household typically lose status when male privilege and authority here is not what it was in the home country, and they may have to take jobs at a much lower status level. Women, who usually enter the labor force after coming to the United States, begin to experience an independence and autonomy that carry over into the negotiation of new roles and patterns of family life (Hondagneu-Sotelo and Messner 1994; Jo 2002; Kibria 2007).

Immigrants bring many strengths to this country: the "immigrant ethos" of strong family ties and high aspirations, hard work, and achievement. Children in immigrant families spend more time on homework and have higher GPAs than the U.S. average, and they adjust well to school. Immigrants exhibit a strong devotion to family and community, respect for work and education, good health, and spirituality (D. Brooks 2006; Lewin 2001a; Reardon-Anderson, Capps, and Fix 2002).

Immigrant families pay payroll, Social Security, property, and sales taxes while having very limited access to government benefits; taxes paid outweigh services consumed.[b] The latest research suggests that there is a small uptick to the gross domestic product and a moderation of prices due to lower wages paid to immigrant workers (Martin and Midgley 2006). But costs and benefits are not evenly distributed. Most immigrant family tax dollars go to the federal government, whereas the costs of immigrants' schooling or emergency health care are largely paid by local governments. Whether immigration is a net gain or loss regarding taxes and benefits depends on the age and occupational level of the immigrant. Probably more is spent to aid elderly and less educated immigrants, while younger, more educated immigrants will pay more in taxes than they gain in benefits (Martin and Midgley 2003).

Whatever our views on immigration policy, immigrants are generally responsible family members doing what they can to improve their family lives.

Critical Thinking

What are some strengths exhibited by immigrant families? What are some challenges they face?

At the societal level, what benefits does recent increased immigration offer the United States? What challenges does it bring?

a. An estimated 3.1 million children have had their undocumented parents deported (Preston 2007).

b. Undocumented immigrants are not entitled to most government services; emergency health care and children's elementary and secondary education are the exceptions. There are also some limitations on benefits for legal residents who are not citizens.

children in tribal homes . . . so that they can learn the customs, values, and traditions that make them separate and distinctive cultures in the United States" (Madrigal 2001, pp. 1505–6).[11]

At $35,345, American Indian households have a median income that is significantly lower than that of whites or Asians (U.S. Census Bureau 2008e, Table 1). However, one-third of Native American families have incomes of more than $50,000 (DeNavas-Walt, Proctor, and Lee 2006, Table 2; U.S. Census Bureau 2006c,

[11] In situations where children need to be removed from the home or a biological mother wishes to relinquish a child for adoption, the parent or the state welfare authorities cannot make those arrangements on their own; the tribe must agree and may, in fact, wish to place the child on the reservation rather than in a white adoptive home. Where agreement cannot be reached among the parties, the issue may be referred to tribal or state courts (B. J. Jones 1995).

Table 37). Yet the poverty rate is high—25.3 percent on average for the years 2006–2008. The poverty rate for children under five reaches 52.1 percent, and the childhood poverty rate for all children under the age of eighteen is 46.1 percent in female-headed families (Lugaila and Overturf 2004; U.S. Census Bureau 2009a, Table S0201). Native Americans are one of the poorest racial/ethnic groups in the United States (Snipp 2005), with the poverty rate highest among those living on reservations (DeVoe and Darling-Churchill 2008). In common with other economically disadvantaged groups, American Indians have high rates of adolescent births and nonmarital births; births to American Indian teens rose 12 percent in the last two years to 59 percent, and 65.2 percent of all births are to unmarried women. Native American birthrates, however, are less than the U.S. average (Hamilton, Martin, and Ventura

Phil Schermeister/Corbis

Many social factors condition people's options and choices. One such factor is an individual's place within our culturally diverse society. These rural Navajo reservation children are learning to weave baskets to sell to tourists. Even within a racial/ethnic group, however, families and individuals may differ in the degree to which they retain their original culture. Many Navajo live in urban settings off the reservation or go back and forth between the reservation and towns or cities.

2009; Martin et al. 2006, Tables 2, 4, 18). The American Indian infant mortality rate is higher than the overall U.S. rate (MacDorman and Mathews 2008, Figure 3), as is the childhood mortality rate (Heron et al. 2009).

Still, the American Indian population saw a tremendous rate of growth in recent censuses, doubling between 1990 and 2000. This surge is ascribed to "ethnic shifting." Individuals who might in the past have hidden their heritage, fearing discrimination, no longer feel they need to. Others, who did not have a clear American Indian identity or heritage, have investigated their background, found some Native American ancestry, and so claimed an American Indian identity (Hitt 2005).

American Indians have a higher rate of cohabitation and a lower percentage of married couples than the U.S. average. More than half of married American Indians have spouses who are not Native Americans. Still, Native Americans are more likely to live in family households than the U.S. average (68.4 percent compared to 66.6 percent) (Ogunwole 2006; Snipp 2005; U.S. Census Bureau 2009a, Table S0201). American Indian tribes are now beginning to deal with issues of gay lifestyles and same-sex marriage (Duncan 2005; Leland 2006).

In 2006, 38 percent of American Indian and Alaska Native children lived with mother only and 11 percent with father only, whereas 51 percent lived with two parents (DeVoe and Darling-Churchill 2008, Table 1.5). Children and youth often move between households of extended family members, and, given the high rates of alcoholism on the reservation, children may be placed with foster families, Indian or non-Indian (Lobo 2001). Children often live with grandparents on the reservation when their parents go to the city to find work since jobs are scarce on the reservation (Snipp 2005). Officials of one tribe estimated that only half of the reservation children lived with a parent year-round (Kershaw 2005).

Native American culture gives great respect to elders as leaders and mentors. Older women may also be relied upon for care of grandchildren. In return, Native American families take care of the elderly, although they are finding that more difficult to do when so many adults live off the reservation.

Acculturation seems to have lessened the dominant position of men in the family. The increase in female-headed families has been "the most significant role change in recent times," and one that has enhanced female authority and status in the family and the tribe (Yellowbird and Snipp 2002, p. 243). The number of women tribal leaders has doubled in the last twenty-five years, and it seems that one result is greater attention to child welfare, other social services, and education. At the same time, some women tribal leaders believe the resistance they have often encountered is due to their gender (Davey 2006).

How American Indian families live depends on the tribe and on whether they live on or off the reservation. Only one-third live on reservations, which is surprising, given the historic and symbolic importance of the reservation (Ogunwole 2006). City dwellers often return to the reservation on ceremonial occasions; to visit friends and family; as a refuge in times of hardship; and to expose their children to tribal traditions. Both on and off the reservation, Native American adults may move around frequently, as they stay in the homes of family members or friends (Lobo 2001).

Because of their high rates of intermarriage and mobility of residence between the reservation and the city, American Indians have complex racial/cultural identities. Self-identity does not always match tribal registration (Ray 2006; Snipp 2002). American Indian identity for all but those living on a reservation has become rather fluid and uncertain (Etheridge 2007). Many American Indians have more ancestors who are white than Indian and so appear white. Further identity confusion results from the fact that non-Indians have taken up certain Indian symbols and practices—for example the 2010 Olympic Games logo featured the Inuit symbol "inuksuit" (a stone landmark)—leaving "real" American Indians wondering what markers do distinguish them ("The Aboriginal Aesthetic" 2009; Hitt 2005).

Dorothy Miller developed a typology to explore the relative influence of Indian and mainstream American culture on urban Indian families. **Miller's typology of urban Native American families** posits a continuum from traditional to bicultural to transitional to marginal families. *Traditional families* retain Indian ways, with minimal influence from the urban settings they live in. *Bicultural families* develop a successful blend of native beliefs and the adaptations necessary to live in urban settings. *Transitional families* have lost Native American culture and are becoming assimilated to the white working class. *Marginal families* have become alienated from both American Indian and mainstream cultures. In her empirical research, Miller found bicultural and marginal families to be most common, as transitional and traditional families move toward the bicultural model (D. Miller 1979; cited in Yellowbird and Snipp 2002, pp. 239–43).

Even on reservations or in Alaskan tribal areas, modern culture has penetrated due to the Internet and television (Kershaw 2004). Some predict that as the cultures of individual tribes fade, eventually a "new urban Pan-Indian tribe" will emerge ("Pow Wow Culture" 2006).

Arab American Families

The Arab American population is a little over 1.5 million, with Lebanese Americans making up the largest single ethnic group (U.S. Census Bureau 2010a, Table 52). Although Arab Americans are a relatively small portion of the population, they have found themselves the subject of media stereotyping and government suspicion—and this was before what has become known as "9/11."

Social theorist Professor Edward W. Said, in his book *Orientalism* (1979), noted that European, then American, scholars have long presented people from the Middle East in ways which stereotyped Arabs as exotic, mysterious, and dangerous. Following in Said's footsteps, media scholar Professor Jack Shaheen examined over one thousand American studio films depicting Arabs or Arab Americans (2009/2001). He found an unchanging and rigid stereotype that presents an image of "barbarism" and "buffoonery." Additionally, much of the scholarly marriage and family literature in the United States focusing on Arab families tends to view this ethnically and religiously diverse group as monolithic and through the lens of Euro-American superiority (Beitin, Allen, and Bekheet 2010).

These scholars join countless others in documenting the negative presentation of people and families of Arabic descent—a negative frame with devastating consequences. After the terrorist attacks of September 11, 2001, and continuing through today, Arab American families have been the subject of harassment, intimidation, vandalism, physical attacks, discrimination, and murder by their fellow Americans (American-Arab Anti-Discrimination Committee Research Institute 2008), making it extremely difficult to lead a normal family life.

Arab American families are made up of parents, children, and extended relatives. Family is very important to Arab Americans. Arab American families are often extended beyond American borders (Beitin, Allen, and Bekheet 2010). Like other immigrant families, there are important differences between immigrants and subsequent generations.

Religion is an important factor in Arab American families, but not in the way American media and cynical politicians have portrayed it. Religion is important to Arab Americans, just as it is to the majority of Americans. Sixty-five percent of Arab Americans are Christian, and most are second or third generation American citizens (M. S. Lee 2005, p. 30). "Those immigrating since the 1950s and most Muslim families are likely to relate less with the white majority culture and more with subcultures in which religious, national-origin, and language traditions are preserved. For those who live in ethnic enclaves, intra-group marriages, and family businesses often limit outside social interaction" (Samhan 2005, p. 2).

Arab American women are employed at lower rates than other women. Scholars suggest the reasons for this are varied, but traditional gender roles are emphasized in Arab American families, and women are considered the "bases of security and stability" for family members (Gold and Bozorgmehr 2007, p. 526).

White Families

Non-Hispanic whites continue to be the numerical majority in the United States, comprising 66 percent of the population. Yet, in talking about family cultures, we typically see nothing distinctive about white families or consider them a part of racial/ethnic diversity. In recent years, however, academics and other scholars have begun to devote conscious attention to whether "white" as a racial category indicates a distinct culture and identity.

White families are largely of European descent and so are sometimes termed **Euro-American families**. There have been many studies of family life in specific European-American settings. Studies of working-class families and rural families are usually based on whites. Much that is written about "the family" or "the American family" is grounded in patterns common among middle-class whites. But the concept of "white families" has not really been considered except for the presentation of government statistical data.

The Demographics of White Families According to data and research, the non-Hispanic white family household, compared to those of most other racial/ethnic groups, appears more likely to be headed by a married couple and less likely to include family members beyond the nuclear family. Whites are older than other groups, on average, and have lower fertility rates, so white families are less likely than Hispanic or black families to have children under eighteen living at home. White families have higher incomes than all groups but Asians and lower poverty rates than all other racial/ethnic groups. White women are less likely than black and Hispanic women to bear children as teenagers or to have nonmarital births; however, such births have become more common, with one in seven white women born after 1962 having nonmarital births (Wildeman and Percheski 2009, pp. 1298–99). In 2008, almost 73 percent of non-Hispanic white children lived with two parents (DeNavas-Walt, Proctor, and Smith 2009; Fields 2004; Hamilton, Martin, and Ventura 2006, Table 1; Sarkisian, Gerena, and Gerstel 2006).

In terms of family structure and economic resources, white children and families are more advantaged. Yet the ties that provide mutual support and care of younger and older members are not as strong. White respondents reported less caregiving to aging family members; they are also less likely to rely on family members as child care providers (American Association of Retired Persons 2004; Uttal 1999). Residential separation of whites from most other racial/ethnic groups continues even in suburban settings (Frey 2002).

Whiteness Studies As part of academic interest in whether there is something distinctive about being white, some universities have developed Whiteness Studies programs—analogous to Black Studies or Latino Studies, but with some important differences. Scholars and students consider what it means to be white. For example, Steve Garner (2007) notes that the concept of "whiteness" comes from being socialized in nations with racial regimes (i.e., racial hierarchies), and exists in opposition to other racially defined groups. Importantly, whiteness, as a concept, is used to define what is normal in society (which is why, when we speak of race, we tend to ignore—make invisible—people who are "white"), and thus create what Bourdieu termed **cultural capital** (Bourdieu and Passeron 1979). Cultural capital is a form of cultural competence in which one has "ownership or control over social goods, such as mannerisms and practices that have recognizable high status value" (Gerbrandt 2007, p. 61).

One theme found in whiteness studies is "privilege"— the idea that non-Hispanic whites have advantages in our society that go unnoticed by them (McDermott and Samson 2005). Like blacks, Latinos, Asians, and American Indians, whose opportunities and living conditions may be somewhat determined by their racial/ethnic background, whites' lives are also strongly shaped by their European ethnic heritage. Whites in the United States generally maintain an advantage in terms of cultural capital. Historically it has been whites who have had the most access to education, were allowed to own land, had access to jobs that were off-limits to people of color, and so forth. This history of advantage continues to present opportunities for whites, while presenting disadvantage for people of color. For example, consider educational attainment—a higher percentage of the white population earn degrees. Most Euro-Americans have been assimilated into American society to such a degree that European ethnic identities are "voluntary" and "symbolic." Individuals can choose to highlight them or not, depending on the occasion and the pleasure certain cultural practices (holidays, food) may give them (Waters 2007).

At Arlington National Cemetery, Buddhist monks escort the coffin of an American soldier killed in Iraq. Immigration has contributed to increasing religious diversity in the United States. There has been a Buddhist presence in the United States since at least the nineteenth century, and Buddhist practices have been followed by many Americans of non-Asian backgrounds. But the number of Buddhists more than doubled from 1990 to 2001 as the Asian American population increased through immigration.

In the future, more white Americans may begin to think about identity in racial/ethnic terms, as high levels of immigration and the increasing visibility of Latinos, Asians, and Native Americans as well as African Americans challenge an unconscious assumption that "American" equals "white."

At the same time, there is as much diversity among white families as there is within other broad racial/ethnic groups. To consider "white" to be the same as middle-class is to ignore the existence of some marginalized identities for whites: regional identities such as "redneck" and "hillbilly," pejorative class identities such as "trailer trash," as well as well-recognized class differences and gay/lesbian identities (McDermott and Samson 2005; Royster 2005). In fact, Garner points out that such white identity promotes an association with dominant ideologies, even at the detriment of one's own self-interest (2007). For example, a voter might vote against a politician because he or she is a person of color (or gay/lesbian), even though that candidate's political policies would be beneficial for the voter.

Whiteness Studies programs and scholars have examined militant and racist "white power" movements, which advocate white superiority and racial separatism, sometimes engaging in violence. These acts of violence are known as *hate crimes*. A hate crime is a "criminal act motivated by the victim's personal characteristics, such as race, national origin, or religion" (Gerstenfeld 2010, p. 258). Whiteness Studies programs go beyond scholarship to mobilize *antiracist* attitudes and behaviors of students and the public. Antiracist attitudes and behaviors are those which actively oppose racism and racist behaviors.

Multi-Ethnic Families[12]

As noted in Chapter 1, Barack Obama's campaign and election to the presidency of the United States helped to promote a national dialogue about interracial and international relationships. President Obama is the product of a marriage between a white American and a black Kenyan. Born in Hawaii, President Obama spent some of his childhood in different parts of the world, giving this biracial American a more global identity. Previously, golfer Tiger Woods's emergence as a celebrity had made interracial and interethnic families visible. Multiracial and multiethnic families are created by marriage or establishment of an unmarried-couple household (often followed by the birth of children), and/or by adoption of children who are of a different race than

Image copyright Rob Marmion, 2010. Used under license from Shutterstock.com

Multiracial families are formed through interracial marriage or formation of a nonmarital partnership and also by the adoption of children across racial lines.

their new parents. Since colonial times there has been racial mixing in the United States in marital and other sexual relationships (Maillard 2008). Now the former dichotomy of black and white has expanded into a multiplicity of racial/ethnic identities, including multiracial or multiethnic identities and families.

According to Michael J. Rosenfeld, a Stanford University demographer, approximately 7 percent of married-couple households include spouses whose racial/ethnic identities (regarding racial self-identification and Hispanic heritage) differ. This pattern was even more common in households of unmarried couples. Fifteen percent of opposite-sex partners and male same-sex partners and 13 percent of female same-sex partners reported different racial/ethnic identities (U.S. Census Bureau 2010b, Table 60).

However, only a small percentage of individuals claim a multiracial identity. Only 2.6 percent of the population checked more than one race in the 2000 census. The self-identified two-race population is younger than the overall U.S. population; 25 percent are under ten (N. Jones 2005). The proportion of multiracial children in the population is likely to grow with increasing intermarriage and perhaps a greater tendency to acknowledge a mixed racial/ethnic heritage. There are seven million people who self-identify as bi- or multiracial. "Five percent of Blacks, 6% of Latinos, 14% of Asians, and 2.5% of whites identified themselves as members of at least two races" (Leong 2006, p. 4). One estimate is that by the end of the century, 37 percent of African Americans, 40 percent of Asians, and two-thirds of Latinos will claim a multiracial identity (Rodriguez 2003). "Claim" is the key word here, and the future depends on how individuals come to see their racial/ethnic identity.

How individuals define their multiracial/multiethnic identity when their heritage is mixed varies by age,

[12] Measuring multiracial identity is difficult. Sample surveys taken before the 2000 census indicated that a straightforward question would not work. A decision was made to allow respondents to check more than one race in the 2000 census. However, only around 2.6 percent of the population did so (N. Jones 2005; Jones and Smith 2001).

gender, and the racial/ethnic combination (Jayson 2006b; N. Jones 2005; Rockquemore 2002). Most blacks are well aware that typically they would have a mixed race heritage, but thus far most have maintained a black self-identity. Studies used in this text are the most recent available, and they report findings from the 2000 census. The 2010 Census will provide researchers with newer data. The available data shows that less than half of parents of multiracial children report them as multiracial (Olvera 2009; Tafoya, Johnson, and Hill 2004). As immigration brings new racial/ethnic identities to the fore and intermarriage increases, American patterns of racial/ethnic self-identification are likely to become more fluid and to reflect a multicultural heritage (Bean et al. 2004).

As yet, we have had few studies focused on multiracial/multiethnic families in all their complexity. The U.S. Census Bureau reports fifty-seven combinations of race and ethnicity, and the studies that exist have varying combinations of race and ethnicity. Research on multiracial/multiethnic families has essentially just begun.

Census data indicate that multiracial couples are more apt to cohabitate than couples from the same racial category (Rosenfeld 2008), and another study reports that multiracial children are no more likely to live in unmarried or unstable families than are single-race children, although this varies with the particular racial/ethnic combination (Goldstein and Harknett 2006). Class indicators—occupation, income, education—also vary considerably by racial/ethnic combination (N. Jones 2005). A study of high school students found that multiple ethnic and racial identifications is correlated with high levels of psychological well-being. In fact, the greater the self-identification, the greater the positive outcomes (Binning et al. 2009).

Tensions may arise out of cultural differences within families, and issues may need to be worked out before a couple and their children can reap the benefit of their rich cultural mix. However, a *Washington Post* national survey of 540 interracial married or cohabitating couples tells us that families have been accepting, on the whole. African American/white couples have encountered more difficulty in marriage or with parents than have Asian/white or Latino/non-Hispanic white couples. Recent research suggests that the black husband/white wife marriage experiences the greatest instability, most likely related to the differences that remain between the two groups because of the history of slavery in this country (Bratter and King 2008). Yet by and large, the *Washington Post* survey indicates that interracial couples are positive about the benefits of diversity. They believe that their children are more advantaged than disadvantaged by their multicultural heritage. (Interracial unions are discussed further in Chapter 6.)

Religion

Religion is increasingly analyzed as an important element of family life. Indeed, religious affiliation and practice is a significant influence on family life, ranging from what holidays are celebrated to the placement of family relations into a moral framework. For example, recent research shows that when children and adolescents have deeper religious connections, they tend to have less premarital sex and to be older when they have their first sexual experience (Eggebeen and Dew 2009; Wildeman and Percheski 2009).

Looking at religion over the life course, family scientist Elizabeth Miller (2000) found that religion offers rituals to mark such important family milestones as birth, coming of age, marriage, and death. Religious affiliation provides families with a sense of community, support in times of crisis, and a set of values that give meaning to life. Membership in religious congregations is associated with age and life cycle; young people who have not been actively religious tend to become so as they marry and have children. At the same time, family disruption—divorce, separation, and remarriage—seems to lead to a renewed sense of religion's importance. What seems to be important is not which religion or religions family members belong to, but the fact that family members hold religious beliefs (Vaaler, Ellison, and Powers 2009).

The United States is among the most religious of modern industrial nations. Eighty-three percent of American adults surveyed in 2007 indicated a religious identification (Pew Forum on Religious Life 2008), although in 2008 that percentage dropped to about 75 percent (U.S. Census Bureau 2010b, Table 75).[13] Relating religion to family life, recent research suggests that "religious couples are less prone to divorce because, on average, they enjoy higher marital satisfaction, face a lower likelihood of domestic violence, and perceive fewer attractive options outside the marriage than their less religious counterparts" (Vaaler, Ellison, and Powers 2009, p. 930).

The historically dominant religion in the United States has been Protestantism, especially "mainstream" denominations such as Presbyterianism and Methodism. But mainstream Protestantism has been in decline. Presently, about 17 percent of the U.S. population belongs to a mainstream Protestant denomination, and evangelical Protestantism now encompasses 7 percent

---

[13] The Census Bureau does not collect data on religion, but it does publish survey results in its *Statistical Abstract*. These data are from the American Religious Identification Survey in 2001 over 50,000 households (Kosmin, Mayer, and Keysar 2001), as reported in the 2007 *Statistical Abstract* (U.S. Census Bureau 2007a, Table 73), and from the Pew Forum on Religion & Public Life (2009), which surveyed 35,556 adults.

of Americans. Roman Catholics make up 18.8 percent, Mormons 1 percent, and Jews .08 percent of the population (U.S. Census Bureau 2010b, Table 75).

The U.S. Muslim population is estimated at around 2.5 million (Pew Forum on Religion & Public Life 2009, p. 24). With immigration from Asia, Hindus and Buddhists have increased in numbers. The extent of religious diversity in the contemporary United States is illustrated by the fact that a midwestern city like Omaha, Nebraska, has a Buddhist center, a Hindu temple, and a mosque. Some other religions have become increasingly visible, notably Native American religion and neo-paganism or Wicca—a nature-based religion drawing on European pre-Christian traditions, whereas "[r]oughly one-quarter of adults express belief in tenets of certain Eastern religions" (Pew Forum on Religion & Public Life 2009a, p. 7).

Earlier research in the sociology of religion focused on how Catholics and Jews differed from Protestants in their family patterns. The formalities of doctrine have not always had the effect we might assume. Catholics, for example, appear to have shifted from traditional church teachings to modern conceptualizations of family and sexuality, and their views do not differ much from those of the general population on such issues as homosexuality, birth control, and family size (D. Moore 2005). For example, 61 percent of American Catholics say that it should be left to individuals to decide whether to use birth control, regardless of Church doctrine (Simmons and Eckstrom 2009). In fact, 50 percent of people who have left the Catholic faith have done so because of the Church's stance on birth control (Pew Forum on Religion & Public Life 2009b).

Today, social scientists find certain other religious groups more interesting to study, notably Latter-day Saints (popularly, but incorrectly, termed *Mormons*) and conservative Christians. Large families are encouraged by the LDS Church and is reflected in Utah's distinctively high birthrates (Hamilton, Martin, and Ventura 2006, Table 8). Latter-day Saints, Evangelical Christians, Jehovah's Witnesses, and Muslims reject homosexuality perhaps more strongly than some others. Conservative Protestant Christians and Latter-day Saints are strongly opposed to abortion, whereas Muslims and Catholics remain almost evenly split (Pew Forum on Religion & Public Life 2008, p. 135). Members are not always in conformity with their church's teachings. But religious influences tend to be powerful enough to produce differences in responses to surveys about attitudes and values in many family-related areas.

The theology of conservative Christian religions endorses traditional gender roles, specifically the concept of "headship"—the man as head of the family. But there is a diversity of viewpoints on gender roles *within* the evangelical community (Bartkowski 2001; S. Gallagher 2004). Researchers find a similar diversity among Latter-day Saints women (Beaman 2001). For some, headship is a symbolic value denoting a separation from aspects of American culture thought immoral or threatening. It is not necessarily an everyday reality. A "servant-head" and "mutual submission" interpretation of headship seems to lead to egalitarian decision making in day-to-day family life. "Headship has been reorganized along expressive lines, emptying the concept of virtually all of its authoritativeness" (Wilcox 2004, p. 173).

Evangelical women believe they benefit from the headship concept because they perceive it as providing them with love, respect, and security. Conservative Christian men, more than mainstream Protestant men, seem more emotionally expressive with their wives and children, and more committed to their marriages. Conservative Christian husbands also have lower rates of domestic violence than average (S. Gallagher 2004; Wilcox 2004).

Marital sharing and equality do not seem to extend to household labor. Both men and women see a woman's role as giving priority to the domestic sphere, while husbands do not expect to do many household tasks. This unequal division of labor appears to be a site of conflict in households where the husbands and wives differ in terms of their religiosity. For example, in households where the husband is more religious than the wife, particularly in Islam and evangelical Protestantism, such strictly gendered divisions lead to family tension and decreased marital satisfaction, particularly for the wife, leading to increased risk for divorce (Duba

AP Photo/East Valley Tribune, Heidi Huber

Two volunteers at the American Muslim Women's Association work on a craft project to benefit poorer immigrants and refugees. This organization, with many professional members, also works to reshape the roles of women in Islam. One young woman wears a headscarf; the other does not. Some modern young Muslim women have recently adopted the head scarf to express an intensified identification with Islam in the context of experiences of discrimination or challenge to their religious community.

and Watts 2009; Vaaler, Ellison, and Powers 2009; Zink 2008). Economic need shapes a pragmatic approach to the employment of women, but conservative Christian women are more likely to scale back their work hours and job level upon marriage and/or childbirth (Glass and Nath 2006).

Cohabitation is strongly associated with a decline in religiosity (Uecker, Regnerus, and Vaaler 2007). The groups with the highest rates of nonmarital cohabitation are those who are unaffiliated with any religious organization. Hindus have the highest marriage rates (at 78 percent), followed closely, at 71 percent, by Latter-day Saints (Pew Forum on Religion & Public Life 2008, p. 67). With regard to marital stability, "born-again" Christians are far less likely than other Americans to enter cohabiting relationships, but their divorce rate does not significantly differ from others (Myers 2006).

Religion may be associated with child-rearing practices. Recent studies suggest that regardless of the religious faith (and regardless of the level of religiosity), children in families who adhere to a religious belief tend to be better adjusted. The thinking is that it is less about religious beliefs and more about being a member of a community that is important for family life (Good and Willoughby 2006; Lees and Horwath 2009). These spiritual benefits seem in line with social theorist Pierre Bourdieu's previously discussed *social capital*. In fact, one social researcher, Gerald Grace (2002), has recently added to Bourdieu's capital typology by suggesting religious beliefs offer *spiritual capital*. Spiritual capital includes "resources of faith and values derived from commitment to a religious tradition" (p. 236). Those religions such as Islam or Judaism that depart from a Christian tradition have the added burden of raising children in a society that does not support their faith; this could be said of conservative Protestant groups as well.

Conservative Protestant fathers and Chinese American mothers appear more likely than the general population to be authoritarian, emphasize obedience, and to use physical punishment. At the same time, they seem more emotionally expressive in their child rearing and more likely to praise and hug their children (Lindner Gunnoe, Hetherington, and Reiss 2006). It also appears that conservative Protestant fathers are more likely than mainstream Protestant or unaffiliated fathers to engage in one-on-one interaction with children in leisure activities, projects, homework help, and just talking. They are more likely than religiously unaffiliated fathers to have dinner with children and to participate in youth activities as coaches or leaders. Catholic fathers follow the same pattern of involvement (Wilcox 1998, 2002).

Interfaith marriage provides different types of challenges. There are a good number of interfaith marriages in America today. For example, 23 percent of Catholics, 33 percent of Protestants, 27 percent of Jews, and 21 percent of Muslims are married to people of different faiths (Duba and Watts 2009; Lara and Duba Onedera 2008). Holidays may be difficult because outsiders may expect Muslim or Jewish children to join in Christian celebrations. Some religious groups consider Halloween to be satanic and do not permit their children to celebrate it. Religiously mixed couples may experience tension over how to celebrate the holidays as a family (Haddad and Smith 1996; Horowitz 1999). Issues other than holiday celebrations may be involved. Conservative Christians may not permit their children to date (Goodstein 2001). Regardless of the difficulties interfaith families may have, research shows that couples and families that adhere to religious beliefs, regardless of how those beliefs might differ, have much healthier and well-adjusted, warm, and communicative relationships (Brimhall and Butler 2007).

Islamic families provide an example of the difficulty of maintaining a religiously appropriate family life in the context of a culture that does not share their beliefs. For Muslims, dating, marital choice, child rearing, employment of women, dress, and marital decision making are all religious issues. As one Muslim mother stated to a researcher:

> I think that integration into the non-Muslim environment has to be done with the sense that we have to preserve our Islamic identity. As long as the activity or whatever the children are doing is not in conflict with Islamic values or ways, it is permissible. But when we see it is going to be something against Islamic values, we try to teach our children that this is not correct to our beliefs and practices. They understand it and they are trying to cope with that. (Haddad and Smith 1996, p. 19)

Muslim families now have the added burden of facing suspicion and hostility in the wake of 9/11, but those experiences have also fostered a stronger identity. Younger Muslims seem more consciously and conservatively religious (Pew Forum on Religion & Public Life 2008).

For all religions, finding a balance between participating in the larger society and preserving unique values and behaviors and a sense of community is a challenge in a society characterized by religious freedom rather than a religious establishment. Yet that freedom seems to be cherished by virtually all religious groups in the United States.

As we explore various aspects of American families in greater detail throughout the remainder of this text, it may help you to recall that families differ according to social context—religion, race/ethnicity, social class, age structure, and the historical time in which they live. In the next chapter, we examine gender as a major determinant of experiences in the family.

Summary

- Families exist in a social context that affects many aspects of family life.

- Historical events and trends have affected family life over the last century. These include economic and cultural trends as well as wars and other national crises.

- The age structure affects family patterns and social policy regarding families. The proportion of older Americans in the population is increasing, so we may anticipate a growing responsibility for them. The proportion of children in the population is decreasing, leading to questions about the future of society's commitment to children.

- The economy has a strong impact on family life. Americans do not like to acknowledge class differences, but economic resources affect family options. So do differences in values and preferences that characterize blue-collar and white-collar sectors of society.

- Race and ethnicity shape family life because African American, Latino, Asian, American Indian, Pacific Islander, and white families have some differences in structure, resources, and cultures. The increasing rate of racial/ethnic intermarriage suggests that more families in the future will be multicultural.

- Immigration has risen to a level that contributes visibly to the diversity of family life.

- Religious traditions and prescriptions shape family life. In studying the family, we are apt to take more notice of religions that maintain distinctive family norms.

Questions for Review and Reflection

1. In the everyday lives of families (yours or those you observe), what economic pressures, opportunities, and choices do you see?

2. Describe one specific social context of family life as presented in the text. Does what you read match what you see in everyday life?

3. Do you see the lives of military families as basically similar to or different from those of other families?

4. What are some significant aspects of the family lives of immigrant families? Do you think immigrant families will change over time?

5. **Policy Question.** Which age group is increasing as a proportion of the U.S. population, children or the elderly? What social changes might occur as a result? What social policies do we need to maintain or develop to care for children and the elderly?

Key Terms

binational families 64
cultural capital 72
ethnicity, ethnic identity 57
Euro-American families 71
habitus 56
infant mortality rate 65
life chances 53

Miller's typology of urban Native American families 71
race 57
segmented assimilation 64
sex ratio 62
social class 56
transnational families 67
undocumented immigrant 64

Online Resources

Sociology CourseMate

www.CengageBrain.com

Access an integrated eBook, chapter-specific interactive learning tools, including flashcards, quizzes, videos, and more in your Sociology CourseMate, accessed through CengageBrain.com.

www.CengageBrain.com

Want to maximize your online study time? Take this easy-to-use study system's diagnostic pre-test, and it will create a personalized study plan for you. By helping you identify the topics that you need to understand better and then directing you to valuable online resources, it can speed up your chapter review. CengageNOW even provides a post-test so you can confirm that you are ready for an exam.

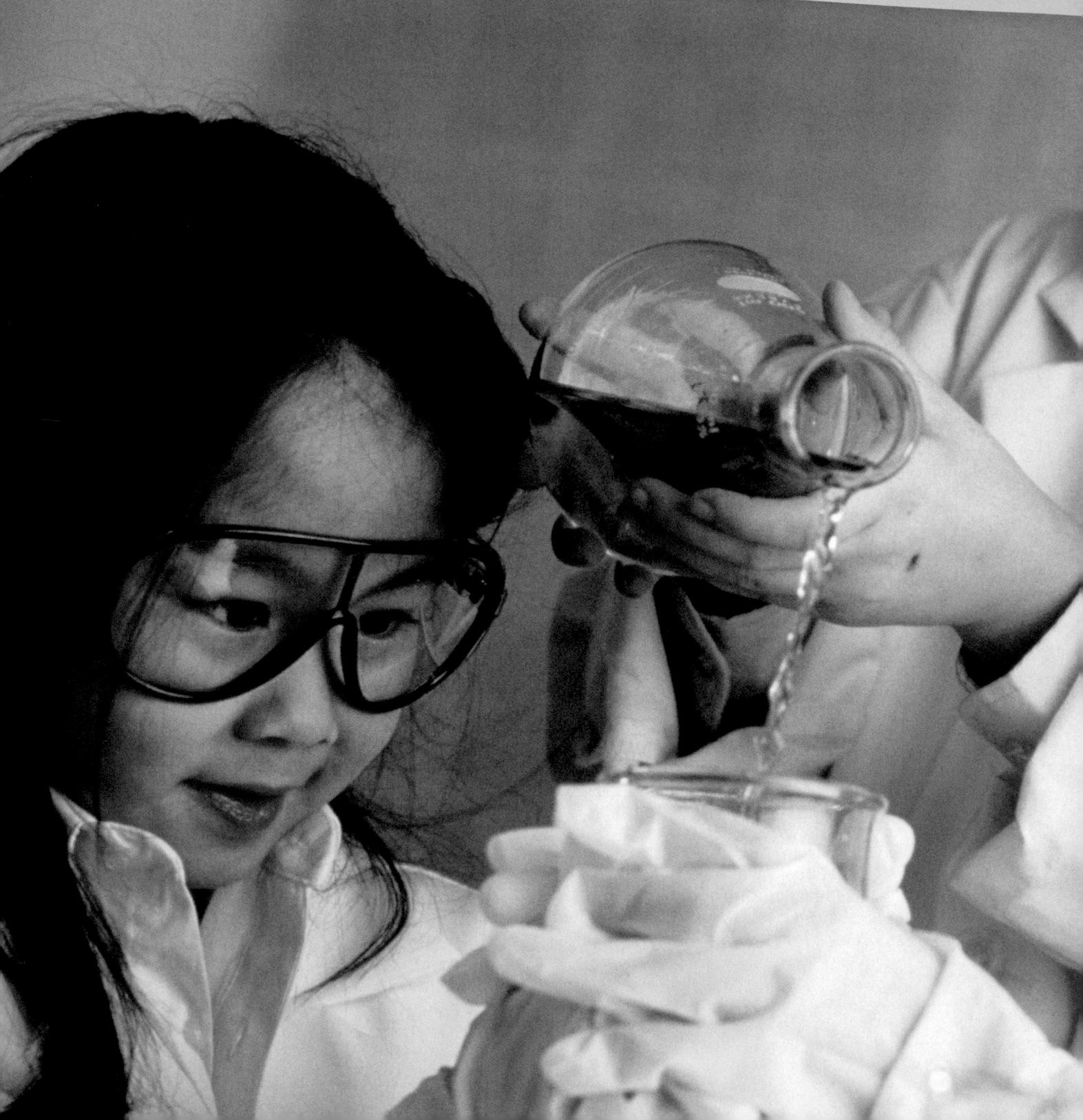

4

Our Gendered Identities

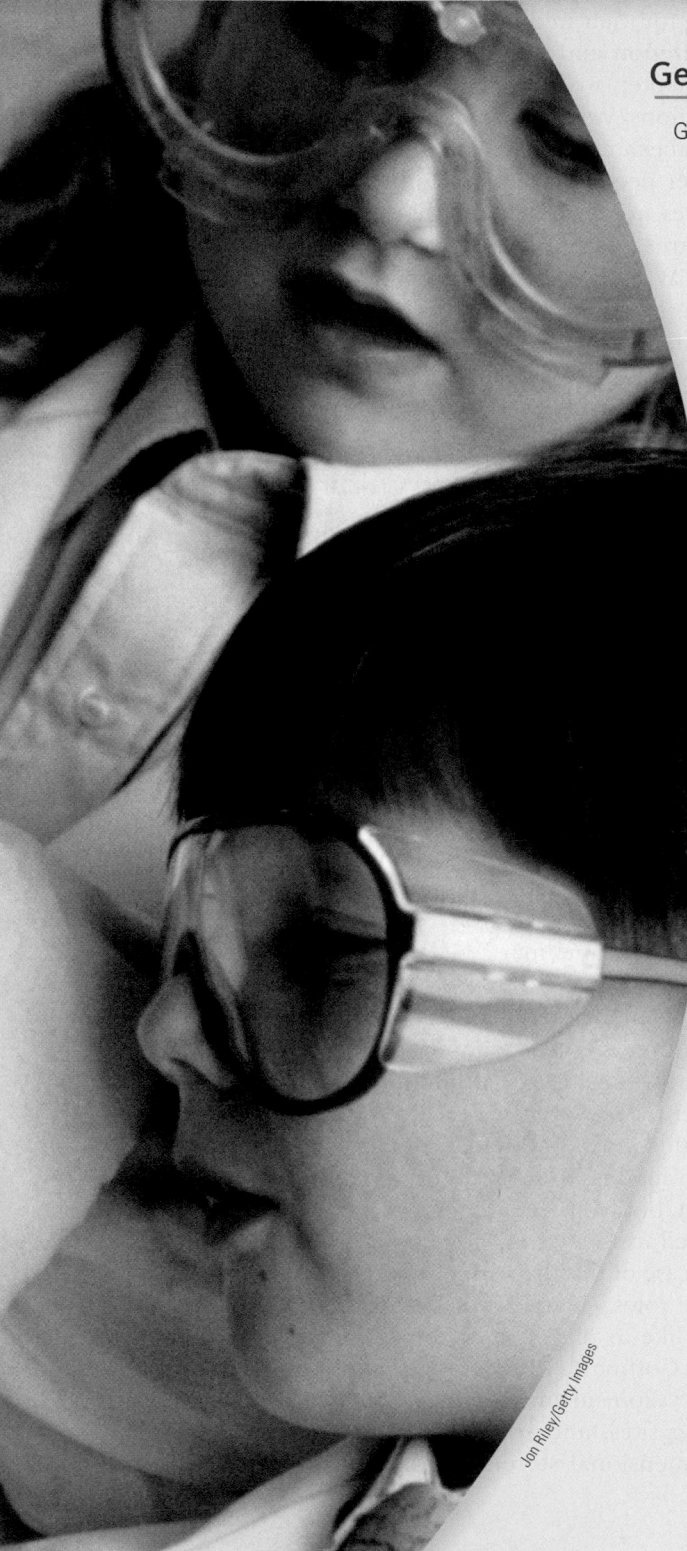

Jon Riley/Getty Images

As we think about family life, we need to consider gender issues. Cultural expectations about how boys and girls, men and women should behave and relate to each other are important influences on personal identities, family roles, and life choices. But to what extent have traditional expectations changed?

A statement from earlier editions of this textbook, "gender influences virtually every aspect of people's lives and relationships," is now being challenged. Some scholars assert that the significance of gender is declining. Others, however, see gender identity and gender inequality as continuing to be enormously important, but point to how our *thinking* about gender has changed.

From an earlier 1970s perspective, in which all women were seen to be disadvantaged compared to all men, we now look at gender in relation to its structural linkages to race, class, and sexual orientation, as well as to global interconnections (Andersen and Collins 2007b; Baca Zinn, Hondagneu-Sotelo, and Messner 2007; Theidon 2009). For example, a black woman immigrant from Haiti who is a single mother working as a domestic has a very different life from that of her employer, a white woman lawyer who is an Ivy League graduate with a professional husband. The husband's life is much different from that of a white male high school dropout who remains single into his forties because he finds himself financially unable to marry (Porter and O'Donnell 2006). Additionally, a women in another country whose soldier-husband has laid down his weapons may have domestic violence issues similar to those of a military wife in the United States (Meneses 2008; Sontag 2008). As noted in our analysis of race/ethnicity in Chapter 3, *within-group* differences have become an important theme in the study of gender.

Gendered Identities

We are using the term *gender* rather than *sex* for an important reason. The word **sex** is used in reference to male or female anatomy and physiology. We use the term **gender** (or **gender role**) far more broadly—to describe societal attitudes and behaviors expected of and associated with the two sexes (Duck and Wood 2006, p. 170).[1] Another concept, **gender identity**, refers to the degree to which an individual sees herself or himself as feminine or masculine.

A further complication occurs because a small number of people are born with ambiguous sexual

characteristics or do not feel at ease with their sex as recorded at birth. "Issues for Thought: Challenges to Gender Boundaries" addresses variations in biology that affect sex and gender identification.

In this chapter, we will examine various aspects of gender. In doing so, we'll consider personality traits and cultural scripts typically associated with masculinity and femininity. For example, note the differences between women's and men's perceptions of fairness and equity as they divide the tasks associated with preparing for their weddings in "Issues for Thought: 'Wife' Socialization and the Heterosexual Wedding." Also in this chapter, we'll analyze gender inequality in social institutions. We'll discuss the possible influence of biology and examine the socialization process as we explore whether people are taught to behave as either females or males, are born that way, or simply adapt to the social structures and opportunities they find as they become adults. We'll discuss the lives of adults as they select from options available to them. And we'll examine the social movements that have arisen around gender issues. We'll speculate about what the future may hold in terms of gender equality.

Gender Expectations and Cultural Messages

We live in an ambiguous time regarding gendered attitudes and behavior. On the one hand, it is now taken for granted that women have careers and that most will work regardless of motherhood. Examples of women in nontraditional roles abound: women as astronauts, CEOs, and officers and enlisted personnel in the military. On the other hand, the media and research explore the continuing disadvantages faced by women and the uncertainty that many women have about their choices and the ability to realize them.

Men have begun to consider where they stand as well—in relationships, in the family, and at work. Some men have begun to move into traditionally female occupations, and married fathers are doing more at home than they used to (Bianchi, Robinson, and Milkie 2006). Women and men are asserting their individuality and joining as equals many areas of society that were, at one time, considered "off-limits" to the other gender. Frequently, however, we receive mixed messages from institutions such as the mass media.

Our actions, thoughts, and feelings come not from instinct, but from social messages, and often those messages tell us the wrong story about each other, leading, in the case of this discussion, to difficulties trying to sort out the roles we are expected to play and the roles we would like to play as men and women. Social theorist Erving Goffman (1967) notes that the messages we see on television, in magazines, online via our social networking sites, and so on, often present an idealized, one-dimensional stereotype of what women and men

[1] The distinction between *sex* and *gender*, first made by sociologist Ann Oakley (1972), is a dominant perspective in social science (Laner 2003). But not all social theorists agree with this conceptualization. Judith Butler (1990) argues that the two are essentially one—gender—and others agree that the biological as well as behavioral aspects of sex are socially constructed (Cresswell 2003).

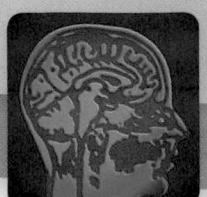
Planning a wedding is more than picking a wedding gown, ordering flowers, and choosing a caterer. Research reveals that heterosexual wedding planning serves as a tool to prepare brides for their future roles as wives. The work involved in planning a wedding highly resembles the domestic and family work that married women have long performed. Weddings, thus, give brides a taste of, and practice for, their future roles as family and relationship managers.

Brides and grooms have a tendency to view themselves as equal participants in wedding planning work. However, when a closer look is taken, it becomes clear that wedding planning work is grossly unequal. For example, in a recent study, every single bride-to-be completed disproportionate amounts of the behind-the-scenes work, such as information gathering. It was found, for example, that brides had a greater familiarity with informational resources such as popular wedding guides and magazines. Yet not one groom had purchased, read, or mentioned wedding media when interviewed for this study. Additionally, every bride kept track of planning resources and information such as telephone numbers, business cards, and appointments in an organizer. No grooms, however, created or maintained any such organizer. The grooms, though, were not left out of the loop. The typical pattern of action included brides gathering information about a wedding location or caterer, and narrowing that information down to a set of choices which were then presented to the groom.

Brides also complete a disproportionate amount of kin work. This work includes anticipating family needs and facilitating ongoing family ties. In wedding planning, kin work ranges from anticipating special meal requirements of family and guests to making sure all appropriate family are invited, included, and acknowledged. Brides, for example, are much more likely to assist family members and guests with rides to and from the airport, hotel accommodations, and entertainment during their stay. Wedding planning kin work is the same sort of kin work married women have long been performing.

Parallel to the managerial role women play in performing household labor, brides take responsibility for wedding planning—becoming wedding managers. Brides take it upon themselves to oversee, follow up on, and supervise wedding activities. Brides, for example, are the ones who send letters, e-mails, or make telephone calls to the bridal party informing them of the schedule for the photographs, rehearsals, and so forth that the wedding party is expected to attend.

Wedding planning is a microcosm of the family work that takes place within heterosexual marriages. Brides are introduced to the world of kin work and the behind-the-scenes work it takes to construct and maintain family through planning a wedding. This is probably not the first time brides have experienced this work. Growing up, many girls are given kin work as chores, and they have probably seen their mothers, sisters, grandmothers, and aunts perform it. The wedding, however, often becomes her first major project in family work and the time when she transitions to being a major player in the game—the wife.

Critical Thinking

Think about the ways you have been socialized into a particular gender role. Discuss the household chores you did as a young person—were they different from those of your siblings or friends of the opposite sex? Did those chores have the effect of preparing you for any particular adult roles?

Source: Tamara Sniezek. You can read more of Professor Sniezek's research into weddings in her 2005 journal article: "Is It Our Day or the Bride's Day? The Division of Wedding Labor and Its Meaning for Couples," *Qualitative Sociology* 28(3):215–234.

"are." When our lives, lifestyles, and behaviors do not match up with these media-driven stereotypes, we sometimes feel *stigmatized*. To be stigmatized is to have others act disapprovingly toward us in such a way that we feel badly about ourselves.

American attitudes have grown more liberal regarding men's and women's roles. Few agree, for example, that "sons in a family should be given more encouragement to go to college than daughters," an expression of **traditional sexism** (Sherman and Spence 1997). Traditional sexism is the belief that women's roles should be confined to the family and that women are not as fit as men for certain tasks or for leadership positions. Such beliefs have declined since the 1970s (Twenge 1997a, 1997b). Some social theorists believe, however, that traditional sexism has not gone completely away. Research into transgendered identity (see "Issues for Thought: Challenges to Gender Boundaries") finds that discriminatory attitudes toward transgendered people are rooted in traditional sexism (Serano 2007).

A more subtle **modern sexism** has replaced traditional sexism. It takes the form of agreement with statements such as the following: "Discrimination in the labor force is no longer a problem" and "in order not to appear sexist, many men are inclined to overcompensate women" (Campbell, Schellenberg, and Senn 1997; Tougas et al. 1995). Yet, data for 2008 shows us that being a secretary was the top job for women, and the median wage for women was $36,000 (Douglas 2009, p. 289). Modern sexism denies that gender discrimination

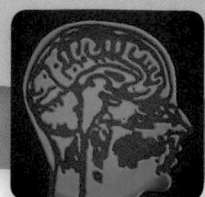
We take for granted that sex is a dichotomy: You are either male or female. Yet somewhere between 1 and 4 percent of live births are **intersexual**—that is, the children have some anatomical, chromosomal, or hormonal variation from the male or female biology that is considered "typical." "Chromosomes, hormones, the internal sex structures, the gonads and the external genitalia all vary more than most people realize" (Fausto-Sterling 2000, p. 20; Gough et al. 2008).

In the 1950s intersex babies (then termed *hermaphrodites*) were assigned a gender identity by doctors, and parents were advised to treat them accordingly. The children typically underwent surgery to give them genitals more closely approximating the assigned gender.

Intersexuality emerged as an area of political activism with the formation of the Intersex Society of North America (ISNA) in 1993 (Gough et al. 2008). Members have demonstrated against arbitrary gender assignment and the surgical "correction" of intersexed infants, demanding the acceptance of gender ambiguity (Preves 2002). Some medical ethicists take the position that "the various forms of intersexuality should be defined as normal" (Lawrence McCullough, quoted in Fausto-Sterling 2000, p. 21; see also Ghosh 2009), and that surgical "corrections" are not only unethical, but serve to "reinforce the stigma through degradation and shame" (Gough et al. 2008, p. 494; Warne and Bhatia 2006).

Although some **transsexuals** (who have been raised as one sex while emotionally identifying with the other) still wish surgery to conform their bodies to their gender identity, others may simply adopt the dress and demeanor of the sex with which they identify, vary their appearance and self-presentation, or adopt a style that is not gender identified.

The term **transgendered** describes an identity adopted by those who are uncomfortable in the gender of their birth. They may be in transition to a new gender or simply wish to continue to occupy a middle ground. Some universities have established gender-neutral housing at the request of transgendered students (F. Bernstein 2004; Raymond and Gordon 2008). Some bureaucratic forms now include a "transgender" box as well as those for "male" and "female," suggesting the beginning of societal accommodation of a more complex sex/gender system.

Chaz Bono (formerly Chastity Bono) was raised as the daughter of entertainers Sonny Bono and Cher. At age forty, she decided to have gender reassignment. *Gender* (or *sex*) *reassignment* is a surgical procedure in which a person's primary and secondary sex organs are changed to that of the other sex. Although the public transformation from Chastity to Chaz has provided a positive treatment of the issue, such transformations prove difficult for the people undergoing reassignment, not because of the physical and psychological transformations, but because of the responses to the changes by friends and family—and even the public.

In 1998, classical pianist David Buechner became Sara Buechner. Buechner was, prior to 1998, a famous musician. After her gender reassignment, the public, friends, and family did not support the changes, and her career stalled. Like Chaz Bono's mother, Sara Buechner's mother had a great deal of difficulty accepting her child's desires and subsequent change to a woman. In both cases, like many others, eventually family, friends, and even the public become accustomed to the new person. Not every transgender story is as positive, but as more people choose to redefine themselves in this way, the rest of society will follow with relative levels of acceptance (Marikar 2009; Winerip 2009).

The biological, psychological, and social realities presented by intersexed or transgendered individuals are challenges to the notion that there are clearly demarcated masculine and feminine genders and gender roles. In turn, parents, physicians, mental health professionals, and educators are coping with ethical decisions regarding surgery and socialization for those born with ambiguous sex characteristics or who appear in childhood to have uncertain sexual identities (P. L. Brown 2006; Lerner 2003; Weil 2006).

Critical Thinking

Have transgendered individuals been politically visible in your campus or community? What are your own thoughts as to whether gender is a dichotomy or a continuum along which individuals may vary?

---

persists and includes the belief that women are asking for too much (Swim et al. 1995). Moreover, though work, family, and civic roles have changed and modernized, there is still a sense on the part of the average person that men and women are different in personality and aptitudes (Begley 2009).

Modern sexism is endemic in the mass media. The mass news and entertainment media present images and stories that suggest "full equality for women is real—that now…[women] can be or do anything they want—but then simultaneously suggest that most women prefer domesticity over the workplace. This reinforces the notion that women and men together no longer need to pursue greater gender equality at work and at home" (Douglas 2009, pp. 283–284). The repetition of such distorted messages leads to unhelpful and sometimes harmful outcomes. For example, males are made to worry that they should resemble *hypermasculine* (i.e.,

Image courtesy of The Advertising Archives

Mass media present us with an idealized, one-dimensional stereotype of ourselves. This has important connotations for our relationships, how we see ourselves, and how we see each other.

characterized by distorted or exaggerated masculine traits) media images, causing some to act aggressively towards other males in an effort to "prove" themselves (Cooper 2008; Kane 2009; C. Lee 2008).

You can probably think of some characteristics typically associated with being feminine or masculine. Stereotypically masculine people are often thought to have **agentic** (from the root word *agent*) or **instrumental character traits**—confidence, assertiveness, and ambition—that enable them to accomplish difficult tasks or goals. A relative absence of agency characterizes our expectations of women, who are thought to embody **communal** or **expressive character traits**: warmth, sensitivity, the ability to express tender feelings, and placing concern about others' welfare above self-interest.

The ways in which men are expected to show agency and women expressiveness are embedded in the culture around us. Let's examine some of our cultural messages about masculinity and femininity.

Masculinities We need to state the obvious: Men are not all alike. Recognizing this, scholars have begun to analyze **masculinities** in the plural, rather than the singular—a recent and subtle change meant to promote our appreciation for the differences among men.

An important cultural expectation is that a man should be occupationally or financially successful, or at least should be working to support his family—which should include children sired in marriage. A man is also expected to be confident and self-reliant, even aggressive. An alternative cultural message emphasizes adventure, sometimes coupled with violence and/or the need to outwit, humiliate, and defeat other men in barroom brawls, contact

sports, and war (Katz 2006; Sullivan and McHugh 2009).

Scholars have expanded on this last idea. If a male finds that legitimate avenues to occupational success are blocked to him because of, for example, lack of education or racial/ethnic status, he might "make it" through an alternative route, through physical aggression or striking a "cool pose." The latter involves dress and postures manifesting fearlessness and detachment, adapted by some racial/ethnic minority males for emotional survival in a discriminatory and hostile society (Crook, Thomas, and Cobia 2009; Martin and Harris 2006; Smith and Beal 2007; Wester et al. 2006). White supremacy movements can be seen as a similar response to status disadvantage for less advantaged white males (Swain 2002, pp. 1–2 and Chapter 3).

During the 1980s, a new cultural message emerged. According to this message, the "liberated" male or "new man" is emotionally sensitive and expressive, valuing tenderness and equal relationships with women (Messner 1997, pp. 36–38, 41). Since then, another transformation of the ideal male image has occurred in response to the terrorist attacks of 9/11. Now, an image of unafraid "can-do" men who tackle fear and traumatic events head-on, providing "unambiguous and uncomplicated performances of masculinity," while at the same time publicly shedding tears after they deal with traumatic issues, is becoming a more prevalent image of masculinity in American society (Adelman 2009, p. 279).

Femininities There are a variety of ways of being a woman, according to cultural messages of **femininities**. The pivotal expectation for a woman requires her to offer emotional support. Traditionally, the ideal woman was physically attractive, not too competitive, a good listener, and adaptable. She served as a man's helpmate, aiding and cheering his accomplishments. She was further expected to be a "good" mother, putting her family's and children's needs before her own. This cultural message continues to be promoted in our mass media.

Communications theorist Sut Jhally (2007), using Goffman's analysis of *gender display* (gender display are the behaviors we exhibit because of our socialization as men or women), examined a variety of mass media products, such as television, film, and magazines. One element he examined was people's hands. He noted that female "hands are shown not as assertive or controlling

Davis Factor/CORBIS

More and more women are entering nontraditional occupations, such as the military.

of their environment but as letting the environment control them" (Jhally 2009, p. 6).

An expectation that emerged as more women entered the workforce and the feminist movement arrived is that of the "professional woman:" independent, ambitious, self-confident. This cultural model may combine with the traditional one to form the "superwoman" message, according to which a good wife and/or mother also efficiently attains career success and/or supports her children by herself. An emerging female expectation is the "satisfied single:" a woman (either heterosexual or lesbian, usually employed, and perhaps a parent) who is quite happy not to be in a serious relationship with a male.

Gender Expectations and Diversity The view of men as instrumental and women as expressive was based primarily on images of white, middle-class heterosexuals. The use of European American, middle-class heterosexual people and families as "normal" in social science research often causes social scientists and policy makers to interpret differences as deficiencies, leading to the stereotyping of different groups (Leidy et al. 2009; Parke and Buriel 2006).

The "strong black woman" cultural messages for black women range from "bitches and bad (black) mothers to modern mammies . . . black women are [stereotyped as] either extremely educated or a high school dropout, ambitious or listless, sexy or ugly" (Boylorn 2008, pp.

417–418; Collins 2004; see also Andersen and Collins 2007a; Squires 2007). Latinas and Asian women are stereotyped as being more submissive than non-Hispanic white women (Andersen and Collins 2007a; Covert and Dixon 2008; Gewertz 2009). Latino men are stereotyped as patriarchal, following a *machismo* cultural ideal of extreme masculinity and male dominance (Hondagneu-Sotelo 1996; McLoyd et al. 2000; Ramos-Sánchez and Atkinson 2009). Some Latinos engaging in this stereotypical machismo behavior may do so because of the history of Latinos in the United States. According to William Carrigan and Clive Webb (2009), racial prejudice by whites against Mexicans in the late 1800s through early 1900s emphasized Mexican men as having more feminine attributes such as cowardice and a preference for wearing "fancy" clothing. This demeaning legacy is an important part of the psychological and cultural history of Latinos that has contemporary ramifications.

Other research indicates that racial/ethnic differences in role expectations and behaviors, particularly for males, are actually not as strong as either stereotypes or sociohistorical perspectives have suggested. The preeminence of the ideology of the male provider role is a powerful theme in *all* racial/ethnic groups (e.g., Taylor, Tucker, and Mitchell-Kernan 1999). Yet this view has changed significantly over the course of contemporary American labor history, especially with the necessity of dual incomes (Cunningham 2008).

African American families have "flexible family roles," but the male's involvement in child care and other expressive roles does not have the same priority as the provider role, despite the difficulty encountered by African American males in fulfilling this role. Black men and women express preferences for egalitarian relationships (Cowdery et al. 2009; Furdyna, Tucker, and James 2008). African American men are more supportive of employed wives than white men are. Yet, gender ideology among African Americans does differentiate the sexes by the importance of the male provider role.

Women are likely to perform more of the household labor than men (though African American men do more than white men) (Cowdery et al. 2009). African American men show up as more conservative than white men in other ways—for example, in a stronger conviction that men and women are essentially different: Men are "manly," and women are "womanly," or soft and feminine (Haynes 2000, p. 834). Interestingly, this may be related to the pursuit of the traditional family structure that has long eluded most African Americans. In many African American families, the ability to have traditional gender relationships is evidence of economic success (Cowdery et al. 2009; Furdyna, Tucker, and James 2008).

Similarly complex gender patterns are observed in Latino families. For example, some Mexican American women engage in stereotyped *marianismo* where they carry the primary responsibility for housework and child care (Pinto and Coltrane 2008). But roles have been modified in the migration process and with women's entry into the labor force. Mexican American women do more housework and child care than men, but less than their counterparts who still live in Mexico or who are immigrants (Pinto and Coltrane 2008).

(Asian) Indo American women, studies show, typically find themselves on similar trajectories. Like Mexican American women, as they obtain greater levels of education and develop their own careers, these wives are making greater demands on their husbands to help out with the burden of domestic chores and child care (Bhalla 2008; Kallivayalil 2004). These are but a few examples of the complexity of role expectations and behavior in real-life families.

To What Extent Do Women and Men Follow Cultural Expectations?

It is one thing to recognize cultural images but another to follow them. Consequently, we continue to ask: To what extent do individual men and women, boys and girls, exhibit gender-differentiated behaviors?

In adult life, women seem to have greater connectedness in interpersonal relations and, perhaps due to gender stereotypes, find themselves pushed into the caregiving professions in greater numbers than men, whereas men tend to be in more socially dominant, competitive, and achievement-oriented occupations (Beutel and Marini 1995; Webster and Rashotte 2009). But there is great individual variation, and the situational context accounts for much of the apparent difference between men and women (Gormley and Lopez 2010; Meier, Hull, and Ortyl 2010; Rosenfeld 2007; Webster and Rashotte 2009).

Moreover, behavior vis-à-vis the gendered expectations we've discussed generally fits an overlapping pattern (Basow 1992). We can visualize this as two overlapping distribution curves (see Figure 4.1). For example, although the majority of men are taller than the majority of women, the area of overlap in men's and women's heights is considerable, and some men are

"It's her first bench–clearing brawl."

Traditional stereotypes of children define males as aggressive and competitive and girls as sensitive and concerned for others. Real behavior is far more varied than these stereotypes and depends very much on the situation.

shorter than some women. It is also true that differences among women or among men (*within-group variation*) are usually greater than the average difference between men and women (*between-group variation*).

Although males and females differ little on basic traits and abilities, the opportunities available in the social structure affect the options of men and women and, ultimately, their behaviors as they adapt to those options. The "deceptive differences" (Epstein 1988) we observe or think we observe typically involve men and women assigned to different social roles. A woman secretary, for example, is expected to be compliant and supportive of her male boss's decisions. To observers, she seems to have a gentle and submissive personality, while he is seen to have leadership qualities (Eagly, Wood, and Diekman 2000; Meier, Hull, and Ortyl 2010; Ridgeway and Smith-Lovin 1999; Webster and Rashotte 2009).

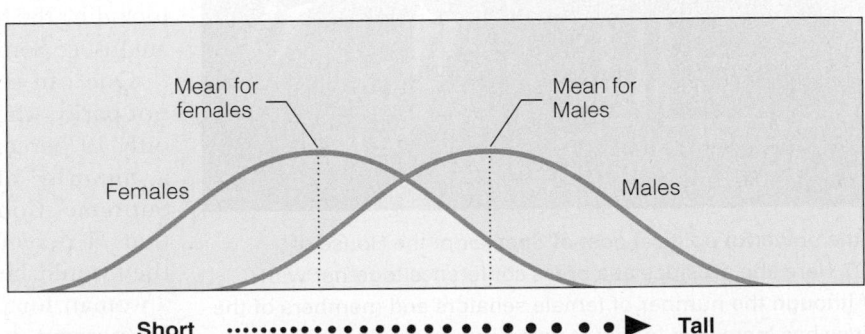

Figure 4.1 How females and males differ on one trait, height, conceptualized as overlapping normal distribution curves. Means (averages) may differ by sex, but trait distributions of men and women occupy much common ground.

The Gender Similarities Hypothesis

The preponderance of research on gender differences suggests that in fact they are few. Psychologist Janet Hyde (2005) offers a **gender similarities hypothesis** to replace the usual assumption of gender differences. The gender similarities hypothesis "holds that males and females are similar on most, but not all, psychological variables. . . . That is, men and women, as well as boys and girls, are more alike than they are different" (p. 581).

Hyde finds virtually no difference on most traits, a few moderate differences, and very few large differences. These conclusions apply even to math and verbal ability, self-esteem, and tendency toward aggression, areas where gender differences were thought to be pronounced. She did find evidence of gender differences (1) in motor performance, especially in throwing distance and speed; (2) in sexuality, especially male's greater incidence of masturbation and acceptance of casual sex; and (3) in physical aggressiveness. She did not find clear differences in relational aggression. Other research that fails to find confirmation of most stereotypical gender differences in emotions and emotional expression supports the gender similarities hypothesis (Else-Quest, Hyde, and Linn 2010; Hyde 2007).

Hyde goes on to argue that mistaken assumptions about gender differences have serious costs, hurting women's opportunities in the workplace, men's confidence in nontraditional family roles, and both sexes' confidence in their ability to communicate with each other. Despite the validity of Hyde's gender similarities hypothesis, virtually all societies, including our own, are structured around some degree of gender inequality.

Gender Inequality

Male dominance describes a situation in which the male(s) in a dyad or group assume authority over the female(s). On the societal level, male dominance is the assignment to men of greater control and influence over society's institutions and, usually, greater benefits. In this section we address gender difference and gender inequality in certain major social institutions: politics and government, religion, health, education, and the economy.

Male Dominance in Politics

As of 2009, in the U.S. Congress there were seventeen women in the Senate and seventy-six in the House of Representatives (U.S. House of Representatives 2009). In 2007, Democratic congresswoman Nancy Pelosi became Speaker of the House, the highest position in the House of Representatives and second after the vice president in the line of succession. A recent *New York Times* article quotes political strategists as saying that "[v]oters have grown more accustomed to women in powerful positions" (Toner 2007).

Beginning with the Clinton administration and continuing under President Barack Obama, women have been more visible in the executive branch of government as well. Hillary Clinton, the former first lady and a presidential contender herself, serves as secretary of state. Women justices have served on the Supreme Court, including the newest justice, Associate Justice Sonia Sotomayor.

One can see progress, though not parity, when women comprise only 19 percent of Congress and a minority of the cabinet and Supreme Court. Surveys report that 71 percent of the public say they would be willing to vote for a woman for president—but only 56 percent believe their family, friends, and coworkers are willing to do so (Rasmussen Reports 2008).

AP Photo/Paul Sakuma

Nancy Pelosi attained the powerful political post of Speaker of the House of Representatives in 2007. Here she presides at a press conference together with other House leaders. Although the number of female senators and members of the House of Representatives has increased in recent years and women have had some senior appointments in the executive branch in addition to Pelosi's congressional position, women remain a minority in positions of political power.

Male Dominance in Religion

Religion as an institution evidences male dominance as well. Although most U.S. congregations have more female than male participants, men more often hold positions of authority, while women perform secretarial, housekeeping, and low-level administrative chores (Chaves, Anderson, and Byassee 2009). As with politics, there are some striking exceptions that suggest future change. Women have been elected as bishops and denomination leaders in the African Methodist Episcopal, Anglican, United Methodist, and Presbyterian churches, but recent surveys suggest that women lead only 8 percent of congregations in the United States (Banerjee 2006a, 2006b; Butt 2009; Chaves, Anderson, and Byassee 2009; Gott 2010; "Professor Says" 2007).

The effects of personal religious involvement on women's daily lives are complex. The growth of evangelical Protestantism, Islamic fundamentalism, and the Latter-day Saints religion, along with the charismatic renewal in the Catholic Church, has fostered a traditional family ideal of male headship and a corresponding rejection of feminist-inspired redefinitions of family roles. We note that strict gender divisions in the home lead to family tension and decreased marital satisfaction, particularly for the wife. Such dissatisfaction increases the risk of divorce even in the most traditional of couples (Duba and Watts 2009; Vaaler, Ellison, and Powers 2009; Zink 2008).

On the other hand, actual practice among evangelicals seems more egalitarian than formal doctrine (Bartkowski 2001, Chapter 7). There is also a growing feminist movement among American Muslim women who seek to combine their religio-cultural heritage with equal rights for females. This feminism "signifies a commitment to faith that challenges secular feminist notions" about Islam and women (Karim 2009, p. 93).

Gender and Health

When it comes to health, we can no longer speak of male advantage. From birth onward, indeed prior to birth (fetal loss), males have higher death rates. Male infants have higher rates of infant mortality and adverse conditions (Heron et al. 2009, Table D). Nature recognizes this disparity, and in the United States, around 105 boys are born for every 100 girls, with boys outnumbering girls under age eighteen (U.S. Census Bureau 2009j).

Life expectancy for the total population dropped slightly to 77.7 years in 2006—80.2 years for females and 75.1 years for males, a difference of 5.1 years (Heron et al. 2009). The life expectancy gap between males and females peaked in 1979 at almost eight years and has been decreasing since (see Figure 4.2).

Gender difference in longevity has been attributed to greater risk factors for boys and men, including smoking and drinking, accidents, suicide, and murder victimization, as well as some not-well-understood vulnerability to infection and stress (Heron et al. 2009). Men also have far fewer doctor visits (i.e., checkups) than women, and men who are middle-aged and more traditional visit the doctor even less (Dotinga 2009; Painter 2006; Rabin 2006).

For years, medical researchers paid women little attention. But partly due to the feminist movement, "women's health has been a national priority" (Rabin 2006). Now men's advocacy groups are calling for increased attention to men's health and greater investment in research on their unique health conditions—funding for breast cancer research, for example, exceeds that for prostate cancer by 40 percent. In any case, we can applaud the increasing attention to gender equality in medical research.

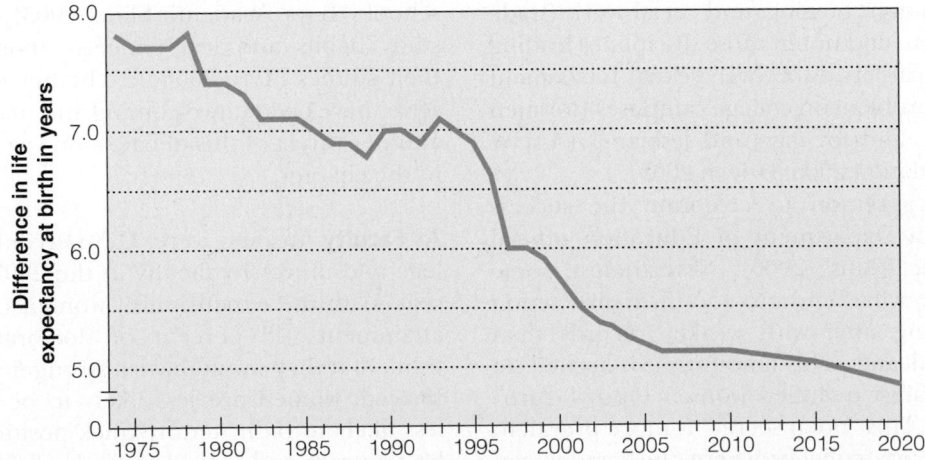

Figure 4.2 The difference in life expectancy between males and females, United States, 1975–2020. From a difference of almost eight years in 1979, this gap has been steadily decreasing. The years post-2006 are projections by the U.S. Census Bureau.

Gender and Education

Whether women or men are relatively disadvantaged in higher education is in dispute. (The relative circumstances of boys and girls in K–12 schooling will be discussed in the Gender and Socialization section later in this chapter.)

As Students Women have been the majority of college students since 1979 and now surpass men in the proportion of the total population that are college graduates (Lewin 2006c; "Women Catching Up" 2008). In 2007, women earned 57 percent of bachelor's and 60.5 percent of master's degrees, 50 percent of first professional degrees, and 50 percent of doctorates (U.S. Census Bureau 2010b, Table 288).

The changing gender balance in higher education—indeed, in high school completion, with 72 percent of girls but only 65 percent of boys graduating high school (Lewin 2006b)—has led to cries of alarm (e.g., "leaving men in the dust," Lewin 2006a, p. A1; "stagnation for men," "Study: Academic Gains" 2006). Advocates of attention to boys argue that with the advances made by women (attributable to feminism), it is now the "boys' turn." Otherwise, "many will needlessly miss out on success in life" ("Big [Lack of] Men" 2005).

In response to the assumption that women "have it made" and men are the disadvantaged category, counterarguments and data have been offered (see, e.g., Gandy 2005). First of all, men's college enrollments have *not* declined; they have simply not increased as rapidly as women's (Corbett, Hill, and St. Rose 2008; U.S. National Center for Education Statistics 2006). Second, any disadvantage to males depends on age and class. Among upper-income students of traditional age, *males*—white, black, Hispanic, and Asian—remain a majority of college enrollees (Cataldi, Laird, and KewalRamani 2009; Lewin 2006c), though barely so. Women's master's degrees are primarily in education, nursing, and social work (traditional female areas), and not in those disciplines leading to the most elite careers. Moreover, sexual harassment continues to be a problem on college campuses (for men as well as women, and for gays and lesbians) (AAUW Educational Foundation 2006; Dziech 2003).

"There is every reason to celebrate the success of women," says a Department of Education official ("Study: Academic Gains" 2006). Nevertheless, some colleges now have what amounts to affirmative action for men, admitting men with weaker records than some women applicants. This amounts, it is argued, to discrimination against qualified women ("Boys' Turn" 2006; Britz 2006). This "open secret" is more prevalent at smaller liberal arts colleges where the gender gap is more obvious (Jaschik 2009; Meyer 2009). The U.S. Commission on Civil Rights has begun to research the issue because, it says, ". . . some college administrators argue that they must discriminate against women or the gender balance at their institutions will become . . . off-kilter" (Heriot 2009).

Some education scholars believe that "the new emphasis on young men's problems . . . is misguided in a world where men still dominate the math–science axis, earn more money, and wield more power than women" (Lewin 2006a, p. 18). Such concerns may be misguided given the recent data released by the U.S. Department of Education. A new report suggests that reading scores for males at ages nine, thirteen, and seventeen have improved since 2004, and the gaps between boys and girls is decreasing (Rampey, Dion, and Donahue 2009). Additionally, recent research also points out the continued disparity in earnings between the sexes, with women, on average, earning just 77.1 cents for every dollar a man earns, which is down from 77.8 cents in 2007 (Institute for Women's Policy Research 2009).

The data have made visible two patterns. First, it becomes clear that the college achievement gap is greater among racial/ethnic groups *within* gender categories and especially points to black and Hispanic male disadvantage (Aronson 2003; Corbett, Hill, and St. Rose 2008). This has set off a debate over whether these trends show a worrisome achievement gap between men and women or whether the concern should be directed toward the educational difficulties of poor boys, whether they are black, white, or Hispanic (Cataldi, Laird, and KewalRamani 2009; Corbett, Hill, and St. Rose 2008; Lewin 2006a, p. 18).

A second and alarming pattern is the apparent difference between males and females in goals and attitudes toward schooling. Such attitudes have long-run implications for males' educational attainment, placement in the workforce, and ability to maintain a marriage that has a stable financial underpinning. A *USA Today* study found that 84 percent of girls but only 67 percent of boys think it is important to continue beyond high school ("Boys' Academic Slide" 2003). Boys have poorer study habits and less concern about doing well in their studies (Tyre 2006c). Whether the schools themselves have been unwelcoming and unadapted to boys is another thrust of this debate—one we will take up later in the chapter.

As Faculty Women were 41.8 percent of full-time college and university faculty in the 2007–2008 academic year. With the expansion of women's graduate degree attainment (48 percent of doctorates), one would think that they would have a stronger presence by now. Instead, women are less likely to be full-time faculty; less likely to be in tenure-track positions; less likely to be tenured; and less likely to be full professors (U.S. Department of Education 2009, Table 245; West and Curtis 2006). Disadvantage to women is especially strong at the more elite schools. Sixty-two percent of

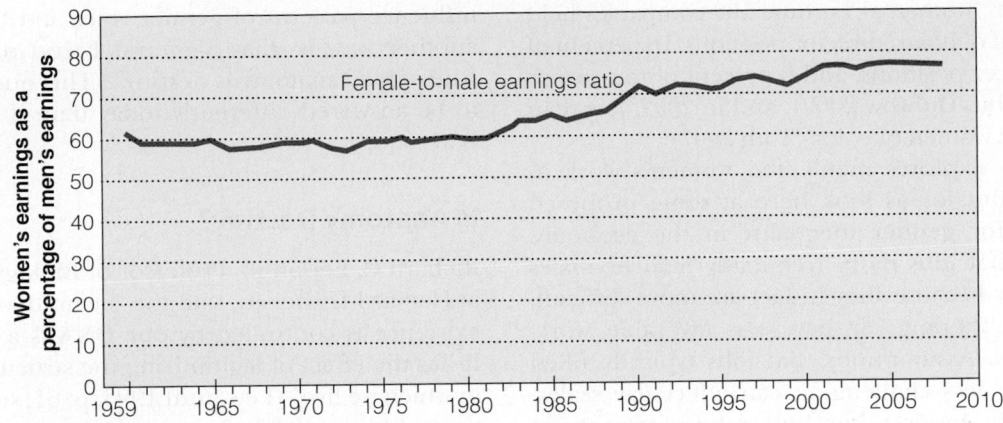

Figure 4.3 Female-to-male earnings ratio of full-time, year-round workers by sex: 1955–2008.

those hired by Ivy League universities in 2005 were men (Arenson 2005).

Little gain is seen in faculty racial/ethnic diversity. In 2007, 44.6 percent of college faculty were white males; 32 percent, white females; 2.8 percent, black females; 2.7 percent, black males; 2.5 percent Asian/Pacific Islander females; 4.8 percent, Asian/Pacific Islander males; 1.6 percent, Hispanic females; 1.9 percent Hispanic males; and less than 1 percent each, Native American males and females (Snyder, Dillow, and Hoffman 2009).

Male Dominance in the Economy

It is on gender inequality in the economy that most attention has been focused. Although the situation is changing, as you can see from Figure 4.3, men have been and continue to be dominant economically. Before examining the gendered pay differences, we would first like to discuss what appears to be economic role reversals for some married couples. According to a recent poll by the Pew Research Center (see Figure 4.6 later in the chapter), "only 4% of husbands had wives who brought home more income than they did in 1970, a share that rose to 22% in 2007" (Fry and Cohn 2010, p. 2). The reasons for this are many, but it is important to note that there are fewer married couples than in 1970, more women than men obtain college degrees, and in 2008 men were unemployed at three times the rate of women. Males accounted for about 75 percent of the 2008 unemployment rate (Borbely 2009; Fry and Cohn 2010, p. 2).

In 2008, women who were employed full time earned 80 percent of what men employed full time did (U.S. Bureau of Labor Statistics 2009a); this is down a full percentage point from 2006. Younger women between the ages of sixteen and twenty-four and between the ages of twenty-five and thirty-four earn a larger proportion (91 percent and 89 percent, respectively) of what men do

(U.S. Bureau of Labor Statistics 2009a). Non-Hispanic white and Asian women both earned just under 80 percent of what males earned. Black and Hispanic female/male comparisons are more favorable to women, at 90 percent, but this is primarily because black and Hispanic men have much lower earnings than do white men (U.S. Bureau of Labor Statistics 2009a).

Even in the same occupational categories, women earn less than men. For instance, in 2008, in the highest-paying occupation, that of chief executive, women made $83,356, whereas men earned $103,948 on average (U.S. Bureau of Labor Statistics 2009a). The lowest paid occupational category, food preparation worker, also had a difference in pay, with women earning $17,576 and men earning $19,136 on average. Some analysts suggest the difference is related to choice of occupational specialty and practice setting; however, as evidenced by the differences in pay for the same profession at both ends of occupational earnings, such a simplistic explanation provides little clarity to a problem that has haunted women since the Industrial Revolution and that has devastating impacts on poverty, hunger, health, and quality of life.

Overall, the earnings gap between men and women narrowed in recent decades, but, as Figure 4.3 indicates, that gap is widening slightly. Some proportion of the convergence is due to falling wages for men or their increased time out of the labor force, but by and large the narrowing gap is due to rising wages for women as they have increased their human capital (education and skills) and labor force participation (England 2006). Yet, as discussed above, the pay gap between male and female college graduates has widened a bit since the 1990s.

Men also continue to dominate corporate America. "A decade ago it was possible to imagine that men and women with similar qualifications might one day soon be making identical salaries. Today that is harder to imagine" (Leonhardt 2006b, p. A1; see also "Women Still Lag"

2004). In 2008, "women at Fortune 500 companies held only 15 percent of board director positions, 16 percent of corporate officer positions, and 6 percent of top earner positions" (Tuhus-Dubrow 2009); and in 2007, less than 2 percent were women of color (Toth 2007).

Chapter 11 explores men's and women's work in more detail. But let us look here at some proposed explanations for gender inequality in the economy. Some argue that jobs more frequently held by males may pay better because they in fact are more difficult, require more training, or have less favorable working conditions. Assumptions that jobs typically filled by women are less challenging can affect pay scales. Women may themselves buy into gender stereotypes and believe they are less competent to achieve certain jobs or occupational levels. And as women enter the job pool for certain professional jobs (e.g., journalism), the field becomes more crowded and wages and salaries are less competitive (Blau, Brinton, and Grusky 2006).

Women also contend with employers' *assumptions* that women will opt out, which may affect their careers. This is where the costs of gender stereotyping come in. Employers may be less likely to select women for advanced training and positions with upward mobility potential even though the women affected may be highly ambitious, with a commitment to continuous employment (Laff 2007; Pinto 2009). This form of discrimination ultimately leads to less advancement and earnings for women. For this reason, it is more likely a woman's income will be seen as less essential to the family than a man's, and the family may favor his career over hers in terms of residential mobility, the taking of time-pressured jobs and promotions, and the like.

But not all of the wage gap is explainable by objective characteristics of men's and women's employment histories, skills, or women's own choices. It is likely that discrimination against women as employees continues to some degree. After various factors associated with earnings are taken into account, research finds a varying (by perhaps as much as a 22.9 percent) differential in earnings that is usually attributed to discrimination, albeit of a nonobvious sort (Blau and Kahn 2006; Institute for Women's Policy Research 2009; National Women's Law Center 2007).

We have been discussing gender inequality. The state of affairs can be nicely summed up by Figure 4.3, which illustrates the gains women have made economically *and* the gap that still hasn't closed. Male dominance in the United States today is, of course, more moderate than in our past or compared to contemporary societies that are more traditional or overtly oppressive of women. Still, cross-culturally and historically, it appears that virtually all societies have been characterized by some degree of male dominance.

This leads us to ask whether male dominance might be anchored in biology. To what extent does biology influence patterns of gender roles and behaviors? Put another way, is what Sigmund Freud once proposed true—that "anatomy is destiny"? This question is likely to be answered differently today than it was just a few years ago.

Is Anatomy Destiny?

Richard C. Lewontin, Professor of Biology and Zoology at Harvard University, said the "claim that all of human existence is controlled by our DNA is a popular one. It has the effect of legitimizing the structures of society in which we live" (Lewontin 1993, p. 61; see also Lewontin and Levins 2007). Lewontin is suggesting that some researchers attribute socially constructed human behavior to inherited genetics. In particular, these researchers look to race and gender differences, trying to find a genetic explanation for what most social scientists recognize as structurally created forms of inequality, often relying heavily on functionalist explanations of social organization—hence the term sociobiology (MacCallum and Hill 2006; Nielsen 2009). In other words, these researchers have privileged simplistic explanations for sociohistorical complexity.

Biological theories of gender difference were initially offered by ethologists, who study humans as an evolved animal species (e.g., Tiger 1969). Tiger, who primarily studied baboons, found males to be dominant and argued that *Homo sapiens* inherited this condition through natural selection. Newer data on nonhuman primates have challenged these conclusions as socially constructed myths. Primate species vary in their behavior, and within species there is some environmentally shaped variation (Bartlett 2009; Haraway 1989; Vogel et al. 2009; Wood and Eagly 2002, p. 721). The facts suggest that male and female behavior is not differentiated in a consistent way in the animal species most closely related to humans.

A subsequent sociobiology theory of gendered behavior focused on genes. This theoretical model is most often found in the realm of *evolutionary psychology*, where human behavior is thought to be associated with biological adaptation. In this view, in order to continue their genes, individuals act to maximize their reproduction or that of close kin. Different strategies characterize males (who seek to impregnate many females) and females (who seek the best conditions in which to nurture their small number of children) (Buss 2009; Buss and Shackelford 2008; Dawkins 1976). Other researchers reverse the order, suggesting that rather than viewing biology as impacting human behavior and society, we should instead examine the ways in which society and social interactions impact our biology (McIntyre and Edwards 2009).

Other evolutionary perspectives focus on the hunting–gathering era of early human evolution. Men, because of their greater physical strength and their freedom from reproductive responsibility, were able to hunt. But women, who might be pregnant or breastfeeding, gathered food that was naturally available close to home while they also cared for children. According to these evolutionary theories, circumstances elicited different adaptive strategies and skills from men and women that then became encoded in the genes—greater aggression and spatial skills for men, nurturance and domesticity for women. According to this perspective, these traits remain part of our genetic heritage and are today the foundation of gender differences in personality traits, abilities, and behavior (Maccoby 1998, Chapter 5). Newer research suggests that such traits merely reflect cultural differences (i.e., such traits vary from culture to culture).

The notions that current behaviors stem from our adaptation during the hunting–gathering era, discussed above, have been called into question, along with the conflation of evolution with adaptation. For example, physical size must be taken into account when analyzing the difference between males and females in such activities as throwing and spatial reasoning (Hooven et al. 2004; Yang et al. 2007). Another set of traits brought into question are empathy (female) and aggression (male). Recent research shows that basal testosterone levels (found in both males and females) has significant impact on the level of empathy or aggression in either sex. In other words, aggressive or empathetic behavior (or the lack thereof) has less to do with one's gender and more to do with levels of basal testosterone in the body at any given time (Carré and McCormick 2008; Mehta, Jones, and Josephs 2008; van Honk et al. 2004).

The genetic heritage is expressed through **hormonal processes** (as noted above).[2] However, the relationship between hormones and behavior goes in both directions: What's happening in one's environment may influence hormone secretion levels. Several studies have found, for example, that the hormonal levels of males in romantic relationships and those of new fathers undergo changes parallel to those associated with maternal behavior (e.g., lower testosterone and cortisol and detectable levels of estradiol) (Gray 2003;

Gray, Ellison, and Campbell 2007; McIntyre et al. 2006; van Anders and Watson 2006, 2007). Testosterone rises in men in response to an athletic or other competition or in response to insults (Carré and McCormick 2008; Mehta, Jones, and Josephs 2008).

Biologists have relinquished deterministic models in their thinking about gender and family.[3] They present an evolutionary theory that acknowledges strong environmental effects on animal behavior. Much current biological theorizing leaves plenty of room for culture (e.g., Emlen 1995; Geary and Flinn 2001; Lewontin and Levins 2007; McIntyre and Edwards 2009; Nielsen 2009). In fact, cognitive psychologists continue to debate the existence and significance of gender differences; however, a growing body of important research shows that *stereotype threat* may have a greater impact on gendered differences in cognitive testing results. A stereotype threat is "the sense of threat that can arise when one knows that he or she can possibly be judged or treated negatively on the basis of a negative stereotype about one's group" (Goff, Steele, and Davies 2008, p. 92; see also Campbell and Collaer 2009; Harvard University 2005; Pinker 2005; Spelke 2005).

Sociologists who work from a biosocial perspective (e.g., Alan Booth and colleagues) are finding complex interactions among gender, social roles, and biological indicators rather than categorical gender differences (see Figure 4.4 for differing ideas of gender role development). It is safe to say that there is convergence on the opinion that in gender, as well as other behavior, biology interacts with culture in complex and constantly changing ways that cannot be reduced to biological determinism.

Although adult men and women seem to be converging in social roles and personal qualities, gender differences and separation still seem rather powerful in the younger years. We look now at the various theories and practices related to the socialization of boys and girls.

---

[2] **Hormones** are chemical substances secreted into the bloodstream by the endocrine glands; they influence the activities of cells, tissues, and body organs. The primary male sex hormones are androgens. Testosterone, produced in the male testes, is an androgen. Testosterone levels in males peak in adolescence and early adulthood, then slowly decline throughout the rest of a man's life. Females also secrete testosterone and other androgens, but in smaller amounts. The primary female hormones are estrogen and progesterone, secreted by the female ovaries.

Sex hormones influence *sexual dimorphism*—that is, differences between the sexes in body structure and size, muscle development, fat distribution, hair growth, voice quality, and the like. The degree to which hormones produce gender-differentiated behavior is disputed.

[3] Another area of biological research and theorizing on gender differences has to do with brain organization and functioning. *Brain lateralization* refers to the relative dominance and the synchronization of the two hemispheres of the brain. The argument was that male and female brains differ due to greater amounts of testosterone secreted by a male fetus. As a result, researchers believed different sides of the brain may be dominant in males and females, or males and females may differ in the degree to which the two brain halves work together (Blum 1997; Liu et al 2009; Springer and Deutsch 1994).

Overall, brain lateralization studies have produced unconvincing evidence concerning sex differences in brain organization or a connection to verbal, spatial, or mathematical abilities. Recent research finds no significant difference between the genders in either mathematical or language skills, instead noting that the brain functioning is asymmetrical in both sexes. Furthermore, gender differences in measured math ability and achievement have declined dramatically, arguing against a biological explanation (Hyde, Fenema, and Lamon 1990; Kimmel 2000, pp. 30–33; Liu et al. 2009; Pinel and Dehaene 2009; Rogers 2001; Sun and Walsh 2006; Wood and Eagly 2002, p. 720).

Gender as a Functional Role	Gender as a Situational Role
1. Gender role behaviors developed to meet functional system needs of the family; functional needs of nuclear family system create differentiation into an instrumental leader (father) and an expressive leader (mother); the two roles describe division of labor and other aspects of interaction within families.	1. Gender role differences depend up on cultural definitions; interaction requires both influential, agentic, proactive behaviors and acquiescent, expressive, reactive behaviors depending on the nature of the risk, characteristics of other individuals, and other situational facts.
2. Instrumental and expressive roles are equally valuable because they have equal functional importance; individuals recognize that, so the two roles are equally valued and rewarded; also, each role has its own ranking of assumed competence.	2. Instrumental and expressive roles are seen as unequal in importance, social esteem, and perceived value; instrumental and agentic attitudes and behaviors are favored; social-emotional and responsive attitudes and behaviors are less favored; also the instrumental role is associated with high general competence, whereas the expressive role is associated with low general competence; thus the two kinds of roles are associated with power and prestige structures in face-to-face interaction.
3. Childhood gender socialization entails learning either the instrumental (male) or expressive (female) role; by late adolescence, gender-appropriate attitudes and behaviors usually have been thoroughly internalized.	3. Socialization includes learning both superordinate and subordinate behaviors and attitudes; it also includes learning social cues to tell a person's place within situational social hierarchies.
4. Because of the functional importance of filling the two gender roles, socialization is repetitive and is so intense that the roles tend to be overlearned and individuals become encased in them; a fully socialized individual typically is only able to display one role set; fully socialized adults have difficulty understanding or displaying the complementary role to their own.	4. Because of the importance of situation-appropriate behavior, individuals learn considerable flexibility in displaying both instrumental and social-emotional attitudes and behaviors; an individual can display different role sets in different circumstances.
5. Changing gender role behaviors and attitudes (presumably in the direction of greater quality) would require changing the gender role socialization experiences of children from earliest childhood through late adolescence; such changes would need to be made in all or most families to change society; however, that would endanger the functioning of future families.	5. Changing gender role behaviors and attitudes (presumably in the direction of greater equality) would require devising situation-specific interventions to change structural characteristics of particular situations.

Figure 4.4 The functional role vision compared to the situational vision of gender.

Gender and Socialization

Psychologist Eleanor Maccoby, earlier associated with a review of research on sex differences that found them few and mostly unimportant (Maccoby and Jacklin 1974), has now come to see biology as grounding some *childhood* differences between the sexes, as well as children's tendency to prefer sex-segregated play.

Maccoby sees boys' rough play, earlier separation from adults, poor impulse control, and competition-seeking behavior, and girls' interest in young infants, earlier verbal fluency, and earlier self-regulation as biologically based. She attributes these differences to *prenatal* hormonal priming (see also Hines et al. 2002), noting that hormonal levels *during childhood* do *not* match these observed patterns. Nor do hormonal levels vary much by sex until adolescence. It is worth noting that she sees the biological influence as very specifically *not* marking the existence of generalized sex differences. However, an important question arises in regard to biological bases for sex differences concerning the relationship and interactions each child has with its same-gendered parent and the development of gender

roles (Caldera, Huston, and O'Brien 1989; Feldman 2003). For example, if a female who conducts herself according to the expected gender roles has more interaction with a female child, would we not see a more pronounced development of those roles in that child?

Such questions bring us to look at biology and childhood in a different way. In contrast to other species, much of the behavior of humans involves behavior that is learned, not programmed as instinct. There is a lengthy period of dependency on parents or other adults during which this learning takes place (Geary and Flinn 2001). This process, termed **socialization**, is "[a] process by which people develop their human capacities and acquire a unique personality and identity and by which culture is passed from generation to generation" (Ferrante 2000, p. 521).

In this section we look specifically at *gender* socialization, but the socialization processes discussed here are applicable to other aspects of cultural values and behavioral expectations as well. First, we examine some theories of socialization, then we look at specific settings for gender socialization, and finally we address some current issues regarding boys and girls.

Children learn much about gender roles from their parents, whether they are taught consciously or unconsciously. Parents may model roles and reinforce expectations of appropriate behavior. Children also internalize messages from available cultural influences and materials surrounding them.

Theories of Socialization

There are a number of competing theories of gender socialization, each with some supporting evidence.

Social Learning Theory According to **social learning theory** (Bandura and Walters 1963), children learn gender roles as they are taught by parents, schools, and the media. Children observe and imitate *models* of gender behavior, such as parents, and/or they are *rewarded (or punished)* by parents and others for gender-appropriate or -inappropriate behavior. Fathers seem to have stronger expectations for gender-appropriate behavior than do mothers. Although this theory makes intuitive sense, researchers have found little association between children's personalities and parents' characteristics (Andersen 1988; Losh-Hesselbart 1987). Other research, especially on the realm of familial violence, suggests there is some association between children's aggressive behaviors and observed parental behavior (Button and Gealt 2010; Dunn 2005; Hoffman, Kiecolt, and Edwards 2005).

Self-Identification Theory Some psychologists think that what comes first is not rules about what boys and girls should do but rather the child's awareness of being a boy or a girl. In this **self-identification theory** (also termed *cognitive-developmental theory*), children categorize themselves as male or female, typically by age three. They then identify behaviors in their families, in the media, or elsewhere appropriate to their sex and adopt those behaviors. In effect, children socialize themselves from available cultural materials (Kohlberg 1966; Signorella and Frieze 2008).

Gender Schema Theory Similar to self-identification theory, gender schema theory posits that children develop a framework of knowledge (a **gender schema**) about what girls and boys typically do (Bem 1981). Children then use this framework to organize how they interpret new information and think about gender. Once a child has developed a gender schema, the schema influences how she or he processes new information, with gender-consistent information remembered better than gender-inconsistent information. For example, a child with a traditional gender schema might generalize that physicians are men even though the child has sometimes had appointments with female physicians. Overall, gender schema theorists see gender schema as maintaining traditional stereotypes. This theoretical framework continues to be tested and continues to show that younger children maintain more rigid gender stereotypes than do adolescents (Crouter et al. 2007; Signorella and Frieze 2008).

Symbolic Interaction Theory In **symbolic interaction theory** (Cooley 1902, 1909; Mead 1934), children develop self-concepts based on social feedback—the *looking-glass self* (see Chapter 2). Also important is their *role taking*, as they play out roles in interaction with significant others such as parents and peers. As children grow, they take on roles representing wider social networks, and eventually *internalize* norms of the community (termed by Mead *the generalized other*). Although this is a general theory of socialization, you can see how it can be applied to gender. Little girls play "mommy" with their dolls and kitchen sets, whereas little boys play with cars or hypermasculine action figures. But things are changing, and it is now likely that little girls as well as little boys play "going to work."

All the socialization theories presented here seem plausible, but none has conclusive research support. Self-identification theory and gender schema theory, as well as biologically deterministic theories, are especially lacking. It is also the case that gender socialization is a moving target in a rapidly changing social world. When noting the difficulties of these theories in trying to fully explain gendered behavior, keep in mind that human behavior is based on many factors. Our genes, our socialization, our interactions, our environment, and our cultural and social encounters are just a few of the things that impact who each individual "is."

The complexity and informational intake capabilities of our brains may never be fully known. But each encounter we have with another person, another object, with visual stimuli, with auditory stimuli, and so forth all work together to impact each of us in different ways. Thus, the social sciences have the most difficult of tasks—which is to try and make sense of who we are and how we got to where we are now. Sometimes, along the way, researchers bring their own biases into the mix, and we thus find ourselves combating deterministic theoretical models.

Regardless of the long-term value of each theoretical construct, all are ultimately valuable because they serve to move the scientific disciplines forward, allowing researchers to develop more sophisticated explanations of who we are.

Settings for Socialization

We'll turn now to some empirical findings regarding gender socialization in concrete and specific settings: the family, play and games among peers, media influences, and the schools. We saw in Chapter 3 that religious groups often specify appropriate gender roles, and they are also a setting for gender socialization.

Boys and Girls in the Family From the 1970s on, parents have reported treating their sons and daughters similarly.[4] "[T]he specialization of men for dominance and women for subordination that emerged [as a socialization pattern] in patriarchal societies has eroded with the weakening of gender hierarchies in postindustrial societies" (Wood and Eagly 2002, p. 717). Differential socialization still exists, but it is typically not conscious. Instead, it "reflects the fact that the parents themselves accept the general societal roles for men and women," though this is no longer universal (Kimmel 2000, p. 123). New analysis also reinforces parental roles as important in the socialization process, suggesting that low levels of parental involvement as well as negative and coercive parental behaviors are likely to contribute to negative behaviors in children—particularly in female adolescents (Kroneman et al. 2009; Svensson 2003).

Still, encouragement of gender-typed interests and activities continues. A study of 120 babies' and toddlers' rooms found that girls had more dolls, fictional characters, children's furniture, and the color pink; boys had more sports equipment, tools, toy vehicles, and the colors blue, red, and white. Fathers, more than mothers, enforce gender stereotypes, especially for sons; it is more acceptable, for example, for girls to be tomboys (Adams and Coltrane 2004; Bussey and Bandura 1999; Feldman 2003, p. 207; Kimmel 2000, Chapter 6; Marks, Lam, and McHale 2009; Pomerleau et al. 1990).

Exploratory behavior is encouraged more in boys than in girls (Feldman 2003, p. 207). Toys considered appropriate for boys encourage physical activity and independent play, whereas "girl toys" elicit closer physical proximity and more talk between child and

caregiver (Athenstaedt, Mikula, and Bredt 2009; Caldera, Huston, and O'Brien 1989; Zahn-Waxler and Polanichka 2004). Even parents who support nonsexist child rearing for their daughters are often concerned if their sons are not aggressive or competitive "enough"—or are "too" sensitive (Blakemore and Hill 2008; Pleck 1992). Girls are increasingly allowed or encouraged to develop instrumental attitudes and skills. Meanwhile, boys are still discouraged from, or encounter parental ambivalence about, developing attitudes and skills of tenderness or nurturance (Blakemore and Hill 2008; Chaplin, Cole, and Zahn-Waxler 2005; Kindlon and Thompson 1999; Martin and Ross 2005; Pollack 1998).

Beginning when children are about five and increasing through adolescence, parents allocate household chores—both the number and kinds—to their children differentially, according to the child's sex. With African American children often an exception, Patricia will more likely be assigned cooking and laundry tasks; Paul will find himself painting and mowing (Burns and

Courtesy Deb Glover and Celeste Wheeler

Toys send messages about gender roles. What does this toy say?

---

[4] Because most of the classic research presented in this section has focused on middle-class whites, findings may or may not apply to other racial/ethnic or class groups. Racial/ethnic and class variation in child rearing is discussed in Chapter 10.

Homel 1989; McHale et al. 1990). Because girls' chores typically must be done daily, whereas boys' are sporadic, girls spend more time doing chores—a fact that "may convey a message about male privilege" (Basow 1992, p. 131). African American girls, however, are raised to be more independent and less passive. Research also indicates that African American boys as well as girls are socialized for roles that include employment and child care (Brown et al. 2009; Hale-Benson 1986; Staples and Boulin Johnson 1993; Theran 2009).

Although relations in the family provide early feedback and help shape a child's developing identity, play and peer groups become important as children try out identities and adult behaviors. In fact, the author of one review of psychological research argues that peers have much more influence on child and adolescent development in general than do parents (J. Harris 1998; Rose and Rudolph 2006).

Play and Games The role of **play** is an important concept in the interactionist perspective. In G. H. Mead's theory (1934), play is not idle time, but a significant vehicle through which children develop appropriate concepts of adult roles as well as images of themselves.

Boys and girls tend to play separately and differently (Aydt and Corsaro 2003; Maccoby 1998, 2002; McIntyre and Edwards 2009). Girls play in one-to-one relationships or in small groups of two and three; their play is relatively cooperative, emphasizes turn taking, requires little competition, and has relatively few rules. In "feminine" games such as jump rope or hopscotch, the goal is skill rather than winning (Basow 1992; Munroe and Romney 2006). Boys more often play in fairly large groups, characterized by more fighting and attempts to effect a hierarchical pecking order. Boys also seem to exhibit high spirits and having fun (Maccoby 1998; Munroe and Romney 2006). From preschool through adolescence, children who play according to traditional gender roles are more popular with their peers; this is more true for boys (Martin 1989).

Especially in elementary schools, many cross-sexual interaction rituals such as playground games are based on and reaffirm boundaries and differences between girls and boys. Sociologist Barrie Thorne (1992), who spent eleven months doing naturalistic observation at two elementary schools, calls these rituals **borderwork** (Aydt and Corsaro 2003).

Sports play a role, both the informal and organized sports of childhood and the images presented in the media (Hardin and Greer 2009; Messner 2002). Now girls have more organized sports available to them, as well as more media models of women athletes. Girls who take part in sports have greater self-esteem and self-confidence (Andersen and Taylor 2002; Dworkin and Messner 1999).

The Power of Cultural Images Media images often convey gender expectations. These images are called *media frames*. A media frame is the way a story has been created for the consumer (this includes television shows, commercials, magazine articles, photographs, radio shows, music, video games, etc.). A media frame is the way that the writers of a story make sense of the stories and events we are viewing, hearing, interacting with, and/or reading about. In other words, the media frame guides us through what the subject is and its meaningful qualities. Such framing has important implications for our gender expectations.

Children's programming more often depicts boys than girls in dominant, agentic roles (B. Carter 1991). On music videos, females are likely to be shown trying to get a man's attention. Some videos broadcast shockingly violent misogynistic (i.e., hatred of women) messages (Jhally 2007). In TV commercials, men predominate by about nine to one as the authoritative narrators or voiceovers, even when the products are aimed at women (Craig 1992; Kilbourne 1994). Cultural images in the media indicate to the audience what is "normal." In studying media frames, researchers "focus on *how* issues and other objects of interest are reported by news media as well as *what* is emphasized in such reporting" (Weaver, McCombs, and Shaw 2004, p. 257, emphasis in original).

The media coverage of the 2008 presidential election is an excellent recent example of the continuation of gender stereotypes that has haunted women politicians since the Suffragette movement (Baird 2008). Women who are government officials or running for political office are more likely to have reporters, television shows, and even opponents emphasize their physical appearance (how their hair looks, what kind of clothing they wear) and to discuss their children and marital status. Communication scholars have noted that "Palin's attractiveness resulted in frequent and varied references to her 'sexiness,' whereas Clinton was viewed as not feminine enough in pantsuits that covered her 'cankles'" (thick ankles) (Carlin and Winfrey 2009, p. 330; see also Bystrom 2006; Heldman, Carroll, and Olson 2005).

Socialization in Schools There is considerable evidence that the way girls and boys are treated differently in school is detrimental to both genders (Corbett, Hill, and St. Rose 2008; Sullivan, Riccio, and Reynolds 2009). School organization, classroom teachers, and textbooks all convey the message that boys are more important than girls. At the same time, some critics posit that school expectations are unreasonably difficult for the typical boy to meet.

Teachers' Practices Research shows that teachers pay more attention to males than to females, and males tend to dominate learning environments from nursery school

Are boys behaving differently than girls in this photo? How does this fit with the discussion of gender and socialization in school? What does that discussion say about girls in school? Boys in school?

through college (Lips 2004; Zaman 2008). Researchers who observed more than 100 fourth-, sixth-, and eighth-grade classes over a three-year period found that boys consistently and clearly dominated classrooms. Teachers called on and encouraged boys more often than girls. If a girl gave an incorrect answer, the teacher was likely to call on another student, but if a boy was incorrect, the teacher was more likely to encourage him to learn by helping him discover his error and correct it (Sadker and Sadker 1994). Compared to girls, boys are more likely to receive a teacher's attention, to call out in class, to demand help or attention from the teacher, to be seen as model students, or to be praised by teachers. Boys are also more likely to be disciplined harshly by teachers (Epstein 2007; Kindlon and Thompson 1999).

In subtle ways, teachers may reinforce the idea that males and females are more different than similar. There are times when boys and girls interact together relatively comfortably—in the school band, for example. But in classrooms, teachers often pit girls and boys against each other in spelling bees or math contests (Thorne 1992).

African American Girls, Latinas, and Asian American Girls in Middle and High School Journalist Peggy Orenstein (1994) spent one year observing pupils and teachers in two California middle schools, one mostly white and middle class and the other predominantly African American and Hispanic and of lower socioeconomic status. Orenstein found that in both schools, girls were

subtly encouraged to be quiet and nonassertive whereas boys were rewarded for boisterous and even aggressive behaviors.

However, African American girls were louder and less unassuming than non-Hispanic white girls were when they began school—and some continued along this path (Theran 2009). In fact, they called out in the classroom as often as boys did. But Orenstein noted that teachers' reactions differed. The participation, and even antics, of white boys in the classroom was considered inevitable and rewarded with extra attention and instruction, whereas the assertiveness of African American girls was defined as "menacing, something that, for the sake of order in the classroom, must be squelched" (p. 181). Orenstein further found that Latinas, along with Asian American girls, had special difficulty being heard or even noticed. Probably socialized into quiet demeanor at home and often having language difficulties, these girls' scholastic or leadership abilities largely went unseen. In some cases, their classroom teachers did not even know who the girls were when Orenstein mentioned their names.

In high school, Latina girls experience cross-pressures when the desire to succeed in school and move on to a career is in tension with the traditional assumption of wife and mother roles at a young age. Eighteen percent of sixteen- to seventeen-year-olds do not finish high school, compared to 8.4 percent of black and 4.5 percent of white girls (Cataldi, Laird, and KewalRamani 2009, Table 3). Factors such as poverty and language barriers, as well as the pressure to contribute to the family, affect the educational attainment of both boys and girls. But young Latinas, especially recent immigrants, seem torn between newer models for women as mothers *and* career women and the traditional model of marriage and homemaking (Canedy 2001). Nevertheless, the majority of Latinas and Latinos finish high school.

School Organization In 2007–2008 school year, 75.6 percent of all school employees were women (Coopersmith 2009, Table 3), although only a little over half (51 percent) of the principals were women (Battle 2009, Table 3). Eighty-five percent of elementary teachers and 59 percent of secondary teachers were women (Coopersmith 2009, Table 3). These numbers represent a change toward greater balance since 1982, when only 21 percent of principals, 24 percent of officials and administrators, and 49 percent of secondary teachers were women (U.S. Census Bureau 2003a, Table 252).

Programs and Outcomes One concern related to schooling has been whether girls are channeled into or themselves avoid the traditionally masculine areas in high school study. We don't really know the answer to this question. A recent review of 1,000 research studies (AAUW Educational Foundation 1999) reports that high school boys and girls now take similar numbers of science courses, but boys are more likely to take all three core courses: biology, chemistry, and physics. Girls enroll in advanced placement (AP) courses in greater numbers than boys, including AP biology. But fewer girls than boys get high enough scores on AP tests to get college credit. Girls take fewer computer courses, and they cluster in traditional female occupations in career-oriented programs (Halpern et al. 2007).

Girls versus Boys?

Girls have long been the primary focus of attention in examining the possible bias of educational institutions. The previous sections make a good case for such concern, and the 1994 Women's Equal Education Act declares girls an "under-served population." Despite the litany of difficulties girls and women may face in educational settings, the intention is to identify problems that need continued attention. In fact, girls are doing well on the whole. Women and girls "are on a tear through the educational system. . . . In the past 30 years, nearly every inch of progress . . . has gone to them" (Thor Mortenson of the Dell Institute for the Study of Opportunity in Higher Education, in Conlin 2003, p. 76).

In recent years, attention has turned to boys. Some writers attack the "myth of girls in crisis" (Sommers 2000a, p. 61). A resident scholar at the conservative think-tank American Enterprise Institute, Sommers in her critique goes beyond a concern for balance to argue that there is a "war on boys" (2000b). In Sommers's view, boys are actively discriminated against by the educational establishment (2000b, p. 60; see also 2000a, p. 23). "[B]oys are resented, both as the unfairly privileged sex and as obstacles on the path to gender justice" (2000b, p. 60; see also 2000a, p. 23).

Sommers and others point to the declining male share of college enrollments and note that on a number of indicators, girls do better in school: better grades, higher educational aspirations, and greater enrollment in AP and other demanding academic programs.

Currently, girls are even more likely to outnumber boys in higher-level math and science courses; however, in 2007 boys outscored girls in math in all grades tested (Snyder, Dillow, and Hoffman 2009). Girls also outnumber boys in student government, honor society, and student newspaper staffs. More boys fall behind grade level and more are suspended, and they are far more likely to be shunted into special education classes or to have their inattentive and restless behavior defined as deviant, and medicated. Indicators of deviant behavior—crime, alcohol, and drugs—show more involvement by boys. Other research also shows that boys have a greater incidence of diagnosis of emotional disorders, learning disorders, attention-deficit disorders, and teen deaths (Conlin 2003; Goldberg 1998; Sommers 2000a, 2000b).

To the argument that boys do better on SAT and other standardized tests, Sommers responds that the pool of girls taking the test is more apt to include disadvantaged and/or marginal students, whereas their male counterparts do not take these exams. Sommers's concerns may be misplaced, however. Other scholars note that when examining sex differences, socioeconomic and racial differences should be taken into account as well. When these are taken into account, the sex differences are no longer significant, but there is a correlation between race and class and a higher SAT score (Zwick 2001). In fact, in an examination of the most recent SAT data (see Table 4.1), we can see that the higher the test taker's family income, the better the scores (Total Group Report 2009, p. 4; see also Rampell 2009).

Other analysts do not necessarily share Sommers's allegations of active discrimination against boys. But they argue that attention to girls' educational needs and the success of men in the work world tended to obscure boys' problems in school. They see a mismatch between typical boy behavior—high levels of physical activity and more challenges to teachers and school rules—and school expectations about sitting still, following rules, and concentrating (Poe 2004). Moreover, a survey by the Public Education Network (Metropolitan Life Insurance Company 1997) found that 31 percent of boys in grades 7–12 felt that teachers do not listen to them, as compared with 19 percent of girls.

A recent research study (Meadows, Land, and Lamb 2005) sought to examine and compare the situations of boys and girls. The researchers took note of Sommers's critique regarding boys, as well as the research cited earlier on the disadvantaged situation of girls in schools. They noted Carol Gilligan's influential books (1982; Gilligan, Lyons, and Hanmer 1990), which argue that girls' strengths in relationships and emotional expressions are devalued in an individualistic and competitive American society and that girls become discouraged as they arrive on the threshold of adolescence (see also Pipher 1995).[5]

---

[5] Meadows et al. (2005) note that Gilligan's conclusions about gender differences are based on small numbers of interviews with girls (no boys) and on anecdotes and that she has never been willing to make her data available for review by other scholars.

Table 4.1 SAT Score by Family Income, 2009

SAT All Test Takers (N = 1,530,128)	Test Takers Percent 100	Critical Reading Mean 501	Mathematics Mean 515	Writing Mean 493
Family Income				
$0-$20,000	10	434	457	430
$20,000-$40,000	15	462	475	453
$40,000-$60,000	15	488	497	476
$60,000-$80,000	15	503	512	491
$80,000-$100,000	13	517	528	505
$100,000-$120,000	11	525	538	516
$120,000-$140,000	5	529	542	520
$140,000-$160,000	4	536	550	527
$160,000-$200,000	5	542	554	535
More than $200,000	7	563	579	560

Source: Adapted from The College Board (http://professionals.collegeboard.com, Table 11).

Essentially, Meadows and her colleagues asked: How goes it with boys and girls? For answers they turned to the status of boys and girls on twenty-eight social indicators of well-being, which are also combined into an index. These indicators measure seven life domains: material well-being, health, safety, productive activity, intimacy, place in the community, and emotional well-being (Meadows, Land, and Lamb 2005, p. 5). Meadows and her colleagues concluded that "gender differences in well-being, when they do exist, are very slight and that overall both boys and girls in the United States currently enjoy a higher quality of life than they did in 1985" (p. 1).

Other approaches to concerns about boys are exactly opposite to the approach of Sommers. Psychiatrist William Pollack's perspective is that "what we call . . . normal boy development . . . not only isn't normal, but it's traumatic and that trauma has major consequences" (quoted in Goldberg 1998, p. A-12; see also Kimmel 2001). To express vulnerability is to run the risk of victimization. Boys, particularly racial/ethnic minority youth, face a dilemma in that acting tough to protect themselves is threatening to adults (psychologist Dan Kindlon, cited in Goldberg 1998). One effort suggested by this line of thought is to take measures to decrease bullying (Kimmel 2001; Totura et al. 2009). Another is to encourage boys to express their emotions and to redefine the male role to include emotional expression (Espelage and Swearer, 2004; Zaman 2008).

Other proposals include the following: (1) accepting a certain level of boys' rowdy play as not deviant, (2) implementing more active learning-by-doing to permit physical movement in classroom settings, and (3) encouraging activities shared by boys and girls and boy–girl dialogues about gender (Kindlon and Thompson 1999). The bottom line may be the observation by Marie C. Wilson, president of the Ms. Foundation for Women: "We'd be so naive to think we could change the lives of girls without boys' lives changing" (quoted in Goldberg 1998, p. A-12). It may be that "girls and boys are on the same side in this issue."

In this section on socialization, we have examined how socialization shapes gender identities and gendered behavior. Socialization continues throughout adulthood as we negotiate and learn new roles—or as those already learned are renegotiated or reinforced. The varied opportunities we encounter as adults influence the adult roles we choose and play out, and the qualities and skills we develop. And those have changed in recent decades, in response to the Women's Movement and the men's movements that followed.

"We don't believe in pressuring the children. When the time is right, they'll choose the appropriate gender."

Social Change and Gender

The increasing convergence of men's and women's social roles, though incomplete, reflects a dramatic change from the more gender-differentiated world of the mid-twentieth century. Such changes are due not only to structural forces (especially economic) that led to women's increased entry into the labor force but also to active change efforts by women and their allies in the Women's Movement. Men's movements followed.

The Women's Movement

The nineteenth century saw a feminist movement develop, but from around 1920 until the mid-1960s there was virtually no activism regarding women's rights and women's roles.[6] Women did make some gains in the 1920s and 1930s in education and occupational level, but these were eroded during the more familistic post–World War II era.

Media glorification of housewife and breadwinner roles made them seem natural despite the reality of increased women's employment. But contradictions between what women were actually doing and the roles prescribed for them became increasingly apparent. Higher levels of education for women left college-educated women with a significant gap between their abilities and the housewife role assigned to them. Employed women chafed at the unequal pay and working conditions in which they labored and began to think that their interest lay in increasing equal opportunity. Betty Friedan's book *The Feminine Mystique* (1963) captured this dissatisfaction and made it a topic of public discussion. Further, the Civil Rights Movement of the 1960s provided a model of activism. In a climate in which social change seemed possible, dissatisfaction with traditional roles precipitated a social movement—the Second Wave of the Women's Movement. This movement challenged theretofore accepted traditional roles and strove to increase gender equality.

In 1961 President Kennedy set up a Presidential Commission on the Status of Women, and some state commissions were established subsequently. The National Organization for Women (NOW) was founded in 1966. Meanwhile, in Congress, the Civil Rights Act of 1964 had been amended by opponents (as a political tactic) to include sex—and it passed! Title VII of the Civil Rights Act gradually began to be enforced. Grassroots feminist groups with a variety of agendas and political postures developed across the country.

NOW had multiple goals—opening educational and occupational opportunities to women, along with establishing support services such as child care. As well, NOW recognized the commitment of a majority of women to marriage and motherhood and spoke to the possibility of "real choice" (National Organization for Women 1966). Although supporting traditional heterosexual marriage for those who chose it, the organization came to support the more controversial choice (in those times) of a lesbian lifestyle, as well as reproductive rights, including abortion.

Women vary in their attitudes toward the Women's Movement and in what issues are important to them. Figure 4.5 indicates "top priority" issues that were evident in a survey of women conducted in 2003.

Some women of color and white working-class women may find the Women's Movement irrelevant to the extent that it focuses on psychological oppression or on professional women's opportunities rather than on "the daily struggle to make ends meet that is faced by working class women" (Aronson 2003, p. 907; Langston 2007). Black women have always labored in the productive economy under duress or out of financial necessity and did not experience the enforced delicacy of women in the Victorian period. Nor were they ever housewives in large numbers, so the feminist critique of that role may seem irrelevant (Lessane 2007).

Chicano/Chicana (Mexican American) activism gave *la familia* a central place as a distinctive cultural value. Latinos of both sexes placed a high value on family solidarity, with individual family members' needs and desires subsumed to the collective good, so that Chicana feminists' critiques of unequal gender relations in *la familia* often met with hostility (Manago, Brown, and Leaper 2009).

Muslim women and Arabic women also view feminism through the lens of culture, religion, and community. Strict adherence to religious and cultural traditions are often at odds with western feminist views, yet many women are challenging sharia law and using the Qur'an to question patriarchy and demand women's rights (Karim 2009; Sayeed 2007).

African American and Latino women consider racial/ethnic as well as gender discrimination in setting their priorities (Arnott and Matthaei 2007). In fact, it is more precise to say their feminist views are characterized by *intersectionality*—structural connections among race, class, and gender (Baca Zinn, Hondagneu-Sotelo, and Messner 2007; Labaton and Martin 2004; Roth 2004; Yuval-Davis 2006).

In some ways, such as their experience with racial/ethnic discrimination and their relatively low wages, Chicanas are more like Mexican American men, who are also subordinated, than they are like non-Hispanic

---

[6] The "First Wave" of feminism began with a convention on women's rights that produced the Seneca Falls Declaration in 1848. Nineteenth-century women were also active in abolitionist and temperance movements. The First Wave of feminism came to an end when a major goal, voting rights for women, was achieved in 1920 (Rossi 1973).

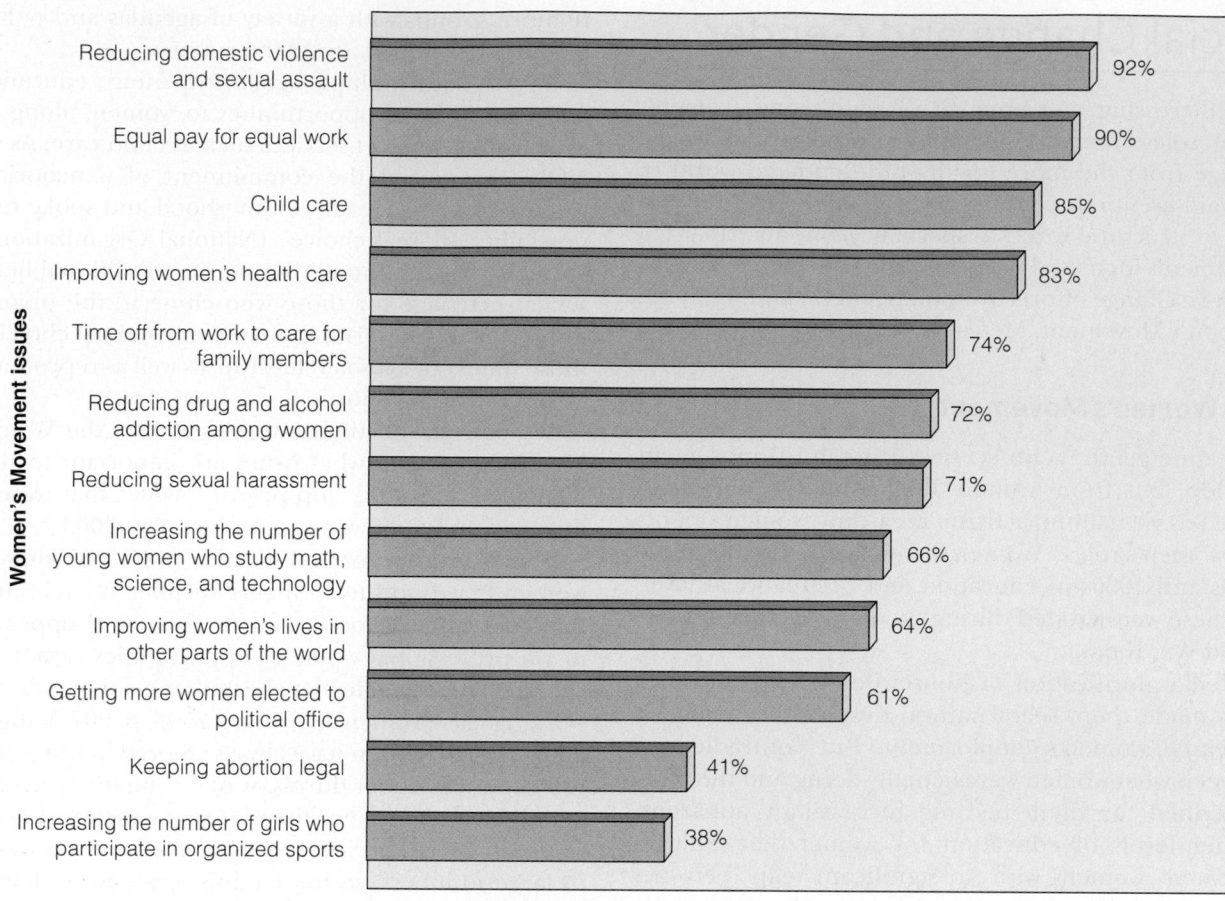

Percentage of women surveyed who rated the issue a "top priority"

Figure 4.5 "Top priority" issues for the Women's Movement. More than 3,300 women were asked to indicate which issues they felt the Women's Movement should focus on. The 2003 survey was conducted for the Center for the Advancement of Women by Princeton Survey Research Associates.

Source: Center for the Advancement of Women 2003, p. 11 (www.advancewomen.org).

white women. Nevertheless, a Chicana feminism emerged during the 1960s and 1970s. Generally, Chicanas support women's economic issues, such as equal employment and day care, while showing less support for abortion rights than do Anglo women. Latinas formed some grassroots community organizations of their own to offer social services such as job training, community-based alternatives to juvenile incarceration, and bilingual child development centers. The Mexican American Women's National Association (now MANA) was established in 1974 (Segura and Pesquera 1995).

African American women are more critical of gender inequality than are white women (Kane 2000). A National Urban League report states that "a feminist perspective has much to offer Black America" (West 2003). African American women and men are more likely than whites to endorse political organizing for women's issues (Hunter and Sellers 1998). Sixty-eight percent of Latina women ($n = 354$) and 63 percent of African American women ($n = 352$) surveyed in 2001 as

part of a national sample of 2,329 "strongly agree" that there is a need for a women's movement today (Center for the Advancement of Women 2003). Post Colonial, Third World, and Transnational Feminisms are led by women of color and Third World women. These movements developed out of western feminism's lack of attention to issues of racism, imperialism, colonial oppression, power, sexuality, and resistance to hegemony. These feminisms seek to create justice and equality across national borders (Gurel 2009).

There are other variations in attitudes toward the Women's Movement. Some women deplore the rise of feminism and encourage traditional marriage and motherhood as the best path to women's self-fulfillment (Enda 1998; Passno 2000). Many feminists would define the movement as one that advances the interests and status of women as mothers and caregivers, as well as workers (Showden 2009). Surveys in the 1980s indicate that large majorities reject the notion that the Women's Movement is antifamily (Hall and Rodriguez 2003).

Dan Koeck/The New York Times/Redux

Native Americans, members of what were once hunting and gathering and hoe cultures, have a complex heritage that varies by tribe but may include a matrilineal tradition in which women have owned (and may still own) houses, tools, and land. Native American women's political power declined with the spread of Europeans into their territories and the subsequent reorganization of Indian life by federal legislation in the 1920s. Recently, Native American women have begun to regain their power. Erma J. Vizenor is chairwoman of the White Earth Nation, the largest tribe in Minnesota. Dr. Vizenor, who holds a doctorate from Harvard, is one of 133 women tribal leaders.

The media often assert that a younger "post-feminist" generation does not support a women's movement. The assumption is that they may have a negative image of feminism or be latently feminist, but believe that women's rights' goals have already been achieved. Sometimes such articles assert that younger women are simply too busy with work and family to have the time to be active (Showden 2009).

Differences of opinion among women on issues related to sexuality and reproduction are undoubtedly divisive. Most recent research finds a complex array of definitions of feminism (Aronson 2003; Center for the Advancement of Women 2003), and cultural meanings of *feminism* do seem to vary by age cohort (Peltola, Milkie, and Presser 2004; Showden 2009).

Nevertheless, research suggests that "post-feminism" is a myth. Hall and Rodriguez, who did an extensive review of survey data, found an increase in support for the Women's Movement from the 1970s and 1980s to the middle or late 1990s (see also Bolzendahl and Myers 2004). Young adults age eighteen to twenty-nine reported more favorable attitudes than older cohorts (Hall and Rodriguez 2003, p. 895; see also Aronson 2003). "One might note that many of the ideologies associated with feminism have become relatively common place and speak to the success of feminism in attaining much

broader acceptance of gender equality" (Schnittker, Freese, and Powell 2003, pp. 619–20). Although not all supporters of the Women's Movement self-identify as feminists, a majority (54 percent) "say that being a feminist is an important part of who they are" (Center for the Advancement of Women 2003).

Men's Movements

As the Women's Movement encouraged changes in gender expectations and social roles, some men responded by initiating a men's movement. The first National Conference on Men and Masculinity was held in 1975 and has been held almost annually since. The focus of this men's movement is on changes that men want in their lives and how best to get them. One goal has been to give men a forum—in consciousness-raising groups, in men's studies college courses, and, increasingly, on the Internet—in which to air their feelings about gender and think about their life goals and their relationships with others.

Kimmel (1995) divides today's men's movement into three fairly distinct camps: antifeminists, profeminists, and masculinists. *Antifeminists* believe that the Women's Movement has caused the collapse of the natural order, one that guaranteed male dominance, and they work to reverse this trend. The National Organization for Men (NOM) opposes feminism, which it claims is "designed to denigrate men, exempt women from the draft and to encourage the disintegration of the family" (Siller 1984, quoted in Kimmel 1995, p. 564).

According to Mark Kann, men's self-interest may lead to an antifeminist response even among men who wish women well in an abstract sense:

> I would suggest . . . that men's immediate self-interest rarely coincides with feminist opposition to patriarchy. Consider that men need money and leisure to carry out their experiments in self-fulfillment. Is it not their immediate interest to monopolize the few jobs that promise affluence and autonomy by continuing to deny women equal access to them? . . . Why should they commit themselves to those aspects of feminism that reduce men's social space? It is one thing to try out the joys of parenting, for example, but quite another to assume sacrificial responsibility for the pains of parenting. (Kann 1986, p. 32)

Profeminists support feminists in their opposition to patriarchy. They analyze men's problems as stemming from a patriarchal system that privileges white heterosexual men while forcing all males into restrictive gender roles (Ashe 2004). In 1983, profeminist men formed the National Organization for Changing Men (changed in 1990 to the National Organization for Men Against Sexism, or NOMAS), whose purposes are to transcend gender stereotypes while supporting

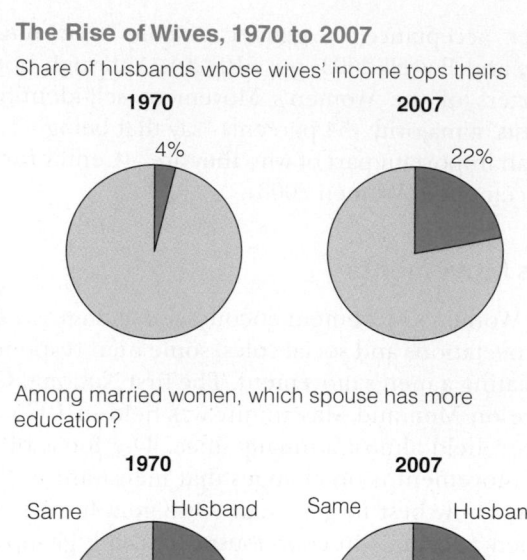

The Rise of Wives, 1970 to 2007

Share of husbands whose wives' income tops theirs

1970 — 4%

2007 — 22%

Among married women, which spouse has more education?

1970 — Same 52%, Husband 28%, Wife 20%

2007 — Same 53%, Husband 19%, Wife 28%

Figure 4.6 The Rise of Wives.

women's and gays' struggles for respect and equality ("Who We Are" 2009).

The newer *masculinists*, who emerged in the early 1990s, tend not to focus on patriarchy as problematic (although they might agree that it is). Instead, masculinists work to develop a positive image of masculinity, one combining strength with tenderness. Their path to this is through therapy, consciousness-raising groups, and rituals. Through rituals, men are to get in touch with their feelings and heal the buried rage and grief caused by the oppressive nature of corporate culture, the psychological and/or physical absence of their fathers, and men's general isolation due to a socialized reluctance to share their feelings (Kimmel 1995). Robert Bly's *Iron John* (1990) is a prominent example of these ideas.

In examining men's movements and their goals, it is important to appreciate that men's social situations vis-à-vis traditional roles are as diverse as women's. The idea of a universal patriarchy and male dominance is challenged by the obvious point that all men are not privileged in the larger society (Connell 2005; Hebert 2007), whether or not they are in gender relations. "When race, social class, sexual orientation, physical abilities, and immigrant or national status are taken into account, we can see that in some circumstances 'male privilege' is partly—sometimes substantially—muted" (Baca Zinn, Hondagneu-Sotelo, and Messner 2004, p. 170, citing Kimmel and Messner 1998).

Personal and Family Change

Sometimes, in response to available options, adults reconsider earlier choices regarding gender roles. For example, a small proportion of men choose to be full-time fathers and/or househusbands. Others may effect more subtle changes, such as breaking through previously learned isolating habits to form more intimate friendships and deeper family relationships.

Ironically, that very expansion in the range of people's opportunities may lead to mixed feelings and conflicts, both within oneself and between men and women as we confront the "lived messiness" of gender in contemporary life (Heywood and Drake 1997, p. 8). Stay-at-home moms may worry about the family budget and about their options if their marriages fail or if they desire to work when children are older. They may feel others consider them uninteresting or incompetent. Women who are employed may wish they could stay home full time with their families or at least have less hectic days and more family time. Moreover, a wife's career success and work demands may lead her into renegotiating gender boundaries at home, and that may produce domestic tension.

Modern men may be torn between egalitarian principles and the advantages of male privilege. For one thing, husbands are still expected to succeed as principal family breadwinners (see Figure 4.6 to see how some wives are now earning more than their husbands). The "new man" is expected to succeed economically *and* to value relationships and emotional openness. Although women want men to be sensitive and emotionally expressive, they also want them to be self-assured and confident.

Men often face prejudice when they take jobs traditionally considered women's (Simpson 2005; Snyder and Green 2008). They may also encounter more resistance than women when they try to exercise "family friendly" options in the workplace (Neal and Hammer 2007; Peterson 2004). This resistance men encounter, however, may also be a self-imposed perception in which they believe they will be viewed as "less committed to their company" should they access such options (Neal and Hammer 2007, p. 159). Such difficulties, coupled with technological advancements, have allowed some fathers to take the "daddy track" option. This is where more men work from home or opt for flexible work hours (Jarrell 2007; Shellenbarger 2007).

These conflicts are more than psychological. They are in part a consequence of our society's failure to provide support for employed parents in the form of adequate maternity and paternity leave or day care, for example. American families continue to deal individually with problems of pregnancy, recovery from childbirth, and early child care as best they can. Adequate job performance, let alone career achievement, is difficult for women under such conditions regardless of ability. Declining economic opportunities for non–college-educated men, coupled with criticisms of male privilege sparked by the Women's Movement, lead some men to feel unfairly picked on (Hebert 2007).

Today's men, like today's women, find it difficult to have it all (Jarrell 2007). If women find it difficult to combine a sustained work career with motherhood, men face a conflict between maintaining their privileges and enjoying supportive relationships. But many "will find that equality and sharing offer compensations to offset their attendant loss of power and privilege" (Gerson 1993, p. 274).

Chapter 11 discusses work and family in more detail, and Chapter 13 addresses family power.

The Future of Gender

"Gender is a ruling idea in people's lives—even where egalitarian ideology is common, as among young, affluent, educated Americans. . . . Despite the extraordinary improvements in women's status over the past two centuries, some aspects of gender have seemed exceptionally resistant to erosion in recent decades"—notably child raising and occupational achievement and pay. "[P]eople still think about women and men differently and . . . men still occupy most of the highest positions of political and economic power" (Jackson 2006, pp. 215, 229).

If men and women are seen through a lens of differential competence, if men and women continue to interact in situations in which they (men) have greater power and status, then assumptions about inherent status differences are supported. Moreover, gender differences may be maintained because of the power of gendered self-concepts. Men and women may have a psychological stake in the maintenance of these differences as important aspects of personal identity (Ridgeway 2006).

Yet sociologist Robert Max Jackson is convinced that the forces of history will sweep gender inequality aside. Two hundred years of change suggest that, indeed, gender is not embedded so deeply in society that it cannot change. "It is fated to end because essential organizational characteristics and consequences of a modern, industrial, market-oriented, electorally governed society are inherently inconsistent with the conditions needed to sustain gender inequality" (Jackson 2006, p. 241).

Looking at marriage and the family, sociologist Steven Nock expects marriages of "equally dependent spouses" to represent the future toward which American marriage is evolving. By this he means a model in which both spouses are earners as well as family caretakers, and they are dependent on their common earnings as the basis of their family life. In the short run, women's increased earnings have enabled them to leave a marriage of poor emotional quality. Yet the falling divorce rate suggests that in the long run this will not be the case because, as Nock says,

> . . . we are returning to a more traditional form of marriage than we have had in the last century. Marriage was historically based on extensive dependency by both partners. . . . I believe that the recent increases in marital stability reflect the gradual working out of the gender issues first confronted in the 1960s. If so, this implies that young men and women are forming new types of marriages that are based on a new understanding of gender ideals. And if so, the growth of equal dependency that we are likely to see in the next decade bodes well. (Nock 2001, pp. 773–74)

Gendered expectations and behaviors—both as they have and have not changed—underpin many of the topics explored throughout this text. For example, gender is important to discussions of family power, communication, and parental roles, as well as to work and family roles. In future chapters we will explore these topics, keeping in mind gendered expectations and social change.

Reprinted by permission of Anne Gibbons.

Summary

- Roles of men and women have changed over time, but living in our society remains a somewhat different experience for women and for men. Gendered cultural messages and social structure influence people's behavior, attitudes, and options. But in many respects, men and women are more alike than different.

- Generally, traditional masculine expectations require that men be confident, self-reliant, and occupationally successful—and engage in "no sissy stuff." During the 1980s, the "liberated male" cultural message emerged, according to which men are expected to value tenderness and equal relationships with women. We have seen the return of respect to the physically tough and protective "manly man."

- Traditional feminine expectations involve a woman's being a man's helpmate and a "good mother." An emergent feminine role is the successful "professional woman." When coupled with the more traditional roles, this results in the "superwoman."

- Individuals vary in the degree to which they follow cultural models for gendered behavior. The extent to which men and women differ from each other and follow these cultural messages can be visualized as two overlapping normal distribution curves.

- Although there are significant changes, male dominance remains evident in politics, in religion, and in the economy.

- There are racial/ethnic and class differences in stereotypes, as well as some differences in actual gender and family patterns. This and other diversity has come to be expressed in responses to the women's and men's movements.

- Biology interacts with culture to produce human behavior, and the two influences are not really separable. Sociologists give considerable attention to the socialization process, for which there are several theoretical explanations. Advocates have expressed concern about barriers to the opportunities and achievements of both boys and girls.

- Underlying both socialization and adult behavior are the social structural pressures and constraints that shape men's and women's choices and behaviors. These have changed in recent decades, in part as a consequence of the Women's Movement and subsequent men's movements.

- Turning our attention to the actual lives of adults in contemporary society, we find women and men negotiating gendered expectations and making choices in a context of change at work and in the family. New cultural ideals are far from realization, and efforts to create lives balancing love and work involve conflict and struggle, but promise fulfillment as well.

- Whether gender will continue to be a moderator of economic opportunities and life choices in the future is uncertain, as men's and women's roles and activities converge, but men remain more advantaged. It is likely that gender identity will continue to be important to both men and women.

- Some social scientists predict that the shared economic and family roles emerging in younger marriages will produce more stable marriages.

Questions for Review and Reflection

1. What are some characteristics generally associated with males in our society? What traits are associated with females? How might these affect our expectations about the ways that men and women should behave? Are these images still influential?

2. Do you think men are dominant in major social institutions such as politics, religion, education, and the economy? Or are they no longer dominant? Give evidence to support your opinion.

3. Which theory of gender socialization presented in this chapter seems most applicable to what you see in the real world? Can you give some examples from your own experience of gender socialization?

4. Women and men may renegotiate and change their gendered attitudes and behaviors as they progress through life. What evidence do you see of this in your own life or in others' lives?

5. **Policy Question.** What family law and policy changes of recent years do you think are related to the women's and men's movements? What policies do you think would be needed to promote greater gender equality and/or more satisfying lives for men and women?

Key Terms

agentic (instrumental) character traits 83
borderwork 95
communal (expressive) character traits 83
femininities 84
gender 80
gender identity 80
gender role 80
gender schema theory of gender socialization 93
gender similarities hypothesis 86
hormonal processes 91
hormones 91
intersexual 82

male dominance 86
masculinities 83
modern sexism 81
play 95
self-identification theory of gender socialization 93
sex 80
social learning theory of gender socialization 93
socialization 92
symbolic interaction theory of gender socialization 93
traditional sexism 81
transgendered 82
transsexual 82

Online Resources

Sociology CourseMate

www.CengageBrain.com

Access an integrated eBook, chapter-specific interactive learning tools, including flashcards, quizzes, videos, and more in your Sociology CourseMate, accessed through CengageBrain.com.

www.CengageBrain.com

Want to maximize your online study time? Take this easy-to-use study system's diagnostic pre-test, and it will create a personalized study plan for you. By helping you identify the topics that you need to understand better and then directing you to valuable online resources, it can speed up your chapter review. CengageNOW even provides a post-test so you can confirm that you are ready for an exam.

5

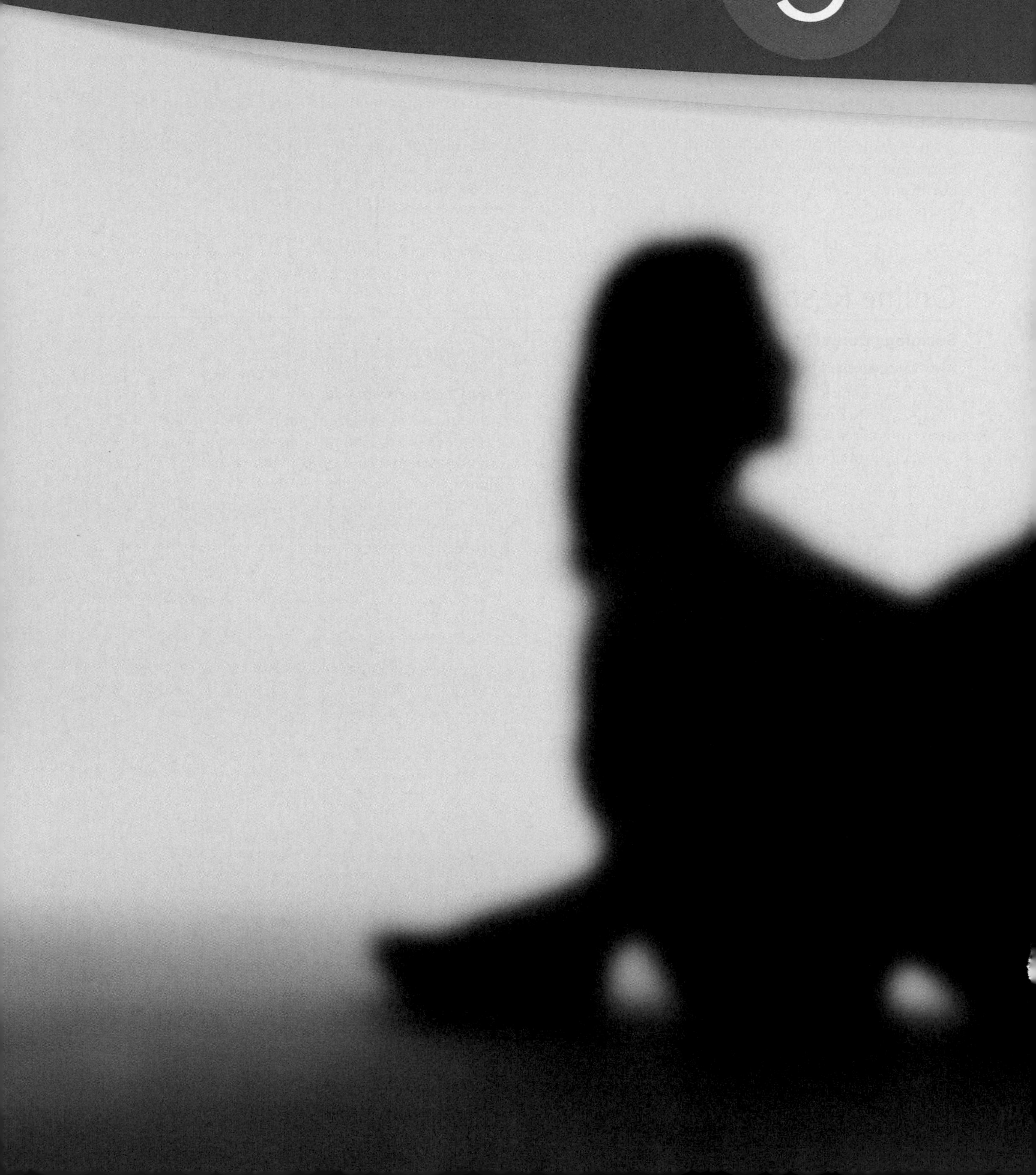

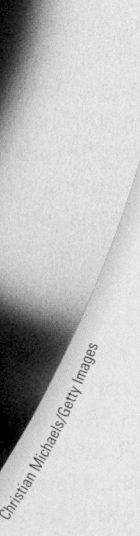

Our Sexual Selves

Christian Michaels/Getty Images

From childhood to old age, people are sexual beings.[1] Sexuality has a lot to do with the way we think about ourselves and how we relate to others. It goes without saying that sex plays a vital role in marriages and other intimate partner relationships.

Despite the pleasure it may give, sexuality may be one of the most baffling aspects of our selves. Finding mutually satisfying ways of expressing sexual feelings can be a challenge.

In this chapter, we will look briefly at children's sexual development. We define sexual orientation and examine the situation of gay men, lesbians, and bisexual individuals in today's society. We will review the changing cultural meanings of sexuality through our history, as well as the varied sexual standards present in contemporary culture. We will discuss sex as a pleasure bond that requires open, honest, and supportive communication, and then look at the role sex plays throughout marriage.

We will look at some challenges that are associated with sexual expression. What happens when one or both partners has an affair? How has the emergence of HIV/AIDS as a pandemic disease affected sexual relationships and families? Finally, we will examine ways that sexuality, research on sex, and sex education have become political issues in our society.

Before we discuss sexuality in detail, we want to point out our society's tendency to reinforce the differences between women and men and to ignore the common feelings, problems, and joys that make us all human. The truth is, men and women aren't really so different. Many physiological parts of the male and female genital systems are either alike or analogous, and sexual response patterns are similar in men and women (Masters and Johnson 1966).

Space limits our ability to present much detail on the various possibilities of sexual expression, which include kissing, fondling, cuddling, and even holding hands, as well as more overtly sexual activities. We consider that you might have concerns about sexuality that are difficult to address in the limited space of this textbook.

Sexual Development and Orientation

Knowledge about children's sexual development and the emergence of sexual orientation is not as extensive as we would like, but we do know some things.

[1] Five online appendices give information on sexual and reproductive topics: Appendix A: Human Sexual Anatomy, Appendix B: Human Sexual Response, Appendix C: Sexually Transmitted Diseases, Appendix D: Sexual Dysfunctions and Therapy, and Appendix E: Conception, Pregnancy, and Childbirth. These appendices can be accessed at the book website.

Children's Sexual Development

"Human beings are sexual beings throughout their entire lives" (DeLamater and Friedrich 2002, p. 10). As early as twenty-four hours after birth, male newborns get erections, and infants may touch their genitals. In a study of almost one thousand children in Minneapolis and Los Angeles, pediatric researchers sought to establish a baseline of "normative" sexual behavior—that is, to indicate the normality of children's sexual interest to parents, social workers, and others. These researchers found that young children often exhibit overtly sexual behaviors.

Reports by "primary female caregivers," using the Child Behavior Checklist and Child Sexual Behavior Inventory, indicate that between the ages of two and five, a substantial number of children engage in "rhythmic manipulation" of their genitals, which the researchers term a "natural form of sexual expression" (DeLamater and Friedrich 2002, p. 10; see also Kellogg 2009; "Sexual Development" 2009). Children may also try to look at others who are nude or undressing or try to touch their mother's breasts or genitals. Sixty percent of boys and 44 percent of girls in this age group touched their own sex organs (Friedrich et al. 1998, Tables 3 and 4). Children also "play doctor," examining one another's genitals. There were few sex differences overall.

Researchers are interested in these physical manifestations of childhood sexual development. However, they place this in context, noting that overall sensual experiences from infancy onward shape later sexual expression, while attachment to parents in infancy and childhood provides the emotional security essential to later sexual relationships (DeLamater and Friedrich 2002; Kellogg 2009; "Sexual Development" 2009).

Early sexual behavior peaks at age five, declining thereafter until sexual attraction first manifests itself around age eleven or twelve. Children are maturing about two years earlier than they were one hundred years ago and much earlier than in 1500, when the average age of puberty was nineteen in England (Brink 2008; Sanghavi 2006; Steingraber 2007). As the age of puberty has declined, the age at marriage has risen, leaving a more extended period during which sexual activity may occur among adolescents and unmarried adults.

Sexual Orientation

As we develop into sexually expressive individuals, we manifest a sexual orientation. **Sexual orientation** refers to whether an individual is drawn to a partner of the same sex or the opposite sex. **Heterosexuals** are attracted to opposite-sex partners and **homosexuals** to same-sex partners.[2] **Bisexuals** are attracted to people of both sexes. A person's sexual orientation does not necessarily predict his or her sexual behavior; abstinence is a behavioral choice, as is sexual expression with partners of the nonpreferred sex. All these terms designate one's choice of sex partner only, not general masculinity or femininity or other aspects of personality.

We tend to think of sexual orientation as a dichotomy: One is either "gay" or "straight." Actually, sexual orientation may be a continuum. Freud, Kinsey, and many present-day psychologists and biologists maintain that humans are inherently bisexual; that is, we all have the latent physiological and emotional structures necessary for responding sexually to either sex.

From the interactionist point of view (see Chapter 2), the very concepts "bisexual," "heterosexual," and "homosexual" are social inventions. They emerged in scientific and medical literature in the late nineteenth century (Katz 2007, pp. 10–12; Seidman 2003, pp. 46–49, 56–58). Although same-sex sexual relations existed all along, the conceptual categories and the notion of sexual orientation itself were cultural creations. Developing a sexual orientation today may be influenced by the resultant tendency to think in dichotomous terms: Individuals may sort themselves into the available categories and behave accordingly. In time, social pressures to view oneself as either straight or gay may inhibit latent bisexuality or inconsistencies (Gagnon and Simon 2005; Katz 2007, pp. 25–27; Rosario et al. 2006; Schwartz 2007). In this light, the recent assertion of **asexuality** as a sexual orientation represents a challenge to the traditional dichotomy. *Asexuality* is described in "A Closer Look at Diversity: Is It Okay to Be Asexual?"

The reason some individuals develop a gay sexual orientation has not been definitively established—nor do we yet understand the development of heterosexuality. The American Psychological Association (APA) takes the position that a variety of factors impact a person's sexuality. The most recent literature from the APA says that "sexual orientation is most likely the result of a complex interaction of environmental, cognitive and biological factors . . . is shaped at an early age . . . [and evidence suggests] biology, including genetic or inborn hormonal factors, play a significant role in a person's sexuality," and that sexual orientation is not a choice and cannot be changed at will (American Psychological Association 2010).

Among gays and lesbians, sexual identity through a sense of being different might have been felt in childhood. Sexual attraction to same-sex people occurs as early as ten for boys and fourteen for girls. Same-sex sexual activity typically begins around age fourteen for males, whereas women's initial experiences tend to occur around age sixteen. "Coming out"—identifying oneself as gay to others—occurs on average just before or after high school graduation (Drasin et al. 2008; Savin-Williams 2006). Recent empirical research reaffirms these findings. Of 2,560 California high school students surveyed over a three-year period, 11 percent reported a gay or lesbian identification, approximately 12 percent identified as bisexual, and nearly 5 percent noted that they were questioning their sexuality (Russell, Clarke, and Clarey 2009).

Individuals vary in this process. A study of 156 lesbian, gay, and bisexual youths recruited from gay organizations and public colleges in New York City found that 57 percent consistently identified as gay or lesbian and were more certain and accepting of their same-sex sexuality, involved in gay social activities, and more comfortable with others knowing. Fifteen percent consistently identified as bisexual. Another 18 percent experienced a gradual transition to the establishment of a gay self-identity. Contrary to stereotype, female youth were more likely to have a consistent gay sexual orientation earlier than male youth (Rosario et al. 2006; see also Rosario, Scrimshaw, and Hunter 2008).

Deciding who is to be categorized as **gay**, **lesbian**, or bisexual for research purposes is not easy—How much experience? How exclusively homosexual? With survey respondents possibly concealing sexual orientation, this precludes any certain calculation of how many gay and lesbian individuals are in our society. Until fairly recently, it was stated that about 10 percent of adult individuals are gay or lesbian. However, the National Health and Social Life Survey (NHSLS; Laumann et al. 1994) suggested that the proportion of homosexual individuals in the population is probably lower. An analysis of combined NHSLS data and University of Chicago National Opinion Research Center (NORC) survey data found that 4.7 percent of men have had some same-sex experiences since age eighteen, whereas 2.5 percent had exclusively same-sex experiences over the last year. Of women, 3.5 percent report some adult same-sex experiences, whereas 1.4 percent had exclusively same-sex experiences over the last year (Black et al. 2000).

In terms of self-identification, the National Survey of Family Growth, conducted in 2002 and 2003 by the U.S. Centers for Disease Control and Prevention, found that approximately 4.1 percent of each sex in the eighteen to forty-four age range reported a gay, lesbian, or bisexual self-identification, whereas 90 percent identified as heterosexual (Mosher, Chandra, and Jones 2005, pp. 1–3, Table B; Gates 2006).

---

[2] Everyday terms are *straight* (heterosexual), *gay*, and *lesbian*. The term *gay* is synonymous with *homosexual* and refers to males or females. Often *gay* or *gay male* is used in reference to men, while *lesbian* is used to refer to gay women. The Committee on Lesbian, Gay, Bisexual, and Transgender Concerns of the American Psychological Association (1991) prefers the terms *gay male* or *gay man* and *lesbian* to *homosexual* because the committee thinks that the latter term may be associated with negative stereotypes.

A Closer Look at Diversity

Is It Okay to Be Asexual?

A majority of Americans are heterosexual—that is, attracted to potential partners of the opposite sex. Some Americans are **GLBT**: gay male, lesbian, bisexual, or transgendered, attracted to same-sex partners or those of either sex, as the case may be. A newly identified sexual orientation is asexuality. An unknown but probably small number of Americans simply do not experience sexual attraction to others. This is different from *celibacy* or *abstinence*, which is a *decision* not to have sexual relations, at least for a time, but not from a lack of desire.

Asexual individuals have emotional feelings and may desire intimate relationships with others, just not sexual ones. Some asexuals experience physical arousal or even masturbate, "but feel no desire for partnered sexuality" (Jay 2005).

Little research on asexuals exists; asexuals have been virtually invisible. In fact, by age forty-four, 97 percent of men and 98 percent of women have had

sexual intercourse (Mosher, Chandra, and Jones 2005, p. 1). Such sexual activity, however, does not always mean that feelings of sexual desire exist.

In an exploratory study of asexuality conducted in 1994 and based on a national probability sample of British residents age eighteen through fifty-nine (Bogaert 2004), 1.05 percent reported themselves to be asexual even though 44 percent were or had been married or cohabitating. Women were more likely to report an asexual orientation than men, but age was not related to asexuality. An interesting finding of this study was that there were large differences between sexual and asexual people in education and social class, with asexuals more likely to have less education and lower social origins. Bogaert speculates that asexuality may be related to a less-advantaged social environment. Needless to say, this and other findings of the study need to be tested against further research conducted in other societies with diverse class structures.

The Asexual Visibility and Education Network (AVEN) was founded in 2001, as a networking and information resource (www.asexuality.org). This group would like to see *asexuality* become a recognized sexual orientation so that absence of sexual desire is not treated as dysfunctional but, rather, as a "normal" alternative. Clinicians vary in whether they agree, but as one example, Dr. Irwin Goldstein, director of the Boston University Center for Sexual Medicine, considers that "[l]ack of interest in sex is not necessarily a disorder or even a problem . . . unless it causes distress" (Duenwald 2005, p. 2).

Critical Thinking

Had you been aware of the concept of asexuality before reading about it here?

In your opinion, does our society define asexuality as dysfunctional? Can you think of examples that support your viewpoint? Might asexuality become recognized as a legitimate sexual orientation?

The existence of a fairly constant proportion of gays and lesbians in virtually every society—in societies that treat homosexuality harshly as well as those that treat it permissively—suggests a biological imperative (Bell, Weinberg, and Hammersmith 1981), as does the fact that 450 mammal and bird species engage in same-sex sexual activity (Roughgarden 2009a, 2009b; Mackay 2000, p. 22). No specific genetic differences between heterosexuals and gays have been conclusively established (Greenberg, Bruess, and Haffner 2002, p. 367; Roughgarden 2009a). Meanwhile, a biological imperative does not equal predetermined sexuality and means instead that a variety of factors influences sexual orientation. Research on biological influences on sexual orientation continues (Abrams 2007; Roughgarden 2009a).

Whether a same-sex sexual orientation finds expression is clearly affected by environment, apart from or in conjunction with any genetic dispositions. A study using data from the General Social Survey and the National Health and Social Life Survey found increases in same-sex partnering between 1998 and 2002, especially for women (Butler 2005). Butler points to changes in social norms and in the legal climate, as well as increasing

economic opportunities for women, as likely factors shaping this change. Butler (along with other researchers) notes that the social climate impacts not only the public's perspectives on sexuality but also the kinds of research questions that are being explored regarding sexual orientation. Anne Fausto-Sterling, a noted professor of biology, put it best when she said, "we should debate what it is we want to understand about human sexuality, argue about the forms of knowledge we seek, and consider what the best ways of pursuing such knowledge might be" (2007, p. 55).

Theoretical Perspectives on Human Sexuality

There are various theoretical perspectives concerning marriage and families, as we saw in Chapter 2. Many of these have been applied to human sexuality. We can, for example, look at sexuality using a structure–functional perspective. In this case, we see sex as a focus of norms designed to regulate sexuality so that it serves the societal function of responsible reproduction.

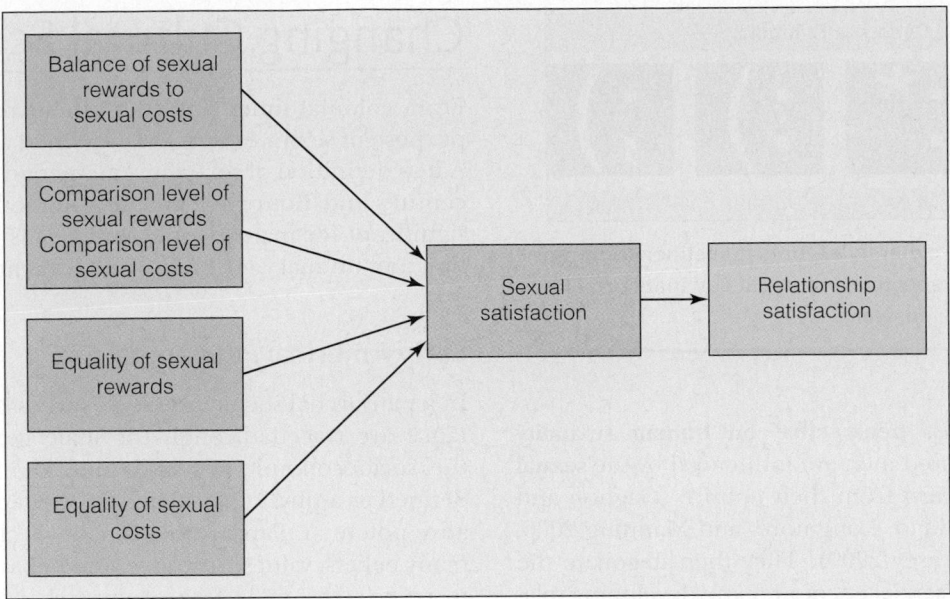

Figure 5.1 Model of factors associated with sexual and relationship satisfaction.

Those looking at sexuality from a biosocial perspective consider that humans—like the species from which they evolved—are designed so as to enable them to efficiently transmit their genes to the next generation. According to this biosocial perspective, men are naturally promiscuous, seeking multiple partners so as to distribute their genes widely, whereas women, who can generally have only one offspring a year, are inclined to be selective and monogamous (Dawkins 1976).

Both these perspectives have their limits. The structure–functional perspective tells us little about the emotions and pleasures of sexual relationships, whereas the biosocial perspective argues a genetic determinism that is contradicted by historical and cross-cultural variation in sexual behavior and relationships. Two more useful ways of looking at sexual relationships in a sociological perspective are exchange theory and interaction theory, both introduced in Chapter 2.

The Exchange Perspective: Rewards, Costs, and Equality in Sexual Relationships

From a general *exchange theory* perspective, women's sexuality and associated fertility are resources that can be exchanged for economic support, protection, and status in society. However, an exchange theory perspective that brings sex closer to our human experience is the **interpersonal exchange model of sexual satisfaction** (Lawrance and Byers 1995; Kisler and Christopher 2008).

Figure 5.1 shows us that in the interpersonal exchange model of sexual satisfaction, satisfaction depends on the *costs* and *rewards* of a sexual relationship, as well as the participant's *comparison level*—what the person expects

out of the relationship. Also important is the *comparison level for alternatives*—what other options are available, and how good are they compared to the present relationship? Finally, in this day and age, expectations are likely to include some degree of *equality*. Research to test this model found that these elements of the relationship did indeed predict sexual satisfaction in married and cohabiting couples (Byers 2005; Lawrance and Byers 1995; Kisler and Christopher 2008); however, social class remained a necessary variable to take into consideration within this exchange model of sexual satisfaction (Neff and Harter 2003).

The Interactionist Perspective: Negotiating Cultural Messages

The *interactionist* perspective emphasizes the interpersonal negotiation of relationships in the context of sexual scripts: "*That* we are sexual is determined by a biological imperative toward reproduction, but *how* we are sexual—where, when, how often, with whom, and why—has to do with cultural learning, with meaning transmitted in a cultural setting" (Fracher and Kimmel 1992). Cultural messages give us legitimate reasons for having sex, as well as who should take the sexual initiative, how long a sexual encounter should last, how important experiencing orgasm is, what positions are acceptable, and whether masturbating is appropriate, among other things. Recently, cultural messages have concerned what sexual interaction or relationships are appropriately conducted over the Internet, as well as with the newer phenomenon of "sexting" via cell phones with video capabilities.

Cybersex Symbols			
Hug	Kiss	Smile	Wink

Cybersex. Is it sex—cyberstyle—or is it abstinence? From an interactionist perspective, we might say that society is still constructing the answer.

An **interactionist perspective on human sexuality** holds that women and men are influenced by the **sexual scripts** that they learn from their culture (Gagnon and Simon 2005; Giordano, Longmore, and Manning 2006; VanderLaan and Vasey 2009). They then negotiate the particulars of their sexual encounter and developing relationship (A. Stein 1989, p. 7; MacNeil and Byers 2009).

Sex partners assign meaning to their sexual activity—that is, sex is symbolic of something, which might be affection, communication, recreation, or play, for example. Whether each gives their sexual relationship the same meaning has a lot to do with satisfaction and outcomes. For example, if one is only playing while the other is expressing deep affection, trouble is likely. A relationship goal for couples becoming committed is to establish a joint meaning for their sexual relationship.

Sex has different cultural meanings in different social settings. In the United States (and elsewhere), messages about sex have changed over time.

Self-disclosure and physical pleasure are key qualities in building sexually intimate relationships. Tenderness is a form of sexual expression valued not just as a prelude to sex but also as an end in itself.

Changing Cultural Scripts

From colonial times until the nineteenth century, the purpose of sex in America was defined as reproductive. A new definition of sexuality emerged in the nineteenth century and flourished in the twentieth. Sex became significant for many people as a means of communication and intimacy (D'Emilio and Freedman 1988).

Early America: Patriarchal Sex

In a patriarchal society, descent, succession, and inheritance are traced through the male genetic line, and the socioeconomic system is male-dominated. Sex is defined as a physiological activity, valued for its procreative potential. **Patriarchal sexuality** is characterized by many beliefs, values, attitudes, and behaviors developed to protect the male line of descent. Men are to control women's sexuality. Exclusive sexual possession of a woman by a man in monogamous marriage ensures that her children will be legitimately his. Men are thought to be born with an urgent sex drive, whereas women are seen as naturally sexually passive; orgasm is expected for men but not for women. Unmarried men and husbands whose wives do not meet their sexual needs may gratify those needs outside marriage. Sex outside marriage is considered wrong for women, however.

Although the patriarchal sexual script has been significantly challenged, it persists to some extent and corresponds to traditional gender expectations. If masculinity is a quality that must be achieved or proven, one arena for doing so is sexual accomplishment or conquest. A 1992 national survey by the NORC (National Opinion Research Center) at the University of Chicago, based on a representative sample of 3,432 Americans age eighteen to fifty-nine, found that men were considerably more likely than women to perform, or "do," sex. For example, more than three times as many men as women reported masturbating at least once a week. Three-quarters of the men reported always reaching orgasm in intercourse, whereas the fraction for women was nearer to one-quarter. Men are also much more likely to think about sex (54 percent of men and 19 percent of women said they think about it at least once a day) and to have multiple partners. Men are also more excited by the prospect of group sex (Laumann et al. 1994). One might

© Paul Fusco/ Magnum Photos

return to a biosocial perspective to explain these differences except that they are less pronounced among the youngest cohorts.

The Twentieth Century: The Emergence of Expressive Sexuality

A different sexual message has emerged as the result of societal changes that include the decreasing economic dependence of women and the availability of new methods of birth control. Because of the increasing emphasis on couple intimacy, women's sexual expression is more encouraged than it had been earlier (D'Emilio and Freedman 1988). With **expressive sexuality,** sexuality is seen as basic to the humanness of both women and men; there is no one-sided sense of ownership. Orgasm is important for women as well as for men. Sex is not only, or even primarily, for reproduction, but is an important means of enhancing human intimacy. Hence, all forms of sexual activity between consenting adults are acceptable.

The 1960s Sexual Revolution: Sex for Pleasure

Although the view of sex as intimacy continues to predominate, in the 1920s, an alternative message began to emerge wherein sex was seen as a legitimate means to individual pleasure, whether or not it was embedded in a serious couple relationship. Probably as a result, the generation of women born in the first decade of the twentieth century showed twice the incidence of nonmarital intercourse[3] as those born earlier (D'Emilio and Freedman 1988). This probably occurred mostly in relationships that anticipated marriage (Zeitz 2003). Further liberalization of attitudes and behaviors characterized the sexual revolution of the 1960s.

What was so revolutionary about the sixties? For one thing, the birth control pill became widely available; as a result, people were freer to have intercourse with more certainty of avoiding pregnancy.[4] At least for heterosexuals, laws regarding sexuality became more liberal. Until the U.S. Supreme Court decision in *Griswold v. Connecticut* (1965) recognized a right of "marital privacy," the sale or provision of contraception was illegal

in some states. The idea that sexual and reproductive decision making belonged to the couple, not the state, was extended to single individuals and minors by subsequent decisions (*Eisenstadt v. Baird* 1972; *Carey v. Population Services International* 1977).

Americans' attitudes and behavior regarding sex changed during this period. In 1959, about four-fifths of Americans surveyed said they disapproved of sex outside marriage. In 2006, only 25 percent said it was "always wrong" (Smith 1999; Schott 2007). Not only did attitudes become more liberal, but behaviors (particularly women's behaviors) changed as well. The rate of nonmarital sex and the number of partners rose, while age at first intercourse dropped (Wells and Twenge 2005)—7.6 percent of young people were having sexual intercourse before the age of 13 (Eaton et al. 2008). The trend toward higher rates of nonmarital sex has continued. Now "almost all Americans have sex before marrying" (Finer 2007, p. 73).

Today, sexual activity often begins in the teen years. Table 5.1 shows the percentages of sexually experienced teens in each of the major racial/ethnic groups according to the Youth Risk Behavior Surveillance System, a national high school–based survey conducted by the U.S. Centers for Disease Control and Prevention (Eaton et al. 2008). Not surprisingly, sexual experience increases with age. In 2007, almost half (47.8 percent) of high school students had had sexual experience (49.8 percent of males and 45.9 percent of females), and 35 percent are currently sexually active. For the vast majority (85 percent) of teens who had experienced sex, their first experience was with a romantic partner, although some 7.8 percent of students were forced to have sex (Eaton et al. 2008; Ryan, Manlove, and Franzetta 2003).

Perhaps the most significant change the sexual revolution ushered in, among heterosexuals at least, has been in marital sex. "Today's married couples have sexual intercourse more often, experience more sexual pleasure, and engage in a greater variety of sexual activities and techniques than people surveyed in the 1950s" (Greenberg, Bruess, and Haffner 2002, p. 437). In the NORC study, 88 percent of married partners said they enjoy great sexual pleasure (Laumann et al. 1994).

---

[3] *Nonmarital sex* refers to sexual activity by people who are not married to each other, whether they have never married or are divorced, widowed, or currently married (although we usually use the term *extramarital sex* for this last situation). *Nonmarital sex* replaces the previously common term *premarital sex. Premarital* connotes the anticipation of marriage, reflecting the fact that before the sexual revolution, a substantial portion of nonmarital sexual activity took place between people who were formally engaged or informally pledged to marry or who would subsequently marry.

[4] It is important to note that both sexual liberation and the use of birth control were already in progress. "The pill did not create America's sexual revolution so much as it accelerated it" (Zeitz 2003).

Table 5.1 Sexual Experience of High School Students by Race/Ethnicity and Gender, 2005

| | Percentage who . . . | | | |
| | "Ever had Sexual Intercourse" | | "Are Currently Sexually Active" | |
Ethnicity	Males	Females	Males	Females
White	42%	44%	31%	34%
Black	75%	61%	51%	44%
Hispanic	58%	44%	36%	34%

Source: Eaton et al. 2006, Tables 44, 46.

The 1980s and 1990s: Challenges to Heterosexism

If the sexual revolution of the 1960s focused on freer attitudes and behaviors among heterosexuals, recent decades have expanded that liberalism to encompass lesbian and gay male sexuality. Until several decades ago, most people thought about sexuality almost exclusively as between men and women. In other words, our thinking was characterized by **heterosexism**—the taken-for-granted system of beliefs, values, and customs that places superior value on heterosexual behavior and that denies or stigmatizes nonheterosexual relations. However, since "Stonewall" (a 1969 police raid on a U.S. gay bar) galvanized the gay community into advocacy, gay males and lesbians have not only become increasingly visible but have also challenged the notion that heterosexuality is the one proper form of sexual expression.

Gay men and lesbians have won legal victories, new tolerance by some religious denominations, greater understanding on the part of some heterosexuals, and sometimes positive action by government. Some states and communities have passed sexual orientation anti-discrimination laws. Corporations also are increasingly likely to enact antidiscrimination policies.

The public's attitudes toward homosexuality, though never as favorable as toward nonmarital sex generally, became more favorable in the 1990s, after earlier high rates of disapproval. In the early 1970s, about 70 percent thought homosexual relations were "always wrong." In 1986, the Supreme Court decision in *Bowers v. Hardwick* declined to extend privacy protection to gay male or lesbian relationships, and homosexual conduct remained criminalized in some states. Then in a 2003 case (*Lawrence et al. v. Texas*), the Supreme Court reversed its earlier decision, striking down a Texas law criminalizing homosexual acts, thus legalizing same-sex sexual relations. Figure 5.2 shows that 56 percent of those surveyed in a Gallup poll agreed that "homosexual

relations should be legal" ("Gay and Lesbian Rights" 2009).

Americans are divided over whether gay men and lesbians choose their sexual orientation, a split that shapes attitudes. People who see being gay as a choice are less sympathetic to lesbians or gay men regarding jobs and other rights (Loftus 2001). Americans are more likely to approve of civil rights protections for gays and lesbians than of a gay or lesbian lifestyle, and that approval has continued to strengthen since the 1970s. In 2006, 89 percent of Americans surveyed in a Gallup poll agreed that gays "should have equal job opportunities" (Saad 2006a). The workplace is "becoming friendlier" to gay, lesbian, and bisexual employees, and the vast majority of Forbes 500 companies include sexual orientation in their antidiscrimination policies (Fidas and Luther 2010). Gay employees are increasingly open about their sexual orientation ("The Office Closet Empties Out" 2006).

Homophobia—viewing homosexuals with fear, dread, aversion, or hatred—is still present in American society. A recent poll found that "most Americans oppose legalizing marriage between same-sex couples by 57 percent to 40 percent." What is interesting about these numbers, however, is that when those polled knew someone who is gay, lesbian, or bisexual, their support for gay and lesbian rights increased. In other words, the polls reflect what studies show: The more familiar we are with sexual minorities, the more likely we are to support sexual equality ("Gay and Lesbian Rights" 2009; Herek 2009a, 2009b).

As discussed in the previous paragraph, knowing a gay or lesbian person tends to make us more supportive of sexual rights. It seems, however, that semantics are also important when Americans are contemplating gay rights. For example, CBS and *The New York Times* polled Americans in 2009, and found that a majority (59 percent) favored allowing homosexuals to serve in the U.S. military, while the same poll showed that an

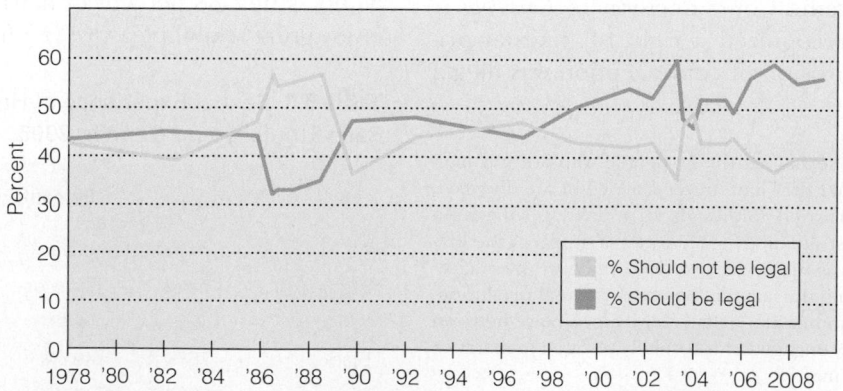

Figure 5.2 Do you think gay or lesbian relations between consenting adults should or should not be legal?

Joseph Sohm/Visions of America/Corbis

Lesbian and gay male unions and families have become increasingly visible over the past decade. Meanwhile, discrimination and controversy persist.

even stronger majority (71 percent) favored allowing gay men and lesbians to serve in the U.S. military. As readers can see, the pollsters used two different types of wording to ask the same question. In one question, they used the term "homosexuals," whereas in another question, they used the term "gay men and lesbians."

Comparing Gay Male and Lesbian Sexual Behaviors Sexual conduct, experience, and satisfaction have much to do with cultural trends. Leonore Tiefer notes that how we develop our desires and our expectations have very much to do with our "socially produced expectations [which] affect meaning and satisfaction" (Tiefer 2007, p. 246).

Philip Blumstein and Pepper Schwartz (1983), who studied a large national sample of 12,000 volunteers from the Seattle, San Francisco, and Washington, DC areas prior to the HIV/AIDS epidemic,[5] described lesbian relationships as the "least sexualized" of four kinds of couples: heterosexual cohabiting and married couples, lesbians, and gay men. Lesbians have sex less frequently than gay men, although it may be difficult to make comparisons because lesbians' physical

relationship can take the form of hugging, cuddling, and kissing, not only genital contact (Peplau, Fingerhut, and Beals 2004; Frye 1992; Tracy and Junginger 2007). Nevertheless, lesbians report greater sexual satisfaction than do heterosexual women: "Their greater tenderness, patience, and knowledge of the female body are said to be the reasons" (Konner 1990, p. 26; see also Holmberg, Blair, and Phillips 2010). Gay male sexuality is more often "body-centered" (Ruefli, Yu, and Barton 1992), whereas lesbian sexuality is more person-centered. The conventional wisdom is that gay men may be more accepting of nonmonagamous relationships than are lesbians—or heterosexuals (Adam 2007; Christopher and Sprecher 2000). Some studies, especially those done prior to the HIV/AIDS epidemic, show that gay men have more transitory sex than lesbians. Other studies point out that even though gay men may say they are in open relationships, "they did not act on this option." At the same time, significant percentages of men in both homo- and heterosexual monogamous relationships had slept with someone other than their own partner since "their primary relationships began" (Adam 2007, pp. 123, 125). Casual sex among lesbians appears to be relatively rare.

We have discussed differences, but patterns of sexual frequency and satisfaction in gay and lesbian relationships resemble those of heterosexual marriage and cohabitation in some ways. In all couple types studied by Kurdek (1991)—gay, lesbian, heterosexual cohabitants, and married couples—sexual satisfaction within

---

[5] Although the Blumstein and Schwartz study was published in 1983, and is not a random sample, it "continues to be the most extensive study on the sexuality of gay and lesbian couples to date" (Christopher and Sprecher 2000, p. 1,007). These data are still being used by well-respected researchers (e.g., Kurdek 2006). Still, one must keep in mind that these data predate the AIDS epidemic and other societal changes that have an impact on sexuality and sexual expression.

1. Don't assume anything you send or post is going to remain private.

2. There is no changing your mind in cyberspace—anything you send or post will never truly go away.

3. Don't give into the pressure to do something that makes you uncomfortable, even in cyberspace.

4. Just because a message is meant to be fun doesn't mean the person who gets it will see it that way . . . many teen boys (29 percent) agree that girls who send such content "are expected to date or hook up in real life."

5. Nothing is truly anonymous.

Source: Adapted from *Sex and Tech: Results from a Survey of Teens and Young Adults,* by the National Campaign to Prevent Teen and Unplanned Pregnancy and CosmoGirl.com (2009), p. 2.

---

each group was associated with general relationship satisfaction and with sexual frequency. "Despite variability in structure, close dyadic relationships work in similar ways" (Kurdek 2006, p. 509; see also Holmberg and Blair 2009).

The Twenty-First Century: Risk, Caution—and Intimacy

Although pleasure seeking was the icon of sixties sexuality, warnings to be cautious in the face of risk characterizes contemporary times. Some heterosexual young adults see AIDS as a threat only for other people. Others do acknowledge the risk, but decide to take their chances. "Most emerging adults . . . say that fear of AIDS has become the framework for their sexual consciousness, deeply affecting their attitudes toward sex and the way they approach sex with potential partners," perhaps asking for a test result or insisting on condom use (Arnett 2004, p. 91). (HIV/AIDS is discussed in more detail later in this chapter.) Meanwhile, a number of singles have multiple partners over time; males tend to report higher rates of multiple partners, with 29 percent reporting "fifteen or more female sexual partners" compared to 9 percent of females stating they have had "fifteen or more male sexual partners in a lifetime" (Fryar et al. 2007, p. 3).

Sexting may be a new word to many older generations, but is a well-known and growing phenomenon among young people of all genders, sexualities, classes, and belief systems in our modern, technological era. According to a new survey, many teens and young adults have sent sexually provocative photographs and text messages over their cell phones to people they don't know, or that they want to date or hook up with. In addition, 71 percent of teen girls, 67 percent of teen boys, 83 percent of young adult women, and 75 percent of young adult men admit to sending sexually suggestive photos and text messages of themselves to people with whom they are in a romantic relationship. Many believe that their sexual images or text will remain with their romantic interest, yet 48 percent of young adult women and 46 percent of young adult men say "it is common for nude or seminude photos to get shared with people other than the intended recipient" (Sex and Tech 2009, pp. 2–3).

Today, there is risk in a sexual encounter. Critics of sexual liberation argue that this is especially disadvantageous for women (Shalit 1999). At the same time, a more liberal sexual environment offers the potential for expressive sexuality and true sexual intimacy. People now have more knowledge of the principles of building good relationships (whether or not they always succeed

"Should I come out to everyone all at once or one cubicle at a time?"

in putting them into practice). Now that the possibility of satisfying sexual relationships seems more attainable, how do men and women negotiate those sexual relationships inside and outside of marriage?

Negotiating (Hetero)sexual Expression[6]

Although relationships between the sexes are more equal today than in the past, many—though assuredly not all—women and men today may have internalized divergent sexual scripts, or messages. Today's heterosexuals negotiate sexual relationships in a context in which new expectations of equality and similarity coexist within a heritage of gender-related difference. Men may be somewhat more accepting of recreational sex than women are, and studies continue to show that women are more interested than men in romantic preliminaries (Purnine and Carey 1998).

The "pressure on men to be more sensitive, less predatory, and less macho has been mounting for several decades" (Schwartz and Rutter 1998, p. 48; see also Seal and Ehrhardt 2003). After the sexual revolution of the 1960s, men became more interested in communicating intimately through sex, whereas women showed more interest than before in physical pleasure (Pietropinto and Simenauer 1977; Seal, O'Sullivan, and Ehrhardt 2007). Masters and Johnson (1976) had argued that more equal gender expectations lead to better sex.

This discussion points again to the fact that cultural messages vary and that sexual relationships are negotiated in a social context. "Since early in the twentieth century the bonds between marriage and sexual activity have been unraveling" (Smith 2006, p. 26). In the next sections, we discuss nonmarital—outside of marriage/committed relationships—sexual activity. After that, we will examine sexuality within marriage/committed relationships—where most sexual activity takes place—and then look at what is known about racial/ethnic diversity in sexual expression. Keep in mind that sexual expression may include more than intercourse—activities ranging from genital intercourse, to oral and anal sex, to kissing and cuddling.

Four Standards for Sex Outside Committed Relationships

In the 1970s, sociologist Ira Reiss (1976) developed a fourfold classification of societal standards that illustrates the varied cultural messages about nonmarital sex.

Today we can apply his typology to sex outside of all committed relationships. Reiss's four standards—abstinence, permissiveness with affection, permissiveness without affection, and the double standard—were originally developed to apply to premarital sex among heterosexual couples. However, they have since been applied to the sexual activities of unmarried people generally. (Later we'll discuss extramarital sex—that is, sexual relations of married people or cohabitants outside of their committed relationship.)

Abstinence The standard of **abstinence** maintains that, regardless of the circumstances, nonmarital intercourse is wrong for both women and men. Many contemporary religious groups, especially the conservative Christian and Islamic communities, encourage abstinence as a moral imperative.

The several years before 2008 saw an increase in rates of teen pregnancy (3 percent), births (4 percent), and abortions (1 percent) (Kost, Henshaw, and Carlin 2010, p. 2). That reversed a downward trend, as teen birth rates had declined 34 percent from 1991 to 2005 (Hamilton, Martin, and Ventura 2010), with accompanying declines in teen pregnancy rates (Dyess 2009; Gavin et al. 2009; Schlesinger 2010). The trend has reversed again: In 2008, the teen birth rate dropped, along with birth rates in most other age groups, a decline attributed in part to the recession (Hamilton, Martin, and Ventura 2010).

Generally, teens who do not engage in sexual activity give conservative values or fear of pregnancy, disease, or parents as their reasons (Blinn-Pike 1999; Rasberry and Goodson 2009, pp. 79, 81). A study of college students suggests that women who refrain from sexual activity do so because of an absence of love, a fear of pregnancy or sexually transmitted diseases (STDs), a belief that use of contraception will cause infections or cancer, or a belief system that endorses nonmarital virginity (Kaye, Sullentrop, and Sloup 2009). Men's hesitancy to pursue sexual involvement comes more often from feeling "inadequate or insecure" (Christopher and Sprecher 2000, p. 1,009).

Individuals adopt the abstinence standard for other reasons as well. Some women and men have withdrawn from nonmarital sexual relationships entirely to avoid bad experiences. Some withdraw from sexual risk, at least for a time, rather than feeling vulnerable in the open sexual climate of the sexual revolution (Rasberry and Goodson 2009). The feminist movement is cited as empowering women to be abstinent if they wish (Ali and Scelfo 2002).

Advocates of abstinence claim that it is "a new sexual revolution" (Laub 2005, p. 103; see also Herzog 2008) in a "campus life [that] has become so drenched in sexuality, from the flavored condoms handed out by a resident adviser to the social pressure of the hook-up scene."

[6] The idea that sexual expression is negotiated was developed in regard to heterosexual relationships; this principle may be applied to gay and lesbian relationships.

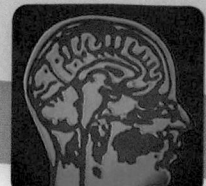
Two similar new patterns of sexual behavior—"hooking up" and "friends with benefits"—made their appearance at the turn of the twenty-first century. Researchers believe that "hooking up" and "friends with benefits" have been around for a while, but are just now becoming visible to researchers and others not immersed in youth culture.

The "demise of the date and the rise of the hookup is a national trend," with dating coming about only after the hookup has turned into a lasting emotional attachment (England and Thomas 2007, p. 152; Bogle 2008, pp. 44, 48). Indeed, a growing body of research suggests that, at least among a majority of U.S. students, *dating*—pre-planned couple outings—has virtually disappeared (Epstein et al. 2009; B. Wilson 2009). Now that same impro-vised sociability has been extended to "hooking up" or finding a "friend with benefits"—that is, pairing off from the group for a sexual encounter (Bogle 2008). In fact, the idea of the hookup has moved beyond the college to people who have diseases, disfigure-ments, and even psychological issues, and who have difficulty finding others who can understand their needs. Take, for example, cancer survivors. Many cancer survivors, once they divulge that history to someone they've been dating, often find themselves quickly abandoned by the love interest. As a result, new organizations and websites have emerged, creating a social space

where people can "hook up" with oth-ers dealing with similar issues (Alexan-der 2010; Durham 2010).

The basic idea in hooking up and friends with benefits is that a sexual encounter means nothing more than just that—sexual activity. Hooking up can occur with no prior acquaintance between the parties and no further contact afterward. *Friends with benefits* is described as sex that takes place with friends, but without expectations for romantic love or commitment. At least that is the initial premise.

These sexual scripts may have emerged because many of today's young adults—who are delaying mar-riage, going to college, and developing careers—still want to have sex. They want some intimate connection with-out risking romantic disappointment and emotional loss. When hooking up or finding friends with benefits occurs among high school students, similar rea-sons may be in place. As one or both sex partners anticipate leaving for college, they may want to keep their options open (Edwards 2006; B. Wilson 2009).

What does this development mean for young lives and future marriage prospects? Reactions of social scientists range from alarm at the disappearance of a courtship path to marriage (Glenn and Marquardt 2001; Stepp 2007), to the thought that hooking up is simply the new way to begin a relationship that may become serious despite its ini-tial intent (England and Thomas 2007;

Epstein et al. 2009). By carefully outlin-ing the positives and negatives of both dating and hooking up, one observer reminds us that dating also has draw-backs as a relationship development process (Bogle 2004/2008).

Meanwhile, some observers note that men continue to have greater relation-ship power in hookup settings, as indeed they did in the dating era (Bogle 2008). Now that there are more women than men on most college campuses, it is sug-gested that male power is reinforced inasmuch as competition among het-erosexual women for the fewer available men is heightened. However, this is not necessarily the case, as men must nego-tiate the sexual scripts just as much as women (Manning, Giordano, and Long-more 2006; Pollack 2006; Smiler 2008).

There is considerable research on hooking up for such a newly recognized phenomenon. Maybe academics, shut up in their offices analyzing data, find vicari-ous enjoyment in the topic. More seri-ously, though, due to researchers' interest, we should in time learn more about how the gap between "hooking up" and "mar-ried with children" is to be bridged.

Critical Thinking

What do you think about the advisabil-ity of hooking up or having friends with benefits? How might a researcher from the interactionist perspective investigate the way that couples bridge the gap between "hooking up" and "married with children"?

At Princeton, a pro-abstinence organization, the Anscombe Society, was formed to support abstinence outside of tra-ditional marriage and to provide a setting for exploration and discussion of the abstinence alternative (Peterson 2005b, p. B3). Other organizations and books (e.g., Doan and Williams 2008; Moffett 2007; Winner 2006) offer intellectually sophisticated and sexually frank discussions of this alternative.

Permissiveness with Affection The standard of **permis-siveness with affection** permits nonmarital intercourse

for both men and women equally, provided they have a fairly stable, affectionate relationship. This standard may be the most widespread sexual norm among unmar-ried individuals. A Gallup poll taken in 2008 found that 57 percent of Americans find sex between an unmar-ried man and woman "morally acceptable" ("Marriage" 2009).

In a 1997 national poll by *U.S. News & World Report* magazine, a majority of respondents under the age of forty-five said that adult nonmarital sex "generally ben-efits people" in addition to offering sexual pleasure.

A majority also felt that having had a few sexual partners makes it easier for a person to choose a sexually compatible spouse (Whitman 1997). The previously mentioned NORC survey concluded that we have sex mainly with people we know and care about. Seventy-one percent of Americans have only one sexual partner in the course of a year (Laumann et al. 1994).

Permissiveness without Affection Sometimes called recreational sex, **permissiveness without affection** allows intercourse for women and men regardless of how much stability or affection is in their relationship. Casual sex—intercourse between partners only briefly acquainted—is permitted.

About fifteen years ago, a *New York Times* article surprised many by describing a pattern among adolescents termed "**friends with benefits**" (Denizet-Lewis 2004). This article focused on teens "**hooking up**" casually for sexual encounters with friends and acquaintances, completely outside a romantic relationship context. Today, sociologists see this as a growing trend among adolescents and young adults. The point of the "hookup" seems to be that teens or young adults, who feel themselves to be unready for romance and commitment, are able to explore their sexuality in what is intended to be an emotionally neutral context. According to Dr. Kathleen Bogle, a sociologist who studies this phenomenon, dating used to be something that led up to sex, but in "the hookup era, something sexual happens, even though it may be less than sexual intercourse, that may or may not ever lead to dating" or romantic involvement (B. Wilson 2009; see also Bogle 2008, pp. 24–49). Of course, it doesn't always work out that way.

Sexual activity does not necessarily mean intercourse; oral sex is quite common (Bogle 2008, p. 25). "Issues for Thought: 'Hooking Up' and 'Friends with Benefits'" discusses this new sexual script.

The Double Standard According to the **double standard**, women's sexual behavior must be more conservative than men's. In its original form, the double standard meant that women should not have sex before or outside marriage, whereas men could. Within the context of marriage and committed relationships, femininity "is typically framed in terms of being sexually desirable rather than sexually desiring whereas masculinity connotes sexual aggression and prowess" (Elliott and Umberson 2008, p. 392). More recently, the double standard has required that women be in love to have sex, or at least have fewer partners than men have. Now it appears that there are even different expectations for males and females in "hooking up."

An exploratory study of this phenomenon among undergraduates at two Eastern colleges found some informal rules: Women should be less aggressive, should not hang around fraternity houses, and should have fewer partners than men. It was quite difficult for these college women to maintain reputation and self-esteem while engaging in hooking up—"the only game in town" (Bogle 2004, p. 99; see also Stepp 2007, p. 117).

At first glance, it would seem that men have greater sexual freedom and greater power in these relationships. Bogle, in fact, concludes: "[T]here is one crucial commonality . . . [of *hooking up* and dating, its predecessor sexual script]. . . . [*Men*] *have a greater share of power in both eras*" (p. 229; see also Bogle 2008, pp. 23, 173). Other research notes, however, that this is not necessarily the case; in fact, many men find these ever-transforming sexual scripts difficult to negotiate. Furthermore, young men seem to be as emotionally vulnerable as women and are, frequently, using the "hookup" as a means to find lasting and meaningful relationships (Epstein et al. 2009; Manning, Giordano, and Longmore 2006; Pollack 2006; Smiler 2008).

In our society generally, men and women may have different expectations, with men exposed to cultural conditioning that encourages them to separate sex from intimacy, whereas among women, sexual expression more often symbolizes connection with a partner and communicates intimacy. One observer also argues that gender differences in permissiveness may reflect differences in social power and vulnerability, prompting a woman's strategy of self-protection through adherence to conventional cultural expectations (Howard 1988).

Sexual Infidelity

Up to this point, we've been examining scripts for sex among noncommitted relationships. Now we will look at a different form of sex outside marriage and committed relationships—sexual infidelity, or "affairs."

As we will see in Chapter 7, marriage typically involves promises of sexual exclusivity—that spouses and committed partners will have sexual relations only with each other. Cohabiting and other committed relationships also involve expectations of fidelity, although to a somewhat lesser degree (Treas and Giesen 2000). In this era of expressive sexuality, "people still feel that the self-disclosure involved in sexuality symbolizes the love relationship and therefore sexuality should not be shared with extramarital partners" (Reiss 1986, pp. 56–57). Americans believe in fidelity and sexual exclusivity regardless of a legally binding commitment to one another (see Chapter 7), but it seems that beliefs and actions to not always quite match up, and some find themselves unable to completely adhere to their own expectations.

Although infidelity is found in virtually any society and throughout our known history, the proscription against extramarital sex is stronger in the United States than in many other parts of the world. Ninety-two percent of Americans consider extramarital affairs "morally wrong" (Newport 2009). Cohabiting couples also generally expect each other to be sexually faithful (94 percent, compared to 99 percent of married couples). However, the rate of sexual infidelity is higher among cohabiting couples than among married couples (Treas and Giesen 2000).

As you will see in Chapters 7 and 8 of this text, long-term relationships (including marriages and cohabitations) are generally founded on the agreements, by both parties, of fidelity. Some researchers distinguish among emotional infidelity, sexual infidelity, and combined emotional and sexual infidelity. The latter is most disapproved. Emotional (without sexual) infidelity is least disapproved (Blow and Hartnett 2005). The impacts of any form of infidelity, however, are lasting. Sexual infidelity is engaging in sexual relations with someone who is not one's own marriage or committed partner. Some researchers define emotional infidelity as an "intense, primarily emotional, nonsexual relationship" with someone who is not one's own marriage or committed partner (Potter-Efron and Potter-Efron 2008, p. 2). Both of these forms of infidelity have long-term, sustained negative impacts on the marriage or committed relationship (Meier, Hull, and Ortyl 2009).

Statistics on sexual infidelity are based on what people report: Some spouses or committed partners hesitate to admit an affair; others boast about affairs that didn't really happen. Social researcher Pepper Schwartz (2009) says that "over a lifetime approximately 25 to 50 percent of married men and women are going to cheat on their partner. Make that 50 percent plus of cohabiters" (see also Chapters 7 and 8 of this text). Although these rates are lower than those found in earlier research, "these percentages translate into a significant number of Americans who have experienced sex with someone other than their spouse at least once" (Christopher and Sprecher 2000, p. 1,006). A survey conducted in 2002 by the Centers for Disease Control and Prevention found that 92 percent of married men and 93 percent of married women had sexual contact with only one opposite-sex partner (the spouse) during the past year. This compares to 80 percent of male and female cohabitants whose sexual relations were limited to their cohabiting partner. New research out of the University of Washington finds that "the lifetime rate of infidelity for men over 60 increased to 28 percent in 2006, up from 20 percent in 1991. For women over 60, the increase is more striking: to 15 percent, up from 5 percent in 1991" (Parker-Pope 2008, p. D1), which suggests that women's infidelity is on the rise.

Of married men, 3.4 percent reported a lifetime experience of affairs with other men; 5.3 percent of cohabiting men reported such experiences. Of married women, 7.2 percent reported some sexual experience with other women, and 10.8 percent of cohabiting women did so. (The higher rates for women are probably due to the much looser definition of "sexual

"Hey look, before this goes any further, I should probably tell you we're married."

experience" for women (Mosher, Chandra, and Jones 2005, Tables 1, 2, 8.)

Risk Factors Sociologists Judith Treas and Deirdre Giesen (2000) developed a conceptual model of risk factors for extramarital sex. They tested this model with data from the 1992 National Health and Social Life Survey, a national probability sample, which included 2,870 married or cohabiting individuals ages eighteen through fifty-nine.[7]

Treas and Giesen found that entering an extramarital affair is a rational decision. That is, affairs are generally *not* spontaneous (the result of too much alcohol, for example), nor are they the consequence of overwhelming romantic passion. Rather, "[p]eople contemplating sexual infidelity described considered decisions" (Treas and Giesen 2000, p. 49). In a recent study, researchers also found that loneliness is an important factor in one's decision to be unfaithful. This study found that the conditions of many undocumented workers in the United States is such that loneliness, as well as fear of deportation, led to unsafe sexual practices and an increased risk of sexually transmitted disease (Hirsch et al. 2009).

Although the book you are reading focuses on the United States, there is some important research examining sexual infidelity in other countries. This research provides evidence of cross-cultural similarities when it comes to love and sexual infidelity. For example, in all parts of the world, the decision to engage in sexual

[7] This study draws on the National Health and Social Life Survey data set (Laumann et al. 1994), but analyzes only data from married and cohabiting respondents.

infidelity appears to be unrelated to feelings of love the adulterer has for his or her partner, but instead comes from a complex interplay of sexuality, identity, ideology, ego, access, and even peer pressure (Hirsch et al. 2009/2010).

Not surprisingly, individuals who have a stronger sex interest and who have more permissive sexual values and more past sex partners are more likely than others to engage in sexual infidelity (Treas and Giesen 2000). Lower satisfaction with a marital or cohabiting relationship is another unsurprising risk factor. Relationship dissatisfaction is a motive more important to women. Sexual dissatisfaction and declines in frequency are also associated with affairs, especially for men (Blow and Hartnett 2005). In fact, research shows that the risk of extramarital affairs for both genders is "significantly higher among marriages characterized by spousal violence, divorce proneness, a past experience of marital separation, or the practice of spending relatively little time together," but marital and sexual satisfaction seemed to be less important to someone's decision to have an extramarital affair (DeMaris 2009, p. 605).

Opportunity plays a role. Couples who lead separate lives and who have jobs requiring travel are more likely to have extramarital affairs (Treas and Giesen 2000). Workplace opportunity per se was not a significant factor for lifetime rates of extramarital affairs, but it was associated with having an affair in the last twelve months. The researchers speculate that if an opportunity came along at a low point in the marriage, it might be taken advantage of. Additionally, women's economic independence may have something to do with their increasing rates of extramarital sex. Women are working longer hours, traveling more, and have the same access to cell phones, text messaging, and so on that men have to create and nurture intimate connections outside of marriage or committed relationship (Parker-Pope 2008).

The 1990s saw the emergence of a new brand of marital infidelity—adultery on the net, or **cyberadultery**. The Internet has created new opportunities for individuals to develop secret relationships (Jayson 2008a; Potter-Efron and Potter-Efron 2008; Whitty and Quigley 2008). The emotional connection may lead to a meeting—and then perhaps to a sexual relationship. At the same time, the Internet makes it more likely that a partner or employer will be able to find out about an affair (Cooper 2004; Crooks and Baur 2005). Such a discovery often triggers a couple's move into therapy.

Historically, societies have depended on community pressures to control disapproved sexual activity of any sort. Shared social networks of family and friends, as well as church attendance, seemed to operate as social controls discouraging infidelity (Hutson 2009).

Union duration of marriage or cohabitation, which can be a measure of both *investment* in the relationship

and *habituation,* showed a positive relationship with likelihood of extramarital sex during the union. This provides some support for the **habituation hypothesis**—that is, that familiarity reduces the reward power of a sexual encounter with a spouse or partner compared to a new relationship (Liu 2000). At the same time, union duration is also a simple measure of exposure to the risk of an extramarital affair (Treas and Giesen 2000).

Previous researchers have found gender differences evident in the analysis of patterns of extramarital sex (Harris 2003a), with more husbands than wives having had an affair sometime during their marriage (Laumann et al. 1994; Schwartz 2009). If a wife has an affair, she is more likely to do so because she feels emotionally distanced by her husband. Men who have affairs are far more likely to do so for the sexual excitement and variety they hope to find. Moreover, "men feel more betrayed by their wives having sex with someone else; women feel more betrayed by their husbands being emotionally involved with someone else" (Glass 1998, p. 35; Begley 2009; Blow and Hartnett 2005). However, when other risk factors are controlled, gender differences may be reduced or even eliminated (Treas and Giesen 2000).

Effects of Sexual Infidelity The secrecy required by an affair erodes the connection between partners. When discovered, the betrayal may spark jealousy—or it may create a crisis that motivates a search for the resolution of more general relationship problems (Crooks and Baur 2005; Snyder, Baucom, and Gordon 2008).

An affair *can* have positive effects such as encouraging closer relationships, paying greater attention to couple communication, and placing a higher value on the family (Olson et al. 2002). But only a small percentage of couples see an improved relationship (Blow and Hartnett 2005). Not only has trust been eroded and feelings been hurt, but the uninvolved spouse may also have been exposed to various sexually transmitted diseases—not a rare occurrence (Crooks and Baur 2005). For many partners, concern about HIV/AIDS heightens anger and turmoil over affairs. The uninvolved partner may feel exploited financially as well, because money has been unilaterally spent on an intimate outside the relationship (T. Smith 2006).

Research among married individuals is mixed as to whether infidelity "causes" divorce. That seems to depend on the previous level of marital satisfaction, the motives attributed to the unfaithful spouse, attitudes toward infidelity in general, and the efforts of both spouses to work things out (Blow and Hartnett 2005).

Recovering from an Affair Given that affairs do occur, people will have much to think about if they discover that their spouse has had (or is having) one. The uninvolved mate will need to consider how important the affair is relative to the marital relationship as a whole.

Can she or he regain trust? In some cases, the answer is no; trust never gets reestablished, and the heightened suspicion gets incorporated into other problems the couple might have (Baucom, Snyder, and Gordon 2009, p. 325).

Whether trust can be reestablished depends on several factors. One is how much trust there is in the first place. For this reason, new relationships may be especially vulnerable to breaking up after an affair. Many couples do recover from an affair; however, "it's hard to do without a therapist" (S. Glass 1998, p. 44; see also Baucom et al. 2006). Therapists suggest that doing so requires that the offending partner:

- apologize sincerely and without defending her or his behavior.

- allow and hear the verbally vented anger and rage of the offended partner (but not permit physical abuse).

- allow for trust to rebuild gradually and to realize that this may take a long time—up to two years or more.

- do things to help the offended partner to regain trust—keep agreements, for example, and call if he or she is running late.

Meanwhile, the offended partner needs to decide whether she or he is committed to the relationship and, if so, needs to be willing to let go of resentments as much as is possible (Snyder, Baucom, and Gordon 2008). Finally, the couple should consider relationship counseling (described in Chapter 12). According to Shirley Glass, "The affair creates a loss of innocence and some scar tissue. I tell couples things will never be the same. But the relationship may be stronger" (1998, p. 44; 2003).

Sexuality throughout Marriage and Committed Relationships

It might surprise you that various aspects of nonmarital sex are more likely to be studied than are those within marriage and that sexual activities of teens receive more research attention than those of adults (L. Davis 2006). "More is known about sexuality in marriage at this time than has ever been true in the past. But we still have only a limited view of how sexuality is integrated into the normal flow of married life" (Christopher and Sprecher 2000, p. 1,013).

Research in recent years has been much better methodologically, but has tended to focus on sexual frequency: How often do married couples have sex, and what factors affect this frequency? Before we get to the answers, we need to say something about how the information is gathered. "Facts About Families: How Do We Know What We Do? A Look at Sex Surveys" discusses the history and progress of research on sexuality.

How Often?

Social scientists are interested in sexual frequency because they like to examine trends over time and to relate these to other aspects of intimate relationships. For the rest of us, "How often?" is typically a question motivated by curiosity about our own sexual behavior compared to that of others. Either way, what do we know?

Married couples have sex more often than single individuals, though less often than cohabiting couples (Christopher and Sprecher 2000; T. Smith 2006; Yabiku and Gager 2009). In the NORC survey, described earlier in this chapter, the average frequency of sex for sexually active, married respondents under age sixty was seven times a month. About 40 percent of married individuals said they had intercourse at least twice a week (Laumann et al. 1994). Of course, these figures are averages: "People don't have sex every week; they have good weeks and bad weeks" (Pepper Schwartz, quoted in Adler 1993).

So does the ratio of good to bad weeks change over the course of a marriage? Yes: You have fewer good weeks (sorry).

Fewer Good Weeks To examine sexual frequency throughout marriage, Call, Sprecher, and Schwartz (1995) looked at the responses of 6,785 marrieds with a spouse in the household (and 678 respondents who were cohabiting) in the NSFH data set described earlier. Like researchers before them and since (T. Smith 2006), they found that sexual activity is highest among young marrieds. About 96 percent of spouses under age twenty-five reported having had sex at least once during the previous month. The proportion of sexually active spouses gradually diminished until about age fifty, when sharp declines were evident. Among those fifty to fifty-four years old, 83 percent said they had sex within the previous month; for those between sixty-five and sixty-nine, the figure was 57 percent; 27 percent of respondents over age seventy-four reported having had sex within the previous month.

The average number of times that married people under age twenty-five had sex is about twelve a month. That number drops to about eight times a month at ages thirty through thirty-four, then to about six times monthly at about age fifty. After that, frequency of intercourse drops more sharply; spouses over age seventy-four average having sex less than once each month.

It used to be that describing sexuality over the course of a marriage would be nearly the same as discussing sex as people grow older. Today, this is not the case. Many

How do we know what Americans do sexually? In serious social science, researchers strive for *representative samples* that reflect, or represent, all the people about whom they want to know something.

Pioneering Research

The pioneer surveys on sex in the United States were the Kinsey reports on male and female sexuality (Kinsey, Pomeroy, and Martin 1948; 1953). Kinsey used volunteers; he believed that a statistically representative survey of sexual behavior would be impossible because many of the randomly selected respondents would refuse to answer or would lie.

Recent Surveys

More recent scientific studies on sexual behavior have used random samples. In 1992, the National Opinion Research Center (NORC) at the University of Chicago conducted interviews with a representative sample of 3,432 Americans, age eighteen to fifty-nine—the National Health and Social Life Survey (Laumann et al. 1994). Eighty percent agreed to be interviewed—an impressively high response rate.

Respondents were questioned in ninety-minute face-to-face interviews. To provide some anonymity for the more sensitive part of the interview—questions about oral and anal sex, for example—specific sexual behavior questions were asked by means of a questionnaire. The respondent wrote answers and sealed them in an unlabeled envelope.

Findings of the National Health and Social Life Survey may be generalized to

the U.S. population under age sixty with a high degree of confidence. Indeed, the results have been welcomed as the first-ever truly scientific nationwide survey of sex in the United States. Because the NORC sample included only people under sixty, however, it cannot tell us anything about the sexual activities of older Americans.

Another study of sex among married individuals (Call, Sprecher, and Schwartz 1995) sought to compensate for the NORC study's deficiencies by using another national data set, the National Survey of Families and Households (NSFH). Between 1987 and 1988, the NSFH staff, affiliated with the University of Wisconsin, did in-person interviews with a representative national sample of 13,000 respondents age eighteen and over (Sweet, Bumpass, and Call 1988), and the survey was repeated between 1992 and 1994. Considered very reliable, the NSFH data are used as a basis for analysis regarding many topics discussed in this text. Some analyses have combined the National Survey of Families and Households and NORC data (Black et al. 2000).

A more recent survey, the National Survey of Family Growth, was conducted in 2002 and early 2003, by the U.S. Centers for Disease Control and Prevention. This survey also had an almost 80 percent response rate. It consists of in-person, in-home interviews with 12,571 people—4,928 men and 7,642 women. Measures of sexual behavior were collected by means of computer-assisted

self-interviewing. This survey was limited to those fifteen through forty-four years of age, with different analyses involving various age combinations within this range (Mosher, Chandra, and Jones 2005). A new edition of this survey is currently taking place, with data collection ongoing. To see for yourself the process that takes place in this kind of research, explore the *Planning and Development of the Continuous National Survey of Family Growth* that is available online.[a]

NORC continues to conduct a biennial General Social Survey that includes questions about sexual behavior. It publishes extensive reports on sexual behavior (e.g., Smith 2006), as well as providing current data on attitudes of the general public toward sexual activity. Additionally, in a study titled, National Children's Study, NORC is conducting a longitudinal study of American children from prior to birth to the age of twenty-one in an effort to study environmental influences on children's health and development.

Conclusions based on survey research on sensitive matters such as sexuality must always be qualified by an awareness of their limitations—the possibility that respondents have minimized or exaggerated their sexual activity or that people willing to answer a survey on sex are not representative of the public. Nevertheless, with data from these national sample surveys, we have far more reliable information than ever before.

a. http://www.cdc.gov/nchs/data/series/ sr_01/sr01_048.pdf

couples are remarried, so that at age forty-five, or even seventy, a person may be newly married. Nonetheless, we may logically assume that young spouses are in the early years of marriage.

Young Spouses

Young spouses have sexual intercourse more frequently than do older mates. Young married partners, as a rule, have fewer distractions and worries. The high frequency

of intercourse in this age group may also reflect a self-fulfilling prophecy: These couples may have sex more often partly because society expects them to.

After the first few years, sexual frequency declines (T. Smith 2006). Why so? The sexual intensity of the honeymoon period subsides, and "from then on almost everything—children, jobs, commuting, housework, financial worries—that happens to a couple conspires to reduce the degree of sexual interaction while almost nothing leads to increasing it" (Greenblatt 1983, p. 294).

Indeed, later research does indicate that pregnancy, the presence of small children, and a less than certain birth control method are factors that reduce sexual activity in young marriages.

Researchers have begun to wonder how sexual relations in early marriage might differ between couples who have established a sexual relationship before marriage and those who did not (Sprecher 2002), but there has been little examination of the transition from premarital to marital sex.

Spouses in Middle Age

On average, as people get older, they have sex less often. Physical aging is not the only explanation for the decline of sexual activity over time, although it appears to be the most important one. Marital satisfaction was the second largest predictor of sexual frequency.

Sexual satisfaction, marital satisfaction, and sexual frequency are interrelated throughout marriage (Byers 2005; Crooks and Baur 2005; Sprecher and Schwartz 1994; T. Smith 2006; Elliott and Umberson 2008). However, "even among couples who rate their marriages as very happy and among those who say they are still 'in love,' frequency of intercourse declines with age," and some of the decline is gendered with more women reporting lower levels of desire then men as they age (T. Smith 2006, p. 13; Elliott and Umberson 2008).

Despite the declining frequency of sexual intercourse, respondents in a variety of small studies emphasized the continuing importance of sexuality (Greenblatt 1983; Elliott and Umberson 2008). They pointed to the total marital relationship rather than just to intercourse—to such aspects as "closeness, tenderness, love, companionship and affection" (Greenblatt 1983, p. 298)—as well as other forms of physical closeness such as cuddling or lying in bed together. In other words, with time, sex may become more broadly based in the couple's relationship. During this period, sexual relating may also become more sophisticated, as the partners become more experienced and secure (Purnine and Carey 1998).

Older Partners

In our society, images of sex tend to be associated with youth, beauty, and romance; to many young people, sex seems out of place in the lives of older adults. Not too many years ago, public opinion was virtually uniform in seeing sex as unlikely—even inappropriate—for older people. With Masters and Johnson's work in the 1970s, indicating that many older people are sexually active, public opinion began to swing the other way. Then, in the 1980s, researchers began to caution against the romanticized notion that biological aging could be abolished (Cole 1983, pp. 35, 39). Of course, physical changes associated with aging do affect sexuality (Christopher and Sprecher 2000, p. 1,002).

In a nationally representative sample surveyed by the American Association of Retired Persons, a majority (56 percent) of those individuals forty-five and older agreed that a satisfying sexual relationship is important to one's quality of life. But they rated family and friends, health, being in good spirits, financial security, spiritual well-being, and a good relationship with a partner as more important than a fulfilling sexual connection (Jacoby 2005). Thus, "it stands to reason that individuals and couples…who have developed the capacity over the years to experience optimal sexuality have much to teach the rest of us" (Kleinplatz et al. 2009, p. 15). For example, when asked what they would tell younger generations about sexual enjoyment, elderly respondents who had been in a committed relationship twenty-five years or longer told researchers that good sex over the long term includes patience and practice. To illustrate this metaphorically, one respondent told the researchers, "instead of rushing by the windows in a train, one watches the scenery." Reminiscing during lovemaking also enhanced both the sexual desire and the sexual experience in elderly lovemaking (Melby 2010, p. 4; Kleinplatz et al. 2009).

Men and women in their late forties both placed an equal and high priority on sex. Additionally, 73 percent of people aged fifty-seven to sixty-four and 53 percent of people aged sixty-five to seventy-four remain sexually active, but by age sixty, a gender gap becomes evident. Sixty-two percent of men but only 27 percent of women gave sex a high priority (DeLamater and Sill 2005; Lindau et al. 2007; Melby 2010, p. 4).

Some older partners shift from intercourse to petting as a preferred sexual activity. On the other hand, sexual intercourse does not necessarily cease with age. Among the sexually active, 90 percent said they found their mates "very attractive physically" (Greeley 1991). Sexually active spouses over age seventy-four have sex about four times a month. Indeed, retirement "creates the possibility for more erotic spontaneity, because leisure time increases" (Allgeier 1983, p. 146). New research warns, however, that although maintaining a healthy sex drive into the "golden years" is normal, care needs to be taken to ensure that people who choose to accept a decreasing sex drive as they age are not treated as "victims of a pathology" (Marshall 2009, p. 219).

When health problems do not interfere, both women's and men's emotional and psychological outlooks are as important as age in determining sexual functioning. Factors such as monotony, lack of an understanding partner, mental or physical fatigue, and overindulgence in food or alcohol may all have a profound negative effect on a person's capacity for sexual expression. Another important factor is regular sexual activity—as in "use it or lose it" (Marshall 2009, p. 218).

What about Boredom?

Jokes about sex in marriage are often about boredom. Among social scientists, one explanation often offered for the decline in marital sexual frequency is **habituation**—the decreased interest in sex that results from the increased accessibility of a sexual partner and the predictability in sexual behavior with that partner over time. Decreases due to habituation seem to occur early in the marriage; sexual frequency declines sharply after about the first year of marriage no matter how old (or young) the partners are. The reason for "this rather quick loss of intensity of interest and performance" appears to have two components: "a reduction in the novelty of the physical pleasure provided by sex with a particular partner and a reduction in the perceived need to maintain high levels of sexual behavior" (Call, Sprecher, and Schwartz 1995, p. 649; see also Liu 2000).

However, research by Erickson (2005) and by Elliott and Umberson (2008) suggests that the decline in sexual frequency may be more complex than the previous explanation. They use social theorist Arlie Hochschild's (1983) concept, **emotion labor**, in which women, through their gendered work at home, display certain emotions that they believe are expected of them—in other words, it's a gendered management of emotions. This emotion labor includes all of the pressures of work, running the household, dealing with children's needs, and so on—the burdens of women's daily lives. Such work is exhausting, oftentimes leading to, the authors suggest, reduced sex drive on the part of women. The authors suggest that the complexities found in reduced sex drive among women may have quite a bit to do with the fact that women are exhausted from their emotion work that is endemic to running a household, as well as the exhaustion that comes from displaying emotions that are expected but that they don't feel in the course of their daily and family lives.

"Owen, look—the good sex fairy."

Given that the decline in marital frequency occurs most sharply early in marriage and only gradually after that, these researchers reasoned that "it is difficult to determine ... whether habituation to sex actually occurs" throughout the marriage (Call, Sprecher, and Schwartz 1995, p. 647). Comparisons of first-married and remarried couples can shed some light on this. Remarried respondents reported somewhat higher rates of sex frequency compared to people in first marriages who were the same age, and this was particularly true for those under age forty. Because people who remarry do renew the novelty of marital sex with a new partner, this finding is evidence for the habituation hypothesis (Call, Sprecher, and Schwartz 1995).

Sexual Satisfaction in Marriage and Other Partnerships

All this discussion of the frequency of intercourse may tempt us to forget that committed partners' sexuality is essentially about intimacy and self-disclosure. In other words, sex between partners—heterosexual partners and gay and lesbian partners as well—both gives pleasure and reinforces the relationship. A relatively recent study found that those who "reported both the greatest emotional satisfaction and the greatest physical pleasure in their intimate relationships were those who were partnered in a monogamous relationship" (Hendrick 2000, p. 4). Another study comparing cohabiting, married, and single individuals found that cohabiting and married individuals had the highest—and equal—levels of physical pleasure, with emotional satisfaction with sex greatest among married people (Waite and Joyner 2001).

Despite declining sexual frequency, sexual satisfaction remains high in marriages over the life course (of course, the less satisfied may have opted for divorce); 88 percent report that they are "extremely" or "very physically pleased" (Laumann et al. 1994; see also Christopher and Sprecher 2000, p. 1,003). General satisfaction with sexual relationships was also characteristic of gay and lesbian couples.

Race/Ethnicity and Sexual Activity

Table 5.1 reports differences in the sexual experience of high school students. As discussed previously in this chapter, there is variation between racial and gender groups when it comes to sexual activities. Additionally, a study in the *American Journal of Public Health* noted that, in 2006 (the most recent year that data was collected), the vast majority of men from all racial groups had only one sex partner. The prevalence, however, "of multiple sexual partnerships varied substantially by race/ethnicity. Non-Hispanic black and Hispanic men (28 percent and 18 percent, respectively) were more likely to have had multiple sexual partners than were non-Hispanic white men (13 percent) and men of other racial/ethnic groups (9 percent)" (Adimora, Schoenbach, and Doherty 2007, p. 2,232). Comparisons between white males and females and their black and Hispanic counterparts about experience with oral and anal sex vary with the particular item. Overall, whites are more likely to have had "unconventional sex" (Mosher, Chandra, and Jones 2005, Tables 1, 2, p. 12).

African Americans and non-Hispanic whites are more similar than dissimilar in at least some aspects of their sexual behavior, and any dissimilarities are related to socioeconomic status, not race (Knox and Zusman 2009). Asians report fewer sexual partners over a lifetime and tend to have their first sexual experiences later than whites and Hispanics. For Hispanic men, the more acculturated into the mainstream American culture they are, the more casual sexual encounters they have (Meston and Ahrold 2010). Research on *married couples* suggested that sexual frequency does not vary significantly with race, social class, or religion (Christopher and Sprecher 2000).

Gay/lesbian sexuality, like heterosexual behavior, has been explored among African Americans mostly in the context of problems (e.g., AIDS) and at the lower end of the social scale. An exception is found in the analysis of the 2000 census data on black same-sex households, which make up 14 percent of all such households. Black same-sex households tend to be less well off economically than other same-sex households. They are more likely to be raising children (Dang and Frazer 2004).

Social scientists writing about gay black male sexuality believe it is not as visible as among whites because blacks may find white gay subcultures alien, and they may tend to be integrated into heterosexual communities and extended families that strongly disapprove of homosexuality (Bowleg et al. 2003; Mays, Cochran, and Zamudio 2004). Similar issues are found in the more traditional Latino and Asian cultures where heterosexuality is emphasized. For example, the "machismo" image is a strong component of Latino culture, and assertions of virility are an important part of traditional Asian cultures as well (Calzo and Ward 2010, p. 1,104). As part of that social integration, as well as the racism that men of color perceive in the gay community (Han 2008), black and Latino gay men may be more likely to be bisexual than exclusively homosexual and less likely to assert a gay identity even when engaged primarily in same-sex relations (Sandfort and Dodge 2008). This pattern of engaging in sex with other men while maintaining a straight masculine identity has been labeled "the down low" (Denizet-Lewis 2003; Malebranche 2007).

Black lesbians are relatively invisible due to their smaller numbers and integration into extended family relationships. The issues faced by black gay and bisexual men tend to be relatively invisible as well. The black lesbians in a qualitative study of 530 lesbians and 66 bisexual women were well-educated, middle-class women in their thirties, who first became conscious of their attraction to women at around age fourteen, with first same-sex sexual experiences at age nineteen. Their adult relationships have been generally satisfying and close (Mays and Cochran 1999).

Now let us turn to a more general discussion of sexual expression as conceptualized by sex researchers William Masters and Virginia Johnson, who initiated contemporary sex therapy.

Sex as a Pleasure Bond: Making the Time for Intimacy

The convergence of sexual satisfaction with general satisfaction serves to support Masters and Johnson's (1976) view of sex as a **pleasure bond**, by which partners commit themselves to expressing their sexual feelings with each other.

In sharing sexual pleasure, partners realize that sex is something partners do with each other, not to or for each other. Each partner participates actively, as an equal in the sexual union. Further, each partner assumes **sexual responsibility**—that is, responsibility for his or her own sexual response. When this happens, the stage is set for conscious, mutual cooperation. Partners feel freer to express themselves sexually.

Just as it is important for families to arrange their schedules so that they may spend time together, it's also important for couples to plan time to be alone and intimate (Marano 2010; Masters, Johnson, and Kolodny 1994). Planning time for intimacy involves making conscious choices. Boredom with sex after many years in a marriage may be at least partly the consequence of a decision by default.

Therapists suggest that couples might create romantic settings at home or—if they can afford it—take a weekend retreat. Partners may choose to set aside at least one night a week for themselves alone where they can cuddle and watch movies, for example. Another idea is leisurely going out together for a cup of coffee together. Partners do not have to have intercourse during these times: They should do only what they feel like doing. But scheduling time alone together does mean mutually agreeing to exclude other preoccupations and devote full attention to each other. The point is to have "us time," so this time together should not be spent discussing finances, family, or work.

Sexual Expression, Family Relations, and HIV/AIDS

This suggestion may be easier for parents with young children who are put to bed fairly early. A common complaint from parents of older children is that the children stay up later and, by the time they are teenagers, the parents no longer have any private evening time together. One woman's solution to this problem:

> Our house shuts down at 9:30 now. That doesn't mean we say "It's your bedtime, kids. You're tired and you need your sleep." It means we say, "Your dad (or your mom) and I need some time alone." The children go to their rooms at 9:30. Help with homework, lunch money, decisions about what they'll wear tomorrow—all those things get taken care of by 9:30 or they don't get taken care of. (Monestero 1990)

The important thing, these therapists stress, is that partners don't lose touch with either their sexuality or their ability to share it with each other. In other words, as Patti Newbold, who lost her husband, sadly notes, "marriage isn't about my needs or his needs or about how well we communicate about our needs. It's about loving and being loved" (Marano 2010, p. 71).

We have been talking about human sexual expression as a pleasure bond. It is terribly unfortunate that sexuality can also be associated with disease and death. Indeed, the fact that it is so difficult to make a transition to the next topic points to the multifaceted, even contradictory, nature of contemporary human sexual expression.

HIV/AIDS has now been known for about thirty years. The HIV, or "human immunodeficiency virus," which produces AIDS, has existed longer than that, but it was only in 1981 that AIDS was recognized as the cause of a rapidly increasing number of deaths.

An HIV infection eventually progresses to full-blown AIDS. AIDS stands for "acquired immunodeficiency syndrome"; it is a viral disease that destroys the immune system. With a lowered resistance to disease, a person with AIDS becomes vulnerable to infections and other diseases that other people easily fight off. "Facts About Families: Who Has HIV/AIDS?" presents some details on the demographics and transmission modes of HIV/AIDS.

The rates of new HIV diagnoses increased 15 percent between 2004 and 2007 (U.S. Centers for Disease Control and Prevention 2009, p. 6), with about 56,000 people becoming infected each year (Kates et al. 2009, p. 8). The rise of new HIV cases comes at the same time as county and state, and federal budgets for HIV/AIDS/STD prevention funding were cut or remained flat (Kates et al. 2009, pp. 6–7, 15). Because of its lethal character—over 583,298 deaths in the United States through 2007 (U.S. Centers for Disease Control and Prevention 2009)—we focus here on HIV/AIDS. Appendix C describes various STDs and presents information on transmission modes, prevention, and treatment.

A theme of this text is that social, political, economic, and cultural conditions affect people's choices. We examine here the impact of HIV/AIDS as a societal phenomenon that has changed the consequences of decisions about sexual activity and one that intersects with other social characteristics.[8]

HIV/AIDS and Heterosexuals

Some heterosexual adults have responded to the threat of AIDS with changed behavior, but many others have not. As you can see from Figure 5.3, the incidence of AIDS transmission is growing among heterosexuals, making up 31 percent of all new cases by 2007. Heterosexuals report increased use of condoms and fewer partners than in the past (T. Smith 2006).

---

[8] We discussed, in Chapter 3, that poor people (including a substantial portion of people of color) have lower rates of education, and when they do have access to education, it is often of poorer quality. What this combination of factors suggests to many contemporary researchers examining this issue is that poverty denies access to education about diseases such as HIV/AIDS (as well as knowledge of prevention). Furthermore, poverty denies access to medical care and resources, which also may be of help at preventing this and other transmittable (but preventable) diseases.

Who Has HIV/AIDS?[a]

Over one million people are living with HIV or full-blown AIDS: 40 percent black, 38 percent white, 16 percent Hispanic, and less than one percent each Asian/Pacific Islander and Native American/Alaska Native. The cumulative total of AIDS cases reported through 2007 is over a million, with around 56,000 new cases diagnosed each year. There have been over 580,000 deaths since AIDS was first identified in 1981 (Altman 2005; U.S. Centers for Disease Control and Prevention 2009, Tables 4, 8; "HIV/AIDS Epidemic in the United States" 2009, p. 1).

On the positive side, infections are being caught in the early stages, and new treatments have enabled longer lives for those with the virus (Antiretroviral Therapy Cohort Collaboration 2008).[b] On the other hand, the increase in infections suggests a growing sense of complacency among groups at risk of contracting the disease (Cooter and Stein 2009). Estimates are that 21 percent of those with HIV have not been tested and are unaware of their condition ("HIV/AIDS Epidemic in the United States" 2009, p. 1).

Age and HIV/AIDS

HIV/AIDS has most affected young and middle-aged adults. As of 2007, around 70 percent of AIDS cases were diagnosed in people in the twenty-five through forty-four age range, with the greatest increase in diagnoses going to those age forty to forty-four, who accounted for 15 percent of all new cases. The proportion of HIV/AIDS cases among teenagers is small, under one percent (U.S. Centers for Disease Control and Prevention 2009, p. 6, Tables 1, 3), but keep in mind that individuals who are older at diagnosis may have been infected as adolescents. Expanded testing and treatment of HIV-infected pregnant women have lowered the rate of new cases of prenatal transmission to fewer than 2 percent of births to infected women ("Pregnancy and Childbirth" 2007, n.p.). There were only twenty-eight new cases of AIDS in children in 2007 (U.S. Centers for Disease Control and Prevention 2009, Table 5).

We seldom think of AIDS as affecting older individuals, but around 10 percent of cases are found among those age fifty and up. Only 1.5 percent of AIDS cases are reported for individuals age sixty-five

and older, but that is over 15,853 cases among senior citizens (U.S. Centers for Disease Control and Prevention 2009, Table 3). Currently, there are efforts to create programs to educate older Americans about risks and precautions concerning HIV/AIDS (Villarosa 2003).

Gender and HIV/AIDS

Men accounted for 74 percent of AIDS cases diagnosed in 2007, among adolescents and adults. The dominant source of AIDS among males is having sex with other men (53 percent), with intravenous drug use and heterosexual contacts also significant causes of infection. Cumulatively, 80 percent of AIDS cases among women arose from heterosexual contact, and 20 percent from intravenous drug use (U.S. Centers for Disease Control and Prevention 2007, p. 7, Table 3).[c]

Critical Thinking

Pick one of the previously mentioned demographic categories—for example, teen women. What ideas can you think of for an HIV/AIDS prevention program for this group? You may want to

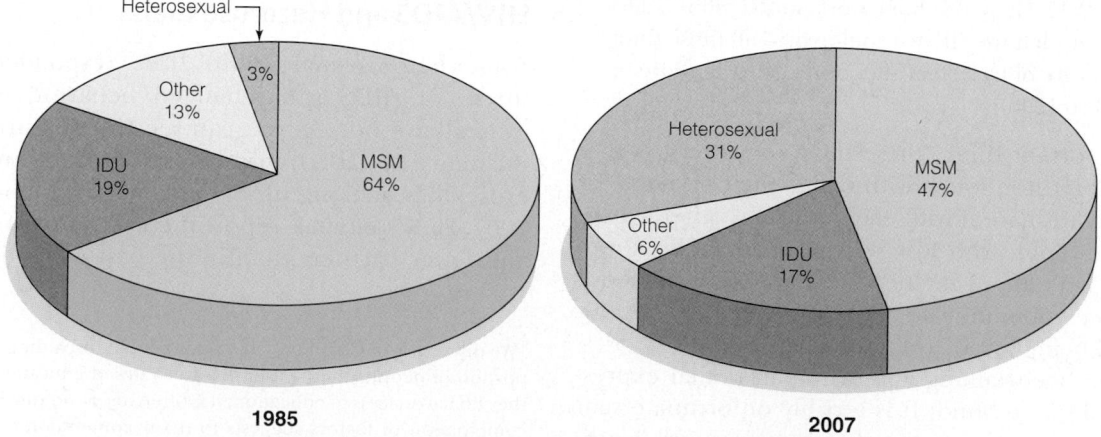

AIDS Diagnoses by Transmission Category, United States, 1985 & 2007

1985
Heterosexual 3%
Other 13%
IDU 19%
MSM 64%

2007
Heterosexual 31%
Other 6%
IDU 17%
MSM 47%

MSM=Men who have sex with men (gay and bisexual men)
IDU=Injection drug use

Figure 5.3 The transformation of AIDS diagnoses. What was once considered a "gay disease" has become more prevalent in the heterosexual population.

consult Appendix C, "Sexually Transmitted Diseases," on the website.

a. The term HIV/AIDS is used in general references to this sexually transmitted disease. When speaking about numbers of cases, HIV (the virus that causes AIDS) and AIDS (the active disease) are often distinguished. Most of those who become infected with the HIV virus will progress to full-blown AIDS.

The incidence (number of new cases) and prevalence (current cases) of HIV infection are only estimates, as there is no population-wide screening program. Many people who may be HIV-positive are not tested, and test results are not always reported consistently. Consequently, most of the data in these sections on the social distribution of HIV/AIDS are based on AIDS cases, as those are more definite in diagnosis and are reported more accurately.

b. Initially, many cases of AIDS arose through infection from blood transfusions, but this mode of transmission declined after 1985, when donated blood began to be rigorously screened for HIV. Blood transfusion accounted for less than one percent of cases in 2007 (U.S. Centers for Disease Control and Prevention 2009, Table 4).

c. Infection from woman-to-woman sexual contact is rare (U.S. Centers for Disease Control and Prevention 2007).

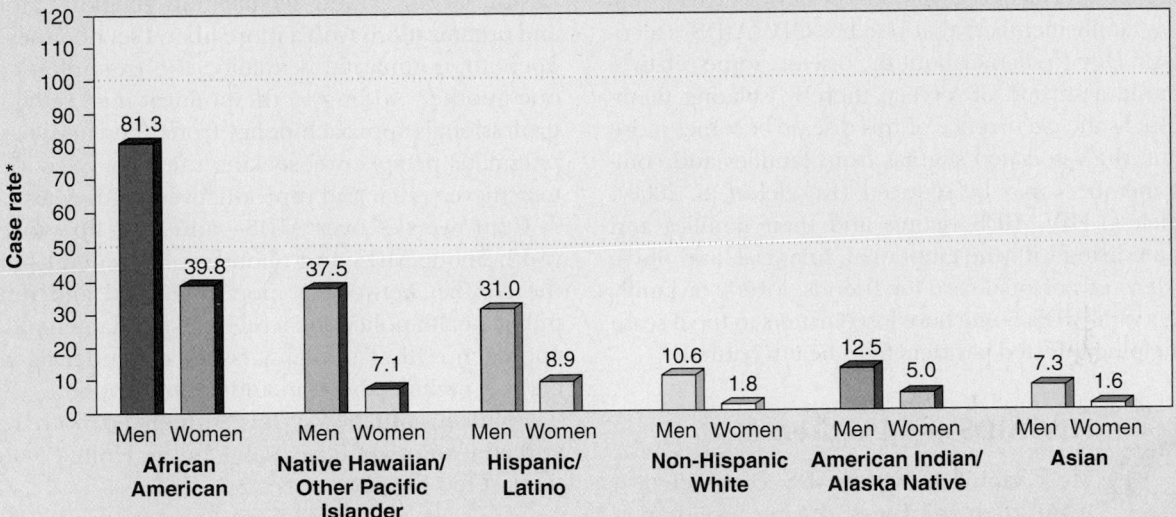

Figure 5.4 Estimated rates of AIDS cases reported among adults and adolescents by race/ethnicity, 2005.*

*Number of cases per 100,000 in respective racial/ethnic and gender group.

Number of cases per 100,000 in respective racial/ethnic and gender group. Source: U.S. Centers for Disease Control and Prevention 2009, Table 6b.

They may insist on a recent blood test verification of HIV status or, more often, condom use. "Even for many [young adults]…who have been on the conservative side in their behavior, AIDS is part of their consciousness. …They realize that even if they have been careful in their sexual behavior, their partners may not have been, and that puts them at risk" (Arnett 2004, p. 91).

In an effort to reduce potential contact with the disease, some heterosexuals opt for periods of celibacy. Given the increasing heterosexual transmission of HIV/AIDS (see Figure 5.3), "women should embrace a philosophy of always protecting themselves from HIV" ("Third of New HIV Cases" 2004). Perhaps 20 percent of gay men marry at least once (K. Butler 2006b; J. Gross 2006d), and upwards of 70 percent of "straight-identified men having sex with men are married" (DeNoon 2006). Consequently, some heterosexual women may be regularly exposed to the virus if their husbands are sexually active with men.

HIV/AIDS and Gay Men

Many gay men modified their sexual behavior in the 1980s. Multiple, frequent, and anonymous sexual contacts had been common elements of lifestyle and sexual ideology for many gay men (Blumstein and Schwartz 1983). However, attitudes and behavior changed enough, at least among men in their thirties and over, to have dramatically reduced the incidence of new cases among gay males for a time, but we saw a 25 percent increase in new cases of male-to-male transmission of HIV/AIDS from 2004 to 2007 (U.S. Centers for Disease Control and Prevention 2009, Table 1).

Meanwhile, life expectancy (average number of years remaining to those treated between age twenty and thirty-five) is now between thirty and forty years with antiretroviral therapy (Antiretroviral Therapy Cohort Collaboration 2008, Table 2). As the mood in the gay community lightened since the early days of HIV/AIDS in the 1980s, some gay men have returned to unprotected sex (barebacking) with many and anonymous partners, and we have seen a surge of HIV infections among younger gay men ("HIV/AIDS Epidemic in the United States" 2009). Gay activists and public health professionals have expressed concern that drug ads with pictures of relatively hearty gay men convey a misleading message about the difficulties of living with AIDS, a message that may reduce caution and prevention (Cooter and Stein 2010, pp. 14–15; M. Gross 2006).

HIV/AIDS and Family Crises

Some families will face crises and loss because of AIDS. Telling one's family members that one has HIV/AIDS is a crisis in itself. Due to shame about the disease, some relatives grieve amid a shroud of secrecy, thereby isolating themselves. But as the occurrence of this disease becomes more prevalent, the associated stigmas from families and community members may be reduced (Knodel et al. 2009). Nevertheless, HIV/AIDS victims and their families and friends are living with the emotional, financial, and physical burdens of personal care for friends, lovers, or family members with AIDS. Some have lost partners to the disease or are helping infected partners fight health battles.

HIV/AIDS and Children

Most children with AIDS contracted it from their mothers during pregnancy, at birth, or through breast milk. This form of AIDS transmission has declined dramatically due to voluntary prenatal HIV testing of pregnant women and the subsequent administration of prenatal drug therapies.

Women who have tested positive for HIV/AIDS are not always willing to give up the prospect of motherhood; some are deciding to have children after diagnosis now that the risk of transmission may be drastically reduced by medication. In fact, if "women take these drugs before and during birth, and their babies are given drugs after birth, HIV transmission is reduced from 25 percent to less than 2 percent (fewer than 2 in 100)" ("Pregnancy and Childbirth" 2007, n.p.). Men with HIV are beginning to hope for parenthood also, with a procedure called "sperm washing" designed to minimize transmission to a female partner (Cichocki 2009; Kolata 2002a).

Children with AIDS have a unique array of treatment requirements and are often in the hospital. Some are abandoned by their parents to hospital care. Others are raised by grandparents or foster parents. A new set of problems has arisen as babies with AIDS enter their teens. Clinical professionals report behavioral, emotional, and cognitive problems among some of these babies who have survived to adolescence. Public health workers have begun to take note of the teens' needs for services (Dee 2005; Hazra, Siberry, and Mofenson 2010).

HIV/AIDS and other sexually transmitted diseases are more than a medical or a family problem—they are conditions imbued with social meanings and consequences. Politics enters into decisions about policy related to HIV/AIDS—and policy regarding sexuality in general.

The Politics of Sex

One of the most striking changes over the past several decades has been the emergence of sexual and reproductive issues as political controversies. Religious and political conservatives and secular and religious individuals and organizations with a more liberal set of values—more open to nonmarital sexuality, for example—confront one another. Adding to the political mix, public health professionals approach policy from a research-based and pragmatic perspective, seeking the most effective means to achieve sexual and reproductive health goals.

Controversies over AIDS—and over how to educate youth about AIDS and about sex in general—illustrate the conflict between a morally neutral and pragmatic public health policy and a religious fundamentalist moral approach. Other sexuality issues engendering political conflict include abortion and contraception.

Political controversy has influenced both research and education about sexuality in the United States over the last few decades.

Politics and Research

At first, the emergence of AIDS seemed to legitimate sex research and lead to the funding of research on sexual behavior because of the implications for controlling the AIDS epidemic (Christopher and Sprecher 2000). The need for more comprehensive and current data on sexual behavior twice led to efforts to mount federally funded national sample surveys to be conducted by teams of well-respected social scientists. The studies were initially funded, but then Congress canceled a pilot study on the grounds that a sex survey would be too controversial. NORC conducted a much smaller, though national, sample survey without federal funds through the support of grants from private agencies.

A planned survey of 24,000 teens in grades seven to eleven was also canceled ("U.S. Scraps" 1991). Ironically, comparisons with other countries suggest that our society's tendency to deny sexuality at the same time we encourage it sends mixed messages that, among other negative consequences, probably help account for the unusually high rates of teen pregnancy and abortion in the United States (Dyess 2009; Risman and Schwartz 2002). The climate for the scientific study of sexuality has been hostile enough that "only the brave study sex" (Carey 2004). Whether this trend holds in the near future remains to be seen, although there have been some signs of change via the actions on the part of the Obama administration.

The politicization of research has taken other forms. Some reports of research that do not support the government position on an issue have been removed from government websites or changed after initial posting (Lewontin 2004; Rest and Halpern 2007; Simoncelli 2005). For example, a review of many studies which concluded that abortion does *not* cause breast cancer (contradicting the position of the National Right to Life movement) was removed from the National Cancer Institute website (Shulman 2008). On January 12, 2010, a new page was uploaded onto the National Cancer Institute website that noted scientists "concluded that having an abortion or miscarriage does not increase a woman's subsequent risk of developing breast cancer" (U.S. National Cancer Institute 2010). Research on sex education which found that providing information about contraception to teens does *not* increase their sexual activity was removed from the U.S. Centers for Disease Control and Prevention (CDC) website. The CDC fact sheet on condoms was changed to de-emphasize the protective value of condoms vis-à-vis HIV infection (Shulman 2008). On February 8, 2010, the U.S. Center for Disease Control and Prevention website uploaded a new page that said, "Consistent and correct use of male latex condoms can reduce (though not eliminate) the risk of STD transmission" ("Condoms and STDs: Fact Sheet for Public Health Personnel" 2010).

Adolescent Sexuality and Sex Education

Indeed, adolescents are sexually active; 47.8 percent of high school students responding to the 2007 Youth Risk Behavior Surveillance (Eaton et al. 2008) have had sexual intercourse. What may surprise you is that teen sexual intercourse has declined since 1991; in that year, 54 percent of high school youth had had intercourse. Studies suggest that males have changed more than females; a slim majority of high school males (50.2 percent) were virgins in 2007 (Eaton et al. 2008; Risman and Schwartz 2002). The downward trend in adolescent sexual intercourse (and births) predates the emphasis on "abstinence-only" sex education programs (Dailard 2003; "Improvements" 2006).

Experts attribute the decline in sexual intercourse to comprehensive sex education and to fear of sexual disease. Some have argued that sociosexual values have simply become more conservative generally, but that possibility has not been adequately researched (N. Bernstein 2004a; Risman and Schwartz 2002; Santelli et al. 2007). Data from the National Survey of Family Growth (1995 and 2002 waves) indicate that 86 percent of the decline in pregnancy risk is due to improved contraception (regular use, better methods), whereas only 14 percent is due to delayed sex (Santelli et al. 2007). These declines in pregnancy risk reversed

in 2005, but the downward trend in teen births resumed in 2008 (Hamilton, Martin, and Ventura 2010).

Sociologist Frank Furstenberg, who has long researched teen sex, reproduction, and parenthood, as well as the transition to adulthood, believes that young people have observed how difficult it is in today's economy to establish a satisfying life if parenthood comes too early. He thinks that teenagers "getting the picture" might be a key element, as they strive to delay sex and prevent pregnancy to avoid the difficult lives they have seen around them (in N. Bernstein 2004a).

The decline in teen sexual activity has slowed in the twenty-first century, virtually "flat-lining" since 2001, except for declines in current sexual activity of black youth (Brenner et al. 2006; Feijoo 2004; "Improvements" 2006). Moreover, in assessing the decline of teen sexual activity, it is important to note that a lot depends on one's definition of sexual activity. Yes, sexual intercourse is less frequent among adolescents than in the past. It is also true that teens exhibit surprisingly high rates of oral sex: 55 percent of males ages fifteen through nineteen and 54 percent of females engage in oral sex (Mosher, Chandra, and Jones 2005, Tables 3 and 4). Many adolescents do not consider oral sex to be "sex" so they can still consider themselves virgins. They also seem to be attracted to oral sex because it does not present a risk of pregnancy (true) or of sexually transmitted disease (not true).

Sex Education Current controversy centers on whether sex education should be "abstinence only" or "abstinence plus" (also termed "comprehensive"). Since 1996, the federal government has taken the official position that abstention from sexual relations unless in a monogamous marriage is the only protection against sexually transmitted disease and pregnancy—and that abstinence is the only morally and rationally appropriate principle of sexual conduct. "Abstinence-only" programs may mention contraception, if at all, only to cite allegedly high failure rates. Programs are urged to convey to students that nonmarital sex for people *of any age* is likely to have harmful physical and psychological effects (J. Brody 2004; Dailard 2002).

Surveys of parents indicate that they prefer that an "abstinence-plus" sex education program be presented in the schools, one that would include contraception and AIDS prevention as well as promotion of abstinence. As Figure 5.5 indicates, more than 74 percent of adults "approve of health education classes that teach about sex and abstinence," and 49 percent believe abstinence-only classes have some impact on preventing teen pregnancy (Rasmussen Reports 2009). Yet, a substantial and growing number of school sex education programs are "abstinence only," as government funding is limited to such programs (Lindberg, Santelli, and Singh 2006).

Research indicates that comprehensive sex education programs (those that include contraception) do not lead to any earlier commencement of sexual activity; in fact,

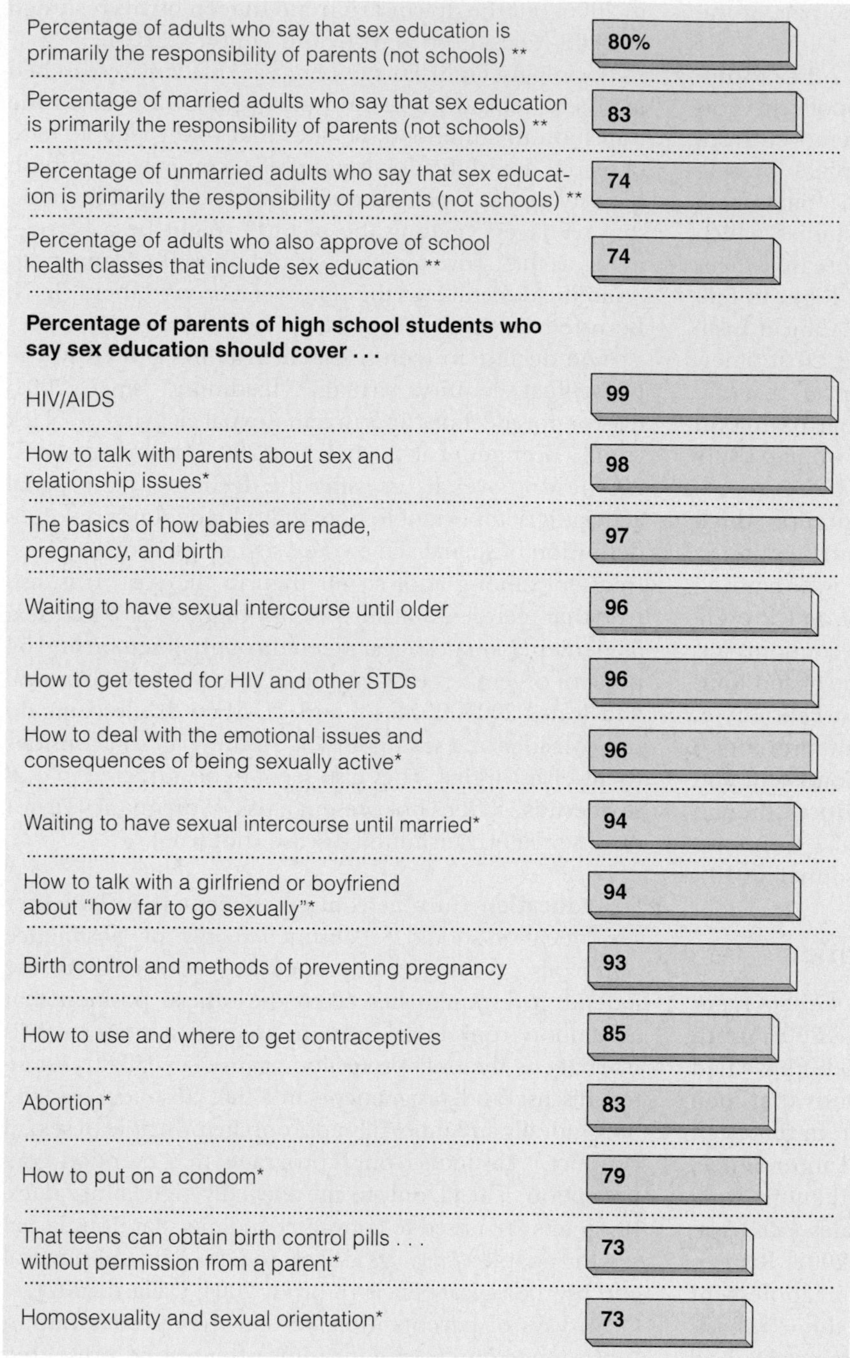

Percentage of adults who say that sex education is primarily the responsibility of parents (not schools) ** | 80%

Percentage of married adults who say that sex education is primarily the responsibility of parents (not schools) ** | 83

Percentage of unmarried adults who say that sex education is primarily the responsibility of parents (not schools) ** | 74

Percentage of adults who also approve of school health classes that include sex education ** | 74

Percentage of parents of high school students who say sex education should cover . . .

HIV/AIDS | 99

How to talk with parents about sex and relationship issues* | 98

The basics of how babies are made, pregnancy, and birth | 97

Waiting to have sexual intercourse until older | 96

How to get tested for HIV and other STDs | 96

How to deal with the emotional issues and consequences of being sexually active* | 96

Waiting to have sexual intercourse until married* | 94

How to talk with a girlfriend or boyfriend about "how far to go sexually"* | 94

Birth control and methods of preventing pregnancy | 93

How to use and where to get contraceptives | 85

Abortion* | 83

How to put on a condom* | 79

That teens can obtain birth control pills . . . without permission from a parent* | 73

Homosexuality and sexual orientation* | 73

Figure 5.5 Who should teach children about sex, and what parents want sex education to teach their children.

* Questions marked with an asterisk were asked of only half the sample.

**Questions marked with a double asterisk were asked in a 2009 telephone survey by Rasmussen Reports.

Source: Survey of 1,001 parents of children in seventh to twelfth grade sponsored by National Public Radio (NPR)/Kaiser Family Foundation/Kennedy School of Government (2004). The survey was conducted in September/October 2003. *The high school parent subsample = 450.

** Rasmussen Reports National Survey of 1,000 Adults, conducted January 12-13, 2009.

Reports National Survey of 1,000 Adults, conducted January 12–13, 2009.

research indicates that comprehensive sex education delays the start of sexual activity. "Encouraging abstinence and urging better use of contraception are compatible goals" (Kirby 2001, p. 18). There is as yet no evidence that "abstinence-only" programs are effective in delaying sex or preventing pregnancy, nor are they scientifically accurate. The federal government's own study of the effectiveness and accuracy of its federally funded abstinence-only education programs found them wanting ("The Abstinence-Only Delusion" 2007; Begley 2007; J. Brody 2004; "Conclusions Are Reported" 2007; Kirby, Laris, and Rolleri 2006; Crosse 2008, p. 5). In fact, the only abstinence-based education program that was effective (although only slightly) at delaying sex would not have qualified to be taught in middle schools because it did not meet the federal mandated guidelines. This recent study examined abstinence-focused education and found a link between such education and a reduction in sexual activity in twelve-year-old African American children. This study of 662 children found that, two years after the abstinence-focused course, the children who took it had slightly reduced sexual activity, compared with those who had taken a traditional sex education course (Guttmacher Advisory 2010; Jemmott, Jemmott, and Fong 2010; Schlesinger 2010).

While abstinence-only education is ineffective in delaying sex, early research found that virginity pledges taken in certain circumstances seemed to delay adolescent sexual activity (Bearman and Brückner 2001), but later research has found no difference in delay of sexual activity between those who take the pledge and those who do not (Rosenbaum 2009; Tanne 2009; Thomas 2009). Recent research also found that once these teens become sexually active, they do so without precautions and have higher rates of pregnancy and STDs than other teens (Altman 2004; Rosenbaum 2009, p. e114).

Those results compare adolescents who are still minors. A study based on the National Longitudinal Study of Adolescent Health, which compared young adults (nineteen through twenty-five) who had made a virginity pledge and those who had not, found that contraceptive use did not differ at this later point. Both groups had high rates of intercourse (89.7 percent and 75 percent, respectively) and other sexual activity. In other words, "by the time they become young adults, some 81 percent of pledgers have engaged in some type of sexual activity" (Rector and Johnson 2005, p. 13).

Sex education needs to take into account the teen propensity to engage in oral sex and to consider it risk-free despite high rates of STDs among youth (Halpern-Felsher et al. 2005). Moreover, the first experience for a small proportion of teens—7.8 percent—is forced sex (Eaton et al. 2008). Some others' experiences are ambiguous as to wantedness (Houts 2005).

Sex education programs may emphasize peer pressure, troubled families and neighborhoods, or hormonal processes and rarely consider the broad array of teen motivations for sexual activity. It seems that both adolescent males and females seek sex because they expect it to meet needs for intimacy, sexual pleasure, and social status (Ott et al. 2006). These issues need to be taken into account in sex education programs. Long-time sex education researcher Douglas Kirby and his colleagues have identified some programs that seem effective in discouraging early sexual activity and sexual risks (Kirby, Laris, and Rolleri 2006).

Sexual Responsibility

People today are making decisions about sex in a climate characterized by political conflict over sexual issues. Premarital and other nonmarital sex, homosexuality, abortion, and contraception represent political issues as well as personal choices. Public and private communication must rise to new levels. The AIDS epidemic has brought the importance of sexual responsibility to our attention in a dramatic way.

Making knowledgeable choices is a must. Because there are various standards today concerning sex, all individuals must determine what sexual standard they value, which is not always easy. Today's adults may be exposed to several different standards throughout the course of their lives. Even when individuals feel that they have clear values, applications to particular situations may be difficult. A person who believes in the standard of sexual permissiveness with affection, for example, must determine when a particular relationship is affectionate enough.

Making these choices and feeling comfortable with them requires recognizing and respecting your own values, instead of just being influenced by others when in a sexual situation. Anxiety may accompany the choice to develop a sexual relationship, and there is considerable potential for misunderstanding between partners. This

Intimacy and sexuality require communication—physical as well as verbal—from both partners. Sexuality has become more expressive and less patriarchal in the United States, and each generation finds itself reevaluating sexual assumptions, behaviors, and standards.

Eric K. K. Yu/CORBIS

section addresses some principles of sexual responsibility that may serve as guidelines for sexual decision making.

One obvious responsibility concerns the possibility of pregnancy. Partners should plan responsibly whether, when, and how they will conceive children and then use effective birth control methods accordingly.

A second responsibility concerns the possibility of contracting STDs or transmitting them to someone else. Individuals should be aware of the facts concerning HIV/AIDS and other STDs. They need to assume responsibility for protecting themselves and their partners. They need to know how to recognize the symptoms of an STD and what to do if they get one (see Appendix C).

A third responsibility concerns communicating with partners or potential sexual partners. As we've seen in this chapter, sex may mean many different things to different people. A sexual encounter may mean love and intimacy to one partner and be a source of achievement or relaxation to the other. Honesty lessens the potential for misunderstanding and hurt between partners.

A fourth responsibility is to oneself. In expressing sexuality today, each of us must make decisions according to our own values. A person may choose to follow values held on the basis of religious commitment or put forth by ethicists or by psychologists or counselors. People's values change over the course of their lives, and what's right at one time may not appear so later. Despite the confusion caused both by internal changes as our personalities develop and by the social changes going on around us, it is important for individuals to make thoughtful decisions about sexual relationships.

Summary

- Social attitudes and values play an important role in the forms of sexual expression that people find appropriate and enjoyable.

- Despite decades of conjecture and research, it is still unclear just how sexual orientation develops and whether it is genetic or socially shaped. Recent decades have witnessed increased acceptance of gay, lesbian, bisexual, and transgender (GLBT) individuals, though some disapproval, discrimination, and hostility remain.

- Whatever one's sexual orientation, sexual expression is negotiated amid cultural messages about what is sexually permissible or desirable. In the United States, these cultural messages have moved from patriarchal sex, based on male dominance and on reproduction as its principal purpose, to a message that encourages sexual expressiveness in myriad ways for both genders equally.

- Four standards of nonmarital sex are abstinence, permissiveness with affection, permissiveness without affection, and the double standard—the latter diminished since the 1960s, but is still alive. Extramarital sex is not approved, but it does occur and represents a challenge to marital trust.

- Marital sex changes throughout the life course. Young spouses have sex more often than do older mates. Although the frequency of sexual intercourse declines over time and through the length of a marriage, some 27 percent of married people over age seventy-four are sexually active.

- Making sex a pleasure bond, whether a couple is married or not, involves cooperation in a nurturing, caring relationship. To fully cooperate sexually, partners need to develop high self-esteem, to break free from restrictive gendered stereotypes, and to communicate openly.

- HIV/AIDS has had an impact on relationships, marriages, and families.

- Sexuality, sexual expression, and sex education are public issues at present, and different segments of American society have divergent views.

- Whatever the philosophical or religious grounding of one's perspective on sexuality, there are certain guidelines for personal sexual responsibility that we should all heed.

Questions for Review and Reflection

1. Give some examples to illustrate changes in sexual behavior and social attitudes about sex. What might change in the future?

2. Do you think that sex is changing from "his and hers" to "theirs"? What do you see as some difficulties in making this transition?

3. Discuss what you've learned about the nature of sexual relationships. What did you find useful and relevant to everyday life? What seems remote from real-world experience?

4. How do you think HIV/AIDS affects sex and sex relationships—or does it?

5. **Policy Question.** What role, if any, has government policy played in sex education, research and information on sex, and sexual regulation?

Key Terms

abstinence 117	HIV/AIDS 127
asexual 110	homophobia 114
asexuality 109	homosexuals 108
bisexual 109	hooking up 119
cyberadultery 121	interactionist perspective on human sexuality 112
double standard 119	interpersonal exchange model of sexual satisfaction 111
emotion labor 125	lesbian 109
expressive sexuality 113	patriarchal sexuality 112
friends with benefits 119	permissiveness with affection 118
gay 109	permissiveness without affection 119
GLBT 110	pleasure bond 126
habituation 125	sexting 116
habituation hypothesis 121	sexual orientation 108
heterosexism 114	sexual responsibility 126
heterosexuals 108	sexual scripts 112

Online Resources

Sociology CourseMate

www.CengageBrain.com

Access an integrated eBook, chapter-specific interactive learning tools, including flash cards, quizzes, videos, and more in your Sociology CourseMate, accessed through CengageBrain.com.

www.CengageBrain.com

Want to maximize your online study time? Take this easy-to-use study system's diagnostic pre-test, and it will create a personalized study plan for you. By helping you identify the topics that you need to understand better and then directing you to valuable online resources, it can speed up your chapter review. CengageNOW even provides a post-test so you can confirm that you are ready for an exam.

6

Love and Choosing a Life Partner

Almost three-quarters of young adults believe in "one true love," and more than 90 percent would like a "soul mate" (Robison 2003; Whitehead and Popenoe 2001). We all want to be loved, and most of us expect to be in a committed relationship—if not now, then in the future. Being in a committed relationship involves selecting someone with whom to become emotionally and sexually intimate and, often, with whom to raise children. Accordingly, the choice of a partner is a major life course decision. In Chapter 1, Figure 1.2, "The Cycle of Knowledgeable Decision Making," illustrates that making knowledgeable decisions requires an awareness of one's personal beliefs and values, as well as conscious consideration of alternatives and serious thought about the probable consequences. You may want to refer to Figure 1.2 as you study this chapter.

Research suggests that the best way to choose a life partner is to look for someone to love who is socially responsible, respectful, and emotionally supportive. It is also important that the person is committed both to the relationship and to the value of staying together. Lastly, it helps if that person also demonstrates good communication and problem-solving skills (Bradbury and Karney 2004; Hetherington 2003).

Equally important, research findings suggest looking for a mate with values that resemble one's own, because similar values and attitudes are strong predictors of ongoing happiness and relationship stability. Although romantic love is usually a very important ingredient (Amato 2007), successful life partnerships are also based on partners' maturity, common goals, qualities of friendship, and the soundness of their reasons for getting together (Gaunt 2006; Lacey et al. 2004).

In this chapter, we'll look at some things that influence the choice of a life partner and subsequent relationship satisfaction. We will examine how a relationship develops and proceeds from first meeting to commitment. We will also discuss interreligious and interracial unions. We'll examine research on how cohabiting before marriage affects marital stability. To begin, we explore some things that we know about love.

Nearly three-quarters of women and almost two-thirds of men marry by age thirty; by age forty, more than 80 percent of Americans have married (Goodwin, McGill, and Chandra 2009). By age twenty-four, more Americans are married than cohabiting (Saad 2008b). Therefore, the topic of choosing a marriage partner is critically important. Meanwhile, we need to note here that research and published counseling advice on choosing a committed life partner have focused almost solely on heterosexual, *marital* mate selection. One reason for this situation is that choosing a marriage partner is more easily identifiable for researchers—one is either married or not—whereas the existence and nature of committed, lifelong relationships outside marriage are harder to identify and research.

Furthermore, although more than three-quarters of Americans who have divorced remarry within ten years (Bramlett and Mosher 2001), research tends to focus on choosing a spouse for a first marriage. However, much of what is said in this chapter can probably be applied to choosing a partner for a committed nonmarital relationship and for remarriages as well. Research specifically related to choosing a spouse for remarriage is discussed in Chapter 16.

Love and Commitment

"With respect to love, the gap between everyday people and family scholars is surprisingly wide.... [M]ost researchers have avoided the topic.... Yet attitude surveys reveal that the great majority of Americans view love as the primary reason for getting and staying married" (Amato 2007, p. 306).

> Romantic love can be defined in multiple ways. For example, one definition might be that romantic love is a strong emotional bond with another person that involves sexual desire, a longing to be with the person, a preference to put the other person's interests ahead of one's own, and a willingness to forgive the other person's transgressions. (Amato 2007, p. 206)

With the obvious exceptions of physical and emotional abuse, loving involves the acceptance of partners for themselves and "not for their ability to change themselves or to meet another's requirements to play a role" (Dahms 1976, p. 100). People are free to be themselves in a loving relationship, and to expose their feelings, frailties, and strengths (Armstrong 2003). Related to this acceptance is caring, or empathy—the concern a person has for the partner's growth and the willingness to "affirm [the partner's] potentialities" (May 1975, p. 116; Jaksch 2002).

Psychologist Erich Fromm (1956) chastises Americans for their emphasis on wanting to *be loved* rather than on learning to *love*. Many of the ways we make ourselves lovable, Fromm writes, "are the same as those used to make oneself successful, 'to win friends and influence people.' As a matter of fact, what most people in our culture mean by being lovable is essentially a mixture between being popular and having sex appeal" (p. 2). Showing empathy, of course, is something very different and is essential in loving relationships (Ciaramigoli and Ketcham 2000). Rollo May defines empathy, or caring, as a state "composed of the recognition of another; a fellow human being like one's self; of identification of one's self with the pain or joy of the other" (1969, p. 289). In addition to empathy, loving someone requires commitment.

Maintaining a loving relationship requires commitment of both partners (Dixon 2007). Committing oneself to another person involves working to develop a relationship "where experiences cover many areas

Jack Hollingsworth/Photolibrary

Marriages between individuals with a relatively secure attachment style that take place after about age twenty-five and are between partners who grew up in intact (nondivorced) families are the most likely to be satisfying and stable. But having grown up as a child of divorce does *not* mean that an individual will *necessarily* have an unhappy or unstable marriage.

of personality; where problems are worked through; where conflict is expected and seen as a normal part of the growth process; and where there is an expectation that the relationship is basically viable and worthwhile" (Altman and Taylor 1973, pp. 184–87; Amato 2007).

Committed lovers have fun together; they also share more tedious times. They express themselves freely and authentically (Smalley 2000). Committed partners view their relationship as worth keeping, and they work to maintain it despite difficulties or disagreements (Amato 2007; Love 2001). **Commitment** is characterized by this willingness to work through problems and conflicts as opposed to calling it quits when problems arise. In this view, commitment involves consciously investing in the relationship (Etcheverry and Le 2005). Then too, committed partners "regularly, routinely, and predictably attend to each other and their relationship no matter how they feel" (Peck 1978, p. 118). Psychological research expands upon these ideas; the following section is an example.

Sternberg's Triangular Theory of Love

In research on relationships varying in length from one month to thirty-six years, psychologist Robert Sternberg (1988a, 1988b, 2006) found three components necessary to authentic love: intimacy, passion, and commitment (see also Overbeek et al. 2007). According to **Sternberg's triangular theory of love**, *intimacy* "refers to close, connected, and bonded feelings in a loving relationship. It includes feelings that create the experience of warmth in a loving relationship…[such as] experiencing happiness with the loved one;…sharing one's self and one's possessions with the loved one; receiving…and giving emotional support to the loved one; [and] having intimate communication with the loved one" (Sternberg 1988a, pp. 120–21).

Passion "refers to the drives that lead to romance, physical attraction, sexual consummation, and the like in a loving relationship" (Sternberg 1988a, pp. 120–21). *Commitment*—the "decision/commitment component of love"—consists of not only deciding to love someone, but also deciding to maintain that love. **Consummate love** (see Figure 6.1), composed of all three components, is "complete love,… a kind of love toward which many of us strive, especially in romantic relationships" (Sternberg 1988a, pp. 120–21).

The three components of consummate love develop at different times, as love grows and changes: "Passion is the quickest to develop, and the quickest to fade.… Intimacy develops more slowly, and commitment more gradually still" (Sternberg, quoted in Goleman 1985). Passion, or "chemistry," peaks early in the relationship but generally continues at a stable, although fluctuating, level and remains important both to our good health

Relationships evidence different characteristics or personalities. John Alan Lee (1973) classified six love styles, initially based on interviews with 120 white, heterosexual respondents of both genders. Lee subsequently applied his typology to same-sex relationships (Lee 1981). Researchers then developed a Love Attitudes Scale (LAS): eighteen to twenty-four questions that measure Lee's typology (Hendrick, Hendrick, and Dicke 1998). Although not all subsequent research has found all six dimensions, this typology of love styles has withstood the test of time, has proven to be more than hypothetical, and may even have cross-cultural relevance (Lacey et al. 2004; Le 2005; Masanori, Daibo, and Kanemasa 2004).

Love styles are sets of distinctive characteristics that loving or lovelike relationships take. The word *lovelike* is included in this definition because not all love styles amount to genuine loving as this chapter defines it. People may incorporate different aspects of several styles into their relationships. What are Lee's six love styles?

1. **Eros** (AIR-ohs) is a Greek word meaning "love"; it forms the root of our word *erotic*. This love style is characterized by intense emotional attachment and powerful sexual feelings or desires. When erotic couples establish sustained relationships, these are characterized by continued emotionally intense sexual interest. A sample question on the Love Attitudes Scale (LAS) designed to measure eros asks respondents to agree or disagree with the following: "My partner and I have the right chemistry between us" (Hendrick, Hendrick, and Dicke 1998).

2. **Storge** (STOR-gay) is an affectionate, companionate style of loving. This love style focuses on deepening mutual commitment, respect, friendship over time, and common goals.

Storgic lovers' basic attitudes to their partners are one of familiarity: "I've known you a long time, seen you in many moods" (Lee 1973, p. 87). Storgic lovers are likely to agree that "I always expect to be friends with the one I love" (Hendrick, Hendrick, and Dicke 1998).

3. **Pragma** (PRAG-mah) is the root word for *pragmatic*. Pragmatic love emphasizes the practical element in human relationships and rational assessment of a potential partner's assets and liabilities. Arranged marriages are often examples of pragma. So is a person who decides very rationally to get married to a suitable partner. The following is one LAS statement that measures pragma: "A main consideration in choosing a partner is/was how he/she would reflect on my family" (Hendrick, Hendrick, and Dicke 1998).

4. **Agape** (ah-GAH-pay) is a Greek word meaning "love feast." Agape

---

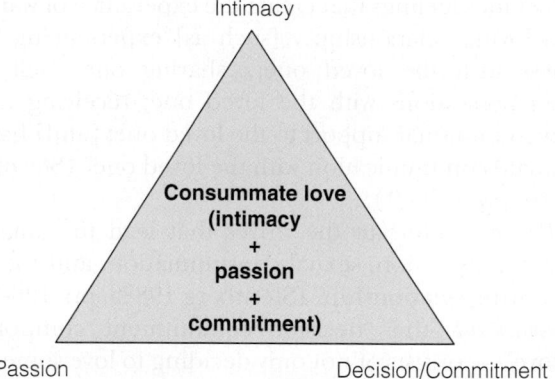

Figure 6.1 The three components of love: triangular theory.

Source: Adapted from "Triangular Love," by Robert J. Sternberg (1988a), Figure 6.1, p. 121. In Robert J. Steinberg and Michael L. Barnes (eds.), *The Psychology of Love.* Copyright © 1988 Yale University Press. Adapted by permission.

(Kluger 2004) and to the long-term maintenance of the relationship (Love 2001). Intimacy, which includes conveying and understanding each other's needs, listening to and supporting each other, and sharing common values, becomes increasingly important as time goes on. In fact, psychologist and marriage counselor Gary

Smalley (2000) argues that a couple is typically together for about six years before the two feel safe enough to share their deepest relational needs with one another.

Commitment is essential; however, commitment without intimacy and some level of passion is hollow. In other words, all these elements of love are important. Because these components not only develop at different rates but also exist in various combinations of intensity, a relationship is always changing, if only subtly (Sternberg 1988b). In addition to Sternberg's model, attachment-theory scholars (see Chapter 2) offer insight into loving relationships.

Attachment Theory and Loving Relationships

Applying attachment theory to loving relationships, we can presume that people with more secure attachment styles would have less ambivalence about emotional closeness and commitment. "Secure attachment, in part, depends on regarding the relationship partner as being available in times of need and as trustworthy" (Kurdek 2006, p. 510). In Figure 6.2, attachment style is incorporated in *beliefs and attitudes about the partner or the relationship.*

We might therefore conclude that those with a *secure* attachment style are better prospects for a committed relationship (Rauer and Volling 2007). An *insecure/anxious*

emphasizes unselfish concern for a beloved's needs even when that requires personal sacrifice. Often called *altruistic love*, agape emphasizes nurturing others with little conscious desire for a return other than the intrinsic satisfaction of having loved and cared for someone else. Agapic lovers would likely agree that, "I try to always help my partner through difficult times" (Hendrick, Hendrick, and Dicke 1998).

5. **Ludus** (LEWD-us) focuses on love as play or fun. Ludus emphasizes the recreational aspects of sexuality and enjoying many sexual partners rather than searching for one serious relationship. Of course, ludic flirtation and playful sexuality may be part of a more committed relationship based on one of the other love styles. LAS questions designed to measure ludus include the following: "I enjoy playing the game of love with a number of different partners" (Hendrick, Hendrick, and Dicke 1998).

6. **Mania**, a Greek word, designates a wild or violent mental disorder, an obsession, or a craze. Mania involves strong sexual attraction and emotional intensity, as does eros. However, mania differs from eros in that manic partners are extremely jealous and moody, and their need for attention and affection is insatiable. Manic lovers alternate between euphoria and depression. The slightest lack of response from a love partner causes anxiety and resentment. Manic lovers would be likely to say, "When my partner doesn't pay attention to me, I feel sick all over" or "I cannot relax if I feel my partner is with someone else" (Hendrick, Hendrick, and Dicke 1998). Because one of its principal characteristics is extreme jealousy, we may learn of manic love in the news when a relationship ends violently. Of Lee's six love styles, mania least fits our definition of love, described earlier.

How do these love styles influence relationship satisfaction and continuity? Psychologists Marilyn Montgomery and Gwendolyn Sorell (1997) administered the LAS to 250 single college students and married adults of all ages. They found that eros can last throughout marriage and is related to high satisfaction. Agape is also positively associated with relationship satisfaction (Neimark 2003). Interestingly, Montgomery and Sorell found storge to be important only in marriages with children. Ludus did not necessarily diminish relationship satisfaction among those who are mutually uncommitted. However, ludic attitudes have been empirically associated with diminished long-term relationship and marital satisfaction (Le 2005; Montgomery and Sorell 1997).

attachment style entails "fear of abandonment" with consequent possible negative behaviors such as unwarranted jealousy or attempts to control one's partner. An *avoidant* attachment style leads one to pass up or shun closeness and intimacy either by evading relationships altogether or demonstrating ambivalence, seeming preoccupied, or otherwise distancing oneself (Benoit and Parker 1994; Fletcher 2002; Hazen and Shaver 1994).

The attachment style of one's partner can either magnify or lessen the effects of one's own attachment style. For example, if both individuals are insecure and anxious, the relationship will be characterized that way as well. On the other hand, a person with an insecure attachment style who is in a committed relationship with someone having a secure attachment style may gradually learn to feel more secure (Banse 2004). Individual or relationship therapy may help people change their attachment style.

Attachment theory and Sternberg's triangular theory of love are not the only ways of looking at love, of course. "Facts about Families: Six Love Styles" analyzes love in yet a third way. Still another way to better understand love is to think about what love is *not*. We turn now to an examination of three things love isn't: martyring, manipulating, and limerence.

Three Things Love Isn't

Love is not inordinate self-sacrifice. And, loving is not the continual attempt to get others to feel or do what we want them to—although each of these ideas is frequently mistaken for love. Nor is love all those crazy feelings you get when you can't get someone out of your mind. We'll examine these misconceptions in some detail.

Martyring Martyring involves maintaining relationships by consistently minimizing one's own needs while trying to satisfy those of one's partner. Periods of self-sacrifice are necessary through difficult times. However, as a premise of a relationship, excessive self-sacrifice or martyring is unworkable. Martyrs may have good intentions, believing that love involves doing unselfishly for others without voicing their own needs in return. Consequently, however, martyrs seldom feel that they receive genuine affection. A martyr's reluctance to express his or her needs is damaging to a relationship, for it prevents openness and intimacy.

Manipulating Manipulators follow this maxim: If I can get her [or him] to do what I want done, then I'll be sure she [or he] loves me. **Manipulating** means seeking to control the feelings, attitudes, and behavior of your partner or partners in underhanded ways rather than by directly (not abusively!) stating your case. When not getting their way, manipulators are likely to find fault with a partner, sometimes with verbal abuse. "You don't really love me," they may accuse. Manipulating, like martyring, can destroy a relationship.

Limerence Have you ever been so taken with someone that you couldn't get him or her out of your mind? Although the object of your attention may be unaware of your feelings, you review every detail of the last time you saw him or her and fantasize about how you might actually develop a relationship. Psychologist Dorothy Tennov (1999 [1979]) named this situation limerence (LIM-er-ence). She makes the following points: First, limerence is not just "lust" or sexual attraction. People in limerence fantasize about being with the limerent object in all kinds of situations— not just sexual ones. Second, many of us have experienced limerence. Third, limerence can possibly turn into genuine love, but more often than not, it doesn't.

People discover love; they don't simply find it. The term *discovering* implies a process—developing and maintaining a loving relationship require seeing the relationship as valuable, committing to mutual needs, satisfaction, and self-disclosure, engaging in supportive communication,

and spending time together. We now turn to factors that affect how that love plays into the selection of a life partner.

Mate Selection and Relationship Stability

Designed to apply to same-sex unions as well as to heterosexual marriages, Figure 6.2 depicts a model of factors that affect relationship stability—whether partners remain together over time (Kurdek 2006). Relationship stability, happiness, and satisfaction depend upon how the partners interact with each other as well as on the perceived social support the couple receives from family members, friends, and the community in general. How partners interact with each other, in turn, depends upon a person's ideas about the partner and the relationship. These beliefs and attitudes depend, in turn at least partly, upon the personality traits that each partner brings to the union (Kurdek 2006, p. 510).

As an example, let's say that Fran and Maria are considering marriage. Each wonders about the odds of

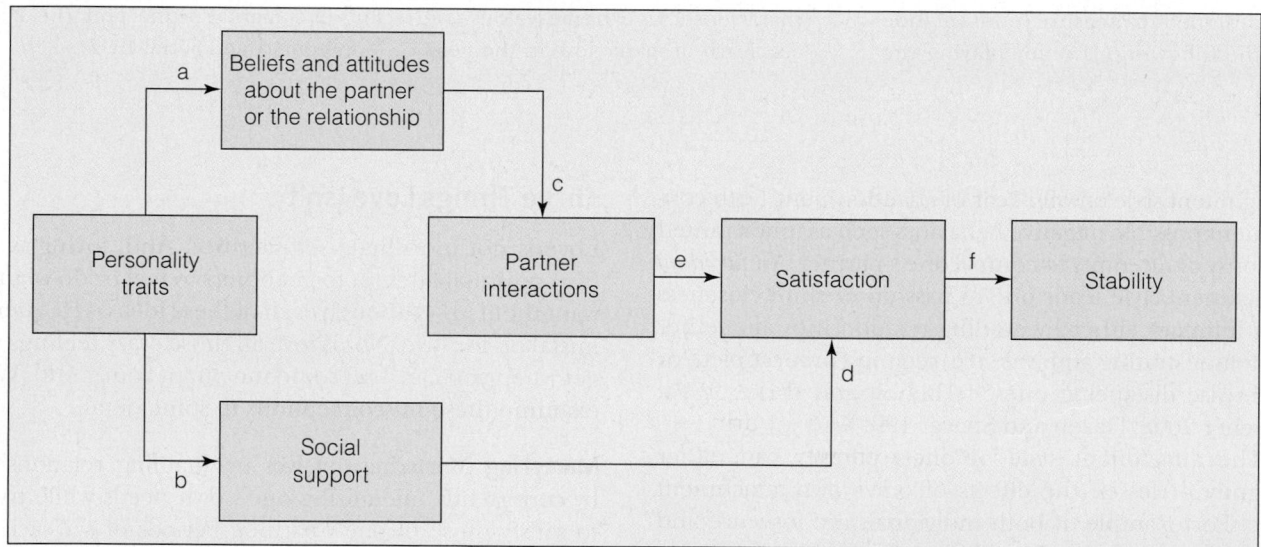

Figure 6.2 A time-ordered sequential model of relationship outcomes.

"The model has six components that form a time-ordered sequence of six linkages (letters a through f). The [personality traits] component refers to personality traits partners bring to their relationships that affect both the manner in which the relationship events are appraised (...[beliefs and attitudes about the relationship]; Link a) and the quality of perceived social support that is received (Link b)....The [beliefs and attitudes about the partner or the relationship] component refers to beliefs and attitudes about the partner or the relationship that affect how partners interact with each other (Link c)....

The component social support underscores the view that intimate relationships coexist with other personal relationships, particularly those involving friends and family members....The partner interactions component represents how partners behave toward one another and [along with social support, Link d] forms another basis for overall satisfaction with the relationship (Link e)....Relationship satisfaction refers to the overall level of positive affect experienced in the relationship and the extent to which important personal needs are being met in the relationship and is one determinant of relationship stability (Link f)" (adapted from Kurdek 2006, pp.510–11).

staying married (**marital stability**). First, Maria and Fran need to take their personality traits into account. Is Fran thoughtful, dependable, reliable, and honest? Is Maria? Is one or both prepared to support a family? What beliefs and attitudes do they have about each other and about their relationship? Does Fran believe that marriage to Maria is likely to result in marital satisfaction? Or is Fran marrying Maria despite misgivings about her or the relationship? Does Maria believe that marital stability is likely for her? Or does she see this marriage as likely to end in divorce but "worth a try" anyway? Positive attitudes about the relationship, coupled with realistically positive assessments of a prospective spouse's personality traits, are important to marital stability.

According to this model, Maria and Fran's respective personality traits influence the degree of social support that they will receive—and believe that they receive—from family members and friends. More perceived social support will result in greater marital satisfaction. Then too, Fran and Maria's beliefs and attitudes about each other and about their relationship will affect how they interact with each other. Will they—do they now—interact primarily in supportive ways? Do they handle conflict well? Supportive interaction results in greater marital satisfaction. Greater marital satisfaction, in turn, results in the greater likelihood of marital stability.

Other chapters in this text focus on various aspects of partner interactions and social support. Here we focus on choosing a partner who is best predisposed psychologically to maintain a stable and committed relationship.

Psychologists and counselors advise choosing a partner who is integrated into society by means of school, employment, a network of friends, and who fairly consistently demonstrates supportive communication and problem-solving skills (Cotton, Burton, and Rushing 2003). We're reminded that, "Heavy or risky drinking is associated with a host of marital difficulties including infidelity, divorce, violence and conflict" (L. Roberts 2005, p. F13). The same can be said for other forms of substance abuse (Kaye 2005, F15). Furthermore, research shows that relationships are more likely to be stable when partners' parents have not been divorced.

The Intergenerational Transmission of Divorce Risk

Either because they know the statistics, have divorced friends, or have experienced their parents' divorce, many young adults are cautious about getting married and possibly going through the pain and economic upheaval of divorce, especially if they plan to have children (Arnett 2004; Wallerstein 2008). It is important to remember that assuredly not all children of divorced parents will themselves divorce. No one is suggesting that a child of divorced parents be automatically rejected as a future spouse, and not all researchers agree that there is an intergenerational transmission of divorce risk (Li and Wu 2008). However, "[s]tudies based on large national samples consistently show that parental divorce increases the risk of marital instability in offspring" (Hetherington 2003, p. 325; Teachman 2004). Cross-national research has also found this to be true in industrialized countries besides the United States (Dronkers and Harkonen 2008). When both spouses come from divorced families, the probability of their own divorce is still higher (Amato and DeBoer 2001; Wolfinger 2005).

Family scholars refer to this phenomenon as the **intergenerational transmission of divorce risk**: A divorced parental family transmits to its children a heightened risk of getting divorced. Noting that "apparently, there is something in the divorce experience beyond that of parental conflict that exacerbates problems in stability in intimate relations in offspring" (Hetherington 2003, p. 326), researchers have suggested the following four hypotheses to explain the intergenerational transmission of divorce risk. Children of divorce are themselves more likely to get divorced because they have:

1. more—and more serious—personality problems
2. neither been exposed to nor learned supportive communication or problem-solving skills
3. less commitment to the relationship
4. more accepting attitudes toward divorce (Dunne, Hudgins, and Babcock 2000; Hetherington 2003; Hetherington and Kelly 2002; Wolfinger 2005)

We might also hypothesize that children of divorce, as a category, are less likely to have developed a secure attachment style, discussed previously.

After what we've said here, it is important to emphasize that children of divorce will themselves not *necessarily* divorce (Zimmerman and Thayer 2003). Research shows that a supportive, well-adjusted partner "can play a protective role" in minimizing the intergenerational transmission of divorce risk. Prominent researcher on children of divorce, E. Mavis Hetherington (2003) has described her findings on this point:

> Under conditions of low stress with a supportive partner, there was no difference in couple instability between the offspring of divorced and nondivorced parents. For these well-married youths in a benign environment, no intergenerational transmission of marital instability was found. Under conditions of high stress, there was a marginally significant trend for the offspring of divorced parents, even with a supportive partner, to show somewhat more marital instability than those from nondivorced families. (p. 328)

Minimizing Mate Selection Risk

Mate selection plays a part in the intergenerational transmission of divorce risk because individuals from divorced families are themselves more inclined to have the characteristics described earlier and to choose partners who have them. Hetherington (2003) reports her findings concerning **mate selection risk** as follows:

> Youths from divorced families were more likely to select high-risk partners who were also from divorced families and who were impulsive, socially irresponsible, and had a history of antisocial behaviors such as alcohol and drug abuse, minor misdemeanors, troubles with the law, problems in school and at work, fighting, and an unstable job history. (p. 328)

Other research has found that *mate selection risk* may apply to adult children of alcoholics as well as to those of divorce (Olmsted, Crowell, and Waters 2003). What can one do to minimize mate selection risk?

A first step in minimizing mate selection risk is to let go of misconceptions we might have about love and choosing a partner. Selecting a partner wisely involves balancing any insistence on perfection against the need to be mindful of one's real needs and desires. In the absence of adequate role models for maintaining a supportive relationship, many of us may embrace misconceptions about finding a partner. For instance, we might believe that "I can be happy with anyone I choose, if I work hard enough," or that "Falling in love with someone is sufficient" (Cobb, Larson, and Watson 2003, p. 223). However, working things out requires both partners' willingness and ability to do so; just one's own willingness to work hard at a marriage is not enough. Furthermore, if having fallen in love is assumed to be enough to make a union last, then other, potentially detrimental partner characteristics may be minimized. Generally, long-term relationships built on "respect, mutual support and affirmation of each other's worth are more likely to survive" (Hetherington 2003, p. 322).

Later in this chapter, the section "Some Things to Talk About" gives other ideas on assessing your own and a prospective partner's values and attitudes. Of course, it's important to be truthful when relating with a potential partner, as well as to ascertain how truthful a partner is being (S. Campbell 2004). Some couples go to counseling to assess their future compatibility and commitment (Marech 2004). Others may access marital compatibility tests on the Internet. Although we, your authors, are unable to attest to the efficacy of these, they do stimulate couple discussions about important topics. At this point, we turn to the social science analogy of choosing a mate in a marketplace.

The Marriage Market

Imagine a large marketplace in which people come with goods to exchange for other items. In nonindustrialized societies, a person may go to market with a few chickens to trade for some vegetables. In more industrialized societies, people attend hockey equipment swaps, for example, trading outgrown skates for larger ones. People choose partners in much the same way: They enter the *market* (traditionally called the **marriage market**) armed with resources—personal and social characteristics—and then they bargain for the best "buy" that they can get.

Arranged and Free-Choice Marriages

In much of the world, particularly in parts of Asia and Africa that are less Westernized, parents have traditionally arranged their children's marriages. In **arranged marriage**, future spouses can be brought together in various ways. For example, in India, parents typically check prospective partners' astrological charts to assure future compatibility. Traditionally, the parents of both prospective partners (often with other relatives' or a paid matchmaker's help) worked out the details and then announced the upcoming marriage to their children. The children may have had little or no say in the matter, and they may not have met their future spouse until the wedding. However, today it is more common for the children to marry only when they themselves accept their parents' choice. Unions like these, sometimes called "assisted marriages," can be found among some Muslim groups and other recent immigrants to the United States (Ingoldsby 2006b; MacFarquhar 2006). Research shows that—at least at first—couples who have had more input report greater marital satisfaction (Madathil and Benshoff 2008).

Arranged marriage is still practiced, but the custom is waning (Zang 2008). Arranged marriage was observed throughout most of the world into the twentieth century—and well into the eighteenth century in Western Europe. As pointed out in "My Family: An Asian Indian American Student's Essay on Arranged Marriages," we can still find arranged marriages in many parts of the less industrialized world today. The majority of young couples in cultures that have traditionally practiced arranged marriage continue to heed extended family members' opinions about a prospective mate (Zhang and Kline 2009).

The fact that marriages are arranged doesn't mean that love is ignored by parents. Indeed, marital love may be highly valued. However, couples in arranged marriages are expected to develop a loving relationship *after* the marriage, not before (Tepperman and Wilson 1993). A study that compared marital satisfaction among arranged marriages in India to those more

My Family

An Asian Indian American Student's Essay on Arranged Marriages

The college student who wrote the following essay is the daughter of Asian Indian immigrants to California. Her essay points out that some people in the United States experience arranged marriages today and that arranged marriage is changing.

My parents had an arranged marriage twenty-three years ago and are still married to this day. Their wedding day was the first time they saw each other, and even then it was just a quick glance. A mutual friend of both families mentioned…that their kids would be a good match. So my mom's parents…interviewed my dad. My grandparents took into consideration various things like my dad's height, weight, education, family background, age, income, house, and health….

My dad's parents did the same thing to my mom. They went to see…whether they liked her. One of their main concerns was what their grandchildren would look like. So they wanted a tall, pretty, healthy wife for their son. Both sets of parents agreed to the other's child, and the deal was made. Both sets of my grandparents went home to tell their son and daughter that they were getting married. There were no questions asked by my parents. Out of respect for their parents, they agreed to be married. Actually, this is incorrect, because they didn't really agree; they couldn't have agreed [because] they were never asked. Basically they weren't given an option. However, neither of my parents argued; they just went along with their parents' wishes.

When I asked my parents if they were disappointed when they first saw each other on their wedding day, they said no. They did not fall in love at first sight either, but they thought their parents did a good job in finding them a spouse…. Neither of my parents regret[s] marrying each other even though they really didn't have much of a choice. They are happy with their lives, their children, and each other.

Arranged marriages in the East Indian culture are a lot different today than they were [twenty-three years ago]. I have friends my own age who have had arranged marriages. Now they get the opportunity to actually meet the person whom their parents have in mind for them. Better yet, they get to decide whether or not they want to marry that person.

A friend of mine just went to India with her parents to get married. Once she got there she discovered that her family members had already picked out about fifteen different men from whom she could choose. She interviewed all of them with her parents. She crossed off the names of the ones she wasn't interested in and had a second interview with the remaining men on her list. With three men left on her list after the second interview, her parents decided that she could meet these men alone without any parent chaperons….

She ended up liking one of them. She said she was very attracted to him. He was a dentist, tall, had a nice body, very polite, and treated her like a princess. She married this guy and brought him over to California. She couldn't be happier. She told me that she doesn't think that she would have ever been able to find such a wonderful husband without the help of her family. She is very grateful to them.

As for my two brothers and me, my parents have not necessarily expected us to have an arranged marriage. But before we do get married they want us to possess the best qualities, mainly a high education, good manners, respect for others, and high self-esteem, which will enable us to choose a partner of equal qualities….I did not have an arranged marriage. I married my husband because I truly loved him. However, before I made the decision to get married, I did try to make sure that we would be compatible and have similar goals.

I sometimes wonder what it would be like to have an arranged marriage. What would the wedding night be like? How could two people who don't really know each other or love each other make love? I did not feel comfortable enough to ask my parents or my friend about this. But it is something that I wonder about.

Critical Thinking

What are some advantages of arranged marriage? Some disadvantages? In what ways might personal choice be involved in arranged marriages today?

---

freely chosen in the United States found no differences in marital satisfaction between the two groups. According to the authors, "Although this is not a case in favor of arranged marriages, it provides no support for a position opposing this tradition" (Myers, Madathil, and Tingle 2005, p. 189). Meanwhile, with global Westernization, arranged marriages are less and less common, especially among those with higher education (D. Jones 2006; Hoelter, Axinn, and Ghimire 2004).

The United States is an example of what cross-cultural researchers call a **free-choice culture**: People choose their own mates, although often they seek parents' and other family members' support for their decision. Immigrants who come to the United States from more collectivist cultures, in which arranged marriages have been the tradition, face the situation of living with a divergent set of expectations for selecting a mate. Some immigrant parents from India, Pakistan, and other countries arrange for spouses from their home country to marry their offspring. Either the future spouse comes to the United States to marry the young person, or the young person travels to the home country for a wedding ceremony, after which the newlyweds usually live in the United States (Dugger 1998). In this case, the marriage is typically characterized

Although the arranged marriage of this couple in Northern India (right) may seem to be a world apart from the more freely chosen marriage of this couple in the United States (left), bargaining has occurred in both of these unions. In arranged marriages, families and community do the bargaining, based on assets such as status, possessions, and dowry. In freely chosen marriages, the individuals perform a more subtle form of bargaining, weighing the costs and benefits of personal characteristics, economic status, and education.

by the greater Westernization of one partner (the young person who has lived in the United States) and the spouse's simultaneous need to adjust not only to marriage but also to an entirely new culture.

The incidence of **cross-national marriages** like these may decline as the required visa for an immigrating spouse is far harder to come by now than prior to September 11, 2001. Furthermore, American-born children of immigrants may see themselves as too Americanized for this approach to finding a partner (MacFarquhar 2006). Whether unions are arranged or not, we can think of choosing a marital partner as taking place in a market.

Regarding arranged marriages, parents go through a bargaining process not unlike what takes place at a traditional village market. They make rationally calculated choices after determining the social status or position, health, temperament, and, sometimes, physical attractiveness of their prospective son- or daughter-in-law. Professional matchmakers often serve as investigators and go-betweens, just as we might engage an attorney or a stockbroker in an important business deal.[1]

With arranged marriage, the bargaining is obvious. The difference between arranged marriages and marriages in free-choice cultures may seem so great that we are inclined to overlook an important similarity: *Both*

[1] Sometimes, as in the Hmong culture, the exchange involves a *bride price*, money or property that the future groom pays the future bride's family so that he can marry her. More often, the exchange is accompanied by a *dowry*, a sum of money or property the female brings to the marriage. As one example, Asian Indians have traditionally practiced the dowry system, now illegal there but still widespread (Self and Grabowski 2009; Srinivasan and Lee 2004). A woman with a large dowry can expect to marry into a higher-ranking family than can a woman with a small dowry, and dowries are often increased to make up for qualities considered undesirable (M. Kaplan 1985, pp. 1–13). For instance, parents in eighteenth-century England increased the dowries of daughters who were pockmarked.

The man on the decorated horse is an investment banker from New York who has traveled to Jaipur, India, to marry a native Asian Indian woman. The marriage had been arranged in India by the groom's mother. The couple will return to New York to live. "I always knew I'd probably end up getting married to someone who wasn't very American, because I'm not myself in some ways," he said.

involve bargaining. What has changed in free-choice societies is that individuals, not family members, do the bargaining.

Social Exchange

The ideas of bargaining, market, and resources used to describe relationships come to us from exchange theory, discussed in Chapter 2. A basic idea of exchange theory is that whether relationships form or continue depends on the rewards and costs they provide to the partners. Individuals, it is presumed, want to maximize their rewards and avoid costs, so when they have choices, they will pick the relationship that is most rewarding or least costly.

This analogy is to economics, but in relationships, individuals are thought to have other sorts of resources to bargain besides money: physical attractiveness, intelligence, educational attainment, earning potential, personality characteristics, family status, the ability to be emotionally supportive, and so on. Individuals may also have costly attributes, such as belonging to the "wrong" social class, religion, or racial/ethnic group, being irritable or demanding, and being geographically

inaccessible (a major consideration in modern society). The increasing number of single people with children in today's market may find parenthood to be a costly attribute (Goldscheider, Kaufman, and Sassler 2009).

The Traditional Exchange Historically, women have traded their ability to bear and raise children, coupled with domestic duties, sexual accessibility, and physical attractiveness, for a man's protection, status, and economic support. This traditional exchange characterizes marriages in many immigrant groups that have recently arrived in the United States (Hill, Ramirez, and Dumka 2003). Evidence from classified personal ads shows that the traditional exchange still influences heterosexual relationships in general. Men are more likely to advertise for a physically attractive woman; women, for an economically stable man. Although women increasingly have their own employment and income, women continue to expect greater financial success as a category from prospective husbands than vice versa (Buss et al. 2001; Fitzpatrick, Sharp, and Reifman 2009). National data that looked at black, Hispanic, and white males found that the probability of a man's getting married largely depends on his earning power (Edin and Reed 2005; Lichter, Qian, and Mellott 2006; Schoen and Cheng 2006).

Bargaining in a Changing Society "As gender differences in work and family roles blur, individuals' criteria for an acceptable mate are likely to change" (Raley and Bratter 2004, p. 179). For instance, research that looked at heterosexual mate preferences in the United States over the past sixty years showed that men and women—but especially men—have increased the importance that they put on potential financial success in a mate, while a woman's domestic skills have declined in importance (Siegel 2004; Sweeney and Cancian 2004). One study indicates that, for today's young man, a woman's high socioeconomic status increases her sexiness (Martin 2005).

As gender roles become more alike, exchange between partners may increasingly include "expressive, affective, sexual, and companionship resources" for both partners. In fact, a high-earning woman might bargain for a nurturing, housework-sharing husband, even if his earning potential appears to be lower than hers (Press 2004; Sprecher and Toro-Morn 2002). As college-educated young women approach occupational and economic equality with potential mates, the exchange becomes more symmetrical than in the past, with both genders increasingly looking for physical attractiveness, emotional sensitivity, and earning potential in one another (Buss et al. 2001; Montoya 2008). Marriages based on both partners' contributing roughly equal economic and status resources are more egalitarian. Changes in men's roles toward greater emotional expressiveness may improve relationship communication and satisfaction.

Desiring wives who can make good money, college-educated men are now much more likely to marry college-educated women than a few decades ago. In fact, today *both* men and women are likely to want a spouse with more education or who earns more than they do (King and Allen 2009; Raley and Bratter 2004).[2] The fact that college-educated women and men tend to marry one another points to another concept associated with mate selection—assortative mating.

Assortative Mating—A Filtering Process Individuals gradually filter, or sort out, those who they think would not make the best life partner or spouse. Research has consistently shown that people are willing to date a wider range of individuals than they would live with or become engaged to, and they are willing to live with a wider range of people than they would marry (Jepsen and Jepsen 2002). For instance, one study has found that women are less likely to consider the economic prospects of their male partners when deciding whether to cohabit than when deciding about marriage (Manning and Smock 2002). Social psychologists call this process **assortative mating** (or, sometimes, *assortive mating*). Assortative mating raises another factor shaping partner choice—the tendency of people to form committed, and especially marital, relationships with others with whom they share certain social characteristics. Social scientists term this phenomenon **homogamy**.

Homogamy: Narrowing the Pool of Eligibles

Individuals tend to make relationship choices in socially patterned ways, viewing only certain others as potentially suitable. The market analogy would be to choose only certain stores or websites at which to shop. Each shopper has a socially defined **pool of eligibles**: a group of individuals who, by virtue of background or birth, are considered most likely to make compatible mates.

Americans tend to choose partners who are like themselves in many ways. People tend to form committed relationships with people of similar race, age, education, religious background, and social class (Kossinets and Watts 2009). As an example, the Protestant, Catholic, and Jewish religions, as well as the Muslim and Hindu religions, have all traditionally encouraged **endogamy**: marrying within one's own social group.

The opposite of endogamy is **exogamy**, marrying outside one's group, or **heterogamy**—that is, choosing someone dissimilar in race, age, education, religion, or social class. For example, age and educational heterogamy have been more pronounced among black individuals than among whites, partly because an "undersupply" of educated black men prompts black women to partner with considerably older or younger men with less education (Surra 1990). Overall, however, with "the loosening of relationship conventions," more older women are dating or marrying men at least five years younger—the media-hyped "cougar" phenomenon (Kershaw 2009b). This "loosening of relationship conventions" evidences itself in other types of heterogamy as well.

As young adults experience increased independence from family influence, we can expect a rise in interracial and interethnic unions (Rosenfeld 2008). In spite of a trend toward less religious homogamy and a lessened tendency of Asian, Hispanic, and European Americans—such as Irish, Italians, or Poles—to marry within their own ethnic groups, homogamy is still a strong force (Fu and Heaton 2008).[3] About 7.5 percent of U.S. marriages involve spouses of different races or a Hispanic married to a non-Hispanic (U.S. Census Bureau 2010b, Table 60). The growth in race/ethnic intermarriage *rates* for Asians and Hispanics has declined some since the 1990s (although their numbers continued to rise). More than 90 percent of non-Hispanic and of black couples are racially homogeneous. About 60 percent of Asian Americans and 75 percent of Hispanics marry within their group (Lee and Edmonston 2005). However, nearly 54 percent of Native Americans marry outside their race, and more than 80 percent of Arab Americans marry outside their ethnicity (Kulczycki and Lobo 2002). Interracial and interethnic marriage and cohabitation rates involving African Americans have continued to increase significantly (Qian and Lichter 2007).

With regard to socioeconomic class and education, although people today are marrying across small class distinctions, they still are not doing so across large ones. For instance, individuals of established wealth or high

---

[2] As a result of this situation, finding an acceptable spouse could prove problematic for both women and men. "If both sexes are looking to marry ['up,' or] hypergamously, there is a mismatch between the preferences of men and women" (Raley and Bratter 2004, p. 168).

[3] "It often comes as a surprise to whites born after 1980 that crossing the ethnic boundaries to date or marry had social consequences in recent American ethnic history. The dating and marriage of, for example, an Italian American and an Irish American [in the first half of the twentieth century] not only raised eyebrows in each community but often brought disappointment and even estrangement from family members" (C. Gallagher 2006, p. 143).

These Hmong immigrants in St. Paul, Minnesota, are celebrating the Hmong New Year, which also serves as a courting ritual. As in Laos, teenagers line up—boys on one side, girls on the other—and play catch with desirable potential mates. Catching the ball begins conversation. Tossing the ball gives girls a chance to meet boys under conditions approved by their parents. In Minnesota, however, the traditional Laotian black cloth ball is often replaced with a fluorescent tennis ball (Hopfensperger 1990, p. 1B). Because virtually all participants are Hmong, the ritual helps to ensure racial/ethnic homogamy.

education levels seldom marry those who are poor or who have low educational achievement (Fu and Heaton 2008). We know relatively little about religious, racial, or socioeconomic homogamy in cohabiting relationships, but we would presume—given the filtering process of assortative mating—that cohabiting couples exhibit less homogamy than married couples. All in all, at least with regard to marriage, an individual is likely to choose someone who is similar in basic social characteristics. We'll look at a hypothetical case to see why this is so.

Reasons for Homogamy

Andrea is attracted to Alex (and vice versa), who is a college student like herself. Andrea's parents are upper-middle class. They live in the expensive section of her hometown, have a housekeeper, and frequently have parties by their pool. Catholic, they go to Mass every Sunday. Alex's parents are working class. They are separated. His mother lives in an apartment and works as a checker in a supermarket. The family believes in "being good people," but do not belong to any organized religion.

How likely is it that Andrea and Alex will marry? If they do marry, what sources of conflict might occur? We can begin to answer these questions by exploring

four related elements that influence both initial attraction and long-term happiness. For one thing, people often find it easier to communicate and feel more at home with others from similar education, social class, and racial or ethnic backgrounds (Lewin 2005). Alex is likely to have attitudes, mannerisms, and vocabulary different from those of Andrea. Each may feel out of place in surroundings that the other considers natural. Two other factors—geographic availability and social pressure—are important reasons that many relationships are generally homogamous.

Geographic Availability Geographic availability (traditionally known in the marriage and family literature as *propinquity* or *proximity*) has historically been a reason that people meet others who are like themselves (Harmanci 2006; Travis 2006). For instance, as the size of various immigrant communities in the United States grows, the geographic availability of eligibles in the same ethnicity increases, resulting in ethnic homogamy (Gowan 2009; Qian and Lichter 2007). Geographic segregation, which can result from either discrimination or strong community ties, contributes to homogamous marriages (C. Gallagher 2006; Iceland and Nelson 2008; Lichter et al. 2007). Intermarriage patterns within the American Jewish community are an example. Only about 6 percent of Jews married non-Jews in the late 1950s. Now that the barriers that used to exclude Jews from certain residential areas and colleges are gone, about half marry gentiles (Sussman 2006).

Geographic availability also helps to account for educational and social class homogamy. Middle-class people tend to socialize together and send their children to the same schools; upper- and lower-class people do the same. Unless they had met in a large, public university or online, it is unlikely that Alex and Andrea would have become acquainted at all.

Today, first encounters may occur in cyberspace, and people meet others as far away as other continents. Websites such as InterRacialMatcher.com facilitate heterogamy. However, the Internet may actually encourage endogamy among religious or racial/ethnic groups, who can advertise online for homogamous dating partners (e.g., allhispanicdating.com; asianromance.com; blacksingles.com; christian.com; jewishconnect.com; meetmuslimsingles.com; see also Desmond-Harris 2010). Illustrating these points, Russel K. Robinson, a black gay man writing in the *Fordham Law Review*, describes his experience as follows:

> Although I lived on the wealthy, predominantly white west side of [Los Angeles], the Internet created opportunities for me to interact with men in [less wealthy areas of the city]—men I almost certainly would not meet randomly while going through my daily routine.....Even as the Internet increases romantic opportunity, it also channels interactions....Like many dating websites, Match.com

prompts the user to indicate which races he will and will not date....[I]f a white user is interested only in white romantic partners, he can easily structure his screen so that he never even has to view nonwhite profiles. (Robinson 2008, 2791–92)

The Internet also allows us to meet people with similar values and attitudes if not similar in socioeconomic characteristics, such as vegetarians or humanitarian social activists (K. Baker 2008). Then too, online daters' mutual ability to access the Internet and then to travel, if necessary, to meet each other face-to-face assures some degree of educational and/or financial homogamy. We know of no research on the actual effect of the Internet on marital homogamy, and we'd like to suggest that this question would be a good one for future research—perhaps yours.

Social Pressure A second reason for homogamy is social pressure. Interethnic relationships are more likely to develop when young adults are relatively independent of parental influence and/or when one's parents have an ethnically diverse network of friends (Rosenfeld and Byung-Soo 2005). Meanwhile, for the majority of us, cultural values encourage marrying someone who is socially similar to ourselves. Andrea's parents, friends, and siblings may not approve of Alex because he doesn't exhibit the social skills and behavior of their social class. Meanwhile, Alex's mother and friends may say to him, "Andrea thinks she is too good for us. Find a girl more like our kind."

Sometimes, social pressure results from a group's concern for preserving its ethnic or cultural identity. Arab, Asian, or Hispanic immigrants may pressure their children to marry within their own ethnic group to preserve the culture (Kitano and Daniels 1995; S. M. Lee 1998). Whether blatant or subtle, social pressure toward homogamy can be forceful. Making knowledgeable choices involves recognizing the strength of social pressure and deciding whether to act in accordance with others' expectations.

We've discussed some reasons for homogamy, but not all marriages are homogamous, of course. Heterogamy refers to marriage between those who are different in race, age, education, religious background, or social class.

Examples of Heterogamy

How does marrying someone from a different religion, social class, or race/ethnicity affect a person's chances for a happy union? In general, marriages that are homogamous are more likely to be stable because partners are more likely to share the same values and attitudes when they come from similar backgrounds (Furstenberg 2005; Gaunt 2006). As an example, Islamic marriage counselor Aneesah Nadir suggests that, in a homogamous Muslim marriage, both spouses—not just one—are likely to find value in referring to the Quran for answers to disagreements (Nadir 2009).

In this section, however, we examine the relationship between heterogamy and marital success. We will discuss interfaith marriages, and then look at interracial and interethnic unions. First, however, it is worth noting that partners can experience cultural differences even when they share a religious, racial, or ethnic category. For instance, Muslims (as well as those of other religions) may not only practice their religion to varying degrees but also be of different ethnicities (Nadir 2009). As a second example, with regard to race, the U.S. black population is itself culturally diverse because it includes black individuals whose ancestors have been in this country for generations as well as recent immigrants from Africa or the Caribbean (Kent 2007). Similarly, the category "Asian" includes individuals from a variety of nations and cultures.

Interfaith Marriages It is estimated that between 30 percent and 40 percent of Jewish, Catholic, Mormon, Muslim, and a higher percentage of Protestant adults and children in the United States live in interfaith or interdenominational households (D'Antonio et al. 1999). Being highly educated seems to lessen individuals' commitment to religious homogamy (Petersen 1994). Religions that see themselves as the one true faith and people who adhere to a religion as an integral component of their ethnic/cultural identity (for example, some Catholics, Jews, and Muslims) are more likely to encourage homogamy, sometimes by pressing a prospective spouse to convert (Bukhari 2004). Often, religious bodies are concerned that children born into the marriage will not be raised in their religion (Sussman 2006).

Some switching, no doubt, also takes place because partners agree with the widely held belief that interreligious marriages tend to be more stressful and less stable than homogamous ones—a belief supported by research (Mahoney 2005). One probable reason that religious homogamy improves chances for marital success involves value consensus. Religion-based values and attitudes may come into play when negotiating leisure activities, child-raising methods, investments and expenditures of money, and appropriate spousal roles (Curtis and Ellison 2002). Meanwhile, analysis of data from a national random telephone survey of Protestant and Catholic households concluded that although marital satisfaction was less for interdenominational couples, the difference disappeared when the interdenominational couple had generally similar religious orientations, good communication skills, and similar beliefs about child raising (Hughes and Dickson 2005; Williams and Lawler 2003).

One study conducted with homogamously married Christian, Jewish, and Islamic couples found strong religious beliefs to be associated with less couple conflict. Shared religiosity gave them a shared sense of purpose and commitment to permanence, coupled with a willingness to forgive the spouse when conflicts emerged (Lambert and Dollahite 2006). One research team has attributed the higher rate of marital happiness associated with religious homogeneity almost entirely to the positive effect of church attendance: Homogamously married partners go to church more often and at similar rates and, as a result, show higher marital satisfaction and stability (Heaton and Pratt 1990).

Some recent research shows a declining effect of religious differences on marital satisfaction over the past several decades due to the greater effect of couples' gender, work, and co-parenting concerns (Myers 2006; Williams and Lawler 2003).

Interracial/Interethnic Marriages **Interracial marriages** include unions between partners of the white, black, Asian, or Native American races with a spouse outside their own race. As defined by the U.S. Census Bureau, Hispanics are not a separate race but, rather, an ethnic group. Unions between Hispanics and others, as well as between different Asian/Pacific Islander, Hispanic, or black ethnic groups (such as Thai–Chinese, Puerto Rican–Cuban, or African American—black Caribbean) are considered **interethnic marriages**.

Interracial unions have existed in the United States throughout our history (Maillard 2008). However, not until June 1967 (*Loving v. Virginia*) did the U.S. Supreme Court declare that interracial marriages must be considered legally valid in all states. At about the same time, it became impossible to gather accurate statistics on interracial marriages. Many states no longer require race information on marriage registration forms, so these data are incomplete at best.

Available statistics show that the proportion of interracial and interethnic marriages is fairly small (between 7 and 8 percent of the adult U.S. population, or 4.5 million couples) (U.S. Census Bureau 2010b, Table 60). Although this proportion has steadily increased since 1970, when the proportion was just one percent, growth in interracial and interethnic marriage may recede somewhat in the future as immigrant groups become large enough to provide an ample pool of eligibles within their own ethnic categories (Lee and Edmonston 2005; Qian and Lichter 2007). If we count cohabiting couples, the percentage today of racially or ethnically heterogeneous couples would be somewhat higher than the statistics for married couples because (due to the assortative mating process) cohabiting couples are less homogamous than married couples (Batson, Qian, and Lichter 2006; Joyner and Kao 2005).

As shown in Figure 6.3, of all interracial marriages in 2008 (this does not count Hispanic–non-Hispanic unions), about 20.5 percent (481,000 couples) were black–white (U.S. Census Bureau 2010b, Table 60).[4] The vast majority of the remainder were combinations of whites with Asians, Native Americans, and others. Two-thirds of black-white marriages involved black men married to white women. Among Hispanic marrieds, about one-third have a spouse of non-Hispanic origin (U.S. Census Bureau 2010b, Table 60).

[4] Native-born African Americans are significantly more likely to intermarry racially than are black recent immigrants from the West Indies or Africa (Batson, Qian, and Lichter 2006).

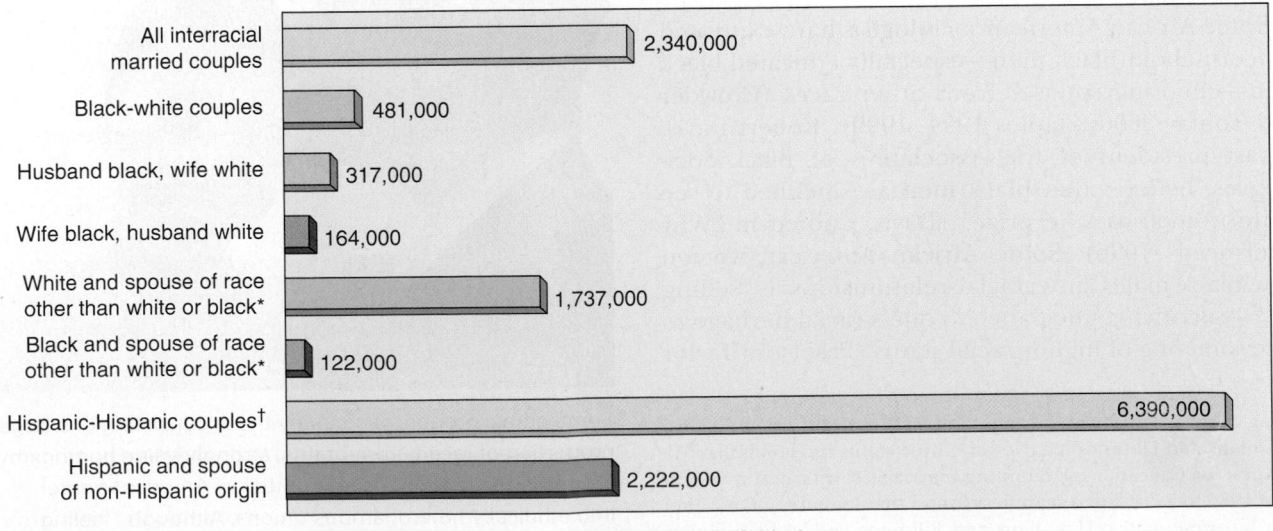

*Neither white nor black, but Asian, Native American, Aleut, Pacific Islander.

†Persons of Hispanic origin may be of any race.

Figure 6.3 Number of interracial and Hispanic–non-Hispanic married couples, 2008.

Source: U.S. Census Bureau 2010b, Table 60.

Reasons for Interracial and Interethnic Unions Much attention has been devoted to why people marry or form other romantic unions interracially. One apparent reason among racial/ethnic groups that are relatively small in number is simply that they have a smaller pool of eligibles in their own race/ethnicity and are also more likely than larger racial/ethnic groups to interact with others of different races (Qian and Lichter 2007; Robinson 2008).

Another explanation is the **status exchange hypothesis**—the argument that an individual might trade his or her socially defined superior racial/ethnic status for the economically or educationally superior status of a partner in a less-privileged racial/ethnic group (Kalmijn 1998). In this regard, racial stereotypes may play a part:

> [A] society dominated by Euro-Americans will unsurprisingly privilege a standard of beauty and cultural styles that is a mirror image of itself, even if that image is a media distortion.... [I]nterviews with Asian men and women [found] that a sizable minority of respondents preferred whites as potential or current mates because of their preference for "European" traits including tallness, round eyes, buffness for men and more ample breasts for women. (C. Gallagher 2006, p. 150)

Applying the status exchange hypothesis to black–white intermarriage would suggest marrying "up" socioeconomically on the part of a white person who, in effect, trades socially defined superior racial status for the economically superior status of a middle- or upper-middle-class black partner. Little research has been done to test this hypothesis, but a recent study of intermarriage among native Hawaiians, Japanese, Filipinos, and Caucasians in Hawaii supports the hypothesis (Fu and Heaton 2000).[5]

Some African American sociologists have expressed concern about black men—especially educated black men—choosing spouses from other races (Crowder and Tolnay 2000; Staples 1994, 1999). Robert Davis, a past president of the Association of Black Sociologists, believes that black men are inclined to see white women as "the prize" (Davis, quoted in "Why Interracial" 1996). Some African American women view black males' interracial relationships as "selling out"—sacrificing allegiance to one's racial heritage to date someone of higher racial status (Paset and Taylor

1991). In a paper on this topic, a Latina student wrote the following:

> It is not just in the African American community that this is happening....I have noticed when a Mexican American man gets educated, he usually ends up dating and marrying an Anglo female. Being with an Anglo female is more of a trophy to the Mexican man....I would like to meet an educated Mexican male and date him but there are not that many around. I am not totally set on just dating a Mexican male, but you hardly see white males dating other races. Sometimes it seems there is little hope for Hispanic and African American females to ever find a good partner. (Torres 1997)

Having said this, we note that research on interracially married couples has generally found that "with few exceptions, this group's motives for marriage do not appear to be any different from those of individuals marrying...within their own race" (Porterfield 1982, p. 23). As with homogamous couples, those in heterogamous relationships find their partners by means of social exchange in a marriage market (Fryer 2007). When asked about their motives for marrying heterogamously, the most common answers they give are love and compatibility (Porterfield 1982).

Interracial/Interethnic Heterogamy and Marital Stability

Marital success can be measured in terms of two related, but different factors: (1) stability—whether or how long the union lasts, and (2) the happiness of the partners.

Some ethnic groups, particularly those consisting of a large proportion of recent immigrants, strongly value homogamy. Nevertheless, an increasing number of Americans enter into ethnically heterogamous unions. Although "feeling at home"—a factor that encourages homogamy—may be difficult at first, some individuals thrive on the cultural variety characteristic of interracial or interethnic relationships.

---

[5] As a result of historical events and cultural definitions, native Hawaiians and Filipinos have lower ethnic status in Hawaii than do Japanese or Caucasians. Examining marriage certificates in Hawaii from 1983 to 1994, the researchers found that to marry a Caucasian or a Japanese, native Hawaiians and Filipinos had to have higher economic or educational status than those who married within their own ethnic group. At the same time, Japanese and Caucasians who married native Hawaiians or Filipinos were "of lower status in their own group" (Fu and Heaton 2000, p. 53).

Marital *stability* is not synonymous with marital *happiness* because, in some instances, unhappy spouses remain married, whereas less unhappy partners may choose to separate. Just as information on interracial/interethnic marriages is incomplete, so is information on their divorces. Only about half the states and the District of Columbia report race/ethnicity on divorce records. What evidence we have on whether interracial/interethnic marriages are more or less stable than intraracial/intraethnic unions comes from national survey data.

One study that analyzed data from the National Survey of Families and Households (NSFH) showed that partners in interethnic unions—both married and cohabiting—reported lower relationship quality than did those in same-ethnic unions (Hohmann-Marriott and Amato 2008). In general, research suggests that marriages that are homogamous in age, education, religion, and race are the most stable (Bratter and King 2008; Larson and Hickman 2004).

However, a recent study based on analysis of data from 23,139 married couples "failed to provide evidence that interracial marriage *per se* is associated with an elevated risk of marital dissolution." Instead, the risk of divorce or separation among the interracial couples sampled was similar to that of the race of the spouse from the more divorce-prone race. Accordingly, "Mixed marriages involving Blacks were the least stable followed by Hispanics, whereas mixed marriages involving Asians were even more stable than endogamous White marriages" (Zhang and Van Hook 2009, p. 104).

Additionally, these researchers found black husband–white wife pairings to be the least stable of all the marriage types they looked at:

> One plausible interpretation of these results is that they reflect persistent racism and distrust directed toward Blacks, particularly Black men in the United States. The qualitative findings from Yancey (2007) indicated that Whites who married Blacks experienced more firsthand racism as compared to Whites who married other non-Black minorities. Specifically, White women reported encountering more racial incidents with their Black husbands (e.g., inferior service, racial profiling, and racism against their children) and more hostilities from families and cohorts as compared to other interracial pairings. Research in communication and cultural studies echoed Yancey's findings and found that the above-mentioned social pressures tend to increase social isolation of Black-White unions, especially from the White community, and consequently negatively impact the survival of these marriages. (Zhang and Van Hook 2009, p. 105)

We can offer at least three explanations for any differences in the marital stability that may exist among interracial or interethnic couples. First, significant differences in values and interests between partners can create a lack of mutual understanding, resulting in emotional gaps and increased couple conflict (Durodoye and Coker 2008; Lincoln, Taylor, and Jackson 2008). Second, such marriages may create conflict between the partners and other groups, such as parents, relatives, and friends. Continual discriminatory pressure from the broader society may create undue psychological and marital distress (Bratter and Eschbach 2006; Childs 2008). If they lack a supporting social network, partners may find maintaining their union in times of crisis more difficult. Also, a higher divorce rate among heterogamous marriages may reflect the fact that these partners are likely to be less conventional in their values and behavior, and unconventional people may divorce more readily than others (see Hohmann-Marriott and Amato 2008).

Interracial/Interethnic Heterogamy and Human Values

One recent study of unmarried interracial couples in college found *higher* relationship satisfaction compared to same-race couples (Troy, Lewis-Smith, and Laurenceau 2006). A comparison of Mexican American–non-Hispanic white marriages with those of homogamous white and homogamous Mexican American couples found little difference in marital satisfaction among the three groups (Negy and Snyder 2000). Whether these findings would apply to all interethnic or interracially married couples is unknown.

In any case, it is important to note the significance of human values. Many people do not want to limit their social contacts—including their life partners—to socially similar people. Although many people may retain a warm attachment to their racial or ethnic community, and some ethnic groups strongly value homogamy, social and political change has been in the direction of breaking down racial and ethnic barriers. People committed to an open society find intermarriage to be an important symbol, whether or not it is a personal choice, and do not wish to discourage this option (Dunleavy 2004; Moran 2001).

The data on homogamy may be interpreted to mean that, regardless of differences in race or ethnicity, common values and lifestyles contribute to relationship stability. A heterogamous pair may have common values that transcend their differences in background. Some problems of interracial (or other heterogamous) marriages have to do with social disapproval and lack of social support from either race (Felmlee 2001; Gullickson 2006). However, individuals can choose to work to change the society into one in which heterogamous marriage will be more accepted and hence will pose fewer problems.

Moreover, opinion polls and other research show that Americans are becoming less disapproving of interracial dating and marriage (Carroll 2007a; J. Jones

2005). Also, Americans are now more likely to have a close confidant of another race, suggesting that interracial bridges among people are increasing (Hulbert 2006). Still, among both blacks and whites, a minority of individuals strongly disapprove of interracial marriage (Jacobson and Johnson 2006). To the degree that racially, religiously, or economically heterogamous marriages increase in number, they are less likely to be troubled by the reactions of society.

Again, we see that private troubles—or choices—are intertwined with public issues. Some ethnic groups strongly value homogamy. Meanwhile, it is also true that if people are able to cross racial, class, or religious boundaries and at the same time share important values, they may open doors to a varied and exciting relationship. Chapter 10 explores raising children in interracial families. In our society, choice of life partners, whether homogamous or heterogamous, typically involves developing an intimate relationship and establishing mutual commitment. The next section examines these processes.

Developing the Relationship and Moving Toward Commitment

Social scientists have been interested in the process through which a couple develops their relationship and mutual commitment. What first brings people together? What keeps them together?

Meandering Toward Marriage and First Meetings

Young people today "meander toward marriage," feeling that they'll be ready to marry when they reach their late twenties or so (Arnett 2004, p. 197). Experiencing unprecedented freedom, today's young adults often express the need to explore as many options as possible before "settling down." As one young woman explained:

> I think everyone should experience everything they want to experience before they get tied down, because if you wanted to date a black person, a white person, an Asian person, a tall person, short, fat, whatever, as long as you know you've accomplished all that, and you are happy with who you are with, then I think everything would be OK. I want to experience life and know that when that right person comes, I won't have any regrets. (Quoted in Arnett 2004, p. 113)

Some young couples "hook up" for nonrelationship or "recreational" sex (see Chapter 5). They may find sexual pleasure in "friends with benefits" as they

experiment with many relationships before they think about looking for a spouse (Kan and Cares 2006; Manning, Giordano, and Longmore 2006). According to sociologist Kathleen Bogle (2008), the emergence of the "hookup" is "a major shift in the culture over the past few decades"—a shift from dating with a focus on developing a long-term, possibly marital relationship to getting together only for a sexual encounter. "Bogle says the hookup is what happens when high school seniors and college freshmen suddenly begin to realize they won't be marrying for five, ten, or fifteen years" (Wilson 2009; see also Meier and Allen 2009; Wallace 2007).

Despite its growing appeal among college students, hooking up has unfortunately been empirically linked to known risky behaviors such as alcohol abuse and engaging in sexual intercourse without using a condom (Gute and Eshbaugh 2008). As discussed in Chapter 5, hooking up can also evidence the sexual double standard as men more often see the hookup primarily as providing sexual gratification whereas women are more likely to hope that a hookup could be the beginning of a relationship (Bogle 2008). Then too, casual dating such as this can be associated with date or acquaintance rape, as discussed in "Issues for Thought: Date or Acquaintance Rape."

First Meetings Americans tend to believe that they find more socially desirable personality traits in those who are physically attractive. An examination of research studies on mate preferences since 1939 shows that physical attractiveness increased as a value over the past century and is especially important upon first meeting and in the early stages of a relationship (Malakh-Pines 2005; Tender 2008).

The majority of couples meet for the first time in face-to-face encounters, such as at school, work, a game, or a party. However, more couples today, especially those who are older, meet through singles' ads or online (Ellin 2009). Development of a face-to-face romantic relationship moves from initial encounter to discovery of similarities and self-disclosure (Knobloch, Solomon, and Theiss 2006). However, meeting for the first time online is a bit different.

Internet relationships—sometimes coupled with Internet background checks—progress through "an inverted developmental sequence." That is, without first seeing one another, two people who find each other intriguing gradually get to know one another through keyboard discussions. Then, too, as noted earlier, emerging niche websites introduce people who share specific characteristics, such as not wanting children, or being dog owners or vegetarians (Kirby 2005). This development makes it possible to establish the groundwork for rapport from the beginning, rather than having rapport develop as people gradually discover more about each other. Over time, emails become more intimate, and a powerful connection may be established (Merkle and

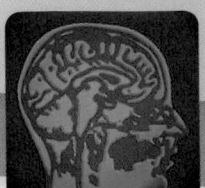
Contrary to the impression that we are likely to get from the media, most rape victims know their rapists (G. Cowan 2000). **Date** or **acquaintance rape**—being involved in a coercive sexual encounter with a date or other acquaintance—emerged as an issue on college campuses over the past two decades, but date rape no doubt plagued the dating scene for a long time before that (Friedman, Boumil, and Taylor 1992).

Often, excessive use of alcohol is involved (Foran and O'Leary 2008; Peralta and Cruz 2006). Findings from various research studies over the past decade show that sexually coercive men tend to dismiss women's rejection messages regarding unwanted sex and differ from noncoercive men in their approach to relationships and sexuality: They date more frequently; have higher numbers of sexual partners, especially uncommitted dating relationships; prefer casual encounters; and may "take a predatory approach to their sexual interactions with women."

Closely related to the concept of date rape is *sexual coercion* (Ryan and Mohr 2005). Some researchers have noted the existence of female-initiated sexual coercion—although significantly fewer women than men are sexually coercive, and when they are coercive, they use less forceful techniques. Men's experiences with being coerced most often do not advance beyond kissing or fondling,

whereas women's experiences most often result in unwanted, sometimes violent intercourse (Christopher and Sprecher 2000).

Many female victims blamed themselves, at least partially—a situation that can result in still greater psychological distress (Breitenbecher 2006). One reason victims blame themselves has to do with **rape myths**: beliefs about rape that function to blame the victim and exonerate the rapist (G. Cowan 2000). Rape myths include the ideas that (1) the rape was somehow provoked by the victim (for example, she "led him on" or wore provocative clothes); (2) men cannot control their sexual urges, a belief that consequently holds women responsible for preventing rape; and (3) rapists are mentally ill, a belief that encourages potential victims to feel safe with someone they know, no matter what (G. Cowan 2000). Increasingly, college men report that they recognize the male's responsibility for rape; this finding may be evidence that campus rape-prevention information and workshops make a difference (Domitrz 2003).

Critical Thinking

What can you do to help prevent date rape? What should you do if you or a friend is raped by an acquaintance? What would or should you do if a friend or acquaintance of yours was known to be the perpetrator of a date or acquaintance rape?

At its 1985 national convention, members of Pi Kappa Phi fraternity unanimously adopted a resolution not to tolerate any form of sexually abusive behavior on the part of their members. The fraternity also produced this poster and distributed it to all its chapters. The illustration is a detail from the painting *The Rape of the Sabine Women*. Beneath the large message, a smaller one reads, "Just a reminder from Pi Kappa Phi. Against her will is against the law."

Richardson 2000; Tender 2008). However individuals meet, what is it that draws and keeps them together? One way to explore this process involves the idea of a developing relationship as moving around a wheel.

The Wheel of Love

According to this theory, the development of love has four stages, which is a circular process—a **wheel of love**—capable of continuing indefinitely. The four stages—rapport, self-revelation, mutual dependency, and personality need fulfillment—are shown in Figure 6.4, and they describe the span from attraction to love.

Rapport Feelings of rapport rest on mutual trust and respect. A principal factor that makes people more likely to establish rapport is similarity of values, interests, and background (Gottlieb 2006)—social class, religion, and so forth. The outside circle in Figure 6.4 is meant to convey this point. However, rapport can also be established between people of different backgrounds, who may perceive one another as an interesting contrast to themselves or see qualities in one another that they admire.

Self-Revelation Self-revelation, or *self-disclosure*, involves gradually sharing intimate information about oneself. People have internalized different views about how

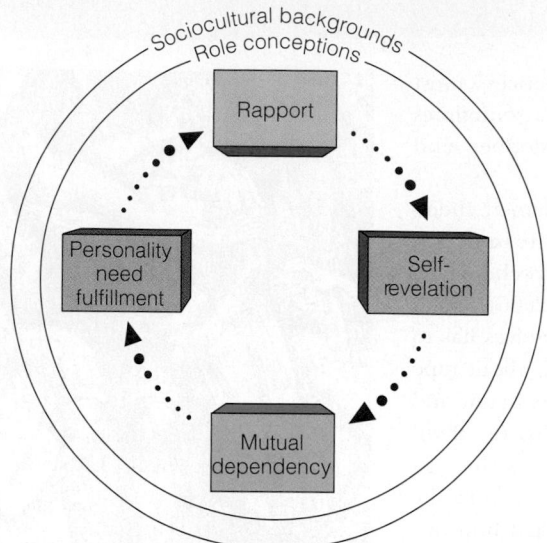

Figure 6.4 Reiss's wheel theory of the development of love.

Source: From *Family Systems in America,* 3rd ed., by I. Reiss © 1980 Wadsworth, a division of Cengage Learning.

much self-revelation is proper. The middle circle of Figure 6.4, "Role conceptions," signifies that ideas about social class-, ethnic-, or gender-appropriate behaviors influence how partners self-disclose and respond to each other's self-revelations and other activities.

For most of us, love's early stages produce anxiety. We may fear that our love won't be returned. Maybe we worry about being exploited or are afraid of becoming too dependent. One way of dealing with these anxieties is, ironically, to let others see us as we really are and to share our motives, beliefs, and feelings (Peck and Peck 2006). As reciprocal self-revelation continues, an intimate relationship may develop while a couple progresses to the third stage in the wheel of love: developing interdependence, or mutual dependency.

Mutual Dependency In this stage of a relationship, the two people desire to spend more time together and thereby develop interdependence or, in Reiss's terminology, *mutual dependency.* Partners develop habits that require the presence of both partners. Consequently, they begin to depend on or need each other. For example, watching a good DVD may now seem lonely without the other person because enjoyment has come to depend not only on the movie but on sharing it with the other. Interdependency leads to the fourth stage: a degree of mutual personality need fulfillment.

Need Fulfillment As their relationship develops, two people find that they satisfy a majority of each other's emotional needs. As a result, rapport increases, leading to deeper self-revelation, more mutually dependent habits, and still greater need satisfaction. The relationship is one of ongoing emotional exchange and mutual

support. Along this line, social scientist Robert Winch (1958) once proposed a theory of complementary needs, whereby we are attracted to partners whose needs complement our own (Malakh-Pines 2005). Sometimes people take this idea to mean that "opposites attract." This may make intuitive sense to some of us, but needs theorists more often argue that we are attracted to others whose strengths are *harmonious* with our own (Klohnen and Mendelsohn 1998; Schwartz 2006). Couples also tend to be matched on sex drive and attitudes about sex (Lally and Maddock 1994; Murstein 1980). "As We Make Choices: Harmonious Needs in Mate Selection" further explores these ideas. As partners develop mutual need satisfaction and interdependence, they gradually define their relationship. Part of defining a relationship should involve, according to counselors, talking about serious questions (Pendley 2006). What are some things that deserve discussing?

Some Things to Talk About

Talking about the relationship that you and your partner want may bring up differences, many of which can be worked out. If value differences are uncovered and cannot be worked out—for example, about whether or not to have children—it might be better to end the relationship before committing to a union that cannot satisfy either partner. Openly and honestly discussing matters like the ones that follow is important to successful mate selection (Pendley 2006):

1. When should a relationship be dissolved and under what circumstances? How long and in what ways would you work on an unsatisfactory relationship before dissolving it?

2. What are your expectations, attitudes, and preferences regarding sex?

3. Do you want children? If so, how many? Whose responsibility is birth control?

4. If you have children, how will you allocate child-rearing responsibilities? Do the two of you agree on child-raising practices, such as whether to spank a child?

5. What is the financial situation of each partner as he or she comes into the union? How will the couple manage any prior debt, credit problems, and their existing financial situations in general?

6. Will partners equally share breadwinning and home-making responsibilities, or not? How will money be allocated? Who will be the owner of family property, such as family businesses, farms, or other partnerships?

7. Do you expect your partner to share your religion? Will you attend religious services together? If you are of a religion different from that of your mate, where will you worship? What about the children's religion?

Finding a spouse with needs that are in harmony with one's own means matching different (but complementary) needs in some cases. In other cases, finding the "right" partner involves matching similar needs. The following three areas in which couples' needs should be similar are suggested by prominent sociologist Pepper Schwartz (2006) as important for a happy, long-term match:

1. **Personal Energy.** Your marriage may have more chance for success when your general energy level matches your partner's. "Whenever a couple spends time together their energy levels come into play.... While lovers may be willing to accommodate a leisurely stroll along the beach or a speed walk up the nearest mountain, at the end of the day, constant accommodation can be taxing and frustrating" (Schwartz 2006, p. 17).

2. **Outlook.** "People's attitudes and mood ranges from cheerful and upbeat to serious and earnest.... There is nothing wrong with either of these two emotional approaches to the world, but [it can be frustrating] if one person always feels the other is 'raining on her parade.'...Meanwhile, the partner's take on this may be to see that kind of optimism as simplistic or even scary...[and to] stop trusting their partners' instinctive impulse to see the brightest side of everything" (Schwartz 2006, p. 18).

3. **Predictability.** "If you draw comfort from surrounding yourself with familiar patterns and places, you are not going to be happy with someone for whom the very thought of predictable days, weeks, and places fills them with an urge to run.... The opposite of this need for predictability is the passion for variety.... [I]f the person who craves variety and the person who craves predictability find themselves together, they are going to feel betrayed and angry and worst of all, trapped" (Schwartz 2006, p. 21; see also Smithson and Baker 2008).

Critical Thinking

Do you agree with Pepper Schwartz that couples' needs should be similar in these three areas? Can you think of examples that would support Schwartz's points? Can you think of exceptions? Can you think of other situations in which the best match would be between partners with similar needs? Can you think of areas in which partners' different needs might complement one another?

---

8. What are your educational goals? How about your prospective partner's?

9. How will each of you relate to your own and to your partner's relatives?

10. What is your attitude toward friendships with people of the opposite sex? How about cyberfriends? Would you ever consider having sex with someone other than your mate? How would you react if your partner were to have sex with another person?

11. How much time alone do you need? How much are you willing to allow your partner?

12. Will you purposely set aside time for each other? If communication becomes difficult, will you go to a marriage counselor?

13. What are your own and your partner's personal definitions of *intimacy, commitment,* and *responsibility*?

Discussing topics such as these is an important part of defining a couple's relationships.

Defining the Relationship

From an interaction-constructionist perspective (see Chapter 2), qualitative research with serious dating couples shows that they pass through a series of fairly predictable stages (Sniezek 2002, 2007) by means of which they further define their relationship. Hinting, testing, negotiating, joking, and scrutinizing the partner's words and behavior characterize this process. As one example, a thirty-two-year-old emergency room worker described reading her partner's joking about marriage—and his use of the word *yet*—as a possible sign that he had considered marrying her:

> It was right here in our kitchen I put it (the food) down on his plate and I'm like "prison food." And he said "Um gee and we're not even married yet." And that was like the first joke. It stuck in my mind. (Sniezek 2007)

As the relationship progresses toward an eventual wedding, the "marriage conversation" is introduced. In the research being described here, women were more likely to initiate marriage talk—cautiously and indirectly: "Yeah it's like so what are you thinking? Where is this relationship heading?" In other cases, the marriage conversation began more directly. One woman raised the question of marriage when her partner suggested that they live together:

> When he asked me to move in with him, I told him I felt uncomfortable living with someone and not being married— a moral issue for me. So we talked about [the probability of getting married] at that time. (Sniezek 2007)

Once marriage talk is initiated, the couple faces negotiating a joint definition of the relationship as premarital. If one partner rejects the idea that the relationship should lead to marriage, several responses can occur. In some cases, one partner's marriage hopes may be relinquished, although the relationship continues. In other cases, a partner may deliver an ultimatum. Sometimes an ultimatum results in marriage; in other cases, stating an ultimatum causes an irreparable rift in the relationship.

Finally, most couples do not define themselves as "really" engaged until one or more ritualized practices take place—buying rings, setting a wedding date, public announcements to family and friends, holding engagement parties, and so on. These practices make the redefinition of the relationship increasingly public and "hardened" (Sniezek 2007).

In another qualitative study on this subject, two social scientists conducted lengthy interviews with 116 individuals in premarital relationships. They examined the process by which these partners gradually committed to marriage (Surra and Hughes 1997). From the interviews, the researchers classified the respondents' relationships in two categories: *relationship-driven* and *event-driven*.

Relationship-driven couples followed the evolving, wheel-like pattern described earlier. However, in event-driven relationships, partners vacillated between commitment and ambivalence. Often they disagreed on how committed they were as well as why they had become committed in the first place. The researchers called this relationship type *event-driven* because events—fighting, discussing the relationship with one's own friends, and making up—punctuated each partner's account.

It's probably no surprise that event-driven couples' satisfaction with the relationship fluctuated over time. Although often recognizing their relationships as rocky, they do not necessarily break up, because positive events (for example, a discussion about getting married or an expression of approval of the relationship from others) typically follow negative ones. At least some event-driven couples would probably be better off not getting married. We turn now to an even more serious issue—that of dating violence.

Dating Violence—A Serious Sign of Trouble

Sometimes we need to make decisions about continuing or ending a relationship that is characterized by physical violence and/or verbal abuse. Physical violence occurs in 20 percent to 40 percent of dating relationships (Luthra and Gidycz 2006)—a high, "most surprising incidence" (Johnson and Ferraro 2000, p. 951). Most incidents of aggression involve pushing, grabbing, or slapping. Between 1 percent and 3 percent of college students have reported experiencing severe violence, such as beatings or assault with an object.

Researchers are concerned that dating violence among teens is widespread and that many teens—as well as others—apparently minimize violence or view it as to be expected in certain situations (Hoffman 2009; Prospero 2006; Sears et al. 2006).

Both genders engage in physical aggression (Ryan, Weikel, and Sprechini 2008). However, by far, the more serious injuries result from male-to-female violence (Johnson and Ferraro 2000). Furthermore, women are more inclined to "hit back" once a partner has precipitated the violence, rather than to physically strike out first (Luthra and Gidycz 2006).

Dating violence typically begins with and is accompanied by verbal or psychological abuse (Lento 2006) and tends to occur over jealousy, with a refusal of sex, after illegal drug use or excessive drinking of alcohol, or upon disagreement about drinking behavior (Cogan and Ballinger 2006; Ryan, Weikel, and Sprechini 2008).

Researchers have found it discouraging that about half of abusive dating relationships continue rather than being broken off (Few and Rosen 2005). Given that the economic and social constraints of marriage are not usually applicable to dating, researchers have wondered why violent dating relationships persist. Evidence suggests that having experienced domestic violence in one's family of origin—even chronic verbal abuse in the absence of physical violence—is significantly related to both being abusive and to accepting abuse as normal (Cyr, McDuff, and Wright 2006; Tshann et al. 2009). A recent qualitative study of twenty-eight female undergraduates in abusive dating relationships found that some of these women felt "stuck" with their partner (Few and Rosen 2005). A majority had assumed a "caretaker identity," similar to martyring. As one explained:

> I always was a rescuer in my family. I felt that I was rescuing him [boyfriend] and taking care of him. He never knew what it was like to have a good, positive home environment, so I was working hard to create that for him. (p. 272)

Others felt stuck because they wanted to be married, and their dating partner appeared to be their only prospect: "I think near the end, one of the reasons I was scared to let go was: 'Oh, my God, I'm twenty-seven.' I was worried that I was going to be like some lonely old maid" (p. 274).

What are some early indicators that a dating partner is likely to become violent eventually? A date who is likely to someday become physically violent often exhibits one or more of the following characteristics:

1. Handles ordinary disagreements or disappointments with inappropriate anger or rage

2. Has to struggle to retain self-control when some little thing triggers anger

3. Goes into tirades

4. Is quick to criticize or to be verbally mean

5. Appears unduly jealous, restricting, and controlling

6. Has been violent in previous relationships (Island and Letellier 1991, pp. 158–66)

Dating violence is never acceptable. Making conscious decisions about whether to marry a certain person raises the possibility of not marrying him or her. Letting go of a relationship can be painful. Next, we'll look at the possibility of breaking up.

The Possibility of Breaking Up

Returning to Reiss's wheel theory of love, we note that once people fall in love, they may not necessarily stay in love. Relationships may "keep turning," or they may slow down or reverse themselves. Sometimes love's reversal, and eventual breakup, is a good thing: "Perhaps the hardest part of a relationship is knowing when to salvage things and when not to" (Sternberg 1988a, p. 242). Being committed is not always noble, as in cases of relationships characterized by violence or consistent verbal abuse, for example (partner abuse is discussed in Chapter 13). Committed love will require some sacrifices over the course of time. However, as one therapist put it, "Love should not hurt" (Doble 2006).

According to the exchange perspective, dating couples choose either to stay committed or to break up by weighing the rewards of their relationship against its costs. As partners go through this process, they also consider how well their relationship matches an imagined, ideal one. Partners also contemplate alternatives to the relationship, the investments they've made in it, and barriers to breaking up. (This perspective is also used when examining people's decisions about divorce, as discussed in Chapter 15.)

When a partner's rewards are higher than the costs, when there are few desirable alternatives to the relationship, when the relationship comes close to one's ideal, when one has invested a great deal in the relationship, and when the barriers to breaking up are perceived as high, an individual is likely to remain committed. However, when costs outweigh rewards, when there are desirable alternatives to the relationship, when one's relationship does not match one's ideal, when little has been invested in the relationship in comparison to rewards, and when there are fewer barriers to breaking up, couples are more likely to do so.

Even when a couple does not break up, recent research on dating couples has found support for the *principle of least interest* whereby the less involved partner wields more power in and control over the continuation or ending of the relationship (Waller

"Put me on your do-not-call list."

1951). Relatedly, some research has found that the lesser valued partner in a relationship is more inclined toward jealousy and yet also more willing to forgive the more highly valued partner for relationship indiscretions (Sidelinger and Booth-Butterfield 2007). However, high relationship satisfaction and stability are associated with equal emotional involvement (Crawford, Feng, and Fischer 2003; Sprecher, Schmeeckle, and Felmlee 2006). Having examined the processes through which individuals and couples move as they select a spouse, we turn to a discussion of cohabitation as a step in choosing a marriage partner.

Cohabitation and Marital Quality and Stability

Cohabitation serves different purposes for different couples: Living together "may be a precursor to marriage, a trial marriage, a substitute for marriage, or simply a serious boyfriend–girlfriend relationship" (Bianchi and Casper 2000, p. 17). Since the 1970s, the proportion of marriages preceded by cohabitation has grown steadily, and by 1995, a majority of marriages followed this pattern (Bumpass and Lu 2000). Cohabitation as a "substitute for marriage" is discussed at length in Chapter 8. Here we address cohabitation as a stage in choosing a spouse. Specifically, we will explore this question: How does cohabiting affect subsequent marital quality and stability?

Since about 1990, the proportion of cohabitors who eventually married their partners has declined (Seltzer 2000, p. 1,252). This situation is largely due to the fact that cohabiting has become more socially acceptable, a cultural change that "contributes to a decline in cohabiting partners' expectations about whether marriage is the 'next step' in their own relationship" (Seltzer 2000, p. 1,249). Another reason that fewer cohabitors are marrying has to do with economics: Poor cohabiting couples are less likely to marry (Gibson-Davis 2009; Lichter, Qian, and Mellott 2006).

Nevertheless, at least half of today's married couples between ages eighteen and forty-nine report having lived together before their wedding (Saad 2008b). Many of them began cohabiting with definite plans to marry their partner. "Thus, first-time cohabitors often believe their union is part of the marriage process" (Guzzo 2009a, p. 198). On the other hand, cohabitors may gradually come to believe that they'll marry eventually. One study found that cohabitors who had talked about future marriage had "generally been living with their partners for about two years, indicating that the issue of greater permanence in their relationships surfaces over time" (Sassler 2004, p. 501). All else being equal, cohabiting couples who identify as conservative Protestants are more likely than other cohabitors to marry (Eggebeen and Dew 2009).

Many young people today follow the intuitive belief that "cohabitation is a worthwhile experiment for evaluating the compatibility of a potential spouse, [and therefore] one would expect those who cohabit first to have even more stable marriages than those who marry without cohabiting" (Seltzer 2000, p. 1,252; see also Manning 2009; "Marriage" 2008). Among high school seniors, about two-thirds agreed that "[i]t is usually a good idea for a couple to live together before getting married in order to find out whether they really get along" (The National Marriage Project 2009, Figure 18).

At this point, research in inconclusive on whether doing so is really a good idea. Interestingly, however, a study of 120 heterosexuals cohabiting for about one year found that those who lived together to test their compatibility had more negative couple communication and generally poorer quality relationships than those who reported cohabiting for other reasons (Rhoades, Stanley, and Markman 2009).

So far, there has not been much research on whether cohabitation that is limited only to one's future spouse increases or decreases the odds of marital success. We can report that findings from a national representative sample show the divorce rate for serial cohabitors to be twice that for women who cohabited only with their eventual husbands (Lichter and Qian 2008). A second study that also used a national representative sample (of 6,577 women) found that premarital cohabitation that was limited to the woman's future husband did not increase the couple's likelihood of divorce (Teachman 2003). This situation also appears to be true for remarriages: Cohabiting with only one's future second spouse has been found not to increase divorce likelihood (Teachman 2008a).

Meanwhile, research over the past twenty years has consistently shown that marriages which are preceded by more than one instance of cohabitation are *more* likely to end in separation or divorce than are marriages in which the spouses had not previously cohabited at all (Xu, Hudspeth, and Bartkowski 2006). However, one study shows that these findings apply to non-Hispanic whites but not to African Americans or Mexican Americans, for whom cohabiting may be a more normative life course event, as discussed in Chapter 8 (Phillips and Sweeney 2005). Why might serial cohabitation before marriage be related to lower marital stability? Hypotheses to answer this question can be divided into two categories—*experience* and *selection*—both of which are supported to some degree by research.

First, the **experience hypothesis** posits that cohabiting experiences themselves affect individuals so that, once married, they are more likely to divorce (Seltzer 2000). For example, serial cohabitation may adversely affect subsequent marital quality and stability inasmuch as the experiences actually weaken commitment because "'successful' cohabitation demonstrates that reasonable alternatives to marriage exist" (Thomson and Colella 1992, p. 377). There is also evidence that "young adults become more tolerant of divorce as a result of cohabiting, whatever their initial views were," possibly because "cohabiting exposes people to a wider range of attitudes about family arrangements than those who marry without first living together" (Seltzer 2000, p. 1,253; see also Dush, Cohan, and Amato 2003; Popenoe and Whitehead 2000).

A related hypothesis suggests that some cohabiting couples, who would not have married if they had been simply dating but not living together, do end up marrying just because getting married seems to be the expected next thing to do. We can assume that it is less difficult to end an unsatisfactory dating relationship than a cohabiting one. Furthermore, research has found that cohabitors who marry after having a nonmarital birth experience lower marital relationship quality than do nonparent cohabitors who eventually marry (Tach and Halpern-Meekin 2009). This finding may result from the fact that cohabiting parents are more likely to marry mainly because they feel that they should. Choosing by default, couples may "slide" from cohabiting into marrying, rather than making more deliberative decisions (Stanley, Rhoades, and Markman 2006; Stanley 2009).

Second, the **selection hypothesis** assumes that individuals who choose serial cohabitation (or who "select" themselves into cohabitating situations) are different from those who do not; these differences translate into higher divorce rates. Serial cohabitors are more likely to have low relative education and income as well as less effective problem-solving and communication skills—factors related to divorce (Amato et al. 2008; Lichter and Qian 2008; Thornton, Axinn, and Xie 2007). Furthermore, those who choose serial cohabitation may have more negative attitudes about marriage in general and more accepting attitudes toward divorce.

An important international study lends considerable support to the selection hypothesis (Liefbroer and Dourleijn 2006). This study looked at the effects of cohabitation on marital stability in several countries and found that cohabiting had no negative effect on marital stability in countries such as Norway, where cohabiting is more common than in the United States. The researchers reasoned that in societies where cohabitation is about as common as marriage, those who live together before marrying would not be significantly different from those who do not. Therefore, no selection effect would be operating. The fact that this research found no negative effect of cohabitation on marital stability in societies where there would be little selection effect supports the selection hypothesis. Other support

© Digital Vision/Getty Images

Love is a process of discovery that involves continual exploration and sharing. Choosing a supportive partner is an important factor in developing a satisfying long-term relationship. Creating and maintaining that union involves recognizing that challenges will arise and committing to face and overcome them.

for the selection hypothesis is the finding that negative effects of cohabitation apply more strongly for non-Hispanic whites than for blacks or Mexican Americans, the latter two racial/ethnic groups having a higher percentage of cohabitors (Phillips and Sweeney 2005).

Maintaining a satisfying long-term relationship is challenging, if only because two people, two imaginations, and two sets of needs are involved. Differences *will* arise because no two individuals have exactly the same points of view. Relationships can more often be permanently satisfying, counselors advise, when spouses learn to care for the "unvarnished" other, not a "splendid image" (Van den Haag 1974, p. 142). In this regard, sociologist Judith Wallerstein, reflecting on her own marriage of fifty years, writes:

> I certainly have not been happy all through each year of my marriage. There have been good times and bad, angry and joyful moments, times of ecstasy and times of quiet contentment. But I would never trade my husband, Robert, for another man. I would not swap my marriage for any other. This does not mean that I find other men unattractive, but there is all the difference in the world between a passing fancy and a life plan. For me, there has always been only one life plan, the one I have lived with my husband. (Wallerstein and Blakeslee 1995, p. 8)

Choosing a supportive partner is an important factor in developing this kind of long-term love and relationship satisfaction.

Summary

- Loving is a caring, responsible, and sharing relationship involving deep feelings, and it is a commitment to intimacy. Genuine loving in our competitive society is possible and can be learned.

- Love should not be confused with martyring, manipulating, or limerence.

- People discover love; they don't simply find it. The term *discovering* implies a process—developing and maintaining a loving relationship require seeing the relationship as valuable, committing to mutual needs satisfaction and self-disclosure, engaging in supportive communication, and spending time together.

- Historically in Western cultures, marriages were often arranged in the marriage market, as business deals. In some of the world that is less Westernized, some marriages are still arranged. Some immigrant groups in the United States (and other Westernized societies) today practice arranged or "assisted" marriage.

- Whether marriage partners are arranged, "assisted," or more freely chosen, social scientists typically view people as choosing marriage partners in a marriage market; armed with resources (personal and social characteristics), they bargain for the best deal they can get.

- Although gender roles and expectations are certainly changing, some aspects of the traditional marriage exchange (a man's providing financial support in exchange for the woman's childbearing and child-raising capabilities, domestic services, and sexual availability) remain. Nevertheless, couples today are increasingly likely to value both partners' potential for financial contribution to the union.

- An important factor shaping partner choice is homogamy, the tendency of people to select others with whom they share certain social characteristics. Despite the trend toward declining homogamy, it is still a strong force, encouraged by geographical availability, social pressure, and feeling at home with people like ourselves.

- Committed relationships develop through building rapport and gradually negotiating the relationship as premarital, leading to marriage.

- Serial cohabiting (although not necessarily cohabiting before marriage only with one's future spouse) has been shown to increase the likelihood of divorce. The suggested reasons for this involve the *selection hypothesis* and the *experience hypothesis*.

Questions for Review and Reflection

1. Sternberg offers the triangular theory of love (Figure 6.1). What are its components? Are they useful concepts in analyzing any love experience(s) you have had?

2. Explain reasons why marriages are likely to be homogamous. Why do you think homogamous unions are more stable than heterogamous ones? How might the stability of interracial or interethnic relationships change as society becomes more tolerant of these?

3. If possible, talk to a few married couples you know who lived together before marrying, and ask them how their cohabiting experience influenced their transition to marriage. How do their answers compare with the research findings presented in this chapter?

4. This chapter lists topics that are important to discuss before and throughout one's marriage. Which do you think are the most important? Which do you think are the least important? Why?

5. **Policy Question.** What social policies, if any, presently exist to discourage couples who are experiencing dating violence from getting married? What new policies might be enacted to further discourage dating violence?

Key Terms

arranged marriage 144
assortative mating 148
commitment 139
commitment (Sternberg's triangular
 theory of love) 139
consummate love 139
cross-national marriages 146
date rape (acquaintance rape) 155
endogamy 148
exogamy 148
experience hypothesis 160
free-choice culture 145
geographic availability 149
heterogamy 148
homogamy 148
interethnic marriages 151

intergenerational transmission of divorce risk 143
interracial marriages 151
intimacy (Sternberg's triangular theory of love) 139
manipulating 142
marital stability 143
marriage market 144
martyring 141
mate selection risk 144
passion (Sternberg's triangular theory of love) 139
pool of eligibles 148
rape myths 155
selection hypothesis 160
self-revelation 155
status exchange hypothesis 152
Sternberg's triangular theory of love 139
wheel of love 155

Online Resources

Sociology CourseMate

www.CengageBrain.com

Access an integrated eBook, chapter-specific interactive learning tools, including flash cards, quizzes, videos, and more in your Sociology CourseMate, accessed through CengageBrain.com.

www.CengageBrain.com

Want to maximize your online study time? Take this easy-to-use study system's diagnostic pre-test, and it will create a personalized study plan for you. By helping you identify the topics that you need to understand better and then directing you to valuable online resources, it can speed up your chapter review. CengageNOW even provides a post-test so you can confirm that you are ready for an exam.

7

Marriage: From Social Institution to Private Relationship

Stockbyte/Jupiterimages

Between 80 and 90 percent of American adults today are, have been, or will be married for at least part of their lives (Stevenson and Wolfers 2007). Seventy-five percent of those in their twenties plan to marry someday (Bergman 2006a). Consistently, surveys show married people to be happier and healthier than unmarrieds. In research reported by sociologist Cherlin (2005), "One question asked of adults was whether they agreed with the statement, 'Marriage is an outdated institution.' Only 10 percent of Americans agreed—a lower share than any developed nation except Iceland" (Cherlin 2005, p. 44).

Nevertheless, according to a national Gallup poll, fewer than two-thirds of American adults under age fifty think that it's very important for a committed couple to marry—even when they plan to spend the rest of their lives together (Saad 2006b). That's a major change from sixty years ago, when marriage seemed the only option for the vast majority of committed (heterosexual) couples. Nevertheless, although the situation is much less clear-cut than now, marriage in the United States continues to be the most socially acceptable—and stable— gateway to family life.

In this chapter and the one that follows, we explore marriage as a changing institution, along with other ways to fashion family life. This chapter describes what distinguishes marriage from other couple relationships, then examines the changing nature of marriage, and ends with an exploration of recently married couples' relationships. We will see that getting married announces a personal life course decision to one's relatives, to the community, and, yes, to the state. Despite wide variations, marriages today have an important element in common: the commitment that partners make publicly—to each other and to the institution of marriage itself (Cherlin 2004; Goode 2007 [1982]). Put another way, getting married—as opposed to cohabiting, for instance—is not only a private relationship but also a publicly proclaimed commitment. We'll further explore research on the benefits of marriage for adults and children, and examine government initiatives to strengthen marriage. We begin by looking at marital status in the United States today.

Marital Status: The Changing Picture

Do you have friends who are "living together"? Maybe they are raising children. Maybe you know someone who says that she or he is "happily divorced." Do you know married couples who don't have children, either because they're putting it off or they don't want children at all? If you're in your twenties or thirties, you may have trouble believing that these situations were far from ordinary not long ago. However, marriage is different now than it was in the days of our parents and grandparents. Figure 7.1 shows the marriage, divorce, and birth rates in the United States from 1950 to between 2007 and 2008. As you can see from that figure:

1. The marriage rate has generally declined—from 11.1 marriages (that is, weddings) per 1,000 population in 1950, to 7.1 in 2008 (Tejada-Vera and Sutton 2009, Table A). In 1960, nearly 90 percent of women and men between ages thirty-five and forty-four were

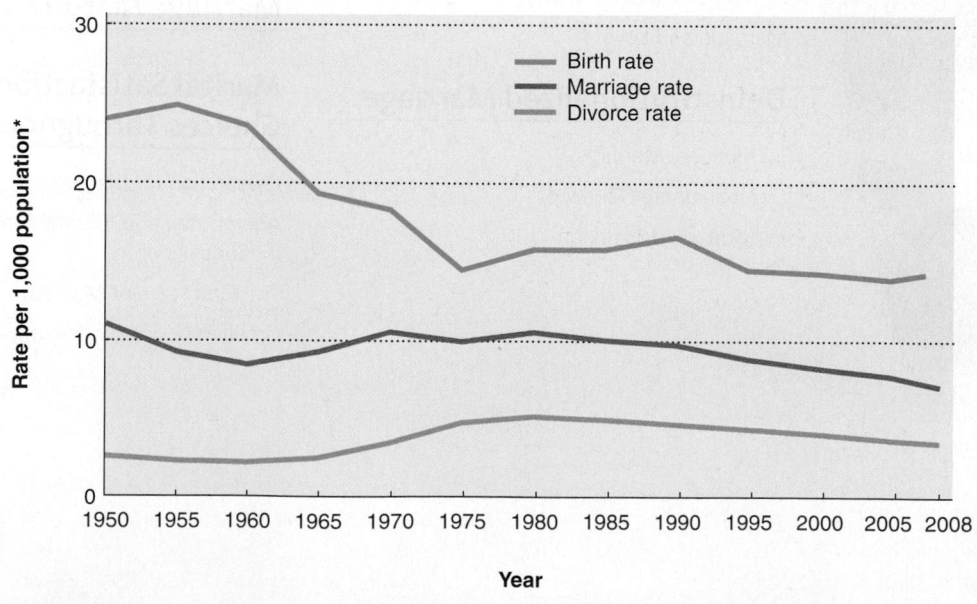

Figure 7.1 U.S. marriage, divorce, and birth rates, 1950 to 2007–08

Sources: Hamilton, Martin, and Ventura 2009, p. 2; Tejada-Vera and Sutton 2009, Table A; U.S. Census Bureau 2007a, Table 72.

married, compared with about 70 percent today (The National Marriage Project 2009, Figure 3).

2. The divorce rate is higher today than in 1960, when there were 2.2 divorces per 1,000 population. In 2008, that figure was 3.5 (Tejada-Vera and Sutton 2009, Table A). With ups and downs, the U.S. divorce rate has climbed since the nineteenth century. In the 1920s, the divorce rate accelerated somewhat. Then between about 1965 and 1975, the divorce rate doubled—a phenomenon that some policy makers call the *divorce revolution* (Popenoe 2007). By the mid-1970s, divorces reached an all-time high. Although the rate has slowly but steadily declined since then, it remains significantly higher than it was fifty years ago.

3. The birth rate has steadily declined since 1950—from 24.1 births per 1,000 population in 1950, to 14.3 in 2007 (Hamilton, Martin, and Ventura 2009, p. 2).[1]

Marking a couple's commitment, weddings are public events because the community has a stake in marriage as a social institution. Publicly proclaiming commitment to the marriage premise can help to enforce a couple's mutual trust in the permanence of their union. More and more, however, marriage seems to be reserved for the middle and upper classes—those who feel that they can afford this component of the American dream.

These three indicators—marriage, divorce, and birth rates—present a changing picture of marriage over the past sixty years. Today, 57 percent of U.S. adults are currently married (U.S. Census Bureau 2009a, Table A1; 2010b, Table 56)—a proportion that has been slowly declining over the past several decades. Although this situation results partly from a continuing trend of generally rising divorce rates since the early twentieth century, it also marks a fairly dramatic change from a few decades ago (Coontz 1992, 2005b). Throughout the first half of the twentieth century, the trend was for more people to marry and at increasingly younger ages.[2] Moreover, about 80 percent of those unions lasted until the children left home (Scanzoni 1972). In the 1960s, that trend reversed, and since then, the tendency has been for smaller and smaller proportions of Americans to be married. "Facts About Families:

Marital Status—The Increasing Proportion of Unmarrieds" further explores marital status in the United States today.

One reason for these changes—perhaps seeming ironic at first glance—is that we increasingly expect to find love in marriage. How would expecting to find love in marriage be associated with fewer of us being married? The following sections answer this question. To begin, we examine the time-honored marriage premise, with its expectations for permanence and sexual exclusivity.

The Time-Honored Marriage Premise: Permanence and Sexual Exclusivity

Why does a marriage today require a wedding, witnesses, and a license from the state? Around four hundred years ago in Western Europe, the government, representing the community, officially became involved in marriage (House 2002; Thornton 2009). For about one century before that, Roman Catholic Canon Law included rules, or canons, that regulated European marriage—although the canons, difficult to enforce in widely separated rural villages, were often ignored (Halsall 2001; House 2002; Therborn

---

[1] Rates per 1,000 population are not the ideal way to present these data, because population characteristics, such as the fact that the U.S. population has aged, are not taken into account. However, rates per 1,000 population are the only data available for presenting this long-term, historical comparison. The fertility rate for women of childbearing age (15 to 44) evidences the same trend as in Figure 7.1 (Hamilton, Martin, and Ventura 2009, Figure 1).

[2] For men, median age at first marriage in 1890—the year when the government first began to calculate and report this statistic—was 26.1. For both women and men, median ages at first marriage fell from 1890 until 1960, when they began to rise again. Around 1950, family sociologists described a standard pattern of marriage at about age twenty for women and twenty-two for men (Aldous 1978).

The proportion of unmarrieds age eighteen and over climbed from 28 percent of the total population in 1970, to 43 percent in 2008 (U.S. Census Bureau 2010b, Table 56). Figure 7.2 compares marital status proportions for non-Hispanic white, Hispanic, African American, and Asian women and men. As you can see in Figure 7.2, for instance, Asian Americans are most likely to be married and least likely to be divorced. African Americans are likely to be never married, followed by Hispanics. Cohabitors, more fully explored in Chapter 8, may be never married, divorced, or widowed.

The Never-Married

There is a growing tendency for young adults to postpone marriage until they are older. By 2009, the median age at first marriage for both men and women had risen to 25.9 for women and 28.1 for men (U.S. Census Bureau 2010a, Table MS-2).

As a consequence of postponing marriage, the proportion of singles in their twenties has risen dramatically. In 1970, 36 percent of women age twenty through twenty-four were never-married; by 2008, that figure had risen to 79 percent. The ranks of never-married men age twenty through twenty-four have increased from 55 percent in 1970 to 87 percent in 2008 (Saluter and Lugaila 1998; U.S. Census Bureau 2010b, Table 57).

This proportion of unmarrieds is striking when compared with the 1970s—and all the more striking when compared with the 1950s. However, this situation is not so unusual in a broader time frame. That is, the percentage of never-married men and women age twenty through twenty-four today is comparable to the proportion of young adults never married at the turn of the twentieth century (Arnett 2004).

The Divorced

The growing divorce rate has contributed to the increased number of singles. In 2008, 9 percent of men and 12 percent of women age eighteen and over were divorced. These proportions show a sharp increase from 1980, when 4 percent of men and 6 percent of women were divorced (U.S. Census Bureau 2000, Table 53; U.S. Census Bureau 2010b, Table 56). Although the divorce rate is no longer rising, it is stable at a high level, and the divorced will continue to be a substantial component of the unmarried population. Chapter 15 addresses divorce.

The Widowed

Unlike the other unmarried categories, the proportion of widowed women and men has remained about the same over the past several decades—between 2 and 3 percent for men, and between 9 and 12 percent for women (U.S. Census Bureau 2010b, Table 56). Death rates declined throughout the twentieth century, reducing the chances of widowhood for the young and middle-aged—although the wars in Afghanistan and Iraq unfortunately remind us that the widowed can be young as well. Meanwhile, the proportion of older people in the population has increased, and an older person has a greater risk of losing a spouse. Furthermore, widows (though not widowers) find it difficult to remarry, due to the significantly higher number of older women than older men, a situation discussed in Chapter 17.

Critical Thinking

How do you think the decreasing proportion of marrieds has affected American society in general and child raising in particular? What changes in cultural attitudes have helped to cause the high proportion of unmarrieds today? What structural factors have helped to cause the high proportion of unmarrieds?

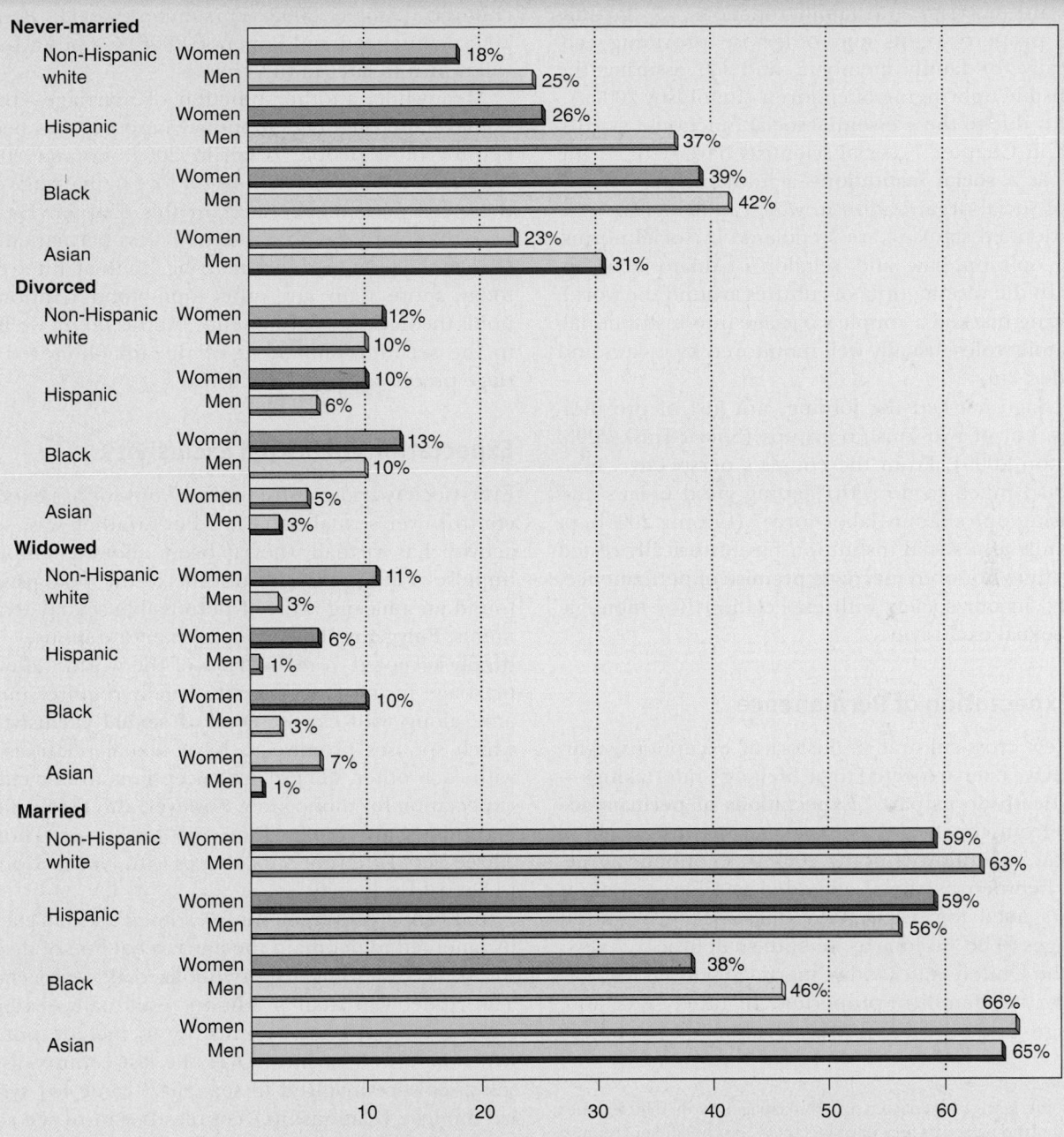

NOTE: Hispanics can be of any race. Unmarried cohabitors can be never-married, divorced, or widowed.

Figure 7.2 Marital status of the U.S. population, age 18 and over, 2008, by race/ethnicity and age

Source: U.S. Census Bureau 2010b, Table 57.

2004).[3] Even in the absence of Canon Law, communities throughout the world, represented by kinship groups or extended families, had always claimed a stake in two important marriage and family functions: (1) guaranteeing property rights and otherwise providing economically for family members, and (2) assuring the responsible upbringing of children (Ingoldsby 2006a).

Partly due to these essential social functions, also discussed in Chapter 1, social scientists have defined the family as a **social institution**—a fundamental component of social organization in which individuals, occupying defined statuses, are "regulated by social norms, public opinion, law and religion" (Amato 2004, p. 961).[4] In the vast majority of cultures around the world, a wedding marked a couple's passage into institutionalized family roles, usually well monitored by in-laws and extended kin.

Marriage marked the joining, not just of two individuals, but of two kinship groups (Sherif-Trask 2003; Thornton 2009). From the couple's perspective, marriage had much to do with "getting good in-laws and increasing one's family labor force" (Coontz 2005b, p. 6). Family as a social institution has historically rested on the time-honored **marriage premise** of permanence, coupled in our society with expectations for monogamous sexual exclusivity.

The Expectation of Permanence

With few cross-cultural or historical exceptions, marriages have been expected to be lifelong undertakings—"until death do us part." **Expectations of permanence** derive from the fact that marriage was historically a practical institution (Coontz 2005b). Economic agreements between partners' extended families, as well as society's need for responsible child raising, required marriages to be "so long as we both shall live."

In the United States today, marriage seldom involves merging two families' properties. In other ways, too, marriage is less critically important for economic security. Furthermore, marriage today is less decisively associated with raising children, although marriage remains significantly related to better outcomes for children (Amato 2005; Furstenberg 2003; Popenoe 2008; Whitehead and Popenoe 2006)—a point that we will return to later in this chapter.

Meanwhile, another function of marriage—providing love and ongoing emotional support—has become key for most people (Cherlin 2004; Coontz 2005b). We explore how expectations for love in marriage affect those for permanence later in this chapter. Here we note that marriage is considerably less permanent now than in the past. However, marriage in the United States today, more than any other non-blood relationship, holds the hope for permanence. At this point, we'll turn to the second component of the time-honored marriage premise—sexual exclusivity.

Expectations of Sexual Exclusivity

Every society and culture that we know of has exercised control over sexual behavior. Put another way, sexual activity has virtually never been allowed simply on impulse or at random. Meanwhile, anthropologists have found an amazing array of permissible sexual arrangements. **Polygamy** (having more than one spouse) is culturally accepted in many parts of the world.[5] However, marriage in the United States legally requires monogamy, along with **expectations of sexual exclusivity**, in which spouses promise to have sexual relations only with each other. (There are exceptions to our cultural expectation for monogamy, however, and three of these exceptions are touched on in "Issues for Thought: Three Very Different Subcultures with Norms Contrary to Sexual Exclusivity.")

In Europe, requirements for women's sexual exclusivity emerged to maintain the patriarchal line of descent; the bride's wedding ring symbolized this expectation. The Judeo-Christian tradition eventually extended expectations of sexual exclusivity to include not only wives but also husbands. Over the last century, as "the self-disclosure involved in sexuality [came to] symbolize the love relationship," couples began to see sexual exclusivity as a mark of romantic commitment (Reiss 1986, p. 56).

Today, expectations of sexual exclusivity have broadened from the purely physical to include expectations of emotional centrality, or putting one's partner first. Indeed, some marriage counselors now speak of "emotional affairs" (Herring 2005; Meier, Hull, and Ortyl

---

[3] The Netherlands first enacted a civil marriage law in 1590 (Gomes 2004). England passed its first Marriage Act in 1653 but did not require a legal marriage license until 1754 (House 2002). Shortly after Europeans established colonies in the United States, they enacted rules for marriage similar to those that they had known in Europe (Cott 2000). In the four hundred years since then, our federal and state governments have generated a massive number of marriage-related laws and court decisions. For instance, polygamy has been illegal in the United States since 1878, and due to laws enacted at the turn of the twentieth century, unmarried cohabitation is still illegal in some states, although the laws are seldom enforced (Hartsoe 2005). Also, before issuing a marriage license, some states require blood tests for various communicable diseases. Many states have waiting periods, ranging from seventy-two hours to six days, between the license application date and the wedding ("Chart: State Marriage License" 2006).

[4] Social scientists typically point to five major social institutions: family, religion, government or politics, the economy, and education.

[5] Polygamy can be divided into two types. *Polygyny*, a form of polygamy whereby a man can have multiple wives, "is a marriage form found in more places and at more times than any other" (Coontz 2005b, p. 10). However, polygyny is not always that frequent, because many men cannot afford multiple wives. Polyandry is rare (Stephens 1963). *Polyandry*—a woman's having many husbands—is still less frequent.

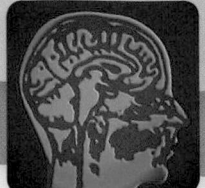

Although a very substantial majority of Americans value monogamy as a cultural standard, there are subcultural exceptions. This box looks at three of these subcultural exceptions, each one very different from the others—polygamy, polyamory, and swinging.

Polygamy

Polygamy has been illegal in the United States since 1878, when the U.S. Supreme Court ruled that freedom to practice the Mormon religion did not extend to having multiple wives (*Reynolds v. United States* 1878). Today, some activists are pursuing U.S. legalization of polygamy (Stacey and Meadow 2009).

Although a 2006 Gallup poll found that one-quarter of Americans think that most Mormons endorse polygamy (Carroll 2006), this is not the case. The Church of Jesus Christ of Latter-day Saints (LDS) no longer permits polygamy. Nevertheless, there are dissident Mormons (not recognized as LDS by the mainstream church) who follow the traditional teachings and take multiple wives (Woodward 2001, p. 50). Some multiple wives have argued that polygyny is a feminist arrangement because the sharing of domestic responsibilities benefits working women (D. Johnson 1991; Joseph 1991).

Federal law prohibits prospective immigrants who practice polygamy from entering the United States. However, polygamy has been found in New York among some immigrants from countries in which it is practiced (Bernstein 2007). Civil libertarians argue that the Supreme Court should rescind its *Reynolds* decision on the grounds that the right to privacy permits this choice of domestic lifestyle as much as any other (Slark 2004).

Polyamory

Polyamory means "many loves" and refers to marriages in which one or both spouses retain the option to sexually love others in addition to their spouse (Polyamory Society n.d.).

Deriving their philosophy from the sexually open marriage movement, which received considerable publicity in the late 1960s and 1970s, polyamorous spouses agree that each may have openly acknowledged sexual relationships with others while keeping the marriage relationship primary. Unlike in swinging, outside relationships can be emotional as well as sexual. Couples usually establish limits on the degree of sexual and/or emotional involvement of the outside relationship, along with ground rules concerning honesty and what details to tell each other (Macklin 1987, p. 335; Rubin 2001). "Polyamorists are more committed to emotional fulfillment and family building than recreational swingers" (Rubin 2001, p. 721).

Some polyamorous couples are raising children. The Polyamory Society's Children Educational Branch offers advice for polyamorous parents and maintains a PolyFamily scholarship fund, as well as the Internet-based "PolyKids Zine," and "PolyTeens Zine," both designed to present "uplifting PolyFamily stories and lessons about PolyFamily ethical living" (Polyamory Society n.d.). Like polygamy, polyamory has received some media attention in the last several years, and polyamorists are working toward greater social acceptance. Some polyamorists want to establish legally sanctioned group marriages and have begun to organize in that direction (Anderlini-D'Onofrio 2004). Conservative groups, such as the Institute for American Values, see such moves as evidence of an emergent "radical sensibility" that threatens American values and harms children (Marquardt n.d., p. 30; Kurtz 2006).

Swinging

Swinging is a marriage arrangement in which couples exchange partners to engage in purely recreational sex. Swinging gained media and research attention as one of several "alternative lifestyles" in the late 1960s and early 1970s (Rubin 2001). At that time, it was estimated that about 2 percent of adults in the United States had participated in swinging at least once (Gilmartin 1977).

Although little research has been done on swinging in the past few decades, "lifestyle practitioners," as some swingers now prefer to be called, still exist as a minority subculture. It has been estimated that there are now about 3 million married swingers in the United States, an increase of about one million since 1990. Some of this growth is probably due to the Internet, which helps to link potential swingers (Rubin 2001).

Interestingly, as a category, swingers tend to be middle-aged, middle-class, and more socially and politically conservative than one might expect (Jenks 1998; Rubin 2001). Although they often face the challenge of managing jealousy, swingers emphasize the lifestyle's positive effects—variety, for example (deVisser and McDonald 2007). Former swingers who have given up the lifestyle point to problems with jealousy, guilt, competing emotional attachments, and fear of being discovered by other family members, friends, or neighbors (Macklin 1987).

A couple considering a sexually nonexclusive marriage must take into account not only personal values and relationship management challenges but also the increased risk of being infected with HIV/AIDS. However, condoms are typically available at swing clubs, and "the fear of disease has apparently not inhibited the recent growth of swinging" (Rubin 2001, p. 723).

Critical Thinking

What do you think about these exceptions to monogamy? Do you see them as threatening to American values? If so, why? If not, why not? Does one or more of them seem reasonable to you while others do not? If so, why? If not, why not?

2009). With over 90 percent of us believing an affair is morally wrong ("Marriage" 2008), Americans are less accepting of extramarital sex than are people in many monogamous societies. (Although the vast majority of Americans say that they disapprove of extramarital sex, the picture is somewhat different in practice. Sexual infidelity is explored in Chapter 5.)

To summarize, the marriage premise has changed somewhat over the past century. Expectations for permanence have diminished while those for sexual exclusivity have been extended to include not just physical sex but also emotional centrality. The following section explores how these changes came about.

From "Yoke Mates" to "Soul Mates"—A Changing Marriage Premise

Chapter 1 points to an individualist orientation in our society. In eighteenth-century Europe, **individualism** emerged as a way to think about ourselves. No longer were we necessarily governed by rules of community. Societies changed from **communal**, or **collectivist**, to **individualistic**. In individualistic societies, one's own self-actualization and interests are a valid concern. In collectivist societies, people identify with and conform to the expectations of their extended kin. Western societies are characterized as individualistic, and individualism is positively associated with valuing romantic love (Dion and Dion 1991; Goode 2007 [1982]). (By *Western*, we mean the culture that developed in Western Europe and now characterizes that region and Canada, the United States, Australia, New Zealand, and some other societies.)

The Industrial Revolution and its opportunities for paid work outside the home, particularly in the growing cities and independent of one's kinship group, gave people opportunities for jobs and lives separate from the family. In Europe and the North American colonies, people increasingly entertained thoughts of equality, independence, and even the radically new idea that individuals had a birthright to "the pursuit of happiness" (Coontz 2005b). These ideas were manifested in dramatically unprecedented political events of the late 1700s, such as the U.S. Declaration of Independence and the French Revolution.

The emergent individualistic orientation meant a generally diminished obedience to group authority, because people increasingly saw themselves as separate individuals, rather than as intrinsic members of a group or collective. Individuals began to expect self-fulfillment and satisfaction, personal achievement, and happiness. With regard to marriage, an emergent

individualist orientation resulted in three interrelated developments:

1. The authority of kin and extended family weakened.
2. Individuals began to find their own marriage partners.
3. Romantic love came to be associated with marriage.

Weakened Kinship Authority

Kin, or extended family, include parents and other relatives, such as in-laws, grandparents, aunts and uncles, and cousins. Some groups, such as Italian Americans, African Americans, Hispanics, and gay male and lesbian families, also have "fictive" or "virtual" kin—friends who are so close that they are hardly distinguished from actual relatives (Furstenberg 2005; Sarkisian, Gerena, and Gerstel 2006). In collectivist, or communal, cultures, kin have exercised considerable authority over a married couple. For instance, in traditional African societies, a mother-in-law may have more to say about how many children her daughter-in-law should bear than does the daughter-in-law herself (Caldwell 1982).

In Westernized societies, however, kinship authority is weaker. By the 1940s in the United States, at least among white, middle-class Americans, the husband–wife dyad was expected to take precedence over other family relationships. Sociologist Talcott Parsons noted that the American kinship system was not based on extended family ties (1943). Instead, he saw U.S. kinship as comprised of "interlocking conjugal families" in which married people are members of both their **family of orientation** (the family they grew up in) and their **family of procreation** (the one formed by marrying and having children). Parsons viewed the husband–wife bond and the resulting family of procreation as the most meaningful "inner circle" of Americans' kin relations, surrounded by decreasingly important outer circles. However, Parsons pointed out that his model mainly characterized the American middle class. Recent immigrants and lower socioeconomic classes, as well as upper-class families, still relied on meaningful ties to their extended kin.

Although the situation is changing, the extended family (as opposed to the married couple or nuclear family) has been the basic family unit in the majority of non-European countries (Ingoldsby and Smith 2006). In the United States, extended families continue to be important for various European ethnic families, such as Italians, and for Native Americans, blacks, Hispanics, and Asian Americans, as well as other immigrant families (Gowan 2009; Kent 2007; Mather 2009; Richardson 2009).

Norms about extended-family ties derive both from cultural influences and from economic or other practical circumstances (Hamon and Ingoldsby 2003; Wong, Yoo, and Stewart 2006). Immigrants from many

"The Dinner Quilt," *Faith Ringgold, 1986*

less-developed nations work in the United States and send money to extended kin in their home countries (Ha 2006). Among Hispanics, *la familia* ("the family") means the extended as well as the nuclear family.

More and more Hispanics today value the primacy of the conjugal bond (Hirsch 2003). Meanwhile, like the Italians that Gans (1982 [1962]) studied in the 1960s, many Hispanics live in comparatively large, reciprocally supportive kinship networks (Sarkisian, Gerena, and Gerstel 2006; Lugo Steidel and Contreras 2003). For example, many Puerto Rican families have lived in "ethnically specific enclaves" and may rely as much on extended kin as on conjugal ties (Wilkinson 1993). Asian immigrants are also likely to emphasize extended kin ties over the marital relationship (Glick, Bean, and Van Hook 1997).

We can sometimes get a glimpse of mainstream American individualism through the eyes of fairly recent immigrants from more collectivist societies. For example, a Vietnamese refugee describes his reaction to U.S. housing patterns, which reflect nuclear, rather than extended-family, norms:

> Before I left Vietnam, three generations lived together in the same group. My mom, my family including wife and seven children, my elder brother, his wife and three children, my little brother and two sisters—we live in a big house. So when we came here we are thinking of being united in one place. But there is no way. However, we try to live as close as possible. (quoted in Gold 1993, p. 303)

American housing architecture similarly discourages many Muslim families—from India, Pakistan, or Bangladesh, for example—who would prefer to live in extended-family households (Nanji 1993).

All this is not to say that extended-family members are irrelevant to non-Hispanic white families in the United States. Nuclear families maintain significant emotional and practical ties with extended kin and parents-in-law (Lee, Spitze, and Logan 2003). A qualitative study with a sample that was 95 percent white showed that uncles often mentor nephews or nieces (Milardo 2005). Extensive data from the Longitudinal Study of Generations show that young adults today highly value their parents and extended families (Bengston, Biblarz, and Roberts 2007). However, as individuals and couples increasingly become more urban—and more geographically mobile—the power of kin to exercise social control over family members declines. If an individualist orientation has weakened kinship authority, it has also led to the desire to find one's own spouse.

Finding One's Own Marriage Partner

Arranged marriage has characterized collectivist societies (Hamon and Ingoldsby 2003; Ingoldsby 2006b; MacFarquhar 2006; Sherif-Trask 2003). Because a marriage joined extended families, selecting a suitable mate was a "huge responsibility" not to be left to the young people themselves (Tepperman and Wilson 1993, p. 73).

Analyzing arranged marriage in contemporary Bangladesh, sociologist Ashraf Uddin Ahmed notes that an individual's finding his or her own spouse "is thought to be disruptive to family ties, and is viewed as a child's transference of the loyalty from a family orientation to a single person, ignoring obligations to the family and kin group for personal goals" (Ahmed, quoted in Tepperman and Wilson 1993, p. 76). Moreover, there is concern that an infatuated young person might choose a partner who would make a poor spouse.

Ahmed argues that the arranged marriage system has functioned not only to consolidate family property but also to keep the family's traditions and values intact. But as urban economies developed in eighteenth-century Europe and more young people worked away from home, arranged marriages gave way to those in which individuals selected their own mates. Love rather than property became the basis for unions (Coontz 2005b).

Marriage and Love

Throughout the first five thousand years of human history in all the world's cultures that we know of, people probably fell in love, but they weren't *expected* to do so with their spouses. Marriage was thought to be "too vital an economic and political institution to be entered into solely on the basis of something as irrational as love" (Coontz 2005b, p. 7). Love—an intense, often unpredictable, and possibly transitory emotion—was viewed as threatening to the practical institution of marriage. Valuing romance could lead individuals to ignore or challenge their social responsibilities.[6]

However, with time, the ideology of romantic love came to be expected of marriage (Meier, Hull, and Ortyl 2009). In family historian Stephanie Coontz's words, basing marriage on love and companionship

> represented a break with thousands of years of tradition. . . . Critics of the love match argued . . . that the values of free choice and egalitarianism could easily spin out of control. If the choice of a marriage partner was a personal decision, . . . what would prevent young people . . . from choosing unwisely? If people were encouraged to expect marriage to be the best and happiest experience of their lives, what would hold a marriage together if things were "for worse" rather than "for better"? (2005b, pp. 149–50)

To use Coontz's metaphor, couples were no longer yoked together (like field oxen). "Where once marriage had been seen as the fundamental unit of work and politics, it was now viewed as a place of refuge from work, politics, and community obligations—a haven in a heartless world" (Coontz 2005b, p. 146; Lasch 1977). A successful marriage came to be measured by how well the union met its members' emotional needs.

To summarize this section, emergent individualism in eighteenth-century Europe meant that people, increasingly valuing personal satisfaction and happiness, began to associate romantic love with marriage and, hence, to want to find their own marriage partners, a practice that both resulted from and further caused weakened kinship authority. Couples were no longer bound by the yoke of kin control. As you might guess, the nature of marriage changed. We'll explore that change next.

[6] An interesting way that Europe's twelfth- and thirteenth-century noblemen and women managed love's threat to marriage as a social institution was the practice of *courtly love*. As we have seen, most marriages in the upper levels of society during this period were based on pragmatic considerations, not love. But, as the saying went then, "marriage is no real excuse for not loving" (quoted in Coontz 2005b, p. 6). Among Europe's noblemen and women, romantic love was expressed in relationships outside marriage in which a knight worshipped his lady, and ladies had their favorites. These relationships involved a great deal of idealization and could be adulterous but were not necessarily sexually consummated (Stone 1980). The distinction between romance and marriage was also evident in the lower classes (Coontz 2005b, p. 7).

Deinstitutionalized Marriage

Coontz asserts that love and expectations for intimacy have "conquered marriage" (2005b). What does she mean? Coontz is talking about what family sociologist Andrew Cherlin (2004) has called the **deinstitutionalization of marriage**—a situation in which time-honored family definitions and social norms "count for far less" than in the past (p. 853). For instance, childbearing outside of marriage, once severely stigmatized, now "carries little stigma" (Cherlin et al. 2009a, p. 919; but also see Usdansky 2009a, 2009b).

The following sections present and expand upon Cherlin's analysis of the shift from *institutional* to *companionate* to *individualized* marriage. As we discuss these three kinds of marriage, we need to remember that they are abstractions, or *ideal types*.[7] In reality, marriages approximate these types to varying degrees.

Institutional Marriage

We have witnessed a gradual historical change in Western and Westernized societies away from **institutional marriage**—that is, marriage as a social institution based on dutiful adherence to the time-honored marriage premise, particularly the norm of permanence (Cherlin 2004, 2009a; Coontz 2005b; Thornton 2009).

> Once ensconced in societal mandates for permanence and monogamous sexual exclusivity, the institutionalized marriage in the United States represented the age-old tradition of a family organized around economic production, kinship network, community connections, the father's authority, and marriage as a functional partnership rather than a romantic relationship. . . . Family tradition, loyalty, and solidarity were more important than individual goals and romantic interest. (Doherty 1992, p. 33)

Institutional marriage generally offered practical and economic security, along with the rewards that we often associate with custom and tradition (knowing what to expect in almost any situation, for example). With few exceptions over the past five thousand years, institutional marriage was organized according to patriarchal authority, requiring a wife's obedience to her husband and the kinship group. It is also true that, legally, institutional marriage could involve what today we define as wife and child abuse or neglect. Child and wife abuse were not recognized as social problems in this country until the 1960s and 1970s respectively.

Across cultures, the strength and scope of patriarchal authority varied, however. As an extreme example, in ancient Rome, the *paterfamilias* (family father), having

[7] In this context, the word *ideal* indicates that a type exists as an idea, not that it is necessarily good or preferable.

absolute authority over his wife and children, could legally kill them or sell them into slavery.[8] No matter how old they were, sons were subject to the authority of the *paterfamilias* until he died. A daughter lived under her father's rule until she married, when her father's authority over her was legally transferred to her husband (Long 1875; S. Thompson 2006). In the United States, of course, patriarchal authority never approached anything near that of the ancient Roman *paterfamilias*.

Companionate Marriage

By the 1920s in the United States, family sociologists had begun to note a shift away from institutional marriage, and, in 1945, the first sociology textbook on the American family (by Ernest Burgess and Harvey Locke) was titled *The Family: From Institution to Companionship*. By **companionate marriage**,

> Burgess was referring to the single-earner, breadwinner-homemaker marriage that flourished in the 1950s. Although husbands and wives in the companionate marriage usually adhered to a sharp division of labor, they were supposed to be each other's companions—friends, lovers—to an extent not imagined by the spouses in the institutional marriages of the previous era. . . . Much more so than in the 19th century, the emotional satisfaction of the spouses became an important criterion for marital success. However, through the 1950s, wives and husbands tended to derive satisfaction from their participation in a marriage-based nuclear family. . . . That is to say, they based their gratification on playing marital roles well; being good providers, good homemakers, and responsible parents. (Cherlin 2004, p. 851)

With companionate marriage, middle-class Americans often dreamed of attaining "the white picket fence." That is, they saw marriage as an opportunity for idealized domesticity within the "haven" of their own single-family home.[9] (This is why we have drawn a picket fence to symbolize the companionate marriage bond

The married couple embedded in this family of Eastern European immigrants who arrived in New York City in 1832 may be in love, but they were not *expected* to find love in marriage. Instead, their union is held together by strong expectations of permanence, bolstered by the social control of the kinship group.

New York Public Library Digital Image Collection

---

in Figure 7.3.) Meanwhile, women's increasing educational and work options, coupled with their expectations for marital love, sowed the seeds for the demise of companionate marriage (Cherlin 2004; Coontz 2005c).

An individualistic orientation views each person (both husband *and* wife) as having talents that deserve to be actualized. In this climate, women in companionate marriages began to pursue opportunities for self-actualization, as well as to expect a husband's expressive support for their doing so (Jackson 2007). Furthermore, women challenged centuries of previously ignored domestic violence. Given the tension between gender inequality and expectations for emotionally supported self-actualization, the companionate marriage "lost ground" (Cherlin 2004, p. 852).

By the 1970s, observers noted a movement away from people's finding of personal satisfaction primarily in acceptable role performance—for example, in the role of husband/breadwinner or wife/homemaker. Research on college students showed a shift in self-orientation away from defining themselves according to the roles they played. More and more, they identified themselves in terms of their individual personality traits. But individuals' appreciation for the esteem they get from playing their roles well "buttresses the institutional structure" (Turner 1976, p. 1,011; Babbitt and Burbach 1990). As one result, critics began to warn that American culture was becoming "narcissistic": Individuals appeared less focused on commitment or concern for future generations (Bellah et al. 1985; Lasch 1980).

---

[8] The occasions on which the *paterfamilias* actually exercised his authority to kill family members were uncommon, however (S. Thompson 2006).

[9] Companionate marriages of the 1950s "were exceptional in many ways. Until that decade, relying on a single breadwinner had been rare. For thousands of years, most women and children had shared the tasks of breadwinning with men. . . . Also new in the 1950s was the cultural consensus that everyone should marry, and that people should do so at a young age. The baby boom of the 1950s was likewise a departure from the past, because birthrates in Western Europe and North America had fallen steadily during the previous 100 years" (Coontz 2005c).

Feminists defined this situation somewhat differently: Attention to domestic abuse, unequal couple decision making, and unfair division of household labor—as well as a wife's ability to more easily leave an intolerable situation through divorce—could be good things (Hackstaff 2007). Some celebrated the fact that American culture would finally begin to make room for "thinking beyond the heteronormative family" (Roseneil and Budgeon 2004, p. 136; Stacey 1996). Coontz summarizes the situation more neutrally: "For better or worse," over the past thirty years, "all the precedents established by the love-based male breadwinner family were . . . thrown into question" (2005b, p. 11; 2005c). However one saw it, by the late 1980s, companionate marriage—which had lasted for but a minute in the long hours of human history—had largely given way to its successor, individualized marriage.

As an ideal type, the *companionate marriage* that characterized most of the twentieth century emphasized love and compatibility, as well as separate gender roles. However, in reality, couples represent this ideal type to varying degrees. Although this Russian immigrant couple, who own and operate a small Los Angeles grocery store, illustrate companionate marriage in *some* ways, they do not fit the definition of companionate marriage in at least one important way: They share the family provider role.

Individualized Marriage

Four interrelated characteristics distinguish **individualized marriage**:

1. It is optional.
2. Spouses' roles are flexible—negotiable and renegotiable.
3. Its expected rewards involve love, communication, and emotional intimacy.
4. It exists in conjunction with a vast diversity of family forms.

Partly because marriage is optional today, brides, grooms, and long-married couples have come to expect different rewards from marriage than people did in the past. They continue to value being good partners and, perhaps, parents. However, today's spouses are less likely to find their only, or definitive, rewards in performing these roles well (Byrd 2009). More than in companionate marriages, partners now expect love and emotional intimacy, open communication, role flexibility, gender equality, and personal growth (Cherlin 2004, 2009a; Meier, Hull, and Ortyl 2009). Over the course of about three centuries, couples have moved "from yoke mates to soul mates" (Coontz 2005b, p. 124).

Intense romantic feelings have been associated with greater marital happiness and may serve to get a married couple through bad times (Udry 1974; Wallerstein and Blakeslee 1995). There can be a downside to all this

though. The idealization and unrealistic expectations implicit in individualized marriage can cause problems. Social theorist Anthony Giddens argues that expectations for a relationship based on intimate communication to the extent that "the rewards derived from such communication are the main basis for the relationship to continue" often lead to disappointment: "[M]ost ordinary relationships don't come even close" (Giddens 2007, p. 30). Giddens may be overstating the case. Probably many marriages do come close. However, the fact remains that such high expectations may be associated with the following results:

1. A person's deciding not to marry, because she or he can't find a "soul mate" who can promise this level of togetherness;
2. A high divorce rate (although assuredly there are other reasons for divorce, too, as described in Chapter 15);
3. A lower birth rate as individuals focus on options in addition to raising children, a topic addressed in Chapter 9.

One theme of this text is that society influences people's options and thereby impacts their decisions. To the extent that they are legally, financially, and otherwise able, people today organize their personal, romantic, and family lives as they see fit (Byrd 2009). Some

Figure 7.3a The institutional marriage bond. Couples are "yoked" together by high expectations for permanence, bolstered by the strong social control of extended kin and community.

Figure 7.3b The companionate marriage bond. Couples are bound together by companionship, coupled with a gendered division of labor, pride in performing spousal and parenting roles, and hopes for "the American dream"—a home of their own and a comfortable domestic life together.

Figure 7.3c The individualized marriage bond. Spouses in individualized marriages remain together because they find self-actualization, intimacy, and expressively communicated emotional support in their unions.

engage in "dyadic innovation" (Green 2006, p. 182)—that is, they fashion their relationships with little regard to traditional norms. As a twenty-eight-year-old woman told an interviewer:

> Marriage, just because it's a piece of paper, doesn't necessarily mean it's a relationship or a long-standing relationship. A long-standing relationship can be a boyfriend. If you're with somebody and you love them, I don't really care about the piece of paper. So marriage really never enters my mind. (in Byrd 2009, p. 324)

To summarize, "How good is your relationship?" is often a question equal in importance to "Are you married?" (Giddens 2007).

In this climate, a wide variety of family forms emerge. What today we call the *postmodern* family (see Chapter 1), characterized by "tolerance and diversity, rather than a single-family ideal," takes many forms (Doherty 1992, p. 35). As noted in Chapter 1, some observers view the deinstitutionalization of marriage as a loss for society, a "decline" that hopefully can be turned around (e.g., Whitehead and Popenoe 2006). Others see the deinstitutionalization of marriage simply as an inevitable historical change (e.g., Coontz 2005b, 2005c).

Individualized Marriage and the Postmodern Family— *Decline* or Inevitable *Change*?

Those who view individualized marriage as a *decline* assert that our culture's unchecked individualism has caused widespread moral weakening and self-indulgence. They say that Americans, more self-centered today, are less likely than in the past to choose marriage, are more likely to divorce, and are less child-centered (Blankenhorn 1995; Popenoe 2007, 2008; Stanton 2004a, 2004b; Whitehead and Popenoe 2008). From this point of view, the American family has broken down.

Others, in contrast, see the deinstitutionalization of marriage as resulting from inevitable social *change*. These thinkers point out that, for one thing, people who look back with nostalgia to the good old days may be imagining incorrectly the situation that characterized marriage throughout most of the nineteenth and twentieth centuries. For instance, large families with many children and higher death rates for parents with young children meant that many children were not raised in two-parent households (Coontz 1992). Moreover, we cannot go back:

> [J]ust as we cannot organize modern political alliances through kinship ties or put the farmers' and skilled craftsmen's households back as the centerpiece of the modern economy, we can never reinstate marriage as the primary source of commitment and caregiving in the

modern world. For better or worse, we must adjust our personal expectations and social support systems to this new reality. (Coontz 2005c)

In a climate characterized by debate between spokespersons from these opposing perspectives, researchers and policy makers examine the social consequences of deinstitutionalized marriage.

Deinstitutionalized Marriage: Examining the Consequences

In her seminal 1995 presidential address to the Population Association of America, family demographer Linda Waite (1995) asked rhetorically, "Does Marriage Matter?" She concluded that indeed it does, for both adults and children. After thoroughly reviewing prior research that compared the well-being of family members in married unions with that of those in unmarried households, Waite reported that, as a category, spouses:

- had greater wealth and assets.
- earned higher wages.
- had more frequent and better sex.
- had overall better health.
- were less likely to engage in dangerous risk taking.
- had lower rates of substance abuse.
- were more likely to engage in generally healthy behaviors.

Comparing children's well-being in married families with that of those in one-parent families, Waite found that, as a category, children in married families:

- were about half as likely to drop out of high school.
- reported more frequent contact and better-quality relationships with their parents.
- were significantly less likely to live in poverty.

Since her address, many sociologists and policy makers have further researched and debated Waite's findings. In the section following this one, we will examine the responses of policy makers. Here we review a

sampling of demographic data and research findings on the question, "Does marriage matter?"

National income and poverty data apparently support Waite's argument that marrieds are financially better off. The median income for married-couple families in 2007 was $72,589, compared with just $44,358 and $30,296 for unmarried, male- and female-headed households, respectively (U.S. Census Bureau 2010b, Table 683). As Table 7.1 indicates, even when a wife is not in the labor force, married-couple households earn from $2,000 to $4,000 more annually than do unmarried male householders. This income gap is dramatically higher when marrieds are compared with unmarried female householders (U.S. Census Bureau 2010b, Table 683). Clearly, these data support the argument that higher income is positively associated with marriage. Furthermore, since Waite's address, studies have continued to find that, compared to unmarrieds, spouses in enduring marriages generally have better physical and mental health (Dush, Taylor, and Kroeger 2008; Liu 2009; Popenoe 2008; Williams, Sassler, and Nicholson 2008).

However, research also suggests that the association between marriage and positive outcomes is more complex than Waite indicated. For example, marrieds, on average, are less often depressed than the widowed and the divorced. But those in first marriages are not necessarily less depressed than either the remarried or the never married (Bierman, Fazio, and Milkie 2006; LaPierre 2009). Then, too, in addition to being married, education, a comfortable income, and (among blacks) not having to suffer from society-wide racism improve mental health (Bierman, Fazio, and Milkie 2006; Mandara et al. 2008). Finally, marrieds have more frequent sex than unmarrieds when all unmarrieds are categorized together, but they do not have more frequent sex than cohabiting couples (Waite 1995).

An early criticism of Waite's claims was that much—although not all—of the association between marriage and positive outcomes was due to *selection effects*. In researchers' language, people may "select" themselves into a category being investigated—in this case, marriage—and this self-selection can yield the results for which the researcher was testing. Increasingly,

Table 7.1 Median Income of Families by Types of Family in Constant (2007) Dollars: 1990 to 2007

	All Married-Couple Families	Married-Couple Families, Wife in Paid Labor Force	Married-Couple Families, Wife Not in Paid Labor Force	Unmarried Male Family Householder	Unmarried Female Family Householder
1990	$61,354	$71,937	$46,544	$44,669	$26,039
2000	$71,157	$83,361	$48,140	$45,425	$30,963
2007	$72,589	$86,435	$47,329	$44,358	$30,296

Source: U.S. Census Bureau 2010b, Table 683.

individuals with superior education, incomes, and physical and mental health are more likely to marry (Bierman, Fazio, and Milkie 2006; England and Edin 2007; Goodwin, McGill, and Chandra 2009, Figure 6; Schoen and Cheng 2006; Teitler and Reichman 2008). The **selection hypothesis** posits that many of the benefits associated with marriage—for example, higher income and wealth, along with better health—are therefore actually due to the personal characteristics of those who choose to marry (Cherlin 2003). For example, married women are more likely than those who are cohabiting or heading single-family households to inherit wealth (Ozawa and Lee 2006). Being positioned to inherit wealth from one's family of origin is a personal characteristic that *precedes* getting married.

Nevertheless, not all the benefits associated with marriage are accounted for by selection effects. In contrast to the selection hypothesis, the **experience hypothesis** holds that something about the *experience* of being married itself causes these benefits—a point that we will return to at the end of this chapter. Figure 7.4 illustrates the selection and the experience hypotheses. Meanwhile, considerable research has focused on examining the relationships between marriage and the consequences for children.

Child Outcomes and Marital Status: Does Marriage Matter?

The proportion of children under age eighteen living with two married parents declined steadily over the past forty years—from 85 percent in 1970, to 77 percent in 1980, to 67 percent in 2008 (U.S. Federal Interagency Forum on Child and Family Statistics 2006, 2009). About 20 million children under age eighteen (27 percent of all U.S. children) live in single-parent households (U.S. Federal Interagency Forum on Child and Family Statistics

2009). Twenty-three percent of all U.S. children reside in single-mother households, with another 4 percent living with single fathers. Nearly half (49 percent) of all single-parent families are non-Hispanic white. Blacks comprise 29 percent of single-parent families; Hispanics, 19 percent; and Asians, 2 percent (U.S. Census Bureau 2010b, Tables 66, 69).

Some "single" parents have cohabiting partners. Of all U.S. children, about 4.6 million (6 percent) live with a parents or parents who are cohabiting. Of children who live with cohabiting couples, about half (2.3 million) live with both of their unmarried biological or adoptive parents (U.S. Census Bureau 2009a, Table S0901; U.S. Census Bureau 2009d, Table C3; U.S. Federal Interagency Forum on Child and Family Statistics 2009).

Considerable research supports Waite's overall conclusion that growing up with married parents is better for children (Kreider and Elliott 2009a, 2009b; Magnuson and Berger 2009; Popenoe 2008). For instance, when compared with teens in homes with two married biological parents, those in single-parent and cohabiting families are more likely to experience earlier premarital intercourse, lower academic achievement, and lower expectations for college, together with higher rates of school suspension and delinquency (Carlson 2006; Manning and Lamb 2003; VanDorn, Bowen, and Blau 2006). In addition, studies that compared economically disadvantaged six- and seven-year-olds from families of various types found fewer problem behaviors among children in married families (Ackerman et al. 2001). Other research has found that, among couples with comparable incomes, married parents spend more on their children's education (and less on alcohol and tobacco) than do cohabiting parents (DeLeire and Kalil 2005). Furthermore, a recent analysis of national data found that, when compared to those in other family forms, married mothers exhibited the healthiest prenatal behaviors (Kimbro 2008).

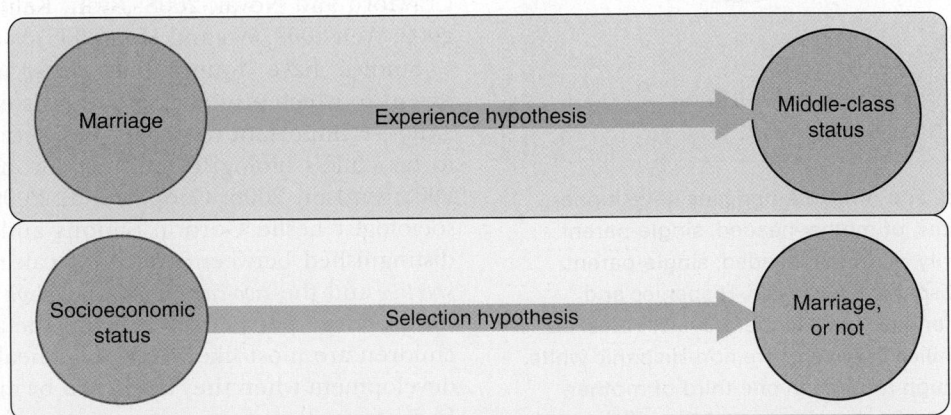

Figure 7.4 Causal order: Experience hypothesis, selection hypothesis
Source: Adapted from Marsh et al. 2007, p. 739.

As the experience hypothesis would suggest, one reason that, as a category, children with married parents evidence better outcomes may be the *experience* of growing up in a married-couple household. With its presumption of permanent commitment to the family as a whole, marriage "allows caregivers to make relationship-specific investments in the couple's children—investments of time and effort that, unlike strengthening one's job skills, would not be easily portable to another relationship" (Cherlin 2004, p. 855; Popenoe 2008).

However, as with the benefits of marriage for adults, researchers, working to unravel the statistical correlation between marriage and positive child outcomes, have uncovered complexities. For instance, findings differ according to how the variable *marriage* is defined.

Rhoberazzi/Getty Images

Although Hispanics and African Americans have higher percentages, or rates, of mother-headed, single-parent families, the majority of mother-headed, single-parent families are non-Hispanic white. Also, Hispanics and African American families have higher poverty rates, but the majority of families in poverty are non-Hispanic white. Furthermore, although more than one-third of mother-headed, single-parent families live below the official poverty level, nearly two-thirds do not.

Results differ when the variable *marriage* allows an investigator to compare the effects of having two biological, continuously married parents with those having a remarried stepparent. Using data from the National Longitudinal Survey of Youth, one study (Carlson 2006) compared outcomes for adolescents in several family structures. Similar to prior research, this study found fewer behavior problems among teens who lived with their continuously married biological parents. However, adolescents born outside of marriage but whose biological parents later married or were cohabiting, or whose mother married a stepfather, had more behavior problems than teens whose biological parents were continuously married (Carlson 2006). Furthermore, transitions to and from various family structures have been found to result in poorer outcomes for children (Bures 2009; Magnuson and Berger 2009).

In addition to refining the marriage variable, researchers have proposed supplementary or alternative causes for children's marriage-associated benefits. For one thing, children raised by married families are less likely to live in poverty—a situation that has serious negative effects on child outcomes (Moore et al. 2009). ("Facts about Families: Marriage and Children in Poverty" describes various effects of growing up in poverty.) We know that married parents are less likely to live below poverty level, but factors in addition to marital status are related to poverty as well. For example, in a study focused on Hispanic children, researchers showed that, for Mexican American children, poverty is related to a combination of marital status and the number of children in the household, the latter "an important predictor of poverty regardless of marital status" (Crowley, Lichter, and Qian 2006).

Beside marital status and poverty, still other factors correlate with children's outcomes—the child's neighborhood and peers, family conflict, parental nurturance and involvement in the child's school activities, parents' participation in religious services, and parents' available social support (Broman, Li, and Reckase 2008; Crawford and Novak 2008; Ryan, Kalil, and Leininger 2009; Wen 2008; Wu and Hou 2008).

Studies have found that *father involvement*—the extent to which a biological father is engaged with his child—is important regardless of whether he is married to his child's biological mother (Bronte-Tinkew et al. 2008; Carlson 2006; Cooper et al. 2009). Accordingly, sociologist Leslie Gordon Simons and her colleagues distinguished between what they call the *marriage perspective* and the *two-caregivers perspective*: "What we label the marriage perspective rests on the assumption that children are most likely to display healthy growth and development when they are raised by married parents." In contrast, the two-caregivers perspective contends that children do best when raised by two caregivers rather than by a single caregiver (Simons et al. 2006, p. 805).

More than 13.3 million U.S. children under age eighteen live at or below poverty; children comprise 36 percent of this nation's poor (U.S. Census Bureau 2010b, Table 697). More than 17 million children live in "near poor" conditions—that is, at less than 125 percent of poverty level (U.S. Census Bureau 2009a, Table S1703). In recent years, poverty rates have declined for African Americans and Hispanics. Nevertheless, 14 percent of white, 12 percent of Asian, 34 percent of African American, and 28 percent of Hispanic children live in poverty (U.S. Census Bureau 2010b, Table 696; see also Moore et al. 2009). Figures such as these may lead us to think of poverty in terms of black or Hispanic families, but the majority of poor children are non-Hispanic white. Despite lower *rates* of poverty, non-Hispanic whites predominate in sheer numbers, comprising nearly two-thirds of all poor children in the United States (U.S. Census Bureau 2010b, Table 696).

Regardless of their parents' marital status, children growing up in poverty often do not have enough nutritious food; are more likely to live in environmentally unhealthy neighborhoods; have more physical health, socioemotional, and behavioral problems; must travel farther to attain health care; attend poorly financed schools; do less well academically; and are more likely to drop out (Goosby 2007; Moore et al. 2009; Teachman 2008b). Furthermore, the rate of severe violence toward children is about 105 per 1,000 in families below the poverty line, compared to about 30 per 1,000 children in other families (Gelles and Cavanaugh 2005).

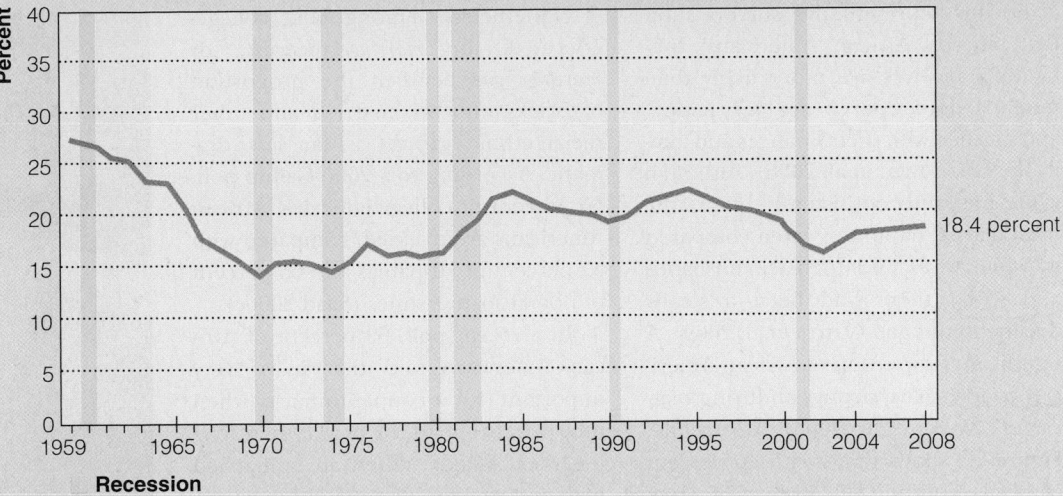

Figure 7.5 U.S. poverty rate for children under age 18, 1959 to 2008

Sources: Proctor and Dalaker 2003; U.S. Census Bureau 2009f, Table 690.

Economic hardship in childhood, even in a two-parent married family, particularly when it lasts for a long time or occurs in adolescence, is related to lowered emotional well-being in early adulthood (Sobolewski and Amato 2005; Vandewater and Lansford 2005). Not having enough money causes stress, which often leads to mothers' depression, parental conflict, and general household turbulence. Household turbulence and mothers' depression, in turn, are associated with lower parent–child (especially teen) relationship quality, a situation that results in a child's lower psychological well-being (Goosby 2007; Jackson, Choi, and Bentler 2009; Sobolewski and Amato 2005; Teachman 2008b).

As Figure 7.5 illustrates, the child poverty rate for all races calculated together was about 27 percent in 1959, but beginning with President Lyndon Johnson's War on Poverty[a] in the 1960s, it dropped consistently during the 1970s, to a low of about 14 percent. A decade later, a series of economic recessions occurred along with the phasing out of many War on Poverty measures. As a result, the child poverty rate began to rise in the late 1970s, then fell again after about 1993. However, the rate began to rise again in 2000. In 2008, the child poverty rate was 18.4 percent—up from 15.6 percent at the turn of the twenty-first century (Proctor and Dalaker 2003; U.S. Census Bureau 2010b, Table 697).

The War on Poverty offered *structural* strategies to decrease poverty, such as community meal programs and health centers, legal services, summer youth programs, senior centers, neighborhood development, adult education, job training, and family planning (Garson n.d.). Commitment to the War on Poverty diminished after the 1970s, with national rhetoric shifting to debates focused on individual responsibility. Today, however, scholars and some policy makers are again insisting that the United States must pay attention to ecological and structural supports for children and families regardless of—or in addition to—concerns about changing family structure (Cherlin 2009a; Moore et al. 2009; Popenoe 2009).

a. War on Poverty measures, first proposed in 1964 by President Lyndon Johnson and enacted by Congress in the subsequent Economic Opportunity Act, allocated federal funds to reduce poverty. You may have heard of War on Poverty programs, such as the Job Corps or the Neighborhood Youth Corps, Head Start, or Adult Basic Education. Although the majority of War on Poverty measures have ended, Head Start and the Job Corps continue to exist.

A Closer Look at Diversity

African Americans and "Jumping the Broom"

Nationally representative surveys show that, among African Americans, husbands and wives are more likely than unmarrieds to report being "very happy" and satisfied with their finances and family life (Blackman et al. 2006). Although white husbands consistently report the most marital happiness when compared to white wives and to black husbands and wives, there had been a steady decline in this gap (Corra et al. 2009). A significant proportion of African American couples have strong, enduring marriages (Marks et al. 2008). Meanwhile, Figure 7.2 shows that—with 46 percent of black men and 38 percent of black women currently married—African Americans are considerably less likely to be wed than are other U.S. racial/ethnic groups. A large body of literature, written by both blacks and whites, is accumulating on the structural–cultural reasons for this situation (McAdoo 2007; Saad 2006c; Wilson 2002). Chapter 3 explores this literature.

Nonetheless, among the college-educated, the *marriage disparity,* or *marriage gap* between the proportion of blacks who are married and other racial/ethnic groups is far less dramatic. According to a 2006 Gallup poll, 55 percent of college-educated African Americans are married, compared with 57 percent of Hispanics and 65 percent of non-Hispanic whites (Saad 2006c).

In a recent poll, 69 percent of African Americans said that it is "very important" for a couple to marry when they plan to spend the rest of their lives together. Asked, "When an unmarried man and woman have a child together, how important is it to you that they legally marry?" college-educated African Americans were *more* inclined than either Hispanics or non-Hispanic whites to say that marrying in this situation is "very important." The figures were 55 percent of blacks, 46 percent of Hispanics, and 37 percent of non-Hispanic whites (Saad 2006c). Furthermore,

Although somewhat controversial because it can be a reminder of slavery, jumping the broom at African American weddings is going through some revival as black couples plan wedding celebration rituals designed to incorporate their cultural heritage.

Analyzing data on 867 African American children from the Family and Community Health Study, Simons and her colleagues found that "child behavior problems were no greater in either mother-grandmother or mother-relative families than in those in intact nuclear families." At least among blacks, these researchers found mother-grandmother families to be "functionally equivalent" (Simons et al. 2006, p. 818). Then too, other extended kin in black families—uncles, for example—may be involved in child care (Richardson 2009).

Interestingly, research based on a national representative sample of more than 10,000 U.S. teens found that the negative effects of time lived with a single mother were less serious for black and Hispanic adolescents than they were for whites. Why might this be? First, although nonmarried parenthood remains somewhat stigmatized in the United States (Usdansky 2009a, 2009b), it is noteworthy that this stigma may be minimal within the black community where single-parent families are more normative (Heard 2007). Accordingly, "The common history among blacks allows for the emergence and primacy of social supports, such as women-centered kinship networks, coresidence with extended family, and strong ties to the church, which can buffer the negative effects of stress caused by family

instability" (Heard 2007, p. 336). (For more on blacks and marriage, see "A Closer Look at Diversity: African Americans and 'Jumping the Broom.'")

As with the benefits of marriage for adults, researchers hypothesize that selection effects explain much—although not all—of marriage's advantage for children. We've seen that, on average, individuals who marry are better educated and have higher incomes. As parents, they live in neighborhoods more conducive to successful child raising. Less likely to be stressed due to financial problems, they are more likely to practice effective parenting skills (Manning and Brown 2006; also see Teachman 2008b). In her address to the Population Association of America, Waite acknowledged the contribution of selection effects to child outcomes. She added, however, that "we have been too quick to assign *all* the responsibility to selectivity here, and not quick enough to consider the possibility that marriage *causes* some of the better outcomes we see . . ." (1995, p. 497, italics in original).

To summarize, a large body of research shows that marriage is associated with benefits for adults and children. However, this relationship is complex, and much of it may be due to variables other than marital status as well as to selection effects (Teachman 2008b).

attitude surveys consistently show that African Americans value marriage, perhaps more than non-Hispanic whites do (Bramlett and Mosher 2001; Johnson and Staples 2005; Saad 2006c).

The news media have focused so frequently on poverty-level African Americans and on the relatively low proportion of married blacks that we may forget about the 10.6 million (40 percent of) African Americans who *are* married (U.S. Census Bureau 2010b, Table 56). Google "African American marriage" and you'll find several websites marketing wedding products and services to middle-class blacks. One book, *Jumping the Broom* (Cole 1993) is a wedding planner for American blacks. If you're not African American, there's a fairly good chance that you have not heard of jumping the broom. What is it?

For African Americans, the significance of the broom originated among the Asante in what is now the West African country of Ghana. Used to sweep courtyards, the handmade Asante broom was also symbolic of sweeping away past wrongs or warding off evil. Brooms played a part in Asante weddings as well. To culminate their wedding ceremony, a couple might jump over a broom lying on the ground or leaning across a doorway. Jumping the broom symbolized the wife's commitment to her new household, and it was sometimes said that whoever jumped higher over the broom would be the family decision maker (DiStefano 2001; Prahlad 2006).

Among slaves brought to the Americas, jumping the broom continued. Not allowed to marry legally, slaves

sometimes jumped a broom as an alternative ceremony to mark their marital commitment. The association of jumping the broom with slavery has stigmatized the tradition for some African Americans. However, the ritual is coming back as more middle-class blacks seek culturally relevant wedding celebrations (African Wedding Guide n.d.; Anyiam 2002; DiStefano 2001).

Critical Thinking

How do you think that jumping the broom might be used to symbolize the time-honored marriage premise? Why would it be important to incorporate traditions that are relevant to one's own culture into a wedding ceremony? Why do you think we hear relatively little about African Americans' weddings or marriages?

A theme of this text is that research findings expand our knowledge so that we can better make decisions knowledgeably. One thing that the research implies is that, although generally marriage is advantaged, additional factors affect individual outcomes, and the disadvantages for some ethnic groups may not be as severe as for the population as a whole.

Just as researchers have responded in various ways to Waite's address, policy makers have had conflicting reactions. More conservative policy leaders, associated with the *decline* and "family breakdown" perspective, once hoped to effect a "family turnaround" (Whitehead and Popenoe 2003). They noted "a greater emphasis on short-term gratification and on adults' desires rather than on what is good for children" that they attributed to government welfare programs and to decreased attention to religious principles (Giele 2007, p. 76).

Today, these policy makers appear to have given up on a broad "family turnaround." In the words of sociologist David Popenoe (2008):

> [Realistically] there is probably not much that government policies or social action can do to change the situation. If major change is to come about it will have to occur through a broad cultural shift, reflected in the hearts

and minds of the citizenry, in the direction of stronger interpersonal commitments and families. . . . Still, there surely are actions that societies can take to try to improve the situation and not make it worse; actions that discourage cohabitation and encourage marriage, at least when children are involved. (pp. 15–16)

On the other hand, policy makers who see marriage simply as *changing* recognize that many families are struggling but criticize the solutions offered by conservatives and propose their own. The following section explores this policy debate.

Valuing Marriage—The Policy Debate

Policy advocates from a marital *change* perspective are mainly concerned about the high number and proportions of parents and children living in poverty. They view poverty as causing difficult child-raising environments with resulting negative outcomes for some—though not all—of America's children. From this viewpoint, family struggle results from structural conditions, such as recession. Accordingly, these spokespersons argue for structural, or ecological (such as neighborhood-level), solutions.

From the *decline* perspective, on the other hand, concerns about "family breakdown" include the high number of federal dollars spent on "welfare" for poverty-level single mothers, coupled with the irresponsible socialization of children (Giele 2007). They define the causes for these concerns as primarily cultural, such as changes in individuals' values and attitudes regarding marriage. Therefore, they offer motivational and educational programs to effect a family "turnaround."

Policies from the Family Decline Perspective

An important goal from the *decline* perspective is to return to a society more in line with the values and norms of companionate marriage. As means to that end, advocates have established programs to encourage marital permanence. Many religions insist on premarital counseling as a way to dissuade couples in inappropriate matches from marrying and to encourage those who do marry to stay together (Nock 2005; Nadir 2009; Ooms 2005).

Covenant Marriage Some conservative Christian organizations and legislators advocate **covenant marriage**. In covenant marriage, partners agree to be bound by a marriage "covenant" (stronger than an ordinary contract) that will not let them get divorced easily ("Covenant Marriages Ministry" 1998). Between ten and twenty years ago, three states—Louisiana, Arizona, and Arkansas—enacted covenant marriage laws. Twenty other states have considered, but failed to pass, covenant marriage laws (Leon 2009).

How does covenant marriage work? Before their wedding, couples choose between two marital contracts, conventional or covenant. If legally bound by a marriage covenant, couples are required to get premarital counseling and may divorce only after being separated for at least two years or if imprisonment, desertion for one year, adultery, or domestic abuse is proved in court. In addition, a covenant couple must submit to counseling before a divorce (Brown and Waugh 2004).

Typically, fundamentalist Christian religions are enthusiastic about covenant marriage, whereas feminists and other critics are not ("Couple Support" 2006; Leon 2009; Sanchez et al. 2002). For instance, critics point out that proving adultery or domestic abuse in court may be difficult and expensive, while living in a violent household can be deadly (Gelles 1996).

Despite promoters' early enthusiasm, covenant marriage has failed to become a serious social movement, never having spread beyond the original three enacting states. Relatively few couples in the states where it is available have opted for covenant marriage ("More Binding Marriage" 2004). In addition to covenant

marriage, the federal Healthy Marriage Initiative encourages marital stability (Chaney 2009; Graefe and Lichter 2007).

Government Initiatives States have promoted marriage education, some offering money incentives for couples to participate. Other state initiatives include home visitation programs for families that might be targeted for government assistance because of a variety of reasons, such as a birth to a teenager or an unstable marriage; mentoring, marriage counseling, communication skills, and anger management workshops; state-funded resource centers that provide information on marriage; and state websites that include marriage enrichment information and links to service-related sites (Dion 2005; Ooms 2005).

As part of the federal Healthy Marriage Initiative (HMI), these programs largely began after 2004, when Congress reauthorized the Temporary Assistance for Needy Families (TANF), or "welfare reform" program.[10] Although this program has been less emphasized after the presidency of George W. Bush, HMI's family-related goals continue to include, among other things, ending the dependence of single parents on government benefits by promoting not only job preparation but also marriage (Carlton et al. 2009; Healthy Marriage Initiative 2009). Proponents argued that giving single mothers "accurate information on the value of marriage in the lives of men, women, and children," along with marriage skills education, encourage marriage and reduce divorce (Ooms 2005; Rector and Pardue 2004).

The disparity in marriage rates between the poor and those who are not poor has become significant enough that social scientists have coined a term for this situation—the *marriage gap*. Meanwhile, critics of programs specifically designed to motivate people to marry argue that low-income Americans value marriage and would like to marry, but marriage is difficult to achieve for many of them. Low-income single mothers want trustworthy, steadily employed husbands who will help with both finances and child care (Burton et al. 2009; Joshi, Quane, and Cherlin 2009). As a young, college-educated woman in a qualitative study of African American single mothers explained, "I realized that when I do decide to enter marriage, my partner must have the same ambitions as me or similar to [mine]. . . . Not too many men my age or older have the ambition that I have" (in Holland 2009, p. 173). For rich and

[10] The 1996 Personal Responsibility and Work Opportunity Reconciliation Act, or "welfare reform bill," effectively ended the federal government's sixty-year guarantee of assisting low-income mothers and children. The federal Aid to Families with Dependent Children (AFDC) program ended in 1997, and a different federal program, Temporary Assistance for Needy Families (TANF), ensued. TANF limits assistance to five years for most families, with most adult recipients required to find work within two years.

Evelyn Hockstein/MCT/Landov

The disparity in marriage rates between the poor and those who are not poor has become significant enough that social scientists have coined a term for this situation—the *marriage gap*. Low-income Americans value marriage and would like to marry, but marriage appears difficult to achieve for many, although certainly not all, of them.

poor alike, a wedding symbolizes personal achievement (Cherlin 2004; Mandara et al. 2008). After interviewing women in low-income neighborhoods, two researchers concluded that, for poor women, marriage

> has become an elusive goal—one they feel ought to be reserved for those who can support . . . a mortgage on a modest row home, a car and some furniture, some savings in the bank, and enough money left over to pay for a "decent" wedding. (Edin and Kefalas 2007, p. 508; also see Gibson-Davis 2009; King and Allen 2009)

Due to declining work opportunities for the less well educated and consequent high unemployment rates for men in poor neighborhoods, many potential husbands in these communities cannot promise a steady income (Burton and Tucker 2009; Harris and Parisi 2008; Huston and Melz 2004).

Relieving poverty will require solutions other than—or at least in addition to—promoting marriage.

The Relationship between Marriage and Poverty Data that relate child poverty rates to children's living arrangements show that residing with married parents does significantly lessen the likelihood of growing up in

poverty (Kreider and Fields 2005, Table 2). As you can see from Table 7.2, when all races are taken together, 4.6 percent of married-couple families live below the official poverty line. This figure compares with 13.3 percent of single male-householder families and 28.3 percent of single female-householder families. We might conclude that encouraging people to get married would work *somewhat* to lessen poverty (Amato 2005; Thomas and Sawhill 2005).

However, the association between marriage and poverty is hardly the whole story. Table 7.2 shows that—despite the fact that they are married—4.5 percent of white, nearly 8 percent of Black, about 13 percent of Hispanic, 6.5 percent of Asian, and about 10 percent of American Indians or Alaska Natives live in poverty. Obviously, marriage alone is not sufficient to alleviate poverty. For one thing, as shown in Table 7.2, female householders with no spouse present are more than twice as likely as their male counterparts to live in poverty (28.3 percent, compared with 13.2 percent). In addition to marital status, low wages for women contribute to poverty (Ozawa and Lee 2006).

Moreover, a majority of unmarried families is not living in poverty. We must conclude that marriage contributes to a family's economic well-being, but a child's

Table 7.2 U.S. Families Below Poverty Level*

Family Type	Married couple	Male householder, no spouse present	Female householder, no spouse present
All races, number	2,910,000	671,000	4,087,000
White, number	2,278,000	440,000	2,400,000
Black, number	346,000	177,000	1,484,000
Hispanic, number	903,000	139,000	881,000
Asian, number	178,000	26,000	57,000
American Indian/Alaska Native, number	32,785	**	62,062
All races, %	4.6	13.2	28.3
White, %	4.5	11.5	25.1
Black, %	7.9	20.4	36.6
Hispanic, %	13.3	14.7	36.0
Asian, %	6.5	11.6	15.4
American Indian/ Alaska Native, %	10.2	**	37.6

*Data are for 2006, except American Indian/Alaska Native data, which are 2008. Percentages in poverty for all races for married couples with children and for female householders with children in 2008 were 6.5 and 36.3, respectively.
**Data unavailable.

Sources: U.S. Census Bureau 2009a, Table S1702; U.S. Census Bureau 2009f, Table 694.

having married parents is not absolutely necessary to grow up above the poverty line.

Policies from the Family Change Perspective

Many policy makers maintain that Americans are struggling with economic and time pressures that get in the way of their ability to realize family values (Ozawa and Lee 2006; Teitler et al. 2009). As remedies for poverty, policy leaders in this camp propose structural solutions such as support for education, job training, drug rehabilitation, improved job opportunities, neighborhood improvements, small business development, and parenting skills education (Amato 2005; S. Brown 2004; Ozawa and Lee 2006). Indeed, "[l]ow-income communities have been neglected for so long that the resources needed to rebuild them will require a major shift in public priorities over an extended period of time, possibly generations" (Huston and Melz 2004, p. 956).

Andrew Cherlin (2009a) argues that in a climate of "contradiction" between the American values of commitment to marital stability and individual freedom and happiness, it is a bit naïve today to think that encouraging people to get or stay married will work to better facilitate raising children responsibly. He finds marital stability virtually impossible to enhance in our American values climate, and therefore argues that it is not pragmatic to continue to insist on legal marriage

as a public policy goal. Instead, he argues for *family stability*—supporting children and therefore their parents in whatever family form they find themselves (Cherlin 2009a).

Unfortunately, finding resources for ecological and structural support for families is even more difficult today than prior to the recession that began in 2008. Not only are resources more scarce, but also politicians and others debate whether (a) "welfare" encourages single parenthood while lessening the motivation to work, or (b) some form of family "safety net" is necessary and should not be stigmatized (DeParle 2009).

Having looked at research and policy on the question of whether marriage matters, we can conclude that marriage does matter, at least for those who can afford to get married. We end this chapter with a discussion of how marriage contributes to spouses' happiness and life satisfaction.

Happiness and Life Satisfaction: How Does Marriage Matter?

No longer a "marker of conformity," a wedding today marks a couple's public announcement that they have chosen marriage, among other available options, as a way to define and live their lives (Cherlin 2004, p. 856). Marriage, more than any other relationship, promises to shore up that love, helping partners to keep it, once discovered.

Academic research (e.g., Wienke and Hill 2009) and opinion polls show that both husbands and wives are far more likely than others to say that they are "very happy." Nearly two-thirds (62 percent) of marrieds say they are "very happy," compared to less than half (45 percent) of unmarrieds (Carroll 2005; Taylor, Funk, and Craighill 2006). What is it about the experience of being married that works to create this difference?

For one thing, there are some pragmatic reasons that spouses (and cohabitors, but to a lesser degree) benefit from an *economy of scale*. Think of the saying, "Two can live as cheaply as one." Although this principle is not entirely true, some expenses, such as rent, do not necessarily increase when a second adult joins the household (Thomas and Sawhill 2005; Goode 2007 [1982]; Waite 1995). Then, too, the promise of permanence associated with the marriage premise accords spouses the security to develop some skills and to neglect others because they can count on working in complementary ways with their partners (Goode 2007 [1982]; Nock 2005). Furthermore, "[s]pouses act as a sort of small insurance pool against life's uncertainties, reducing their need to protect themselves *by themselves* from unexpected events" (Waite 1995, p. 498).

In addition, marriage offers enhanced social support (Manning and Brown 2006). Marriage can connect people to in-laws and a widened extended family, who may be able to help when needed—for instance, with child care, transportation, a down payment on a house, or just an emotionally supportive phone call. The enhanced social support that often accompanies marriage works to encourage the union's permanence (Giddens 2007). For example, family and friends send anniversary cards, celebrations of the years the couple has spent together and reminders of the couple's vow of commitment. Beginning with a public ceremony, marriage makes for what sociologist Andrew Cherlin (2004) calls *enforced trust*:

> Marriage still requires a public commitment to a long-term, possibly lifelong relationship. This commitment is usually expressed in front of relatives, friends, and religious congregants. . . . Therefore, marriage . . . lowers the risk that one's partner will renege on agreements that have been made. . . . It allows individuals to invest in the partnership with less fear of abandonment. (p. 854)

Furthermore, marriage offers *continuity*, the experience of building a relationship over time and resulting in a uniquely shared history. And, finally, marriage provides individuals with a sense of obligation to others, not only to their families but also to the broader community (Goode 2007 [1982]; Wolfinger and Wolfinger 2008). This, in and of itself, gives life meaning (Waite 1995, p. 498).

Marital Satisfaction and Choices Throughout Life

Our theme of making choices throughout life surely applies both to couples anticipating marriage as well as to decisions made during the early years of marriage. We'll examine these topics now.

Preparation for Marriage

Given today's high divorce rate, clergy, teachers, parents, policy makers, and others have grown increasingly concerned that individuals be better prepared for marriage. High school and college family life education courses are designed to prepare individuals of various racial/ethnic groups for marriage (Coalition for Marriage, Family and Couples Education 2009; DeMaria 2005; Fincham, Hall, and Beach 2006). Premarital counseling, which often takes place at churches or with private counselors, is specifically oriented to couples who plan to marry. For example, many Catholic dioceses require premarital counseling before a couple may be married by a priest. Christians of other denominations, and Islam, Jewish, and other religions offer premarital

AP Images/Erik S. Lesser

As an effort to improve family life, this neighborhood restoration project involves the community in effecting structural, as opposed to cultural or attitudinal, change.

counseling as well. Illustrating the connection between private lives and public interest, a few states now require premarital counseling for engaged couples under age eighteen (Holloway 2008; Murray 2006).

Premarital counseling goals involve helping the couple to evaluate whether their relationship should lead to marriage; develop a realistic, yet hopeful and positive vision of their future marriage; recognize potential problems; and learn positive problem-solving and other communication skills (Dinkmeyer 2007; Holloway 2008). Some programs have been developed especially for those contemplating post-divorce cohabitation or remarriage, particularly when children are involved (Gonzales 2009).

Research designed to assess how effective these programs are shows that they do improve a couple's communication skills and relationship quality, at least in the short term (Blanchard et al. 2009). Unfortunately, however, we have little data on the relationship between premarital counseling and relationship stability. Furthermore, "a lack of racial/ethnic and economic diversity in the samples prevented reliable conclusions about the effectiveness of [premarital counseling] for disadvantaged couples, a crucial deficit in the body of research" (Hawkins et al. 2008, p. 723).

Among other factors, success depends upon the personality characteristics of each partner, as well as on couple characteristics, such as the interactional styles with which they begin the program, influences from their families of origin, and their motivation to learn from the program (Murray 2004). One study found that those who become actively involved in premarital counseling are more likely to value marriage and to be kind and considerate to begin with (Duncan, Holman, and Yang 2007). Overall, however, family experts see these programs as important, especially for those under age eighteen and for adult children of troubled or divorced families (Dinkmeyer 2007; Holloway 2008). Psychologist Scott Stanley identifies four benefits of premarital education:

(a) it can slow couples down to foster deliberation,
(b) it sends a message that marriage matters, (c) it can help couples learn of options if they need help later, and (d) there is evidence that providing some couples with some types of premarital training . . . can lower their risks for subsequent marital distress or termination. (Markman, Stanley, and Blumberg 2001, p. 272)

The idea of "slowing couples down" prompts questions about the relationship between age at marriage and the union's stability.

Age at Marriage, Marital Stability and Satisfaction

In the 1950s, men tended to marry at about age twenty-four and women at about age twenty-two (Aldous 1978). We've seen that the median age at first marriage today is about

twenty-six for women and twenty-eight for men—considerably older than at mid-twentieth century. Nevertheless, "Although much attention has been paid to the increasing age at first marriage in the United States, many Americans continue to marry at young ages. More than one-quarter of young women and more than 15 percent of young men marry before their twenty-third birthday" (Uecker and Stokes 2008, p. 844). Is there a "right" age to marry?

Over the past several decades, researchers have given considerable attention to the relationship between age at first marriage and marital stability. Findings consistently show that the odds of marital stability increase with age at marriage (Amato et al. 2007). Those who marry young are, on average, more emotionally immature and impulsive and less apt to be educationally, financially, or psychologically prepared to responsibly choose a partner or perform marital roles (Clements, Stanley, and Markman 2004; Martino, Collins, and Ellickson 2004). Low socioeconomic origins, poor communication and problem-solving skills, premarital pregnancy, lack of interest in school, and financial struggles are associated with marrying early (Larson and Hickman 2004; Uecker and Stokes 2008). Teen marriages (the majority between eighteen- and nineteen-year-olds) are the least stable (Bramlett and Mosher 2001; McGinn 2006b).

Typically, policy makers have developed programs encouraging teens to postpone marriage. Meanwhile, when they do occur, early marriages would benefit by recognition and support:

Given early marriage's known association with marital dissolution, it is important to pay adequate attention to . . . individuals who marry in early adulthood. . . . Early marriage comes with its own set of difficulties, however, and if understanding and supporting all marriages—be they early, normative, or late—is a goal of scholarship and policy, this population should garner more attention from both researchers and policy makers. (Uecker and Stokes 2008, p. 844, 845)

Until recently, research on age at first marriage focused solely on marital stability. However, a 2009 analysis of findings from several major national surveys examined the relationship between age at first marriage and marital happiness and satisfaction (Glenn, Uecker, and Love 2009). Findings show that marriages occurring today, when spouses are between ages twenty-two and twenty-five, are most likely to be not only stable but also happy. Spouses who first married after age thirty reported lower marital satisfaction even as they were likely to stay married.

Based on a thorough review of prior research, Glenn, Uecker, and Love (2009) offer possible explanations. For one thing, more "set-in-their-ways" older spouses may find it more difficult to fashion a compatible life together. Also, marrying after about age thirty may mean selecting a spouse from a market in which "lots of the good ones are gone." Or it may suggest that an individual has been searching for the perfect partner—a situation that can only lead to later disappointment. Despite lower satisfaction levels, however, age at first marriage does act as a deterrent to divorce. If they are older, mildly unhappy partners may feel hesitant to reenter singlehood, the dating game, or the marriage market. Some advice from the study:

> The findings of this study *do* indicate that for most persons, little or nothing in the way of marital success is likely to be gained by deliberately delaying marriage beyond the mid twenties. For instance, a 25 year old person who meets an excellent marriage prospect would be ill-advised to pass up that opportunity only because he/she feels not yet at the ideal age for marriage. Furthermore, delaying marriage beyond the mid twenties will lead to the loss during a portion of young adulthood of any emotional and health benefits that a good marriage would bring. . . . On the other hand, it is extremely important to stress that the findings of this study should not lead anyone of any age to panic and thus make a bad choice of a spouse. (Glenn, Uecker, and Love 2009, pp. 42–43)

The First Years of Marriage

The first years of marriage tend to be the happiest, with gradual declines in marital satisfaction afterward (Dush, Taylor, and Kroeger 2008; Tach and Halpern-Meekin 2009). Why this is true is not clear. One explanation points to life cycle stresses as children arrive and economic pressures intensify; others argue that falling in love and new marriage are periods of emotional intensity from which there is an inevitable decline (Glenn 1998; Whyte 1990). We do know something about the structural advantages of the early years of marriage, and it is likely that these contribute to high levels of satisfaction. For one thing, partners' roles are relatively similar or unsegregated in early marriage. Spouses tend to share household tasks and, because of similar experiences, are better able to empathize with each other.

In the 1950s, marriage and family texts characteristically referred to the first months and years of marriage as a period of adjustment, after which, presumably, spouses had learned to play traditional marital roles. Today we view the first months and years of marriage more as a time of role-*making* than of role-*taking*.

From the interaction-constructionist theoretical perspective (see Chapter 2), **role-making** refers to modifying or adjusting the expectations and obligations traditionally associated with a role. Role-making involves issues explored more fully in other chapters of this text. Newlyweds negotiate expectations for sex and intimacy (Chapter 5), establish communication (Chapter 12) and decision-making patterns (Chapter 13), balance expectations about marital and job or school responsibilities (Chapter 11), and come to some agreement about becoming parents (Chapter 9) and how they will handle and budget their money. When children are present, role-making involves negotiation about parenting roles (Chapter 10). Role-making issues peculiar to remarriages are addressed in Chapter 16. Generally, role-making in new marriages involves creating, by means of communication and negotiation, identities as married people (Rotenberg, Schaut, and O'Connor 1993). The time of role-making is not a clearly demarcated period but continues throughout marriage.

Couples must also accomplish certain tasks during this period. In general, "the solidarity of the new couple relation must be established and competing interpersonal ties modified" (Aldous 1978, p. 141; Rotenberg, Schaut, and O'Connor 1993). Getting through this stage requires making requests for change and negotiating resolutions, along with renewed acceptance of each other. Indeed, research by psychoanalyst John Gottman shows that communication as newlyweds tends to influence the later happiness—and even the permanence—of the marriage (Gottman et al. 1998; Gottman and Levenson 2000, 2002). The couple constructs relationships and interprets events in a way that reinforces their sense of themselves as a couple (Wallerstein and Blakeslee 1995). Relationships with parents change (Sarkisian and Gerstel 2008). Perhaps expectedly, a recent study in Switzerland found that newlyweds were happier when relatives were supportive but not interfering (Widmer et al. 2009).

A national study undertaken by family researchers at Creighton University in Omaha identified three main, potentially problematic topics for couples in first marriages: (1) money—balancing job and family, dealing with financial debt brought into the marriage by one or both spouses, and what to do with money income; (2) sexual frequency; and (3) agreeing on how much time to spend together—and finding it! Challenges associated with learning to balance work and/or college courses and a marital relationship are real (Christopherson 2006). Feeling supported by parents and extended kin helps (Kurdek 2005). Other issues the couples in the Creighton study mentioned above involved expectations about who would do household tasks (and how well), communication challenges, and problems with in-laws (Brennan 2003; Risch, Riley, and Lawler 2004). A more general, necessary goal for couples in early marriage is to create couple connection.

Creating Couple Connection

Partners who desire enduring emotional relationships must keep their relationship as a high priority. Some research suggests that, on average, today's marriages are happier when both spouses are employed (Schoen, Rogers, and Amato 2006). Also, husbands and wives who engage in supportive communication and who together pursue leisure activities that they both enjoy are more

© Corbis/ Jupiter Images

Because, more than others, a married couple can count on continuity, bolstered by enforced trust, spouses are freer to plan for a future together.

compatible and satisfied with their marriage (Crawford et al. 2002; Johnson et al. 2005; Kurdek 2005).

Research on marital satisfaction also suggests that couples who make time for shared new experiences are more happily married (Burpee and Langer 2005). Assuredly for wives, time spent together is important to marital happiness (Gager and Sanchez 2003). An important psychologist and expert on marital communication, John Gottman, offers this advice:

> Happy, solid couples nourish their marriages with plenty of positive moments together. . . . Too often, families lead complex—even grueling—lives in which they sacrifice the happy times for more materialistic, fleeting goals. . . . Sundays at the office take the place of Sundays at the park. But if you want to keep your marriage alive, it's essential to rediscover—or perhaps simply make time for—those experiences that make you feel good about your spouse and your marriage. (1994, p. 223)

Comparative analysis of data collected from national samples in 1980 and 2000 revealed that spouses spend less time interacting with each other now than they did twenty-five years ago. (However, their reported marital satisfaction had not declined significantly, partly because they were more satisfied with the decision-making equality in their marriage [Amato et al. 2003].) Nevertheless, increased emphasis on other matters, such as managing debt, job pressures and long work hours, or children's needs, can result in exhaustion, increased conflict, and slow emotional erosion (Dew 2008; Roberts and Levenson 2001). Noting that "[l]ove is not an express lane concept,"

observers suggest creating daily "connecting moments" when you can be alone together and pay attention to your relationship (Brennan 2003; Brotherson 2003).

Keeping one's marriage vital requires that partners consciously and continuously strive to maintain intimacy. An emotionally meaningful relationship does not develop long term "by drift or default" (Cuber and Harroff 1965, p. 145). Satisfaction with the marital relationship has a great deal to do with the choices that partners make. One important set of decisions involves practicing positive communication skills. Research that followed 135 Denver couples over thirteen years, from the time they were engaged and into their marriages, concluded that "the seeds of marital distress and divorce are sown for many couples before they say 'I do.' . . . [N]egative premarital and early marital interactions . . . prime a marriage for the erosion of positivity over time" (Clements, Stanley, and Markman 2004, p. 621). Building better communication skills is addressed in Chapter 12.

Summary

- The marriage premise involves permanence and—for Western societies—monogamous sexual exclusivity.

- New norms for love-based marriage gradually became prevalent in Europe throughout the 1700s and 1800s. Expectations for personal happiness and love in marriage have changed the marriage premise over the past three hundred years.

- Marriages and families have become deinstitutionalized. Marriage has changed from institutionalized, to companionate, to individualistic.

- Overall, researchers consistently find a significant correlation between marriage and many positive outcomes for both adults and children, but the relationship is more complicated than it first appears because some of the benefits associated with marriage are due to selection effects.

- Scholars and policy makers who view individualized marriage and the postmodern family as indications that the institution of marriage and family is in "decline" and "breakdown" have proposed ways to effect a turnaround—such as covenant marriage and the Healthy Marriage Initiative.

- Scholars and policy makers who view individualized marriage and the postmodern family as results of inevitable historical change, resulting in struggle for some families more than for others, have proposed structural solutions to poverty and family struggle—such as higher wages for women, neighborhood development, an adequate minimum wage, and more employment opportunities.

- Optional and less permanent than in the past, marriage continues to offer benefits; married adults are more likely than others to say that they are happy with their lives. Being married continues to help bolster the marriage premise, due largely to family and community social support that results in enforced trust.

- There is society-wide concern about preparation for marriage. Premarital counseling and family life education are two approaches that have been developed, but we need more research data on their effectiveness.

- Partners change over the course of a marriage, so a relationship needs to be adaptable if it is to continue to be emotionally satisfying.

- Spouses in the first years of marriage engage in role-making, a process that includes—among other things—negotiating issues surrounding money, sexual frequency, and time together.

Questions for Review and Reflection

1. Discuss the marriage premise, with its expectations of permanence and sexual exclusivity. Describe how love has changed the marriage premise over the past three centuries. What does family demographer Stephanie Coontz mean when she says that "love conquered marriage"?

2. Pointing out the pros and cons of each, compare and contrast companionate marriage and individualized marriage.

3. Explain the connection between federal "welfare reform" and the National Marriage Initiative.

4. Recognizing the possibility of selection effects, what are some ways that the experience of being married can enhance happiness and life satisfaction?

5. **Policy Question.** Are you more inclined to agree with policy leaders from the family decline or from the family change perspective? Using research evidence from this chapter, explain the reasons for your answer.

Key Terms

collectivist society 172
communal society 172
companionate marriage 175
covenant marriage 184
deinstitutionalization of marriage 174
expectations of permanence 170
expectations of sexual exclusivity 170
experience hypothesis 179
family of orientation 172
family of procreation 172
individualism 172

individualistic society 172
individualized marriage 176
institutional marriage 174
kin 172
marriage premise 170
polyamory 171
polygamy 170
role-making 189
selection hypothesis 179
social institution 170
swinging 171

Online Resources

Sociology CourseMate

www.CengageBrain.com

Access an integrated eBook, chapter-specific interactive learning tools, including flash cards, quizzes, videos and more in your Sociology CourseMate, accessed through CengageBrain.com.

CENGAGENOW™

www.CengageBrain.com

Want to maximize your online study time? Take this easy-to-use study system's diagnostic pre-test, and it will create a personalized study plan for you. By helping you identify the topics that you need to understand better and then directing you to valuable online resources, it can speed up your chapter review. CengageNOW even provides a post-test so you can confirm that you are ready for an exam.

8

Living Alone, Cohabiting, Same-Sex Unions, and Other Intimate Relationships

Jon Riley/Getty Images

When answering national polls, nearly all Americans say that our families are very important to us. Our families are essential to us, yet they are likely to be different from the "traditional" family of the mid-twentieth century—as we saw in Chapters 1 and 7. Today's "postmodern" family is characterized by a diversity of family forms. As we look at various living arrangements in this chapter, we will examine some of these family forms. In the process, we will find that the distinction between being single and being married is not that sharp today. Many people who are legally single are embedded in families of one form or another.

Many college students today think of "being single" as not being in a romantic relationship. By this way of thinking, a person in a dating relationship or cohabiting would not be single. To the U.S. Census Bureau, however, *single* means *unmarried*, and many researchers tend to use these terms interchangeably. In this chapter, we will examine what social scientists know about the large and growing number of singles: the never-married, the divorced, and the widowed.

Marriage as a social institution is explored in Chapter 7. In this chapter, we'll discuss a variety of living arrangements other than marriage. We will explore cohabitation, as well as same-sex families. We'll also look at living alone, residing with one's parents, and living communally, or in groups. To begin, we'll examine some reasons for the increasing proportion of unmarrieds in our society today.

Reasons for the Increasing Proportion of Unmarrieds

Figure 7.2, in Chapter 7, shows the proportions of never-married, divorced, and widowed individuals by various racial/ethnic categories. These percentages are considerably higher than in past decades. In 1970, less than 28 percent of U.S. adults were single. Today that number is about 44 percent (Saluter and Lugaila 1998, Figure 1; U.S. Census Bureau 2010b, Table 56). Much of this change is due to a growing proportion of widowed elderly. However, the high proportion of singles also results from a high divorce rate and young adults' postponing marriage, together with a dramatically escalating cohabitation rate. Several social factors—demographic, economic, technological, and cultural—encourage Americans to postpone marriage, not to marry at all, or to choose to divorce rather than stay married.

Demographic, Economic, and Technological Changes

One reason for the growing proportion of singles is *demographic*, or related to population numbers. A high rate of heterosexual marriage presumes that there are matching numbers of marriage-age males and females in the population. Therefore, the **sex ratio**—the number of men to women in a given society or subgroup—influences marital options and singlehood.[1] Throughout the nineteenth and early twentieth centuries, the United States had more men than women, mainly because more men than women migrated to this country and, to a lesser extent, because a considerable number of young women died in childbirth. Today this situation is reversed due to changes in immigration patterns and greater improvement in women's health. In 1910, there were nearly 106 men for every 100 women. The sex ratio was 100, or "even," a few years after the end of World War II, in about 1948. In 2008, there were about 97 men for every 100 women (U.S. Census Bureau 2010b, Table 7).

Beginning with middle age, there are increasingly fewer men than women in every racial/ethnic category. Sex ratios differ somewhat for various racial/ethnic categories, however. For instance, at younger ages, the sex ratio is lower for blacks and Native Americans than for Hispanics or non-Hispanic whites. Incarceration rates as well as employment prospects also affect individuals' odds of marrying (Dixon 2009; King and Allen 2009).

In addition to demographics, *economic* factors have increased the proportion of nonmarrieds. For one thing, expanded educational and career options for college-educated women over the past several decades have encouraged many of them to postpone marriage:

> With so many options open to them, and with so little pressure on them to marry by their early twenties, the lives of young American women today have changed almost beyond recognition from what they were 50 years ago. And most of them take on their new freedoms with alacrity, making the most of their emerging adult years before they enter marriage and parenthood. (Arnett 2004, p. 7)

In addition, many middle-aged and older career and/or divorced women tend to look on marriage skeptically, viewing it as a bad bargain once they have gained financial and sexual independence (Levaro 2009; Swartz 2004; Zernike 2006).

Moreover, research shows that people today view marriage as a status that needs to be financially affordable. The fact that many men's earning potential has declined, relative to women's, may make marriage less attractive to both genders (Raley and Bratter 2004). Growing economic disadvantage and uncertainty

---

[1] The sex ratio is expressed in one number: the number of males for every 100 females. Thus, a sex ratio of 105 means that there are 105 men for every 100 women in a given population. More specialized sex ratios may be calculated—for example, the sex ratio for specific racial/ethnic categories at various ages or the sex ratio for unmarried people only.

appear to have made marriage less available to many who might want to marry but feel that they can't financially afford it (Gibson-Davis 2009; King and Allen 2009; Roberts 2009).

In addition to economics, *technological* changes over the past sixty years have affected the proportion of singles. Beginning with the introduction of the birth control pill in the 1960s, improved contraception has contributed to the decision to delay or forego marriage. With effective contraception, sexual relationships outside marriage and without great risk of unwanted pregnancy became possible (Gaughan 2002; Coontz 2005b). Moreover, as discussed in Chapter 1, reproductive technologies such as artificial insemination, offer the possibility for planned pregnancy to unpartnered heterosexual women as well as to same-sex couples. In addition to these structural reasons for the increasing proportion of nonmarrieds, cultural changes have played a part.

The increase in the number and proportion of unmarrieds in our society is a result of many factors, including unfavorable sex ratios, especially in older age categories; economic constrictions; improved contraception; and changing attitudes toward marriage and singlehood, which have resulted in more young adults postponing marriage.

Cultural Changes

As discussed in Chapter 2, social scientists note a fairly new life cycle stage called *emerging adulthood:* Young people today spend more time in higher education and/or exploring options regarding work, career, and family making than in the past (Arnett 2000; Furstenberg 2008). Although Americans of all ages have helped to increase the proportion of singles, emerging adulthood accounts for much of the greater proportion of unmarrieds today.

Several other cultural changes over the past few decades also account for the growing proportion of nonmarrieds. First, attitudes toward nonmarital sex have changed dramatically over past decades. With about two-thirds of adults of all ages approving, sexual intercourse outside marriage has become widely accepted ("Marriage" 2008). Currently, "hooking up"—discussed in Chapter 6 and called "recreational sex" in the 1960s—has gained attention (Wilson 2009).

Second, as American culture gives greater weight to personal autonomy, many find that—at least "for now"—singlehood is more desirable than marriage (Furstenberg 2008; Meier and Allen 2009). As one young man explained:

> It would kind of bum me out to be married. One day I was at work and my friend called me up from Florida and said, "What are you doing?" I'm like, "Just working," and he said, "Can you come down?" I'm like, "When?" and he's like, "Tomorrow," and I'm like, "Well, let me see what I can do." So I took a week off all of a sudden and went down to Florida. And I know I'd never be able to do that if I was married. (in Arnett 2004, p. 101)

A young woman evidences a similar attitude:

> I hope to be married by the time I'm 30. I mean, I don't see it being any time before that. I just think I have a lot of life left in me, and I want to enjoy it. There's so much out there, not that you couldn't see it with your husband, but why have to worry "Is he going to get mad at this?" Just go out and enjoy life and then settle down, and you'll know you've done everything possible that you wanted to do, and you won't regret getting married. (in Arnett 2004, p. 103)

Being single has become an acceptable option, rather than the deviant lifestyle that it was once thought to be. During the 1950s, people (including social scientists) tended to characterize the never-married as selfish, neurotic, or unattractive. The divorced were also stigmatized. These views have changed so much that the popular press today enthusiastically runs stories about those who are "embracing the solo life" (Hurwitt 2004; Sanders 2004).

Then too, getting married is no longer virtually the only way to gain adult status. Before about 1940, the most legitimate reason for leaving home, at least for

© Asia Images/Jupiter Images

women, was to get married. Today, 45 percent of young men and 39 percent of young women say that they first left home for other reasons, often to attend college and/or "to gain independence." The nationwide General Social Survey (GSS), conducted by the National Opinion Research Center (NORC), found that people now see becoming self-supporting as the first transition to adulthood, followed by no longer living with one's parents, having a full-time job, completing school, being able to support a family financially, and—sixth on the list—getting married (T. Smith 2003). As more young adults choose to claim independence simply by moving, marriage has lost its monopoly as the way to claim adulthood (Arnett 2004; Furstenberg et al. 2004).

Moreover, cohabitation is emerging as a socially accepted alternative to marriage (Cherlin 2009a). Among teens, about 70 percent say it's okay for couples to live together before they get married, and fewer than half (45 percent) of adults today think that it is "morally wrong" to have a baby outside of marriage (Lyons 2004; "Marriage" 2008). Forty-two percent of U.S. adults believe that an unmarried couple that has lived together for one year is just as committed as a couple that has been married for one year ("Marriage" 2008). Although beliefs such as these are considerably less likely among many recent immigrants and members of some religions, young adults in general experience greater independence and less parental pressure to marry than in the past (Arnett 2004; Rosenfeld 2008).

Finally, the changing nature of marriage itself may render marriage less desirable now than in the past (Cherlin 2009a). Marriage has become less strongly defined as permanent, and high divorce rates have led at least some singles to fear a potential divorce of their own. This fear reduces the likelihood that they will marry (Waller and Peters 2008). Historically, the expectation of permanence offered a significant benefit to getting married (Amato et al. 2007). If marriage is losing its permanent status, then

> [i]ndividuals, as a result, have less faith that a successful marriage is possible, and they transfer support for marriage into support for other coupling arrangements, such as cohabitation—arrangements that are easier to dissolve if (and when) problems arise. (Willetts 2006, p. 125)

To summarize, it appears that much of the increase in singlehood results from (1) low sex ratios, particularly in certain regions and among specific age and racial/ethnic groups; (2) increasing educational and economic options for some, coupled with growing financial disadvantage for others; (3) technological changes regarding pregnancy; and (4) changing cultural attitudes toward marriage and singlehood: greater acceptance of premarital sex and emphasis on personal autonomy, development of singlehood and cohabitation as acceptable lifestyles, marriage having lost its monopoly as a way to

claim adulthood, and the diminished permanence of marriage. We can apply the exchange theoretical perspective (see Chapter 2) to this issue of less compelling reasons to get married. Overall, as people weigh the costs against the benefits of being married, marriage offers fewer benefits now, relative to being single, than in the past. If fewer Americans are married today, what are singles' various living arrangements?

Singles—Their Various Living Arrangements

As a category, singles make a variety of choices about how to live. Some live alone; others, with parents; still others, in groups or communally. Some unmarrieds cohabit with partners of the same or opposite sex. This section explores these living arrangements.

Living Alone

The number of one-person households has increased dramatically over past decades. Individuals living alone now make up over 28 percent of U.S. households—up from just 8 percent in 1940 (U.S. Census Bureau 1989, Table 61; U.S. Census Bureau 2008a). Figure 8.1 gives the percentage of U.S. adults living alone, by age. As you can see from the figure, the likelihood of living alone increases with age. This is true for all racial/ethnic groups and is markedly higher for older women than for older men (U.S. Census Bureau 2010b, Table 58).

Asians and Hispanics of all ages are less likely to live alone than are blacks or non-Hispanic whites. Although the social science literature tends to focus on middle-class blacks as married, researchers have pointed to a growing number of middle-class, never-married black singles who live alone (Marsh et al. 2007). Significantly less likely to be married than other racial/ethnic groups (see Figure 7.2), blacks are more likely than others to be living by themselves, particularly in older age groups. More collectivist Asian and Hispanic cultures help to discourage living alone in these ethnic groups. Of course, some who live alone are actually involved in long-term committed relationships.

Living Alone Together

An emerging lifestyle choice is *living alone together (LAT)*. Here a couple is engaged in a long-term relationship, but each partner also maintains a separate dwelling (A. Roberts 2005). The number of these relationships is difficult to ascertain, because the U.S. Census Bureau does not measure them. However, European social scientists have noted this family form. According to David Popenoe, codirector of the National Marriage Project at

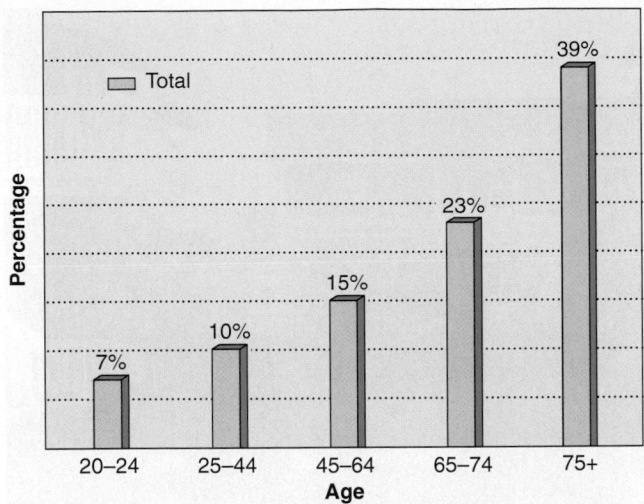

Figure 8.1 Percentage of people over 19 years old living alone, by age, 2008

Source: Calculated from U.S. Census Bureau 2010b, Table 58.

Table 8.1 Percentage Living with their Parents, Sex and Year

	Age 18–24	Age 25–34
Men		
1960	52	11
1995	58	15
2002	55	14
2008	56	15
Women		
1960	35	7
1995	47	8
2002	46	8
2008	48	10

Source: U.S. Census Bureau 2009e, Table AD-1.

Rutgers University, LAT is clearly an emerging trend in the United States. We know this partly from anecdotal evidence and "partly [from] the fact that every other significant European trend in family life has turned out to be happening in America" (Popenoe, quoted in Brooke 2006). Although we know relatively little about LAT, it is apparently at least partly motivated by a desire to retain autonomy. As one woman said, "I like my own life, my own identity and want to keep it. I like having the things I love around me." As one man put it, "I am as devoted as any husband to her, . . . but I like my alone time and being around my stuff, not [hers]" (in Brooke 2006).

For an older adult, living alone together "allows for unencumbered contact with adult children from previous relationships while protecting their inheritance and offering freedom from caregiving as a prescribed duty. . . Separate homes also allow a tangible line of demarcation in terms of gender equity and the distribution of household labor" (Levaro 2009, p. F10). Some young adults in LAT relationships reside with their parents (S. Smith 2006).

Living with Parents

A large proportion of young adults today are living with one or both parents. In Table 8.1, we see that the percentage of young adults living at home has increased moderately since 1960. In 2008, 56 percent of men and 48 percent of women age eighteen through twenty-four lived with their parents. For men and women age twenty-five to thirty-four, the proportions are 15 percent and 10 percent, respectively (U.S. Census Bureau 2009e, Table AD-1). Overall, about 11 percent of all adults over age eighteen live with their parents. Some adults who live with their parents have never moved out, but others—called boomerangers—have left home and then returned (Arnett 2004; Wang and Morin 2009).

Back in 1940, the proportion of adults under age thirty living with their parents was quite high. Sociologists Paul Glick and Sung-Ling Lin suggest why:

> The economic depression of the 1930s had made it difficult for young men and women to obtain employment on a regular basis, and this must have discouraged many of them from establishing new homes. Also, the birth rate had been low for several years; this means that fewer homes were crowded with numerous young children, and that left more space for young adult sons and daughters to occupy. (Glick and Lin 1986, p. 108)

These same reasons apply to many young people today.

Looking at Table 8.1, you will note that the proportion of young adults living with parents declined slightly between 1995 and 2002. As Figure 3.1 illustrates, median household income rose between 1995 and 2002. We can argue that a better economy during that period helped to reduce the percentage of adults living with their parents.

Not surprisingly, the economic recession that began in 2008 led to an increase in boomerangers (Palmer 2008). In one national poll, 10 percent of adults between ages eighteen and thirty-four said they had recently moved back home due to the poor economy (Wang and Morin 2009). A high unemployment rate adds to singles' job-finding difficulties. Then too, "in recent decades, young adults [have been] taking longer to find secure, well-compensated employment" (Danziger and Rouse 2008, F8). Moreover, even before the recent recession, housing in urban areas was too expensive for many singles to maintain their own apartments.

Although many observers focus on economic reasons as primary causes for living with parents, some others see this trend as resulting from "an indulgent parenting style" that demands little of young adult offspring (Pisano 2005). However, it is also true that parents may appreciate the companionship and help of their live-in adult children (Arnett 2004; Straus 2009). Researchers and journalists who have interviewed parents and adult children note that the "generation gap," which in past decades might have made financial dependency and close living annoying to one or both generations, seems to have vanished. According to sociologist Barbara Risman, parents and adolescents or emerging adults today have more in common in lifestyle and values than did baby boomers and their depression/World War II–era parents (in Jayson 2006a). On the other hand, conflict with parents can precipitate the decision to move out and take up residence with a romantic partner (Sassler 2004).

We need to note here that living with parents can occur in a variety of circumstances. Some ethnic groups, such as the Hmong, expect single women to reside with their parents until marriage. Unmarried women who have babies, especially those who became mothers in their teens, may be living with parents. Formerly married young men and women may return to their parental home after divorce. Just as economic considerations, desire for emotional support, or need for help with child raising may lead young singles to live with parents, similar pressures may encourage singles to fashion group or communal living arrangements.

Group or Communal Living

Groups of single adults and perhaps children may live together. Often these are simple roommate arrangements. But some group houses purposefully share aspects of life in common. **Communes**—that is, situations or places characterized by group living—have existed in American society throughout its history and have widely varied in their structure and family arrangements.[2]

Living communally has declined in the United States since its highly visible status in the 1960s, when many communes were established as ideological retreats from what their founders saw as the misguided American life characteristic of the 1950s. However, some communes that were established then still exist. Furthermore, small-scale and nonideological versions of group living have more recently surfaced (Jacobs 2006). For example, the recession that began in 2008 prompted 12 percent of single adults between ages eighteen and thirty-four to

Courtesy of Matt Kramer

Cohousing started in Denmark and spread to the United States in the early 1980s. Residents own their own homes, with residences clustered closely to leave open space, which is community-owned. Cohousing complexes, which typically combine private areas with communal kitchens—and, often with community gardens—offer alternative living arrangements and can be a way to cope with some of the problems of aging, unattached singlehood, or single parenthood.

acquire a roommate, and another 2 percent took in a boarder (Wang and Morin 2009).

Communal living, either in single houses or in cohousing complexes that combine private areas with communal kitchens and "family rooms," may be one way to cope with some of the problems of aging, unattached singlehood, or single parenthood ("Cohousing" 2006). In a small but growing number of cohousing complexes, people of diverse races, ethnicities, and ages choose to reside together, sharing some meals and recreational activities. Communal living is designed to provide enhanced opportunities for social support and companionship. More commonly, financial considerations and the desire for companionship encourage romantically involved singles to share households. We turn now to a discussion of "living together," or cohabitation.

---

[2] In some communes, such as the traditional Israeli kibbutzim (Spiro 1956) and nineteenth-century American groups such as the Shakers and the Oneida colony (Kephart 1971; Kern 1981), all economic resources are shared. Work is organized by the commune, and commune members are fed, housed, and clothed by the community. Other communes may have some private property. Sexual arrangements also vary among communes, ranging from celibacy to monogamous couples (the kibbutzim and some communes in the United States) to the open sexual sharing found in both the Oneida colony and some modern American groups. Children may be under the control and supervision of a parent, or they may be raised more communally, with a de-emphasis on biological relationships and responsibility for discipline and care vested in the entire community.

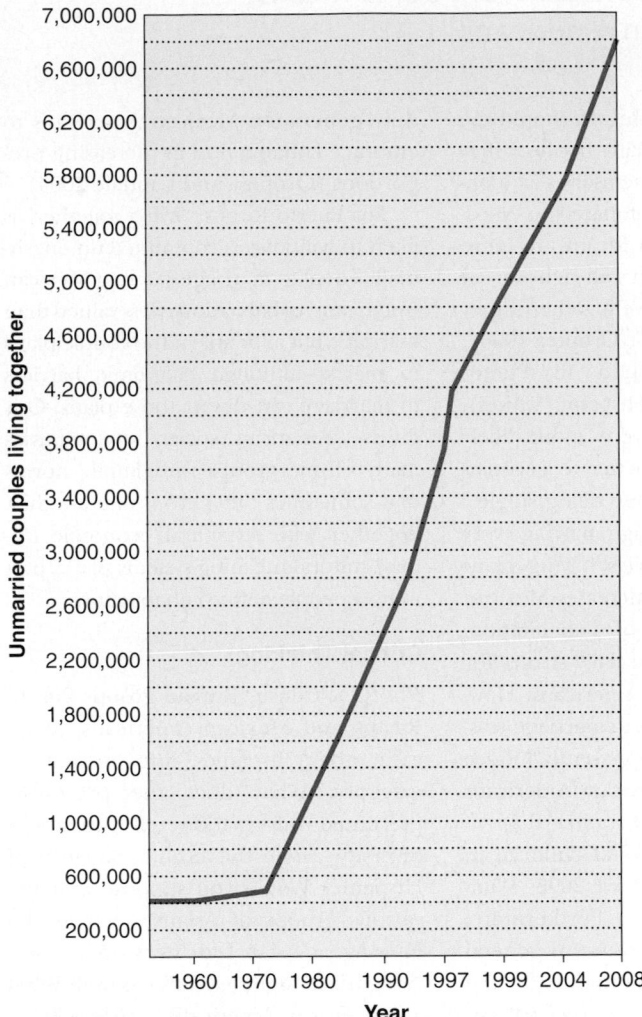

Figure 8.2 Unmarried heterosexual couples living together in the United States, 1960–2008

Sources: Edwards 2009; Glick and Norton 1979; Schneider 2003; Simmons and O'Connell 2003; U.S. Census Bureau 2007a, Table 61.

Cohabitation and Family Life

Cohabitation, or nonmarrieds living together, gained widespread acceptance over the past several decades and is "widely viewed as one of the most important changes in family life in the past 40 years, dramatically altering the marital life course by offering a prelude to or a replacement for marriage" (S. Smith 2006, p. 7). Not only in the United States but in other industrialized nations as well, "living together" has dramatically increased. In this country, the cohabitation trend spread widely in the 1960s, took off sharply in the 1970s, and has risen steadily ever since, as Figure 8.2 illustrates. Today, about 6.8 million U.S. heterosexual couples cohabit (Edwards 2009). This number may be an undercount, because cohabitors do not necessarily move into separate housing. Instead,

they may live together in a parental home or reside with roommates and therefore would not be included in a census count (Manning and Smock 2005).

By 2005, an estimated one-half of U.S. women age fifteen through forty-five had cohabited with a male partner at some time in their lives—an increase of nearly 10 percent since 1995 (Chandra et al. 2005, Table 47). The proportion of teenage cohabitors is fairly small; nevertheless, the likelihood of a fifteen- to nineteen-year-old's currently cohabiting nearly tripled in the twenty-five years between 1980 and 2005 (Houseknecht and Lewis 2005).

By 2008, opposite-sex, unmarried-couple households were 5 percent of all households (U.S. Census Bureau 2009b, Table S1101). Six percent of all households with children are headed by cohabiting partners (Kreider and Elliott 2009b, pp. 6 and 8). About 10 percent of infants under age one—but just 1 percent of children age twelve to seventeen—live with cohabiting parents. "This [age] difference may indicate both the fact that the prevalence of cohabitation has risen over the last 10 to 15 years and the fact that cohabiting couples have high rates of dissolution, so they may not remain together for 12 to 17 years after the child's birth" (Kreider and Elliott 2009b, p. 16). The incidence of cohabitation is expected to escalate further as future generations become still more accepting of this family form (Lyons 2004). Furthermore, new generations are growing up in cohabiting families and thereby may be socialized to take cohabiting for granted (Seltzer 2004).

Approximately three-quarters of cohabitants are under age forty-five (U.S. Census Bureau 2008b, Table UC4). Nevertheless, the proportion of middle-aged cohabitors has increased over the past two decades. Middle-aged and older cohabitors are generally in relationships of longer duration, and—as you might expect—they are more likely to be divorced (King and Scott 2005). Approximately 5 percent of cohabitants are age sixty-five and over—a 50 percent increase since 1990 (U.S. Census Bureau 2000, Table 57; 2008b Table UC4). Older singles express less desire to marry (or remarry) than do younger singles (Mahay and Lewin 2007). Older couples have found that living together absent legal marriage may be economically advantageous, because they can retain some financial benefits that are contingent on not being married (King and Scott 2005). With regard to older couples,

> Financial considerations are often at the forefront of decisions to forgo marriage. New couples are aware of their children's fear for their inheritance and for that reason may. . . refrain from marriage. . . . When it comes to marriage and money, widows often say straight out that they are unwilling to risk the financial security of a departed husband's pension. (Levaro 2009, p. F10)

Comparing marrieds to cohabitors, analyses of data from several sources find cohabitors to be younger, less

You probably have an idea of what cohabitation means to you, and you may assume that living together signifies the same thing to all of us. But researchers who have analyzed national survey data to study cohabitation among various racial/ethnic groups have uncovered interesting differences (Castro Martin 2002; Manning 2004). Cohabiting means different things to different people—and to different categories of people (Landale, Schoen, and Daniels 2009).

For instance, Puerto Rican women have a long history of **consensual marriages** (heterosexual, conjugal unions that have not gone through a legal marriage ceremony). The tradition of consensual marriages probably began among Puerto Ricans due to lack of economic resources necessary for marriage licenses and weddings: "Although nonmarital unions were never considered the cultural ideal, they were recognized as a form of marriage and they typically produced children" (Manning and Landale 1996, p. 65). Therefore, for Puerto Ricans in the continental United States, cohabitation symbolizes a committed union much like marriage, and they don't necessarily

feel the need to marry legally should the woman become pregnant because they have already defined themselves as (consensually) married. Compared to Mexican Americans, Puerto Ricans are more likely to agree that "[i]t's all right for an unmarried couple to live together if they have no plans to marry" (Oropesa 1996).

Meanwhile, compared to Puerto Ricans (and to non-Hispanic whites), Mexican Americans were more likely to agree that "[i]t is better to get married than go through life being single." Mexican Americans weigh marriage very positively, and a couple's having plans to marry significantly increases Mexican Americans' approval of cohabitation. These findings are especially strong for foreign-born Mexican Americans. However, economic barriers to marriage (discussed in Chapter 7) apparently induce many low-income Mexican Americans to continue to cohabit, rather than to marry, and to raise several children in cohabiting families (Lloyd 2006; Wildsmith and Raley 2006). Furthermore, as a result of their exposure to general cultural values and attitudes in the United States, we can expect second- and

third-generation Mexican Americans to embrace cohabitation in increasing proportions (Oropesa and Landale 2004).

For Puerto Ricans, "living together" is likely to symbolize a committed union, virtually equal to marriage. Among Mexican Americans, cohabitation is less valued than marriage but allowable if the couple plans to marry—although economic barriers to marriage can thwart those plans. Our diverse American society encompasses many ethnic groups with family norms that sometimes differ from one another. Together with structural/economic factors, cultural meaning systems play a part in how people define cohabitation.

Critical Thinking

The U.S. Census Bureau groups Puerto Ricans and Mexican Americans, along with other Hispanics, into one ethnic category. What does the previously presented information tell you about diversity *within* the ethnic category of Hispanic? Would you suppose that the various groups of Asians, such as the Hmong or Asian Indians, who are also categorized together, differ as well? What about African Americans, or whites?

educated, earning less income, less likely to own their own homes, more likely to be nonwhite, and likely to have experienced more transitions in living arrangements as children (S. Ryan et al. 2009; Schoen et al. 2009; Thornton, Axinn, and Xie 2007; U.S. Census Bureau 2008b, Table UC4). Research shows that some of these trends begin in adolescence: Cohabiting women were found to have had lower academic achievement and fewer parental resources in adolescence (Amato et al. 2008). Relatively conservative religious affiliation, attendance, and fervor are negatively related to cohabitation (Eggebeen and Dew 2009). Nevertheless, people from all social classes, educational categories, and religious persuasions have cohabited.

Cohabitation as an Acceptable Living Arrangement

As "A Closer Look at Diversity: The Meaning of Cohabitation" suggests, cohabiting means different things to different people. Generally, however, "[c]ohabitation is

very much a family status, but one in which the levels of certainty about commitment are less than in marriage" (Bumpass, Sweet, and Cherlin 1991, p. 913; see also Thornton, Axinn, and Xie 2007). British demographer Kathleen Kiernan (2002) has described a society-wide, four-stage process through which cohabitation becomes a socially acceptable living arrangement, equal in status to marriage.

In the first stage, the vast majority of heterosexuals marry without living together first. We saw this stage in the United States until the late 1970s. In the second stage, more people live together but mainly as a form of courtship before marriage, and almost all of them marry with pregnancy. Today, some cohabitants consider their lifestyle a means of courtship, or *premarital cohabitation*. As one young woman explained, "We wanted to try it out and see how we got along, because I've had so many long-term relationships. I just wanted to make sure we were compatible. And he's been married before, and he felt the same way" (in Arnett 2004, p. 108). Living together as a means of selecting a committed life partner is explored in Chapter 6.

In the third stage, cohabiting becomes a socially acceptable alternative to marriage. A couple no longer feels it necessary to marry with pregnancy or childbirth, and people routinely take an unmarried partner to work or family get-togethers. Nevertheless, in stage three, legal and social differences remain between marriage and "just" living together. In the fourth stage, cohabitation and marriage become virtually indistinguishable, both socially and legally. In this stage, the numbers of married and cohabiting couples are about equal, and a cohabiting couple may have several children (Kiernan 2002).

Social historian Stephanie Coontz (2005b) characterized the United States as "transitioning from stage two to stage three at the end of the twentieth century" (p. 272).[3] Perhaps in some large metropolitan areas of this country—where "cohabitation has replaced marriage as a first union experience for a growing majority of young adults" (Lichter and Carmalt 2009, p. F12)—cohabitation has fully reached stage three. According to Coontz, Sweden is an example of a society in stage four. Norway is a second example. Interestingly, in Norway where cohabitation is common and virtually institutionalized, cohabitors' relationships are more like those of marrieds than in the United States (Hansen, Moum, and Shapiro 2007).

Will the United States ever get to stage four? Coontz is skeptical, because "people [in the United States] still place much more importance on getting married than Swedes do" (2005b, p. 272). "As We Make Choices: Some Things to Know about the Legal Side of Living Together" discusses the legal implications of cohabiting in the United States today.

Cohabitation as an Alternative to Unattached Singlehood and to Marriage

In a qualitative study of 120 heterosexual cohabitors, nearly two-thirds ranked, "I wanted to spend more time with my partner" as their first reason for moving in together (Rhoades, Stanley, and Markman 2009). Some couples begin to live together shortly after their first date; others wait for months or longer (Sassler 2004). Accounts of how cohabitation begins suggest that cohabiting does not always result from a well-considered decision. As one twenty-three-year-old woman who had been living with her parents explained,

I was looking for my own apartment at this time. . . . He was like, "Why don't you just move in with me?" I was like, "Let's give it some time," or whatever. So I dated him for like a month and then finally all my stuff ended up in his house. (in Sassler 2004, p. 496)

Some cohabitants view living together as an alternative to dating or unattached singlehood (Manning and Smock 2005). As one respondent told her interviewer:

Um, he had came over, and we had talked and. . . he had spent the night and then from then on he had stayed the night, so basically . . . he just honestly never went home. I guess he had just got out of a relationship, the person he was living with before, he was staying with an uncle and then once we met, it was like love at first sight or whatever and um, he never went home, he stayed with me. (in Manning and Smock 2005, p. 995)

Psychologist Jeffrey Arnett (2004) has dubbed those who live together as an alternative to being single *uncommitted cohabitors*.

Other cohabitors view living together as an alternative to marriage (Cherlin 2009a; Willetts 2006). As they construct their own definitions of commitment, we can think of these couples as *committed cohabitors* (Arnett 2004; Byrd 2009). As one cohabitor explained:

We've been together 10 years. We met at college. . . . We graduated and started living together. . . . We never say never, but we certainly don't have any plans to [marry]. We're very happy being unmarried to each other. (in Sachs, Solot, and Miller 2003)

And as a thirty-two-year old woman who has been in a monogamous relationship for ten years explained:

I have friends who have been married and divorced already in the time that we have been together. . . . And I think I like the luxury of the fact that every day that we are together I know we are together because we both choose to be and not because we feel some artificial obligation to be together. (in Byrd 2009)

People's reasons for living together as an alternative to marrying often include the belief that marriage signifies loss of identity or stifles partners' equality and communication (Moore, McCabe, and Brink 2001; Willetts 2003). In the United States today, this view is unusual among young adults, and committed cohabitors are generally older (Arnett 2004).

The Cohabiting Relationship

As a category, cohabiting couples differ from married couples in several ways. First, cohabitors are less homogamous, or alike in social characteristics, than are marrieds. At 12 percent, cohabiting couples are about twice

---

[3] In the early twentieth century, living together outside marriage was illegal in every state. Due to laws enacted at the turn of the twentieth century, unmarried cohabitation has remained illegal in a handful of states, although the laws are seldom enforced. The American Civil Liberties Union has sued to overturn anticohabitation laws in states where they still exist (Jonsson 2006).

When unmarried partners move in together, they may encounter regulations, customs, and laws that cause them problems, especially if they're not prepared. Consulting a lawyer is strongly advised. Some potential trouble spots:

Domestic Partners

- In many areas and employment sites, opposite- and same-sex unmarried couples may register their partnership and then enjoy some rights, benefits, and entitlements that have traditionally been reserved for marrieds, such as access to joint health insurance.

- Registering as **domestic partners** usually requires joint residence and finances, plus a statement of loyalty and commitment.

Residence

When two unmarried people are renting, landlords may ask each to sign the lease—a legally binding contract—so that each is held responsible for all rent and associated costs.

Bank Accounts

Any couple can open a joint bank account, but one partner can then withdraw money without the other's approval.

Power of Attorney for Finances

Important when one partner becomes incapacitated, a document establishing power of attorney for finances allows the authorized partner to pay bills, run the partner's business, file taxes, and so on.

Credit Cards

If an unmarried couple shares a credit card, both partners are legally responsible for all charges made by either of them, even after the relationship ends. Creditors generally will not remove one person's name from an account until it is paid in full.

Property

Have a written agreement about what happens to property that was purchased together should the relationship end.

Insurance

- The routine extension of auto and home insurance policies to "residents of the household" cannot be presumed to include nonrelatives.

- Anyone may name anyone else as the beneficiary on a life insurance policy. However, insurance companies sometimes require an "insurable interest," generally interpreted to mean a conventional family tie.

as likely as marrieds to be interracial (Gates 2009b, p. ii). Compared with married women, cohabiting women are more likely to earn more and be several years older than their partners (Fields 2004). Cohabitors have been more likely than marrieds to be nontraditional in many ways, including attitudes about gender roles, and to have parents with nontraditional attitudes and/or who have divorced (Baxter 2005; Cunningham 2005; Davis, Greenstein, and Gertelsen Marks 2007; Teachman 2004).

On average, cohabiting relationships are relatively short-term. Half last less than one year, because the couple either break up or marry (Bumpass and Lu 2000). However, one national survey found that 39 percent of unmarried couples were still together after five to seven years (Bianchi and Casper 2000, p. 17). Still, "[c]ompared with married couples, cohabitors are much more likely to break up" (Seltzer 2000, p. 1,252). Reasons include the fact that, for the most part, cohabiting partners are not committed to their relationship in the same way that married partners are. Then, too, cohabitation may not include widely held norms to guide behavior to the degree that marriage does. As a result, the

relationship may suffer as partners struggle to define their situation. Finally, lack of social support may negatively impact the stability of cohabitation "as members of the [cohabiter's acquaintance] network . . . provide the partners possibilities for other intimate relationships" (Willetts 2006, p. 114).

Uncertainty about commitment, together with less well-defined norms for the relationship, may be reasons that, compared to marrieds, cohabitants pool their finances to a lesser extent (Kenney 2004; King and Scott 2005); are less likely to say that they are happy with their relationships and find them less fair (Skinner et al. 2002); report a higher incidence of depression than marrieds (Kim and McKenry 2002; Lamb, Lee, and DeMaris 2003); place greater importance on sexual frequency (Yabiku and Gager 2009); and have more sex outside the relationship than marrieds do (Treas and Giesen 2000).

However, research that analyzed data from the National Survey of Families and Households (NSFH) has found that the relationship quality of "*long-term*" cohabiting couples (who were together for at least four years) differed little from marrieds in conflict

Wills and Living Trusts

If you have no will or living trust when you die, your property will pass to individuals designated by state law—usually legal spouses and blood relatives. A surviving partner could inherit nothing, not even property that he or she paid for.

Health Care Decision Making

Anyone too ill to be legally competent should have an agent to act for him or her in medical decision making. Many cohabitants want their partners to play this role. To be sure that medical personnel honor this desire, designate your partner as decision maker through a "durable power of attorney for health care" document.

Children

With the 1972 case of *Stanley v. Illinois* (405 U.S. 645), an unmarried mother is no longer entitled to sole disposition of the child in many states. Although courts have placement discretion, unmarried couples should stipulate in writing that custody is to go to the partner if the other parent dies. Note also that financial obligations for child support do not depend on marital status.

Some—but not all—courts grant visitation rights to a nonbiological, same-sex co-parent should the relationship end. Unmarried parents to a partner's child should consider three documents:

- Co-parenting Agreement spelling out the rights and responsibilities of each partner
- Nomination of Guardianship that adds language to a will or living trust
- Consent to Medical Treatment allowing the co-parent to authorize the child's medical procedures

Breaking Up

Ending a cohabiting relationship is not to be taken lightly:

- Couples who do not stipulate in writing—and preferably with an attorney's assistance—paternity, property, and other agreements can expect legal hassles. See an attorney about the laws in your state.
- Ending a registered domestic partner agreement in California and some other states requires a formal property settlement agreement and a dissolution proceeding in court (Clifford, Hertz, and Doskow 2007).

Critical Thinking

Does having to worry about the legal aspects of cohabitation lessen what appear to be some of the advantages of living together? Why, or why not?

---

levels, amount of interaction together, or relationship satisfaction. One thing did differ, however: For both marrieds and long-term cohabitors, relationship satisfaction declined with the addition of children to the household, but this decline was more pronounced for cohabitors (Willetts 2006). Other research has found that, compared with younger cohabitors, older cohabiting couples generally report higher relationship quality. Among younger cohabitors, lack of plans for marriage is associated with lower relationship satisfaction (King and Scott 2005). One study found that cohabiting men with intentions to marry their partner do more housework than do other cohabiting males (Ciabattari 2004).

Cohabitation and Intimate Partner Violence Evidence also exists of considerable domestic, or intimate partner violence (IVP), in cohabiting relationships (Brownridge and Halli 2002; DeMaris 2001)—more than among marrieds or dating partners. This situation may also be due to a combination of relatively low commitment (Johnson and Ferraro 2000) and conflict over "rights, duties, and obligations" (Magdol et al. 1998, p. 52). In addition,

selection effects—the situation (discussed more thoroughly in Chapter 7) in which individuals "select" themselves into a category being investigated—in this case, cohabitation—probably help to account for these findings. As we have seen, individuals who live together without marrying tend to be less well educated and poorer than marrieds, and—although domestic violence occurs at all economic and education levels—low income and education are statistically associated with higher levels of domestic violence (Schumacher et al. 2001).

Although not as often as one might expect, IVP can result in the dissolution of the relationship (DeMaris 2001). Research on the economic consequences of cohabitors' breaking up finds similarities to getting divorced. On average, men experience moderate financial decline, whereas women's economic decline is more pronounced (Avellar and Smock 2005). Counselors stress the importance of being fairly independent before deciding to cohabit, understanding one's motives, having clear goals and expectations, and being honest with and sensitive to the needs of both oneself and one's partner. This is especially necessary when children are involved.

© Royalty-Free/CORBIS

According to Pamela J. Smock, associate director at the Institute for Research at the University of Michigan, Ann Arbor, cohabiting "has become the typical and, increasingly, the majority experience of persons before marriage and after marriage" (quoted in "What Happened? . . ." 2003). On average, cohabiting relationships are relatively short-term. Half last less than one year, because the couple either break up or marry. Cohabiting men with intentions to marry their partner are likely to do more housework than other cohabiting males.

Cohabiting Parents and Outcomes for Children

Today, between 10 and 20 percent of all births occur to a cohabiting mother (Chandra et al. 2005, Table 16; Lichter and Carmalt 2009). About half of all nonmarital births occur to cohabiting couples (Dye 2005; Hamilton et al. 2006). Perhaps half or more of births to cohabitors are planned (Lichter and Carmalt 2009; Manning 2001). As shown in Figure 8.3, 38 percent of cohabiting heterosexual households contain children under age eighteen—a proportion only five percentage points lower than that of married-couple

households with children (U.S. Census Bureau 2009c, Table UC3; U.S. Census Bureau 2010b, Table 65). Six percent of children under age eighteen (about 4.6 million) live with a parent or parents who are cohabiting. Of children who live with cohabiting couples, half live with both of their unmarried biological or adoptive parents (U.S. Census Bureau 2009a, Table S0901; U.S. Census Bureau 2009d, Table C3; U.S. Federal Interagency Forum on Child and Family Statistics 2009).

These statistics describe a situation at one point in time. However, "it is important to keep in mind that as children age, they may spend time in several [living] arrangements" (Kreider and Elliott 2009a, p. 16). More than ten years ago, demographers estimated that about one in four U.S. children would live in a family headed by a cohabiting couple "at some point during childhood" (Graefe and Lichter 1999, p. 215). Today—due to the increasing incidence of cohabiting and of childbearing among unmarried couples—we estimate this proportion to be higher (Lichter and Carmalt 2009).

Although a large majority of cohabiting couples with children have one child, a significant number (about 1.5 million cohabiting couples) are raising two or more children (Chandra et al. 2005, Table 9). Research from at least three national samples has found this situation to be more characteristic of black and Hispanic cohabitors than of non-Hispanic whites, who are more likely to marry upon becoming pregnant (Chandra et al. 2005, Table 18; Manning 2004).

A qualitative study with thirty cohabiting working-class childfree couples found that most intended to defer having children—many until after they marry, if they do (Sassler, Miller, and Favinger 2009). In another qualitative study, in-depth interviews with twenty-four childless cohabitors who had had some college found that they associated their desire to be in the middle class with marrying in the event that they had children. For these couples, cohabitation "appears to serve as a staging area, a time when couples can be together, complete schooling, and get fiscally established prior to marrying and beginning families" (Sassler and Cunningham 2008, p. 22).

Having a child while cohabiting does not necessarily increase a couple's odds of staying together, but conceiving a child during cohabitation and then marrying before the baby is born apparently does increase union stability. Why would this be? "Although birth in cohabitation indicates a decision to remain together during pregnancy, it also represents a decision not to commit to marriage" (Manning 2004, p. 677). Cohabiting parents who see a father's involvement in parenting as very important are more likely to stay together (Hohmann-Marriott 2009). Perhaps ironically, the fear of divorce among unmarried parents

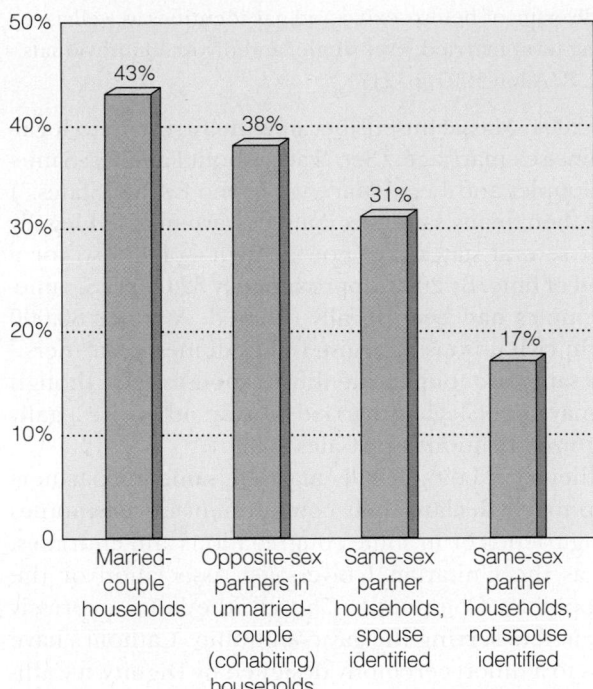

Figure 8.3 Percentage of children under age 18 in four U.S. household types, 2008

Note: Unmarried partners' children refers to at least one biological child under age 18 of either parent. Percentages for opposite-sex cohabiting households are for 2007.

Sources: Gates 2009b, Figure 13; U.S. Census Bureau 2009c, Table UC3; U.S. Census Bureau 2010b, Table 69.

may reduce their likelihood of marrying (Waller and Peters 2008).

Children's Outcomes For many families, "[c]ohabitation may not be an ideal childrearing context precisely because of the stress associated with the uncertainty of the future of the union" (S. Brown 2004, p. 353). Because cohabiting couples are significantly less likely to stay together than marrieds, it has been noted that many children in cohabiting-couple families will experience a series of changes, or transitions in their family's living arrangements (Lichter and Carmalt 2009; Raley and Wildsmith 2004). This cumulative instability is related to problematic outcomes for children (Cavanagh 2008). "Residential and other household changes associated with the formation of new partnerships may disrupt well-established patterns of [parental] supervision" (Thomson et al. 2001, p. 378; see also Heard 2007; Magnuson and Berger 2009). Other research has found a relationship between a mother's overall stress (which negatively affects parenting) and her forming a coresidential relationship with a nonbiological father to her child (Cooper et al. 2009).

Accordingly, many family scholars have expressed concern regarding outcomes for adolescents and

younger children living in opposite-sex, cohabiting families (Booth and Crouter 2002; S. Brown 2004). For example, research that compared economically disadvantaged children from families of various forms found more problem behaviors among children in various types of unmarried families, including cohabiting unions (Ackerman et al. 2001; M. J. Carlson 2006). Other research has found that, among couples with comparable incomes, cohabiting parents spend less on their children's education than do marrieds (DeLeire and Kalil 2005).

When compared to those living with two married biological parents, adolescents who have lived with a cohabiting parent are more likely to experience earlier premarital intercourse, higher rates of school suspension, and antisocial and delinquent behaviors, coupled with lower academic achievement and expectations for college (Albrecht and Teachman 2003; S. Brown 2004; M. J. Carlson 2006). Having a cohabiting male in the household who is not the biological father appears not to enhance adolescents' outcomes when compared with living in a single-mother household (Manning and Lamb 2003). Nevertheless, research also shows that, compared to growing up in a single-parent home, children do benefit economically from living with a cohabiting partner, provided that the partner's financial resources are shared with family members (Manning and Brown 2006).

A relatively new area for court determination of child custody concerns children of cohabiting parents who separate. Some courts treat nonmarital relationships as "sufficiently marriage-like" for marital law to apply (Judge Heather Van Nuys in "Court Treats" 2002). Some proponents of the idea that only legally married couples should be treated as married oppose court involvement in custody issues of unmarried parents. However, "Courts aren't trying to contribute to the demise of traditional families. But they recognize the reality of families today and functional parents" (Duke University Law School Dean, Katherine Bartlett, quoted in Biskupic 2003, p. 2A).

As cohabitation becomes increasingly common, we become more aware that today's pluralistic family is comprised of many forms. We turn now to another relationship type that exemplifies the pluralistic family— same-sex couples.[4]

---

[4] Not only cohabiting and same-sex couples add diversity to the postmodern family. Transgendered identity—physically changing surgically and/or with the use of hormones from male to female or vice versa—further complicates a traditional view of family life. As an example, a man who married a woman in 1988 thereafter gradually changed gender. The couple stayed together. Is their marriage "gay"? On another note, some states do not recognize sex changes. If the former male, now a female, were to remarry now in those states, she would be allowed only to marry a female (Boylan 2009).

Same-Sex Couples and Family Life

Google GLBT (gay, lesbian, bisexual, and transgendered) organizations, and you'll find websites for blacks, Latinos/as, Jews, and Muslims, among others. Lesbian and gay singles make up a diverse category of all ages and racial/ethnic groups. In 2008, there were approximately 565,000 same-sex couple households in the United States, about evenly divided between gay male and lesbian couples (Gates 2009b).[5, 6]

Families of same-sex couples include lesbian co-parents, as well as an array of combinations of lesbian mothers and biological fathers, surrogate mothers, and gay biological fathers (less frequent) (Goldberg 2010; C. Patterson 2000). A family studies professor describes the diversity apparent in her own lesbian family as follows:

> My partner and I live with our two sons. Our older son was conceived in my former heterosexual marriage. At first, our blended family consisted of a lesbian couple and a child from one partner's previous marriage. After several years, our circumstances changed. My brother's life partner became the donor and father to our second son, who is my partner's biological child. My partner and I draw a boundary around our lesbian-headed family in which we share a household consisting of two moms and two sons, but our extended family consists of additional kin groups. For example, my former husband and his wife have an infant son, who is my biological son's second brother. All four sets of grandparents and extended kin related to our sons' biological parents are involved in all our lives to varying degrees. These kin comprise

a diversity of heterosexual and gay identities as well as long-term married, ever-single, and divorced individuals. (K. R. Allen 1997, p. 213)

In 2004, Massachusetts became the first state to legalize same-sex marriage. (See "Facts About Families: Same-Sex Couples and Legal Marriage in the United States.") Since then, many same-sex couples have married legally in the several states that now allow it—or did so for a period of time. By 2009, approximately 32,000 U.S. same-sex couples had been legally married. Another 80,000 were in civil unions or registered as domestic partners.[7] Many same-sex couples identify as spouses even though they may not be legally married or in an otherwise legally recognized relationship (Gates 2009b).[8]

Other than being legally married, same-sex partners may publicly declare their commitment in ceremonies among friends or in some congregations and churches, such as the Unitarian Universalist Association or the Metropolitan Community Church, the latter expressly dedicated to serving the gay community. Catholics have access to a union ceremony designed by Dignity, a Catholic support association, although the Catholic Church does not recognize these unions ("Registration of Holy Union . . ." 2008).

Secular commitment ceremonies for gay men and lesbians are common enough to have sparked a number of wedding-planning businesses for same-sex couples (e.g., twobrides.com, twogrooms.com). Same-sex couples also use other commitment markers, such as joint estate planning, buying a house together, wearing rings, or hyphenating their last names (Porche and Purvin 2008; Reczek, Elliott, and Umberson 2009; Suter, Daas, and Bergen 2008). Registering as domestic partners (described in "As We Make Choices: Some Things to Know about the

---

[5] The Census Bureau conducts an annual American Community Survey (ACS) between decennial censuses. The number of same-sex couples reported by the 2008 ACS was considerably lower than the 780,000 reported in 2006. "This is likely a result of changes in the format and processing of the 2008 ACS that reduced the probability of respondents making errors" (Gates 2009b, p. 2). For a detailed account of how this error originally occurred—and how it was corrected in the 2008 ACS—see Gates 2009a.

[6] Until the 2000 census, calculations of unmarried couples did not include same-sex partners. Therefore, comparing estimates from before 2000 with census data thereafter is inappropriate.

Beginning with the 2000 census, unmarried same-sex couples were counted as such. Because Massachusetts legalized same-sex marriage in 2004, there was hope that the 2010 census would calculate same-sex married, as well as unmarried, couples. However, the 2010 census does not count married same-sex couples as married; instead it counts them as unmarried. This situation is largely due to the Census Bureau's interpretation of the federal Defense of Marriage Act (which defines marriage as necessarily "between one man and one woman") as not permitting an enumeration of same-sex couples as married. However, the Census Bureau has announced a large-scale project to improve future data collection on same-sex couples, and it is anticipated that questions on their marital status will appear in the future (O'Connell and Lofquist 2009; Quinn 2009; "Same-Sex Couples . . ." 2009).

[7] The terms *domestic partnership* and *civil union* both refer to officially recognized unions according to which unmarried couples enjoy some (although not all) rights and benefits ordinarily reserved for marrieds. However, the term *civil union* is used to refer to various states' granting same-sex couples a legal status virtually equivalent to marriage. The term domestic partnership is typically used to refer to same- or opposite-sex unmarried partner benefits offered by some employers, cities, counties, and states. Generally, domestic partnerships grant couples lesser status and fewer benefits than do civil unions. However, California, Oregon, and Washington use the term *domestic partnership* for legislation similar or equivalent to civil union laws in other states ("Marriage, Domestic Partnerships, and Civil Unions . . ." 2009).

[8] Same-sex couples living together in long-term, committed relationships are not a recent development. According to social historian Samuel Kader (1999), same-sex committed couples date back to the Old Testament, and commitment ceremonies between same-sex partners were not unknown in early Christianity. More recently, scholars "have uncovered a long and complicated history of gay relationships in nineteenth-century America. Sometimes women passed as men to form straight-seeming relationships; sometimes men or women lived together as housemates but were really lovers; sometimes individuals would marry but still carry on romantic, sometimes lifelong same-sex intimate relationships" (Seidman 2003, p. 124).

Legal Side of Living Together") can have emotional significance for same-sex couples who may do so partly as a way of publicly expressing their commitment.

Demographically, same-sex couples who identify as spouses (even if not legally married) are much like heterosexual marrieds in several respects. Same-sex couples who identify as spouses have an average age of fifty-two, average household incomes of $91,558, and 31 percent are raising children. "That compares with an average age of fifty, household income of $95,075, and 43 percent raising children for married heterosexual couples" ("Report: Gay Couples Similar . . ." 2009). Additionally, in both couple types, a little over 20 percent have a college degree, about 80 percent own their own homes, and 6 to 7 percent are interracial (Gates 2009b).[9]

Same-sex couples live in virtually every county of every state (Gates 2009b). Often they find community in urban areas that have high concentrations of gays and lesbians, along with strong activist organizations. But lesbians and gays also live in the suburbs and smaller towns. The Internet has changed life for many gay men and lesbians, especially those living in rural areas. Accessing websites such as PlanetOut.com, homosexuals from all over the world can meet and interact online regardless of geographical boundaries (Gudelunas 2006).

Although discrimination assuredly persists, attitudes toward GLBT rights generally have become more accepting over the past twenty-five years. Gallup polls have shown a gradual increase—from 34 percent in 1983 to 57 percent in 2008—in agreement with the idea that being gay or lesbian is an "acceptable alternative lifestyle" (Saad 2008a; see also Newman 2007). Then too, more employers, cities, and states are extending various benefits to same-sex couples—health insurance for an employee's partner, for example (Surdin 2009).

The Same-Sex Couple's Relationship

In many respects, same-sex relationships are similar to heterosexual ones. Like heterosexuals, same-sex partners highly value love, faithfulness, and commitment (Meier, Hull, and Ortyl 2009). Research indicates that the need to resolve issues of sexual exclusivity, and power and decision making is not much different in same-sex pairings than among heterosexual partners (Kurdek 2006). However, research shows that same-sex partners of both genders are likely to evidence more equality and role sharing than couples in heterosexual marriages (Kurdek 2007; Parker-Pope 2008).

Nevertheless, a recent study of thirty-two New York City black lesbians in stepfamilies (that is, one partner brought her child or children into a same-sex cohabiting relationship) found that for these women constructing "family" meant following traditional heterosexual gender roles to some degree. Specifically, biological mothers in these stepfamilies were more involved not only in child care but also in housework. The author argues that following heterosexually "appropriate" marital roles may be especially important for lesbian mothers who previously bore their children in heterosexual contexts, because "they willingly moved from validated relationships with men to . . . often stigmatized, same-sex unions. Given this status change, these mothers may find 'appropriate' gender construction that much more important" (Moore 2008, p. 353).

Same-sex couples must daily negotiate their private relationship within a heterosexual—often, heterosexist—world (Oswald and Masciadrelli 2008; Suter et al. 2006). A comparison of same-sex couples with married heterosexual couples found that the former spent more time discussing the state of their relationship. The researchers suggest that this difference may reflect the absence of a legal bond: "It appears that to some degree heterosexual married couples may take for granted that they are bound together through legal marriage, whereas gays and lesbians must frequently 'take the pulse' of the relationship to assess its status" (Haas and Stafford 2005, p. 56).

A traumatic example of living out a private relationship amid heterosexism occurred in 2009, when a Florida hospital prevented a distraught lesbian from being at her dying partner's bedside ("Federal Court Dismisses Lawsuit . . ." 2009). "This is an anti-gay city and state," the hospital receptionist reportedly explained (Parker-Pope 2009). One year later, President Obama ordered that any hospital that received federal funds through Medicare or Medicaid must grant hospital visiting rights to same-sex partners (Stolberg 2010). Nevertheless, discrimination against same-sex couples persists in many areas of their lives.

Discrimination adds stress for same-sex couples and may result in lowered mental health and relationship quality (Otis et al. 2006). In addition, the potential for discrimination gives partners unique avenues for dealing negatively with couple conflict. "For example, 'outing' one's partner is not an issue for heterosexuals but is a surprisingly common weapon for gay people in an abusive relationship" (Burke and Owen 2006, p. 6; Sorenson and Thomas 2009).

Same-Sex Intimate Partner Violence Stress resulting from discrimination was once thought to be one reason that statistics indicated higher rates of intimate partner violence (IVP) among same-sex couples versus among heterosexual couples. However, recent analysis has uncovered methodological deficiencies in earlier reports of higher IPV among same-sex couples, and current research suggests that the rates are

---

[9] Interestingly, same-sex unmarried partners and different-sex cohabiting partners have similar rates of being interracial—13 percent of same-sex unmarried partners and 12 percent of different-sex cohabiting partners (Gates 2009b, p. 11).

The 1974 U.S. Supreme Court decision in *Singer v. Hara* defined marriage as a union between one man and one woman (*Singer v. Hara* 1974). Nevertheless, the federal government has traditionally recognized the right of individual states to create, interpret, and enforce laws regarding marriage and families. Consequently, the battle over legal marriage for same-sex couples has largely been fought within individual state courts and legislatures.

Lawsuits Claiming Discrimination and Varied State Responses

Beginning in the late 1990s, same-sex couples in several states filed lawsuits claiming that barring lesbians and gays from legal marriage is unconstitutional because it discriminates against same-sex couples. Some (although not all) courts have agreed and ordered their state legislatures to address this problem by passing new, nondiscriminatory laws. Results have been varied. In 2004, Massachusetts became the first state to allow gay and lesbian couples to marry legally. About 10 percent of all states now allow same-sex marriage. However, marriage in these states remains "an incomplete legal status, because same-sex marriages are denied federal recognition of their relationship by virtue of the 1996 federal [Defense of Marriage Act, or] DOMA" (Oswald and Kuvalanka 2008, p. 1060). (DOMA is described later in this boxed material.) Other states have passed *civil union* laws.

Civil unions allow same-sex partners access to virtually all marriage rights and benefits on the state level, but none on the federal level. For instance, a couple would have state-regulated rights to joint property and tenancy, inheritance without a will, and hospital visitation and health care decisions for their partners. However, they cannot collect federal Social Security benefits upon a partner's death, nor can a non-U.S. partner become a U.S. citizen upon joining a civil union (Oswald and Kuvalanka 2008).

Meanwhile, some states have amended their state constitutions to stipulate that marriage in that state is to be defined only as heterosexual. As an example, in 2008, California voters (by a margin of 52 percent to 48 percent) passed the California Marriage Protection Act (Proposition 8), a state constitutional amendment declaring that "only marriage between a man and a woman is valid or recognized in California." Proposition 8 overturned a prior California court ruling which said that same-sex couples have a constitutional right to marry.

However, a state constitution does not outrank a federal court ruling, and in August 2010, a federal court ruled that California's Proposition 8 violated the U.S. constitution because it discriminated against a category of U.S. citizens without a rational basis for such discrimination. Proponents of Proposition 8 had argued that to allow same-sex marriage would erode traditional marriage, but the federal judge who struck down Proposition 8 ruled that, "Tradition alone…cannot form the rational basis for law" (McKinley and Schwartz 2010).

This decision, however, does not definitively determine the legality of same-sex marriage because it applies only to one jurisdiction and because the case is expected to advance to the U.S. Ninth Circuit Court of Appeals and from there to the United States Supreme Court (Schwartz 2010). How a U.S. Supreme Court ruling might affect the existing Federal Defense of Marriage Act and the various state laws permitting or prohibiting same-sex marriage (and domestic partnership) remains to be seen.

The Federal Defense of Marriage Act (DOMA)

States usually recognize one another's legal decisions. This *principle of reciprocity* would require a state to recognize a legal marriage performed in another state. However, to allow states *not* to follow the principle of reciprocity regarding same-sex marriages, the United States passed the 1996 **Defense of Marriage Act (DOMA)**. The Defense of Marriage Act is a federal statute declaring marriage to be a "legal union of one man and one

about the same for both couple types (Sorenson and Thomas 2009). Meanwhile, society-wide responses to same-sex violence "can be described as neglectful at best" (Sorenson and Thomas 2009, p. 349; Kulkin et al. 2007):

> Restraining orders, a legal remedy that is widely used when a victim is trying to end the relationship that requires the abuser to have either no or only peaceful contact with the victim . . . are not available to gay men and lesbians in [some] states. In addition, social services are not always welcoming; lesbian victims of IPV report not feeling comfortable seeking domestic violence services that generally are geared to heterosexual females. And most shelters do not accept male clients, making such services off limits to gay male IPV victims. (Sorenson and Thomas 2009, p. 349)

Gay activists argue that domestic violence laws need to specifically include lesbian and gay partners, and police must be trained to more effectively address intimate partner violence among same-sex couples (Kulkin et al. 2007). In sum, the problem of same-sex IVP "is exacerbated" by a political climate that often "treats gays and lesbians as a marginalized population" (Burke and Owen 2006, p. 7).

"*There's nothing wrong with our marriage, but the spectre of gay marriage has hopelessly eroded the institution.*"

Robert Mankoff/Bios/Cartoonbank

woman," withholding federal recognition of legal same-sex marriages in any state, and relieving states of the obligation to grant reciprocity to marriages performed in another state. As a result of DOMA, a large majority of states have passed laws or state constitutional amendments that refuse to allow legal marriage in that state and/or recognize a marriage obtained by same-sex couples in another state (Oswald and Kuvalanka 2008).

In 2009 a bill, named the Respect for Marriage Act, was introduced into Congress. If passed, the act will repeal DOMA ("Legislation Introduced . . ." 2009).

A Proposed Federal Amendment to the U.S. Constitution

A proposed federal amendment to the U.S. Constitution would define marriage as between one man and one woman and ban same-sex marriage in the United States while allowing states to create civil unions or domestic partnerships (Allen and Cooperman 2004; Page and Benedetto 2004). Because the originators of the U.S. Constitution intended for amendments not to be undertaken lightly, they made them very difficult to pass. Passing a U.S. constitutional amendment requires a two-thirds majority in both the U.S. House of Representatives and Senate. Then the amendment must be approved by three-fourths of the states. A constitutional amendment that would define marriage as between one man and one woman has not passed in Congress.

This range—from legal same-sex marriage in several states to a possible federal constitutional amendment that would ban same-sex marriages across the country—points up the serious public divide in the United States regarding same-sex marriage. For further details on legal marriage for same-sex couples in the United States and throughout the world, visit the following websites: American Civil Liberties Union (aclu. org); Partners Task Force for Gay and Lesbian Couples (buddybuddy.com); and DOMA Watch (domawatch.org).

Critical Thinking

Today it would be difficult to escape the public debate over whether, on the one hand, the family as institution is inappropriately threatened or, on the other hand, tolerance for diversity is fitting when the issue is legal marriage for same-sex couples. What do you think? Is the institution of marriage and family threatened by same-sex marriage? Why, or why not? Can you back up your opinion with facts?

Same-Sex Parents and Outcomes for Children

Enough same-sex couples are establishing families with children that, by the early 2000s, observers noted a "gay baby boom" (Johnson and O'Connor 2002). A 2008 census bureau survey found that 31 percent of same-sex couples who identified themselves as married, and 17 percent of other same-sex households now include children under age eighteen (Gates 2009b, Figure 13). (See also Figure 8.3.) These children were born to the union, were adopted, or were born in prior heterosexual relationships.

Same-sex couples become parents through adoption, foster care, planned sexual intercourse, or artificial insemination (Bell 2003; Gomes 2003; Mundy 2007). Some—but certainly not all—courts permit a lesbian partner to adopt the biological child of the other partner or will grant joint adoption to same-sex couples, ensuring legal parenthood to both members of the couple raising a child (Human Rights Campaign 2007).

Religions vary in their policies regarding such families. For example, the Catholic Church officially opposes both legal marriage and adoption by same-sex couples (Buchanan 2006; Egelko 2008; LeBlanc 2006a). Courts also vary in their receptiveness to same-sex families.

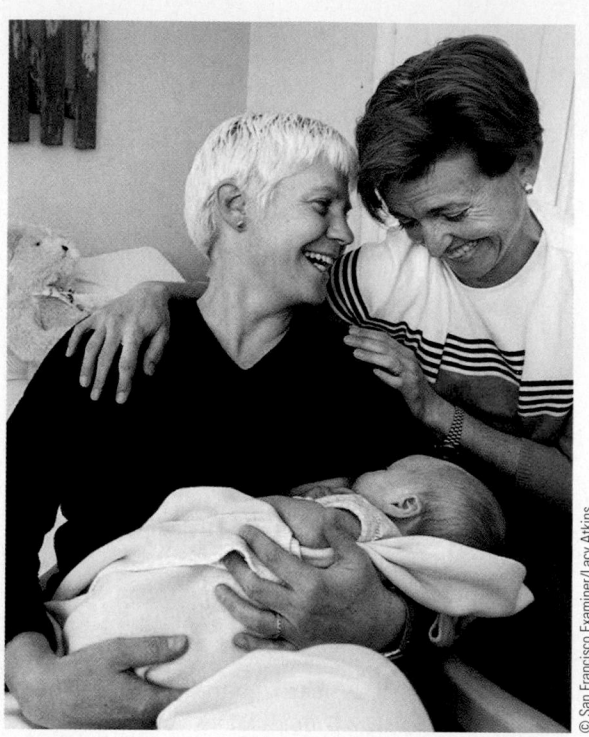

Lesbian couples may take advantage of AID (artificial insemination by donor) technology so that one partner gives birth to a baby they both want. Research indicates that children of lesbian or gay male parents are generally well adjusted and have no noticeable differences from children of heterosexual parents.

Some courts have permitted a lesbian co-parent to adopt a biological child born to her partner, ensuring legal parenthood to both members of the couple raising a child (Human Rights Campaign 2007). Many courts grant joint adoption to gay male couples. However, some states prohibit same-sex couples from adopting children. Florida, for example, prohibits same-sex couple adoption, although it does allow them to serve as foster parents (Waddell 2005).[10]

Among lesbian couples, one partner may give birth to a child that both partners parent. When a couple decides to follow this course, the women face a series of decisions: Who will be the biological mother? How will a sperm donor be chosen? What will they call themselves as parents? How will they negotiate parenthood within a heterosexual society? Where and from whom will they find support? (Chabot and Ames 2004; Goldberg and Smith 2008).

[10] The Williams Institute on Sexual Orientation, Law and Public Policy, affiliated with the UCLA School of Law, estimates that Florida's prohibiting adoption by GLB individuals and same-sex couples costs the state $2.5 million annually in unnecessary foster care expenses (Goldberg, Badgett, and Cooper 2009).

Compared to that regarding heterosexuals, the amount of research on same-sex parents is small but growing. One study has found that lesbian partners who become parents are more likely than heterosexual wives to remain committed to full-time work as well as to motherhood (Peplau and Fingerhut 2004). Another study focused on the relationship quality of twenty-nine lesbian couples who gave birth to their first child by means of artificial insemination. These researchers found that, similar to heterosexual couples, lesbian partners' relationship satisfaction declined with the transition to parenthood. This situation was largely due to having less time to be alone as a couple after the baby was born (Goldberg and Sayer 2006).

"Living in a society fixated on labels and family terminology," lesbian couples are often asked, "Who is the real mom?" (Chabot and Ames 2004, p. 354). One non-biological mother illustrates this point:

I think a lot of people have issue with that . . . you're not *really* the mother if you're not the biological mother. You're just sort of playing this role, or something. Maybe you're just the one who's also responsible, but you're not "*the mom.*" We don't care what anybody else thinks. We both are the moms. (in Chabot and Ames 2004, p. 354)

Then, too, issues of support from the couple's extended families may cause tension:

Biological mothers' families may undermine the nonbiological mother's relationship to the child, seeing her as "less of a mother." . . . Another possibility is that biological mothers' families . . . meet or even surpass nonbiological mothers' expectations for support, but their frequent presence or greater involvement ultimately causes conflict between the partners. (Goldberg and Sayer 2006, p. 97)

Meanwhile, same-sex parents emphasize their similarity to heterosexual parents: "We go to story time at the library and worry about all the same food groups" (in M. Bell 2003). "Contrary to stereotypes of these families as isolated from families of origin, most report that children had regular (i.e., at least monthly) contact with one or more grandparents, as well as with other adult friends and relatives of both genders" (Patterson 2000, p. 1,062). A wide network of adult friends and relatives may even include ex-husbands. A lesbian mother who, like her partner, brought a daughter into the relationship from a heterosexual previous marriage, explains:

[B]oth of [the girls'] fathers live very close. We stayed right within the same school district that I was in with the little one. Little Monica's father being just in the next school district over. So, the fathers were always there visiting and taking care of [the girls], especially my ex-husband when . . . I was back in school, so . . . he always had the responsibility of being there when they got home [from school]. (in Hequembourg 2007, p. 169)

Children's Outcomes Research from an accumulation of more than one hundred studies finds children of gay male and lesbian parents to be generally well adjusted, with no noticeable differences from children of heterosexual parents in cognitive abilities, behavior, or emotional development. There is no evidence that children of same-sex parents are confused about their gender identity, either in childhood or adulthood, or that they are more likely to be homosexual (Goldberg 2010; Meezan and Rauch 2005).

Although not necessarily refuting these findings, sociologists note that the research methodologies of many of these studies are not rigorous, largely because it is very difficult to locate representative samples of gay male and/or lesbian parents. Sociologist Tim Biblarz argues that insufficient long-term, large-scale research exists to determine whether being raised by same-sex parents affects sexual identity (Jayson 2009). The research that we do have—which is largely on lesbian, white, and middle- or upper-middle class parents—concludes that same-sex parents, especially those who identify as spouses, are much like married heterosexuals in their parenting practices (Goldberg 2010). Its members having themselves reviewed the literature, the American Academy of Pediatrics officially supports gay male and lesbian couples' adopting, bearing, and raising children (Perrin 2002).

Meanwhile, like children of other minority groups, those in same-sex families may experience prejudice from friends, classmates, or teachers. Regarding relationships with schools, the Family Pride Coalition urges same-sex parents to

> [t]ell the teachers who is in your family and names your children use to identify them, and provide a glossary of correct terms for lesbian and gay families. Give the library a list of books, videos and other resource materials . . . , and encourage school administrators and librarians to purchase these materials for the school. (Brickley et al. 1999)

In a heterosexist society, adult children of same-sex parents face questions about coming out about their parents (Goldberg 2007, 2010). Children of same-sex parents have formed a support group called COLAGE (Children of Lesbians and Gays Everywhere) and maintain a website (www.colage.org). Their purpose is to "engage, connect, and empower people to make the world a better place for children of lesbian, gay, bisexual, and/or transgender parents and families." Being allowed to marry legally might be a benefit to children being raised in same-sex households, because marriage is associated with increased "durability and stability of the parental relationship" as well as enhanced in-law, grandparent, and other extended-family investment (Meezan and Rauch 2005, p. 108; Goldberg 2010; Wildman 2010). We turn now to the debate over legal marriage for same-sex couples.

The Debate over Legal Marriage for Same-Sex Couples

In 2000, the Netherlands became the first country to allow same-sex partners to marry.[11] At the time of this writing, several other European countries including Portugal, as well as Argentina, Canada, Iceland, and South Africa, now allow same-sex marriage (Barrioneuvo 2010; Partners Task Force for Gay and Lesbian Couples 2009).[12] Meanwhile, the United Nations Commission on Human Rights has been unable to pass a resolution to add sexual preference as a reason that people's human rights must not be violated. The motion was dropped "in the midst of intense pressure" from the Vatican and the Conference of Islamic States ("United Nations Drops" 2004). We can conclude that the **culture war**—deep conflict over matters concerning human sexuality and gender—is global.

The conservative Family Research Council website (www.frc.org), one of several that speak out against same-sex marriage, urges people to "Take a Stand for Marriage!" Other websites, such as Gay and Lesbian Advocates and Defenders (GLAD), the Partners Task Force for Gay and Lesbian Couples, or the National Black Justice Coalition advocate for the other side (www .glad.org; www.buddybuddy.com; www.nbjc.org). Having first emerged as a remote possibility in the 1970s, legal marriage for gay and lesbian couples "became a frontline issue" after 1991, when gay activists formed the Equal Rights Marriage Fund (Seidman 2003; Taylor et al. 2009). "Facts about Families: Same-Sex Couples and Legal Marriage in the United States" outlines political developments regarding legal marriage for same-sex couples.

As shown in Figure 8.4, 57 percent of Americans favor allowing same-sex couples to enter into legal agreements, such as civil unions and domestic partnerships, that would give them many of the same rights as married couples. Our country is about evenly split between those who favor (46 percent) and who oppose (48 percent) allowing gays and lesbians to adopt children. Just as attitudes have become more accepting about GLBT rights generally, public opposition to legal same-sex

---

[11] For a detailed account of developments regarding same-sex legal marriage around the world, see the following: Human Rights Campaign (www.hrc.org); Lambda Legal (www.lambdalegal.org); American Civil Liberties Union (www.aclu.org); and "Legal Marriage Report: Global Status of Legal Marriage," Partners Task Force for Gay and Lesbian Couples 2009, at buddybuddy.com.

[12] U.S. citizens are allowed to marry in Canada. However, at this writing, their unions are not recognized by either the United States government or the vast majority of state governments. "Another complication arises if a couple wishes to divorce. They would not be able to do so in their resident state if their state did not recognize the marriage in the first place. To get a divorce, one of the partners would need to reside in Canada for a year" (Partners Task Force for Gay and Lesbian Couples 2006c).

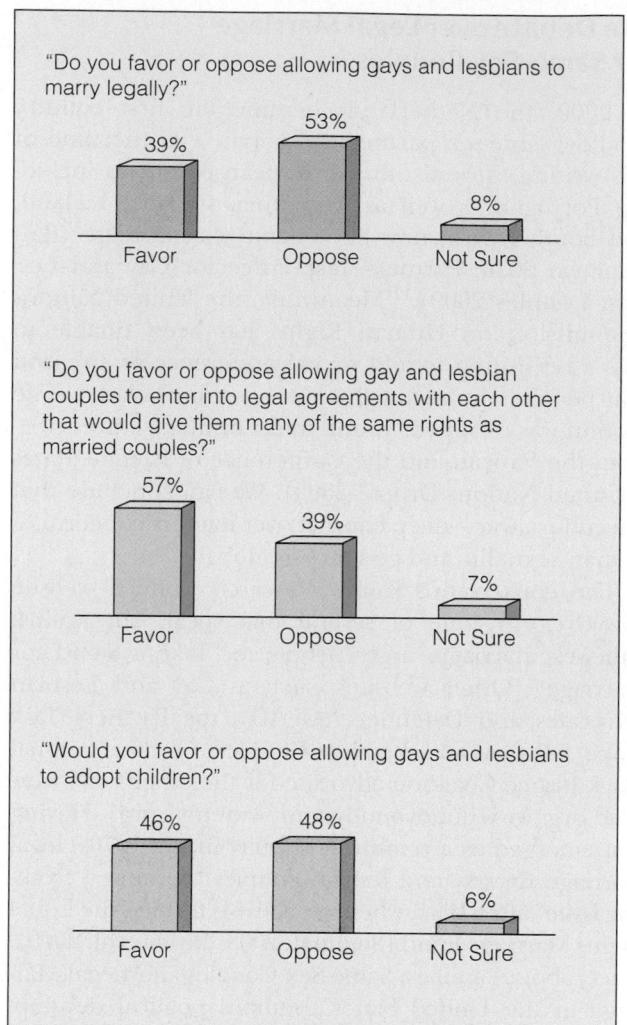

Figure 8.4 American public opinion regarding legal marriage, domestic partnerships/civil unions, and adoption for same-sex couples, 2008–2009.

Source: Pew Research Center 2008, 2009c.

marriage has softened somewhat since the mid-1990s, when about only 25 percent of Americans believed that same-sex marriage should be legal (Vestal 2009). Thirty-nine percent of Americans say that they favor legal marriage for same-sex couples, while 53 percent oppose (Pew Research Center 2009c). Scholars see the national divide over legal same-sex marriage as a mark of cultural ambivalence resulting from conflicting core values: the sanctity of marriage versus personal freedom and civil rights (Brumbaugh et al. 2008). What are the arguments against and for legal marriage for same-sex partners?

Arguments for Legal Marriage as Heterosexual Only Not all religions oppose legal same-sex marriage. The Mormon and Catholic churches, the National Association of Evangelicals, and Islam and Orthodox Jewish congregations oppose legal same-sex marriage. "On the other side are the Unitarians, the United Church

of Christ, the Union for Reform Judaism, the Soka Gakkai branch of Buddhism, and dissident groups of Mormons, Catholics, and Muslims" (Egelko 2008, p. A10). Christian social psychologist David Myers, making a "Christian Case for Gay Marriage," argues that if marriage is good for people and society, as discussed in Chapter 7, then marriage should be an option for everyone, including lesbians and gays (Myers and Scanzoni 2006). However, for religious fundamentalists and other conservative groups, the move to legalize marriage is an "attempt to deconstruct traditional morality" (Smolowe 1996a; J. Wilson 2001).

Those who favor defining legal marriage as only heterosexual are more likely to value traditional gender roles (as well as traditional family structure) and to live in communities in which those values are reinforced through daily communication with like-minded neighbors (McVeigh and Diaz 2009). Proponents of legal marriage as only heterosexual argue that heterosexual marriage alone has deep roots in history, as well as in the Judeo-Christian and other religious traditions (Hartocollis 2006; McKinley and Schwartz 2010). They further claim that only heterosexual married parents can provide the optimum family environment for raising children and that legalizing same-sex marriage would weaken an institution already threatened by single-parent families, cohabitation, and divorce (Blankenhorn 2007). Some contend that "[e]ven more ominously, permitting gays to marry would open the door for all sorts of people to demand the right to marry—polygamists, children, friends, kin—even more than two partners" (Seidman 2003, p. 128).[13] Finally, they argue that legal marriage for same-sex couples is unnecessary, given legislation, such as civil unions in some states, which give virtually all the rights of marriage to same-sex couples without the title "married" or "spouse" (Seidman 2003).

Arguments for Legal Same-Sex Marriage In her book, *Beyond (Straight and Gay) Marriage* (2008), American University law professor Nancy Polikoff argues that all family forms need to be valued under the law. Same-sex families comprise a family form that is not going to disappear; if marriage is thought to be good for spouses and children emotionally, financially, and healthwise, then to deny these benefits to a significant number of individuals is unethical and socially costly (Rauch 2004).

---

[13] This fear is not entirely unfounded. In 2003, the U.S. Supreme Court struck down laws criminalizing sodomy (interpersonal sexual acts that do not allow for procreation, such as oral sex or anal intercourse). At that time, the Supreme Court ruled that individuals have "the full right to engage in private conduct without government intervention." Today some advocates seek the same considerations for proponents of bigamy (Kurtz 2006; Soukup 2006). Moreover, changing the legal definition of marriage as necessarily "between one man and one woman" could affect not only the gender stipulation but also the requirement that marriage take place between only two individuals (Kurtz 2003; Stacey and Meadow 2009).

Those who favor legalized same-sex marriage further argue that denying lesbians and gays the right to marry legally violates the U.S. Constitution because it discriminates against a category of citizens (Schwartz 2010). Legal marriage yields economic and other advantages. For instance, hospital visitation rights would be guaranteed to a same-sex spouse. As another example,

> The right to divorce is a benefit associated with marriage that often goes unmentioned. Relationships end, and marriage provides the opportunity for a legal chaperon when partners are unable or unwilling to manage their separation, dissolution, and post-divorce parenting in a productive manner. Legal marriage for LGBT partners would protect the interest of all members of the family upon dissolution, just as it does for heterosexual partners. (Allen 2007, p. 181)[14]

Allowing same-sex legal marriage would facilitate child custody cases in the event of dissolution (Polikoff 2008; Hare and Skinner 2008). Chapter 15 further explores dissolution of same-sex unions and of unmarried heterosexual relationships.

However, due to the federal Defense of Marriage Act (DOMA), which defines marriage as only between "one man and one woman," even same-sex couples that are legally married in states that allow it cannot receive federal benefits designated for married couples (Clifford, Hertz, and Doskow 2007; Oswald and Kuvalanka 2008). There are more than one thousand federal laws in which marital status is a factor, including the rights to veterans' benefits, for example ("Federal Marriage Benefits Denied…" n.d.; U.S. General Accounting Office 1997). Other federal laws that apply only to legally married partners include those granting Social Security benefits to a widowed or disabled spouse, the right of legally married partners to inherit from one another without a will, or laws making it possible for the immigrant spouse of a U.S. citizen to also become a citizen.[15]

Proponents of gay and lesbian rights further argue that creating domestic partnerships or civil unions, instead of allowing marriage for same-sex partners, creates second-class citizens (Allen 2007; Leff 2006; Seidman 2003). As one lesbian spouse, legally married in Massachusetts, said:

[14] Couples who marry in states where same-sex marriage is legal, then return to their own states where it is not legal, can expect to encounter difficulties in the unforeseen event of a later desire to divorce. Same-sex divorce is impossible in a state that does not recognize the marriage, and getting divorced in the state where a couple was married can be challenging. Most states have a residency requirement for divorce, some as long as one year (Clifford, Hertz, and Doskow 2007).

[15] Under federal law, legally married spouses can petition for immigration and citizenship status for their foreign-born husbands or wives. In 2009, a bill was introduced in the U.S. Congress that would similarly allow American citizens and legal immigrants to pursue U.S. residency for their same-sex partners (Preston 2009). At this writing, the outcome of this proposed legislation has not been determined.

> Same-sex marriage not only makes me take my relationship more seriously, it's making me take my country more seriously. I always felt oppressed and not a part of America, not really. But this seems like finally there is a light in the dark, like finally . . . the government is saying that my relationship counts and I count, too. (in Lannutti 2007, p. 141)

Many children of gay male or lesbian couples also view the legalization of same-sex marriage as giving them security and comfort (P. L. Brown 2004; Goldberg 2007). Relatedly, one mother described her daughter's school experience:

> She would be quite open about [our family] in school and answer, "I have two moms." And kids would say to her, "Well, they can't be married." And she would say we're married because that's how we always represented ourselves. So she had a boy in her class that would say, "They're not married, they can't be married!" So I figured it was such a thrill for her to be able to say, with confidence, "They're married, and yes it is legal," and "I'm just like you." (in Porche and Purvin 2008, p. 153)

Interestingly, however, although many lesbians and gays support the claim for same-sex marriage in principle, in Massachusetts where same-sex marriage has been legal since 2004, a minority of same-sex couples has chosen to marry. A recent qualitative study of Massachusetts same-sex couples who had been together for twenty years or more sheds light on this development. One reason that committed same-sex couples may not marry when given the option is that they have already "spent thousands of dollars instituting all the legal protections they felt they needed and did not see a need to complicate those arrangements by changing their status through a marriage recognized only by Massachusetts." Attitudes of other participants in this study resembled those of some heterosexual cohabiting couples. They said that "marriage is made 'not by some sort of legal sanctions' but by the commitment of the people in the relationship to each other" (Porche and Purvis 2008, p. 155). Gay men and lesbians themselves have been divided somewhat on the desirability of legalized same-sex marriage, at least for themselves ("Gays Want the Right" 2004).

Dissenting Arguments among Lesbians and Gay Men With the majority of states offering *no* legal recognition at all for same-sex unions, some GLBT spokespersons argue that too much emphasis is placed on advocating for same-sex marriage when activist resources would be better spent working for wider enactment of civil unions (Leff 2009). Furthermore, some gays and lesbians have opposed legal same-sex marriage in principle. Generally they have objected to mimicking a traditionally patriarchal institution based on property rights and institutionalized husband–wife roles and characterized by a high divorce rate.[16] Opponents have also objected to giving the state power to regulate primary adult relationships (Peele 2006).

Moreover, they have stressed that legalizing same-sex unions would further stigmatize any sex outside marriage, with unmarried lesbians and gay men facing heightened discrimination ("Monogamy: Is It for Us?" 1998; Seidman 2003). We can't say how representative the following statement is, but it does illustrate the viewpoint of at least one lesbian when contemplating the possibility of legalized same-sex marriage in her state:

> I don't want to get married so this marriage thing is going to make it harder for me to find a person to be in a relationship with. I know that because I don't want to get married, women will think I'm not a good potential partner, and move on. . . . It sucks, because . . . now I have to limit myself to the other non-marrying kinds out there—like finding a great girl wasn't hard enough already! (in Lannutti 2007, p. 145)

Whether and to what extent allowing same-sex couples to legally marry increases their personal life satisfaction is a matter for future research. In the following section, we turn to a discussion of life satisfaction among the unmarried.

Although there is undeniable evidence for the physical and psychological benefits of marriage, unattached singles do point to benefits of their lifestyle. Among these are less irritation with coresident family members and a greater sense of control over their lives. Moreover, when we think of singlehood as a continuum, we realize that not all singles—even those who live alone—are socially unattached, disconnected, or isolated. Maintaining close relationships with family and friends is associated with positive adjustment and satisfaction among singles.

Maintaining Supportive Social Networks and Life Satisfaction

Perhaps not surprisingly, life satisfaction is associated with income as well as marital status (Pelham 2008). As discussed in Chapter 7 and illustrated in Table 7.1, many singles and single parents, particularly women, just do not make enough money (U.S. Census Bureau 2010b, Table 700). Many work more than one low-paying job, and then take care of their homes and children (Huston and Melz 2004). For them, "career advancement" means hoping for a small raise or just hanging on to a job in the face of growing economic insecurity. Pursuing higher educational opportunities means rushing to class one evening a week after working all day. Moreover, research shows that poor women have less-effective private safety nets than do others, because their families and friends are also very likely to be poor and overburdened financially and emotionally (Harknett

2006). These women are dealing with work, parenting, and low-income issues, enjoying neither the stereotypical "swinging singles" lifestyle characterized by personal freedom and consumerism nor the commitment of marriage.

As pointed out in Chapter 7, research and polls consistently find that, as a group, marrieds are happier than singles (Carroll 2005; Taylor, Funk, and Craighill 2006; Wienke and Hill 2009). Research also shows that, regardless of whether they are legally married, people in secure interpersonal heterosexual or same-sex relationships—and those who socialize often with friends and family—are happier than are those who spend considerable time alone (Harter and Arora 2008; Pelham 2008; Wienke and Hill 2009).

If we think of the various living arrangements of unmarrieds as forming a *continuum of social attachment* (Ross 1995), we realize that not all singles are socially unattached, disconnected, or isolated. In sociologist Catherine Ross's (1995) research with a nationally representative sample of about two thousand adults who

---

[16] With irony, San Francisco columnist Mark Morford questions why same-sex couples would *want* to marry: "Show me a single scientific experiment where fully 50 percent of the results turn out negative and induce collapse and emotional breakdown and childhood therapy and Xanax and alcoholism and screaming, and I'll show you a scientist who will quickly scrap the whole thing and start over" (Morford 2006).

For singles, it's important to develop and maintain supportive social networks of friends and family. Single people place high value on friendships, and they are also major contributors to community services and volunteer work.

were interviewed in 1990, people in close relationships—whether or not married and whether or not living alone—were significantly less depressed than those with no intimate partner at all (see also Pelham 2008). Furthermore, the relationship between being involved and not being depressed held *only* for those in happy, or supportive, arrangements.

One young woman, single by choice, actually planned her own wedding ceremony—to herself. She wore white, carried a bouquet, and invited about twenty friends. She "chose to join herself in matrimony" a few days after her thirtieth birthday. Her friend, serving as officiator, asked, "Do you promise to love, honor, and respect yourself from this day forward for as long as you live?" "I do," answered the woman. She was subsequently declared "wedded to life" (Seligman 2006). Those who view themselves as "wedded to life"—by choice—are probably more satisfied with their lives than are those who are single against their wishes.

Meanwhile, for unattached singles, living alone can be lonesome. However, living alone does not necessarily imply a lack of social integration or meaningful connections with others (Trimberger 2005). Nevertheless, unattached singles have tended to report feeling lonely more often than have marrieds (Harter and Arora 2008; Pelham 2008). Poor and older singles are especially likely to be lonely, perhaps because the low incomes and ill health that tend to accompany old age make socializing very difficult. Besides age and income, being single as a result of divorce apparently affects loneliness (Kim and McKenry 2002).

Sociologist E. Kay Trimberger (2005) argues that the "heaviest thing" for unattached, middle-aged women is the "idea of the couple, and that's so internalized." Trimberger identifies the following "pillars of support" for unattached single women: a nurturing home, satisfying work, satisfaction with their sexuality, connections to the next generation, a network of friends and possibly family members, and a feeling of community.

Some research has found cohabitants to be midway between unattached singles and marrieds in mental and physical well-being (Kurdek 1991), while other studies have shown no difference between cohabitants and other singles, "suggesting that the protection effects of marriage are not as applicable to cohabitation" (Kim and McKenry 2002, p. 905). However, marriage also involves a set of obligations and the responsibility of coping with both the burdens of other family members and the disappointments that come with family life. Valuing personal autonomy, Americans may find these obligations emotionally stressful (Gove, Style, and Hughes 1990). There are some areas in which nonmarrieds may feel better off than the married. Less irritation and a greater sense of control over one's life can be among the advantages of being single (Hughes and Gove 1989).

All of us need support from people whom we are close to and who care about us. Isolation increases feelings of unhappiness, depression, and anxiety (Umberson et al. 1996), whereas being socially connected "seems to keep stress responses . . . from running amok," according to UCLA psychologist Shelley Taylor (quoted in "Save the Date" 2004). Maintaining close relationships with parents, siblings, and friends is associated with positive adjustment and life satisfaction (Kurdek 2006; Soons and Liefbroer 2009; Spitze and Trent 2006). A social network that can function as

"a parent safety net" is important to positive parenting practices among unmarried mothers (Ryan, Kalil, and Leininger 2009).

A crucial part of one's support network involves valued friendships. Despite changing gender roles, men remain less likely than women to cultivate psychologically intimate relationships with siblings or same-sex friends (Levy 2005; Weaver, Coleman, and Ganong 2003). Indeed, a man may be more open and disclosing with a woman friend (Wagner-Raphael, Seal, and Ehrhardt 2001). One study of men in the construction industry found that many of them, rather than building truly supportive friendships, talked instead about horseplay, alcohol consumption, risk taking, and physical prowess, and generally engaged in one-upmanship (Iacuone 2005). Men (as well as women) who do not establish friendships based on emotional honesty run the risk of feeling socially isolated. In addition to friendships, other sources of support for singles include group living situations, religious fellowships, and volunteer work (Lyons 2003; Mustillo, Wilson, and Lynch 2004). Singles also reach out to their families of origin (Arnett 2004; Bengston, Biblarz, and Roberts 2002). However a person chooses to live the single life, establishing a sense of belonging by maintaining supportive social networks is important.

Summary

Vicky Kasala

Besides a variety of living arrangements, factors such as age, sex, residence, religion, and economic status contribute to the diversity and complexity of single life. An elderly man or woman existing on Social Security payments and meager savings has a vastly different lifestyle from two unmarried professionals living together in an urban area. Also, the experience of being unmarried differs according to whether one is single by choice or involuntarily.

- Since the 1960s, the number of unmarrieds has risen dramatically. Much of this increase is due to young adults' postponing of marriage, coupled with the rise in the incidence of cohabitation.

- One reason people are postponing marriage today is that increased job and lifestyle opportunities may make marriage less attractive.

- The low sex ratio—fewer men for women of marriageable age—has also caused some women to postpone marriage or put it off entirely.

- Attitudes toward marriage and singlehood have changed, so that being unmarried is more often viewed as preferable, at least "for now."

- More and more young unmarrieds are living in their parents' homes, usually at least partly as a result of economic constraints.

- Some unmarrieds have chosen to live in communal or group homes.

- A substantial number and growing percentage of heterosexual unmarrieds are cohabiting.

- As heterosexual cohabitation becomes more acceptable, more and more cohabiting households include children either born to the union or from a previous relationship.

- The relative instability of heterosexual cohabiting unions has led to some, apparently warranted, concern for the outcomes of children living in cohabiting families.

- Some couples live together in gay male or lesbian unions; a little more than one-third of lesbian and nearly one-quarter of gay male households include children either from the same-sex union or from a previous (often heterosexual) relationship.

- Same-sex couples must daily negotiate their private relationship within a heterosexual—and often, heterosexist—world.

- Research finds that children raised by same-sex couples are not significantly different from those raised by heterosexual parents.

- Congruent with the emergence of the pluralistic family, we are witnessing a national and global debate over whether legal marriage should be extended to include lesbians and gay men.

- However one chooses to live the single life, it is important to maintain supportive social networks.

Questions for Review and Reflection

1. Individual choices take place within a broader social spectrum—that is, within society. How do social factors influence an unmarried individual's decision regarding his or her living arrangements?

2. What do you see as the advantages and disadvantages of cohabitation compared to marriage?

3. What does current research tell us about the outcomes generally of children raised in homes with two married biological parents, compared to those raised in cohabiting families?

4. On average, do the outcomes of children raised by heterosexual parents differ from the outcomes of those raised by same-sex couples?

5. **Policy Question.** Do you think that legalizing same-sex marriage is a good idea? Give arguments based on facts to support your opinion.

Key Terms

cohabitation 199
communes 198
consensual marriages 200
culture war 211

Defense of Mrriage Act (DOMA) 208
domestic partners 202
sex ratio 194

Online Resources

Sociology CourseMate

www.CengageBrain.com

Access an integrated eBook, chapter-specific interactive learning tools, including flash cards, quizzes, videos, and more in your Sociology CourseMate, accessed through CengageBrain.com.

www.CengageBrain.com

Want to maximize your online study time? Take this easy-to-use study system's diagnostic pre-test, and it will create a personalized study plan for you. By helping you identify the topics that you need to understand better and then directing you to valuable online resources, it can speed up your chapter review. CengageNOW even provides a post-test so you can confirm that you are ready for an exam.

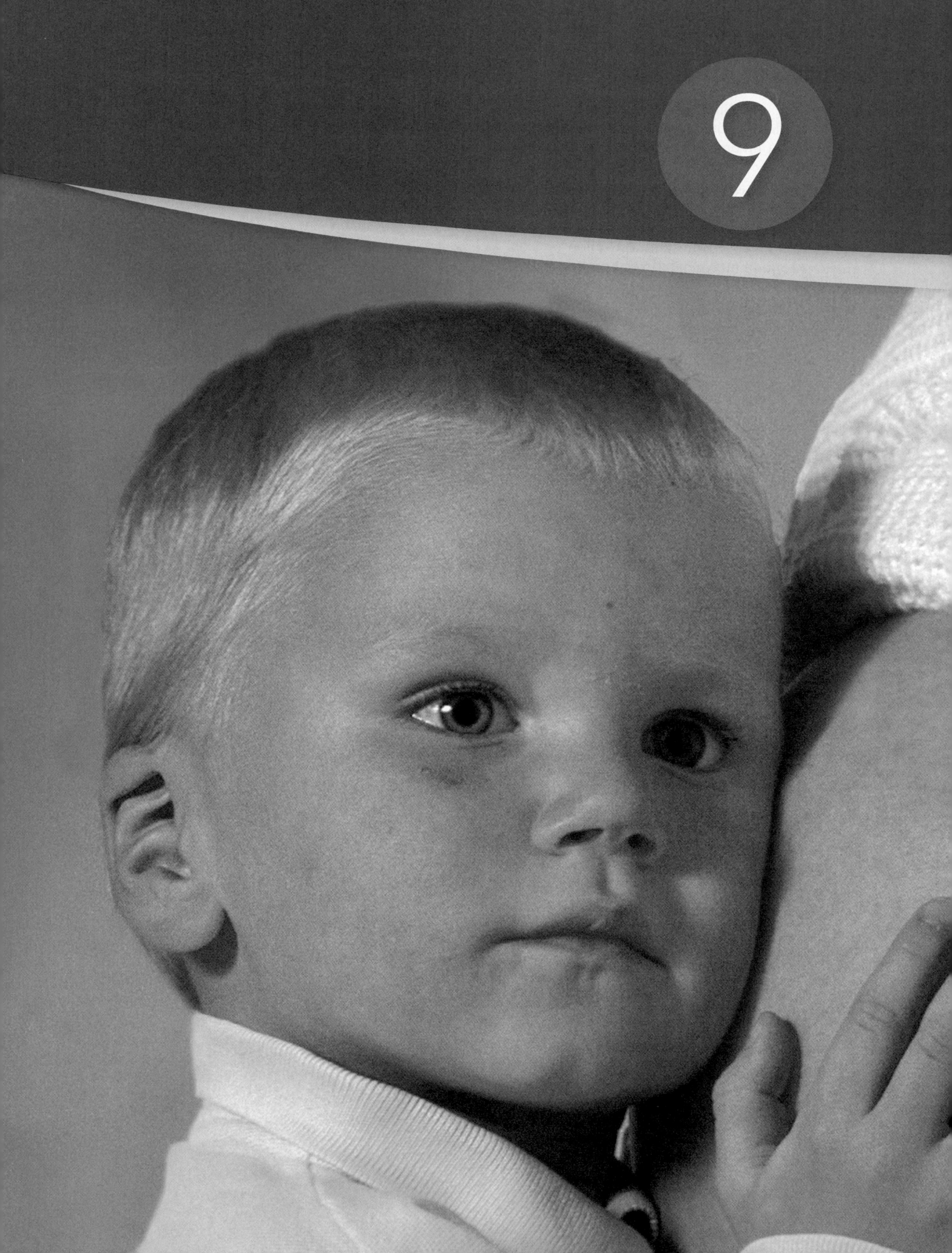

To Parent or Not to Parent

© IT Stock International/Creatas

There may be someone in your class who has been adopted, perhaps by parents of another race. There may be someone in your class who is thinking about infertility treatment. Or who is thinking about having an abortion. Or about having a first child. Or about whether to ever have children. Or about having and raising an only child, or a larger family. All these decisions focus on some aspect of whether (or how) to become a parent. They are very personal choices, but in this chapter we'll see that they are nevertheless influenced by the society around us.

Significant changes have taken place in American childbearing patterns in the decades since World War II. For one thing, the average number of children an American woman bears has declined. For another, women are having children at later ages. And finally, childlessness—by choice or circumstance—is more common today.

The **total fertility rate (TFR)** is the number of births a typical woman will have during her lifetime.[1] In the United States, the TFR dropped sharply from a high of 3.5 at the peak of the baby boom to the lowest level ever recorded: 1.738 in 1976. In recent years, the total fertility rate has fluctuated around 2.0; on average, American women are now having around two children each (J. Martin et al. 2005, Table 4; Hamilton, Martin, and Ventura 2009, 2010, Table 1; and see Figure 9.1). At the same time, choosing not to be a parent is more acceptable today.

Family Group by Henry Spencer Moore, 1947

Even with some recent fluctuations, overall fertility[2] levels have dropped. Although 2007 showed a slight increase in teen birthrates at 1 percent, we continue to see childbearing increasingly shifted to later ages—in

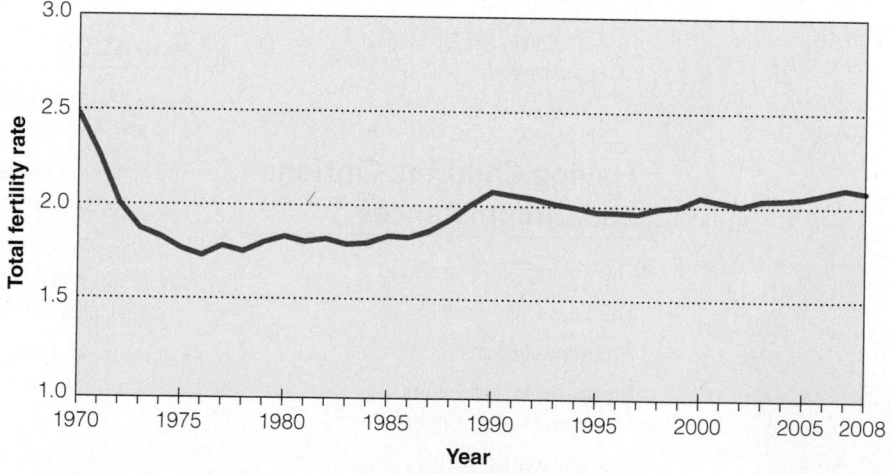

Figure 9.1 Total fertility rates, United States, 1970–2008

[1] The total fertility rate (TFR) for a given year is an artificial figure arrived at through complex mathematical calculations. In common-sense terms, the TFR indicates how many children an average woman would have if present trends continue. It is the figure most used in this textbook to grasp trends in fertility and family size. The total fertility rate and other birth rates may be computed for various sectors of the population (for example, unmarried women, white women, adolescent women, and so forth).

[2] The term *fertility* is used by demographers to refer to actual births. In everyday language we use the term *fertility* to mean ability to reproduce. However, the technical social science term for reproductive capacity is *fecundity*.

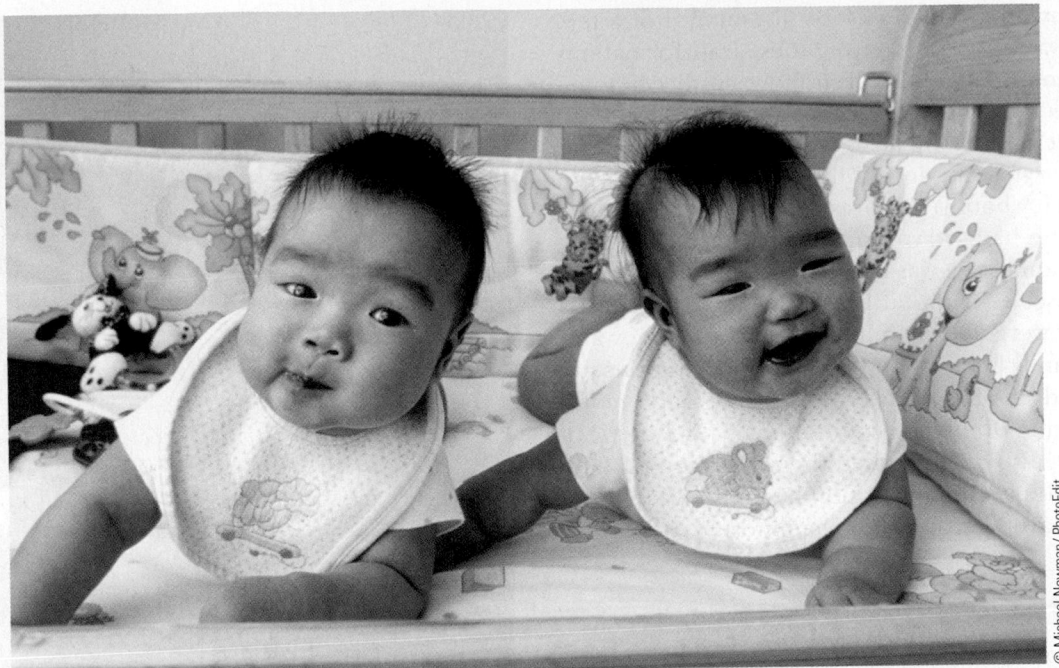

The numbers of twins, triplets, and higher-order births have increased dramatically since 1980. Families raising two or more children who are the same age gain the attention of onlookers when they are out and about—and they face challenges at home. As their numbers have grown, so have organizations to bring them together to share common concerns and joys.

fact, 2008 showed a 2 percent drop in teen birthrates and a 3 percent drop in birthrates to women in their twenties. Married women had been waiting longer to have their first babies. For example, women in their thirties who had postponed parenthood are now having first, second, and, in some cases, third children, and some are becoming mothers for the first time in their forties. The 2008 birthrate for women in their thirties has dropped around 1 percent from its forty-year high in 2005, whereas the birthrate for women in their early forties has reached a forty-three-year high (Mathews and Hamilton 2009; Hamilton, Martin, and Ventura 2009, 2010). As for multiple birthrates, the birthrate for twins remains at 32.1 for every 1,000 births, while triplets and higher-order births are down 5 percent (Martin, Hamilton, Sutton, Ventura, Menacker, and Kirmeyer, Mathews 2009, pp. 2, 20).

At the same time, childlessness is higher for women now than in the recent past. In 2006, 20 percent of women age forty through forty-four were childless, twice the percentage of childless women in that age group in 1976 (Dye 2008, Table 2). Twenty-two percent of men age forty through forty-four reported that they were childless (in 2002; Martinez et al. 2006, p. 1).

These points describe the sum total of many couple and individual decisions. Throughout this chapter, we'll be looking at the choices that individuals and couples have to make about whether to have children and, if so, how many. Among other things, we'll see that modern scientific and technological advances have both increased people's options and added new wrinkles to their decision making. We'll see, too, that technological progress does not mean that people can exercise complete control over their fertility. To begin, we'll review fertility trends in the United States in more detail. Then we'll examine the decision whether or not to become a parent.

Fertility Trends in the United States

Lower U.S. fertility appears to be a major change when we compare current birthrates to those of the 1950s. But the decline in fertility is actually a continuation of a long-term pattern dating to about 1800. Alternatives to the motherhood role began to open up with the Industrial Revolution and the resulting creation of a labor force that worked in production outside the home. Previously, in a preindustrial economy, women could combine productive work on the farm or home artisan shop with motherhood. But when work moved from home to factory, the roles of worker and mother were not so compatible. Consequently, as women's employment increased, fertility declined.

Declining infant mortality, another change affecting fertility over time, is a result of improved health and living conditions. Gradually, it became unnecessary to bear so many children to ensure the survival of a few. Changes in values accompanying these transformations made large numbers of children more costly economically and less satisfying to parents.

In the face of the long-term decline over the past two centuries, the upswing in fertility in the late 1940s and 1950s (the baby boom) requires explanation. It appears that those who had grown up during the Great Depression, when family size was limited by economic factors, found themselves in an affluent postwar economy as adults. They were able to fulfill dreams of a happy and abundant family life to compensate for deprivations suffered as children (Easterlin 1987). Marriage and motherhood became dominant cultural goals for American women; men also concentrated their attention on family life. Couples in this generation had more than three children, on average, and some had more.

Family Size

Today, approximately 2.5 children constitute Americans' ideal family size (Carroll 2007). In 2006, only 6.4 percent of women who had completed their childbearing had four children and 3.5 percent had five or more children, this compared to more than three times that in 1976 (Dye 2008, Table 1). Although support for those who choose to have four or more children appears to be declining, other important factors in declining family size are found in a lack of social support for women's childbearing. For example, contemporary pressures on women in their childbearing years such as extended educational pursuits, "demanding work schedules," along with postponement of marriage, etcetera, appear to serve as barriers to increased family size (Hagewen and Morgan 2005; Morgan and Rackin 2010). See "A Closer Look at Diversity: Choosing Large Families in a Small-Family Era" for a discussion of this viewpoint.

U.S. fertility tends to hover slightly below or slightly above **replacement level** (some studies put it below at 1.9; other studies put it above at 2.12).[3] The total fertility rate in the United States has never dropped as low as those of some European and Asian countries (Dye 2008; Goldstein, Sobotka, and Jasilioniene 2009, Table 1; Hamilton, Martin, and Ventura 2010). So a current question regarding American fertility is this: Why does the United States have higher fertility than other countries with parallel levels of economic development and better family support policies?

Historically, America's fertility was higher than that of Europe (Goldstein, Sobotka, and Jasilioniene 2009;

The ideal family size in the United States is now two children.

Weeks 2002, p. 236), so cultural factors may be at work that make the United States exceptional. The United States appears to have strong fertility norms encouraging at least two children and discouraging childlessness and one-child families. The pattern of two-child preference has been consistent for over thirty years. Looking back, demographers now interpret the drop in fertility between the 1970s and the 1980s as not really a fertility decline but, rather, a postponement of births, whereas the slightly stronger fertility norms may have something to do with optimism and a general level of prosperity found in the United States (Goldstein, Sobotka, and Jasilioniene 2009; Hagewen and Morgan 2005; Myrskylä, Kohler, and Billari 2009).

Differential Fertility Rates

Not surprisingly, fertility rates vary among segments of the U.S. population. Usually, more highly educated and well-off families have fewer children, and that is true of current fertility rates (Morgan and Rackin 2010). Although better-off families have more money, their

[3] *Replacement level* is the level of fertility necessary for a society to replenish its population. For each two adults, two children are needed to replace them. The total fertility rate is pegged at 2.1, to take into account that some women will die before reaching reproductive age, some will be biologically unable to reproduce or may choose not to, and some will be institutionalized or in religious orders mandating celibacy.

Large families of four or more children are a distinct minority in the United States today—and have been for some time. In 2006, only 6.4 percent of women who had completed their childbearing had four children, and 3.5 percent of women had five or more children (Dye 2008, Table 1).

The U.S. family size ideal for thirty years has been two children, with the three-child family being the next most popular. Families of four or more are thought better than childless or only-child families. But favorable opinion of large family size has been declining, dropping by half (20 percent to 10 percent) between 1970 and 1974 and between 2000 and 2002 (Hagewen and Morgan 2005). Current polls are consistent with these findings. A 2007 Gallup poll found 52 percent saying two children is ideal, followed by 25 percent believing that three children is ideal (Carroll 2007d). "Being a good parent is now [perceived as] largely inconsistent with having more than a small number of children" (Morgan 2003, p. 593).

A qualitative study of sixty women who have chosen various family sizes—childless, only one child, the "normative" family size of two or three, and "supernormative" families of four or more—suggests that views of large families have become negative and involve significant stereotyping. Mothers of large families report that they are stigmatized, seen as uneducated, insufficiently attentive to their children, messy housekeepers, ignorant of birth control, and experiencing unintended pregnancies (Hagewen and Morgan 2005; Zernike 2009). Environmental scientists are also researching the impact multiple childbirths have on the planet. Their argument is that children born to American parents tend to have a greater impact on the sustainability of the planet (because Americans consume more of the world's resources); thus, the more children Americans have, the greater the long-term impact (Murtaugh and Schlax 2008; Weisman 2008). Although experiencing pressures to limit family size that began, in some cases, after their second child was born, the majority of these mothers of large families also received some positive feedback. Many larger families are religious and find acceptance in their religious communities (Zernike 2009). Online communities (interactive chat sites, blogs, and so on) have also been created to provide resources and support for members of large families.

As discussed previously in this chapter, as well as in Chapter 3, women fifteen to forty-four years old who had a bachelor's degree or higher bore fewer children on average (1.6 births) than those without a high school diploma (2.4 births) (Dye 2008). So where do we find large families? Women with no high school diploma are most likely (21 percent) to have borne four or more children, dropping off to 11 percent for high school graduates, 8 percent of those women with some college, and only 3 percent of women who hold a bachelor's degree or more (Chandra et al. 2005, Table 2). Mothers of large families are less likely to be in the labor force. In a recent study of young urban males, researchers found similarities to the previous discussion of mothers. The lower the levels of education in men, the greater likelihood to have fathered multiple children outside of marriage or committed relationships. As the levels of education went up, the rates of fatherhood dropped, and the rates of marriage and income increased (Bronte-Tinkew et al. 2009, Table 1).

"The acceptable/normative family may very well vary from culture to culture, such that women having family sizes considered deviant in some segments of the U.S. population may experience acceptance within their own culture" (Mueller and Yoder 1999, p. 918). Some commentators note higher fertility rates in the Great Plains and Southwest regions (D. Brooks 2004). The Southeast Asian Hmong ethnic group has an average family size of 5.34 (Fadiman 1998; U.S. Census Bureau 2009a, Table S0201). Some Pacific Islander groups and the Amish, among others, also favor large families (Greksa 2002; Harris and Jones 2005; Hurd 2006; U.S. Census Bureau 1993).

There is in fact a pro–large family movement termed "Quiverfull," whose families aim for six or more children. "Quiverfull mothers think of their children as . . . an army they're building for God" (A. Joyce 2006, p. St-1; see also Zernike 2009). This movement is situated in fundamentalist churches. Quiverfull mothers homeschool their children, accept male headship, and make no attempt to control family size or timing.

A desire for large families, though not common, is not limited to conservative religious sectors of American society:

> We come from all different faiths and some of us are not religious at all. We are well-to-do and of modest means. We are home schoolers, public schoolers, and private schoolers. We are stepparents, adoptive parents, and foster parents. . . . We are a diverse bunch, but the one thing we at larger families.com all have in common is that we love our kids. (Francis 2007)

Critical Thinking

What do you see as the advantages and disadvantages of large families? Small families? What is *your* ideal family size? Why?

As Figure 9.2 shows, women in the various U.S. racial/ethnic groups vary quite a bit in the number of children they have. Here we look at fertility patterns among the major American racial/ethnic groups.

Fertility Rates among Non-Hispanic Whites

According to 2008 preliminary data from National Vital Statistics Reports, fertility patterns of the non-Hispanic white population are similar to those described for the total population, although slightly lower. The total fertility rate for whites in 2008 was 1.835 compared to 2.085 for the total population (Hamilton, Martin, and Ventura 2010, Table 1). Because the white population was historically such a large part of the total population,

explanations offered for historical changes in fertility apply to changes in fertility among non-Hispanic whites.

Fertility Rates among African Americans

Some racial/ethnic minority populations in the United States have fertility rates that are higher than those of non-Hispanic whites, and that is true of African American women at the present time. Earlier in our history, in the late eighteenth century, white and black women appeared to have borne children at approximately the same rates. At about that time, the white population began to reduce its fertility, and by the end of the nineteenth century, childbearing among whites had declined significantly. The birthrate among

blacks did not decline much until after 1880, when it began to drop rapidly.

When individuals have satisfying options other than parenthood, they typically choose to limit their childbearing. The differences in timing of the fertility decline of white and black populations in this country suggest that although education and other opportunities opened up for whites with the Industrial Revolution earlier in the nineteenth century, they did not do so for blacks until well after the Civil War. By the 1930s, the black fertility rate was close to that of whites. Through the 1940s, 1950s, and after, trends in fertility among blacks generally paralleled those among whites, though at a higher level.

With regard to the current fertility rate among black Americans, the same differential opportunity explanation holds true as an explanation for blacks' higher fertility. Nevertheless, African American fertility rates declined by almost 25 percent in the 1990s and into the current century. The total fertility rate for non-Hispanic African Americans was 2.110 in 2008 (Hamilton, Martin, and Ventura 2010, Table 1).

Fertility Rates among Latinos

Latinos (termed *Hispanics* in government statistical documents) have the highest fertility rate of any U.S. racial/ethnic group. Their total fertility rate (TFR) of 2.905 in 2008 was over 58 percent higher than that of non-Hispanic white women (Hamilton, Martin, and Ventura 2010, Table 1). The TFR for Mexican American women is the highest among Hispanics (3.107), whereas women of Puerto Rican (2.167) and

Figure 9.2 Total fertility rate by race/ethnicity, United States, 2008

Source: Hamilton, Martin, and Ventura 2010, Table 1.

children are also more costly, for these parents expect to send their children to college and to provide them with expensive experiences and possessions. Moreover, people with high education or income have options other than parenting. They may be involved in demanding careers or enjoy travel, activities that they weigh

against the investment of time and money required in parenting more children (Weeks 2007).

Women who are not in the labor force have higher birthrates and a larger completed family size on average than employed women. This may be intentional; women may shape their employment commitments

Cuban background (1.601) have moderate or low total fertility rates. At a TFR of 3.014, women of other Central American and South American background have fertility rates just slightly lower than those of Mexican American women (Martin et al. 2009, Table 15).

Reasons for the high birthrates include the fact that Hispanics migrate from nations that have high birthrates and Catholic and rural traditions that value large families. Large families may serve important functions, especially in poorer families. Children might be an insurance policy against a parent's old age in a society that lacks adequate welfare or retirement systems. Even while children are growing up, their earnings might be an important part of the family income. Large-family norms based on these needs may be carried over to the United States (P. H. Collins 1999, p. 202).

Moreover, the lifetime fertility of Latinas varies strongly with their educational attainment, and Latinas are relatively more concentrated in the lower educational categories. Similarly, on average (Cubans excepted), Latino families have lower incomes and higher rates of poverty than the non-Hispanic white population, also factors associated with higher fertility.

Finally, Latinos are younger, with more concentration in childbearing ages. Latinas, especially Mexican Americans, typically begin having children at younger ages. Latinas in their early twenties have a much higher fertility rate than women this age in other racial/ethnic groups.

© Bill Bachmann/ PhotoEdit

Native American women who live on reservations have significantly higher fertility than those who do not. Differential birthrates reflect the fact that people in various cultures have different beliefs and values about having children.

Fertility Rates among Asian Americans/Pacific Islanders

Asian American/Pacific Islander and Native American births are a relatively small proportion of U.S. births. Asian/Pacific Islander women had a comparatively low total fertility rate (2.056) in 2008 (Hamilton, Martin, and Ventura 2010, Table 1), but there is wide variation by country of origin. As Asian/Pacific Islander immigrants assimilate, their birthrates tend to converge with those of whites (Hwang and Saenz 1997).

Fertility Rates among Native Americans/Alaska Natives

Fertility rates of Native Americans/Alaska Natives have declined by over 20 percent since 1990, to a total fertility rate of 1.843 in 2008 (Hamilton, Martin, and Ventura 2010, Table 1). Native American women who live on reservations have significantly higher fertility than those who do not (Taffel 1987), probably because of the limited educational and economic opportunities noted earlier.

Critical Thinking

The total fertility rate, which is an approximation of average family size, is lower in all racial/ethnic groups than it was during the baby boom era (1946 to 1964). Why do you think this is so? Does it have to do with economic pressures? Changing attitudes toward children? Or something else?

to their birth intentions and vice versa (Dye 2008, Table 2).

Differential birthrates also reflect the fact that beliefs and values about having children vary among cultures—see "Facts about Families: Race/Ethnicity and Differential Fertility Rates" for a discussion of fertility among the diverse racial/ethnic groups of the United States.

Decision making about having children now takes place at a time of more reproductive options than ever before. This observation highlights the point that in the early twenty-first century, parenthood is a choice, made in a social context. We now focus on people as they make their decisions about becoming parents.

The Decision to Parent or Not to Parent

The variations in birthrates just described reflect decisions shaped by values and attitudes about having children. In traditional society, having children was not viewed as a matter for decision; couples didn't decide to have children. Children just came, and preferring not to have any was unthinkable. Now couples and individuals intentionally consider becoming parents.

Early family planning efforts focused on the timing of children and family size. Now choices include "whether" to have children, not only "when," but also "how many," and "how." Although social change and technology provide more choices, they also present dilemmas. It is not always easy to choose whether to have children, how many to have, when to have them, and when to use reproductive technology. Not all choices can be realized, whether they reflect a desire to have children or to avoid having children.

The extent to which people today consciously choose (or reject) parenthood or experience it as something that simply happens to them is uncertain. Because people have more control over their lives generally, many now approach parenthood as a conscious choice. Among others—teenagers, for example—parenthood is often less thought out. Moreover, some people may be philosophically disinclined to plan their lives (Luker 1984). For whatever reason, in a 2002 survey, women respondents reported that over one-third of their recent births were unintended—14 percent unwanted and 21 percent mistimed. Additional research suggests that as many as half of all births to American mothers may be unintended (Chandra et al. 2005, p. 1; Finer and Henshaw 2006; Trussell and Wynn 2008).[4] Nevertheless, more so than in the past, our society presents the possibility of choice and decision making about parenthood.

In the following pages, we'll look more closely at some of the factors involved in an individual's or couple's decision making about whether to become a parent: first, the social pressures, and then the personal pros and cons.

Social Pressures to Have Children

Social pressures to have children exist in our society, as "strong norms against childlessness persist" (Hagewen and Morgan 2005, p. 512). Social scientists refer to this cultural phenomenon as a **pronatalist bias**. Having children is taken for granted, whereas not having children seems to need a justification. Eighty-three percent of American women say being or becoming a mother is important to their identity (Center for the Advancement of Women 2003, p. 8). Negative stereotypes of the voluntarily childfree were prevalent at least through the 1970s, and continue in some lesser form through our contemporary era (Abma and Martinez 2006; Kelly 2009; Mueller and Yoder 1999).

Some scholars believe pronatal pressures are becoming stronger now that the countercultural trends of the 1960s and 1970s have been replaced by an emphasis on "family values" (Bulcroft and Teachman 2004; Dowland 2009). But demographers who have reviewed survey data on this point argue that the expectation for married couples to have children is less pronounced than in the past. "Negative views toward voluntary childlessness . . . may be changing" (Hagewen and Morgan 2005, p. 513; see also Koropeckyj-Cox and Pendell 2007). Although some contemporary researchers are uncomfortable with the term *childfree,* it is often used now instead of the more negative-sounding term *childless.*[5]

Is American Society Antinatalist?

Some observers, in fact, argue that U.S. society has become antinatalist—that is, slanted against having children or, at least, not doing all it can to support parents and their children. These family policy scholars view American society as characterized by **structural antinatalism** (Huber 1980), insufficiently supportive of parents and children. Given the previous discussion in this chapter on increasing numbers of childfree adults, this viewpoint may have some merit. Critics of American family policy point out that nutrition, social service, financial aid, and education programs directly affecting the welfare of children are not adequate compared to other nations at our economic level (Children's Defense Fund 2008). Nor do we provide paid parental leave or other support for parents of young children as many other countries do. Children in the United States are more likely to be poor than children in comparable countries, and the United States ranks twentieth out of twenty-one in overall child well-being among advanced industrial nations (Taubman 2009; UNICEF 2007, p. 1).

Other features of our society make parenthood difficult. Some municipalities keep taxes down by restricting housing size to discourage families with children from living there, meanwhile providing tax breaks and housing preferences for the elderly (C. Jones 2004; Monaghan 2010; Peterson 2005a). There has been a

---

[4] See Santelli et al. 2003 for a discussion of how *unintended* has been defined and the limitations of research on unintended pregnancy.

[5] Each term conveys an inherent bias. For that reason and because there are no easy-to-use substitute terms, we use both *childless* and *childfree* in this text.

push back against "family-friendly" policies in the workplace, as workers who do not have children challenge policies that support parents (Burkett 2000). As fewer people are active parents, some advocates anticipate political reverberations that will disadvantage child-rearing families—less willingness of communities to support good schools, for example. As a numerical minority, parents may experience "a growing sense of isolation from the American mainstream" (Whitehead and Popenoe 2006).

At the same time, nonparents may perceive that *they* are the less-favored group and resent what they see as assertions of privilege on the part of parents. Sidewalk standoffs, where parents pushing super-sized strollers meet adult pedestrians who decline to give way, represent the "latest fissure in a long-standing divide between parents and nonparents over who has made the right choice in life" (Rosenbloom 2005, pp. ST1–2).

Of course, those choices that people make about becoming parents reflect not only external social pressures or cultural moments but also their own needs, values, and attitudes (Lundquist, Budig, and Curtis 2009). Still, given "strong antinatalist forces" that include direct and indirect costs and responsibilities for parents, some scholars have asked this question: "Why do people choose to have any children?" (Hagewen and Morgan 2005, p. 513). In the next sections, we will look at some of the rewards and costs associated with parenthood.

Motivation for Parenthood

Traditionally, children were viewed as economic assets—more hands for the work in the fields and kitchens. The shift from an agricultural to an industrial society and the development of compulsory education transformed children from economic assets to economic liabilities. But as their economic value declined, children's emotional significance to parents increased, partly because declining infant mortality rates made it safe to become attached to children. Parents' desire was for "a child to love" (Zelizer 1985, p. 190).

Children bring many emotional benefits and other satisfaction to parents. Although less true for today's adults (see Chapter 2), becoming a parent can certify one's attainment of adulthood. For men as well as women, parenthood represents an important personal identity (Games-Evans 2009; Hagewen and Morgan 2005). When parents are interviewed, they often express their desire "to have a child to love" and talk about the "joy that comes from watching a child grow"

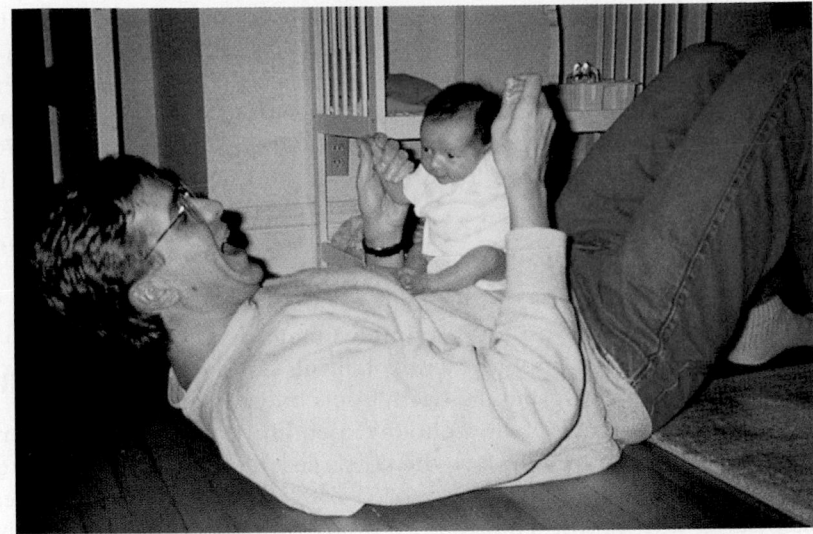

© Beth Huber

Children can bring vitality and a sense of purpose into a household. Having a child also broadens a parent's role in the world: Mothers and fathers become nurturers, advocates, authority figures, counselors, caregivers, and playmates.

(Morgan and King 2001, p. 11). Parenthood can give one a sense of commitment and meaning in an uncertain social world (McMahon 1995 in Morgan and King 2001, p. 11).

In having children, parents can find a satisfaction that is lacking in their jobs. Family life also offers an opportunity to exercise a kind of authority and influence that one may not have at work. Children add considerable liveliness to a household, and they have fresh and novel responses to the joys and vexations of life. In sum, children "are capable of bringing profound meaning and purpose into people's lives" (Groat et al. 1997, p. 571). This idea that children bring unique benefits to parents has been termed the **value of children perspective** (Hoffman and Manis 1979; Lamanna 1977).

The *value of children perspective* has recently been supplemented by a **social capital perspective**[6] on the benefits of parenthood:

> Parenthood intensifies interaction with, and assistance from, other family members. It facilitates exchanges with neighbors and other community members. . . . [P]arenthood can bring parents into an extensive and supportive network. (Schoen and Tufis 2003, p. 1032)

Research suggests that the anticipated social capital benefits of parenthood may be one motivation for childbearing, not only for married but also for unmarried

---

[6] We usually think of *capital* as money. But more generally, the term refers to a resource that can be used to one's benefit. Social capital, then, refers to social ties that are or can be helpful resources.

prospective parents. Analysis of responses from 1,155 unmarried women in the National Survey of Families and Households (NSFH) found that nonmarital conceptions occurred more often to women who anticipated social capital benefits from children (Schoen and Tufis 2003; South and Crowder 2010).

Costs of Having Children

Although the benefits of having children can be immeasurable, children are costly. On a purely financial basis, children decrease a couple's level of living considerably. In husband–wife families with two children, an estimated 42 percent of household expenditures are attributable to children. In fact, the yearly cost of child-care for an infant can run upwards of $14,480, and costs associated with raising children are rising faster than our incomes. For example, social researcher Pamela Paul notes that our household incomes rose 24 percent over the last decade, whereas our costs associated with child raising rose 66 percent (Paul 2009, p. 5, 7). The average cost of raising a child born in 2007 to age eighteen is estimated at $269,040 for middle-income families (Lino 2008, Tables 12).

Added to the direct costs of parenting are **opportunity costs**: the economic opportunities for wage earning and investments that parents forgo when rearing children. These costs are more often felt by mothers. In particular, this cost more keenly impacts white married, divorced, and never-married mothers. This cost also impacts African American married mothers of more than two children. Opportunity costs are not, interestingly, associated with Hispanic motherhood (Glauber 2007). Although Glauber's research did not go into detail about the reasons for opportunity costs or the lack thereof, it does acknowledge a differential between income levels and race (see Chapter 3 for more discussion on gender, race, and income). A woman's career advancement may suffer as a consequence of becoming a mother, especially in a society that does not provide adequate day care or a flexible workplace. A couple in which one partner quits work to stay home with a child or children faces loss of up to half or more of its family income (Longman 1998). The spouse (more often the woman) who quits work also faces lost pension and Social Security benefits later. All in all, in our society, there is "a heavy financial penalty on anyone who chooses to spend any serious amount of time with children" (Crittenden 2001, p. 6).

Conversely, loss of free time and increased stress are two important costs of trying to lead two lives, as a family person and as a career person. Parents generally experience a loss of freedom of activity and schedule flexibility with the arrival of a first child (Beck et al. 2010; Hagewen and Morgan 2005).

All in all, "from the day children are born they become a source of joy and a source of burdens for their parents" (Nomaguchi and Milkie 2003, p. 372). Still, a large proportion of men (68 percent) and women (70 percent) who have children and who were surveyed in 2002 answered "strongly agree" to this statement: "The rewards of being a parent are worth it despite the cost and work it takes"; only 2 to 3 percent disagreed (Martinez et al. 2006, Figure 25; see also Schindler 2010).

How Children Affect Marital Happiness

Marital strain is considered a common cost of having children. Evidence shows that children, especially young ones, stabilize marriage; that is, parents are less likely to divorce. But a stable marriage is not necessarily a happy one: "[C]hildren have the paradoxical effect of increasing the stability of the marriage while decreasing its quality" (Bradbury, Fincham, and Beach 2000, p. 969). A major review of the research in this area finds that not only do parents report lower marital satisfaction than nonparents, but also, the more children there are, the lower marital satisfaction is. Additionally the drop in marital satisfaction is "sudden and persists over time" (Doss et al. 2009, p. 617; Twenge, Campbell, and Foster 2003; see also Coontz 2009). Additional research involving 1,000 families suggests that part of the issue is that even in the most well-planned pregnancies, "most couples become much more traditional in their approach to housework and childcare. No matter how much they think the tasks will be shared, most women wind up doing more housework than they did before the birth, and more of the childcare than they expected. The discrepancy between what the couples hoped for and the reality of wives having to take on a 'second shift' at home leads to feelings of tension, depression, and sometimes anger in both partners" (Cowan and Cowan 2009, n.p.). Parents are also more likely to experience depression than are nonparents (Evenson and Simon 2005).

Spouses' reported marital satisfaction tends to decline over time whether or not they have children. But serious conflicts over work, identity, and domestic responsibilities can erupt with the arrival of children, especially in "soul-mate" relationships that have to be "nurtured and coddled in order to thrive" (Whitehead and Popenoe 2008, p. 8). In particular, fathers tend to find a reduced sense of confidence in their place in the family after the birth of a child, and that confidence tends to continue to deteriorate, whereas mothers find themselves having greater difficulty managing conflict in the family (Doss et al. 2009, pp. 614, 615). A study that followed Swedish couples through the parenting years found that, though at any one time the majority of parents described their marriages favorably, the

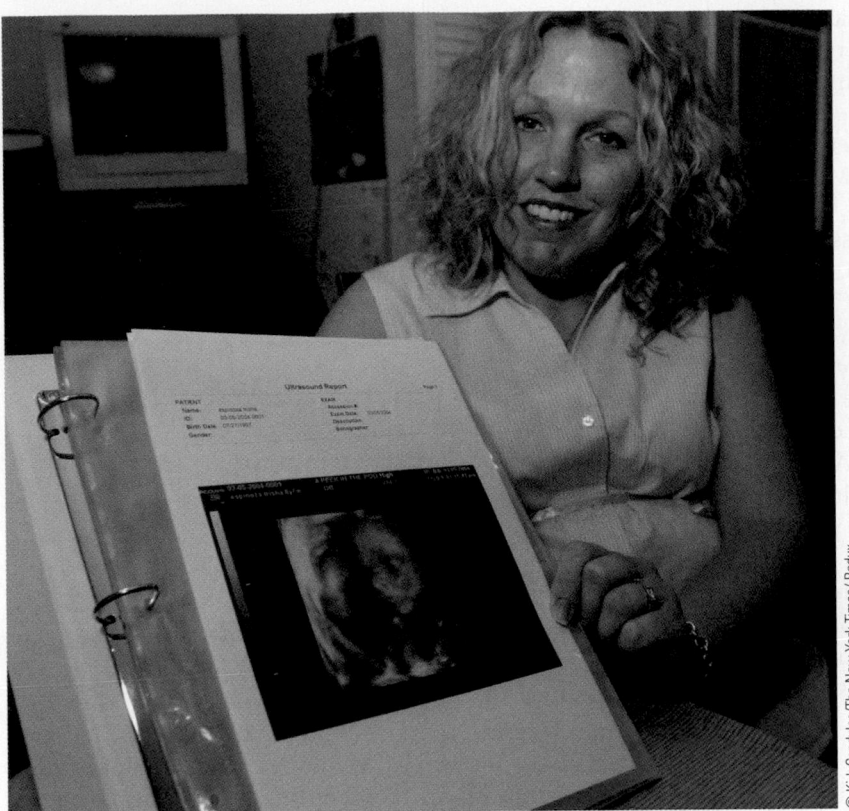

A military wife displays the ultrasound image she has sent to her husband in Iraq so he can keep a photo of their baby-to-be with him.

marital relationship became less harmonious over time, disharmony peaking at the child's ages of ten through twelve (Stattin and Klackenberg 1992). The couple's relationship is especially affected if one or both partners are not cooperative in their parenting (Belsky and Hsieh 1998).

When they have children, spouses may find that they begin responding to each other in terms of more traditional role obligations, and that in turn affects marital happiness negatively (Coltrane 1990; Cowan and Cowan 2009, n.p.; Nock 1998). Spouses, who now are not only busier as parents but also more dissimilar in their dominant roles, begin to do fewer things together and to share decision making less (Bird 1997). Although parenthood is viewed positively and increases life satisfaction, research indicates that positive feelings about children are not sufficient to offset the negative effects on marital happiness of changes in marital structure brought about by the arrival of children (Tsang et al. 2003). Dissatisfaction with one's marriage after the arrival of the first child seems more pronounced and longer lasting for wives than for husbands (Glenn 1990, p. 825; Coleman 2006). Some 38 percent of mothers of infants have high marital satisfaction, whereas 62 percent of childless women do (Twenge, Campbell, and Foster 2003).

Twenge, Campbell, and Foster's comprehensive review of the research (2003) noted that the negative effects of children on marital satisfaction seem to be stronger for younger cohorts. Perhaps couples today experience a greater "before–after" contrast when children arrive. They have often married and become parents later in their lives, and so experienced a great deal of personal freedom and a career focus for many years. Women's roles, especially, change with parenthood, leaving a big gap between the child-free workingwoman's lifestyle and that of a new mother. Moreover, the increased individualism of our culture may make day-to-day responsibility for the care of young children seem less natural than in the 1950s, when social obligations were culturally dominant (Turner 1976).

Even though the addition of a child necessarily influences a household, the arrival of a child is less disruptive when the parents get along well and have a strong commitment to parenting (Whitehead and Popenoe 2008). One longitudinal study shows that the drop in marital satisfaction is less for couples who were happy before the birth and actively planned for the infant (Cowan and Cowan 1992). New friendship networks, such as with other parents, may provide some of the social support previously given by one's spouse. (Chapter 10 looks more closely at the relationships between parents and their children.)

Remaining Childfree

We have been discussing factors that influence the decision whether or not to have children. Involuntary childlessness, the result of infertility or other adverse circumstances, is discussed later in this chapter. Here we examine **voluntary childlessness**, the choice reported by 7 percent of American women in a 2002 survey (Abma and Martinez 2006; Dye 2008; Kelly 2009, p. 157). Although voluntary childlessness has long been seen as the purview of whites, recent studies indicate there is little difference between racial groups—particularly between African Americans and whites—when it comes to the decision to remain childless (Lundquist, Budig, and Curtis 2009, Figure 1).

There is often some ambiguity about the "decision" to remain childfree. For many, it is a gradual decision over time. For others, it is a decision by default, as age or relationship status lead eventually to realization that one will not have children (Whitehead and Popenoe 2008, p. 14). For an increasing number of younger women, it represents an early commitment not to have children. "Firm choices to have no children may signal an increasing proportion of women who see the costs of childbearing as too high" (Hagewen and Morgan 2005, p. 522; see also Kelly 2009).

An increase in voluntary childlessness is ascribed to the social changes of recent decades. The rise of feminism challenged the inevitability of the mother role. More than 70 percent of women surveyed in 2001 said no to the question of whether "a woman need[s] the experience of motherhood to have a complete life," including 69 percent of mothers (Center for the Advancement of Women 2003, p. 8). According to demographer David Foot, a greater ability to control fertility, greater participation of women in paid employment, concern about overpopulation and the environment, and/or an ideological rejection of the traditional family provide the social context for some people's decisions—these reasons and decisions are consistent across all cultures (Kingston 2009, p. 39; see also Rowland 2007).

The Lives of the Childfree The voluntarily childless have more education and are more likely to have managerial or professional employment and higher incomes. They are more urban, less traditional in gender roles, less likely to have a religious affiliation, and less conventional than their counterparts (Kingston 2009; Lundquist 2009; Rowland 2007).

Childfree women tend to be attached to a satisfying career. Childless couples value their relative freedom to change jobs or careers, move around the country, and pursue any endeavor they might find interesting (Majumdar 2004; Park 2005). Most studies have found childfree couples to be more satisfied with their relationship than parenting couples are. The childless elderly are as satisfied with their lives and less stressed than parents—in fact, studies show the childless elderly have lower levels of depression than their counterparts who did have children (Bures, Koropeckyj-Cox, and Loree 2009). They seem to have developed social support networks in lieu of children (Dykstra and Hagestad 2007; Park 2005).

Men's and Women's Motives for Childlessness An earlier review of the literature by Sharon Houseknecht (1987), summarized by Park (2005), found the most important motive for voluntary childlessness to be "freedom from child care responsibilities and greater opportunity for self-fulfillment and spontaneous mobility," reported in 79 percent of the studies and true of both men and women. "Higher marital satisfaction" was reported as a motive for remaining childless in 62 percent of the studies and was important to both sexes.

Men were more affected by "monetary advantages," reported in 55 percent of studies, whereas women stated that "female career considerations" (55 percent) shaped their decisions to be childless (Park 2005, p. 379). Noting the connection between childlessness and career commitment, Abma and Martinez (2006) also hypothesize that voluntarily childless women are simply satisfied with their lives as they are and are not necessarily driven by a need to sacrifice for their careers (also see Kelly 2009; Majumdar 2004).

Having Children: Options and Circumstances

Discussions about having children often evoke images of a young, newly married couple. More and more, however, as the discussion of the changing life course in Chapter 2 suggests, decisions about becoming parents are being made in a much wider variety of circumstances. In this section, we address childbearing with reference to postponing parenthood; the one-child family; nonmarital childbearing; decisions about having children in stepfamilies; and multipartnered fertility. Gay and lesbian parenthood is discussed in Chapter 8.

The Timing of Parenthood

Births to women in their twenties, the primary ages for childbearing, constitute just over half of all births in the United States (Dye 2008), but the age of first birth increased from twenty-one years old to twenty-five (Mathews and Hamilton 2009). After a recent two-year increase, teen birthrates have declined to the lowest ever recorded in sixty-five years of record keeping, including a historic low for Hispanic teen birthrates. Meanwhile, birthrates for women in their thirties has also dropped, whereas women in their forties see their birthrates continue to increase dramatically; the rate for women forty through forty-four is the highest recorded since 1967 (Hamilton, Martin, and Ventura 2010). Birthrates for men forty-five through forty-nine—that is, the rate at which older men have fathered children—have increased by over 20 percent since 1980 (Martin et al. 2009). What are the factors producing this change, and how do early and late parenthood look as choices at the present time?

Postponing Parenthood Later age at marriage and the desire of many women to complete their education and become established in a career appear to be important factors in the high levels of postponed childbearing. Both sexes remain longer in the "emerging adulthood" stage of the life course, enjoying a greater degree of personal freedom and ability to concentrate on career than is possible after assuming family responsibilities. In fact, the highest level of births are by women with graduate or professional degrees (Dye 2008; p. 2). Moreover, with the availability of reliable contraception and the promise of assisted reproduction technology, people can now plan their parenthood for earlier or later in their adult lives. In addition to delayed first-time parenthood, some births to older women (and men) may follow the breakup of marriage or other relationships, and be followed by new pairings and the desire to have children with the new partner.

But fertility declines with age, for men as well as women, although less dramatically for men. Older male age also increases the likelihood of having children with genetic abnormalities or other conditions. "I think what we're saying is that men, too, need to be concerned about their aging," says Dr. Brenda Estenazi of the University of California School of Public Health (Rabin 2007, p. 6).

Many couples today are postponing parenthood into their thirties, sometimes later.

It has been known for some time that older mothers have higher rates of premature or low-weight babies and multiple births—all risks for learning disabilities and health problems. Older mothers also have higher rates of miscarriage, as well as health problems such as diabetes and hypertension (MacDorman and Kirmeyer 2009). Physicians nevertheless advise that pregnancy risk factors should not deter women who want children from having them at older ages: "The take-home message is that while a lot of complications of labor and pregnancy are increased . . . the vast majority of [older mothers] do perfectly fine" (Dr. William Gilbert, quoted in "Older Moms" 1999). In fact, given the scientific advancements in reproductive technology, younger women can be assured of at least the possibility of pregnancy into their forties through cryopreservation of healthy eggs (Lehmann-Haupt 2009).

Still, a more intense public concern about the dangers of postponing parenthood for women emerged with the publication of economist Sylvia Ann Hewlett's book *Creating a Life* (2002) based on her survey of 1,168 older "high-achieving career women" (women in the top 10 percent of earners). Hewlett found a high rate of childlessness among successful managerial and professional career women, most of whom had not intended to be childless. Hewlett faults women for focusing on careers based on an assumption that it would be easy enough to have children later in life. To avoid childlessness, Hewlett goes so far as to suggest that women start their families earlier by intentionally seeking a husband while in their twenties, "even if this involves surrendering part of one's ego" (p. 199).

The implication that women need to minimize their career interests is surprising in this era of generally advancing gender equality, and although Hewlett's advice is questionable, her caution about the limits of reproductive technology are valid. In fact, a recent article in the *Journal of Family and Reproductive Health* noted, "As women delay childbearing, there is now an unrealistic expectation that medical science can undo the effects of aging" (Karimzadeh and Ghandi 2008, p. 62). But critics note that Hewlett has overgeneralized from a small segment of women at a particular point in their lives. Although her description of high-achieving women at ages twenty-eight through thirty-five is accurate, by age forty, high-achieving women are *more* apt to be married and mothers than are other employed women (Boushey 2005b). Boushey and other critics (e.g., Pollitt 2002; Mishel, Bernstein, and Shierholz 2009; Whitehead and Popenoe 2008, see pg. 31) argue that the real problem is the failure of policy support for working families.

Early and Late Parenthood Now that postponing parenthood to the thirties is increasingly common, early parenthood tends to be seen as the more difficult path (Jayson 2010). Choosing early parenthood means more certainty of having children, but young parents may have to forgo some education and get a slower start up the career ladder. Early parenthood can create strains on a marriage if the breadwinner's need to support the family means little time to spend at home or if young parents lack the maturity needed to cope with family responsibility (Gillmore et al. 2008). Moreover, couples who have children early usually start late on saving for college or retirement and must work harder and longer to meet family needs if they have low incomes (Jayson 2010; Joshi, Quane, and Cherlin 2009; Tyre 2004). Early mothers' identities seemed much more dominated by their maternal role, whereas later mothers' identities were more variegated (Walter 1986).

Parenting early means greater freedom later. The Brewers, a young college couple, were twenty-two and twenty when they married, and they had two children in the next few years. "Believe it or not, the couple planned all this. Three months after their first date, they both knew they wanted the same thing. . . . Have some kids in their twenties and happily wave them off to college in their early forties" (Christopherson 2006).

Women who postponed parenthood found that combining established careers with parenting created unforeseen problems. Career commitments may ripen just at the peak of parental responsibilities. On the other hand, late mothers had more confidence in their ability to manage their changed lives because of the organizational skills they had developed in their work. They also had more money with which to arrange support services, and they felt confident of their ability as parents (Jayson 2010). Psychiatrists speak of the maturity, patience, and good parenting skills of later-life parents (Tyre 2004).

A book based on interviews with a nonrandom set of older fathers, mostly white and middle class, found that men who had children in later life expressed a great deal of joy in parenthood, particularly if they had given priority to jobs with earlier-born children (Carnoy and Carnoy 1995). They saw themselves as more patient with children (Vinciguerra 2007).

Early mothers felt that they had had more spontaneity as youthful parents (Walter 1986). "'We wanted to be young parents [said one mother]. . . . We didn't want to be sixty when they got out of high school'" (Poniewozik 2002, pp. 56–57). In a study based on interviews with 114 Canadian expectant mothers, the younger pregnant women (in their twenties) spoke of their physical health as an asset, as well as the health of their parents—they expected to rely on parents' help with the children. They were also pleased to think that as younger parents they would have less of a "generation gap" between themselves and their children (Dion 1995).

Older expectant mothers in this study (in their thirties) looked to friendship networks for support, including a sense of being "on time" in attaining parenthood—since they had friends who had also delayed parenthood (Dion 1995). They spoke of having needed

a period of time for personal development—not just career development—for themselves and their spouses. They felt that delaying parenthood meant greater maturity and preparation for parenthood.

Older parents worry about their physical limitations. And for older parents, there is a sense of limited time with children that both increases pleasure in parenting—"Everything is more precious"—and creates anxiety about the future—"That he could die before his daughter reaches adulthood 'is a reality that I live with,'" said one father who was fifty-nine when his daughter was born (Vinciguerra 2007, p. ST-1). That he may not live to see grandchildren is another reality.

Being born to older parents affects children's lives as well. They usually benefit from the financial and emotional stability that older parents can provide and the attention given by parents who have waited a long time to have children. But children of older parents often experience anxiety about their parents' health and mortality (Vinciguerra 2007). Their parents may become frail before they have established themselves in their adult lives. Not only does having children later in life put the burden of elderly parent care on their children at a younger age, but it also limits their children's children's years with grandparents.

For prospective parents who seek to time their parenthood to be early or later in life, it's important to have an awareness of the trade-offs—plus an understanding that having children is a challenge at any age!

The One-Child Family

Some prospective parents consider the challenges of parenthood daunting, but also reject the idea of childlessness. For them, the solution is the one-child family. In 2006, 16.9 percent of women age forty through forty-four had just one child (Dye 2008). The number of one-child families continues a steady increase, making up 19 percent of American families (U.S. Census Bureau 2010b, Table 64).

The proportion of one-child families in America appears to be growing due to at least three factors: (1) women's increasing career opportunities and aspirations in a context of inadequate domestic support; (2) the high cost of raising a child through college; and (3) peer support: the choice to have just one child becomes easier to make as more couples do so. Divorced people who do not remarry or form a new reproductive partnership may end up with a one-child family because the marriage ended before more children were born.

Negative stereotypes present only children as "socially unskilled, dependent, anxious, and generally maladjusted" (Hagewen and Morgan 2005, p. 514). To find out whether there was any basis for this image, psychologists in the 1970s produced a staggering number

© Fancy / Alamy

Some families choose to have only one child, a decision that can ease time, energy, and economic concerns. There may be extra pressure on only children, and they do not experience sibling relationships. But only children tend to receive more personal attention from parents, and parents may enjoy their child more when they do not feel so overwhelmed as they might with more offspring to care for.

of studies that generally concluded that no negative effects of being an only child could be found (Pines 1981, p. 15; see also Falbo 1976 and Hawke and Knox 1978). Recent studies of kindergartners and adolescents have concluded that sibling relationships foster the development of interpersonal skills and reduce incidences of depression, particularly in young girls, although the differences between children with siblings and only children is small (Downey and Condron 2004; Kim et al. 2007).

Research reports only children to be more intelligent and mature, with more leadership skills, and better health and life satisfaction both as children and into adulthood (Deveny 2008; Hagewen and Morgan 2005). In a 1998 survey using a national sample of more than 24,000 eighth graders, sociologist Douglas Downey (1995) found that only children were significantly more likely to talk frequently with their parents; to have attended art, music, or dance classes outside of school; and to have visited art, science, or history museums.

Advantages Parents with only one child report that they can enjoy parenthood without feeling overwhelmed and tied down. They have more free time and are better off financially than they would have been with more children (Deveny 2008; Downey 1995). Researchers have found that family members shared decisions more equally and could afford to do more things together (Hawke and Knox 1978).

Research shows that the child in a one-child family has some advantages over children with siblings. Parents of only children had higher educational expectations for their child, were more likely to know their child's friends and the friends' parents, and had more money saved for their child's college education.

Disadvantages There are disadvantages, too, in a one-child family. For the children, these include the obvious lack of opportunity to experience sibling relationships, not only in childhood but also as adults. The 2000 census indicated that some 700,000 siblings live together. Siblings may provide social support, as well as exchanges

If these sisters get along well—as they appear to—they can provide companionship and support for each other as they go through life. Over 700,000 siblings shared a residence in 2000 (C. Lee 2006).

of material assistance and someone to rely on in emergencies (C. Lee 2006; Riedmann and White 1996).

Only children may face extra pressure from parents to succeed, and they are sometimes under an uncomfortable amount of parental scrutiny. As adults, they have no help in caring for their aging parents. Disadvantages for parents include the fear that the only child might be seriously hurt or might die and the feeling, in some cases, that they have only one chance to prove themselves good parents.

Nonmarital Births

In 2008, 40.6 percent of all births were to unmarried women (Hamilton, Martin, and Ventura 2010, p. 5). Nonmarital births take place in many different contexts in terms of parents' relationship status, age, financial resources, and so forth. We will first look at general trends in nonmarital births, as well as at racial/ethnic variation. We will then touch on births in cohabiting families, in "fragile families," to older single mothers, and to adolescent women.

After declining during the 1990s, nonmarital birthrates have risen again to an all-time high (Hamilton, Martin, and Ventura 2010). Meanwhile, childbearing in marriage has declined, leaving births outside of marriage a larger proportion of total births (Martin et al. 2009, Table D). From 1940 until the early 1960s, only 4 to 5 percent of all births were to unmarried women; as recently as 1980, 18 percent were. Although there was a slight decline in nonmarital births from the mid-1990s to 2002, there has been a marked increase and we are now seeing "the highest number ever recorded in the United States" (Martin et al. 2009, Table D, p. 11).

The current figure of 1.6 million nonmarital births represents a profound change in our society of the context of parenthood. Public attitudes correspond to these behavioral trends. Although 67 percent of those responding to a Gallup poll in 2006 thought it "very important" for a couple to marry if they planned to spend their lives together, only 37 percent felt marriage was "very important" "when a couple has a child together" (Saad 2006b).

Biologically, women mature earlier today, but they marry later and are more likely to divorce than in the past, so they spend more years at risk of a nonmarital pregnancy. They are much less likely now to marry upon the discovery of a nonmarital pregnancy. In 2006, 64.5 percent of women experiencing a birth in the previous twelve months were married to the father, and another 4.8 percent were living with their unmarried partner. The remainder of first-time mothers were not married or cohabiting with the father (Dye 2008, Table 3, p. 8).

Thirty-two percent of births to unmarried women in 2006 were to non-Hispanic white mothers (Martin et al. 2009, p. 11). The proportion of nonmarital births in each racial/ethnic category is somewhat different. As Figure 9.3 shows, in 2008, 72.3 percent of African American births, 65.8 percent of American Indian/ Alaska Native births, 52.5 percent of Hispanic births,

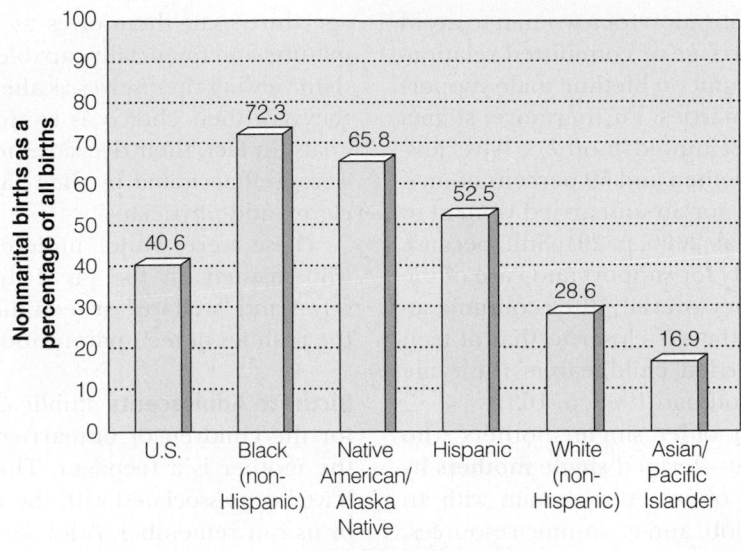

Figure 9.3 Births to unmarried women as a percentage of all births, by race and ethnicity, United States, 2008

Source: Hamilton, Martin, and Ventura 2010, Table 1

28.6 percent of non-Hispanic white births, and 16.9 percent of Asian/Pacific Islander births occurred outside marriage (Hamilton, Martin, and Ventura 2010, Table 1).

The nonmarital birthrate of African American women has declined substantially and remains significantly lower in 2008 than at its 1989 peak (Hamilton, Martin, and Ventura 2010; Martin et al. 2009, Table 19). While the nonmarital birthrates for African Americans remain lower, there were markedly steep increases for both white, non-Hispanic, and Hispanic women age twenty and older through 2006 (Martin et al. 2009, Table 19).

Births to Cohabitants "Fertility during cohabitation continues to account for almost all of the recent increases in nonmarital childbearing. . . . Cohabitation . . . has increasingly become . . . a two-parent family union in which to have and raise children outside of marriage" (Manning 2001, p. 217). Forty percent of nonmarital births are to heterosexual cohabiting women, and birthrates for never-married cohabitants are virtually the same as those for married women (Whitehead and Popenoe 2008, p. 6; Chandra et al. 2005). Cohabiting families (heterosexual and same-sex) are discussed in Chapter 8.

Gays and lesbians are included in this section because they are not legally allowed to marry in most states. Through the use of artificial insemination and surrogacy, these families are increasingly able to become biological parents (Berkowitz and Marsiglio 2007).

Births to "Fragile Families" It is now realized that less visibly attached unmarried parents, including those not living together, may have a more regular relationship than previously thought. The Fragile Families study (McLanahan et al. 2001; McLanahan and Carlson 2002) is a national longitudinal study conducted in twenty large U.S. cities. Initially, it comprised a sample of 3,700 children, born between 1998 and 2000, to unmarried parents and 1,200 born to married parents. This analysis, based on 1,764 new mothers in seven cities, found that the vast majority of new parents described themselves as "romantically involved on a steady basis," with 50 percent cohabiting and 33 percent "visiting." These fathers helped the mother during pregnancy and/or visited the hospital. Nearly 100 percent expressed a desire to be involved in the child's life, and 93 percent of mothers agreed. "The myth that unwed fathers are not around at the time of the birth could not be further from the truth" (McLanahan et al. 2001, p. 217). Nevertheless, involvement of these fathers is likely to decline over time, given their limited resources, and this has important implications for their offspring—particularly for daughters (McLanahan et al. 2001; Mitchell, Booth, and King 2009; Wu and Wolfe 2001). Many researchers encourage policy support for these "fragile families," especially in African American and Hispanic communities where father absence is commonplace (Whitehead and Popenoe 2008, p. 17; Robbers 2009).

Births to Older Single Mothers Although unwed birthrates are highest among young women in their twenties, they have increased dramatically for older women in recent years (Martin et al. 2009, Table 19; Ventura 2009, Figure 2). As opportunities grow for women to support themselves and as the permanence of marriage becomes less certain, there is less motivation for a woman to avoid giving birth outside of marriage or committed relationship because she cannot count on lifetime male support for the child even if she marries. Furthermore, stigma and discrimination against unwed mothers have lessened. Seventy percent of women and 59 percent of men surveyed believe it is "okay for an unmarried woman to have a child" (Martinez et al. 2006, p. 29). Still, because the burden of responsibility for support and care of the child remains on the mother, overall, "the economic situation of older, single mothers is closer to that of teen mothers than that of married childbearers the same age" (Foster, Jones, and Hoffman 1998, p. 163).

There is a category of older single mothers who may be better off than that—termed **single mothers by choice**. The image is that of an older woman with an education, an established job, and economic resources, who has made a choice to become a single mother. Not having found a stable life partner, yet wanting to parent, a woman makes this choice as she sees time running out on her "biological clock" (Lehmann-Haupt 2009).

Whether this is a significant development in terms of numbers is uncertain (Musick 2002). But it has drawn the attention of researchers. Sociologist Rosanna Hertz (2006a) interviewed sixty-five single mothers who had their first child at age twenty or older and who, more significantly, are self-sufficient economically. Not having the "chance" to be in stable, child-rearing marriages, they became mothers through various routes: accidental biological pregnancy, artificial insemination by known or unknown donor, or adoption.

"For the women in this study, single motherhood was never a snap decision" (Hertz 2006a, p. 26):

> I always had in the back of my mind that if I was thirty and not married, then I'd have children on my own. Then it was when I was thirty-two. Then it was when I went back to school at Princeton to get my master's degree. Then it was thirty-six and I had just broken up with another man. (p. 26)

Women were often surprised to find themselves taking what they saw as an unconventional step:

> Daring to consider getting pregnant on my own just seemed like such an outrageous thing to do. And from that point of thinking about it, to doing it, was the longest stretch because I was kind of shocked that I would think that way, and I wasn't sure of what I really wanted to do. (p. 27)

Once a mother, the parenting practices of single mothers by choice and their sense of family were very traditional. In fact, they saw themselves as exemplifying family values by having chosen parenthood.

Similarly, in two smaller studies of single mothers by choice (Bock 2000; Mannis 1999), researchers interviewed women who adopted children or who purposefully became pregnant. These mothers, usually over age thirty, saw themselves as responsible, emotionally mature, and financially capable of raising a child. Rather than viewing themselves as alternative lifestyle pioneers, they saw their choice as conforming to normal family goals. In fact, their decisions to become single mothers were well accepted by their family, friends, employers, clergy, and physicians.

These were white, middle-class, educated women who insisted on the great difference between themselves and "welfare" or teen mothers. What, in fact, are the realities of teen parenthood today?

Births to Adolescents Public concerns about outcomes for the children of unmarried parents intensify when the mother is a teenager. The words *teenage pregnancy* have been associated with the word *problem* since most of us can remember. Adolescent birthrates rose in the late 1960s, as sexual behavior liberalized. However, by the time a "teen pregnancy epidemic" was identified, adolescent birthrates had already begun to decline (see Figure 9.4). Declines in the adolescent birthrates have been especially large for young black women. Teens are using contraception more regularly, and sexual activity has leveled off. The teen abortion rate has dropped also. Teen pregnancies remain at a historic low for the nation and account for just 23 percent of nonmarital births (Ventura 2009; Ventura et al. 2006; Martin et al. 2006).

Nevertheless, the United States still has by far the highest teen pregnancy, abortion, and birthrates of any industrialized country (Abma et al. 2004), and teen pregnancy is still problematic. In the 1950s, when teen birthrates were actually higher, most teen mothers were already married or they married before the child's birth, and a strong economy provided young fathers with jobs that could support a family.

Now, as Figure 9.4 indicates, most teen women giving birth are not married, and so they lack the economic support of a spouse and the support of a co-parent. Women as well as men need more education in today's world, and women are expected to seek employment. Teenage parents, especially those with more than one child, face a bleak educational future, limited job prospects, and a very good chance of living in poverty, compared to peers who do not become parents as teenagers (Perper, Peterson, and Manlove 2010; Logan et al. 2007). As described earlier in this chapter, the costs associated with raising children can be daunting for even the most well-planned family, so it is intuitive, then, that the issues teen parents face are magnified dramatically for these young people who are just beginning their transition into adult lives. Prospects for the children of teen parents have included lower academic achievement and, because of the lack of resources related to poverty, a trend toward a cycle of early unmarried pregnancy themselves (South and Crowder 2010; Bronte-Tinkew et al. 2009).

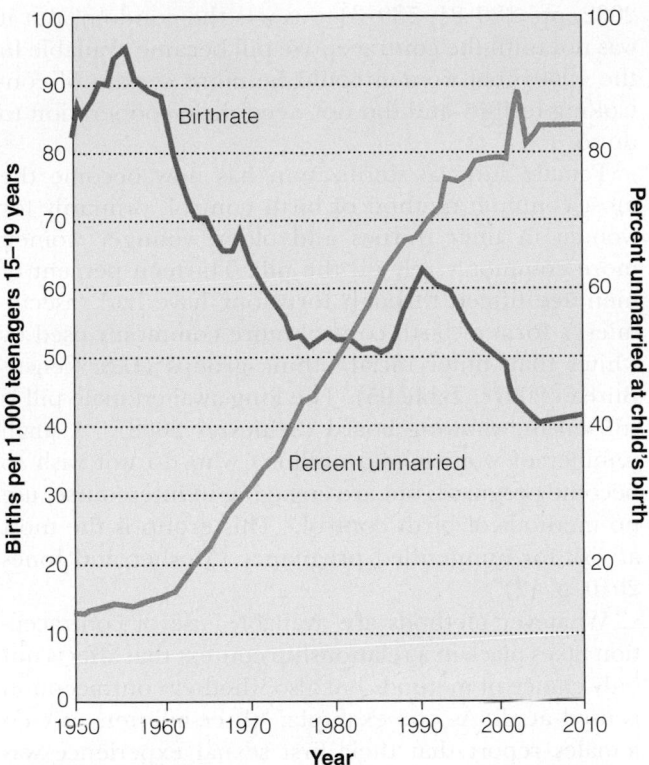

Figure 9.4 Birthrate for teen women fifteen through nineteen years and percentage of teen births that are to unmarried teen women, 1950 to 2010

Sources: Ventura, Mathews, and Hamilton 2001, Figure 1; Downs 2003, Figure 1; Martin et al. 2009, Tables A, 17; Hamilton, Martin, and Ventura 2010, Tables 1, 2; Dye 2008, p. 8; Hamilton et al. 2005.

Yet we have begun to recognize that economic and/or racial/ethnic disadvantage may be playing a larger role than age in shaping a teen mother's limited future (Geronimus 1991; Gueorguieva et al. 2001; Mauldon 2003; South and Crowder 2010, Turley 2003). Moreover, outcomes of teen parenthood vary and are not by any means uniformly negative. One longitudinal study of black teen mothers from low-income families in Baltimore concluded that although early childbearing increases the risk of ill effects for mother and child, it is unclear that the risk is so high as to justify the popular image of the adolescent mother as an unemployed woman living on welfare with a number of poorly cared-for children. To be sure, teenage mothers do not manage as well as women who delay childbearing, but most studies have shown that there is great variation in the effects of teenage childbearing (Furstenberg, Brooks-Gunn, and Morgan 1987, p. 142).

Similarly, a careful study based on national sample data sets finds that "teen childbearing plays no causal role in children's test scores and in some behavioral outcomes of adolescence." Research on other outcomes

is inconclusive. "We . . . suggest caution in drawing conclusions about early parenthood's overarching effects" (Levine, Emery, and Pollack 2007, p. 105).

Stepparents' Decisions about Having Children

When people remarry or form a new committed partnership, they have decisions to make about having children together. Does it make a difference whether one or both partners already have children? The answer to that question is yes.

A study of more than two thousand couples drawn from a national sample—the National Survey of Families and Households—found that individuals living with a second spouse or partner were most likely to want to have a child if there were no stepchildren of either partner (S. Stewart 2002). Desire for another child was lower for cohabiting couples than for married ones. If *both* partners already had children, an intention to have another child was especially low, with one exception. Because of the symbolic importance of joint parenthood, if the couple did not have a biological child, they were very likely to intend to have one.

"Increasing numbers of children are being born into complex living arrangements" (S. Stewart 2005b, p. 470). The motivation of the parents is often to integrate the stepfamily around a new child who is biologically related to all family members. In fact, there is little firm knowledge about the impact of a new biological child on a stepfamily. Stewart urges more research on the long-term effects of adding children to a stepfamily.

Multipartnered Fertility

Multipartnered fertility is a new interest and a very new area of research for family social scientists. Researchers participating in the Fragile Families and Child Well-Being Study of urban parents at the time of their first birth realized in follow-up that some of those parents, particularly those unmarried at the birth, went on to have children with new partners.

How frequent is multipartnered fertility? What are the implications for family life and, especially, for the well-being of children? Research has begun, and we offer some information that is just the beginning of what is likely to become an extensive area of research.

The Fragile Families study follow-up (discussed previously in the chapter), and reviews of fertility history, found that three-fourths of the mothers had children by only one father. Most of the others had children by two fathers, a few by three or four fathers.

Multipartnered fertility is most common in nonmarital families as these have a high rate of breakup; moreover, the participants tend to be younger with more of their lives ahead of them. Black (non-Hispanic) men and women are more likely than other racial/ethnic groups to have children by more than one partner (Carlson and Furstenberg 2006; "Urban Parents" 2006).

Multipartnered fertility seems likely to lead to very complex family systems and weaker ties with extended families. Indeed, multipartnered fertility is associated with less financial, housing, and child care support from kin networks (Carlson and Furstenberg 2006; Harknett and Knab 2007).

Another study using data from the National Longitudinal Study of Adolescent Health (Guzzo and Furstenberg 2007) paid particular attention to the policy implications of multipartnered fertility. "On the one hand, having children by different fathers can present daunting challenges for young mothers. Having to negotiate paternal support and involvement with different men is stressful and may result in different levels of involvement for children who live in the same household but do not share the same father" (p. 37). On the other hand, social welfare authorities can hardly expect these very young women not to have additional children if their first relationship breaks up after only one child.

Much of the foregoing discussion of fertility issues, and especially reports of declining birthrates in some sectors, leads us to the question of preventing unwanted pregnancies.

Preventing Pregnancy

Falling birthrates from the nineteenth century onward indicate that people did not always want to have as many children as nature would make possible. As early as 1832, a book describing birth control techniques and devices was published in the United States. The diaphragm was invented in 1883, and was a common method of birth control for married couples (Weeks 2002, pp. 180–81, 530–31), as was the condom. But it was not until the contraceptive pill became available in the 1960s that women could be more certain of controlling fertility and did not need male cooperation to do so.

Female surgical sterilization has now become the most common method of birth control, primarily for women in their thirties and older; younger women more commonly rely on the pill. Thirteen percent of men age fifteen through forty-four have had vasectomies, a form of birth control more commonly used by whites than other racial/ethnic groups (U.S. Census Bureau 2007c, Table 95). The long-awaited male pill is still on the drawing board (Schieszer 2008). A small number of women (4.5 million) who do not wish to become pregnant, but are engaging in intercourse, use no methods of birth control. This group is the most at risk for unintended pregnancy (Mosher and Jones 2010, p. 12).

Whatever methods are available, use of contraception takes place in a relationship context that affects not only choice of methods but also whether contraception is used at all. As one example, "three-quarters of teen females report that their first sexual experience was with a steady boyfriend, a fiancé, a husband or a cohabiting partner," and a vast majority of them (74 percent) used contraceptives the first time they had sex ("Facts on American Teens" 2010, p. 1).

The physical and opportunity costs of children tend to be higher for women than for men, whether married or unmarried, and family planning services have always been oriented to women as a clientele. More recently, family planning organizations have realized that they need to reach out to men to provide them with contraceptive and health information and so influence couple decisions. So far there are few such programs, and men are typically unaware

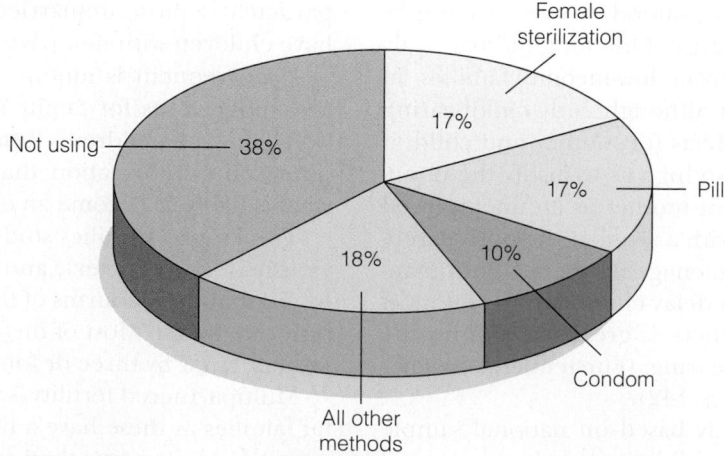

Figure 9.5 Percent distribution of women aged 15–44 years, by current contraceptive status: United States, 2006-2008

In our society, sexuality and reproduction have become increasingly politicized. Nowhere is this more apparent than in the intensely heated pro-life/pro-choice debate over abortion, one of the most polarizing issues in America today.

of their availability where they do exist (Finer, Darroch, and Frost 2003). New research and services have targeted adolescent young men as an approach to adolescent pregnancy prevention (Ball and Moore 2008).

Abortion

Effective contraception[7] prevents the potential problems associated with unwanted pregnancy. When contraception is not used or fails, however, many women who do not want to remain pregnant decide to have an abortion. We will look next at this option, a controversial social issue.

The term **abortion** is used for the expulsion of the embryo or fetus from the uterus either naturally (*spontaneous abortion* or *miscarriage*) or medically (surgically or drug-induced abortion). This section addresses **induced abortion**—that is, purposefully obtained

abortion, which is what we usually mean in discussions of "abortion."

Approximately 40 percent of American women have had an induced abortion at some point in their lives (Ehrenreich 2004a; Henshaw and Kost 2008; Kliff 2010). Abortion decisions are primarily made within the context of unmarried, accidental pregnancy or a failed relationship. However, some married couples may consider aborting an unwanted pregnancy if, for example, they feel that they have already completed their family or could not manage or afford to raise another child. About 40 percent of unintended pregnancies are aborted (Guttmacher Institute 2006). The question of abortion can also arise for couples who, through prenatal diagnosis techniques, find out that a fetus has a serious defect.

Less than a quarter of pregnancies ended in abortion in 2005. Around 1.22 million abortions were performed in the United States in that year, down from a peak of about 1.6 million in 1990 (Hesnshaw and Kost 2008, p. 7). The rate of abortions per 1,000 women of childbearing age has been falling since 1980 (U.S. Census Bureau 2007a, Tables 96 and 97). Abortions take place much earlier in pregnancy than in the past—more than half occur within the first nine weeks of pregnancy, and only

[7] For more up to date information on contraception, please visit Planned Parenthood at www.plannedparenthood.org and the Guttmacher Institute at www.guttmacher.org

11 percent after thirteen weeks or more (U.S. Census Bureau 2007a, Table 97).[8]

Reasons for abortion reported in surveys and interviews at abortion sites include the following: having a child would interfere with the woman's education, work, or ability to care for dependents (74 percent); not being able to afford a baby at this time (73 percent); not wishing to be a single mother or having relationship problems (48 percent); and the woman or couple had completed child-bearing (38 percent) or were not ready to have a child (almost one-third; Finer et al. 2005; Boonstra et al. 2006, pp. 8–9). A Guttmacher Institute report concluded that "[a]lthough women who have abortions and women who have children are often perceived as two distinct groups, in reality they are the same women at different points in their lives" (2006, p. 9; see also Henshaw and Kost 2008).

Less than one percent report that an abortion decision was the result of pressure from parents or partners. Sixty percent had consulted someone, usually a husband or partner, in making the decision (Boonstra et al. 2006; Finer et al. 2005). The "right" of the male partner to compel completion of the pregnancy arises occasionally as an issue in the media or the courts, lacking in prominence outside the debates between advocacy groups. U.S. Supreme Court decisions (e.g., *Planned Parenthood v. Danforth* 1976) have made clear that an abortion decision is the woman's, not that of her husband, her parents, or her reproductive partner.

The Politics of Abortion

Throughout world history, abortion has been a way of preventing birth. The practice was not legally prohibited in the United States until the mid-nineteenth century. Laws prohibiting abortion stood relatively unchallenged until the 1960s, when an abortion reform movement succeeded in modifying some state laws to permit abortions approved by physicians on a case-by-case basis. The movement culminated in the 1973 U.S. Supreme Court decision *Roe v. Wade,* which legalized abortion throughout the United States.[9] This legal right has been tested frequently, with the most recent challenge coming from Nebraska governor Dave Heineman, making it illegal, in that state, to abort a fetus after twenty weeks of gestation (Davey 2010).

As virtually everyone is aware, pro-choice and pro-life activists—those who favor or oppose legal abortion—have made abortion a major political issue. Legislation and other public policy responses to abortion have been shaped by this struggle, as has been the availability of abortion services. Abortion was never widely available outside urban areas. Now states have placed various restrictions on access to abortion including a ban on a certain procedure commonly termed *partial birth abortion*, used later in a pregnancy (*Gonzales v. Carhart* 2007; Stout 2007). Moreover, the scaling back of training in abortion procedures in medical education as a consequence of political pressures means that as current providers age out or become discouraged by social pressures and threats to their lives, the practical effect is to reduce abortion options. Although clinic violence is down, women arriving for abortion appointments must usually pass a gauntlet of picketers.

Although the Supreme Court has upheld many state restrictions on abortion, it has not outlawed the procedure—at least not yet. Nor has a constitutional amendment to criminalize abortion made it through Congress. The result is that abortion continues to be legally available, as pro-choice advocates wish, while the goals of pro-life advocates have been partially reached through legal and practical restrictions on abortion availability. This has some correspondence with the centrist position of the American public, which favors abortion under certain circumstances.

According to 2005–2006 Gallup polls, 53 percent of respondents described themselves as pro-choice, while 42 percent chose the pro-life label. The majority of Americans (68 percent) believe that *Roe v. Wade* should remain the law of the land. Support for abortion is heavily qualified, however, with only 21 percent believing that abortion should be legal in all circumstances, 57 percent in some circumstances, and 18 percent believing abortion should be illegal in all circumstances (Gallup Poll News Service 2009). Table 9.1 shows the particular circumstances that influence people's attitudes about abortion.

Public opinion of abortion differs sharply by the trimester of pregnancy. Two-thirds of poll respondents approved of first-trimester abortion, whereas more than two-thirds disapproved of later-stage abortions (Saad 2003). Few abortions (0.02 percent) take place in the third trimester of pregnancy (Henshaw and Kost 2008).

Approval of abortion seems to be decreasing among younger people. A 2003 national survey of college

---

[8] In 2002, more than four-fifths of abortions were obtained by unmarried women. More than half (56 percent) were obtained by women in their twenties, 17 percent by teens. White women account for the greatest number of abortions (41 percent of total abortions, compared to 32 percent for blacks, 20 percent for Hispanics, 6 percent for Asian/Pacific Islanders, and 1 percent for Native Americans (Jones, Darroch, and Henshaw 2002, Table 1)). But the percentage of pregnancies that are terminated by abortion (the *abortion ratio*) is highest among black and Asian/Pacific Islander women. Researchers conclude that "women who have abortions are diverse, and unintended pregnancy leading to abortion is common in all population subgroups" (Jones, Darroch, and Henshaw 2002, p. 232). Still, women in poverty account for a disproportionate share of abortions (Boonstra et al. 2006). Just under half (46 percent) have had a previous abortion (U.S. Census Bureau 2007a, Table 97).

[9] *Roe v. Wade* did not legalize any and all abortions in any and all situations. *Roe v. Wade* allows abortion to be obtained without question in the first trimester of pregnancy. But abortion is subject to regulation of providers and procedure in the second trimester and may be outlawed by states after fetal viability (the point at which the fetus is able to live outside the womb), which occurs in the third trimester.

freshmen conducted by UCLA found that 55 percent support legal abortion, compared to 64 percent ten years earlier: "We're the first generation to be more pro-life than our parents," said one freshman (in Rosenberg 2004).

The Safety of Abortions

National Right to Life claims that abortion is a threat to both medical and physical health, including threats to future reproductive capacity and an increased risk of breast cancer (National Right to Life 2005, 2006).

Abortions and Physical Health The evidence indicates that abortion is a safe medical procedure. According to the American College of Obstetricians and Gynecologists (n.d.), a surgical abortion "is a low risk procedure. . . . An early abortion has less risk than carrying a pregnancy to term." The safety of the less common RU-486 (pill) method of abortion remains under investigation (U.S. Food and Drug Administration 2006).

Research indicates that abortion has no impact on the ability to become pregnant—sterility following abortion is very uncommon—and there is virtually no risk to future pregnancies from a first-trimester abortion of a first pregnancy (Boston Women's Health Book Collective 1998, p. 407).

A panel assembled by the National Cancer Institute to review the research concluded that there is no association between induced abortion and breast cancer (Collaborative Group on Hormonal Factors in Breast Cancer 2004; U.S. National Cancer Institute 2003; see also Bakalar 2007).

Abortions and Psychosocial Outcomes It is safe to say that for most women (and for many of their male partners), abortion is an emotionally charged, often upsetting, experience. Some women report feeling guilty or frightened, a situation that can be heightened by demonstrators outside abortion clinics. Emotional stress is more pronounced for second-trimester than earlier

abortions and for women who are uncertain about their decision (Boston Women's Health Book Collective 1998, pp. 406–08). Women from religious denominations or ethnic cultures that strongly oppose abortion may have more negative and mixed feelings after an abortion, but outcomes are more complex than that, as they seem to depend also on the woman's emotional well-being prior to having the abortion (Major et al. 2009).

Some women have reported that the decision to abort enhanced their sense of personal empowerment (Boston Women's Health Book Collective 1998, pp. 406–7; Kliff 2010). Research has found positive educational, economic, and social outcomes for young women who resolve pregnancies by abortion rather than giving birth. In one study, "those who obtained abortions did better economically and educationally and had fewer subsequent pregnancies than those who chose to bear children" (S. Holmes 1990). A more recent study, done in New Zealand, also reports these outcomes, but finds most of these benefits of abortion to be related to preexisting circumstances of the women. Choosing abortion did result in enhanced educational outcomes (Fergusson, Boden, and Horwood 2007).

The decision to abort is often very difficult to make and act on. But it is important to make a distinction between negative feelings and psychiatric problems. "Having an emotion is not the same as having a mental disorder" (Rubin and Russo 2004, p. 74). The consensus at present is that there is no clear relationship of abortion to mental health (Major et al. 2009; Sit et al. 2007; Steinberg and Russo 2008).

According to a longitudinal study of almost 5,000 black and white women, the emotional distress involved in making the decision and having the abortion does not typically lead to severe or long-lasting psychological problems (Adler et al. 1992; Boonstra et al. 2006, pp. 22–24; O'Malley 2002; Russo and Dabul 1997; Russo and Zierk 1992; Rubin and Russo 2004). The American Psychological Association (2005a) has taken the position that "[a]bortion is a safe procedure that carries few . . . psychological risks."[10]

Women (and men) making decisions about abortions are most likely to make them in accordance with their values. A detailed review of religiously or philosophically grounded moral and ethical perspectives on abortion is

Table 9.1 Percentage of U.S. Adults Approving of Abortion under Certain Circumstances

ABORTION SHOULD BE LEGAL...	
when the woman's life is endangered.	85%
when the woman's physical health is endangered.	77
when the pregnancy was caused by rape or incest.	76
when the woman's mental health is endangered.	63
when there is evidence that the baby may be physically impaired.	56
when there is evidence that the baby may be mentally impaired.	55
when the woman or family cannot afford to raise the child.	35

Source: Gallup poll, January 10–12, 2003 (Saad 2003).

[10] The New Zealand study of women who had an abortion prior to age twenty-one and were followed to age twenty-five did find some mental health impact (Fergusson, Horword, and Ridder 2006). It is difficult to evaluate this study vis-à-vis American comparisons because, in New Zealand, permission for an abortion requires demonstrating a physical or mental health need for an abortion in the first place. The age range is also somewhat limited. Fergusson et al. simply note that this study contradicts the American Psychological Association (APA) conclusion and advises that the issue remain open for further research. Rubin and Russo (2004) meanwhile note that a pro-life framing of the issue and other movement tactics could induce guilt and other negative feelings, and they provide advice to therapists who are working with women who have had abortions.

outside the scope of this text, but as we note in Chapter 1, values provide the context for such decisions.

Involuntary Infertility and Reproductive Technology

For some, concern about fertility means avoiding unwanted births. Other couples and individuals face a different problem. They want to have a child, but either they cannot conceive or they cannot sustain a full-term pregnancy. We turn now to the issue of involuntary infertility.

The Social and Biological Context of Infertility[11]

As medically defined, **involuntary infertility** is the condition of wanting to conceive and bear a child but being physically unable to do so. It is usually diagnosed in terms of unsuccessful efforts to conceive for at least twelve months. A related concept, **impaired fertility**, describes the situation of a woman or a couple who has a physical barrier to pregnancy or who has not been able to carry a pregnancy to full term. *Subfecundity* or *secondary infertility* describes a situation in which a woman or a couple has had children previously, but now cannot. We will use the term *infertility* for all the situations in which a woman or a couple is not able to have a desired child. Around 12 percent of American women have impaired fertility, while 7 percent of married women are infertile (Chandra et al. 2005, pp. 21–22).

Infertility has become more visible because the present tendency to postpone childbearing until one's thirties or even forties creates a class of infertile potential parents who are intensely hopeful and financially able to seek treatment. "For most women and their partners, infertility is a major life crisis" (Boston Women's Health Book Collective 1998, p. 532).

Many couples only gradually become aware that, unlike other couples they know who are happily planning their pregnancies without apparent difficulty, they themselves remain without desired children. At this point, couples are likely to seek a medical solution to their problem (Matthews and Matthews 1986).

Infertility Services and Reproductive Technology

Louise Brown, the first "test-tube" baby, gave birth to *her* first child in 2006 ("World's 1st" 2007). This look back reminds us of how astonishing were the first developments

in reproductive technology. Now, more and more, **assisted reproductive technology (ART)** has become an accepted reproductive option. In 2004, there were 128,000 ART procedures, which resulted in almost 50,000 infants (U.S. Centers for Disease Control and Prevention 2006a).

This quote from a six-year-old "dreaming of motherhood" suggests just how normalized ART has become: "Mommy, Mommy, when I grow up, I want to be a mommy just like you. I want to go to the sperm bank just like you and get some sperm and have a baby just like me" (in Ehrensaft 2005, p. 1).

American women of childbearing age increased their use of fertility services from 9 percent in 1982 to 15 percent in 1995, and then reduced their use to 12 percent by 2002 (detailed information on use of infertility services can be found in Figure 9.6). Only 1.4 percent of women who have ever used an infertility service have used assisted reproductive technology (Chandra and Stephen 2010). Almost half (44 percent) of women with fertility problems sought medical help in 1995 (the latest year for which detailed data are available). They were more often white (75 percent) and typically older (twenty-five through forty-four), married, better educated, and with higher family incomes than those not seeking treatment. But 20 percent of those using infertility services were not married, and 11 percent were poor (Chandra and Stephen 1998). With regard to race and ethnicity, little social science data exist. What limited available data does show, however, is that women identifying as Hispanic had the highest use of infertility services in 2002 at 50.6 percent. That same year, 44.7 percent of women identified as non-Hispanic white, and 43.1 percent of women identified as non-Hispanic black accessed infertility services (Chandra and Stephen 2010, Table 1).

Medical procedures involving drug therapies, donor insemination, in vitro fertilization, and related techniques are discussed in Chapter 1. Fertility treatment is stressful. The procedures can be uncomfortable. Scheduling sex for the main purpose of reproduction can feel depersonalizing and can add conflict to a relationship (Becker 1990, 2000). As one wife explained:

> All the things you read about—that men feel like they are just a tool. You have to have an erection and ejaculate at a certain time whether you want to or not. He has said to me in times out of genuine anger, "I feel like all you want me for is to make a baby. You don't really want me, you just want me to do it." (Becker 1990, p. 94)

Anthropologist Gay Becker undertook an ethnographic study of infertility treatment, conducting five hundred interviews and doing field observation over a four-year period. The study included 143 women and 134 men and is a good source of information about the experience of infertility and its treatment.

When faced with involuntary infertility, an individual or a couple experiences a loss of control over life plans

---

[11] The physiology of conception, pregnancy, and childbirth is described in Appendix E, "Conception, Pregnancy, and Childbirth." For a more detailed technical discussion of infertility, see Becker 2000; Greenberg, Bruess, and Haffner 2002; or current books on the subject addressed to a lay (nonmedical) audience.

Number (in millions) and percentage of women in the United States, age 22-44 years, with current fertility problems who have ever used infertility services, by type of service, 1995 and 2002.							
	Women aged 22-44 years				Nulliparous* women aged 22-44 years		
	Number in millions		Percent		Number in millions		Percent	
Infertility Services	**1995**	**2002**	**1995**	**2002**	**1995**	**2002**	**1995**	**2002**
Total	2.7	2.7	44.5	38.5	1.1	1.2	43.0	44.4
Medical help to get pregnant	2.1	2.3	35.2	32.0	1.0	1.0	42.1	37.5
Advice	1.7	1.8	27.5	25.1	0.8	0.8	32.5	30.4
Infertility testing (male or female)	1.4	1.5	23.1	21.5	0.8	0.7	31.1	28.3
Female testing	1.3	1.4	21.2	19.9	0.7	0.7	28.3	27.1
Male testing	1.0	1.1	16.8	16.0	0.6	0.6	22.2	21.2
Ovulation drugs	1.0	1.2	16.1	16.2	0.5	0.5	18.3	20.3
Surgery or treatment for blocked tubes	0.5	0.3	7.4	3.6	0.2	0.1	8.3	4.3
Artificial insemination (including intrauterine)	0.4	0.4	5.9	6.1	0.2	0.2	9.1	8.5
Assisted reproductive technology	0.1	0.1	0.8	1.1	0.0	0.0	1.4	1.4
Medical help to prevent miscarriage	1.2	0.9	19.2	12.9	0.2	0.3	7.6	12.6

Figure 9.6 Women seek a wide variety of infertility services. Advice is the largest single form of infertility assistance women seek, with 1.8 million women asking their doctor for advice in 2002.

* Never having given birth to a child.

and may feel helpless, defective, angry, and often guilty. Although men and women may differ in their specific responses to infertility, both are very affected emotionally by the challenge to taken-for-granted life plans and their sense of manhood or womanhood. For example, upwards of one-half of women undergoing infertility treatments and 15 percent of men say that it "was the most upsetting experience of their lives" ("Psychological Impact" 2009, p. 1). The psychological burden of infertility may fall especially hard on professional, goal-oriented individuals. These people "have learned to focus all their energies on a particular goal. When that goal becomes a pregnancy that they cannot achieve, they see themselves as failures in a global sense," and this situation can hurt their relationship (Berg 1984, p. 164).

Going through infertility treatment is costly, with the average cost for one in vitro fertilization running at $8,158, and an additional $3,000 to $5,000 per cycle for fertility drugs. A current controversy involves whether employee health insurance should cover it. As of this writing, only fifteen states require insurance companies to offer policies that cover infertility treatment ("Psychological Impact" 2009).

Infertility treatment can be successful, but often it is not. When it is not, couples are faced with yet another decision—whether, or when, to quit trying. Said one husband, "The technology . . . has given us so many options that it is hard to say no" (quoted in Stolberg 1997, p. A1).

Reproductive Technology: Social and Ethical Issues

Reproductive technologies enhance choices and can reward infertile couples with much-desired parenthood. They enable same-sex couples or uncoupled individuals to become biological parents. But reproductive technologies have tremendous social implications for the family as an institution and raise serious ethical questions as well.

The following sections explore the commercialization of reproduction, inequality of access to reproductive technology, and the sometimes confusing and ambiguous parent–child relationships created by reproductive technology.

Commercialization of Reproduction A general concern is that the new techniques, when performed for profit, commercialize reproduction. Prospective parents and their bodies are treated as products and thereby dehumanized (Rothman 1999). Examples include the selling of eggs or sperm to for-profit fertility clinics and the marketing of sperm or eggs with certain donor characteristics such as intelligence, physical attractiveness, and athletic ability.

Reports of fraud, overstatement of positive outcomes, failure to warn about the risk of multiple births, and other professional violations (Leigh 2004) make it important to understand that an individual seeking treatment is in fact a consumer and should interview the doctor and

"So you're having trouble conceiving. Have you tried sex?"

Michael Shaw/New Yorker/Cartoonbank

investigate the facility. There has been little attempt by the government to regulate assisted reproductive technology in the same way that adoption, for example, is regulated. The federal government does require annual reports of procedures and success rates, but no licensing is required (Skloot 2003). Clients may not realize that the average success rate (a baby) is around 27 percent. That varies with the type of procedure and with age; chances are much less for women over forty and as low as one percent for women age forty-six or older (Spar 2006, Table 2.2).

Inequality Issues Reproductive technologies raise social class and other inequality issues. Assisted reproductive technology is usually not affordable by those with low incomes. By and large, after initial diagnosis, lower-income couples do not go on to more advanced (and expensive) treatment (Chandra and Stephen 2010). Said one woman, "There need to be some options for people like us who don't have money sitting in the bank" (Becker 2000, p. 20).

Who Is a Parent? Reproductive technology creates "family" relationships that depart considerably from what is possible through unassisted biology. Surrogacy, along with embryo transfer, creates the possibility that a child could have three mothers (the genetic mother, the gestational mother, and the social mother), as well as two fathers (genetic and social). In such a situation, how do courts define the "real" parents? A recent Florida court decision suggests that a male who donates sperm, even if he is present in the child's life, without prior agreement has no rights to the resulting child (O'Neill 2009).

An interesting recent development is the emergence of sperm donors as putative fathers, sought out by their "children" as they enter adolescence or young adulthood (Harmon 2005b, 2007b). Many states have laws by which sperm donors, with the exception of the husband, have no parental rights, but this barrier between sperm donors and their biological children is gradually being broken.[12]

What Kind of Child? As technology advances, the potential to create a child with certain traits expands. Embryo screening—a technology for examining fertilized eggs before implantation to choose or eliminate certain ones—is a boon for prospective parents whose family heritage includes disabling genetic conditions (Harmon 2006). But embryo screening also raises the possibility of sex selection (Grady 2007; Marchione and Tanner 2006) and perhaps selection for other traits. Those using sperm donors are already scanning the records to find evidence of traits they would like in their children. Philosophers ponder the implications for parents and children when children are made to order!

We should note that these concerns about reproductive technology are primarily articulated by medical and public health professionals, academics, policy analysts, and ethicists. For the most part, prospective parents themselves are more focused on their desire for a child and not so inclined to view ART with a critical eye—at least not initially.

Reproductive Technology: Making Personal Choices

Choosing to use reproductive technology depends on one's values and circumstances. Religious beliefs and cultural values influence decisions.[13] Fertility treatment can be financially, physically, and emotionally draining. The need for frequent physician's visits can interfere with job obligations, and infertility treatment can lead to tensions in a marriage.

There are certain situations in which the need for reproductive technology can be anticipated. Men or women undergoing medical treatments that will leave them infertile may bank sperm or eggs, and couples may take similar action regarding freezing embryos. In the last few years, men going off to war in Iraq have banked sperm, anticipating contact with hazardous materials—or death. Indeed, a baby was recently born to a father who was recently killed in Iraq (Oppenheim 2007).

People who become parents through successful assisted reproductive therapy are euphoric. The vast majority of children born by means of in vitro fertilization or donor sperm are thoroughly normal. There are, however, slightly elevated incidences of birth defects in children born through the use of ART. Although these slight increases intrigue researchers who are curious

---

[12] Legislation in Sweden and court decisions in the United Kingdom have given children in those countries the right to obtain identifying information about a donor. Some American sperm clinics have responded to the *identity release movement* by developing *open sperm donor* programs, which agree to make information available to the child at eighteen; offer photos of the donor; or, at a minimum, offer genetic/health information. Some previously anonymous donors and their biological offspring have met, arranging meetings through the clinic when all parties mutually agree to do so (Villarosa 2002a; Talbot 2001).

The American Society for Reproductive Medicine recommends against secrecy: "It's no longer possible to think of sperm donation without thinking of what the child it produces may someday want" (Talbot 2001, p. 88).

[13] The Catholic Church prohibits all forms of reproductive technology, including artificial insemination by the husband (Congregation for the Doctrine of the Faith 1988; McCormick 1992). The Jewish tradition requires physical union for adultery and so does not define donor insemination (DI) as adulterous. But Judaism does view masturbation as sinful. Hence, a man's obtaining sperm either to sell or to artificially inseminate his wife is morally problematic; this is true of Catholic teaching as well (Newman 1992). Some interpretations of Protestantism, on the other hand, note that the Bible sees infertility as cause for sorrow and exalts increasing human freedom beyond natural barriers (Meilander 1992).

about the reasons why these birth defects occur at greater rates, the chances of bearing children with birth defects (regardless of using ART or traditional methods to become pregnant) remains at approximately 3 percent (Kolata 2009, D1).

For others who had hoped to become parents, infertility treatment eventually became the problem instead of the solution. Coming to terms with infertility has been likened to the grief process, in which initial denial is followed by anger, depression, and usually ultimate acceptance: "When I finally found out that I absolutely could not have children . . . it was a tremendous relief. I could get on with my life" (Bouton 1987, p. 92). Some people gradually choose to define themselves as permanently and comfortably childfree (Koropeckyj-Cox and Pendell 2007).

A second way to get on with life yet retain the hope of parenthood is through adoption. Indeed, some couples explore adoption options even as they continue infertility treatments (Ford 2009).

The comment of author Debora Spar about her study of the reproductive technology business (2006) can apply equally well to adoption. When asked what most surprised her in research for her book, *The Baby Business* (2006), Spar reflected: "Everyone I spoke to who had gone through these difficult processes came out with a child that they were convinced was the only child that they were ever destined to have. . . . To me it shows that there's something in humans that connects us to our children and it goes even deeper than genetics alone" (quoted in Dreifus 2006).

Adoption

The U.S. census looked at adoption for the first time in 2000. In that year, there were more than two million adopted children in U.S. households, about 2.5 percent of all children. In terms of *numbers,* there are more adopted children in non-Hispanic white families (more than 70 percent of all adopted children). But Asian/Pacific Islander families have the highest *rate* of adoption relative to their population. More girls than boys are adopted. Women, especially single women, prefer to adopt girls, and girls are more likely to be available for adoption. Ninety-five percent of Chinese babies available for adoption, for example, are girls (Fields 2001; Kreider 2003; U.S. Census Bureau 2007a, Table 65).

Census data do not distinguish adoptions by biological relatives or stepparents from nonrelative adoptions. But earlier research found that a majority of adopted children were related to their adoptive parents by blood or marriage. Most commonly, those who adopted unrelated children have no other children, have impaired fertility, and have used infertility

services. They are more likely to be older and highly educated and to have higher incomes (Bausch 2006; Kreider 2003; U.S. Census Bureau 2007a, Table 65). A study in one U.S. county found adoption applicants drawn to adoption by their pronatalist beliefs (the importance of children and parenting) and their exposure to adoption (through friends and family members) (Bausch 2006).

To encourage adoption, there is now a federal tax credit of $10,000 toward adoption expenses for low- and middle-income parents. Corporations sometimes subsidize employees' adoptions. One survey of one thousand companies of various sizes found that 44 percent offered paid leave to newly adoptive parents, and 83 percent assisted with finances (Clemetson 2006b).

Some children are adopted informally—that is, the children are taken into a parent's home, but the adoption is not legally formalized. **Informal adoption** is most common among Alaska Natives, African Americans, and Hispanics (Kreider 2003).

Adoptions increased through much of the twentieth century, reaching a peak in 1970, but the number has declined since, with only 1.4 percent of American families adopting in 2002 (the latest data available) (J. Jones 2008). Fewer infants are available due to more effective contraception and legalized abortion. And white unmarried mothers, those most likely to relinquish their infants in the past, are now likely to keep their babies (J. Jones 2008, Table 16).

Some couples pursue international or transracial adoption, whereas others adopt "special needs" children—those who are older, come with siblings, and/or are disabled.

© Matt Miller/The Omaha World-Herald

In this photo, taken in Plainview, Nebraska, four-year-old adopted daughter Natalie has just been sworn in as a new U.S. citizen. Since 2001, children adopted internationally by U.S. citizens receive their American citizenship automatically.

Facts about Families

Through the Lens of One Transracial Adoptee
By Cathy L. Wong,[a] PhD

In history today we are studying about China. I do not know anyone there, but they might know of me. The teacher seemed to think that I should know about "my people." I tell her I am adopted and she mumbles cautiously that it was "probably for the best." She continues, "at least you will fit in if you were to visit the homeland someday." I'm confused, but wonder if she had stated that because I don't "fit" into America? Or was she implying I did not fit in to this community? I'm confused and saddened by what I think I understand. I look around and see no one who looks like me. I find myself staring intensely at the pictures of all those people climbing the Great Wall and wonder if anyone misses me. I come to realize that I cannot speak Chinese—I feel like a foreigner within both worlds. (Cathy L. Wong, personal journal, 1972, age 11)

Despite the love, attention, and caring of two well-meaning adoptive parents, as long as I can recall, I was left with many unanswered questions. These questions seemed to be fueled by my need to understand a past that was lost but not forgotten upon my entrance into the United States.

If one makes meaning from the stories of one's life, how do transracial adoptees make sense of their world?

In the midst of all of my unanswered questions and the voids in my own narrative, I have constructed meaning, finding a place for myself within the larger context of the society in which I live.

Using autoethnography, I add my own voice to a growing body stories in professional and academic fields. Autoethnography seeks (1) a deeper understanding one's experience, (2) a deeper understanding of how lives are intricately intertwined and influenced by cultural structures and social interactions with others in society, and (3) to understand how meaning evolves from the social interactions with others—in this case, the focus is on the life of the transracial adoptee.

Transracial Adoption Today

Transracial adoptions occur both internationally and domestically. Transracial or interracial adoptions (TRA) occur when children are adopted by people who are of a race that is different from their own, and/or where the children are from a different country. This is best exampled through my own American family: I am Chinese, my mother is of Swiss and Italian decent, and my father is of Greek origins. In my case, like those of many other transracial adoptees, history plays an important role. For example, war, economic strife, and perhaps even poverty led to the abandonment of thousands of children in Hong Kong—mostly girls like me—at the end of World War II. This is how I ended up here, in the United States.

Adoption of children from overseas has been steadily increasing since the 1980s. Between 1985 and 2003, American families adopted 40,496 children from China (Grice 2005, p. 124). The rate of adoptions from China has increased since that time, with 27,748 more children from China being placed in American homes since 2004—including 3,852 in 2008 (Yearbook of Immigration Statistics: 2008, Table 12).

The Debate

In 1972, debates arose in the United States regarding adoptions of children across racial and ethnic lines. At the heart of the debate are white, predominately middle- or upper-middle-class households in first-world countries who are adopting children from second- and third-world countries. Questions were raised such as, "Can white people properly raise children of color?" and "Is this in the best interest of adopted children?" The debate still continues today. Missing from the debate, however, have been the voices and stories of these transracially adopted children. Missing from all of this were our voices—our voices that allowed us to demand visibility and claim our space, and that allowed others

The Adoption Process

The experience of legal adoption varies widely across the country, partly because it is subject to differing state laws. Adoptions may be public or private. *Public adoptions* take place through licensed agencies. *Private adoptions* (also called *independent adoptions*) are arranged between the adoptive parent(s) and the birth mother, usually through an attorney. Legal fees and the birth mother's medical costs are usually paid by the adopting couple.

More and more, adoptions are open; that is, the birth and adoptive parents meet or have some knowledge of each other's identities. Even when an adoption is closed, as adoptions used to be, some states now have laws permitting the adoptee access to records at a certain age or under specified conditions.

A concern that arose in recent decades because of some high-profile cases is whether birth parents can claim rights to a biological child after the child has been adopted. In those cases, a nonmarital biological father had not given consent or even been notified, and he was able to assert his parental rights ("Biological Fathers' Rights" 2007; Burbach and Lamanna 2000). States have begun to reexamine laws requiring birth fathers to register with a "putative father registry" if they wish to have a role in decisions about the child's future (that is, with regard to relinquishment for adoption). Often the time window is very short, and the registries are not well publicized; legal challenges by biological fathers are in the works (Lewin 2006d; Markon 2010). However, of all domestic adoptions in the recent past, fewer than one percent have been contested by biological parents (Stewart 2007).

to understand our struggles and challenges—the barriers we adoptees face. It is not my intent to speak for my fellow transracial adoptees, but to add my voice and story to theirs. The research I am conducting is my means of cultivating my own understanding of how TRA has influenced my personal and professional life.

What I Found

Since my birth in 1962, adoption laws and policies have changed dramatically. Only a very limited number of adoptees lack access to their original birth records; open adoptions have continued to grow; and transracial adoptees have developed support systems that allow them to come together and share their experiences with each other.

My transracial adoption story is my own and not necessarily reflective of other transracial adoptee experiences. What my research, however, does is illuminate the unique social context in which I was raised: the small rural town where I grew up in the 1960s, the era of secrecy that left me and other adoptees with many unanswered questions, and a society that was struggling to embrace a burgeoning societal trend—transracial adoptions and their multiracial families in the United States.

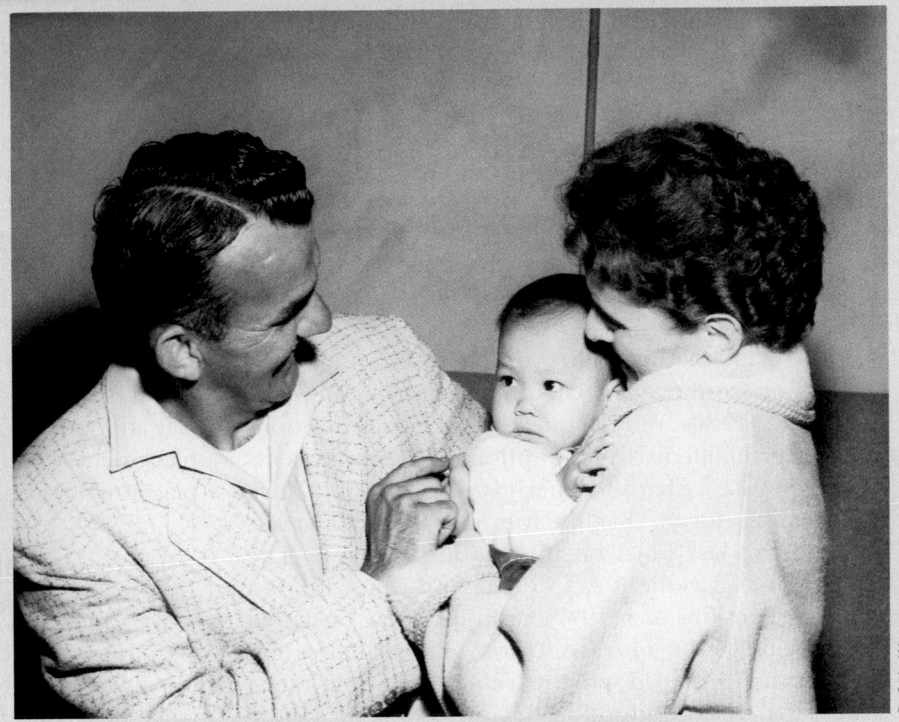

Special delivery 1962. Cathy Wong, the author as an infant, arriving in the United States from Hong Kong after being adopted by her white American parents.

a. This work was adapted from Dr. Cathy L. Wong's unpublished dissertation: "Filling the Void: An Autoethnographic Study of a Transracial Adoptee." Dr. Wong is a member of the Department of Sociology and Gerontology at California State University, Stanislaus. Her published work includes a chapter entitled, "Intergenerational Pathways: Getting Along in Space Takes on New Meaning," in Living In Space: Cultural and Social Dynamics, Opportunities, and Challenges in Permanent Space Habitats, edited by Sherry Bell and Langdon Morris.

Another concern of prospective adoptive parents has to do with the adjustment of adopted children—are they likely to have more problems than other children? Research suggests that adopted children, especially males, are at higher risk of problems in school achievement and behavior, psychological well-being, and substance use. A recent careful study, based on a large, nationally representative sample, confirms earlier findings of small to moderate differences between adopted and nonadopted children (Simmel, Barth, and Brooks 2007). As can be the case in social science, a different research review concluded that the overall body of research supports "the view that most adoptive families are resilient" and that positives outweigh negatives (O'Brien and Zamostny 2003, p. 679).

Adoption of Racial/Ethnic Minority Children

Today, "40 percent of adopted children are of a different race, culture, or ethnicity than both of their adoptive parents" (Vandivere, Malm, and Radel 2009, p. 9). The family diversity created by transracial adoption seems in tune with the increasing diversity of American society, and now includes four out of ten adoptions in the United States (Vandivere, Malm, and Radel 2009). Yet it has been controversial.

In 1971, agencies placed more than one-third of their black infants with white parents (Nazario 1990). At that time, the number of black adoptive homes was much smaller than the number of available children, whereas the reverse was true for whites. But interracial adoptions, having increased rapidly in the 1960s and early 1970s, were much curtailed after 1972, when the National Association of Black Social Workers strongly objected. Suggesting that transracial adoption amounted to cultural genocide, racial/ethnic minority advocates expressed concern about identity problems and the loss of children from the black community ("Preserving Families of African Ancestry" 2003). Native American activists have successfully asserted tribal rights and collective interest in Indian children. In addition to identity concerns, they expressed the fear that coercive pressures might be put on parents to relinquish their children to provide adoptable children to white parents. Indeed, this practice was pervasive through the 1960s (Fanshel 1972).[14]

As a result of this controversy, adoption agencies shied away from transracial adoption for many years. In the late 1980s, only about 8 percent of adoptions were interracial, usually adoption by white parents of mixed-race, African American, Asian, or Native American children (Bachrach et al. 1990). Congress has had the last word on this matter, however. The Multiethnic Placement Act (1994) and the Adoption and Safe Families Act (1997) prohibit delay or denial of adoption based on race, color, or national origin of the prospective adoptive parents. But some racial issues still arise in adoption decisions. Joseph Crumble typifies the concerns of some black social workers: "For blacks, it's about how confident whites can be with the issues of race when their race is in conflict with the race of the child" (in Clemetson and Nixon 2006, p. A18).

Long-term studies suggest that transracial adoption has proven successful for most parents and children, including with regard to racial issues. Sociologist Rita Simon and social work professor Howard Altstein followed interracial adoptees from their infancy in 1972 to adulthood. They were able to locate eighty-eight of the ninety-six families from the 1984 phase of the study for their latest book (2002). They concluded that, as adolescents and later, transracially adopted children "clearly were aware of and comfortable with their racial identity" (p. 222).

Another longitudinal study of transracial (white parents and African American, Asian, and Latino children) and in-race adoptions (white parents, white

children) followed the children from the mid-1970s to 1993, when they were in their early twenties. There were no differences in general adjustment or problem behavior between the two groups. Such adjustment difficulties as did exist among the transracially adopted children tended to be connected to racial issues—discrimination and "differentness" of appearance. Not surprisingly then, researchers found that neighborhood made a difference within the transracial adoptee group; those who were reared in mixed-race neighborhoods were more confident in their racial identity (Feigelman 2000).

Some researchers have suggested that, rather than causing serious problems, transracial adoptions may produce individuals with heightened skills at bridging cultures. "The message of our findings is that transracial adoption should not be excluded as a permanent placement when no appropriate permanent inracial placement is available" (Simon 1990).

Adoption of Older Children and Children with Disabilities

Together with certain racial/ethnic minorities, children who are no longer infants and children with disabilities make up the large majority of youngsters now handled by adoption agencies (Finley 2000). Special needs adoptions are pursued not only by couples who are infertile but also for altruistic motives. Gay men have adopted infants with HIV/AIDS, for example (Morrow 1992). In some cases, lesbian and gay male or older couples adopt such hard-to-place children because law or adoption agency policy denies them the ability to adopt other children.

The majority of adoptions of older children and children with disabilities work out well. Disruption and dissolution rates rise with the child's age at adoption. Among adoptions generally, only about 2 percent of agency adoptions end up being *disrupted adoptions* (the child is returned to the agency before the adoption is legally final) or *dissolved adoptions* (the child is returned after the adoption is final). But 4.7 percent of adoptions of children age three to five at adoption, 10 percent of those age six to eight years, and perhaps as high as 40 percent of children adopted between the ages of twelve and seventeen are disrupted or dissolved (Festinger 2005).

What causes these disrupted and dissolved adoptions? For one thing, some children available for adoption may be emotionally damaged or developmentally impaired due to drug- or alcohol-addicted biological parents, physical abuse from biological or foster parents, or previous broken attachments as they have been moved from one foster home to another. Some develop **attachment disorder**, defensively unwilling or unable to make future attachments (Barth and Berry 1988;

[14] The Indian Child Welfare Act of 1978 requires that "adoptive placement be made with (1) members of the child's extended family, (2) other members of the same tribe, or (3) other Indian families" so as "to protect the rights of the Indian child as an Indian and the rights of the Indian community and tribe in retaining its children in its society." In practice, outcomes of contested adoption cases have depended on the parents' attachment to the reservation and other circumstances. Tribes have also agreed to placements with white guardians or adoptive parents when they have believed it to be in the child's best interest.

Vandivere, Malm, and Radel 2009). Observers have seen attachment disorder among adoptees from Romania and other eastern European orphanages (Mainemer, Gilman, and Ames 1998).

Adoption professionals point out that parents are willing to adopt children with problems as long as they know what they are getting into (Groze 1996: Vandivere, Malm, and Radel 2009). Agencies have increasingly tried to gain information about the circumstances of the pregnancy and the child's early life and to match children's backgrounds with couples who know how to help them (Ward 1997).

International Adoptions

International adoption grew dramatically through 2003, but has slowed down to below 18,000 adoptions in 2004. In 2007, 60 percent of all children adopted from overseas by American parents were from Asia, especially from China; 15 percent from India, Kazakhstan, Colombia, Ukraine, Philippines, and Ethiopia; 13 percent from Russia; and 11 percent each from Guatemala and Korea (Vandivere, Malm, and Radel 2009, Table 3).

Parents who have adopted internationally have encountered all kinds of difficulties: the expense of travel to a foreign country—and getting time off from work to go to the child's country for an extended stay; difficulty with negotiations and paperwork in a foreign language, and the need to rely on translators and brokers; the uncertainty about being able to choose a child, as opposed to having one thrust upon the parent; the occasional unexpected expansion of adoption fees or expected charitable contributions; the ambivalence and reluctance of a nation to place its children abroad; and the complete failure to bring home a child.

The biological mother's consent is an issue in overseas adoptions because it is more difficult to be sure that the mother has willingly placed her child for adoption rather than being coerced or misled by a baby broker. Romania placed a moratorium on adoptions, fearing corruption of their entire system. Russia has also recently placed a temporary moratorium on adoption applications, and Guatemala is revising its process to comply with the Hague Convention on Intercountry Adoption. (The United States ratified this treaty in 2008.) China is moving to impose new rules on foreign adoptions, including not only establishing an age requirement for adoptive parents (under fifty) and stable marriage specifications but also ruling out obese prospective parents (Belluck and Yardley 2006; Clemetson 2007; "Hague Convention" 2010; J. Gross 2007; N. Knox 2004; Lacey 2006; Yin 2007).

International adoptions can pose some of the same problems as the adoption of older children. Conditions in homes and institutions overseas may not be ideal beginnings, and children may have health problems or suffer from attachment disorder (Elias 2005; J. Gross 2006c; Vandivere, Malm, and Radel 2009). But the vast majority of international adoptions are successful (Tanner 2005). A meta-analysis of around one hundred studies found that adopted children are referred to mental health services more often than nonadoptees are, perhaps a function of adjustment concerns and high-income parents more than troubling behavior. "Most international adoptees are well adjusted" (Juffer and van Uzendoorn 2005, p. 2,501). They "are underrepresented in juvenile court and adult mental health placements," according to Dr. Laurie C. Miller, editor of the *Handbook of International Adoptive Medicine* (Miller 2005b, p. 2,533; see also Miller 2005a).

Those who adopt internationally say they made this choice for several reasons. They are more apt to be able to adopt a healthy infant, with a shorter wait and often fewer limits in terms of age or marital status. The adoption is perceived to be less risky in that there is little likelihood of a birth mother seeking to reclaim the child (Clemetson 2006a; Zuang 2004). To what degree racial preferences enter into the choice of international adoption is difficult to determine.

Today, there are not only more agencies for arranging international adoptions but also more resources for coping with any postadoption difficulties. There are now specialists in "adoption medicine" who can address medical and cognitive problems of children adopted overseas, as well as psychologists who are prepared to address international or transracial adoption issues (J. Gross 2006c; Tuller 2001). There are "culture camps" (Chappell 1996), schools (Zhao 2002), parent groups (Clemetson 2006a), and other resources for bridging the cultural gap for a child raised in America but conscious of having started life in another country. Most parents try very hard to maintain a bicultural identity for the child (Brooke 2004), and some undertake travel to the child's country of origin. Sometimes, though, internationally adopted children just want to simply be the American child that they also are.

International adoption produces more and more multicultural families in an increasingly multicultural America. The many media photos of happy adoptive parents and children tell a story of hopes for parenthood that are realized.

Contemporary society offers many choices about whether to have children and how many. In addition, modern technology has increased people's options about *how* to have children. New trends in adoption, such as international adoption, have added still more options and precipitated more decisions. As this text has often suggested, the best way to make decisions about whether or not to parent is to make them knowledgeably.

Summary

- Today, individuals have more choices than ever about whether and when to have children and how many to have.

- Although parenthood has become a choice, the majority of Americans continue to value parenthood. Only a small percentage expects to be childless by choice.

- Nevertheless, it is likely that changing values concerning parenthood, the weakening of social norms prescribing marriage and parenthood, a wider range of alternatives for women, the desire to postpone marriage and childbearing, and the availability of modern contraceptives and legal abortion will result in a higher proportion of Americans remaining childless or having only one child in the future.

- Some observers believe that societal support for children is so lacking in the United States that it amounts to *structural antinatalism*. They point to the absence of a society-wide program of health insurance and

health care for children, to workplace inflexibility, to the lack of affordable quality day care, and to the absence of paid maternal or paternal leave, as is provided in Europe and elsewhere.

- Children can add a fulfilling and highly rewarding experience to people's lives, but they also impose complications and stresses, both financial and emotional.

- Birthrates have declined for married women, and many women are waiting longer to have their first child. Although other nonmarital birthrates have risen in recent decades, teen birthrates have declined. Pregnancy outside of marriage has become increasingly acceptable, but some unmarried pregnant women choose abortion.

- Deciding about parenthood today can include consideration of postponing parenthood, having a one-child family, engaging in nonmarital births, having new biological children in stepfamilies, adopting, and taking advantage of infertility treatment.

Questions for Review and Reflection

1. What are some reasons that there aren't as many large families now as there used to be?

2. Discuss the advantages and disadvantages of having children. Which do you think are the strongest reasons for having children? Which do you think are the strongest reasons for *not* having children?

3. How would you react to becoming the parent of twins? Triplets? More? If your choice is to take fertility treatments that pose a risk of multiple births or

to not have children at all, what would you do—and why?

4. Which reproductive technology would you be willing to use? In what circumstances?

5. **Policy Question.** How is a pronatalist bias shown in our society? Are there antinatalist pressures? What policies might be developed to support parents? Are there any special policy needs of nonparents? Why might a society's social policies favor parents over nonparents?

Key Terms

abortion 239
assisted reproductive technology (ART) 242
attachment disorder 249
fecundity 220
fertility 220
impaired fertility 242
induced abortion 239
informal adoption 246
involuntary infertility 242
multipartnered fertility 237

opportunity costs (of children) 228
pronatalist bias 226
replacement level (of fertility) 222
single mothers by choice 236
social capital perspective (on parenthood) 227
structural antinatalism 226
total fertility rate (TFR) 220
value of children perspective (on parenthood) 227
voluntary childlessness 229

Online Resources

Sociology CourseMate

www.CengageBrain.com

Access an integrated eBook, chapter-specific interactive learning tools, including flash cards, quizzes, videos, and more in your Sociology CourseMate, accessed through CengageBrain.com.

www.CengageBrain.com

Want to maximize your online study time? Take this easy-to-use study system's diagnostic pre-test, and it will create a personalized study plan for you. By helping you identify the topics that you need to understand better and then directing you to valuable online resources, it can speed up your chapter review. CengageNOW even provides a post-test so you can confirm that you are ready for an exam.

Raising Children in a Diverse Society

"Whoever came up with the Peace Corps motto, 'The toughest job you'll ever love,' probably wasn't a parent" (Picker 2005, p. 46). Although raising children may be a joyful and fulfilling enterprise, parenting today takes place in a social context that makes child rearing an enormously difficult task. For most of human history, adults raised children simply by living with them and thereby providing examples and socialization into adult roles. From an early age, children shared the everyday world of adults, working beside them, dressing like them, sleeping near them.

At least in Europe, the concept of childhood as different from adulthood did not emerge until about the seventeenth century, according to historian Phillipe Ariès (1962). As education became available to all children, not just those of the wealthy, and as they spent more of their time in school, children gradually spent less time participating in the everyday lives of adults. One result is that today we regard children as people who need special training, guidance, and care (Apple 2006). Nevertheless, compared to sixty years ago, U.S. society can seem indifferent to the needs of parents and their children. For instance, the rate of child poverty in the United States exceeds that of the nation as a whole and is considerably higher than in other wealthy industrialized nations (Moore et al. 2009; U.S. Census Bureau 2010b, Table 697).

In this chapter, we will discuss the parenting process in the United States, a society that is diverse economically, by race/ethnicity, and in terms of family structure. As you study this chapter, we encourage you to note the *intersection* of these circumstances: how the parenting process is influenced by ways that gender, race/ethnicity, and social class interconnect, or overlap within a family structure, or form. As just one example, within the single-parent family form, parenting is a different experience for a low-income father of color than for a middle-class, non Hispanic white mother.

We'll begin by looking at some general characteristics of the parenting process today. Next we'll examine how gender affects parenting. We will then describe parenting styles, noting that the authoritative parenting style is advised by child development experts. We'll address ways that parenting differs according to race/ethnicity. We'll describe grandparents who serve as parents.

Other issues related to children appear throughout this text. Child outcomes related to cohabitation and same-sex couples are addressed in Chapter 8. Combining work and parenting roles is explored in Chapter 11. Suggestions about how best to communicate with children appear in Chapter 12. Violence against children is discussed in Chapter 13. The economic concerns of divorced parents, as well as their children's outcomes, are addressed in Chapter 15. Issues unique to stepparents are considered in Chapter 16. Here we address the parenting *process* in a diversity of social circumstances.

Parents in Twenty-First Century America

As shown in Figure 10.1, married couples comprise just under two-thirds (64 percent) of families with a joint child under age eighteen. Single-mother families represent almost one-quarter (24.6 percent) of parenting family groups. The remaining parental family groups

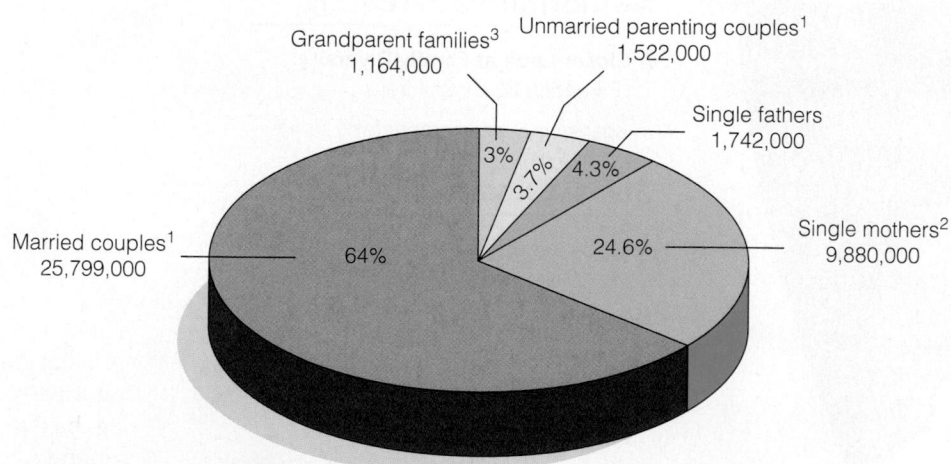

[1]Have at least one joint never-married child under age eighteen in the home.
[2]Parent may have a partner, but none of the children is also the child of the cohabiting partner.
[3]Grandparent householder with grandchildren for whom the grandparent is responsible.

Figure 10.1 Family Groups with Children under Age 18, 2009

Source: Calculated from U.S. Census Bureau 2010a, Table FG10. Percentages do not total exactly 100 percent due to rounding errors.

include single fathers, unmarried cohabiting couples with at least one joint child under age eighteen, and grandparent families (U.S. Census Bureau 2010a, Table FG10). According to Census Bureau definitions, a single parent can be either cohabiting or not.

In addition to variations in family form, many parents display a marked fluidity in living arrangements, resulting in a degree of complexity not remarked upon by researchers until fairly recently. For example, a single mother may move from her own apartment to live with her mother, then move once more to reside with other relatives or a romantic partner. Then too, **multipartnered fertility** (a person's having children with more than one partner) can mean that a father resides, perhaps temporarily, with one or more of his children but not with others (Harknett and Knab 2007). As a result, the parenting situations discussed in this chapter should be understood as changeable.

Regardless of their living arrangements or family structure, parents today face a myriad of questions that would not have been imagined just a few decades ago: How much fast food is too much? Should I let my child walk to school without a chaperone? Should I believe the teacher who says my child needs medication? Does my teenager spend too much time on Facebook? What should I tell my child about terrorism?

Parenting Challenges and Resilience

We would not want to point out the difficulties facing today's parents without first noting some positives. In general, parents now have higher levels of education and are likely to have had some exposure to formal knowledge about child development and child-raising techniques. Many fathers are more emotionally involved than several decades ago (Bianchi, Robinson, and Milkie 2006). In many respects, technology has improved our quality of life. The Internet offers countless sources of information for parents dealing with just about any situation. New communication technologies allow parents and children to keep in virtually continual contact and make being in touch with other family members far easier and more likely than a generation ago (Devitt and Roker 2009).

Nevertheless, parents face difficulties and make mistakes. It helps to know that children can be remarkably **resilient**: Children (and adults) can demonstrate the capacity to recover from or rise above adverse situations and events (Coyle et al. 2009; Goldman 2006; Werner and Smith 2001). Furthermore, research indicates that "[a]dults who acknowledge and seem to have worked through difficulties of their childhood are apparently protected against inflicting them on their children" (Belsky 1991, p. 124). There is also evidence that one caring, conscientious adult can generate a resilient child (Johnson 2000; Soukhanov 1996). Meanwhile,

Raising a child with disabilities reminds us that parents and children can evidence resilience, which is enhanced by strong familial bonds.

the family ecology perspective (see Chapter 2) leads us to look at ways that the larger environment challenges parents today.

Some Ways That the Social Environment Makes Parenting Difficult Here we list six societal features that make parenting difficult:

1. In our society, the parenting role conflicts with the working role, and employers typically place work demands first (Barnett et al. 2009; Bass et al. 2009; Marshall and Tracy 2009). A majority of American parents say they worry about juggling the demands of work and family and wish they could spend more time with their children (Erickson and Aird 2005; Snyder 2007).

2. Today's parents raise their children in a pluralistic society characterized by diverse and conflicting values. Parents are but one of several influences on children. Among others are schools, television, movies, music, and the Internet. Concerns about outside influences

may be greater for immigrant parents whose cultural values differ from some that they encounter in the United States (Driscoll, Russell, and Crockett 2008). But high percentages of all American parents worry about negative messages in the media, protecting their children from drugs and alcohol, or about the possibly problematic influences of other kids on their child (Farkas, Johnson, and Duffett 2002).

3. Various experts have publicized the fact that parents influence their children's health, weight, eating habits, math and language abilities, behaviors, and self-esteem. Although much parenting advice is useful, the emphasis on how parents influence their children can lead us to feel anxious about our performance as parents.[1]

4. Today's parents often find themselves sandwiched between simultaneously caring for children and elderly parents. Although caregiving can increase life satisfaction, stress can build as family members juggle employment, housework, child care, and parent care (Cullen et al. 2009).

5. In many areas of the United States, community appreciation for and assistance to parents has diminished over the past fifty years. The proportion of children under eighteen in our population—about one-quarter—represents a substantial drop from the 1960s, when more than one-third of Americans were children (U.S. Census Bureau 2010b, Table 8). Accordingly, the proportion of households with children has dropped from almost half in 1960 to fewer than one-third. Therefore, the parenting role is less predominant today. As parenting has become one lifestyle choice among many, society-wide support has diminished for the child-rearing role, once taken for granted as central (Whitehead and Popenoe 2008).

6. Today's parents are given full responsibility for successfully raising "good" children, but their authority is often questioned. For example, the state may intervene in parental decisions about schooling, discipline and punishment, medical care, and children's safety as automobile passengers ("Home School Laws" 2010; "HPV Policy" 2007; Jervey 2004).

As a result of these factors, being a parent today can be far more challenging than many nonparents realize. It's no wonder that parents, especially when employed, are considerably more stressed than nonparents (Carroll 2007c). Then too, a parent's physical illness, as well as raising a child with special needs, can add stresses peculiar to these situations, and recent government budget cuts have meant diminished resources for children with special needs (Firmin and Phillips 2009; Ontai 2008).

A Stress Model of Parental Effectiveness

As we saw in Chapters 7 and 8, considerable research shows that growing up with married parents is statistically related to better child outcomes (Kreider and Elliott 2009a; Magnuson and Berger 2009). Researchers attribute much of this finding to differential stress levels. Rather than family structure itself, stresses that are peculiar to family forms other than marriage account for divergent child outcomes. Being raised in a supportive family atmosphere is statistically related to more desirable outcomes for children, regardless of family structure (Doohan et al. 2009; Schoppe-Sullivan, Schermerhorn, and Cummings 2007).

According to a **stress model of parental effectiveness** (see Figure 10.2), stress that parents experience—from sources such as job demands, financial worries, concerns about neighborhood safety, feeling stigmatized due to negative stereotypes associated with living in a nonmarital family form, or racial/ethnic discrimination—cause parental frustration, anger, and depression, increasing the likelihood of household conflict. Parental depression and household conflict, in turn, lead to poorer parenting practices—inconsistent discipline, limited parental warmth or involvement, and lower levels of parent-child trust and communication. Poorer child outcomes result (Benner and Kim 2010; Burrell and Roosa 2009; Jackson, Choi, and Bentler 2009; Teachman 2009; White et al. 2009). Having social support mediates, or diminishes, this adverse relationship (Lee et al. 2009).

The Transition to Parenthood

More than forty years ago, in what has become a classic analysis, social scientist Alice Rossi asserted that the **transition to parenthood** is difficult for several reasons. Many first-time parents approach child rearing with little experience. Moreover, new parents abruptly assume

---

[1] Today you can read about how to raise "respectful" (Cartmell 2006; Rigby 2006), "happy" (Adkins 2007; Biddulph and Biddulph 2007), optimistic and "depression-proofed" (Murray and Fortinberry 2006), "successful" (Brodkin 2006; Burt and Perlis 2006), "well-adjusted" (Gangstad 2006), "confident" (Apter 2007), "socially skilled" (Markway and Markway 2006), "charitable" (Weisman 2006), "kind" (Siegel 2006), "resourceful" (Nelsen, Erwin, and Duffy 2007), "generous" (Gallo and Gallo 2005), and "balanced" kids (Campbell and Suggs 2006). You can read about how to raise kids to "lead change" (Gustafson 2009), or "make a change" (Tim Smith 2006), kids destined for "true greatness" (Kimmel 2006)—even "athletic stars" (Dance and Place 2006). Advice may be specifically directed to parents who are raising boys (Cox 2006; M. Jones 2006; Lewis 2007), girls (Preuschoff 2006; Trevathan and Goff 2007), twins (Gottesman 2006; Heim 2007), children who are deaf (Marschark 2007), children with special needs (Winter 2006), and "gifted" children (Klein 2007), as well as those who are "strong-willed" (Pickhardt 2005), shy (Markway and Markway 2006), or "spirited" (Kurcinka 2006). You can learn to raise great children properly by using "6 keys" (Leman 2006), "8 steps" (Gallo and Gallo 2005), "12 secrets" (Wright 2006a), "13 dynamics" (Inman and Koenig 2006), "52 brilliant ideas" (Dosani and Cross 2007), "101 truths" (Scott 2006), or "135 tools" (Arnall and Elicksen 2007).

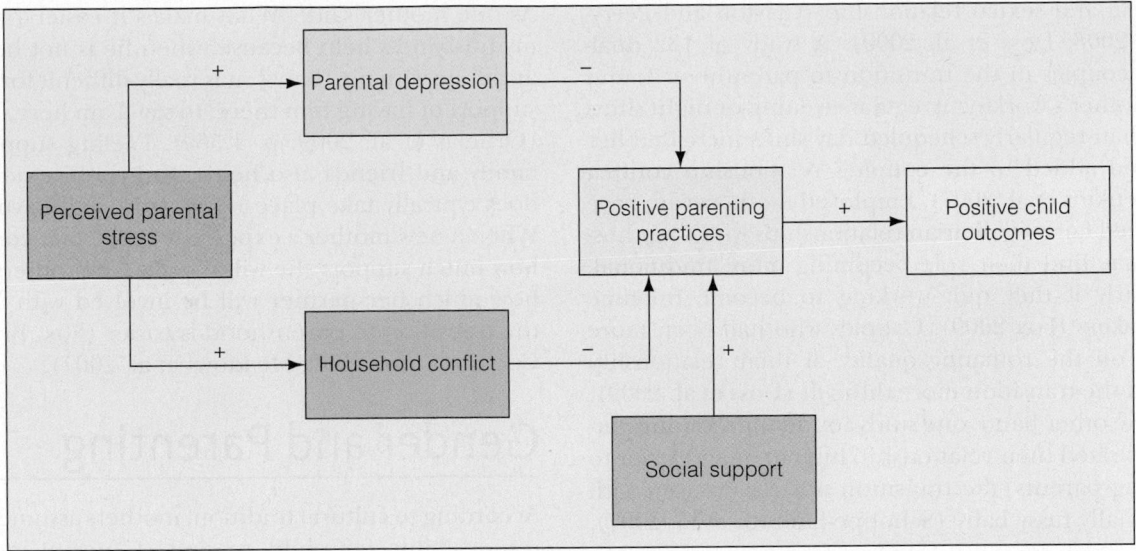

Figure 10.2 Stress Model of Effective Parenting. In this figure, plus signs (+) depict positive relationships between variables, and minus signs (-) depict negative ones. Greater use of positive parenting practices results in more positive outcomes for children. However, higher stress levels result in more (+) parental depression and more (+) household conflict. Parents might feel stressed due to job or educational demands; financial difficulties; concerns about neighborhood safety; or feeling stigmatized as a result of racial/ethnic discrimination or negative stereotyping associated with nonmarital family forms. Increased parental depression and household conflict result in diminished (-) use of positive parenting practices. Meanwhile, higher levels of perceived social support—due to high levels of family cohesion, private safety nets, or policies and programs that support parents—are positively related (+) to effective parenting practices, hence to positive child outcomes.

Sources: This figure was designed by Agnes Riedmann and derived from research findings from the following: Baxter 1989; Benner and Kim 2010; Broman, Li, and Reckase 2008; Bronte-Tinkew, Horowitz, and Scott 2009; Brush 2008; Burrell and Roosa 2009; Goosby 2007; Jackson, Choi, and Bentler 2009; Joshi and Bogen 2007; Lee et al.

twenty-four-hour duty, caring for a dependent and fragile infant (Rossi 1968). Often surprised at how disruptive an infant can be, new parents report being bothered by the baby's interrupting their sleep, work, and leisure time. In the words of one new mother, "[Y]ou have to run to bathe yourself because the child is going to wake up" (in Ornelas et al. 2009, p. 1,564).

Moreover, parents are more likely than in the past to be geographically distant from their own parents and other relatives who might give advice and help. For example, a Mexican immigrant mother told an interviewer:

> In Mexico, when you have a baby, well your mother is always there, or your family is there. . . . They help you like to pick him up, to hold him, to change him. All of that. So, you come here, and you find yourself alone and with a little baby that you don't even know how to pick up. (in Ornelas et al. 2009, p. 1,568)

More disconnected from friends and others than before the baby's arrival, nonimmigrant new mothers also report feeling isolated (Paris and Dubus 2005). Employed mothers of infants, especially those who have jobs with inflexible hours and little opportunity for advancement, are more likely to feel stressed (Marshall and Tracy 2009). Along with other factors, the difficulties associated with a new baby result in postpartum depression in about 10 percent of new mothers (Formichelli 2001).[2]

Meanwhile, a study of low-income black families found that mothers who were more pleased about their pregnancy were less likely later to view parenting as burdensome (Ispa et al. 2007). Becoming a parent typically involves what one researcher has called the *paradox of parenting:* New parents feel overwhelmed, but the motivation to overcome their stress and do their best proceeds from the stressor itself—the child as a source of love, joy, and satisfaction (Coles 2009).

For couples, the transition to parenthood means less time spent relaxing together and declines in their

---

[2] It helps to know that babies differ even at birth; the fact that a baby cries a lot does not necessarily mean that she or he is receiving the wrong kind of care (Rankin 2005). Infants may have different "readabilities"—that is, varying clarity in the messages or cues they give to tell caregivers how they feel or what they want (R. Bell 1974). Although new parents' attitudes, overall mood, and self-esteem influence how they view their babies, it also appears that babies have varied temperaments at birth. Some are "easy," responding positively to new foods, people, and situations, and transmitting consistent cues (such as tired cry or hungry cry). Other infants are more "difficult." They have irregular habits of sleeping or eating that sometimes extend into childhood; they may adapt slowly to new situations and stimuli; and they may cry endlessly, for no apparent reason (Komsi et al. 2006; Roisman and Fraley 2006; Thomas, Chess, and Birch 1968).

emotional and sexual relationship (Clayton and Perry-Jenkins 2008; Doss et al. 2009). A study of 132 dual-income couples in the transition to parenthood found that a mother's working irregular, evening or night shifts rather than regularly scheduled day shifts increased her stress and added to the couple's relationship conflict (Perry-Jenkins et al. 2007). Employed mothers who have established fairly egalitarian relationships with their husbands may find their role becoming more traditional, particularly if they quit working to become full-time homemakers (Fox 2009). Couples who had been more focused on the romantic quality of their relationship may find the transition more difficult (Doss et al. 2009).

On the other hand, one study found that, among parents who rated their relationship high in quality prior to becoming parents, the transition was easier, even with an unusually fussy baby (Schoppe-Sullivan et al. 2007).

Tetra Images / Getty Images

Transition to parenthood can be difficult for a number of reasons, including upset schedules and lack of sleep. It's a paradox that (1) new parents feel overwhelmed, while (2) inspiration to overcome their stress and do their best is provided by the stressor itself—the child as a source of profound delight.

As one mother said, "What makes it easier [is] having my husband's help because, when he is not here and I am alone with my [baby], it is really difficult for me. The support of having him there, to say 'I am here,' [helps]" (Ornelas et al. 2009, p. 1,569). Feeling supported by family and friends also helps, and positive adjustment does typically take place (Bost, Cox, and Payne 2002). When a new mother's expectations are met concerning how much support she will receive from others and/or how much her partner will be involved with the baby, the transition to parenthood is easier (Fox, Bruce, and Coombs-Orme 2000; Meadows et al. 2007).

Gender and Parenting

According to cultural tradition, mothers assume primary responsibility for child rearing (Cancian and Oliker 2000). Whether employed or not, a mother is generally expected to be the child's primary **psychological parent**, assuming—with self-sacrifice when necessary—major emotional responsibility for the safety and upbringing of her children (Springer, Parker, and Leviten-Reid 2009). Historically, fathers have been expected to be breadwinners and not necessarily competent in or desirous of nurturing children on a day-to-day basis (Gerson 1997). Today, however, our culture prescribes that "good" fathers not only assume considerable (usually primary) financial responsibility but also actively participate in the child's care (Troilo and Coleman 2008).

The theoretical perspective that emphasizes social reality as constructed (Chapter 2) reminds us that family members adapt culturally understood roles to their own situations.[3] Put another way, individuals "do" family, especially in a postmodern society characterized by fluidity in family forms (Hertz 2006b). How do cultural expectations regarding mother- and fatherhood correspond with the daily experiences of mothers and fathers? Put another way, how does each gender "do" parenthood?

Doing Motherhood

Whether single, cohabiting, or married, mothers typically engage in more hands-on parenting than do fathers, and they take primary responsibility for their children's upbringing (Hook and Chalasani 2008; Newport 2008).[4] Perhaps not surprisingly, adolescents of

[3] As discussed in Chapter 7, from the social constructionist perspective, individuals engage in role making: We adjust, or "make," rather than simply "take" our roles.

[4] An obvious exception involves gay men who choose to parent. "Gay men who choose to parent, either as a couple or alone, must cope with the fact that they will be challenging societal notions regarding the absence of a woman as the primary caregiver. Under this assumption, many men, both gay and nongay, will struggle with questions concerning their ability to parent based solely on their exposure to traditional gender scripts" (Berkowitz and Marsiglio 2007, p. 367).

both genders are likely to name their mother as their closest family confidant (Nomaguchi 2008). A study of employed, coupled heterosexual parents found that mothers often define "quality time" differently from fathers. Fathers are more likely to see quality time with their children as being home and available if needed. Mothers more often see quality time as having heart-to-heart talks with their children or engaging in child-centered activities (Snyder 2007).

Moreover, heterosexual fathers see themselves as more involved with their children than their partners do (Mikelson 2008). A mother may try to manage her child(ren)'s relationship with their father by encouraging father–child activities and constructing for them a positive image of him (Seery and Crowley 2000). "Despite much attention in recent years to the . . . 'new, nurturing father,' . . . women still do most child raising and homemaking" (Arendell 2000, p. 1,198; Dermott 2008). This intense daily contact with children is viewed ambivalently by many mothers—as a source of great life satisfaction but causing frustration and stress (Warner 2006).

Some women quit successful careers to accommodate their mothering role (Stone 2007; Tyre 2006b). As mothers have entered the labor force in greater numbers, many men have been encouraged by their family's need and the redefinition of male roles to want to play a larger part in the day-to-day care of their families (Bass et al. 2009; Bianchi, Robinson, and Milkie 2006). When mothers see fathers as competent parents—and when fathers believe that their child's mother has confidence in them—fathers are more likely to be highly involved (Fagan and Barnett 2003). Stressful for virtually all parents, mothering as a single parent is generally even more so.

Single Mothers

About 38 percent of all births occur to unmarried women (U.S. Census Bureau 2010b, Tables 80, 85). As discussed in Chapter 8, some of these births are to women in same-sex couples. In addition, between one-quarter and one-half of all nonmarital births occur to cohabiting heterosexual couples (Carter 2009; Dye 2005; Hamilton et al. 2006). We can therefore conclude that between 15 and 20 percent of all births occur to uncoupled, single mothers. The intersection of gender with family form is evident in the fact that single women, dramatically more often than single men, assume responsibility for child rearing (Brush 2008).

The category *single mother* is diverse by race/ethnicity, immigration experience, education, and socioeconomic class. Many single mothers never intended to raise their children without a partner. On the other hand, some women have purposefully decided to raise a child as a single parent.

Single Mothers by Choice Some single women adopt or, using donor sperm and artificial insemination, purposefully conceive and intend to raise their child without a partner. Called **single mothers by choice**, these women tend to be in their thirties or early forties, middle- or upper-middle class, European American, and relatively highly educated (Bock 2000).

Sociologist Rosanna Hertz (2006a) describes several stages in the decision to become a single mother by choice. First, a woman begins to realize that finding a partner first, then becoming a parent is unlikely to pan out for her. Next, she begins to investigate options regarding adoption or nonstandard insemination procedures. She also mobilizes support from family and friends. After the baby has entered her life, she continues to construct her roles around providing financial support and caring for her child. Researchers have given relatively little attention to the parenting practices of single mothers by choice. More numerous today—and receiving far more research attention—are single mothers by circumstance.

Single Mothers by Circumstance For the majority of women, the decision to be a single mother is less than deliberate (Hertz 2006a). **Single mothers by circumstance** arrive at this status in a variety of ways. Some women, becoming mothers while married or cohabiting, believed that their relationship would last but have since divorced or separated. Others realized that their relationship was not permanent, and the pregnancy may have been unplanned, but they chose to bear the child rather than have an abortion (Carter 2009). As one never-married college student and single mother told an interviewer, "My heart felt ready for the baby. I knew it was something I could do with or without a spouse" (in Holland 2009, p. 173).

As a category, single mothers' median family incomes are considerably lower than either married mothers' or single fathers, and single mothers are more likely to live in poverty, as shown in Tables 7.1 and 7.2. Some single mothers fantasize about marrying a responsible partner: "Wouldn't that Prince Charming dream be nice just once in awhile?" (in Gemelli 2008, p. 112). Others may have wanted to wed but did not find a man whom they considered acceptable (Holland 2009). Some single mothers keep the fathers of their children at a distance due to poor relationships with them, safety concerns for their children, apprehension about the father's illegal activities, or their seeing him as generally unreliable (England and Edin 2007).

Nevertheless, single mothers are well aware that to be married is the cultural ideal. One told an interviewer, with reference to her never-married status:

> Even though I have come so far, I have a car, I have a house, not just an apartment, and I'm a CNA [certified nursing assistant] and I will never have to work a minimum wage job again ever. My kids aren't in want for anything really important, but . . . I'm nowhere. I'm at the bottom, you know. (in Gemelli 2008, p. 112)

In a society that strongly advocates a two-parent child-rearing model,[5] single mothers report feeling stigmatized (LaRossa 2009; Thornton 2009; Usdansky 2009a, 2009b). A low-income single mother told her interviewer, "I've had to do motherhood in a hostile environment" (in Gemelli 2008, p. 106). As pointed out in Chapter 9, disgrace attached to unwed motherhood has diminished since the mid-twentieth century. Nevertheless, popular culture continues to shame or dishonor single mothers, particularly those whose children were born out of wedlock (e.g., Coulter 2009). Negative attitudes about unmarried parenthood encourage society-wide reluctance to provide resources for single mothers and their children (Mollborn 2009).

Meanwhile, single mothers evidence creativity and resilience as they construct support networks to help with finances, housing, child care, and other needs (Chu-Yuan Lee et al. 2009; King, Mitchell, and Hawkins 2010; Levine 2009). For instance, an association called CoAbode facilitates house sharing for single mothers (coabode.com). As another example, single mothers maintain the Internet-based M.O.M.S. support group that offers practical assistance, such as providing business clothes for those in need (www.singlemoms.org). In addition, single mothers may rely on brothers, brothers-in-law, grandfathers, uncles, or male cousins to serve as father figures for their children (Juffer 2006; Richardson 2009).

A **private safety net**, or social support from family and friends, is associated with children's better adjustment (Ryan, Kalil, and Leininger 2009). But social support from extended family is not always without cost. For example, a single mother told of her father's offering to pay for her family's medical insurance but only on the unspoken condition that she listen while he regularly criticized her. Single mothers, especially when low-income, undertake an "ungainly balancing act . . . as they walk a tightrope of reciprocity, social isolation, and material support frequently coupled with humiliating condemnation" (Brush 2008, p. 128).

To improve life for themselves and their children, many single mothers choose further education. This decision is not without added stress, however (Duquaine-Watson 2007). Given their work and parenting obligations, finding time to attend assigned off-campus activities or meetings to plan group projects poses problems. Moreover, some instructors make unappreciated, stereotypic assumptions. As one student explained, "We were going to be starting to talk about

welfare laws and programs the following week and [my instructor] wanted to know if I would be comfortable sharing my experiences with the rest of the class. I never even got welfare" (in Duquaine-Watson 2007, p. 234).

For single mothers who do receive welfare, legislative changes have meant added stress (Gemelli 2008). As noted in Chapter 7, the 1996 "welfare reform bill" dismantled the federal Aid to Families with Dependent Children (AFDC) program. A new federal program, Temporary Assistance for Needy Families (TANF), replaced AFDC. Intended to encourage both marriage and the work ethic, TANF limits to five years what had previously been open-ended assistance and requires that most recipients be employed.[6] For many single mothers, TANF has meant unsatisfying work at poverty-level wages, new day care struggles, and less time with their children (Cook et al. 2009; Neblett 2007).

In accordance with the stress model of parental effectiveness, single mothers' time constraints, generally poorer economic resources, and resultant higher depression levels—not family structure per se—result, on average, in less effective parenting behaviors (Guzzo and Lee 2008; Teachman 2009). Instead of between-group comparisons of children raised by single mothers with those raised in two-parent homes, researchers sometimes make within-group comparisons of children raised in various single-mother families. In these latter

Given work and other demands, more than 50 percent of mothers tell pollsters that they wish they had more time with their children. One way that these middle-class mothers cope with time pressures is by taking their toddlers to their yoga class. One thing that may be sacrificed, however, is mom's time for personal relaxation.

---

[5] The two-parent model is culturally potent enough that, in one instance, a Vermont judge threatened to rescind custody from a lesbian biological mother because she refused to allow her former partner access to their daughter (Ring 2009).

[6] As one woman said in response to welfare changes meant to encourage the work ethic, "It's funny that low-income women or women living in poverty can stay at home and be a bad role model, but a middle class mother or wealthy mother is not a bad role model when they stay home. It's like America has two standards and it's based on class" (in Gemelli 2008, p. 101).

cases, variations in income, education, stress, and depression levels largely explain child-outcome differences (Brush 2008; Crawford and Novak 2008). Not surprisingly, stress is less pronounced among single mothers with relatively higher education, fewer children, better jobs, and more personal resources (Lleras 2008).

Doing Fatherhood

Over the past decade, research on fathers has increased dramatically as policy makers and social scientists have concerned themselves with the importance of fathers to mothers' and children's lives. In general, these studies show that a father's involvement in his child's upbringing is related to positive cognitive, emotional, and behavioral outcomes from infancy into adolescence.[7] On the flip side of the coin, father absence has generally been associated with adverse effects on children's cognitive, moral, and social development (Bronte-Tinkew et al. 2008; Dermott 2008; J. Jones 2008; Mitchell, Booth, and King 2009).[8]

However, this situation is more complicated than suggested by these generalizations (Pan and Farrell 2006). Incidences of a father's substance abuse and of father-perpetrated partner violence and child abuse remind us that encouraging father contact is not always best for children (Blazei, Iacono, and McGue 2008; Lee, Bellamy, and Guterman 2009; Osborne and Berger 2009; Salisbury, Henning, and Holdford 2009). Furthermore, **social fathers** (nonbiological fathers in the role of father, such as stepfathers) do not seem to improve *adolescents'* outcomes when compared with living in a single-mother household (Bzostek 2008). Indeed, a mother's male relatives—for example, the child's uncle or grandfather—may be better and more reliable parent figures than a romantic partner (Jayakody and Kalil 2002). Nevertheless, research does show that, compared to growing up in a single-mother home, younger children benefit economically from living with a social father, provided that he shares his financial resources with the family (Manning and Brown 2006).

---

[7] Not only is father involvement often good for children, but it can also benefit the broader community. There is evidence that fatherhood changes a man toward greater altruism for the rest of his life. Based on data from the National Survey of Families and Households (NSFH), researchers found that, compared with men who have never been fathers, middle-aged men of all social classes who at some point in their lives had become fathers and were highly engaged with children were significantly more likely to have altruistically oriented social relationships and to be involved in service organizations (Eggebeen, Dew, and Knoester 2009).

[8] We note here that "father absence" can be other than simply residential absence and can, even among married fathers, include psychological absence, or indifference with minimal positive father-child interaction. Of course, residential fathers have more opportunity to develop psychological presence than do nonresident fathers (Krampe 2009).

Married fathers are increasingly invested in their children's daily lives as they engage in breadwinning, planning, sharing activities, and teaching their children (Gaertner et al. 2007; Whitehead and Popenoe 2008). Children's interaction with fathers often differs from that with their mothers as fathers more typically play with or engage in leisure activities with their children than do mothers (Cancian and Oliker 2000). Although we tend to stereotype low-income fathers of color as unmarried and absent, interviews with young African American and Latino fathers in New York City uncovered married fathers who were actively involved in their children's upbringing. Better educated fathers with more satisfying jobs showed higher levels of parental engagement (Wilkinson et al. 2009). Experiencing high levels of workplace stressors, including low levels of employee self-direction, adds to fathers' stress, resulting in less effective parenting (Goodman et al. 2008).

Married-Couple Families with a Stay-at-Home Father In 2008, about 140,000 married-couple families had a stay-at-home father (U.S. Census Bureau 2010b, Table 68)—a situation that has since increased with men's job losses during the recession that began in 2008 (Bar 2009; O'Reilly 2009). Some of these men have been laid off or closed family-owned businesses and remain out of the workforce (Aasen 2009; Kershaw 2009a). Others have wives who earn more than they could and question sending their children to day care when the father could stay home (St. George 2010; J. A. Smith 2009).

Fathers' responses to these situations tell us something about the joys of daily experiencing the little things that growing children say and do (Swager 2009). Their responses also point out the relative lack of status associated with parenting, at least in some circles. For example, one stay-at-home father, recently laid off from a Fortune 500 company, told a *New York Times* reporter, "To go from the 24-7, high-end, deal-making prestige of working for places that are written about in newspapers to this, it took a long time to get comfortable. . . . It's humbling " (Kershaw 2009a).

Single Fathers

Compared to mothers, the proportion of fathers who serve as the principal parent is dramatically small. Among families with children under age eighteen, about 4 percent are single-father families—5 percent for blacks and non-Hispanic whites, 3 percent for Hispanics and Asians. About half of these nearly 2 million fathers are never married. A significant proportion are divorced, and a much smaller fraction, widowed (U.S. Census Bureau 2010b, Tables 66 and 67). The majority of single fathers care for just one child, but some are parenting three or more (U.S. Census Bureau 2010b, Table 64).

Single fathers typically assumed their role because they "stepped up" in difficult and unforeseen circumstances: "You gotta do what you gotta do" (in Coles 2009). In some cases, single fathers have considered their relationship with their own fathers, learned from their past, and want to do things differently. As one single, black father said, "A lot of people take their father not being there when they were young as a bad thing. But I just took the good out of it and took what he did do and took what I'm not going to do like him" (in Coles 2009, p. 1,328). Single fathers often know that extended family support is available, but may not rely on it: "I call my sister occasionally for advice, but I have a strong autonomous streak in me. I'd rather do it myself" (in Coles 2009, p. 1,329).

Whether single or married, poor or financially better off, fathers as primary parents report fighting stereotypes as odd, unmasculine, or weak (Troilo and Coleman 2008). Watched and evaluated as parents, they feel that they have to prove themselves capable (Coles 2009). In response, single fathers have organized various support groups (e.g., www.athomedad.org).

Nonresident Fathers

Nonresident fathers are biological or, much less often, adoptive fathers who do not live with one or more of their children. Less true for divorced fathers, never-married nonresidential fathers move in and out of their children's lives (Amato, Meyers, and Emery 2009; Roy, Buckmiller, and McDowell 2008). Due to multipartnered fatherhood, a father may be living with one or more of his biological children but "nonresident" with regard to others. Then too, a nonresident father may be serving as a social father to one or more children whom he did not conceive, usually because he lives with a woman who had at least one child from a previous relationship (McMahon et al. 2007).

Although often stereotyped as "absent" and disinterested (Troilo and Coleman 2008), many nonresident fathers express love and genuine concern for their children: "These two guys . . . are the reason I live, you know" (in Coles 2009, p. 1,334; also see Wilkinson et al. 2009). A study of fifty nonresident fathers who had previously been arrested for drug problems found that many saw their children daily, several times a week, or weekly (McMahon et al. 2007). Indeed, the majority of nonresident fathers maintain some presence in their children's lives and provide them with various kinds of practical support, at least while the children are young. Some economically disadvantaged fathers take on significant child care responsibility as a vehicle for expressing their contribution to the family's well-being (Amato, Meyers, and Every 2009; England and Edin 2007). Researchers have found that a nonresident father's being involved depends on his employment status, age, education,

religious participation, and substance abuse history as well as on his family background (Goldscheider et al. 2009; Knoester, Petts, and Eggebeen 2007; McMahon et al. 2007). One recent study shows that nonresident fathers are more involved when their child is male (Bronte-Tinkew and Horowitz 2010).

In addition, a nonresident father's involvement largely depends upon his relationship with his child's mother and, to a lesser extent, her extended family (Guzzo 2009b; Marsiglio 2008; Ryan, Kalil, and Ziol-Guest 2008). He tends to be more highly involved when his relationship with his child's mother is generally without conflict (Jackson, Choi, and Franke 2009). Involvement is enhanced when the father has been involved prenatally, possibly because he assumes an identity as father during the prenatal period (Cabrera, Fagan, and Farrie 2008a; Doherty 2008; Marsiglio 2008). The next section explores what children need, regardless of the family form in which they reside.

What Do Children Need?

Children of all ages need encouragement, adequate nutrition and shelter, parental interest in their schooling, and consistency in rules and expectations. Parental guidance should be congruent with the child's age or development level (Barnes 2006; Mental Health America 2009).

Children's Needs Differ According to Age

Children's needs differ according to age. Infants need to bond with a consistent and dependable caregiver. To develop emotionally and intellectually, they need affectionate, intimate relationships as well as conversation and variety in their environment. Discipline is never appropriate for babies because they cannot understand its purpose and are unable to change their behavior in response (Brazelton and Greenspan 2000; Hall 2008).

Preschool children need opportunities to practice motor development as well as wide exposure to language, especially when people talk directly to them (Cowley 2000). They also need consistent, clear definitions of what behavior is unacceptable (Del Vecchio and O'Leary 2006; Dorman 2006). School-age children need to practice accomplishing goals appropriate to their abilities and to learn how to get along with others. To better accept criticism as they get older, they need realistic feedback regarding task performance—neither exaggerated praise nor aggressive criticism. They also need to feel that they are contributing family members by being assigned tasks and taught how to do them (Dinkmeyer, McKay, and Dinkmeyer 1997; Hall 2008; Jayson 2005b).

Although the majority of teenagers do not cause familial "storm and stress" (Kantrowitz and Springen 2005), the teen years do have the special potential for reducing marital quality and sparking parent-child conflict (Cui and Donnellan 2009; Whiteman, McHale, and Crouter 2007). As they search for identity and begin to define who they are and will be as adults, adolescents need firm guidance, coupled with parental accessibility and emotional support (Guilamo-Ramos et al. 2006; D. Walsh 2007). Teens also need to learn effective methods for resolving conflict (Tucker, McHale, and Crouter 2003).

Despite stereotypical ideas to the contrary, influences from teens' peers are not necessarily negative and can, in fact, be positive (Hall 2008). Furthermore, parents can and do influence their teenagers' behavior (Dillon et al. 2008; Bersamin et al. 2008; Longmore et al. 2009). It's important for parents to remember "the obvious fact that most adolescents make it to adulthood relatively unscathed and prepared to accept and assume adult roles" (Furstenberg 2000, p. 903). Regardless of age, children have been shown to benefit from an authoritative parenting style (Junn and Boyatzis 2005).

Experts Advise Authoritative Parenting

Parents gradually establish a *parenting style*—a general manner of relating to and disciplining their children. As shown in Table 10.1, we can distinguish among authoritarian, permissive, and authoritative parenting styles (Baumrind 1978; Maccoby and Martin 1983). The **authoritarian parenting style** is low on emotional warmth and nurturing but high on parental direction and control. The authoritarian parent's attitude is, "I am in charge and set/enforce the rules, no matter what" (Gaertner et al. 2007). Parents who employ this style are more likely to spank their children or use otherwise harsh punishment (Grogan-Kaylor and Otis 2007). Unnecessarily high parental direction or control has been associated with a child's decreased sense of personal effectiveness or mastery over a situation, even among children as young as four (Moorman and Pomerantz 2008).

The children in this family have different needs that correspond with their varied ages. Meanwhile, all children need encouragement along with consistent parental expectations and rules. Authoritative parents are emotionally involved with their children, setting limits while encouraging them to develop and practice their talents.

Karan Kapoor / Getty Images

Table 10.1 Parenting styles

		Parental Warmth	
		Low	High
Parental Monitoring	**High**	Authoritarian	Authoritative/Positive
	Low	Permissive-Emotional Neglect	Permissive-Indulgent

Parenting styles combine two dimensions: (1) parental warmth, and (2) parental expectations, coupled with monitoring of their children. When both warmth and monitoring are high, parents are said to exhibit an authoritative parenting style. At least for white, middle-class children, research consistently shows that an authoritative parenting style is the most effective of the four possible styles.

The **permissive parenting style** gives children little parental guidance. Although low on parental direction or control, permissive parenting may be high on emotional nurturing—a situation, characterized as *indulgent*, that leads to the classic "spoiled child." A second variant of the permissive style is low on *both* parental direction *and* emotional support—a situation of *emotional neglect*. Authoritarian and permissive parenting styles are associated with children's and adolescents' depression and otherwise poor mental health, low school performance, behavior problems, high rates of teen sexuality and pregnancy, and juvenile delinquency (Hall 2008; Waldfogel 2006).

Child psychologists prefer the **authoritative parenting style**, sometimes called *positive parenting*. This style, characterized as warm, firm, and fair, combines emotional nurturing and support with conscientious parental direction (although not excessive control). Authoritative parents would agree with the statements, "I consider my child's wishes and opinions along with my own when making decisions," "I value my child's school achievement and support my child's efforts," and "I expect my child to act independently at an age-appropriate level" (Manisses Communications Group 2000, p. S1). Authoritative parenting involves encouraging the child's individuality, talents, and emerging independence, while also consciously setting limits and clearly communicating and enforcing rules (Brooks and Goldstein 2001; Ginott, Ginott, and Goddard 2003).[9] Authoritative parents monitor their children's

activities and whereabouts while giving appropriate consequences for misbehavior when warranted (Waldfogel 2006).

Regardless of family structure, authoritative parents are more likely than others to have children who do better in school and are socially competent, with relatively high self-esteem and cooperative, yet independent, personalities (Crawford and Novak 2008; Fivush et al. 2009; Jackson-Newsom, Buchanan, and McDonald 2008). Positive effects of authoritative parenting last into adulthood (Schwartz et al. 2009). "A Closer Look at Family Diversity: Parenting LGBT Children" asks you to consider how an authoritative parenting style would apply to this situation.

When two parents are involved, their collaboration or working together renders them more effective, especially when both parents use the authoritative parenting style (Feinberg, Kan, and Hetherington 2007; Kjobli and Hagen 2009; Simons and Conger 2009). An interesting study of two-parent Mexican American families found that children and their parents were happier when parents agreed with and supported one another (Formoso et al. 2007). Collaborative parenting reduces stress and enhances parents' feelings of competence (Jackson, Choi, and Franke 2009), as well as partners' relationship satisfaction (Ehrenberg et al. 2001; Kluwer, Heesink, and Van de Vliert 2002). A sample of married and cohabitating working-class parents of first graders found that families in which both parents practiced an authoritative parenting style were most effective in raising well-adjusted children. Families in which only one parent used an authoritative parenting style were more effective than those in which neither parent did (Martin, Ryan, and Brooks-Gunn 2007; Meteyer and Perry-Jenkins 2009).

Psychological Control versus Authoritative Parenting As opposed to direct forms of parental control, such as asking direct questions, explicitly stating expectations, and giving time-outs or denying privileges, **psychological control** involves using manipulative strategies such as inducing guilt or withdrawing signs of affection. Rather than conveying unconditional love for the child while correcting misbehavior, psychological control relies on negative cues, such as refusing to acknowledge the child.

The underlying message is that the child's behavior has hurt the parent's feelings. To reduce tension, the child is expected to comply with the parent's wishes. At least among Latino and European American adolescents, and among African American girls, psychological control has been found to hinder children's emergent sense of agency or mastery over their behavior and future goal attainment. Depression can result (Bean and Northrup 2009; Mandara and Pikes 2008; Soenens, Vansteenkiste, and Sierens 2009).

[9] Limits are best set as house rules and stated objectively in third-person terms. A parent may say, for example, "The time to be home is 10 o'clock." With preschoolers, limits need to be set and stated very clearly: A parent who says, "Don't go too far from home" leaves "too far" to the child's interpretation. "Don't go out of the yard at all" is a wiser rule.

A Closer Look at Family Diversity

Parenting LGBT Children

Parenting a child who is lesbian, gay, bisexual, or transsexual (LGBT) can be not only pleasurable and broadening but also emotionally challenging ("For Parents . . ." 2004). When a lesbian, gay male, or bisexual child "comes out," or discloses his or her identity to family members, the parent may feel confused, ambivalent, alone, embarrassed, and/or angry (Martin et al. 2009). Even parents who view themselves as progressive, readily accepting GLBT friends and acquaintances, may be surprised at their feelings of disappointment and grief upon finding out that their child is lesbian, gay, or bisexual. If homosexuality is against the parent's religion, a child's coming out can be even more disconcerting. It helps to recognize the following facts:

- All cultures and historical periods include individuals who have identified themselves as gay, lesbian, bisexual, or transgender.

- The American Psychological Association and the American Pediatric Association do not consider being GLBT as a psychological disorder.

- GLBT adolescents and young adults may feel guilty about their sexual orientation, worried about responses from their families and friends, and fearful of discrimination in clubs, sports, college, or the workplace ("Gay and Lesbian Adolescents" 2006).

- Hiding one's sexual orientation can be extremely stressful and isolating. Adolescents and adult children come out because they want to live their lives openly and honestly without deception ("Questions and Answers" n.d.; Wright and Perry 2006).

- The physical and mental health of GLBT youth is better when they feel social support (Wright and Perry 2006).

- Parents' acceptance of their child's sexual identity allows for open discussions about the child's dating relationships and related issues, such as ways to deal with prejudice and discrimination or how to reduce risks associated with HIV/AIDS ("For Parents . . ." 2004).

Experts advise that, whatever a parent's feelings when a child comes out, the child needs assurance that she or he is loved just as much as before:

Can you imagine the feelings of a youngster who bravely tells his or her parents that they are gay only to be confronted by an anger which may be so severe that they are put out of the house, or told that he or she has brought shame to the family? Yet this happens and it is a fact that gay people occasionally commit suicide because they have been so badly ostracized and made to feel alienated. ("If Your Child . . ." 2008).

As parents work through their feelings, they may want to talk their situation over with others. Professionals urge parents not share the information without their child's consent. An exception involves talking with a counselor. Other resources are available as well. PFLAG (Parents, Families, and Friends of Lesbians and Gays) is a national organization comprised of local educational and support groups, as well as Internet resources.

Critical Thinking

How might the stress model of effective parenting be applied to this situation of a GLBT child's coming out to her or his parents? How might an authoritarian parent's reaction to their child's coming out differ from the response of an authoritative parent?

Before leaving this discussion, we need to note that some scholars view the authoritarian/permissive/authoritative model as biased. For instance, one study shows that authoritative parenting is a more important prediction of behavior for children of European descent than for the Hmong (Supple and Small 2006). Some scholars argue that the model is ethnocentric, or Eurocentric—that is, it uses European, white, middle-class beliefs about parenting as the standard to which all others are compared—often unfavorably (Farver et al. 2007). This point is developed throughout the section "Racial/Ethnic Diversity and Parenting," later in this chapter. The next section focuses on the question of whether spanking is ever appropriate.

Is Spanking Ever Appropriate?

Spanking refers to hitting a child with an open hand without causing physical injury. In 2007, a California state legislator proposed legislation that would have outlawed spanking children under three years old. Had the proposal become law, California would have joined the approximately fifteen European nations in which spanking children is illegal (Straus 2007). However, the California-based suggestion met enough negative media response that the idea was dropped, and the proposed bill was revised to involve only more serious forms of corporal punishment, such as kicking or hitting a child with an implement (Steinhauer 2007). Whether spanking children is ever a good idea is controversial (Coombs-Orme and Cain 2008; Larzelere 2008).

Analysis of data from the 13,000 respondents in the National Survey of Families and Households found that about one-third of fathers and 44 percent of mothers had spanked their children during the week prior to being interviewed. Boys, especially those under age two, are spanked the most often. Children over age six are spanked less often, but some parents spank their children during early adolescence (Guzzo and Lee 2008).

Mothers spank more often than fathers. Younger, less-educated parents in households with more children and less social support, parents who argue a lot with their children, sociopolitical conservatives, those with a fundamentalist religious orientation, and parents who live in relatively violent neighborhoods are more likely to spank (Button 2008; Ellison and Bradshaw 2009). The stress of first-time parenthood is associated with spanking (Guzzo and Lee 2008). One study found that single mothers who become more seriously involved in a romantic relationship, whether with the biological father of their child or not, are more likely to spank their children. The authors speculated that the mothers experienced increased strains as they incorporated the child into their developing relationship (Guzzo and Lee 2008).

A leading domestic violence researcher, sociologist Murray Straus (2007) advises parents never to hit children of any age under any circumstances. At least among European American children, being frequently spanked in childhood is linked to later behavior problems, as well as to depression, suicide, alcohol or drug abuse, physical aggression against one's parents in adolescence, and later to abusing one's own children and to intimate partner violence (Berlin et al. 2009; Lansford and Dodge 2008). Straus has argued that spanking teaches children a "hidden agenda"—that it is all right to hit someone and that those who love you hit you. This confusion of love with violence sets the stage for domestic violence. Then, too, especially when a parent spanks in anger—which is never advised—"spanking can escalate and apparently does mix in with more severe hitting" (Kazdin and Benjet 2003, p. 102; Roberto, Carlyle, and Goodall 2007). Infants and babies under two years old should never be spanked (Coombs-Orme and Cain 2008). Spanking or vigorously shaking an infant can lead to permanent damage, even death.

Meanwhile, some researchers contend that Straus and his colleagues may be overstating the case (Saadeh, Rizzo, and Roberts 2002). For instance, psychologist Marjorie Gunnoe (cited in S. Gilbert 1997) theorizes that spanking is most likely to have negative results for children only when they perceive being spanked as an aggressive act. She hypothesizes that children under age eight tend to think it is their parent's right to spank them. However, recent research does show that corporal punishment has generally negative effects (Christie-Mizell, Pryor, and Grossman 2008; Mulvaney and Mebert 2007).

The American Academy of Pediatrics advises that children under two years old and adolescents should *never* be spanked and recommends that parents learn disciplinary methods other than spanking (American Academy of Pediatrics 1998). Straus has argued that spanking usually accompanies other, more effective discipline methods such as explaining or denying privileges. These nonspanking discipline methods are effective by themselves, and parents should be encouraged to follow the principle, "just leave out the spanking part" (Straus 1999a, p. 8; Straus 2007). In general, middle- and upper-middle-class parents are less likely than low-income parents to spank. Low-income parents "may be unaware of current American Academy of Pediatrics policy recommendations about spanking. Or they may consciously disagree with them" (Guzzo and Lee 2008).

Social Class and Parenting

There are effective and ineffective parents in all social classes (Jackson, Choi, and Bentler 2009). Meanwhile, this section examines some ways that social class impacts parental alternatives and choices. You'll recall a theme of this text: Decisions are influenced by social conditions that expand or limit one's options. Virtually all opportunities and experiences, or *life chances*, are influenced by **socioeconomic status (SES)**—one's position in society, measured by educational achievement, occupation, and/or income. Parenting is no exception (Furstenberg 2006; Lareau 2006).

Research shows that family education and income have more influence on parenting behaviors and children's outcomes than do race/ethnicity or family structure in and of itself (Gibson-Davis 2008). We have seen that parents who are less stressed and relatively content practice more positive child-rearing behaviors (Burrell and Roosa 2009; Gibson-Davis 2008). Reduced stress and emotional well-being, in turn, are statistically correlated with higher socioeconomic status.

Middle- and Upper-Middle-Class Parents

In this climate of economic uncertainty, even middle- and upper-middle-class parents with relatively high education have suffered layoffs, salary reductions, and reduced health and retirement benefits (Coy, Conlin, and Herbst 2010). Having already tightened their belts, some have trouble paying their bills (Pew Research Center 2009b). Nevertheless, compared with lower-SES parents, those with higher income can better afford to provide for their children's needs and wants. For instance, more than 90 percent of households with annual incomes of $100,000 have Internet access, compared with 50 percent of those making between

Parents' and children's life experiences are significantly related to their socioeconomic status. Middle- and upper-middle-class parents tend to emphasize concerted cultivation of their children's development and talents.

$25,000 and $35,000 (U.S. Census Bureau 2010b, Table 1118). Then too, higher-SES parents have the resources to hire household help or purchase devices such as baby-monitoring equipment that might help with parenting (Knoester, Haynie, and Stephens 2006; Nelson 2008).[10] Furthermore, they reside in neighborhoods conducive to successfully raising and educating their children.

More highly educated parents have fewer children on average (U.S. Census Bureau 2010b, Table 92) and are likely to emphasize **concerted cultivation** of their child's talents and overall development. According to this parenting model, they more often praise their children; play with or talk to them "just for fun"; read to them; create and enforce rules about watching television; engage their children in extracurricular lessons, clubs, and sports; take them on outings; enroll them in private or charter schools; and say that there are people in the neighborhood whom they can count on (U.S. Census Bureau 2009h, Tables D5-D30).

More so than in low-income neighborhoods, middle- and upper-class children are likely to have neighborhood and school friends who share their parents' values and can therefore serve as parallel socialization agents (Hall 2008). Then too, volunteering at their children's schools, and monitoring other students' and even teachers' behavior, highly educated parents secure educational advantages for their children. Should problems arise at school, higher-SES parents are likely to have network contacts with community professionals who can help and may even challenge school officials' decisions (Hassrick and Schneider 2009).

Higher-SES parents are likely to get parenting information from professional sources such as books or the Internet (Radey and Randolph 2009). Often using the authoritative parenting style, they negotiate with their children in ways meant to foster language and critical thinking skills, self-direction, initiative, and self-advocacy (Lareau 2006; Shinn and O'Brien 2008). This parenting model well prepares children for success in the broader society because schools and professions value the self-direction, critical thinking, and self-advocacy that these children learn at home. But can parents take concerted cultivation too far?

Hyperparenting—The "Hurried Child" and "Helicopter Parents" According to some observers, many higher-SES parents engage in **hyperparenting**. Dubbed "helicopter parents," they hover above, meddling excessively in their children's lives.[11] Typically, they scramble to be "perfect," providing their children with more than is either necessary or beneficial (Warner 2006). Some parents have become vulnerable to marketing that prods "good" parents to spend unnecessarily large amounts of money on baby gear and toys, among other things (Deveny 2007; Purcell 2007). For many, excessive spending on their children cuts into their ability to save for emergencies, retirement, or college (Paul 2008).

Critics warn that many higher-income parents not only give their children too much but also engage

---

[10] Ironically, although possibly enhancing parental freedom, baby-monitoring devices can also increase anxiety levels because they encourage defining the infant as extremely fragile (Nelson 2008). Can you think of other parental aids that might have similar effects?

[11] For example, some nutritionists are concerned that many parents have become overzealous, even obsessive, in efforts to engender good eating habits in their children—not allowing a child to eat cake at a birthday party, for instance ("What's Eating Our Kids?" 2009). And camp counselors report being contacted by anxious parents because their child looked sleepy or unhappy in a group photo ("Helicopter Parents" 2007). College personnel report being emailed by irate parents because there was too much salt in the cafeteria chicken or their undergraduate's roommate ate their ramen noodles ("Mom Needs an A" 2007).

them in too many scheduled activities—private lessons, extracurricular activities associated with school or church, and organized recreational programs. "If the hearth was the center of the home in the 19th century, the calendar is really the center of many middle class homes" (Lareau, in "Class in America" 2003). Expecting achievement in these endeavors, parents may place too many demands on their children, even encouraging them to compete for places in the most preferred preschools, for example (Saulny 2006). Filling their own needs while believing that they are acting in the best interest of their children, these parents may be determined to raise "trophy kids" (Perrow 2009).

Although there is some evidence that tightly scheduling a child may not be harmful (Cloud 2007), developmental psychologist David Elkind (1988) warned a generation ago that "scheduled hyperactivity" (Kantrowitz 2000) can produce the "hurried child," not to mention frazzled parents (Elkind 2007a, 2010; Warner 2006). The over-scheduled, or "hurried child" is denied free playtime while encouraged to assume too many challenges and responsibilities too soon (Elkind 2007a, 2007b).[12] Hurried children suffer the stress induced by the pressure to accomplish (Anderson and Doherty 2005). Or they may abandon goal-directed academic and/or extracurricular activities (Su 2007).

As the hurried child enters young adulthood, some helicopter parents meddle inappropriately in their college student's educational experiences or attempt to negotiate job offers, salary, and benefits for their offspring (Armour 2007; Blanck 2007; "Mom Needs an A" 2007). Some individuals who see themselves in negative descriptions of helicopter parents find the criticism unfair, especially in this unsettling economy and because they pay considerably for their children's education.

Nevertheless, psychologists warn that helicopter parents risk turning their children against them. Furthermore, excessive hovering denies youth opportunities to develop self-confidence or problem-solving skills (Su 2007). Some parents have created an organization called "Putting Family First" (www.puttingfamilyfirst.org) that supports parents who want to reduce their children's involvement in organized activities and have more unscheduled family time (Kantrowitz 2000).

Compared with life in higher-SES families, "childhood looks different" in families of lower socioeconomic status where children grow up with less supervised play, fewer scheduled activities, and more

"So many toys—so little unstructured time."

unplanned interaction with extended family members and friends (Lareau 2003a). Sociologists Elliot Weininger and Annette Lareau (2009) have noted an inconsistency. Higher-SES parents determine to promote self-direction in their children but exercise subtle forms of control that can undermine their intent. Conversely, lower-SES parents tend to espouse obedience but often grant children considerable autonomy (as they play outside unsupervised, for example), thereby limiting emphasis on conformity.

Working-Class Parents

Working-class parents work at construction, manufacturing, repair, installation, and service jobs such as health care assistants, that require at least a high school education and pay higher than minimum wage. More so than with higher-SES parents, working-class parents have suffered the negative effects of declining factory work and union power; decreased wages and benefits; insecurities associated with temporary work; and escalating housing, utilities, and transportation costs.

Working-class parents do not necessarily view the concerted development parenting model as good parenting. In fact, they may view this model as negative, creating demanding children (Guzzo and Lee 2008). Instead, they tend to follow the **facilitation of natural growth parenting model**, according to which children's abilities are allowed to develop naturally (Lareau 2006). Some working-class parents employ the authoritative parenting style. Nevertheless, much parent-child communication tends to be authoritarian, emphasizing obedience and conformity and less often eliciting children's feelings or opinions (Lareau 2003a, 2006). Working-class parents are more likely to tell their children what to do rather than trying to persuade them with reasoning. When dealing with professionals (e.g., doctors, religious clergy, teachers, public officials), working-class

---

[12] A related issue involves schools' "hurrying" of children with the increased use of standardized testing and decreased emphasis on music, art, or time for spontaneity and play (Cloud 2007; Trudeau 2006; Tyre 2006a).

parents are likely to encourage their children to keep their thoughts and questions to themselves (Shinn and O'Brien 2008).

Many working-class parents are involved in their children's schools and do promote academic success in their children (Cooper, Crosnoe, et al. 2009; Woolley and Grogan-Kaylor 2006). However, the natural growth parenting model, coupled with the authoritarian parenting style, does not correspond well with the middle-class culture and expectations of schools and professions. Although higher-SES children appear to gain a sense of entitlement, working-class children are likely to grow up with feelings of discomfort, constraint, and distrust regarding their school and work experiences (Lareau 2003a; Lucas 2007). For children from working-class families who embark upon professional careers, this sense of not fitting in can persist (Lubrano 2003).

Low-Income and Poverty-Level Parents

The majority of low-income and poverty-level parents work at minimum- or less-than-minimum-wage jobs with irregular and unpredictable hours and no employer-subsidized medical insurance or other benefits. Because more and more low-income jobs are part-time and also because working even full-time at minimum wage does not pay enough to live above poverty level, many low-income parents have two or three jobs (Ames, Brosi, and Damiano-Teixeira 2006). Analysis of data from the National Longitudinal Survey of Youth (NLSY) shows that single mothers who work part-time in low-wage jobs with nonstandard hours generally raise their children in poorer-quality home environments (Lleras 2008).[13] Irregular work schedules in low-wage jobs with little autonomy and high supervisor surveillance, coupled with housing or neighborhood troubles as well as financial worries, cause stress.

[13] Specifically, what does it mean to say that a low-income single mother's home environment is less than desirable? In the National Longitudinal Survey of Youth (NLSY), *home environment* is operationally defined (see Chapter 2) using measures of cognitive stimulation, maternal responsiveness, and safety of the physical environment. For three- to six-year-olds, cognitive stimulation involves eight measures that are whether the parent helps the child learn (1) numbers, (2) shapes, (3) colors, and (4) the alphabet; (5) whether the mother reads to the child at least three times a week; (6) whether the child has ten or more books; (7) whether the family gets at least one magazine regularly; and (8) whether there is a record or tape player in the home. Maternal responsiveness involves four measures that are whether the mother (1) talks to the child, (2) kisses or hugs the child, (3) answers the child's questions verbally, and (4) voices positive feelings about the child. Physical environment has four items that are whether (1) the play environment appears safe, and whether the rooms are reasonably (2) clean, (3) lighted, and (4) uncluttered (Lleras 2008).

Living in rented homes, apartments, or motel rooms, they struggle with rent burdens, utility payments, and housing instability (Berger, Heintze, et al. 2008; Ehrenreich 2001). Moreover, many poor families move from city to city to live with relatives or to search for jobs. This situation makes it difficult for a parent to establish support systems and hinders children's chances for school continuity and success (Molyneux 1995). In sharp contrast to higher-SES parents, many low-income parents struggle to give their children a few "extras," such as a "respectable" birthday party, school field trips, or a high school class ring (Lee, Katras, and Bauer 2009; Mistry et al. 2008).

With fewer resources, lower-income parents are less likely to live in neighborhoods that value education or encourage high achievement (Coles 2009; England and Edin 2007; Henry et al. 2008). In fact, items that middle-class Americans take for granted, such as relatively safe, gang-free neighborhoods, are often unavailable (Loukas et al. 2008), and parental control is more difficult to achieve in neighborhoods characterized by antisocial behavior (Gayles et al. 2009; Moore et al. 2009).

Furthermore, poverty-level families are more likely than others to live with air pollution, and to have poorer nutrition, more illnesses such as asthma, schools that are less safe, and limited access to quality medical care (Downey and Hawkins 2008; Seccombe 2007). Children living in poverty—more often disabled or chronically ill than other children—have expensive health care needs that welfare or other social services do not always or completely cover (Cohen and Bloom 2005; Levine 2009).

About 8 percent of children who are raised in poverty (compared with about 5 percent of children raised in families that are not poor) have emotional or behavioral difficulties (Simons et al. 2006, p. 3; Teachman 2008b). "Facts about Families: Marriage and Children in Poverty," in Chapter 7, further describes children's outcomes that result from growing up in poverty.

Homeless Families Over the past three decades, extreme poverty, a shortage of affordable housing, job erosion, home foreclosures, declining public assistance, lack of affordable health care, domestic violence, substance addiction, and mental illness have helped to create a significant number of homeless families—a phenomenon that would have been unthinkable forty years ago (National Coalition for the Homeless 2009c).

Families with children are among the fastest-growing segments of the homeless population, a situation that has become more pronounced since the beginning of the recession that began in 2008. Partly as a result of changes to welfare laws, described earlier in this chapter, approximately 40 percent of the homeless are mothers and children, with children under age eighteen making up about one-quarter of the homeless (Seccombe 2007). Fathers, some of whom are single parents, are also found among the homeless (National Coalition for the Homeless 2009a, 2009c).

Homeless parents, especially those who have been without housing for a longer time, move often and have little in the way of a helpful social network ("Homeless Families with Children" 2001). Getting children to school and supervising their homework—not to mention being actively involved in a child's classroom—become extremely difficult for homeless parents. Although families benefit from entering shelters, life in a homeless shelter is itself stressful. Some shelters require that the family leave during the day, regardless of the weather: "How can this mother go out and look for a job or even look for a place to live when she's got three kids, and it's raining, or it's cold?" Problematic rules involve bedtimes, mealtimes, keeping children quiet, and the requirement that children be with their parents at all times. Other stressors occur as well. For example, one mother told of a single male resident's "getting fresh with my older girl" (Lindsey 1998, p. 248).

Social class may be more important than race/ethnicity in terms of parental values and interactions with children (Lareau 2003b, 2006). Middle-class parents of all racial/ethnic groups are more alike than different, and so are poverty-level parents. Upper-middle-class black parents perform their role differently than do working-class black parents or those living below

poverty level (Peters 2007). At the same time, social scientists do look at how various ethnic groups evidence culturally specific parenting behaviors (Cohen, Tran, and Rhee 2007). The major focus of the following section is on parenting behaviors and challenges that are specific to various racial, ethnic, and religious minorities.

Racial/Ethnic Diversity and Parenting

As a beginning, we need to note two factors. First, there is considerable overlap among class and racial/ethnic categories. For instance, although the upper middle class now includes substantial numbers of people of color, particularly Asians, it is still largely non-Hispanic white. Many African American and Hispanic families are now solidly middle class, but these race/ethnic groups remain overrepresented in low-income and poverty categories. A second factor to note is that there is considerable ethnic diversity *within* the following groups. For instance, Asian Americans include a broad range of ethnicities, including Chinese, Japanese, Korean, Vietnamese, and Asian Indians, among others.

African American Parents

Evidence suggests that African American parents' attitudes, behaviors, and hopes for their children are similar to those of other parents in their social class (Peters 2007). Nevertheless, the impact of race remains important. For instance, even when social class is taken into account, it appears that African American mothers (but not fathers) are more likely than European Americans to spank their children. However, spanking may not have the same negative effects on black children as it does on European American children. Among blacks, physical punishment is more acceptable and hence more likely to be viewed by both the parent and the child as an appropriate display of maternal warmth and positive parenting (Jackson-Newsom, Buchanan, and McDonald 2008). It follows that comparing African American parents to other ethnic groups can be seen as Eurocentric (Dodson 2007).

John Moore/Getty Images

Families benefit from homeless shelters, but living there can be stressful in itself. Stress-inducing regulations involve bedtimes and mealtimes, along with the expectation that children be quiet or with their parents at all times.

Chris Hondros/Getty Images

As with other race/ethnic minorities, African Americans' parental attitudes and behaviors are similar to other parents in their socioeconomic status (SES). Nevertheless, the intersection of gender and race with SES means that this father, culturally expected to be an effective breadwinner, also risks race discrimination as he navigates a job search in this recession economy. He is pictured here at a New York State employment services office in Brooklyn.

Besides putting up with research findings that are possibly biased against them, even higher-SES African Americans remain vulnerable to discrimination (Lacy 2007; Welborn 2006). So simple a matter as buying toys becomes problematic. Black dolls only? Should the child choose? What if the choice is a white Barbie doll? For middle-class African American parents, forging a unique *black middle-class* identity and then instilling this identity into their children is a major undertaking (Lacy 2007).

Native American Parents

Native American parents have been described as exercising a permissive parenting style that some critics have viewed as bordering on neglectful. However, describing Native American parenting in this way smacks of Eurocentrism (Seideman et al. 1994). Traditionally, Native American culture has emphasized personal autonomy and individual choice for children as well as for adults. Before the arrival of Europeans and for some time thereafter, Native Americans successfully raised their children by using nonverbal teaching examples and "light discipline," possibly coupled with "persuasion, ridicule, or shaming in opposition to corporal punishment or coercion." Native Americans continue to respond with warmth to their children's needs and also to "respect children enough to allow them to work things out in their own manner" (R. John 1998, p. 400; Seideman et al. 1994).

Given the problems of substance abuse and high teenage suicide rates documented among Native American youth, we might conclude that the traditional method of raising Native American children is no longer effective, due to changes in the broader society. However, valuing their cultural heritage, many Native Americans have been reluctant to assimilate into the broader society—and this reluctance may mean a rejection of the authoritative parenting style advised by European American psychologists. Meanwhile, the use of tribal elders "to help mitigate the loss of parental involvement and early nurturant figures in the lives of Native American adolescents" is important to many tribes (R. John 1998, p. 404).

Meanwhile, researchers have noted that Native American parents and children demonstrate resilience. For example, a longitudinal study of twenty-nine Navajo Native American mothers who as teenagers bore infants found that, twelve to fifteen years later, many had completed or gone beyond high school (Dalla et al. 2009). Moreover, although single-parent households do occur in Native American communities, the extended family serves as an instrument of group solidarity by reinforcing cultural standards and expectations, and lending practical assistance. With symbolic and actual leadership status in family communities, Native American grandparents often monitor grandchildren and may fully adopt the parenting role when necessary (Letiecq, Bailey, and Kurtz 2008).[14]

---

[14] However, despite the idea that the extended family may be strongly institutionalized among Native Americans and the resultant possibility that raising grandchildren might be less stressful for Native American grandparents, one study has found that Native American grandmothers raising their grandchildren were *more* depressed than were their European American counterparts (Letiecq, Bailey, and Kurtz 2008).

Hispanic Parents

Hispanic parents have been described as more authoritarian than their non-Hispanic counterparts. However, as with other racial/ethnic minorities, it may be that this description is Eurocentric and therefore inaccurate. The concept of **hierarchical parenting**, which combines warm emotional support for children with a demand for significant respect for parents, older extended-family members, and other authority figures, may more aptly apply to Hispanic parents. Hierarchical parenting is designed to instill in children a more collective value system rather than the relatively high individualism that European Americans favor (McLoyd et al. 2000).

This collectivism has been found to be functional. For instance, a recent study that compared Mexican American with European American parents in Southern California found that family cohesion (*familismo*) lessened the relationship between economic stress and negative parenting (Behnke et al. 2008; see also Martyn et al. 2009). Research shows similar positive effects of family cohesion for Vietnamese American as well as for families of other ethnicities, including European Americans (Lam 2005; Vandeleur et al. 2009).

Hispanic parents teach their children the traditions and values of their cultures of origin while often coping with a generation gap that includes differential fluency and different attitudes toward speaking Spanish (Gonzales et al. 2006; Pew Research Center 2009a; Smokowski, Rose, and Bacallao 2008). As in other bicultural families, intergenerational conflicts may extend into many matters of everyday life as the younger generation becomes more assimilated into U.S. culture. For example, "My mother would give me these silly dresses to wear to school, not jeans," complained a fifteen-year-old Mexican American female (Suro 1992, p. A-11). As Hispanic immigrant parents adjust to U.S. culture, they are likely to place less emphasis on *familismo* and to become increasingly permissive, the consequences of which can be detrimental to adolescents' behavior (Baer and Schmitz 2007; Driscoll, Russell, and Crockett 2008).

Asian American Parents

Despite often being portrayed as affluent, Asian Americans are found in lower socioeconomic strata and may experience economic hardship (Ishii-Kuntz et al. 2009). Nevertheless, compared to the average of 29.4 percent for all Americans over twenty-four years old, 52.6 percent of Asian Americans have completed four years or more of college (U.S. Census Bureau 2010b, Table 224). Ironically, meanwhile, the Asian American parenting style is often characterized as authoritarian (Greenfield and Suzuki 2001) and even hostile (McBride-Chang and Chang 1998). Research findings are mixed regarding the success of the expert-preferred authoritative parenting style among Asian Americans, again suggesting that the model may be Eurocentric (Pong, Johnston, and Chen 2010).

Social scientists have offered an alternative parenting concept, the **Confucian training doctrine**. This parenting model is named after the sixth-century Chinese social philosopher Confucius, who stressed (among other things) honesty, sacrifice, familial loyalty, and respect for parents and all elders. The Confucian training doctrine blends parental love, concern, involvement, and physical closeness with strict and firm control, or "training" (Chao 1994; McBride-Chang and Chang 1998). "Training" may involve parents' use of guilt, shame, and moral obligation to control their children's behavior (Farver et al. 2007). However, the extremely strong ties between Asian American parents and their children have been found to lessen parent-child conflict, along with potential negative effects of shaming (Benner and Kim 2009; Park, Vo, and Tsong 2008).

Like other ethnic minorities, Asian Americans have suffered from discrimination (Lau, Takeuchi, and Alegria 2006; Tong 2004). Moreover, Asian immigrant parents may face conflicts with their children when expecting traditional behavior characteristic of the homeland while their children assimilate into the American culture and no longer adhere to traditional expectations regarding dating, for example, or marital monogamy (Ahn, Kim, and Park 2008; Farver et al. 2007). Then too, Asian American youth must contend with high expectations created by the stereotyping of Asians as a "model," or "super" minority (Abboud 2006).

Parents of Multiracial Children

According to the 2000 census, which was the first to offer citizens the option of identifying themselves as more than one race, nearly 7 million Americans are of mixed race ("Mixed Race Americans" 2009; Roth 2005). Of those, more than 40 percent are children under age eighteen. As multiracial children reach childbearing ages and as racial heterogamy loses its taboo, the number of multiracial births is expected to climb (Qian and Lichter 2007).

Raising biracial or multiracial children has unique challenges, although it is not without rewards as well (Rockquemore and Laszloffy 2005). One challenge may be tension between parents and children—and between the parents themselves—over cultural values and attitudes. For instance, Euro-American parents, particularly white single mothers who are living in mostly white communities, may find that "[y]ou have to seek, to go out of your way to give them that African

At this festival marking Eid, the end of Ramadan, this Muslim community in central Texas gathers for afternoon prayers. Muslim parents hope that their children will remain true to their religious tradition. Meanwhile, like parents of other minority religions in the United States, they must help their children face fear of ridicule and actual discrimination.

American side" (respondent, quoted in O'Donoghue 2005, p. 148).

A small, qualitative study of eleven East Coast white women who were raising biracial black-white teens unexpectedly found that—because their physical characteristics made it possible—some adolescents (three females) chose to embrace a Latina identity as a way to deal with racial ambiguity (O'Donoghue 2005). One psychologist surveyed multiracial adults and asked whether they thought that their parents had been prepared to raise children of mixed race. The majority did not believe so (Dunnewind 2003). Today, however, many schools are more sensitive to the needs of multiracial children (Wardle 2000), and more resources are available for parents raising multiracial children. According to one recent study, multiracial and multi-ethnic families that foster an explicit family identity as multicultural, multiracial, or multiethnic have happier, better-adjusted children (Soliz, Thorson, and Rittenour 2009). Chapter 9 also addresses these issues.

Religious Minority Parents

Ethnicity is often associated with religious belief. For example, Chinese Americans are likely to be Buddhists; Asian Indian Americans are likely to be Hindu or Sikh. In a Christian dominant culture, diverse ethno-religious affiliations affect parenting for many Americans. For instance, Muslims have their own holy days, such as Ramadan, which may not be taken into account in public schools' scheduling. Wearing flowing robes and, more often, head scarves or veils (called *hijab*), Muslims report that they fear ridicule and face discrimination from employers and others (American Moslem Society 2010; "Muslim Parents Seek" 2005).

American parents of religious minorities generally hope that their children will remain true to their religious heritage amid a majority culture that seldom understands and is sometimes threatening ("Muslim Parents Seek" 2005; F. R. Lee 2001).[15] One solution has been the emergence of religion-based summer camps for children of Bahia, Buddhist, Catholic, Hindu, Jewish, Mormon, Mennonite, Muslim, and Sikh parents, among others (www.mysummercamps.com).

Raising Children of Racial/Ethnic Identity in a Racist and Discriminatory Society

A parent's feeling victimized by racism adds stress to an already stressful parenting process (Benner and Kim 2009; Brody et al. 2008). Families of color (or religious minorities) attempt to serve as an insulating environment, shielding children from and/or confronting injustices (Brody et al. 2008; Brown et al. 2007; Cohen, Tran, and Rhee 2007). Most engage in **race socialization**—developing children's pride in their cultural heritage while warning and preparing them about the possibility of encountering discrimination (Scottham and Smalls 2009; Umana-Taylor et al. 2009). Higher levels of race socialization are associated with parents' having been discriminated against, their higher sense of personal efficacy, and greater concern that their children will actually encounter racism (Benner and Kim 2009; Brody et al. 2008; Crouter et al. 2008).

Valuing one's cultural heritage while simultaneously being required to deny or "rise above" it to advance poses problems for individuals and between parents and their children. For instance, Native Americans must often choose between the reservation and an urban life that is perhaps alienating but may present greater economic opportunity (Seccombe 2007). Latinos may see a threat to deeply cherished values of family and community in the competitive individualism of the mainstream American achievement path (McLoyd

---

[15] This desire that their children maintain their ethno-religious heritage is a principal reason for some immigrant parents' preference that their children marry homogamously, sometimes in arranged marriages (see Chapter 6).

et al. 2000). Asian Americans may follow the "model minority" route to success but experience emotional estrangement from their culturally traditional parents (Kibria 2000). We turn now to a discussion of another variation in the parenting experience—grandparents as parents.

Grandparents as Parents

About 11 percent of U.S. grandparents are raising grandchildren (Lumpkin 2008). More than 3.6 million children under age eighteen are living in a grandparent's household, a few with only their grandfather, many with two grandparents, and many more with only their grandmother (King, Mitchell, and Hawkins 2010; U.S. Census Bureau 2010b, Table 70). Having risen dramatically over the past ten years, the number of grandparents raising children or residing with and helping to raise their grandchildren is expected to rise, especially in today's recession economy (Spratling 2009).

Taken together, unmarried parenthood, divorce or separation, poverty, substance abuse, HIV/AIDS, domestic violence, abandonment, and incarceration account for a very large majority of grandparent families, or **grandfamilies** (Henderson et al. 2009). Unsurprisingly, the family often views a grandparent's assuming the role of primary parent as a crisis. One study found that grandparents' coping strategies involved relying on their religious faith as well as imagining that the situation would somehow "just go away" (Lumpkin 2008). Handling family crises is addressed in Chapter 14.

Sylvie de Toledo is a social worker whose nephew was raised by her mother after her sister's suicide. As a result of this experience, Toledo founded a support group called Grandparents as Parents (GAP) "to meet the urgent and ongoing needs of grandparents and other relative caregivers raising at-risk children" (www .grandparentsasparents.org). As Toledo writes,

> Sometimes the call comes at night, sometimes on a bright morning. It may be your child, the police, or child protective services. "Mama, I've messed up. . . ." "We're sorry. There has been an accident. . . ." "Mrs. Smith, we have your grandchild. Can you take him?" Sometimes you make the call yourself—reporting your own child to the authorities in a desperate attempt to protect your grandchild from abuse or neglect. Often the change is gradual. At first your grandchild is with you for a day, then four days, a month, and then two months as the parents slowly lose control of their lives. You start out baby-sitting. You think the arrangement is temporary. You put off buying a crib or moving to a bigger apartment. Then you get a collect call from jail—or no call at all. (Toledo and Brown 1995, p. 9)

At other times, the change is more sudden, as when a grandchild's parents are killed in an auto accident (Landry-Meyer and Newman 2004).

Becoming a primary parent requires considerable adjustment for grandparents (Lumpkin 2008). Living with children in the house is a significant change after years of not doing so (Dolbin-MacNab 2006). A grandparent's circle of friends and work life may change. He or she may retire early, reduce work hours, or try to negotiate more flexible ones. On the other hand, a grandparent may return to work to finance raising the child(ren). In either case, a grandparent's finances may suffer while paying for items such as additional beds, food, and clothing. To help, under the **formal kinship care** system, some states offer financial compensation to grandparents (or other relatives, such as aunts) who raise their grandchildren as state-licensed foster parents.[16] "Facts about Families: Foster Parenting" further discusses foster parenting.

The Center for Law and Social Policy finds formal kinship care to be generally good for children. Compared with children in nonrelative foster care, those fostered by relatives are less likely to have tried to run away, and more likely to say that they feel loved, like those with whom they live, and want their current placement to be their permanent home (Conway and Hutson 2007). However, a few critics argue that, at least with regard to black children, formal kinship care is overused and detrimental to families in the long run because it fails to emphasize reunification with the children's parents (Harris and Skyles 2008).

Grandparents' raising grandchildren is typically characterized by family members' ambivalence (Letiecq, Bailey, and Kurtz 2008). Unsure whether or when their grandchildren will return to the parental home, grandparents may "learn a . . . stance of detachment to cope with the shifts they are sure to experience and probably even applaud" (Nelson 2006, p. 822). Furthermore, there are often questions about the possible legal termination of the parent's parental rights (McWey, Henderson, and Alexander 2008). When parental rights are not terminated, grandparents who are responsible for the children in their care lack legal rights over them (Letiecq, Bailey, and Porterfield 2008).

Some grandchildren being raised by a grandparent see one or both parents either regularly or sporadically; but generally these relationships are complex, often marked with difficulties. In a qualitative study with white, black, and mixed-race children being raised by grandparents, some children hoped for reunification

---

[16] However, some grandparents report having trouble navigating their state's kinship care system due to, among other reasons, fear and distrust of the child welfare system and daunting bureaucratic regulations (Letiecq, Bailey, and Porterfield 2008).

Every state government has a department that monitors parents' treatment of their children. An example is California's Department of Child Protective Services. When government officials determine that an individual under age eighteen is being abused or neglected, they can take temporary or permanent custody of children and remove them from the parental home to be placed in **foster care**. As wards of the court, foster children are financially supported by the state.

About 460,000 children are in foster care in the United States. There would be more, but there are not enough foster parents or other facilities to fill current needs ("Foster Care" 2005; Wingett 2007). Seventy-one percent of foster care takes place in a licensed foster parent's home—47 percent with nonrelatives, and 24 percent with relatives. The remainder of children in foster care live in various arrangements, including group homes (6 percent) or institutional settings (10 percent), such as Nebraska's Boystown (which also accepts girls) (U.S. Department of Health and Human Services 2009a; www.boystown.org).

The mean age of children in foster care is 9.7 years. Children stay in foster care for an average of about two years, but 20 percent stay for only one to five months, and nearly 10 percent remain until age eighteen when they "age out." Although the rate of children in foster care is higher for African Americans than for other race/ethnic groups, the highest percentage of children in foster care are white (40 percent), followed by 31 percent for blacks, 20 percent of Hispanics, and 5 percent of mixed-race children (U.S. Department of Health and Human Services 2009a).

Very often without family support, those who "age out" of the system—for the most part, youths who were older upon becoming foster children and the developmentally neediest—face serious challenges as they work toward assuming adult roles (Osgood et al. 2005). They are more likely than other young adults to become imprisoned, homeless, unemployed, or pregnant outside marriage (Koch 2009). In 2009, Congress passed a bill encouraging states to extend foster care services until age twenty-one. According to this legislation, the federal monies pay for half of these services, with states and counties responsible for the remainder (Koch 2009). About half of the states extend Medicaid health insurance coverage beyond age eighteen for former foster children and/or offer them scholarships or free public college tuition.

Among others, motivations for becoming a foster parent include fulfilling religious principles, wanting to help fill the community's need for foster homes, enjoying children and hoping to help them, providing a companion for one's only child or for oneself, and earning money. Technically not salaried, foster parents are "reimbursed" in regular monthly stipends by the government.

Although the ultimate goal in half of the cases is reunification of foster children with their parents or principal caretakers, about one-quarter of foster children are available for adoption (U.S. Department of Health and Human Services 2009a). Some foster parents see fostering as a step toward adopting (Baum, Crase, and Crase 2001). We end with the words of Jo Ann Wentzel, senior editor of the magazine *Parenting Today's Teen* and foster mother to more than seventy-five children over the course of her career:

> I don't regret anything I've ever done for any of my [foster] kids. . . . Every once in a while, a kid will track me down and leave a cryptic message on my answering machine, which says, I know I was a pain-in-the-butt when I lived with you but I really learned a lot from you. . . . Or maybe they will tell me about their successes and claim it was because of something we did or said. They tell me they called because they wanted us to know they turned out good [sic] or because they respected our opinion on something. (Wentzel 2001, p. 2)

Critical Thinking

From the structure-functional perspective, discussed in Chapter 2, foster parents are functional alternatives to biological or adoptive parents. What are some ways, do you think, that the foster parent system is functional? What are some instances in which it could be dysfunctional?

---

with their parents while the majority had accepted the situation (Dolbin-MacNab and Keiley 2009).

The extent to which children benefit from a grandparent's intervention has just begun to be studied, and results are mixed (Dunifon and Kowaleski-Jones 2007). On the one hand, a grandparent's living in the home of a poor single mother is advantageous inasmuch as it adds income, from Social Security benefits, for example (Barnett 2008; Mutchier and Baker 2009). In addition, researchers and social workers generally maintain that grandparents provide stability, family cohesiveness, and solidarity while often enhancing young children's cognitive development (Dunifon and Kowaleski-Jones 2007; Spratling 2009). On the other hand, not all grandparents who are raising their grandchildren employ effective parenting practices (Barnett 2008). Grandmothers have been found to be most sensitive and beneficial to infants (Dunifon and Kowaleski-Jones 2007).

Research is also just beginning on the responses and feelings of adults who were raised by their grandparents. One study found that some adult children were grateful and felt a strong bond with their grandmothers whereas others evidenced distance and distrust (Dolbin-MacNab and Keiley 2009). Social service agencies have initiated educational and coping programs for grandfamilies (Dolbin-MacNab 2006; Ross and Aday 2006). The National Center on Grandfamilies promotes awareness of grandfamilies and gives advice on how to help grandparents meet their various needs (www.grandfamilies.org).

Parenting Young Adult Children

Children benefit from parents' emotional support throughout their twenties and after (Kantrowitz and Tyre 2006). As adolescents make the transition to adult roles, parent-child relations often grow closer and less conflicted (Arnett 20004, Chapter 3; Straus 2009).[17] At the same time, young adults may be angry or depressed over lingering childhood issues or difficulties with assuming adult roles (Arnett 2004; Galambos and Krahn 2008). Meanwhile, concerns over the young adult's sometimes faltering transition to adulthood can cause parental ambivalence and parent-child conflict (Hay, Fingerman, and Lefkowitz 2007).

Polls show that a significant majority of higher-SES parents lend or give their children money to repay student loans, buy a car, help with rent or credit card debt, or to put a down payment on a house (Harris 2008). One interesting study found that parents tend to provide money not only to their neediest but also to their most successful children, the latter in anticipation of help from the child as the parent grows older (Fingerman et al. 2009).

Recession, unemployment, and underemployment, along with a decline in affordable housing, make launching oneself into independent adulthood especially difficult today (Danziger and Rouse 2008, F8; Wadler 2009; Wang and Morin 2009). As one result, more and more young adult children either do not leave the family home, or return to it as "boomerangers" after college, divorce, or upon finding first jobs unsatisfactory (Bodnar 2007; Trumbull 2009). As reported in Chapter 8,

more than half of men and almost half of women age eighteen through twenty-four live with one or both parents, and a significant fraction of older adults do too (U.S. Census Bureau 2009e, Table AD-1).

Parents should feel comfortable in setting reasonable household expectations. One way to do this is to negotiate a parent-adult child residence-sharing agreement (Wadler 2009). Following are some issues to negotiate:

1. How much money will the adult child be expected to contribute to the household? When will it be paid?

2. What benefits will the child receive?

3. Who will have authority over utility usage?

4. What are the standards for neatness?

5. Who is responsible for cleaning what and when?

6. Who is responsible for cooking what and when?

7. How will laundry tasks be divided?

8. What about noise levels?

9. What about guests? When are they welcome, and in what rooms of the house? Will the home be used for parties?

10. What are the expectations about informing other family members of one's whereabouts?

11. What will be the rules about using the personal possessions of others?

12. If the adult child has returned home with children, who is responsible for their care?

More detailed residence-sharing agreements are available on the Internet, some for sale (e.g., "Boomerang Kids Contract" 2009). Although a residence-sharing agreement can help temporarily, the goal of the majority of parents is for their adult children to move on. Accomplishing this may be complicated by differing ideas on just what a parent owes an adult child. Our culture offers few guidelines about when parental responsibility ends or how to withdraw it.

Toward Better Parenting

What are some steps that we can take to improve parenting in the United States? Studies show that optimal parenting involves the following factors:

- Supportive family communication (Leidy et al. 2009; Lindsey et al. 2009)

- Involvement in a child's life and school (Cooper, Crosnoe, Suizzo, and Pituch 2009)

- Private safety nets—that is, support from family and/or friends (Lee et al. 2009; Ryan, Kalil, and Leininger 2009)

---

[17] A possible exception involves offspring from non- or nominally religious families who in young adulthood embrace Orthodox Judaism or Christian or Muslim fundamentalism. In these cases, the children may be concerned that their parents lack appropriate religious fervor and/or—at least in the case of fundamentalist Christianity—may not be saved. Meanwhile, some parents can find their child's new fundamentalism "appalling" ("Religion's Generation Gap" 2007).

White Packert/Getty Images

Good parenting involves having adequate economic resources, being involved with the child, using supportive communication and having support from family and/or friends, along with workplace and broader social policies that bolster all families.

- Adequate economic resources (Guzzo and Lee 2008)

- Workplace policies that facilitate a healthy work-family balance and support parenting in other ways as well (Aber 2007; Bass et al. 2009)

- Safe and healthy neighborhoods that encourage positive parenting and school achievement (Dudley 2007)

- Society-wide policies that bolster all parents (Marshall and Tracy 2009)

Chapter 12 explores the first factor listed, supportive family communication. Here we note some existing programs designed to improve child rearing, and then discuss further ways to promote better parenting.

Over the past several decades, many national organizations have emerged to help parents. Some programs serve parents in general. One of these is Thomas Gordon's Parent Effectiveness Training (PET) (Gordon 2000). Another is Systematic Training for Effective Parenting (STEP) (Center for the Improvement of Child Caring n.d.). Both STEP and PET combine instruction on effective communication techniques with emotional support for parents. Some other parent education classes incorporate anger management training (Fetsch, Yang, and Pettit 2008). However, arguing that there remains a "fundamental lack of knowledge about infant cognitive skills and alternative strategies for dealing with troublesome behavior," advocates increased promotion of basic knowledge about child development (Coombs-Orme and Cain 2008).

Meanwhile, programs have been designed to improve parenting in specific situations. For example, a variety of intervention programs aim to help teenage and/or substance-abusing parents (Tolan, Szapocznik, and Sambrano 2007). Other programs have been developed for low-income parents and for grandparents (Dolbin-MacNab 2006; Ross and Aday 2006). Some curricula are designed for particular racial/ethnic groups (Center for the Improvement of Child Caring n.d.; Kumpfer and Tait 2000).

Moreover, there are programs intended to increase fathers' involvement, some fairly successful (Cowan et al. 2009; Fagan 2008; Hawkins et al. 2008). Because still more children could benefit from fathers' economic support and additional contact, policy makers advise further interventions that facilitate father involvement (Amato, Meyers, and Emery 2009; Teachman 2009).

In line with the family decline perspective, discussed in Chapters 1 and 7, some policy makers promote marriage as the most effective way to enhance father involvement. According to this perspective,

[I]t is the institution of marriage that helps men to "sign on" to fatherhood. By choosing to make a legal, social and public commitment to a spouse, a man voluntarily agrees—often well ahead of the actual arrival of a child—to take on the legal and social role of a father. (Whitehead and Popenoe 2008, p. 18; see also Doherty 2008 and Gillmore et al. 2008)

From the family change perspective, on the other hand, there is need for "shifting the paradigm in support of multiple family forms" (Jones 2008, p. 208).

What are some ways that social policy could better support all parents, regardless of family structure? The stress model of effective parenting suggests that reducing parents' stress would improve parenting. Accordingly, "Improving the socioeconomic conditions of parents, particularly among the most vulnerable, might improve parenting outcomes across all relationship types" (Guzzo and Lee 2008, p. 58).

Moreover, because free time to engage in leisurely social interaction and activities is crucial to psychological well-being (Harter and Arora 2008), work and society-wide policies aimed at freeing up time for mothers and fathers would improve parenting. More concerted attention on the part of employers to parental child-raising needs and responsibilities would help, along with their greater recognition that good parenting is essential to a civil society.

Then too, elementary school administrators might schedule children's performances and parent-teacher conferences so that they are less likely to conflict with parents' work schedules (Barnett et al. 2009). And single parents in particular would benefit from more—and more affordable—day care services (Ornelas et al. 2009). Also, college regulations might be widened to better accommodate student parents. An example would be an institution's explicitly stating that exams can be made up in the case of a child's illness (Duquaine-Watson 2007).

We've seen that informal social support has been found to mitigate stress and hence to be related to more positive parenting (Lee et al. 2009; Ryan, Kalil, and Leininger 2009). But policy makers also urge greater civic and community activism on the part of parents (Dudley 2007; Whitehead and Popenoe 2008). Some parent-education programs include instruction on how to become more civically involved or engage in community activism (Doherty, Jacob, and Cutting 2009). "Pediatrics is politics," the late pediatrician Benjamin Spock once said (quoted in Maier 1998). He meant that good parenting involves working for better neighborhoods, communities, and family-centered social policies—and these, in turn, result in better parenting.

Summary

- Family form, gender, socioeconomic class, and race/ethnicity intersect to result in parenting experiences for individuals.

- The family ecology theoretical perspective reminds us that society-wide conditions influence the parent-child relationship, and these factors can place emotional and financial strains on parents.

- We began by considering reasons why parenting can be difficult today. We noted that work and parent roles often conflict.

- The stress model of effective parenting posits that stressors of various sorts lead to parental depression

and household conflict, which in turn result in less positive parenting practices and ultimately in poorer child outcomes.

- Although more fathers are involved in child care today, mothers are the primary parent in the vast majority of cases and continue to do the majority of day-to-day child care.

- Not only mothers' but fathers' roles can be difficult, especially in a society like ours, in which attitudes have changed so rapidly and in which there is no consensus about how to raise children and how mothers and fathers should parent.

- Child psychologists prefer the authoritative parenting style, although some scholars describe the authoritarian/permissive/authoritative parenting style typology as ethnocentric or Eurocentric.

- The need for supportive—and socially supported—parenting transcends social class and race or ethnicity. At the same time, we have seen that parenting differs in some important ways, according to economic resources, social class, and whether parent and child suffer discrimination due to religion, racial/ethnic status, or sexual orientation of the parents. Raising children in lower socioeconomic strata is a very different experience from parenting in wealthier social classes.

- Higher-SES parents tend to follow the *concerted cultivation parenting model*, whereas working-class parents are more likely to adhere to the *facilitation of natural growth model*.

- Besides concerns for basic necessities, such as food, clothing, shelter, and health care, poverty-level parents may live in depressingly blighted neighborhoods.

- A trend over the past several decades has been for an increasing number of grandparents to serve as primary parents, often as a result of some crisis in the child's immediate family.

- More than 460,000 children are in foster care today, many of them in formal *kinship care*.

- To have better relationships with their children, parents are encouraged to accept help from others (friends and the community at large as well as professional caregivers), to build and maintain supportive family relationships (the subject matter for Chapter 12), and to engage in community or civic activism.

Questions for Review and Reflection

1. Describe reasons why parenting can be difficult today. Can you think of others besides those presented in this chapter?

2. Compare these three parenting styles: authoritarian, authoritative, and permissive. What are some empirical outcomes of each? Which one is recommended by most experts? Why?

3. How does parenting differ according to social class? Use the family ecology theoretical perspective to explain some of these differences.

4. What unique challenges do African American, Native American, Hispanic, and/or Asian American parents face today, regardless of their social class? How would *you* prepare an immigrant child or a child of color to face possible discrimination?

5. **Policy Question.** Describe some social policies that could benefit all low-income parents, regardless of their gender, race/ethnicity, or family structure.

Key Terms

authoritarian parenting style 263

authoritative parenting style (also known as positive parenting) 264

concerted cultivation 267

Confucian training doctrine 272

facilitation of natural growth parenting model 268

formal kinship care 274

foster care 275

grandfamilies 274

hierarchical parenting 272

hyperparenting 267

multipartnered fertility 255

permissive parenting style 264

private safety net 260

psychological control 264

psychological parent 258

race socialization 273

resilient 255

single mothers by choice 259

single mothers by circumstance 259

social fathers 261

socioeconomic status (SES) 266

stress model of parental effectiveness 256

transition to parenthood 256

Online Resources

Sociology CourseMate

www.CengageBrain.com

Access an integrated eBook, chapter-specific interactive learning tools, including flash cards, quizzes, videos, and more in your Sociology CourseMate, accessed through CengageBrain.com.

www.CengageBrain.com

Want to maximize your online study time? Take this easy-to-use study system's diagnostic pre-test, and it will create a personalized study plan for you. By helping you identify the topics that you need to understand better and then directing you to valuable online resources, it can speed up your chapter review. CengageNOW even provides a post-test so you can confirm that you are ready for an exam.

Work and Family

Providing and caring for all family household members, including dependents and the elderly, is integral to our definition of families. Until recently, historically speaking, cooperative labor for survival was the dominant purpose of marriage. Women as well as men engaged in economically productive labor not limited to the personal care of family members.

"Where do you work?" is a new question in human history. Only since the Industrial Revolution has working been considered separate from family living, and only since then have the concepts "employed" and "unemployed" emerged. With the Industrial Revolution, economic production moved outside the household to factories, shops, and offices. Although human beings have always worked, it was not until the industrialization of the workplace in the nineteenth century that people characteristically became wage earners, hiring out their labor to someone else and joining a **labor force.**[1]

First men and then women have become workers in the labor force. How has that affected family life? In this chapter, we'll look at the interrelationship of work and family roles for both women and men. We'll look at paid employment and unpaid household work. We'll consider how people use their time to meet work and family responsibilities and consider whether time spent with children has been cut short. And we'll look at the strategies and choices partners use to manage their work and their family relations.

Women in the Labor Force

As the Industrial Revolution got under way, women by and large remained in the home, depending on social class, of course. Women of lower social classes, immigrant women, women of color, and widowed women often supported themselves and their families by taking in laundry, marketing baked goods, working as domestic labor in other people's homes, and housing boarders; before they married, they may have worked in factories. Still, it was largely men who held "jobs" and were visible in economic production.

Women's Entry into the Labor Force

As family size declined and especially as the need for clerical workers and light factory workers expanded, women began to enter the labor force. Industrialization gave rise to bureaucratic corporations, which depended heavily on paperwork. Clerical workers were needed, and not enough men were available. Textile industries sought workers with a dexterity thought to be possessed by women. The expanding economy needed

more workers, and women were drawn into the labor force in significant numbers beginning around 1890. As Figure 11.1 shows, women's participation in the labor force has increased greatly since the beginning of the nineteenth century.

This trend accelerated during World War I, the Great Depression, and World War II, and then slowed following the war. As soldiers came home, the government encouraged women to return to their kitchens. Despite these cultural pressures, the number of wage-earning women rose again. Material expectations increased for housing and consumer goods, and families became more likely to think of college education for the kids.

Beginning in about 1960, the number of employed women began to increase rapidly. Stagnant and declining earnings for men and economic uncertainty for previously successful industries led more families to require a second earner. The growth in the divorce rate left women uncertain about the wisdom of remaining out of the labor force and, hence, dependent on a husband's earnings. The women's movement emerged and was a strong force for anti–sex discrimination laws that opened formerly male occupations to women. The movement also altered attitudes about careers for women. By 1979, a majority of married women were employed outside the home.

At first, the largest group of wage-earning women consisted of young unmarried women; relatively few women worked during child-rearing years. Although many mothers remained at home while children were small, by 1970, half of wives with children between ages six and seventeen earned wages, and that figure increased to 76.2 percent in 2007 (U.S. Census Bureau 2010b, Table 585; U.S. Bureau of Labor Statistics 2009c, Bulletin 2307).

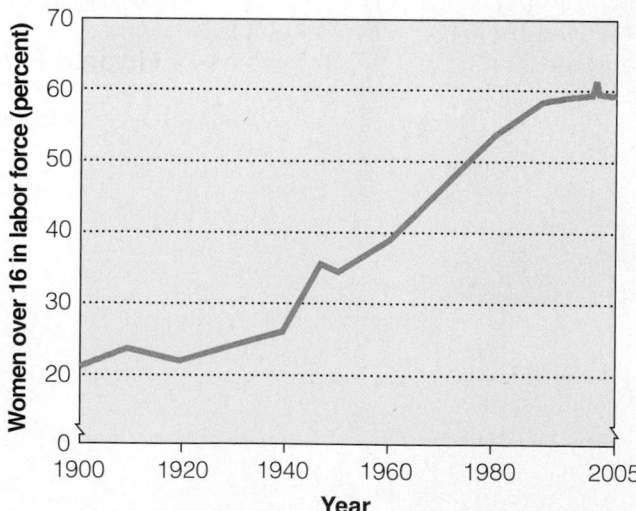

Figure 11.1 The participation of women over age sixteen in the labor force, 1900–2005

Source: Thornton and Freedman 1983; U.S. Census Bureau 1998, 2010b, Table 575.

---

[1] The term *labor force* refers to those people who are employed or who are looking for a paid job.

Keith Brofsky/Jupiterimages

Women have entered the labor force in greater and greater numbers since the 1960s.

of Hispanic women are employed (U.S. Census Bureau 2010b, Table 575). Attitudes changed along with behavior. By the late 1990s, fewer than 20 percent of women and men disapproved of married women working (Sayer, Cohen, and Casper 2004, Table 1). Urban and rural women carry a particularly heavy burden, as both types of communities have been hit tremendously hard by the long-term changes in the American economy. Nearly "one in five rural married women contribute the majority of the couple's earnings, representing a 56 percent increase since 1970. The proportion of urban married women as primary and sole breadwinners increased by 65 percent over the same time period" (Smith 2008, p. 23).

Women's Occupations

When women are in the labor force and work for pay (sometimes termed "market work"), what kinds of jobs do they hold? Occupational distribution of women differs from that of men, as Figure 11.2 indicates.

The pronounced tendency for men and women to be employed in different types of jobs is termed **occupational segregation**. Figure 11.2 depicts the major occupational categories of employed women for 2008. As you can see, 24.5 percent of all employed women were office or sales workers. Only 15.2 percent of employed women were in management, business, or finance positions, whereas 21.1 percent were in professional work (U.S. Bureau of Labor Statistics 2009c, Table 10). Asian American (46 percent) and white women (40.6 percent) were the most likely to hold managerial or professional jobs, compared to black (31.3 percent) and

Mothers of young children were the last women to move into employment outside the home. In 2007, 61.5 percent of wives with children under age six were paid employees. In fact, 57.8 percent of married mothers of children under age one had joined the labor force. Even larger proportions of single women were employed: 78 percent of those with children age six to seventeen, and 67.4 percent of those with children under six (U.S. Census Bureau 2010b, Tables 585, 586).

The *rate* of increase in employment has been greater for white women than black women, who historically had been more likely to work for wages (England, Garcia-Beaulieu, and Ross 2004). Now, white women—with a labor force participation rate of 59.2 percent—are catching up to black women at 61.3 percent. Fifty-nine percent of Asian women and 56.2 percent

© James Marshall/The Image Works

Women in blue-collar jobs are still a minority, although more women are entering these jobs, which tend to pay better than traditional women's jobs in service or clerical work.

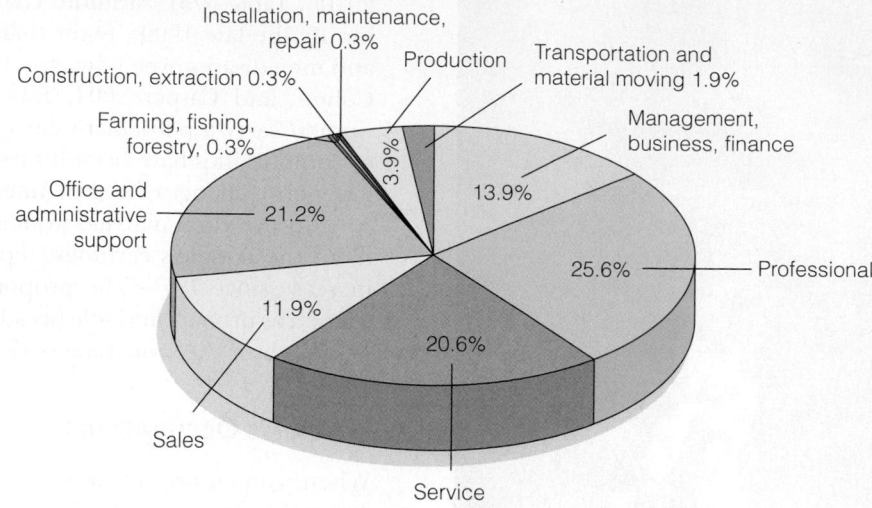

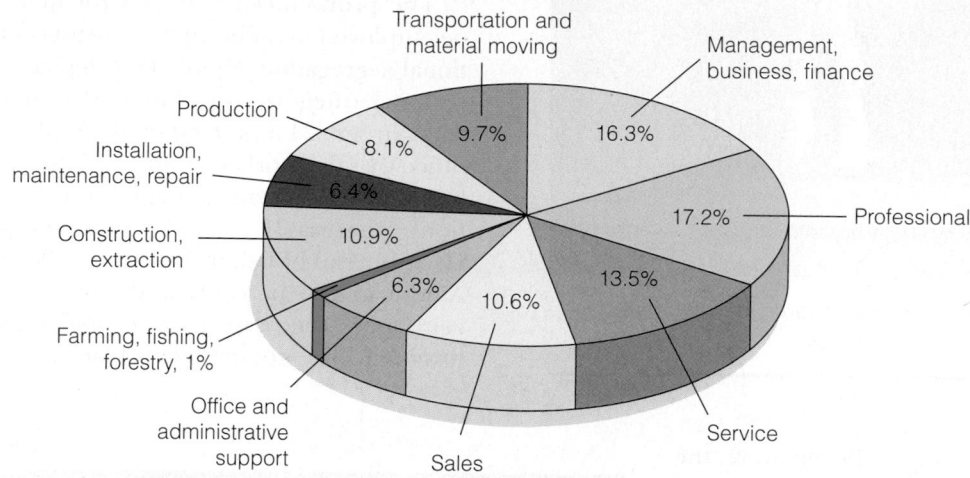

Figure 11.2 The jobs held by women and men, 2005. The percentages in each sector of the pie charts tell us what percentages of women and what percentages of men hold certain jobs. For example, 25 percent of women have professional employment whereas 20 percent are in service occupations. Seventeen percent of men are in professional jobs, whereas 13 percent are service workers.

ª Percentages may not add to 100 percent due to rounding.

Source: U.S. Bureau of Labor Statistics 2009c. *Women in the Labor Force: A Databook,* Table 10.

Hispanic women (23.5 percent) (U.S. Bureau of Labor Statistics 2009c, Table 12).

Jobs typically held by men and women differ *within* major occupational categories, with men more likely to hold the upper-level jobs within each sector. Even though women are proportionately more likely to be professionals than men, they occupy the lower-paying ranks. For example, women are 27.2 percent of the country's dentists, but 96.3 percent of the dental assistants; 34.4 percent of lawyers, but 87.7 percent of paralegals and legal assistants; 30.5 percent of physicians, but 91.7 percent of registered nurses (U.S. Bureau of

Labor Statistics 2009c, Table 11). This occupational segregation contributes to the difference between men's and women's average earnings.

The Wage Gap

Differences in earnings persist in comparisons of employed women and men. Women who worked full time in 2008 earned eighty-eight cents for every dollar men earned (U.S. Bureau of Labor Statistics 2009c, Table 16). The **wage gap** (the difference in earnings between men and women) varies considerably depending on occupation and tends to be greater in the more elite, higher-paying occupations (Weinberg 2004). For instance, the starkest difference in pay occurs in the highest-paying occupation, that of physician. In this occupation in 2008, women on average earned $95,766 or just 60.8 percent of men's earnings ($157,510) (Cheeseman and Rosenthal 2008, Table B). The difference was thought to be related to choice of occupational specialty and practice setting (for example, pediatrics, which pays less, is a popular choice of female physicians); however, U.S. Census researchers Cheeseman and Downs (2007) show that the problem is endemic throughout all occupations. For example, males who enter the nursing profession make higher wages than women who work in the same profession. To some extent, men's and women's employment remains segmented into dual labor markets, with women in a narrower range of jobs offering fewer benefits and advancement opportunities.

Men continue to dominate corporate America. In 2010, less than 3 percent of the highest-earning executives in Fortune 500 companies were women ("Women CEOs" 2010). Although racism blocks the path to management for nonwhite or Hispanic men, both racism and sexism block the path for nonwhite and Hispanic women, who hold only two executive positions and who make up 3.2 percent of the boards of directors positions in the top 500 American corporations (Angelo 2010; Soares, Carter, and Combopiano 2009). In 2010, only one (and the first) African American woman held a CEO position in the *Fortune 500* list of top companies (Angelo 2010).

Occupational segregation (that is, differentiation in jobs of men and women) declined from 1960 to 1990, at all levels of the occupational range. The decline was most pronounced for the college-educated. "But it is also clear that men and women continue to occupy separate spheres in the world of work. It is also clear that the pace of change has slowed" (Cotter, Hermsen, and Vanneman 2004, pp. 13–14). Another important point is the growing divide among women (and men) who have or do not have a college education. Increasingly, the earnings of less-educated men and women, especially men, are falling behind those of more highly educated individuals (Wessel 2010).

Whether the wage gap is due to discrimination or represents men's and women's differing job choices (some of which are impacted by family obligations) is disputed (Boraas and Rodgers 2003, p. 14). Do women's family responsibilities lower their career achievement, and, if so, is the pay gap a result of personal choice? We have discussed the wage gap more generally in Chapter 4. Here, we focus on the relationship between lesser earnings and women's motherhood role.

The concept of the **motherhood penalty** describes the fact that motherhood has a tremendous negative lifetime impact on earnings, a long-term gender-earnings gap. "Women still earn a small proportion of what men earn [over a lifetime] . . . and remain financially dependent on men for income during the child-rearing years and indeed throughout much of their adult lives" (Hartmann, Rose, and Lovell 2009, p. 125). Furthermore, the motherhood penalty has not declined over time despite women's increasing education, involvement in the labor force, somewhat less discrimination, and more opportunities to advance their careers (National Women's Law Center 2010; Goodman 2009; Hegewisch and Liepmann 2010).

A study conducted by the National Women's Law Center (NWLC) found that women were simply paid less than men for the same amount and kind of work. This accounted for much of the difference between the earnings of women and men. Married mothers work approximately one hour less per day in the labor market than married fathers; although this impacts the wages of married mothers, it is not enough to explain the differences between all women's and all men's pay (U.S. Bureau of Labor Statistics 2008). Citing the 2003 U.S. General Accounting Office research on the gender gap, the NWLC found that when factors such as "marital status, race, number and age of children, and income, as well as work patterns such as years of work, hours worked, and job tenure" are controlled for, women make 80 percent of what men do (National Women's Law Center 2010, p. 3). A recent class-action lawsuit against Wal-Mart exemplifies this pattern at the lower end of the pay scale. In 2010, Wal-Mart Corporation settled the largest sex-discrimination lawsuit in history when it agreed to pay damages in the amount of $86 million to over 200,000 female employees in California for paying them less than it paid its male employees (Stempel 2010). Things appear even worse for women at the higher end of the pay scale. Here, women are paid even less than their male counterparts, averaging seventy-five cents to every dollar earned by a man. For example, female CEOs make on average $1,500 per week whereas their male counterparts average $2,000 weekly (Censky 2010).

The Great Recession was impacting the global economy at the time of this writing. Regardless of whether or not the economic outlook has improved between then and now, both women and men will see long-term impacts on job and income prospects. Many commentators examining the Great Recession offer an additional, tongue-in-cheek term for the current economic woes: mancession (a merging of the words man and *recession*). The reason for this is because the hardest hit in this most recent economic downturn have been men— particularly the lesser educated, those who are poorer, and men of color (Elsby, Hobjin, and Şahin 2010; Hartmann, English, and Hayes 2010; Wessel 2010). At the same time, however, women's participation in the labor force continues to increase. Although 75 percent of those unemployed in 2008 were men, women compose an ever larger share of the employed, increasing from 46.4 percent at the end of 2007 to nearly half (47.4 percent) in December 2009 (Borbely 2009; Hartmann, English, and Hayes 2010; Taylor et al. 2010, p. 2). This has important ramifications for familial relationships—especially around the division of labor where women typically bear the majority of child care and domestic duties.

Opting Out, Stay-at-Home Moms, and Neotraditional Families

"The housewife" has vanished, more or less. That would be a woman who views her adult role as one devoted to the home, while she remains economically dependent on the earnings of her husband, the breadwinner. Today, though, 67.7 percent of women with children under eighteen are in the labor force, as are 69.4 percent of married mothers (U.S. Census Bureau 2010b, Table 586; U.S. Bureau of Labor Statistics 2009c, Table 6). Are there no traces left of the housewife?

The movement of women into the labor force has dropped a percentage point or so from its 2000 peak (see Figure 11.1), but some of this may be due to the changing demographics of the American population, rather than a retreat back into the house. The percentage of mothers who return to work within the year after a birth has dropped slightly from a 1998 peak of 59 percent to 57.3 in 2006 (Dye 2008, Figure 3), but has remained relatively steady over the years. These data suggest that workingwomen generally continue to do so, even after giving birth to a child. As has been discussed previously, and later in this chapter, economic trends in the United States are such that mothers will have increasing pressure to enter or remain in the labor force.

In 2009, 22 percent of married women with children under fifteen were "stay-at-home" mothers, while 7 percent of fathers were. These parents gave "to care for home and family" as their reason for not participating in the labor force. Each of these "stay-at-home" mothers and fathers had employed husbands or wives who worked fifty-two weeks in the last year (U.S. Census Bureau 2009e, Table SHP-1).

We don't know from these bare-bones data whether these mothers (and fathers) plan to remain out of the labor force, having made a commitment to being a full-time parent, or whether they plan to return to work. We also don't know what their occupations are, although highly educated mothers and those with higher-level occupations are more likely to be employed, and they return to the labor force more quickly after giving birth (Johnson and Downs 2005). The long view of the American economy is such that the trend of stay-at-home

fathers may actually increase, whereas many mothers are likely to become the new family breadwinners (Belkin 2009). Research from the United States Federal Reserve shows that males in the United States bear the largest brunt of unemployment. Since 1980, following each recession, men have lost ground by an average of 1.2 percent, whereas women have lost an average of .73 percent. This kind of chronic job loss suggests that American males will increasingly have difficulty finding and keeping their jobs (Elsby, Hobijn, and Şahin 2010, Table 1).

Opting Out Opting out was conceived of in more limited terms than a complete withdrawal from the labor force, as the young women interviewed or informally surveyed spoke of part-time jobs or leaving the labor force for a few years. Little evidence exists that substantial "opting out" has in fact occurred (see Gerson 2010, p. 240, note 21 for additional sources). Economist Heather Boushey of the Center for Economic and Policy Research (2005a; 2006) suggests, based on her analysis, that a weak labor market since 2001 had led to a very slight downturn in labor force participation for both women and men, whereas labor force participation of single women and of high school dropouts continued to grow (Porter 2006). Since 2001, we have seen a trend in which women are becoming the majority of workers in the American labor market (Belkin 2009, p. O11).

A look at the choices of more affluent women, such as graduates of elite universities, found that these women also did not leave the workforce after having children, at least not for very long—58 percent were never out of the job market for more than six months, and on average, the women spent 1.6 years out of the labor force. Most were married and had children (Goldin 2006). For example, interviews with African American women lawyers, technology experts, corporate managers, and entrepreneurs indicate that they are concerned about the need to build financial security for their families and, often, the need to help extended-family members. "Among highly educated women aged 25 to 45, the effect of having children on women's labor force participation has been negligible since 1984, and remains so today" (Boushey 2006). In fact, the job losses felt by men during the current economic downturn has many highly educated and affluent mothers returning to the labor force, many of them sooner than they had intended (Greenhouse 2009; Belkin 2009). Moreover, some companies are beginning to develop reentry programs for women who have taken time out of the labor force (Joyce 2007b; McGinn 2006a).

Stay-at-Home Moms In 2009, 26 percent of mothers of children under fifteen in married-couple families were stay-at-home mothers, wives of steadily employed men who remained out of the labor force for the entire year, giving as their reason "taking care of home and family" (U.S. Census Bureau 2009e, Table SHP-1; Jayson 2009). These data give no indication of whether this arrangement is temporary or permanent. Given that most women leave the labor force for relatively short periods over a lifetime, "stay-at-home mom" is a status that is temporary for the majority of women. But what would women prefer? The Gallup Poll has been asking that question for many years. In 1978, a decisive majority favoring employment over the traditional homemaker role of women appeared in poll data for the first time. "Since then, no clear consensus in either direction has emerged, with small majorities of women sometimes opting for working outside the home and, . . . [at other times] small majorities favoring the traditional role of family caretaker" (Moore 2005). In 2007, the poll showed that a slight majority of American women preferred paid employment over the traditional role of family caretaker (Saad 2007).

Neotraditional Families There are families, termed **neotraditional families** (Wilcox 2004, pp. 209–11), for whom a traditional division of labor is the ideal:

> This [neotraditional] order is appealing to men and women who are discontented with . . . family modernization, the lack of clarity in gender roles . . . , and the pressures associated with combining two full-time careers. It is also appealing to women who continue to identify with the domestic sphere, who wish to see homemaking and nurturing accorded high value, and who wish to have husbands who share their commitment to family life. . . . Men who continue to seek status as domestic patriarchs who have the primary earning responsibility and at least titular authority over their families are also attracted to this order. (Wilcox 2004, p. 209)

Wilcox associates this family model with evangelical Christianity, as well as Orthodox Judaism, traditional Catholicism, and Mormonism; and suggests it is most likely to be found in the middle and working classes of the outer suburbs and in rural areas (Wilcox 2004, p. 210). Kathleen Gerson, interestingly, finds a heightened sense of egalitarianism coming from men and women who were raised in neotraditionalist families. She finds that although terms such as *equality* and *egalitarian* are elusive concepts to define, nonetheless these men and women are hoping "to find a lifelong partner, to balance work and family, and to share breadwinning and caretaking" (2010, p. 107).

Wilcox found active conservative Protestants more likely than mainline Protestants to agree that men should be breadwinners and women homemakers. Interestingly, all groups showed a decline in this viewpoint from the 1970s to the 1990s, though "active conservative Protestants" remain at almost 60 percent support for this model (Wilcox 2004, Figure 3.3). It stands to reason that economic pressures force many neotraditional women into the labor force, though they are likely to organize that work as much as possible around part-time or in-home work, or take substantial time out of the labor force when children are small (Eikhof, Warhurst, and Haunschild 2007; Cordes 2009).

Some women have chosen to opt out of the labor force to raise their children at home. This former executive may return to the labor force eventually.

Noah Berger/The New York Times/Redux

Men's Occupations

The work situations of men are many and varied as Figure 11.2 shows us. For example, many of the blue-collar jobs that paid good wages to earlier generations of men have vanished, and men without college degrees have experienced eroding incomes. "'In the past guys could drop out of school after finishing high school, or even without finishing, and go into a factory and get a steady job with benefits. ... But there has been a deterioration in young men's economic position'" (sociologist Valerie Oppenheimer, quoted in Porter and O'Donnell 2006). Between 1979 and 2003, there was no gain for those with some college but no degree, while high school grads' earnings declined 8 percent (Mincy 2006; Wessel 2010). Hartmann, English, and Hayes (2010) research suggests this trend continues. For example, men without a high school degree in 2009 have a 7.8 percent unemployment rate, whereas men with college degrees have a 2.6 percent unemployment rate (Hartmann, English, and Hayes 2010, p. 33). As discussed elsewhere in this book, the unemployment rates are even higher for men of color.

Yet, the provider role is an important one for men of all social classes. What is the present state of the provider?

The Provider Role

What sociologist Jessie Bernard terms the **good provider role** for men emerged in this country during the 1830s. Before then, a man was expected to be "a good steady worker," but "the idea that he was *the* provider would hardly ring true" (Bernard 1986, p. 126), because in a farm economy both husband and wife had roles in producing the family's income. The provider role (and its counterpart, the housewife role) lasted into the late 1970s. The proportion of married-couple families in which only the husband worked gradually declined from 42 percent in 1960 to 17.8 percent in 2007 (Wilkie 1991; U.S. Bureau of Labor Statistics 2009c, Table 23; and see Figure 11.3).

As Figure 11.3 indicates, in the vast majority (57.6 percent) of married couples, both husband and wife are now employed. In 5.6 percent of married couples, only the wife is employed. Although the role of family wage earner is no longer reserved for husbands, many Americans still believe that the man should be the principal provider for his family, and it works out that way in practice to some degree.

Overall, whether single or married, parent or not, men work more hours than women and are more likely

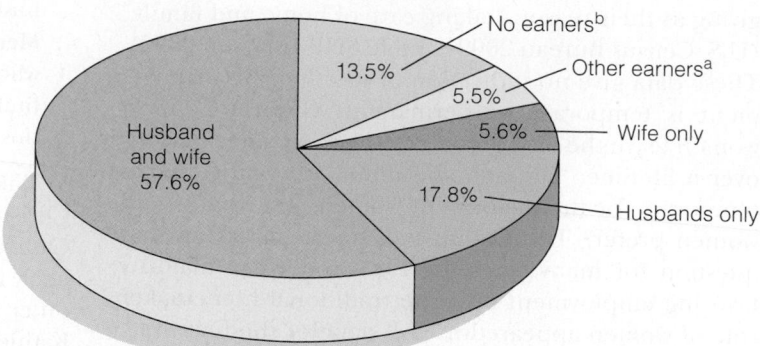

Figure 11.3 Married-couple families by number and relationship of earners, 2007

Source: U.S. Bureau of Labor Statistics 2006d, Table 23.

[a] Includes husband and other family member(s); wife and other family member(s); other earners, neither husband nor wife.

[b] The spouses may be unemployed, retired, disabled, institutionalized, or imprisoned. (The term *unemployed* refers to people who are in the labor force and are looking for work but who presently have no job.)

(89.5 percent) to work full time than are women (75.4 percent). Employed wives contribute over a third (36 percent) of a family's income (U.S. Bureau of Labor Statistics 2009c, Tables 20, 24). Men continue to be primary breadwinners in the majority of couples, and most men (in all racial/ethnic groups) identify with this role (Coltrane 2000).

> [S]ocietal notions of the meaning of work for men and women are still quite distinctive. Both men and women may view working as a choice for women, even when the woman has no real alternative to being employed. In contrast, there is a strong societal imperative for men to be employed outside the home, and those who choose not to do so are viewed skeptically. (Taylor, Tucker, and Mitchell-Kernan 1999, p. 756)

In fact, men's success—as measured in terms of employment and higher earnings—still seems to be important in "facilitating marriage and enhancing marital stability" (Bianchi and Casper 2000, p. 31). Although social pressure may push men to view their lives through the lens of job status and earnings, Kathleen Gerson notes that, "when fathers found themselves doing all of the breadwinning despite a preference for sharing [domestic duties], their ambivalence spilled into daily domestic life" (2010, p. 78).

It is also difficult to live up to societal expectations that may not mesh with the reality of economic opportunities. This situation is especially applicable to blue-collar and racial/ethnic minority husbands in the twenty-first-century economy where "basic economic shifts ... leave men with shrinking opportunities for secure, well-paid, and unionized work" (Gerson 2010, p. 203). Moreover, husbands who want to share household work and child care will not find it easy to do so while continuing as the primary breadwinner (Cordes 2009). For this reason, partners who want to create new options for themselves need to work for changes in the public and corporate spheres, an option explored later in this chapter.

Some husbands today are rejecting the idea that dedication to one's job or occupational achievement is the ultimate indicator of success; in fact, 73 percent of men say they want a job or career that is fulfilling (*Time Opinion Poll on Gender* 2009). Some are choosing less-competitive careers and are spending more time with their families. Four-fifths of men age twenty through thirty-nine who were interviewed in 2000 rated a work schedule that would give them more family time as a more desirable job quality than challenging work or high income. Seventy percent of these younger men said they would exchange money for time with their families—compared to 26 percent of men over sixty-five. This suggests an important generational change, and one that might be happening because a substantial majority of the younger men (70 percent) had working mothers (Grimsley 2000). In a recent poll, nearly half of American men said that companies should provide more flexible work schedules to both men and women (Halpin and Teixeira 2009, p. 412 Table 3). Surprisingly, even during an economic recession, a Harris poll shows that 37 percent of fathers with children under the age of eighteen would leave their job if their significant other made enough to support the family, and 38 percent would accept less pay in exchange for more time to spend with their children (Careerbuilder.com 2007). Although this is down from the previously referenced 2000 poll, it shows a significant level of dissatisfaction

with the *work–life*[2] balance for men, even in the face of economic insecurity.

Meanwhile, there is an effort on the part of some social scientists (e.g., Christiansen and Palkovitz 2001) to change the *meaning* of the standard male provider role so that it is seen to be as much a form of family work and fathering as "hands-on" parenting. Men with children work increased hours compared to childless men, on average.

Anthropologist Nicholas Townsend (2002) interviewed thirty-nine men who graduated from the same northern California high school in the 1970s; thirty were non-Hispanic white, six were Hispanic, and three were Asian American. Regardless of ethnicity, the men described their lives and goals in terms of "the package deal," which was composed of marriage, children, home ownership, and a steady job. Work was seen to be part of being a good father: "Everybody has a purpose in life. It's the same basic, mundane thing. You get up, you go to work, you come home. Your purpose is to provide for your family" (Skip, quoted in Townsend 2002, p. 117). Although these men desired to spend more time with their children and thought that important, in reality their time was devoted to paid work—many had two jobs or put in extensive overtime.

Still, it appears that there are two distinct models for the father-as-provider role. Some fathers (*good providers*) work *more* hours than childless men, whereas others (*involved fathers*) work *fewer* hours. A man's ideological commitment to one or the other role makes a difference.

[2] Work–life balance is the attempt men and women make to balance their work (the demands of job/career) and life (family, leisure, personal activities) in such a way as to find enjoyment and satisfaction in both.

Many men today expect to work at home doing child care or domestic work, as well as to hold a job.

And "it seems clear that a shift away from the provider role and toward the involved father role [has occurred] in recent years" (Kaufman and Uhlenberg 2000, p. 934). Thus, some fathers try to decrease the demands of the workplace in order to participate more at home.

Why Do Men Leave the Labor Force?

Men may relinquish employment as a positive choice: the desire to spend more time with their children. But they may also not be employed because of poor health or disability, or their loss of a job may have developed into long-term unemployment.

Men may be dissatisfied with the competitive grind or the nature of their work, and find themselves in a situation—a working wife who earns enough to support the family or an early retirement package—that permits them to seek new options (J. Smith 2009; Tyre and McGinn 2003). Some couples may size up the situation and recognize that the woman is more desirous of pursuing a career, or has a higher-earning career, and/or is more successful than the man, and they can decide as a couple to reverse roles. In nearly 26 percent of couples, wives earn more than their husbands (U.S. Bureau of Labor Statistics 2009c, Table 25). For a variety of reasons, including the demise of secure employment at all economic levels, labor force participation rates have fallen for men (Krueger 2004).

Although they are a small minority, some men have relinquished breadwinning to become **househusbands**: men who stay home to care for the house and family while their wives work. About 158,000 fathers with children under fifteen remained out of the labor force for that purpose in 2009 (U.S. Census Bureau 2010c, Table FG8). The Census Bureau only considers those whose wives worked full time as **stay-at-home dads.** Although few men—0.4 percent—are stay-at-home dads by this definition (Fields 2004, Table 5), some 20 percent of fathers of preschool children whose mothers were employed were principle caregivers. This number, however, fluctuates between the regular school year and the summer months, as well as by income level (J. Johnson 2005, Table 2; Laughlin and Rukus 2009, slide 21).

In 26 percent of gay male couples with children, one parent often stays at home: "To some gay men, the idea of entrusting the care of a hard-won child to someone else seems to defeat the purpose of parenthood" (Bellafante 2004). These couples will, of course, have a male earner, which underscores the fact that the options for men who would like to give more time to their families are limited because men in our society typically earn more than women do. Consequently, whatever their preferences, many heterosexual couples find themselves needing to encourage the man's dedication to his job or career in the interests of the family's overall financial well-being (Lewin 2009; Smith 2009).

Although fathers who are primary parents express more sense of isolation than stay-at-home mothers and may experience the loss of a career-based identity, being a househusband is not the lonely choice it once was. Local groups, national organizations, and Internet chat rooms bring househusbands together, and mothers at home are more welcoming of their male counterparts than they used to be.

As with many aspects of family life, choice is the key to a man's satisfaction with the househusband role, as is mutual understanding by the couple about the specifics of their division of labor (Cordes 2009; Gerson 2010; Lewin 2009; Smith 2009; Tyre 2009).

Two-Earner Marriages— Work/Family Options

As recently as 1968, there were equal proportions of dual-earner and provider–housewife couples: 45 percent of each (Hayghe 1982). Today, **two-earner marriages**, in which both partners are in the labor force, are the statistical norm among married couples.

Even though we may tend to think of two-earner couples as ones in which both partners are employed nine to five, spouses display considerable flexibility in how they design their two-earner unions. (Households headed by single parents, of course, have more constraints on their choices.) These arrangements are ever-changing and flexible, varying with the arrival and ages of children and with both spouses' job opportunities and working conditions, and involving experimentation with different solutions for managing work and family commitments.

Although these work–family arrangements are fluid, we have observed certain patterns. In this section, we examine some ways in which couples choose to structure their work commitments and family life: the two-career marriage, part-time employment, shift work, working at home, and temporarily leaving the labor force.

Two-Career Marriages

Careers differ from *jobs* in that they hold the promise of advancement, are considered important in themselves—not just a source of money—and demand a high degree of commitment. Career men and women work in occupations that usually require education beyond the bachelor's degree, such as medicine, law, academia, financial services, and corporation management.[3]

---

[3] Higher-income men tend to be married to higher-income women, as the tendency to marry homogamously (see Chapter 6) would suggest. One effect of the trend toward dual-earner families is increasing inequality between families with two high-status, high-paying careers and those with two poorly paid jobs (Paul 2006; Schwartz and Mare 2005). Families depending on one woman's income fare even worse.

Some dual-earner couples choose to work together in a joint business.

Two-career families often outsource domestic work and are likely to employ an in-home caregiver, a **nanny.**

Part-Time Employment

A little over 25 percent of women worked part-time in 2008 (compared to 10.5 percent of men; U.S. Bureau of Labor Statistics 2009c, Table 20). Mothers try to scale back their employment while children are preschoolers (children under six years old); however, only 6.3 percent of mothers with preschool-aged children were unemployed in 2008. By the time a child is nine months old, nearly 60 percent of all mothers are back at work—the majority working full-time, with only 22 percent of these mothers working only part-time (Han et al. 2008, p. 17).

Greater family and personal time is a clear benefit of part-time employment, but there are costs. As it exists now, part-time work seldom offers job security or benefits such as health insurance. And part-time pay is rarely proportionate to that of full-time jobs. For example, a part-time teacher or secretary usually earns well below the wage paid to regular staff. In higher-level professional/managerial jobs, a different problem appears. To work "part-time" as an attorney, accountant, or aspiring manager is to forgo the salary, status, and security of a full-time position and still to put in forty hours a week.

Shift Work

Sometimes one or both spouses engage in **shift work**, defined by the Bureau of Labor Statistics as any work schedule in which more than half an employee's hours

The vast majority of two-earner marriages would not be classified as *dual career* because the wife's or the husband's employment does not have the features of a *career*. Nevertheless, the dual-career couple is a powerful image. Most of today's college students view the **two-career marriage** as an available and workable option.

For two-career couples with children, family life can be hectic, as partners juggle schedules, chores, and child care. Career wives, in particular, often find themselves in a paradoxical situation. The career world tends to view the person who splits time between work and family as less than professional, yet society encourages working women to do exactly that (Hochschild 1997).

A San Francisco choreographer goes back to work, taking her new baby to a ballet rehearsal—another way to combine work and family.

A Closer Look at Family Diversity

Diversity and Child Care

Extreme Child-Care Maneuvers

By Sue Shellenbarger
for *The Wall Street Journal* May 20, 2009

It was a hand-off reminiscent of a spy movie.

Rushing from a client meeting earlier this month, advertising consultant Ted Villa wheeled his Jetta into an office-building parking lot, whipped out his cellphone and reported his location to his business partner and wife, Nancy Snow Villa. Moments later, Ms. Villa piloted their SUV into a nearby spot, their three small children riding in back. After a brief bathroom break inside the building for the kids, Mr. Villa hopped into the SUV and drove away with the children. Leaving the Jetta where he parked it, Ms. Villa raced into the building for her own client meeting.

Ted Villa and Nancy Snow Villa swap work and child-care duties throughout the day in order to spend as much time with their three children as possible.

Time elapsed: less than 10 minutes. Money spent on child-care help: zero.

In a shift that is speeding a change in marital roles, the complex dance of the dual-earner couple is escalating to new extremes. Forced by the recession to cut costs while snapping up every opportunity to work, husbands and wives are swapping roles and bending work schedules at levels never seen before. Layoff victims are squeezing in freelance work amid family duties. Couples are coordinating their calendars down to the minute.

This new marital choreography in some ways "is a throwback to the kind of family that prevailed for most of the 20th century," when families tended to do everything for themselves without hiring child care or housekeeping help, says Stephanie Coontz, an author on the history of marriage and research director of the Council on Contemporary Families, Chicago. In other ways, the trend is historically unprecedented—accelerating a move toward more equitable sharing of responsibility and power between husbands and wives. "We're moving into uncharted territory here," Ms. Coontz says. "It would be a fascinating social experiment—if it weren't so painful."

Marriage expert Thomas Bradbury likens the pressures to the treadmill stress tests used to screen cardiac patients.

"Some couples will pass the test, a few will not," says Dr. Bradbury, a psychology professor and co-director of the Relationship Institute at the University of California, Los Angeles. Healthy marriages will grow stronger, but the hardships will bring others' weaknesses "into sharp relief," requiring mindful effort to survive.

Some scenes from the new dual-earner two-step:

Tag-team parents: The Villas started exUrban, a marketing and advertising concern, after he was laid off late last year. To hold down child-care costs and spend as much time as possible with Jane, age 5; Sam, 3; and Ben, 7 months, the Needham, Mass., couple swap roles throughout the day. They plan client meetings week-by-week to avoid conflicts, then fit writing, creative work and planning "in the gaps" between appointments, time with the children and housework, calling only occasionally on sitters or family members for help, Ms. Villa says.

On the day of the parking-lot hand-off, "Ted had meetings all morning. I had meetings all afternoon, and the kids needed to be picked up at noon" at the Montessori school, where they're all enrolled, she says. She picked them up, Mr. Villa met her at her client's office, and he took over family duties from there. "Some of our colleagues say, 'How sustainable is this?'" Ms. Villa says. But for now, she adds, it's what both she and her husband want.

With extensive planning, the Villas find time for family and work.

The tag-team act gets tougher for people who lack control over their work hours. Sabrina Holmes, mother of a 2-year-old son, can't afford to pay for his $700-a-month child-care center with the office jobs she is able to get.

The Green Valley, Ill., mother stopped using paid child care and found flexible work at home instead, as a call-center agent for LiveOps.com. Ms. Holmes has begun splicing her work in every waking hour her husband is home to help with their son.

The downside: Randy, her steelworker husband, works rotating shifts, moving from daytime to evening to graveyard hours every few weeks. That are before 8 a.m. or after 4 p.m. It has been estimated that in one-quarter of all two-earner couples, at least one spouse does shift work; one in three if they have children (Presser 2000). Some spouses use shift work for higher wages or to ease child care arrangements. In 2004 (the most recent data available), 17.7 percent of all workers worked in shift work, with men making up the preponderance of shift workers (19.1 percent), while 16.1 percent of women worked in shift work. Fathers tend to work more weekend days than mothers (McMenamin 2007, pp. 9, 11).

"Tag team" and "split shift" are just a couple of colloquial terms used for dual-income earners who do shift work, as well as single parents who juggle multiple friend and family child care resources in a given day (Gornick, Presser, and Batzdorf 2009). The terms represent the clockwork timing necessary to exchange childcare duties from one working parent to the other one or more times a day. These kinds of handoffs are typically scheduled weeks in advance and are more often used by parents who have some control over their work schedules. For example, one parent drives to a

means Ms. Holmes's workday moves too, interspersed with the naps and meals of her rambunctious son.

Not surprisingly, she sometimes finds it hard to "keep my eyes open," Ms. Holmes says. Also, "you kind of give up your free time with each other." On many days, "when my husband gets home from work, I go in my office, shut the door and that's it," she says. Meanwhile, her husband is doing more laundry, cleaning and yardwork. The result: Both of them are "getting spread pretty thin."

Dueling dockets: Melissa and Joel Selcher, both managers in high-pressure jobs, want to make sure at least one of them is available at all times in case one of their children, Lily, 9 months, or Jackson, 3, needs to be picked up at their child-care center. They also want to stay on top of their jobs amid layoffs and the recent acquisition of Mr. Selcher's company.

To coordinate their calendars, they've developed "a complex bartering system," Ms. Selcher says.

At the beginning of each month, the Burlingame, Calif., couple send each other meeting lists, then decide which commitments take precedence. One partner's meeting with an executive vice president beats out the other's meeting with a vice president.

If one spouse is making a presentation or running a meeting, "that person gets trump power," Ms. Selcher says. And each has the power of the veto; "we don't accept or agree" to a new meeting "unless we check with the other one," she says. They also have developed an emergency-alert code for getting in touch: If one partner rings the other's cellphone three times without leaving a message, that means "step out of your meeting and call me," she says. They used it recently when Jackson threw up at day care and had to be picked up midday.

Ms. Selcher acknowledges that the carefully orchestrated system "feels like a constant juggle." But for them, it works. "We're trying to make the right call for both of us," she says. "It's not just what's best for me or what's best for him, but what's best for the whole family unit."

Pooling assets: Eva Silva Travers, a Studio City, Calif., mother of two, and her husband, Joe Travers, were accustomed to separate checking accounts when they married, and kept them that way for a while. But now, coping with her layoff last October as a creative-department manager, and his reduced income as a touring musician, they've pooled their assets for the first time in a joint account.

"It's easier and less stressful to pay the bills out of joint money," she says. "Not having those little nickel-and-diming conversations is such a stress reliever."

She also has begun tracking Mr. Travers's schedule closely on tours, so they can plan Internet calls rather than relying on costly world cellphones for spontaneous calls, as they did in the past.

"Good things can come from hard times," says Ms. Travers, who is currently a freelancer on Elance.com. "This has taught us to be more of a couple, really. Just managing our work and money and schedules is a skill. . . . It takes a conscientious effort. The recession has forced our hand in making that change, which is ultimately really good."

To weather the strain, couples need consciously to acknowledge the stress and "make allowances for it," Dr. Bradbury says. Get help if needed, such as taking a class on relationships, committing to time alone together or seeing a therapist.

Second, cut your partner some slack. If you see him or her becoming overwhelmed, "do more: drive the kids, go to the grocery, cook dinner, pay the bills, whatever," he says. But "don't crow about how helpful you are," because that "can make your partner feel worse."

Third, "set up a firewall" against frustration, to prevent it from spilling over into your relationship and eroding the good feelings you have for each other, he says.

Instead of attacking or unloading on your partner, try to offer gestures of support or affection. And "even in those moments when he or she is driving you nuts," Dr. Bradbury says, remind yourself of why you fell in love—and focus on those qualities.

* Write to Sue Shellenbarger at sue. shellenbarger@wsj.com.

Critical Thinking

What family theory or theories could you use to analyze this situation?

Source: Shellenbarger, Sue. 2009. "Extreme Child-Care Maneuvers." *The Wall Street Journal* May 20, D1.

predetermined location with the children he's taken care of during the morning hours; he gets out and takes his wife's car to work, while she gets into his car and takes the children home where she will stay with them for the afternoon (Shellenbarger 2009).

Shift workers not only face physical stress with night work or frequently changing schedules, but shift work also reduces the overlap of family members' leisure time, and that can affect the marriage: "To the extent that social interaction among family members provides the 'glue' that binds them together, we would expect that the more time spouses have with one another, the more likely they are to develop a strong commitment to their marriage and feel happy with it" (Presser 2000, p. 94). Unsurprisingly, then, shift work is associated with a decrease in marital stability. For single parents the results of shift work can be devastating. Because of the increase in women's shift work (especially at night), an increasing number of single mothers find themselves forced to leave their children unattended, which has tragic consequences for their children (Gornick, Presser, and Batzdorf 2009).

Doing Paid Work at Home

Home-based work (working from home, either for oneself or for an employer) has increased dramatically over the past decades—a 55 percent increase between 1990 and 2000 (Bergman 2004). The number of self-employed has increased 25 percent from around 3.47 million in 1999 to 4.34 million in 2005, while the number of people working from home increased nearly 20 percent (from 9.48 to 11.33 million) between those same years (Tozzi 2010; U.S. Census Bureau 2005, Table 1).

Home-based work used to involve *piecework*, sewing or flower making, for example. This mode of home production is declining due to competition from low-wage workers overseas. It still exists, particularly in the assembly of medical kits, circuit boards, jewelry, and some textile work, but many home-based workers are educated and are engaged in professional services such as law, accounting, computer programming, consulting, marketing, finance, and so on (Tozzi 2010). Other home-based businesses include the direct selling of cosmetics, kitchenware, and other products, as well as working as an independent contractor to handle customer service calls (Armour 2006). In fact, in 2005 (the most recent data available), 42 percent of people working exclusively from home had family incomes of $75,000 or greater, and 46.5 percent held a bachelor's degree or higher (U.S. Census Bureau 2005, Table 1).

Home-based work now includes working from home for an employer, perhaps through telecommuting—connecting to the office, customers, clients, or others by the Internet, telephone, videoconferencing, or other means. In 2008, 21.1 percent of workers worked at home as part of their primary job, and 55 percent of them were self-employed. Two-thirds were managerial or professional employees. A little over half of home-based workers are women, and they make up 37.8 percent of all self-employed people (U.S. Bureau of Labor Statistics 2009c, Table 36/2009d, Tables 6 and 7; U.S. Census Bureau 2007a, Table 592/2005, Table 1).

The reason women give most often for working at home is to catch up with work, but 32 percent of women with children under the age of six say it is to "coordinate work with personal/family needs" (Wight and Raley 2009, Table 1). Remarking on the advantages of flexibility, mothers of young children were the most likely see telecommuting favorably: "I can take care of the sick child and get my work done. A win-win situation" (Hill, Hawkins, and Miller 1996, p. 297).

As the author of a study of women in a home-based direct-selling business noted, however, "many women soon discovered . . . that they had exchanged one set of challenges for another. Mothers employed at home report problems with interruptions . . . ; they are often asked to . . . run errands for relatives, to watch neighbors' children when bad weather closes the school, or to keep an eye out for the older kids" (Kutner 1988; see also Gudmunson et al. 2009).

A study comparing office-based employees to teleworkers found that teleworkers were no more likely than the office workers to feel they had enough time for family life. Some said that they tended to work more hours than they would otherwise (Hill, Hawkins, and Miller 1996, p. 297). Indeed, work–family flexibility may be a double-edged sword. The families of some teleworkers "struggled because workplace and schedule flexibility blurred the boundaries between work and family life" (p. 293). Home-based workers faced the same tension between career advancement—which required putting in long work hours—and family time as did employees working in a more conventional setting (Berke 2003; Gudmunson et al. 2009).

Unpaid Family Work

Unpaid family work involves the necessary tasks of attending both to the emotional needs of all family members and to the practical needs of dependent members (such as children or elderly parents), as well as maintaining the family domicile.

Caring for Dependent Family Members

Our cultural tradition and social institutions give women principal responsibility for raising children. Moreover, our culture designates women as "kinkeepers" (Salari and Zhang 2006), whose job it is to keep in touch with—and, if necessary, care for—parents, adult siblings, and other relatives. The vast majority of informal elderly care is provided by female relatives, usually daughters and (albeit less often) daughters-in-law (Piercy 2007).

Family responsibilities and resources in meeting the needs of elderly, ill, or disabled family members are topics included in Chapters 14 and 17, although many of the chapters deal with the emotional aspects of family life. In this chapter, we look more closely at housework and child care.

Housework

Utopians and social engineers alike once shared a hope that advancing technology and changed social arrangements would make obsolete the need for families to cook, clean, or mind children (D. Hayden 1981). But collective arrangements proposed by utopians and early feminists never caught on. Servants, who had done much of the work for earlier middle-class housewives, entered factory work or took other, better jobs, and middle-class women were left to do their own housework (Cowan 1983). Technology seems merely to have raised

Doris Lee (1905–1983), Thanksgiving, 1935. Oil on canvas.

Thanksgiving © The Art Institute of Chicago

the standards rather than making housework less time-consuming. For example, instead of changing clothes at infrequent intervals, we now do so daily, although it will most likely be a woman washing those dirty clothes (Cowan 1983; Newport 2008).

The Second Shift Housework—even with the decline—remains substantial. Including child care, many employed wives (and some husbands) put in what sociologist Arlie Hochschild calls a **second shift** of unpaid family work that amounts to an extra month of work each year (Hochschild 1989).

Increased immigration has provided a class of women who will do child care and cleaning for affluent, dual-career families. But despite changing attitudes among couples and media portrayals of two-earner couples who share housework, women in fact continue to do more of it. Although the gap has lessened (Artis and Pavalko 2003), data from about 8,500 participants in a University of Michigan study showed that women, on average, spend twenty-seven hours a week on housework (compared to forty hours in 1965), whereas men increased their housework time from twelve hours in 1965 to sixteen hours in 1999, but this dropped down to thirteen hours in 2005 (Institute for Social Research 2002; Swanbrow 2008).

Women's revolutionary entry into the labor force would seem to require a concurrent restructuring of household labor. Husbands *are* doing somewhat more around the house than they did twenty years ago. But "women continue to feel responsible for family members' well-being and are more likely than men to adjust their work and home schedules to accommodate others" (Coltrane 2000, p. 1212).

Who Does the Housework? A researcher commenting on the Michigan study said: "Women have shown a massive decline in the time spent in housework and a massive increase in paid work. Men have picked up a bit of

the slack at home, but at some point have said, 'I've put the dishes in the dishwasher five nights this week. What else do you want from me?'" (K. Peterson 2002, p. D-06). Husbands are typically more willing to do child care—especially "fun" activities—than housework, although nearly as many women take care of the children on a daily basis as well as do the housework (S. Berk 1985; Hochschild 1989; Newport 2008). According to the U.S. Bureau of Labor Statistics *American Time Use Survey*, women spend 2.1 hours daily engaged in domestic chores, while men spend only 1.3 hours. Women spend 1.73 hours caring for children (such as bathing, feeding, reading to, playing with, and so on), whereas men spend 0.84 hours in these activities (2009d, Table 9; Hartmann, English, and Hayes 2010, p. 4). This imbalance exists regardless of employment status (Galinsky, Aumann, and Bond 2009).

Over the long history of the United States, men's participation in household labor has been consistently less than that of women, and it has long been thought to be generally related to the degree of equality of earnings between the spouses and the proportionate share of those earnings produced by the wife. Where the disagreements amongst social scientists occur is the impact that men's unemployment has on their level of participation in domestic chores and child care. For example, some researchers thought that when men are unemployed they may actually do less (Coltrane 2000)—that perhaps being a breadwinner is so symbolically important that unemployed family men are reluctant to do anything that might seem to undermine their manhood, such as labor traditionally considered women's work (Shelton and John 1993).

Recent research into this issue suggests a more complex picture. For example, in 2008, it was found that employed women spent an average of 2.52 hours in housework and caregiving per day, whereas an employed man spent 1.59 hours doing the same housework and caregiving. Unemployed women, in 2008, spent an average of 3.96 hours engaged in domestic duties and child care, whereas unemployed men spent an average of 2.87 hours per day (Hartmann, English, and Hayes 2010, Table 5; U.S. Bureau of Labor Statistics 2009d). These data appear to reinforce what authors and journalists are suggesting: that, after the initial trauma (or blow to the ego) of job loss, unemployed men do increase their participation in housework and child care duties (Cooke 2006; della Cava 2009, Eckel 2010). A longitudinal study of women's and men's participation in household duties from 1976 to 2005 shows that women are engaged in fewer hours of domestic chores while men are doing more. For example, the University of Michigan's Panel Study of Income Dynamics, conducted since 1968, shows that women spent twenty-six hours per week doing housework in 1976, whereas they spent approximately seventeen hours per week in

2005. During this same time frame, men's housework doubled from six hours per week to thirteen hours per week; men's child care duties have also tripled since the mid-1960s (Galinsky, Aumann, and Bond 2009; Kelleher 2007; Sullivan and Coltrane 2008; Swanbrow 2008).

This pattern of less housework time also was thought to characterize those men whose wives earn more than they do (Brines 1994; Hochschild 1989; Tichenor 1999; see also Kroska 1997), but other research suggests quite the opposite. According to social researcher Stephanie Coontz, married women who outearn their partners have husbands that contribute more to household duties and are more likely to have greater marital stability (Coontz 2007; see also Bianchi, Robinson, and Milkie 2006; Cooke 2006; Crary 2008; Sullivan and Coltrane 2008).

Ways in which individual families manage vary. Some two-earner couples hire household help, especially in upper-income white families, and purchase the services of immigrant, racial/ethnic minority, and working-class people for housekeeping and child care work or other chores (Coontz 2007; Ehrenreich and Hochschild 2002; Kelleher 2007). Researchers note that women more often coordinate paid services as well as do more housework themselves.

Another housework option might appear to be help from children. Some studies find that children, especially in single-parent families, help significantly with housework, while others find that children in married-couple families do more work. In any case, "while many children do some household labor . . . their contribution is typically occasional and their time investment small" (Shelton and John 1996, p. 311; see also Coltrane 2000, pp. 1225–26).

Another issue is that housework can have a meaning beyond simple household maintenance. Performing certain household tasks considered traditionally feminine (or masculine) may reinforce a masculine or feminine gender identity. "Housework is not just the performance of basic household tasks but it is also a symbolic expression of gender relations, particularly between wives and husbands" (Artis and Pavalko 2003, p. 748; see also Gilbert 2008). Recent polls show that even with the increased egalitarian division of household labor, there remains a tendency to split the tasks along gender lines. For example, men say their household tasks tend to include car upkeep, yard work, and investment decisions, whereas women tend to do more meal preparation and dish washing, grocery shopping, cleaning the house and laundry, and child care (Newport 2008). That said, however, this division of labor does not seem to incorporate the "moral quality" that it was once thought to have.

Researchers at the Families and Work Institute note that regardless of the "precise objective degree of responsibility men are assuming for various aspects of family work, it has clearly become more socially acceptable for men to be and to say they are involved in child care, cooking and cleaning over the past three decades than it was in the past" (Galinsky, Aumann, and Bond 2009, p. 18). American couples who divide the household duties more equitably have greater marital satisfaction and lower incidences of divorce than couples with more traditional divisions of household labor (Cooke 2006; Hall and MacDermid 2009; Sullivan and Coltrane 2008). Psychologist Joshua Coleman suggests that men's participation in domestic chores offers increased opportunities for intimacy amongst couples, noting "if a woman feels stressed out because the house is a mess and the guy's sitting on the couch while she's vacuuming, that's not going to put her in the mood" (Crary 2008).

Race/Ethnicity and Other Factors In some ethnic groups, such as Vietnamese and Laotian, housework is significantly shared, if not by husbands, by household members other than the wife/mother (P. Johnson 1998). Among African Americans, adult children living at home, extended kin, and nonresident fathers are likely to share housework and child care (Gerson 2010). The latter may provide child care or help with repairs.

Children do some household labor, but it is more often a socialization device or family group activity than a substantial sharing of parents' household tasks.

Anderson Ross/Jupiterimages

Research on racial/ethnic differences finds that the pattern of men's spending less time than women in housework occurs in white, black, Asian Indian, and Hispanic families. However, black men spend more time in unpaid family work than do white men (Bhalla 2008; Barajas and Ramirez 2007; Gerson 2010). One explanation offered for black men's greater participation in housework is that they have more egalitarian attitudes, at least in this domain, and that African American wives are more likely to be employed and to have earnings that are closer to equality compared to their husbands than is true for other groups (Forry, Leslie, and Letiecq 2007; Gerson 2010). However, when factors other than race/ethnicity that affect men's household labor were taken into account—such as age, number of children, sex-role attitudes, and wives' sex-role attitudes—race/ethnicity was no longer so significantly associated with household labor time. In other words, the differences among white, black, and Hispanic men's household labor time may reflect other differences among them, as well (Coontz 2007; Kelleher 2007; Sullivan and Coltrane 2008).

Is Housework Vanishing? One of the ways in which families have adjusted to women's entry into the labor force is to scale down what is thought necessary—assisted by microwaves, fast food, and so forth, and sometimes by paid services. The University of Michigan researchers use the term *vanishing housework* in noting that as men and women are both putting in more hours of employment, the total amount of time a couple spends on housework has declined. "This may mean that women who work, and especially those who work in high-paying jobs, cut back the amount of time they spend on cooking and cleaning by living with a little more dust and baking fewer homemade cookies, or simply that, because both spouses work, these families are more likely to be able to hire someone to do housework while both spouses maintain careers" (Achen and Stafford 2005, p. 12).

Another reason for the decline in housework may be a related change in culture. A study that looked at different cohorts of women found that younger women do less housework, suggesting that "socialization about family life, gender, and household labor may have been substantially different for newer cohorts" (Artis and Pavalko 2003, p. 758).

The Leisure Gap Women interviewed by sociologist Arlie Hochschild

> tended to talk more intently about being overtired, sick, and "emotionally drained." . . . They talked about how much [sleep] they could "get by on.". . . These women talked about sleep the way a hungry person talks about food. (Hochschild 1989, p. 9)

She and other researchers concluded that the second shift for women means a **"leisure gap"** between

After a long day on the job, Cabral and Denys get some sleep on the seventeen-mile shuttle bus trip from the plant to Moline, Illinois, where they live. Longer hours of employment mean that time families spend on housework is "vanishing."

husbands and wives, as women sacrifice leisure—and sleep—to accomplish unpaid family work.

But according to recent research, the leisure gap seems to have vanished, at least so far as work demands are concerned. In their research based on time diaries, Bianchi, Robinson, and Milkie (2006) add employment hours and household work hours to get total time spent in work for men and women. Men, it is true, spend fewer hours in housework, but they spend more in paid employment. North American women spent an average of fifty-two to fifty-seven hours on employment plus domestic work (van der Lippe 2010, p. 51; Treas and Drobnič 2010).

Still, women have half an hour less than men of leisure time. Moreover, a lot depends on what is meant by "leisure." "Because women tend to be the coordinators of family life, it is often difficult for them to take time for themselves independent of household responsibilities" (Mattingly and Bianchi 2003, p. 1001). What counts as leisure time for women often involves their organizing of family activities for others. For example, while a mother is enjoying a child's birthday party, she is simultaneously managing the occasion. Mothers' ostensibly "leisure" time includes time spent with children and a great deal of multitasking, or "contaminated" leisure, as they do household tasks or supervise children while engaging in recreational activities. Free time away from the household is less available to women than men (Mattingly and Bianchi 2003).

Another component of leisure time is not related to a division between genders, but to a national division. The United States has the highest annual average of weeks worked than any other country. Americans worked an average of 46.7 weeks in 2005 (most recent data available), whereas the average for workers in the other industrialized nations was 42.6 weeks. Because the United States does not have federally mandated minimum vacation time nor federally mandated paid holidays (as opposed to all other members of the Organization for Economic Cooperation and Development—OECD), Americans work more during the year than all other workers in industrialized nations (Mishel, Bernstein, and Shierholz 2009, p. 367, Table 8.6). The implications of this for American families is important because the inability to escape work-life demands, even for a short time, are correlated with depression, anxiety, marital conflict, and so on.

Fairness and Marital Happiness A conclusion easily drawn from research is that employed women are carrying an unfair share of domestic tasks. But do couples themselves see it that way? That depends on the meaning of household work to the couple and what they consider "fair."

Although, overall, unequal shares of household labor are associated with marital dissatisfaction, this relationship is altered by perceptions of fairness. Citing a number of studies, Michelle Frisco and Kristi Williams (2003) found perceived fairness to be more strongly associated with marital happiness (and, in their own study, with the likelihood of divorce) than differences in actual hours spent in domestic work. To a wife or woman partner, a man's taking up *some*, if not an equal share of, household tasks may signify caring.

Interesting is that men and women perceive "fair share" differently. Of those men in dual-earner families who perceived that they were doing *more than their fair share*, 43 percent were actually doing *less than half* the housework. In other words, they did not think it would be fair for them to do as much as half the housework. Meanwhile, of women who perceived themselves to be doing a *fair share*, almost two-thirds were doing *all or more* of the housework. An uneven split seemed fair to them (Frisco and Williams 2003 [the Frisco and Williams study used the *Marital Instability Over the Life Course* data set]; White and Booth 1991; see also Greenstein 2009 for international comparisons). Probably both men and women have in mind as a standard of comparison the breadwinner–housewife model. Men, then, are doing more, whereas women are unloading some of their former responsibility. That seems "fair." And men may lump housework and employment together and add that up to feel the total burden of family responsibility is a fair one (Lavee and Katz 2002). The relatively comparable men's and women's total hours of paid and household work reported by Bianchi, Robinson, and Milkie (2006) would support this conclusion.

Gay Men, Lesbians, and Housework Given that gay couples are composed of people of the same gender, how does their household division of labor work out and what impact does it have on the relationship?

A small qualitative study of forty-three gay male and thirty-six lesbian couples explored these questions. Each partner was employed full-time, and there were no children residing with the couples (which is the majority pattern among gay/lesbian couples). The study looked at who performed some traditionally female tasks—it provides some interesting insights, though given the small sample, it cannot be conclusive.

Partners were asked how often they performed six tasks compared to how often their partner did. Generally, lesbian couples' division of labor was more egalitarian than that of gay male couples. The researcher was most interested in the impact on the relationship. Perceived equality was closely tied to relationship satisfaction and that, in turn, to relationship stability (Kurdek 2007). As with heterosexual couples, research shows that the amount of hours individual members of a

The second shift is probably more enjoyable when shared by both partners.

couple work in paid labor has the greatest influence on the division of household labor (Sutphin 2006).

In the next section, we will examine how partners juggle household labor demands, along with employment.

Juggling Employment and Family Work

The concept of juggling implies a hectic and stressful situation. A great deal of research and other writings on the subject suggest that today's typical dual-earner family or a family with a working single parent is a hectic one (e.g., Hochschild 1989, 1997). This is particularly true when children are in the home, and more so for single men and married and single women than for married men due to the greater "role overload" of the first three groups (Kiecolt 2003, p. 34).

Work, Family, and Leisure: Attitudes and Time Allocation

An influential study of changes in time at work concluded that working people are spending significantly more hours at work than in the recent past (Schor 1991). In their book based on data from the Current Population Survey, sociologists Jerry Jacobs and Kathleen Gerson affirmed this conclusion. They point to an "increasing mismatch between our economic system and the needs of American families" (Jacobs and Gerson 2004, back cover)

American workers lead the industrial world in the number of hours worked—with the average worker working 8.44 hours a day or 42.5 hours per week, and producing "$63,885 of wealth per year, more than their counterparts in all other countries" ("U.N.: U.S. Workers" 2007; U.S. Bureau of Labor Statistics 2009d, Table 4; U.S. Census Bureau 2010b, Table 589). A little over 26 percent of all employees now work more than forty hours per week, with 10 percent working between fifty and sixty hours a week. Just over 5 percent of the labor force held two or more paid jobs in 2008, with slightly higher proportions of women than men (U.S. Census Bureau 2010b, Tables 588, 589, 596).

Virtually every researcher studying work–family time hears expressions of time pressure, and of feeling rushed and stressed (e.g., Jacobs and Gerson 2004; Bianchi, Robinson, and Milkie 2006). Yet Bianchi and her colleagues argue that such stress is concentrated in the children's early years and especially for women with demanding careers, as well as for single mothers, whereas Jacobs and Gerson see more pervasive problems for working families. These different conclusions as to whether work hours have increased are difficult to resolve, but they seem to reflect methodological

differences. Bianchi and her colleagues argue that their time diary methodology is more accurate because it is specific and timely—study participants are "walked" through activities of the preceding twenty-four hours. Jacobs and Gerson point to the small size of time diary samples and to other methodological issues and argue the merits of their Current Population Survey data.

Jacobs and Gerson do a careful analysis of theirs and the time diary studies and conclude that each is measuring different things. The hours worked per week by each employee have not changed much over recent decades. But the weeks devoted to work by family members have increased dramatically because of women's entry into the labor force. Total hours of work are expanded by the increasing tendency for women to work full-time and not leave the labor force for an extended period.

Some interesting research has been designed to assess the impact of women's entry into the labor force on health, marital quality, and marital stability. One such study has found women to have increasingly good (self-reported) health as labor force participation and working hours have increased. Both women's increased education and their employment have contributed to better health, contrary, perhaps, to expectation. Although there was a short-term diminishment of health attributed to the stress of coping with work and family in children's younger years, once children entered school, superior health rebounded. Overall, women have gained in health as employment has become the norm (Schnittker 2007).

A Gallup poll asked individuals whether they have enough time to do what they want. About 55 percent of those age eighteen through thirty-four and 58 percent of those aged thirty-five to fifty-four—the ages of employment and active parenting—say they do not have enough time (Carroll 2008). For a majority of Americans, rest and relaxation time, time for friends and hobbies, and even time for sleep is not what they would

"Hey, Baby, I just dropped the kids off at school, and now I'm going to the grocery store, and then I'm going home and unloading the car—am I making you hot?"

We've talked about employment and household labor. What do people do with the rest of their time?

An American Time Use Survey was conducted in 2008 by the U.S. Bureau of Labor Statistics (2009d). Some 21,000 people were asked to keep time diaries, recording what activities they engaged in and for how much time. The reports of these many individuals were averaged to come up with typical days for different groups. Let's look at employed parents of children under eighteen and see what happens in an average day.[a]

First there are the basics. Just over an hour (1.14 hours) was spent in *eating and drinking*. Around nine hours (9.15) were spent in such *personal care* activities as sleeping, bathing, dressing, and health care.

Work averaged just under six hours (5.77) for men, and just over three hours (3.3) for women (remember that some people work part-time and that all days are not workdays). On average, employed women spent almost two and a half hours (2.36) on *household activities*—the domestic labor—while these tasks occupied an hour and a quarter (1.25) of men's time.

These totals did not include child care, which fell into the category of *caring for and helping household children.*

Women spent an hour and a half (1.51) on caring for household children, while men spent less than an hour (.85).

Shopping—*consumer goods purchases,* as the survey termed it—took almost an hour of an employed mother's daily time (.49), while fathers devoted a half an hour to shopping (.28). Men were able to devote more of their time to *leisure and sports* (4.42 hours) than were women (3.93 hours). The most common use of leisure time for all was watching television.

Men and women participated in *organizational, civic, and religious activities* at about the same rate, each spending about a half of an hour on an average day. Men spent an average of .12 and women spent an average of .16 hour in *caring for and helping nonhousehold* members; and in *educational activities* women spent nearly double the amount men did. *Telephone calls, mail, and email* communication rounded out the day (at .08 hour for men and .22 hour for women).

So what does it all mean? Despite amounts of time in some categories so small as to seem trivial, we can see some interesting things in these figures. The American Time Use Survey shows that women spend more than double the time than men do in the care of the household

and its members, adding one more study to those that show a gender disparity. However, men spent a great deal more time in paid employment than did women.

In other areas, there is little difference in time use between men and women. The time men and women devote to organizational, civic, religious, and educational activities and to helping nonhousehold members is similar. Still, men have more leisure time, whereas women do more shopping (more likely to be grocery shopping than "fun" shopping). Women spend twice as much time as men in communication activities and more time on personal care, a large portion of which is sleeping.

Critical Thinking

How do you spend *your* time? Has the time you spend in various activities changed throughout your life? These data are for employed people who have children at home. If your situation is different, is your time use different as well?

Source: U.S. Bureau of Labor Statistics 2009d, Table 8.

a. This survey divides people by those with children under six years and those with children between the ages of six and seventeen. To simplify the table, we added the percentages from both and divided by two.

---

like it to be (Saad 2004). For more on these and similar issues, see the "Facts About Families: Where Does the Time Go?" box, which reports data from a major government survey on how employed parents spend their time on an average day.

Another study using a national data set found that "wives' full-time employment is associated with greater marital stability" while not affecting quality one way or the other (Schoen, Rogers, and Amato 2006). Still another study looked at gender changes and marital quality and found that most gender-related changes had no negative impact on marital quality, but there was increased marital conflict attributable to "work-family demands based on the combination of wives' employment and preschool-age children" (Rogers and Amato 2000, p. 747). This suggests, once again, that perhaps the negatives for families with employed women are focused on the preschool years of those with children.

There is some indication that younger workers have different attitudes toward work–family balance than did their predecessors. Social scientists see a "gender convergence" in attitudes and values regarding work and family roles. Both men and women want a balance of work and family in their lives (Cohen 2007b, p. A13; Monahan Lang, and Risman 2007).

A 2002 study sponsored by the American Business Collaboration (composed of such prominent corporations as IBM and Johnson and Johnson, with additional support from the Ford Foundation) and conducted by the Families and Work Institute surveyed some 2,800 adults from four generations of workers: (1) "matures" (born 1945 and earlier); (2) "baby boomers" (born between 1946 and 1964); (3) "Generation X" (born between 1965 and 79); and (4) "Generation Y" (born between 1980 and 1994). Respondents were asked if they put work before family ("work-centric"), put family before work ("family-centric"), or prioritized both equally ("dual-centric").

As Figure 11.4 indicates, a majority of the two youngest generations described themselves as family-centric, some as dual-centric, with few likely to style themselves as work-centric. Their predecessors, the baby boomers, were comparatively more work oriented and less family oriented, although a majority of boomers selected the dual-centric and family-centric orientations together. Matures (not included in Figure 11.4) were similar to Generations X and Y in being less work-centric, but were also less family-centric than all other generational groups. A majority were dual-centric (Families and Work Institute 2004).

We will look now at how children in two-earner marriages are doing, and then at parents.

How Are Children Faring?

Before women with children entered the work force in large numbers, working mothers were considered problematic by child development experts and the public. Now they are taken for granted. A 2001 survey of women (not all of them mothers) found more than 90 percent in agreement with the statement that a woman can be a good mother and have a successful career (Center for the Advancement of Women 2003). More recent

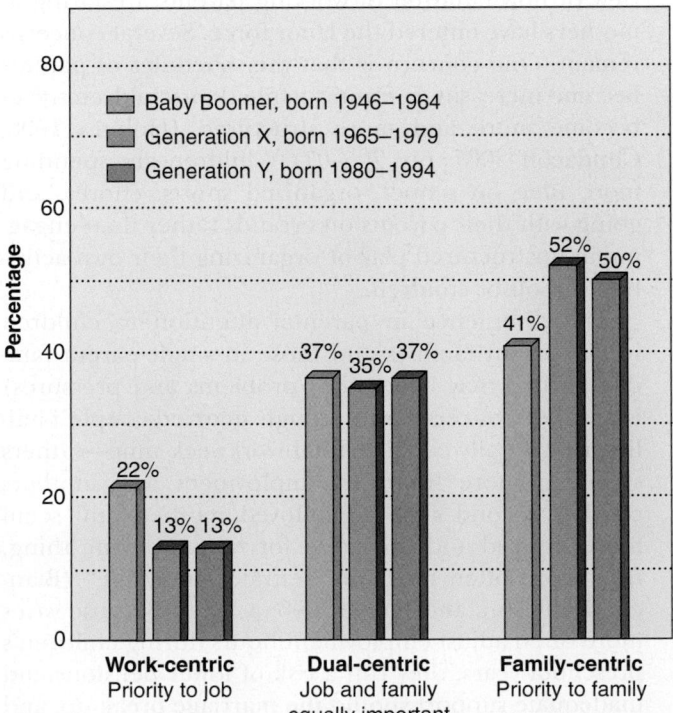

Figure 11.4 Priority given to work, family, or both by 2,800 workers surveyed in 2002: Generational differences

Source: Families and Work Institute 2004.

"Quality time? Do I have to?"

studies conclude that maternal employment does not cause behavior problems in children (Vander Ven et al. 2001; see also Agee, Atkinson, and Crocker 2008), and another study of more than 6,000 children studied at age twelve found no difference between the children whose mothers were employed or not employed during the child's first three years (E. Harvey 1999). What is notable in much of the research, however, is the startling correlations between low family income and childhood problems (Jackson, Choi, and Bentler 2009).

Overall, this continues to be the prevailing view. Furthermore, the economic benefit to children of working mothers cannot be overlooked. Family income tends to be favorably associated with various child outcome measures. Important for parents, though, is keeping their child's needs in the forefront in the face of daily pressures. Recent studies have found that mothers who work part-time are better at this than those who work full-time—and may indeed spend more time helping their children with homework than even full-time homemakers. Before the era of working mothers, so-called full-time mothers did not spend all their time with children, but devoted more time than today's mothers to household work or volunteer work. And some of those mothers—of larger families, especially—made use of paid help in caring for children.

"The puzzling thing about the reallocation of mothers' time to market work [employment] is that it appears to have been accomplished with little effect on children's well-being," noted sociologist Suzanne Bianchi in her presidential address to the Population

Association of America (Bianchi 2000). A variety of studies indicate that parents today spend as much or more time with children as in the past (Milkie et al. 2004; Bianchi, Robinson, and Milkie 2006; and see Figures 11.5 and 11.6). Figure 11.5 presents total weekly hours parents spent with children in 1975 and 2000. That time has increased for married fathers and married mothers. It has, however, decreased for single mothers, though it remains substantial.

The data on "total time" reflect the time a parent spends in the presence of children. Figure 11.6 presents the time parents spend in "primary child care," that is, active caretaking, whether that takes the form of routine care or enrichment activities. Mothers and fathers spent more time in child care in 2000 than did parents in previous measured years going back to 1965. (Time spent doing child care was measured in time diary studies done by various universities using samples ranging from 1,200 to over 5,000; Bianchi, Robinson, and Milkie 2006, Chapter 2).

How did mothers, especially, accomplish an increase in time with children while at the same time dramatically increasing employment hours? (Incidentally, time spent in personal care [sleeping, grooming and eating] and free time remained relatively stable [Bianchi, Robinson, and Milkie 2006, Figure 5.1]). They cut back on housework

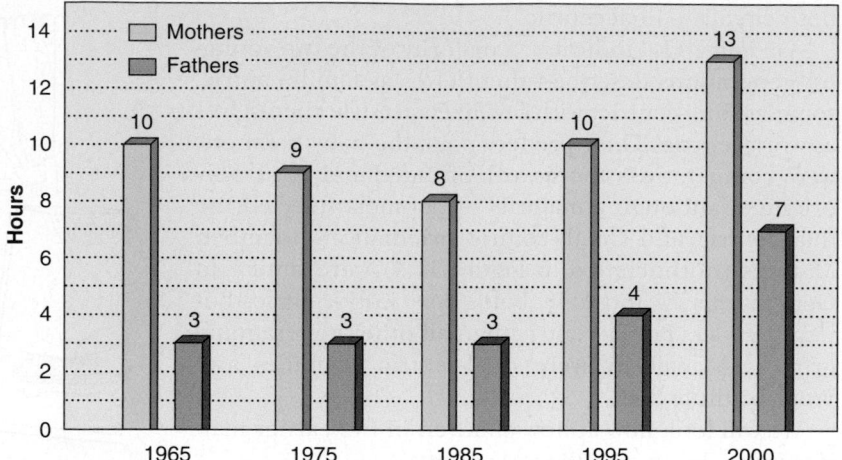

Primary Child Care, Average Weekly Hours

Figure 11.6 Primary child care, average weekly hours, mothers and fathers, 1965–2000

Source: From Bianchi, Robinson, and Milkie, *The Changing Rhythms of American Family Life*, 2006, Figure 4.3, p. 72. Reprinted by permission of the Russell Sage Foundation.

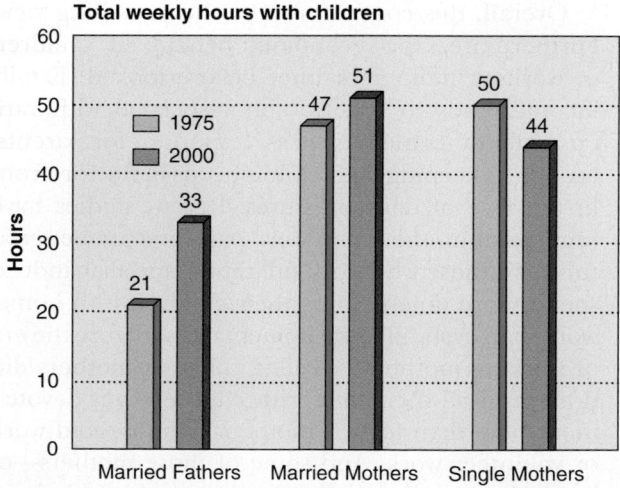

Figure 11.5 Total weekly hours spent with children for married fathers, married mothers, and single mothers, 1975 and 2000

Source: From Bianchi, Robinson, and Milkie, *The Changing Rhythms of American Family Life*, 2006, Figure 4.1, p. 63. Reprinted by permission of the Russell Sage Foundation.

and spent somewhat less time doing things just with spouses. In part, they multitasked; parents spent time with children (and often each other) in children's activities or those of the parents. Today's families are smaller, so that parental attention is less divided; moreover, the increase in father's time with children (for married parents) means increased total attention for children.

These results present an optimistic and reassuring view of how children of working parents are faring as mothers have entered the labor force. Several concerns remain. One concern is that the schedules of parents become increasingly frantic, while their children's lives become more and more structured (Holmes 1998; Chudacoff 2007, pp. 205–07). Children are spending more time on school, organized sports, chores, and going with their parents on errands rather than engaging in unstructured play or organizing their own activities with other children.

The divergence in parental attention to children in two-parent families and those in single-parent families (which may have other problems and pressures) is another concern. And though married-couple families are virtually equal in total workweek time—fathers spending more hours in employment and mothers on the "second shift"—employed mothers still seem more stressed and pressured for time. For one thing, they more often must "orchestrate family life" (Bianchi, Robinson, and Milkie 2006, p. 171). Because wives more often adjust employment hours during children's preschool years, they run a risk of lower pensions and inadequate support should the marriage break up, and they may not have the careers they might have had otherwise. This leads us to the issue of how parents are faring as they juggle paid and unpaid work.

How Are Parents Faring?

This chapter focuses on work and family in marriages (rather than other family forms) for two reasons. First, the vast majority of research on the interface between paid employment and family labor concerns marrieds. Second, single parenting is addressed in some detail in Chapter 15.

> Although the rough edges of the work–family conflict may be particularly sharp for single parents, two-earner marriages assuredly have them also. Whether one is single or married, "career and family involvement have never been combined easily in the same person." (Hunt and Hunt 1986)

An ideal for modern family is to share wage-earning and family responsibilities on an equal basis and, for men, to be an involved father. Indeed, more fathers are taking off work following the birth of a child, and they are more visible in parenting classes, in pediatricians' offices, and dropping off and picking up children in day care centers.

A man, regardless of parental status, who gives family priority may have to deal with challenges to his masculinity or resentment from coworkers. Employers may not see unpaid family work as important or believe that employees, especially males, should allow family responsibilities to interfere with labor force involvement (Hochschild 1997). As a result, workers report that they are reluctant to take advantage of family benefits that are theoretically available (Jacobs and Gerson 2004, p. 6). Some husbands report having lied to bosses or taken other evasive steps at work to hide conflicts between job and family. One man told his boss that he has "another meeting" so that he can leave the office each day at 6 p.m.: "I never say it's a meeting with my family."

The previous discussion applies to working parents generally. In the next section, we will look at some stresses peculiar to *two-career* marriages—keeping in mind the distinction between *two-earner* couples and *two-career* couples made earlier in the chapter.

Two-Career Marriages Some decades ago, as the two-career marriage was emerging as an ideal lifestyle available to all young couples, Hunt and Hunt (1977) noted that dual-career families require a support system of child care providers and household help that depends heavily on ability to pay. That means it is inherently limited to a small number of families. Moreover, the success of today's two-career union is premised on the existence of a labor pool of low-paid, but highly dependable, household help. The vast majority of such help is provided by women, many of whom have their own families to worry about (Romero 1992). Even parents who can afford to pay for it find that locating such help may be difficult.

Two-career partners need the dexterity to balance not only career and family life but also her and his careers so that both spouses prosper professionally in what they see as a fair way. Two careers requiring travel may present added problems as parents "scramble to patch things together" for overnight child care (Shellenbarger 1991). The balance between partners may be upset by career fluctuations as well as family time allocations. The contrast between one career that is going well and one that is not may be hard on the partner on the down side. But the marriage may "operate as a buffer, cushioning the negative impacts of failures or reversals in one or the other career" (Hertz 1986, p. 59). When the marriage is rewarding, compromises, such as turning down opportunities that would require relocation, are acceptable because of the importance given to marriage as well as career.

Sociologist Rosanna Hertz (1986) found that the two-career couples she studied were realistic, though sometimes regretful, about some benefits of the traditional relationships they are giving up. Although men acknowledged that their wives provided less support, they appreciated the excitement, and the status, associated with an achieving wife. Some of them had considered or made career changes that would not have been possible if wives had not been successful wage earners. Both partners claimed fulfillment from and assigned emotional meaning to an egalitarian dual-career marriage: "She has a sense of a full partnership and she should" (in Hertz 1986, p. 75).

Commitment was perceived as truer: "Working . . . has decreased my dependence. . . . That makes it into much more of a voluntary relationship" (in Hertz 1986, p. 75). Very visible to Hertz was the way in which communication was enhanced by similar lives, making possible a higher level of mutual support than in conventional couples: "These couples . . . [had] a different level of understanding about each other's lives, a level that is intimate and empathic" (p. 77).

The couples Hertz studied did report conflict over balancing time, commitment, and career moves. Indeed, the geography of two careers presents a significant challenge to couples.

The Geography of Two Careers Because career advancement often requires geographic mobility—and even international transfers—juggling two careers may prove difficult for married or committed partners. A career move for one may make the other a **trailing spouse** who relocates to accommodate the partner's career. Increasingly, couples turn down transfers because of two-career issues. As a result, some large companies now offer career-opportunity assistance to a trailing spouse, such as hiring a job search firm, facilitating intercompany networking, attempting to locate a position for the spouse in the same institution, or providing career counseling (Jio 2008).

Although wives still move for their husband's career more often than the reverse, the number of trailing husbands has increased, and studies show that the benefits are much better for the trailing husband than for the trailing wife (Shauman and Noonan 2007). More two-career marriages today are based on a conscious mutuality to which partners have become accustomed by the time a career move presents itself. Such couples are less likely to have problems with a female-led relocation than are more traditional marrieds. For many spouses, trailing is preferable to commuting, another solution to the problem of career opportunities in two locations.

To Commute or Not to Commute? Social scientists have called marriages in which spouses live apart **commuter marriages**. The vast majority of commuting couples would rather not do so, but endure the separation for the sake of career or other goals. Since research began on commuter marriages in the early 1970s, social scientists have drawn different conclusions. Some studies suggest that the benefits of such marriages—greater economic and emotional equality between spouses—counter their drawbacks. Other research focuses on difficulties in managing the lifestyle. One conclusion to be drawn from the research is that commuters who are able to have frequent reunions are happier with the lifestyle than those who cannot.

In 2008, just under 3.5 percent of married Americans lived apart (U.S. Census 2010b, Table 66). One study of commuter couples compared life satisfaction for 90 commuting and 133 single-resident, two-career couples. Almost three-fourths of the commuters saw their partner weekly. The researchers were surprised to find that the commuters experienced less stress and overload than the single-residence couples: "Perhaps there is some restructuring in the commuting two-residence couple that simplifies life or perceptions of it. Perhaps short separations facilitate compartmentalization, allowing commuters to keep work life and family life in well-separated spheres, and to confront the demands of each role in alternation rather than simultaneously" (p. 405). Then, too, the commuter couples had significantly fewer babies and young children than did single-residence couples; commuter marriages probably work better in the absence of dependent children (G. Stern 1991).

Commuter marriages are not new or novel—American men have long worked in transient professions where they are gone for extended periods of time (that is, truck drivers, traveling sales, soldiers, and so on). What is different about modern commuter couples is that it is increasingly common for the wife to be the commuter in the couple, instead of the husband. The introduction of children increases the stress in commuter marriages—especially for a spouse who is the primary caregiver, because that spouse becomes a de facto single parent for the duration the commuting spouse is absent. It is suggested that approximately 800,000 children in the United States live in commuter families.

"Every Monday, Jaime Cangas, 40, kisses his wife Karen, 36, goodbye as she leaves their Plano, Texas, home and heads toward the airport. As a consultant for Accenture, she will be gone until late Thursday night, working with clients in faraway cities. Jaime, who sells and markets security software, will drop off their children, Caroline, 7, at school, and Mitchell, 3, at day care. He shops for groceries during his lunch break, then picks them both up at 6. When they get home, the kids blow kisses at Mommy through the webcam" (Cullen 2007). Although this arrangement has its stressors, research suggests more serious strains come about when the commuting ends and the couple or family reunites under the same household. Each member in the relationship has become used to the living arrangement, so an adjustment period becomes necessary after the commuting spouse moves back into the family home (Tessina 2008).

Couples who have been married for shorter periods seem to have more difficulties with commuter marriages. Perhaps because of their history of shared time, more established couples in commuter marriages have a greater "commitment to the unit" (H. Gross 1980; Rhodes 2002).

Social Policy, Work, and Family

Despite the benefits of employment to women and their families, and despite societal pressures and gender-role changes leading to high female employment, neither public policy nor families have fully adapted to this change. This section examines policy issues regarding work and family. Policy issues center on two questions: "What is needed?" and "Who will provide it?"

What Is Needed to Resolve Work–Family Issues?

Researchers and other work–family experts are in general agreement that single-parent and two-earner families are in need of more adequate provisions for child and elder care, family leave, and flexible employment scheduling.

Child Care Policy researchers define **child care** as the full-time care and education of children under age six, care before and after school and during school vacations for older children, and overnight care when employed parents must travel. Child care may be paid or unpaid and provided by relatives or others, including one of the parents.

In her study of dual-earner couples, sociologist Rosanna Hertz (1997) explored parents' approaches to child care and found they fell into three categories. One is the **mothering approach to child care**, whereby the couple prefer that the wife care for the children. An initial strategy of overtime or a second job for the husband often proves to be unworkable, so the wife does have to enter the labor force. But the couple maintains as traditional a division of labor as they can, with the mother working as much as possible during hours the children are sleeping or in school.

In the **parenting approach to child care**, parents share family care, and structure their work to this end. They accept part-time work, for example, and the lower incomes that go with it. But primarily, these are "labor force elites" (Hertz 1997, p. 370), who can be sure of commanding a full-time job when they want to, or whose part-time earnings produce substantial income. In blue-collar or lower-income families, shift care or the periodic unemployment of men produces a parenting approach. In 2005, 18 percent of fathers were principle sources of child care for preschoolers during mothers' day shift (U.S. Census Bureau 2008c).

In the **market approach to child care**, career-oriented couples hire other people to care for their children. We now look at child care in this sense. There are essentially three types of nonrelative child care. Paid care may be provided in the child's home by a nanny, an **in-home caregiver** who lives in or comes to the house daily. The term **family child care** refers to care provided in a caregiver's home, often by an older woman or a mother who has chosen to remain out of the labor force to care for her own children. Parents who prefer family day care seem to be seeking a family-like atmosphere, with a smaller-scale, less routinized setting. Perhaps they also desire social similarity of caregiver and parent to better ensure that their children are socialized according to their own values.

Center care provides group care for a larger number of children in child care centers. The use of child care centers has increased rapidly, partly because of the growing scarcity of in-home caregivers, as relatives or neighbors who formerly cared for children now join the labor force themselves. Increased use of center care is also due to the perception that it offers greater safety and a strong preschool curriculum.[4]

By the time they enter school, an estimated 44 percent of children have been in a nonrelative child care

arrangement. This is more common for black and non-Hispanic white children, a little less so for Asian or Pacific Islander children, and least common among Hispanics (Dye and Johnson 2007, Table 2). Black mothers, who relied heavily on kin networks in the past, saw that option decline by the 1990s, as grandmothers and other relatives entered the labor force themselves (Brewster and Padavic 2002). Nonrelative care is also more common for children in families above the poverty level than for those in lower-income families (Dye and Johnson 2007, Table 3).

Of children of employed mothers in 2006, 28.8 percent were in organized care—day care centers, Head Start, or preschool. Slightly more than 4 percent had in-home care; 6.2 percent had family day care; and the remainder had other arrangements, no care, or multiple arrangements. Relative care was heavily used—25.6 percent of children were in the care of grandparents, with fathers caring for 24 percent; sibs or other relatives cared for 11 percent; and mothers themselves cared for 5.5 percent (by bringing children to the office or other work site, including self-employment work at home; U.S. Census Bureau 2008b, Table 1B).

Now there is extensive research on the developmental outcomes of various child care arrangements. See "As We Make Choices: Child Care and Children's Outcomes" for a discussion of this research.

About 20.8 percent of children age nine to eleven and 31.3 percent of those age twelve to fourteen whose mothers were employed full time were in **self-care**—that is, without adult supervision—for an average of seven hours a week (U.S. Census Bureau 2008b, Table 4). Self-care is more common in white upper-middle- and middle-class families than in black, Latino, or low-income settings, perhaps because of differences in neighborhood safety (U.S. Census Bureau 2008b, Table 4).

Low-income, single-parent, rural, and Hispanic parents are especially likely to have relatives take care of their children (Capizzano, Adams, and Ost 2006). Hispanic parents seem to prefer either relative or family day care rather than center care, a choice attributed to wanting a "warm and family-like atmosphere" rather than a "formal and cold" child care center. Family day care may also be seen as providing a personal relationship between the parent and the caregiver and, perhaps, a bilingual setting (Chira 1994).

African American parents prefer a center for its perceived educational benefits, whereas white parents' preference is more likely to be for the social interaction experiences a center provides for children. Some families may seek a provider of their own racial/ethnic group who will maintain the cultural context children have at home or, at a minimum, a white caregiver or center that will provide "racial safety"—that is, will not

---

[4] With regard to safety, it is important to note that despite a smattering of confirmed cases, concerns about abuse of children in day care have largely proved unfounded; studies indicate that children are at greater risk of abuse in their own homes (Finkelhor, Hotaling, and Sedlak 1991). "Overall, child care is quite safe" (Wrigley and Dreby 2005, p. 729), and center care is safer than care in private homes.

Facts about Families

Child Care and Children's Outcomes

Parents who have to make decisions about child care want to know two things: What are the characteristics of quality child care? And what effect does being in child care have on children? We address child care quality in "As We Make Choices: Selecting a Child Care Facility." Here we look at outcomes for children who have spent time in child care in their early years.

Psychologist Jay Belsky drew considerable attention when he reported an early finding that infants in their first year who are in nonparental care for twenty or more hours per week "are at elevated risk of being classified as insecure in their *attachments* to their mothers at 12 or 18 months of age" (Belsky 1990, p. 895; 2008, p. 9).[a] This set off the "day care wars" (Carey 2007), in which Belsky has continued to engage in dialogue with other child care researchers about whether time in child care is harmful to children and in what circumstances (2002).

A multiple-site longitudinal study, organized by the National Institute of the Child Health and Human Development (NICHD), began in 1991, and has now followed more than 1,300 children from shortly after birth through sixth grade (Belsky et al. 2007; Belsky 2008). The study looks at the impact of various types of child care compared to maternal care. (Commentators on the NICHD research find it noteworthy that the focus of the NICHD study is on *mothers*. A child is considered to be in care when not with the mother—that is, care by the father is considered "child care" [U.S. National Institute of Child Health and Human Development 2002].) Professor Belsky was one of the thirty researchers initially involved in the study, and has continued to participate. He is one of the authors of the latest reports on the NICHD study.

The conclusion first drawn from NICHD research was that children in nonrelative care and children cared for by their own parents differed little in development and emotional stability at fifteen months and three years (U.S. National Institute of Child Health and Human Development 1999). Yet, at the three-year point, child care did have a negative relationship to maternal sensitivity ("how attuned the mother is to the child's wants and needs") and child engagement ("how connected or involved a child appeared to be when relating to his or her mother"). But that finding did not hold for children in *quality care* (NICHD Early Child Care Research Network 1999a; U.S. National Institute of Child Health and Human Development 1999; Belsky 2008).

Moreover, favorable outcomes in terms of cognitive and linguistic skills were associated with *quality* of care, described as "when child care providers talk to children, encourage them to ask questions, respond to children's questions, read to them, challenge them to attend to others' feelings, and to different ways of thinking" (U.S. National Institute of Child Health and Human Development 1999; NICHD Early Child Care Research Network 1999b; Belsky 2008).

As the children approached age five and entered kindergarten, those who had spent longer hours in child care over time were found to have more behavior problems and conflicts with adults (as reported by parents, teachers, and the children themselves). That was true even when quality and type of care were taken into account (NICHD Early Child Care Research Network 2003a).

Although Professor Belsky made much of the results, his co-researchers argued that these were not serious problems—that, in fact, more serious behavior problems were evidenced by children who had not been in day care at all. They also pointed to the fact that problem behavior was confined to a minority of day care children—more than 80 percent of children long in care did *not* exhibit any behavior problems ("Day-Care Researchers" 2001). In this view, the findings had a "lack of clinical significance" (Dworkin 2002, p. 167), meaning that they did not signal a level of trouble that should cause concern. New research upholds these findings but does suggest a slight correlation between length of day care attendance and impulsiveness and risk taking in adolescence, particularly in children who were placed in low-quality day care where the caregivers were more authoritarian and children's free play was limited (Fox 2010; Stein 2010; Vandell et al. 2010).

Moreover, children in high-quality center care outperformed children not in care in measures of cognitive skills and language development (NICHD Early Child Care Research Network 2000b). Family background factors and maternal sensitivity were more important in their impact on children's adjustment than was time in child care (NICHD Early Child Care Research Network 2003a). The small size of the negative effects, the good adjustment of the preponderance of children, and the greater importance of parental influence in terms of the effects of extended child care were reassuring (Vandell et al. 2010).

Research continued through the first 4½ years of the children's lives, their day care years. The latest report

act in a racist way with their children (Uttal 2004). Black children are "more likely than the other groups to be in center-based care across most categories of children examined. The same can be said for the low use of center-based care and the high use of relative care among Hispanic children: these findings persist regardless of the child or family characteristic examined" (Capizzano, Adams, and Ost 2006, p. x).

Average monthly child care costs for infants exceed the costs a family would spend in food. "A family in the United States with one infant faced average prices in 2008 of $4,560 to $15,895 a year for center-based child

on the NICHD study (Belsky et al. 2007; Vandell et al. 2010) assesses the situation of these children since entering school and draws the following conclusions:

1. Change over time has seen some earlier negative or positive effects vanish, while others take their place. The research, too, captures only one point in time in an ongoing process.

2. Children with more experience in child care center settings continued to evidence more behavior problems; however, these problems are more strongly correlated with low-quality child care and poverty. Also, length of time in nonfamilial child care is associated with impulsiveness and risk taking in adolescence. Why this is so remains a mystery, but the researchers speculate that it may be peer interaction and authoritarian child care providers in a center group setting that elicits disruptive behavior.

3. Nonrelative care, mostly center care, is associated with negative effects; however, high-quality care is a protective factor for children of fragile or at-risk families (Vandell et al. 2010, p. 739; see also Burchinal et al. 2010).

4. High-quality care of any kind is associated with better vocabulary.

5. "[P]arenting quality proved to be a far stronger and more consistent predictor of tested achievement and teacher-reported social functioning than was child-care experience" (Belsky et al. 2007, p. 696).

6. The developmental impact of child care on individual children may thus not be so significant, because child development outcomes of child care are "smaller in size and less pervasive than those associated with families and parenting" (Belsky et al. 2007, p. 698). But the "*collective consequences*" of a certain amount of problem behavior associated with children's child care experience can affect "classrooms, schools, communities, and society at large" (Belsky et al. 2007, p. 698).

7. "The current findings suggest that the quality of early child care experiences can have long-lasting (albeit small) effects on middle-class and affluent children as well as those who are economically disadvantaged" (Vandell et al. 2010, p. 750).

There continue to be criticisms of the study, most notably, the question of a "selection effect." The researchers were not, of course, able to assign families randomly to each type of child care; the choices of the parents may reflect something distinctive about the family that is the key factor affecting outcomes. A different and later study took advantage of developments in methodology to counter the effect of selection bias. The focus of this study was on only the first year of the child's life. Results indicated that the mother's working full-time in the first year was associated with negative cognitive and behavioral outcomes; these negative outcomes did not occur when mothers worked part-time in the first year, postponed work, or did not work for the first three years (J. Hill et al. 2005).

Child care researchers consider the policy implications of the research. Belsky (2002) argues for tax or other policies to support full-time parental care in the home, especially during the first year. NICHD research suggests that intervention programs might be effective in enhancing the parenting skills that their research suggests is more important for development than whether or not the child spends time in day care, especially for those children who did not have high-quality child care in their first years. "Experimental studies of high-quality early intervention programs have demonstrated that these programs can enhance social, cognitive, and academic development of economically disadvantaged children" (Belsky et al. 2007; Vandell et al. 2010, p. 738).

Other child care scholars agree with the NICHD researchers that subsidies should be available to permit parents to cut back work hours. At the same time, they urge attention to improving the quality of out-of-home care. They also believe child care can make a positive contribution to social development if done well (Maccoby and Lewis 2003).

Critical Thinking

If you were the mother of a new baby, would you find this research useful in making your decision about returning to work? Or would you be more inclined to rely on the advice of family members or other parents—or your child's reactions to child care? What would you like researchers to find out about children in child care?

a. "**Attachment** represents an active, affective, enduring, and reciprocal bond between two individuals that is believed to be established through repeated interaction over time" (Coleman and Watson 2000, p. 297, citing Ainsworth et al. 1978).

care" while parents of four-year-old children averaged "$4,055 to $11,680 a year in child care fees in 2008" ("Parents and the High Price" 2009, p. 1). Many parents using paid care change their arrangements each year because a caregiver quits, the cost is too high, the hours or location are inconvenient, the child is unhappy, or the parent dislikes the caregiver. As they struggle to find quality, affordable child care, many parents must make more than one arrangement for each child. As they patch together a series of child care arrangements, the system becomes increasingly unpredictable ("Parents and the High Price" 2009, p. 2).

As We Make Choices

Selecting a Child Care Facility

FOCUS ON CHILDREN

Universal, comprehensive, government-funded day care does not exist in the United States today. Although some parents have access to child care facilities through government programs or their employers, many parents are on their own in selecting a child care facility.

Some parents arrange their work schedules to care for their children, while others hire a nanny or recruit relatives into this role. Here we make some suggestions to parents who are choosing from commercially available child care.

State laws, which vary in both provisions and enforcement, establish minimal standards, and professional organizations like the American Academy of Pediatrics have developed guidelines for quality child care. We outline some of the things we think parents should consider when exploring and choosing child care for their children, drawing on the American Academy of Pediatrics guidelines as well as other sources.

Some criteria are very tangible and specific, like the ratio of children to adults. Some are more qualitative and can best be judged by the parent during visits, including post-placement visits to the child care facility. Some are only applicable to center care, whereas others are relevant to family day care as well.

- *Low child-to-staff ratio.* Positive caregiving is associated with a low child-to-staff ratio, especially for very young children. State guidelines vary, but experts believe they tend to be too minimal. Best would be six to eight infants per two caregivers, six to twelve one- to two-year-olds per three teachers, and fourteen to twenty older preschoolers per two teachers.

- *Stable staff.* Some staff turnover is inevitable, but it should not exceed 25 percent a year. If children must

Child care centers and preschools provide care for many children during the workday. Although children may not receive as much adult attention as they might with a single caregiver or with family day care, they benefit from greater interaction with other children and a preschool curriculum.

constantly adjust to changes in personnel, they cannot build the warm and trusting relationships that they need with caregivers. It is also important to learn how much attention is given to preparing children for a caretaker's departure and to helping them adjust to new staff.

- *A well-trained staff.* Trained staff members are likely to be more responsive, more stimulating, and more creative in their activities with children. Because child care workers are poorly paid, it is difficult to find centers with staff members who are highly educated or trained in early childhood education. The ideal situation is for staff to be knowledgeable about child development and to participate in workshops or other ongoing training in best practices. Ask about staff education and plans for further training. In family day care settings, ask whether other family members or others who are not formally "staff" are nevertheless involved in caring for the children.

- *Cultural sensitivity.* Caregivers should be knowledgeable about the diverse racial/ethnic, religious, and social class cultures of this society and should be aware that children may come from various types of families, such as traditional nuclear, dual-earner, gay/lesbian, single-parent, divorced, or remarried.

- *Other staff qualities.* A warm personality and interpersonal sensitivity are essential. Caregivers who let children express their feelings and who will take their views into account are desirable. Because staff members will have an influence on the child's language acquisition, being verbally fluent and well-spoken is an asset. Some parents may have specific preferences, such as male or female caregivers or both, or minority or bilingual staff. Parents seeking family child care may have a specific type of home environment in mind and should consider how well their values and lifestyle match those of the caretaker.

- *Age-appropriate attention.* Babies need a responsive adult who coos and talks to them. One-year-olds need a staff member who will name things for them. Two-year olds need someone who reads to them. Older children can profit from social interaction and activities with other children as well as with adult caregivers.

- Adults should be responsive to children and interact with them, not limit themselves to a directive, organizing role. Do they greet the child warmly? Do they seem interested in what the child is doing or saying? They should make eye contact and perhaps bend to their level when speaking with children, not brush the children off or have a ho-hum attitude. How staff members interact with children can best be ascertained by observation in visits to the center.

- *Age-appropriate and stimulating activities and play spaces.* Experts differ on how academic a preschool program should be, and parents differ in how "educational" a program they are looking for. Look for a facility that also fosters play and community activities such as trips to the zoo or fire station—and one that prepares children for learning rather than offering a first-grade program in preschool. In any case, parents should pick a child care facility that is a good match for their values in this regard. You should find that staff members have a well-thought-out rationale for their program that they can easily describe.

- On the negative side, avoid child care centers or family day care environments that seem to provide only custodial care or allow lots of TV watching. What kinds of indoor and outdoor spaces are there for constructive and imaginative child play? What toys, books, and games are available? What are the ages of the other children who will be with your child in care?

- *Discipline.* Inquire about how staff handle the minor behavior problems that inevitably arise with children. Child experts typically recommend "time-outs," with physical discipline to be avoided. States vary in their laws regarding whether child caretakers are permitted to spank children. Where this is legally permissible, there may be centers or family caretakers who are indeed committed to the use of physical discipline—of course, parents may vary in terms of whether this is acceptable to them. You should both inquire and observe how "incidents" are handled, and ascertain whether the child care facility's policy and practice match what you want for your children. Use of physical discipline suggests that the caretakers are not well trained in handling problems and may create a somewhat fearful atmosphere for children as well.

- A *relationship with parents.* Parent–caregiver relationships will vary depending on whether the child is in family day care or a center. Any child care facility should welcome parental involvement, in the form of visits at a minimum; be wary of facilities that do not allow unannounced visits. You should feel supported in your parental role by the family caretaker or center staff (rather than distanced or unduly criticized). You should feel included in the child's daily life in child care. Especially important is how and how well family caretakers or center staff members communicate with you about problems.

- *Practical and financial considerations.* You will be told the basic hours and fees, but you also need to know what happens when the child is sick or the family leaves town and the child does not attend as usual. Can arrangements be made to have children arrive earlier or leave later than normal center hours on occasion? Regularly? Is transportation provided? If so, how costly is it, and how reliable?

- *Recommendations from other parents.* Talk to other parents about the facility. If you don't know any parents with children in the center, ask for names and phone numbers of parents who have children enrolled there, and talk to them about the facility. If a center declines to give you this information, try to determine whether the reason is that the facility's board adopted a privacy policy to protect parents or whether the management is being elusive and defensive.

- *Visits.* Visit the day care center as often as you can—and, if possible, unannounced—both before and after selecting a facility.

- *Accreditation.* The National Association for the Education of Young Children (NAEYC) is an accrediting agency for child care centers. If you plan to use center care, you might want to check the association's website (www.naeyc.org) for listings of accredited centers in your state. Although not all good child care centers have taken this step, accreditation by the NAEYC is a good sign.

Critical Thinking

What qualities do you think are most important in choosing a child care center? How would you compare in-home care, family day care, or center care on the qualities you think are important?

Sources: American Academy of Pediatrics 1992; Coordinated Access for Child Care 2001; Find Care 2002; Galinsky 2001; NICHD Early Child Care Research Network 2000a, 2003b; U.S. National Institute of Child Health and Human Development 2002; Watson 1984; Working Moms Refuge 2001.

Family day care and many child care centers are usually open weekdays only and close by 7 p.m. Some parents, such as single mothers on shift work or those who travel, need access to twenty-four-hour care centers. Child care is difficult to find for mildly ill youngsters too sick to go to their regular day care facility, although there are now centers beginning to fill this need (National Association for Sick Child Day Care 2010). Adding to the difficulty of finding day care is parents' need for *quality child care.* "As We Make Choices: Selecting a Child Care Facility" offers guidelines for evaluating the quality of a child care setting.

Elder Care There are some parallels between workers' responsibility for child care and for elder care. **Elder care** involves providing assistance with daily living activities to an elderly relative who is chronically frail, ill, disabled, or just in need of assistance. Many parents of the large baby boom generation are in their eighties and may live far away from their adult children. An estimated 34 million Americans are taking care of their aging parents. Some workers have retired early or just quit to care for parents, whereas others have turned down promotions, switched to part-time work, taken leaves of absence, or simply taken time off from work (Piercy 2007).

The need for companies to offer employees help with elderly dependents beyond unpaid family leave is becoming more recognized. Given that 90 percent of U.S. companies recently surveyed had a workforce of fifty-five years and older, many American companies are now offering elder care benefits as a recruiting tool (Woldt 2010). Supervisors may offer flexibility on an individual basis, but formal programs of assistance for elder care are in the beginning stages. Some 27 percent of companies now offer elder care benefits (Joyce 2007a). In addition to a graying population needing some assistance, a new phenomenon is taking root—older Americans finding new careers in taking care of the elderly. According to Paraprofessional Healthcare Institute, in 2008, some 28 percent of professional elder care workers were over the age of fifty-five (Leland 2010, p. 14; *Older Direct-Care Workers* 2010, p. 2). Care of the elderly is discussed in more detail in Chapter 17.

Family Leave **Family leave** involves an employee being able to take an extended period of time from work, either paid or unpaid, for the purpose of caring for a newborn, for a newly adopted or seriously ill child, for an elderly parent, or for their own health needs, with the guarantee of a job upon returning. The concept of family leave incorporates maternity, paternity, ill-child, and elder care leaves.

The 1993 Family and Medical Leave Act mandates up to twelve weeks of unpaid family leave for workers in companies with at least fifty employees. But unpaid leave will not solve the problem for a vast majority of employees, as most working parents need the income.

More employers are now offering paid maternity leave. In 1981, only 37 percent of first-time mothers who had worked during pregnancy took paid leave (maternity leave, sick leave, or vacation). By 2003 (most recent data available), that number had increased to 49 percent. Women's unpaid leave reached an all-time high of 45 percent by 1996, where it remained until 2000, declining since that time to 39 percent. Another 8.5 percent took disability leave in 2003. Twenty-five percent of these first-time mothers quit their jobs, while 3.8 percent were let go. This last percentage—first-time mothers being fired from their jobs—has increased from a low of 2 percent in 1996 (Johnson 2007).

According to the Institute for Women's Policy Research, approximately 93 percent of the top 100 American companies offer one or more weeks of paid maternity leave, with the majority offering between five and eight weeks (Lovell, O'Neill, and Olsen 2007, p. 1). As for the remainder of the private sector, parental leave was far more dismal, with some 8 percent finding paid family leave available. Not surprisingly, the workers with the highest rates of paid family leave available were those in management, professional, and related fields (14 percent), whereas just 4 percent of those who worked in production, transportation, and material moving had paid family leave available ("National Compensation Survey" 2007, Table 19) "'Gen X and Gen Y men [men born later than 1964] are demanding to have the ability to play a larger role in family life than their fathers did," states Joan Williams, director of Work/Life Law at American University ("More New Dads" 2005, p. Bus. 1).

Flexible Scheduling About 27.5 percent of full-time workers have flexible schedules (U.S. Bureau of Labor Statistics 2009c, Table 30). **Flexible scheduling** includes such options as **job sharing** (two people share one position), working at home or telecommuting, compressed workweeks, flextime, and personal days (days off for the purpose of attending to a personal matter such as a doctor's appointment or a child's school program). Compressed workweeks allow an employee to concentrate the workweek into three or four or sometimes slightly longer days. **Flextime** involves flexible starting and ending times, with required core hours.

Flexible scheduling, although not a panacea, can help parents share child care or be at home before and after an older child's school hours. Some types of work do not lend themselves to flexible scheduling (see "A Closer Look at Family Diversity: Diversity and Child Care" earlier in this chapter), but the practice has been adopted by the federal government and by some companies because it offers employee-recruiting advantages, prevents turnover, and frees up office space when some employees work at home. Even when not formally offered, it may be possible.

Economist Edith Josten noted that employees who are allowed some flexibility in deciding when to begin and end work have reduced incidences of dissatisfaction, and those who choose to work longer hours in a day (allowing for a shorter workweek) complain less about job-related fatigue (2002, pp. 87-88). Employees who have flexible hours report enhanced job satisfaction and loyalty to the employer, but they find that flextime does not alleviate all or even most family–work conflicts. For one thing, women are slightly less likely to have this option than men are, though they are more in need of it given the typical division of labor in the home (U.S. Bureau of Labor Statistics 2009c, Table 30).

Who Will Provide What Is Needed to Resolve Work–Family Issues?

Policy experts, lawmakers, employers, parents, and citizens disagree over who has the responsibility to provide what is needed regarding various work–family solutions. A principal conflict concerns whether such solutions as child care or family leave should be government policy or constitute privileges for which a worker must negotiate.

The countries of northwestern Europe, which have a more pronatalist and social-welfare orientation than the United States, tend to view family benefits as a right (Lewis 2008). There is "the pervasive belief . . . that children are a precious national resource for which society has collective responsibility" (Clinton 1990, p. 25). Putting this belief into practice, most European countries are committed to *paid* maternity (or parental) leave for up to at least six months and usually much longer (Lewis 2008). Accustomed to a lack of family policy at the federal level, American parents sometimes turn their attention to local schools as a source of help for the care of older children in after-school programs and younger children in preschool programs and all-day kindergarten.

As you've read elsewhere in this chapter, some large corporations demonstrate interest in effecting **family-friendly workplace policies** that are supportive of employee efforts to combine family and work commitments. Such policies include on-site child care centers, sick-child care, subsidies for child care services or child care locator services, flexible schedules, parental or family leaves, workplace seminars and counseling programs, and support groups for employed parents. Such research as exists on outcomes for employers suggests that these policies help in recruitment, reduce employee stress and turnover, enhance morale, and thus increase productivity ("Balancing Work and Life" 2009).

But family-friendly policies are hardly available to all American workers (Heymann, Earle, and Hayes 2007). Professionals and managers are much more likely than technical and clerical workers to have access to leave policies, telecommuting, or flexible scheduling

("National Compensation Survey" 2007, Table 19). "At the high end, the big corporations are stepping up to provide benefits to help families, and at the lower end, as women leave welfare, there's now much more support for the idea that they deserve help with child care. But the blue-collar families, the K-Mart cashier, get nothing" (work–family policy expert Kathleen Sylvester, quoted in Lewin 2001b).

An estimated 40 percent of the workforce is made up of unmarried people. Single individuals or childless workers have begun to complain about what they see as the privileging of parents of young children when they themselves may have family caregiving needs: for elderly parents, siblings, or friends with whom they maintain caregiving relationships. They may find it onerous to cover for coworkers who are on leave or out of the office. They may feel that simple fairness should permit some flexibility in their schedules as well, for personal needs. Some companies have begun to accommodate these workers by instituting sabbaticals, "flexible culture," and "employee-friendly" policies, redefining policies previously characterized as "family-friendly" (Joyce 2006; 2007a).

We have devoted attention to work–family policies because these issues so strongly influence the options and choices of individual families. We would like to think that family-friendly companies represent the future of work. After all, "children . . . are 'public goods'; society profits greatly from future generations as stable, well-adjusted adults, as well as future employees and tax payers" (Avellar and Smock 2003, p. 605). Nevertheless, these voluntary programs and benefits do depend on cost constraints and corporate self-interest and are not likely to be so available during economic downturns or restructuring. Moreover, family-friendly programs need to be more comprehensive in terms of benefits and more widely available to all echelons of workers. However, keep in mind that most workers need extensive family support only during the period in which they are parenting young children. From that perspective, the challenge looks less daunting.

Entering the political arena to work toward the kinds of changes families want is one aspect of creating satisfying marriages and families. But employed couples also want to know what *they* can do themselves to maintain happy marriages. We now turn to that topic.

The Two-Earner Marriage and the Relationship

As you have been reading throughout this chapter, there are many challenges associated with two-earner marriages. But research shows that multiple roles (such as employee, spouse, parent) does not add to stress

© Stewart Cohen/Index Stock Imagery

This dual-earner couple have common job experiences—something traditional spouses do not have an opportunity to share.

(provided there is enough time to accomplish things), and in fact may enhance personal happiness. Research also points to the heightened satisfaction, excitement, and vitality that two-earner couples can have because these partners are more likely to have common experiences and shared worldviews than do traditional spouses, who often lead very different everyday lives. At the same time, conflict may arise in two-earner marriages as couples negotiate the division of household labor and more generally adjust to changing roles.

Gender Strategies

How a couple allocates paid and unpaid work and then justifies that allocation can be thought of as a *gender strategy*, a way of working through everyday situations that takes into account an individual's beliefs and deep feelings about gender roles, as well as her or his employment commitments (Hochschild 1989). In today's changing society, conscious beliefs and deeper feelings

about gender may conflict. For example, a number of men in Hochschild's study of working couples articulated egalitarian sentiments, but had clearly retained gut-level traditional feelings about sex differences. Tensions exhibited by many of Hochschild's respondents were a consequence of "faster-changing women and slower-changing men" (p. 11).

Even when spouses share similar attitudes about gender, circumstances may not allow them to act accordingly. In one couple interviewed by Hochschild, both partners held the traditional belief that a wife should be a full-time homemaker. Yet because the couple needed the wife's income, she was employed, and they shared housework on a nearly equal basis. How couples manage their everyday lives in the face of contradictions reflects a consciously or unconsciously negotiated gender strategy.

One gender strategy used by wives who would like their husband to do more, but know he won't and are reluctant to insist, is to compare their husbands to other men "out there" who apparently are doing even less. A common gender strategy, according to Hochschild (1989), is to develop *family myths*—"versions of reality that obscure a core truth in order to manage a family tension." For example, when a husband shares housework in a way that contradicts his traditional beliefs and/or feelings, couples may develop a myth alleging the wife's poor health or incompetence to protect the man's image of himself. A common family myth defines the wife as an organized and energetic superwoman who has few needs of her own, requires little from her husband, and congratulates herself on how much she can accomplish.

Sociologist Bradford Wilcox (2004) uses the term *"enchanted" economies of gratitude* (p. 137, referencing Hochschild) in his study of evangelical families (to explain evangelical husbands' greater-than-average expressions of appreciation for their wives' household work). The commitment of the couple to a religiously based traditional division of labor is an anomaly in the context of today's ideal of egalitarian sharing. The evangelical wife's greater household labor is a "gift" that has symbolic significance for their religious and family world, and the husband reciprocates with "emotion work" (Wilcox 2004, Chapter 5).

Maintaining Intimacy While Negotiating Provider Roles and the Second Shift

Two kinds of changes are involved in moving toward more egalitarian family roles: Women come to share the provider role, while men take greater responsibility for household work. In considering the provider role, we turn to the notion of *meaning* again: Is women's sharing of the provider role a *threat*, so that men fear losing masculine identity, women's domestic services, and power?

Or is a woman's sharing the provider role a *benefit,* because men benefit materially from wives' employment and earnings and from a partner's enthusiasm for the wider world? Recent research suggests that men are more apt to see women's employment as a benefit. As a result, there is an ideological shift of men toward egalitarianism (Gerson 2010, pp. 107, 117).

Household work seems to be the greater arena for stress and conflict as roles change. Study after study shows that marital satisfaction is greater when wives feel that husbands share fairly in the household work. But a woman's employment does not necessarily lead to a husband's sharing of household work (Greenstein 2009).

Although husbands may now carry a greater share of the family work than in the past, getting comfortable with transitions in marital roles is not a quick and easy process. But when the transition proceeds from a mutual commitment to achieve an equitable relationship, the result may be greater intimacy. It follows from the general principle articulated in Chapter 1 that initial choices may need to be revisited over the life cycle. Gender issues may be revisited as partners adjust to changing work–family realities and as children enter the picture and then grow older. A first step is to address conflict.

Accept Conflict as a Reality The idea that marital partners may sometimes have competing interests departs from the more romanticized view that sees marriages and families as integrated units with shared desires and goals. As a first step toward maintaining intimacy during role changes, partners need to recognize their possibly competing interests and to expect conflict (Gerson 2010, p. 123).

Accept Ambivalence After accepting conflict as a reality, the next step in maintaining intimacy as spouses adjust to two-earner marriages is for both to recognize that each may have ambivalent feelings. The following excerpt from one young husband's essay for his English composition class is illustrative of a man's dilemma in assessing fairness in the division of labor: "I'm in school six days a week. My wife works between 40 to 50 hours a week. So I do the majority of the cooking, cleaning, and laundry. To me this is not right. But am I wrong to think so?" Women may also be ambivalent. They want their husbands to be happy, they want their husbands to help and support them, they feel angry about any past inequalities, and they feel guilty about their declining interest in housekeeping and their decreasing willingness to accommodate their husbands' preferences. Furthermore, men who participate have opinions about how child rearing or housework should be done. As a husband begins to pitch in, his wife may resent his intrusion into her traditional domain.

Empathize A next step is to empathize. This may be difficult, for it is tempting instead to point out where a partner falls short. But if couples are to maintain intimacy, they must make sure that *both* partners "win." Wives are often irritated by observing that husbands may underestimate the number of hours that household labor takes (Galinsky, Aumann, and Bond 2009; Kelleher 2007; Swanbrow 2008). It is never easy to adjust to new roles, and men especially may feel they have a lot to lose. Men can gain, too, of course: They develop domestic skills, their marriage is enhanced, there is more money, and they benefit from spending time with their children. In Hochschild's study (1989), some fathers who felt they had been emotionally deprived in relationships with their own fathers took great pleasure in creating more satisfying family relationships with and for their children.

As husbands empathize, they need to be aware that their willingness to participate in household tasks is vitally important to wives, especially to employed wives. A husband's sharing carries a symbolic meaning for a wife, indicating that her work is recognized and appreciated and that her husband cares (Crary 2008).

Strike an Equitable Rebalance Researchers who studied 153 Pennsylvania couples with children in school concluded the following: "Our data imply that the adjustment of individual family members, as well as harmonious family relationships, requires a *balance* among the very different and often conflicting needs and goals of different family members" (McHale and Crouter 1992, pp. 545–46, italics in original). Once equity is habitual, calculation and constant comparison are no longer necessary; some observers point out that the balance need not be an exactly calculated fifty-fifty split.

Show Mutual Appreciation Once partners have committed themselves to striking a balance, they need to create ways to let each know the other is loved. Traditional role expectations are relatively rigid and limiting, but they can be a way of expressing love and caring. When a wife cooks her husband's favorite meal or a husband can pay for family travel, each feels cared about. As spouses relinquish some traditional behaviors, they need to create new ways of letting each other know they care. Many people have noted the potential of shared work and of shared provider and caregiving roles for enriching a marriage (Galinsky, Aumann, and Bond 2009; van der Lippe 2010).

This discussion of the second shift has been framed in terms of marriage, the relationships of husbands and wives as they negotiate this marital challenge. Marriage *is* most likely to draw on cultural expectations of a traditional division of labor. But the second shift exists in other family forms. In heterosexual cohabiting couples, the woman does less household labor and the man more than in marriage, whereas the domestic division of labor is rather egalitarian in gay and lesbian couples. Single women and men also have work to do to maintain their households, especially if they are parents. Single men tend to do more than married men, whereas single women do less than married women. Interestingly, remarried couples are more likely to share housework than are men and women in a first marriage (Coltrane 2000; Patterson 2000), as are couples who cohabited before marriage (Rhoades et al. 2006). We should keep in mind that employees are embedded in diverse families and that partners may come up with a variety of ways of accomplishing providing and caregiving.

Despite the unresolved tensions of the second shift, research by sociologist K. Jill Kiecolt (2003) suggests that employed men and women are largely happy with their home lives. She set out to explore a thesis developed by Arlie Hochschild (1997) in her study of workers at one company. Hochschild concluded that family life for employed people is so hectic that work becomes a refuge, a place where individuals would prefer to be. Hochschild's was a case study, so no statistical conclusions could be drawn.

Testing this thesis with General Social Survey data from National Opinion Research Center (NORC) over the period between 1973 and 1994, Kiecolt found that only 13 percent of workers saw it that way. In the most recent year she studied, over 40 percent of respondents had "high work–home satisfaction," whereas home was viewed as a haven for another 40 percent plus.

The next chapter examines communication and managing conflict in families, skills that can smooth the negotiation of work–family roles.

Summary

- We look at men's and women's participation in the labor force. Traditionally, the husband's job was as provider, the wife's as homemaker. These roles changed as more and more women entered the workforce. Women remain segregated occupationally, and they earn lower incomes than men, on average.

- We have seen that paid work is not usually structured to allow time for household responsibilities and that women, more than men, continue to adjust their time to accomplish both paid and unpaid work. Many wives would prefer shared roles, and negotiation and tension over this issue can cast a shadow on a marriage. An incomplete transition to equality at work and at home affects family life profoundly. However, in recent years, men have been increasing their share of the housework, and men and women now have a balance in total work hours.

- Household work and child care are pressure points as women enter the labor force and the two-earner marriage becomes the norm. To make it work, either the structure of work must be changed, social policy must support working families, or women and men must change their household role patterns—very probably all three.

- We have emphasized that both cultural expectations and public policy affect people's options. As individuals come to realize this, we can expect pressure on public officials and corporations to meet the needs of working families by providing supportive policies: parental leave, child care, and flextime.

- To be successful, two-earner marriages will require social policy support and workplace flexibility. But there are some things couples themselves can do to better manage a working-couple family. Recognition of both positive and negative feelings and open communication between partners can help working couples cope with an imperfect social world.

Questions for Review and Reflection

1. Discuss to what extent distinctions between husbands' and wives' work are disappearing.

2. What do you see as the advantages and disadvantages of men being househusbands? Discuss this from the points of view of both men and women.

3. What are some advantages and disadvantages of home-based work?

4. What work–family conflicts do you see around you? Interview some married or single-parent friends of yours for concrete examples and for some suggestions for resolving such conflicts.

5. **Policy Question.** What family-friendly workplace policies would you like to see instituted? Which would you be likely to take advantage of?

Key Terms

attachment 307
center care 305
child care 304
commuter marriage 304
elder care 310
family child care 305
family-friendly workplace policies 311
family leave 310
flexible scheduling 310
flextime 310
good provider role 288
househusband 290
in-home caregiver 305
job sharing 310
labor force 282
leisure gap 297
market approach to child care 305

motherhood penalty 285
mothering approach to child care 305
nanny 291
neotraditional families 287
occupational segregation 283
opting out 286
parenting approach to child care 305
second shift 295
self-care 305
shift work 291
stay-at-home dad 290
trailing spouse 303
two-career marriage 291
two-earner marriage 290
unpaid family work 294
wage gap 284

Online Resources

Sociology CourseMate

www.CengageBrain.com

Access an integrated eBook, chapter-specific interactive learning tools, including flash cards, quizzes, videos, and more in your Sociology CourseMate, accessed through CengageBrain.com.

www.CengageBrain.com

Want to maximize your online study time? Take this easy-to-use study system's diagnostic pre-test, and it will create a personalized study plan for you. By helping you identify the topics that you need to understand better and then directing you to valuable online resources, it can speed up your chapter review. CengageNOW even provides a post-test so you can confirm that you are ready for an exam.

12

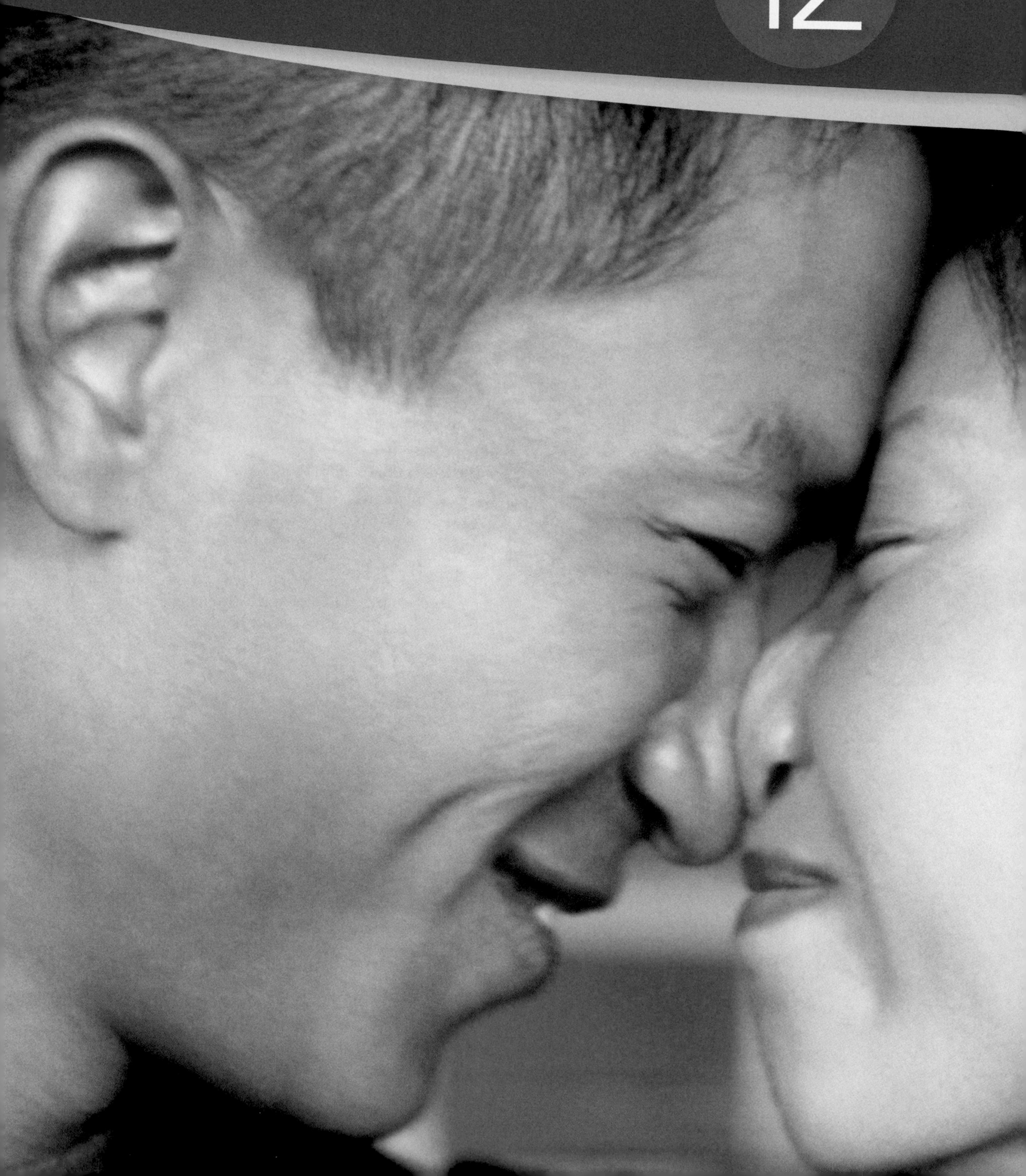

Communication in Relationships, Marriages, and Families

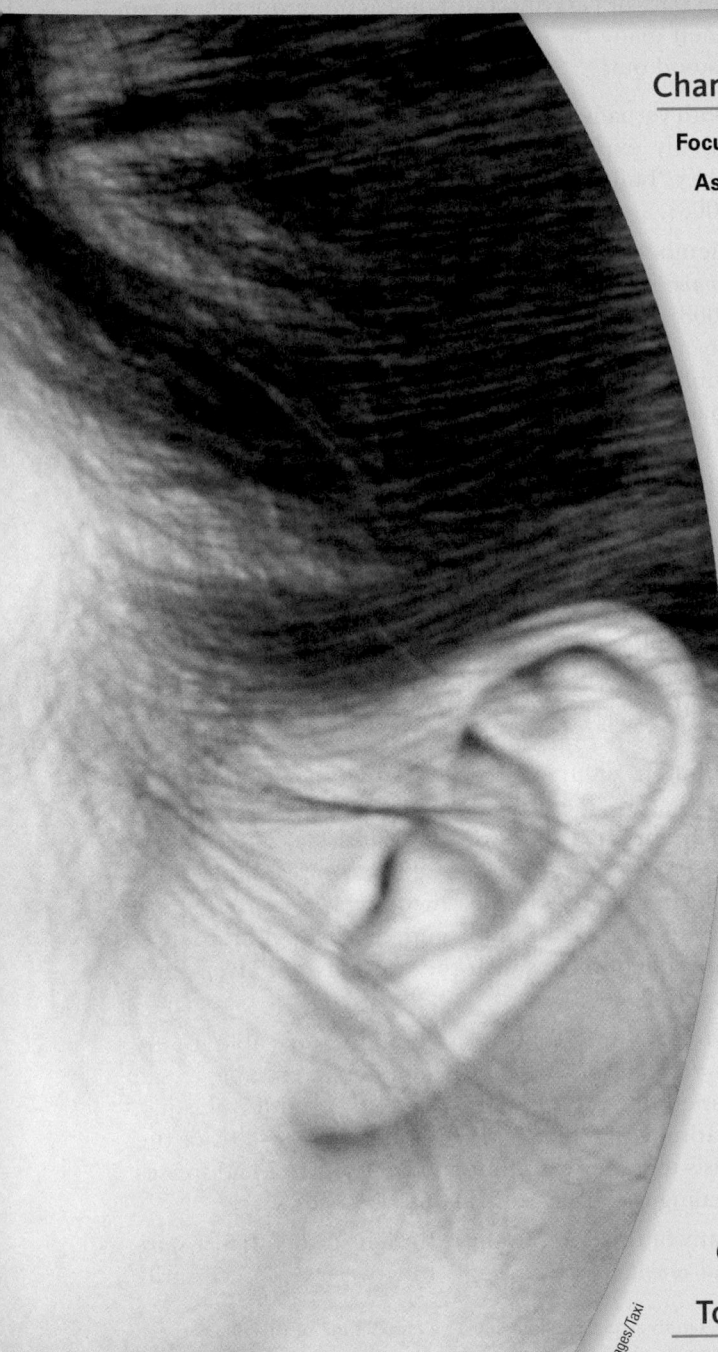

Characteristics of Cohesive Families

Focus on Children: Children, Family Cohesion, and Unresolved Conflict

As We Make Choices: Communicating with Children—How to Talk So Kids Will Listen and Listen So Kids Will Talk

Communication and Couple Satisfaction

As We Make Choices: Ten Rules for Successful Relationships

Conflict in Relationships

Indirect Expressions of Anger

John Gottman's Research on Couple Communication and Conflict Management

Issues for Thought: A Look Behind the Scenes at Communication Research

The Four Horsemen of the Apocalypse

Gender Differences and Communication

Working Through Conflicts in Positive Ways—Ten Guidelines

Guideline 1: Express Anger Directly and with Kindness

Guideline 2: Check Out Your Interpretation of Others' Behaviors

Guideline 3: To Avoid Attacks, Use "I" Statements

Guideline 4: Avoid Mixed, or Double Messages

Guideline 5: When You Can, Choose the Time and Place Carefully

Guideline 6: Address a Specific Issue, Ask for a Specific Change, and Be Open to Compromise

Guideline 7: Be Willing to Change Yourself

Guideline 8: Don't Try to Win

Guideline 9: Be Willing to Forgive

Guideline 10: End the Argument

Toward Better Couple and Family Communication

Facts about Families: Relationship and Family Counseling

The Myth of Conflict-Free Conflict

© Ron Chapple/Getty Images/Taxi

Providing emotional security is an important function of today's families. Moreover, families are powerful environments. Virtually nowhere else in our society is there such capacity to support, hurt, comfort, denigrate, reassure, ridicule, hate, and love. Research from a variety of samples and pertaining to a variety of family situations overwhelmingly supports what may be intuitively obvious: Conveying affection for one's partner and other family members is a very important determinant of relationship and family happiness, as well as each family member's psychological well-being (Fagan 2009; Soliz, Thorson, and Rittenour 2009). And although conflict is a natural part of every relationship, developing positive communication skills can help family members to resolve conflicts in positive ways.

This chapter will address the importance of communicating affection as well as addressing conflict in positive ways. We'll examine the relationship between communication and relationship satisfaction. We'll discuss gender differences with regard to couple and family communication. We will review ten guidelines suggested for addressing family and couple conflicts. To begin, we'll look at characteristics of cohesive families.

Characteristics of Cohesive Families

Family cohesion, or "togetherness," is defined as "the emotional bonding that couples and family members have toward one another" (Olson and Gorall 2003, p. 516). A couple or family can have too much cohesion (an *enmeshed* couple or family) or too little (a *disengaged* or *disconnected* couple or family). Experts advise a *balanced level of cohesion*—one that combines a reasonable and mutually satisfying degree of emotional bonding with individual family members' need for autonomy. In this chapter, we will use the term *family cohesion* to refer to a balanced degree of cohesion—neither enmeshed nor disengaged.

Before going further, we should recognize that for different families—and for families of different race/ethnicities—the definition of *balance* with regard to family cohesion varies. "If a couple's/family's expectations or subcultural group norms support more extreme [cohesion levels], families can function well as long as all family members desire the family to function [at that level]" (Olson and Gorall 2003, p. 522). For instance, in Mexican American families, a relatively high level of family cohesion has been related to positive outcomes for adolescents (Behnke et al. 2008; Martyn et al. 2009).

To find out what makes families cohesive, social scientist Nick Stinnett researched 130 "strong families" in rural and urban areas throughout Oklahoma (Stinnett 1985, 2008). Obviously, this limited sample, selected with help from home economics extension agents, has

no claim to representativeness. Furthermore, the concept *strong family* is subjective. Various individuals or groups have their own ideas about just what a strong family is. But Stinnett's research helped to advance ideas about what makes for couple and/or family cohesion. In general, Stinnett's families constructed their lives in ways that enhance family relationships. Instead of drifting into relationship habits by default, they made knowledgeable choices, each member playing an active part in carrying out family commitments. When Stinnett made his observations, the following six qualities stood out:

1. Both verbally and nonverbally, family members often openly expressed their *appreciation for one another.* They "built each other up psychologically" (Stinnett 2008).

2. Members of cohesive families had a *high degree of commitment* to the family group as a whole (Stinnett 2008). Families like this create a shared family identity and reality (Rueter and Koerner 2008).[1] From a symbolic interactionist perspective, families create a shared reality through frequent, spontaneous, and unconstrained conversations that allow family members to participate together in defining family beliefs, values, situations, events, and rituals.

3. Stinnett found that, on a regular basis, family members *arranged their personal schedules* so that they could do things together. They invested in their family. When life got so hectic that members didn't have enough time for their families, they listed the activities they were involved in, found those that weren't worth their time, and scratched them off their lists (Stinnett 2008). Leisure time together is important for family members (Clayton and Perry-Jenkins 2008; Smith, Freeman, and Zabriskie 2009; Smith, Freeman, and Zabriskie 2009). For example, in a college-student sample, those who reported more shared time with their grandparents had more satisfying relationships with them (Mansson, Myers, and Turner 2010).

4. Stinnett found that strong families were able to *deal positively with crises.* Family members were able to see something good in bad situations, even if it was just gratitude that they had each other and could face the crisis together (Stinnett 2008). Chapter 14 addresses dealing creatively with stress and crises.

5. Many of the families that Stinnett studied had a *spiritual orientation.* Although they were not necessarily

---

[1] Some family structures may have an easier time of this than others. For instance, many same-sex couples and their children must create a family identity in the absence of full cultural legitimacy. Then too, "Genetically related family members likely share a sense of belonging based on physical appearance, blood ties, and shared social attitudes or cognitions based in genetic inheritance. All these shared characteristics facilitate their ability to create a shared reality" (Rueter and Koerner 2008).

members of any organized religion, they did have a sense of some power and purpose greater than themselves and typically evidenced a "hopeful attitude toward life" (DeFrain 2002).

6. These families had *positive communication* patterns. Members of families like these talk with and listen to one another, conveying respect and interest (Schrodt 2009). They confirm, validate, and accept each other (Dailey 2009). One study of college-aged daughters found that they were happier with their father-daughter relationship when it involved higher levels of ordinary and reciprocal conversation (Punyanunt-Carter 2008).

In addition to the previously mentioned, cohesive families and supportive couple relationships involve some "old-fashioned" virtues such as prudence, humility, tolerance, gratitude, justice, charity, and forgiveness (Fincham, Stanley, and Beach 2007). How do children benefit from family cohesiveness?

Children, Family Cohesion, and Unresolved Conflict

Regardless of family structure, a family characterized by warmth, cohesion, and generally supportive communication is better for children (Hillaker et al. 2008; Lindsey et al. 2009; Matiasko, Grunden, and Ernst 2007). Furthermore, parental values are more readily passed on to children when the family atmosphere is generally cohesive (Roest, Dubas, and Gerris 2009).

Conversely, a home characterized by significant, unresolved, and ongoing conflict negatively impacts children (Schoppe-Sullivan, Schermerhorn, and Cummings 2007). Meanwhile, contrary to the idea that couple conflicts in the home are necessarily detrimental to children, conflicts can end in constructive ways from the children's perspective.

> We posit that children's positive response to conflict resolution is an indication of enhanced emotional security. That is, children feel an increased sense of well-being resulting from the confidence of knowing that although their parents disagree, their relationship is safe and will endure. (Goeke-Morey, Cummings, and Papp 2007, p. 751)

However, a climate of *unresolved* marital conflict, especially when accompanied by parental depression, which it often is, correlates with children's emotional insecurity (Kouros, Merrilees, and Cummings 2008). A Hong Kong study of children's responses to ongoing parental conflict found that the children felt anxious over the future of their parents' relationship as well as feeling that they had to mediate the conflict (Lee et al. 2010).

Research shows a link between unresolved parental conflict and children's behavior problems (Feinberg, Kan, and Hetherington 2007; Teachman 2009). One study (Buehler et al. 1998) sampled 337 sixth through eighth graders. Three-quarters were non-Hispanic white, 12 percent were Hispanic, and 13 percent represented other racial/ethnic groups. The parents of 87 percent of the children in the sample were married. The parents' average education level was somewhere between high school graduate and some college.

The students were asked to fill out questionnaires that assessed their behavior and any conflict between their parents. *Externalizing behavior problems* (associated with "acting out," aggression toward others, or rule breaking) were measured by students' agreeing or disagreeing with statements such as "I cheat a lot" or "I tease others a lot." *Internalizing behavior problems* (those associated with emotional or psychological problems) were measured by students' agreeing or disagreeing with statements such as "I am unhappy a lot" or "I worry a lot." The children were also asked about their parents' conflict. Negative *overt parental conflict styles* involved such things as the parents' calling each other names, telling each other to shut up, or threatening each other in front of the child. Negative *covert parental conflict styles* included such things as trying to get the child to side with one parent and asking the child to relay a message from one parent to the other because the parents refused to speak to each other.

The researchers found that conflict between parents was not the only cause of children's behavior problems. Nevertheless, for both girls and boys, a strong correlation existed between interparental

Among other things, cohesive families have high levels of commitment and positive communication patterns. Making time to be together, they build one another up psychologically.

© MBI/Alamy

There are more and less effective ways to communicate with children, and a knowledgeable choice would involve more effective ways. What are some of these methods?

Helping Children Deal with Their Feelings

Children—including adult children—need to have their feelings accepted and respected.

1. *You can listen quietly and attentively.*

2. *You can acknowledge their feelings with a word.* "Oh . . . mmm . . . I see. . . ."

3. *You can give the feeling a name.* "That sounds frustrating!"

4. *You can note that all feelings are accepted, but certain actions must be limited.* "I can see how angry you are at your brother. Tell him what you want with words, not fists."

Engaging a Child's Cooperation

1. *Describe what you see, or describe the problem.* "There's a wet towel on the bed."

2. *Give information.* "The towel is getting my blanket wet."

3. *Describe what you feel.* "I don't like sleeping in a wet bed!"

4. *Write a note* (above towel rack): Please put me back so I can dry. Thanks! Your Towel

Instead of Punishment

1. *Express your feelings strongly—without attacking character.* "I'm furious that my saw was left outside to rust in the rain!"

2. *State your expectations.* "I expect my tools to be returned after they've been borrowed."

3. *Show the child how to make amends.* "What this saw needs now is a little steel wool and a lot of elbow grease."

4. *Give the child a choice.* "You can borrow my tools and return them, or you can give up the privilege of using them. You decide."

Encouraging Autonomy

1. *Let children make choices.* "Are you in the mood for your gray pants today or your red pants?"

2. *Show respect for a child's struggle.* "A jar can be hard to open."

3. *Don't ask too many questions.* "Glad to see you. Welcome home."

---

conflict and behavior problems. When parents used an overtly negative style, the youth were more likely to report externalizing behavior problems. When parents used a covert, negative style, the youth were more likely to report internalizing behavior problems.

In another study of fifty-five Caucasian middle- and upper-middle-class five-year-olds and their married mothers, the mothers completed questionnaires on parent–child relations and interparental arguing that were mailed to them at home. Later, the mothers took their children to be observed in a university laboratory setting. The researchers found that marital discord was positively related to children's externalizing and internalizing behavior problems. However, this research also showed that the interparental conflict influenced a child's behavior *indirectly:* Marital discord negatively affected parental discipline and the parent-child relationship more generally. This situation then negatively affected the child's behavior.[2]

The researchers concluded that "if parents are able to maintain good relations with children in the face of marital conflict, the children may be buffered from the potential emotional fallout of the conflict" (Harrist and Ainslie 1998, p. 156; see also Lindsey, Caldera, and Tankersley 2009; Schoppe-Sullivan, Schermerhorn, and Cummings 2007; Schrodt, Witt, and Messersmith 2008). Recent research has investigated the effects of parental and general family conflict on sibling relationships.

Unresolved Family Conflict and Sibling Relationships One interesting study of mothers, fathers, and adolescents from 200 middle- and working-class, mostly European American families found parental conflict to be causally associated with parents' differential treatment of their children. A possible reason is because ongoing, unresolved parental conflict encourages a parent to form a parent-child alliance with one sibling—a situation that leaves other siblings out. Perceived differential treatment among siblings leads to sibling conflict and underlying resentments that can linger into adulthood (Kan, McHale, and Crouter 2008).

Another study looked at a racially and ethnically diverse sample of elderly mothers with at least two adult children and found that a mother's perceived favoritism while the children were growing up reduced siblings' closeness in adulthood. "Further, mothers' favoritism appeared to reduce closeness regardless of which child

---

[2] A recent study from the biosocial theoretical perspective (see Chapter 2) sampled mothers and found that a mother's high level of cortisol, a chemical associated with being stressed, "spilled over" after having been secreted during parental conflict to subsequently have a negative effect on her parenting behaviors (Sturge-Apple et al. 2009). Interestingly, other research shows that an individual's ingesting the hormone oxytocin—see footnote #11—reduces cortisol levels during couple conflict and also increases positive communication behaviors (Ditzen et al. 2009; see also Ellison and Gray 2009; Priem, McLaren, and Solomon 2010).

4. *Don't rush to answer questions.* "That's an interesting question. What do you think?"

5. *Encourage children to use sources outside the home.* "Maybe the pet shop owner would have a suggestion."

6. *Don't take away hope.* "So you're thinking of trying out for the play! That should be an experience."

Praise and Self-Esteem

Instead of evaluating, describe.

1. *Describe what you see.* "I see your car parked exactly where we agreed it would be."

2. *Describe what you feel.* "It's a pleasure to walk into this room!"

3. *Sum up the child's praiseworthy behavior with a word.* "You sorted out your pencils, crayons, and pens, and put them in separate boxes. That's what I call *organization*!"

Freeing Children from Playing Roles

1. *Look for opportunities to show the child a new picture of himself or herself.* "You've had that toy since you were three, and it looks almost like new!"

2. *Put children in situations in which they can see themselves differently.* "Sara, would you take the screwdriver and tighten the pulls on these drawers?"

3. *Let children overhear you say something positive about them.* "He held his arm steady even though the shot hurt."

4. *Model the behavior you'd like to see.* "It's hard to lose, but I'll try to be a sport about it. Congratulations!"

5. *Be a storehouse for your child's special moments.* "I remember the time you. . . ."

Critical Thinking

What bit of advice given here might you choose to practice when communicating with the child(ren) in your life? Why is it important to encourage children to talk? Why is it important to listen to children? Why does how we talk to children matter?

Source: Excerpts from Rawson Associates/Scribner, an imprint of Simon & Schuster, from *How to Talk So Kids Will Listen and Listen So Kids Will Talk*, by Adele Faber and Elaine Mazlish. Copyright © 1980 by Adele Faber and Elaine Mazlish. Also see Faber and Mazlish (2006) as well as the Faber Mazlish website, www.fabermazlish.com, and the Mental Health America website, www.mentalhealthamerica.net.

---

was favored, suggesting that siblings' relationships are shaped . . . by principles of equity" (Suitor et al. 2009, p. 1032).[3] However, a different study found that, in childhood and adolescence, the sibling who felt slighted was likely to be depressed (Shanahan et al. 2008). Meanwhile a study of 246 two-parent Mexican American families found that adolescents in families with a solution-oriented conflict management style, rather than ongoing unresolved conflict, had better sibling relationships (Killoren, Thayer, and Updegraff 2008).

Of course, children's behavior also depends on how the parents communicate with the children themselves, even when family conflict exists (Schrodt et al. 2009). "As We Make Choices: Communicating with Children— How to Talk So Kids Will Listen and Listen So Kids Will Talk" describes some effective ways to communicate with children.

By now you may have surmised that in families headed by coupled partners, family communication tends to be influenced by the degree of supportiveness or negativity in the couple relationship itself (Doohan et al. 2009). Distressed couples tend toward negative exchanges that put family relationships on a downward spiral (Driver and Gottman 2004). Partners who communicate mutual affection create a positive "spiraling effect" so that the atmosphere becomes one of emotional support (L. White 1999).

Communication and Couple Satisfaction

Couples demonstrate different **relationship ideologies**—expectations for closeness and/or distance as well as ideas about how partners should play their roles. A pivotal task for all couples is to balance each partner's need for autonomy with the simultaneous need for intimacy, togetherness, and support (Brock and Lawrence 2009; Lavy et al. 2009).

Couples also differ in their attitudes toward conflict. Some expect to engage in conflict only over big issues. Others argue more often. Still others expect a relationship that largely avoids not only conflict but also demonstrations of affection (Fitzpatrick 1995). All of these couple types can be happy with their relationship. What matters is whether the partners' actual interaction matches their ideology.

Meanwhile, unhappy relationships have some common features: less positive and more negative verbal and nonverbal communication, together with more reciprocity of negative—but not of positive—messages (Noller and Fitzpatrick 1991; Gottman and Levenson 2000).

---

[3] According to linguist Deborah Tannen (2006), many grown daughters continue to feel rejected because their mothers persist in showing preference to their brothers.

An important characteristic of happy couples involves disclosure of feelings and showing affection for one another. Meanwhile, even the happiest couples experience conflicts. Whether and how a couple resolves interpersonal conflicts creates a "spiraling effect" that positively or negatively influences communication throughout the family.

Regarding married couples, Huston and Melz (2004) found that after the honeymoon stage, there is a "coming down to earth" stage in a marriage. Interestingly, however, in the years after "the honeymoon's over," couples did not necessarily argue more. Instead, their marriages showed a decline in signs of love and affection. "One year into marriage, the average spouse says 'I love you,' hugs and kisses their partner, makes their partner laugh, and has sexual intercourse about half as often as when they were newly wed." Although marriages do not necessarily "become more antagonistic as time passes, the unpleasant exchanges that *do* occur are embedded in a less affectionate context, and thus, the spouses are likely to come to feel that their marriage is less of 'a haven in a heartless world'" (p. 951).

Having gathered data on married couples, researchers Ted Huston and Heidi Melz (2004) classified relationships into four types: *warm*, or friendly; *tempestuous*, or stormy; *bland*, or empty shell; and *hostile*, or distressed (p. 951).[4] As indicated in Figure 12.1, warm relationships are high on showing signs of love and affection while low on antagonism. Tempestuous unions are high on both affection and antagonism. Bland marriages are low on showing signs of affection as well as on antagonism. Hostile marriages are low on love and affection but high on antagonism.

We can assume that warm and friendly relationships best fill the family function of providing emotional security. We can also conclude that hostile ones are undesirable. Huston and Melz called both bland and tempestuous unions "mixed blessing" relationships because these two types evidenced only one of two desirable attributes. Although bland relationships have little antagonism, they lack displays of affection. And although tempestuous couples intermittently show affection, they deal with conflicts in aggressive, or antagonistic, ways.

If our goal is to identify the early signs of a marital rupture, our research suggests that we look to the loss of love and affection early in marriage as symptomatic. . . . This loss of good feelings, rather than the emergence of conflict early in marriage, seems to be what sends relationships into a downward spiral, no doubt eventually leading to increased bickering and fighting and, ultimately, to the collapse of the union. (Huston and Melz 2004, pp. 951–52)

This situation helps to explain the general finding, discussed in Chapter 7, that for many couples the early years of marriage are the happiest. Of course, partners can change this by making knowledgeable decisions about communicating intimacy.

Other research, conducted by widely recognized communication psychologist, John Gottman and his colleagues, found that "[t]he absence of positive affect and not the presence of negative affect . . . was most predictive of later divorcing" (Gottman and Levenson 2000, p. 743). **Positive affect** involves the verbal or nonverbal expression of affection. Gottman further argued that, at least for the middle-class couples in his sample, he could predict a married couple's later divorce by examining how well the spouses showed that they were interested in each other:

In a careful viewing of the videotapes, we noticed that there were critical moments during the events-of-the-day conversation that could be called either "requited"

[4] This research by Huston and Metz (2004) studied heterosexual, married couples only. The extent to which their findings and conclusions apply to otherwise committed couples, such as cohabiting or same-sex partners, is unknown. Increasingly, researchers are making the point that correlates of relationship satisfaction need to be studied among other than heterosexually married couples, and researchers are beginning to do this (e.g., Lincoln, Taylor, and Jackson 2008).

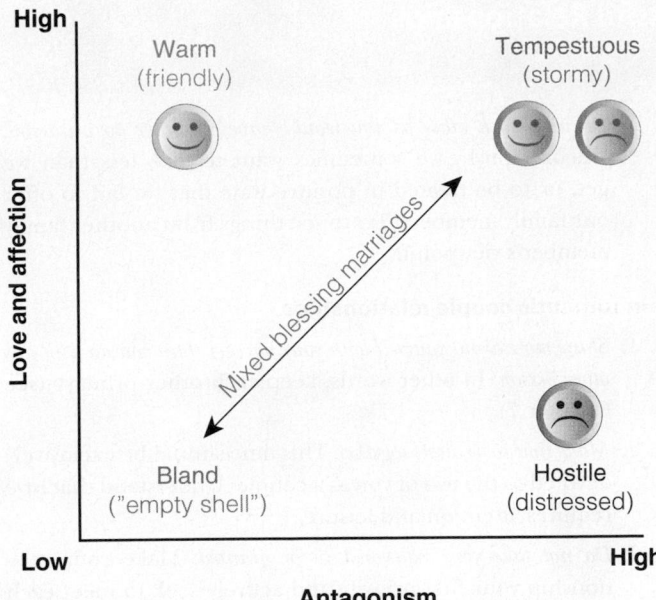

High

Warm
(friendly)

Tempestuous
(stormy)

Love and affection

Mixed blessing marriages

Bland
("empty shell")

Hostile
(distressed)

Low   **High**

Antagonism

Figure 12.1 The emotional climates of committed relationships. This figure depicts the classification of emotional climates along two dimensions: (1) love and affection, and (2) antagonism. Warm relationships are high on love and affection while low on antagonism. Tempestuous unions are high on love and affection but also on antagonism. Bland relationships are low on love and affection as well as on antagonism. Hostile relationships are low on love and affection but high on antagonism. Why do you think that the tempestuous and bland types are called "mixed blessings"?

Source: From Huston and Melz 2004. Reprinted by permission of Blackwell Publishing. See footnote #2 in this chapter.

[returned, acknowledged, or reciprocated] or "unrequited" interest and excitement. For example, in one couple, the wife reported excitedly about something their young son had done that day, but she was met with her husband's disinterest. After a time of talking about errands that needed doing, he talked excitedly about something important that happened to him that day at work, but she responded with disinterest and irritation. No doubt this kind of interaction pattern carried over into the rest of their interaction, forming a pattern for "turning away" from one another. (Gottman and Levenson 2000, p. 744)

Relatedly, UCLA psychologist Shelly Gable studied how one partner responds when something positive happens to the other one, such as a promotion at work (Gable et al. 2004). A partner might respond enthusiastically ("That's wonderful, and it's because you've had so many good ideas in the past few months."). But he or she could instead respond in a less-than-enthusiastic manner ("Hmmm, that's nice."), seem uninterested ("Did you see the score of the Yankees game?"), or point out the downsides ("I suppose it's good news, but it wasn't much of a raise."). According to Gable's research, the only "correct" reaction—the response that's correlated with intimacy, satisfaction, trust, and continued commitment—is the first response: the enthusiastic, active one (Lawson 2004a).

"As We Make Choices: Ten Rules for Successful Relationships" presents ideas on how to show positive affect. But even the happiest couples have conflicts, and how they are addressed has much to do with maintaining supportive relationships.

Conflict in Relationships

"It worries me that you keep referring to our honeymoon as our 'honeymoon period.'"

Psychologists Nathaniel Branden and Robert Sternberg have developed some rules for nourishing relationships. Here are ten. The first seven can be applied to all family relationships. The final three pertain to romantic couple relationships.

In all family relationships

1. *Express your love verbally.* Say "I love you."

2. *Be physically affectionate.* Offer (and accept) a touch or hug that says, "I care," "I'm sorry," or "I understand."

3. *Express your appreciation.* Tell your loved ones what you like, enjoy, and admire about one another. Listen with interest.

4. *Help the relationship or family to become an emotional support system.* Be there for each other in times of illness, difficulty, and crisis; be generally helpful and nurturing—devoted to each other's well-being.

5. *Express your affection in material ways.* Send cards or give presents on more than just expected occasions. Lighten a family member's burden once in a while by doing more than your agreed-upon share of the chores.

6. *Accept your family members' shortcomings.* We are not talking about putting up with physical or verbal abuse here. But harmless shortcomings are part of every relationship. Love your family members, not an unattainable idealization of them.

7. *Do unto each other as you would have the other do unto you.* Unconsciously, we sometimes want to give less than we get, or to be treated in positive ways that we fail to offer our family members. Try to see things from another family member's viewpoint.

In romantic couple relationships

1. *Share more about yourself with your partner than you do with any other person.* In other words, keep each other primary (see Chapter 7).

2. *Make time to be alone together.* This time should be exclusively devoted to the two of you as a couple. Understand that love requires attention and leisure.

3. *Do not take your relationship for granted.* Make your relationship your first priority and actively seek to meet each other's needs.

Critical Thinking

Often, we read a list like the previous one and think about whether our partner or other family members are doing them, not whether we ourselves are. How many of the items on this list do you yourself do? Which two or three items might you begin to incorporate into a relationship?

Sources: Branden 1988, pp. 225–28; Sternberg 1988b, pp. 272–77; see also Gottman and Silver 1999; Gottman and DeClaire 2001; Mackey, Diemer, and O'Brien 2000; Markman, Stanley, and Blumberg 2001.

Conflict in Relationships

Couples argue about their children, money, household chores, in-laws, how to allocate their time, other relatives, and irritating little habits that one of them has. Another common topic for arguments involves the couple's communication itself, often with each partner feeling that the other is not paying attention or understanding (Papp, Cummings, and Goeke-Morey 2009). Anger and conflict are challenges to be met rather than avoided (Schechtman and Schechtman 2003).

Sociologist Judith Wallerstein (Wallerstein and Blakeslee 1995) conducted lengthy interviews with fifty predominantly white, middle-class married couples in northern California. The shortest marriage was ten years and the longest forty years. To participate, both husband and wife had to define their marriage as happy. When discussing what she found, Wallerstein wrote this:

> [E]very married person knows that "conflict-free marriage" is an oxymoron. In reality it is neither possible nor desirable. . . . [I]n a contemporary marriage it is expected that husbands and wives will have different opinions. More important, they can't avoid having serious collisions on big issues that defy compromise. (p. 143)

The couples in Wallerstein's research quarreled:

> In one marriage the husband and wife sat in the car to argue, to avoid upsetting the children. She told him that passive smoke was a proven carcinogen, and while the children were young he could not smoke in their home. He could do what he wanted outside. The man admitted that the request was reasonable, but he was furious. He punished her by not talking to her except when absolutely necessary for three months. Then he accepted the injunction on his smoking and they resumed their customary relationship. (p. 148)

Wallerstein concluded that

> [t]he happily married couples I spoke with were frank in acknowledging their serious differences over the years. . . . What emerged from these interviews was not only that conflict is ubiquitous but that these couples considered learning to disagree and to stand one's ground one of the gifts of a good marriage. (p. 144)

Counselors generally advise that an important aspect of learning to disagree involves expressing anger directly, a skill explored later in this chapter. Here we turn to an examination of *indirect* expressions of anger.

Indirect Expressions of Anger

Many of us may feel uncomfortable about expressing our anger directly. As a result, we can find ourselves engaging in **passive-aggression**—that is, expressing anger indirectly. Chronic criticism, nagging, nitpicking, and sarcasm are all forms of passive-aggression. Procrastination, especially when you have promised a partner that you will do something, may be a form of passive-aggression (Ferrari and Emmons 1994). These behaviors create unnecessary distance and pain in relationships. Most people who use sarcasm do so unthinkingly, unaware of its hurtful consequences. But being the target of sarcastic or otherwise hurtful remarks can result in partners feeling alienated from each other (Murphy and Oberlin 2006). Then too, sex and other expressions of intimacy become arenas for ongoing conflict when partners passive-aggressively withhold them. For example, a partner makes a disparaging comment in front of company. The hurt spouse says nothing at the time but rejects the other's sexual advances later.

Other forms of indirect anger include sabotage and displacement. **Sabotage,** a means of getting revenge, or "payback" (Boon, Deveau, and Alibhal 2009), involves one partner's attempts to spoil or undermine some activity that the other has planned. For example, the partner who is angry because the other invited friends over when he or she wanted to relax may sabotage the evening by acting bored. In **displacement**, a person directs anger at people or things that the other cherishes. An individual who is angry with a partner for spending too much time on a career may hate the partner's expensive car.

Typically, individuals express anger indirectly because they are afraid of conflict, either generally or with reference to a specific person or persons (Murphy and Oberlin 2006; Oyamot, Fuglestad, and Snyder 2010). Advising partners to express anger directly rests on assumptions of equitable power and feelings of security in a relationship (Knobloch and Knobloch-Fedders 2010; Knudson-Martin and Mahoney 2009). Nevertheless, partners and family members who do not express their anger directly risk emotional and/or sexual detachment (Gottman and Levenson 2000). Of course, it is important to recognize that direct expressions of anger can go too far, resulting in domestic violence (Rehman et al. 2009), discussed in Chapter 13. We turn now to what an important research team has to say about conflict management.

© Myrleen Ferguson/PhotoEdit

Learning to express anger and dealing with conflict early in a relationship are challenges to be met rather than avoided. Acknowledging and resolving conflict is painful, but it often strengthens the couple's union in the long run. A key to effective conflict management is to share everyday— and positive—events in friendly, supportive ways so that arguments occur within an overall context of couple satisfaction and mutual trust.

John Gottman's Research on Couple Communication and Conflict Management

Social psychologist John Gottman (1979, 1994, 1996; Gottman et al. 1998; Gottman and DeClaire 2001; Gottman and Notarius 2000, 2003) has made his reputation in the field of marital communication.[5] In the 1970s, applying an interactionist perspective to partner communication, he began studying newly married couples in a university lab while they talked casually, discussed issues that they disagreed about, or tried to solve problems. Video cameras recorded the spouses' gestures, facial expressions, and verbal pitch and tone. After he began this research, Gottman kept in contact with more than 650 of the couples, some for as many as fourteen years. Typically, the couples were videotaped intermittently. Some couples volunteered for laboratory observation that monitored shifts in their heart rate and chemical stress indicators in their blood and urine as a result of their communicating with each other (Gottman 1996).

Studying marital communication in this detail, Gottman and his colleagues were able to chart the

[5] As explored in "Issues for Thought: A Look Behind the Scenes at Communication Research," the extent to which Gottman and colleagues' findings apply to other than heterosexually married couples has been questioned.

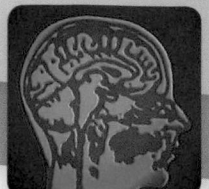

An important scientific norm requires that research findings be critically reviewed by others to help ensure that they are accurate. One way to follow this norm is for subsequent studies to try to reproduce, or *replicate,* the findings of an earlier researcher—that is, follow the first researcher's design and methods over again to discern whether the same findings emerge. If a study replicates well (the same findings show up), we can be better assured that the research is to be taken seriously. If the study does not replicate (the same findings do not emerge in both studies), then we cannot be certain what to think until even more studies are done.

In 2007, a team of researchers (Kim, Capaldi, and Crosby 2007) sought to reproduce findings from the much publicized earlier research of well-known and highly respected social psychologist John Gottman and his colleagues. The research that sought to reproduce Gottman's findings used observation methods similar to Gottman's. However, the sample Kim and colleagues used was purposefully different. Gottman had used a sample of newlyweds who had been married for the first time within the previous six months and who answered media ads that requested their participation. Most of these couples were middle-class college graduates. Kim, Capaldi, and Crosby (2007) used a sample of young adults who had grown up in poor neighborhoods and who, by their twenties, were unlikely to have graduated from college. Furthermore, Kim and colleagues' sample included both married

and cohabiting couples. Kim and colleagues used a different sample to discern whether Gottman's findings could be applied to couples other than middle-class marrieds.

One thing that Gottman and his colleagues had found was that a serious communication problem faced by couples occurs when the wife raises complaints to her husband in ways that he perceives as abrasive or attacking. The husband then withdraws and apparently ignores her—the *female-demand/male-withdraw communication* pattern. Based on their findings, Gottman and his colleagues advised therapists to focus on gender differences when counseling (heterosexual) couples' communication. Therapists were encouraged to help wives to raise issues more gently and husbands to be more willing to be influenced by their wives (Gottman et al. 1998).

The later researchers pointed out that Gottman and his colleagues had published advice to therapists without the qualification that it might only apply to middle-class, married couples. Therefore, they argued that Gottman's research needed to be further examined to see whether his findings applied to couples who were not middle class and/or were not married, but cohabiting.

This story might not be news if Kim and colleagues had successfully reproduced Gottman's findings. However, this was not the case. For one thing, the later researchers' findings did not support

the female-demand/male-withdraw pattern. In a published response, Gottman argued, among other things, that his research was not necessarily meant to apply to cohabiting couples (Coan and Gottman 2007). Nevertheless, Kim and colleagues agued that their failure to replicate Gottman's findings

calls into question the extent to which [Gottman and his colleagues'] findings . . . should be used as a basis for recommendation for therapy and interventions with young couples and in general provides a caution against translating empirical findings to treatment recommendations without replication. (Kim, Capaldi, and Crosby 2007, p. 66)

When researchers fail to replicate another's findings, "it is often difficult for readers to know what to conclude." Therefore, we need to be "cautious with initial findings until they are replicated" (Heyman and Hunt 2007, p. 84) and tested in many different samples.

Critical Thinking

Given Kim, Capaldi, and Crosby's (2007) failure to reproduce the findings of Gottman et al. (1998) in a sample that was not primarily middle-class and included cohabiting couples, what do you think about the generalizability of Gottman's suggestions to therapists? How might yet another research team further investigate Gottman's findings?

---

effects of small gestures. For example, early in his career he reported that when a spouse—particularly the wife—rolled her eyes while the other was talking, divorce was likely to follow sometime in the future, even if the couple was not thinking about divorce at the time (Gottman and Krotkoff 1989). Gottman's research has been challenged and may apply only to middle-class, married couples (see "Issues for Thought: A Look Behind the Scenes at Communication Research"). Recognizing the need for continued research in this area, we present Gottman's highly influential findings here.

The Four Horsemen of the Apocalypse

Gottman's research (1994) showed that conflict and anger themselves did not predict divorce, but processes that he called the **Four Horsemen of the Apocalypse** did.[6] The Four Horsemen of the Apocalypse are contempt, criticism, defensiveness, and stonewalling. Rolling

[6] The word *apocalypse* refers to the biblical idea that the world is soon to end, being destroyed by fire. The Four Horsemen are allegorical figures representing war, famine, and death, with the fourth uncertain (*Concise Columbia Encyclopedia* 1994, p. 309). Gottman used the phrase to indicate attitudes and behaviors that foreshadow impending divorce.

one's eyes indicates **contempt**, a feeling that one's spouse is inferior or undesirable. **Criticism** involves making disapproving judgments or evaluations of one's partner. **Defensiveness** means preparing to defend oneself against what one presumes is an upcoming attack. **Stonewalling** involves resistance—refusing to take a partner's complaints seriously.[7] In several of Gottman's studies, these behaviors identified those who would divorce, with an unusually high accuracy of about 90 percent.

Later, after more research, Gottman added **belligerence**, a behavior that challenges the other's power or authority (for example, "What can you do if I do go drinking with Dave? What are you going to do about it?") (Gottman et al. 1998, p. 6). Still later, Gottman and his colleagues identified similar patterns among gay and lesbian couples (Gottman et al. 2003; see also Houts and Horne 2008). To illustrate how Gottman's horsemen make their ways into communication, consider the following exchange:

> PARTNER A: I can't find my cell phone.
>
> PARTNER B: Don't accuse me of taking it. It's never on when I call you anyway.
>
> PARTNER A: What's that got to do with anything?
>
> PARTNER B: So look for it.
>
> PARTNER A: So help me.
>
> PARTNER B: You're just like your dad—always expecting somebody to do things for you.
>
> PARTNER A: Jerk!

In this scenario, *Partner A* mentions having misplaced a cell phone, and an argument develops, illustrating contempt, defensiveness, and belligerence. When *Partner A* announces the need for the cell phone, *Partner B* becomes defensive: "Don't accuse me of taking it." At this point, *B* raises a complaint: "It's never on when I try to call you anyway." In a less-distressed couple, *A* might respond to *B*'s complaint. However, *A* fails to de-escalate the interchange and does not acknowledge *B*'s complaint. Less distressed couples might stop this negative spiral with shared humor or some sign of affection. However, *Partner A* subsequently requests that *B* help look for the cell phone. *Partner B*'s reply is contemptuous and critical: "You're just like your dad—always expecting somebody to do things for you." Again, the couple fails

to de-escalate the negative affect. This time *A* calls *B* a jerk. Name-calling is contemptuous.

It appears the couple has forgotten what the fight is about. In fact, one wonders whether they ever knew what the fight was about. Counselors point out that distressed couples, like the couple depicted here, may unconsciously allow trivial issues to become decoys so that they evade the real area of conflict and leave it unresolved. In sum, contempt, criticism, defensiveness, stonewalling, and belligerence characterize unhappy marriages and may signal impending divorce (Gottman and Levenson 2002). Gottman, like other researchers, found that communicating positive feelings for a partner characterized happier, more stable unions.

Positive versus Negative Affect Gottman and his colleagues videotaped 130 newlywed couples as they discussed a problem that caused ongoing disagreement in their marriage for fifteen minutes (Gottman et al. 1998). Each couple's communication was coded in one-second sequences, and then synchronized with each spouse's heart rate data, which was being collected at the same time. The heart rate data would indicate each partner's physiological stress.

The researchers examined all the interaction sequences in which one partner first expressed **negative affect**: anger, sadness, whining, disgust, tension and fear, belligerence, contempt, or defensiveness. Belligerence, contempt, and defensiveness (three of Gottman's indicators of impending divorce) were coded as *high-intensity, negative affect*. The other emotions listed previously (anger, sadness, whining, and so on) were coded as *low-intensity negative affect*.

Next, the researchers watched what happened immediately after a partner had expressed negative affect or raised a complaint. Sometimes the partner reciprocated with negative affect in kind, either low or high intensity. As examples, *Partner A* whines, and *Partner B* whines back; *A* expresses anger, and *B* responds with tension and fear; or *A* is contemptuous, and *B* immediately becomes defensive.

At other times, one partner's first negative expression was reciprocated with an escalation of the negativity. As examples, *Partner A* whines, and *Partner B* grows belligerent; or *A* expresses anger, and *B* becomes defensive. Gottman and his colleagues called this kind of interchange *refusing-to-accept influence*, because the spouse on the receiving end of the other's complaint refuses to consider it and, instead, escalates the fight. Negative escalation is evidenced in the previous cell phone conflict.

Meanwhile, still other couples were likely to communicate with positive affect, responding to each other warmly with interest, affection, or shared (not mean or contemptuous) humor. Positive affect typically de-escalated conflict (Gottman and Levenson 2000, 2002). Gottman and his colleagues found that "[t]he only variable that predicted both marital stability and marital

---

[7] Stonewallers react to their partner's attempts to raise tension-producing issues by refusing to entertain them. Avoiding or evading an argument is an example of stonewalling. Argument evaders use several tactics to avoid fighting, such as vacating the scene when an argument threatens; turning sullen and refusing to talk; declaring, "I can't take it when you yell at me"; using the hit-and-run tactic of filing a complaint, then leaving no time for an answer or resolution; saying "OK, you win" without meaning it.

happiness among stable couples was the amount of positive affect in the conflict" (1998, p. 17). In stable, happy couples, shared humor and expressions of warmth, interest, and affection were apparent even in conflict situations and, therefore, de-escalated the argument.

The researchers "found no evidence . . . to support the [idea that] anger is the destructive emotion in marriages" (1998, p. 16). Instead, they found that contempt, belligerence, and defensiveness were the destructive attitudes and behaviors. Furthermore, Gottman and his colleagues concluded that the interaction pattern best predicting (heterosexual) divorce was a wife's raising a complaint, followed by her husband's refusing-to-accept influence, followed, in turn, by the wife's reciprocating her husband's escalated negativity, and the absence of any de-escalation by means of positive affect. Despite changing gender roles (see Chapter 4), researchers continue to observe gender differences in communication patterns.

Gender Differences and Communication

Before the nineteenth century, men's and women's domestic activities involved economic production, not personal intimacy. With the development of separate gender spheres in industrializing societies during the nineteenth century, expressions of emotion became the domain of middle-class women, whereas work was defined as more appropriate to masculinity. As a result of this historical legacy, we have come to see men as less well equipped than women for emotional relatedness (Real 2002).

Sociologist Francesca Cancian (1987) has expanded on these points to argue that men are equally loving, but women, not men, are made to feel primarily responsible for love's endurance or success. Furthermore, expressions of love are defined and perceived mostly on feminine terms—that is, verbally—and women are the more verbal sex. Expressions of love that men may make, such as doing favors or reducing their partners' burdens, are not credited as love (Cancian 1985). Recent research appears to support Cancian's analysis. For instance, a study with 453 heterosexual couples drawn from a national representative survey looked at changes that women would like in their partners, compared with changes that men would like. The women were more likely than the men to want increases in a partner's demonstrations of positive emo-tion (Heyman et al. 2009).

In Cancian's analysis, "The consequences of love would be more positive if love were the responsibility of men as well as women and if love were defined more broadly to include instrumental help as well as emotional expression" (1985, p. 262). Cancian has also argued that a more balanced view of how love is to be expressed—one that includes masculine as well as feminine elements—would find men equally loving and emotionally profound.

Meanwhile, Deborah Tannen's book *You Just Don't Understand* (1990) suggests that men typically engage in **report talk**, conversation aimed mainly at conveying information. Women, on the other hand, are likely to engage in **rapport talk**, speaking to gain or reinforce intimacy or connection with others. Men are likely to bring up problems, for instance, only when hoping to trigger suggestions for solution. Women, on the other hand, are likely to talk about problems simply to share or foster rapport. These gendered differences lead to an imbalance in many families. If the mother is telling about troubles she confronted during her day but the father is not, the result is that mothers come across as more problem-ridden and insecure than fathers. And many men, because they don't tend to talk in this way, understandably assume that a woman who recounts a problem must be seeking help solving it; why else would she talk about

"Sometimes I wonder what life would be like with you."

it? That's why they generously provide solutions. So the woman's conversational gambit ends up being refracted through the man's point of view. This misunderstanding of women's rapport-talk often results in mothers appearing to their families as less confident, or even less competent, than their husbands. (Tannen 2006, pp. 83–84)

Moreover, some researchers speculate that women, being more expressive and attuned to the emotional quality of a relationship, are more likely than men to bring conflict into the open, sometimes in an attention-getting negative tone (Cui et al. 2005). Men try to minimize the impending conflict either by conciliatory gestures or by stonewalling. The male's minimization of conflict may appear to the female as failure to recognize her emotional needs (Noller and Fitzpatrick 1991; Canary and Dindia 1998). In the following, a husband describes this situation:

> The more I try to be cool and calm her the worse it gets. I swear, I can't figure her out, I'll keep trying to tell her not to get so excited, but there's nothing I can do. Anything I say just makes it worse. So then I try to keep quiet, but . . . wow the explosion is like crazy, just nuts. (in Rubin 2007, p. 323)

We might compare this to a wife, who told her interviewers that,

> I can't stand that he's so damned unemotional and expects me to be the same. He lives in his head all of [the] time, and he acts like anything that's emotional isn't worth dealing with. (in Rubin 2007, pv. 322)

Reviews of research on couple communication in the 1990s (Gottman and Notarius 2000; Bradbury, Fincham, and Beach 2000) concluded that men and women differ in their responses to negative affect in close relationships. When faced with a complaint from a partner, men tend to withdraw emotionally whereas women do not. Researchers have found this pattern to be common enough that some call it the *female-demand/male-withdraw communication pattern* (Gottman and Levenson 2000). In distressed marriages, this pattern becomes a repeated cycle of negative verbal expression by one partner and withdrawal by the other (Bradbury, Fincham, and Beach 2000).

Many researchers and therapists agree that generally there is a female-demand/male-withdraw pattern (Miller and Roloff 2005; Weger 2005). However, an alternative view argues that "it is not gender per se but the nature of the marital discussion—for example, whether it is the wife or the husband who desires a change—that may determine who is demanding and who is withdrawing" (Roberts 2000, p. 702; also see Kim, Capaldi, and Crosby 2007). Research on same-sex couples has found the same pattern—that is, one partner demands while the other withdraws (Parker-Pope 2008). And research in stepfamilies suggests that neither partner is likely to

demand as much as in high-risk, first marriages, and both are more likely to withdraw from a conflict (Halford, Nicholson, and Sanders 2007).

Obviously, the **female-demand/male-withdraw interaction pattern** leads to both partners feeling misunderstood, thereby decreasing marital satisfaction (Weger 2005). Gottman and his colleagues (1998) concluded that wives and husbands have different goals when they disagree:

> The wife wants to resolve the disagreement so that she feels closer to the husband and respected by him. The husband, though, just wants to avoid a blowup. The husband doesn't see the disagreement as an opportunity for closeness, but for trouble. (p. 17)

In one husband's words, "I just got mad and I'd take off—go out with the guys and have a few beers or something. When I'd get back, things would be even worse." From his wife's perspective, "The more I screamed, the more he'd withdraw, until finally I'd go kind of crazy. Then he'd leave and not come back until two or three in the morning sometimes" (Rubin 1976, pp. 77, 79).

Gottman and his colleagues sought to better understand this pattern. You'll recall that the researchers monitored spouses' heart rates as indicators of physiological stress during conflict. They hypothesized that, "it is likely that the biological, stress-related response of men is more rapid and recovery is slower than that of women, and that this response is related to the greater emotional withdrawal of men than women in distressed families" (Gottman et al. 1998, p. 19). That is, when confronted with conflict from an intimate, men may experience more intense and uncomfortable physical symptoms of stress than women do. Therefore, men are more likely than women to withdraw emotionally and/or physically.

An alternative—or complementary—view is that men have been socialized to withdraw. The cultural options for masculinity include "no 'sissy' stuff," according to which men are expected to distance themselves from anything considered feminine. We guess this could include a wife's complaints. In two books on men and communication, the first called *I Don't Want to Talk About It*, therapist Terrence Real (1997, 2002) attributes males' withdrawal to a "secret legacy of depression," brought on by men's traditional socialization, particularly society's refusal to let them grieve over losses (e.g., "Don't cry over nothing"). It is possible that physiology and culture interact to create a female-demand/male-withdraw pattern.

What Couples Can Do The general conclusion of Gottman's research on couple communication and conflict management is as follows. Both partners:

1. need to try to be gentle when they raise complaints;
2. can help to reduce anxiety in their mate by communicating care and affection, hence reducing physical stress symptoms;

3. can learn techniques for reducing anxiety in one-self—for example, taking a time-out or saying that "I just can't talk about it now, but I will later" (and meaning it);

4. need to be willing to accept influence from each other;

5. need to do what they can—perhaps using authentic, shared humor, kindness, and other signs of affection—to de-escalate the argument. It is important to recognize that this does not mean avoiding the issue altogether.

Finally, Gottman and his colleagues (1998) suggest that, as we have already seen, it is important for couples to think about communicating with positive affect more often in their daily living and not just during times of conflict (Gottman and DeClaire 2001). "As We Make Choices: Ten Rules for Successful Relationships" suggests ways to do this. Then too, as we have seen, cohesive families have arguments. But the argument ends; conflicts are resolved. We turn to some guidelines for accomplishing this.

Working Through Conflicts in Positive Ways—Ten Guidelines

Different cultural groups vary in their endorsement of openly expressing emotion and directly expressing conflicts (see, for example, Hirsch 2003). We need to recognize that preferred standards for communication vary culturally (Matsunaga and Imahori 2009) and that the guidelines suggested in this chapter, which accent direct communication styles, may be ethno- or Eurocentric.[8] Nevertheless, counselors do advise that there are better (and not-so-good) ways that virtually all couples and family members can resolve differences.

Before going further, we want to point out that not all negative facts and feelings need to be communicated. Before voicing a complaint, we might ask ourselves, "How important is it?" (Sanford 2006). Counselors suggest that if, after giving it some time and thought, we believe that raising a particular grievance is important, then we should do so. Similarly, when offering negative information, it is important to ask ourselves why we

want to do so and whether the other person really needs to know. We turn now to ten specific guidelines for constructive conflict management.

Guideline 1: Express Anger Directly and with Kindness

Family members may have the false belief that their intimates automatically know—or should know—what they think and how they feel. This incorrect idea is detrimental to relationships (Hamamci 2005). When complaints are not addressed directly, conflict goes unresolved, with lingering grievances sparked again and again by "subtle triggers." Consider the following family situation:

> An ongoing point of contention in this family is the mother's belief that her teenage daughter, Joyce, spends too much money on clothes and makeup, which she buys in upscale stores rather than more economical stores, like Wal-Mart. So when the father, who is scanning a newspaper, remarks, "I see Wal-Mart set a record for sales yesterday," the seed is planted for an argument to sprout. (Tannen 2006, p. 123)

The underlying conflict is voiced as follows:

Mom: So? We don't shop at Wal-Mart, so what's the point?

Dad: Okay.

Joyce: What does that have to do with anything?

Mom: Okay, I'm just saying—

Joyce: Saying what?

Mom: Yeah, so what's the point?

Joyce: What point, Mom? You don't shop there either, Mom.

Mom: Yes, I do. You could shop there for toiletries.

Joyce: For clothes you shop there, Mom?

Mom: No.

Joyce: See, so why should we go shopping there for toiletries? . . . I don't go shopping for toiletries anywhere because you buy them for me.

Mom: No, but you buy makeup.

Dad: Well, this year we can do all our Christmas shopping at Wal-Mart. (Tannen 2006, pp. 123–25)

Tension and conflict go unresolved.

Counselors advise expressing anger directly because doing so makes way for resolution (Bernstein and Magee 2004). For example, the mother might say, "I feel that you've been spending more than we can afford on makeup." Counselors further advise that a grievance will be less threatening to the receiver when positive feelings are conveyed at the same time that the grievance is voiced.

---

[8] Deborah Tannen's (1990) book *You Just Don't Understand*, which drew wide attention for its comparison of men's and women's communication styles, also points out cultural communication differences among, for example, New Yorkers, Californians, New Englanders, and Midwesterners, and among Scandinavians, Canada's native peoples, and Greeks. Interpersonal communication differences are evidenced by other race/ethnicities as well (Matsunaga and Imahori 2009).

Chronic stonewallers may fear rejection or retaliation and therefore hesitate to acknowledge their own or their partner's angry emotions. Examples of stonewalling include saying things like, "I can't take it when you yell at me," or turning sullen and refusing to talk. It may sound impossible to fight more fairly when you're angry, but "practice makes better." Using "I" statements, avoiding mixed messages, focusing your anger on specific issues, and being willing to change are some guidelines worth trying.

Family Member B: No, I'm irritated because I was tied up in traffic an extra half hour on my way home.

Guideline 3: To Avoid Attacks, Use "I" Statements

Attacks, sometimes interpreted as blame, involve insults or assaults on another's character or self-esteem. Needless to say, attacks do not help to bond a couple (Sinclair and Monk 2004). A rule in avoiding attack is to use the word, *I* rather than *you* or *why*. For example, instead of declaring, "You're late," or asking "Why are you late?"—both of which can smack of blame—a statement such as, "I was worried because you hadn't arrived" may allow for more positive dialogue. The receiver is more likely to perceive "I" statements as an attempt to recognize and communicate feelings; "you" and "why" statements are more likely to be perceived as attacks, even when not intended as such.

Of course, making "I" statements may be too much to ask in the heat of an argument. One social psychologist has admitted what many of us may have experienced: "It is impossible to make an 'I-statement' when you are in the . . . 'wanting-revenge, feeling-stung-and-needing-to-sting-back' state of mind" (quoted in Gottman et al. 1998, p. 18). Of course, this is partly the point. Keeping in mind the possibility of expressing a complaint—at least *beginning* a confrontation—with an "I" statement can discourage family members from getting to that wanting-revenge state of mind in the first place.

If you're angry and resentful, requests for change will be met with resistance and countercharge efforts: "It's not my problem; it's your problem." But if you learn to approach each other with acceptance and empathy, you can create a collaborative context, and often people will make spontaneous changes. ("Loving Your Partner" 2000)

So, even better, the mother might say, "You always look nice, and I like the way that you choose to wear your makeup, but I feel that you're spending more than we can afford on it." Being direct is not the same as being unnecessarily critical.

Guideline 2: Check Out Your Interpretation of Others' Behaviors

Because family members and partners in distressed relationships seldom understand each other as well as they think they do, a good habit is to ask for feedback by a process of *checking it out:* asking the other person whether your perception of her or his feelings or of the present situation is accurate. Checking it out often helps to avoid unnecessary hurt feelings or imagining trouble that may not exist, as the following example illustrates:

Family Member A: I think you're mad about something. *(checking it out)* Is it because it's my class night and I haven't made dinner?

Guideline 4: Avoid Mixed, or Double Messages

Mixed, or double messages contradict each other. Contradictory messages may be verbal, or one may be verbal and one nonverbal. For example, a family member offers to take the family to a movie yet sighs and says that he or she is exhausted after a really hard day at work. Or a partner insists, "Of course I love you" while picking an invisible speck from his or her sleeve in a gesture of indifference.

Senders of mixed messages may not be aware of what they are doing, and mixed messages can be very subtle. They sometimes result from simultaneously wanting to recognize and to deny conflict or tension. A classic example is the *silent treatment.* One partner becomes aware that she or he has said or done something upsetting and asks what's wrong. "Oh, nothing," the other replies without much feeling, but everything about the partner's face, body, attitude, and posture suggests that something is indeed wrong (Lerner 2001).

Moreover, communication involves both a sender and a receiver. Just as the sender gives both an overt message and an underlying *meta-message*,[9] so also does a receiver give cues about how seriously she or he is taking the message. For example, listening while continuing to do chores sends the nonverbal message that what is being heard is not very important.

Guideline 5: When You Can, Choose the Time and Place Carefully

Arguments are less likely to be constructive if the complainant raises grievances at the wrong time. One partner may be ready to argue about an issue when the other is almost asleep or working on an important assignment, for instance. At such times, the person who picked the fight may get more—or less—than he or she had expected.

Family members might negotiate a time and place for addressing issues. Arguing "by appointment" may sound silly and be difficult to arrange, but doing so has advantages. For one thing, complainants can organize their thoughts and feelings more calmly and deliberately, increasing the likelihood that they will be heard. Also, recipients of complaints have time before the argument to prepare themselves to hear some criticism.[10]

Guideline 6: Address a Specific Issue, Ask for a Specific Change, and Be Open to Compromise

Constructive relationships aim at resolving current, specific problems. Recipients of complaints need to feel that they can do something specific to help resolve the problem raised. This will be difficult if they feel overwhelmed by old gripes. Furthermore, complainants should be ready to propose one or more solutions. Recipients might come up with possible solutions themselves. When family members can entertain potential solutions to a definite problem at hand, they are better able to negotiate alternatives.

John Gottman found that happily married couples reached agreement rather quickly. Either one partner gave in to the other without resentment, or the two compromised. Unhappily married couples continued in a cycle of stubbornness and hostility (Gottman and Krotkoff 1989; see also Busby and Holman 2009).

Guideline 7: Be Willing to Change Yourself

The principle that couples or family members should accept each other as they are sometimes merges with the idea that individuals should be exactly what they choose to be. The result is an erroneous assumption that if someone loves you, he or she will accept you just as you are and not ask for even minor changes. In truth, partners need to be willing to be influenced by their loved ones and to change themselves (Lerner 2001).

Therapists note that, in some relationships, each person expects the other one to do the changing: "You have to understand, she's [or he's] impossible to live with" (Ball and Kivisto 2006, p. 155). One counselor team (Christensen and Jacobson 1999) has suggested "acceptance therapy," helping individuals accept their partners and other family members as they are instead of demanding change—although these counselors also suggest that, paradoxically, showing acceptance can lead to a partner's changing behavior. We need to balance acceptance of another against not being a doormat, but being willing to change ourselves is key.

Guideline 8: Don't Try to Win

Counselors encourage us to recognize that there are probably several ways to solve a particular problem, and backing others into a corner with ultimatums and counter-ultimatums is not negotiation but attack. Moreover, wanting to win a dispute with a loved one typically encourages us to use unnecessarily hurtful language, which nonproductively increases the recipient's stress (Priem, McLaren, and Solomon 2010). We're reminded that recipients of painful messages typically see them as more hurtful than do the senders (Zhang 2009). How we say things impacts how others perceive them (Young 2010). Even hurtful information can be conveyed with sensitivity.

Societies that emphasize competition, such as ours does, encourage people to see almost everything they do in terms of winning or losing (Fromm 1956). Yet research clearly indicates that for same-sex and heterosexual couples, the tactics associated with winning in a particular conflict are also those associated with lower relationship satisfaction (Clunis and Green 2005; Heene, Buysse, and Van Oost 2007; Houts and Horne 2008). Losing lessens a person's self-esteem, increases resentment, and adds strain to the relationship. On the

---

[9] Communication scholars and counselors point out that there are two major aspects of any communication: *what* is said (the verbal message) and *how* it is said or interpreted (the nonverbal "meta-message"). The meta-message depends on tone of voice, inflection, and body language, as well as on the receiver (Nierenberg and Calero 1973). In a mixed message, the verbal message does not correspond with the meta-message.

[10] A qualitative British study of teenagers and their parents found that some parents and teens use mobile phones to raise sensitive issues that they intend to later pursue face to face. One teen said, "Well, I think mobiles can be really good if you've got something you don't wanna tell straight away," and another young respondent said, "I'd maybe text if it's something that I can't, I dunno, something I can't get across [face to face] and stuff" (in Devitt and Roker 2009, p. 192).

other hand, everyone wins when family members mutually agree on solutions to their differences (Carroll, Badger, and Yang 2006).

Guideline 9: Be Willing to Forgive

A growing number of therapists suggest that being willing to forgive is critical to ongoing happy relationships (Fincham, Hall, and Beach 2006). Forgiveness "is the idea of a change whereby one becomes less motivated to think, feel, and behave negatively (e.g., retaliate, withdraw) in regard to the offender." Forgiveness is not something to which the offender is necessarily entitled, but it is granted nevertheless.

Contrary to what many individuals believe, however, forgiveness does not require that the offended partner minimize or condone the offense. Rather, "an individual forgives despite the wrongful nature of the offense and the fact that the offender is not entitled to forgiveness." Further, "forgiveness is distinct from denial (an unwillingness to perceive the injury) . . . or forgetting (removes awareness of offence from consciousness)" (Fincham, Hall, and Beach 2006, p. 416).

Forgiveness is often a process that takes time, rather than one specific decision or act of the will. Being willing to forgive has been associated in research with marital satisfaction, lessened ambivalence toward a partner, conflict resolution, enhanced commitment, and greater empathy (Fincham, Hall, and Beach 2006).

Guideline 10: End the Argument

Ending the argument is important. Sometimes when individuals are too hurt to continue, they need to stop arguing before they reach a resolution. A family member may signal that he or she feels too distressed to go on by calling for a time-out. Or it could help to bargain about whether the fight should continue at all.

As pointed out earlier in this chapter, the happily married couples that Gottman and his colleagues, as well as Wallerstein, interviewed knew how and when to stop fighting. Arguments can end with compromise, apology, submission, or agreement to disagree (Goeke-Morey, Cummings, and Papp 2007). Ideally, a fight ends when there has been a mutually satisfactory airing of each partner's views.

Toward Better Couple and Family Communication

Keeping a loving relationship or creating a cohesive family is not automatic. Doing so requires working on ourselves as well as on our relationships. A first step involves consciously recognizing how important the relationship is to us. A second step is to set realistic expectations about the relationship (Cloud and Townsend 2005). As one married woman put it,

> You just have this idealized version of getting married, you know, everybody plays it up as so romantic and so wonderful and sweet. Now that I am married and now that I have gotten older and hit the real world I'm kind of like . . . It's a lot more hands-on, you know, getting stuff done . . . than it is that idealized romantic notion that you get as a girl. (in Fairchild 2006, p. 13)

A third step involves improving our own (1) **emotional intelligence**—awareness of what we're feeling so that we can express our feelings more authentically; (2) ability and willingness to repair our moods, not unnecessarily nursing our hurt feelings; (3) healthy balance between controlling rash impulses and being candid and spontaneous; and (4) sensitivity to the feelings and needs of others (Keaten and Kelly 2008). We can develop greater flexibility of thought, learning to think of several alternative workable solutions to problems and to have several ways of responding to a situation, not just one that habitually comes up by default (Koesten, Schrodt, and Ford 2009). Support is mutually reinforcing. When we can support others, they are more likely to be supportive (Priem, Solomon, and Steuber 2009).

With regard to the relationship itself, counselors encourage making time for play and incorporating new activities into relationships (Lawson 2004b; Smith, Freeman, and Zabriskie 2009).[11] Social psychologist John

---

[11] An intriguing area of research from the biosocial theoretical perspective (see Chapter 2) suggests that brain chemistry helps to explain why shared new activities enhance romantic relationships (Gottlieb 2006; Slater 2006). This research points to two hormones. The first, dopamine, is a chemical naturally produced in our brains. Although dopamine has many functions, its importance to love is that it acts upon the pleasure center in our brains, giving us powerful feelings of enjoyment and motivating us to do whatever we're doing that is so pleasurable over and over again (Berridge and Robinson 1998). Dopamine helps to explain why we have a second helping of a really tasty dessert, for example. Furthermore, dopamine is associated with new or novel pleasurable experiences and activities. Research shows that when people are newly in love, they tend to have higher brain levels of dopamine (Slater 2006). Dopamine makes you "high on" your partner. The second hormone relevant here is oxytocin, also produced naturally in our brains. Some researchers have nicknamed oxytocin the "love" or "cuddle" hormone (Barker n.d.; Bosse 1999).

Like dopamine, oxytocin has several functions (inducing labor and stimulating breast milk production in females, for example). Research in mammals has long demonstrated that oxytocin facilitates more general maternal, nurturing behaviors ("Oxytocin" 1997). In addition, oxytocin seems to be related to human feelings of deep friendship, trust, sexuality, love, bonding, and commitment ("Biology of Social Bonds" 1999; Bosse 1999; see also Ditzen et al. 2009). Hormones affect feelings and behavior, but the reverse is also true: Behaviors can stimulate hormone production. Doing novel things together and engaging in supportive touch, including sex, stimulate the production of dopamine and oxytocin, respectively (Slater 2006; see also Ellison and Gray 2009). All this gives a somewhat new slant to the phrase "making love," doesn't it?

Crosby points out that people may misinterpret the idea of "working at" committed relationships: Instead of working *at* relationships, "we may, with all good intentions, end up making work *of*" them (Crosby 1991, p. 287).

Some of us have grown up with poor role modeling on the part of our parents (Ledbetter 2009; Rovers 2006; Schrodt et al. 2009; Zimmerman and Thayer 2003). Regardless of how our parents behaved, we can choose to change how we communicate (Braithwaite and Baxter 2006; Turner and West 2006; Wright 2006b).

Training programs in couple and family communication, often conducted by counseling psychologists, have proven effective in helping to change negative communication patterns (Blanchard et al. 2009; Bodenmann, Bradbury, and Pihet 2009; Sevier et al. 2008; Yalcin and Karaban 2007). One program for married and cohabiting couples is ENRICH, originally developed by social psychologist David Olson at the University of Minnesota. A similar program is PREP (the Prevention and Relationship Enhancement Program), developed by marital communication psychologists Scott Stanley and Howard Markman, with the overall aim of strengthening marriages and preventing divorce (Markman, Stanley, and Blumberg 2001; Schilling et al. 2003). Marriage Encounter and similar organizations offer weekend workshops, designed for mostly satisfied marrieds who want to improve their relationship (Yalcin and Karaban 2007). Advertising "psychological care for the whole family," the Family Success Consortium offers programs for all couples, whether or not married (www.familysuccessconsortium.com).

Some programs have been designed for same-sex couples (Heffner, 2003; Unitarian Universalist Association nd.). Men's groups aimed at encouraging their expressions of emotional intimacy have been shown to enhance couple and family relationships (Garfield 2010). Some conflict management programs have been developed for young and adolescent siblings (Kennedy and Kramer 2008; Thomas and Roberts 2009) and/or for families of particular race/ethnicities (e.g., Soll, McHale, and Feinberg 2009). As mentioned in Chapter 10, some parenting enhancement programs incorporate anger management components (Fetsch, Yang, and Pettit 2008).

Couples or family members who want to work for change on their own might practice the previously mentioned guidelines for conflict resolution. As partners and family members grow accustomed to voicing grievances regularly and in more respectful or caring ways, their disagreements less often become full-fledged fights: Family members gradually learn to incorporate many irritations and requests into their normal conversations, arguing in normal tones of voice and even with humor.

Although these suggestions may help, learning to fight fair is not easy. Sometimes couples and families feel that they need outside help, and they may decide to engage a counselor. See "Facts about Families: Relationship and Family Counseling" for a discussion of this alternative.

Couples can change their fighting habits. The key to staying happily together is to make knowledgeable choices—about not avoiding conflict but dealing with it openly, or directly, and in supportive ways. Doing so involves listening—without judgment, without formulating a response while the other talks, and without interrupting. The goal isn't necessarily agreement, but acknowledgment, insight, and understanding.

Then too, a number of books and Internet resources are available that could help. As examples, there are books on overcoming passive-aggressive behavior (Murphy and Oberlin 2006), recognizing how we sabotage our relationships (Matta 2006), and changing habits that can thwart a satisfying life in general (Kagan and Einbund 2008). Some books, such as *Person to Person: Positive Relationships Don't Just Happen* (Hanna, Suggett, and Radtke 2008) focus on both individual self-improvement and couple communication. Susan Halpern's *Finding the Words: Candid Conversations with Loved Ones* (2009) covers topics such as cultivating conscious conversations as a couple, communicating in ways that might lessen the disruptive effects of divorce, and improving communication between parents and their adult children. There are books on communication designed specifically for same-sex couples (Clunis and Green 2005). And, of course, there are university courses and textbooks

Relationship and family counseling is a professional service having two goals: (1) helping individuals, couples, and families gain insight into the actually or potentially troublesome dynamics of their relationship(s); and (2) teaching clients more effective and supportive communication techniques. According to the American Association for Marriage and Family Therapy (AAMFT), this type of counseling is meant to be "solution-focused; specific, with attainable therapeutic goals; [and] designed with the 'end in mind'" ("What Is Marriage and Family Therapy?" 2005; Clinton and Trent 2009).

Experts advise couples or families to visit a counselor when communication is typically hostile or conflict goes unresolved, when they cannot figure out how to resolve a family problem themselves, or when a partner is thinking of leaving a committed relationship. However, counseling is also appropriate—and perhaps more effective—as a preventive technique, undertaken at the onset of family stress or when a couple or family sees a potentially troublesome transition ahead.

People go to counselors for help in working through premarital and engagement issues, as well as cultural clashes, same-sex couple, cohabitation, infidelity, divorce, substance abuse, finances, unemployment, co-parenting conflict, infertility, sexual difficulties, and changing roles such as with retirement, remarriage, and stepfamily issues, among others (Clinton and Laaser 2010; Mayo Clinic Staff 2005).

Qualifications of Counselors

The qualifications of counselors vary. A counselor who is a member of the American Association for Marriage and Family Therapy (AAMFT) has a graduate degree and at least three years of clinical training under a senior counselor's supervision. The safest way to choose a qualified counselor is to select one who belongs to the AAMFT. To do so, check the organization's website, www.aamft.org. Personal recommendations from family members or friends or both may also be helpful.

It is important to have a counselor whom you like and trust and who empathizes with you. It is also important that the counselor respect your religious and personal values. Even well-trained counselors can be capable of unintentional bias that may get in the way of productive therapy (Charles, Thomas, and Thornton 2005; Knudson-Martin and Laughlin 2005). If after three or four sessions you do not feel comfortable with the counselor or don't believe the counselor is effective, it might be a good idea to try someone else. Experts advise interviewing a prospective counselor before beginning therapy. The Mayo Clinic Staff (2005) advises asking lots of questions, including the following:

- Are you a clinical member of the AAMFT or licensed by the state, or both?
- What is your educational and training background?
- What is your experience with my type of problem?
- How much do you charge?
- Are your services covered by my health insurance?
- Where is your office and what are your hours?
- How long is each session?
- How often are sessions scheduled?
- How many sessions should I expect to have?
- What is your policy on canceled sessions?
- How can I contact you if I have an emergency?

Will Counseling Save a Relationship?

Despite its substantiated benefits, the extent to which counseling "saves" a relationship is difficult to measure (Corliss and Steptoe 2004; Sprenkle 2003). Counseling is based on the presumption that partners are willing to cooperate, and it is possible that one's partner may not be willing. No counselor can or will attempt to change a person to a partner's liking without active cooperation from all involved (Rasheed, Rasheed, and Marley 2010). To read—or participate in—an online discussion of ordinary people's opinions on whether counseling "saved" their relationship, you might want to visit the Berkeley Parents Network webpage, "Does Couples Counseling Work?" on the University of California, Berkeley website (parents.berkeley.edu).

Critical Thinking

Can you think of a specific example from your own experience when couple or family counseling was helpful? When it could have been helpful? Can you think of examples when couple or family counseling might be less than helpful?

on interpersonal communication and relationships (e.g., Knapp and Vangelisti 2009; Verderber, Verderber, and Berryman-Fink 2010). In addition, there is a vast number of good (and perhaps not-so-good, so be selective) Internet resources (e.g., Robinson 2009; Von Rosenvinge n.d.; and see Gilkey, Carey, and Wade 2009).

Family relationships are dynamic and can change for the better. For example, an adult woman told

Many observers strongly criticize the way that American culture tends to equate love with infatuation, or chemistry. "Every pop-cultural medium portrays the heights of adult intimacy as the moment when two attractive people who don't know a thing about each other tumble into bed and have passionate sex." But infatuation "merely brings the players together" (Lewis, Amini, and Lannon 2000, pp. 206–07). "Relationships live on time" (Lewis, Amini, and Lannon 2000, p. 205). We need to move from infatuation to "the deep connection that is the hallmark and destination of true love" (Love 2001, p. xi). Positive communication is critical to this process.

this story about her improving relationship with her sister:

> [We] spent some time together. . . . We hadn't done that in 3 or 4 years. . . . It was . . . getting to the point where . . . we could just continue to stick our head in the sand or we could . . . try this again. Because this is the only family. . . . So [now, after beginning to repair the relationship], it's sort of inching along like that. A little better, a little better. (in Connidis 2007, p. 489)

The Myth of Conflict-Free Conflict

By now, enough attention has been devoted to conflict resolution techniques that it may seem as if conflict itself can be free of conflict. It can't. Even the fairest fighters hit below the belt once in a while, and just about all fighting involves some degree of frustration and hurt feelings. Moreover, some individuals have a partner who chooses not to learn to face conflict positively. In relationships where one wants to change and the other doesn't, sometimes much can be gained if just one partner begins to communicate more positively. Other times, however, positive changes in one individual do not spur growth in the other. Situations like this may end in alienation, separation, or divorce.

Then too, even when both partners develop constructive habits, all their problems will not necessarily be resolved (Booth, Crouter, and Clements 2001; Driver and Gottman 2004). Although a complainant may feel that he or she is being fair in bringing up a grievance and discussing it openly and calmly, the recipient may view the complaint as critical and punitive, and may not want to bargain about the issue.

Not every conflict can be resolved, even between the fairest and most mature individuals. If an unresolved conflict is not crucial, then the two may simply have to accept their inability to resolve that particular issue. Family cohesiveness, as well as supportive couple relationships, has much to do with commitment, gentleness, and humor, and on letting our loved ones know how much we care about and appreciate them—a task largely accomplished by little gestures such as a touch or hug, and also by sharing ourselves and listening with genuine interest (Love and Stosny 2007).

Summary

- Members of cohesive families express their appreciation for each other, have a high level of commitment to the family group as a whole, do things together, know how to deal positively with stress or crises, and evidence positive communication patterns.

- There is evidence that a family's having a spiritual orientation is positively related to cohesiveness.

- Research on couple communication indicates the importance to relationships of both positive communication and the avoidance of a spiral of negativity.

- Although some family communication patterns may reach the point of pathology, family conflict itself is an inevitable part of normal family life.

- Although arguing is a normal part of the most loving relationships, there are better and worse ways of managing conflict.

- Alienating practices, such as belligerence and the Four Horsemen of the Apocalypse—contempt, criticism, defensiveness, and stonewalling—should be avoided.

- Constructive arguing habits may not only resolve issues but also bring participants closer together.

- Constructive arguments are characterized by efforts to be gentle and by de-escalation of negativity. No one loses.

- There is no such thing as conflict-free conflict.

Questions for Review and Reflection

1. Explain why families are powerful environments. What are the advantages and disadvantages of such power in family interaction?

2. Explain the interactionist theoretical perspective on families, and show how John Gottman's research illustrates this perspective.

3. Describe the Four Horsemen of the Apocalypse. If someone you care for treated you this way in a disagreement, how would you feel? Do you ever treat others with one or more of the "four horsemen"?

4. Discuss your reactions to each of the ten guidelines proposed in this chapter for constructive arguing. What would you add—or subtract?

5. **Policy Question.** Besides the suggestions in "As We Make Choices: Ten Rules for a Successful Relationship," what *society-wide* ideas might you offer for maintaining loving relationships?

Key Terms

belligerence 327
contempt 326
criticism 326
defensiveness 326
demand/withdraw interaction pattern 329
displacement 325
emotional intelligence 333
family cohesion 318
Four Horsemen of the Apocalypse 326

mixed, or double, messages 331
negative affect 327
passive-aggression 324
positive affect 327
rapport talk 328
relationship ideologies 321
report talk 328
sabotage 325
stonewalling 326

Online Resources

Sociology CourseMate

www.CengageBrain.com

Access an integrated eBook, chapter-specific interactive learning tools, including flash cards, quizzes, videos, and more in your Sociology CourseMate, accessed through CengageBrain.com.

www.CengageBrain.com

Want to maximize your online study time? Take this easy-to-use study system's diagnostic pre-test, and it will create a personalized study plan for you. By helping you identify the topics that you need to understand better and then directing you to valuable online resources, it can speed up your chapter review. CengageNOW even provides a post-test so you can confirm that you are ready for an exam.

13

Power and Violence in Families

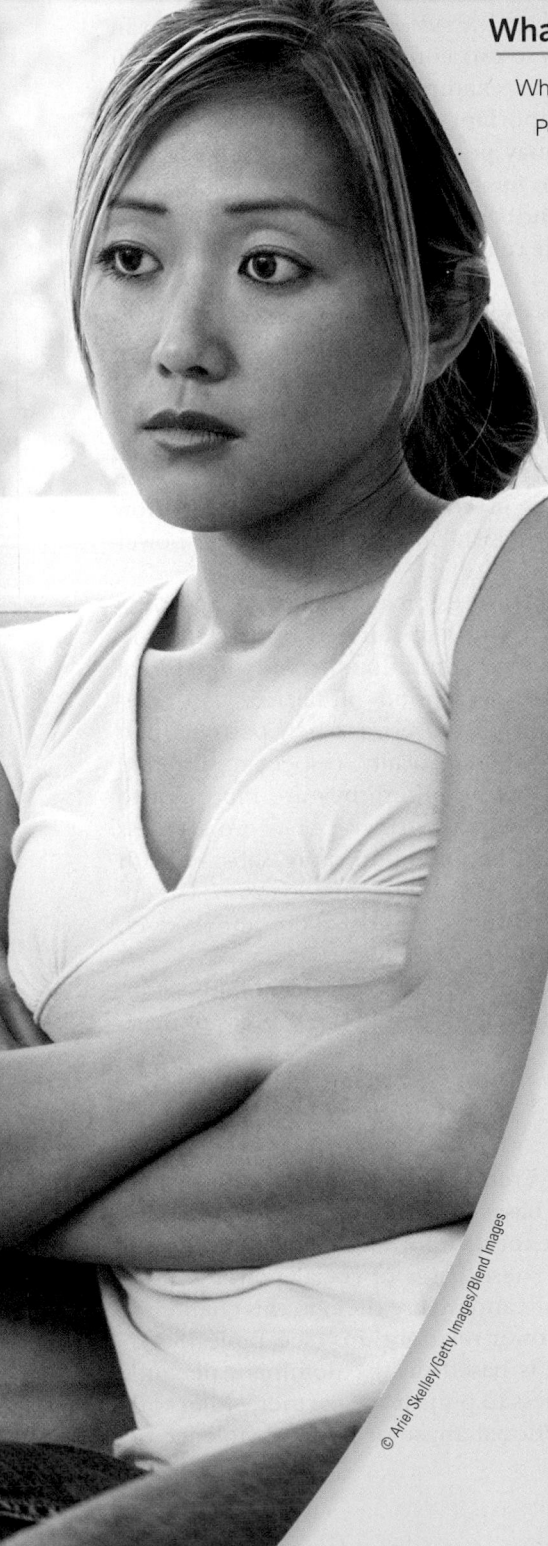

© Ariel Skelley/Getty Images/Blend Images

- Sarah gets a chance for a promotion at work, but accepting it will mean moving to another city; Sarah's spouse does not want to relocate.

- Antonio wants a new stereo for his truck; his partner would prefer to spend the money on ski equipment.

- Marietta would like to talk to her husband about what he does (and doesn't do) around the house, but he is always too busy to discuss the issue.

- Greg feels that he gives more and is more committed to his marriage than his wife is.

This chapter examines power in relationships, particularly marriage and intimate partner relationships. We will discuss some classic studies of marital decision making and look at what contemporary social scientists say about marital power. We will discuss why playing power politics is harmful to intimacy and explore an alternative. Finally, we will explore one tragic result of the abuse of power in families—family violence between intimate partners and violence involving children. We begin by defining power.

What Is Power?

Power may be defined as the ability to exercise one's will. There are many kinds of power. Power exercised over oneself is *personal power*, or autonomy. Having a comfortable degree of personal power is important to self-development. *Social power* is the ability of people to exercise their wills over the wills of others. Social power may be exerted in different realms, including within the family. Parental power, for instance, operates between parents and children. (In this chapter, we focus more on the general dynamics of power in *couple* relationships, rather than parent–child relationships, but we do include child abuse and neglect in the section on family violence.)

The analysis of power in couples originally focused on marriage, but it has been extended to include couples who are not married, both heterosexual cohabitors and same-sex partners. We will use the term **intimate partner power** in referring to unmarried couples or to unmarried *and* married couples, when discussing both together. But because much research on relationship power still focuses on married couples, we will often be discussing **marital power,** with the partners described as husbands and wives.

What Does Marital Power Involve?

Marital power is complex and has several components. First, marital power involves *decision making:* Who gets to make decisions about everything from where the couple will live to how they will spend their leisure time? Second, marital power involves the *division of labor:* Who

earns money? Who does the work around the house? A third arena of marital power is the *allocation of money* earned by either or both partners. Who controls spending for the household? Who has access to personal spending money? Finally, marital power involves a partner's *sense of empowerment, being able to influence* one's partner and feeling free to raise complaints to one's spouse about the relationship.

In addition to the components of marital power, the concept involves both *objective measures of power* (who actually makes more—or more important—decisions, and so on) and a *subjective measure of fairness* in the marriage. These two concepts may be related, but not necessarily. For example, a husband who makes virtually all the important decisions and does relatively little housework may perceive the relationship as fair, whereas a wife who has a larger role in decision making and whose husband shares the housework may nevertheless feel that the relationship is unfair, depending on her expectations.

Judging fairness can be grounded in an **equality** standard—both partners should share equally in the rights and responsibilities of the relationship. Or fairness can be thought of in terms of **equity**—are the rewards and privileges of the relationship proportional to the contributions of the partners? The difference between these concepts will be discussed later, when we analyze how gender plays out in marital and intimate partner power relationships.

Both objective measures of *actual* equality and partners' subjective *perceptions* of fairness influence marital satisfaction, marital commitment, and the risk of disruption, but the perception of fairness is generally more powerful. Furthermore, when partners perceive themselves as reciprocally respected, listened to, and supported by the other, they are more apt to define themselves as equal partners. They are also less depressed, generally happier, and more satisfied with their marriage (Corra et al. 2009; DeMaris 2007; Greenstein 2009; Sullivan and Coltrane 2008; and Weigel, Bennett, and Ballard-Reisch 2006b).

Understanding that marital power is a complex concept, we turn to an examination of the sources of marital power.

Power Bases

Two psychologists (French and Raven 1959) developed a typology of six bases, or sources, of social power: coercive, reward, expert, informational, referent, and legitimate power. These bases of social power can be applied to the family, and we use them in this chapter in analyzing couple power relationships (see Table 13.1).

Coercive power is based on the dominant person's ability and willingness to punish the partner, either with psychological–emotional abuse or physical violence or,

Table 13.1 Bases of Social Power as Applied to the Family

Type of Power	Source of Power	Example
Coercive Power	Ability and willingness to punish the partner	Partner sulks, refuses to talk, and withholds sex; physical violence
Reward Power	Ability and willingness to give partner material or nonmaterial gifts and favors	Partner gives affection, attention, praise, and respect to partner, and assists him or her in realizing goals; takes over unpleasant tasks; gives material gifts
Expert Power	Knowledge, ability, judgment	Savings and investment decisions shaped by partner with more education or experience in financial matters
Informational Power	Knows more about a consumer item, child rearing, travel destination, housing market, health issue	Persuades other parent about most effective mode of child discipline, citing experts' books
Referent Power	Emotional identification with partner	Partner agrees to purchase of house or travel plans preferred by the other because she or he wants to make partner happy
Legitimate Power	Society and culture authorize the power of one or the other partner, or both	In traditional marriage, husband has final authority as "head" of household; current ideal is that of equal partners

Source: Typology of power concepts from French and Raven (1959). Specific wording of definitions and the examples are the authors' (Lamanna/Riedmann).

based on the persuasive content of what the dominant person tells another individual. A husband may be persuaded to stop smoking by his wife's giving him information on smoking's health dangers.

Referent power is based on a person's emotional identification with the partner. In feeling part of a couple or group, such as a family, whose members share a common identity, an individual gets emotional satisfaction from thinking as the more dominant person does. Alternatively, *referent power* might be a source of influence for the partner who is generally less dominant, for the dominant partner may gain satisfaction in behaving as the "referent" to individual wishes. A husband who attends a social function when he'd rather not "because my wife wanted to go and so I wanted to go too" has been swayed by *referent power.* In happy relationships, *referent power* increases as partners grow older together (Raven, Centers, and Rodrigues 1975; Pyke and Adams 2010).

Finally, **legitimate power** stems from the dominant individual's ability to claim authority, or the right to request compliance. *Legitimate power* in traditional marriages involves acceptance by both partners of the husband's role as head of the family. Although this is not the case for all families in the United States, the current ideal in mainstream culture is an egalitarian couple partnership (Gerson 2010, pp. 106–07).

Throughout this chapter, we will see the various power bases at work. The consistent research finding, for instance, that the economic dependence of one partner on the other results in the dependent partner's being less powerful may be explained by understanding the interplay of both *reward power* and *coercive power.* If I can reward you with financial support—or threaten to take it away—then I am more able to exert power over you.

more subtly, by withholding favors or affection (Davies, Ford-Gilboe, and Hammerton 2009, p. 28). Slapping a mate and spanking a child are examples of *coercive power,* so is refusing to talk to the other person—the silent treatment. **Reward power** is based on an individual's ability to give material or nonmaterial gifts and favors, ranging from emotional support and attention to financial support or recreational travel.

Expert power stems from the dominant person's superior judgment, knowledge, or ability. Although this is certainly changing, our society traditionally attributed expertise in such important matters as finances to men, while women were attributed special knowledge of children and expertise in the domestic sphere. **Informational power** is

Native Man and Woman by unidentified Native artist, Chukotka Peninsula, Russia. Artist workshop, Uelen.

The Dynamics of Marital Power

We turn now to look more specifically at research on marital power and the theoretical perspectives used to explain couple power relationships.

Classical Perspectives on Marital Power

Research on marital power began in the 1950s. At that time—before the feminist movement of the 1970s—interest in marital power was more academic than political. Social scientists Robert Blood and Donald Wolfe were curious about how married couples made decisions. Their book *Husbands and Wives: The Dynamics of Married Living* (1960) was based on interviews with wives only. Nevertheless, it was a significant piece of research and shaped thinking on marital power for many years.

Egalitarian Power and the Resource Hypothesis Blood and Wolfe began with the assumption that although the American family's forebears were patriarchal, "the predominance of the male has been so thoroughly undermined that we no longer live in a patriarchal system" (pp. 18–19). They reasoned that the relative power of wives and husbands results from their relative resources. The **resource hypothesis** holds that the spouse with more resources has more power in marriage. Resources include education and earnings; within marriage, a spouse's most valuable resource would be the ability to provide money. Another resource would be good judgment, probably enhanced by education and experience. (Note that the resource hypothesis is a variation on *exchange theory*, presented in Chapter 2.)

To test their resource hypothesis, Blood and Wolfe interviewed about 900 wives in Greater Detroit and asked who made the final decision in eight areas, such as what job the husband should take, what car to buy, whether the wife should work, and how much money the family could afford to spend per week on food. From their interviews, they drew the conclusion that most families (72 percent) had a "relatively egalitarian" decision-making structure (that is, the spouses held roughly equal power, whether that involved separate areas of decision making or joint decisions). However, there were families in which the husband made the most decisions (25 percent), and a few wife-dominated families (3 percent).

The resource hypothesis was supported by the finding that the relative resources of wives and husbands were important in determining which partner made more decisions. Wage and salary earnings or other individual income was a major source of decision-making power. Older spouses and those with more education also made more decisions. Blood and Wolfe also found the relative power of a wife to be greater after she no longer had young children (and was less dependent on her husband) or when she worked outside the home and thereby gained wage-earning resources for herself.

The Blood and Wolfe study had the effect of encouraging people to see marital power as shared rather than patriarchal and resting on their individual attributes or resources rather than on social roles. But this study was strongly criticized, particularly because it tended to ignore the nuances of hidden power—the power of gender expectations and roles that most of us are socialized into.

Criticism of the Resource Hypothesis One criticism concerns Blood and Wolfe's criteria for attributing power to husbands or wives. The decisions wives made (such as how much to spend on food) were generally less important than those that husbands typically made (such as which city the couple should live in): "Having the power to make trivial decisions is not the same as having the power to make important ones" (Brehm et al. 2002, p. 321). And there were important areas of family life that were not included in the Blood and Wolfe study—such as sexual life, how many children to have, and how much freedom partners might have for same- or opposite-sex friendships.

Critics stated that power between spouses involves far more than which partner makes the most *final* decisions—deciding what *alternatives* are going to be considered may be the real decision. Moreover, the person who seems to be making a decision may in fact be acting on a delegation of power from the other partner (Safilios-Rothschild 1970). For example, a husband might ask his wife to make vacation arrangements, specifying that it be a skiing vacation.

Another criticism of the resource hypothesis concerns its narrow focus—on individuals' background characteristics and abilities—but does not take into account their personalities and the way they interact (Brehm et al. 2002). And finally, marital power is more than decision making; it also implies the relative autonomy of wives and husbands, along with the division of labor in marriages (Safilios-Rothschild 1970; Tichenor 2010, p. 419).

Blood and Wolfe came under heaviest fire for their conclusion that a patriarchal power structure had been replaced by egalitarian marriages.

Resource and Gender Feminist Dair Gillespie (1971) pointed out that power-giving resources tend to be unevenly distributed between the sexes. Husbands usually earn more money even when wives work, so husbands have access to more economic *resources*. Husbands are often older (and at the time of the study were often better educated than their wives). So husbands are more likely to have more status, and they may be more knowledgeable, or seem to be (*expert* or *informational power*).

Even their greater physical strength may be an important resource (*coercive power;* Collins and Coltrane 1995), although it can be a destructive one, as we will see later in this chapter.

Women are likely to have fewer alternatives to the marriage than their husbands, especially if wives cannot support themselves or are responsible for the care of young children. Moreover, men can remarry more readily than women. Consequently, according to Gillespie, the resource hypothesis, which presents resources as neutral and power as gender-free, is simply "rationalizing the preponderance of the male sex." Marriage is hardly a "free contract between equals" (p. 449; see also Komter 1989 and Tichenor 2010).

Current research tends to support Gillespie's insight that American marriages continue to be inegalitarian even though they are no longer traditional, although younger marrieds are working toward more egalitarian relationships (Gerson 2010; Rosenbluth, Steil, and Whitcomb 1998; Spade and Valentine 2010; Wilkie, Ferree, and Ratcliff 1998). True, resources make a difference, and an important factor in marital power is whether or not a wife is employed. Wage-earning wives have more to say in important decisions and in the division of household labor, but gendered divisions remain nonetheless (Johnston 2007).

One way in which women come to have fewer resources is through their reproductive roles and resulting economic dependence. Just after marriage, the relationship is apt to be relatively egalitarian, with the husband only moderately more powerful than the wife—if at all. Often at this point, the wife has considerable economic power in relation to her husband because she is employed and may even have a well-paying career. But relationships tend to become less egalitarian with the first pregnancy and birth (Coltrane and Ishii-Kuntz 1992).

During the childbearing years of the marriage, the practical need to be married is felt especially strongly by women, who are more often than not the primary caregivers as well as bearers of children (Johnson and Huston 1998). A divorce would likely mean that the woman must parent and support small children alone. Women engaged in reproduction and child rearing may have less energy to resist dominance attempts. On the other hand, a mother may exert power over her husband by threatening to leave and take the infant with her (LaRossa 1979).

As we noted, working contributes to marital power. But working for wages or even outearning a husband does not necessarily give a wife full status as an equal partner (Coltrane 2000; Teachman 2010; Tichenor 2005). Even though a working wife is, in theory, less obliged to defer to her husband and has greater authority in making family decisions, she does not necessarily participate equally in decision making in fact, and she is still unequally burdened with housekeeping and child rearing.

Resource Theory Researchers have come to realize that resource theory does not fully explain marital power. Although women's employment rates, occupational status, and income have increased in recent decades, their share of household work has not declined to a similar degree (Cooke 2006; Hall and MacDermid 2009; Sullivan and Coltrane 2009). This "failure of resource and exchange perspectives to explain marital power dynamics in two earner couples" (Tichenor 1999, pp. 638–39; 2005) has led scholars to turn to other theoretical perspectives.

Resources in Cultural Context Studies comparing traditional societies with more modern ones suggest that in a traditional society, norms of patriarchal authority may be so strong that they override personal resources and give considerable power to all husbands (Safilios-Rothschild 1967; Blumberg and Coleman 1989). Put another way, in a traditional society, male authority is *legitimate power*. This perspective, termed **resources in cultural context**, stresses the idea that resources are not effective in conferring marital power in traditional societies that legitimate male dominance with a patriarchal norm.

This situation may be especially true for immigrant families from traditional societies, such as in Asia or Central and South America, at least for those who are newly arrived. However, subsequent generations may be expected to adopt the more common American pattern. The generation born in the United States has moved to a **transitional egalitarian situation** regarding marital power, typical of the rest of the country, in which "husband–wife relationships are more flexible and negotiated . . . [and] socioeconomic achievements become the basis for negotiation within the family" (Cooney et al. 1982, p. 622).

Even among native-born Americans, however, we must recognize the continuing salience of tradition and the assumption that it is legitimate to some degree for husbands to wield authority in the family (Nock, Sanchez, and Wright 2008; Komter 1989). The continued importance of the legitimation of husbands' authority is apparent in religious groups that accept the principle of male headship of the family (Meyers 2005; Wilcox 2004). Although egalitarianism is undoubtedly the most sought after mode among American couples generally, whether an **egalitarian norm** of marital power is fully realized in any sector of American society is a question we will discuss throughout this chapter. Presently it seems those most likely to have attained this idea are lesbian couples, in which the resources, for example, that each brings to the relationship do not affect each person's power (Jeong and Horne 2009).

In sum, the cultural context conditions resource theory explains marital power only when there is no overriding egalitarian norm or **patriarchal norm of marital power**. Put another way, if traditional norms of male authority are strong, husbands will almost inevitably dominate regardless of personal resources. Similarly, if an egalitarian norm of marriage were completely accepted, then a husband's superior economic achievements would be irrelevant to his decision-making power because both spouses would have equal power. It is only in the present transitional egalitarian situation, in which neither patriarchal norms nor egalitarian norms are firmly entrenched, that marital power is negotiated by individual couples, and the power of husbands and wives may be a consequence of their resources (Stevenson and Wolfers 2006).

Love, Need, and Power Some have argued that a primarily economic analysis does not do justice to the complexities of marital power. Perhaps a wife has considerable power through her husband's love for her, what we have termed *referent power*. Generally, however, the wife holds the less-powerful position even in this reckoning. In our society, women value close emotional relationships more than men do (Lois 2010). They are encouraged to express their feelings, whereas men are less likely to articulate their feelings for their partners. Overt dependency affects power: "A woman gains power over her husband if he clearly places a high value on her company or if he expresses a high demand or need for what she supplies. . . . If his need for her and high evaluation of her remain covert and

unexpressed, her power will be low" (Cancian 1985, p. 258).

But another way of looking at it is that men are less relatively powerful in the private, intimate sphere than they are in the public world because the private world is more likely to give priority to *referent power*. Therapists and mass media, and, to an increasing degree, the public, support women's desire for more expression of feelings. Men have been much influenced by the expectation that they should respond favorably to wives' influence (Connell 2005). They are encouraged to engage in "emotion work," "to express emotion to their wives, to be attentive to the dynamics of their relationship and the needs of their wives, [and] . . . to set aside time for activities focused especially on the relationship" (Wilcox and Nock 2006, p. 1,322).

We have spent some time on earlier theory and research because these lead to issues of marital power that are still current.

Current Research on Marital Power

The first research on marital power by Blood and Wolfe focused solely on decision making. A later research project by sociologists Philip Blumstein and Pepper Schwartz (1983) covered other aspects of couple relationships—money and sex, for example—and it compared married couples to heterosexual cohabiting couples and gay male and lesbian couples. As a major step forward in the study of couple relationships, this study is still cited (Gilbert and Rader 2008; Moore 2008; Pinto and Coltrane 2009).

One reason for its longevity may be that social scientists had seemed to lose interest for a time in researching "marital power." Instead, they pursued marital and partner equality issues indirectly, by examining women's expanding entry into the labor force and related issues of who does the unpaid labor of household and child care. Recently, however, the word *power* has begun to reappear in journal articles and books examining couples' allocation of money, their capacity to influence each other and to raise touchy issues, and, still, the question of fairness (or not) in the division of household labor (Vogler, Lyonette, and Wiggins 2008).

Equality, equity, and *gender* and their interrelationship are themes of current thinking about marital and other intimate partner power. Are men and women now equal in their family relationships? If not, why not?

© Don Smetzer/ Getty Images

Although an older generation may hold to traditional patriarchal power, the next generation may renegotiate and consciously change those roles, especially as women assume more autonomy and make gains in the workplace. In this photo, the classic card game of Hwoa-tu and the Chinese vase and screen in the background suggest a world of traditional authority, a man's *legitimate power* as head of the family. The posture and clothing of the younger family members suggest more casual and democratic family relations.

© Steve Skjold/PhotoEdit

Gay and lesbian couples are more likely to share domestic duties than are heterosexual couples, although attainment of an egalitarian ideal eludes many gay/lesbian couples as well. In marriages, men's participation in housework has increased, although wives continue to do more.

Social scientists generally agree that the cultural ideal today is one of spousal and partner equality and of shared work and family roles. According to the resource hypothesis, as wives entered the labor force and began earning substantial income, they would be able to bargain for equality at home based on their resources. By and large, this vision of equality has not been attained in a number of respects:

> Feminists and scholars assumed that women's moving into the labor force and becoming important co-breadwinners would increase their power in the family—especially in terms of control over money management and decision making. However, the marital power literature over the last several decades has not borne this assumption out. Women's power in decision-making has increased somewhat, but not to a degree commensurate with the level of income many of them have been earning. In short, their income does not seem to buy them the same . . .

power that men have typically enjoyed. (Tichenor 2005, p. 91; see also Vogler, Lyonette, and Wiggins 2008)

We examine the complexities of partner equality by looking at current research on couples and their unpaid household work, control over money, decision making, and the expression of grievances and management of emotions. We draw on a number of studies (research reviews, quantitative and qualitative studies) of varying methodologies, sample sizes, and social locations.[1] We then look at where things stand regarding gender equality in the family and consider the future of family power.

Household Work and Leisure Time Virtually all research indicates that women's satisfaction with the fairness of their partners' contributions to household work is strongly associated with women's (and sometimes men's) relationship happiness, marital commitment, and depression and with the risk of marital disruption. Where women have more egalitarian expectations than men fulfill, there is often marital conflict (DeMaris 2007; Gerson 2010; Greenstein 2009; Sullivan and Coltrane 2009).

Fairness of the division of household labor is not usually evaluated on a 50–50 standard. What's "fair" to a man may be less than half, while a woman has to be doing all or almost all the housework to find it "unfair" (Kendall 2007). "[U]nequal divisions of labor are accepted as normal" (Coltrane 2000, p. 1,223).

Women whose husbands work more hours are less apt to see the division of household labor as unfair (Kendall 2007). So are more traditional women, perhaps because their expectations are shaped by a religious doctrine of separate spheres and male headship (Myers 2006; Nock, Sanchez, and Wright 2008). Still, even in evangelical couples, there is an implicit acknowledgment of a norm of equality in the attention given by evangelical men to expressing great appreciation for their wives' doing the preponderant share of housework. In the context of a societal egalitarian ideal, the additional domestic work of evangelical wives becomes a "gift," which is reciprocated by the husband's emotional work of expressed appreciation in an "economy of gratitude" (Vaaler, Ellison, and Powers 2009; Wilcox 2004, p. 154, referencing Hochschild 1989).

In American society more generally, there has been a significant increase in men's share of housework (Eckel 2010; Gilbert 2008; Kelleher 2007; Sullivan and Coltrane 2008). Although that share still does not approach equality, the increase since 1965 is quite dramatic. In 1965, women did seven hours of housework for every hour that men put in; in 2005, it was nearly one and a half hours (see Figure 13.1).

Social scientists now tend to use housework as one criterion of power (on the assumption that no one really

[1] In this limited space, it becomes impossible to discuss the details of the research methodology of each source cited. For more detail, consult the original sources.

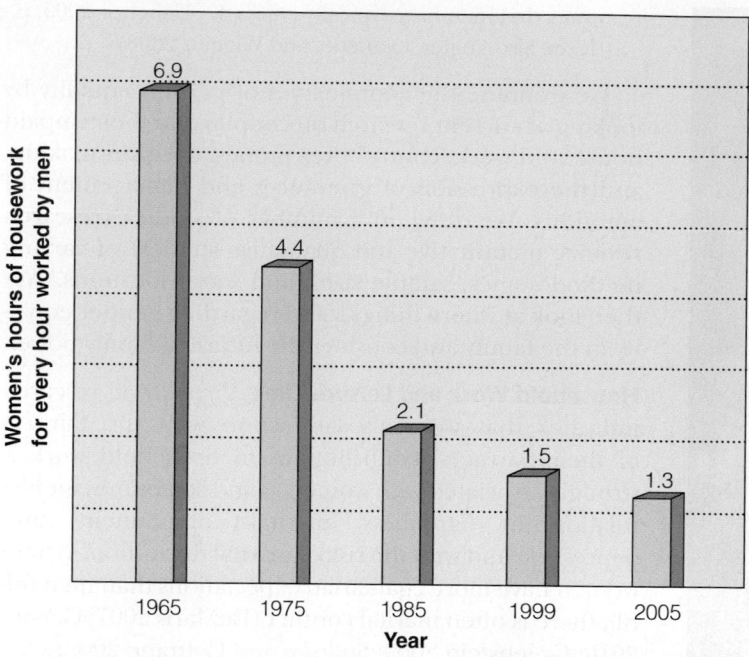

Figure 13.1 Trends in average weekly housework[a] hours for women and men (ages twenty-five through fifty-four), 1965–2005 (ratio of women's hours to men's hours)

[a]"Housework" includes "core housework" (cooking meals, meal cleanup, housecleaning, laundry, and ironing) as well as "other housework" (outdoor chores, repairs, gardening, animal care, bills, other financial).

Source: Adapted from Sayer, Cohen, and Casper 2004, Table 3; Mixon 2008.

wants to do it). The current situation—women do one and one half times what men do in housework—can be seen as a metaphor for their relative marital power. Women have gained men's participation in housework—and gained in power—but have not achieved absolute equality.

There also continues to be a "leisure gap." Although in formal terms, women have only one-half hour less of leisure time than men have, what is labeled "leisure" for women is often indirectly child care and household management (Bianchi, Robinson, and Milkie 2006; Gerson 2010).

Some women see their greater responsibility in household work as enabling the acquisition of some measure of power at the practical level (Tichenor 2005):

As wives gather information in preparation for having some kind of discussion, they often form opinions about what . . . they would prefer to happen. They are then free to present the information . . . in such a way that makes what they want to do seem like the . . . most reasonable course of action. . . .

[This] suggests that information is an important source of power in these relationships. While women carry a tremendous burden in terms of managing household responsibilities, that work . . . gives women access to knowledge that either gives them direct control or sometimes allows them an extra measure of influence in joint decision

making. Most important is that women are often completely aware of this power and use it consciously to their advantage. (Tichenor 2005, p. 96)

Thus, women's *informational power* can offset men's *resource* or *legitimate power* or enable them to have the influence that their own *resource power* apparently does not. Of course, it must be exercised clandestinely.

Control Over Money Research on couples' **allocation systems**—whether they pool their money and who controls pooled or separate money—is relatively recent. British social policy scholar Jan Pahl (1989; see also Kenney 2006 and Vogler 2005 for recent research using this typology) developed a typology of allocation systems that subsequent researchers have used or adapted.

In the industrial era, a family's allocation system was typically one of complete control of his earnings by the male breadwinner, who doled out a housekeeping allowance to his homemaker wife. The allowance was often rather skimpy, while the husband was privileged to take money "off the top" for personal spending and recreation.

Even before the emergence of feminism, this system began to be seen as inappropriate to a companionate model of marriage in which men and women were seen as equal, though with different family roles. To resolve the tension between a theoretical equality of men and women in the family, but their much different worth in the market economy, the typical marital allocation system became one of pooled resources. The husband's earnings were deposited into an account maintained in both names and controlled jointly by the spouses—at least, that was the theory. Given male dominance in decision making in this era, the husband usually controlled the pooled account. Moreover, nonearning women typically felt uncomfortable making decisions about "his" earnings. In reality, the joint account was not jointly controlled.

Feminists began to criticize the joint pool system. Women were now earning money as well. Separate financial accounts and control began to be seen as a favored alternative, with each spouse or partner making equal contributions to running the household. But in another iteration, scholars and feminists have pointed out that equal household contributions may not be equitable when women typically earn less than men—their contribution to household expenses represents a larger proportion of their earnings (Kenney 2006; Tichenor 2005; Vogler, Lyonette, and Wiggins 2008).

A variety of allocation systems operate in American (and European) marriages at present, involving two dimensions: whether to pool and who controls—man,

This couple seems to be sharing control over their money on an equal basis.

woman, or both?. There continues to be a tension in many marriages and other partner relationships between the communal values of the couple relationship and the individualism of the market, in which each person may have a very different level of earned income (Vogler 2005). Cohabitants and those who have been previously divorced are especially likely to maintain separate money (Kenney 2006).

Gender still plays a strong role, and men seem to retain more control over the family's income no matter who earns it. They are especially likely to retain personal spending money and/or to feel free to spend the family's income on personal and recreational desires without consulting their partner. Women may spend some of their "personal" money on household needs. And even otherwise egalitarian men may assume they have "veto power" over major decisions (Tichenor 2005; Vogler, Lyonette, and Wiggins 2008; Vogler, Brockmann, and Wiggins 2008). This would seem to be an example of traditional *legitimate power* ascribed to a male, overriding the wife's *resource power.*

Power and Decision Making Here, we look at decision making and marital power in general terms rather than in the specific domains of spending and housework. Using a national sample survey to compare decision making in 1980 and 2000, Amato, Johnson, Booth, and Rogers found that, in 2000, "respondents were significantly more likely to report equal-decision-making" (2003, p. 9). Even wives in evangelical families often have more decision-making power than their formal submission to the male family head would indicate. In fact, some research shows that "co-parenting and joint decision-making are more common in evangelical homes than in secular and mainline religious households" (Bartkowski and Read 2003, p. 88; Bartkowski 2001; Vaaler, Ellison, and Powers 2009).

Power asymmetry was found more often among dissatisfied couples. Men's power may not be visible, as they may have the ability to suppress issues so that they never arise overtly (Tichenor 2005, p. 25). "The spouse with less power [usually the wife] typically spends more time aligning emotions with [the spouse's] expectations" (Coltrane 1998) rather than risking confrontation. This increases emotional pressure. Coltrane notes that some men may *feel* powerless despite their greater power:

> Men's subjective sense of powerlessness—of lost or slipping privilege—is often a precursor to wife-beating or sexual abuse. . . . This does not mean that these men are less powerful than their wives. . . . This contradictory co-existence of felt powerlessness and actual (if latent) power is quite common for men. (Coltrane 1998, p. 201)

Women, on the other hand, may fear appearing too powerful. "For some women, expressing or exercising power seems threatening either to their relationships or to their gender identities. Some wives speak openly of the danger that power poses for them" (Tichenor 2005, p. 110). They are concerned about their husbands' sense of masculinity, as well as their gender identity (p. 114). In fact, many women attempt to "preserve their husbands' masculinity by backing away from power . . . deferring to husbands and adopting various strategies to make it look as if husbands were in control" (Vogler, Lyonette, and Wiggins 2008, p. 131).

Bases of Marital Power To sum up, in these discussions of marital power and decision making, we see the interplay of three bases of social power: *resource power, legitimate power,* and *informational power.*

Resource power traditionally gave provider husbands greatest power in marital decision making, including the capacity to keep troubling issues and decisions from even arising. But the equation of resources (that is, earnings) with power hasn't worked in the same way for women. A study of wives who earn more than their husbands suggests that "the gender structure exerts an influence that is independent of breadwinning or relative financial contributions" (Tichenor 2005, p. 117). A residual sense of the propriety of traditional male privilege—that is, *legitimate power*—ascribed more authority to men even in situations where they lacked *resource power.* "Just as women's income does not buy them either relief from domestic labor or greater financial power . . . , it does not give them dominion in decision making" (p. 117).

In an interesting twist, women can sometimes gain power from their greater knowledge of the household. They can use this *informational power* to shape decisions about purchases and household arrangements, as we noted earlier.

Equality, Equity, and Gender Spouses are usually aware of imbalances in marital power. Wives typically have a greater sense of unfairness because they are so often disadvantaged (DeMaris 2007, p. 192). From 1980 to 2000, there has been a "shift toward more egalitarian marital relations" (Amato et al. 2003, p. 9; see also Gerson 2010, pp. 117, 122–23). At the same time, wives seem to have gained in average marital happiness. "Wives were less happy than husbands in both surveys, but the gap between husbands and wives grew smaller between 1980–2000. The narrowing of the gender gap in happiness may be attributable to more egalitarian marital relations" (p. 11; see also Corra et al. 2009).

The Future of Marital Power

Power disparities discourage intimacy, which is based on honesty, sharing, and mutual respect. For most, therefore, attainment of the American ideal of equality in marriage would seem to support the development of intimacy in marital relationships.

Yet the United States is a pluralistic society, and so we may expect to find varied visions of the future of marital power. Whether or not they reflect an egalitarian ideal, they generally take into account an egalitarian norm of marital power. We first sketch out this diversity, and then look at some of the specific marriage models scholars envision as they ponder the direction of change.

The Road to the Future One way to bring varied visions of the future together is to imagine couples driving along an interstate highway that leads to gender equality. Some couples are committed to getting there fast, and they take the express lane directly to shared and equal roles and power.

Most of the couples remain on the main interstate and are not completely sure if they want to travel all the way to the end. They keep consulting their maps to see if there is a stopping point they would like better, a mix of equality and gender identity.

A few couples exit the interstate, looking for a setting, perhaps a small town, where they can reproduce the more differentiated gender roles of the mid-twentieth century, though with some respect for equality in the relationship.

Some couples pull into a rest stop to continue the quarrel that has sprung up—she prefers a greater degree of sharing of power and household work than he does. After some negotiation, they reach at least a temporary compromise and continue on their way.

Finally, some couples have a complete breakdown and are waiting along the road for assistance, to return home and then move into separate lives.

Mutually Economically Dependent Spouses or Other Egalitarian Relationships Sociologist Steven Nock sees the future as one of *mutually economically dependent spouses*. "What I propose [as] . . . the emerging

This family is having breakfast in a household where roles may be somewhat differentiated by gender, but there is no sharp difference in status and power between the adults.

form of American marriage [is] a relationship in which couples are equally dependent on one another's earnings" (2001, p. 755). He defines *MEDS,* or **mutually economically dependent spouses**, as dual-earner couples in which each spouse earns between 40 and 59 percent of the family's income. Examining data from the 1999 Current Population Survey, he finds that presently just under one-third of all dual earners (some 20 percent of all couples) are mutually economically dependent. This pattern occurs at all economic levels (Nock 2001; see also Amato and Hohmann-Marriott 2007).

Nock not only sees equality in this arrangement but also finds it less threatening to marriage than one might think. True, an independent income increases the risk of divorce on the part of wives who are dissatisfied with household contributions of their husbands (Stevenson and Wolfers 2006). At the same time, in a MEDS marriage, the two would become equally dependent financially, which could strengthen commitment if husbands change to contribute more household work.

This is a complex argument and rests on the thought that a marriage would be grounded in "extensive dependencies by both partners," as was the case in traditional society. As Nock sees it, the decline in divorce rates since the early 1980s may reflect "the gradual working out of the gender issues first confronted in the 1960s. If so, this implies that young men and women are forming new types of marriages that are based on a new understanding of gender ideals" (Nock 2001, p. 774). Men growing up today, often with working mothers, are more likely to adapt to changing gender roles. "We are still at the early stages of a fundamental realignment of gender in our society" (p. 773).

Another possibility is that *norms of equality* may come to be so strong that men and women will have equal power in marriage regardless of resources. *Legitimate power* would endorse women's equality with men in the family. Pepper Schwartz's research on *peer marriage* (also referred to in the literature as "post-gender" or "equal sharer") offers an example of a strong equality norm at work (1994; 2001; see also Hall and MacDermid 2009). She studied couples who attempt to make their marriage according to this ideal. "As We Make Choices: Peer Marriage" describes this research.

Neotraditional Families In a pluralistic society, there are alternative visions of the family model. Among evangelical Christians and other conservative religious sectors, a gendered division of labor, formal male dominance in decision making, and an egalitarian spirit combine in the **neotraditional family** (see Chapter 11 for more discussion on this type of family).

Although a husband's dominant power is legitimated in this milieu, marital power in practice is often negotiated. First "articulated by evangelical feminists,"

the "mutual submission" (of husband and wife to each other) has become increasingly popular because it justifies the shared decision making that characterizes many evangelical marriages (Dolan 2008; p. 32; Bartkowski and Read 2003; Bartkowski 2001; Vaaler, Ellison, and Powers 2009). Another way in which a norm of equality is represented in these ostensibly husband-dominant marriages is in the emotional "economy of gratitude" (Pugh 2009, p. 6; Wilcox 2004, p. 154), as husbands display appreciation for their wives' "gift" of household work. Evangelical couples *are* committed to a headship model of marital power, yet this has an "enchanted" quality, symbolic of the religious commitment of the couple (Wilkins 2009, p. 363). In many ways, then, the edges of difference and dominance are softened in the neotraditional family, as the title of sociologist Bradford Wilcox's book—*Soft Patriarchs*—suggests.

A Gender Model of Marriage Writing from a perspective somewhat different than Nock's MEDS vision, Wilcox and Nock (2006) suggest that the egalitarian, equal-resource model of marriage does not represent what most couples want, at least at present, or what makes them happy with their marriage. Instead, they construct a **gender model of marriage**.

Wilcox and Nock see the egalitarian definition of marital power as circumscribed by the symbolic importance of maintaining gender boundaries. Couples want to construct conventional relationships and marriages in which they are comfortable. Compromise with an egalitarian ideal occurs "as spouses work together to construct appropriate gender identities and maintain viable marriages" (Tichenor 2005, p. 32).

> Because wives—even wives with egalitarian attitudes—have been socialized to value gender-typical patterns of behavior, wives will be happier in marriages with gender-typical practices in the division of household labor, work outside the home, and earnings. Because husbands—even husbands with egalitarian attitudes—have been socialized to value gender-typical patterns of behavior, husbands will be happier in marriages that produce gender-typical patterns and will be more inclined to invest themselves emotionally in their marriages than husbands organized along more egalitarian lines. (Wilcox and Nock 2006, p. 1,328)

In their study of over 5,000 couples drawn from the second wave of the National Survey of Families and Households (1992–94), Wilcox and Nock did find support for the hypothesis that "the gendered character of marriage seems to remain sufficiently powerful as a tacit ideal among women to impact women's marital quality" (pp. 1,339–40). Nevertheless, when men's household work departed from the expectations of more egalitarian wives, marital quality was also affected. "'Her' marriage is happiest when it combines elements of the new and old" (p. 1,321).

A piece of research by Pepper Schwartz paints a picture of couples who have developed egalitarian marriages, or tried to. Schwartz followed up her earlier research on couples with an exploration of the factors that facilitate **peer marriage** (1994, 2001): "I began looking for couples who had worked out no worse than a 60–40 split on child rearing, housework, and control of discretionary funds, and who considered themselves to have 'equal' status or standing in the relationship" (2001, p. 182). The study was based on fifty-seven egalitarian couples, with some additional interviews with couples considered **near peers** and **traditionals** for comparison.

Near peers believed in equality, but the combination of the arrival of children and the desire to maximize income meant that the husband did not participate as much as the couple's egalitarian ideals required. *Traditionals* were those marriages in which males dominated decision making except regarding children, but both parties were OK with this—the wife did not seek equality.

Peer marriages did not necessarily stem from a feminist ideology. Only 40 percent of women and 20 percent of men in peer marriages cited feminism as a motive. The rest gave other reasons for wanting a peer marriage: a rejection of negative parental models (women resented their father's dominance of their mothers; men wished for more involvement as parents); a desire to undertake co-parenting; and, in some cases, a period of serious tension in the marriage that required renegotiation

of roles. However, for men especially, partner's preference was what led their marriage in an egalitarian direction. In an interesting twist on the expressive role of women, "[m]any of these men told me they had always expected a woman to be the emotional architect of a relationship and were predisposed to let her set the rules" (p. 183).

There was "no single blueprint" (p. 189), nor were these peer marriage couples high-earning "yuppies" or academics with flexible schedules. Peer marriage seemed, in the long run, to require an intense desire to have such a marriage and a persistent willingness to forgo male career advancement and income. Everyday responsibilities also had to be constantly monitored and renegotiated. Over time, the peer marriage couples evolved strong egalitarian norms that overrode the surrounding structure and typical power processes, much as Blumstein and Schwartz had earlier found in their lesbian couples (and sociologist Barbara Risman [1998] found in her research on "fair" marriages). The couple's respect for each other as described in Schwartz's study of peer marriage is essentially a "no-power" relationship.

Critical Thinking

Do you see peer marriages as an ideal for yourself, or not? How common do you think peer marriages will become in the future?

Source: Pepper Schwartz 1994.

Power Politics versus No-Power Relationships

Marriage and other intimate partner relationships that partners find fair and equitable are generally more apt to be stable and satisfying. Social scientist Peter Blau terms this situation *no-power*. **No-power** does not mean that one partner exerts little or no power; it means that each partner has the ability to mutually and reciprocally influence and be influenced by the other (Gottman et al. 1998; Schwartz 1994). We turn now to a discussion of the process of changing power relationships in marriage.

A fair division of household labor is not the only standard by which women and men judge the equality or lack of equality in their marriage. The respect one has for the other's views is extremely important; that is a central element in a no-power conceptualization of marriage and other partner relations. As we use the term, *no-power* also implies that partners do not seek to exercise their relative power over each other.

No-power partners seek to negotiate and compromise, not to win (see Chapter 12). They are able to avoid **power politics**.

Power Politics in Marriage

As gender norms move from traditional toward egalitarian, all family members' interests and preferences gain legitimacy, not only or primarily those of the husband or husband/father. For example, the man's occupation is no longer the sole determining factor in where the family will live or how the wife will spend her time. This means that decisions formerly made automatically, or by spontaneous consensus, must now be consciously negotiated. A possible outcome of such conscious negotiating, of course, is greater intimacy; another is locking into power politics and conflict.

In the worst case, both equal and unequal partners may engage in a cycle of devitalizing power politics. Partners come to know where their own power lies, along with the particular weaknesses of the other. They may alternate in acting sulky, sloppy, critical, or distant, or even hint at leaving the marriage (Blumberg and Coleman 1989;

"I'm trying to look at it from my point of view."

Chafetz 1989). The sulking partner carries on this behavior until she or he fears the mate will "stop dancing" if it goes on much longer; then it's the other partner's turn. This kind of seesawing may continue indefinitely, with partners taking turns manipulating each other. However, the cumulative effect of such power politics is to create distance and loneliness for both spouses.

New research into marital power politics in immigrants examines the changes that take place in immigrant marriages once these couples integrate into the American culture. The findings suggest similar patterns of tension, resistance, and acceptance to shifts in power as seen in marriages of native-born Americans. The reason for this, the researchers suggest, is that patterns of gendered power evolve over time—no couple's relationship is static. Because human beings and their relationships are dynamic, patterns of power will also change, regardless of consciousness and social context (Falicov 2007; Maciel, Van Putten, and Knudson-Martin 2009). Few couples knowingly choose power politics, but this is an aspect of marriage in which choosing by default may occur. Our discussion of power in marriage is designed to help partners become sensitive to these issues so that they can avoid such a power spiral, or reverse one if it has already started.

Alternatives to Power Politics

There are alternatives to this kind of power struggle. Robert Blood and Donald Wolfe (1960) proposed one in which partners grow increasingly separate in their decision making; that is, they take charge of separate domains: one buying the car, perhaps the other taking charge of disciplining their children. This alternative is a poor one for partners who seek intimacy, however, for it enforces separateness rather than the sharing of important decisions.

A second, more viable alternative to perpetuating an endless cycle of power politics is for one partner to disengage from power struggles, as described in "As We Make Choices: Disengaging from Power Struggles." This includes a third, perhaps best, alternative, which is for the more powerful partner to consciously relinquish some power in order to save or enhance the marriage. We saw in Chapter 12, for instance, that marriage communication expert John Gottman and his colleagues (Gottman et al. 1998) advise husbands to be willing to share power with their wives if they want happy, stable marriages, something he says can be difficult for men who had difficult childhoods (pp. 18–19; Coan and Gottman 2007).

The Importance of Communication As we've noted earlier, partners who see themselves as mutually respected, equally committed, and listened to when they raise concerns are more likely to see their relationship as egalitarian and are more satisfied overall with their relationship.

Meanwhile, unequal relationships discourage closeness between partners: Exchange of confidences between unequals may be difficult, especially when self-disclosure is seen to indicate weakness. Men more than women have been socialized not to reveal their emotions (Hoyt 2009; Panayiotou and Papageorgiou 2007; Rudman and Glick 2008; Spade and Valentine 2010). Women, feeling less powerful and more vulnerable, may resort to pretense and the withholding of sexual and emotional response. In fact, because the female partner in a heterosexual relationship is the only socially acceptable outlet for a male to express his vulnerabilities, when the woman leaves the relationship, the effect is a "greater emotional cost" for the man than for the woman (Rudman and Glick 2008, p. 223).

Nevertheless, trying to change the balance of marital power may bring the risk of devitalizing a relationship, depending on how partners go about it. Mates who try to disengage from power struggles, without explaining what they are doing and why, risk estrangement. The reason is that dominant partners may have taken a mate's compliance as evidence of love rather than fear. If this deference is withdrawn, a dominant partner may conclude that "she (or he) doesn't love me anymore" and escalate efforts at control, contributing to a spiral of alienation and estrangement that is not acknowledged or discussed openly.

The Change at a No-Power Relationship Changing power patterns can be difficult, even for couples who talk about it, because these patterns usually have been established from the earliest days of the relationship. From the interactionist perspective, certain behaviors come not only to be expected but also to have symbolic meaning. In addition, sociologist William Goode had an insight that continues to be relevant for many couples. He wrote about the important change in men's position as women gain equality in society and in the family. According to Goode:

The late Carlfred Broderick, sociologist and marriage counselor, offered the following exercise to help people disengage from power struggles. The object of this exercise is to get you out of the business of monitoring everyone else's behavior and free you from the unrewarding power struggles resulting from that assignment. Here is the exercise:

1. Think of as many things as you can that your spouse or children should do, ought to do, and would do if they really cared, but don't do (or do only grudgingly because you are always after them). Write them down in a list.

2. From your list, choose three or four items that are especially troublesome right now. Write each one at the head of a sheet of blank paper. These are the issues that you, considerably more than your spouse, want to resolve (even though he or she, by rights, should be the one to see the need for resolution). Right now you are locked in a power struggle over each one, leading to more resentment and less satisfaction all around.

3. In this step you'll consider, one by one, optional ways of dealing with these issues without provoking a power struggle. Place an A, B, C, and D on each sheet of paper at appropriate intervals to represent the four options listed below. Depending on the nature of the issue, some of these options will work better than others, but for a start, write a sentence or paragraph indicating how each one might be applied in your case. Even if you feel like rejecting a particular approach out of hand, be sure to write something as positive as possible about it.

Option A: Resign the Crown

Swallow your pride and cut your losses by delegating to the other person full control and responsibility for his or her own life in this area. Let your partner reap his or her own harvest, whatever it is. In many cases, your partner will rise to the occasion, but if this doesn't happen, resign yourself to suffering the consequences.

Option B: Do It Yourself

There's an old saying: "If you want something done right, do it yourself." Accordingly, if you want something done, and if the person you feel should do it doesn't want to, it makes sense to do it yourself the way you'd like to have it done. After all, who ever said someone should do something he or she doesn't want to do just because you want him or her to do it?

Option C: Make an Offer Your Partner Can't Refuse

Too many interpret this, at first, as including threats of what will happen if the partner doesn't shape up. The real point, however, if you select this approach, is to find out what your partner would really like and then offer it in exchange for what you want him or her to do. After all, it's your want, not your spouse's, that is involved. Why shouldn't you take the responsibility for making it worth your spouse's while?

Option D: Join with Joy

Often the most resisted task can become pleasant if one's partner shares in it, especially if an atmosphere of play or warmth can be established. This calls for imagination and goodwill, but it can also be effective in putting an end to established power struggles.

Critical Thinking

Have you ever tried any of these in a couple situation? If so, how did it work out? Would these principles be useful, do you think, in other relationships, such as with children, extended family, or coworkers?

Source: Broderick 1979a, pp. 117–23.

Men have always taken for granted that what they were doing was more important than what the other sex was doing, that where they were, was where the action was. Men occupied the center of the stage, and women's attention was focused on them. . . . [But] the center of attention shifts to women more now than in the past. I believe that this shift troubles men far more, and creates more of their resistance, than the women's demand for equal opportunity and pay in employment. (Goode 1982, p. 140)

One small study of twelve fairly equal newlywed couples found that some of them either consciously or unconsciously avoided issues about marital power and developed a "myth of equality" (Knudson-Martin and Mahoney 1998). Sometimes this seems to work—but only for a while. The best way to work through power changes is to openly discuss power and to fight about it fairly, using the techniques and cooperative attitudes we describe in Chapter 12. The partner who feels more uncomfortable can bring up the subject, sharing his or her anger and desire for change but also stressing that he or she still loves the other. Indeed, research suggests that spouses think of their marital relationship as fair when they feel listened to and emotionally supported (Gudmunson 2009; Rudman and Glick 2008; Soliz, Thorson, and Rittenour 2009).

Meanwhile, partners need to remember that managing conflict about power in a positive way is easier said than done. Attempts at communication—and open communication itself—do not solve all marital

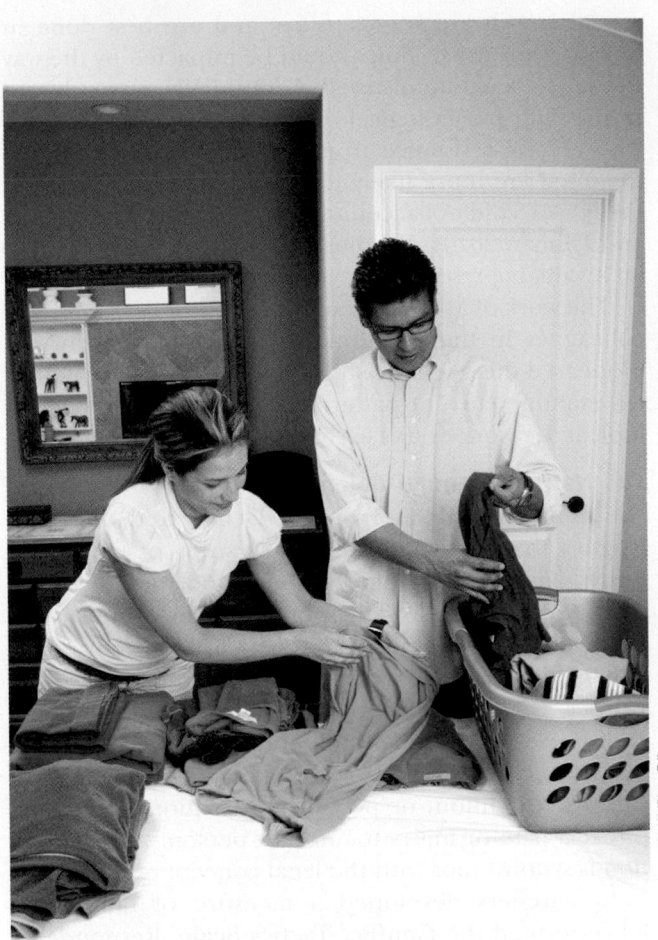

© David Young-Wolff/PhotoEdit

Doing laundry does not seem to have become the site of a power struggle for this couple, but rather it is just something that needs to be done. Partners in a no-power relationship work at doing things on equal terms and seek to negotiate and compromise, thus avoiding deadly power games.

problems. Changing a power relationship is a challenge to any marriage. It can be painful for both partners, though promising a more rewarding relationship in the long run. One option for handling power and gender role change is to seek the help of a qualified marriage counselor or counselor team.

The Role That Marriage Counselors Can Play

Today, many marriage counselors are committed to viewing couples as two human beings who need to relate to each other as equals. In other words, they are committed to helping couples develop no-power relationships. They realize that once both spouses admit—to themselves and to each other—that they do in fact love and need each other, the basis for power politics is gone. On this assumption, counselors help spouses

learn to respect each other as people and not to engage in coercive withdrawal.

Couples need to be aware that, like everybody in society, marriage counselors have internalized their own perspectives on gender roles—and these may not match the goals of the couple. There may be issues concerning potential racial and cultural bias on the part of therapists (Sluzki 2008) or simple lack of awareness of cultural differences in communication style or other matters. Choosing (or retaining) an appropriate counselor should involve an assessment of the counselor's sensitivity to the values, goals, and needs of the couple (Soliz, Thorson, and Rittenour 2009).

The counselor's gender may be an issue for some couples as they explore power and gender issues. A dominant husband, fearful that "it's going to be two against one," may feel threatened by a female counselor. On the other hand, a wife may fear that a male counselor will be too traditional or unable to relate to her. In this situation, counselors sometimes work as a team, woman and man. In any case, it is important that both partners feel comfortable with a counselor or counseling team from the beginning.

Whether on their own or with the help of counselors, partners can choose to emphasize no-power over the politics of power. No marriage—indeed, no relationship of any kind—is entirely free of power politics. But as Chapter 12 points out, the politics of love requires managing conflict in such a way that both partners win.

When, on the other hand, power politics triumphs over no-power, one result may be family violence—psychological (emotional) and/or physical.

Family Violence

The use of physical violence to gain or demonstrate power in a family relationship has occurred throughout history, but only in the last fifty years has violence been labeled a social problem.[2]

The identification of child abuse as a social problem in the 1960s was followed in the 1970s by attention to wife abuse. With the 1980s came concern about elder abuse, as well as husband abuse. More recently, attention has been given to youth dating violence, to violence in adult dating and cohabiting relationships, including

[2] Dr. C. Henry Kempe and his colleagues are credited with the "discovery" of child abuse. They published an article on the "battered child syndrome" based on their observation of hidden injuries to children revealed by X-rays (Kempe et al. 1962). Social scientists took note, and then pursued their interest in child abuse and other forms of family violence. The discovery of child abuse was somewhat like Columbus's discovery of America in that the phenomenon was always there but had not been noticed or taken seriously by academics or authorities.

same-sex relationships, sexual coercion in marital and nonmarital relationships, sibling violence, and child-to-parent violence. We discuss many of these forms of violence in this chapter; dating violence and acquaintance rape are discussed in Chapter 6 and elder abuse and neglect in Chapter 17.

Major Sources of Data on Family Violence

There are several major sources of current data on family violence. Probably the best for our purposes is the National Crime Victimization Survey (NCVS), conducted every two years by the Bureau of Justice Statistics (BJS). This is a national sample survey that asks respondents about all violence they have experienced, their relationship if any to the perpetrator, and whether the violence was reported to the police. Violent acts covered by the survey include assault and rape/sexual assault, as well as other crimes not relevant to family violence. Although murder is obviously not included in the victimization survey, other data on homicides are included in Bureau of Justice Statistics reports on **intimate partner violence**. Spouses, ex-spouses, and current or former boyfriends or girlfriends, including same-sex partners, are considered *intimate partners*.

Other relevant government data include the Uniform Crime Reports of the FBI based on the National Incident-Based Reporting System (NIBRS). Data on criminal incidents reported to the police are compiled from records submitted by many (though not all) local law enforcement agencies (Durose et al. 2005; Catalano 2007). A weakness of these data is that many crimes are not reported to the police, including an estimated one-half of intimate partner violent crimes (Groves and Cork 2008). However, because most homicides are reported, the homicide data are more valid (e.g., Fox and Zawitz 2004; Loftin, McDowall, and Fetzer 2008). The BJS uses these data to supplement the NCVS in producing its report series, *Intimate Partner Violence in the United States*.

A third major source of data on family violence is the National Violence Against Women Survey (NVAWS) (Chen and Ullman 2010), commissioned by the National Institute of Justice and the Centers for Disease Control and Prevention and conducted between 1995 and 1996. The survey employed a modified version of Murray Straus's Conflict Tactics Scale rather than asking about "crimes." While this survey remains the most important tool in understanding violence against women, it is not problem-free, and it is important to acknowledge those problems. Social researchers Yingyu Chen and Sarah E. Ullman recently analyzed the survey and noted that the NVAWS used "random-digit dialing methods to select the sample, which excluded women without telephones and those who were homeless or institutionalized" (2010, p. 275). Also, the survey interviews were done over the telephone, but topics such as rape and

sexual assault are very sensitive and are best done in person, thus the findings could be impacted by the way the survey was administered. Additionally, research into post-assault psychological outcomes shows that women who receive assistance from others after they've been assaulted are more likely to report sexual assaults, but the NVAWS did not account for those differences (Chen and Ullman 2010). Despite the limitations, the study is an invaluable resource on violence against women.

The work of Murray Straus, Richard Gelles, and their colleagues in their National Family Violence Surveys pioneered the scientific study of family violence. Before we examine current patterns of family violence, let us look at the work of this early research group.[3]

The National Family Violence Surveys The early and continuing research of Straus, Gelles, and their colleagues shaped the social science study of family violence. This research group undertook a household survey in 1975, followed by a 1985 telephone survey; together, the surveys produced data from more than 8,000 husbands, wives, and cohabiting individuals (Straus, Gelles, and Steinmetz 1980; Straus and Gelles 1986, 1988, 1995; Gelles and Straus 1988).

The authors defined violence as "an act carried out with the intention, or perceived intention, of causing physical pain or injury to another person." This definition is synonymous with the legal concept of assault.

Researchers developed a measure of family violence termed the **Conflict Tactics Scale**. Respondents were asked about the following acts: threw something at the other; pushed, grabbed, or shoved; slapped or spanked; kicked, bit, or hit with a fist; hit or tried to hit with something; beat up the other; burned or scalded (children) or choked (spouses); threatened with a knife or gun; and used a knife or gun (Straus and Gelles 1988, p. 15). *Severe violence* was defined as acts that have a relatively high probability of causing an injury: kicking, biting, punching, hitting with an object, choking, beating, threatening with a knife or gun, using a knife or gun—and, for violence by parents against children, burning or scalding the child (Straus and Gelles 1988, p. 16). Later modified somewhat (Straus et al. 1996), the Conflict Tactics Scale is different from and broader than the crime categories of assault and homicide that form the basis of criminal justice system statistics. Many other family violence researchers have used the scale.

The 1975 and 1985 National Family Violence Surveys found that, in 16 percent of the couples surveyed, at least one of the partners had engaged in a violent act

---

[3] Murray Straus still leads the Family Violence Research Program at the University of New Hampshire's Family Research Laboratory. Many of the colleagues he has copublished with are there. Richard Gelles is now at the University of Pennsylvania.

against the other during the previous year. Considering the entire length of the marriage rather than just the previous year, respondent reports indicated that a violent act occurred in 28 percent of couples. National Family Violence Survey data also yielded information on violence directed toward children by parents and siblings. The National Family Violence Surveys explored social variation in family violence and some of the circumstances thought to be associated with family violence, such as stress and alcohol use.

Having presented some major sources of data on family violence, we examine the circumstances and outcomes of spouse or partner abuse in more detail in the next sections.

Intimate Partner Violence

Intimate partner violence—the physical or emotional abuse of spouses, cohabiting or noncohabiting relationship partners, or former spouses or intimate partners—is a serious and significant problem. First identified in terms of *wife abuse*, the growing practice of cohabitation places many unmarried women in similar situations. Husbands or male partners may also be subject to abuse from intimate partners as may same-sex partners of gay males and lesbians. Employing the term *intimate partner violence,* the federal government now includes all these forms of couple violence in its reports on domestic violence.

We will focus primarily on marital violence in analyzing the dynamics of intimate partner violence, but it is worth noting that the rate of violence between cohabiting partners is higher than that of spouses (Magdol et al. 1998). As the proportion of cohabiting couples in the population increases, this setting becomes of greater importance in an overall perspective on domestic violence. That is, if cohabiting couples have higher rates of domestic violence and if there come to be more of them, theories of domestic violence may need to be modified to take this group into greater account.

We focus on physical abuse, but verbal abuse (such as name-calling, demeaning verbal attacks), and other kinds of emotional abuse (such as threats to take away children, threats to the victim's extended family or friends, and threats or attacks on pets) virtually always occur along with physical aggression (Johnson 2008, p. 88; McCue 2008) and may be part of a pattern of control and domination.

Intimate partner violence can indeed result in serious injuries. National Crime Victimization Survey data indicate that, for the period between 1993 and 2005, 5 percent of female victims of intimate partner abuse and 4 percent of male victims were seriously injured. More women (44 percent) than men (36 percent) had minor injuries (Catalano 2007). Another source of injury data is the Study of Injured Victims of Violence (SIVV), which is a count of emergency room admissions

Local advocacy groups draw attention to efforts to prevent domestic violence and to the need for more resources. In many localities, there are still not enough shelters to meet the needs of battered women and their children.

attributable to family or nonfamily violence. In 1994, there were more emergency room admissions due to serious family violence (not only intimate partner violence) than reported in the National Crime Victimization Survey. However, the emergency room study was small and the two data sources are not truly comparable (Durose et al. 2005, p. 72).

Who Are the Victims of Intimate Partner Violence?
According to a report based on the National Crime Victimization Survey, there were over 615,000 "victimizations" by intimate partners in 2005. One-third of these were serious violent crimes: rapes, sexual assaults, and aggravated assaults, and/or crimes involving serious injuries, weapons, or sexual offenses. The other two-thirds were lesser offenses, mostly simple assaults (Catalano 2007).

Women are the primary victims of intimate partner violence reported in the National Crime Victimization Survey. A much larger proportion of non-fatal violent victimizations of women are perpetrated by intimate partners: 22 percent compared with only 4 percent of the violent victimizations of men, who are more at risk of violence from others than from intimate partners.

The rate of non-fatal violent victimization of women by intimate partners is over four times that of men. In every racial/ethnic category (see Figure 13.2), women have higher rates of victimization than men. Although intimate partners commit only 5 percent of murders of men, they account for 30 percent of homicides of women (Catalano 2007).

Younger women (twenty through twenty-four) are more likely to have experienced intimate partner victimization, as are those women who are separated or divorced (Catalano 2007), although some experts question the lower rates of marital violence, arguing that married women are simply less likely to acknowledge their victimization ("Domestic Violence Decline" 2006).

Although the data suggests that the rates of intimate partner violence have dropped for all racial groups since 2004, as Figure 13.2 indicates, rates of intimate partner violence vary greatly by race/ethnicity. Victimization rates are strikingly higher (11.1 per 1,000) for Native American women. Black women's rates are high (5.0), but still less than one-half the rate for Native American women. White and Hispanic females have moderate rates (4.0 and 4.3, respectively), whereas Asian females, as well as Asian males, have very low rates of intimate partner violence victimization (1.4 and > 0.1, respectively). White, black, and Hispanic male victimization rates are also low, whereas those of Native American men are relatively high (Catalano 2007).

Multiple studies have demonstrated that cohabiting partners have higher rates of *situational couple* violence (violence that arises out of a quarrel and is often mutual), but married women are more likely to be the target of *intimate terrorism* (a systematic pattern of violence employed by a man to intimidate and control his partner) (Brownridge and Halli 2002; Johnson 2010; Leone, Johnson, Cohan 2007). These concepts will be developed in more detail later in the chapter. A variety of explanations have been offered, but none conclusively

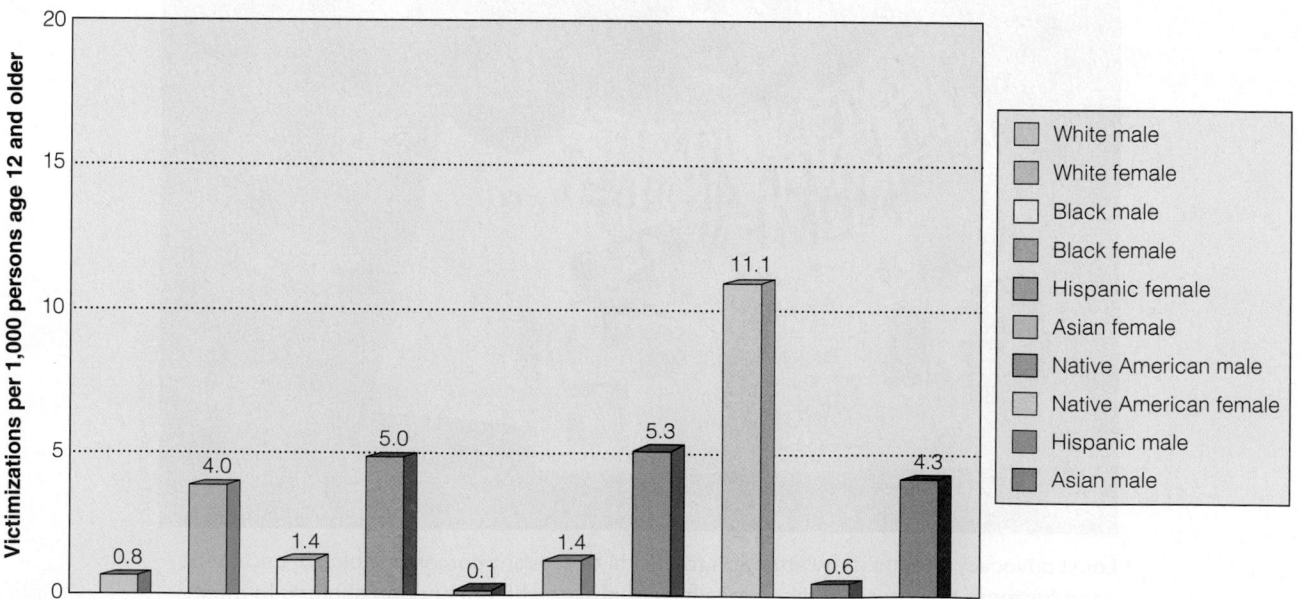

Figure 13.2 Intimate partner victimization rate (nonfatal) by gender, race, and Hispanic origin, 1993–2005
Source: Catalano 2007.

proven. Overall, cohabitors are younger, less integrated into family and community, and more likely to have psychobehavioral problems such as depression and alcohol abuse—all factors associated with family violence (Stets 1991). Another possibility is that there is less institutional control over cohabiting relationships than over marriage (Ellis 2006; Nock 1995). Still another thesis is that the less-violent cohabiting couples end up getting married whereas more-violent married couples get divorced, sharpening the difference between the two groups (Kenney and McLanahan 2006).

Several recent studies have reported that pregnancy increased the likelihood of physical violence by intimate partners (Cox 2008; Burch and Gallup 2004; S. Martin et al. 2004). None of these studies used a representative sample. Earlier studies using national samples found that when age was controlled, there was no increased risk with pregnancy. Still, the many studies, however imperfect, that have found an association between pregnancy and violence have kept this hypothesis alive, along with a possible explanation—jealousy, specifically that the new baby would interfere with the wife's attention to and care of the man. What is known with more certainty is that those pregnant women who are abused seek medical care later in pregnancy and are more likely to have preterm and low-birth-weight babies (Janssen et al. 2003; Sarkar 2008).

Substance abuse, especially of alcohol, is often cited as a factor in male violence against women. That seems to be true of heavy use of alcohol and binge drinking, though not necessarily for other patterns of alcohol use. Alcohol is implicated in violence through cognitive impairment, impulsivity, and a tendency to perceive threats, particularly in married-couple relationships where research finds an association between alcohol and "more severe forms of intimate partner violence" (Catalano 2007; Kaukinen 2004; Wiersma et al. 2010, p. 372). Drinking may also serve as a rationalization and excuse for violence that would have occurred in any case (Gelles 1974).

Marital Rape Wife and female partner abuse may take the form of sexual abuse and rape. Estimates are that between 10 and 14 percent of women experience marital rape (Ferro, Cermele, and Saltzman 2008, p. 765; Bergen 2006). These sexual assaults often involve other violence as well.

The issue of **marital rape** arose as a feminist one in the 1970s, and as such was conceptualized in terms of the law of *marriage*. Under traditional common law, a husband's sexual assault or forceful coercion of his wife was not considered rape because marriage meant the husband was entitled to unlimited sexual access. The legal situation has improved since the 1970s, as a result of feminist

political activity. As of 1993, all states have provisions against marital rape in their legal codes[4] (McMahon-Howard, Clay-Warner, and Renzulli 2009, p. 507). More research is needed in this area; the "lack of empirical and theoretical attention to sexual assault and coercion in marriage . . . is striking" (Christopher and Sprecher 2000, p. 1,007). Data on sexual assault of women intimate partners, married or nonmarried, are collected as part of the federal government's documentation of intimate partner violence.

Intimate Partner Violence is Declining The preceding data show us how significant the problem of intimate partner violence still is. Yet, the direction of change gives us some indication that efforts to combat domestic violence are paying off. Intimate partner violence declined dramatically from 1993 to 2005, as Figures 13.3 and 13.4 indicate. The rate of nonfatal victimization declined nearly 44 percent for men and over 63 percent for women. Intimate partner homicide rates declined 48 percent for men and 24 percent for women (Catalano 2007; Pastore and Maguire 2007, Tables 3.131.2005, 3.132.2005).

Shannan Catalano, author of the Bureau of Justice Statistics intimate partner violence study, cites experts' opinions that stronger law enforcement, increased education, and expanded services for battered partners have led to this decline. A cautious note is also sounded: "[T]he apparent decline could [also] mean that women are choosing to suffer in silence rather than seek help ("Domestic Violence Decline" 2006).

As we go on to explore gender issues in intimate partner violence, we will consider why women may not seek

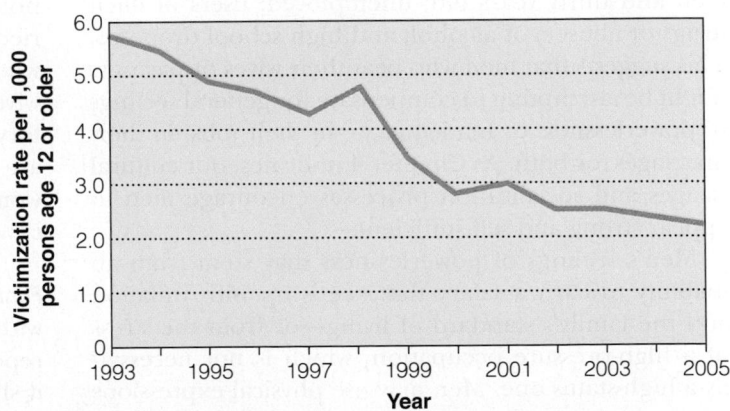

Figure 13.3 Intimate partner victimization rate (nonfatal), 1993–2005
Source: Catalano 2007.

[4] There are some exemptions in the laws of thirty of the states. One common example is that if a wife is asleep or unconscious, thus legally unable to consent, a husband may be exempt from prosecution (National Clearinghouse on Marital and Date Rape 2005).

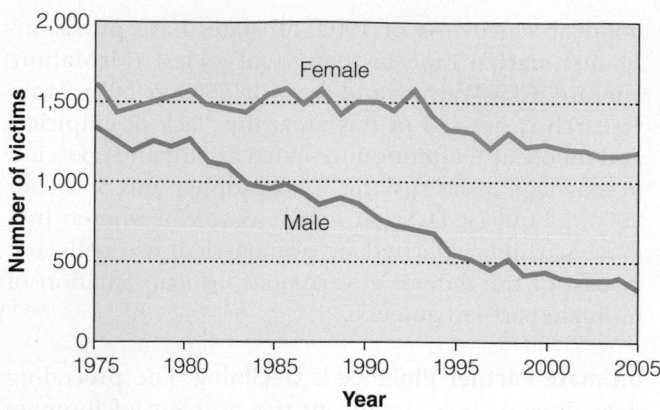

Figure 13.4 Homicides of intimates by gender of victim, 1976–2005

Source: Catalano 2007.

help or leave their marital or other relationships—and why and how they do. We'll also consider whether men are equally victims of domestic violence.

Gender Issues in Intimate Partner Violence

Three questions arise regarding gender and intimate partner violence: Why do men beat their wives and partners? Why do women live with it? And what about husband or male partner abuse?

Why Do Men Do It? From data in the National Crime Victimization Survey, we find that about 96 percent of females experiencing nonfatal intimate partner violence were victimized by a male (Catalano 2007).

Richard Gelles (1994, 1997) lists "risk factors" for men who abuse women: those who are between eighteen and thirty years old; unemployed; users of illicit drugs or abusers of alcohol; and high school dropouts. This suggests that men who beat their wives or partners might be attempting to compensate for general feelings of powerlessness or inadequacy—in their jobs, in their marriages, or both. As Chapter 4 indicates, our cultural images and socialization processes encourage men to appear strong and self-sufficient.

Men's feelings of powerlessness may stem from an inability to earn a salary that keeps up with inflation and the family's standard of living—or from the stress of a high-pressure occupation, which is not necessarily a high-status one. Men may use physical expressions of supremacy to compensate for their lack of occupational success, prestige, or satisfaction (K. Anderson 1997). Research using the National Survey of Families and Households found that financial adequacy reduced the risk of couple violence. Employment in low-status and unpleasant jobs that increased irritability, on the other hand, was associated with man-to-woman violence, a stress explanation of family violence (Fox et al. 2002; "Promoting Respectful" 2009). The husband's

unemployment is also associated with domestic violence (Condon 2010; Lauby and Else 2008).

In terms of relative status, a woman's risk of experiencing severe violence is greatest when she is employed and her husband is not. Much research has found violence associated more generally with status reversal, where the woman is superior in some way to the man in terms of employment, earnings, or education (Kaukinen 2004). A man's loss of status upon immigration—when jobs commensurate with education or expectations do not measure up, economic hardship is the family's lot, and wives, children, and people in general do not accord a male the respect he is accustomed to in a more hierarchical society—can lead to family conflict and violence (Min 2002).

Absent a *reward power* base for family power, some men resort to *coercive power:* "[V]iolence will be invoked by a person who lacks other resources to serve as a basis for power—it is the "ultimate resource" (Goode 1971, p. 628; see also Allen and Straus 1980, p. 190, in Fox et al. 2002). Men may use violence to attempt to maintain control over wives or partners trying to become independent of the relationship (Dutton and Browning 1988). Figure 13.5, developed by staff members of a program for male batterers in Duluth, Minnesota (Pence and Paymar 1993), illustrates how a male partner's need for power and control may result in both psychoemotional and physical violence. This type of family violence has been called *intimate terrorism* (M. Johnson 2008) and will be discussed along with other types of intimate partner violence in a subsequent section of this chapter.

Why Do Women Continue to Live With It? Women do not like to get beaten up. However, they may stay married to husbands or remain with violent male partners who beat them repeatedly. For the most part, battered wives leave and/or seek divorce only after a long history of severe violence and repeated conciliation. There are several reasons for this, and they all point to those women's lack of personal resources with which to take control of their own lives.

Fear Battered women's lack of personal power begins with fear (DeMaris and Swinford 1996). "First of all," reports social scientist Richard Gelles, "the wife figures if she calls police or files for divorce, her husband will kill her—literally" (Gelles, quoted in C. Booth 1977, p. 7). This fear is not unfounded. An estimated 74 percent of murders of women by their male partners occurred in response to the woman's attempt to leave the relationship (Seager 2009, p. 2287). Husbands or ex-husbands have shown enormous persistence in stalking, pursuing, and beating or killing women who try to leave an abusive situation (Snider et al. 2009). Fear of reprisals by the batterer continues to be a barrier to

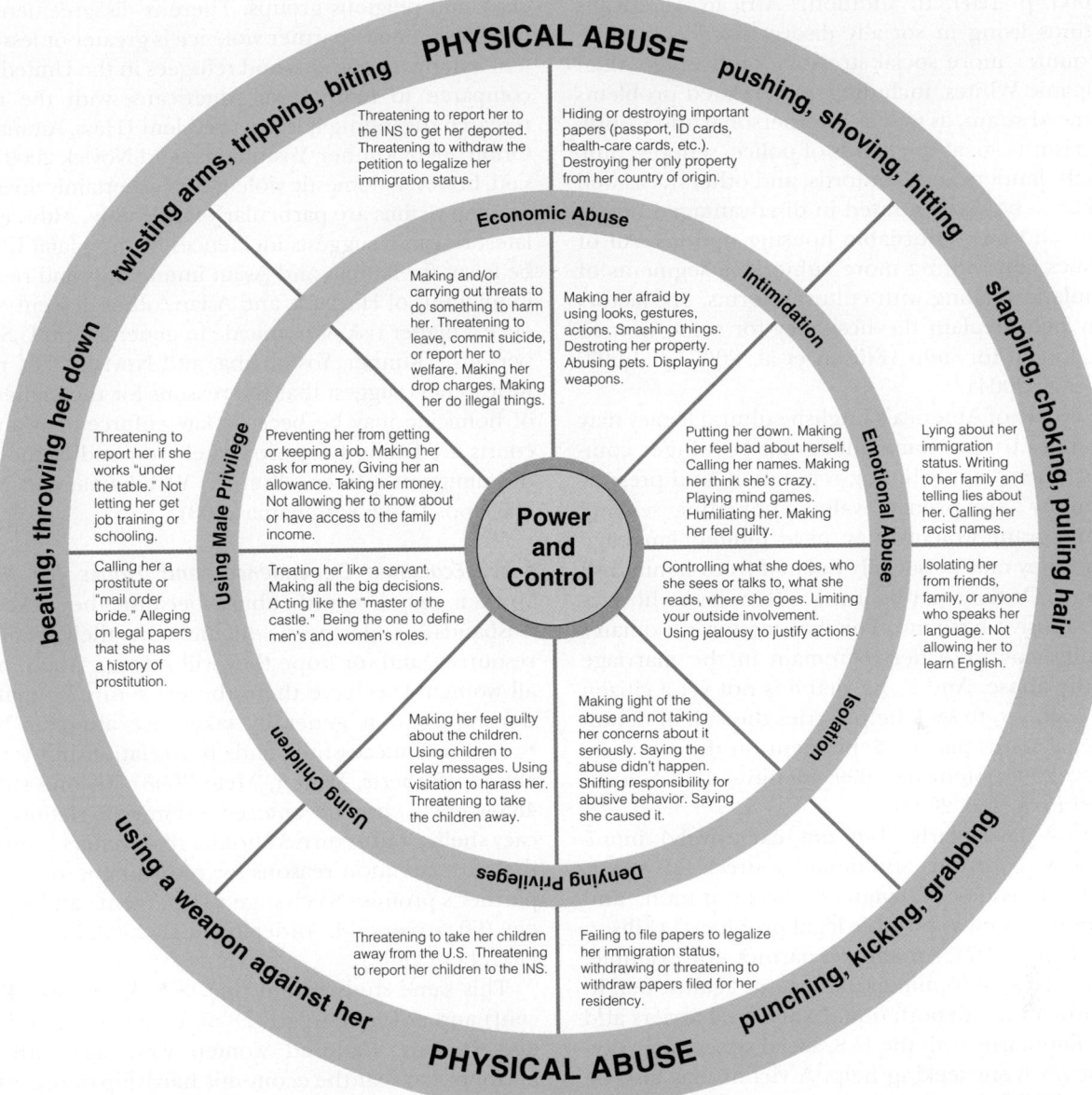

Figure 13.5 The power and control wheel for immigrant and native born: behaviors that some male partners use for coercive power and control

Source: Immigrant Battered Women Power and Control Wheel, produced and distributed by National Center on Domestic and Sexual Violence, Austin, TX, available at www.endingviolence.org/files/uploads/ImmigrantWomenPCwheel.pdf and adapted from the original wheel by Domestic Abuse Intervention Project, Duluth, MN.

seeking police intervention, according to recent studies (M. Anderson et al. 2003; Wolf et al. 2003). In addition to the immediate concerns of injury and death, women who live with violent partners exhibit greater "adverse health conditions and health risk behaviors" than the general population (Black and Breiding 2008, p. 116 Table 2; Breiding, Black, and Ryan 2008). Women with disabilities suffer greater rates of intimate partner violence than the general population (Black and Breiding 2008; Brownridge 2006).

Cultural Norms Historically, women were encouraged to put up with abuse. English common law, the basis of the American legal structure, asserted that a husband had the right to physically "chastise" an errant wife. Although the legal right to physically abuse women has long since disappeared, our cultural heritage continues to have an influence on seeking help and getting it (Ellison et al. 2007). For some women of color, hesitancy to call the police may derive from historic tensions between racial/ethnic communities and the police force (Wolf et al. 2003, p. 124). In addition, "African Americans and Latinos living in socially disorganized communities encounter more social stressors, on average, than non-Hispanic Whites, including work-related problems and financial strain, as well as interpersonal and institutional racism (e.g., at the hands of police, schools, public officials, lenders and landlords, and others)," as well as tending to be concentrated in disadvantaged neighborhoods and lack affordable housing options. All of these issues confronting more vulnerable segments of our population, along with cultural norms, are factors that may help explain the hesitancy for some women of color to call for help (Ellison et al. 2007, p. 1,097; Benson et al. 2004).

What is true of America's English cultural legacy may be even more true of some immigrant or refugee communities, when "family honor, reputation, and preserving harmony" are primary values, impeding seeking help. Immigrant women may have limited language skills, and they may be socially isolated from family and community. Or they may be living with in-laws who support the abusive husband. For that matter, a woman's own family may urge her to remain in the marriage despite the abuse. And if the victim is not yet a citizen or legal resident, to seek help carries the risk of deportation; legal status may be dependent on the marriage (Childress 2003; Mehrota 1999; Menjívar and Salcido 2002; Yoshioka et al. 2003).

"Latinos—particularly (but not exclusively) immigrants—may confront additional sources of stress, centering on issues of language, acculturation, and assimilation, as well as possible legal problems" (Ellison et al. 2007, p. 1,097). An abusive partner, for example, "may use the victim's immigration status against her, in effect, threatening deportation. Language barriers and a lack of familiarity with the U.S. social system may prevent a victim from seeking help. A victim may also be afraid that if she reports violence to the authorities, she and/or her partner will be treated with insensitivity, hostility, and/or discrimination. That fear may be justified; mainstream organizations may lack sociocultural understanding and/or may have discriminatory or insensitive attitudes toward immigrants and refugees," as can be seen in newspaper reports of immigrant domestic violence issues (Runner, Yoshihama, and Novick 2009, pp. 4, 11; Yoshihama 2008).

Programs are beginning to emerge to assist immigrant women (Abraham 1995, 2000). Moreover, there is evidence for changed attitudes in some settings. An article about Jewish immigrants from Central Asia living in New York reported that the "word has spread" that wife abuse (and harsh physical discipline of children) is not the way things are "done in America" ("Old Ways" 2003).

Of course, domestic violence issues are clearly an American problem that crosses all generational, social, cultural, class, and religious groups. There is disagreement as to whether intimate partner violence is greater or less prevalent among immigrants and refugees in the United States compared to native-born Americans, with the newest research suggesting it is less prevalent (Hass, Ammar, and Orloff 2006; Runner, Yoshihama, and Novick 2009). This said, however, domestic violence most certainly does exist, and the victims are particularly vulnerable. Although the latest research suggests incidences of "non-fatal IPV may be lower for Latinas and Asian immigrants and refugees, immigrants of Hispanic and Asian/other descent experience a higher risk of homicide in general than U.S.-born persons" (Runner, Yoshihama, and Novick 2009, p. 11). Researchers suggest that the reasons for the higher rates of homicide may be because law enforcement and the courts are failing to appropriately respond to the needs of immigrant women (Runner, Yoshihama, and Novick 2009, pp. 41–42; Yoshihama 2008).

Love, Economic Dependence, and Hopes for Reform

Women may live with abuse because they love their husbands or partners, depend on their economic resources, and/or hope they will reform. About half of all women who leave their abusers return to that relationship, and it generally takes five attempts before the woman successfully ends her relationship with her abuser (Roberts, Wolfer, Mele 2008). In one study of 485 women who had entered a domestic victims' advocacy shelter but returned home, researchers found that the most common reasons for returning were the male partner's promise to change (71 percent) and his apology (60 percent; M. Anderson et al. 2003; Enander and Holmberg 2008).

This same study found that lack of money (40 percent) and nowhere to go (28 percent) were also important reasons. Battered women who stay with their partners may fear the economic hardship or uncertainty that will result if they leave. They hesitate to summon police or to press charges not only out of fear of retaliation but also because of the loss of income or damage to a husband's professional reputation that could result from his incarceration. Fear of economic hardship is heightened when children are involved. For a mother, leaving requires being financially able to take along her children and support them—or leaving them behind, where they may also be in danger.

A new wrinkle in economic dependency has emerged with the passage of welfare reform legislation in 1996. Studies show that 20 to 30 percent of women on welfare are in situations of risk for domestic violence. Some men become abusive when the woman gets a job, which may threaten his control. Abusive men inhibit a woman's economic independence in many ways, for example, physical restraint, harassment on the job (including calling a woman's supervisor or coworkers), stalking, and destroying her work clothes (Kimerling et al. 2009).

Compliance with the requirement to report paternity of children may also trigger retaliation by a man, who now will be pursued for child support. Although the law contains a waiver provision directed at exactly these problems, it is not certain that women are being informed or that the provision is implemented. Some women will find it difficult to comply with welfare requirements because of the objections and control tactics of men in their lives (Riger, Staggs, and Schewe 2004, p. 812; see also Handler and Hasenfeld 2006, p. 304). Cut off from welfare, they become even more dependent on violent men.

Apart from such special circumstances, it is possible that though a dramatic rise in women's employment and earnings may prove threatening to low-earning husbands in the short run, in the long run, mutual awareness of a woman's potential economic independence may deter wife abuse by changing the family power dynamic (Yakushko and Espin 2010).

Gendered Socialization Another factor that helps perpetuate abuse is the cultural mandate that it is primarily a woman's responsibility to keep a marriage or relationship from failing. Believing this, wives are often convinced that their emotional support may lead husbands to reform. Thus, wives often return to violent mates after leaving them (Roberts, Wolfer, and Mele 2008).

Childhood Experiences Research suggests that people who experience violence in their parents' home while growing up may regard beatings as part of married life, and this is another factor associated with women's living with abuse. Men, as well as women, are more likely to be victims of intimate partner violence as adults if they were exposed to child abuse or witnessed parental interpersonal violence as children (Brown and Bulanda 2008; Pettit et al. 2010).

Low Self-Esteem Finally, unusually low self-esteem interacts with fear, depression, confusion, anxiety, feelings of self-blame (Walker 2009, pp. 155–165), and loss of a sense of personal control (Umberson et al. 1998) to create the *battered woman syndrome*, in which a wife cannot see a way out of her situation (Walker et al. 2009).

A Way Out: Shelters and Domestic Violence Programs
A woman in such a position needs to redefine her situation before she can deal with her problem, and she needs to forge some links with the outside world to alter her circumstances. This usually occurs over time, with some unsuccessful attempts to leave as part of the process.

Although there are not enough of them, a network of shelters for battered women provides a woman and her children with temporary housing, food, and clothing to alleviate the problems of economic dependency and physical safety. These organizations also provide counseling to encourage a stronger self-concept so that the woman can view herself as worthy of better

treatment and capable of making her way alone in the outside world if need be. Finally, shelters provide guidance in obtaining employment, legal assistance, family counseling, or whatever practical assistance is required for a more permanent solution.

This last service provided by shelters—obtaining help toward longer-range solutions—is important. Two face-to-face interviews with the same 155 wife-battery victims (a "two-wave panel study") were conducted within eighteen months during 1982 and 1983 in Santa Barbara, California. Each of the women interviewed had sought refuge in a shelter. Findings showed that victims who were also taking other measures (for example, calling the police, trying to get a restraining order, seeking personal counseling or legal help) were more likely to benefit from their shelter experience: "Otherwise, shelters may have no impact or perhaps even trigger retaliation (from husbands) for disobedience" (Berk, Newton, and Berk 1986, p. 488). As the researchers conclude,

> The possibility of perverse shelter effects for certain kinds of women poses a troubling policy dilemma. On the one hand, it is difficult to be enthusiastic about an intervention that places battered victims at further risk. On the other hand, a shelter stay may for many women be one important step in a lengthy process toward freedom, even though there may also be genuine short-run dangers. (p. 488)

As with some other decisions discussed in this text, social scientists have applied *exchange theory* to an abused woman's decision to stay or leave (McDonough 2010). As Figure 13.6 illustrates, an abused wife weighs such things as her investment in the relationship, her (dis)satisfaction with the relationship, the quality of her alternatives, and her beliefs about whether it is appropriate for her to leave ("subjective norm") against such questions as whether she will be better off if she leaves (might her husband retaliate, for example?) and whether she can actually do it. The woman's personal resources along with community (structural) resources, such as whether shelters or other forms of assistance are available, further affect her decision. Personal barriers might involve not having either a job with adequate pay or an extended family that could help. Structural barriers might include the lack of community systems for practical help.

Michael Johnson and Kathleen Ferraro (2000) answer the question "Why do they stay?" with: "The truth is, they don't stay." Instead, abused women went "through a process of leaving and returning, each time gaining more psychological and social resources . . . until they escaped from the web" (pp. 956–57).

Men as Victims of Intimate Partner Abuse Both women and men sometimes resort to violence. A major question regarding family violence is whether female-to-male violence is trivial in numbers and effects or should be regarded as a serious social problem.

The early National Family Violence Surveys reported approximately equal amounts of both minor and serious partner violence on the part of men and women. Researcher Murray Straus and his colleagues continue to point to data indicating comparable levels of male and female intimate partner violence. In fact, "IPV by men, but not by women has been decreasing since the mid 1970s but assaults by women on male partners have stayed about the same" (Straus 1999b; 2005; 2008; Straus and Ramirez 2007, p. 9). A number of other studies and research reviews also find that comparable numbers of males and females have engaged in physical violence (Dutton and Nicholls 2005; Medeiros and Straus 2006).

As with most social research, there are debates as to the methodological validity of the sample and the study itself. This holds true for research into intimate partner violence—particularly the research into intimate partner violence against males. Although the National Family Violence Survey is a national sample survey, many of the other studies cited as evidence for gender symmetry are convenience samples, studies of college students, or clinical samples of couples who have sought help for their marital problems—in other words, some might argue these other studies are not so strong methodologically. On the other hand, meta-data analysis of seventeen large-scale studies suggests gender symmetry in intimate partner violence (Cook 2009, pp. 14–15, Tables 1.3, 1.4). In addition, critiques of relevant studies (e.g., Were women asked about perpetration of violence, or only of victimization?) suggest there has been a bias toward an assumption of women as victims of male violence, thus skewing the surveys themselves in favor of the preconception (Straus and Scott, forthcoming).

Crime victimization data, including large-sample surveys, indicate that women are overwhelmingly the victims of intimate partner violence. These conflicting reports set off a dispute, still not resolved, as to whether intimate partner violence is *asymmetrical*—with women primarily victims of male aggression—or whether couple violence is *symmetrical*—both men and women engage in intimate partner violence and at similar rates.

The first assumption—that males primarily perpetrate violence against partners—underlies policy directed toward providing resources for women victims; it is a strongly held feminist perspective. If, on the contrary, women are as likely as men to perpetrate intimate partner violence, then they begin to look less like victims and more like aggressors.

Differences in conclusions may simply arise from methodological differences. Critics of the National Crime Victimization Survey (which finds *asymmetric* violence) point out that "crime" terminology may dampen reports of less-serious female-to-male violence because those acts may not seem to be crimes to those interviewed. Questions on the Conflict Tactics Scale (CTS) include a broader range of actions that may be characterized by the survey respondent as family conflict and violence, but not criminal victimization (Dutton and Nicholls 2005; Medeiros and Straus 2006; Straus 1999b, 2005; 2008; Straus and Ramirez 2007). A key problem with the Conflict Tactics Scale (used in many studies that find *symmetric violence*), however, is that sexual assault, a substantial part of male-to-female violence, was not included in the first version of the CTS used in the 1975/1985 National Family Violence Surveys. (The later modification of the CTS does include sexual assault [Straus et al. 1996].)

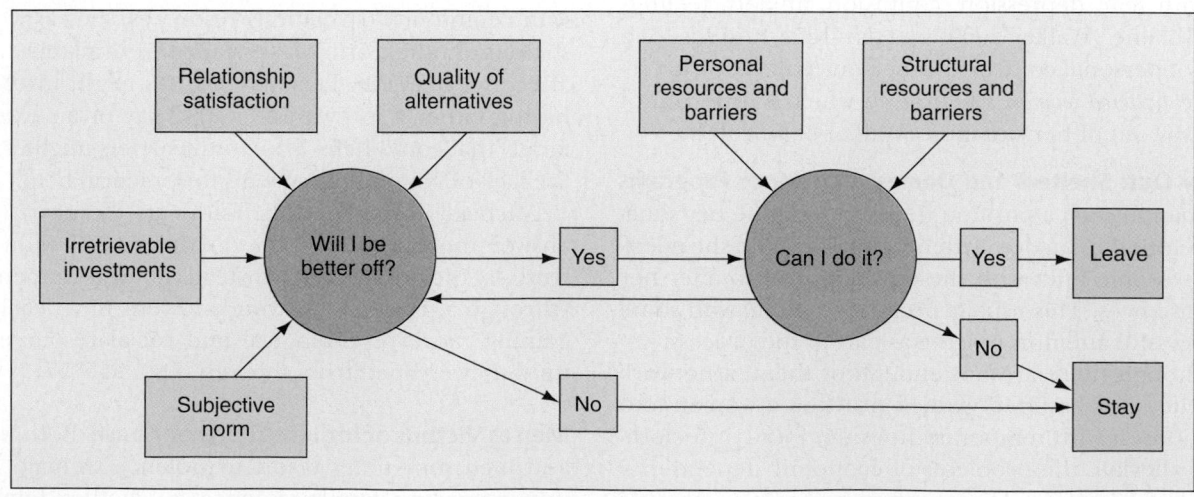

Figure 13.6 Conceptual model of abused women's stay/leave decision-making process
Source: Choice and Lamke 1997, p. 295.

The Conflict Tactics Scale, used by many researchers and most associated with findings of gender-balanced violence, has been widely criticized for lack of context. In reporting lifetime or annual incidence of violence, a single, never-repeated act could be equated with a marriage-long pattern of abuse. Feminist critics assert that the context of violence is ignored in simple counts of male and female violence: Where does a particular violent act fit into the couple relationship? Who initiated the violence? Was it in self-defense? (Carney, Buttell, and Dutton 2007; Hines, Brown, and Dunning 2007; Loseke and Kurz 2005); at the same time, other critics assert that all violence against partners must be examined fully through the lens of science rather than through the lens of ideology (Hines and Douglas 2009).

Recent reviewers of the literature have tried to make distinctions that might explain the contradictory conclusions about who is violent. Definitions and measurements of violence continue to be relevant, as does whether a survey asks about victimization only or also asks whether the respondent has been a perpetrator of violence (Hines and Douglas 2009; Straus and Scott in press). The most convincing explanations for contradictory findings—that men are the more violent sex (asymmetrical violence) or that men and women are both violent (symmetrical violence)—are (1) sample differences, (2) measures, and (3) typologies of intimate partner violence (to be discussed momentarily).

Studies and research reviews that are dominated by samples of younger people show "that among violent adolescent relationships, the percentage of relationships in which there was reciprocal partner violence ranged from 45% to 75%." Moreover, a "recent meta-analysis found that a woman's perpetration of violence in a youthful relationship was the strongest predictor of her being a victim of partner violence." The focus on research into adolescent and young adult relationship violence is important because youthful relationship violence is a strong predictor of adult relationship violence (Stith et al. 2004; Whitaker et al. 2007, p. 941).

The question of whether wives' violence toward husbands is mostly in self-defense, as feminist violence researchers argue, is part of the debate about gender differences in domestic violence perpetration. Demie Kurz (Kurz 1993; Loseke and Kurz 2005) offers evidence that intimate partner violence by women is largely in self-defense or at least retaliation rather than the initiation of a violent attack. Sociologist Murray Straus claims that better data indicate that wives often strike out first and that the data do "not support the hypothesis that assaults by wives are primarily acts of self-defense or retaliation" (Straus 1993, p. 76; 2008; Straus and Ramirez 2007). Moreover, he argues that though women's violence produces fewer serious injuries and deaths

than men's, these are substantial enough in numbers to challenge any minimization of women's intimate partner violence (Dutton and Nicholls 2005; Medeiros and Straus 2006; Straus 2005; 2008). A substantial number of researchers have argued that couple violence is gender symmetrical, despite the criticism they have received from feminists who argue that male intimate partner violence is simply a "means for men to systematically dominate, control, and devalue women" (Dutton and Nicholls 2005, p. 684).

Other researchers examining the issue suggest that the root explanation for contradictory research results may be that there are two primary forms of heterosexual violence against women—"intimate terrorism" (formerly termed "patriarchal terrorism") and "situational couple violence" (formerly termed "common couple violence") (M. Johnson 1995; 2008).[5] **Intimate terrorism** refers specifically to abuse that is almost entirely male and that is oriented to controlling the partner through fear and intimidation. Physical abuse is but one of the tools a terrorist uses; emotional abuse is frequent as well. Intimate terroristic violence is not focused on a particular matter of dispute between the partners, but is intended to establish a general pattern of dominance in the relationship. This form of intimate partner violence occurs more often in marriage than in cohabitation. It includes more incidents, is likely to escalate, and is more likely to produce serious injury.

Situational couple violence refers to mutual violence between partners that often occurs in conjunction with a specific argument. It involves fewer instances, is not likely to escalate, and tends to be less severe in terms of injuries (M. Johnson 2008; Johnson and Ferraro 2000). Situational couple violence appears to be perpetrated by women as well as men and may be more common than intimate terrorism, producing the gender-balanced rates found in some studies (M. Johnson 2008, pp. 60–71).

Some researchers argue that there is "compelling evidence that men's and women's experiences with violence at the hands of marital and cohabiting partners differ greatly" (Tjaden and Thoennes 2000, p. 156). The problem with such analysis is that other analyses of the same data suggest it contains serious problems. For example, psychologist John Archer (2000) takes issue with Tjaden and Thoennes (2000) because the NWAWS "was presented to respondents as a study of victimization

---

[5] Johnson's typology of domestic violence includes additional types that reflect a pattern of violent resistance to a partner's violence, sometimes mutual violent resistance to each other's violence. Although an important consideration, Johnson's conceptualization of violent resistance has changed somewhat in a short time (2006; 2008). For the moment it seems most useful to emphasize his two primary types of violence, while including some research on the topic of violent resistance in the chapter.

of women, it contained 'filters' or demand characteristics that would make men less likely to report their own victimization" (cited in Dutton and Nicholls 2005). Also, other studies show that, as with immigrant women abuse discussed earlier in the chapter, the police appear ill-equipped to deal with female abuse of males, going so far as to downplay violence against men even when they're called to the scene (Brown 2004).

A good deal of research shows that there is an overwhelming victimization of women shown by crime victimization data, including data on homicides, which suggests that women are victims of the most serious violence. Unfortunately, however, when research shows comparable rates of violence between the genders, oftentimes "female-perpetrated abuse is minimized and understood as either defensive or situational in nature, an isolated expression of frustration in communicating with an unsympathetic partner, in contrast to the presumably intentional, pervasive, and generally controlling behaviors exhibited by men" (Hamel and Nicholls 2006, p. xxxix; M. Johnson & Leone 2005).

American society socializes men to believe they are the "strong" ones. In fact, the image of men (discussed in Chapter 4) that is presented to society is that of a successful and strong "man," who is confident, self-reliant, and even aggressive; and who can physically outwit and defeat any opponent (Sullivan and McHugh 2009; Katz 2006). Given the socialization of both men and women in our modern society, some researchers question if it is scientifically appropriate to assume that males socialized into this cultural milieu would acknowledge on surveys or in interviews that they had been battered by the "weaker" sex—their female companions, given the internalized self-concepts males in the United States are socialized to believe (Brown 2004; Dutton and Nicholls 2006).

There is a need for programmatic support for male victims of spouse abuse: "Compassion for victims of violence is not a zero sum game. . . . Reasonable people would rationally want to extend compassion, support, and intervention to all victims of violence" (Kimmel 2002, p. 1,354). Indeed, the male victim of violence has few resources and often little sympathy (Cook 2009, p. 41).

Abuse among Same Gender, Bisexual, and Transgender Couples

We discuss gay male, lesbian, and bisexual intimate partner violence apart from our discussion of married-couple and heterosexual cohabitants violence because the analysis of heterosexual intimate partner violence is largely based on gender *difference*. In fact, it was initially assumed that the likely greater similarity in power of same-sex couples would deter couple violence—unfortunately not. Rates of same-sex intimate partner violence (SSIPV) are "comparable to rates of heterosexual

domestic violence, with approximately one quarter to one half of all same-sex intimate relationships demonstrating abusive dynamics," and upwards of half of all transgendered people reporting intimate partner victimization (Murray and Mobley 2009, p. 361; Bornstein et al. 2006, p. 162).

Same-sex intimate partner violence has long been shrouded in silence in lesbian and gay communities (Murray and Mobley 2009). Lesbians may have effectively denied the issue through a lack of conceptualizing lesbian intimate partner violence as domestic violence (particularly because domestic violence has long had a connotation of male-perpetrated violence in the research community). As for violence in the relationships of gay men, there may be even more silence and denial, as there is a very real fear (as is for immigrant women and some women of color discussed earlier in this chapter) of being victimized by the system (Island and Letellier 1991).

Research on lesbian, gay male, bisexual, and transgendered relationship violence was initially scanty, but more research and service-oriented articles and books have appeared in the last few years. Studies done to date suggest that violence between same-sex partners occurs at the same or greater rate as in heterosexual relationships (Kulkin et al. 2008; Little and Terrance 2010; Murray and Mobley 2009). One large-sample study of 499 couples found that 9 percent reported physical violence in current couple relationships, and 32 percent in past relationships (Brown 2008; Kulkin et al. 2008; Little and Terrance 2010; Turell 2000).

As is also true for straights, domestic violence may be found in all racial/ethnic categories, social classes, education levels, and age groups. Among lesbians, neither "butch/femme" roles nor the women's physical size has been found to figure into violence (Obejas 1994).

Some of the relationship dynamics in same-sex abusive partnerships are similar to those in abusive straight relationships (Kurdek 1994), and because of this, the Violence Against Women Act was expanded in April 2010, to include same-gender couples under its criminal provisions (Barron 2010; Savage 2010). As with heterosexual intimate partner violence, batterers may use drugs or alcohol or have a history of childhood exposure to violence, justifying and excusing attacks on his or her partner. The abusive partner uses violence or threats of violence to keep the partner from leaving. Also like heterosexuals involved in domestic violence, the gay or lesbian couple is likely to deny or minimize the violence, along with believing that the violence is at least partly the victim's fault (Kulkin et al. 2008).

A special problem for lesbian and gay male domestic-violence victims is that few resources exist to serve their needs. The availability of legal protection is problematic in many states, although as of 2007, seventeen states and many municipalities now have legal protections for sexual minorities (Saxe 2007, p. 58). Although some states specifically exclude same-sex couples from domestic violence laws, the federal government has expanded the criminal provisions under the Violence Against Women Act to include sexual minority intimate partner violence (Barron 2010; Savage 2010). Still, a real lack of resources are available for lesbian and gay male domestic-violence victims, often because of the stigmatization of sexual minorities.

Such social stigmatization exacerbates the problem because it sometimes leads sexual minorities to suffer from **minority stress**. Lesbians and bisexual women, in particular, are subject to "a double risk of minority stress due to their social status as both women and sexual minorities. Lesbian women of color often experience what is known as 'triple jeopardy,' that is, they experience minority stress threefold, gender, race, and sexual preference" (Balsam and Szymanski 2005; Brown 2008, p. 459). Gay men of color also face a double risk of minority stress because they are both sexual and racial minorities.

Such external stressors are an important component in domestic abuse issues. As has been discussed elsewhere in this text, vulnerable families, such as those living in economic deprivation, who face discrimination, and so on, tend to have greater rates of domestic distress. Hate crimes, discrimination, internalized homophobia, fear of being "outed," are all stressors that can impact homosexual and bisexual relationships. For sexual minorities, these minority stresses

are directly correlated with intimate partner violence. Because gays, lesbians, bisexuals, and transgendered people are sexual minorities, there is a real fear of a negative response by police, domestic abuse shelters, courts, and other services responding to homosexuals and bisexuals involved in domestic abuse (Brown 2008; Fountain et al. 2009).

Gay/lesbian/bisexual individuals may be afraid to go to the police—or to use any domestic violence intervention services—for fear of having their gay identity revealed or receiving a hostile response. In fact, although sexual minorities experience all of the same threats as heterosexual victims, they have an additional concern—the abusive partner can threaten to "out" a person to employers, family members, and friends (Brown 2008). When victims finally do call the police, some lesbians, for example, will describe their batterers as male because they fear being "outed," as well as discrimination on the part of first responders, hospitals, and the courts (Simpson and Helfrich, 2005).

Reporting by member organizations of The National Coalition of Anti-Violence Programs (see Figure 13.7) show that women callers to their domestic violence hotlines were 51 percent, which is "similar to the percentage reported in 2007. Those identifying as male represented the next largest category (42%), signifying a slight decrease in reports (10%). Callers identifying as intersex rose from 7 to 19 (171% increase). Reports from transgender men dropped 14% and those from transgender women rose 1%" (Fountain et al. 2009, p. 20). The reduction in calls from transgendered men is of concern because transgendered men experience the greatest amounts of social stigma from the broader society, and have the greatest risk of all sexual minority groups for hostility and discrimination from police and service providers when facing intimate partner violence (IPV).

An important obstacle for gay men is the fear of being feminized if they seek help. The dominant ideology in American society is that men are strong, so, just like heterosexual male IPV victims, to acknowledge abuse is to acknowledge weakness. This means gay men, like heterosexual men, not only suffer the abuse, but struggle with their own masculine identities. Although lesbians do seek help from the same services as heterosexual women, they find friends, counselors, and relatives the most helpful sources of support. Gay men find friends, counselors, and support groups of greatest help. Domestic violence services oriented to gay men, lesbians, bisexuals, and transgendered people are now somewhat available in larger cities with substantial gay/lesbian communities (Fountain et al. 2009).

Immigrants include gay men, lesbians, bisexuals, and transgendered people, but little research has been done on these particular segments of the

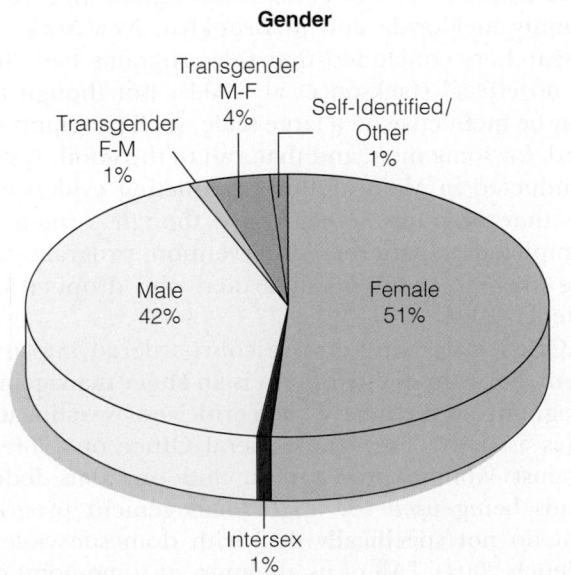

Gender

Transgender M-F 4%

Transgender F-M 1%

Self-Identified/ Other 1%

Male 42%

Female 51%

Intersex 1%

Figure 13.7 Gender Identity of Victims and Survivors. Data are from 2989 cases in which sexual orientation was known.

Source: Fountain et al. 2009.

immigrant population who, like heterosexual immigrant spouses, face a far different set of difficulties than native-born citizens when dealing with IPV. Sexual minority immigrants are particularly vulnerable in American society (because of immigration policies that favor heterosexuals), and tend to remain invisible to the broader society. Even still, according to research done by the National Coalition of Anti-Violence Programs, intimate partner violence in sexual minority immigrant communities is being reported at an increasing rate. "Striking increases [in domestic violence calls] were seen in a relatively new category of immigration, especially people with recent visas (1700%), refugees and asylees (900%), and people who are undocumented (250%)," particularly high are immigrants from East Africa (Fountain et al. 2009, pp. 24, 57).

The wide variety of research into the issues raised by domestic violence tells us is that "perpetrators and survivors of abuse, regardless of sexual orientation were more aggressive, hostile, more distressed, had more substance abuse problems, and reported being less satisfied with their relationships than those not in abusive relationships" (Brown 2008, p. 459).

Stopping Relationship Violence

The debate over whether women are as violent as men—or not—may be resolved at the practical level by noting that the interests of men as a group converge with those of women in curbing spousal violence—not only in hopes of having viable relationships but simply for survival. Even though men kill wives and girlfriends at a much higher rate than women kill husbands and boyfriends, some victimized women do murder or seriously injure their male partners. Progress in stopping intimate partner violence will benefit both sexes.

We have already discussed the shelter movement. Other approaches involve (1) counseling and group therapy directed toward abusive male partners (or the couple), and (2) the criminal justice system.

Counseling and Group Therapy Counseling and group therapy were earlier thought to be ineffective for male abusers. That may have been partly due to the inapplicability of general programs to this specific problem.

A number of male batterer intervention (or treatment) programs have now been developed. Many abusing husbands and male partners have difficulty controlling their response to anger and frustration, dealing with problems, and relinquishing their excessive control over their partner. Many men have a sincere desire to stop abusing their partners, even though abusers are difficult to reach and may drop out of treatment. However, data from therapists suggest that supportive and therapeutic-style group therapy reduces dropout rates because it promotes change in "abuse-supporting attitudes" in such a way as to reduce stigma and provide a setting in which abusers can learn more-constructive ways of both coping with anger and balancing autonomy and intimacy. In other words, supportive group therapy focuses on "psychological targets" such as emotion regulation (that is, anger management and stress tolerance), sobriety, and psychopathology (depression, anxiety, and so on), as opposed to "re-education" (Dutton 2006; Scott 2004; Straus and Scott, forthcoming, p. 34).

Although there is ongoing debate about the number of women who batter their spouses, the debate does not reduce the importance of treatment for these offenders. Research suggests that, like male abusers, women are at risk for re-offending unless provided appropriate and adequate therapy and counseling. Findings suggest that, like male abusers, female abusers must learn to regulate their emotions, deal with substance abuse issues, take responsibility for their behavior, and learn how to peacefully resolve familial conflicts (Carney and Buttell 2006; Carney, Buttell, and Dutton 2007; Dowd, Leisring, and Rosenbaum 2005).

Approximately 25 percent of male abusers engage in repeat intimate partner violence (Straus and Scott, forthcoming, p. 35). It is difficult to evaluate the level of success of male batterer intervention programs because of design problems, low response rates, and high program dropout rates and because programs do not always follow the research protocol (Jackson et al. 2003). Two recent and rigorously designed evaluations sponsored by the National Institute of Justice were conducted on intervention programs in Broward County in Florida and in Brooklyn, New York. The researchers concluded that the programs had "little or no effect" (Jackson et al. 2003). But though they may be ineffective on a large scale, such programs may work for some men, and that's all to the good. A study conducted in Maine found "conflicting evidence on whether programs were effective, though . . . men who completed a batterers' intervention program were less likely to re-offend than men who dropped out" (Hench 2004).

Often male batterers are court-ordered into treatment. Sometimes one option is an anger management program, but these have been criticized by victim advocates as ineffective. The Federal Office on Violence Against Women now agrees, and prohibits federal funds being used for anger management programs that do not specifically deal with domestic violence (Hench 2004). "All of us are angry at some point during our day; violence, on the other hand, is very different. It's an action," says one director of a program that offers both kinds of treatment to varied clients. Another says, "The issue regarding domestic violence is power and control. The offender is likely to beat or abuse the victim whether or not he or she is angry" (in Hench 2004).

In the past two decades, some couples' therapy programs have emerged to treat wife abuse. Typically, these programs counsel husbands and wives—or just husbands—separately over a period of up to six months. After this first treatment phase, couples are counseled together and are taught anger management techniques, along with communication, problem-solving, and conflict-resolution skills. Research has shown that spirituality is important to many people who have experienced IPV, and religious communities are strengthening their domestic violence counseling for couples in an effort to reduce incidences of re-battering (Ellison et al. 2007; Gillum, Sullivan, and Bybee 2006).

Couples' therapy programs designed to stop domestic violence are somewhat controversial because they proceed from the premise that a couple's staying together without violence after an abusive past is possible. Feminist scholars have expressed concern that therapists underestimate the danger that women face in violent relationships. There is some evidence that negative social sanctions from either partner's relatives or friends may help stop wife abuse.

Same-gender couple IPV also includes many of the components found in heterosexual couple violence counseling and therapy, but this poses some problems as well as solutions. For example, individual therapists may not take IPV as seriously and begin the couple in counseling (rather than suggesting police intervention and shelter services) where abuse can become even more pronounced (Helfrich and Simpson 2005; Kulkin et al. 2008).

The Criminal Justice Response There was little legal protection for battered women in the past. The street wisdom among police, as well as those who worked with battered women, was that calling the police was an ineffective strategy and posed some risk to the woman. Arresting an abusive partner or pressing charges would only aggravate the situation and result in escalating violence later. Officers also felt themselves to be at risk in responding to domestic violence calls.

Police officers typically avoided making arrests for assault that would be automatic if the man and woman involved were not married. The laws themselves contributed to police reluctance: Statutes might require a police officer to witness the act before making an arrest at the scene, or more severe injury might be required for prosecution for battery. In some cases, restraining orders required additional court action before they could be enforced.

A sociological experiment in Minneapolis in the 1980s obtained results indicating that mandatory arrest could be an effective deterrent to future violence (Sherman and Berk 1984),[6] although caution must be used when assigning causality to such laws because it is possible that the changes could also be the result of "increased efforts of the battered women's and shelter movement over the past several decades which have aided women in safely exiting violent relationships" (Leisenring 2008, p. 462). As a consequence of this experiment, laws have been changed to make arrests for domestic violence more feasible, and some states or jurisdictions have policies that mandate arrest in certain situations involving family violence.

Most subsequent replications of the arrest experiment did not get the same result. In fact, according to the National Institute of Justice, "policies and services designed to help victims of domestic violence appear to have two possible and opposing effects: either they decrease the abuse and risk of homicide, or they have the unintended consequence of increasing them." (Dugan, Nagin, and Rosenfeld 2004, p. 21; Leisenring 2008). It now appears that arrest will deter future violence only on the part of men who are employed and married, men with a "stake in conformity." Other men, those who are unemployed and/or not married to the woman they abused, may react to arrest by *increased* violence (Dugan, Nagin, and Rosenfeld 2004).

An even more serious problem of the arrest strategy has been that a literal reading of a mandatory arrest law has resulted in the arrest of victims, along with perpetrators, when the victim has resisted with violent force. Women also fear that reporting domestic violence to the police will risk contact with Child Protective Services and the removal of their children from the home. Some women who did contact police reported that the batterer was not arrested, as they had expected, and that the police sometimes trivialized their situation. In some cases, women claimed that the exchange between the perpetrator and the police officer was characterized by "male bonding," in which the perpetrator's story overrode the woman's complaint of violence.

Some women reported positive and protective experiences to researchers: "[S]o when the police did intervene that night, they made it pretty clear that I didn't deserve it (the abuse). . . . [T]hey talked to me and I filed a report . . . and that's the last I saw of my husband" (M. Wolf et al. 2003).

This seems a good time to remind ourselves of the good news that appears in recent reports. Both fatal

---

[6] In the Minneapolis experiment, officers were randomly assigned to respond by arresting the (presumably male) perpetrator, by counseling the parties, or simply by separating them for a cooling-off period. A six-month follow-up by telephone and an examination of police call records indicated that arrest was the most effective response in deterring subsequent violence (Sherman and Berk 1984).

and nonfatal violence against intimate partners has declined since 1993, and that may well be a consequence of the support and treatment programs we have described.

The drop in the male homicide rate is attributed to the greater availability of options for abused women. When women kill a partner, it is usually out of desperation to exit a violent relationship. The shelter programs and other resources that now exist have given battered women escape routes so that they are less likely to kill spouses or partners in an attempt to stop the violence (Dugan, Nagin, and Rosenfeld 2004; Leisenring 2008).

Shelter options, women's increased ability to support themselves, a cultural change that takes domestic violence seriously and endorses women's taking self-protective actions, and increased interest in and understanding of domestic violence on the part of law enforcement agencies are all developments that may account for the decrease in fatal and nonfatal violence against women by their intimate partners.

We turn now to another type of family violence in which the more powerful abuse the less powerful—child abuse.

Violence Against Children

Perceptions of what constitutes child abuse or neglect have differed throughout history and in various cultures.[7] Practices that we now consider abusive were accepted in the past as the normal exercise of parental rights or as appropriate discipline.

Even today, standards of acceptable child care vary according to culture and social class. What some groups consider mild abuse, others consider right and proper discipline. In 1974, however, Congress provided a legal definition of *child maltreatment* in the Child Abuse Prevention and Treatment Act. (The federal government and some researchers use the umbrella term *child maltreatment* to cover both abuse and neglect.)

The act defines child abuse and neglect as the "physical or mental injury, sexual abuse, or negligent

treatment of a child under the age of 18 by a person who is responsible for the child's welfare under circumstances that indicate that the child's health or welfare is harmed or threatened" (U.S. Department of Health, Education, and Welfare 1975, p. 3).

Child Abuse and Neglect People use the term **child abuse** to refer to overt acts of aggression—excessive verbal derogation (emotional child abuse) or physical child abuse such as beating, whipping, punching, kicking, hitting with a heavy object, burning or scalding, or threatening with or using a knife or gun. (By current American standards, spanking or hitting a child with a paddle, stick, or hairbrush is not "abuse," although it is in Sweden and several other countries [Straus and Donnelly 2001]; and see Chapter 10 of this text.) Data collected by the federal government between the years of 1992 and 2007 show a 52 percent decline in physical abuse. Fatalities of children from abuse and neglect, however, increased 15 percent from 2006 to 2007 (U.S. Department of Health and Human Services 2010, pp. 55, 94). The number of child fatalities due to neglect and abuse has been increasing in the past few years.

Child abuse is not specific to the United States. This pamphlet was produced by social service agencies in Hong Kong.

---

[7] A dramatic example of cultural difference in defining child abuse is the controversy surrounding *female genital mutilation (FGM)*. Some sub-Saharan African and Muslim cultures practice FGM, which is the surgical removal of the clitoris and other external female genital organs, and suturing of the vaginal opening until marriage. In those cultures, FGM is an important rite of passage for young girls and considered necessary to make them eligible to marry. (It does not seem to be a Muslim religious teaching, however.) FGM has been brought to the United States by some immigrants as part of their cultural heritage. It has been outlawed in the United States since 1996, and is now prohibited in some African countries. FGM is still practiced clandestinely here (Renteln 2004, pp. 51–53).

Some of the increase can be attributed to improvements in data collection.

Child Neglect In 2008, nearly 24 percent of investigations or assessments by child protective services agencies "determined at least one child to be a victim of abuse or neglect," with 70 percent of them suffering neglect (U.S. Department of Health and Human Services 2010, pp. xii, xiii). Child neglect includes acts of omission—failing to provide adequate physical or emotional care. Physically neglected children often show signs of malnutrition, lack immunization against childhood disease, lack proper clothing, attend school irregularly, and need medical attention for such conditions as poor eyesight or bad teeth. Often these conditions are grounded in parents' or guardians' economic problems, mental health issues, lack of parenting skills, a history of childhood abuse, and so on (Currie and Widom 2010; Daniel, Taylor, and Scott 2010). Research into the differences between child neglect and child abuse suggests that the largest cases of child neglect are found in children from birth to age three, and are associated with larger families with low incomes. Children twelve and older are the most frequent victims of physical and sexual abuse (Bundy-Fazioli, Winokur, and DeLong-Hamilton 2009; U.S. Department of Health and Human Services 2010, p. 47, Table 3-12).

Emotional child abuse or neglect involves a parent's often being overly harsh and critical, failing to provide guidance, or being uninterested in a child's needs. Emotional child abuse might also include allowing children to witness violence between parents—there were children in residence in 35.2 percent of households where intimate partner violence took place (Catalano 2007). Although emotional abuse may occur without physical abuse, physical abuse results in emotional abuse as well.

Child Sexual Abuse Another form of child abuse is **sexual abuse**: a child's being forced, tricked, or coerced, by an older person, into sexual behavior—exposure, unwanted kissing, fondling of sexual organs, intercourse, rape, incest, prostitution, and pornography—for purposes of sexual gratification or financial gain (Goldman et al. 2003; U.S. Department of Health and Human Services 2010, p. 115). Nine percent of abused children (of all ages) in a national sample survey were sexually abused in 2008 (U.S. Department of Health and Human Services 2010, p. 45, Table 3-10). Data collected between 1992 and 2007 show that child sexual abuse has declined 53 percent (U.S. Department of Health and Human Services 2010, p. 94).

Incest involves sexual relations between related individuals. The most common forms are sibling incest followed by father–daughter incest. The definition of child sexual abuse *excludes* mutually desired sex play between or among siblings close in age, but coerced sex by strong and/or older brothers is sexual abuse and is more widespread than parent–child incest (Bass et al. 2006; Carlson, Maciol, and Schneider 2006; Kiselica and Morrill-Richards 2007; Thompson 2009). Incest is the most emotionally charged form of sexual abuse; it is also the most difficult to detect. Incest appears to be in the background of a variety of sexual, emotional, and physical problems among adults who were abused as children (Brand and Alexander 2003; Carlson, Maciol, and Schneider 2006; Schlesinger 2006).

We see occasional media stories about female sex abusers, but research indicates that sexual abuse is almost entirely perpetrated by males (Peter 2009). Data on sexual exploitation indicate that 47 percent of sexual assaults on children were by relatives; 49 percent by others such as teachers, coaches, or neighbors; and only 4 percent by strangers (Hernandez 2001). Sexual abuse by paid caregivers and by mentors such as teachers, coaches, youth program directors, and clergy is a problem being addressed by policy makers and child care professionals. Sexual exploitation of homeless children is yet another abuse problem (Tyler and Cauce 2002).

Research by social psychologists finds lower self-esteem and greater incidences of depression among adults who have been victims of child abuse (Banyard et al. 2009; Sachs-Ericsson et al. 2010). A study of nearly 43,000 adolescents found that those who had been physically and/or sexually abused were more prone to binge drinking and thoughts of suicide. However, high levels of supportive interest and monitoring from at least one parent decreased the risk for these outcomes among sexually abused adolescents (Luster and Small 1997).

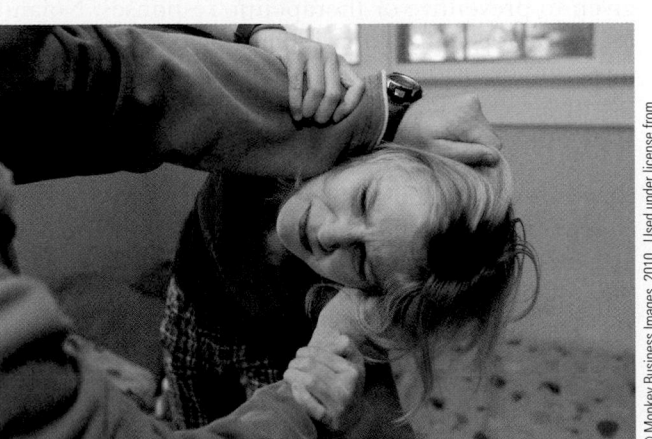

Sibling violence is not "kid stuff." This under-the-radar form of violence is rather frequent and can be quite injurious.

Sibling Violence Sibling violence is often overlooked and rarely studied (K. Butler 2006a; Finkelhor et al. 2005), even though the early National Family Violence Survey found it to be the most pervasive form of family violence (Straus, Gelles, and Steinmetz 1980).

Nor is it only of the "harmless" teasing variety ("UF Study" 2004). A national sample study found that 35 percent of children had been hit or attacked by siblings in the previous year. Fourteen percent were repeatedly attacked, 5 percent hard enough to have injuries such as bruises, cuts, chipped teeth, and sometimes broken bones. Two percent were hit with rocks, toys, broom handles, shovels, or knives (K. Butler 2006a).

Child psychologist John Caffaro (in K. Butler 2006a) sees sibling abuse as situational, not personality driven. When parents are frequently physically or emotionally absent from the home, or when they have their own problems, sibling violence is more apt to occur. Failure to intervene effectively also plays a part, as does parental favoritism of one child over another. Trauma, anxiety, and depression are likely to result from experiencing sibling violence, as well as an increased likelihood to perpetrate violence as an adult and to have relationship problems (K. Butler 2006a; Hoffman and Edwards 2004; Noland et al. 2004).

Perpetrators of sibling violence are more likely than others to become perpetrators of dating violence, according to a study of more than 500 men and women at a Florida community college. "Siblings learn violence as a form of sibling manipulation and control as they compete with each other for family resources. . . . They carry these bullying behaviors into dating, the next peer relationship in which they have an emotional investment" (researcher Virginia Noland in "UF Study" 2004; Noland et al. 2004). Yet sibling violence has received comparatively little research attention and even less attention has been given to preventive or therapeutic responses. Noland recommends that sibling violence be taken more seriously and that anger management programs be implemented while these violent individuals are still kids ("UF Study" 2004).

How Extensive is Child Abuse?

As noted earlier in this chapter, rates of child physical abuse and child sexual abuse declined in the 1990s, and in the early years of the twenty-first century (Finkelhor and Jones 2004, 2006; U.S. Department of Health and Human Services 2010).

Current estimates from the federal report *Child Maltreatment 2010* are based on state reports of child abuse. Of the reported cases of child maltreatment, 71.1 percent are of neglect; 16.1 percent, of physical abuse; 9.1 percent, of sexual abuse; 7.3 percent, of psychological mistreatment; and 2.2 percent, of medical neglect.

(The remainder are cases that include multiple factors or unspecified maltreatment.)[8] An estimated 1,740 children died from abuse or neglect in 2008, with nearly 80 percent under the age of four (U.S. Department of Health and Human Services 2010, pp. 27, 55, Figure 3-5).

Abused children live in families of all socioeconomic levels, races, nationalities, and religious groups, although child abuse is reported more frequently among poor and nonwhite families than among middle- and upper-class white families. Families below the poverty line have three times the rate of severe violence to children. Differences in rates may be partly due to differences in reporting—children of the poor are more apt to be seen in the emergency room or by social welfare authorities (Gelles and Cavanaugh 2005). Another reason may be unconscious racial discrimination on the part of physicians and others who report abuse and neglect (Lane et al. 2002). Experts believe there are also real differences, however (Gelles and Cavanaugh 2005), and a stress explanation is often offered.

In 2008, according to government data, 45.1 percent of the victims of child maltreatment were white, 21.9 percent were African American, 20.8 percent were Hispanic, 1.2 percent were American Indian/Alaska Native, and less than 1 percent were Asian. When the size of each racial/ethnic group is taken into account, it appears that African American, American Indian, and Pacific Islander children had the highest victimization rates; Hispanics and whites had moderate levels of victimization; and Asian American children had low rates of child maltreatment (U.S. Department of Health and Human Services 2010, pp. 42–43, Table 3-9).

The percentages of male (48.3 percent) and female (51.3 percent) victims were not very different. The youngest children (through age three) were more

---

[8] A controversy has arisen over accusations of child sexual and other abuse in the context of a child custody dispute. On the one hand, parents alleging child abuse (usually mothers) have been accused of fabricating the charge to gain an advantage in the custody determination (Childress 2006). On the other, those parents have claimed that their intent is to protect the child from real abuse by the other parent (usually fathers).

A study of over 9,000 contested divorces found that only 1 to 8 percent involved allegations of child abuse (Goldstein and Tyler 1998; McDonald 1998; citing Thoennes and Tjaden 1990). Half the allegations were considered "founded," that is, found to be true. In 33 percent of the cases, no abuse was found, and in 17 percent, it could not be determined if child abuse had occurred or not.

The key point about fraudulent reports is that only 14 percent were "deliberate false accusations" (Goldstein and Tyler 1998, p. 1). The remainder of the "unfounded" cases were sincerely made reports that were later found to be in error, typically due to misunderstandings of children's behavior or statements.

vulnerable than older children (U.S. Department of Health and Human Services 2010, p. 26). A child faces the greatest risk of becoming a victim of homicide during the first year of life (Collymore 2002).

Eighty-four percent of abused children were mistreated by at least one parent: by mother only in 38.3 percent of cases; by father only in 18 percent of cases; and by both mother and father in nearly 18 percent of cases. In 10 percent of cases, children were mistreated by other caregivers: foster parents or legal guardians, day care workers, or unmarried partners of a parent (U.S. Department of Health and Human Services 2010, pp. 28, 51, Table 3-15).

Abuse versus "Normal" Child Rearing It is too easy for parents to go beyond reasonable limits when angry or distraught or to include as "discipline" what most observers would define as abuse (Baumrind, Larzelere, and Cowan 2002; Feigelman et al. 2009). Hence, child abuse must be seen as a potential behavior in many families—even those we think of as "normal" (Feigelman et al. 2009).

Immigrant families may come from cultures where rather severe physical punishment is considered necessary for good child rearing. Those parents may not be aware that what they are doing by way of parental discipline is illegal in this country. They may instead view themselves as very responsible parents (Renteln 2004, pp. 54–57).[9]

Risk Factors for Child Abuse Consider the following society-wide beliefs and conditions that, when exaggerated, may encourage even well-intentioned parents to mistreat their children:

- A belief in physical punishment is a contributing (but not sufficient) factor in child abuse. Abusive parents have learned—probably in their own childhood—to view children as requiring physical punishment (Milner et al. 2010, p. 335).

- Parents may have unrealistic expectations about what the child is capable of; often, they lack awareness and knowledge of the child's physical and emotional needs and abilities (Letarte, Normandeau, and Allard 2010, p. 254). For example, slapping a bawling

toddler to stop her or his crying is completely unrealistic, as is too-early toilet training.

- Parents who abuse their children were often abused or neglected themselves as children. Violent parents are likely to have experienced and thereby learned violence as children. Whether victims of child abuse or witnesses of adult interpersonal violence, those exposed to family violence in childhood are more likely than others to abuse their own children, and their spouses as well (Heyman and Slep 2002; Milner et al. 2010; Zielinski 2009). This does not mean that abused children are predestined to be abusive parents or partners. Gelles and Cavanaugh (2005) report an intergenerational transmission rate of 30 percent. That is much higher than the general average of 2 to 4 percent; nevertheless, "[t]he most typical outcome for individuals exposed to violence in their families of origin is to be nonviolent in their adult families. This is the case for both men and women" (Heyman and Slep 2002, p. 870).

- Parental stress and feelings of helplessness play a significant part in child abuse (Milner 2010). "Economic adversity and worries about money pervade the typical violent home" (Gelles and Straus 1988, p. 85; Zielinski 2009). Overload, often related to family problems, also creates stress that may lead to child abuse. Other causes of parental stress are children's misbehavior, changing lifestyles and standards of living, and a parent's feeling pressure to do a good job but being perplexed about how to do it.

- Families have become more private and less dependent on kinship and neighborhood relationships. Hence, parents are alone with their children, shut off at home from the "watchful eyes and sharp tongues that regulate parent–child relations in other cultures" (Skolnick 1978, p. 82). In neighborhoods that have support systems and tight social networks of community-related friends—where other adults are somewhat involved in the activities of the family—child abuse and neglect are much more likely to be noticed and stopped (Falconer et al. 2008).

- Other circumstances that are statistically related to child maltreatment include parental youth and inexperience, marital discord and divorce, and unusually demanding or otherwise difficult children (Milner et al. 2010). Other risk factors involve parental abuse of alcohol or other substances (Milner et al. 2010); a mother's cohabiting with her boyfriend (who could potentially abuse the child; Crary 2007); and having a stepfather (because stepfathers are as likely as biological fathers to abuse their children; Crary 2007). Still, it is important to remember that 80 percent of people who commit child abuse

---

[9] Immigrant parents may be mistakenly identified as having abused children because of certain cultural practices not initially understood in this country. There are healing practices in certain cultures that can produce what looks like evidence of injuries to an American doctor or social service worker. Southeast Asians employ a practice known as "coining" whereby they rub the edge of a coin along the skin. This leaves marks that can appear to be those of a whip (Child Abuse Prevention Council of Sacramento n.d.). Similarly, Asian children may have "Mongolian spots" on their skin, a natural phenomenon, but one that appears as bruising to an unaware health practitioner (Families with Children from China 1999).

are the children's own biological parents, whereas cohabitants and stepparents make up just 4.4 percent each (U.S. Department of Health and Human Services 2010, p. 66).

Combating Child Abuse Three major approaches to combating child abuse and willful neglect are the punitive approach, which views abuse and neglect as crimes for which parents should be punished; the therapeutic approach, which views abuse as a family problem requiring treatment; and the social welfare approach, which looks to stress factors and the family's social context.

The Criminal Justice Approach Those who favor the punitive approach believe that one or both parents should be held legally responsible for abusing a child.

A complicated issue emerging with regard to this approach involves holding battered women criminally responsible for "failing to act" to prevent such abuse at the hands of their male partners. Feminist legal advocates have begun to question whether the law should hold a battered woman responsible for failing to prevent harm to her children when, as a battered woman, she cannot even defend herself: "When the law punishes a battered woman for failing to protect her child against a batterer, it may be punishing her for failing to do something she was incapable of doing. . . . She is then being punished for the crime of the person who has victimized her" (Erickson 1991, pp. 208–9). Legal experts note that fathers are typically *not* held accountable for child abuse committed by female partners (Liptak 2002).

The courts are starting to recognize this paradox. The Illinois Supreme Court overturned such a mother's conviction in 2002, and courts have ruled in favor of mothers who lost custody or had children removed from the home, citing the mothers' domestic violence victimization (Liptak 2002; *Nicholson v. Scopetta* 2004; Nordwall and Leavitt 2004).

The Therapeutic Approach All states have criminal laws against child abuse. But the approach to child protection has gradually shifted from punitive to therapeutic. Not all who work with abused children are happy with this shift. These critics prefer to hold one or both parents clearly responsible. They reject the family system approach to therapy because it implies distribution of responsibility for change to all family members (M. Stewart 1984). Nevertheless, social workers and clinicians—rather than the police and the court system—increasingly investigate and treat abusive or neglectful parents.

The therapeutic approach involves two interrelated strategies: (1) increasing parents' self-esteem and their knowledge about children, and (2) involving the community in child rearing (Goldstein, Keller, and Erne

1985). A typical voluntary program holds regular meetings to enhance self-esteem and educate abusive parents. Programs may attempt to reach stressed parents before they hurt their children, and many operate a twenty-four-hour hotline for parents under stress. High school classes on family life, child development, and parenting are now virtually universal and are thought to reduce child abuse by giving future parents an understanding of what they can expect from children at different ages.

Involving the community means getting people other than parents to help with child rearing. There are options such as *supplemental mothers,* who are available to babysit regularly with potentially abused children. Another community resource is the *crisis nursery,* where parents may take their children when they need to get away for a few hours. Ideally, crisis nurseries are open twenty-four hours a day and accept children at any hour without prearrangement.

One form of protection for abused or neglected children is to remove them from their parents' homes and place them in foster care. This practice is controversial, as foster parents have been abusive in some cases, and there are not enough foster parents to go around in many regions of the country. Moreover, removal from the home can be quite traumatic to children, who are often very attached to their parents despite the abuse (Kaufman 2006). They may blame themselves for the breaking up of the family (Gelles and Cavanaugh 2005).

An alternative is **family preservation**, whereby a Child Protective Services worker is able to "leave the child with an impoverished or troubled family and provide support in the form of housekeeping help or drug treatment, and then visit frequently to monitor progress" (Kaufman 2006, p. A12). The family preservation approach would not be appropriate if harm to the child appears imminent. Family preservation is a controversial strategy (Gelles 2005; Wexler, 2005), but both removal of the child from the home and a family preservation approach carry risk.[10]

The Social Welfare Approach The social welfare approach overlaps with the therapeutic approach but takes note of the social, cultural, and economic context of child maltreatment to provide services and parent education that may make child abuse less likely. Housing assistance and subsidized child care, for example, might prevent a low-income or socially isolated parent from taking the risk of leaving children alone while working.

---

[10] For detailed discussions of this controversy see articles by Richard Wexler (2005; "Family Preservation Is the Safest Way to Protect Most Children") and Richard Gelles (2005; "Protecting Children Is More Important Than Preserving Families") in Loseke, Gelles, and Cavanaugh (2005).

Parent education directed toward new immigrant parents might mitigate the development of situations that end in removal of children from the home. For example, if some immigrant families do not realize that their traditional disciplinary practices constitute criminal child abuse in the United States, parent education offered through refugee service centers could anticipate that problem. The same is true regarding leaving children alone at home. This may be customary and perfectly safe in a small tribal village but not so safe in the United States; moreover, it is illegal (Gonzalez and O'Connor 2002; Renteln 2004, pp. 54–58).

Commercial Sexual Exploitation of Children We close this section on child maltreatment with a look at a form of child abuse that is not, strictly speaking, family violence, but which is often set in motion by developments in seriously troubled families—that is the commercial sexual exploitation of children. Researchers Richard J. Estes and Neil Weiner of the University of Pennsylvania go beyond family violence per se to look at society-wide organized sexual exploitation of children. Their study was based on interviews with victims and child welfare workers in twenty-eight cities in the United States, Mexico, and Canada (Estes and Weiner 2002; Hernandez 2001; Memmott 2001). Based on this research, they estimate that as many as 300,000 to 400,000 children a year are molested or used in pornography or prostitution. The National Institute of Justice supports this estimate, noting as many hotline tips regarding child exploitation (National Institute of Justice 2007).

Family dynamics often place children in harm's way. Typically, victims of organized sexual exploitation are runaways, "throwaways" (children who have been kicked out of the home by parents), or other homeless children who trade or sell sex to meet their basic survival needs. Some sexually exploited children live at home but are offered for sexual purposes by their families in exchange for money, drugs, or other benefits (*Domestic Sex Trafficking of Minors* n.d.; Harris 2009).

We are much more aware of parents' abuse of children than we are of children's abuse of parents, but it does happen. Sometimes it is an outgrowth of earlier child abuse. We turn now to the topic of child-to-parent abuse.

Child-to-Parent Abuse

The discussion of **child-to-parent abuse** is brief because not much research has been done (Cottrell and Monk 2004). Yet, like other forms of family violence, child-to-parent abuse has been there all along.

This section relies heavily on a review article by Cottrell and Monk (2004). Data suggest that 9 to 14 percent of parents have been abused by adolescent children, with injuries that include bruises, cuts, and broken bones. Types of assaults have included kicking, punching, biting, and weapons. Mothers, especially single mothers, and elderly parents of youth are the most frequent victims.

Adolescent boys are the most frequent perpetrators, and their growth in size and strength is associated with increases in violence. Although there are no clear findings of differences in race/ethnicity or social class, poverty and other family stressors are associated with this form of violence.

Abusive children may exhibit diminished emotional attachments to parents. The child may have been abused by the parent or witnessed intimate partner abuse in the household. Overly permissive parents and those who abandon their authority in response to the violence tend to see more of it. Parents whose child-rearing styles contradict each other are also at risk. Drug use by the adolescent may play a role.

Parents who are victims of assaults by their adolescent children often engage in denial. Unfortunately, at the moment, few if any support services exist, and the criminal justice system has not responded systematically (Cottrell and Monk 2004).

Generally, we see any form of family violence as more likely to occur in situations of unequal rather than equal power. We close this chapter with a reminder of everyone's basic right to be respected—and not to be physically, emotionally, or sexually abused—in any relationship. And we end on an optimistic note, as most forms of family violence show evidence of declining rather than increasing.

Summary

- Power, the ability to exercise one's will, may rest on cultural authority, on economic and personal resources that are gender-based and/or involve love and emotional dependence, on interpersonal dynamics, or on physical violence.

- Marital power or power in other intimate partner relationships includes decision making, control over money, the division of household labor, and a sense of empowerment in the relationship. American marriages experience a tension between egalitarianism on the one hand and, on the other, gender identities that in effect preserve male authority.

- The relative power of a husband and wife within a marriage or other intimate partnership varies by education, social class, religion, race/ethnicity, age, immigration status, and other factors. It varies by whether or not the woman works and with the presence and age of children. Studies of married couples, cohabiting couples, and gay and lesbian couples illustrate the significance of economically based power and of norms about who should have power.

- Couples can consciously work toward more egalitarian marriages or intimate partner relationships and relinquish "power politics." Changing gender roles, as they affect marital and intimate relationship power, necessitate negotiation and communication.

- Physical violence is most commonly used in the absence of other resources.

- Researchers do not agree on whether intimate partner violence is primarily perpetrated by males or whether males and females are equally likely to abuse their partners. The effects of intimate partner violence indicate that victimization of women is the more crucial social problem, and it has received the most programmatic attention. Recently, some programs have been developed for male abusers. Studies suggesting that arrest is *sometimes* a deterrent to

further wife abuse illustrate the importance of public policies in this area.

- Economic hardships and other stress factors (among parents of all social classes and races) can lead to physical and/or emotional child abuse as can lack of understanding of children's developmental needs and abilities. One difficulty in eliminating child abuse is drawing a clear distinction between "normal" child rearing and abuse.

- Physical, verbal, and emotional abuse, as well as sexual abuse and child neglect, are forms of violence against children. Sibling violence is an often overlooked form of child abuse.

- Criminal justice, therapeutic, and social welfare approaches are ways of addressing the problem of child maltreatment.

- Child-to-parent abuse is a recently "discovered" form of family violence. It may grow out of previous abuse of a child.

Questions for Review and Reflection

1. How is gender related to power in marriage? How do you think ongoing social change will affect power in marriage?

2. Do you think that power in a marriage or other couple relationship depends on who earns how much money? Or does it depend on emotions? Is it possible for a couple to develop a no-power relationship?

3. Looking at domestic violence, why might women remain with the men who batter them? Do you think that shelters provide an adequate way out for these

women? What about arresting the abuser? Should intimate partner violence against men receive more attention in the form of social programs? Why or why not?

4. What factors might play a role when well-intentioned parents abuse their children?

5. **Policy Question.** What can we as a society do to combat child neglect that is really due to family poverty?

Key Terms

allocation systems 346
child abuse 368
child-to-parent abuse 373
coercive power 340
Conflict Tactics Scale 354
egalitarian norm (of marital power) 343
emotional child abuse or neglect 369
equality 340
equity 340
expert power 341
family preservation 372
gender model of marriage 349
incest 369

informational power 341
intimate partner power 340
intimate partner violence 354
intimate terrorism 363
legitimate power 341
marital power 340
marital rape 357
minority stress 365
mutually economically dependent spouses (MEDS) 349
near peers (Schwartz's typology) 350
neotraditional family 349
no-power 350
patriarchal norm (of marital power) 343

Online Resources

Sociology CourseMate

www.CengageBrain.com

Access an integrated eBook, chapter-specific interactive learning tools, including flash cards, quizzes, videos, and more in your Sociology CourseMate, accessed through CengageBrain.com.

www.CengageBrain.com

Want to maximize your online study time? Take this easy-to-use study system's diagnostic pre-test, and it will create a personalized study plan for you. By helping you identify the topics that you need to understand better and then directing you to valuable online resources, it can speed up your chapter review. CengageNOW even provides a post-test so you can confirm that you are ready for an exam.

14

Family Stress, Crisis, and Resilience

© Gideon Mendel/Corbis

Americans say they're stressed. In a national poll taken just before the onset of the recession that began in 2008, only 16 percent of eighteen- to twenty-nine-year-olds and 12 percent of thirty- to forty-nine-year-olds said that they rarely experience stress (Carroll 2007c). Since the recession's onset, Americans' stress levels have climbed (Elias 2009; Shugrue and Robison 2009). In a recent poll by the American Psychological Association, nearly half of U.S. adults reported that they had lain awake at night with worry at least once during the previous month (American Psychological Association 2009).

Children are worried as well. A poll by the national organization, KidsHealth, asked children ages nine to thirteen whether they worry and, if so, what about. Results showed that children as young as nine worry about the health of family members and getting into automobile accidents, among other concerns ("Kids-Poll" 2008). Unfortunately, but a sign of the times, some children as young as in kindergarten worry about their body image ("Body Image" 2008). Children with a deployed military parent worry about whether their parent will come home and when, or about whether the parent will be injured: "The worst time is when the phone rings because you don't know who is calling. They could be calling, telling you that he got shot or something" (in Huebner et al. 2007, p. 117).[1]

We can think of the family as continually balancing the demands put upon it against its capacity to meet those demands. This chapter addresses family stress, crisis, and resilience. We will review various theoretical perspectives on the family and discuss how these can be applied to family stress and crises. We'll discuss what precipitates family stress or crisis, and then look at how families define or interpret stressful situations and how their definitions affect the course of a family crisis.

In several places throughout this text, we point out that families are more likely to be happy when they work toward mutually supportive relationships—and

when they have the resources to do so. Nowhere does this become more apparent than in a discussion of how families manage stress and crises. To begin, we'll define the concepts of *stress*, *crisis*, and *resilience*.

Defining Family Stress, Crisis, and Resilience

As sociologist Pauline Boss (1997) reminds us,

> Perhaps the first thing to realize about stress is that it's not always a bad thing to have in families. In fact it can make family life exciting—being busy, working, playing hard, competing in contests, being involved in community activities, and even arguing when you don't agree with other family members. Stress means change. It is the force exerted on a family by demands. (p. 1)

Family stress is a state of tension that arises when demands test or tax a family's capabilities. Situations that we think of as good, as well as those that we think of as bad, are all capable of creating stress in our families. Moving to a different neighborhood, taking on a new job, and bringing a baby home might be examples of "good" situations that create family stress.

Family stress might be also be caused by more potentially harmful, ambiguous, or difficult situations such as finding adequate housing on a poverty budget, financing children's education on a middle-class income, being laid off in a recession economy, or losing one's home due to the recent mortgage crisis. A family member's injury or a death in the family is a source of family stress. Responding to the needs of aging parents is stressful for a family (see Chapter 17). Undergoing infertility treatments is a stressor. Living as a cancer survivor can be a stressor (Marshall 2010).

Family stress calls for family adjustment (Patterson 2002b). As an example of adjustment, more and more older parents are moving into the homes of their grown children in response to recession-related financial pressures (M. Alvarez 2009). When adjustments are not easy to come by, family stress can lead to a **family crisis**: "a situation in which the usual behavior patterns are ineffective and new ones are called for immediately" (National Ag Safety Database n.d., p. 1; Patterson 2002b). We can think of a family crisis as a sharper jolt to a family than more ordinary family stress. The definition of *crisis* encompasses three interrelated ideas:

1. Crises necessarily involve change.

2. A crisis is a turning point with the potential for positive effects, negative effects, or both.

3. A crisis is a time of relative instability.

Family crises are turning points that require some change in the way family members think and act to meet a new situation (Hansen and Hill 1964; McCubbin

---

[1] We can never completely shield our children from natural disasters, news of terrorism, or other dangers. And it's appropriate for children to feel anxious about parents and other relatives serving in the military, being hospitalized, imprisoned, addicted to drugs or alcohol, or going through anxiety-provoking situations. Various websites offer information addressed to children and stress. Among them are the following:

- American Academy of Pediatrics has developed a guide for pediatricians (www.aap.org/terrorism/index.html);

- Department of Defense Education Activity on school safety, children of military personnel, and other programs (www.dodea.edu);

- FEMA (Federal Emergency Management Agency), on preparing children for natural disasters and national security emergencies (www.ready.gov/kids);

- The KidsHealth organization has material designed for parents and children regarding things that children worry about and how parents can help (kidshealth.org);

- The National Child Traumatic Stress Network, sponsored by UCLA and Duke University, is another good resource (www.nctsnet.org).

and McCubbin 1991; Patterson 2002b). In the words of social worker and crisis researcher Ronald Pitzer:

> Crisis occurs when you or your family face an important problem or task that you cannot easily solve. A crisis consists of the problem and your reaction to it. It's a turning point for better or worse. Things will never be quite the same again. They may not necessarily be worse; perhaps they will be better, but they will definitely be different. (Pitzer 1997a, p. 1)

In part, what makes the difference between whether things get better depends on a family's level of resilience—the ability to recover from challenging situations. We return to the topic of resilience later in this chapter. Meanwhile, we note that social science theory gives insight into family stress, crisis, and resilience.

Theoretical Perspectives on Family Stress and Crises

We saw in Chapter 2 that there are various theoretical perspectives concerning marriages and families. Throughout this chapter, we will apply several of these theoretical perspectives to family stress and crises. Here we give a brief review of several theoretical perspectives that are typically used when examining family stress and crises.

You may recall that the *structure–functional* perspective views the family as a social institution that performs essential functions for society—raising children responsibly and providing economic and emotional security to family members. From this point of view, a family crisis threatens to disrupt the family's ability to perform these critical functions (Patterson 2002b).

The *family development,* or *family life course,* perspective sees a family as changing in predictable ways over time. This perspective typically analyzes **family transitions**—expected or *predictable* changes in the course of family life—as family stressors that can precipitate a family crisis (Carter and McGoldrick 1988). For example, having a first baby or sending the youngest child off to college taxes a family's resources and brings about significant changes in family relationships and expectations. Over the course of family living, people may form cohabiting relationships, marry, become parents, break up or divorce, remarry, and make transitions to retirement and widowhood or widowerhood. All these transitions are stressors (Cooper, McLanahan, et al. 2009).

In addition, the family development perspective focuses on the fact that predictable family transitions, such as an adult child's becoming financially independent, are expected to occur within an appropriate time period, even though the window of acceptable time has lengthened over that past several decades (Arnett 2004; Furstenberg et al. 2004). As discussed in Chapter 2, transitions that are "outside of expected time" create greater stress than those that are "on time" (Hagestad 1996; Rogers and Hogan 2003). Partly for this reason, teenage pregnancy is often a family stressor. Another example, explored in Chapter 10, involves a grandparent's filling the parent role. "A Closer Look at Family Diversity: Young Caregivers" provides a third example.

The *family ecology* perspective explores how a family influences and is influenced by the environments that surround it. From this point of view, many causes of

"Sergeant Michael Buyas, with sons Jaiden (left) and Justin, says that in his dreams he still has legs. In waking moments, he worries about how he'll teach his three boys wrestling, his favorite sport in high school. Michael's legs were blown off by an improvised explosive device just before Christmas 2004 in Iraq" (Ryan 2006). A crisis necessarily involves change. The *family ecology* perspective focuses on how factors external to the family, such as war in Iraq or Afghanistan, can result in family crisis. From a *family systems* perspective, all the members in this family must adapt to their father's injuries.

Chronicle photo by Michael Macor

We tend to think of caregivers as middle age or older, and a large majority of them are. But caregivers are diverse in age, some of them children. For young caregivers, the role involves additional stress inasmuch as it does not take place "on time." Here's what one thirty-something caregiver wrote in her blog:

> Being thrust into a caregiver role at a younger age, when my mom at the age of 57 had a debilitating stroke, I was faced with all the "common" caregiver challenges but at a time in my life when it was least expected and with absolutely no warning. I immediately left my career, my home, my friends to move back home (2,000 miles away) to do everything that was humanly and sometimes inhumanly possible to help my mom. . . . Being a caregiver, especially at a young age, is a huge sacrifice. I don't regret it, but sometimes I can't help but feel that I am missing out on some of the best years of my life.

> During my 20s, I mostly focused on my career. I was always a very driven person and while I had one or two serious relationships during that time, I was not ready to "settle down." In my mind, I felt like that's what my 30s would be for. Had I been able to predict the future, I would've married my college boyfriend and started having babies immediately. OK, maybe not, but the idea of it sure sounds

good now (laugh). So, here I am, one year into caregiving and I just started working again (my career had to be redefined too). . . .

> Meanwhile, my friends and acquaintances are getting married, having babies, buying houses, etc. Sometimes I feel like everyone is moving forward, and I am frozen in time. . . . I wonder if and when will I have the opportunity to fulfill my own hopes and dreams. As a young caregiver, and in my particular situation, this is my biggest challenge and fear. . . . So to all the young caregivers out there—whether you are caring for your spouse/significant other, a sibling, or a parent—You are not alone. (Caregiver Support Blog, caregiversupport.wordpress.com).

In addition to caregivers in their twenties and thirties, an estimated 1.5 million U.S. children under age eighteen serve as caregiver to a family member (Shifren 2009; "Young Caregivers" 2009). Experts expect the numbers to grow as chronically ill patients leave hospitals sooner and live longer, as the recession compels patients to forgo paid help, and as more returning veterans need home care (Belluck 2009).

Child and teen caregivers are often responsible for keeping the care recipient company. In addition, they shop, do household chores, and help with meal preparation. Some assist the care recipient with eating, getting in and out of

bed, getting dressed, taking a bath, or going to the bathroom. Some administer medications; help the care recipient communicate with doctors, nurses, or other medical professionals; make appointments; or arrange for others to help the care recipient (Hunt, Levine, and Naiditch 2005; see also Champion et al. 2009).

Caretaking can give purpose to a young person's life (Shapiro 2006a). Although some child caregivers do well, others grow depressed and/or angry as they sacrifice social and extracurricular activities. Some miss—or even quit—school (Belluck 2009). Policy makers urge further research on questions, such as how to improve support groups for young caregivers, how teachers and schools can assist them, and how educational, social, and career opportunities can be fostered within the context of caregiving (Hunt, Levine, and Naiditch 2005).

Support organizations for young caregivers include the American Association of Caregiving Youth (AACY), the National Alliance for Caregiving (NAC) (www.caregiving.org), the National Family Caregivers Association (NFCA) (www.thefamilycaregiver.org), the Family Caregiver Alliance (FCA) (www. caregiver.org), the Children of Aging Parents (CAPS) (www.caps4caregivers .org), and the Caregiving Youth Project, sponsored by AACY (Belluck 2009).

Critical Thinking

From a policy point of view, what might be done to assist young caregivers? What might your local community do to help?

---

family stress originate outside the family—in the family's neighborhood, workplace, and national or international environment (Boss 2002; Socha and Stamp 2009). Living in a violent neighborhood causes family stress and has potential for sparking family crises (Bertram and Dartt 2009). Conflict between work and family roles that is largely created by workplace demands is another example of an environmental factor that may cause family stress (Barnett et al. 2009; Bass et al.

2009). Natural disasters, such as tornadoes, hurricanes, or earthquakes, create family stress and crises (Sattler 2006; Taft et al. 2009). Moreover, as we'll see later in this chapter, our family's external environment offers or denies us resources for dealing with stressors.

The *family system* theoretical framework looks at the family as a system—like a computer system or an organic system, such as a living plant or the human body. In a system, each component or part influences all the other parts. When one family member changes a role, all the family members must adapt and change as well. As an example, when a family member becomes addicted to alcohol or other mind-altering drugs, the entire family system is affected (El-Sheikh and Flanagan 2001; Foster and Brooks-Gunn 2009).

Finally, exploring the discussions, gestures, and actions that go on in families, the *interactionist perspective* views families as shaping family traditions and family members' self-concepts and identities. By interacting with one another, family members struggle to create shared family meanings that define stressful or potentially stressful situations—for example, as good or bad, disaster or challenge, someone's fault or no one's fault. As we will explore later in this chapter, "a family's shared meanings about the demands they are experiencing can render them more or less vulnerable in how they respond" (Patterson 2002b, p. 355).

What Precipitates a Family Crisis?

Demands put upon a family cause stress and sometimes precipitate a family crisis. Social scientists call such demands **stressors**—a precipitating event or events that create stress. Stressors vary in both kind and degree, and their nature is one factor that affects how a family responds. In general, stressors are less difficult to cope with when they are expected, are brief, and gradually improve over time.

Types of Stressors

There are several types of stressors, as Figure 14.1 shows. We will briefly examine nine of them here.

1. Addition of a Family Member Adding a member to the family—for example, through birth, adoption (Bird, Peterson, and Miller 2002), marriage, remarriage, or the onset of cohabitation—is a stressor. You may recall the discussion on why the transition to parenthood is stressful in Chapter 10. The addition of adult family members may bring people who are very different from one another in values and life experience into intimate social contact. Furthermore, not only are in-laws (and increasingly stepparents, step-grandparents, and step-siblings) added through marriage or cohabitation but also a whole array of their kin come into the family. Then too, having an adult child return home or having a member of one's extended family move into the household because of financial problems are stressors.

Moreover, like any system, a family has boundaries. Family members need to know "who is in and who is outside the family" (Boss 1997, p. 4). Adding a family member is stressful because doing so involves family boundary changes; that is, family boundaries have to shift to include or "make room for" new people or to adapt to the loss of a family member (Boss 1980). This situation applies to the addition of a cohabiting partner to the family, as well as his or her departure from the family (Cherlin 2009a).

2. Loss of a Family Member The death of a family member is, of course, a stressor. Family systems theory reminds us that children as well as adults grieve the absence of a family member, and their grief needs to be addressed (Boss 1980; Monroe and Kraus 2010).

Meanwhile, the likelihood of death in a society can influence how people define a death in the family. For instance, under the mortality conditions that existed in this country in 1900, half of all families with three children could expect to have one die before reaching age fifteen. Social historians have argued that parents defined the loss of a child as almost natural or predictable and, consequently, may have suffered less emotionally than do parents today (Wells 1985, pp. 1–2). Family members who lose a child of any age today do so "outside of expected time," a situation that exacerbates, or adds to, their grief. The long-term effects of grieving such a loss may negatively affect a couple's intimacy (Gottlieb, Lang, and Amsel 1996).

Loss of potential children through miscarriage or stillbirth has the possible added strain of family disorientation. Attachment to the fetus may vary substantially so that the loss may be grieved greatly or little. Add to

| Addition of a family member | Loss of a family member | Ambiguous loss | Sudden, unexpected change | Ongoing family conflict | Caring for a dependent, ill, or disabled family member | Demoralizing event | Daily family hassles | Anxieties about children in a culture of fear |

Figure 14.1 Types of stressors

that the generally minimal display of bereavement customary in the United States and the omission of funerals or support rituals for perinatal (birth process) loss, and "all these ambiguities mean that a family may have to cope with sharply different feelings among family members . . . [and] the family as a whole may have to cope with the fact that they as a family have a very different reaction to loss than do the people around them" (Rosenblatt and Burns 1986, p. 238).

In addition to permanent loss, the temporary loss of a family member, such as through an older sibling's going away to college or a parent's leaving for long periods due to work demands, is a stressor. Temporary losses that not only create change in family structure but also introduce fear of the unknown are a form of ambiguous loss (Huebner et al. 2007; Whealin and Pivar 2006).

3. Ambiguous Loss

The loss of a family member is ambiguous when it is uncertain whether the family member is "really" gone (Boss 2007):

> Ambiguous loss is a loss that remains unclear. . . . [U]ncertainty or a lack of information about the where-abouts or status of a loved one as absent or present, as dead or alive, is traumatizing for most individuals, couples, and families. The ambiguity freezes the grief process and prevents cognition, thus blocking coping and decision-making processes. Closure is impossible. (Boss 2007, p. 105)

Having a family member who has been called to war or who is missing in action are situations of ambiguous loss (Pittman, Kerpelman, and McFadyen 2004; Boss 2007).

In addition, a family member may be physically present but psychologically absent, as in the case of family members with alcoholism or mental illness, those suffering from Alzheimer's disease or who have experienced brain injury, or children with cognitive impairment or severe disabilities (Blieszner et al. 2007; Roper and Jackson 2007). The ambiguity of post-separation or postdivorce family boundaries can be stressful. A nonresident father whose relationship with the child's mother—and hence with his future child—is uncertain can experience ambiguous loss (Leite 2007).

From the family systems perspective, ambiguous loss is uniquely difficult to deal with because it creates family **boundary ambiguity** (see Figure 14.2)—"confused perceptions about who is in or out of a particular family" (Boss 2004, p. 553; Carroll, Olson, and Buckmiller 2007):

> With a clear-cut loss, there is more clarity—a death certificate, mourning rituals, and the opportunity to honor and dispose [of the] remains. With ambiguous loss, none of these markers exists. The clarity needed for boundary maintenance (in the sociological sense) or closure (in the psychological sense) is unattainable. . . . [P]arenting roles are ignored, decisions are put on hold, daily tasks are undone, family members are ignored or cut off, and

rituals and celebrations are canceled even though they are the glue of family life. (Boss 2004, p. 553)

4. Sudden, Unexpected Change

A sudden, unexpected change in the family's income or social status may also be a stressor. Having a child run away is an example (Cohen 2008). Sudden job loss is another example. Natural disasters, mentioned earlier, cause sudden change. Most people think of stressors as being negative, and some sudden changes are. But positive changes, such as winning the lottery (don't you wish?) or getting a significant promotion, can cause stress too.

5. Ongoing Family Conflict

Ongoing, unresolved conflict among family members is a stressor (Hammen, Brennan, and Shih 2004). Deciding how children should be disciplined can bring to the surface divisive differences over parenting roles, for example. The role of an adult child living with parents is often unclear and can be a source of unresolved conflict. Watching an adult grandchild go through family conflict can be a stressor for a grandparent. If children of teenagers or of divorced adult children are involved, the situation becomes even more challenging (Hall and Cummings 1997).

6. Caring for a Dependent, Ill, or Disabled Family Member

Caring for a dependent or disabled family member is a stressor (Berge and Holm 2007; Patterson 2002a). An example involves being responsible for an adult child or a sibling with mental illness and/or physical or developmental disabilities (Fields 2010; Levine 2009; Lowe and Cohen 2010). Due mainly to advancing medical technology, the number of dependent people and the severity of their disabilities have steadily increased over recent decades. For instance, more babies today survive low birth weight and birth defects.[2] Also, more people now survive serious accidents, and many seriously injured soldiers in Iraq and Afghanistan have survived. These family members may require ongoing care and medical attention. "Issues for Thought: Caring for Patients at Home—A Family Stressor" discusses how recent technological advances, coupled with the goal of containing medical care costs, have created new stressors for families who are increasingly expected to care for very ill patients at home.

In addition, parents may be raising children with chronic physical conditions, such as asthma, diabetes, epilepsy, or autism (O'Brien 2007; Rao and Beidel 2009). Families may need to see their children through bone marrow, kidney, or liver transplants, sometimes requiring several months' residence at a medical

---

[2] Caring for a disabled child can be stressful enough for parents that they decide not to have another. Analysis of national data found that mothers of firstborn children with a disability were statistically less likely to have a second child (MacInnes 2008).

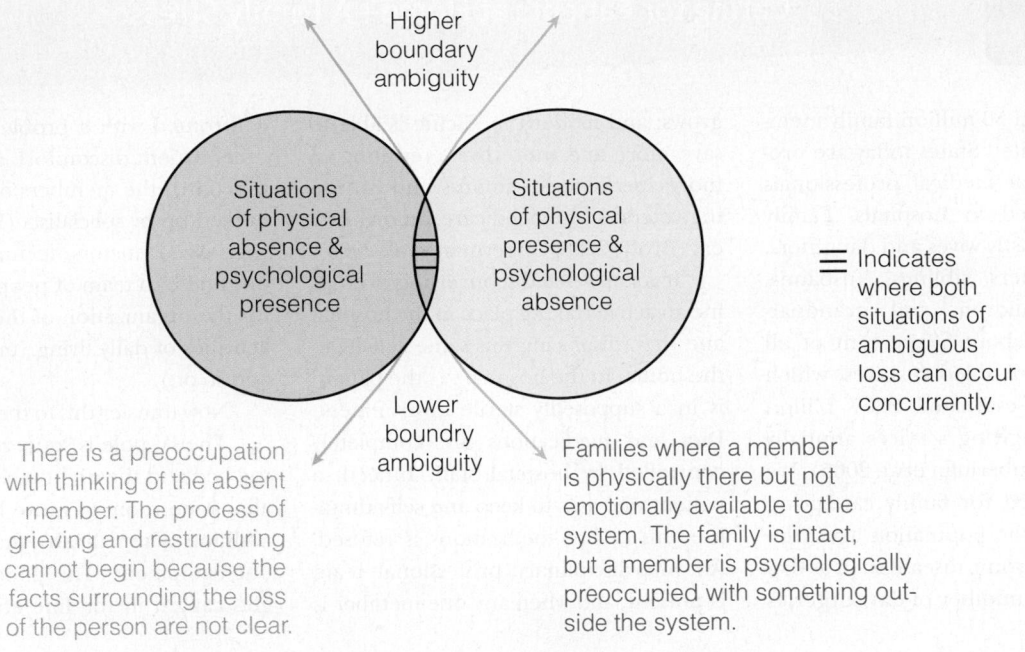

Higher
boundary
ambiguity

Situations
of physical
absence &
psychological
presence

Situations
of physical
presence &
psychological
absence

≡ Indicates
where both
situations of
ambiguous
loss can occur
concurrently.

Lower
boundry
ambiguity

There is a preoccupation
with thinking of the absent
member. The process of
grieving and restructuring
cannot begin because the
facts surrounding the loss
of the person are not clear.

Families where a member
is physically there but not
emotionally available to the
system. The family is intact,
but a member is psychologically
preoccupied with something outside the system.

Catastrophic and unexpected situations

- war (missing soldiers)
- natural disaster (missing persons)
- kidnapping, hostage-taking, terrorism
- incarceration
- desertion, mysterious disappearance
- missing body (murder, plane crash, etc.)

- Alzheimer's disease and other dementias
- chronic mental illness
- addictions (alcohol, drugs, gambling, etc.)
- traumatic head injury, brain injury
- coma, unconsciousness

More common situations

- divorce
- military deployment
- young adults leaving home
- elderly mate moving to a nursing home

- preoccupation with work
- obsession with computer
  games, Internet, TV

Figure 14.2 High boundary ambiguity—two forms: (1) a family member's physical absence coupled with psychological presence, and (2) a family member's physical presence coupled with psychological absence. "Sometimes a family experiences an event or situation that makes it difficult—or even impossible—for them to determine precisely who is in their family system" (Boss 1997, pp. 2–3).

Sources: Adapted from Boss 1997, pp. 2–3; and Boss 2004, p. 555. Used by permission of Blackwell Publishing.

center away from home (LoBiondo-Wood, Williams, and McGhee 2004). Adults with advanced AIDS may return home to be taken care of by family members. A family's caring for a terminally ill member can be a stressor for young children, who may exhibit behavior problems as a response, as well as for the adults in the household (Seltzer and Heller 1997). Chapter 17

addresses the issue of the sandwich generation of caring for one's own children as well as for aging parents.

7. Demoralizing Events Stressors may be demoralizing events—those that signal some loss of family morale. Demoralization can accompany the stressors already described (see, for example, Early, Gregoire, and

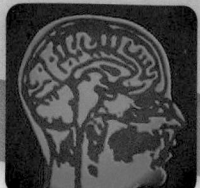
Between 20 and 50 million family members in the United States today are providing care that medical professionals once performed in hospitals. Family caregivers—mostly wives and daughters, but also partners, siblings, husbands, sons, grandchildren, and grandparents—provide about 80 percent of all care for ill or disabled relatives, which represents an estimated $237 billion in unpaid caregiving services annually (Brody 2008; Guberman et al. 2005). We can expect need for family caregiving to increase as the population ages; the incidence of chronic disease such as diabetes rises; the number of day surgeries grows; and modern medicine is able to save more and more lives, resulting in more special-needs infants and returning veterans who need care, among others (Brody 2008; Guberman et al. 2005).

"It is somewhat disconcerting to imagine an activity taking place in the hospital and then displacing this same activity to the home. In the hospital . . . the patient is in a supposedly sterile environment. Diet and medications are completely controlled by hospital staff. Indeed, a patient who asks to keep and self-administer his or her medications is refused. An interdisciplinary professional team is present, and when any one member is confronted with a problem (leaking IV tube, patient discomfort, apparatus malfunction), the members of the team are backed up by specialists (IV technicians, specialized doctors, technicians, and so on) and by a team of people responsible for the organization of the instrumental activities of daily living (meals, toileting, and so on). . . .

"Now transfer this to the home setting. . . . The IV pole is squeezed in between the bed and the night table, and there is almost no room to move because of the addition of a small table that is used to lay out equipment. The patient frequently gets caught in the line and loosens the

McDonald 2002). But, among other things, this category also includes poverty, homelessness, having one's child placed in foster care, juvenile delinquency or criminal prosecution, scandal, family violence, mental illness, incarceration, or suicide (Burgess 2008; Grekin, Brennan, and Hammen 2005; McNamara 2008; Taft et al. 2009; Wadsworth and Berger 2006). Being the brunt of racist treatment is potentially demoralizing (Murry et al. 2001). Grandparents' raising grandchildren is a situation that is often—although not always—associated with demoralizing events (Henderson et al. 2009).

Physical, mental, or emotional illnesses or disorders can be demoralizing. Alzheimer's disease or brain injury, in which a beloved family member seems to have become a different person, can be heartbreaking. In military personnel who have served during wartime, post-traumatic stress disorder (PTSD) can be demoralizing, causing "family members [to] feel hurt, alienated, or discouraged, and then become angry or distant toward the partner" ("PTSD and Relationships" 2006). Some illnesses can be especially demoralizing when they are associated with the possibility of being socially stigmatized. Being homeless, learning disabilities, HIV/AIDS, attention deficit/hyperactivity disorder (ADHD), anorexia nervosa, and bulimia are examples (Burgess 2008; Lerner and Johns 2009; Odom 2009; McNamara 2008; Rumney 2009; Shannon 2009).

© Tom Stewart/CORBIS

Sometimes a situation may be classified as more than one type of stressor. Due to advancing medical technology, for instance, more newborns today survive low birth weight or birth defects but may need ongoing remedial attention. Therefore, adding a baby to the family may also mean caring for a medically fragile child.

8. Daily Family Hassles Daily family hassles are stressors. Examples involve balancing work against family demands, working odd hours, being regularly stuck in traffic on long commutes to work, or arranging child care or transportation (Evans and

catheter in his or her vein. When it starts bleeding at the site, the home care nurse has already come and gone. What to do? The caregiver makes an adjustment. He or she is abused for hurting the patient, but the IV starts to flow again and the bleeding stops. There is another dispute between patient and caregiver concerning hygiene around the IV. What does keeping a sterile area mean? Can the dog sit on the bed? Does the caregiver have to wear gloves? Are these questions important enough to disturb medical personnel for answers? Who should be called—the hospital, the home care nurse, or the 24-hour medical-information line? . . .

"[P]erhaps the most unsettling aspect of the transfer of care responsibilities to patients and their families is the anxiety and insecurity of assuming this care without sufficient supervision and emergency backup. In the hospital, you have an emergency call button if something goes wrong. But what replaces this button when you are being cared for at home? Indeed, the home is psychologically, and sometimes physically, very far from immediate help in the case of an emergency or an unforeseen development. [In this study, the] majority of patients and caregivers assuming complex care felt alone and abandoned, causing high

levels of stress and anguish and conflicts within couples and within families. . . .

"Based on our study, we raise serious questions about the legitimacy of the transfer of high-tech care to the family."

Critical Thinking

Can you apply the family ecology theoretical perspective to this situation? What are some creative ways that a family might deal with high-tech caregiving at home? In what ways might community activism play a part in addressing this situation?

Source: Largely Excerpted from Guberman et al. 2005, pp. 247–72; also Brody 2008).

Wachs 2010). Another example involves protecting children from danger, especially in neighborhoods characterized by violence (Bertram and Dartt 2009; Foster and Brooks-Gunn 2009). A recent study found that, compared to married mothers, single mothers were more likely to feel greater stress and feelings of inadequacy, evidenced by migraines or chronic back pain, when experiencing daily hassles associated with their children's allergies or frequent colds (Ontai et al. 2008). "Facts about Families: ADHD, Stigma, and Stress" explores this point further.

Some scholars have investigated everyday stressors that are unique to certain professions. For instance, especially in recent years, military families "are subjected to unique stressors, such as repeated relocations that often include international sites, frequent separations of service members from families, and subsequent reorganizations of family life during reunions" (Drummet, Coleman, and Cable 2003, p. 279; see also Bowen et al. 2003). And families of Protestant clergy experience not only the stressors of ministry demands but also family criticism and situations in which members of the congregation "intrusively assume that the minister will fulfill their expectations without due consideration of the minister's priorities" (Lee and Iverson-Gilbert 2003, p. 251).

9. Anxieties about Children in a "Culture of Fear" A final stressor involves living in a situation of chronic anxiety with regard to children's safety. Increased media portrayal of various dangers seems to have led to a general "culture of fear" (Glassner 1999), which makes "anxiety about children . . . a central matter in twentieth-century American culture" (Fass 2003a; Stearns 2003). High-profile kidnappings and school shootings certainly

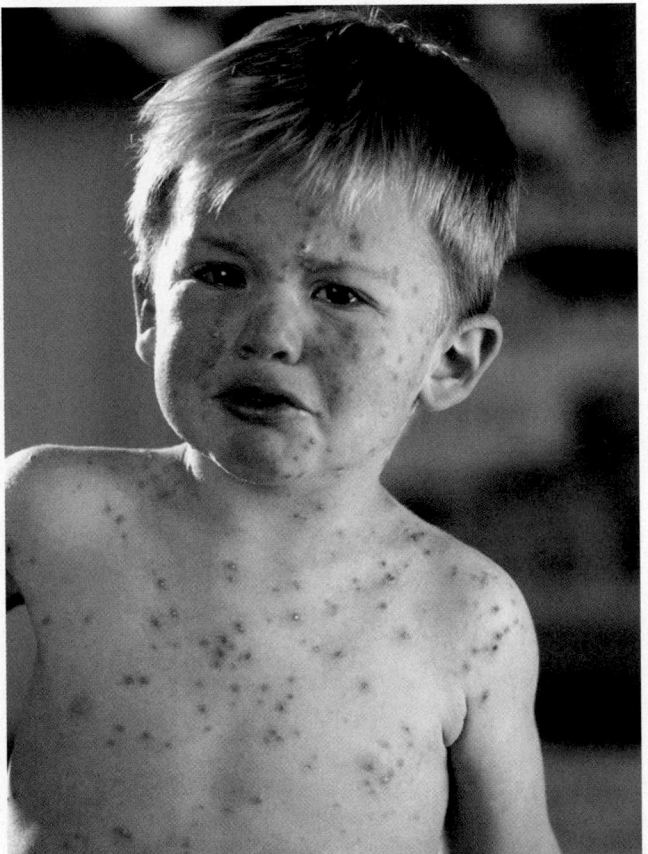

© SW Productions/Getty Images

Daily family hassles, such as a child's coming down with chicken pox, put demands on a family. Sometimes everyday hassles pile up to result in what social scientists call "stressor overload." This is especially true when a new stressor is added to already difficult daily family life.

According to estimates by the American Psychiatric Association, attention deficit/hyperactivity disorder (ADHD) has been diagnosed in 3 to 7 percent of U.S. school children, more often in boys than in girls (Firmin and Phillips 2009). ADHD can also be an adult diagnosis (Retz and Klein 2010).

Families with children diagnosed with ADHD face ongoing stressors that can be understood as daily hassles, some severe and demoralizing. Parents report that, due to the child's behavior, they often feel interrupted; miss social events because they are hesitant to leave their child with a babysitter; are anxious about taking the child out in public; must deal with other parents', neighbors', teachers', and/or school bus drivers' complaints; spend excessive amounts of time with the child's homework; worry that the child will get into trouble or be injured; have difficulty finding adequate after-school placement for their child and are unable to find or afford other professional or school services for the child; lose patience due to especially trying morning routines; must address siblings' resentment of parents' extra time and attention spent with the child diagnosed with ADHD; miss work; face lack of sleep due to disrupted bedtimes; and do not have enough time for themselves (Firmin and Phillips 2009; Reader, Stewart, and Johnson 2009).

Raising a child diagnosed with ADHD can involve feeling embarrassed as a result of specific instances of misbehavior and also as a consequence of being stigmatized. Stigmatizing others involves prejudice or discrimination based on others' perceived negative characteristics, status, or behaviors (Goffman 1963). Courtesy stigma refers to a situation in which not only the initially stigmatized individual but also her or his intimates are stigmatized by association (Goffman 1963; Koro-Ljungberg and Bussing 2009).

Focus groups (see Chapter 2) with thirty parents of children diagnosed with ADHD revealed that the parents often received unsolicited advice and felt negatively judged by both extended family members and strangers. Indeed, public debate over whether the diagnosis itself is legitimate or simply a convenient label for bad behavior increases the possibility of stigma (Koro-Ljungberg and Bussing 2009).

Some parents informed others of the diagnosis to ward off potential criticism of their child's behavior. Parents who were able to resist or shrug off negativity from neighbors, extended kin, or community were better able to cope. Some parents mentioned spirituality as a resource. As one mother said, "I just leave it in God's hand because the only thing I can do is just pray for him. Just pray and ask God to shield and protect him" (in Koro-Ljungberg and Bussing 2009, p. 1,192).

For many parents, however, management of courtesy stigma may involve withdrawal on the one hand, coupled with activism on the other hand. The majority of the parents in the focus group research managed stigma mainly by avoiding potentially stressful situations. They kept the diagnosis to themselves, interacting primarily or only with families of children who demonstrated behaviors similar to their child's. Many admitted doing homework and school projects for their children to reduce the possibility of being stigmatized by their child's teacher or by other parents (Koro-Ljungberg and Bussing 2009).

Parents also engaged in activism. Some volunteered at school to advocate for their child. They pressed for special school services for children diagnosed with ADHD, and they politicized ADHD by demanding more public education and increased awareness that could help reduce the stigma associated with ADHD (Koro-Ljungberg and Bussing 2009).

A variation of the STEPP (Strategies to Enhance Positive Parenting) program has been developed specifically to address parenting children diagnosed with ADHD (Chacko et al. 2008). Although sponsored by the pharmaceutical company Shire, the website ADHDactionguide.com includes nonpharmaceutically based tips for managing adult ADHD (see also Retz and Klein 2010).

Critical Thinking

Do you know an adult, child, or parents of a child who has been diagnosed with ADHD? Could you have added to their feelings of being stigmatized? If you are parenting a child diagnosed with ADHD, have you experienced courtesy stigma? If so, how have you handled it? How might you handle it in the future?

---

inspire worry, but parental fear may exceed the reality of the risk. Misrepresented by the media as high and/or rising, many perceived threats to children are statistically low or have actually declined. In 2008, the U.S. murder rate was at its lowest level since the 1960s (Von Drehle 2010). Crimes against children at school have declined in the past several years (Dinkes, Kemp, and Baum 2009). And although one wouldn't know it by watching televised news, the odds that a stranger will kidnap a child are extremely low.[3]

[3] We don't mean to imply that kidnappings, crimes at school, or other feared threats to children never occur. A realistic analysis of dangers, collecting information on strategies appropriate to living safely in our neighborhoods, a plan for talking with children about protection from actual risks, and a "check it out" attitude toward frightening media stories are good parental approaches.

As you read this section, you may have noted that sometimes a single event can be classified as more than one type of stressor or combine stressor types. Adopting a child with special needs often involves both adding a family member and caring for a disabled child (Schweiger and O'Brien 2005). As another example, raising a child with emotional or behavior problems adds to everyday family hassles, can be demoralizing, and may precipitate family conflict (Chacko et al. 2008; Talan 2009).

The September 11, 2001, attack on New York City and Washington, DC, was an event that can be classified as a sudden change in our family environments—a change that, among other things, sparked parents' need to consider how to talk with their children about terrorism (Myers-Walls 2002; Walsh 2002). For many, the attack was a demoralizing event. For others, the event was not only sudden and demoralizing but also sadly marked the loss of one or more family members. Many who had a family member in the Twin Towers on September 11 experienced ambiguous loss as they searched for missing relatives (Boss 2004).

Stressor Overload

A family may be stressed not just by one serious, chronic problem but also by a series of large or small, related or unrelated stressors that build on one another too rapidly for the family members to cope effectively (McCubbin, Thompson, and McCubbin 1996). This situation is called **stressor overload**, or **pileup**:

> Even small events, not enough by themselves to cause any real stress, can take a toll when they come one after another. First an unplanned pregnancy, then a move, then a financial problem that results in having to borrow several thousand dollars, then the big row with the new neighbors over keeping the dog tied up, and finally little Jimmy breaking his arm in a bicycle accident, all in three months, finally becomes too much. (Broderick 1979b, p. 352)

Characteristically, stressor overload creeps up on people without their realizing it. Even though it may be difficult to point to any single precipitating factor, an unrelenting series of relatively small stressors can add up to a crisis. In today's economy, characterized by longer working hours, two-paycheck marriages, fewer high-paying jobs, fewer benefits, and little job security, stressor overload may be more common than in the past. A second example of stressor overload is the addition of depression to an earlier stressor, such as chronic poverty or an adolescent family member's living with epilepsy (Seaton and Taylor 2003). A third example might involve the ambiguous loss of a family member deployed in Afghanistan or Iraq, followed by the stressors associated with the family member's return to the family, possibly compounded by the soldier's

serious physical injuries and/or post-traumatic stress disorder (England 2009; Hoge 2010; Johnson 2010). We'll return to the idea of stressor pileup shortly. Now, however, with an understanding of the various kinds of events that cause family stress and can precipitate a family crisis, we turn to a discussion of the course of a family crisis.

The Course of a Family Crisis

Family stress "is simply pressure put on the family"; in a family crisis, there is an "imbalance between pressure and supports" (Boss 1997, p. 1). A family crisis ordinarily follows a fairly predictable course, similar to the truncated roller coaster shown in Figure 14.3. Three distinct phases can be identified: the event that causes the crisis, the period of disorganization that follows, and the reorganizing or recovery phase after the family reaches a low point. Families have a certain level of organization before a crisis; that is, they function at a certain level of effectiveness—higher for some families, lower for others. Families that are having difficulties or functioning less than effectively before the onset of additional stressors or demands are said to be **vulnerable families**; families capable of "doing well in the face of adversity" are called **resilient families** (Patterson 2002b, p. 350).

In the period of disorganization following the crisis, family functioning declines from its initial level. Families reorganize, and after the reorganization is complete, (1) they may function at about the same level as before; (2) they may have been so weakened by the crisis that they function only at a reduced level—more

Resilient families do well in the face of adversity. Greater financial resources are advantageous in coping with family stress and crises, but low income families are often creatively resilient in locating resources.

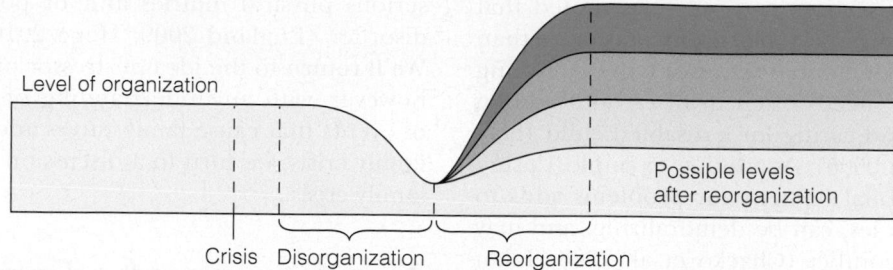

Figure 14.3 Patterns of family adaptation to crisis
Source: Adapted from Hansen and Hill 1964, p. 810.

often the case with vulnerable families; or (3) they may have been stimulated by the crisis to reorganize in a way that makes them more effective—a characteristic of resilient families.

At the onset of a crisis, it may seem that no adjustment is required at all. A family may be confused by a member's alcoholism or numbed by the new or sudden stress and, in a process of denial, go about their business as if the event had not occurred. Gradually, however, the family begins to assimilate the reality of the crisis and to appraise the situation. Then the **period of family disorganization** sets in.

The Period of Disorganization

At this time, family organization slumps, habitual roles and routines become nebulous and confused, and members carry out their responsibilities with less enthusiasm. Although not always, this period of disorganization may be "so severe that the family structure collapses and is immobilized for a time. The family can no longer function. For a time no one goes to work; no one cooks or even wants to eat; and no one performs the usual family tasks" (Boss 1997, p. 1). Typically, and legitimately, family members may begin to feel angry and resentful.

Expressive relationships within the family change, some growing stronger and more supportive perhaps, and others more distant. Sexual activity, one of the most sensitive aspects of a relationship, often changes sharply and may temporarily cease. Parent–child relations may also change. As one example, a child described life with his mother during the period of disorganization after his father was deployed:

> I could tell my mom was getting like really depressed and since she wouldn't talk, I wouldn't talk. And so around the house everyone was just kind of depressed for a little while and you could tell because they didn't speak a lot. (in Huebner et al. 2007, p. 117)

Relations between family members and their outside friends, as well as the extended kin network, may also change during this phase. Some families withdraw from all outside activities until the crisis is over; as a result, they may become more private or isolated than before the crisis began. As we shall see, withdrawing from friends and kin often weakens rather than strengthens a family's ability to meet a crisis.

At the **nadir**, or low point, of family disorganization, conflicts may develop over how the situation should be handled. For example, in families with a seriously ill member, the healthy members are likely either to overestimate or to underestimate the sick person's incapacitation and, accordingly, to act either more sympathetically or less tolerantly than the ill member wants (Conner 2000; Pyke and Bengston 1996). Reaching the optimal balance between nurturance and encouragement of the ill person's self-sufficiency may take time, sensitivity, and judgment.

During the period of disorganization, family members face the decision of whether to express or to smother any angry feelings they may have. Expressing anger as blame will usually sharpen hostilities; laying blame on a family member for the difficulties being faced will not help to solve the problem and will only make things worse (Stratton 2003). At the same time, when family members opt to repress their anger, they risk allowing it to smolder, thus creating tension and increasingly strained relations. How members cope with conflict at this point will greatly influence the family's overall level of recovery.

Recovery

Once the crisis hits bottom, things often begin to improve. Either by trial and error or by thoughtful planning, family members usually arrive at new routines and reciprocal expectations. They are able to look past the time of crisis to envision a return to some state of normalcy and to reach some agreements about the future. Some families do not recover intact, as today's high divorce and separation rates illustrate. Divorce, as well as the separation of a cohabiting or same-sex relationship, can be seen both as an adjustment to family crisis and as a family crisis in itself (see Figure 14.4).

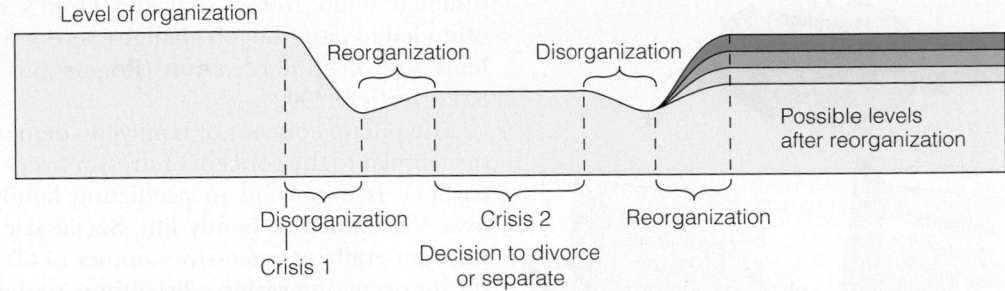

Figure 14.4 Divorce or separation as a family adjustment to crisis and as a crisis in itself

Other families stay together, although at lower levels of organization or mutual support than before the crisis. As Figure 14.3 shows, some families remain at a very low level of recovery, with members continuing to interact much as they did at the low point of disorganization. This interaction often involves a series of circles in which one member is viewed as deliberately causing the trouble and the others blame that individual and nag him or her to stop. This is true of many families in which one member is alcoholic or is otherwise chemically dependent, an overeater, or a chronic gambler, for example. Rather than directly expressing anger about being blamed and nagged, the offending member persists in the unwanted behavior.

Some families match the level of organization they had maintained before the onset of the crisis, whereas others rise to levels above what they experienced before the crisis (M. McCubbin 1995). For example, a family member's attempted suicide might motivate all family members to reexamine their relationships.

Reorganization at higher levels of mutual support may also result from less dramatic crises. For instance, partners in midlife might view boredom with their relationship as a challenge and revise their lifestyle to add some zest—by traveling more or planning to spend more time together rather than in activities with the whole family, for example.

Now that we have examined the course of family crises, we will turn our attention to a theoretical model specifically designed to explain family stress, crisis, adjustment, and adaptation.

Family Stress, Crisis, Adjustment, and Adaptation: A Theoretical Model

Some decades ago, sociologist Reuben Hill proposed the ABC-X family crisis model, and much of what we've already noted about stressors is based on the research of Hill, his colleagues, and his successors (R. Hill 1958; Hansen and Hill 1964). The **ABC-X model** states that **A** (the stressor event) interacting with **B** (the family's ability to cope with a crisis; their crisis-meeting resources) interacting with **C** (the family's appraisal of the stressor event) produces **X** (the crisis) (see Sussman, Steinmetz, and Peterson 1999). In Figure 14.5, *A* would be the demands put upon a family, *B* would be the family's capabilities—resources and coping behaviors—and *C* would be the meanings that the family creates to explain the demands.

As Figure 14.5 illustrates, families continuously balance the demands put upon them against their capabilities to meet those demands. When demands become heavy, families engage their resources to meet them while also appraising their situation—that is, they create meanings to explain and address their demands. When demands outweigh resources, family adjustment is in jeopardy, and a family crisis may develop. Through the course of a family crisis, some level of adaptation occurs (Patterson 2002b).

Stressor Pileup

Building on the ABC-X model, Hamilton McCubbin and Joan Patterson (1983) advanced the *double* ABC-X model to better describe family adjustment to crises. In Hill's original model, the *A* factor was the stressor event; in the double ABC-X model, *A* becomes *Aa*, or "family pileup." *Pileup* includes not just the stressor but also previously existing family strains and future hardships induced by the stressor event.

When a family experiences a new stressor, prior strains that may have gone unnoticed—or been barely managed—come to the fore. Prior strains might be any residual family tensions that linger from unresolved stressors or are inherent in ongoing family roles, such as being a single parent or a partner in a two-career family. For example, ongoing but ignored family conflict may intensify when parents or stepparents must deal with a child who is underachieving in school, has joined a criminal gang, or is abusing drugs. As another example, financial and time constraints typical of single-parent

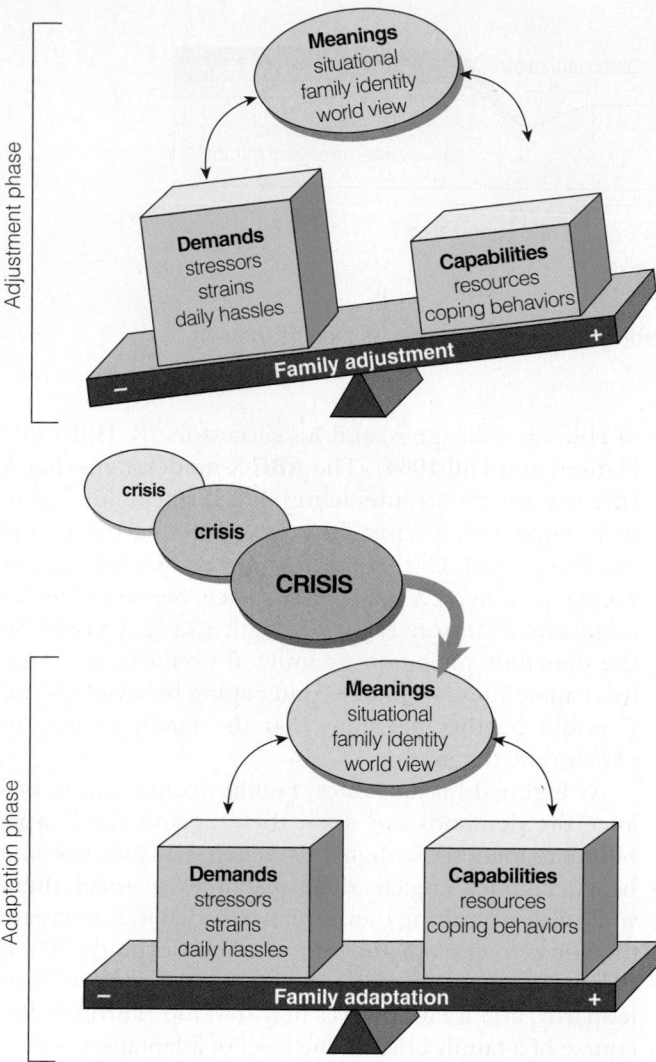

Figure 14.5 Family Stress, Crisis, Adjustment, and Adaptation. Families continuously balance the demands put upon them against their capabilities to meet those demands. When demands become heavy, families engage their resources to meet them while also appraising their situation—that is, they create meanings to explain and address their demands. When demands outweigh resources, family adjustment is in jeopardy, and a family crisis may develop. Through the course of a family crisis, some level of adaptation occurs (Patterson 2002b).

Source: From Patterson 2002b, p. 351. Reprinted by permission of Blackwell Publishing.

Note: Previously adapted from J. M. Patterson 1988, pp. 202–37. Copyright © 1988 by Families, Systems & Health, Inc.

families may assume crisis-inducing importance with the addition of a stressor, such as caring for an injured child.

An example of future demands precipitated by the stressor event would be a parent's losing a job, a stressor followed by unpaid bills. A study about parenting a

disabled child found that the child's rehabilitation often led to parental job changes, severe financial problems, and sleep deprivation (Rogers and Hogan 2003; S. Porterfield 2002).

The pileup concept of family-life demands, or stressors (similar to the concept of stressor overload described earlier), is important in predicting family adjustment over the course of family life. Social scientists believe that, generally, an excessive number of life changes and strains occurring within a brief time, perhaps a year, are more likely to disrupt a family.[4] Put another way, pileup renders a family more vulnerable to emerging from a crisis at a lower level of effectiveness (McCubbin and McCubbin 1989).

We have examined various characteristics of stressor demands put upon a family. Next we will look at how the family makes meaning of, defines, or appraises those demands. We'll look at crisis-meeting resources and coping behaviors after that.

Appraising the Situation

From an interactionist perspective, the meaning that a family gives to a situation—how family members appraise, define, or interpret a crisis-precipitating event—can have as much or more to do with the family's ability to cope as does the character of the event itself (McCubbin and McCubbin 1991; Patterson 2002a, 2002b). For example, a study of families faced with caring for an aging family member found that some families felt more ambivalent or negative about having to provide care than did others, who saw caregiving as one more chance to bring the family together (Pyke and Bengston 1996; see also Roscoe et al. 2009).

Several factors influence how family members define a stressful situation.[5] One is the nature of the stressor itself.

---

[4] The Holmes–Rahe Life Stress Inventory/Scale, developed in 1967, provides a way to measure an individual adult's or child's stress level. Many of the scale items, such as death of a spouse, death of a parent, divorce, divorce of parents, marital separation, death of a close family member, marriage, marital reconciliation, change in health of a family member, pregnancy, fathering a pregnancy out of marriage, gaining a new family member, a child leaving home, and trouble with in-laws, are actually family stressors and result not only in individual stress but also in family stress or crisis.

The authors rank various life events according to how difficult they are to cope with. For instance, death of a spouse, the most stressful life event on the adult scale, is equivalent to 100 *life change units.* Divorce is equivalent to 73 life change units, while trouble with in-laws is equivalent to 29. For children, divorce of parents is equivalent to 77 life change units, while hospitalization of a parent is equivalent to 55 life change units and loss of a job by a parent is equivalent to 46. You can find the Holmes–Rahe Stress Inventory Scale on the Internet.

[5] Although we are discussing the family's definition of the situation, it is important to remember the possibility that each family member experiences a stressful event in a unique way: "These unique meanings may enable family members to work together toward crisis resolution or they may prevent resolution from being achieved. That is, an

For instance, sometimes in the case of ambiguous loss, families do not know whether a missing family member will ever return or whether a chronically ill or a chemically dependent family member will ever recover. Being in limbo this way is very difficult.

In addition to the nature of the stressor itself, a second factor is the degree of hardship or the kind of problems the stressor creates. Temporary unemployment at age seventeen is less a hardship than is a layoff at age fifty-five when there are few job prospects. Being victimized by a crime is always a stressful event, but coming home to find one's house burglarized may be less traumatic than being robbed at gunpoint. Caregiving for a short time after a family member's surgery differs from the long-term caregiving of indefinite length associated with serious chronic illness.

A third factor is the family's previous successful experience with crises, particularly those of a similar nature. If family members have had experience in nursing a sick member back to health, they will feel less bewildered and more capable of handling a new, similar situation. Believing from the start that demands are surmountable, and that the family has the ability to cope collectively, may make adjustment somewhat easier (Pitzer 1997b; Wells, Widmer, and McCoy 2004). Family members' interpretations of a crisis event shape their responses in subsequent stages of the crisis. Meanwhile, the family's crisis-meeting resources affect its appraisal of the situation.

A fourth, related factor that influences a family's appraisal of a stressor involves the adult family members' legacies from their childhoods (Carter and McGoldrick 1988).[6] For example, growing up in a family that tended to define anything that went wrong as a catastrophe or a "punishment" from God might lead the family to define the current stressor more negatively. On the other hand, growing up in a family that tended to define demands simply as problems to be solved or as challenges might mean defining the current stressor more positively.

A positive outlook, spiritual values, supportive communication, adaptability, public services, and informal social support—all these, along with an extended family and community resources, are factors in family resilience, or meeting a crisis creatively.

Crisis-Meeting Resources

A family's crisis-meeting capabilities—resources and coping behaviors—constitute its ability to prevent a stressor from creating severe disharmony or disruption. We might categorize a family's crisis-meeting resources into three types: personal/individual, family, and community.

The personal resources of each family member (for example, intelligence, problem-solving skills, and physical and emotional health) are important. At the same time, the family *as family* or family system has a level of resources, including bonds of trust, appreciation, and support (family harmony); sound finances and financial management and health practices; positive communication patterns; healthy leisure activities; and overall satisfaction with the family and quality of life (Boss 2002; Patterson 2002b).

Family rituals are resources (Boss 2004; Oswald and Masciadrelli 2008). A study of families with alcoholism found that adult children of alcoholics who came from families that had maintained family dinner and other rituals (or who married into families that did) were less likely to become alcoholics themselves (Bennett, Wolin, and Reiss 1988; Goleman 1992).

And, of course, *money* is a family resource. For instance, a breadwinner's losing his or her job is less difficult to deal with when the family has substantial savings. In a qualitative study among U.S. working-poor rural families, one respondent explained that "I had absolutely nothing after I paid my bills to feed my kids. I scrounged just so that they could eat something, and I had to shortchange my landlord so that I could feed

individual's response to a stressor may enhance or impede the family's progress toward common goals, may embellish or reduce family cohesion, may encourage or interfere with collective efficacy" (Walker 1985, pp. 832–33).

[6] In their model of family stress and crisis, social workers Betty Carter and Monica McGoldrick (1988) see "family patterns, myths, secrets [and] legacies" as *vertical stressors*—because they come down from the previous generations. These authors call the type of stressors that we have been discussing in this chapter *horizontal stressors* (p. 9).

them, too, which put me behind in rent." Another said, "I felt overwhelmed and stressed because every time I get paid, I just don't have money for everything . . . because I have two . . . children [with medical problems]" (Dolan, Braun, and Murphy 2003, p. F14). At the other end of the financial spectrum are families who can afford to send troubled adolescents to costly "wilderness camps," for example, or other residential treatment facilities for illegal drug use or otherwise negative behaviors. Parents have told evaluation researchers that facilities such as these help to abate or relieve a family crisis and also to stabilize the family (Harper 2009).

The family ecology perspective alerts us to the fact that *community resources* are consequential as well (Socha and Stamp 2009). Increasingly aware of this, medical and family practice professionals have in recent years designed a wide variety of community-based programs to help families adapt to medically related family demands, such as a partner's cancer or a child's diabetes, congenital heart disease, and other illnesses (Peck 2001; Marshall 2010; Tak and McCubbin 2002). In fact, in many instances, family members have become community activists, working to create community resources to aid them in dealing with a particular family stressor or crisis. Parents have been "a driving force" in shaping services and laws related to individuals with mental challenges (Lustig 1999). As a second example, parent groups and adults with disabilities worked together to help pass the Americans with Disabilities Act (Bryan 2010; Turnbull and Turnbull 1997).

Vulnerable versus Resilient Families Ultimately, the family either successfully adapts or becomes exhausted and vulnerable to continuing crises. Family systems may be high or low in vulnerability, a situation that affects how positively the family faces demands; this enables us to predict or explain the family's poor or good adjustment to stressor events (Patterson 2002b).

More prone to poor adjustment from crisis-provoking events, *vulnerable families* evidence a lower sense of common purpose and feel less in control of what happens to them. They may cope with problems by showing diminished respect or understanding for one another. Vulnerable families are also less experienced in shifting responsibilities among family members and are more resistant to compromise. There is little emphasis on family routines or predictable time together (McCubbin and McCubbin 1991).

From a social psychological point of view, *resilient families* tend to emphasize mutual acceptance, respect, and shared values. Family members rely on one another for support. Generally accepting difficulties, they work together to solve problems with members, feeling that they have input into major decisions (McCubbin et al. 2001). It may be apparent that these behaviors are less

difficult to foster when a family has sufficient economic resources. The next section discusses factors that help families to meet crises creatively.

Meeting Crises Creatively

Meeting crises creatively means that after reaching the nadir in the course of the crisis, the family rises to a level of reorganization and emotional support that is equal to or higher than that which preceded the crisis. For some families—for example, those experiencing the crisis of domestic violence—breaking up may be the most beneficial (and perhaps the only workable) way to reorganize. Other families stay together and find ways to meet crises effectively. What factors differentiate resilient families that reorganize creatively from those that do not?

A Positive Outlook

In times of crisis, family members make many choices, one of the most significant of which is whether to blame one member for the hardship. Casting blame, even when it is deserved, is less productive than viewing the crisis primarily as a challenge (Stratton 2003).

Put another way, the more that family members can strive to maintain a positive outlook, the more it helps a person or a family to meet a crisis constructively (Burns 2010; Thomason 2005). Electing to work toward developing more open, supportive family communication—especially in times of conflict—also helps individuals and families meet crises constructively (Stinnett, Hilliard, and Stinnett 2000). Families that meet a crisis with an accepting attitude, focusing on the positive aspects of their lives, do better than those that feel they have been singled out for misfortune (Burns 2010). For example, many chronic illnesses have downward trajectories, so both partners may realistically expect that the ill mate's health will only grow worse (Marshall 2010). Some couples are remarkably able to adjust to this, "either because of immense closeness to each other or because they are grateful for what little life and relationship remains" (Strauss and Glaser 1975, p. 64).

Spiritual Values and Support Groups

"Spirituality, however the family defines it, can be a strong comfort during crisis" (Thomason 2005, p. F11). Some authors have argued that strong religious faith is related to high family cohesiveness (Lepper 2009) and helps people manage demands or crises, partly because it provides a positive way of looking at suffering (Wiley, Warren, and Montanelli 2002). A spiritual outlook may be fostered in many ways, including through Buddhist, Christian, Muslim, and other religious or philosophical traditions. However, a sense of spirituality—that is, a conviction that there is some power or entity greater than oneself—need not be associated with membership in any organized religion. Self-help groups, such as Alcoholics Anonymous or Al-Anon for families of alcoholics, incorporate a "higher power" and can help people take a positive, spiritual approach to family crises.

Open, Supportive Communication

Families whose members interact openly and supportively meet crises more creatively (Olson and Gorall 2009; Orthner, Jones-Sanpei, and Williamson 2004). For one thing, free-flowing communication opens the way to understanding (Thomason 2005). As an example, research shows that expressions of support from parents help children to cope with daily stress (Valiente et al. 2004). As another example, the better-adjusted husbands with multiple sclerosis believed that even though they were embarrassed when they fell in public or were incontinent, they could freely discuss these situations with their families and feel confident that their families understood (Power 1979). And as a final example, talking openly and supportively with an elderly parent who is dying about what that parent wants—in terms of medical treatment, hospice, and burial—can help (Fein 1997).

Knowing how to indicate the specific kind of support that one needs is important at stressful times. For example, differentiating between—and knowing how to request—just listening as opposed to problem-solving discussion can help reduce misunderstandings among family members—and between family members and others as well (Stinnett, Hilliard, and Stinnett 2000; Tannen 1990). Families whose communication is characterized by a sense of humor, as well as a sense of family history, togetherness, and common values, evidence greater resilience in the face of stress or crisis (Thomason 2005).

Adaptability

Adaptable families are better able to respond effectively to crises (Boss 2002; Uruk, Sayger, and Cogdal 2007). And families are more adaptable when they are more democratic and when conjugal power is fairly egalitarian. In families in which one member wields authoritarian power, the whole family suffers if the authoritarian leader does not make effective decisions during a crisis—and allows no one else to move into a position of leadership (McCubbin and McCubbin 1994). A partner who feels comfortable only as the family leader may resent his or her loss of power, and this resentment may continue to cause problems when the crisis is over.

Family adaptability in aspects other than leadership is also important (Burr, Klein, and McCubbin 1995). Families that can adapt their schedules and use of space, their family activities and rituals, and their connections with the outside world to the limitations and possibilities posed by the crisis will cope more effectively than families that are committed to preserving sameness. For example, a study of mothers of children with developmental disabilities found that mothers who worked part-time had less stress than those who worked full-time or were not employed at all (Gottlieb 1997). As another example, a study of married parents caring for a disabled adult child found that when their division of labor was adapted to feel fair, both parents experienced greater marital satisfaction and less stress (Essex and Hong 2005).

Informal Social Support

It's easier to cope with crises when a person doesn't feel alone (see, for example, Bowen et al. 2003; Tak and McCubbin 2002; Wickersham 2008). In fact, polls show that time spent with others is necessary to individuals' emotional well-being (Harter and Arora 2008). Families may find helpful support in times of crisis from kin, good friends, neighbors, and even acquaintances such as work colleagues (Johnson 2010). Analysis of data from the National Survey of Black Americans found that many of them in times of crisis received support from fellow church members (Taylor, Lincoln, and Chatters 2005). These various relationships provide a wide array of help—from lending money in financial emergencies to helping with child care to just being there for emotional support. Research on families in poverty shows that, although the informal social support that they receive rarely helps to lift them out of poverty, it does help them to cope with their economic circumstances (Henly, Danziger, and Offer 2005).

Even continued contact with more casual acquaintances may be helpful, as they often offer useful information, along with enhancing one's sense of community (Orthner, Jones-Sanpei, and Williamson 2004). And, of course, the Internet offers information and support for many, many stressors (Gilkey, Carey, and Wade 2009). A qualitative study that recruited participants by means of Web pages asked the seventy-seven respondents who answered an Internet-based survey about the advantages and disadvantages of Internet support, compared with face-to-face social networks (Colvin et al. 2004). Respondents mentioned two main Internet advantages—anonymity and the ease of connecting with others in the same situation despite geographical distance. Disadvantages related to lack of physical contact: "No one can hold your hand or give you a Kleenex when the tears are flowing" (p. 53).

© Image 100/Jupiter Images

Many—although not all—turn to their extended family for social support in times of stress. Kin may provide emotional support, monetary support, and practical help.

An Extended Family

Sibling relationships and other kin networks can be a valuable source of support in times of crisis (Ryan, Kalil, and Leininger 2009). Grandparents, aunts, uncles, or other relatives may help with health crises or with more common family stressors, like running errands or helping with child care (Milardo 2005). Families going through divorce often fall back on relatives for practical help and financial assistance. In other crises, kin provide a shoulder to lean on—someone who may be asked for help without causing embarrassment—which can make a crucial difference in a family's ability to recover.

Although extended families as residential groupings represent a small proportion of family households, kin ties remain salient (Furstenberg 2005). One aspect of all this that is beginning to get more research attention involves reciprocal friendship and support among adult siblings (White and Riedmann 1992; Kluger 2006; Spitze and Trent 2006). In times of family stress or crisis, new immigrants (as well as African Americans) may rely on **fictive kin**—relationships based not on blood or marriage but rather on "close friendship ties that replicate many of the rights and obligations usually associated with family ties" (Ebaugh and Curry 2000).

We need to be cautious, though, not to overestimate or romanticize the extended family as a resource. For instance, a study that compared mothers who had children with more than one father found that the women received less support from their kin networks than did single mothers who did not have multipartnered births. The researchers concluded that "smaller and denser kin networks seem to be superior to broader but weaker kin ties in terms of perceived instrumental support" (Harknett and Knab 2007). Among the poor, extended kin may not have the resources to offer much practical help (Henly, Danziger, and Offer 2005).

Moreover, along with some previous research, a small study of low-income families living in two trailer parks along the mid-Atlantic coast concluded that "low-income families do not share housing and other resources within a flexible and fluctuating network of extended and fictive kin as regularly as previously assumed." Extended family members may not get along, or individuals may be too embarrassed to ask their kin for help. One woman explained that neither her parents nor any one of her five siblings could help her because "they all have problems of their own." A Hispanic mother told the interviewer, "I know you've probably heard that Hispanic families are close-knit, well, hmmph! No, we take care of ourselves" (Edwards 2004, p. 523). Then, too, among some recent immigrant groups, such as Asians or Hispanics, expectations of the extended family may clash with the more individualistic values of Americanized family members.

Community Resources

In 2008, Nebraska became the last state in the United States to adopt a safe-haven law whereby parents can abandon their children at hospitals without fear of prosecution. Intended as a way to save unwanted newborns from being murdered or left in dumpsters or motel rooms, Nebraska's safe-haven law failed to limit the ages of children who could be legally abandoned. During the month after the legislation passed, more than thirty youngsters were left at Nebraska hospitals. Most of them were older than eleven. Some had extremely severe mental and behavioral problems (Hansen and Spenser 2009). Some had been transported to Nebraska by overwhelmed parents from outside the state. A month later, Nebraska amended its law to require that legally abandoned children had to be younger than thirty days old (Italie 2008; Jenkins 2008). The story became fuel for jokes on late-night television and afternoon talk shows.

But someone other than Maury Povich needs to be paying attention to this. . . . Something is wrong when so many parents are so eager to abandon so many children. . . . Because what happened in Nebraska constitutes a message from overstressed parents, one we ignore at our own peril. It is not a complicated message. On the contrary it is as simple and succinct as a word: Help. (Pitts 2008)

Upon calling a special session of the Nebraska legislature to address this issue, more than one legislator indicated that the state would have to examine the accessibility of social services for older children and their families (Eckholm 2008). Subsequent reviews showed that in some cases the state had failed to help desperate parents who did not receive necessary services for their children until after the children had been dropped off. In other cases, the parents themselves did not seem to know where to turn—although appropriate services were available—until they heard about the safe-haven law (Hansen and Spenser 2009).

The success with which families meet the demands placed upon them depends upon the availability of community resources, coupled with families' knowledge of and ability to access the community resources available to help (Odom 2009; Trask et al. 2005).

Community-based resources are defined as all of those characteristics, competencies and means of persons, groups and institutions outside the family which the family may call upon, access, and use to meet their demands. This includes a whole range of services, such as medical and health care services. The services of other institutions in the family's . . . environment, such as schools, churches, employers, etc.[,] are also resources to the family. At the more macro level, government policies that enhance and support families can be viewed as community resources. (McCubbin and McCubbin 1991, p. 19, boldface added)

Among others, community resources include schools and school personnel; social workers and family welfare agencies; foster child care; church programs that provide food, clothing, or shelter to poor or homeless families; twelve-step and other support programs for substance abusers and their families; programs for crime or abuse victims and their families; support groups for people with serious diseases such as cancer or AIDS, for parents and other relatives of disabled or terminally ill children, or for caregivers of disabled family members or those with cancer or Alzheimer's disease; and community pregnancy prevention and/or parent education programs. An Oregon study of non-Hispanic white and Hispanic teen mothers found that a government-funded home-visitation program increased family functioning, especially for the Hispanics in this sample (Middlemiss and McGuigan 2005).

A unique example of parent education programs, mandated by the U.S. government in 1995, involves federal prison inmates. Parent inmates learn general skills, such as how to talk to their child. They also learn ways to create positive parent–child interaction from prison—such as games they can play with a child through the mail—as well as suggestions on what to do when returning home upon release (Coffman and Markstrom-Adams 1995; see also Comfort 2008). "Issues for Thought: When a Parent Is in Prison" further describes some of these programs.

Another community resource, family counseling (see Chapter 12) can help families after a crisis occurs, such as a family member's suffering from post-traumatic stress disorder (PTSD) (England 2009). Counseling can also help when families foresee a family change or future new demands (Clinton and Trent 2009; Rasheed, Rasheed, and Marley 2010). For instance, a couple might visit a counselor when expecting or adopting a baby, when deciding about work commitments and family needs, when the youngest child is about to leave home, or when a partner is about to retire.

Family counseling is not just for relationships that are in trouble but is also a resource that can help to enhance family dynamics. Increasingly, counselors and social workers emphasize empowering families toward the goal of *enhanced resilience*—that is, emphasizing and building upon a family's strengths (Burns 2010; Power 2004).

In addition to counseling, resources include books on various subjects related to family stress and crises. Some examples: Barbara Monroe and Frances Kraus's *Brief Interventions with Bereaved Children* (2010); Avis Rumney's *Dying to Please* (2009) on eating disorders (see also Siegel, Brisman, and Weinshel 2009); Lynn Adams's (2010) "survival guide" for parenting autistic children (see also Roth and Barson 2010); Wes Burgess's *The Bipolar Handbook for Children, Teens, and Families: Real-Life Questions with Up-To-Date Answers* (2008); Kenneth Talan's *Help Your Child or Teen Get Back on Track* (2009), which addresses emotional and behavior problems (see also Kearney 2010); Chelsea Lowe and Bruce Cohen's *Living with Someone Who's Living with Bipolar Disorder* (2010); and Diane England's *The Post Traumatic Stress Disorder Relationship* (2009). Some books on topics of family stress or crisis are written specifically for children. Julianna Fields's *Families Living with Mental and Physical Challenges* (2010) is one example.

We also note the countless resources available online. Resources on an enormous variety of stressors—from involuntary infertility (www.resolve.org); to having a disabled child (www.supportforfamilies.org); to experiencing the death of a child (compassionatefriends.org); to having a family member in prison (prisontalk.com)—offer Web-based virtual communities and information from experts as well as from others who are experiencing similar family demands. For families who might feel stigmatized by the stressors that they are experiencing

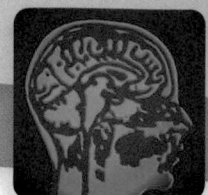
More than two million children have a parent who is in jail or prison (Sabol and West 2009)—a demoralizing family stressor event, coupled with boundary ambiguity. Incarceration rates rose sharply during the 1990s (Arditti 2003). Although rates are not increasing as rapidly today as ten years ago, they do continue to rise (McCarthy 2009). A child's risk of parental incarceration increased by 60 percent between 1978 and 1990. Although the risk remains low for non-Hispanic white children, calculations show that 14 to 15 percent of black children born in 1978 and 25 to 28 percent "of black children born in 1990 had a parent imprisoned by the time the child was 14" (Wildeman 2009, p. 271).

Prior to their imprisonment, 79 percent of mothers and 53 percent of fathers were living with their children (National Resource Center on Children and Families of the Incarcerated 2009a.). While mothers are in prison, about one-quarter of their children live with their fathers. Grandparents care for about half of all children with incarcerated mothers. Non-Hispanic white children are more likely to be in nonfamily foster care

(see Chapter 10) than are African American or Hispanic children (Lee, Genty, and Laver 2005). This may be because black and Hispanic communities have had more of a tradition of shared care of children (e.g., Stack 1974), a situation facilitating making arrangements that place children with adult relatives, often grandparents (Enos 2001; Poehlmann 2005). The children's caregivers often

> feel compelled to lie about their loved one's whereabouts. If the children are young, their mother may explain the father's absence by saying that "Daddy's away on a long trip" or "He's working on a job in another state." One caregiver . . . explained to her nephews that their father was away at "super-hero school." Older children who know the truth may feel that they need to be careful not to discuss it at school or with friends. (Arditti 2003, p. F15)

Children's visiting an incarcerated parent can be expensive and otherwise difficult to arrange, because prisons are often far from their homes (McManus

2006; National Resource Center on Children and Families of the Incarcerated 2009b). One study found that half of children of women prisoners did not visit at all during their mother's incarceration. However, including phone calls and letters, 78 percent of mothers and 62 percent of fathers had at least monthly contact with children (Mumola 2000).

More and more, policy makers have realized that disrupted family ties have a severe and negative impact on the next generation (Arditti 2003; Comfort 2008; Poehlmann 2005). Consequently, a number of correctional systems, including the Federal Bureau of Prisons, have developed visitation programs to facilitate parent–child contact. Many correctional facilities have returned to an earlier practice of permitting babies born in prison to remain with their mothers for a time (Rutgers University School of Criminal Justice and the New Jersey Institute for Social Justice 2006; Comfort 2008). Although visitation programs were initially oriented solely to mothers, prisons have more recently developed programs for fathers as well (Enos 2001; McManus 2006).

---

and are therefore reluctant to seek informal and community support, Web-based resources have the advantage of offering information and support anonymously (Colvin et al. 2004).

Crisis: Disaster or Opportunity?

A family crisis is a turning point in the course of family living that requires members to change how they have been thinking and acting (McCubbin and McCubbin 1991, 1994). We tend to think of *crisis* as synonymous with *disaster*, but the word comes from the Greek for *decision*. Although we cannot control the occurrence of many crises, we can decide how to cope with them.

Most crises—even the most unfortunate ones—have the potential for some positive as well as negative effects. For example, Professor Joan Patterson, a recognized

expert in the field of family stress, has observed that many parents who are raising children with "complex and intense" medical needs

> seem to find new meaning for their life. Having a child with such severe medical needs and such a tenuous hold on life shatters the expectations of most parents for how life is supposed to be. It leads to a search for meaning as a way to accept their circumstances. When families get to this place, they not only accept their child and their family's life, but they often experience a kind of gratitude that those of us who have never faced this level of hardship can't really understand. (2002a, p. F7)

Whether a family emerges from a crisis with a greater capacity for supportive family interaction depends at least partly on how family members choose to define the crisis. A major theme of this text is that, given the opportunities and limitations posed by society, people create their families and relationships based on the choices they make. Families whose members choose to

Having a family member in prison or jail is a crisis a small but growing number of families face today. Family stress and adjustment experts tell us that virtually all family crises have some potential for positive as well as negative effects. Can you think of any possible positive effects in this case? What community supports might help this family? What might be some alternatives to incarcerating parents who have been actively involved in raising their children?

We focus here on children's needs, but imprisonment demoralizes other family members as well and usually has a negative economic impact on the family system, not only when the prisoner has been an essential breadwinner but also due to costs associated with visiting the prisoner and making long-distance family telephone calls, among others (Arditti, Lambert-Shute, and Joest 2003). Moreover, because of stigma associated with incarceration, prisoners' families receive little community support (Arditti 2003, p. F15).

Strong family bonds appear to reduce children's negative behaviors (Poehlmann 2005), although "incarceration can undermine social bonds, [and] strain marital and other family relationships" (Western and McLanahan 2000, p. 323). Policy analysts argue that "[a]n over reliance on incarceration as punishment, particularly for nonviolent offenders, is not good family policy" (Arditti 2003, p. F17; Wildeman 2009). They propose alternatives to incarceration, such as home confinement with work release (Comfort 2008; see also sentencingproject.org).

Critical Thinking

Can you apply the ABC-X model to this situation of having an incarcerated family member?

be flexible in roles and leadership meet crises creatively.

However, even though they have options and choices, family members do not have absolute control over their lives (Coontz 1997; Kleber et al. 1997). Many family troubles are really the results of public issues. For example, the serious family disorganization that results from poverty is as much a social as a private problem (Trask et al. 2005). Also, most American families have some handicaps in meeting crises creatively. The typical American family is under a high level of stress at all times. Providing family members with emotional security in an impersonal and unpredictable society is difficult even when things are

It's important to remember that not all stressors are unhappy ones. Happy events, such as moving into a new house, can be family stressors too.

running smoothly. Family members are trying to do this while holding jobs and managing other activities and relationships.

Moreover, many family crises are more difficult to bear when communities lack adequate resources to help families meet them (Coontz 1997; Pitts 2008). One response to this situation is to engage in community activism (Bryan 2010). For example, one couple, frustrated by the lack of organized community support available to them and their autistic child, founded Autism Speaks. A project of Autism Speaks is "to develop a central database of 10,000-plus children with autism that will provide, for the first time, the standardized medical records that researchers need to conduct accurate clinical trials" (Wright 2005, p. 47; see also autismandactivism.com; Roth and Barson 2010). A second example is the grassroots Disability Rights Movement (Bryan 2010). When families act collectively toward the goal of obtaining needed resources for effectively meeting the demands placed upon them, family adjustment can be expected to improve overall.

Summary

- Throughout the course of family living, *all* families are faced with demands, transitions, and stress.

- Family stress is a state of tension that arises when demands test, or tax, a family's resources.

- A sharper jolt to a family than more ordinary family stress, a family crisis encompasses three interrelated factors: (1) family change, (2) a turning point with the potential for positive and/or negative effects, and (3) a time of relative instability.

- Demands, or stressors, are of various types and have varied characteristics. Generally, stressors that are expected, brief, and improving are less difficult to cope with.

- The predictable changes of individuals and families—parenthood, midlife transitions, post-parenthood, retirement, and widowhood and widowerhood—are all family transitions that may be viewed as stressors.

- A common pattern can be traced in families that are experiencing family crisis. Three distinct phases can be identified: (1) the stressor event that causes the crisis, (2) the period of disorganization that follows, and (3) the reorganizing or recovery phase after the family reaches a low point.

- The eventual level of reorganization a family reaches depends on a number of factors, including the type of stressor, the degree of stress it imposes, whether it is accompanied by other stressors, the family's appraisal or definition of the crisis situation, and the family's available resources.

- Meeting crises creatively means resuming daily functioning at or above the level that existed before the crisis.

- Several factors can help families meet family stress and/or crises more creatively: a positive outlook, spiritual values, the presence of support groups, high self-esteem, open and supportive communication within the family, adaptability, counseling, and the presence of a kin network.

Questions for Review and Reflection

1. Compare the concepts *family stress* and *family crisis*, giving examples and explaining how a family crisis differs from family stress.

2. Differentiate among the types of stressors. How are these single events different from stressor overload? How might economic recession cause stressor overload?

3. Discuss issues addressed in other chapters of this text (e.g., work–family issues, parenting, separation, divorce, and remarriage) in terms of the ABC-X model of family crisis.

4. What factors help some families recover from crisis while others remain in the disorganization phase?

5. **Policy Question.** In your opinion, what, if anything, could/should government do to help families in stress? In crisis?

Key Terms

ABC-X model 389
boundary ambiguity 382
community-based resources 395
family crisis 378
family stress 378
family transitions 379
fictive kin 394

nadir of family disorganization 388
period of family disorganization 388
resilient families 387
stressors 381
stressor overload (pileup) 387
vulnerable families 387

Online Resources

Sociology CourseMate

www.CengageBrain.com

Access an integrated eBook, chapter-specific interactive learning tools, including flash cards, quizzes, videos, and more in your Sociology CourseMate, accessed through CengageBrain.com.

www.CengageBrain.com

Want to maximize your online study time? Take this easy-to-use study system's diagnostic pre-test, and it will create a personalized study plan for you. By helping you identify the topics that you need to understand better and then directing you to valuable online resources, it can speed up your chapter review. CengageNOW even provides a post-test so you can confirm that you are ready for an exam.

Divorce: Before and After

All marriages end—either in death or divorce. Divorce has become a common experience in the United States for all social classes, age categories, and religious and ethnic groups. Over 40 percent of recent first marriages are likely to end in divorce (Teachman, Tedrow, and Hall 2006).[1] In this chapter, we'll examine factors that affect people's decisions to divorce, the experience itself, and ways the experience may be made less painful and become the prelude to the future, alone or in a new marriage. We'll also analyze why so many couples in our society decide to divorce and examine the debate over whether a divorce should be harder to get than it is today. We'll begin by looking at divorce rates in the United States, which are among the highest in the world.

Today's High U.S. Divorce Rate

The divorce rate started its upward swing in the nineteenth century (Amato and Irving 2006; Teachman, Tedrow, and Hall 2006, Figure 4.1).[2] The frequency of divorce increased throughout most of the twentieth century, as Figure 15.1 shows, with dips and upswings

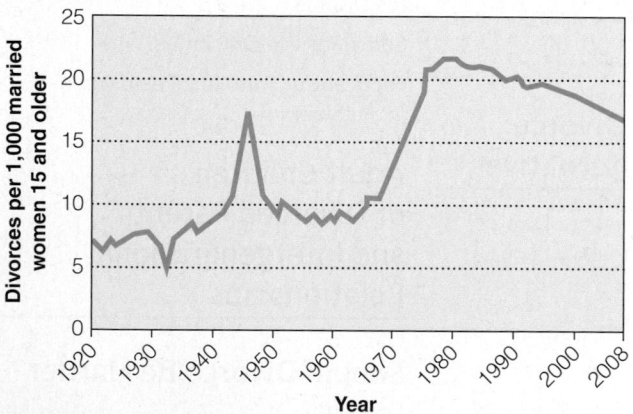

FIGURE 15.1 Divorces per 1,000 married women age fifteen and older in the United States, 1920–2008. This includes the latest data available for the refined divorce rate.

Source: U.S. National Center for Health Statistics 1990a, 1998, p. 3; Wilcox and Marquardt 2009, p. 76, Figure 5.

[1] For years estimates were that as many as 50 percent of first marriages would end in divorce. Experts now think marital dissolutions never reached 50 percent and likely never will (Hurley 2005, citing Rose Kreider of the U.S. Census Bureau). As of 2001 (the most recent available data), the estimated percentages of marriages ending in divorce for men and women now in their fifties were 41 percent for men and 39 percent for women (Kreider 2005).

[2] See Paul. R. Amato and Shelley Irving's chapter in *Handbook of Divorce and Relationship Dissolution* (2006) for a presentation of divorce rates, divorce law, and attitudes toward divorce in various eras of American history.

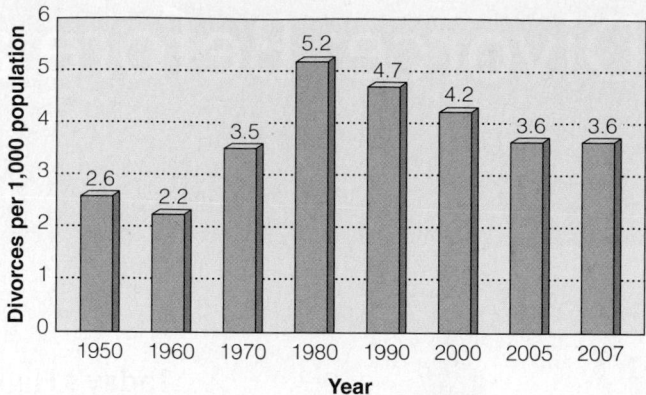

FIGURE 15.2 Divorces per 1,000 population, 1950 to 2007 (crude divorce rate).

Source: Adapted from U.S. Census Bureau 2010b, Table 78; U.S. National Center for Health Statistics 2006, Table A.

surrounding historical events such as the Great Depression, Great Recession of 2007, and major wars. Between 1960 and its peak in 1979, the **refined divorce rate** more than doubled. The *refined divorce rate* declined throughout the nineties (Wilcox and Marquardt 2009, p. 75, Table 5).

We can extend the time line to 2005, if we use the **crude divorce rate** (see Figure 15.2). The crude divorce rate has declined almost 30 percent since 1979, and has not been so low since around 1970 (Stevenson and Wolfers 2007; U.S. Census Bureau 2010b, Table 78).[3]

The decline in divorce rates varies by social category. It has declined dramatically for women college graduates, whereas less-educated women have experienced virtually stable divorce rates (S. Martin 2006; Martin and Parashar 2006). This has produced what sociologist Steven Martin calls the **divorce divide** (in Hurley 2005; Ono 2009). Predictions are that only 25 percent of college women who married in the early 1990s will divorce, whereas over 50 percent of less-educated women will experience a divorce.

Divorces occur relatively early in marriage. The median length of a first marriage that ends in divorce

[3] The *refined divorce rate* is the number of divorces per 1,000 married women. The refined divorce rate compares the number of divorces to the number of women at risk of divorce (that is, married women). It is a more valid indicator of the rate at which marriages are dissolved than the crude divorce rate.

The *crude divorce rate* is the number of divorces per 1,000 population. This rate includes portions of the population—children and the unmarried—who are not at risk for divorce. Despite its limitations, the crude divorce rate is used for comparisons over time because these data are the only long-term annual data available. The federal government discontinued compilation of the refined divorce rate in the mid-nineties (Broome 1995).

In lieu of detailed government data, we also rely on national-sample surveys—including the Census Bureau's Current Population Survey—for individuals' reports of their current or cumulative experience of divorce. These divorce data are *not* collected annually, but less frequently.

They are high-school sweethearts who seemed happily married for 40 years after raising four children. But the announcement that Al and Tipper Gore were splitting up was not a surprise to researchers who study the most divorcing cohort in American history. Al is 61 and Tipper is 62, which means they came of age in the 1960s and '70s, an era when even the most intimate relationships were radically altered by huge social upheaval. "This is the generation that weathered a lot of changes that didn't match their expectations when they walked down the aisle," says Betsey Stevenson, an assistant professor of business and public policy at the University of Pennsylvania who studies marriage trends. "In some respects, it's a miracle how many of them stayed together."

Marriage researchers have long known that the rockiest years are the early ones, and, generally speaking, the longer a couple is married, the less likely they are to divorce. That's still true, says Stevenson, but "silver divorce" is no longer rare, and that's particularly true for the baby-boom generation, born between 1946 and 1964. In the 1950s, many couples married just past age 20, but only 8 percent were divorced at the 10-year mark. After 20 years, just 19 percent had split; after 30 years, 26 percent were divorced; and after 40 years, only 30 percent were no longer together.

The Gores married in 1970, when divorce rates had just begun a dramatic spike upward. Their contemporaries had much less stable marriages. Within 10 years, 27 percent of their unions had broken up. And after 30 years together, more than half were divorced. As those couples now start to hit the 40-year mark, the rate is slowing down, but some continue to divorce.

"If you look at every single year of marriage, they have the highest divorce rates of anyone born before or after them," Stevenson says. In fact, she adds, this group of baby boomers is most responsible for the commonly heard statistic that one out of every two first marriages will eventually end in divorce. Based on their track record to date, however, baby boomers' divorce rates will clearly end up being higher than that. About 4 percent of divorces every year now involve those married 40 years or more, she says.

The increase in silver divorces is not strictly an American phenomenon. The rate of divorce among those 55 and older is also creeping up in Britain, France, Canada, and Japan, even though divorce rates on the whole are inching down. "I keep hearing people say that this is rare or unusual, but it's not," Stevenson says.

Ironically, it's the fact that the '70s cohort was also the most marrying generation that set them up to be the most likely to divorce. After all, you have to be married to get divorced. And because they tended to marry in their early 20s and are expected to live into their 80s or 90s, theirs may be the generation that lives the most years as married couples, especially if you include people who remarry. "A couple like the Gores, if they had stayed together, had the prospect of being married for seventy years before one of them was likely to die," Stevenson says. "That's a long time."

And the years of social turbulence for these couples are far from over, she says. In the early years of their marriages, these couples were the first to confront the challenges of juggling two careers and balancing family and work. Now, as they face retirement, there are likely to be more battles over whether to keep working or start a new, slower-paced way of life. "They're not only trying to figure out what to do with the years ahead, but who they want to do it with," Stevenson says. "They're blazing a new way to live their lives past sixty, and they're figuring it out right now.

"Some will decide they have a lot of living left to do, and they may want to stop and reevaluate whether their marriage will continue to work for them over the next two decades," she says. "Some may make different choices about how to live the rest of their lives than they would if they thought they would die in a few years."

Rather than assuming divorces like the Gores' represent failure, it may make sense to cast their relationship as more than 35 years of success. "There is no way of knowing when things stopped working well for the Gores," Stevenson says, thinking back on the couple's famous kiss at the 2000 Democratic convention. "But none of us has any reason to believe it wasn't a good marriage for thirty or more years. I don't think we should look at a marriage that ends after forty years as a failure because it didn't make it to sixty or seventy years. It doesn't mean that most of those years weren't as special as we thought they were."

In any case, they are definitely a sign of what's to come. As the newlyweds of the '70s continue to age, we'll likely see record numbers of silver anniversaries reached, as well as a record number of silver divorces. Once again, the Gores will be trendsetters.

Source: Wingert and Kantrowitz 2010.

is about eight years. The proportion of divorces for couples married twenty years or more has increased. This "'silver divorce' is no longer rare, and that's particularly true for the baby-boom generation, born between 1946 and 1964," where more than half of all couples married in the 1970s were divorced by the time of their thirty-year anniversary (Kreider 2005; Wingert and Kantrowitz 2010, n.p.). These statistics need to be contextualized. The baby-boom generation is one of the largest in history, and subsequently, is the generation with the largest number of marriages. It is also the generation with the greatest rate of marriage. With that many people getting married, it's not surprising that we would find a high divorce rate as well. A recent *Newsweek* article discusses this in the Facts about Families box.

Most observers (though not all)[4] conclude that the divorce rate has stabilized, and even declined, for the time being (Teachman 2008a). Probably the most significant reason is the rise in the age at marriage. Fewer people are marrying at the vulnerable younger ages. Those who wait are likely to make better choices and to have the maturity and commitment to work through problems (Heaton 2002; Teachman, Tedrow, and Hall 2006).

Another reason is that better-educated and better-off working couples have had the economic tide in their favor. According to public policy professor Andrew Cherlin, "Families with two earners with good jobs have seen an improvement in their standard of living, which leads to less tension at home and lower probability of divorce," so long as the wife does not earn more than her husband. This is particularly true in white marriages, where the greater a woman's income, compared to her husband's, destabilizes the marriage (in "Divorce Rate" 2007; Teachman 2008, p. 16; Wilcox and Marquardt 2009, p. 44). Also, some societal adjustment to women's employment and dual-earner families now seems to have occurred (Teachman, Tedrow, and Hall 2006; Teachman 2008). Spouses "are learning how to negotiate marriages based on less rigid gender roles than in the past" (historian Stephanie Coontz in "Divorce Rate" 2007).

Some credit marriage education programs funded by the federal government for falling divorce rates. For poorer families, who could not afford family counseling on their own, the programs may have helped couples manage their marital relationships (Crary 2007b; "Divorce Rate" 2007). Moreover, some observers noted an increased determination on the part of children of a divorcing generation to make their own marriages work (Crary 2007b; Teachman, Tedrow, and Hall 2006).

[4] See Schoen and Canudas-Romo 2006; Teachman, Tedrow, and Hall 2006; and a comment by sociologist Andrew Cherlin in Hurley 2005, p. A7.

On a less enthusiastic note, demographers point out that divorce rates may have stabilized or declined because cohabitation has increased—if riskier relationships never become marriages, they never become divorces either. The fact remains that, since around 1980, there has been an unanticipated decline in divorce rates. Nevertheless, divorce rates remain high by historical standards.

Historians also point to the fact that marriage can be dissolved by death as well as divorce. The longer life span attained in the twentieth century gives people who remain married more time together. Married couples are now much more likely to reach their fortieth anniversary than they were at the beginning of the twentieth century, and children are more likely to be raised by both parents. In fact, couples who married "in the 1980s were less likely to part ways" than those who married during the 1970s, "and those marrying in the 1990s

Deciding to divorce is difficult. Couples struggle with concerns about the impact on children and feelings about their past hopes and current unhappiness.

and 2000s have been even more reluctant to divorce" (Stevenson 2010, n.p.).

Still, the continuation of a high incidence of divorce contributes to the increased prevalence of single-parent families. Children's living arrangements vary greatly by race and ethnicity, as Figure 15.3 indicates. Based on the most recent available data, we find that Asian and non-Hispanic white children are most apt to be living in two-parent families (with biological parents or a parent and stepparent). A majority of Hispanic, American Indian/ Alaska Native, and Hawaiian/Pacific Islander children live with two parents, whereas just under a majority of black children are living in a single-mother household (Lugaila and Overturf 2004). There are many reasons why so many African American children live in single-mother households. We have discussed these reasons in previous chapters, and will continue to shed light on the issue in the next chapters of this book.

In summing up the statistics, we need to note that a high divorce rate does not mean that Americans have given up on marriage. It means that they find an unhappy marriage intolerable and hope to replace it with a happier one. But a consequence of remarriages—which have higher divorce rates than first marriages—is an emerging trend of **redivorce.** Many who divorce—and their children—can expect several emotionally significant transitions in family structure and lifestyle (Teachman 2008). (The stability of remarriages is addressed in greater detail in Chapter 16.)

In a context of a high though declining divorce rate, along with a positive view of marriage, why is it that married couples do divorce?

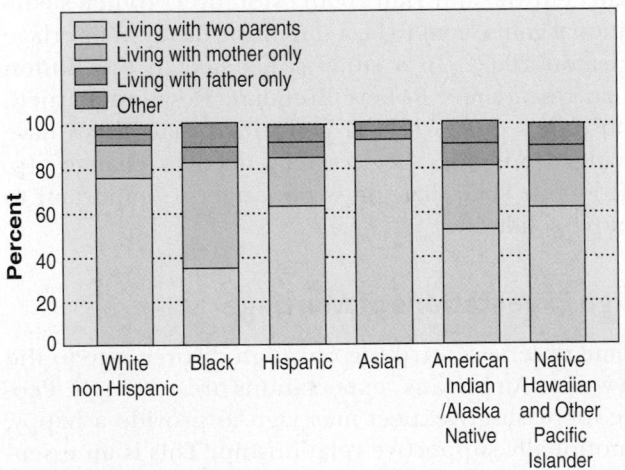

FIGURE 15.3 Living arrangements of children under eighteen by race/ethnicity, 2000. *Other* includes children who are living in the homes of relatives, in foster homes, or with other nonrelatives, or who are heads of their own households.

Source: Adapted from Lugaila and Overturf 2004, Table PHC-T-30b.

Why Are Couples Divorcing or Dissolving Their Unions?[5]

Various factors can bind married couples together: economic interdependence; legal, social, and moral constraints; and the spouses' relationship itself. Yet the binding strength of some of these factors has lessened. "[A]ll Western [and some non-Western] countries have been moving toward a less familistic set of attitudes and toward greater individual investments in self, career, and . . . personal growth and goals" (Goode 1993, p. 81; see also Gubernskaya 2010 for similar research findings).

Economic Factors

Traditionally, as we've seen, the family was a self-sufficient productive unit. Survival was far more difficult outside of families, so members remained economically bound to one another. But today, because family members no longer need one another for basic necessities, they are freer to divorce than they once were (Davis 2010).

Families are still somewhat interdependent economically. Even though marriage "has become less economically necessary . . . it remains economically advantageous in most cases" (Wilcox and Marquardt 2009, p. 42). As long as marriage continues to offer practical benefits, economic interdependence will help hold marriages together. The economic practicality of marriages varies according to several conditions.

Divorce and Social Class The higher the social class as defined in terms of education, income, and home ownership, the less likely a couple is to divorce. Income loss has been found to increase the likelihood of divorce, especially when it is the male who loses his income (Wilcox and Marquardt 2009, pp. 19, 34). Both the stress of living with inadequate finances and the failure to meet expectations for economic or educational attainment seem to contribute to marital instability (Rampell 2009a). This so concerns researchers that the general sense is that "the deep economic downturn of the last two years seems likely to pose a threat to the long-term health of working class marriage" (Wilcox and Marquardt 2009, p. 20). This situation, together with the tendency of low-income groups to marry relatively early, helps explain why less well-off families have the highest rates of marital disruption, including divorce, separation, and desertion; and why more advantaged groups, taking longer to marry, tend to have lower rates of marital disruption (Sassler, Cunningham, and Lichter 2009, p. 772).

---

[5] Although this research is specific to marriage, we think many of the dynamics apply to breakups of committed nonmarital relationships.

Some groups traditionally associated with a lower social class appear to have lower rates of divorce, even though they fit the general patterns of those in the lower strata of society. Latinos, for example, have lower rates of divorce. Researchers, however, have found that Latinos are more likely to separate but not divorce (Umaña-Taylor and Alfaro 2006), thus while Latinos may have divorce rates on par with whites, this hides the fact that many more Latinos may no longer be living as married couples.

Wives in the Labor Force The upward trend of divorce and the upward trend of women in the labor force have accompanied each other historically. But are they causally connected?

Much research, though not all, indicates that wives' employment in itself makes no difference in marital quality (Sayer and Bianchi 2000; Schoen, Rogers, and Amato 2006). As Chapter 11 points out, whether husbands are supportive of their wives' employment and share in housework *does* relate to wives' marital satisfaction. *Conflict theorists* hypothesize and research confirms that marital conflict may increase if women go into the job market but their husbands do not take over an equitable share of the domestic tasks. Symbolic interactionist research suggests this as well. For example, Arlie Hochschild's research on the "second shift" (discussed in Chapter 11) shows that even though women's employment, thus pay, has become an equally important component in the family income, the burden of household duties continues to be borne by working women, leading to marital conflict, and increasing the chances of divorce.

Although it may not affect marital quality, employment might nevertheless contribute to a divorce by giving an unhappily married woman the economic power, the increased independence, and the self-confidence to help her decide on divorce—called the **independence effect** (Sayer and Bianchi 2000; Teachman 2010). Some economists and sociologists posit that marriages are most stable and cohesive when husbands and wives have different and complementary roles—the husband the primary earner, while the wife bears and rears children and is the family's domestic and emotional specialist. Drawing on *exchange theory*, they assert that economic interdependency in marriage is a strong bond holding a marriage together (Becker 1981/1991; Oppenheimer 1997; Springer 2010). But this comes at a cost. New research into this issue finds that men who most strongly internalize the ideology of being the primary earner suffer greater physical and mental health issues because, as you have read elsewhere in this chapter and in this book, it is extremely difficult to thrive as a family on just the male income in our modern American economy. This promotes stress and feelings of impotence in men who adhere strongly to the male breadwinner ideology, ultimately leading, in many cases, to poor health outcomes for those men as they age (Springer 2010).

Most research does *not* support the premise that specialized roles and economic interdependence are necessary to marital stability. Moreover, there is an **income effect** to women's employment. Among low-income couples, a wife's earnings may actually help to hold the marriage together by counteracting the negative effects of poverty and economic insecurity on marital stability (Sayer and Bianchi 2000; Schoen, Rogers, and Amato 2006).

The effects of women's employment on marriage may depend on gender ideology. "[T]he sharp rise in the rates of divorce between 1965 and 1980 may have been at least partly a function of disjuncture between the expectations of spouses [at the time] and the reality of wives' labor market activities" (Teachman, Tedrow, and Hall 2006, p. 71). But for couples today, expectations of role sharing are common. To explore whether wives' employment has positive or negative effects on divorce proneness, researchers Sayer and Bianchi (2000) analyzed a national sample survey based on 3,339 female respondents interviewed around 1988 and again around 1994. When gender ideology and other variables related to likelihood of divorce were taken into account, there was no direct effect of women's employment on divorce. The desirability of the marriage relationship was a much more important factor in predicting divorce.

It is probably the case that there is considerable variety in the impact of women's employment and earnings on a marriage. It is also likely that the effect of wives' employment on marital stability is in transition (Teachman, Tedrow, and Hall 2006). Moreover, women's educational gains seem to be a stabilizing factor in marriage (Heaton 2002). In a study conducted in the Boston area, researchers Robert Brennan, Rosalind Barnett, and Karen Gareis (2001) concluded that "times have changed and the theories may need to change" (p. 179)—role specialization is no longer so important to couple solidarity.

High Expectations of Marriage

Some observers attribute our high divorce rate to the view that Americans' expectations are too high. People increasingly expect marriage to provide a happy, emotionally supportive relationship. This is an essential family function, yet too-high expectations for intimacy between spouses may push the divorce rate upward (Demo and Fine 2010). Research has found that couples whose expectations are more practical are more satisfied with their marriages than are those who expect completely loving and expressive relationships (Demo and Fine 2010; Plotnick 2007; see also

Oberlander et al. 2010 for a discussion of racial differences and marital expectation). Although many couples part for serious and specific reasons, others may do so because of unrealized expectations and general discontent.

Decreased Social, Legal, and Moral Constraints

"**Barriers to divorce** function to keep marriages intact even when attractiveness of the marital relationship is low and the attractiveness of alternatives to the relationship is high" (Knoester and Booth 2000, p. 81). But the social constraints that once kept unhappy partners from separating operate less strongly now.

The official posture of many—though not all—religions in the United States has become less critical of divorce than in the past. **No-fault divorce** laws, which exist in all fifty states,[6] have eliminated legal concepts of guilt and are a symbolic representation of how our society now views divorce. But it does not appear that changes in the law have themselves led to more divorce—in fact, divorces have actually fallen from "23 divorces per 1,000 married couples in 1979 to under 17 per 1,000 in 2005" (Coontz 2010a, p. A29; Wolfers 2006). Rather, legal change seems to have followed the trajectory of cultural attitudes and behavioral practice regarding divorce.[7]

To say that societal constraints against divorce no longer exist would be an overstatement; nevertheless, barriers have weakened. Knoester and Booth (2000) go so far as to conclude that "perhaps the concept of barriers has outlived its usefulness" (p. 98).

Attitudes toward Marriage Virtually no respondents in a study of marital cohesion mentioned stigma or disapproval as a barrier to divorce (Amato and Hohmann-Marriott 2007; Previti and Amato 2003). Emphasis on

the emotional relationship over the institutional benefits of marriage results in marriage being viewed as not necessarily permanent (Cherlin 2004; Demo and Fine 2010). Friedrich Engels, a colleague of Karl Marx and an early family theorist, noted: "If only the marriage based on love is moral, then also only the marriage in which love continues" (1942 [1884], p. 73). The changing nature of marriage is a worldwide phenomenon as far as the industrialized world is concerned (Giddens 2007).

Self-Fulfilling Prophesy Defining marriage as semipermanent can become a self-fulfilling prophecy, says Joshua Goldstein, a Princeton professor of sociology and public affairs: "Expectations of high divorce rates are in some ways self-fulfilling. . . . [T]hat's a partial explanation for why the rates went up in the 1970s." If partners behave as if their marriage could end, it is more likely that it will. But "as word gets out that rates have tempered or even begun to fall, '[i]t could lead to a self-fulfilling prophecy in the other direction'" (in Hurley 2005; see also Goldstein 1999).

Marital Conversation—More Struggle and Less Chitchat If barriers can no longer be counted on to preserve marital stability, the quality of the relationship becomes central to the survival of a marriage (Bodenmann, Ledermann, and Bradbury 2007; Ledermann, et al. 2010). That will "heighten the need for individuals to be committed to the union and the need to make a good marital match—with someone with whom it is possible to negotiate the details of everyday life without relying on the structural constraints generated by a highly gendered division of labor" (Teachman, Tedrow, and Hall 2006, p. 71). No longer are the normative role prescriptions for wives, husbands, or children taken for granted. Consequently, marriage entails continual negotiation and renegotiation among members about trivial matters as well as important ones. As one divorced woman put it,

> It had taken Howard and me only about ten minutes to pronounce the "I do's," but we would spend the next ten years trying to figure out who, exactly, was supposed to do what: Who was responsible for providing child care, finding babysitters and tutors, driving car pools, for which periods and where? (Blakely 1995, p. 37)

Intergenerational Transmission of Divorce

As discussed in Chapter 6, having parents who divorced increases the likelihood of divorcing (Amato and DeBoer 2001; Teachman, Tedrow, and Hall 2006). Researchers are not certain of the reasons for this. It is possible that (1) divorcing parents are models of divorce as a solution to marital problems, or (2) children of divorced parents are more likely to exhibit

---

[6] Some authorities say that "most states" have no-fault divorce laws (Buehler 1995; Stevenson and Wolfers 2007), whereas others cite all fifty states (Nakonezny, Shull, and Rodgers 1995). There is a gray area in that some "no-fault" divorce laws may require a specific period of separation rather than just a declaration by one of the parties that the marriage is over (Hakim 2006). Moreover, some states have retained fault divorce alongside no-fault. In those states, a spouse may choose to file for divorce under a fault provision, alleging that the other partner has committed whatever statutory faults are relevant to the state's marital dissolution laws.

[7] Nakonezny, Shull, and Rodgers (1995) published an article purporting to prove that no-fault divorce laws have played a causative role in increasing divorce rates. Sociologist Norval Glenn (1997) responded with an effective critique of their methodology, concluding that "the adoption of no-fault divorce in itself had very little direct effect on divorce rates" (p. 1,023; and see the response of Rodgers, Nakonezny, and Shull 1997). Further research supports the view that the passage of "unilateral" divorce laws does not account for divorce trends (Wolfers 2006).

personal behaviors that interfere with maintaining a happy marriage. There is also evidence that children of divorced parents marry at younger ages and are more likely to experience premarital cohabitation and births; these factors are associated with higher divorce rates (Heaton 2002; Teachman 2002a).

In a test of the two major hypotheses about **intergenerational transmission of divorce**, using longitudinal data, Amato and DeBoer (2001) found support for the *commitment to marriage* hypothesis. When parents remained married, they served as models of optimism about solving marital problems. A hypothesis about the importance of *parents as models of relationship skills and interpersonal behavior* was not supported in this study (although there is other evidence for it—e.g., Amato 1996). The conclusion of Amato and DeBoer's study is that "it is actual termination of the marriage rather than the disturbed family relationships that affects children. Divorce, rather than conflict, undermines children's faith in marriage" (p. 1,049).

As more parents divorce, more offspring would seem to be vulnerable to the intergenerational transmission of divorce. Yet research spanning the period between 1973 and 1996 finds a decline of almost 50 percent in the rate of intergenerational transmission of divorce. It may be that acceptance of divorce is now so widespread that having parental models is less significant for marital stability (Wolfinger 1999; see Li and Wu 2008 for their critique of this study). It would be nice to conclude on that hopeful note. But analysis of data from the Marital Instability Over the Life Course study suggests that "divorce has consequences for subsequent generations, including individuals not yet born at the time of the original divorce" (Amato and Cheadle 2005, p. 191). Problems evident in the grandchildren of the original divorcing couple include less education, more marital conflict, and poorer relationships with their parents.

On the other hand, research by Li and Wu, analyzing data from the National Survey of Families and Households found no evidence that parent's marital status impacted their children's marital commitment. This study does not deny that children of divorced parents are themselves more likely to divorce, but it demands caution in assuming causality (in the way Wolfinger does). Rather, the findings suggest simply that the longer couples are married, the greater their opportunity for conflict and divorce (Li and Wu 2008). Research on Swedish couples whose parents had divorced and remarried had similar findings, that is, no impact on children's divorce risk (Lyngstad and Engelhardt 2009).

Other Factors Associated with Divorce

Thus far in this section, we have looked at sociohistorical, cultural, and intrafamilial factors that encourage high divorce rates. Another way to think about divorce is to recognize that certain demographic and behavioral factors might be related to divorce rates. These include the following:

- *Remarried mates are more likely to divorce.*

- *Premarital sex and cohabitation before marriage increase the likelihood of divorce, but only when these take place with someone other than the future marital partner* (Heaton 2002; Teachman 2003). "There is evidence that the relationship between premarital cohabitation and divorce is waning" (Teachman, Tedrow, and Hall 2006, p. 74).

- *Premarital pregnancy and childbearing usually increase the risk of divorce in a subsequent marriage* (Heaton 2002; Teachman 2002b). However, if the biological parents marry, and especially if the birth occurred during cohabitation, "it might simply represent the continued evolution of the process of mate selection" (Teachman, Tedrow, and Hall 2006, p. 75).

- Young children stabilize marriage (Hetherington 2003). Hence, *remaining childfree is associated with a higher likelihood of divorce.*

- *Race and ethnicity are differentially associated with the chances of divorcing.* But "racial differences in dissolution are not well understood . . . and we know little about the underlying processes that may generate differences in divorce rates among racial and ethnic groups" (Teachman, Tedrow, and Hall 2006, p. 76).

A government survey reported that, as of 2001 (latest detailed data), the duration of marriage to particular anniversary dates was lower for black, Asian, and Hispanic women than for non-Hispanic whites. If, on the other hand, you look at lifetime experience of divorce for those over age fifteen (see Table 15.1), blacks have a relatively low percentage of "ever divorced," while that of whites is the highest (Kreider 2005, p. 12 and Appendix Table 1).

This discrepancy results from several patterns. Asian and Latino populations are relatively young, so they have not been married and at risk of divorce for as long. Their ultimate divorce rates are difficult to predict.

The black population is also younger than the white non-Hispanic population. But the percentage of blacks who have "ever divorced" is low in large part because of the black "retreat from marriage." "Blacks who marry are an increasingly select subgroup of all blacks . . . who are committed to marriage and therefore less likely to divorce" (Teachman 2002b, p. 345; see also Stevenson and Wolfers 2007). Economic and educational factors seem to play a much more significant role in marital stability among African Americans than in other racial/ethnic groups (Sweeney and Phillips 2004).

- Not surprisingly, *when marital partners are emotionally mature and possess good interpersonal communication skills, "they are better able to deal with the bumps along the road to marital survival"* (Hetherington 2003, p. 322).

The preceding are the connections scholars have made between divorce and other factors. Marital complaints given by the divorced themselves include the partner's infidelity, alcoholism, drug abuse, jealousy, moodiness, violence, low levels of trust, and, much less often, homosexuality, as well as perceived incompatibility and growing apart (Amato 2010; Fincham and Beach 2010). Counselors suggest that some common complaints—about money, sex, and in-laws, for example—are really arenas for acting out deeper conflicts, such as who will be the more powerful partner, how much autonomy each partner should have, and how emotions are expressed (Amato 2010; Amato and Hohmann-Marriot 2007). A general conclusion to be drawn from research is that deficiencies in the emotional quality of the marriage lead to divorce. "In Western cultures, happiness and satisfaction are integral to relationships and are thought to guide decisions regarding their future" (Rodrigues, Hall, and Fincham 2006, p. 97). The personal decision about divorce involves a process of balancing alternatives against the practical and emotional satisfactions of one's present union.

Gay and Lesbian Divorce

In 2005, Carolyn Conrad filed for dissolution of her civil union from Kathleen Peterson—what is noteworthy about this couple, is that they were in the first same-sex marriage in the State of Vermont (Barlow 2005). Like heterosexual relationships that end, the phenomenon of homosexual breakups have been around for a very long time, but some states are now allowing same-sex marriage, and some of those will end in divorce. Same-sex divorce is one of the benefits of same-sex marriage (where it exists) in that it provides for a formal, clearly recognized way to address couple breakup issues regarding property, child custody, and so on.

Currently, six states and the District of Columbia allow same-gender marriage, and three other states recognize the marriages from those six states. The 2007 American Community Survey showed approximately 340,000 married same-sex couples (Sherman 2009). The United States is beginning to see a smattering of divorce in states where same-sex marriage is legal. For example, in 2008, Massachusetts had 10,000 same-sex married couples, and "more than 100 gay divorces had been granted," while around 2 percent of the 8,666 civil unions in Vermont had been dissolved (Henry 2008, n.p.).

Table 15.1 Percentage of Men and Women (Fifteen Years and Older) Ever Divorced, by Race/Ethnicity and Gender, 2001

Race/Ethnicity	Men	Women[a]
White, non-Hispanic	23.3%	25.4%
Black[b]	18.8	20.1
Asian	8.8	10.4
Hispanic	12.7	15.9

[a]Women's rates of divorce are higher because they have usually married younger and so are at greater risk of experiencing a divorce than are men of the same ages.

[b]Over 40 percent of black men and women have never married, so are not at risk of divorce.

Source: Adapted from Kreider 2005, Table 1.

Thinking about Divorce: Weighing the Alternatives

Not everyone who thinks about divorce actually gets one. As divorce becomes a more available option, spouses may compare the benefits of their union to the projected consequences of not being married.

Marital Happiness, Barriers to Divorce, and Alternatives to the Marriage

One model of deciding about divorce, derived from exchange theory (see Chapter 2) by social psychologist George Levinger, posits that spouses assess their marriage in terms of the *rewards* of marriage, *alternatives* to the marriage (possibilities for remarriage or fashioning a satisfying single life), and *barriers* to divorce (Levinger 1965, 1976). Here we look at **Levinger's model of divorce decisions** from the perspective of the person considering divorce.

Respondents to the Marital Instability Over the Life Course surveys named children, along with religion and lack of financial resources, as *barriers* to divorce in open-ended interviews (Previti and Amato 2003). Indeed, another study found that both mothers and fathers anticipated that "divorce would worsen their economic situation and their abilities to fulfill the responsibilities of being a parent" (Poortman and Seltzer 2007, p. 265). However, when researchers Chris Knoester and Alan Booth (2000) examined *quantitative data* from these Marital Instability Over the Life Course surveys, they found that only three of nine barriers studied were associated with a lower likelihood of divorce: (1) when the wife's income was a smaller percentage of the family income, (2) when church attendance was high, or (3) when the couple had a new child.

© Igor Balasanov/iStockphoto

When parents consider divorce, they often think about the potential impact on their children—and that is a barrier to divorce.

Other research evidence shows that young children do serve as a barrier to divorce, especially when one of the children is a boy (Leonhardt 2003). Anticipated economic loss was not as important as parenting concerns. Affection for their children and concern about the children's welfare after divorce discourage some parents from dissolving their marriage. This concern sometimes leads to delaying an intended divorce (Furstenberg and Kiernan 2001; Heaton 2002; Poortman and Seltzer 2007).

Long marriages are less likely to end in divorce. One reason for this, in addition to the marital bond itself, is that common economic interests and friendship networks increase over time and help stabilize the marriage at times of tension (Brown, Orbuch, and Maharaj 2010). When divorce does occur in a longer marriage, it may be partly related to dissatisfaction with one's marital relationship at the onset of the empty nest. After all, people are living longer and are "not only trying to figure out what to do with the years ahead, but who they want to do it with" (Wingert and Kantrowitz 2010, n.p.).

Although some barriers do seem to have an impact on decisions to divorce, it is the *rewards* of marriage—love, respect, friendship, and good communication—that are most effective in keeping marriages together. "Generally, marriages that have built up positive emotional bank accounts through respect, mutual support, and affirmation of each person's worth are more likely to survive" (Hetherington 2003, p. 322).

"Would I Be Happier?"

Alternatives, the third element of Levinger's theory, was found to be the least important in decisions to divorce (Previti and Amato 2003). Yet, some married people may ask themselves whether they would be happier if they were to divorce. This is not an easy question to answer. Some people may prefer to stay single after divorce, but many partners probably weigh their chances for a remarriage.

Some research finds that leaving a bad marriage may have a positive outcome regardless of whether the individual remarries. A British study found people to be less happy one year after separation, but by one year after the divorce, both men and women were happier than they had been while married (Gardner and Oswald 2006).

A study of 1,755 whites in Detroit found that higher levels of depression among the divorced were not apparent among those who saw themselves as escaping marriages with serious, long-term problems (Aseltine and Kessler 1993). Other research (Ross 1995) used data from a national sample of 2,031 adults to compare depression levels among those with no partners and those in relationships of varying quality; still other research, the Health and Retirement Study, shows a wide variety of chronic mental and physical health conditions associated with people who were divorced (Hughes and Waite 2009).

Additionally, researchers compared data from the waves of the National Survey of Families and Households and found, like the other studies, that emotional well-being declines after divorce. What was most interesting about this study, however, are the findings that suggest well-being declines (and improves little) after a divorce regardless of the unhappiness or even violence associated with the now-dissolved marriage. They note,

> our results do not support the hypothesis that disruption of a marriage rated as unhappy, even among those who experienced violence in their marriage, leads to improvements in emotional well-being. In no case did those who divorced or separated show higher well-being than those who remained married, and on some measures they show lower well-being.
>
> We would expect the largest improvements in well-being, should they appear, among those who ended a marriage that they thought was unhappy and entered another. However, on none of the dimensions of well-being that we examined do we see improvements in emotional well-being for those who ended one marriage and formed another, compared to those who remained married. Clearly, if one of the goals of ending a marriage with which one is unhappy is to improve one's emotional well-being, this goal is not typically reached. (Waite, Luo, and Lewin 2009, p. 209)

It is a paradox—people without a partner are likely to be depressed—but those in unhappy relationships are likely to be even more depressed. Marriage can and often does provide emotional support, sexual gratification, companionship, and economic and practical benefits, including better health. But unhappy marriages do not provide these benefits and may be a factor in poorer health (Elias 2004). "[I]t appears that at any particular point in time most marriages are 'good marriages' and that such marriages have a strong positive effect on well-being and that 'bad marriages' have a strong negative effect on well-being" (Gove, Style, and Hughes 1990, p. 14).

Can This Marriage Be Saved? In some cases, partners might be happier trying to improve their relationship rather than divorcing. "Can This Marriage Be Saved?" is the title of a series that ran in the Ladies Home Journal beginning in the 1950s—articles about couples in troubled marriages who were counseled about how to save their marriages. In this spirit, sociologist Linda Waite reports that couples may be in the lowest grouping on marital satisfaction, yet, if they don't divorce, five years later, two-thirds of the unhappily married couples she and her colleagues studied described themselves as "very happy." Those who divorce do not report themselves as very happy later: "If you are playing the odds in favor of happiness, . . . 'staying married is the better bet'" (Waite, quoted in Peterson 2001, p. 8D).

A recently completed ten-year longitudinal study of newly married couples shows that early support in a young couple's life enhanced their marital satisfaction and helped reduce conflict and marital dissolution (Verhofstadt, Ickes, and Buysse 2010; Sullivan et al. 2010). The kinds of support in the form of social support and individual communication behaviors toward one another is key. The researchers note that "how spouses respond to one another's everyday disclosures and requests for support may be more consequential than how they negotiate their differences of opinion in producing behavioral changes that foreshadow later marital satisfaction and stability" (Sullivan et al. 2010, p. 640). The researcher noted specifically that social support is the most important predictor of long-term marital satisfaction. The kinds of support needed for long-term marital satisfaction include, for example, asking for and offering validation of feelings, and asking for and offering understanding and compassion rather than anger and contempt when disagreements arise (Sullivan et al. 2010, p. 641).

Improvements in those marriages came about through the passage of time (children got older, jobs or other problems improved); because partners' efforts to work on problems, make changes, and communicate better were effective; or because individual partners made personal changes (travel, work, hobbies, or emotional disengagement) that enabled them to live relatively happily despite an unsatisfying marriage. In fact, a new study suggests that people who remain in an unsatisfying marriage are less depressed than those who leave their marriages. These unhappily marrieds also tend to have healthier emotional lives than even those who divorce and eventually remarry (Waite et al. 2002; Waite, Luo, and Lewin 2009, p. 205).

One must decide, then, whether divorce represents a healthy step away from an unhappy relationship that cannot be satisfactorily improved or is an illusory way to solve what in reality are personal problems. Going to a marriage counselor may help partners become more aware of the consequences of divorce so that they can make this decision more knowledgeably and not by default.

Marital Separation Nearly two and a half percent of married couples in the United States were separated in 2008 (U.S. Census Bureau 2010b, Table 66). Some marital partners who have separated do make efforts to reconcile. Little research has been done on marital separation. But "each year [it appears] that a substantial number of separated women try to save their marriage" (Wineberg 1996, p. 308). Wineberg's study, using a sample of white women from the 1987 National Survey of Families and Households, found that 44 percent of the separated women attempted reconciliation. Half of the resumptions of marriage that followed took place within a month, suggesting that those separations may have been impulsive and soon regretted. Virtually no marriages were resumed after eight months of separation.

Only one-third of the reconciliations "took"—that is, resulted in a continued marriage (Wineberg 1996). For a majority of individuals, first separation from the spouse denoted permanent dissolution (Binstock and Thornton 2003). Researcher Howard Wineberg cautions that "not all separated couples should be encouraged to reconcile since a reconciliation does not ensure a happy marriage or that the couple will be married for very long" (p. 308).

Stable Unhappy Marriages From time to time, researchers have taken up the question of what happens to couples who are distanced, unhappy, or in conflict if they don't divorce (e.g., Waite, Luo, and Lewin 2009). It is, however, "surprising" that "long-term low-quality marriage . . . has received relatively little attention" (Hawkins and Booth 2005, p. 451). Contradicting other research noted elsewhere in this chapter, Hawkins and Booth's study followed unhappy marriages for twelve years and compared people in unhappy marriages to divorced single and remarried individuals. "Divorced individuals who remarry have greater overall happiness, and those who divorce and remain unmarried have greater levels of life satisfaction, self-esteem, and overall health than unhappily married people. . . . We suggest that unhappily married people who dissolve low-quality marriages likely have greater odds of improving their well-being than those remaining in such unions" (p. 468). The differences in findings you are reading in these sections are useful learning opportunities. What the seemingly contradictory findings remind us of is that social scientists asking different questions to different people at different times in their life may elicit vastly different findings than other social scientists doing the same kind of research. This is good scholarship because it encourages us to critically examine the methodologies used in each study so that we can find where those differences in outcomes might lie.

Is Divorce a Temporary Crisis or a Permanent Stress?

Initially, studies portrayed divorce as a temporary crisis, with adjustment completed in two to four years. Some scholars now consider divorce to be a lifetime chronic stress for both children and adults (Hughes and Waite 2009; Waite, Luo, and Lewin 2009). Although outcomes vary, divorce researcher E. Mavis Hetherington maintains that 70 percent of those who obtain a divorce have a "good enough" postdivorce adjustment (Hetherington and Kelly 2002). In fact, 20 percent of women in this study embraced their new autonomy, developing self-confidence and working toward enhancing their lives (Demo and Fine 2010, p. 114). "Most men, women, and children adapt to their new lives reasonably well within 2–3 years if they are not confronted with continued or additional stresses" (Hetherington 2003, p. 322).

It appears likely that both temporary crisis and chronic stress are outcomes of divorce—some divorced people are rather permanently derailed from an economically and emotionally comfortable life, whereas others are mostly recovered after several years. Of the latter—those who may be considered "recovered"—some have diminished well-being in some respects, whereas others arrive at a higher level of life satisfaction (Amato 2000, 2003; 2010; Hetherington and Kelly 2002). Difficulties and adjustments don't seem to vary much by race or ethnicity (Amato 2000; 2010). They do vary by whether or not the couple are parents. "Generally . . . [for childless couples] recovery is almost always rather swift" (Braver, Shapiro, and Goodman 2006, p. 313).

Getting the Divorce

One of the reasons it feels so good to be engaged and newly married "is the rewarding sensation that out of the whole world, you have been selected. One of the reasons that divorce feels so awful is that you have been de-selected" (Bohannan 1970, p. 33). Anthropologist Paul Bohannan analyzed the divorce experience in terms of six different facets, or "stations": the emotional, the legal, the community, the psychic, the economic, and the co-parental divorce. Experience in each of these realms varies from one individual to

another; some stations, such as the co-parental, do not characterize every divorce. Yet the six stations capture the complexity of the divorce experience. In this section, we will examine the first three stations just listed; we will explore the economic and co-parental aspects of divorce in greater detail later in this chapter, and then make brief mention of the psychic divorce.

The Emotional Divorce

Emotional divorce involves withholding positive emotions and communications from the relationship (Vaughan 1986; Dupuis 2009), typically replacing these with alienating actions and words. Partners no longer reinforce but, rather, undermine each other's self-esteem through endless large and small betrayals: responding with blame rather than comfort to a spouse's disastrous day, for instance, or refusing to go to a party given by the spouse's family, friends, or colleagues. As emotional divorce intensifies, betrayals become greater.

In a failing marriage, both spouses feel profoundly disappointed, misunderstood, and rejected (Brodie 1999). The couple may want the marriage to continue for many reasons—continued attachment, fear of being alone, obligations to children, the determination to be faithful to marriage vows—yet they may hurt each other as they communicate their frustration by look, posture, and tone of voice.

Not all divorced people wanted or were ready to end their marriage, of course. It may have been their spouse's choice. Initiating and noninitiating partners tend to talk about their reasons in different terms. The initiator of the divorce typically invokes a "vocabulary of individual needs" while the noninitiating partner speaks in terms of "familial commitment" (Hopper 1993).[8]

Women are more often the initiators of a divorce, regardless of how long the marriage has endured (Brinig and Allen 2000; Coontz 2010b). The initiator describes many and ongoing complaints that essentially reprise the issues and events described in this chapter and others: conflict over sharing domestic work; infidelity, physical and emotional abuse; alcoholism; political and recreational differences; sexual tensions, disagreements about children; and so on. Whatever the specifics, the initiator decides that the marriage will never be what she or he wants: "'My needs were not being met'; 'I wasn't being fulfilled'" (Hopper 1993, p. 807). The noninitiating partner, who may have been fairly ambivalent

---

[8] This analysis references sociologist C. Wright Mills's (1940) *vocabularies of motive*, a concept that students of sociology or communication may have encountered before. The idea is that individuals construct accounts to justify their actions. These are not necessarily their actual motives, of which they may be somewhat unaware, but narratives of explanation.

and uncertain about the marriage as well, begins to point to the good in the marriage: "'You need to do something to keep this relationship going'; 'you tell someone you're going to be there forever, then you're going to work on it'" (p. 809).

Not surprisingly, research shows that the degree of trauma a divorcing person suffers usually depends on whether that person or the spouse wanted the dissolution because the one feeling "left" experiences a greater loss of control and has much mourning yet to do—the divorce-seeking spouse may have already worked through his or her sadness and distress (Amato 2000; Braver, Shapiro, and Goodman 2006). Even for those who actively choose to divorce, however, divorce and its aftermath may be unexpectedly painful.

The Legal Divorce

A **legal divorce** is the dissolution of the marriage by the state through a court order terminating the marriage. The principal purpose of the legal divorce is to dissolve the marriage contract so that emotionally divorced spouses can conduct economically separate lives and be free to remarry.

Two aspects of the legal divorce make marital breakup painful. First, divorce, like death, creates the need to grieve. But the usual divorce in court is a rational, unceremonial exchange that takes only a few minutes. Divorcing individuals may feel frustrated by their lack of control over a process in which the lawyers are the principals. In one study of divorced women, virtually all had complaints about their lawyers and the legal system (Arendell 1986). A myriad of websites are dedicated to women discussing those complaints (for example, divorcedwomenonline.com; womensenews.org).

A second aspect of the legal divorce that aggravates conflict and misery is the adversary system. Under our judicial system, lawyers advocate their client's interest only and are eager to "get the most for my client" and "protect my client's rights." Opposing attorneys are not trained to and ethically are not even supposed to balance the interests of the parties and strive for the outcome that promises most mutual benefit.

No-Fault Divorce A major change in the legal process of divorce has been the introduction of no-fault divorce. This revision of divorce law was intended to reduce the hostility of the partners and to permit an individual to end a failed marriage readily. Before the 1970s, the fault system predominated. Parties seeking divorce had to prove that they had "grounds" for divorce, such as the spouse's adultery, mental cruelty, or desertion. Obtaining a divorce might require falsifying these facts.

A fault divorce required a legal determination that one party was guilty and the other innocent. The one

judged guilty rarely received custody of the children, and the judgment largely influenced property settlement and alimony awards, as well as the opinions of friends and family (Coontz 2010a; Stevenson and Wolfers 2007; Vlosky and Monroe 2002). Such a protracted legal battle of adversaries increased hostility and diminished chances for a civil postdivorce relationship and successful co-parenting.

Some researchers suggest that the no-fault system of divorce began in Oklahoma in 1953. The reason for concluding 1953 as the date is the way the law was written. The 1953 Oklahoma law allows for a divorce if one party believed there was "irretrievable breakdown and/or irreconcilable differences and/or incompatibility" in the marriage, regardless of what the other person believes or wants (Vlosky and Monroe 2002; Nakonezny, Shull, and Rodgers 1995; Sepler 1981). Other research posits the beginning of no-fault divorce in California as 1970, and continuing until all states passed no-fault legislation. Divorce was redefined as "marital dissolution" and no longer required a legal finding of a "guilty" party and an innocent one. Instead, a marriage became legally dissolvable when one or both partners declared it to be "irretrievably broken" or characterized by "irreconcilable differences." No-fault divorce is sometimes termed **unilateral divorce** because one partner can secure the divorce even if the other wants to continue the marriage.

A final note on the legal divorce is that, by definition, it applies only to marriage. There is no legal forum in which cohabitants, whether heterosexual or gay/lesbian, may obtain a divorce. Some couples may be cohabiting precisely to avoid the prospect of going to court should their relationship sour. However, they are likely to find that the absence of a venue in which to resolve separation-related disputes in a standardized way is also a problem. (The legal side of living together is discussed in more detail in Chapter 8.)

Courts in some states are beginning to grant orders of legal separation or orders of dissolution of civil unions to the nonmarried (Hartocollis 2007; "Iowa Supreme Court" 2007). These may apply the same principles of property arrangements, child custody, and child support to nonmarital cohabiting relationships (Hartocollis 2007). Cohabiting couples may also avail themselves of mediation services.

Divorce Mediation **Divorce mediation** is an alternative, nonadversarial means of dispute resolution by which a couple, with the assistance of a mediator or mediators (frequently a lawyer–therapist team), negotiate the settlement of their custody, support, property, and visitation issues. In the process, it is hoped that they learn a pattern of dealing with each other that will enable them to resolve future disputes. Mediation is recommended or mandatory in all states for child custody and visitation disputes before litigation can be commenced (Comerford 2006).

Early research indicated that couples who use divorce mediation have less relitigation, feel more satisfied with the process and the results, and report better relationships with ex-spouses and children (Bailey and McCarty 2009; Holtzworth-Munroe, Applegate, and D'Onofrio 2009). Recent research summaries are less enthusiastic: "Overall, the corpus of available research does not indicate that mediation (relative to more traditional litigation) serves to either increase or decrease general psychological distress or . . . improve co-parenting relations." However, people were more satisfied with mediation than litigation in child custody cases: "Your feelings were understood" and "Your rights were protected" (Sbarra and Emery 2006, pp. 556–57).

There are arguments for and against mediation in child custody. Women's advocacy groups have claimed that mediation may be biased against females in that they may be less assertive in negotiations. It is also argued that mediators take insufficient account of prior domestic violence (Comerford 2006; Freeman 2008, note 16). Judith Wallerstein believes that the positive effects of divorce mediation for children may be overstated, but new practitioners are attempting to level the field (2003, p. 80; see also Freeman 2008 for a discussion on positive and negative qualities of mediation). Yet it does seem that "[m]ediation produces higher levels of compliance [with court decisions] and lower relitigation rates than litigation or attorney-negotiated settlement." It is less costly and generally less time-consuming than litigation (Comerford 2006; Crary 2007a).

and this does happen. In response, all fifty states have passed grandparent visitation laws. However, a Supreme Court decision struck down Washington's law (*Troxel v. Granville* 2000) because it was considered to interfere with parents' rights to determine how their children are to be raised. The status of other states' laws is uncertain; some courts have allowed grandparents visitation rights in certain circumstances (Dao 2005; Henderson and Moran 2001; Hsia 2002; Stoddard 2006).

In favorable circumstances, grandparents very commonly become closer to grandchildren, as adult children turn to grandparents for help or grandchildren seek emotional support (Henderson et al. 2009; Ruiz and Silverstein 2007). Researchers and therapists have concluded that

> these relationships work best when family members do not take sides in the divorce and make their primary commitment to the children. Grandparents can play a particular role, especially if their marriages are intact: symbolic generational continuity and living proof to children that relationships can be lasting, reliable, and dependable. Grandparents also convey a sense of tradition and a special commitment to the young. . . . Their encouragement, friendship, and affection has special meaning for children of divorce; it specifically counteracts the children's sense that all relationships are unhappy and transient. (Wallerstein and Blakeslee 1989, p. 111; see also Henderson et al. 2009 for a discussion on the importance of grandmother/grandchild relationships and adolescent postdivorce psychological adjustment)

The Community Divorce

Marriage is a public announcement to the community that two individuals have joined their lives. Marriage usually also joins extended families and friendship networks and simultaneously removes individuals from the world of dating and mate seeking. The **community divorce** refers to ruptures of relationships and changes in social networks that come about as a result of divorce. At the same time, divorce provides the opportunity for forming new ties.

Kin No More? Given the frequency of divorce, most extended families find themselves touched by it. Grandparents fear losing touch with grandchildren,

Divorce affects the extended family as well as the nuclear one. In some families, grandparents may lose touch with grandchildren, whereas in others, they may become more central figures of support and stability.

© Paul Barton/ CORBIS

Indeed, children who were close to their grandparents had fewer problems adjusting to their parents' divorce (Connidis 2009; Ruiz and Silverstein 2007).

Of course, more and more grandparents' own marriages are not intact today. Nevertheless, one can assume that even a loving, divorced grandparent could add to the support system of a grandchild of divorce.

Women are more likely than men to retain in-law relationships after divorce, particularly if they had been in close contact before the divorce and if the in-law approves of the divorce (Connidis 2009). Relationships between former in-laws are more likely to continue when children are involved. In any event, grandchildren were most likely to remain closest to maternal grandparents, as mothers typically grew closer to and relied more on their parents after divorce (Connidis 2009; Henderson et al. 2009).

Divorce and remarriage tend to connect chains of people in complex kinship systems. One study looked at the general character of postdivorce extended-kin relationships. This study found that in half the cases, the kinship system included **relatives of divorce** and *relatives of remarriage* (C. Johnson 1988, p. 168). These would be familial connections established through networks of marriage and remarriage: grandparents of half siblings, for example. A photo in this chapter portrays a young boy with his *eight* grandparents, all in attendance at his basketball game (Harmon 2005a).

After divorce, adult children's relationships with their own parents may change. According to one study:

Members of both generations had to revise their expectations of the other, and members of the older generation found themselves in a situation of having to give more of themselves to a child than they had expected to do at their stage of life. They were often forced into a parenting role, and this greater involvement provided more opportunity to observe and comment on their adult child's life. . . . Adult children were more likely to feel that parents should be available to help them with their emotional problems than their parents felt was appropriate. Divorcing children did not want their parents to interfere in child rearing or offer unsolicited advice, while their parents felt they could voice their concerns. (C. Johnson 1988, pp. 190–91)

There was considerable variation in these relationships within the sample of fifty-two adult–child dyads followed over several years in Johnson's study. But most older parents espoused "modern values of personal freedom and self-fulfillment" (p. 191); that is, they did not criticize the decision to divorce from a traditional perspective.

Friends No More? A change in marital status is likely to mean changes in one's community of friends. Divorced people may feel uncomfortable with their friends who are still married because activities are done in pairs; the newly single person may also feel awkward. Couple friends may fear becoming involved in a conflict over allegiances, and they may experience their own sense of loss. Moreover, if married friends have some ambivalence about their own marriages, a divorce in their social circle may cause them to feel anxious and uncomfortable. A common outcome is a mutual withdrawal.

Like many newly married people, those who are newly divorced must find new communities to replace old friendships that are no longer mutually satisfying. The initiative for change may in fact come not only from rejection or awkwardness in old friendships but also from the divorced person's finding friends who share with him or her the new concerns and emotions of the divorce experience. Priority may also go to new relationships with people of the opposite sex; for the majority of divorced and widowed people, building a new community involves dating again.

Deciding knowledgeably whether to divorce means weighing what we know about the consequences of divorce. The next section examines the economic consequences of divorce.

The Economic Consequences of Divorce

Social scientists and policy makers worry about the economic consequences of divorce, especially for children, but also for women and men.

Divorce, Single-Parent Families, and Poverty

Figure 15.4 displays the proportions of children who were living in poverty in 2007, for the largest racial/ethnic groups, comparing poverty rates by family type. As you can see, 43 percent of all children who reside in mother-only, single-parent families live in poverty. This compares to 9 percent of those living in married-couple families. The relationship between family type and poverty is consistent across racial/ethnic categories.

Another consistent research finding is that divorce is related to a woman's and sometimes a man's lowered economic status. "Divorce carr[ies] economic costs" (Sayer 2006, p. 392).

Husbands, Wives, and Economic Divorce

Upon divorce, a couple undergoes an **economic divorce** in which they become distinct economic units, each with its own property, income, control of expenditures, and responsibility for taxes, debts, and so on.

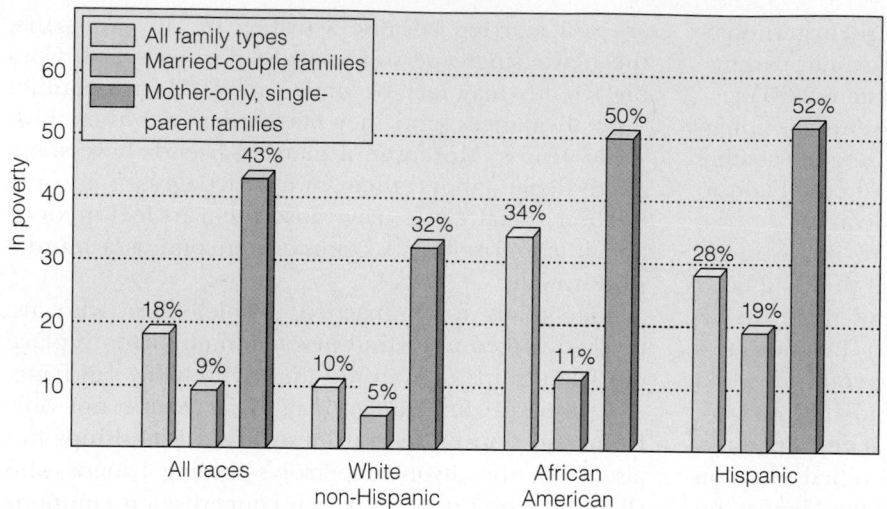

FIGURE 15.4 Poverty status of families with children under eighteen by race/ethnicity and type of family, 2007 (percentage of families with incomes below the poverty level).

Source: Adapted from U.S. Federal Interagency Forum on Child and Family Statistics 2009, Table ECON1A.

Women Lose Financially in Divorce Sociologist Lenore Weitzman addressed the financial plight of divorced women and their children in her landmark book *The Divorce Revolution* (1985). She compared the postdivorce economic decline of women with children to the improved standard of living of ex-husbands.[9] Women and their children experience declines in family income of between 27 percent and 51 percent (depending on the research study). An even more telling statistic is the **income-to-needs ratio**—that is, how well income meets financial needs. Women and their children experience a decline of 20 to 36 percent in their income-to-needs ratio (Meadows, McLanahan, and Knab 2009; Sayer 2006).

A study that compared former spouses in terms of their postdivorce economic situations found that wives with custody of children had only 56 percent of the income relative to needs that noncustodial fathers had (Bianchi, Subaiya, and Kahn 1999). Another study, which differentiated between income quintiles, found that mothers whose income was in the top two quintiles suffered little income loss (with some experiencing economic gains) after the divorce, whereas mothers in the bottom three economic quintiles experienced large losses after the divorce (Ananat and Michaels 2007). Additionally, Braver, Shapiro, and Goodman (2006) contend that the differences in income and

taxation rates between custodial and noncustodial parents generally lead to better financial circumstances for custodial mothers than noncustodial fathers. Regardless of the fineness with which social researchers split the proverbial hair, there remain important differences between male and female outcomes of a divorce.

A fundamental reason for the income disparity between ex-husbands and their former wives is men's and women's unequal wages and different work patterns (Hartmann, English, and Hayes 2010; Oldham 2008a). Despite women's greater participation in the labor force, any reduction in employment during childbearing and child-raising years means they have forgone opportunities for career development. "[A]lthough women are moving toward greater equality with men in the labor market, they remain more economically vulnerable when marriages end" (Bianchi, Subaiya, and Kahn 1999, p. 196).

It is also the case that women who are custodial parents must depend on child support from the other parent to meet their new single-parent family's expenses. Child support amounts are set relatively low, and much child support remains unpaid (as discussed in a later section of this chapter).

A third reason has to do with the typical division of property in divorce. Most state laws require a division of property that is specified as "equitable" (Oldham 2008b). Behind the idea of a fair property settlement run two legal assumptions: The first is that marriage is an economic partnership. A man could not earn the money he earns without the moral support and domestic work of his wife, whether or not she was employed during the marriage. A minority of states have community property laws based on the premise that family property belongs equally to both partners. The remaining states, the majority, have laws promising a divorced wife either an equitable (fair) or an equal (exactly the same) share of the marital property.

A second assumption is that property consists of such tangible items as a house or money in the bank, or other investments. Yet, except for very wealthy people, the valuable "new property" (Glendon 1981) in today's society is the earning power of a professional degree, a business or managerial position, work experience, a skilled trade, or other human capital. When property is divided in divorce, the wife may get an equal share of tangible property, such as a house or savings, but usually that does not put her on an equal footing with

[9] Weitzman's book brought attention to the different economic outcomes of divorce for men and women. Her specific figures were later proven to be erroneous, but a reanalysis of her data still showed a substantial loss for divorced women compared to a slight gain for men (R. Peterson 1996).

Women and their children experience a substantial decline in their standard of living after a divorce. They may need to move to less-expensive—and less-desirable—housing and away from their former neighborhood, school, and friends. Many men also experience a decline in standard of living.

her former husband for the future. An even split of the marital property may not be truly equitable if one partner has stronger earning power and benefits than the other, and if the parent with custody of the children has a heavier child support burden in actuality (even paid-up child support is typically not adequate to meet children's expenses). Put another way, dividing property may be easy compared to ensuring that both partners and their children will have enough to live on comfortably after divorce or at least will be on a similar financial footing. "Most women would have to make heroic leaps in the labor or marriage market to keep their losses as small as the losses experienced by the men from whom they separate" (McManus and DiPrete 2001, p. 266).

In part, this situation results from the assumption of legislators and the courts that women and men would come to have equal earning power. When divorce laws were reformed in the 1970s, self-support for both parties was presumed. Some financially dependent spouses have been awarded short-term **spousal support** or *maintenance*[10] in the form of *rehabilitative alimony*, in which the ex-husband pays his ex-wife maintenance for a few years

[10] These terms have replaced *alimony* to describe a former spouse's support payment to his or her ex-spouse following divorce. Historically, alimony was a payment of husband to wife resting on the assumption that the contract of marriage included a husband's lifetime obligation to support his wife and children. Traditionally, of course, the wife had not been employed, but instead had primary responsibility for making a home and bearing and raising children. Popular myth had it that ex-wives lived comfortably on high alimony awards. But in reality, courts awarded alimony to only a small minority of wives.

while she prepares to reenter the job market. In many, perhaps most, cases, this is not truly enough to enable the woman to reestablish herself financially, and it is not that commonly awarded in any case (Oldham 2008a).

Some activists have argued that wives who left the labor force to raise children or to help with their husband's career deserve not alimony but an *entitlement*, the equivalent of severance pay for the work they did at home during the length of a marriage (Weitzman 1985). (Social Security provisions do allow an ex-wife who had been married at least ten years to collect 50 percent of the amount paid her ex-husband.) Given the risk of divorce, a recent spate of books have warned today's career women not to "opt out" of the labor force to nurture small children, but to maintain employability (Bennetts 2007; Hirschman 2006). In the future, more women who are divorced *will* have a history of almost continual employment and a current job, although they are still likely to have the heavier expenses of a custodial parent.

Anticipating the difference that the stronger labor force attachment and likely higher earnings of younger women may make, one review article takes the position that the long-term divorce disadvantage for women compared to men is, or at least will be, less striking than it now appears (Braver, Shapiro, and Goodman 2006). These authors argue that more research using long-term data is needed and should include custodial and noncustodial parents' differences in tax status and a father's expenses during visitation. They argue that "[i]t is premature to say exactly how the two parents compare in [postdivorce] economic well-being" (p. 324).

Some Men Win, Most Lose Financially in Divorce In fact, the postdivorce economic situation of men has undergone some rethinking. Circumstances have changed since the 1970s, and men and women are both likely to be family earners now. A study covering the years between 1980 and 1993 (McManus and DiPrete 2001) found that most men lose economically in divorce (or in a separation from a cohabiting union). The chief reason for their declining standard of living is the loss of the partner's income. Depending on the study, married men experience a decline in *family income* of from 8 percent to 41 percent; cohabiting men's drop in income is comparable (Sayer 2006). Additional studies examining child support, income, and taxes, suggest that fathers actually suffer significant financial losses in a divorce. The analysis suggests that the financial loss has less to do with a

In 1979, the U.S. Supreme Court determined that laws against sex discrimination should make alimony gender-neutral—in essence, transforming the basis for spousal support from the common-law tradition of husbands' and wives' specialized roles into an economic partnership model. Along with this and other reforms of divorce law came a presumption that spouses should be self-supporting after divorce (Buehler 1995, pp. 102–11).

decline due to loss of a partner's income, and more to do with taxation rates for noncustodial parents—usually fathers. The argument is that the custodial parent is taxed differently (taxed at lower rates), and, coupled with child support, receives benefits that increase financial benefits, whereas the noncustodial parent is now taxed at a much higher rate because (usually) he has no dependents to claim (Braver, Shapiro, and Goodman 2006).

However, though family income drops for a man when he is no longer part of a couple, so do his expenses; his household is now smaller. Consequently, his *income-to-needs ratio* rises anywhere from 8 percent to 41 percent, even taking child support into account (Sayer 2006). But usually, that is not enough to maintain his previous standard of living and quality of material life.

To sum up, though there are no women "winners" in divorce, a majority of men lose, too. Only men who had contributed at least 80 percent of their family's predivorce income in a traditional marriage gain economically in divorce. Still, to continue the gender comparison, "[s]tudies that focus on women's outcomes have yet to unearth any comparable core of women who gain financially following union dissolution" (McManus and DiPrete 2001, p. 266).

Child Support

Child support involves money paid by the noncustodial to the custodial parent to support the children of a now-ended marital, cohabiting, or sexual relationship. Because mothers retain custody in the preponderance of cases, the vast majority of those ordered to pay child support are fathers.

For many years, the child support awarded to the **custodial parent** was often not paid, and states made little effort to collect it on the parent's behalf. Policy makers' concerns about poverty, the economic consequences of divorce for women, and, in particular, welfare and social services costs led to a series of federal laws that have changed that situation considerably.[11] In recent years,

government authorities have been more successful at securing payment, and at standardized amounts that are often higher than previously. With better child support enforcement, the poverty of custodial parents and their children dropped from 33 percent in 1993 to 24.6 percent in 2007.

Nevertheless, child support awards have historically been and continue to be small. In 2008, 54 percent of custodial parents had child support awards, and three-fourths of those received payments. Only 46.8 percent received full payment of what was due, however, and the amounts involved are not very impressive, averaging less than $3,500. Some noncustodial parents do make additional contributions in the form of gifts, clothes, food, medical costs (beyond health insurance), and camp or child care.

Because psychologists and sociologists have understood for some time that noncustodial parents' involvement in their children's lives is beneficial to not only the children, but also to the psychological and emotional health of the parents as well (Schindler 2010), the question is therefore asked: Why do some parents not pay their child support obligations? Some research suggests that the principal reason for a noncustodial parent's failure to pay is unemployment or underemployment (Sorensen 2010). Among families in which the absent parent has been employed during the entire previous year, payment rates are 80 percent or more. Not so when unemployment is involved. "The key to reducing poverty [among single-parent families] thus appears to be the old and unglamorous one, of solving un- and underemployment, both for the fathers and the mothers" (Braver, Fitzpatrick, and Bay 1991, pp. 184–85).

Some noncustodial fathers provide support in ways other than money (Garasky et al. 2007, p. 4,401), such as child care: "In some cases attempts to locate and require payments from such fathers may result in severing these ties" (Peterson and Nord 1990, p. 539). Compliance may be related to the noncustodial parent's involvement in the child's life. Seventy-eight percent are in compliance when they have either joint custody or visitation arrangements; only 67 percent are in compliance when the parent has neither.

Two suggested solutions to the problem of nonpayment of child support are guaranteed child support and a children's allowance. Both are based on the principle of society-wide responsibility for all children. With **guaranteed child support**, a policy adopted in France and Sweden, the government sends to the custodial parent the full amount of support awarded to the child. It then becomes the government's task to collect the money from the parent who owes it. A second alternative, a **children's allowance**, provides a government grant to all families—married or single-parent, regardless of income—based on the number of children they have. All industrialized countries except the United States

[11] The Child Support Enforcement Amendments (1984) to the Social Security Act, the Family Support Act (1988), and the child support provisions of the Personal Responsibility and Work Opportunity Reconciliation Act (1996) did the following: (1) encouraged the establishment of paternity and consequent child support awards, (2) required states to develop numerical guidelines for determining child support amounts, (3) required periodic review of award levels to amend them for inflation and to ensure that the noncustodial parent continued to pay an appropriate share of his or her income, and (4) enforced payments through locator services to find nonpaying noncustodial parents. States were required to implement automatic wage withholding of child support, and some states imposed penalties such as revoking a driver's license, seizing a delinquent payer's assets, or garnishing his or her wages ("Child Support Collected" 1995; Garfinkel, Meyer, and McLanahan 1998; Pirog-Good and Amerson 1997). Perhaps the most interesting collection device is that of the state of Maine, where child support must be paid before worm digging or moose hunting licenses will be issued (Koch 2006).

have some version of a children's allowance. In the present political and economic climate in the United States, it seems unlikely that such measures would be adopted.

As another approach to securing payment of child support, some states have begun experimenting with "responsible fatherhood" programs, often supported by government grants. Low-income fathers are typically expected to pay a higher portion of their income in child support than are middle-class fathers, resulting in a spiral of expanded debt and often withdrawal from their children (Huang, Mincy, and Garfinkle 2005). Recognizing that there are low-income fathers who want to provide support for their children but lack the income to do so, these multifaceted programs provide employment services, family support, and mediation services. Results thus far are only modest. Some increase in child support payments has occurred, though not yet enough to produce the dramatic improvement in the lives of men and their children that program designers had hoped for (Sorenson 2010).

We have been speaking of child support in the context of marriage and divorce, but some courts have awarded child support when same-sex couples who have been raising children together break up (Graham 2008; Kravets 2005; "State Court Orders" 2005). California courts provide an insight into the legal thinking behind child support decisions when same-sex couples break up or dissolve their marriages or civil unions.

The "California Supreme Court confirms the rights of a lesbian mother to have the legal rights of a natural mother" (Graham 2008, p. 1022). It does this using the Uniform Parentage Act to hold the noncustodial parent responsible for child support. According to section 7611 of the Act, the way "a man can be presumed to be the natural father of a child is if he receives the child into his home and openly holds out the child as his natural child. The court construed this statute to apply to mothers who . . . entered into a lesbian partnership and committed to a relationship with children born to their lesbian partner" (Graham 2008, pp. 1,023–24).

We turn now from the economics of postdivorce family support to a broader examination of the aftermath of divorce for children.

Divorce and Children

More than half of all divorces involve children under eighteen, and about 40 percent of children born to married parents will experience marital disruption (Amato 2000). How do separation and divorce affect children? There is strong disagreement on the answer to this question. One research review spoke of the "polemical nature of divorce scholarship" (Amato 2000, p. 1,270).

Outcomes for children depend a great deal on the circumstances before and after the divorce (Amato 2010). Although the divorce experience is psychologically stressful and, in most cases, financially disadvantageous for children, children in high-conflict marriages seem to benefit from a divorce. Living in an intact family characterized by unresolved tension and alienating conflict can cause as great or greater emotional stress and a lower sense of self-worth in children than living in a supportive single-parent family (Barber and Demo 2006). When the conflict level in the home has been low, however, children have poorer postdivorce outcomes. They are likely surprised by a divorce and seem to suffer more emotional damage. Among other things, it is difficult for them to see the divorce as necessary in this situation (Stevenson and Wolfers 2007).

The Various Stresses for Children of Divorce

We begin this section with the research of Judith Wallerstein and her colleagues. This group's research has been very influential in defining the situation of children of divorce for both professionals and the public.

The Wallerstein Research In their longitudinal study of children's postdivorce adjustment, psychologists Judith Wallerstein and Joan Kelly interviewed all of the members of some sixty families with one or more children who had entered counseling at the time of the parents' separation in 1971. Wallerstein and her colleagues reinterviewed children at one year, two years, five years, ten years, and, in some cases, fifteen years later and finally again at the twenty-five-year point (Wallerstein and Lewis 2007, 2008).

In the initial aftermath of the divorce, children appeared worst in terms of their psychological adjustment at one year after separation. By two years postdivorce, households had generally stabilized. At five years, many of the 131 children seemed to have come through the experience fairly well: 34 percent "coped well"; 29 percent were in a middle range of adequate, though uneven, functioning; and 37 percent were not coping well, with anger playing a significant part in the emotional life of many of them (Wallerstein and Kelly 1980). If the middle group is considered to be well enough adjusted, one can say that two-thirds of these children emerged from the divorce intact.

Wallerstein and Lewis performed a twenty-five-year follow-up with fifty-two of the families and found four different types of adult–child and father relations. Their Group A, representing 29 percent of the fathers in the study, had no contact with their adult children. Group B made up 27 percent of the study's fathers. This is the group who seldom or sporadically saw their children, and had very limited contact with them as adults. The third group, Group C, composed 23 percent of the fathers in the study, and this group had good relations with their children postdivorce, which continued on into their children's adulthood. Finally, Group D had varied relationships with their adult children. This group appeared to exhibit some form of preferential treatment to some children—providing economic resources for their preferred adult children but not to the others (Wallerstein and Lewis 2007).

Children whose parents have divorced will more than likely have less money available for their needs. This is especially significant because some of the negative impact of divorce can be attributed to economic deprivation (Strohschein 2005; Wallerstein and Lewis 2007).

Wallerstein found considerable deprivation among even middle-class children compared to what they could have expected had their parents remained married. Because divorce settlements seldom include arrangements to pay for children's college education[12] and family savings are often eroded by divorce, financing higher education is especially problematic. Wallerstein, who followed her sample into young adulthood, was surprised at the extent of their educational downward mobility. Sixty percent of the children in the study were likely to receive less education than their fathers; 45 percent were likely to receive less than their mothers. Even divorced fathers who had retained close ties, who had the money or could save it, and who ascribed importance to education seemed to feel less obligated to support their children through college (Wallerstein and Blakeslee 1989; Wallerstein and Lewis 2007, 2008).

After following their sample of children of divorce for ten years, Wallerstein and her colleagues found the majority to be approaching economic self-sufficiency, to be enrolled in educational programs, and, in general, to be responsible young adults. Even so, the overall impression left by the Wallerstein research is one of loss. In their examination of the impact of divorce on the relationship between children and their fathers, Wallerstein and Lewis eloquently noted that the "twists and turns of the postdivorce father–child relationship, its high vulnerability to change, satisfactions laced with disappointments, and undercurrent of love, longing, and anxiety all call attention to the complexity of building a lasting father-child relationship outside the marriage" (Wallerstein and Lewis 2008, n.p.).

Children may lose fathers, who become uninterested and detached; they may lose mothers, who are overwhelmed by the task of supporting the family and managing a household alone and who either see little chance of happiness for themselves or are busy pursuing their "second chance." Children of divorce experience the loss of daily interaction with one of their parents. Boys, especially, seem to find it difficult to

establish themselves educationally, occupationally, or maritally (Wallerstein and Blakeslee 1989; Wallerstein and Lewis 2007). Wallerstein found that half of the children in her study had experienced a second divorce of one or both parents. ("My Family: How It Feels When Parents Divorce" illustrates many points raised in this section.)

The Wallerstein study has methodological problems: it was a small, unrepresentative sample recruited by offering free counseling to the family; it lacked a control group; and there was difficulty separating family troubles and mental health concerns that predate the separation and divorce from those that might be effects of divorce. It has also been challenged by studies with more representative samples. These studies reach less-negative conclusions. Some critics have also noted that, when the study began in the early 1970s, women were less likely to be in the labor force. The need for an inexperienced mother to enter the labor force created an adjustment problem in the 1970s that would be less of a stress now.

Another methodological problem, although one shared by other longitudinal studies, is that continual interviewing of children about the impact of the divorce might create a mind-set in which any problems are given a divorce-generated interpretation (Ahrons 1994; Cherlin 1999, 2000; Coontz 1997a). As with all research that studies human behavior, the methodological weakness of this study is important, but so too is the qualitative data that a study of this sort provides. True, a large quantitative study provides an important snapshot into various moments in time, but the in-depth interviews found in this study provide rich details and allow for the subjects to be reflective about their own life histories and trajectories. It has allowed social researchers to see something they cannot see in quantitative data—the divergent experiences individuals have, and the impact divorce has had on each of the individual's lives.

Still, other researchers have come to think, like Wallerstein, that divorce has long-term effects (Amato 2010; Conger, Conger, and Martin 2010; Frisco, Muller, and Frank 2007). Divorce is a "risk factor for multiple problems in adulthood" (Amato 2000, p. 1,279). Children of divorce continue to have lower outcomes than children from intact families in the areas of academic success, conduct, psychological adjustment, social competence, and self-concept, and they have more troubled marriages and weaker ties to parents, especially fathers (Amato 2010; Kreider and Elliott 2009b; Magnuson and Berger 2009). See Chapters 7, 8, and 10 for additional discussions on marriage and better child outcomes.

Reasons for Negative Effects of Divorce on Children Researchers and theorists offer a variety of explanations for why and how divorce could adversely affect children. Amato (1993) has summarized five

---

[12] Some divorce agreements provide for support of children through college, provided they are doing well and advancing toward the goal of graduation. It has happened occasionally, but rarely, that courts have ordered such support against the wishes of a parent.

theoretical perspectives concerning the reasons for negative outcomes. We present his typology, along with some relevant research by others. Then we introduce an additional theoretical perspective.

1. The **life stress perspective** assumes that, just as divorce is known to be a stressful life event for adults, it must also be so for children. Furthermore, divorce is not one single event but a process of associated events that may include moving—often to a poorer neighborhood—changing schools, giving up pets, and losing contact with grandparents and other relatives (Benner and Kim 2010; Burrell and Roosa 2009; Jackson, Choi, and Bentler 2009; Teachman 2009; White et al. 2009). This perspective holds that an accumulation of negative stressors results in problems for children of divorce.

2. The **parental loss perspective** assumes that a family with both parents living in the same household is the optimal environment for children's development. Both parents are important resources, providing children love, emotional support, practical assistance, information, guidance, and supervision, as well as modeling social skills such as cooperation, negotiation, and compromise. Accordingly, the absence of a parent from the household is problematic for children's socialization.

3. The **parental adjustment perspective** notes the importance of the custodial parent's psychological adjustment and the quality of parenting. Supportive and appropriately disciplining parents facilitate their children's well-being. However, the stress of divorce and related problems and adjustments may impair a parent's child-raising skills, with probably negative consequences for children. Divorced parents do spend less time with children. Compared to married parents, divorced parents are "less supportive, have fewer rules, dispense harsher discipline, provide less supervision, and engage in more conflict with their children" (Amato 2000, p. 1,279; see also Schoppe-Sullivan, Schermerhorn, and Cummings 2007).

4. The **economic hardship perspective** assumes that economic hardship brought about by marital dissolution is primarily responsible for the problems faced by children whose parents divorce (Amato 2010). Indeed, economic circumstances do condition diverse outcomes for children—perhaps accounting for one-half the differences between children in divorced compared to intact two-parent families. Differences in outcomes exist *within* social class groupings, however. Children in better-off remarried or single-parent families still lag behind children from two-parent families on various outcome indicators (Amato 2010; Kreider and Elliott 2009b; Magnuson and Berger 2009; Waite, Luo, and Lewin 2009).

5. The **interparental conflict perspective** holds that conflict between parents is responsible for the lowered well-being of children of divorce. Many studies, including that of Wallerstein, indicate that some negative results for children may not be simply the result of divorce per se, but are also generated by exposure to parental conflict prior to, during, and subsequent to the divorce (Barber and Demo 2006; Benner and Kim 2010; Burrell and Roosa 2009; Jackson, Choi, and Bentler 2009; Teachman 2009; White et al. 2009).

Visitation is one frequent arena of postdivorce parental disputes. The child isn't ready to go when visitation time starts, or the visiting parent brings the child home late. Child support is another. The Stanford Child Custody Project, which followed over one thousand parents and children, found that a quarter of parents had a conflicted co-parenting relationship three and a half years after the divorce (M-Y. Lee 2002).

The factors that seem to affect co-parenting success are rather straightforward: a previous good co-parenting relationship during the marriage; a mediated rather than hostile divorce process; a reasonably good postdivorce relationship between ex-spouses; and length of time since the divorce. Research on the relationship between good co-parenting and type of custody is inconclusive except that a joint custody arrangement chosen by the parents is much more conducive to good co-parenting than one imposed by the courts. Research on the effect of remarriage on co-parenting is not sufficiently developed for conclusions to be drawn (Adamsons and Pasley 2006).

Multiple Transitions and Children's Outcomes The **family instability perspective** is an additional theory of children's negative outcomes of divorce that has emerged since Amato's original article. The **instability hypothesis** stresses that *transitions* in and out of various family settings are the key to children's adjustment (Barber and Demo 2006; Hetherington 2005). The logic of the instability hypothesis is this:

> Transitions may include parents' separation; a cohabiting romantic partner's move into, or out of, the home of a single parent; the remarriage of a single (noncohabiting) parent[;] or the disruption of a remarriage. The underlying assumption is that children and their parents, whether single or partnered, form a functioning family system and that repeated disruption of this system may be more distressing than its long-term continuation. . . . Stable single-parent households or stepfamilies, in contrast, do not require that children readjust repeatedly to the loss of coresident parents and parent-figures or the introduction of cohabiting parents and stepparents. (Fomby and Cherlin 2007, p. 182)

My Family

In the following excerpts, four children of divorce tell their own stories. As you will see, they talk about issues raised in this chapter.

Zach, Age 13

Even though I live with my Dad and my sister lives with my Mom, my parents have joint custody, which means we can switch around if we feel like it. I think that's the best possible arrangement because if they ever fought over us, I know I would have felt I was like a check in a restaurant—you know, the way it is at the end of a meal when two people are finished eating and they both grab for the check . . . but secretly neither one really wants it, they just go on pretending until someone finally grabs it, and then that one's stuck. . . .

My parents knew they couldn't live together, but they also knew it was nobody's fault. It was as if they were magnets—as if when you turn them the opposite way they can't touch. . . . Neither of them ever blamed the other person, so they worked it out the best they could—for their sakes and ours, too.

Nevertheless, it's very sad and confusing when your parents are divorced. I think I was five when they separated. . . .

When my parents first split up, it affected me a lot. . . . I got real fat and my grades went way down, so I went to a psychologist. She made me do a lot of things which seemed dumb at the time—like draw pictures and answer lots of silly questions. . . . My school work suffered because I was so distracted thinking about my situation that I couldn't listen very well, and for a long time I didn't work nearly as hard as I should have. Everyone told me I was an underachiever, and my parents tell me I still am, but I don't think so. What I do think is that I am a lot more independent—a go-out-and-do-it-yourself person. . . .

I've heard about kids who are having all these problems because their parents are getting divorced, but I can't understand what the big deal is. I mean, it's upsetting, sure, but just because your parents are separated it doesn't mean you're going to lose anybody. . . . It's not something I talk about very much. Most of my friends would rather talk about MTV than talk about divorce.

FIGURE 15.5.a and FIGURE 15.5.b Professional counselors often use art to gain insight into children's feelings about how divorce affects their family. The first drawing reveals the creative coping of a child whose parents are divorcing. She has figured out a way to *include* her father while keeping within the bounds of reality as she knows it. Her sister, on the other hand, used a jagged line to *separate* her father from the rest of the family.

Source: From *The Difficult Divorce: Therapy for Children and Families,* by Marla Beth Isaacs, Braulio Montalvo, and David Abelsohn. Copyright © 1986 by Basic Books, Inc. Reprinted by permission of Basic Books, a member of Perseus Books, L.L.C.

Ari, Age 14

When my parents were married, I hardly ever saw my Dad because he was always busy working. Now that they're divorced, I've gotten to know him more because I'm with him every weekend. And I really look forward to the weekends because it's kind of like a break—it's like going to Disneyland because there's no set schedule, no "Be home by five-thirty" kind of stuff. It's open. It's free. And my father is always buying me presents.

My mom got remarried and divorced again, so I've gone through two divorces so far. And my father's also gotten remarried—to someone I don't get along with all that well. It's all made me feel that people shouldn't get married—they should just live together and make their own agreement. Then, if things get bad, they don't have to get divorced and hire lawyers and sue each other. And, even more important, they don't have to end up hating each other.

I'd say that the worst part of the divorce is the money problem. It's been hard on my Mom because lots of times she can't pay her bills, and it makes her angry when I stay with my father and he buys me things. She gets mad and says things like, "If he can buy you things like this, then he should be able to pay me." And I feel caught in the middle for two reasons: First, I can't really enjoy whatever my Dad does get for me, and second, I don't know who to believe. My Dad's saying, "I don't really owe her any money," and my Mom's saying he does. Sometimes I fight for my Mom and sometimes I fight for my Dad, but I wish they'd leave me out of it completely.

Caleb, Age 7

My parents aren't actually divorced yet. But they're getting one soon. They stopped living together when I was one and a half, and my Dad moved next door. Then, when I was five, he moved to Chicago, and that hurt my feelings because I realized he was really leaving and I wouldn't be able to see him every day. My father's an artist, and when he lived next door to us in New York, I used to go to his studio every day and watch him when he was welding. I had my own goggles and tools, and we would spend many an hour together. I remember when I first heard the bad

news that he was moving away, because I almost flipped my lid. My father said he would be divorcing my Mom but that he wouldn't be divorcing me and we'd still see each other a lot—but not as often. I started crying then and there, and ever since then I've been hoping every single second that he'd move back to New York and we'd all live together again. I don't cry much anymore because I hold it back, but I feel sad all the same.

I get to visit my father quite often. And Shaun. He's my collie. My cat lives in New York with me and Mom. Whenever I talk with Daddy on the phone I can hear Shaun barking in the background. The hardest thing for me about visiting my father is when I have to leave, and that makes me feel bad—and mad—inside. I still wish I could see him every day like I did when I was little. It's hard to live with just one person, because you don't have enough company, though my Mom has lots of great babysitters and that helps a little.

Tito, Age 11

It seems like my parents were always fighting. The biggest fight happened one night when we were at a friend's house. Mommy was inside the house crying, and Daddy was out on the sidewalk yelling and telling my mother to come down, and my little sister, Melinda, and I were outside with a friend of my father's. We were both crying because we were so frightened. Then Daddy tried to break the door down, so Mommy came downstairs. And then the police cars came and Daddy begged Mommy to stay quiet and not say anything and to give him another chance, but she was so unhappy that she got into one of the cars. I was only four but I remember everything. We stayed with our cousin for about two months, and during this time I saw my father whenever he visited us at my grandmother's house. . . . I was always happy to see him, but sometimes it made me feel sad,

too, because I would look forward to our visits so much, and then when we were together it could never be as perfect as I was hoping it would be. He was still so angry at Mommy's leaving him that it was hard for him to feel anything else for anybody. . . .

About the time of the divorce I started to get into fights with other kids, and my mother got worried. She thought I must be feeling very angry and having a hard time expressing my feelings, so she took me to a therapist. . . . We got really close and he'd talk to me about my problems with my Dad. This went on for about two years, and during that time he helped me realize that the divorce was better for me in the long run because our home was more relaxed and there wasn't so much tension in the air.

The other thing that happened around this time was that my mother found out about an organization called Big Brothers, where I could have another male figure in my life. . . . They paired me off with a guy named Pat Kelly, and we've been getting together every weekend for a couple of years. . . . Pat and I do a lot of things like play baseball or video games and eat hot dogs. But the best thing we do is talk—like when I do something good in school I can tell him, and if I feel sad I can talk about that, too. His parents got divorced when he was twelve, and so we have a lot of the same feelings.

Critical Thinking

Overall, do you see these stories as hopeful, dismaying, or both? Why? Were there any particular points that you found surprising or interesting? What do these stories suggest to divorcing parents about how they might help their children cope with divorce?

Source: Excerpts from *How It Feels When Parents Divorce,* by Jill Krementz. Copyright © 1984 by Jill Krementz. Reprinted by permission of Alfred A. Knopf a division of Random House, Inc.

Andrew Cherlin develops this perspective in a book appropriately titled, *The Marriage-Go-Round.* He contrasts the American practice of movement in and out of marriage and relationships with the more stable family patterns of European and Oceanic nations, whether or not these are anchored in marriage (Cherlin 2009a). Cherlin and his colleague Paula Fomby tested the instability hypothesis, as well as a competing *selection hypothesis,* to assess whether the number of transitions produces lower cognitive (academic) outcomes or behavior problems or whether preexisting characteristics of the mother explain both household instability and the effects of that instability on children. They used the National Longitudinal Survey of Youth waves between 1979 and 2000, and a supplemental survey of the original respondents' children.

Multiple transitions did not seem to impact black children. The researchers were not able to determine why from the data they had, but speculate that black children may have more support from the extended family. It may also be the case that other stresses on many black families are so overwhelming as to overshadow changes in family structure.

Multiple transitions did not seem the key to explaining white children's *academic outcomes.* But the number of transitions *did* seem related to white children's *behavioral problems.* Even so, multiple transitions did not seem so powerful a negative influence as living in a single-parent, mother-only family in a child's early years.

A More Optimistic Look at Children in Divorce Having considered reasons for the negative effects of divorce on children, we now try to assess just how important divorce is in the lives of affected children. We've given considerable attention to the research of Wallerstein and her colleagues because it has been very influential. "Judith Wallerstein's research on the long-term effects of divorce on children has had a profound effect on scholarly work, clinical practice, social policy, and the general public's views of divorce" (Amato 2003, p. 332). In fact, this is not a definitive study, and the strongly negative conclusions about divorce that Wallerstein presents seem overstated. "Many of Wallerstein's conclusions about the long-term consequences of dissolution on children are more pessimistic than the evidence warrants" (Amato 2003, p. 332).

Nevertheless, concern that children of divorce are disadvantaged does not rest solely on Wallerstein's research and what many see as her exaggerated presentation of the dangers of divorce (Cherlin 1999). A "persuasive body of evidence supports a moderate version of [her] thesis" (Amato 2003, pp. 338–39; Hetherington 2005). The remaining question, then, is this: *How much does divorce affect children?*

E. Mavis Hetherington has been studying divorcing families for about the same length of time as Judith Wallerstein, and she has a much more optimistic view of the outcomes for children—and adults. Starting in 1974 in Virginia with parents of four-year-olds, forty-eight divorced and forty-eight married-couple families, her research ultimately included 1,400 stable and dissolved marriages and the children of those marriages, some followed for almost thirty years. Hetherington found that 25 percent of these children of divorced parents had long-term social, emotional, or psychological problems, compared to 10 percent of those whose parents had not divorced. However, in assessing the impact of divorce, she would emphasize the 75 to 80 percent of children who are coping reasonably well (Hetherington and Kelly 2002).

> Researchers have clearly demonstrated that, on average, children benefit from being raised in two biological or adoptive parent families rather than separated; divorced; or never married single-parent households.... But ... there is considerable variability, and the differences between groups, while significant, are relatively small. Indeed, despite the well-documented risks associated with separation and divorce, the majority of divorced children as young adults enjoy average or better social and emotional adjustment. (Kelly and Lamb 2003, p. 195, citations omitted; see also Kelly and Emery 2003)

Other scholars agree: "On average, parental divorce and remarriage have only a small negative impact on the well-being of children" (Barber and Demo 2006, p. 291; see also Demo, Aquilino, and Fine 2005, p. 125). All in all, divorce researchers seem to be moving to a middle ground in which they acknowledge that children of divorce are disadvantaged compared to those of married parents—and that those whose parents were not engaged in serious marital conflict have especially lost the advantage of an intact parental home. But many have moved away from simplistic or overly negative views of the outcomes of divorce. A theme that runs through virtually all studies on the impact of marriage and divorce on the well-being of children is socialization. At the end of the day, both negative and positive outcomes for children in both married and divorced households suggest that the behavior of the parents has the greatest impact on their children's well-being. If there is significant discord between married parents, that chronic discord will lead to poor outcomes for their children. If a divorced couple continues to have a relationship that is fraught with conflict, the outcomes for their children will be negative. If children of married, cohabiting, divorced, remarried, or single parents feel nurtured, loved, and supported by parents and families who engage in conflict resolution and work hard at getting along with one another, the outcomes for those children tend to be positive.

In fact, studies that reach different conclusions about overall outcomes show we can still learn much that is potentially useful about what postdivorce circumstances are most beneficial to children's development and what pitfalls to avoid. A good mother–child (or custodial parent–child) bond and competent parenting by the custodial parent seem to be the most significant factors (Amato and Cheadle 2008). Another highly important factor in children's adjustment to divorce is the divorced parents' relationship with each other (Ahrons 2004; Barber and Demo 2006; Freeman 2008). And good nonresident parental relationships are also a positive influence on outcomes (Demo and Fine 2010; Guzzo 2009c).

We now turn to issues of custody, the setting in which children will live after the divorce.

Custody Issues

A basic issue in a divorce of parents is the determination of which parent will take **custody**—that is, assume primary responsibility for caring for the children and making decisions about their upbringing and general welfare.

Custody after Divorce As formalized in divorce decrees, child custody is most commonly an extension of traditional gender roles. Divorced fathers typically have legal responsibility for financial support, while divorced mothers continue the physical, day-to-day care of their children. Eighty-three percent of custodial parents are mothers; 17 percent are fathers. (Not all "custodial mothers" in these government statistics have been divorced; 34 percent were never married.)

Custody patterns and preferences in law have changed over time. As part of a patriarchal legal system, fathers were automatically given custody until the mid-nineteenth century. Then the first wave of the women's movement made mothers' parental rights an issue. Emerging theories of child development also lent support to a presumption that mother custody was virtually always in the child's best interest, the so-called "tender years" doctrine (Artis 2004).

In the 1970s, states' reforms of divorce law incorporated new ideas about men, women, and parenthood; custody criteria were made gender neutral. Under current laws, a father and a mother who want to retain custody have theoretically equal chances. Judges try to assess the relationship between parents and children on a case-by-case basis. But because mothers are typically the ones who have physically cared for the child, and because many judges still have traditional attitudes about gender, some courts continue to give preference to mothers (Demos and Fine 2010). The "best interests of the child" or the more recent "primary caretaker"

"So what's your custody deal?"

standard was often assumed by judges to signal a choice of the mother (Artis 2004).[13]

When both parents seek custody, the odds of father custody are slightly higher when the children are older (Fox and Kelly 1995). Judges may have become more favorable to father custody, but there are no definitive studies on whether fathers are now winning more contested custody cases. One study found that, by 1995, court decisions were almost equalized, with 45 percent of mothers and 42 percent of fathers awarded sole custody and 9 percent sharing custody in disputed cases (Mason and Quirk 1997). Other data, taking as a starting point the original hope of each parent, found that "fathers get

---

[13] Several new legal approaches have appeared in recent child custody cases. One is the *friendly parent* concept, the idea that custody should favor the parent who is more likely to grant access to the child and foster the child's relationship with the other parent. Many states have incorporated the friendly parent doctrine into their statutory standards for custody as one factor among many or as the determining factor.

"There is, however, a small but growing movement to reject the friendly parent statute or limit its application" (Dore 2004, p. 43). This is because in practice it has generated hostility and litigation, as competing parents denigrate each other and sometimes try to provoke the other parent into behavior that will look bad in court. It has also made parents hesitant to raise legitimate child abuse or domestic violence allegations for fear of appearing critical of the other parent. Some courts have rejected the friendly parent doctrine, and some states are modifying their laws (W. Davis 2001; Dore 2004).

A second concept is that of *parental alienation syndrome*, originated by psychiatrist Richard Gardner. This is the idea that one parent has turned the child against the other parent without cause and that a parent may raise false allegations of child abuse (Lavietes 2003). This concept is most often introduced by fathers seeking custody.

Parental alienation syndrome has not received legal acceptance nor is it accepted by the American Psychiatric Association, the American Medical Association, or the American Psychological Association (American Psychological Association 2005b; Lavietes 2003). Courts have not to date permitted it to enter into consideration (e.g., *People v. Michael Fortin* 2000).

the arrangement they prefer less often than mothers do" (Braver, Shapiro, and Goodman 2006, p. 317).

It may be the case that mothers have become less inclined to insist on sole custody. Fatherhood scholar James Levine thinks that "[w]e're seeing some weakening of the constraints on women to feel they can only be successful if they are successful mothers," so they are more willing to concede custody to willing fathers (quoted in Fritsch 2001, p. 4). As with mothers, not all custodial fathers are or were married; in 2003, 21 percent were never married.

Generally, studies have found nothing to preclude father custody or to prefer it (Buchanan, Maccoby, and Dornbusch 1996; Luepnitz 1982). Neither does it seem to make a difference in a child's adjustment whether the custodial parent is the same sex as the child (Powell and Downey 1997).

Noncustodial Mothers With unpromising economic prospects and in the context of changing attitudes about gender roles, some mothers are relinquishing custody. There are more than two million noncustodial mothers (Sousa and Sorensen 2006), concentrated in the twenty-five to forty-five age range and lower- to middle-class economic level. Some of those mothers have lost custody of children due to their abuse or neglect, whereas others have voluntarily surrendered custody or lost a custody contest to the other parent (Eicher-Catt 2004b).

In an earlier study of noncustodial mothers based on interviews with more than 500 mostly white women in forty-four states, only 9 percent reported losing their children in a court battle or ceding custody to avoid a custody fight. The others voluntarily agreed to father custody and gave the following as reasons: money (30 percent); child's choice (21 percent); difficulty in handling the children (12 percent); avoidance of moving the children (11 percent); and self-reported instability or problems (11 percent; Greif and Pabst 1988, p. 88; see also Depner 1993).

More than 90 percent of mothers in the Greif and Pabst study reported that the experience of becoming the noncustodial parent was stressful. So was maintaining a relationship with the child. Deborah Eicher-Catt's qualitative study of noncustodial mothers (2004b) found that the minority of mothers whose custody was abrogated by the courts were restricted in their contact, perhaps permitted only supervised visitation with their children. The larger group who voluntarily relinquished custody found it hard to achieve a workable relationship with the child as well. They were unable to be traditional mothers, but found a "mother-as-friend" role insufficient and uncomfortable. Eicher-Catt advises noncustodial mothers to focus on building a relationship, rather than thinking in terms of the traditional maternal role. "A Closer Look at Family Diversity: A Noncustodial Mother Tells Her Story" describes one

mother's struggle to perform the mother role in this challenging context.

Judith Fisher, who also studied noncustodial mothers, believes that women should not relinquish custody just because they feel inadequate in comparison to their successful husbands. At the same time, she strongly supports the freedom of men and women to make choices about custody—including the woman's choice to live apart from her children—without guilt or stigma. She urges

> the negation of the unflattering stereotypes of noncustody mothers; . . . supportiveness of the woman's choice when it appears to have been well thought out; and . . . [not] blaming the mother when others (the children, the children's father, the courts) decide that the children should live apart from her. (1983, p. 357)

The Visiting Parent To date, most research and discussion on visiting parents has been about fathers, but one study did compare the two sexes. Nonresidential mothers were more apt to telephone and to engage in extended visits with children. But nonresidential mothers and fathers had essentially similar levels of visitation in terms of frequency and activities during the visit. Both were more likely to engage in only leisure activities with their children rather than spending time helping with homework or going to school activities. In reality, less frequent and more recreational visitation seem to be a result of structural factors: distance from the child's home; the difficulty of finding an appropriate setting for the visit; and the wish not to engage in conflict or disciplinary actions in the limited time spent with the child (S. Stewart 1999).[14]

Noncustodial fathers, like noncustodial mothers, find it difficult to construct a satisfying parent–child relationship. During the marriage, a father's authority in the family gave weight to his parental role, but this vanishes in a nonresidential situation. Geographical distance and conflict with the mother may also be barriers to frequent contact. Custodial mothers are effectively gatekeepers, facilitating, or not, the noncustodial father's relationship with his children (Adamsons and Pasley 2006; Sano, Richards, and Zvonkovic 2008).

---

[14] There are some special situations of visitation. In some communities, courts and social workers have developed programs to offer *supervised visitation* between a noncustodial parent and his or her offspring. In this situation, parent–child contact occurs only in the presence of a third party, such as a social worker or a court employee. Supervised visitation is often mandated in situations of alleged domestic violence, drug abuse, long absenteeism, or past imprisonment. Although no doubt a warranted precaution in many cases, supervision is also a hardship on the parent, who must not only visit in a strained situation but also cover the financial cost. Some fathers must forgo visits they desire because they lack money to pay the typical $100 a visit (Kaufman 2007).

A Closer Look at Family Diversity

A Noncustodial Mother Tells Her Story

As I fold the last batch of warm clothes from the dryer, I glance over at the kitchen clock on the wall. Oh dear . . . it's almost four-thirty. . . . It's Sunday and I need to get the guys ready to go home to their dad's. . . .

Nowadays, I live [in a room] in someone else's house, see my kids every other weekend, and pay child support. How can I continue to call myself a mother when I no longer provide their regular care and nurturance? Unlike some mothers . . . who lose custody of their children . . . I voluntarily chose my status. . . . With little money coming in as a full time student, I am unable to provide adequately for them. Like other mothers I've interviewed, making the difficult decision to leave the care of my children to someone else because I deemed it "in their best interest" is not a decision reached lightly. . . . I suddenly realize that it's awfully quiet in the house. "Ty . . . Zachary," I call out, with the basket of clothes now resting on one hip. No response. All I hear is the low drone of a televised basketball game coming from the . . . living room. That's odd, I wonder where they could be. . . . I'm sure they were here just a minute ago.

Have I been too involved in cleaning up our breakfast mess or doing laundry to notice that they'd disappeared? It is a rarity that my boys, Ty, age ten, and Zac, age six, would be unaccounted for during a weekend visit. After all, I consider our time together precious time, although very much punctuated and measured according to planned activities and events. Granted, I have been daydreaming about how good this weekend with them has been. I think I've managed to keep them sufficiently "entertained. . . ." I know that's not my only goal while they're with me, but I do want them to enjoy coming to see me.

Yes, let's see. Friday night we went to the movies. Yesterday afternoon I took them roller skating. . . . We stayed up late last night and watched a rented movie. Dare I say it felt like "family time," if only temporarily? This morning our time together was more improvised. . . . Zac talked me into making waffles for breakfast. How long has it been since I did that? Seldom do I cook anymore.

[After breakfast] I helped Ty finish some homework and talked with him about dealing with his math teacher, whom he hates. I must admit the "down time" with them has been nice. . . . It's the routine patterns of being together, the sense of everydayness, that we miss the most. . . .

Muffled voices outside my window bring me back from my thoughts. They must be outside. . . . "Mom, come here, quick!" Ty yells excitedly . . . [and Zac explains]: "[W]e're building a fort!"

"Well, that sounds pretty good guys, but it's almost time to go and you still haven't packed up yet. Your dad's expecting you for dinner, remember?" "Ah, Mom, can't we stay a little longer," Ty insists, "we just got started." "I know, honey, but this project will be here when you come to visit next time. . . ." I hear myself reluctantly saying. . . . I'm immediately filled with mixed emotions. Although I'm happy to see them finally comfortable enough to make some aspect of this experience their own, I have to begin the departure process. . . . If I don't get them moving now, I won't keep my agreed-on visitation schedule with their dad.

We've entered the kitchen and the door slams behind us, as if to punctuate my words and mark the beginning of our "departure routine." The three of us take the cue and scatter to make preparations.

Source: Adapted from Eicher-Catt 2004a. Deborah Eicher-Catt is assistant professor of Communication Arts and Sciences at Pennsylvania State University.

A related issue is **interference with visitation**, discussed earlier in conjunction with conflict between parents. Such a situation can be very frustrating for fathers who want to maintain close contact with their children (Perrine 2006). In 1998, Congress passed the Visitation Rights Enforcement Act, which requires states to recognize and enforce visitation orders of another state. The earlier Family Support Act of 1988 authorized court intervention programs such as intensive case supervision, mediation, parent education and so on, but so far experimental programs to address interference with visitation have had disappointing results (Pearson and Anhalt 1994; Turkat 1997). As a last resort, courts can order a change of custody.

A new marriage or cohabiting relationship was not itself a factor in decreasing visitation, but the presence of children in a new family, particularly biological children, did lead to a decline. Fathers seemed to find it difficult to parent their children across two families (Guzzo 2009c; Juby et al. 2007; Swiss and Le Bourdais 2009). Be that as it may, the situation of noncustodial fathers seems to have improved since earlier research found that many had detached from their children. "Although the incidence of joint physical custody has remained low, between 35% and 60% of children now have at least weekly contacts with their fathers in many locations, most often a brief midweek visit or overnight" (Kelly 2007). In Ahrons's longitudinal study of postdivorce families, 62 percent of now-adult children reported that their relationships with their father got better or at least stayed the same over the twenty years since the divorce. "Those children whose relationships got better or stayed the same benefited from significantly more father involvement during the first 5 years postdivorce, whereas low father involvement was associated with reports that their relationships with their fathers got worse" (Ahrons 2007, p. 59; Ahrons and Tanner 2003).

A father's minimal or decreased visiting may be painful for children, and so are visits that alienate rather than bond the child and the noncustodial parent. In one example, described in the Wallerstein study,

> [a]lmost always, there would be other adults around or adult activities planned. Carl watched hundreds of hours of television at his father's house, feeling more and more alone and removed from his earlier visions of family life. (Wallerstein and Blakeslee 1989, p. 79)

Children often forgave geographically distant fathers who did not appear frequently, but were very hurt by those nearby fathers who rarely visited. Recent works by Judith Wallerstein and Julia Lewis, and Liam Swiss and Céline Le Bourdais, discuss the complicated nature of these relationships between geographically close fathers who are distant in their relationships with their children.

Wallerstein and Lewis (2008) point out the complex array of hurt feelings, new relationships, trust issues, and other emotional factors that set postdivorce families and their relationships on trajectories that lead to a child's emotional distress. For example, in interviews with Wallerstein and Lewis, fathers spoke of their discomfort with having their adolescent daughters stay overnight with them, while the adolescent daughters of these fathers were themselves anxious about staying with their father if he lived alone. "Several spoke of their embarrassment about experiencing their early menstruation in the father's home" (p. 228). Swiss and Le Bourdais (2009) discuss the material conditions that impact the noncustodial father's relationship negatively. For example, they point out that working-class fathers, especially those who earn "lower incomes are more likely to be working in low-paying, part-time, or shift-oriented work where they may not be available to their children when the latter are free, that is, in the evenings and weekends" (p. 644).

The visitation of fathers does not always affect children positively (Marsiglio et al. 2000). In extreme cases—where there is verbal, physical, or sexual abuse—father contact may actually be damaging to children (King 1994). We've seen that some divorces are precipitated by alcoholism, drug abuse, or domestic violence; in such cases, visitation is not necessarily in the best interest of a child.

When the father enacts an authoritative parenting style (see Chapter 10) and when the visit does not lead to conflict between the parents, it has a favorable impact on the child's adjustment. The most recent studies do show a higher level of paternal parenting skills, so perhaps younger divorced fathers have been more involved with children in the marriage and have a better mastery of parenting during visitation.

Good relationships with noncustodial fathers foster better outcomes for children (Carlson 2006; White and Gilbreth 2001). Contact is a threshold requirement for a father's positive influence. Relationship quality and responsive parenting (that fathers consider the child's point of view and explain decisions) were found to have a positive effect regarding adolescents' "internalizing" (depression) and "externalizing" (aggressive and antisocial behavior) problems. Adolescents' relationships with their mothers were a more powerful influence on well-being, however. Noteworthy is that if an adolescent had a poor relationship with his or her mother, a good one with a nonresidential father seemed to make a difference. Adolescents with poor relationships with both custodial mothers and nonresidential fathers were, not surprisingly, at greatest disadvantage in terms of adjustment (King and Sobolewski 2006; Sobolewski and King 2005).

A custodial parent's moving away has become a significant postdivorce visitation issue, to be discussed later in this section. In some moving cases, as well as others, judges have ordered "electronic communication" or "virtual visits," by video conferencing, instant messaging, webcam setups, and the like. Although they can be costly, these virtual visits allow the noncustodial parent to talk, play chess, view art projects or at-home dance or music performances, and the like over the Internet. On the whole, such electronic communication enhances contact between parents and children. There is some concern that courts will come to rely on electronic visitation rather than the real thing: "'You can't virtually hug your child'" (Clemetson 2006c).

Child Abduction At the other extreme from dissociating from children is kidnapping one's children from the other parent. A study sponsored by the Justice Department reports that over 350,000 children were abducted by family members in 1999. Biological fathers were the most often to abduct their children, but 43 percent were abducted by biological mothers, and 14 percent by grandparents, with the rest by siblings, aunts, uncles, or mothers' boyfriends. Younger children (under six) are most apt to be abducted; there is no difference in the numbers of boys or girls. Most of the children abducted by family members were returned or located (Miller et al. 2008, p. 527).

Child snatching is frightening and confusing for the child, may be physically dangerous, and is usually detrimental to the child's psychological development. Yet for years, the abduction of a child by a biological parent (without custody) was not legally considered kidnapping—or at least not prosecuted as such. Now, however, due to the passage of the Uniform Child-Custody Jurisdiction and Enforcement Act and the Parental Kidnapping Prevention Act, states must recognize out-of-state custody decrees and do more to find the child and prosecute offenders (Fass 2003).

Another form of child snatching is international abduction, increasingly common when cross-cultural marriages fail and a parent has transnational ties. Of the 350,000 child abductions in the United States, it is estimated that 10,000 of them are "held in foreign countries by a parent of a different nationality" (Slota 2009, p. 73). Retrieving a child from another country is very difficult, despite the Hague Convention on the Civil Aspects of International Child Abduction (an international treaty that requires countries to recognize original custody determinations), and is very costly—upwards of $100,000 with no guarantee of retrieval. Advocates argue that the U.S. government has not been sufficiently aggressive in pursuing such cases, and parents often turn to private services that attempt re-abduction of the child back to the parent in the United States (Alanen 2007; Alvarez 2003; Sapone 2000; Slota 2009).

Child abduction is an extreme act, but it points up the frustration involved in arrangements regarding sole custody, or, in some cases, a lack of attention to the noncustodial parent's allegations of child abuse. For example, "50 percent claimed that their children were being abused, neglected or subjected to an unhealthy home environment by the other parent; 42 percent blamed unfair custody terms" (Alanen 2007, p. 12).

Joint Custody California was the first state to enact a statute on joint custody. "Currently joint custody is either recognized, presumed, or mandated in most states [and] provinces of the United States and Canada" (M-Y. Lee 2002, p. 672).

In **joint custody**, both divorced parents continue to take equal responsibility for important decisions regarding the child's general upbringing. When parents live close to each other and when both are committed, joint custody can bring the experiences of the two parents closer together, providing advantages to each. Both parents may feel they have the opportunity to pass their own beliefs and values on to their children. In addition, neither parent is overloaded with sole custodial responsibility. Joint custody gives each parent some downtime from parenting (M-Y. Lee 2002).

Joint custody agreements have two variations. One is *joint legal and physical custody*, in which parents or children move periodically so that the child resides with each parent in turn on a substantially equal basis. The second variation is *joint legal custody*—in which both parents have the right to participate in important decisions and retain a symbolically important legal authority—with physical custody (that is, residential care of the child) going to just one parent (Siebel 2006). Parents with higher incomes and education are more likely to have joint custody, and "[f]athers with higher levels of education and better financial resources were more likely to request, negotiate, and have shared physical custody" (Kelly 2007, p. 37).

Table 15.2 lists advantages and disadvantages of joint custody from a father's perspective. Shared custody gives children the chance for a more realistic and normal relationship with each parent (Arditti and Keith 1993). It results in more father involvement and in closer relationships with both parents (Kelly 2007). The parents of Zach, whose story is included in "My Family: How It Feels When Parents Divorce," have joint custody; this may be one reason for his conclusion that "just because your parents are separated it doesn't mean you're going to lose anybody."

The high rate of geographic mobility in the United States can make joint physical custody difficult. Even without that, some children who have experienced joint custody report feeling "torn apart," particularly as they get older. Although some youngsters appreciate the contact with both parents and even the "change of pace" (Krementz 1984, p. 53), others don't. The following account from eleven-year-old Heather shows both sides of a joint physical custody arrangement:

> The way it works now is we switch houses every seven days—on Friday night at five-thirty. . . . [E]ach of our parents likes to help us with our homework and when we stay with them for a week at a time it's easier for them to keep up with what we're doing. . . . And as long as they're divorced, I don't see any alternative because it wouldn't seem right to live with either parent a hundred percent of the time and only see the other one on weekends. But switching is definitely the biggest drag in my life. . . . My rooms are so ugly because I never take the time to decorate them—I can't afford enough posters and I don't bother to set up my hair stuff in a special way because I know that I'll have to take it right back down and bring it to the next house. Now I'm thinking that I'll try to make one room my real room and have the other one like camping out. I can't buy two of everything, so I might as well have one good room that's really mine. (quoted in Krementz 1984, pp. 76–78)

Joint custody is expensive. Each parent must maintain housing, equipment, toys, and often a separate set of clothes for the children and must sometimes pay for travel between homes if they are geographically distant. Mothers, more than fathers, would find it difficult to maintain a family household without child support, which is often not awarded when custody is shared.

Table 15.2 Advantages and Disadvantages of Joint Custody from a Father's Perspective

Advantages	Disadvantages
Fathers can have more influence on the child's growth and development—a benefit for men and children alike.	Children lack a stable and permanent environment, which can affect them emotionally.
Fathers are more involved and experience more self-satisfaction as parents.	Children are prevented from having a relationship with a "psychological parent" as a result of being shifted from one environment to another.
Parents experience less stress than sole-custody parents.	Children have difficulty gaining control over and understanding their lives.
Parents do not feel as overburdened as sole-custody parents.	Children have trouble forming and maintaining peer relationships.
Generally, fathers and mothers report more friendly and cooperative interaction in joint custody than in visitation arrangements, mostly because the time with children is evenly balanced and agreement exists on the rules of the system.	Long-term consequences of joint custody arrangements have not been systematically studied.
Joint custody provides more free social time for each single parent.	
Relationships with children are stronger and more meaningful for fathers.	
Parental power and decision making are equally divided, so there is less need to use children to barter for more.	

Source: Adapted from Robinson and Barret 1986, p. 89. Copyright © 1986 Guilford Publications. Adapted by permission.

There may be situations—an abusive parent, other domestic violence, or extremely high levels of parental conflict, for example—where sole custody is preferable (Hardesty and Chung 2006).

Research does not consistently support the presumption that joint custody is always best for children of divorced parents. The Stanford Child Custody Study (Maccoby and Mnookin 1992) found that children in mother custody did as well as those in joint custody: "'[T]he welfare of kids following a divorce did not depend on who got custody, but on how the household was managed and how the parents cooperated'" (psychologist Eleanor Maccoby in Kimmel 2000, p. 141; see also Barber and Demo 2006; Freeman 2008). One reviewer of the research literature concluded that there is "no consistent evidence of the superiority of one arrangement over another" (M-Y. Lee 2002, p. 673).

Another review of thirty-three studies of joint and sole custody (Bauserman 2002) did find that children in joint custody arrangements had superior adjustment. In fact, "joint custody and intact family children did not differ in adjustment" in terms of general adjustment, family relationships, self-esteem, emotional and behavioral adjustment, and divorce-specific adjustment (p. 98). This finding of the advantages of joint custody applies to *legal joint custody* as well as to *legal and physical joint custody*. Sole custody was not necessarily bad, but joint custody was simply better in terms of child outcomes.

When a Custodial Parent Wants to Move The desire of a joint custodian or a parent with sole custody to move and take the children "is the hottest issue in the divorce courts at the moment," according to Judith Wallerstein (in Eaton 2004, p. A-1). Some divorce decrees mandate judicial consent and/or the consent of the other parent

for a move to another locality. Other cases have gone to court as a result of a parent's initiating legal action to prevent a move.

In deciding these cases, judges tend to demand that the "custodial parent demonstrate that the move serves the best interests of the child" (Glennon 2008, p. 57). States tend to employ two tests to these cases: The first, the relocation doctrine "require[s] custodial parents who wish to relocate to provide notice of the proposed relocation to the other parent 60 days before the move" (Glennon 2008, p. 58). If the noncustodial parent does not give consent, the custodial parent must ask the court for permission to relocate.

These relocation doctrines prohibit custodial parents from moving if they want to retain custody or share joint custody (Cooper 2004; *Navarro v. LaMusga* 2004). "As a result, women have been torn between their wish to remarry or otherwise rebuild their lives and their wish to have their children reside in their homes at least part-time. . . . In checking with legal experts, I found no instance where the father's wish to move was contested" (Wallerstein and Blakeslee 2003, p. 201).

Some judges do see a constitutional issue in the right to move, especially in this "highly mobile society" (Elrod and Spector 2002, p. 595). This is the custody modification approach in which the courts assume that "as long as the parent has a legitimate reason to relocate . . . [it is the court's duty to simply] determine what custody arrangement is in the best interests of the child upon relocation" (Glennon 2008, p. 59). Courts are now more inclined to permit moves for remarriage or economic reasons than they were in the 1980s and 1990s (Kelly and Lamb 2003).

Judith Wallerstein has appeared in court in support of relocating parents, and her opinion has proven very influential. She takes the position that the well-being of the custodial parent and that parent's relationship with the child are the most important factors in adjustment, trumping the question of contact with the other parent. Another divorce researcher, Richard Warshak, presents the opposite point of view in advocating that attention be paid to the importance of a child's maintaining contact with each parent (Eaton 2004). There is little research to date on the issue of parental relocation and none yet on the adjustment of young children to parental relocation.

Parent Education for Co-Parenting Ex-Spouses

Forty-six states have begun to offer or require parent education for divorcing parents, with twenty-six of them making program attendance mandatory (Kelly 2007; Pollet and Lombreglia 2008, p. 375). In some communities, the children also meet in groups with a teacher or mental health professional (Pollet and Lombreglia 2008). The idea is that parents will continue to raise their children as **co-parents,** and they are likely to need help in meeting this new challenge. Evaluation forms completed after the sessions have shown predominantly positive responses. But "there has been no evaluation of whether the information provided is actively employed by parents in their relationship with each other or with the child or whether, indeed, there is any change in the child's well-being" (Wallerstein 2003, p. 86).

Parents in many states are now required to negotiate a "parenting plan" before their divorce is approved (McGough 2005). Again, we have no research yet as to the effectiveness of this requirement in facilitating cooperative postdivorce parenting.

This discussion of mothers' and fathers' custody, visitation, and child support issues suggests that being divorced is in many ways a very different experience for men and women.

His and Her Divorce

Perhaps nowhere is this difference more evident than in the debate over which partner—the ex-wife or the ex-husband—is the primary victim of divorce. Both are affected by the divorce, but often in different ways.

The first year after divorce is especially stressful for both ex-spouses. Divorce wields a blow to each one's self-esteem. Both feel they have failed as spouses and, if there are children, as parents (Demo and Fine 2010). They may question their ability to get along well in a remarriage. Yet each has particular difficulties that are related to the sometimes different circumstances of men and women. In this discussion, we are primarily speaking of divorced men and women who are parents.

Her Divorce

Women who were married longer, particularly those oriented to traditional gender roles, lose the identity associated with their husband's status. Getting back on their feet may be particularly difficult for older women, who usually have few opportunities for meaningful career development and limited opportunities to remarry (Yin 2008). Women of the baby boom generation and later have usually had significant work experience, so they may find it easier to reenter the work world, if they are not already there.

Divorced mothers who retain sole custody of their children often experience severe overload as they attempt to provide not only for financial self-support but also for the day-to-day care of their children. Monitoring and supervising children as a single parent is especially difficult (Braver, Shapiro, and Goodman 2006). Mothers' difficulties are aggravated by lingering gender discrimination in employment, promotion, and salaries and by the high cost of child care. They may have less education and work experience than their ex-husbands. All in all, custodial mothers frequently feel alone as they struggle with money, scheduling, and discipline problems. Objective difficulties are reflected in decreased psychological well-being (Doherty, Su, and Needle 1989; Ross 1995). An encouraging note, though, is that the poverty rate of single custodial mothers dropped significantly between 1993 and 2007, although at 27 percent, it remains higher than that of custodial fathers.

Although those experiencing marital dissolution are less happy than those who are married, another comparison gives us a picture of "her" divorce that is a bit brighter. A majority of women respondents to the National Survey of Families and Households (1992–1993) who

"Her" divorce often involves financial worries and task and emotional overload as she tries to be the complete parent for the children.

© Rameshwas Das

compared their lives before and after marital separation perceived improvement in overall happiness, home life, social life, and parenting, although not in finances or job opportunities (Furstenberg 2003, p. 172, Figure 1). Women, compared to men, are more likely to have built social support networks, and they do show greater emotional adjustment and recovery than men (Braver, Shapiro, and Goodman 2006).

His Divorce

Divorced men miss having daily contact with their children and are concerned about possible qualitative changes in their parent–child relationships as well (Braver, Shapiro, and Goodman 2006). In some ways, divorced noncustodial fathers have more radical readjustments to make in their lifestyles than do custodial mothers. In return for the responsibilities and loss of freedom associated with single parenthood, custodial mothers escape much of the loneliness that divorce might otherwise cause and are rewarded by social approval for raising their children (Demo and Fine 2010; Wallerstein and Lewis 2008). Many children of divorce, especially daughters, developed closer relationships with their mothers after the divorce (Wallerstein and Lewis 2007).

Custodial fathers, like custodial mothers, are under financial stress. Noncustodial fathers often retain the financial obligations of fatherhood while experiencing few of its joys. Whether it takes place in the children's home, the father's residence, or at some neutral spot, visitation is typically awkward and superficial. The man may worry that if his ex-wife remarries, he will lose even more influence over his children's upbringing. For many individuals, parenthood plays an important role in adult development: "Removed from regular contact with their children after divorce, many men stagnate" (Wallerstein and Blakeslee 1989, p. 143; Swiss and Le Bourdais 2009).

Ex-husbands' anger, grief, and loneliness may be aggravated by the traditional male gender role, which discourages them from sharing their pain with other men. Sociologist Catherine Ross (1995) compared levels of psychological distress for men and women in four different categories: marrieds, cohabitors, those who were dating, and those with no partner. Ross found that divorced men had the lowest levels of emotional support of any group, whereas emotional support among divorced women was "not that much lower than married women's" (p. 138). In situations of isolation and depression, men are more likely than women to be vulnerable to substance abuse and alcoholism (Braver, Shapiro, and Goodman 2006). Yet in most cases, men still hold the keys to economic security, and ex-wives suffer financially more than do ex-husbands.

The fact is that both men's and women's postdivorce situations would be somewhat alleviated by eliminating the economic discrimination faced by women, especially women reentering the labor force, by strong child support enforcement, and by constructing co-parenting relationships that give fathers the sense of continuing involvement as parents that most would like.

"His" divorce involves loss of time with children, as well as a more general loneliness. Being the "visiting parent" is often difficult, but maintaining the father–child bond is significant in a child's adjustment to divorce.

Monkey Business Images/Shutterstock

Some Positive Outcomes?

First came *Creative Divorce.* As divorce rates rose steeply in the 1960s and into the 1970s, many were heartened by Mel Krantzler's 1973 book, which offered the hope that some good would come of this painful experience.

The eighties and nineties saw an accounting of the all-too-real problems of divorce for children and adults. "Creative divorce" seemed not only ironic but almost maliciously misleading to those making the difficult decision of whether to divorce.

Another swing of the pendulum seems to be taking place. Researchers have begun to explore positive outcomes of divorce. Scholars and clinicians have begun to talk about

Not all divorces have the same outcomes, as research by E. Mavis Hetherington and her colleagues demonstrates. These researchers developed a typology to describe the adjustment to divorce of 238 divorced women and 216 men whom they had interviewed regularly over a ten-year period in the Virginia Longitudinal Study of Divorce and Remarriage. The variability of outcomes is striking.

The Virginia researchers developed a typology based on a cluster analysis of eleven adjustment measures: neuroticism, antisocial behavior, social maturity/responsibility, health, achievement, well-being and satisfaction, self-efficacy, autonomy, parenting competence, social relations, and self-esteem. Six patterns of adjustment emerged from the analysis: Figure 16.7 presents the percentages of men and women in each category at the ten-year point.

Enhancers composed 20 percent of the sample, with more women than men in the group. These individuals "grew more competent, well adjusted, and fulfilled" (Hetherington 2003, p. 324) and had good success at work, in social relations, as parents, and often in remarriages. Some had had a good start in terms of their predivorce qualities, while others were "women who looked ordinary until the stresses of divorce and the challenges of being a single parent activated competencies or forced them to seek out additional resources" (p. 324).

Goodenoughs "had some vulnerabilities, some strengths, some successes, and some problems. They fell in the middle on most personal characteristics.... Ten years after divorce, the goodenoughs' postdivorce life looked like their old predivorced life" (p. 324). At 40 percent, the Goodenoughs were the largest group, almost equally divided between men and women.

Seekers were those who "were eager to find a new mate as quickly as possible."

They are hard to quantify because they dropped out of this category once they repartnered. Seekers had less self-esteem and independence, and the men "required a great deal of affirmation.... [If not remarried, Seeker men] succumbed to anxiety, depression, and sometimes sexual dysfunction" (p. 324). They were very dependent on their partners, both old and new. Haste to remarry meant that they sometimes did not make good choices in remarriage partners.

Swingers were a predominantly male group that also declined in numbers over time as the attraction of a libertine lifestyle waned. At the ten-year point, fewer than 10 percent of the divorced people studied were Swingers.

More women than men were *Competent Loners*. "Healthy, well-adjusted, self-sufficient, and socially skilled, competent loners often had gratifying careers, active social lives, and a wide range of hobbies and interests.... [They] were often involved [in] intimate relationships, but these relationships were not enduring ... [as competent loners] had little interest in permanently committing to share

their lives with anyone" (p. 325). Hetherington characterizes Competent Loners, along with Enhancers, as "divorce winners" (p. 325).

Defeated individuals, a little over 10 percent of the divorced group at the ten-year point, had low social responsibility and self-esteem and high depression and antisocial behavior. In essence, they did not have satisfying lives.

Hetherington remarks that in analyzing divorce, many commentators assume that the "defeated" type is the "standard outcome of a marital breakup"—but this is not the case (p. 325). "When marriages dissolve, there is no one adaptive pathway adults follow. Some pathways may be destructive, others may be constructive and enhancing" (p. 329).

Critical Thinking

Hetherington and her colleagues have provided us with a set of types. How well do these types of postdivorce adjustment apply to situations that you may have witnessed or personally experienced?

Source: Hetherington 2003.

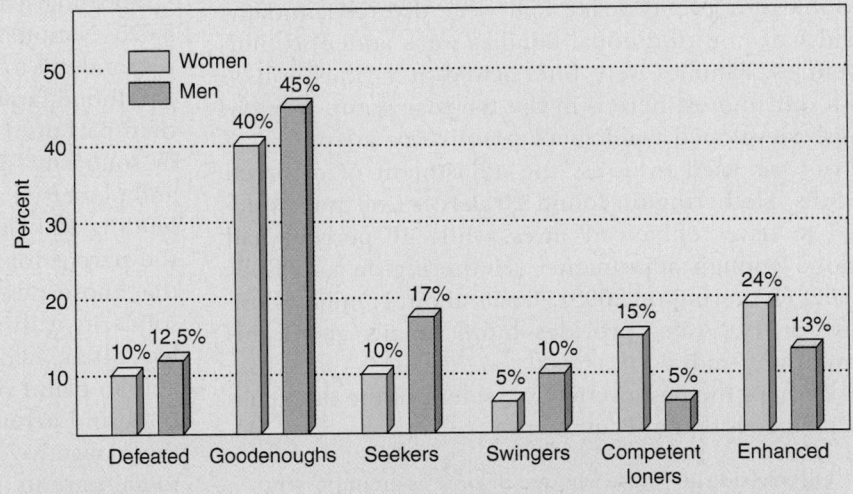

Figure 15.6 Postdivorce adaptive patterns of 216 men and 238 women at the ten-year point: Virginia Longitudinal Study of Divorce and Remarriage.

Source: From Hetherington 2003, pp. 318–331, Figure 3. Reprinted by permission of Blackwell Publishing.

stress-related growth (for children as well as adults). There is now more emphasis on the diversity of outcomes of divorce.

Those studying positive outcomes have connected this line of research to other research on stress-related growth. Stress-related growth can take different paths. A *crisis-related pathway* is marked when a traumatic event generates an ultimate result that makes the person stronger. (See Chapter 14, "Family Stress, Crisis, and Resilience," for a more general discussion of these ideas as applied to the family.) A *stress-relief pathway* occurs, when, for example, the end of a marriage and its problems brings relief to one or both of the partners. Kinds of growth include growth in the self; growth in interpersonal relationships (closer to family and friends); and growth or change in philosophy of life. The specifics are as yet a little vague, and more research is needed. Yet, scholars reviewing the literature conclude that

> [r]esearch on stress-related growth indicates that most individuals who have experienced traumatic events report positive life changes. . . . One thing that is clear from the existing research . . . is that it is at least as common to experience positive outcomes following divorce as negative one[s], and that positive outcomes can coexist with even substantial pain and stress. (Tashiro, Frazier, and Berman 2006, pp. 362, 364)

Another way of looking at stress-related growth—as well as less happy outcomes—comes from E. Mavis Hetherington's Virginia Longitudinal Study of Divorce and Remarriage. She and her colleagues followed 144 couples for twenty years; half were divorced initially, and half not. Additional families were added as time went by. Families were interviewed at various points, but our interest here is in the ten-year point. A previously developed typology of postdivorce adaptive patterns was used to assess the adjustment of divorced adults. Hetherington found 20 percent of those studied to have "enhanced" lives, while 40 percent had "good enough adjustment" (Hetherington and Kelly 2002; Hetherington 2003). "Facts about Families: Postdivorce Pathways" provides more details about this important study.

Perhaps the best overall assessment of the outcomes of divorce is that of Paul Amato:

> On one side are those who see divorce as an important contributor to many social problems. On the other side are those who see divorce as a largely benign force that provides adults with a second chance for happiness and rescues children from dysfunctional and aversive home environments. . . . Based on . . . research . . . it is reasonable to conclude that . . . [d]ivorce benefits some individuals, leads others to experience temporary decrements in well-being that improve over time, and forces others

on a downward cycle from which they might never fully recover. (Amato 2000, p. 1,282)

Adult Children of Divorced Parents and Intergenerational Relationships

We have talked about the general effect of divorce on children, but what do we specifically know about how a parental divorce affects adult children's married and family lives? Marital stability for adult children of divorce is discussed earlier in this chapter, as well as in Chapter 6. Here we address the topic of the quality of intergenerational relationships between adult children and their divorced parents.

There is evidence that adult children of divorced parents have probably come to accept their parents' divorce as a desirable alternative to ongoing family conflict (Ahrons 2004; Ahrons 2007; Wallerstein and Lewis 2007). Nevertheless, a number of studies point to one conclusion: Ties between adult children and their parents are generally weaker (less close, less supportive) when the parents are divorced (Ahrons 2007; Connidis 2009; Kelly 2007). The effect for divorced parents is stronger for fathers, usually the noncustodial parent, but the relationship has been found for mothers as well.

Sociologist Lynn White (1994) analyzed data from 3,625 National Survey of Families and Households respondents to examine the long-term consequences of childhood family divorce for adults' relationships with their parents. Using a broad array of indicators of family solidarity—relationship quality, contact frequency, and perceived and actual social support (doing favors, lending and giving money, feeling that one can call on the parent for help in an emergency)—White found that those raised by single parents reported lowered solidarity with them. As adults, they saw their parents less often, had poorer-quality relationships, felt less able to count on parents for help and emotional support, and actually received less support. White found these negative effects to be stronger regarding noncustodial parents (usually the father) but not limited to them.

In another study, sociologist William Aquilino (1994a) analyzed National Survey of Families and Households data from 3,281 young adults between ages nineteen and thirty-four. All grew up in intact families and had therefore lived with both biological parents from birth to age eighteen, but in about 20 percent of the sample, their parents had subsequently divorced. Aquilino found that even when parents divorce after the child is eighteen, the divorce seems to negatively affect the quality of their relationship. Children of divorced parents were in contact with their parents less often and reported lower relationship quality overall. These findings applied to both mothers and fathers, although the effect was much stronger for fathers.

Generally, evidence suggests that adult children of divorced parents feel less obligation to remain in contact with them and are less likely to receive help from them or to provide help to them. Social scientists (Lye et al. 1995) have posited four reasons for these findings:

1. Children raised in divorced, single-parent families may have received fewer resources from their custodial parent than did their friends in intact families, and thus they may feel less obliged to reciprocate.

2. Strain in single-parent families, deriving from the single parent's emotional stress or economic hardship or both, may weaken subsequent relations between adult children and their parents.

3. The reciprocal obligations of family members in different generations may be less clear in single-parent, postdivorce families.

4. Adult children raised in divorced, single-parent families may still be angry, feeling that their parents failed to provide a stable, two-parent household.

Remarriage and stepfamily relationships may generate similar tensions (remarriage and stepfamilies are discussed in Chapter 16).

Should Divorce Be Harder to Get?

In reviewing the process of divorce and its effects, this question arises: Should divorce be harder to get? Some Americans and some family scholars and policy makers think so. At the same time, the American public holds somewhat ambivalent attitudes about divorce. In one poll, one-half of those surveyed thought divorce should be harder to get (Stokes and Ellison 2010, p. 13, Table 1). It also appears that some unhappily married individuals postponed divorce until their children were older (Foster 2006; Schwartz 2010). Yet one poll found most people saying that they do want divorce laws to be tougher—but not when the divorce is their own ("The Divorce Dilemma" 1996).

Some family scholars argue that our high divorce rate signals the decline of the American family (see Chapter 1). To address this situation, policy makers have proposed changes in state divorce laws so that divorces would be more difficult to get than they have been since the 1970s. As noted earlier, with no-fault divorce laws, a marriage can be dissolved simply by *one* spouse's testifying in court that the couple has "irreconcilable differences" or that the marriage has suffered an "irretrievable breakdown."

Concerned about the sanctity of marriage, the impact on children of marital impermanence, and what seems to some a lack of fairness toward the spouse who would like to preserve the marriage, some states have developed—or at least considered—laws and policies that would make divorce harder to obtain. Three states—Louisiana (in 1997), Arizona (in 1998), and Arkansas (in 2001)—have enacted covenant marriage laws. Such laws have been proposed in a number of other states but not enacted (Nock, Sanchez, and Wright 2008).

Covenant marriage, also discussed in Chapter 7, is an alternative to standard marriage that couples may select at the time of marriage or later. It is essentially a return to fault-based divorce because it requires spouses to prove fault (adultery, physical or sexual abuse, imprisonment for a felony, or abandonment) or to live apart for a substantial length of time to obtain a divorce or to do both. Premarital counseling and counseling directed toward saving the marriage are also required. Although polls indicate that some Americans support covenant marriage (Covenant Marriage: A Fact Sheet 2010), only a small minority of couples have chosen this option—less than 2 percent in Louisiana (Nock, Sanchez, and Wright 2008). "Overall, covenant marriage has not proved as popular as supporters have hoped" (Zurcher 2004, p. 288), although those who have chosen covenant marriage are very satisfied (Nock et al. 2003). Most likely, this is because covenant marriage appeals more to highly religious couples who see it as a "symbol of their belief in a Christian marriage and as a public manifestation of their commitment to God" (Baker et al. 2009, p. 166). Chapter 7 discusses covenant marriage and other marriage support programs in detail.

Those who believe divorce is too readily available have proposed other restrictions on divorce or postdivorce arrangements. These have included the restoration of fault for all divorces; a waiting period of as long as five years; a two-tier divorce process, with a more extensive process for divorces involving children; prioritization of children's needs in postdivorce financial arrangements; requirement of a "parenting plan" to be negotiated prior to granting a divorce; and publicizing research that would convince the public of the risks of divorce (Stokes and Ellison 2010; Wilcox 2009). Many states and cities have established premarital counseling, marriage education, marriage counseling, or some combination of the three as either required or elective for couples planning to marry. Some locales require divorcing parents to attend a parent education program. Few object to such programs when voluntary, though some have objected to their being required. Research is lacking on their effectiveness in preventing divorce; however, research shows these kinds of programs are useful in addressing the needs of separating and divorcing couples, and improving the postdivorce lives of families (Belluck 2000; Pollet and Lombreglia 2008).

Opposition to restrictions on divorce centers around various points. First, divorce is not always or necessarily bad for children. Second, some marriages—those involving physical violence or overt conflict or both—are harmful to children and to one or both spouses. Divorce provides an escape from marital behaviors that may be more harmful than divorce itself, such as a parent's alcoholism or drug abuse (Coontz 1997a). An interesting study by economists Betsey Stevenson and Justin Wolfers (2004; 2007) found that no-fault divorce was associated with a decline in suicide rates for women, as well as a decline in domestic violence against both men and women, and a decline in intimate partner homicides of women. The existence of an escape route seems to change the balance of power and reduce violence.

It also appears that staying married would not necessarily provide economic stability for women and children. Those women who obtain divorces would be better off economically if they remained married, but they would not be as well off as that portion of the population that has not divorced; there are differences between the two groups in economic resources quite apart from divorce (Braver, Shapiro, and Goodman 2006; Stevenson and Wolfers 2007).

Those who oppose restoration of fault divorce point to the fraudulent practices that characterized the divorce process under fault statutes. Seldom was either party truly "innocent" of contributing to the marital difficulties, although one party had to pose as such. "Mental cruelty" was the most easily proven grounds for divorce, but the "evidence" for this was often exaggerated at best and completely trumped up at worst. In New York State—where adultery was the only grounds for divorce—"adultery" was often staged, in a drama organized by lawyers and colluded in by both spouses.

Is Divorce Necessarily Bad for Children?

Having thoroughly reviewed the literature and conducted longitudinal research on the subject, sociologists Paul Amato and Alan Booth (1997) conclude that children whose parents were continuously and happily married are indeed the most successful in adulthood. Children of divorce or those whose parents remained unhappily married were less successful. Amato and

Booth note the significant role that level of parental conflict plays in conclusions about benefit or harm to children from divorce. They divide divorce outcomes into two classes: In one-third of the cases, marital conflict is so serious and so affects children that they are much better off if the parents divorce. But in two-thirds of the cases—in these researchers' opinion—conflict is low-level and not very visible to children. Then the children seem better off if the parents remain married. Amato and Booth present their conclusions as informative, not as advocacy for legal restrictions on divorce.

Is Making Divorce Harder to Get a Realistic Idea?

Noting that divorce is "an American tradition" that began in colonial times and has grown more prevalent with industrialization and urbanization, historian Glenda Riley, among others, argues that making divorce harder to get will not change this trend (Riley 1991; Wolfers 2006).

Given the likelihood that the restrictions on divorce that some reformers would like to see put in place are unlikely to become law, what may be done to address the negative consequences of divorce for children and to help postdivorce families more generally?

Surviving Divorce

Studies in other countries suggest that policy remedies can make divorce less fraught with hardship. And there is research on "the good divorce," one that promotes a workable family in the aftermath of divorce.

Social Policy Support for Children of Divorce

A cross-cultural study gives insight into what the wider U.S. society might do to help. Sociologists Sharon Houseknecht and Jaya Sastry (1996) examined the relationship between "family decline" and child well-being in four industrialized countries: Sweden, the United States, the former West Germany, and Italy. These researchers' measures of family decline included nonmarital birth and divorce rates, the proportion of one-parent households with children, and the percentage of employed mothers with children under three years old. Child well-being was measured by six factors: educational performance, the percentage of children in poverty, infant deaths from child abuse, teenage suicide rates, juvenile delinquency rates, and juvenile drug offense rates. The researchers found that "Sweden, which has the highest family decline score, does not demonstrate a high level of negative outcomes for children compared with other countries with lower levels of family decline. It looks much better than the U.S., which ranks the lowest on child well-being" (p. 736).

What makes the difference between child well-being in the United States and in Sweden? Sweden's welfare policies are "egalitarian and generous" when compared to those of the United States (p. 737). Sweden's society-wide willingness to support children's needs by paying relatively high taxes translates into significantly less child poverty, fewer working hours for parents, and more family support programs, such as paid parental leave. This situation results in relatively high levels of child well-being despite high proportions of nonmarital births, mother employment, divorces, and single-parent households. A review of policies and outcomes in Nordic countries by sociologist William Goode reached similar conclusions (Goode 1993).

The Good Divorce

Is there such a thing as a "good divorce"? That depends on expectations. Against the assumption that divorce is a disaster and solidifies a lasting enmity between the partners to the detriment of their children, one can indeed find a different pattern, whereby couples maintain civility and cooperative parenting, and perhaps even spend some holidays together with their children (Scelfo 2004). "My Family: The Postdivorce Family as a Child-Raising Institution" describes a couple that continued

to parent their children together after divorce, with the assistance of a counselor and after some turmoil.

The Binuclear Family Study The Binuclear Family Study, led by sociologist Constance Ahrons (1994), interviewed ninety-eight divorcing couples approximately one year after their divorce. Ninety percent of them were followed to the five-year point, in a total of three interviews each. These were primarily white, middle-class couples from one Wisconsin county.

At the one-year point, 50 percent of the ex-spouses had amicable relations whereas the other 50 percent did not. In half the cases, the divorce was a bad one and harmful to family members; in the other half, the divorcing spouses had "preserved family ties and provided children with two parents and healthy families" (p. 16):

> In a good divorce a family with children remains a family. The family undergoes dramatic and unsettling changes in structure and size, but its functions remain the same. The parents—as they did when they were married—continue to be responsible for the emotional, economic, and physical needs of their children. (Ahrons 1994, p. 3; see also Kelly 2007 and Ahrons 2007 for additional discussion of positive co-parenting)

The ninety-eight couples represented a broad range of postdivorce relationships. In 12 percent of the cases, couples were what Ahrons termed *perfect pals*—friends who called each other often and brought their common

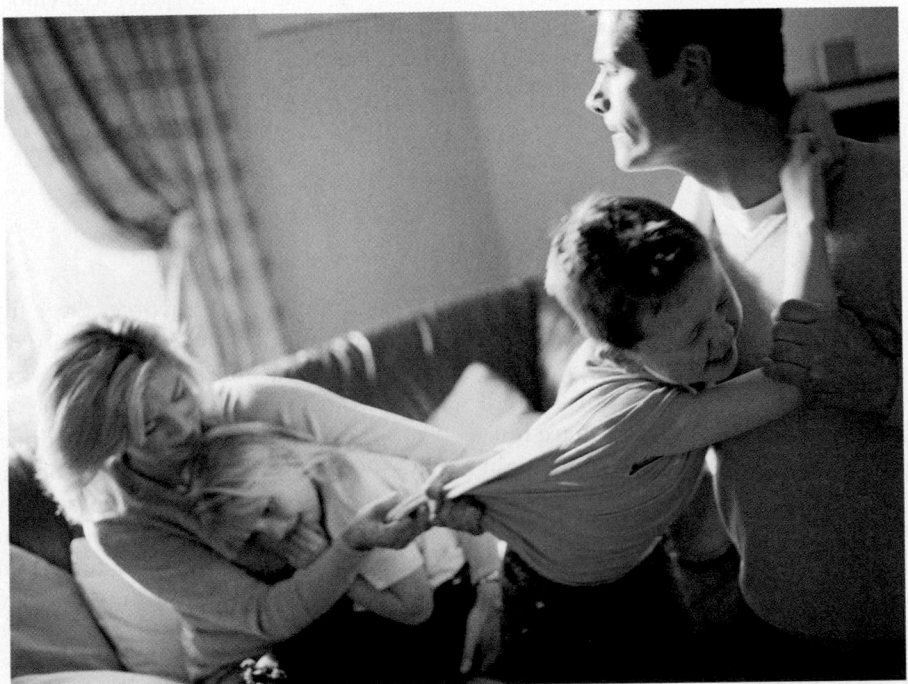

When divorcing parents continue to engage in conflict and especially when children are drawn into it, a child's adjustment is poorer. Interparental conflict does tend to diminish with the passage of time.

My Family

The Postdivorce Family as a Child-Raising Institution

Jo Ann is thirty-eight and has been divorced for six years. She has five children. Gary, nineteen, her oldest, lives with his father, Richard. At the time of this interview, Jo Ann and Richard and their children had recently begun family counseling. The purpose, Jo Ann explained, was to create for their children a more cooperative and supportive atmosphere. Jo Ann and Richard do not want to renew an intimate relationship, but they and their children are still in many ways a family fulfilling traditional family functions.

We've been going to family counseling about twice a month now. The whole family goes—all five kids, Richard, and me. The counselor wants to have a videotaping session. He says it would help us gain insights into how we act together. The two older girls don't want any part of it, but the rest of us decided it might be really good for Joey to see how he acts. [Joey's] the reason we're going in the first place. At the first counseling sessions, he sat with his coat over his head. . . .

Joey's always been a problem. He's used to getting his own way. Some people want all the attention. They will do anything to get it. I guess I never knew how to deal with this. . . . He drives us nuts at home. He calls me and the girls names. . . . He was disrupting class and yelling at the teacher. And finally they expelled him. . . . Joey gets anger and frustration built up in him.

So I took him to a psychologist, and the psychologist said he'd like the whole family to come in, including Richard. Well, Richard still lives in this city and sees all the kids, so I asked him about it. And he said okay. . . .

In between sessions, the counselor wants us to have family conferences with the seven of us together. One day I called Richard and asked him over for supper. In the back of my mind I thought maybe we could get this family conferencing started.

Well, after dinner Joey, our eleven-year-old, started acting up. So I went for a walk with him. We must have walked a mile and a half, and Joey was angry the whole time. He told me I never listen to him; I never spend time with him. Then he started telling me about how he was mad at his dad because his dad won't listen to him.

He said his dad tells all these dumb jokes that are just so old, but he just keeps telling them and telling them. So when we got home, I saw Richard was still there, and I asked Joey, "Would you like to have a family conference? Maybe tell your dad some of the things that are bothering you?" And he said, "Could we?" . . . [During the conference] Joey talked first. Then everybody had a chance to say something. There was one time I was afraid it was going to get out of hand. Everybody was interrupting everybody else. But the counselor had told me you have to set up ground rules. This is where we learned that we got some neat kids because when I said "Let somebody else talk," everybody did! So it went real well. . . . And then finally Richard said, "I think it's time for us to come to a conclusion." I said, "Well, you're right."

Critical Thinking

How might this divorced family differ from the same family before divorce? How is it the same? Even though Jo Ann and Richard's family is no longer intact, what functions does it continue to perform for its members? For society?

---

children and new families together on holidays or for outings or other activities. This was a minority pattern among the "good divorces." More often (38 percent), the couples were *cooperative colleagues,* who worked well together as co-parents but did not attempt to share holidays or be in constant touch—occasionally, they might share children's important events such as birthdays. Ex-spouses might talk about extended family, friends, or work. They still had areas of conflict but were able to compartmentalize them and keep them out of the collaboration that they wanted to maintain for their children (Ahrons 1994; 2007). "As We Make Choices: Ten Keys to Successful Co-Parenting" provides some general guidelines for divorcing parents who want to cooperate in parenting their children.

Other divorcing couples were the *angry associates* (25 percent) or *fiery foes* (25 percent) that we often think of in conjunction with divorce. Over time, one-quarter of the "cooperative colleagues" drifted into one of these more antagonistic categories.[15]

Ahrons's overall point is that the "good divorce" does not end a family but instead produces a **binuclear family**—two households, one family. She argues that we must "recognize families of divorce as legitimate." She notes that it is "important to provide parents with some hope and some goals by informing them that it is never too late to improve their relationship and have a good divorce" (2007, p. 62). To encourage more "good divorces," it is

---

[15] Ahrons (1994; 2007) identified a fifth type of postdivorce couple relationship, the *dissolved duo.* These are couples who have completely lost touch with each other. Because Ahrons's specification of her sample required that a divorced couple have children and be in touch, there were no instances of "dissolved duos" in her sample.

Melinda Blau, author of *Families Apart: Ten Keys to Successful Co-parenting*, observes that "[d]ivorce ends a marriage—it does not end a family; we got divorced; our children didn't" (1993, p. 16). Here are her general guidelines for those who hope to accomplish the "heroic feat" of co-parenting after divorce (pp. 16–17):

- Key #1: . . . get on with your life without leaning on your kids.
- Key #2: . . . care for your kids and . . . act in their best interest.
- Key #3: Listen to your children; understand their needs.
- Key #4: Respect each other's competence as parents and love for the children.
- Key #5: Divide parenting time . . . so that the children feel they still have two parents.

- Key #6: Accept each other's differences. . . .
- Key#7: Communicate about (and with) the children directly, not *through* them.
- Key #8: Step out of traditional gender roles. . . .
- Key #9: Recognize and accept that change is inevitable and therefore can be anticipated.
- Key #10: Know that co-parenting is forever; be prepared to handle holidays, birthdays, graduations, and other milestones in your children's lives with a minimum of stress and encourage your respective extended families to do the same.

Critical Thinking

Do you agree or disagree with the advice presented here?

Source: Blau 1993.

important to dispel the "myth that only in a nuclear family can we raise healthy children" (1994, p. 4). People often find what they expect, and social models of a functional postdivorce family have been lacking.

Ahrons (2004, 2007) reinterviewed 173 children from eighty-nine of the original families in her Binuclear Family Study. Now averaging thirty-one years of age, they had been six through fifteen at the time of the

These divorced parents have both come to meet with their child's teacher. When parents work together to co-parent their children, they continue to have a sense of "family."

marital separation. At present—twenty years later—79 percent think that their parents' decision to divorce was a good one, and 78 percent feel that they are either better off than they would have been, or else not that affected. Twenty percent, however, did not do so well, with "emotional scars that didn't heal" (p. 44). Ahrons judges that the prime factor affecting outcomes was how the parents related to each other in terms of avoiding conflict. The title of Ahrons's most recent book—*We're Still Family* (2004)—characterizes the positive outcomes of divorce that she has found among many of the families she studied.

There is a legal trend that seems to be in tandem with Ahrons's findings and suggestions. This trend is called collaborative divorce. It centers on the idea that divorcing couples will forego litigation and work together—compromising on issues as they proceed—including property divisions and child custody. According to an article by law professor Marsha B. Freeman (2008), collaborative law requires the divorcing couple and their attorneys to "attempt to reach a more amicable, or at least more workable, settlement. Where traditional litigation methods themselves promote anger between the parties and may harm family relationships, collaborative law provides for a more economical and hopefully less hostile way to reach a resolution" (p. 222). In this style of divorce process, the court system does not act as a decision-making body, but instead supports the couple by acting in an advisory capacity, and formalizing the divorce agreement. This style of divorce emphasizes outcomes that are similar to the more collaborative postdivorce relationships Constance Ahrons advocated.

Attachment between Former Partners A more controversial outcome of the "good divorce," and perhaps even some bad ones, is continuing attachment between ex-spouses. One of Paul Bohannon's "stations of divorce" is the psychic divorce, marking the end of the divorce process. **Psychic divorce** refers to the regaining of psychological autonomy through emotional separation from the personality and influence of the former spouse. In Bohannon's view, one must distance oneself from the still-loved aspects of the spouse, from the hated aspects, and "from the baleful presence that led to depression and loss of self-esteem" (Bohannan 1970, p. 53).

By this standard, attachment between former spouses had been seen as a failure of adjustment. But now, clinicians and social scientists are beginning to rethink this stance. "[A]ttachment may be a natural outcome of shared parenting" (Madden-Derdich and Arditti 1999, p. 243). Ahrons and others have written favorably about continuing attachment in the context of co-parenting and maintenance of a family identity and activities (Ahrons 2007).

But what about the postdivorce partner bond as an end in itself? Not only has it been "underacknowledged"

but "the nonparenting aspect of ex-spousal relationships has been considered synonymous to psychological maladjustment" (Masuda 2006, pp. 114, 117).

Taking a positive view of former partners who redefine their relationship as friendship, Masahiro Masuda points out that there can be other than parenting elements in a continued attachment. Exploratory research found some interesting themes that point to good reasons ex-partners might maintain a degree of attachment. What these ex-partners, now friends, talked about was that he or she is "part of my life." They had shared experiences with the former partner that they did not want to consign to limbo. They also claimed to have had a "clean breakup" and that "we're different now." In other words, there were boundary markers that would make sure the past relationship did not prevent formation of new serious relationships despite the continuation of a friendship.

Masuda studied young people and had a sample of limited size—thirty-four college students. But journalists have reported on later-life playing out of postdivorce couple attachments. It appears that it is not unusual for one divorced spouse to provide assistance to the other who needs it at the end of life:

> Hospice workers, academics, and doctors say they are seeing more such cases. . . . Often a person feels deep ties to a former husband or wife or feels a responsibility born of common experience and child rearing. . . . "They are acting more like a brother or sister, or cousin, or extended family member" [said the CEO of a public policy hospice group]. (Richtel 2005, p. ST1-2)

The Divorce-Extended Family A continued postdivorce partner connection can result in a much expanded family. Ahrons, author of *The Good Divorce*, offers a number of suggestions that she believes would contribute to good divorce outcomes and successful binuclear families. One suggestion is that the parent "accept that your child's family will expand to include nonbiological kin" (1994, p. 252). She adds that there is an added benefit to the acceptance of nonbiological kin, saying that the "relationships formed when a parent remarries also tend to be more rewarding for the children as their kinship system expands rather than contracts" (2007, p. 64). Indeed, divorce and the new relationships that follow can produce a **divorce-extended family**.

A surprising phenomenon encountered by those who do research on divorced families is the expansion of the kinship system that is produced by links between ex-spouses and their new spouses and significant others and beyond to *their* extended kin. Sociologist Judith Stacey (1990) speaks of a divorce-extended family (p. 61) and quotes writer Delia Ephron's apt observation:

These eight grandparents, all connected to the young basketball player by marriage, divorce, and remarriage, come together to cheer him on and enjoy his game.

It occurred to me . . . that the extended family is in our lives again. . . . Your basic extended family today includes your ex-husband or -wife, your ex's new mate, your new mate, possibly your new mate's ex, and any new mate that your new mate's ex has acquired. It consists entirely of people who are not related by blood, many of whom can't stand each other. (Ephron 1988, front leaf)

Ephron's version may be out of date in one respect. Some postdivorce extended families find they can enjoy and benefit from connections to one another even when there had earlier been conflict and old tensions sometimes resurface (Kleinfield 2003). "Letting go of old resentments, whether between ex-spouses, parent and child, or stepparent and child, is the most challenging part of" creating divorce-extended families, but to do so presents additional resources, friendship, and love to individuals and families (A. Bernstein 2007, p. 74). Bernstein suggests that in order to create these beneficial connections, families must unlearn the unfavorable conclusions they have made about the members of the new extended family, and create a basis for mutual respect and understanding.

Therapists claim to see "a new norm" of rapport and social contact among the divorce-extended family that is "becoming part of the culture" (Dr. Harvey Ruben, professor of clinical psychiatry at Yale School of Medicine, quoted in Kuczynski 2001, pp. 9-1, 9-6; see also A. Bernstein 2007). In this spirit, "[n]o longer are the names of exes unmentioned at the dinner table. No longer are the details of passing children between homes confined to emotionless email messages. Family therapists,

sociologists, and journalists note that 'family members who 25 years ago might not have had anything to do with one another are finding it desirable to stay connected'" (Kuczynski 2001, p. 9-1).

To the extent that relationships between ex-spouses are cordial and the ties of a divorce-extended family come to seem natural, tension for children moving between families should be reduced. Familial occasions such as graduations and weddings that bring everyone together should be less strained. New research finds that good relationships between children and noncustodial fathers *and* residential stepfathers make independent contributions to good child outcomes (Connidis 2009; Wallerstein and Lewis 2007). If the therapists are right about "new norms," what we could see in the future would be an institutionalization of the "good divorce"—perhaps not attainable for all, but recognized as "normal" for those who do.

That change involves incorporating remarriage bonds into the original family. In Chapter 16, we turn to a consideration of that common step for many divorced people: remarrying.

Summary

- Divorce rates rose sharply in the twentieth century, and divorce rates in the United States are now among the highest in the world. Since around 1980, however, they have declined substantially.

- Among the reasons divorce rates have increased to the present level are changes in society. Economic interdependence and legal, moral, and social constraints are lessening. Expectations for intimacy have risen, while expectations of permanence are declining.

- People's personal decisions to divorce involve weighing the advantages of the marriage against marital complaints in a context of weakening barriers to divorce and an assessment of the possible consequences of divorce.

- Two consequences that receive a great deal of consideration are how a divorce will affect any children and whether it will cause serious financial difficulties.

- Bohannan has identified six "stations of divorce," or aspects of the divorce process and adjustment to divorce. These are the emotional divorce, the legal divorce, the community divorce, the psychic divorce, the economic divorce, and the co-parental divorce.

- The economic divorce is typically more damaging for women than for men, and this is especially so for custodial mothers. Over the past twenty-five years, child support policies have undergone sweeping changes. The results appear to be positive, with more child support being collected. Fathers' chief concern is maintaining a relationship with their children, so visitation, joint custody, and the moving away of custodial mothers are their chief legal and policy issues.

- Researchers have proposed six possible theories to explain negative effects of divorce on children. These include the life stress perspective, the parental loss perspective, the parental adjustment perspective, the economic hardship perspective, the interparental conflict perspective, and the family instability perspective.

- Husbands' and wives' divorce experiences are typically different. Both the task overload and financial decline that characterize the wife's divorce and the loneliness that often accompanies the husband's might be mitigated by less-gender-differentiated postdivorce arrangements. Joint custody offers the opportunity of greater involvement by both parents, although there are some concerns about joint custody as a universal remedy.

- Debate continues among family scholars and policy makers concerning how important a threat of divorce is to children today. Some call for return to a fault system of divorce or other restrictions on divorce. Others see divorce as part of a set of broad social changes, the implications of which must be addressed in ways other than turning back the clock. Now there is also a centrist view of the impact of divorce on children: Yes, there is some disadvantage; no, divorce is not the most powerful influence on children's lives.

- New norms and new forms of the postdivorce family seem to be developing. Some postdivorce families can share family occasions and attachments and work together civilly and realistically to foster a "good divorce" and a binuclear family.

Questions for Review and Reflection

1. What factors bind marriages and families together? How have these factors changed, and how has the divorce rate been affected?

2. How is "his" divorce different from "her" divorce? How are these differences related to society's gender expectations? In your observation, are the descriptions given in this chapter accurate assessments of divorce outcomes for men and women today?

3. In what situation(s), in your opinion, would divorce be the best option for a family and its children?

4. Do you think couples are too quick to divorce? What are your reasons for thinking so?

5. **Policy Question.** Should divorced parents with children be required to remain in the same community? Permitted to move only by court authorization? Be free to choose whether to be geographically mobile?

Key Terms

barriers to divorce 407
binuclear family 438
child support 418
children's allowance 419
community divorce 414
co-parents, co-parenting 431
covenant marriage 435
crude divorce rate 402
custodial parent 418
custody 425

divorce divide 402
divorce-extended family 440
divorce mediation 413
economic divorce 416
economic hardship perspective (on children's adjustment to divorce) 421
emotional divorce 412
family instability perspective (on children's adjustment to divorce) 421
guaranteed child support 419

Online Resources

Sociology CourseMate

www.CengageBrain.com

Access an integrated eBook, chapter-specific interactive learning tools, including flash cards, quizzes, videos, and more in your Sociology CourseMate, accessed through CengageBrain.com.

www.CengageBrain.com

Want to maximize your online study time? Take this easy-to-use study system's diagnostic pre-test, and it will create a personalized study plan for you. By helping you identify the topics that you need to understand better and then directing you to valuable online resources, it can speed up your chapter review. CengageNOW even provides a post-test so you can confirm that you are ready for an exam.

Remarriages and Stepfamilies

Karin Dreyer / Getty Images

- "[My stepdad] is very listening. He's very open to communication. He's willing to talk to you about anything. He's always been there" (in Baxter, Braithwaite, and Bryant 2006, p. 393).

- "I wasn't very happy about [the remarriage] because before this, you know, we got to spend a lot of time with my dad. . . . But then when she came into the picture . . . we kind of got left out" (in Baxter et al. 2009, p. 479).

This chapter explores remarriages and stepfamilies.

Today, approximately 40 percent of all marriages involve a remarriage for one or both partners, and about one-third of all weddings form stepfamilies (Deal 2010). More than half of all U.S. families are re-coupled—either remarried or cohabiting with a subsequent partner (Stewart 2007). Increasingly stepfamilies result from cohabitation as well as from legal remarriage, although the former situation remains a statistical minority. At least one in three Americans are now stepparents, stepchildren, stepsiblings, or living in a stepfamily; and enough remarriages involve stepchildren

that travel packages market the "familymoon"—that is, a honeymoon that includes the entire new stepfamily (D. Miller 2004; Greenberg and Kuchment 2006).

Many remarried people are happy with their relationships and lives. Much of what we have said throughout this book—for example, about good parenting practices, supportive communication, and how best to handle family stress—applies to remarriages and stepfamilies. At the same time, stepfamily members experience unique challenges, because stepfamilies are different from first-marriage families in several important ways. For one thing, forming a cohesive and satisfying stepfamily requires "leaving behind traditional views of [family] belonging" (Sky 2009).

In this chapter, we will discuss happiness and stability in remarriages. We'll look at the various types of stepfamilies. We will note the ambiguity associated with stepfamily living and examine children's well-being in stepfamilies. We'll discuss some challenges typically associated with stepfamilies and explore ideas for creating supportive remarriages and stepfamilies. We'll begin with some basic facts about remarriages and stepfamilies.

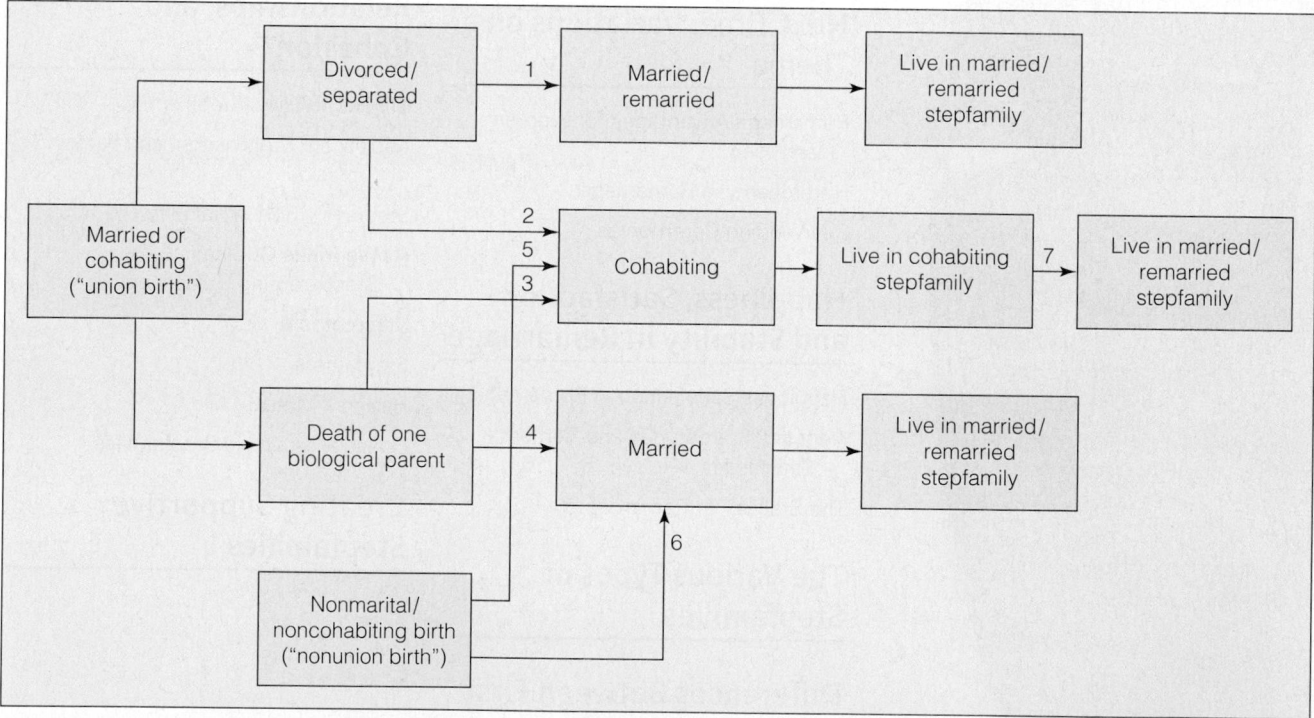

Figure 16.1 Various Pathways to Stepfamily Living. The various possible pathways to stepfamily living begin with a marital or cohabiting union or with nonunion birth or adoption. A union birth/adoption may be followed by divorce, then remarriage, resulting in life in a married or remarried stepfamily (Path 1). A union birth or adoption may also be followed by cohabitation after divorce, resulting in a cohabiting stepfamily (Path 2). When a union birth or adoption precedes the death of one partner, the remaining parent may choose to cohabit (Path 3) or to remarry (Path 4). Following a nonunion birth or adoption, the parent may cohabit (Path 5) or marry (Path 6). In some cases, the partners in a cohabiting stepfamily may marry, resulting in a remarried stepfamily (Path 7). These various paths result in married, remarried, or cohabiting stepfamilies that further differ according to the gender of the biological/adoptive parent and also according to the existence of a living ex-spouse or former cohabiting partner (nonresident parent).

Source: Adapted from Tillman 2007, Figure 1, 383–424.

Remarriages and Stepfamilies: Some Basic Facts

This section describes the various pathways to stepfamily living, and then gives statistics on the prevalence of remarriages and stepfamilies.

Pathways to Stepfamily Living

Sociologist Kathryn Tillman (2007) has described the varied pathways that lead to stepfamily living. As depicted in Figure 16.1, stepfamily formation can begin with a birth to a married or to a cohabiting couple—what Tillman calls a union birth. An alternative is a non-union birth to an uncoupled mother. Children of union births may later experience their parents' divorce or separation, after which a parent marries or remarries, forming a married stepfamily. On the other hand, after a parent's divorce, a child may experience living in a cohabiting stepfamily either permanently or as a transition to a married/remarried stepfamily. Children of nonunion births may later experience a mother's marrying or forming a cohabiting union, resulting in either a married or a cohabiting stepfamily. Multipartnered parenthood (conceiving two or more children with more than one biological partner) adds further complexity to stepfamilies (Manning et al. 2010).

The various pathways to stepfamily living result in different experiences for stepfamily members. For instance, stepfamilies that began with union births—particularly married union births followed by divorce—are likely to include relationships with ex-spouses and relatives of ex-spouses. Although stepfamilies that form after the death of one parent do not have relationships with ex-partners, they may have ongoing ties with relatives of the deceased. The complexity that results from various stepfamily-formation pathways is one factor that influences the diversity of stepfamily experience.[1] We turn now to indicators regarding the prevalence of remarriages and stepfamilies.

Some Remarriage and Stepfamily Statistics

A number of websites are dedicated to remarital wedding planning and services (e.g., www.idotaketwo.com)—an indication that remarriages are fairly common.

Unfortunately, however, the U.S. Census Bureau stopped compiling remarriage statistics in 1988. As a result, we can only estimate today's situation, relying largely on survey data from national random samples (Kreider and Fields 2005; Stewart 2007). We do know that **remarriages** (marriages in which at least one partner had previously been divorced or widowed) are significantly more frequent in the United States today, compared to the middle decades of the twentieth century (Strow and Strow 2006).

Remarriages have always been fairly common in the United States. However, well into the twentieth century, almost all remarriages followed widowhood (Strow and Strow 2006). Indeed, the term stepparent originally meant a person who replaces a dead parent, not an *additional* parent figure (Bray 1999). Today, however, the vast majority of remarrieds have been divorced.

About half of all marriages today are remarriages, a fact that greeting card companies acknowledge.

---

[1] Situations such as living in a multiracial or multiethnic stepfamily (Grady 2009) or raising a disabled stepchild add further diversity to everyday life experiences in stepfamilies. Raising multiracial children is addressed in Chapter 10; raising a disabled child is addressed in Chapter 14.

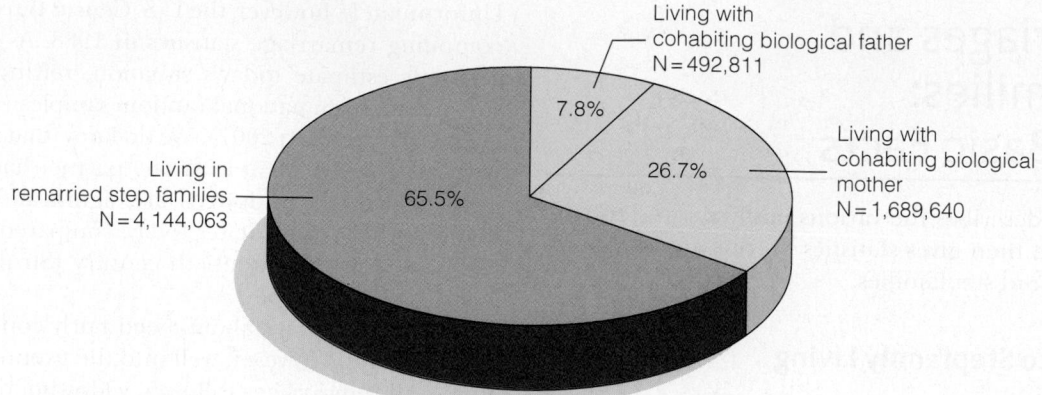

Figure 16.2 U.S. Children Under Age 18 Living in Stepfamilies, 2008
Source: Calculated from U.S. Federal Interagency Forum on Child and Family Statistics 2009 and U.S. Census Bureau 2010b, Table 69.

The remarriage rate rose sharply during World War II, peaking as the war ended. During the 1950s, both the divorce rate and the remarriage rate declined and remained relatively low until the 1960s, when they began to rise again. The remarriage rate peaked again in about 1972, but has declined somewhat since then. One reason for the current decline in remarriage rates is that many divorced people who would have remarried in the past are now cohabiting (Allan, Crow, and Hawker 2008).

A second reason for the decline in remarriage rates may be economic constraints, which discourage divorced individuals who may already be paying child support from assuming even shared financial responsibility for a new family.[2] Nevertheless, about three-quarters of divorced individuals remarry ("Statistics" 2009). Some people divorce, remarry, redivorce, and then remarry again. But a majority of remarriages are second marriages (Stevenson and Wolfers 2007).

Stepfamilies and Children's Living Arrangements

One result of the significant number of remarriages and the growing incidence of cohabitation after separation, divorce, or single parenthood is that more Americans are parenting other people's biological children. In perhaps one-quarter of cohabiting households, "at least one adult brings children from prior relationships, thereby creating cohabiting stepfamily households"

(Coleman, Ganong, and Fine 2000, p. 1290; see also Simmons and O'Connell 2003, Table 4).

About four million, or 8.5 percent of all U.S. children under age eighteen, are stepchildren (calculated from U.S. Federal Interagency Forum on Child and Family Statistics 2009 and U.S. Census Bureau 2010b, Table 69). At least 6 percent of adults over age seventeen who are living with a parent are doing so in stepfamily households (Kreider 2003, Table 1). As you can see in Figure 16.2, 65.5 percent of stepchildren under age eighteen reside in remarried stepfamilies. Another 26.7 percent of stepchildren live with their biological mother and her cohabiting partner, usually a male. In addition, 7.8 percent of stepchildren live in cohabiting households with their biological father and his partner, most often a female.

It is difficult to get accurate statistics on gay male and lesbian (GL) stepfamilies, but the "best guess" of some experts is that "there are millions" in the United States today. The number is expected to increase, given current separation rates and the significant number of GL couples who are raising children (van Eeden-Moorefield, Henley, and Pasley 2005).

The incidence of children's living in stepfamilies is differentially distributed by race and Hispanic origin. Among Hispanics and among non-Hispanic white children who live in two-parent families, about 8 percent live with their biological mother and a stepfather. Among African American children in two-parent families, 13 percent live with their biological mother and a stepfather. Among Asian Americans, the corresponding figure is 2 percent (Kreider and Fields 2005, Table 1). These statistics describe a situation at one point in time. However, "it is important to keep in mind that as children age, they may spend time in several [living] arrangements" (Kreider and Elliott 2009a, p. 16).

---

[2] Some states, concerned about child support, passed legislation in the 1970s, designed to prevent the remarriage of people whose child support was not paid up. But the Supreme Court ruled in *Zablocki v. Redhail* (1978) that marriage—including remarriage—was so fundamental a right that it could not be abridged in this way.

Taken together, Figures 16.1 and 16.2 point to the fact that stepfamily structure is variable. Stepfamilies form as the result of various circumstances, and a stepparent may be a married or cohabiting mother or father (although usually the father). In some stepfamilies, both parents are stepparents to their partner's biological children. Then too, at least 4 percent and perhaps up to 10 percent of households with children are *joint biological stepfamilies*—that is, at least one child is a *mutual child*, the biological child of both parents, and at least one other child is the biological child of one parent and the stepchild of the other parent (Kreider 2003, Table 8). The remainder of this chapter explores what family life is like for stepfamily members. We begin with a look at a couple's dating relationship prior to remarriage.

Choosing Partners the Next Time: Variations on a Theme

Courtship for remarriage has not been a major topic for research. Nevertheless, counselors note that people who ended troubled first marriages through divorce are often still experiencing personal conflicts that they need to resolve before they can expect to fashion a supportive, stable second marriage (Dupuis 2007). Counselors advise waiting until one has worked through grief and anger over the prior divorce before entering into another serious relationship (Marano 2000; see also Bonach 2007).

Meanwhile, dating before remarriage may differ in many respects from that before first marriage. Courtship may proceed much more rapidly, with individuals viewing themselves as mature adults who know what they are looking for—or it may be more cautious, with the partners feeling wary of repeating an unhappy marital experience. Dating may include outings with one or both partners' children and evenings at home as partners seek to recapture their accustomed domesticity. Dating may have a sexual component that is hidden from the children through a series of complex arrangements. "A common courtship pattern is as follows: (a) male partner spends a few nights per week in the mother's household, followed by (b) a brief period of full-time living together, followed by (c) remarriage" (Coleman, Ganong, and Fine 2000, p. 1,290). "Today, a majority of stepfamilies begin as cohabiting relationships" (Cherlin 2009a, p. 22).

The process through which individuals choose a new cohabiting partner whom they do not necessarily intend to marry has been researched even less than has the process of remarital courtship. However, some research suggests that unmarried mothers often

"partner up," forming subsequent cohabiting relationships with men who have better financial stability and fewer behavioral problems than did a prior partner (Bzostek, Carlson, and McLanahan 2007; Graefe and Lichter 2007).

Remarriage Advantages for Women and Men

Many ex-wives, but not necessarily ex-husbands, are thought to gain financially by re-coupling. A longitudinal study by sociologists Donna Morrison and Amy Ritualo (2000) used National Longitudinal Survey of Youth (NLSY) data to track the financial well-being of divorced mothers and their custodial children. Some of the mothers remained single, others remarried, and still others formed cohabiting stepfamilies. The finances of mothers who remarried greatly improved. Mothers who began cohabiting also saw an increase in finances, although not as great.

We can conclude that generally re-coupling is financially advantageous to divorced mothers and their children (Morrison and Ritualo 2000; and see Dewilde and Uunk 2008). This situation changes, of course, when a woman's earnings are relatively high, compared with those of her prospective new partner—and, although still a significant minority, growing numbers of wives today are earning more than their husbands ("Wives Earning More" 2009). Chapter 15 addresses the economic situation of divorced women in more detail.

The fact that remarrying women generally benefit financially more than do remarrying men who, having more income, may be disinclined to pool it, is one reason that women's remarriage rate is considerably lower than men's. This statistic is also a consequence of the very low remarriage rate of widows, for whom few partners are available in later life (Moorman, Booth, and Fingerman 2006). But even comparing only divorced individuals, men's remarriage rates have been substantially higher, particularly after age thirty. A woman's age and her having children work against females' remarriage rates. Chapter 17 addresses the first factor.

Children and the Odds of Remarriage "Children lower the likelihood of remarriage for both men and women, but the impact of children is greater on women's probability of remarriage" (Coleman, Ganong, and Fine 2000, p. 1,289). As discussed in Chapter 15, the mother usually retains custody of children from a previous heterosexual marriage. As a result, a prospective second male partner may look on her children as a financial (or psychological) liability. Interestingly, a study that analyzed data from the National Survey of Families and Households (NSFH) found that divorced fathers with custody of their children were significantly more likely to marry women with children than were men without custody (Goldscheider and Sassler 2006). This situation may exemplify the "traditional exchange," discussed in Chapter 6, according to which women trade child raising for enhanced economic security. We might also define this situation as a form of homogamy in remarriage.

Heterogamy in Remarriage

Unfortunately, there has not been much research on the extent to which remarrying individuals choose partners with social characteristics different from their own (that is, heterogamy, discussed in Chapter 6). Available data suggest that remarriages are somewhat more heterogamous than first unions. Choosing a remarriage partner differs from making a marital choice the first time inasmuch as there is a smaller pool of eligibles with a wider range on any given attribute. As prospective mates move into their thirties and forties, they affiliate in occupational circles and interest groups comprised of diverse backgrounds. As a result, according to the latest government statistics available, remarriages have been less homogamous than first marriages, with partners differing more in age and educational background (U.S. National Center for Health Statistics 1990b). Not only courtship behaviors but also wedding ceremonies differ somewhat among remarrieds.

Re-Wedding Ceremonies

University of Iowa communication studies professor Leslie Baxter and colleagues asked thirty male and fifty female stepchildren in two Midwest colleges to describe their parent's remarriage ceremony (Baxter et al. 2009). The majority of ceremonies the students described ranged from a downsized version of the traditional wedding to a courthouse civil ceremony to an informal event in a backyard or casual restaurant. Sometimes a couple left the area to remarry, and then informed family and friends later.

Many of the students were critical of their stepparent's wedding ceremony. Arguing that "the relationship didn't deserve that" degree of celebration, some criticized even a less elaborate form of the traditional white wedding ceremony. Other students critiqued the ceremony as being too casual: "Well, if you want everyone

Inti St Clair/Getty Images

Remarrying couples differ from first-marrying couples in their degree of homogamy because choosing a remarriage partner differs from making a marital choice the first time inasmuch as there is a smaller pool of eligibles with a wider range on any given attribute. People in their thirties and forties meet others at work or in other settings that bring people together from more varied backgrounds.

to take [your remarriage] seriously, it needs to be a little more than um, a barbeque" (Baxter et al. 2009, pp. 476, 477). A male student said he felt that the ceremony had inadvertently insulted his family of origin:

> The only part that upset me was the pastor was talking about how life's events lead you up to this moment and how there's bumps in the road, and blah, blah, blah, but this is where you're supposed to be. And I got pissed, because I was like, was my mom the bump in the road? (p. 480)

Students whose parent had left the area to remarry and told them later had an especially difficult time.

In a different kind of re-wedding, the bride and groom design a family-centered ceremony. In this case, not only the bride and groom but also all members of the new stepfamily are celebrated. For example, one of Baxter's respondents described how all of the stepchildren as well as the remarrying couple "got little rings to show that we all got married" (p. 475). Although few of the students in this study had experienced a family-centered ceremony, some said they wished they had. For instance, a twenty-five-year-old student whose mother had remarried four years earlier said, "I would have liked to have felt as though . . . a family had been created, or like it was . . . solidifying or memorializing some kind of bond between them and between all of us" (p. 483). Although the family-centered ceremony was the least common wedding type described by the students, it was the one most appreciated:

> Over and over again, participants told us that they had wanted far greater involvement with the remarriage event . . . [whether] it was being granted sufficient time to get to know the stepparent, being informed and consulted about the decision to marry, participating in the planning of the ritual, or creating a ceremony and/or artifacts that celebrated the family. (pp. 481, 485)

Because the re-wedding ceremony can influence how children feel about their future stepfamily, involving them may be more important than a couple realizes.[3] Stepchildren's adjustment is a factor in a remarried couple's overall happiness.

Happiness, Satisfaction, and Stability in Remarriage

As pointed out elsewhere in this text, marital happiness or satisfaction, and marital stability are not the same. *Marital happiness* and *marital satisfaction* are synonymous phrases that refer to the quality of the marital relationship whether or not it is permanent; *marital stability* refers simply to the duration of the union. We'll look at both ways of evaluating remarriages.

Happiness/Satisfaction in Remarriage

Research on remarried partners' happiness and satisfaction was relatively prevalent through the 1980s and 1990s (Shriner 2009). However, since then, scholars have focused on other topics, such as communication in stepfamilies. What research does exist on remarrieds' satisfaction has shown little difference in spouses' overall well-being or in marital happiness between first and later unions (Demo and Acock 1996; Ihinger-Tallman and Pasley 1997; Skinner et al. 2002).

We know that wives' satisfaction with the division of household labor is important to marital satisfaction. Although the evidence is inconclusive, some research suggests that there may be more equity, or fairness, in remarriages than in first marriages because remarried husbands contribute somewhat more to housework than do husbands in first marriages. This situation may be especially true for older remarried couples (Clarke 2005). On the other hand, a small study based on extensive interviews with fifteen adult stepchildren found "the persistence of traditional gender practices in the parenting and stepparenting of children" (Schmeeckle 2007, p. 174). And a recent, small qualitative study of wedding planning among remarrying Canadian brides concluded that their gendered division of labor—at least regarding wedding-planning tasks—was much like that for first marriages (Humble 2009).

Interestingly, research with thirty-two black, New York City partners in lesbian stepfamilies—those in which at least one child was from a mother's prior heterosexual relationship—found that in some ways the women followed traditional gender norms (Moore 2008). Although both women in virtually all of the couples were employed, biological mothers tended to do more child care and household chores. The author speculated that gendered division of labor is defined differently among black lesbian stepmothers than among heterosexuals. Among the lesbian couples, "control over some forms of household labor [resulted in] greater relationship power. Biological mothers want more control over the household because such authority affects the well-being of children—children who biological mothers see as primarily theirs and not their partner's" (p. 344).

Whatever the case regarding gender roles in remarriages and stepfamilies, considerable research shows that remarrieds experience more tension and conflict than do first-marrieds, usually on issues related to stepchildren (Stewart 2007). Nevertheless, supportive

---

[3] A twenty-year-old respondent offered the following advice: "If people are thinking about starting a stepfamily, they should take their time and, you know, keep everyone informed and pay attention to everyone's feelings. . . . Like [a mother should] talk to her daughters and see how we, um, feel, and take those feelings into consideration" (in Baxter et al. 2009, p. 481).

family communication (see Chapter 12) is important to remarital happiness.

Negative Stereotypes and Remarital Satisfaction

In addition to family communication patterns, remarital satisfaction is influenced by the wider society through negative stereotyping of remarriages and stepfamilies (Ganong and Coleman 2004; Stewart 2007). For example, a self-help book, published in 2000, and titled *The Blended Family: Achieving Peace and Harmony in the Christian Home*, compares stepfamilies to "sinners" when it advises that "blended families be given encouragement, support, and teaching *just as the drug addict, murderer, fornicator, adulterer, and other sinners*" (quoted in Coleman and Nickleberry 2009, p. 556, italics added by Coleman and Nickleberry). Some religions, such as Catholicism, do not recognize a remarriage after divorce unless the first marriage has been annulled (Hornik 2001).

Historically (Phillips 1997) and today, potentially harmful myths persist, such as "a stepfamily can never be as good as a family in which children live with both natural parents" (Kurdek and Fine 1991, p. 567; see also Gerlach 2010). In an interesting small study, 211 university students were asked to examine an eight-year-old boy's report card. All the students saw the same report card, but some were told that the child lived with his biological parents whereas others were told that he lived with his mother and stepfather. Asked about their impressions of the boy, male (but not female) students rated the stepchild less positively than the biological child with respect to social and emotional behaviors (Claxton-Oldfield et al. 2002).

Negative stereotypes associated with stepfamilies may influence our appraisal of stepfamily members' functioning (Jones and Galinsky 2003).[4] However, a study of thirty-one white, middle-class spouses in families with stepfathers found that, especially among the wives, believing in none or very few negative myths about stepfamilies and having high optimism about the remarriage were related to high family, marital, and personal satisfaction (Kurdek and Fine 1991). Therapist Anne Bernstein has urged that we work toward

"deconstructing the stories of failure, insufficiency, and neglect" and, instead, "collaboratively reconstruct stories that liberate steprelationships" from this legacy (1999, p. 415).

The Stability of Remarriages

"Remarriages dissolve at higher rates than first marriages, especially for remarried couples with stepchildren" (Coleman, Ganong, and Fine 2000, p. 1,291). According to the Stepfamily Foundation, about 60 percent of remarriages end in divorce. This statistic compares to between 40 and 50 percent of first marriages (Popenoe and Whitehead 2009). Of all remarried or cohabiting stepfamilies with children, about two-thirds break up ("Statistics" 2009; "Stepfamily Fact Sheet" 2010).

There are several reasons for the generally lower stability of remarriages. You may recall the discussion in Chapter 6 about how cohabiting—at least serial cohabiting—before marriage generally increases the odds of divorce. Research that analyzed more than 3,000 remarried respondents from National Survey of Families and Households (NSFH) data similarly found that postdivorce cohabitation is positively associated with remarital instability (Xu, Hudspeth, and Bartkowski 2006). The researchers suggested that the *selection effect* (the idea that divorced people who "select" themselves into cohabitation are different from those who don't) largely explained this situation. For one thing, people who divorce in the first place—and those who cohabit—are disproportionately from the lower-middle- and lower classes, which generally have a higher tendency to divorce or redivorce.

Maybe you've heard people who are about to remarry say that they plan to work harder in their new marriage and not to repeat the mistakes they made in their first one. For many couples, this may be the case (Brimhall, Wampler, and Kimball 2008). After all, a significant proportion of remarriages do not divorce. However, in addition to the selection effect, an ironic second reason that remarriages are more likely to end in divorce may be that remarried partners are reluctant to directly address problems that arise in their relationship. "The experience of destructive conflict that often precedes the breakup of a first marriage can be highly stressful and . . . might prompt avoidance of communication about the difficulties that stepfamily couples have to negotiate" (Halford, Nicholson, and Sanders 2007, p. 480). As pointed out in Chapter 12, researchers and counselors advise directly addressing difficult issues.

A third reason that remarriages are more likely to end in divorce may be that if seemingly irresolvable problems do arise, remarrieds are, as a category, more

---

[4] As a unique form of stepfamilies, many gay and lesbian (GL) families may be "triple-stigmatized" for being (1) gay, (2) gay parents, and (3) stepfamilies (Berger 2000; Erera and Fredriksen 1999; Lynch and Murray 2000). However, in a qualitative study with eleven young adults in GL stepfamilies, no respondent mentioned feeling stigmatized due to living in a stepfamily. All of them said they felt stigmatized due to living in a same-sex family. "Indeed, it is possible that the stigma attributed to stepfamilies is gradually diminishing . . . in keeping with the growing proportion of these families in the population . . . whereas homosexuality and homosexual parents are still judged to be socially 'deviant'" (Robitaille and Saint-Jacques 2009, p. 436).

accepting of divorce; they have already demonstrated their willingness to divorce. As one remarried husband said, "We're not going to tolerate the kind of crap we did the first time around. . . . I don't need it again. She doesn't either" (Brimhall, Wampler, and Kimball 2008, p. 378).

Fourth, although stepfamilies' increasing numbers and visibility have led to their growing social and cultural acceptance, remarried families continue to be stereotyped as "less than" or other than "normal" (Ganong and Coleman 2004). One indication of this situation involves re-weddings. As discussed in Chapter 7, weddings publicly announce a couple's commitment. Indicating their culturally diminished importance, re-weddings are typically less extravagant than first weddings, as we have seen. As a result of the relative devaluing of remarriage, remarrieds may receive less social support from friends or extended kin and be somewhat less integrated with parents and in-laws, thus not experiencing the encouragement or social pressures that can act as barriers to divorce (Coleman, Ganong, and Fine 2000).

Perhaps the most significant factor in the relative instability of remarriages is the presence of stepchildren (Stewart 2005a). Sociologists Lynn White and Alan Booth interviewed a national sample of more than 2,000 married people under age fifty-five in 1980 and reinterviewed four-fifths of them in 1983. White and Booth found that the quality of the remarital relationship itself did not affect the odds of divorce, but the partners' overall satisfaction with family life did:

> [W]e interpret this as evidence that the stepfamily, rather than the marriage, is stressful. . . . These data suggest that . . . if it were not for the children these marriages would be stable. The partners manage to be relatively happy

Oh, that—that's the hard drive from my first marriage."

© Barbara Smaller

despite the presence of stepchildren, but they nevertheless are more apt to divorce because of child-related problems. (White and Booth 1985, p. 696; see also Schrodt, Soliz, and Braithwaite 2008)

The previous discussion suggests that when neither spouse enters a remarriage with children, the couple's union is usually very much like a first marriage. But when at least one spouse has children from a previous marriage, family life often differs sharply from that of first marriages.

The Various Types of Stepfamilies

Remarried families with children are of various types. In the simplest stepfamily type, a divorced or widowed spouse with one child remarries a never-married partner without children. In the most complex stepfamily type, both remarrying partners bring children from previous relationships and also have a mutual child or children together. If a remarriage is followed by redivorce and a subsequent remarriage, the new remarried family structure is even more complex. Moreover,

> [E]x-spouses remarry, too, to persons who have spouses by previous marriages, and who also have mutual children of their own. This produces an extraordinarily complicated network of family relationships in which adults have the roles of parent, stepparent, spouse, and ex-spouse; some adults have the role of custodial parent and others have the role of noncustodial, absent parent. The children all have roles as sons or daughters, siblings, residential stepsiblings, nonresidential stepsiblings, residential half-siblings, and nonresidential half-siblings. There are two subtypes of half-sibling roles: those of children related by blood to only one of the adults, and the half-sibling role of the mutual child. Children also have stepgrandparents and ex-stepgrandparents as well as grandparents. (Beer 1989, p. 8)

"I come from a family that has two sets of stepfamilies," wrote one of our students in a recently assigned essay. "A Closer Look at Family Diversity: Immigrant Stepfamilies" explores an even more complex stepfamily system.

Differences Between First Unions with Children and Stepfamilies

We can point to the following differences between stepfamilies and first marriages with children. In stepfamilies:

A Closer Look at Family Diversity

Immigrant Stepfamilies

Social scientists typically research the topics of remarriage and immigration separately, but social work professor Roni Berger (1997) points out that the two situations can occur simultaneously, making for extra family stress.

In the transition from their country of origin to the new society, immigrants [may] experience loss of a familiar physical, social, and cultural environment, destruction of significant relationships, and a loss of language, belief system, and socioeconomic status. Immigration often means also the bitter loss of a dream because of discrepancies between pre-immigration expectations and the reality of life in the new country. . . .

Immigration and remarriage are similar in that both involve multiple losses, discrepancy between expectations and reality, and integration of two cultures within one unit. Therefore, both processes shake the individual and family foundation of identity and require flexibility in adapting to a totally new situation. . . .

In immigration the family culture may serve as a support and in remarriage the cultural context may do the same. However, when remarriage and immigration coincide, families lose the stability of their anchors and the stresses exacerbate each other. For example, it has been recognized that one source of difficulty in remarriage stems from reactivation of previous losses caused by divorce or death. Immigration is an additional link in the chain of losses that intensifies the already heavy history of losses typical to all stepfamilies. . . .

Case Example

Igor, 15, was referred by the school he attends because of acting-out behaviors in school. At the time of his referral the boy had been in the United States for six months and lived with his divorced and subsequently remarried mother, his stepfather, his four-year-old half-sister, and his maternal grandparents. All these six people are crowded in a one-bedroom apartment.

Igor's biological parents lived in Moscow. They married when both of them were 29, . . . and they eventually divorced when Igor was five years old. However, they continued to live in the same apartment because of housing difficulties. As both parents worked, Igor's maternal grandmother was the main parenting figure, a common practice in Soviet families.

When Igor was nine his father moved in with another woman, a single mother of a boy the same age as Igor. They had together two daughters, married, and emigrated to the United States, where Igor's biological father secured a high engineering position and is financially very successful. For five years Igor had no contact with his biological father and his new family. His mother remarried and had a daughter with her new

---

1. One biological parent is elsewhere.

2. There is a complicated "supra family system," including family members from one or more previous unions.

3. Children may have more than two parenting figures.

4. Children may be members in more than one household.

5. There may be less parental control because there is an influential parent elsewhere.

6. Because relationships between at least one parent and child—or between full siblings—predate the stepfamily formation, there may be preexisting coalitions.

7. There may have been significant relationship losses for all family members.

8. One adult, a stepparent, is not legally related to at least one child in the household.

9. There is a long integration period, prior to which family members must recover from previous transitional stresses.

10. Individuals need validation as members of a "real"—legitimate and worthwhile—family unit.

11. The balance of power is different: Stepparents have relatively little authority initially, and children generally wield more power than in first-married families.

12. There are ambiguous family boundaries and relatively little agreement about family history.

13. At least initially, there is little family loyalty.

14. The first-married, nuclear family model is not a valid guide for stepfamily behaviors (Allan, Crow, and Hawker 2008; Visher and Visher 1996).

Researchers have designed questions for stepfamily members in order to study whether and how factors such as those listed previously affect stepfamily life. "Facts About Families: Measuring Everyday Stepfamily Life" gives some examples. The remainder of this chapter addresses the stepfamily characteristics listed previously and what can be done to meet the challenges they create.

Stepfamilies and Ambiguous Norms

Society offers members of stepfamilies an underdeveloped **cultural script**, or set of socially prescribed and understood guidelines for defining responsibilities

husband, who had never been married before. A year ago the family renewed the contact with Igor's biological father, who sponsored their emigration. Igor's stepfather has been unemployed for most of the last year and Igor's mother works off the books in child care.

His mother and stepfather reported that until the immigration Igor was a "model child." He excelled in school, was popular with friends, involved in extra-curricular activities, played the violin, [was] active in sports and was cooperative and pleasant. During all these years Igor's grandmother remained the major parental figure while practically no relationship developed between him and his stepfather. The troubles started a short time after the family came to the United States. Igor was enrolled in a public school with mostly immigrant black and Hispanic students. He excelled in mathematics and physics which he studied in a bilingual program and in sports.

His language skills were very limited and so were his social relationships. The family's squeezed housing conditions fostered tension and conflicts. Igor's grandmother does not speak any English and could not therefore continue to negotiate with school and social agencies for him anymore, forcing his mother to take on more of a parental role.

The main issues that Igor brought up with the therapist related to his natural father and to his parents' divorce which he returned to time and again. His mother and stepfather were annoyed with his behavior, blamed him for being ungrateful, and used him as a target for all their frustrations and disappointments with the hardships in the new land. Igor felt rejected, idealized his biological father, and blamed his mother and stepfather of being unjust and not understanding. Everybody in the family felt deprived, treated unfairly, disappointed, and angry.

It seemed that while the divorce occurred 15 years earlier and the actual separation four years later, family stresses related to the remarriage of both parents and the birth of the half-sibling gradually piled up. Subsequently the immigration reactivated the experiences of loss and triggered reactions of mourning, accusation, anger, and guilt that have been building up for a long time. . . .

Igor's situation reveals multiple forces operating simultaneously. . . . Systematic research is much needed to study the combined effects of immigration and step-relationships, the issues caused by this combination of stresses . . . and effective strategies to promote the welfare and well-being of immigrant stepfamilies. (pp. 362, 364–369)

Source: Berger 1997, pp. 361–370. Reprinted by permission of Springer Science and Business Media.

and obligations and hence for relating to each other (Ganong and Coleman 2000). Noting the cultural ambiguity of stepfamily relationships, social scientist Andrew Cherlin thirty years ago called the remarried family an **incomplete institution** (1978). For the most part, researchers continue to view the situation this way. Moreover, "Cohabiting stepfamilies are arguably even less institutionalized than married stepfamilies, which are formed through a tie that is legally binding" (Brown and Manning 2009, p. 88).

Professor of social work Irene Levin (1997) has argued that "the nuclear family has a kind of model monopoly when it comes to family forms" (p. 123). According to this **nuclear-family model monopoly**, the first-marriage family is the "real" standard for family living, with all other family forms "seen as deficient alternatives." This situation affects people's understanding of stepfamilies; we incorrectly expect a subsequent union with children to be "more or less the same as the first" (p. 124).

Language as an Illustration of Ambiguous Stepfamily Norms An example of the stepfamily as an incomplete institution can be found in the language we use to address and refer to stepfamily members. Communication researchers interviewed thirty-nine stepchildren at a large Midwestern university (Kellas, LeClaire-Underberg, and Normand 2008). The students described how they purposefully choose language to clarify their family form for others:

> Whenever I talk about [my stepfamily] with people it's always my stepdad, my stepmom, stepsister. . . . I always put those terms in there because I do have a biological, real sister and so I guess, I try to help people out because obviously my family's really confusing. (p. 251)

Note this respondent's reference to her "real" sister.

In addition to clarifying their family situation for others, the students deliberately used language that normalized stepfamily living for outsiders: "If I am outside the family and people ask me where I am going I say I am going to my mom and dad's house. So face-to-face, I call [my stepmother by her name], but with everybody else, it's just my mom" (p. 249).

The students in this study also described the way that language—interestingly, language associated with the nuclear-family model—symbolized and communicated stepfamily members' closeness and solidarity. One participant reported that as a young child she referred to her stepfather as "Daddy" to acknowledge that she felt close to him. Another student reported overhearing his

How do researchers measure aspects of everyday stepfamily life? We can look at two examples.

Measuring Dimensions of Stepfamily Living

Communication scholar Paul Schrodt (2006) developed an instrument to measure various dimensions of stepfamily living. First he reviewed the literature on stepfamilies and uncovered several themes, such as stepfamily cohesiveness, adaptability, communication styles, and conflict. Schrodt then constructed a questionnaire to measure stepfamily members' attitudes and behaviors related to these themes. Respondents were asked to indicate the extent to which they agreed or disagreed with a number of statements.

Some of Schrodt's questions follow. If you live in a stepfamily, you may want to answer them yourself. Answer whether you (1) strongly agree; (2) agree; (3) neither agree nor disagree; (4) disagree; or (5) strongly disagree.

Stepfamily Cohesiveness

1. I am committed to members of my stepfamily.
2. Members of my stepfamily couldn't care less about family traditions.
3. I feel a sense of "family" in my stepfamily.

Stepfamily Adaptability

1. My stepfamily tries new ways of dealing with family problems.
2. In my stepfamily, family meetings are important for discussing problems we have with each other.
3. In my stepfamily, the parents tend to negotiate the rules with the children.

Stepfamily Communication

1. In my stepfamily, family members keep to themselves.
2. I would rather talk about my problems with members of my stepfamily than with people outside of my stepfamily.
3. In my stepfamily, family members tend to speak what is on their minds.

Stepfamily Conflict

1. Overall, we really get along as a stepfamily.
2. When a problem arises in my stepfamily, we have a hard time finding a compromise.
3. I get sick and tired of all the fighting that occurs in my stepfamily.

Measuring Stress Among Married Stepparents

In a second illustration of ways that researchers measure aspects of stepfamily living, social scientists designed a questionnaire to assess the level of difficulty that stepparents experience when faced with certain challenges (Schramm and Higginbotham 2009). Respondents are asked to indicate their level of stress-related difficulty associated with a series of statements. Some examples follow. If you are a stepparent, you may want to answer them yourself. Indicate whether each statement (1) is not at all a current difficulty to (5) is a significant difficulty.

Role of Spouse

1. Working together to solve our problems as a couple.
2. Accepting a different kind of life as a couple than I had imagined.
3. Devoting time to our life as a couple.

Social and Family Dimension

1. Ensuring that the stepparent (me or my spouse) is viewed as a legitimate representative in the children's school environment.
2. Dealing with fiscal problems that arise from living in a stepfamily.
3. Dealing with prejudices regarding stepfamilies.

Role of Parent

1. Knowing how to react when my children express emotions (sadness, anger, and so on) about our stepfamily.
2. Dealing with the fact that my spouse and my children compete for my attention and love.
3. Supporting my spouse when he or she deals with my children.

Role of Stepparent

1. Clearly understanding my spouse's expectations with regards to my role as a stepparent.
2. Establishing a relationship of trust with my spouse's children.
3. Living with children whose values and lifestyles are different from mine.

Critical Thinking

How might questionnaires such as the two above be useful to stepfamily researchers? How might they be useful to family counselors? What question(s) or statement(s) would you add to the ones listed here?

Sources: Schrodt 2006, pp. 427–44; Schramm and Higginbotham 2009, 341–55.

younger stepbrother talking with his friends about how happy he was to have a new big brother: "[H]e called me his brother. . . . After I heard that it went from being a step-family . . . to being an actual family" (p. 249). Similarly, in a different study, a stepfather told an interviewer: "I have never, ever, thought of these two girls as my stepchildren. They're just my daughters, and I've always referred to them as such" (in Hans and Coleman 2009, p. 611).

On the other hand, some students reported strategically using language that communicated separateness: "At the very beginning I wouldn't even call her my stepmom. I would call her my dad's wife. . . . I didn't want that connection" (in Kellas, LeClaire-Underberg, and Normand 2008, p. 250). Other students pointed to consciously negotiated terms by which they referred to or addressed stepfamily members as they balanced relationships in an ambiguous family environment. Fairly common was the decision to call a stepfather by a different term than that reserved for the biological father—referring to a biological father as dad, for example, while calling a stepfather by his first name. One stepdaughter reported that when with her biological father, she "always has to be really careful" to refer to her stepfather as "Paul" although she usually calls him "Dad."

Another student confessed that he consistently addressed his stepfather as "Bill," although he wished he could have called him "Dad," but "it just never came"—even though "he really is the one that raised me" (Kellas, LeClaire-Underberg, and Normand 2008, p. 251). Although society tends to broadly apply the norms of first marriages to stepfamilies, these rules ignore stepfamily complexities. We'll explore two areas in which the stepfamily as an incomplete institution is most apparent: boundary ambiguity and family law.

Stepfamily Boundary Ambiguity

In a stepfamily, "Whose picture goes on the mantle?" (Munroe 2009, p. 168). You may recall that family **boundary ambiguity**, also discussed in Chapter 14, is a "state when family members are uncertain in their perception of who is in or out of the family or who is performing what roles and tasks within the family system" (Boss 1987, quoted in Stewart 2005a, p. 1,003).

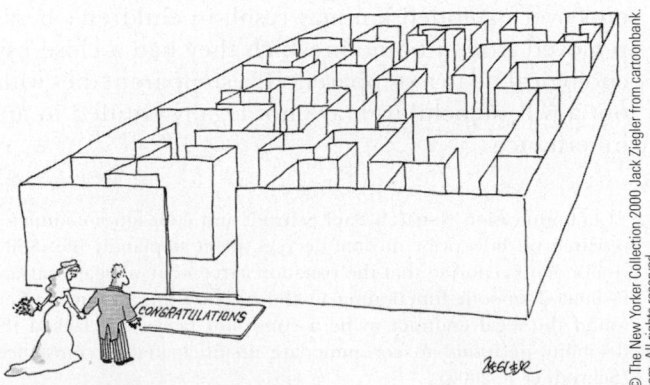

Sociologist Susan Stewart discusses stepfamily boundary ambiguity as follows:

> Family definitions in stepfamilies are dynamic and changeable. One stepfamily member, a wife and mother who complains about never knowing how much to fix for dinner on any given day, describes her family as an "accordion" that "shrinks and expands alternately" [Berger 1998]. . . . Interviews with stepfamily members reveal that definitions of family often differ between parents and children and between siblings. . . . Boundary ambiguity is higher in nontraditional stepfamilies involving cohabitation, part-time residence, and more complex parenting configurations. (Stewart 2007, p. 38)

Stewart (2005a) analyzed data from the National Survey of Families and Households (NSFH). The data included 2,313 stepfamilies, defined as "married or cohabiting couples in which at least one partner has a biological or adopted child from a previous union living inside or outside the household." Stewart operationally defined *boundary ambiguity* as "any discrepancy in partners' reports of shared children (the biological or adopted children of both partners) and/or stepchildren (biological or adopted children from previous unions" (p. 1,009).

Stewart found evidence of boundary ambiguity among 25 percent of parents with stepchildren. When one or more stepchildren lived outside the family household, boundary ambiguity rose to 54 percent. When one or more children of both partners lived outside the family household, boundary ambiguity rose still further—to nearly 80 percent (Stewart 2005a, p. 1,015). Evidence of boundary ambiguity has been found among step- and half siblings (White 1998). Cohabiting stepfamilies are more likely to experience boundary ambiguity than are remarried stepfamilies (Stewart 2005a, p. 1,015).

A more recent study by sociologists Susan Brown and Wendy Manning (2009) compared boundary ambiguity in four family types: (1) families headed by two biological parents, (2) single-mother families, (3) married families in which one parent is a stepparent, and (4) cohabiting families in which one parent is a stepparent. Brown and Manning analyzed data from a national data set consisting of 14,047 interviews with students in grades seven through twelve and their mothers. Separately, adolescents and their mothers were asked to list the members of their families. Adolescent-mother pairs for which lists of family members agreed were characterized as showing no boundary ambiguity. Conversely, adolescent-mother pairs in which one list of family members did not coincide with the other's list were thought to evidence boundary ambiguity.

As shown in Table 16.1, cohabiting stepparent families evidenced greatest boundary ambiguity whereas families with two biological parents evidenced the least. There was near perfect congruence (99 percent) between mother and adolescent reports of living with two biological parents. Among mothers who reported being single mothers, 88 percent of adolescents' reports agreed, with the remaining 12 percent indicating either that their mother was married or (more often) that she was cohabiting. Interestingly, about 20 percent of mothers who reported living in a married stepfamily had a teen who reported living with two biological parents (Brown and Manning 2009).[5] Other research indicates that boundary ambiguity may negatively affect relationship quality and stability (Stewart 2005a, 2007).

Extended Kin Networks in Stepfamilies Relationships with kin outside the immediate stepfamily are complex and uncharted as well (Ganong and Coleman 2004). We have few mutually accepted ways of defining and relating to the new extended and ex-kin relationships that result from remarriage (Stewart 2007). One indication of this situation is that our language has not caught up with the proliferation of new family roles (Kellas, LeClaire-Underberg, and Normand 2008). As partners separate and later form stepfamilies, the new relatives do not so much *replace* as *add to* kin from the first marriage (White and Riedmann 1992). What are the new relatives to be called? There may be stepparents, stepgrandparents, and stepsiblings; but what, for instance, do children call the new wife whom their noncustodial father has married? A further example of ambiguity regarding stepfamilies is the lack of legal definitions for roles and relationships.

Table 16.1 Boundary Ambiguity in Four Family Forms. No boundary ambiguity means that a mother's report of who is in the family coincided with that of her adolescent child. Based on analysis of data from the National Longitudinal Study of Adolescent Health (N=14, 047).

Family Form	No boundary ambiguity %
Two-biological-parent family	99.4
Single-mother family	88.4
Married stepparent family	69.8
Cohabiting stepparent family	34.1

Source: Adapted from Brown and Manning 2009, p. 92.

[5] What might be some reasons that in this study about one-fifth of mothers who said that they lived in a married stepfamily had a teenager who reported living with two biological parents?

Family Law and Stepfamilies

Family law assumes that marriages are first marriages. Therefore, many of the legal provisions that affect stepfamily life proceed from one or both partners' divorce decrees.[6] Meanwhile, few legal provisions exist for several remarried-family challenges—for example, balancing husbands' financial obligations to their spouses and children from current and previous marriages, defining wives' obligations to husbands and children from current and the former marriages, and facilitating legally structured child visitation decrees resulting from a previous divorce. Moreover, in some states, stepparents do not have the authority to see the school records of stepchildren or make medical decisions for them.

Some states have passed legislation that holds stepparents responsible for the support of stepchildren during the marriage (Hans 2002; Malia 2005). However, "in most states, stepparents are not required to support their spouses' children financially, although most voluntarily choose to provide contributions" (Malia 2005, p. 302). Although a stepfamily may come to rely on such financial support, if a remarriage ends in divorce, stepparents are not legally responsible for child support unless they have formally adopted the stepchildren or signed a written promise to pay child support in the event of divorce (Malia 2005). (Interestingly, college financial aid applications require information on a stepparent's income to calculate student need.)

Research suggests that, although remarrieds often share their economic resources, they also take care to protect their individual interests and those of their biological children (Gold 2009; Mason et al. 2002). The preservation of stepparent–stepchild relations when death or divorce severs the marital tie is also a serious issue. Some stepparents continue relationships with their stepchildren after a re-divorce (Dickinson 2002). Visitation rights (and a corresponding support obligation) of stepparents are beginning to be legally clarified (Hans 2002; Mason, Fine, and Carnochan 2001). But outcomes are unpredictable at this point. It's possible that when a custodial, biological parent dies, the absence of custodial preference for stepparents over extended kin may result in children's being removed from a home in which they had a close psychological tie to a stepparent. If a stepparent dies without a will, stepchildren are not legally entitled to any inheritance.

[6] Communication research Paul Schrodt and colleagues conducted research on how prior divorce decrees affect stepfamily life. Some respondents reported that the visitation agreement, a legal contract, facilitated smooth functioning in the co-parenting system. Others found the legal contract to be a constraint because it lacked the flexibility necessary to accommodate unanticipated circumstances (Schrodt et al. 2006).

The only way to be certain that situations like these do not occur is for a stepparent to legally adopt a stepchild, and this situation may be virtually impossible due to the noncustodial, biological parent's objections (Malia 2005; Mason et al. 2002).[7] People who remarry are advised to check with an attorney regarding applicable laws in their state. The vast majority of legislation that is meant to address stepfamilies proceeds primarily from the government's concern over children's well-being.

Children's Well-Being in Stepfamilies

FOCUS ON CHILDREN

How does membership in a stepfamily affect children's well-being? The majority of children in remarried households show few, if any, negative outcomes. However, considerable research has found that, on average, stepchildren of all ages have somewhat higher rates of alcohol abuse and juvenile delinquency, do less well in school, may experience more family conflict, and are somewhat less well-adjusted than children in first-marriage families (Heard 2007; Kirby 2006). "Studies consistently indicate . . . that children in stepfamilies exhibit more problems than do children with continuously married parents and about the same number of problems as do children with single parents" (Amato 2005, p. 80).[8] Sociologist Andrew Cherlin explains these findings:

> the addition of a stepparent increases stress in the family system at least temporarily, as families adjust to new routines, as the biological parent focuses attention on the new partnership, or as stepchildren come into conflict with the stepparent. This increased stress could cause children to have more emotional problems or to perform worse in school, which could counterbalance the positive effects of having a second adult and a second income in the household. (2009a, p. 22)

[7] Stepchild adoption generally requires the waiver of parental rights by the biological parent, who may be actively involved with the child (Malia 2005) and—understandably!—may not want to do so. When the nonresident parent is actively involved, adoption is seldom given serious consideration by either the remarried adults or the children (Pasley 1998a). Children above a certain age, perhaps fourteen, may or must give their consent to stepparent adoption in some states, and even younger children may need to agree to the adoption.

[8] Some research suggests that lowered child well-being in stepfamilies may not apply to African American children, whose culture is more likely to support multiparental models. African American children may be more accustomed to life transitions and stressful life conditions and therefore have greater capacity to adjust to changes, compared with European American children. Analysis of data from a national sample of African American youth found that the presence of a father, whether biological or stepfather, served to increase the likelihood of positive outcomes (Adler-Baeder et al. 2010).

However, because many of the small, negative outcomes for stepchildren are also associated with divorce,

> it is difficult to know the relative contribution of parental remarriage to poor child adjustment. It may well be that most of the negative effects can be attributed to predivorce conditions . . . or postdivorce effects (e.g., reduced income, multiple transitions that accompany divorce), of which parental remarriage is only one. (Ihinger-Tallman and Pasley 1997, p. 31)

Some researchers have found that remarriage lessens some negative effects for children of divorce—but only for those who experienced their parents' divorce at an early age and when the subsequent remarriage remains intact (Arendell 1997; Cherlin 2009a). Additional research shows that younger children adjust better to a parent's remarriage than do older children, especially adolescents (Amato 2005; Carlson 2006).

What about children's well-being in cohabiting stepfamilies? As pointed out in Chapters 7 and 8, compared with growing up in a single-mother home, children benefit economically from living with a cohabiting stepfather, provided that he shares his resources with the family (Manning and Brown 2006). Nevertheless, children in cohabiting stepfamilies generally fare less well than do those in remarried stepfamilies. Cherlin suggests reasons why:

> Single parent families, whether after divorce or in the absence of marriage, create a new family system. Then into that system, with its shared history, intensive relationships, and agreed-upon roles, walks a parent's new live-in partner. . . . Lone mothers may be willing to live with a partner whom they wouldn't necessarily marry. . . . Their partners, in turn, may not be interested in. . . . developing a parentlike relationship with their children. They could even be a net drain on children's resources if the parent becomes preoccupied with the intimate relationship. . . . Moreover, cohabiting partnerships tend to be short-lived, and the departure of a cohabiting partner could once again produce more stress in the household. If the cohabiting stepfamily has established family routines, these would be disrupted again, and the biological parent and the children would have to adjust to the loss. (2009a, pp. 21–23).

Sequential transitions from one family structure to another create upheaval and stress for children (Bures 2009; Cavanagh 2008; Magnuson and Berger 2009).

Although we note the preceding findings, we also recognize other aspects of this story. For one thing, as "My (Step)Family" illustrates, some stepchildren do well and appreciate their stepfamily lives. Moreover, several studies have concluded that stepchildren's well-being and future outcomes largely depend on the quality of the relationships and communication among family members regardless of family structure (Crawford and Novak

© David Young-Wolff/ PhotoEdit

Research consistently shows that stepfamilies are less stable than nuclear families and that, in general, children in stepfamilies have more problems than those in intact, nuclear families. However, research also indicates that the quality of the communication and relationships among family members—and the extent to which children are monitored—may be more important to positive child outcomes than is family structure itself.

2008; Doohan et al. 2009; Schoppe-Sullivan, Schermerhorn, and Cummings 2007).

The extent to which parents or stepparents monitor their children's comings and goings is probably more important to positive child outcomes than is family structure itself (Crawford and Novak 2008). Furthermore, recent research shows that a close, nonconflictual relationship with a stepfather enhances the overall well-being of adolescents, and this is especially true when the child has a similar relationship with the biological mother (Yuan and Hamilton 2006; Booth, Scott, and King 2010). That said, creating a cohesive and supportive stepfamily can be a challenge.

Stepfamily Roles, Relationships, and Cohesion

As we have seen, no cultural script clearly indicates how stepfamily members should play their roles (Ganong and Coleman 2004; Wooding 2008). Given this **role ambiguity**—that is, few clear guidelines regarding what responsibilities, behaviors, and emotions stepfamily members are expected to exhibit—it may not be surprising that relationship and communication patterns in stepfamilies are variable.

In a study somewhat similar to the one on re-weddings, described earlier in this chapter, researchers asked fifty university student stepchildren to describe the communication patterns in their stepfamilies

(Baxter, Braithwaite, and Bryant 2006). As illustrated in Figure 16.3, results showed four different relationship/communication patterns among a biological parent, stepparent, and child.

In a *linked triad*, a child's interaction is connected with the stepparent through the child's biological/adoptive parent: "A lot of stuff I communicate with my stepdad goes through my mom" (p. 389). In the *outsider triad*, the child and the biological/adoptive parent maintain interaction, but the stepparent remains an outsider and pretty much irrelevant to the child's life: "I focus my talking towards my mom. And um like when we're [all] watching the TV, I just always turn to my mom and usually talk to her" (p. 391).

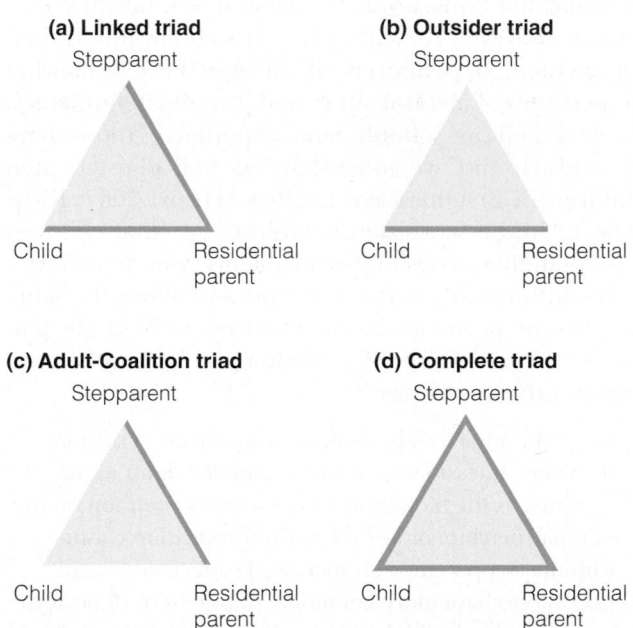

Figure 16.3 Perceived Types of Triadic Communication Structures in Stepfamilies. "A darker line represents the presence of a direct, positive . . . line of communication in a given stepfamily dyad, and a lighter line represents the absence of a direct, positive . . . line of communication" (Baxter, Braithwaite, and Bryart 2006, p. 388). A=Linked Triad. Communication between child and stepparent is linked through the residential biological/adoptive parent. B=Outsider Triad. Child's communication takes place primarily with the residential biological/adoptive parent; child feels little interdependence with the stepparent. In this triad, the stepparent is an outsider. C=Adult-Coalition Triad. Stepparent and residential biological/adoptive parent are viewed as forming a coalition to the relative exclusion of the child. D=Complete Triad. Equal communication and interaction along all three sides of the triad, incorporating all three triadic components equally.

Source: Baxter, Braithwaite, and Bryart 2006, p. 388.

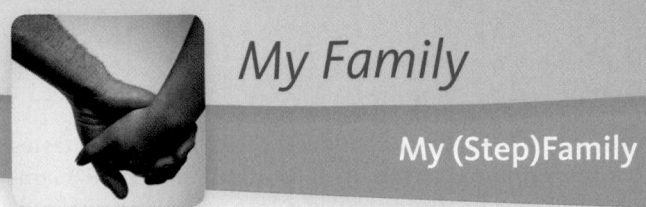

My Family

My (Step)Family

The following essay was written for a marriage and family course by a young college student named David.

I'm writing this paper about my family. My family is made up of my family that I reside with and then my dad, stepmother, and half sister that I visit. My family I reside with is who I consider my real family. We are made up of five girls and three boys, a cat and a dog, my mom and stepfather. The oldest is Harry, then down the line goes Diane, Barbara Ann, Kathy, Debi, Mel, Sharon, and myself. My sister Debi and I are the only kids from my mom's original marriage, so as you can see my mom was taking a big step facing six new kids. (My mom has guts!)

I still consider my dad "family," but I don't come into contact with him that much now. I would like to concentrate on my new "stepfamily," but I really don't like that word for it. My family is my family. My brothers and sisters are *all* my brothers and sisters whether they are step or original. My stepfather, although I don't call him "dad," is my father. My grandparents, step or original, are my grandparents. I can honestly say I love them all the same.

It all began when I was four. I don't remember much about my parents' divorce. The one real memory I have is sleeping with my dad downstairs while my sister slept upstairs with my mom. . . .

My mom met my stepfather, Harry, through mutual friends who went to our church. I was in first grade, and I don't really remember much about their dating. All I knew was either these strange kids came over to my house or I went over to theirs. They were married [when] I was five. At the reception I got to see all my relatives, old and new. . . . After my parents got married, we moved. We didn't move into my stepfather's and his kids' house but to a new house altogether for everyone. I still live there. I love it there and probably will live in that area all my life. . . .

Well, we moved into our new house, and I had to change schools. I had to leave all my old friends and make some new ones. I did make new friends, and what was neat was that our two football teams played each other every year. So I got to play against my old friends with my new friends. When I got into high school, we were all united again. So I had gotten a new family, a new house, and a new school with new friends. I would have to say it was a major life transition. . . .

Both my parents agreed about most issues (discipline, for example), and this led to smooth communication between our parents and us kids. The only thing detrimental I can think of that came out of being in a large family was that I had poor study habits as a young child. The problem was my stepfather can watch TV, listen to the radio, and prepare a balance sheet all at the same time. So he let his kids listen to the radio or watch TV while they did their homework. My mom didn't agree with this, but it was too hard for her to enforce not watching TV with so many kids already used to doing it. . . .

I love all my brothers and sisters very much and would do anything for them. It is neat to see them get married and have kids. I have three nieces and five nephews. It is very exciting to get the whole family together. A lot of my family, nuclear and extended, live in our area. My grandparents, aunts and uncles, cousins, three sisters, and their kids all live within twenty minutes of our house. Having roots and strong family connections are two things I'm very thankful for. These are things I received from my stepfather because if I lived with my dad, I would be on what I consider his nomadic journey: he moves about every three years.

I think I'm very lucky. I had a solid upbringing and relatively few problems. I owe my stepfather a lot. He has given me clothes and food, taught me important lessons, instilled in me a good work ethic, and seen to it that I get a good education—all the way through college. Without him a lot of these things would not be possible; I am grateful for everything he has done for me. I'm happy with the way things turned out, and I love my family dearly.

Critical Thinking

What might be a reason that David does not like to apply the term *stepfamily* to his own family? Divorce and remarriage may be thought of as transitions or crises or both, the subjects of Chapter 14. In what ways does David's essay illustrate meeting crises creatively, as discussed in that chapter? If you live in a stepfamily, how does David's experience compare to your own?

In the *adult-coalition triad*, the child views the biological parent and the stepparent as maintaining a couple relationship that ignores the child's concerns. As an example:

> My sister and I have some difficulties with my stepmom. . . . And [my dad will] say, "Well, honey, you have to understand this and this and this," and like makes excuses for her, and I kind of want to say to him, "Well, we're your *daughters*," you know. (p. 392, italics in original)

In the *complete triad*, communication flows freely, involving all stepfamily members equally: "I'll call home and I'll talk to my stepfather and I'll talk to my mom. . . . He's like a father to me" (p. 393).

Unanimously, the student respondents regarded the adult-coalition pattern as negative and saw the complete triad pattern as ideal. However, the researchers hastened to add that "the outsider triad can be functional for the stepfamily so long as it is compatible with expectations of the stepparent" (Baxter, Braithwaite, and Bryant 2006, p. 395).

Nonetheless, creating stepfamily cohesion is important, although it can be difficult. One obstacle involves the fact that stepparent, stepchild, and stepsibling relationships are often involuntary. Those involved may feel that they had little choice in the matter and hence be disinclined to cooperate (Schrodt, Soliz, and Braithwaite 2008).

Then too, disruptions associated with one or more stepchildren's comings and goings according to a visitation schedule may be stressful (Kheshgi-Genovese and Genovese 1997). Stepchildren in joint custody arrangements (see Chapter 15) may regularly move back and forth between two households with two sets of rules. A child's ties with the noncustodial biological/adoptive parent can make the pre-divorced or separated family seem "more real" than the stepfamily. After returning from visits with the noncustodial parent, a stepchild may unintentionally undermine stepfamily definitions. As an example, a five-year-old visited her biological mother's house where she mentioned a "brother" in her stepfamily. When the girl returned to her stepmother's home, she announced, "My mother says he's not my brother, he's my half-brother" (Bernstein 1997).

Furthermore, stepsiblings may not get along well (Stewart 2007). Especially in the case of multipartnered parenthood, the lives of stepchildren living in one household can vary greatly:

> one child may have a devoted nonresident father who sees her regularly, another child who has no contact with her father jealously watches her half sister go away for weekends with her dad, and a third child—from the new partnership—has both of her parents in the household. . . . The inequalities among children in the same household can be stark. (Cherlin 2009a, p. 195)

In some cases stepsibling rivalry is sparked by actual or perceived inequality of treatment by one parent. As one stepchild explained,

> The way my brothers and I saw it was that my dad treated us the same and he tried to treat her children the same, but we saw a difference in the way she treated her children and the way she treated us. . . . My brothers and I were being treated differently than her children were. (in Baxter, Braithwaite, and Bryant 2006, p. 390)

Moreover, as noted by psychotherapist Susan Pacey:

> Grandparents are powerful figures in the hierarchy of the stepfamily, and can help or hinder the couple in forming a new life together. For example, gifts or bequests made to the biological grandchildren only, when a long established stepfamily home includes step or half siblings, may prove divisive and detrimental to the stepfamily and the couple. (2005, p. 368)

The experience of being a step-grandparent is further discussed in Chapter 17.

Another challenge to fashioning cohesive stepfamilies lies in children's lack of desire to see them work.

After age two or three, children often harbor fantasies that their original parents will reunite (Bray 1999; Gamache 1997). This hope can last into young adulthood. As one college student described hearing of her parent's remarriage intentions,

> It didn't hit me until I hung up the phone and then I remember just crying and I couldn't understand why I was crying. You know? And I think it had just hit me that my parents are never going to get back together. . . . I had never thought that [their] getting divorced was the end of them. (in Baxter et al. 2009, p. 480)

Children and adolescents who want their divorced or separated parents to get together again may feel that sabotaging the new relationship will help to achieve that goal (Kheshgi-Genovese and Genovese 1997).

Furthermore, "[s]tepchildren may feel they are betraying their biological parent of the same sex as the stepparent if they form a friendly relationship with the stepparent" (Kheshgi-Genovese and Genovese 1997, p. 256; see also Kellas, LeClaire-Underberg, and Normand 2008). As one of our students wrote in an essay: "Since

Conflicting expectations concerning a stepfather's— or stepmother's—role may make it stressful. When stepparents can ignore the myths and negative images of the role and maintain optimism about the remarriage, they are more likely to have high family, marital, and personal satisfaction.

I [had] idolized my father for so many years, I didn't want to accept my stepfather. I didn't like the fact that someone else was sleeping with my mother and touching her." (This student gradually changed his attitude, however: "The thing that won me over was [my stepfather's] support of everything that we did. Whenever we went to any of our sporting events he was there to help us.") In the case of remarriage after widowhood, children may have idealized, almost sacred, memories of the parent who died and may not want another to take his or her place (Andersen 2002; Barash 2000).

Furthermore, biological parents may feel caught between loyalties to their biological child and the desire to please their new partner (Bray 1999; Visher and Visher 1996)—and between other loyalties as well. As one stepparent put it,

When you become a stepparent, you find yourself not just playing Piggy in the Middle between your partner and his/her children, but often between your partner and his/her ex, your partner and your ex, your partner and your children, your children and your partner's children. The combinations are endless! (Andersen 2004)

Adolescent Stepchildren and Family Cohesion

Researchers advise that forming stepfamilies when a stepchild-to-be is in adolescence can be especially difficult (Bray and Easling 2005). Many adolescents blame their parents or themselves, or both for the breakup of the first marriage. The stepparent becomes a convenient scapegoat for their hostilities. Stepchildren can prove to be formidable adversaries (Warshak 2000). Our discussion of power in Chapter 13 focuses primarily on marital power, but we're reminded that adolescent stepchildren wield considerable family power (Visher and Visher 1996). Then, too, the desire of other family members to create a cohesive stepfamily may conflict with an adolescent's normal need to express independence and/or with the "substantial degree of autonomy" that they experienced in their former single-parent family (Amato 2005, p. 81). "As We Make Choices: Some Stepparenting Guidelines" contains advice that can make stepparenting easier.

Sociologists Lynn White and Alan Booth's (1985) analysis of stability in remarriages, discussed earlier, considered one additional point: that family tension may be resolved by the child's rather than the partner's exit from the home. Speculating that children might be moved out by sending them to live with the other parent or forcing them to become independent, White and Booth found that older teenage and young adult children in stepfamilies do indeed leave home at significantly lower ages than do those in first-marriage families. Not all teen or young adult stepchildren who leave home early do so for the reasons that White and Booth

suggest (Ganong and Coleman 2005). However, subsequent research supports White and Booth's findings, and today we have a word for this situation—**extrusion**, "defined as individuals' being 'pushed out' of their households earlier than normal for members of their cultural group, either because they are forced to leave or because remaining in their households is so stressful that they 'choose' to leave" (Crosbie-Burnett et al. 2005, p. 213). Based on theory and research findings:

[w]e would conclude that the probability of extrusion in a stepfamily increases (a) when the adolescent cannot communicate effectively about family issues, . . . particularly feelings about a new stepparent entering the family . . . ; (b) when the stepparent and biological parent cannot communicate effectively with the adolescent . . . ; (c) when the biological parent fears losing the relationship with the stepparent if the adolescent remains in the home . . . ; (d) when the biological parent has an insecure, dismissing attachment style, making him or her more likely to "let go of" the adolescent . . . ; (e) when the adolescent reports physical or sexual abuse . . . ; (f) when the stepparent is male, as men tend to have more power than women in conjugal relationships and are more likely to be abusive . . . ; (g) when the stepparent is bringing more resources into the family than the biological parent, giving him or her more power to effect the extrusion . . . ; (h) when the adolescent is not heterosexual . . . ; and (i) when the family is from the mainstream individualist American culture as opposed to a culture strong in familism. (Crosbie-Burnett et al. 2005, pp. 228–29)

Extrusion is an extreme example of a stepfamily's failure to create family cohesion. Among other causes, challenges to stepfamily cohesion result from role ambiguity.

Stepfamily Role Ambiguity

Relatively low role ambiguity has been associated with higher remarital satisfaction, especially for wives, and with greater parenting satisfaction, especially for stepfathers (Kurdek and Fine 1991; Munroe 2009). However, the role of the stepparent is "precarious; the relationship between a stepparent and stepchild only exists in law as long as the biological parent and stepparent are married" (Beer 1989, p. 11). Although some stepchildren certainly do maintain relations with a stepparent after a stepparental divorce, doing so requires forging personalized ways to do this in the absence of commonly understood norms (Dickinson 2002). Stepchildren too must construct their role in the absence of society-wide norms (Speer and Trees 2007).

One result of role ambiguity is that society—and hence the members of the stepfamily itself—seems to expect stepparents and children to love each other in much the same way as biologically related parents and

Preparing to Live in Step

In a stepfamily, at least three (and often more) individuals struggle to form new familial relationships while coping with grief, pain, reminders of the past, or all three. Each family member brings to the situation expectations and attitudes that are as diverse as the personalities involved. The task of creating a successful stepfamily, as with any family, will be easier for all concerned if all members try to understand the feelings and motivations of the others as well as their own.

It is important to discuss the realities of living in a stepfamily before the marriage, when problems that are likely to arise can be foreseen and examined theoretically. If you are contemplating entering a step-relationship, here are some key points to consider:

1. *Plan ahead.* Consider attending an "education for remarriage" workshop, offered by many religious and other community organizations. "Read and understand basic child development so you don't mistake developmentally normal behaviors as inappropriate, uncooperative or as personally against you" (Lavin 2003).

2. *Examine your motives and those of your future partner for marrying.* Get to know your partner as well as possible under all sorts of circumstances. Consider the possible impact of contrasting lifestyles.

3. *Discuss the modifications that will be required in bringing two families together.* Compare similarities and differences in your concepts of child raising.

4. *Explore with your children the changes remarriage will bring:* new living arrangements, new family relationships, and the effect on their relationship with their noncustodial parent.

5. *Give your children ample opportunity to get to know your future partner well.* Consider your children's feelings, but don't allow them to make your decision about remarriage.

6. *Discuss the disposition of family finances with your future partner.* An open and honest review of financial assets and responsibilities may reduce unrealistic expectations and resultant misunderstandings.

7. *Understand that there are bound to be periods of doubt, frustration, and resentment.*

Living in Step

Any marriage is complex and challenging, but the problems of remarriage are more complicated because more people, relationships, feelings, attitudes, and beliefs are involved than in a first marriage. The two families may have differing roles, standards, and goals. Because its members have not shared past experiences, the new family will need to redefine rights and responsibilities to fit both individual and combined needs.

Time and understanding are key allies in negotiating the transition from single-parent to stepfamily status. Consideration of the following points may ease the transition process:

1. *Let your relationship with stepchildren develop gradually.* Don't expect too much too soon—from the children or yourself. Children need time to adjust, accept, and belong. So do parents.

2. *Don't try to replace a lost parent; be an additional parent.* Children need time to mourn the parent lost through divorce or death.

3. *Expect to deal with confusing feelings*—your own, your partner's, and the children's. Anxiety about new roles and relationships may heighten competition among family members for love and attention; loyalties may be questioned. Your children may need to understand that their relationship with you is valued but different from that of your relationship with your partner and that one cannot replace the other. You love and need them both, but in different ways.

4. *Recognize that you may be compared to the absent partner.* Be prepared to be tested, manipulated, and challenged in your new role. Decide, with your mate, what is best for your children, and stand by it.

5. *"Discuss discipline and make sure the biological parent is the one carrying out the discipline of his or her child"* (Lavin 2003)—at least, at first.

6. *Understand that stepparents "do not have the power or authority to 'fix' their stepchildren or the family.* Only a biological parent has that ability" (Lavin 2003). Understand, too, that stepparents need support from biological parents on child-raising issues. Raising children is tough; helping to raise someone else's can seem tougher.

7. *Acknowledge periods of cooperation among stepsiblings.* Try to treat stepchildren and your own with equal fairness. Communicate! Don't pretend that everything is fine when it isn't. Acknowledge problems immediately, and deal with them openly.

8. *Admit that you need help if you need it.* Don't let the situation get out of hand. Everyone needs help sometimes. Join an organization for stepfamilies; seek counseling.

Source: U.S. Department of Health, Education, and Welfare 1978; Lavin 2003; Van Pelt 1985; see also Jeannette Lofas (n.d.).

children do. In reality, this is not often the case, and therapists point out that stepparents and stepchildren should not expect to feel the same as they would if they were biologically related (Barash 2000; Visher and Visher 1996). Therapists advise and research shows that a stepparent's waiting through a period of family adjustment before becoming an active disciplinarian is usually a good idea (Ganong and Coleman 2000). Certain situations are more likely to challenge stepmothers, whereas other difficulties are more common to stepfathers. We'll look at each of these roles and the special challenges associated with them.

Stepmothers

A small study asked 265 stepmothers about their expectations about the stepmother role. The researchers found that stepmothers expected to be included in stepfamily activities but certainly do not see themselves as replacing the stepchild's mother. The more time a stepmother spent with her stepchildren, the more she expected to be included in stepfamily functions and decisions, and the more she behaved as concerned parent, rather than as a friend (Orchard and Solberg 2000). Even into stepchildren's adulthood, the stepmother role is thought by social scientists to be more difficult than is the stepfather role (Coleman and Ganong 1997; Ward, Spitze, and Deane 2009). One important reason for this is a contradiction in expectations for stepmothers:

> Whether mothers or stepmothers, . . . the women's roles are very similar. Irrespectively, they take care of children and housework. [But] . . . the stepparent role expects a certain distance, the female role the opposite. Between the two roles there is a dilemma. One cannot be distant and close at the same time. (Levin 1997, p. 132)

Because of this inherent contradiction, the stepmother role has been described as the **stepmother trap**: On the one hand, society seems to expect almost mythical loving relationships between stepmothers and children. On the other hand, stepmothers are often stigmatized—seen and portrayed as cruel, vain, selfish, competitive, and even abusive (Schrodt 2008; Whiting, Smith, and Barnett 2007). Remember Snow White's, Cinderella's, and Hansel and Gretel's stepmothers?

Maureen McHugh (2007) is a stepmother who writes for the website Second Wives Café: Online Support for Second Wives and Stepmoms (secondwivescafe.com). The following is an excerpt from her online article "The Evil Stepmother":

> My nine-year-old stepson Adam and I were coming home from Kung Fu. "Maureen," Adam said—he calls me "Maureen" because he was seven when Bob and I got married and that was what he had called me before. "Maureen," Adam said, "are we going to have a Christmas Tree?"

> "Yeah," I said, "of course." After thinking a moment, "Adam, why didn't you think we were going to have a Christmas Tree?"

> "Because of the new house," he said, rather matter-of-fact. "I thought you might not let us." It is strange to find that you have become the kind of person who might ban Christmas Trees.

Then too, a stepmother may feel left out by the father's continued relationship with his ex-wife (Barash 2000; Munroe 2009).

Nonresidential Stepmothers Special issues accompany the role of *nonresidential,* or "weekend," *stepmother* when women are married to noncustodial fathers who see their children for visitation periods. Nonresidential stepmothers (remarried women whose stepchildren live in a different household) may try to establish a loving relationship with their husband's children only to be openly rejected, or they may feel left out by the father's ongoing relationship with his offspring. Some nonresidential stepmothers report that their partner's visiting children are bad influences on their own, biological children (Henry and McCue 2009).

Residential Stepmothers Residential stepmothers (women who share the household with stepchildren) may face somewhat different challenges. Social worker and stepmother Emily Bouchard tells her story:

> When I moved in with my husband and his two teenage daughters, he had a real "hands off" approach. . . . Sparks began to fly as soon as I asserted what I needed to be different . . . For example, when I noticed that my car had been "borrowed" (the odometer was different) without my knowledge or permission, I had to show up as a parent the way I needed to parent—setting limits, confronting the greater issues of lying and sneaking, and asserting the natural consequences for unacceptable behavior. This method was foreign to their family, and there were reactions all the way around! Thankfully, my husband supported me in front of his daughter, and then we discussed our differences privately and came to a mutual understanding about how to handle parenting together from then on. (Bouchard n.d.)

One explanation for the difficulties that residential stepmothers face is the fact that stepmother families, more often than stepfather families, begin after heated custody battles subsequent to particularly troubled relations in the pre-divorce family (Schrodt 2008; Stewart 2007).

Stepfathers

Many children have positive relationships with their stepfathers. Psychotherapist Susan Pacey (2005) has noted that there are four different pathways to stepfatherhood: the man may have been single; divorced and childfree; divorced and without custody of his children; or divorced with custody. "Each of these backgrounds is likely to influence the man's approach to his step-parenting role" (Pacey 2005, p. 366). Perhaps not surprisingly, stepfathers who adopt their stepchildren tend to be more involved with them than those who don't (Schwartz and Finley 2006). Then too, whether a stepfather is married or cohabiting tends to influence his role:

> At one end of the continuum are men who should be considered stepparents because they are deeply involved in the children's lives and likely to marry the mothers after a period of living together. At the other end are mothers' short-term romantic liaisons who may have little to do with the children and may be around only for a few months. In between is a gray area of men in the household who are not taking on a parental role but who do spend some time with the children and may be present for a year or two. (Cherlin 2009, p. 101)[9]

In the absence of marriage, father/stepfather involvement drops sharply after relationships between romantic partners end (Tach, Mincy, and Edin 2010).

Men who decide to marry a woman with children come to their new responsibilities with varied emotions, typically quite different from those that motivate a man to assume responsibility for his biological children. "I was really turned on by her," said one stepfather of his second wife. "Then I met her kids." This sequence is a fairly common "situation of many stepparents whose primary focus may be the marriage rather than parenting" (Ceballo et al. 2004, p. 46). Along with feeling positive about what he is undertaking, a new husband may be anxious, fearful, or ambivalent.

To enter a single-mother family, a stepfather must work his way into a closed group—a reason that many stepfathers "tend to be marginalized in households where mothers are regarded as the disciplinarians" (Pacey 2005, p. 366). Furthermore, the mother and children share a common history, one that does not

Many children have positive relationships with their stepfathers. But a stepfather has to integrate himself into a previously established single-mother family that already shares a common history. Children and their stepfathers are more likely to forge positive relationships when the stepfather assumes a parental identity, when his parenting behavior meets his own and other family members' expectations, and when his parental demands are not contested by the child's biological father.

yet include the stepfather. In addition, many stepfathers must construct their stepparenting role parallel to that of the nonresidential, but still involved, biological father (Cheadle, Amato, and King 2010; King 2009). Discipline is likely to be tricky. Children and their stepfathers are more likely to feel positive about their relationship when the stepfather assumes a parental identity, when his parenting behavior meets his own and other family members' expectations, and when his parental demands are not contested by an involved, nonresidential biological father (Coleman, Ganong, and Fine 2000; MacDonald and DeMaris 2002; Marsiglio 2004).

A recent analysis of data from the National Longitudinal Study of Adolescent Health found that, for teens in remarried families, close ties to stepfathers are more likely to develop when the adolescent has close ties to his or her mother before the stepfather entered the family. Prior ties to nonresident fathers were not found to be related to stepfather–stepchild ties (King 2009).

Research shows that both stepmothers and stepfathers play their roles with more distance than do biological parents—more like friends than monitoring parents. Especially when they have biological children of their own, stepfathers tend to be more distant and detached than stepmothers, (Coleman and Ganong 1997). Discipline is likely to be particularly tricky. Stepfathers may react to these difficulties in several different ways:

[9] As a Canadian social worker explained, "when a mother states, 'he's my boyfriend; and the man states, 'She's my girlfriend these days,' I don't tend to consider the man to be a very good father for the child. I am not inclined to include these men in the intervention plan, because I know that they are only passing through" (in Parent et al. 2007, p. 234).

1. The stepfather may be driven away, with the stepfamily ultimately being dissolved.

2. The stepfather may take control, establishing himself as undisputed head of the household and forcing the former single-parent family to accommodate his preferences.

3. The stepfather may be assimilated into a family with a mother at its head and have relatively little influence on the way things are done.

4. The stepfather, his new partner, and her children may all negotiate new ways of doing things (Isaacs, Montalvo, and Abelsohn 1986).

This last possibility is the most positive alternative for everyone, and it is further addressed in the final section of this chapter. First, though, we'll look at what the research has to say about having a mutual child.

Having a Mutual Child

Biological children of both partners in a stepfamily are called *mutual, shared,* or *joint* children (Stewart 2007). Research shows that a principal reason for choosing to have a child together involves hope that the mutual child will "cement" the remarriage bond (Ganong and Coleman 2004). Some women with children feel obliged to give a childless husband a son or daughter of his own. Another cause is perceived social pressure to be like "a normal family."

Some research has found that having a mutual child is associated with increased marital happiness and stability (Pasley and Lipe 1998). However, scholars and therapists have expressed concern about the impact on a stepfamily of having a mutual child early in what is bound to be a complex adjustment. A new child may diminish parental attention to the children already in the stepfamily (Stewart 2005b). Believing that they will now be ignored by the stepparent—or seeing the mutual child as having a privileged place in the family—the stepchildren may feel threatened, jealous, or resentful (Munroe 2009; Pasley and Lipe 1998). Feelings of jealousy about a future mutual child can be set up as early as a parent's re-wedding ceremony. For instance, a nineteen-year-old female whose father had been remarried for two years told an interviewer:

> I think at the wedding it would have been more ideal if like family members had, I don't know, I guess, paid more attention to me and my brother. 'Cause they were kind of like you're married and you're going to have kids and everything, but they kind of forgot that there's already two kids in this family. (in Baxter et al. 2009, p. 482)

A recent study suggests that stepfamily conflict precipitated by stepsiblings increases the odds that the mutual child will have behavior problems (Halpern-Meekin and Tach 2008). Moreover, although joint children

> have both of their biological parents present in the household, they must still deal with the complications of stepfamily life. . . . Qualitative work suggests that children born into stepfamilies face a unique set of challenges. William Beer (1992) reveals that mutual children occupy a privileged yet pressure-filled position [referred to as] the hub. On one hand, the child is related to everyone in the family by blood, so he or she gets more attention and has more control in the family than the other children. On the other hand, this child feels constant pressure to ensure that everyone gets along. (Stewart 2007, pp. 70–71)

For some couples, financial strains associated with stepchildren's expenses are a determining factor in the decision not to have a mutual child (Engel 2000).

Financial Strains in Stepfamilies

Some challenges that characterize stepfamilies often begin with the previous divorce (Gold 2009). This is evident in the case of finances (Henry and McCue 2009; Mason, Fine, and Carnochan 2001). A remarried spouse (usually the husband) generally is financially accountable by law for children from the first union *and* financially responsible—sometimes legally—for stepchildren (Manning, Stewart, and Smock 2003). Not surprisingly, some remarrieds believe that prior child support agreements need to be modified better to accommodate the needs of the stepfamily (Hans 2009).

Moreover, whether legally required to or not, "many stepparents do in fact help to support the stepchildren with whom they reside, either through direct contributions to the child's personal expenses or through payments toward general household expenses such as food and shelter" (Mason et al. 2002; Stewart 2007). Many remarried husbands report feeling caught between what they see as the impossible financial demands of both their former family and their present one (Hans and Coleman 2009).

Some second wives—more often, those without children of their own—feel resentful about the portion of the husband's income that goes to his first partner to help support his children from that union (Hans and Coleman 2009). As an Australian nonresidential stepmother told an interviewer,

> I have always felt very cross that we have to give a large amount of our income [in child support] when I would really like that income to help . . . [my autistic son] . . . with therapy. . . . I am forced to work many hours to pay for [his] therapy. (in Henry and McCue 2009, p. 196)

On the other hand, even nonresidential stepmothers spend their own money on stepchildren—usually for incidentals during visitation periods (Engel 2000).

Meanwhile, a mother may feel guilty about the burden of support that her own children place on their stepfather (Barash 2000). Mothers with children from a former union worry about receiving regular child support from an ex-partner (Manning, Stewart, and Smock 2003). We turn now to a look at what family therapists can tell us more generally about creating supportive stepfamilies.

Creating Supportive Stepfamilies

Creating a supportive stepfamily is not automatic. One stepfamily scholar (Papernow 1993) has suggested a **seven-stage model of stepfamily development**:

1. *Fantasy*—adults expect a smooth and quick adjustment while children expect that the stepparent will disappear and their parents will be reunited.

2. *Immersion*—tension-producing conflict emerges between the stepfamily's two biological "subunits."

3. *Awareness*—family members realize that their early fantasies are not becoming reality.

4. *Mobilization*—family members initiate efforts toward change.

5. *Action*—remarried adults decide to form a solid alliance, family boundaries are better clarified, and there is more positive stepparent–stepchild interaction.

6. *Contact*—the stepparent becomes a significant adult family figure, and the couple assumes more control.

7. *Resolution*—the stepfamily achieves integration and appreciates its unique identity as a stepfamily.

Therapists have condensed this model to four (easier-to-remember) consecutive stages: the fantasy stage, the confusion stage, the conflict stage, and the comfort stage ("Stepfamily Stages: A Thumbnail Sketch" n.d.). Therapists tend to agree that getting from fantasy to comfort takes time—from four to seven years!—and "is one of the most difficult tasks that families can face" (Wark and Jobalia 1998, p. 69; see also Lavin 2003). The transition to successful stepfamily living requires considerable adjustment on the part of everyone involved (Munroe 2009).

It helps to remember that the unrealistic "urge to blend the two biological families as quickly as possible" may lead to disappointment when one or more adult or child members "resist connecting" (Wark and Jobalia 1998, p. 70). You may have noticed that we have not used the once-familiar term *blended family* in this chapter. That's because family therapists and other experts

This family portrait is of a mother and stepfather of two full sisters, along with a baby son from the new union. The remarried family structure, which is complex and has many unique characteristics, has no accepted cultural script. When all members are able to work thoughtfully together, adjustment to a new family life can be easier.

have concluded that stepfamilies do not readily "blend" (Deal, in Kiesbye 2009).[10]

It may be better to think of a stepfamily as a **binuclear family**—a new family type that includes members of the two (or more) families that existed before the divorce and remarriage (Ahrons 2004).

Nevertheless, people can and do create supportive and resilient remarriages and stepfamilies (Ahrons 2004). Counselors remind remarrieds not to forget their couple relationship (Munroe 2009). In many locations, prospective partners can participate in remarriage preparatory courses that alert remarrying couples to expect challenges and help to address them.

In a clever play on words, online columnist and stepmother Dawn Miller (n.d.) titled one of her essays "Don't Go Nuclear—Negotiate." Chapter 6 presents some things for couples to talk about when forging an adaptable, supportive marriage relationship. Many of those questions also apply to remarriages. But there are additional issues to discuss regarding stepfamilies: what will be the household rules, expectations for a stepparent's financial support of stepchildren, how emergency medical care will be handled if the biological parent isn't there to sign a release, questions of inheritance, and perhaps whether there will be a mutual child or children (Lavin 2003).

Chapter 14 points out that life transitions, such as remarriage or the transition to stepparent, are family stressors. That chapter explains that resilient families deal with family transitions creatively by emphasizing mutual acceptance, respect, and shared values (Ahrons 2004). You may also recall that Chapter 12 presents guidelines for constructive family communication—all applicable in stepfamilies.

Moreover, creating supportive and cohesive stepfamilies involves recognizing and building upon some potential family strengths that are unique to stepfamilies. Relationships with new extended kin may be a potential source of new friendships. Beginning a renewed sense of family history is another strength builder. Although holidays often divide the stepfamily because of visitation agreements with a noncustodial parent, it is possible to create new family holidays when the entire stepfamily is sure to be together (Wurzel n.d.)

Today's stepfamilies have access to more resources than in the past. For instance, several online websites by stepfamily counselors and well-respected researchers are designed to give advice and report research findings concerning stepfamilies. Examples are thestepfamilylife .com, and the website of the Stepfamily Foundation (www.stepfamily.org). The website, "I Do! Take Two" (http://www.idotaketwo.com) offers not only purchasable wedding products but also respected counselors' advice on topics ranging from re-wedding etiquette to religious, financial, and legal issues. This website has good suggestions for including children in the re-wedding ceremony, possibly creating a family-forming ritual.

There are also more and more books written by psychologists and others for remarrieds and stepfamily members. Living up to its title, Erin Munroe's *The Everything Guide to Stepparenting* (2009) addresses topics from dating a parent to the logistics of moving in together to questions about maintaining step-relationships after a re-divorce. Increasingly, there are stepfamily books written for children. One of these, directed to teens, is *Stepliving for Teens: Getting along with Step-Parents, Parents and Siblings* (Block and Bartell 2001). Another book for teens and preteens is *The Step-Tween Survival Guide: How To Deal with Life in a Stepfamily* (Cohn, Glasser, and Mark 2008). Sally Hewitt's *My Stepfamily* (2009) is written for younger children. For further suggestions, see Coleman and Nickleberry (2009).

Stepfamily enrichment programs, support groups, and various other group counseling resources for stepfamilies of various ethnicities are increasingly available and have been found to be helpful (Higginbotham, Miller, and Niehuis 2009; Skogrand, Barrios-Bell, and Higginbotham 2009). One example is the Active Parenting for Stepfamilies program (Popkin and Einstein 2006). Another is the Stepfamily Enrichment Program (Michaels 2006). In general, researchers and family therapists tend to agree that:

> it is neither the structural complexity nor the presence/ absence of children in the home *per se* that impacts the marital relationship. Rather, the ways in which couples interact around these issues are the key to understanding marital relationships in general and marital relationships in remarriages specifically. (Ihinger-Tallman and Pasley 1997, p. 25; see also Halford, Nicholson, and Sanders 2007)

Interacting in positive ways in remarriages and stepfamilies involves making knowledgeable choices.

We close this chapter with the paragraph that stepfamily scholar Susan Stewart uses to close her significant book, *Brave New Stepfamilies* (2007):

> One might conclude that Americans can maximize their well-being by getting married, staying married, reproducing their own biological offspring, and toughing it out. Yet an increasing number of Americans live increasing portions of their lives in increasingly diverse families that do not align with this idea. Perhaps Americans might do

---

[10] In fact, Dr. Marjorie Engel (2003), president of the Stepfamily Association of America, warns that "[c]ouples with 'blended' as their objective tend to have the most problematic households and those are the couples most likely to leave the stepfamily because some or all of the members won't buy into the blended concept." Playing with the language, stepmother and online columnist Dawn Miller refers to stepfamily living as "life in a blender" (Miller 2004).

better by admitting the emerging normality of stepfamilies and building institutional support to make their brave new stepfamilies strong. (p. 224)

Summary

- Remarriages have always been fairly common in the United States but are more frequent now than they were early in this century, and they now follow divorce more often than widowhood.

- Remarriages are usually about as happy as first marriages, but they tend to be slightly less stable.

- One reason for relative remarital instability is lack of a widely recognized cultural script for living in remarriages or stepfamilies.

- Remarrieds often unconsciously try to approximate the nuclear-family model, but it does not work well for most stepfamilies because stepfamilies differ from first-marriage families in important ways.

- Relationships within stepfamilies and also with extended kin are often complex, yet there are virtually no social prescriptions and few legal definitions to clarify roles and relationships.

- The lack of cultural guidelines is most apparent in the stepparent role.

- Stepparents are often troubled by financial strains, role ambiguity, and stepchildren's hostility.

- Marital happiness and stability in remarried families are greater when the couple has strong social support, good communication, a positive attitude about the remarriage, low role ambiguity, and little belief in negative stereotypes and myths about remarriages or stepfamilies.

Questions for Review and Reflection

1. Discuss some structural differences between stepfamilies and first-marriage families with children.

2. The remarried family has been called an incomplete institution. What does this mean? How does this affect the people involved in a remarriage? Include a discussion of kin networks and family law. Do you think this situation is changing?

3. What evidence can you gather from observation or your own personal experience or both to show that stepfamilies (a) may be more culturally acceptable today than in the past and (b) remain negatively stereotyped as not as functional or as normal as first-marriage, nuclear families?

4. What are some challenges that stepparents face? What are some challenges faced particularly by stepfathers? Why might the role of stepmother be more difficult than that of stepfather? How might these challenges be confronted?

5. **Policy Question**. In terms of social policy, what might be done to increase the stability of remarried stepfamilies? Of cohabiting stepfamilies?

Key Terms

binuclear family 469
cultural script 455
extrusion 463
boundary ambiguity 457
incomplete institution 455

nuclear-family model monopoly 455
remarriages 447
role ambiguity 460
seven-stage model of stepfamily development 468
stepmother trap 465

Online Resources

Sociology CourseMate

www.CengageBrain.com

Access an integrated eBook, chapter-specific interactive learning tools, including flash cards, quizzes, videos, and more in your Sociology CourseMate, accessed through CengageBrain.com.

www.CengageBrain.com

Want to maximize your online study time? Take this easy-to-use study system's diagnostic pre-test, and it will create a personalized study plan for you. By helping you identify the topics that you need to understand better and then directing you to valuable online resources, it can speed up your chapter review. CengageNOW even provides a post-test so you can confirm that you are ready for an exam.

Aging Families

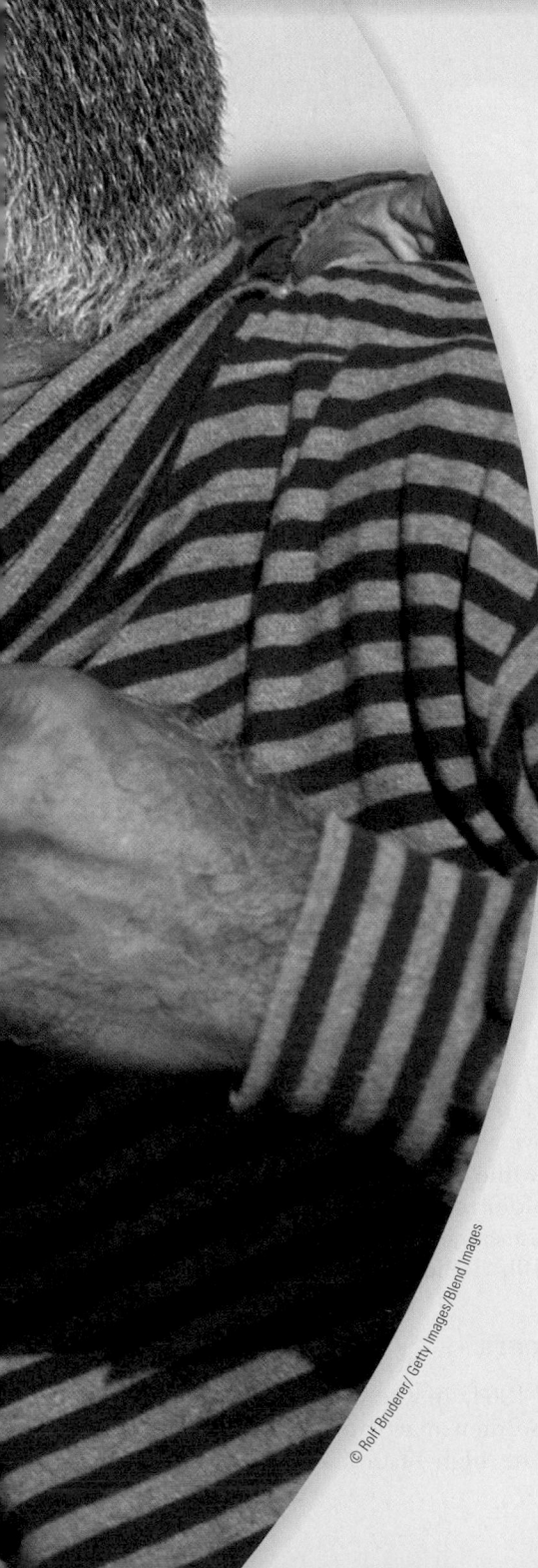

"I feel very young," says an 82-year-old great-grand-mother who lifts weights two to three times a week at a gym, attends watercolor classes and cooks lunch nearly every day for a daughter and a granddaughter who live nearby. She and her husband, Alfred, also 82, . . . go on picnics and attend movies and car shows with their two great-grandchildren. (Rosenbloom 2006)

This chapter examines families in later life.

As a beginning, we note that in recent decades the concept of aging itself has changed. Older Americans behave more youthfully now than in decades past (Sanderson and Scherbov 2008). Americans' happiness level, although highest for those in their early twenties and gradually dropping after that, begins to increase once more at about age sixty and does not drop again until after about age seventy-five. Even into their nineties, 72 percent of Americans tell pollsters that they "experienced happiness, enjoyment, and smiling or laughter during a lot of the day" (Newport and Pelham 2009).

Another phenomenon to note here is the ever-increasing diversity within today's older population—both in race/ethnicity and also in family form. Research is just beginning on family diversity in later life; throughout this chapter, we will focus on the findings.

Many of the topics explored elsewhere in this text apply to aging families. For instance:

- Older families comprise a diversity of family forms, including GLBT couples.
- Older wives—like younger ones—concern themselves with marital equity when it comes to power, decision making, housework, and other/caregiving tasks.
- Older individuals may be engaged in parenting.
- As today's adults age, more and more older families will be stepfamilies.
- Communication is as important in older families as in younger ones.

This chapter focuses on topics specifically related to aging families. We will look at the living arrangements of older Americans and at family relationships in later life. We'll discuss the grandparent role, and then explore issues concerning giving care to older family members. To begin, we'll examine some facts about our aging population.

Our Aging Population

The number of older people in the United States (and all other industrialized nations) is growing remarkably. In 1980, there were 25.5 million Americans age sixty-five or older; today about 40.2 million Americans are age sixty-five or older, and this number is expected to double over

Changing the concept of aging itself, seniors are increasingly active into older ages. According to LeRoy Hanneman of Del Webb Retirement Communities, "Boomers should be called Zoomers" (in "The Demographics of Aging" n.d.).

the next forty years. Americans age seventy-five and older numbered close to 10 million in 1980; by 2010, there were more than 18.7 million. Of those age eighty-five and above, there were 2.2 million in 1980, compared to more than 5.7 million today. Projections are that, by the year 2050, there will be nearly 88.5 million Americans age sixty-five and older, with about 19 million of them age eighty-five and over (U.S. Census Bureau 2010b, Table 8).

Not just the *number* of elderly has increased but also their *proportion* of the total U.S. population. This is especially true for those in the "older-old" (age seventy-five through eighty-four) and the "old-old" (eighty-five and over) age groups. The proportion of Americans age seventy-five and above rose from 4.4 percent in 1980 to 6 percent in 2010, while the proportion of Americans age eighty-five and older rose from 1.0 percent to nearly 2.0 percent over those same years (U.S. Census Bureau 2003a, Table 11; 2010b, Table 8).

Aging Baby Boomers

Between 1946 and 1964, in the aftermath of World War II, more U.S. women married and had children than ever before. The high birthrate created what is

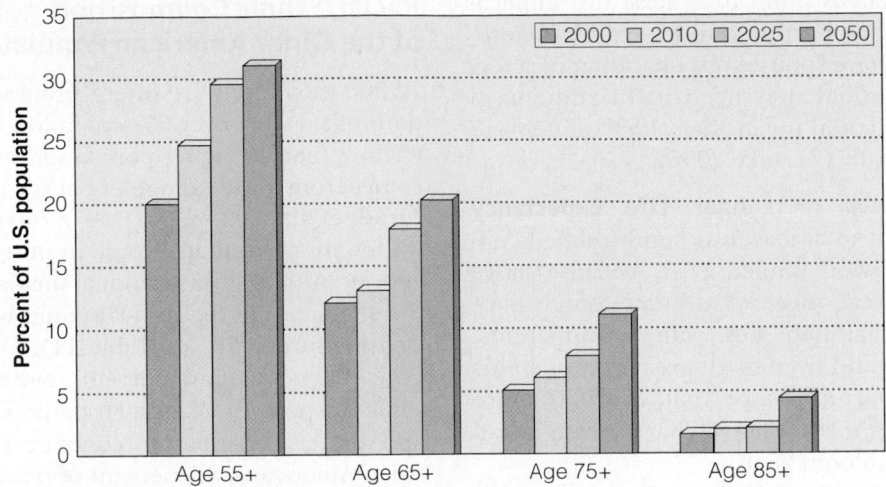

Figure 17.1 Older Americans as a Percentage of the Total U.S. Population, 2000 and 2010, with Projections for 2025 and 2050. Currently, baby boomers are in their fifties and early sixties. As the baby boom cohort grows older, populations over age fifty-five, sixty-five, seventy-five, and eighty-five will increase.

Sources: Calculated from U.S. Census Bureau 2002, Table 12; 2010b, Table 8.

commonly called the **baby boom**. Now baby boomers have begun to retire, and within the next decades, they will generate an unprecedentedly large elderly population (see Figure 17.1). Meanwhile, the number of children under age eighteen is about the same today as it has been for several decades (about 74 million). As a result, children now make up a decreasing proportion—and older Americans a growing proportion—of the population. The changing American age structure is indicated by the nation's median age; it was 36.9 in 2010—up from 30.0 in 1980. Along with the impact of the baby boomers' aging and the declining proportion of children in the population,[1] longer life expectancy has contributed to the fact that, as a whole, our population is growing older.

Longer Life Expectancy

Americans are now living long enough that demographers divide the aging population into three categories: the "young-old" (age sixty-five through seventy-four), the "older-old" (age seventy-five through eighty-five), and the "old-old" (age eighty-five and over). Life expectancy at birth increased from 70.8 years in 1970 (67.1 years for men and 74.7 years for women) to 77.7 years in 2006 (75.1 years for men and 80.2 years for women). By the year 2020, life expectancy at birth is projected to

reach 79.5 (77.1 for men, and 81.9 for women) (U.S. Census Bureau 2010b, Table 102).[2]

Gender and Life Expectancy Women, on average, live about five years longer than men. Consequently, the makeup of the elderly population differs by gender. In 2010, there were 22.9 million women age sixty-five and older, compared to 17.3 million men. For Americans over age eighty-four, there are 3.9 million women and about 1.9 million men (U.S. Census Bureau 2010b, Table 8). This gendered difference in life expectancy means that, among other things, women are more likely than men to be widowed—and poor, for reasons explained later in this chapter—in old age (Federal Interagency Forum on Aging-Related Statistics 2008).

However, trends show that the life-expectancy gap between women and men is slowly narrowing (U.S. Census Bureau 2010b, Tables 102, 103). Should this trend

---

[1] The proportion of the U.S. population under age eighteen was about 36 percent in 1960, compared to about 24 percent in 2010 (U.S. Census Bureau 2003a, Table 11; 2010b, Table 8).

[2] Life expectancy differs by race. The remaining life expectancy at age forty for white males is about thirty-eight years, compared to about thirty-four years for black males. Among women, the figure is about forty-two for whites, compared to thirty-nine for blacks (U.S. Census Bureau 2010b, Table 104). Much of this difference is associated with whites having, on average, higher incomes and lower poverty rates than blacks. Higher incomes, along with higher education levels, are associated with longer life expectancy, largely because people in higher socioeconomic groups have access to better preventive health care and are less likely to work in hazardous environments: "Their educational advantage may also make them more avid consumers of the vast amounts of information available on improving health" (Conner 2000, p. 16). Some of the racial/ethnic difference in life expectancy may also be explained by genetics and by discrimination in health care.

continue, policy analysts point to at least two implications for women. For one thing, more elderly heterosexual women may have spouses or cohabiting partners to care for them should they need it. In addition, a shorter widowhood could mean that older women will be better off financially (Zernike 2006).

Family Consequences of Longer Life Expectancy
Demographers point to at least two family-related consequences of our living longer. First, because more generations are alive at once, we will increasingly have opportunities to maintain ties with grandparents, great-grandparents, and even great-great-grandparents (Bengston 2001). It is estimated that, by 2030, more than two-thirds of eight-year-olds will have a living great-grandparent (Rosenbloom 2006).

A second consequence of longer life expectancy is that, on average, more Americans spend time near the end of their lives with chronic health problems and/or physical disabilities. We can think not just in terms of overall life expectancy but also in terms of **active life expectancy**—the period of life free of disability. After this, a period of being at least partly disabled may follow (Crimmins et al. 2009). As Americans get older, more and more of us will be called upon to provide care for a parent or other relative who is disabled (National Alliance for Caregiving 2009). We will return to issues surrounding giving care to aging family members later in this chapter. We turn now to the racial/ethnic composition of the older American population.

Racial/Ethnic Composition of the Older American Population

"While the elderly are often subsumed under the same umbrella as the 'over 65-generation,' it is important . . . to note that the aging population in the United States comes from a wide range of backgrounds and cultures" (Trask et al. 2009, p. 301). As a category, non-Hispanic whites are older than people in other racial/ethnic categories. Although the national median age is 36.9 years, the median age for non-Hispanic whites is 41.3 (U.S. Census Bureau 2010b, Table 11). Of the total population, 12.9 percent are currently age sixty-five and older, while 16 percent of non-Hispanic whites are sixty-five and above. These figures compare to just 9 percent of Asian Americans, 8.5 percent of blacks, and 5.7 percent of Hispanics (U.S. Census Bureau 2010b, Table 11).

Figure 17.2 looks at the nation's age distribution by race/ethnicity in another way. As you can see from that figure, about 80 percent of the U.S. population over age sixty-four is non-Hispanic white. Another 9 percent is African American, with another 7 percent Hispanic. But the older population is becoming more ethnically and racially diverse as members of racial/ethnic minority groups grow older:

The older population among all racial and ethnic groups will grow; however, the older Hispanic population is projected to grow the fastest, from just over 2 million in 2005 to 15 million in 2050, and to be larger than the older black population by 2028. The older Asian population is

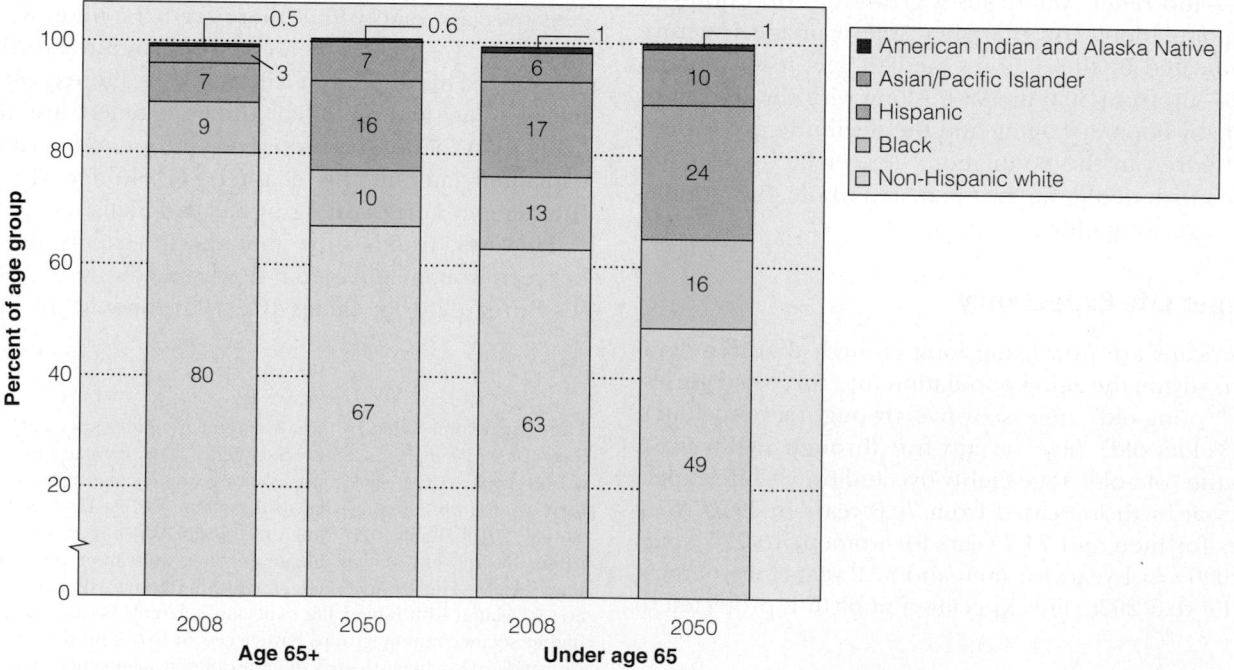

Figure 17.2 U.S. elderly and nonelderly population by race/ethnicity, 2008 and 2050

Note: Hispanic may be of any race.

Sources: del Pinal and Singer 1997; S. Lee 1998; and calculated from U.S. Census Bureau 2010b, Table 9.

also projected to experience a large increase. In 2006, just over 1 million older Asians lived in the United States; by 2050 this population is projected to be almost 7 million. (Federal Interagency Forum on Aging-Related Statistics 2008, p. 4)

Due to immigration and to the relatively high birth-rates of ethnic and racial minority groups, Hispanic, African American, and Asian populations are growing faster than non-Hispanic whites. By 2050, the non-His-panic white share of the population over age sixty-four is projected to fall to 67 percent. Although some senior centers "offer *tai chi* exercise classes or serve tamales for lunch, a reflection of greater ethnic diversity," scholars argue for much more research on the multicultural needs of aging Americans (Treas 1995, p. 8; see also Kershaw 2003; Trask et al. 2009).

Older Americans and the Diversity of Family Forms

We have seen throughout this text that today's families are diverse in form. As the postmodern family (see Chapter 1) grows older, we can expect late-life family forms to exhibit growing diversity. You can see in Figure 17.3 that, in 2008, 72 percent of men and 42 percent of women age sixty-five and older were married. Thirteen percent of women and 10 percent of men in that age category were divorced or separated, with another 4 percent never married (U.S. Administration on Aging 2010, Figure 2). Meanwhile, Figure 7.1, in Chapter 7, shows that the birthrate in the mid 1970s struck a record low while the divorce rate peaked. Americans who were in their twenties and thirties in the 1970s are now moving into later life. Increasingly, therefore, families will enter into older ages with fewer, if any, children and

with histories of cohabitation, separation, divorce, and repartnering. Additionally, now that same-sex families are more visible than in past decades,[3] we can expect researchers and policy makers to pay increased attention to aging same-sex families (Grant 2010). The implications for caregiving of this trend toward increased family diversity are addressed later in this chapter.

Living Arrangements of Older Americans

About 1.44 million or 3.7 percent of Americans age sixty-five and older live in nursing homes (U.S. Administration on Aging 2010, p. 7).[4] Among the vast majority of older adults who are not in institutional settings, some are cohabiting, whereas others live in cohousing arrangements, described in Chapter 8. Still others have established "living alone together" (LAT) relationships, also discussed in Chapter 8. Some older Americans have moved to retirement communities in the "Sun Belt"— Florida and the Southwest—as well as to communities in Mexico and other countries south of the U.S. border (Bjelde and Sanders 2009; Banks 2009). A few retirement communities have begun to emerge specifically for gays and lesbians ("Birds of a Feather" 2010; Grant 2010).

Then too, complementing the term, "boomerang kids," which refers to adult children who move back into their parents' homes, one journalist now writes of "boomerang seniors" (Kluger 2010). Since the onset of the recent economic downturn, a significant and growing number of older Americans have moved into their grown children's or grandchildren's homes (Alvarez 2009). "It's a return to much closer intergenerational ties than we saw through much of the 20th century" (Stephanie Coontz, quoted in Brandon 2008).[5] Some healthy and active older parents move in

[3] The Gay and Lesbian Association of Retired Persons (GLARP) formed in 1999 (www.gaylesbianretiring.org).

[4] The likelihood of living in an institutional setting, such as a nursing home, increases with age: 1.3 percent of Americans between ages sixty-five and seventy-four reside in institutional settings. Among those between ages seventy-five and eighty-four, the percentage is 3.8. Of those age eighty-five and older, 15.4 live in institutional settings (U.S. Administration on Aging 2010, p. 7).

[5] A tip for boomerang seniors and their families: Experts advise that "communication is the key to peaceful multigenerational living. Have regular family conferences to discuss issues before they become problems. Before moving in together, ask family members of all ages to talk about how they expect life to change, including what they want, what they are excited about, and what they're nervous about. Be specific. If grandparents are helping with child care, how much time will they spend babysitting? How do family members want to handle cooking and mealtimes? It's a great way to see where friction may occur and to head it off at the pass" (Alvarez 2s009).

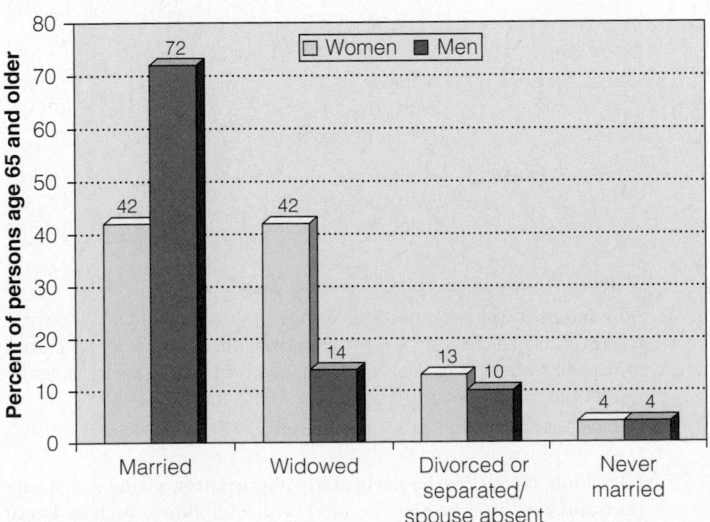

Figure 17.3 Marital Status of Persons Age 65 and Older, 2008
Source: U.S. Administration on Aging 2010, Figure 2, p. 6.

with their adult children to lower expenses ("You Will Be a Parent" 2009). More frail elders reside in their children's homes when the family cannot afford assisted living or other caregiving facilities (Armour 2009).

Nevertheless, historical trends in family living arrangements show a long-term preference for separate households in American society (Bures 2009). Currently, about 27 percent of U.S. households are made up of people living alone. Many of them are older people. This situation represents a growing trend since about 1940. Among Americans age sixty-five and older, approximately 30 percent live alone. Because of the increased likelihood of being widowed as people age, nearly 39 percent of people over age seventy-five live by themselves (U.S. Census Bureau 2010b, Table 58). Despite the recession, demographers project that—due to the increasing number of never-marrieds and to the significant proportion of middle-aged individuals who are divorced—an increasing percentage of older Americans will live alone in the future. Among elderly parents, both they and their adult children generally prefer to live near one another, although not in the same residence (Moody 2006). The elderly who live alone "are usually within a close distance of relatives or only a phone call or email away. Fewer than one out of twenty are socially isolated, and usually are so because they have lived that way most of their lives" (Moody 2006, p. 331).

Gender Differences in Older Americans' Living Arrangements

Due mainly to differences in life expectancy, older heterosexual men are much more likely to be living with their spouse than are older heterosexual women (72 percent of men age sixty-five and older, compared to 42 percent of women). Older women (42 percent) are far more likely than men (14 percent) to be widowed (U.S. Census Bureau 2010b, Table 34). As a result, 50 percent of women age seventy-five and older live alone (U.S. Administration on Aging 2010, p. 7). Of Americans between ages sixty-five and seventy-four who are living alone, 68 percent are women. Of those seventy-five and over who are living alone, 78 percent are women (U.S. Census Bureau 2010b, Table 72).

Moreover, older women are significantly more likely than older men to live with people other than their spouse—a pattern that persists into old-old age (U.S. Administration on Aging 2010, Figure 3). We can conclude that:

> [m]en generally receive companionship and care from their wives in the latter stages of life, while women are more likely to live alone, perhaps with assistance from grown children, to live with other family members, or to enter a nursing home. (Bianchi and Casper 2000, p. 10)

Besides gender, race and ethnicity also affect the living arrangements of older Americans.

Racial/Ethnic Differences in Older Americans' Living Arrangements

Due to economic and cultural differences, the living arrangements of older Americans vary according to race and ethnicity. For instance, Asian Americans and Hispanics are more likely than non-Hispanic whites or blacks to reside in the homes of their adult children (Glick and Hook 2002). Table 17.1 compares the living arrangements of non-Hispanic white, black, Asian, and Hispanic adults age sixty-five and older.

One generalization that we can make from the statistics in Table 17.1 is that African Americans, Asian Americans, and Hispanics are far more likely than non-Hispanic whites to live with people other than their spouse—grown children, siblings, or other relatives (Connidis 2007; Karasik and Hamon 2007; Voorpostel and Blieszner 2008).[6] Partly as a result of economic necessity, coupled with social norms involving family members' obligations to one another, older Asian Americans and Hispanics are less likely than non-Hispanic whites to live alone. This is true for Hispanics even though they are also less likely

Table 17.1 Living Arrangements of People Sixty-Five Years Old and Over, by Race/Ethnicity, 2008

	Living Arrangement	65–74 Years Old %	75 and Older %
Total Population	Alone	23	38
	With spouse	64	44
	With other people[b]	12	19
Non-Hispanic White	Alone	23	40
	With spouse	68	45
	With other people[b]	10	15
Black	Alone	34	41
	With spouse	41	29
	With other people[b]	25	29
Asian American	Alone	14	21
	With spouse	66	47
	With other people[b]	19	31
Hispanic Origin[a]	Alone	19	24
	With spouse	59	42
	With other people[b]	23	34

[a] People of Hispanic origin may be of any race.

[b] The category, with other people, includes relatives other than a spouse, as well as institutional settings, although the proportion of older Americans in institutional settings is relatively small. See Footnote 4 in this chapter for further information on aging individuals in institutional settings.

Source: Calculated from U.S. Census Bureau 2010b, Table 58.

[6] An adult sibling can be particularly important in giving social support and care to a brother or sister with disabilities, such as Down syndrome or mental illness. This situation is particularly true after an aging parent is no longer able to care for a disabled offspring (Hodapp and Urbano 2007; Lohrer, Lukens, and Thorning 2007; Orsmond and Seltzer 2007).

than non-Hispanic whites to live with a spouse (U.S. Census Bureau 2010b, Table 58).[7]

Generally, among older Americans without partners, living arrangements depend on a variety of factors, including the status of one's health, the availability of others with whom to reside, social norms regarding obligations of other family members toward their elderly, personal preferences for privacy and independence, and economics (Bianchi and Casper 2000). Older Americans with better health and higher incomes are more likely to live independently, a situation that suggests strong personal preferences for privacy and independence. However, those in financial need are more likely to live with relatives.

Aging in Today's Economy

Today's older Americans live on a combination of employment income and/or investment income, Social Security benefits, private pensions from employers, personal savings, and social welfare programs designed to meet the needs of the poor and disabled. About 40 percent of the income of Americans age sixty-five and older is from Social Security benefits and related federal programs, such as Medicare, Medicaid, and Supplemental Security Income (SSI).[8] Social Security benefits are the only source of income for one-fifth of Americans over age sixty-four (Federal Interagency Forum on Aging-Related Statistics 2008).

[7] Older African Americans and Hispanics are less likely than non-Hispanic whites to live with a spouse for two reasons. First, due to differences in patterns of marriage and divorce, African Americans are more likely than whites to enter older ages without a spouse. Second, gender differences in life expectancy (with women living longer than men) are slightly higher among blacks and Hispanics (about seven years) than among non-Hispanic whites (about five years) (U.S. Census Bureau 2010b, Table 102).

[8] During the Great Depression of the 1930s, millions of Americans lost their jobs and savings. In response, the federal government passed the 1935 Social Security Act, a dramatically new program designed to assist the elderly. The Social Security Act established the collection of taxes on income from one generation of workers to pay monthly pensions to an older generation of nonworkers. Initially, only those who contributed to Social Security were eligible to receive benefits, but over the years, the U.S. Congress has extended coverage to spouses and to the widowed, as well as to the blind and permanently disabled.

Medicare, begun in 1965, is a compulsory federal program that does not provide money but does offer health care insurance and benefits to the aged, blind, and permanently disabled. Before the program began, just 56 percent of the aged had hospital insurance. In 1992, at least 97 percent of all older people in the United States had coverage because all who qualify for Social Security are eligible for Medicare.

In 1965, intending to provide health care to poor Americans of all ages, Congress created the Medicaid program in conjunction with Medicare. Eligibility for Medicaid is based on having virtually no family assets (saving and checking accounts, stocks, bonds, mutual funds, and any form of property that can be converted to cash) and very little income. In 1972, Congress created the federal Supplemental Security Income program (SSI), a "welfare" program that provides monthly income checks to poverty-level older Americans and the disabled (Meyer and Bellas 2001).

The expansion of Social Security benefits to older Americans resulted in dramatic changes in U.S. poverty rates over the last several decades. Before Social Security was initiated, the elderly were disproportionately poor (Federal Interagency Forum on Aging-Related Statistics 2008). But poverty has declined sharply for those age sixty-five and over—from 36 percent in 1959 to about 10 percent today. Due partly to older Americans' lobbying to protect Social Security benefits, the poverty rate for those over age sixty-four is now about one-half that of children (U.S. Census Bureau 2010b, Table 697). Then too, many of today's older Americans benefit from what was a generally stable or rising economy during most of their working years.

However, having noted that today's older Americans are generally better off than generations preceding them, we need to acknowledge that on average their income declines by up to one-half upon retirement (U.S. Census Bureau 2009b, Table S1903)—a situation that can lead to stress and relationship conflict (Dew and Yorgason 2010). Furthermore, the recession that began in 2008 "marked unprecedented changes in the lives of older Americans"; among those desiring to work, the number of unemployed increased dramatically during 2008 (Mossaad 2010).

Moreover, one-tenth of older adults are living in poverty, and between 6 and 10 percent of the homeless are age sixty-five and older (National Coalition for the Homeless 2009b). Along with the "near poor" (those with incomes at or below 125 percent of the poverty level, who make up another 9 percent of the elderly), these poverty-level older Americans are hardly enjoying the comfortable and leisurely lifestyle that we may imagine when we think of retirement (Employee Benefit Research Institute 2009).[9]

Retirement?

From an historical standpoint, widespread retirement only became possible in the twentieth century, when the industrial economy was productive enough to support sizable numbers of nonworking adults. At the same time, the economy no longer needed so many workers in the labor force, and companies believed that older workers were not as quick or productive as the young. Governments, corporations, labor unions, and older workers themselves saw retirement as a desirable policy, and it soon became the normal practice (Moody 2006). Today, however, an unpredictable economy, together with an aging population, may render the policy of retiring at about age sixty-five outdated (Maestas and Zissimopoulos 2009).

[9] For instance, Medicaid recipients (see footnote 8) living in nursing homes are wards of the state. Hence, their entire monthly income, except for a small personal-needs allowance, goes toward nursing home costs (Meyer and Bellas 2001).

"If we take a late retirement and an early death, we'll just squeak by."

Barbara Smaller

Although most older people eventually retire, some do not—and many of those who don't are employed beyond age seventy (Purcell 2009, Figure 17). Then too, we tend to think of retiring as an abrupt event. However, many people retire gradually by steadily reducing their work hours or intermittently leaving, then returning to the labor force before retiring completely (Kim and DeVaney 2005).

Not wanting to give up the psychological benefits associated with working—that is, feeling that one's life is meaningful and experiencing personal growth—is a reason that people give for not retiring. A less satisfying reason, increasingly relevant in today's economy and particularly applicable to divorced older women, is not being financially able to retire (Taylor 2009). Even before the onset of the recession that began in 2008, there was evidence that the majority of aging baby boomers expected to work at least part-time after retirement age (Brougham and Walsh 2009). By 2009, 38 percent of those over age sixty-two had delayed retirement due to the recession, according to the Pew Research Center. Among those between ages fifty and sixty-one, 63 percent (54 percent of men and 72 percent of women) said that they might need to delay retirement because of the recession (Taylor 2009).

Gender Issues and Older Women's Finances

Older men are considerably better off financially than are older women. In 2008, the median income of American individuals age sixty-five and older was $25,503 for males and $14,559 for females (U.S. Administration on Aging 2010). This dramatically unequal situation is partly due to the fact that, throughout their employment years, men averaged higher earnings than did women (U.S. Census Bureau 2009b, Table S2002; and see Chapter 11). Consequently, older women today have smaller, if any, pensions from employers. Furthermore, older women on average did not begin to save for retirement as early as did men (Even and Macpherson 2004; Herd 2009).

Moreover, although women are much more likely than men to rely on Social Security for at least 90 percent of their income, women's Social Security benefits average only about 76 percent of men's. Maximum Social Security benefits ($2,346 monthly in 2010) are available only to workers with lengthy and continuous labor force participation in higher-paying jobs ("Women and Social Security" 2007; U.S. Social Security Administration 2009). This situation works against older women today, who either did not participate in the labor force at all or are likely to have dropped in and out of the labor force while taking lower-paying jobs (Herd 2009). "Thus, women are penalized for conforming to a role that they are strongly encouraged to assume—unpaid household worker—and their disadvantaged economic position is carried into old age" (Meyer and Bellas 2001, p. 193).

At the time of this writing, the federal Defense of Marriage Act (DOMA) precludes even legally married same-sex couples from receiving federal married-couple benefits, such as Social Security payments to a widowed or disabled spouse (Grant 2010; and see Chapter 8 in this text). However, an older wife married heterosexually for at least ten years to a now-retired worker can receive a spousal "allowance," equal to one-half of her husband's benefits.[10] Ex-spouses also qualify for one-half the amount of their ex's Social Security benefits, provided the marriage lasted at least ten years. Compared to widows, divorced and separated women are worse off financially and many need to work for several years after traditional retirement age (Herd 2009; Ulker 2009). We turn to an examination of couple relationships in later life.

---

[10] Employed women may qualify on the basis of either their own or—if heterosexually married—their husband's work records, although they cannot receive benefits under both categories.

Social Security and Medicare have raised the incomes of older Americans, beginning in 1940, so that the proportion of elderly in the United States living in poverty today has declined and is less than that of children. Nevertheless, about 3.7 million, or 10 percent of older Americans—disproportionately the unmarried and women—are living in poverty. Another 2.4 million of the elderly are classified as "near poor," with income up to 125 percent of the poverty level (U.S. Administration on Aging 2010).

Relationship Satisfaction in Later Life

As shown in Figure 17.3, only about 4 percent of men or women sixty-five and older today have never married (U.S. Administration on Aging 2010, Figure 2). Some later-life marriages are remarriages, and the proportion of repartnered older Americans will increase as those who are now middle-aged grow older. Today, however, the majority of older, heterosexually married couples have been wed for quite some time—either in first or second marriages.

Retirement represents an important and, usually temporarily, stressful change for couples (Dew and Yorgason 2010). For both partners, role flexibility is important to successful adjustment. Also, health is an important factor in morale in later life, and it has a substantial impact on marital quality as well (Connidis 2010).

The majority of older married couples place companionship and intimacy as central to their lives and describe their unions as happy (Szinovacz and Schaffer 2000; Walker et al. 2001). On average, older couples report having fewer disagreements, and marital happiness often increases in later life when couples have the time, energy, and financial resources to invest in their relationship (Hatch and Bulcroft 2004; see also Story et al. 2007). Mothers' marital satisfaction has been shown to increase with age, a finding that is especially true after their grown children leave home (Association for Psychological Science 2008). As a category, aging marrieds are happier and more satisfied with their lives than are their nonmarried counterparts (LaPierre 2009).

This is not to say that all older couples are happy together. One study of national data found that older spouses who felt unfairly treated by their mates were more distressed than were singles of the same age (Hagedoorn et al. 2006). Other research has found that later-life couples who hold more egalitarian attitudes toward gender roles and who experience high levels of warm mutual interaction report significantly greater marital happiness (Kaufman and Taniguchi 2006; Schmitt, Kliegel, and Shapiro 2007).

Sexuality in Later Life

As discussed in Chapter 5, the frequency of sexual intercourse tends to decline with age (Marshall 2009). Nonetheless, many older Americans continue to be sexually active—even into old-old age and even in nursing homes (Lindau and Gavrilova 2010; Purdy 1995; Waite et al. 2009). A 2009 national survey of Americans age forty-five and older found that three-quarters of men and women with partners had had sexual intercourse within the prior six months. Among respondents age seventy and older, 11 percent of women and 22 percent of men reported having sexual intercourse at least once or twice a month (Fisher 2010, Tables 19 and 20). Although researchers found that sexual activity among older adults had declined since 2004—a phenomenon attributed to recession-related worry and stress—80 percent of men and 39 percent of women age seventy and older said that a sexual relationship was important to their quality of life (Fisher 2010, p. 9).

All of this is not to imply that older adults have no sexual problems. Although aging single women may be interested in sex, lack of a partner can be a problem. We have seen that, as they age, women are far more likely than men to be widowed. Moreover, as they grow older, women are adversely affected by the *double standard of aging*—that is, men aren't considered old or sexually ineligible as soon as women are (England and McClintock 2009). Beauty, "identified, as it is for women, with youthfulness, does not stand up well to age" (Sontag 1976, p. 352). For older single women, this situation can exacerbate more general feelings of loneliness (Narayan 2008).

Moreover, stress, dissatisfaction with one's partner, and health-related issues can inhibit sexual desire and activity for both sexes (Laumann, Das, and Waite 2008; Lee 2009). According to psychiatrist Stephen Levine, "Over age 50, the quality of sex depends much more on the overall quality of a relationship than it does for young couples" (quoted in Jacoby 1999, p. 42; see also Elliott and Umberson 2008).

Later-Life Divorce, Widowhood, and Remarriage

Although the majority of couples who divorce do so before their retirement years, some couples do divorce in later life. Little research has been done on the topic, but we might hypothesize that later-life divorces are not necessarily easy on the couple's adult children. Family celebrations and holidays are disrupted. An adult child's graduation or wedding can be difficult when forced to accommodate recently divorced parents.

Furthermore, adult children of divorcing parents may worry about having to become full-time caregivers to an aging parent in the absence of the parent's spouse. "Years after parents split, their children may wind up helping to sustain two households instead of one, and those households can be across town or across the country" (Span 2009a). Nevertheless, although some later-life marriages end in divorce, the vast majority do so with the death of a spouse.

Widowhood and Widowerhood

Adjustment to widowhood or widowerhood is an important common family transition in later life. A spouse's death brings the conjugal unit to an end—often a profoundly painful event. "The stress and emotional trauma of losing a spouse as a confidant might be greater now than in the past, as the average duration of marriage becomes longer with increasing life expectancy" (Liu 2009, p. 1,170).

Because women's life expectancy is longer, and also because older men remarry more often than women do, widowhood is significantly more common than widowerhood. More than half (52 percent) of women between ages seventy-five and eighty-four are widowed, compared with just 17 percent of men. Among those age eighty-five and over, just over three-quarters of women are widowed, compared with about one-third of men (Federal Interagency Forum on Aging-Related Statistics 2008, p. 5).

Typically, widowhood and widowerhood begin with **bereavement**, a period of mourning, followed by gradual adjustment to the new, unmarried status and to the loss. Bereavement manifests itself in physical, emotional, and intellectual symptoms. Recently, widowed

people perceive their health as declining and report feeling depressed. The bereaved experience various emotions—anger, guilt, sadness, anxiety, and preoccupation with thoughts of the dead spouse—but these feelings tend to diminish over time (Connidis 2010; Jin and Chrisatakis 2009). Social support, adult children's help with housework and related tasks, and activities with friends, children, and siblings help (Cornwell and Waite 2009; Ha 2008; Population Reference Bureau 2009).

There is evidence that being single in old age is more physically and emotionally detrimental for men than for women. Wives, as opposed to husbands, tend to be central "in the household production of health" (Jin and Chrisatakis 2009, p. 605). Put another way, females are more likely than males to concern themselves with the health of all household members, so in a wife's absence, a man's health is more likely to decline than vice versa. Then too, as sources of support, women more often have friends in addition to family members. Men are more often dependent solely on family. For some of the widowed, remarriage promises resumed intimacy and companionship.

Age and the Odds of Remarriage

Annually about one-half million Americans over age sixty-five remarry (Belkin 2010). Remarrying elders, particularly those who are widowed, are likely to choose homogamous partners. Sometimes the new partner is someone who reminds them of their first spouse or is someone whom they've known for years.

However, middle-aged and older people may face considerable opposition to remarriage—from restrictive pension and Social Security regulations (Ebeling 2007)[11] and from their adult children. Although some grown children may be supportive of a parent's remarriage, others may find it inappropriate (Sherman and Boss 2007). Adult children may worry about the biological parent's potentially diminished interest in them or about their inheritance (Connidis 2010; Pasley 1998b).[12] Chapter 16 further explores adult children's attitudes about a parent's remarriage.

Especially for women, age reduces the likelihood of remarrying. Although the pattern of remarriage since 1960 has been similar for all age categories,

---

[11] In general, a widow/widower cannot receive Social Security survivors' benefits if he or she remarries (heterosexually) before age sixty. Remarrying after age sixty (or fifty if disabled) does not negatively affect receipt of survivor benefits (Social Security Online n.d.).

[12] Even when a will leaves everything to one's children, a second spouse may have the legal right to claim a share of the estate. Presuming that the remarrying couple desires it, an option may be signing a prenuptial agreement according to which the spouse-to-be relinquishes any claim to the estate (Ebeling 2007; see also Barnes 2009; Palmer 2007).

the remarriage rate for *younger* women is consistently higher than for older women (Moorman, Booth, and Fingerman 2006). An uneven sex ratio (see Chapter 8) decreases the odds that older heterosexual women will repartner: Among Americans age fifty-five and older, there are approximately eighty-two men for every 100 women (U.S. Census Bureau 2010b, Table 7). The *double standard of aging* also works against women in the remarriage market (England and McClintock 2009). Then too, women may be less interested than men in late-life remarriage (Levaro 2009).

Meanwhile, research shows that the myth that widowers are quick to replace a deceased wife is just that—an exaggerated stereotype. "Men with high levels of social support from friends are no more likely than women to report interest in repartnering" (Carr 2004, p. 1,065). During later life, morale and well-being frequently derive from relations with siblings, as well as from friends, neighbors, and other social contacts (Eriksen and Gerstel 2002; Voorpostel and Blieszner 2008). Particularly for females, relationships with adult children and grandchildren continue to be important.

Older Parents, Adult Children, and Grandchildren

More often than spousal relationships, those between parents and their biological children last a lifetime (Kaufman and Uhlenberg 1998). In this section, we examine relations between older parents and their adult children and grandchildren.

Older Parents and Adult Children

Adults' relationships with their parents range from *tight-knit*, to *sociable*, to *obligatory*, to *intimate but distant*, or *detached*. Sociologists Merril Silverstein and Vern Bengston (2001) developed six indicators of relationship solidarity, or connection: geographic proximity, contact between members in a relationship, emotional closeness, similarity of opinions, providing care, and receiving care. Based on survey evidence and using these six indicators, Silverstein and Bengston then developed a typology of these five kinds of parent–adult child relations.

Married older couples may be in first unions or in remarriages. Most older couples describe their marriages as happy. A retired husband may choose to spend more time doing homemaking tasks and give increased attention to being a companionate spouse. Role sharing, feeling that work is fairly shared, and having supportive communication predict good adjustment for retiring couples.

Parent–adult child relations vary depending on how family members combine—or, in the case of the detached relationship style, do not combine—the six indicators. For instance, in tight-knit relations, the parent and the adult child live near each other (geographic proximity), feel emotionally close, share similar opinions, and help each other (give and receive assistance). Sociable relations involve all these characteristics except that the parent and adult child do not exchange assistance. Table 17.2 defines all five relationship types.

Research shows that there is no one typical model for parent–adult child relationships (Arnett 2004; Silverstein and Bengston 2001). Furthermore, parent–adult child relations might change over time, moving from one relationship type to another depending on the parent's and the adult child's respective ages, the parent's changed marital status, and the presence or absence of grandchildren, among other factors. For instance, to be nearer to their aging parents, adult children sometimes return to the area in which they grew up, or retired grandparents may decide to relocate to be near their grandchildren (Lee 2007). Both of these situations could move an intimate-but-distant relationship to a tight-knit one. Then, too, a parent–adult child relationship might change depending only on emotional factors, such as when an adult child chooses to forgive an aging parent for some past transgression, or vice versa.

Table 17.2 Types of Intergenerational Relations

Class	Definition
Tight-knit	Adult children are engaged with their parents based on geographic proximity, frequency of contact, emotional closeness, similarity of opinions, and providing and receiving assistance.
Sociable	Adult children are engaged with their parents based on geographic proximity, frequency of contact, emotional closeness, and similarity of opinions but not based on providing or receiving assistance.
Obligatory	Adult children are engaged with their parents based on geographic proximity and frequency of contact but not based on emotional closeness and similarity of opinions. Adult children are likely to provide or receive assistance or both.
Intimate but distant	Adult children are engaged with their parents based on emotional closeness and similarity of opinions but not based on geographic proximity, frequency of contact, providing assistance, and receiving assistance.
Detached	Adult children are not engaged with their parents based on any of these six indicators of solidarity.

Source: Adapted from Silverstein and Bengston 2001, p. 55.

Using national survey data from a sample of 971 adult children who had at least one surviving noncoresident parent, Silverstein and Bengston (2001) made the following findings (among others):

- The majority of relations were neither tight-knit nor detached, but "variegated"—one of the three relationship styles in between (see Table 17.2). Variegated relations characterized 62 percent of adult children's interaction with their mothers and 53 percent with their fathers.

- Tight-knit relations are more likely to occur among lower socioeconomic groups and racial/ethnic minorities.

- Non-Hispanic whites were more likely than African Americans to have detached relationships with their parents and more likely than blacks or Hispanics to have obligatory relationships with their mothers.

- The most common relationship between a mother and her adult child was tight-knit. The next most common was sociable, followed by intimate but distant, obligatory, and, finally, detached.

- The most common relationship between a father and his adult child was detached, followed by sociable, tight-knit, obligatory, and intimate but distant. Almost four times as many adult children reported being detached from their fathers as from their mothers.

- Daughters were more likely than sons to have tight-knit relations with their mothers.

- Sons were more likely than daughters to have obligatory relations with their mothers.

- Adult children were more likely to have obligatory or detached relations with divorced or separated mothers than with married mothers.

- Adult children were more likely to have detached relations with divorced or separated fathers than with consistently married fathers.

From these findings, we can conclude that daughters are more likely than sons to have close relationships with their parents, especially with their mothers. Even for mothers, a parent's divorce or separation often weakens the bond with adult children (Hans, Ganong, and Coleman 2009). However, having detached relations with one's adult children after divorce is nearly five times greater for fathers than for mothers. Partly, at least, this is true because a divorced father is less likely than either a consistently married father or a divorced mother to live with his biological children and more likely to remarry (Pezzin, Pollak, and Schone 2008; Silverstein and Bengston 2001).

In some families, the reality of past abuse, a conflict-filled divorce, or simply fundamental differences in values or lifestyles makes it seem unlikely that parents and

Financially independent adults' relationships with a parent can be of several types: tight-knit, sociable, obligatory, intimate but distant, or detached. Then too, today's parents can find themselves in the "senior sandwich generation" — paying for a child's college tuition, worrying about the financial burden of elder care for aging parents, while trying to save for their own retirement.

children will spend time together (Arnett 2004). Money matters can also cause tension. This is especially true in stepfamilies where "[a]dult children can feel resentful when they see a stepparent spending what they consider as their rightful inheritance" (Sherman 2006, p. F8). Overall, however, the majority of adult children's relationships with parents, although not necessarily tight-knit, continue to be meaningful.

Grandparenthood

Medical technology and grandparenthood meet as "expectant grandmothers" view ultrasound images of their developing grandchild's fetus (Harpel and Hertzog 2010). Partly due to longer life expectancy, which creates more opportunity for the role, grandparenting—and great-grandparenting—became increasingly important to families throughout the twentieth and into the twenty-first centuries (Bengston 2001; Rosenbloom 2006). Many older Americans are raising grandchildren, as explored in Chapter 10. The discussion in

this section focuses on grandparents who are not primarily responsible for raising their grandchildren.

Young-old grandparents, sometimes called "grand boomers" (Lee 2007), are often employed and often partnered, while old-old grandparents may have physical disabilities. Hence, younger grandparents' experiences with their grandchildren are typically quite different from those of older grandparents (Silverstein and Marenco 2001).

The twentieth century saw increased emphasis on affection and companionship with grandparents. Many grandparents find the role deeply meaningful: Grandchildren give personal pleasure and a sense of immortality. Some elderly grandfathers see the role as an opportunity to be involved with babies and very young children, an activity that may have been discouraged when their own children were young (Cunningham-Burley 2001). Overall, the grandparent role is mediated by the parent. Not getting along with the parent dampens the grandparent's contact and, hence, the relationship with her or his grandchildren (Monserud 2008; Mueller and Elder 2003).

Grandparents often provide practical help (King et al. 2003). They may serve as valuable "family watchdogs," ready to provide assistance when needed (Troll 1985). For instance, the Interactive Autism Network (2010) conducted an online survey of individuals who had a grandchild with autism. This survey was hardly representative of all grandparents in this category, because to be aware of the survey one would have had to be interested enough in the topic to visit the website. However, it is interesting to note that nearly one-third of the grandparent respondents reported being the first in the family to notice anything out of the ordinary regarding their grandchild's development. Some grandparents (14 percent) had moved closer to their grandchild's family to help out. Many grandparents assisted with treatment-related costs, some dipping into retirement savings (Hamilton 2010; Interactive Autism Network 2010).

In low-income and ethnic-minority families, parents and children readily rely on grandparents for child care and other help. Even among middle- and upper-middle-class families, it is not unusual for grandparents to help with child care—or contribute to the cost of a grandchild's schooling, wedding, or first house. A (pre-recession) *New York Times* article featured several upper-middle-class grandparents who commuted by plane weekly to help with child care (Lee 2007). A close relationship with a grandparent can help to facilitate a grandchild's adjustment after parental divorce (Henderson et al. 2009). If an adult child of divorced parents becomes divorced, assistance may be more readily available from a grandparent (Vandell et al. 2003).

Grandparenting Styles Grandparents tend to adopt a grandparenting style similar to one that they themselves experienced as a grandchild (Mueller and

Elder 2003). Of course, grandparenting styles are also shaped by the grandparent's age, health, employment status, and personality (Davey et al. 2009). Among those who are not playing parentlike roles to their grandchildren, three general styles of grandparenting have been identified: *remote, companionate,* and *involved* (Cherlin and Furstenberg 1986). Those who have remote relationships with their grandchildren often, although not always, do so because they live far away. Companionate grandparents do things with their grandchildren but exercise little authority and allow the parent to control access to the youth. Companionate grandparents are often involved in work, leisure, or social activities of their own. Other grandparents are more involved, probably living with or near their grandchildren and frequently initiating interaction with their grandchild.

A grandparent may have different relationship styles with different grandchildren. Although some of them prefer to interact with their teenage grandchildren, grandparents generally are most actively involved with preadolescents, particularly preschoolers (Davey et al. 2009). Preschoolers are more available and respond most enthusiastically to a grandparent's attention. However, after a typically uninterested adolescence, adult grandchildren often renew relationships with grandparents (Cherlin and Furstenberg 1986; see also Mansson, Myers, and Turner 2010).

Race/Ethnicity and Grandparenting Although there is relatively little research on the subject, we do know some ways that race/ethnicity affects grandparenting (Karasik and Hamon 2007). One study found that 87 percent of black grandparents felt free to correct a grandchild's behavior, compared to just 43 percent of white grandparents. As one black grandmother said of her fourteen-year-old grandson, "He can get around his mother, but he can't get around me so well" (quoted in Cherlin and Furstenberg 1986, p. 128).

As another example, Native American elders may serve as *cultural conservator grandparents*—actively seeking contact and temporary coresidence with their grandchildren "for the expressed purpose of exposing them to the American Indian way of life" (Weibel-Orlando 2001, p. 143, quoted in Karasik and Hamon 2007, p. 145). Maintaining an ethnic-minority culture into future generations may be of particular concern for ethnic-minority grandparents.

A study of 112 Asian Indian grandchildren in the United States found that the vast majority maintained regular contact with their grandparents in India. Forty percent had weekly telephone conversations, and another one-third called one or more times a month. Seven percent had weekly email contact. Forty-three percent of the grandchildren visited their grandparent(s) in India every two years. "Regardless

of geographical distance," the grandchildren in this study "depicted their relationships with their grandparent as quite positive and close" (Saxena and Sanders 2009, p. 329).

Divorce, Remarriage, and Grandparenting How does a grandchild's divorce affect the grandparent relationship? Evidence suggests that the news can hit hard, and a grandparent may fret over whether to intervene on behalf of the grandchildren. As might be expected, effects of the divorce are different for the **custodial grandparent** (parent of the custodial parent) than for the **noncustodial grandparent** (parent of the noncustodial parent), with noncustodial grandparents significantly less likely to see their grandchildren as often as they had before the divorce (Connidis 2010). With the current trends in child custody, the most common situation is for maternal grandparent relationships to be maintained or enhanced while paternal ones diminish (Mills, Wakeman, and Fea 2001).

Because of pressure from noncustodial grandparents, states have passed laws giving grandparents the right to seek legalized visitation rights, but courts are reluctant to do so when parents object. "The focus on grandparent visitation rights largely has centered on this legal question: Should the government intrude upon the fundamental rights of parents to allow grandparents to visit their grandchildren?" (Henderson 2005a, p. 640). Grandparents who go to court to seek visitation rights are successful between 30 and 40 percent of the time. When courts do recognize visitation rights for grandparents in spite of parental objections, the reason usually involves the best interests of the child (Henderson 2005a, 2005b).

Remarriages and re-divorces create step-grandparents and ex-step-grandparents. Little information exists on step-grandparents, but available data suggest that they tend to distinguish their "real" grandchildren from those of remarriages. Asked about his step-grandparents, one young man told this story:

> They were pretty good, but again, there was that line. And we knew... . . . I mean I remember a couple of Christmases ago . . . [my grandmother made] a family quilt and everyone was on it except me and my brother. . . . As we got older the fact that they weren't our grandparents became more predominant. (in Kemp 2007, p. 875)

Younger step-grandchildren and those who live with the grandparent's adult child are more likely to develop ties with the step-grandparent (Coleman, Ganong, and Cable 1997; Pacey 2005). "As We Make Choices: Tips for Step-Grandparents" suggests ways to foster positive relationships with step-grandchildren. We turn now to an examination of caregiving to aging family members.

JGI/Blend Images/Corbis

Grandparenting styles differ—shaped by the grandparent and grandchild's ages and personalities, as well as by the grandparent's health and employment status. Ties with a grandchild can be remote, companionate, or involved, and a grandparent may have different ties with different grandchildren. Maintaining one's culture into future generations may be of particular concern for an ethnic-minority, *cultural conservator* grandparent.

Aging Families and Caregiving

When we associate caregiving with the elderly, we may tend to think only in terms of older generations as care *recipients*. However, about one-fifth of Americans age seventy-five and older are engaged in some form of care *giving*, whether child care or caring for other elders (Shapiro 2006b). As well as assisting their adult children and grandchildren financially and otherwise, older Americans give much to their communities. Many are volunteers in their churches, hospitals, schools, and various other settings. In some communities, older people mentor troubled youth (Donahoe 2005).[13] In this section, however, we focus on **elder care**—that is, care provided to the elderly.

---

[13] As another example, you can go online for advice from the Elder Wisdom Circle, a group of "cyber-grandparents" who volunteer to offer guidance to younger people (elderwisdomcircle.org).

Elder care involves emotional support, a variety of services, and, sometimes, financial assistance. A growing number of tax-funded, charity, and for-profit services provide elder care. Nevertheless, the persisting social expectation in the United States is that family members will either care for elderly relatives personally or organize and supervise the care provided by others. About 43.5 million adults age eighteen and over (nearly 19 percent of all American adults) are engaged in **informal caregiving**—unpaid and personally provided care—to a family member or friend who is age fifty or older (National Alliance for Caregiving 2009). In addition, adults responsible for aging family members often pay for some of the care recipient's expenses, such as for groceries, drugs, medical co-payments, or transportation (Gross 2007).

Being concerned about an elderly family member might involve nothing more than making a daily phone call to make sure that he or she is okay or stopping by for a weekly visit. However, **gerontologists**—social scientists who study aging—specifically define **caregiving** as

Step-grandparent relationships are much more likely than biological grandparent ties to be characterized by ambivalence. At the same time, stepchildren can benefit from an older adult's genuine concern and support. The following advice to step-grandparents has been excerpted from the University of Florida Extension website.

"When stepfamilies are formed, many new relationships are created. You may become an instant grandparent with step-grandchildren. You may have both grandchildren and step-grandchildren in the same family. . . . You probably have many thoughts and feelings about this role. You may think:

- I'm not old enough or ready to be a grandparent.

- This interferes with dreams about the birth of my first grandchild.

- Will my step-grandchild like me? Will I like my step-grandchild?

- What expectations do my daughter or son and new son or daughter-in-law have?

- The relationship I have with my other grandchildren is great. I don't want it to change.

- Is it okay to feel differently toward my step-grandchildren than my real grandchildren?

- I feel like I'm expected to treat my step-grandchildren the same as my grandchildren, especially around gift-giving times.

- Will "our" family celebrations and traditions have to change? . . .

"To expect step-grandparents and step-grandchildren to instantly love each other is unrealistic. . . . [but here are some tips for creating supportive step-grandparent relationships]:

"Remember that relationships are built over time. Your relationship and role as a step-grandparent will take time to develop. Communicate and spend time together in order to get to know each other.

"Recognize the vital role of grandparents and step-grandparents in today's families. You can offer children in busy stepfamilies companionship, time, and a listening ear . . . children who are exposed to such contact are less fearful of old age and the elderly. They feel more connoted to their families.

"Create the grandparenting role that is comfortable to you and rewarding for your stepfamily. Step-grandparenting, like other stepfamily roles, is challenging and undefined. It is up to you to carve a role for yourself that fits your son or daughter's new family. Here are some things to consider . . .

[Meanwhile], talk with your step-grandchildren. You may find that all of you are wanting the same things, but have been afraid to communicate. . . . Share these gifts with your grandchildren and step-grandchildren:

- Spend time one-on-one with them.

- Teach them a game or skill.

- Joke and kid with them.

- Listen for their concerns, as well as their joys.

- Talk about family disagreements, but do not criticize the other adults. Use your listening skills.

- Offer companionship for activities they enjoy.

- Share your history and family traditions.

- Show them acceptance."

Critical Thinking

How do these tips correlate with suggestions for supportive family communication, discussed in Chapter 12? How might a step-grandparent benefit from choosing to follow some of these suggestions? How might a step-grandchild benefit?

Source: Excerpted from Ferrer-Chancy 2009.

"assistance provided to persons who cannot, for whatever reason, perform the basic activities or instrumental activities of daily living for themselves" (Uhlenberg 1996, p. 682). Caregiving may be short term (taking care of someone who has recently had joint-replacement surgery, for example) or long term.

The majority of the young-old need almost no help at all, but as older individuals age, they may increasingly require assistance with tasks such as paying bills and, later, bathing or eating (National Alliance for Caregiving 2009, Figure 5). Severely ill or disabled older people, as well as those with dementia, often need a great deal of continuing care. "Facts About Families: Community Resources for Elder Care" describes elder care options.

A number of elder care givers are relatively young—in their thirties, with some in their twenties, partly because children born to older parents begin elder care at younger ages (National Alliance for Caregiving 2009, Figure 3). Then too, grandchildren may be providing elder care (Fruhauf, Jarrott, and Allen 2006). As explored in "A Closer Look at Family Diversity: Young Caregivers," in Chapter 14, some family caregivers are children.

However, the mean age for caregivers is about fifty. Between about 15 percent and one-third of caregivers are age sixty-five or older and may themselves be in poor

More and more older Americans and their care providers are turning to professional elder care service providers for help. With increasing numbers of elderly, there is a growing number of community services and facilities for elder care. The table in this box defines several of these services and facilities. The following are three considerations regarding making decisions about these options:

1. *Don't wait.* Exploring options before they're needed helps all family members know what to expect and begin to prepare for the future (Greenwald 1999, p. 53).

2. *Seek expert advice.* For example, geriatric social workers can help to assess an elderly person's needs and develop action plans: "Such people may be especially helpful in those painful cases when children must take needed steps in spite of the objections of mentally declining parents" (Greenwald 1999, p. 53).

3. *Shop around.* Most providers of senior housing are businesses, not charities, and their products should be scrutinized for cost and quality. Families should visit as many facilities as they can on different days of the week and hours of the day. Ask for references (Shapiro 2001b, p. 60).

Critical Thinking

What kinds of community elder care services can you think of that are not included in this table? How might they be useful to caregivers and to the elderly as well?

The Options	What Is It?
Home care	Wide range of services, including shopping and transportation, health aides who give baths, nurses who provide medical care, and physical therapy brought to the home
Adult day care	A place to get meals and spend the day, usually run by not-for-profit agencies
Congregate housing	A private home within a residential compound, providing shared activities and services; sometimes considered one type of assisted living
Assisted living	Numerous kinds of housing with services for people who do not have severe medical problems but who need help with personal care such as bathing, dressing, grooming, or meals
Continuing-care retirement community (CCRC)	A complex of residences that includes independent living, assisted living and nursing home care, so seniors can stay in the same general location as their housing needs change over time, beginning when they are still healthy and active
Nursing home (skilled nursing facilities)	Residential facilities with twenty-four-hour medical care available for those who need continual attention

Sources: "Choosing Senior Housing" 2007; Greenwald 1999, pp. 54–55; Shapiro 2001a, 2001b; also see "Eldercare Locator" at www.eldercare.gov, and "Glossary of Senior Housing Terms" (2010) at senioroutlook.com/glossary.asp.

health or beginning to suffer from age-related disabilities (Johnson and Wiener 2006; National Alliance for Caregiving 2009, p. 15). About 16 percent of caregivers help out for fewer than six months, but more than one-third provide care for between one and four years, and up to 30 percent assist a disabled family member for five years or more. On average, long-term caregivers spend about twenty hours each week in this role, with another 11 percent spending more than forty hours per week. About 50 percent of caregivers are employed full-time; 11 percent, part-time (National Alliance for Caregiving 2009, Figures 4 and 8, and p. 14).

Which family members provide elder care and how much they provide depend on the family's understanding of who is primarily responsible for giving the care, as well as the care receiver's preference (Connidis and Kemp 2008). The first choice for a caregiver is an available spouse, followed by adult children, siblings, grandchildren, nieces and nephews, friends and neighbors,[14] and, finally, a formal service provider. This system of elderly care receivers' preference for caregivers is termed the **hierarchical compensatory model of caregiving** (Cantor 1979; Horowitz 1985).

[14] A small study of seventeen gay men and twenty lesbians over age sixty-five in Los Angeles concluded that, when relying on neighbors or acquaintances for social support, aging lesbians and gays differ in their approaches (Rosenfeld 1999). Those who formed a gay or lesbian (GL) identity before the 1970s, when homosexuality was highly stigmatized, tended to hide their sexual identity: "Church has become a family. If I need help, they'll help. If they need help, I'll help. Except if they knew what I was, I don't know how they would feel towards me" (p.129). Others—usually baby boomers who formed a GL identity during the 1970s in a "gay liberation atmosphere"—celebrate their GL identity, more often becoming activists for GLBT rights, in elderly housing, health care, and other services.

Should filial responsibility be a law? Consider the following argument by a legal scholar.

"The rapid growth of the senior populations, ever-increasing life expectancy, and a decreasing birth rate—along with instability of government programs like Medicaid, Medicare, and Social Security—place the United States on the verge of having more indigent seniors than ever. More rigidly enforced laws that provide for the care of the elderly from sources other than the government can counter these developments. Prime examples of such legislation are filial responsibility statutes, which impose a duty on children to support their parents. . . .

"A filial responsibility statute is simply a law that 'create[s] a statutory duty for adult children to financially support their parents who are unable to provide for themselves.' Typically, such support includes an obligation to pay for 'food, clothing, shelter, and medical attention.' The rationale behind such laws arises from the reciprocal duty that parents have to care for their children; because parents extended voluntary care to their minor children, it is the filial responsibility of children to return that support to their parents. . . .

"In 1601, filial responsibility was put into statutory form with the enactment [in England] of the Elizabethan Poor Relief Act, from which most modern statutes are derived. . . . Carrying on the historical tradition of supporting indigent parents, [in] the United States . . . prior to the 1960s, federal legislation recognized this obligation as well. However, . . . the establishment of Medicare in the 1960s led to the repeal of the federal statute. Today, there are twenty-two states with filial responsibility statutes, and few—if any—of the states currently enforce these laws. . . .

"[C]hildren are only required to support their parents so long as 'they are of sufficient ability.' . . . Therefore, children should be excused from their filial responsibility if they have no economic means of supporting their parents. . . .

"Children can also avoid filial responsibility if they can demonstrate the parent abandoned them. . . . If the parent did not give the necessary financial support to their child, then it stands to reason that the state should not force the child to give financial support to the parent. . . .

"There are, of course, criticisms of filial responsibility statutes. Among the most prevalent are the administrative difficulties associated with enforcement . . . and the laws' apparent contradiction of America's culture of self reliance. . . . Nonetheless, in recent years, various commentators have advocated the establishment of either a mode filial responsibility law that would make state laws more uniform, or a reenactment of federal filial responsibility law. . . .

"[According to one court decision], 'The selection of the adult children is rational on the ground that the parents, who are now in need, supported and cared for their children during their minority and that such children should in return now support their parents to the extent to which they are capable. . . . [State filial responsibility statutes have been upheld by courts] for the simple reason that they accomplish the goal that they seek—to provide for the indigent elderly and to prevent a strain on our public funds. . . . [Filial responsibility] statutes would be beneficial to our society and provide desperately needed relief for our strained public treasury. Legislatures should give serious consideration to their reinstatement."

Critical Thinking

Other experts in this field find filial responsibility laws problematic. They argue that (1) there is already considerable voluntary assistance from children to parents; (2) filial responsibility legislation can undermine parent–child relationships, creating resentment on the payee's part and guilt on the recipient's part; (3) government programs such as Social Security are preferable. Making Social Security contributions, children are paying for parents, but without the tension created by legislated direct payments (Callahan 1985). What do you think? Is legally requiring filial responsibility a good idea? What would be some benefits to families and to society? What might be some drawbacks?

Source: Excerpted from Lundberg 2009, 534–82.

Older Americans provide a considerable amount of elder care to one another. Up to 40 percent of elder care is provided by the elderly care receiver's spouse (Johnson and Weiner 2006). Caregiving and receiving are expected components of marriage for most of today's older couples, who have developed a relationship of mutual exchange over many years (Machir 2003; Roper and Yorgason 2009). A qualitative study of seventy-five spouse caregivers found that those in longer, emotionally close marriages with little ongoing conflict evidenced better overall well-being (Townsend and Franks 1997; see also Roper and Yorgason 2009). After spouses, adult children are most likely to be providing elder care. Especially for the unmarried and childfree, an older American's siblings are important in mutual caregiving (Eriksen and Gerstel 2002).

Adult Children as Elder Care Providers

Grown children who care for aging parents tend to provide the vast majority of care themselves, although they also seek assistance from formal service providers.

Motivated by **filial responsibility** (a child's obligation to parents), respect, and affection, adult children care for their folks "because they're my parents" (Gans and Silverstein 2006; Klaus 2009). Adult children who are highly religious report higher levels of filial responsibility (Gans, Silverstein, and Lowenstein 2009).

As pointed out in "Issues for Thought: Filial Responsibility Laws," principles of reciprocity are at work regarding filial responsibility. As one caregiver explained to a researcher,

> My mom and I really took care of each other because I had my kids pretty young. I was a teen mom . . . . So I needed her help. . . . My mom's always been there for me, helping me out, so I am always going to be there for her. (in Radina 2007, p. 158)

A study of 387 elderly parents in Florida found that parents expected help from their adult children in proportion to the aid that the parents had once given to their children (Lee, Netzer, and Coward 1994; see also Fingerman et al. 2009). Correspondingly, adult children who have received more financial help from their parents are more likely than their siblings to be engaged in caring for the parent in old age (Henretta et al. 1997). Interestingly, young-adult stepchildren receive less financial and other kinds of practical help than do biological or adopted children (Berry 2008)—a situation that may help to explain the lower level of felt filial responsibility among stepchildren.

Adult Stepchildren and Elder Care In general, parent–adult child ties with stepchildren are not as close as are those with biological or adopted children (Ward, Spitze, and Deane 2009). So it may be no surprise that adult children feel less obligation to aging stepparents than they do to biological parents, especially when the stepparent was acquired later in the adult child's life (Pezzin, Pollak, and Schone 2008). Interestingly, obligation to help one's remarried *biological* parent is often the principal reason for helping the *step*parent. "This type of once-removed family responsibility may reflect the lack of opportunity for building reciprocal relationships between stepparents and stepchildren after later-life remarriages" (Ganong, Coleman, and Rothrauff 2009, p. 176). Then too, reduced filial responsibility to stepparents may reflect the fact that stepfamilies remain incompletely institutionalized, as discussed in Chapter 16.

Shared Elder Care among Siblings There is some evidence that the oldest sibling in a family feels filial responsibility more strongly and thus provides more care to an aging parent than do younger sibs (Fontaine, Gramain, and Wittwer 2009). Today, it's often the case that siblings have geographically moved away from each other and from their aging parents; geographical distance typically excuses a sibling from hands-on, day-to-day caregiving. Even when siblings live near an elderly parent, however, the burden of elder care does not always fall upon each one equally. A study based on forty focus groups (see Chapter 2) asked givers of elder care with siblings to describe how they felt about the caregiving situation:

> Siblings who described an imbalance in caregiving responsibilities reported feeling considerable distress. . . . One participant confessed that she was straddling a "real thin line between just taking her [barely participating sister's] head off some day, because I'm so mad at the inequity of it." (Ingersoll-Dayton et al. 2003, p. 205)

However, the majority of participants in this study sought to define the situation as more or less fair (see also Connidis and Kemp 2008; Kuperminc, Jurkovic, and Casey 2009). They took into account a sibling's geographical distance, employment responsibilities, and other obligations. Some participants (both men and women) called upon gendered expectations to help justify women's inequitable elder care responsibilities. "I guess it's my gender," said a woman whose brother did little to help. "It's just natural" (Ingersoll-Dayton et al. 2003, p. 207).

Gender Differences in Providing Elder Care

Men *are* involved in elder care; about one-third of unpaid caregivers are male. Furthermore, the proportion of sons involved in elder care may increase in the future due to the growing number of only-child sons, smaller sibling groups from which to draw care providers, and changing gender roles that make caregiving an expectation for adult male behavior.

However, women account for about two-thirds of all unpaid caregivers today (National Alliance for Caregiving 2009, p. 14). This statistic partly reflects the fact that, among marrieds, far more caregivers are wives. This situation is mostly due to women's living longer. Feelings of obligation to care for an elderly disabled spouse are fairly equal between husbands and wives (Roper and Yorgason 2009). However, gender makes a significant difference in adult children's caregiving obligations (Cancian and Oliker 2000).

Norms in many Asian American families designate the oldest son as the responsible caregiver to aging parents (Kamo and Zhou 1994; Lin and Liu 1993). Except in this case, the adult child involved in a parent's care is considerably more likely to be a daughter (or even a daughter-in-law) than a son. This situation—one that raises issues of gender equity similar to those of parenting and other unpaid family labor, discussed in Chapters 10 and 11—is partly due to ongoing employment differences between women and men (Sarkisian and Gerstel 2004). Then, too, in accordance with the findings, discussed earlier, that relations with daughters are more often tight-knit than those with sons, a parent may prefer a daughter's help. A son-in-law explained his wife's caring for her mother this way: "Mom just, I think, calls on her more, so that's the way. . . . It's not that the brothers wouldn't help at all, but it's just . . . she gets called on more" (Ingersoll-Dayton et al. 2003, p. 207).

Sons tend to provide elder care only in the absence of available daughters (Lee, Spitze, and Logan 2003). When siblings share in caring for aging parents, sisters do more than brothers, on average, as measured by time spent. Furthermore, men and women tend to provide care differently—in ways that reflect socially gendered expectations (Raschick and Ingersoll-Dayton 2004). Sons, grandsons, and other male caregivers (although not husbands) tend to perform a more limited range of occasional tasks, such as cleaning gutters or mowing the lawn, whereas daughters more often provide continually needed services like housekeeping, cooking, or doing laundry (Hequembourg and Brallier 2005). Sons more often serve as financial managers, as well as organizers, negotiators, supervisors, and intermediaries between the care receiver and formal service providers. Then too, a son is more likely than a daughter to enlist help from his spouse, the elder care receiver's daughter-in-law (Raschick and Ingersoll-Dayton 2004). "Providing intimate, hands-on care is culturally defined as feminine, [and the] dirty parts of care work are mainly women's work" (Isaksen 2002, pp. 806, 809).

The Sandwich Generation

Many daughter-caregivers have children under age eighteen living at home. Indeed, it looks as if "the presence of children in the household connects parents to kin," including aging parents who may need elder care (Gallagher and Gerstel 2001, p. 272). Between 9 and 13 percent of U.S. households with at least one person between ages thirty and sixty have responsibilities for dependent children and elders (Neal and Hammer 2006). About twenty years ago, journalists and social scientists took note of an emergent **sandwich generation**: middle-aged (or older) individuals who are sandwiched between the simultaneous responsibilities of caring for their dependent children and aging parents. More recently, gerontologist Neal Cutler, who studies the effect of aging on finances, coined the term *senior sandwich generation*. These folks are "at least 60 years old and facing the ultimate financial trifecta: college for their kids (either current tuition bills or paying back borrowed money), retirement for themselves and at-home or nursing-home care for one or more parents. All at the same time" (Chatzky 2006).

The sandwich generation experiences all the hectic task juggling discussed in Chapters 9 and 10. For some in this category, work demands take the majority of their attention. Others emphasize child care responsibilities, whereas still others focus more of their energies on care

The majority of the young-old need almost no help at all, but as they grow older they are likely to need more help, sometimes requiring a great deal of continuing care. Often employed, members of the *sandwich generation*—adults sandwiched between the simultaneous responsibilities of caring for their children and aging parents—feel the strains associated with juggling work, child care, and elder care.

for aging parents (Cullen et al. 2009). Regardless of what corner of this triangle predominates—work, children, or aging parents—stress builds as family members maneuver chaotic everyday life (Williams and Boushey 2010).

Elder Care as a Family Process

In accordance with the interactionist theoretical perspective, we can envision elder care as an interactive process during which family members struggle to negotiate various caregiving decisions. Social scientists have noted a **caregiving trajectory** through which the process of elder care proceeds.

First, the caregiver becomes concerned about an aging family member, and often she or he expresses this concern to others, although not necessarily to the older family member. Later, still concerned, the caregiver begins to give advice to the older family member, such as "Don't forget to take your medicine" or "You should get an appointment for new glasses." Still later, the caregiver begins to provide needed services (Cicirelli 2000). Throughout this trajectory, family members may be called upon for advice and counsel in making medical or other significant decisions. A family's decision to move an elderly parent to a nursing home is particularly painful, with concern for the aged parent continuing thereafter (Keefe and Fancey 2000).

Ambivalence, Conflict, and the Need to Set Boundaries

As they make decisions about the elderly family member's condition, disagreements may arise between the caregiver(s) and the receiver as well as among family caregivers (Mills and Wilmoth 2002). Caregiving parent–child relationships might best be characterized by ambivalence (Fingerman, Hay, and Birditt 2004). Older people are often uncomfortable and sometimes stubborn or angry about receiving help. Needing assistance represents a threat to their autonomy and self-esteem, especially if the caregiver is perceived as controlling (Brubaker, Gorman, and Hiestand 1990; Halpern 2009). According to one qualitative study, caregivers may expect a certain amount of deference, or courteous submission to their opinions and decisions, from the aging parent for whom they are caring—and when this does not occur, "intergenerational relations become strained" (Pyke 1999, p. 661).

Then too, parents may become more and more demanding as they get older, a common reaction to loss of bodily and social power with aging and retirement (Bottke 2010). Some conflicts between caregiver and recipient are about finances—when the elderly parent seems to be letting go of money unwisely, for example (Duhigg 2007). Whether an aging family member is able to drive safely or should move into a residential care facility are other potential matters of contention (Dyer, Pickens, and Burnett 2007). More routine issues, such as how urgent is an errand that needs to be run, can also cause caregiver–receiver conflict. Such clashes can be painful; using positive communication skills to set boundaries is important to a caregiver's well-being (Bottke 2010; Halpern 2009).[15]

Disagreements among family caregivers are also to be expected. For instance, conflicts may arise when siblings are "forced to make urgent, complex decisions for loved ones in intensive care units" (Siegel 2004). Adult siblings in the same family may have experienced growing up differently and hold divergent feelings about a parent (Flora 2007). Sometimes old sibling rivalries or perceptions of a parent's favoritism emerge (Suitor et al. 2009) to exacerbate caregiving challenges. All else being equal, the caregiving experience is less stressful and more positive when conflict is low and family members can agree on issues. Put another way, families that have developed a shared understanding of the caregiving situation make more effective caregiving teams (Halpern 2009).

Caregiver Stress Providing elder care may enhance one's sense of purpose and overall life satisfaction due to the self-validating effects of helping another person, enhanced intimacy, and the belief that helping others may result in assistance with one's own needs when the time arrives (Marks, Lambert, and Choi 2002; Shapiro 2006). Nevertheless, "providing help can overwhelm caregivers" (Johnson and Wiener 2006). Caregiving can be physically, financially, and emotionally costly (Christakis and Allison 2006; Parra-Cardona et al. 2007). Caregivers not only spend their own money but (especially women) may also take extended time from work or pass up promotions (National Alliance for Caregiving 2009). Some family caregivers take early retirement or geographically relocate to provide elder care.

Moreover, leisure time with friends and others is important to emotional well-being (Harter and Arora 2008). But older caregivers report losing contact with friends and other family members (National Alliance for Caregiving 2009). Younger caregivers experience limitations on dating and other relationships. As one twenty-six-year-old explained to a research team, "I would like to go camping with my husband once in a while, but I can't just get up and go away, because of taking care of my grandparents" (in Dellmann-Jenkins, Blankemeyer, and Pinkard 2000, p. 181). Providing elder care can be socially isolating, bring on depression, and further strain one's physical health (Clark and Diamond 2010).

Caregiver stress and depression among Americans result partly from the fact that ours is an individualistic culture in which adult children are expected to establish lives apart from their parents and to achieve success as individuals rather than (or as well as) working to benefit the family system (Killian and Ganong 2002). Furthermore, because the chronically ill or disabled of all ages are decreasingly cared for in hospitals, today's informal caregivers are asked to perform complicated care regimens that have traditionally been handled only by health care professionals (Guberman et al. 2005). This situation places added demands on a family caregiver's time, energy, and emotional stamina (Richards 2009). (See "Issues for Thought: Caring for Patients at Home—A Family Stressor," in Chapter 14).

On days when a chronically ill care recipient appears to be doing better, caregivers report being in a better mood (Roper and Yorgason 2009). Also, when the caregiver–recipient relationship feels more reciprocal—with the caregiver feeling that he or she is getting something in return—the caregiver is less often depressed (LeBlanc and Wright 2000). Receiving adequate training and learning specific caregiver skills can lessen caregiver

---

[15] Susan Halpern's *Finding the Words: Candid Conversations with Loved Ones* (2009) is a good resource.

stress, as do various forms of social support—phone calls and other contacts with friends as well as finding information and connecting with other caregivers either in face-to-face support groups or on the Internet (Smerglia, Miller, and Sotnak 2007; Roper and Yorgason 2009; Wilkins, Bruce, and Sirey 2009).[16] However, elder abuse and neglect in families often—although not always—results from caregiver stress (Conner 2000).

Elder Abuse and Neglect

Parallel to child abuse and neglect, discussed in Chapter 13, **elder abuse** involves overt acts of aggression, whereas **elder neglect** involves acts of omission or failure to give adequate care. Elder abuse includes physical assault; verbal abuse and other forms of emotional humiliation; purposeful social isolation (for example, forbidding use of the telephone); and financial exploitation (Hildreth, Burke, and Glass 2009).

The profile of the abused or neglected elderly person is of a female, seventy years old or above, who has physical, mental, and/or emotional impairments and is dependent on the abuser/caregiver for companionship and help with daily activities (Leisey, Kupstas, and Cooper 2009). Studies have found that the *neglected* elderly have more physical and mental difficulties (and, hence, are more burdensome to care for) than are elder *abuse* victims (Pillemer 1986; Whittaker 1995).

At the time of this writing, accurate statistics on the prevalence of elder abuse and neglect do not exist, partly because state agencies cannot count the many unreported instances of elder abuse (Cooper, Selwood, and Livingston 2008; Federal Interagency Forum on Aging-Related Statistics 2008). However, according to the 2010 federal Elder Justice Act, the U.S. government is to gather accurate statistics and encourage research on the prevalence and causes of elder maltreatment, as well as on the effectiveness of programs designed to curb it (American Bar Association 2010). Meanwhile, depending on how broadly a researcher defines elder maltreatment, various studies have concluded that as many as five million—between 1 and over 10 percent—of individuals age sixty and above are abused or neglected annually by professional and family caregivers (Hildreth, Burke, and Glass 2009; Ramnarace 2010). About 38,000 cases of elderly financial abuse are reported every year, and experts believe the actual number of occurrences is much higher (Bendix 2009; Miles 2008; see also Hull 2008).

Current data suggest that *non*family members—typically paid caregivers either in the aged person's home or in an institutional setting—are responsible for more than half of all elder verbal, physical, and financial abuse (Laumann, Leitsch, and Waite 2008; Lee 2009). Education programs for family members and others who work with the elderly—social workers, physicians, nurses, dentists, attorneys—have been designed to facilitate detection and prevention of elder maltreatment in institutional settings (Bendix 2009; Miles 2008; Phelan 2009; Ploeg, Fear, and Hutchison 2009; Rinker 2009; Wagenaar 2009; Wiseman 2008).

Elder Maltreatment by Family Members In 1996, the federal government sponsored the National Elder Abuse Incidence Study, a random sample survey of counties, which combined reports from Adult Protective Services with interviews with "sentinels," people in the community who had contact with the elderly. Estimates from this study—still our best source for statistics—were that about one-half million Americans over sixty and living in family households were abused or neglected (Ramnarace 2010). Neglect was the more common form of elder maltreatment ("Fact Sheet: Elder Abuse Prevalence and Incidence" 2005). A more recent nationally representative survey found that, among those elderly who were abused by an immediate family member, 9 percent experienced verbal abuse whereas 3.5 percent had suffered financial abuse. Less than 1 percent experienced physical abuse at the hands of a family member caregiver (Laumann, Leitsch, and Waite 2008.)

Common to cases of physical elder abuse by family members are shared living arrangements, the abuser's

The majority of elderly Americans maintain their own homes. However, many of the frail elderly depend on, or live with, family members. Although care for an aging parent—most often by daughters— is often given with fondness and love, it can also bring stress, conflicting emotions, and great demands on time, energy, health, and finances.

---

[16] There are many good resources on the Internet. Several examples: American Association of Retired Persons (aarp.org); Center for Retirement Research at Boston College (crr.bc.edu); National Elder Law Network (neln.org); National Institute on Aging (nia.nih.gov); Rand Center for the Study of Aging (rand.org); Social Security Administration (ssa.gov); U.S. Administration on Aging, (aoa.gov); and University of Michigan Retirement Research Center (mrrc.isr.umich.edu).

poor emotional health (often including alcohol or drug problems), and a pathological relationship between victim and abuser (Anetzberger, Korbin, and Austin 1994). Abusive acts may be "carried out by abusers to compensate for their perceived lack or loss of power" (Pillemer 1986, p. 244). "In many instances, both the victim and the perpetrator [are] caught in a web of interdependency and disability, which [make] it difficult for them to seek or accept outside help or to consider separation" (Wolf 1986, p. 221).

Among family members, the largest proportion of abuse is from a spouse or a romantic partner (Laumann, Leitsch, and Waite 2008). In fact, "there is reason to believe that a certain proportion of elder abuse is actually spouse abuse grown old" (Phillips 1986, p. 212; see also Leisey, Kupstas, and Cooper 2009; Shannon 2009). In some cases, marital violence among the elderly involves abuse of the caregiving spouse by a partner who is ill with Alzheimer's disease (Pillemer 1986). Adult children are also responsible for elder abuse (Laumann, Leitsch, and Waite 2008). Data from one study of 300 cases of elder abuse in the Northeast found that an abuser (frequently an adult son) was likely to be financially dependent on the elderly victim (Pillemer 1986).

Researching and combating elder abuse generally proceed from either of two models: the *caregiver model* or the *domestic violence model*. The **caregiver model of elder abuse and neglect** views abusive or neglectful caregivers as individuals who are simply overwhelmed by the requirements of caring for their elderly family members (Abbey 2009). Indeed, abusive or neglectful family caregivers are often stressed by socially structured conditions, such as job conflicts. This model focuses on the fact that burdens associated with caring for an older person may sometimes cause the caregiver to lose control and verbally or physically abuse the receiver (Bainbridge et al. 2009; Wiglesworth, Kemp, and Mosqueda 2008). Professional care providers employed by community agencies are often trained to recognize potentially abusive family situations, and then can work to reduce dangerous levels of caregiver stress that may trigger abuse (Brandl 2007; Thobaben 2008).

In contrast, the **domestic violence model of elder abuse and neglect** views elder abuse as a form of unlawful family violence and focuses on negative personal characteristics of abusers and on a possible criminal justice response (Hagan 2010; Wallace 2008). The criminal justice response is especially likely in the case of financial abuse (Gross 2006b). Now that we have examined providing elder care in general, we turn next to a discussion of racial/ethnic diversity and elder care.

Racial/Ethnic Diversity and Family Elder Care

Adult children of all races and ethnicities feel responsible for their aging parents and to their siblings, with whom they may share elder care obligations. However, racial/ethnic differences do exist regarding elder care.

For instance, Asians have traditionally tended to emphasize the centrality of filial obligations over conjugal relationships (Burr and Mutchler 1999). In addition, older blacks are more likely than non-Hispanic whites to expect their adult children to personally care for them in old age (Lee, Peek, and Coward 1998). As suggested by Figure 17.1, blacks, Hispanics, and Asian Americans are more likely than non-Hispanic whites to expect to share a residence with their grown children if necessary.

Although more than two-thirds (67 percent) of caregivers today are non-Hispanic white, this situation will change as ethnic groups with younger average ages grow older (National Alliance for Caregiving 2009). Research is just beginning to accumulate on how race and ethnicity influence caregivers' emotional and mental health (Skarupski et al. 2009). One study compared eighty-nine Latina with ninety-six non-Hispanic white female caregivers of older relatives with Alzheimer's disease and found that, regardless of ethnicity, lack of financial resources negatively impacted the caregivers' emotional health. However, stress from lack of resources was mitigated, or lessened by Latinas' relatively strong familistic values (Montoro-Rodquez and Gallagher-Thompson 2009). A study of 307 caregivers compared blacks with non-Hispanic whites and found that when exposed to caregiver stress blacks coped better emotionally (Skarupski et al. 2009; see also Wilkins, Bruce, and Sirey 2009). Prior research with African American wife caregivers found that receiving support from their churches lessened their stress and helped their marital relationship (Chadiha, Rafferty, and Pickard 2003).

Meanwhile, racial/ethnic minorities have not been as likely as non-Hispanic whites to use community-based services, such as senior centers. This discrepancy occurs due to language barriers; because individuals don't know that they qualify for government-funded services; or because they are reluctant to include paid service providers as members of their caregiving team (Levine 2008). **Fictive kin** (family-like relationships that are not based on blood or marriage but on close friendship ties) are often resources for elder care exchanges among African Americans (sometimes called "going for sisters"), Hispanics (*compadrazgos*), Italians (*compare*), and other ethnic groups as well (Ebaugh and Curry 2000).

It is a myth that minority ethnic families "take care of their own" and seldom need to rely on community services. However, among many immigrant ethnic groups, **acculturation** (the process whereby immigrant groups adopt the beliefs, values, and norms of their new culture) affects norms of filial obligation. Younger generations are more likely than their elders to become acculturated—a situation that creates the potential for intergenerational conflict (Rudolph, Cornelius-White, and Quintana 2005; Silverstein 2000). A study of older Puerto Ricans found that filial obligation has declined in the younger generations (Zsembik and Bonilla

2000). And research on Chinese immigrant families in California found that sons often outsourced elder care:

> "I told her that I hire you to help me achieve my filial duty," Paul Wang, a 60-year-old Taiwanese immigrant owning a software company in Silicon Valley, California, described . . . his conversation with the in-home care worker he employed for his mother suffering from Alzheimer's disease. (Lan 2002, p. 812)

Acculturation has also meant that, increasingly, aging immigrants live in housing designed for the elderly, rather than with their grown children (Kershaw 2003, p. A10).

The Changing American Family and Elder Care in the Future

In the future, aging individuals may be more and more creative in fashioning caregiving arrangements for themselves and their peers. However, longer life expectancy and more chronically ill Americans in old-old age raise concerns about providing elder care now and in the future (Levine 2008). The American family is changing in form, as we have seen throughout this text. Many of these changes are expected to result in a diminishing caregiver "kin supply" (Bengston 2001, p. 5; Himes 2001).

As noted earlier in this chapter, the current generation of young-old elderly had smaller families as well as higher divorce rates than did generations several decades ago, so fewer adult children will be available for caregiving. Furthermore, females' increased participation in the labor force decreases the time that women, the principal providers of elder care, have available to engage in elder care. Then too, geographical mobility negatively affects face-to-face support for aging relatives.

Today's older Americans are likely themselves to have siblings who can help them, but as more younger parents choose to have fewer or only children, more elderly in the future will have fewer or no siblings. Then too, more siblings in a family can ease each one's burden of care for an aging parent, because they can share necessary tasks (Wolf, Freedman, and Soldo 1997; see also Fontaine, Gramain, and Wittwer 2009). However, increasingly, fewer siblings will be available to share in elder care.

Policy makers' concerns also involve other family-structure changes, such as higher rates of permanent singlehood; remaining childfree; and of cohabitation,

After experiencing it herself, Nancy Edwards, left, wrote a thesis on lesbians coming out later in life. Surrounded by family photos from their prior heterosexual marriages, she and her partner sit in their Minneapolis home. Older individuals who formed a GL identity before the 1970s, when homosexuality was highly stigmatized, may tend to hide their sexual orientation. But baby boomers, who formed a GL identity in a "gay liberation atmosphere" after the 1970s, more often become visible activists for elderly GLBT rights (Rosenfeld 1999).

Tom Wallace, Star Tribune

divorce, and repartnering. Being childfree or without a spouse "eliminates the two most important caregiving resources—spouse and adult child" (Conner 2000, p. 14). Never-married, childfree women are particularly active socially; future childfree and single elderly are likely to fashion fictive-kin families of mutually caring friends (Spencer and Pahl 2006). On the other hand, elderly individuals without children may prove to be "less likely than are parents to have robust network types capable of maintaining independent living " (Wenger et al. 2007, p. 1,419; and see Bures, Koropeckyj-Cox, and Loree 2009).

Moreover, a high sense of filial obligation in a family has been positively related to caregiving (Kuperminec, Jurkovic, and Casey 2009). But parental divorce reduces the younger generation's sense of filial obligation, particularly to divorced fathers (Ganong, Coleman, and Rothrauff 2009; Pezzin and Schone 1999). An elderly parent's daughter's being divorced does *not* decrease her help to her parents—although divorce probably does decrease help from an ex-daughter-in-law (Spitze et al. 1994).

We are gradually learning more about the ramifications of remarriage for elder care. In one qualitative study of late-life remarried caregivers,

> [w]ives revealed how little support or assistance they received from adult stepchildren for their ailing father. Often these caregivers endured a kind of amplified stress, isolation, and conflict in their caregiving role, which they attributed to their remarried, stepmother status. By the same token, adult children and stepchildren are not always granted access to the critical decision-making or caretaking. (Sherman 2006, p. F8; see also Sherman and Boss 2007)

The extent to which one's cohabiting partner participates in elder care is a matter for future research.

Same-sex families and couples confront elder care challenges. "But along with getting older, they also have to face the prejudices of being gay or lesbian. . . . Nursing homes and private retirement centers . . . [often] make assumptions that their residents are heterosexual and structure activities on the basis of these assumptions" (Powell 2004, p. 60). Partly so as not to be separated from their partners by well-meaning relatives who may put them in separate nursing homes, gay men and lesbian couples (who can afford it) have begun to create assisted living communities of their own ("Birds of a Feather" 2010; Grant 2010; Mitchell 2007).

Although many policy analysts express concern about families' ability to effectively provide elder care in the future, some also point to the **latent kin matrix**, defined as "a web of continually shifting linkages that provide the potential for activating and intensifying close kin relationships" (Riley 1983, p. 441; see also Spencer and Pahl 2006). An important feature of the latent matrix is that, although they may remain dormant for long periods, family relations with adult siblings and extended kin emerge as a resource when the need arises (Silverstein and Bengston 2001).

Moreover, while family forms become increasingly diverse, so may ways that members of the postmodern family deal with elder care. For instance, in research involving in-depth interviews with forty-five older respondents, a sixty-seven-year-old separated wife, whose husband had become diabetic and asthmatic and had heart trouble, reported that she still loved him very much: "If something happened to him and this gal [his new romantic partner] didn't take care of him, I would go and take care of him myself" (quoted in Allen et al. 1999, p. 154).

Toward Better Caregiving

Family sociologist and demographer Andrew Cherlin (1996) distinguishes between the "public" and the "private" faces of families. The **private face of family** "provides individuals with intimacy, emotional support, and love" (p. 19). The **public face of family** produces public goods and services by educating children and caring for the ill and elderly: "While serving each other, members of the 'public' family also serve the larger community" (Conner 2000, p. 36; Gross 2004).

Government leaders tend to emphasize the "paramount importance" of "personal responsibility and accountability for planning for one's longevity" (White House Conference on Aging 2005, p. 18). However, the financial and emotional costs of providing good care for ill and disabled family members are often too high for family caregivers to manage without help (Piercy 2010; Span 2009b). Many point to the increasing need for public services to assist family caregivers—for example, transportation, personal care services, and adult day care (Rosen 2007).

However, "this country's family policies lag far behind those of [many nations in] the rest of the world . . . .

> The media constantly reinforce the conventional wisdom that the care crisis is an individual problem. Books, magazines and newspapers offer American women an endless stream of advice about how to maintain their "balancing act," how to be better organized and more efficient or how to meditate, exercise and pamper themselves to relieve their mounting stress. Missing is the very pragmatic proposal that American society needs new policies that will restructure the workplace and reorganize family life. (Rosen 2007, p. 13)

Cancian and Oliker (2000) have proposed the following strategies for moving our society toward better elder care coupled with greater gender equity in providing elder care:

- Provide government funds that support more care outside the family, such as government-funded day care centers for the elderly and respite (time off) services for caregivers.

- Increase social recognition of caregiving—both paid and unpaid—as productive and valuable work.

Fuse

Many older Americans remain active even into old-old age. Nonetheless, longer life span does mean that more and more Americans will spend the last months or years of their lives with chronic health problems and/or disabilities. The aging of the population raises concerns about how a changing family structure will be able to care for tomorrow's elderly and to what extent community and government resources can or will be engaged to help.

- Make caregiving more economically rewarding or, at least, less economically costly to caregivers. (p. 130)

How these policy changes might be accomplished may be difficult to imagine, but this fact negates the utility neither of the vision nor of the political debate that needs to emerge (Gross 2007). Elder care (as well as child care) is indeed a responsibility not only of individual families but also of an entire society.

Summary

- The *number* of elderly, as well as their *proportion* of the total U.S. population, is growing.

- Along with the impact of the baby boomers' aging and the declining proportion of children in the population, longer life expectancy has contributed to the aging of our population.

- For the most part, adult children and their parents prefer to live near each other, although not in the same residence.

- Due mainly to differences in life expectancy, older men are much more likely to be living with their spouse than are older women.

- Among older Americans without partners, living arrangements depend on one's health, the availability of others with whom to reside, social norms regarding obligations of other family members toward their elderly, personal preferences for privacy and independence, and economics.

- Growth in Social Security benefits has resulted in dramatic reductions in U.S. poverty rates for the elderly over the last several decades, although 10 percent of older adults are living in poverty.

- Due to differences in work patterns, wage differentials, and Social Security regulations, older men are considerably better off financially than are older women.

- Most older married couples place intimacy as central to their lives, describe their unions as happy, and continue to be interested in sex, even into old age.

- Even when it is not an abrupt event, retirement represents a great change for individuals and couples, particularly for males who have embraced the traditional masculine gender role.

- Adjustment to widowhood or widowerhood is an important family transition that married couples often must face in later life. Bereavement manifests itself in physical, emotional, and intellectual symptoms.

- Daughters are more likely than sons to have close relationships with their parents, especially with their mothers. However, even among mothers, a parent's divorce, separation, or repartnering often weakens the bond with adult children.

- Partly due to longer life expectancy, grandparenting became increasingly important to families throughout the twentieth century. In the twenty-first century, we will see an increasing number of great-grandparents.

- As members of our families age, elder care is becoming an important feature in family life, with women providing the bulk of it.

- After spouses, adult children (usually daughters) are a preferred choice of an older family member as elder care providers.

- Elder care in families typically follows a caregiving trajectory as the care receiver ages, and it involves not only benefits but also stresses for the caregiver(s).

- Elder maltreatment—that is, abuse and neglect—by family caregivers exists in a small percentage of aging families and is addressed in public policy by either the caregiver model or the domestic violence model.

- There are empirically noted racial/ethnic differences in elder care, but this does not negate that racial/ethnic minorities—as well as non-Hispanic whites—need community and government assistance in providing elder care.

- Changes in the American family lead some policy analysts to be concerned that families will have greater difficulty providing elder care in the future.

However, others point out that family relationships, though latent for long periods, can be activated when needed.

- Better elder care in the future will necessitate involving more men as caregivers and developing public policy that adequately supports expanded community services to assist families in providing elder care.

Questions for Review and Reflection

1. Discuss ways that society's age structure today affects American families.

2. Describe the living arrangements of older Americans today, and give some reasons for these arrangements.

3. Discuss some ways that the growing diversity of family forms among the older population can be expected to affect caregiving.

4. Apply the exchange, interactionist, structure-functionalist, or ecological perspective (see Chapter 2) to the process of providing elder care.

5. **Policy Question.** Describe two suggestions for policy changes that would make family elder care less difficult.

Key Terms

acculturation 495
active life expectancy 475
baby boom 474
bereavement 482
caregiver model of elder abuse and neglect 495
caregiving 487
caregiving trajectory 492
custodial grandparent 486
domestic violence model of elder abuse and neglect 495
elder abuse 494
elder neglect 494

elder care 487
fictive kin 495
filial responsibility 491
gerontologists 487
hierarchical compensatory model of caregiving 489
informal caregiving 487
latent kin matrix 497
noncustodial grandparent 486
private face of family 497
public face of family 497
sandwich generation 492

Online Resources

Sociology CourseMate

www.CengageBrain.com

Access an integrated eBook, chapter-specific interactive learning tools, including flash cards, quizzes, videos, and more in your Sociology CourseMate, accessed through CengageBrain.com.

www.CengageBrain.com

Want to maximize your online study time? Take this easy-to-use study system's diagnostic pre-test, and it will create a personalized study plan for you. By helping you identify the topics that you need to understand better and then directing you to valuable online resources, it can speed up your chapter review. CengageNOW even provides a post-test so you can confirm that you are ready for an exam.

Glossary

ABC-X model A model of family crisis in which A (the stressor event) interacts with B (the family's resources for meeting a crisis) and with C (the definition the family formulates of the event) to produce X (the crisis).

abortion See **induced abortion**.

abstinence The standard that maintains that nonmarital intercourse is wrong or inadvisable for both women and men regardless of the circumstances. Many religions espouse abstinence as a moral imperative, while some individuals are abstinent as a temporary or permanent personal choice.

acculturation The process whereby immigrant groups adopt the beliefs, values, and norms of their new culture and lose their traditional values and practices.

acquaintance rape Forced or unwanted sexual contact between people who know each other, often—although not necessarily—taking place on a date. See also **date rape**.

active life expectancy The period of life free of disability in activities of daily living, after which may follow a period of being at least somewhat disabled.

agape The love style that emphasizes unselfish concern for a beloved, in which one attempts to fulfill the other's needs even when that means some personal sacrifice. See also **eros, ludus, mania, pragma,** and **storge**.

agentic (instrumental) character traits Traits such as confidence, assertiveness, and ambition that enable a person to accomplish difficult tasks or goals.

AIDS See **HIV/AIDS**.

allocation systems The arrangements couples make for handling their income, wealth, and expenditures. Allocation systems may involve pooling partners' resources or keeping them separate. Who controls pooled resources is another dimension of an allocation system.

arranged marriage Unions in which parents choose their children's marriage partners.

asexual, asexuality A person who is asexual does not experience sexual desire. This is different from abstinence or celibacy, which is a choice to not engage in sexual activity despite feelings of sexual desire. Asexuality may be considered a sexual orientation.

assisted reproductive technology (ART) Advanced reproductive technology, such as artificial insemination, in vitro fertilization, or embryo transplantation, that enables infertile couples or individuals, including gay and lesbian couples, to have children.

assortative mating Social psychological filtering process in which individuals gradually filter out those among their pool of eligible individuals they believe would not make the best spouse.

attachment "An active, affective, enduring, and reciprocal bond between two individuals that is believed to be established through repeated action over time" (Coleman and Watson 2000, p. 297, citing Ainsworth et al. 1978).

attachment disorder An emotional disorder in which a person defensively shuts off the willingness or ability to make emotional attachments to anyone.

attachment theory A psychological theory that holds that, during infancy and childhood, a young person develops a general style of attaching to others; once an individual's attachment style is established, she or he unconsciously applies that style to later, adult relationships. The three basic styles are **secure, insecure/anxious,** and **avoidant**.

authoritarian parenting style All decision making is in parents' hands, and the emphasis is on compliance with rules and directives. Parents are more punitive than supportive, and use of physical punishment is likely.

authoritative parenting style Parents accept the child's personality and talents and are emotionally supportive. At the same time, they consciously set and enforce rules and limits, whose rationale is usually explained to the child. Parents provide guidance and direction and state expectations for the child's behavior. Parents are in charge, but the child is given responsibility and must take the initiative in completing schoolwork and other tasks and in solving child-level problems.

avoidant attachment style One of three attachment styles in **attachment theory,** this style avoids intimacy either by evading relationships altogether or by establishing considerable distance in intimate situations.

baby boom The unusually large cohort of U.S. children born after the end of World War II, between 1946 and 1964.

barriers to divorce Impediments to a decision to divorce, such as concern about children, religiously grounded objections to divorce, or financial concerns or dependencies.

belligerence A negative communication/relationship behavior that challenges the partner's power and authority.

bereavement A period of mourning after the death of a loved one.

binational family An immigrant family in which some members are citizens or legal residents of the country they migrate to, while others are **undocumented**—that is, they are not legal residents.

binuclear family One family in two household units. A term created to describe a postdivorce family in which both parents remain involved and children are at home in both households.

biosocial perspective Theoretical perspective based on concepts linking psychosocial factors to anatomy, physiology, genetics, and/or hormones as shaped by evolution.

bisexuals People who are sexually attracted to both males and females.

borderwork Interaction rituals that are based on and reaffirm boundaries and differences between girls and boys.

boundary ambiguity When applied to a family, a situation in which it is unclear who is in and who is out of the family.

bride price Money or property that a future groom pays a future bride's family so that he can marry her.

caregiver model of elder abuse and neglect A view of elder abuse or neglect that highlights stress on the caregiver as important to the understanding of abusive behavior.

caregiving "Assistance provided to persons who cannot, for whatever reason, perform the basic activities or instrumental activities of daily living for themselves" (Cherlin 1996, p. 762).

caregiving trajectory The process through which eldercare proceeds, according to which, first, the caregiver becomes concerned about an aging family member, then later begins to give advice to the older family member, and still later takes action to provide needed services.

case study A written summary and analysis of data obtained by psychologists, psychiatrists, counselors, and social workers when working directly with individuals and families in clinical practice. Case studies may be used as sources in scientific investigation and have played a role in the development of certain family theories.

center care Group child care provided in day-care centers for a relatively large number of children.

child abuse Overt acts of aggression against a child, such as beating or inflicting physical injury or excessive verbal derogation. Sexual abuse is a form of physical child abuse. See also **emotional child abuse or neglect.**

child care The care and education of children by people other than their parents. Child care may include before- and after-school care for older children and overnight care when employed parents must travel, as well as day care for preschool children.

child neglect Failure to provide adequate physical or emotional care for a child. See also **emotional child abuse or neglect.**

child support Money paid by the noncustodial parent to the custodial parent to financially support children of a former marital, cohabiting, or sexual relationship.

child-to-parent abuse A form of family violence involving a child's (especially an adolescent's) physical and emotional abuse of a parent.

children's allowance A type of child support that provides a government grant to all families—married or single-parent, regardless of income—based on the number of children they have.

choosing by default Making semiconscious or unconscious choices when one is not aware of all the possible alternatives or when one pursues the path of least resistance. From this perspective, doing nothing about a problem or issue, or making no choice, is making a choice—the choice to do nothing.

choosing knowledgeably Making choices and decisions after (1) recognizing as many options or alternatives as possible, (2) recognizing the social pressures that can influence personal choices, (3) considering the consequences of each alternative, and (4) becoming aware of one's own values.

civil union Legislation that allows any two single adults—including same-sex partners or blood relatives, such as siblings or a parent and adult child—to have access to virtually all marriage rights and benefits on the state level, but none on the federal level. Designed to give same-sex couples many of the legal benefits of marriage while denying them the right to legally marry.

coercive power One of the six power bases, or sources of power. This power is based on the dominant person's ability and willingness to punish the partner either with psychological–emotional or physical abuse or with more subtle methods of withholding affection.

cohabitation Living together in an intimate, sexual relationship without traditional, legal marriage. Sometimes referred to as *living together* or *marriage without marriage*, cohabitation can be a courtship process or an alternative to legal marriage, depending on how partners view it.

collectivist society A society in which people identify with and conform to the expectations of their relatives or clan, who look after their interests in return for their loyalty. The group has priority over the individual. A synonym is *communal society*.

commitment (to intimacy) The determination to develop relationships in which experiences cover many areas of personality, problems are worked through, conflict is expected and seen as a normal part of the growth process, and there is an expectation that the relationship is basically viable and worthwhile.

commitment (Sternberg's triangular theory of love) The short-term decision that one loves someone and the long-term commitment to maintain that love; one dimension of the triangular theory of love.

common law marriage A legal concept whereby cohabiting partners are considered legally married if certain requirements are met, such as showing intent to enter into a marriage and living together as husband and wife for a certain period. Most states have dropped common law marriage, but cohabiting relationships may sometimes have a similar effect on property ownership and custody rights.

communal (expressive) character traits Traits that foster relationships with others, such as warmth, sensitivity, the ability to express tender feelings, and the desire to place concern about others' welfare above self-interest.

communal society See **collectivist society.**

communes Groups of adults and perhaps children who live together, sharing aspects of their lives. Some communes are group marriages, in which members share sex; others are communal families, with several monogamous couples, who share everything except sexual relations and their children.

community-based resources Characteristics, competencies, and means of people, groups, and institutions outside the family that the family may call upon, access, and use to meet their demands.

community divorce Ruptures of relationships and changes in social networks that come about because of divorce.

commuter marriage A marriage in which the two partners live in different locations and commute to spend time together.

companionate marriage The single-earner, breadwinner–homemaker marriage that flourished in the 1950s. Although husbands and wives in the companionate marriage usually adhered to a sharp division of labor, they were supposed to be each other's companion—friends, lovers—in a realization of trends beginning in the 1920s.

concerted cultivation The parenting model, or style, according to which parents often praise and converse with their children, engage them in extracurricular activities, take them on outings, and so on, with the goal of cultivating their child's talents and abilities.

conflict perspective Theoretical perspective that emphasizes social conflict in a society and within families. Power and dominance are important themes.

Conflict Tactics Scale A scale developed by sociologist Murray Straus to assess how couples handle conflict. Includes detailed items on various forms of physical violence.

Confucian training doctrine Concept used to describe Asian and Asian American parenting philosophy that emphasizes blending parental love, concern, involvement, and physical closeness with strict and firm control.

consensual marriages Heterosexual, conjugal unions that have not gone through a legal marriage ceremony.

consummate love A complete love, in terms of Sternberg's triangular theory of love, in which the components of passion, intimacy, and commitment come together.

contempt One of the **Four Horsemen of the Apocalypse** (which see), in which a partner feels that his or her spouse is inferior or undesirable.

co-parenting, co-parents Shared decision making and parental supervision in such areas as discipline and schoolwork or shared holidays and recreation. Can refer to parents working together in a marriage or other ongoing relationship or after divorce or separation.

courtly love Popular during the twelfth century and later, courtly love is the intense longing for someone other than one's marital partner—a passionate and sexual longing that ideally goes unfulfilled. The assumptions of courtly love influence our modern ideas about romantic love.

courtship The process whereby a couple develops a mutual commitment to marriage.

covenant marriage A type of legal marriage in which the bride and groom agree to be bound by a marriage contract that will not let them get divorced as easily as is allowed under no-fault divorce laws.

criticism One of the **Four Horsemen of the Apocalypse** (which see) that involves making disapproving judgments or evaluations of one's partner.

cross-national marriages Marriages in which spouses are from different countries.

crude divorce rate The number of divorces per 1,000 population. See also **refined divorce rate.**

cultural script Set of socially prescribed and understood guidelines for relating to others or for defining role responsibilities and obligations.

culture war Deep cultural conflict, often buttressed by religious belief systems, over matters concerning human sexuality and gender.

custodial grandparent A parent of a divorced, custodial parent.

custodial parent The parent who has legal responsibility for a child after parents divorce or separate. In sole custody, the child resides with the custodial parent. In joint custody, the child may reside primarily with one parent or may live part of the time with each.

custody Primary responsibility for making decisions about a child's upbringing and general welfare.

cyberadultery Marital infidelity or adultery on the Internet.

data collection techniques Ways that data are gathered when doing research; these include interviews and questionnaires, naturalistic observation, focus groups, experiments and laboratory observation, and case studies, among others.

date rape Forced or unwanted sexual contact between people who are on a date. See also **acquaintance rape.**

Defense of Marriage Act (DOMA) Federal statute declaring marriage to be a "legal union of one man and one woman," denying gay couples many of the civil advantages of marriage and relieving states of the obligation to grant reciprocity, or "full faith and credit," to marriages performed in another state.

defensiveness One of the **Four Horsemen of the Apocalypse** (which see) that means preparing to defend oneself against what one presumes is an upcoming attack.

deinstitutionalization of marriage A situation in which time-honored family definitions are changing and family-related social norms are weakening so that they "count for far less" than in the past.

displacement A passive-aggressive behavior in which a person expresses anger with another by being angry at or damaging people or things the other cherishes. See also **passive-aggression.**

divorce divide The gap in divorce rates between college-educated and less-educated women. The divorce rate has declined substantially for college-educated women, but not for less-educated women.

divorce-extended family Kinship ties that form in the wake of a divorce. May include former in-laws, new spouses of one's ex-spouse, and that person's children and kin as part of one kinship system.

divorce mediation A nonadversarial means of dispute resolution by which the couple, with the assistance of a mediator or mediators (frequently a lawyer–therapist team), negotiate the terms of their settlement of custody, support, property, and visitation issues.

domestic partners Partners in an unmarried couple who have registered their partnership with a civil authority and then enjoy some (although not necessarily all) rights, benefits, and entitlements that have traditionally been reserved for marrieds.

domestic violence model of elder abuse and neglect A model that conceptualizes elder abuse as a form of family violence.

double message See **mixed message.**

double remarriage A remarriage in which both partners were previously married.

double standard The standard according to which nonmarital sex or multiple partners are more acceptable for males than for females.

dowry A sum of money or property brought to the marriage by the female.

economic divorce The aspect of divorce that divides the couple into separate economic units, each with its own property, income, control of expenditures, and responsibility for taxes, debts, and so on.

economic hardship perspective (on children's adjustment to divorce) One of the theoretical perspectives concerning the negative outcomes among children of divorced parents. From this perspective, it is the economic hardship brought about by marital dissolution that is primarily responsible for problems faced by children.

egalitarian norm (of marital power) The norm (cultural rule) that husband and wife should have equal power in a marriage.

elder abuse Overt acts of aggression toward the elderly, in which the victim may be physically assaulted, emotionally humiliated, purposefully isolated, or materially exploited.

elder neglect Acts of omission in the care and treatment of the elderly.

elder care Care provided to older generations.

emerging adulthood The youth and young adult stage of life, which is a period of frequent change and exploration.

emotion A strong feeling arising without conscious mental or rational effort, such as joy, reverence, anger, fear, love, or hate. Emotions are neither bad nor good and should be accepted as natural. People can and should learn to control what they do about their emotions.

emotion labor The display of certain emotions that one believes is expected in a given situation, regardless of whether one feels those emotions.

emotional child abuse or neglect A parent or other caregiver's being overly harsh and critical, failing to provide guidance, or being uninterested in a child's needs.

emotional divorce Withdrawing bonding emotions and communication from the marital or other relationship, typically replacing these with alienating feelings and behavior.

emotional intelligence (1) Awareness of what we're feeling so that we can express our feelings more authentically; (2) ability and willingness to repair our moods, not unnecessarily nursing our hurt feelings; (3) healthy balance between controlling rash impulses and being candid and spontaneous; and (4) sensitivity to the feelings and needs of others.

endogamy Marrying within one's own social group. See also **exogamy.**

equality Power or resources divided between partners so that each has the same amount.

equity A standard for distribution of power or resources of partners according to the contribution each person has made to the unit. Another way of characterizing an equitable result is that it is "fair."

eros The love style characterized by intense emotional attachment and powerful sexual feelings or desires. See also **agape, ludus, mania, pragma,** and **storge.**

ethnicity A group's identity based on a sense of a common culture and language.

Euro-American families Families whose members are of European ethnic background.

evolutionary heritage In the biosocial perspective, human behavior is encoded in genetic or other biological features that come to us as members of a species.

exchange balance Balance of rewards and costs in a relationship.

exchange theory Theoretical perspective that sees relationships as determined by the exchange of resources and the reward–cost balance of that exchange. This theory predicts that people tend to marry others whose social class, education, physical attractiveness, and even self-esteem are similar to their own.

exogamy Marrying a partner from outside one's own social group. See also **endogamy**.

expectations of permanence One component of the marriage premise, according to which individuals enter marriage expecting that mutual affection and commitment will be lasting.

expectations of sexual exclusivity The cultural ideal according to which spouses promise to have sexual relations with only each other.

experience hypothesis The idea that the independent variable in a hypothesis is

responsible for changes to a dependent variable. With regard to marriage, the experience hypothesis holds that something about the experience of being married itself causes certain results for spouses. See also the antonym, **selection hypothesis.**

experiential reality Knowledge based on personal experience.

experiment One tool of scientific investigation, in which behaviors are carefully monitored or measured under controlled conditions. Participants are randomly assigned to treatment or control groups.

expert power One of the six power bases, or sources of power. This power stems from the dominant person's superior judgment, knowledge, or ability.

expressive sexuality The view of human sexuality in which sexuality is basic to the humanness of both women and men, all individuals are free to express their sexual selves, and there is no one-sided sense of ownership.

expressive traits See **communal (expressive) character traits.**

extended family Family including relatives besides parents and children, such as aunts or uncles. See also **nuclear family.**

extrusion "[I]ndividuals' being 'pushed out' of their households earlier than normal for members of their cultural group, either because they are forced to leave or because remaining in their households is so stressful that they 'choose' to leave" (Crosbie-Burnett et al. 2005, p. 213).

facilitation of natural growth parenting model Parenting model, or style, in which the parent defines his/her role as allowing the child's abilities to develop naturally, rather than being consciously cultivated.

familistic (communal) values Values that focus on the family group as a whole and on maintaining family identity and cohesiveness.

family Any sexually expressive or parent-child or other kin relationship in which people live together with a commitment in an intimate interpersonal relationship. Family members see their identity as importantly attached to the group, which has an identity of its own. Families today take several forms: single-parent, remarried, dual-career, communal, homosexual, traditional, and so forth. See also **extended family, nuclear family.**

family boundaries Family members' understandings of who is and who is not in the family. Markers, whether material (e.g., doors, fences, communication devices) or social (e.g., symbols of identity, conversational styles and content, time spent together), indicate the boundaries of the family.

"family change" perspective See **"family decline," "family change" perspectives.**

family child care Child care provided in a caregiver's home.

family cohesion That intangible emotional quality that holds groups together and gives members a sense of common identity.

family crisis A situation (resulting from a stressor) in which the family's usual behavior patterns are ineffective and new ones are called for.

"family decline," "family change" perspectives Some family scholars and policy makers characterize late-twentieth-century developments in the family as "decline," while others describe "change." Those who take the "family decline" perspective view such changes as increases in the age at first marriage, divorce, cohabitation, and nonmarital births and the decline in fertility as disastrous for the family as a major social institution. "Family change" scholars and policy makers consider that the family has varied over time. They argue that the family can adapt to recent changes and continue to play a strong role in society.

family development perspective Theoretical perspective that gives attention to changes in the family over time.

family ecology perspective Theoretical perspective that explores how a family influences and is influenced by the environments that surround it. A family is interdependent first with its neighborhood, then with its social–cultural environment, and ultimately with the human-built and physical–biological environments. All parts of the model are interrelated and influence one another.

family foster care Foster care that takes place in a trained and licensed foster parent's home.

family-friendly workplace policies Workplace policies that are supportive of employee efforts to combine family and work commitments.

family function Activities performed by families for the benefit of society and of family members.

family identity Ideas and feelings about the uniqueness and value of one's family unit.

family instability perspective (on children's adjustment to divorce) The thesis that a negative impact of divorce on children is primarily caused by the number of changes in family structure, not by any particular family form. A stable single-parent family may be less harmful to children than a divorce followed by a single-parent family followed by cohabitation, then remarriage, and perhaps a redivorce.

family leave A leave of absence from work granted to family members to care for new infants, newly adopted children, ill children, or aging parents, or to meet similar family needs or emergencies.

family life course development framework Theoretical perspective that follows families through fairly typical stages in the life course, such as through marriage, childbirth, stages of raising children, adult children's leaving home, retirement, and possible widowhood.

family life cycle Stages of family development defined by the addition and subtraction of family members, children's ages, and changes in the family's connection with other social systems.

family of orientation The family in which an individual grows up. Also called *family of origin.*

family of procreation The family that is formed when an individual marries and has children.

family policy All the actions, procedures, regulations, attitudes, and goals of government that affect families.

family preservation A program of support for families in which children have been abused. The support is intended to enable the child to remain in the home safely rather than being placed in foster care.

family stress State of tension that arises when demands tax a family's resources.

family structure The form a family takes, such as nuclear family, extended family, single-parent family, stepfamily, and the like.

family systems theory An umbrella term for a wide range of specific theories. This theoretical perspective examines the family as a whole. It looks to the patterns of behavior and relationships within the family, in which each member is affected by the behavior of others. Systems tend toward equilibrium and will react to change in one part by seeking equilibrium either by restoring the old system or by creating a new one.

family transitions Expected or predictable changes in the course of family life that often precipitate family stress and can result in a family crisis.

family values See **familistic (communal) values.**

fecundity Reproductive capacity; biological capability to have children.

demand/withdraw interaction pattern A cycle of negative verbal expression by one partner, followed by the other partner's withdrawal in the face of the other's demands.

femininities Culturally defined ways of being a woman. The plural conveys the idea that there are varied models of appropriate behavior.

feminist theory Feminist theories are conflict theories. The primary focus of the feminist perspective is male dominance in families and society as oppressive to women. The mission of this perspective is to end this oppression of women (or related pattern of subordination based on social class, race/ethnicity, age, or sexual orientation) by developing knowledge and action that confront this disparity. See also **conflict perspective.**

fertility Births to a woman or category of women (actual births, not reproductive capacity).

fictive kin Family-like relationships that are not based on blood or marriage but on close friendship ties.

filial responsibility A child's obligation to a parent.

flexible scheduling A type of employment scheduling that includes scheduling options such as **job sharing** or **flextime.**

flextime A policy that permits an employee some flexibility to adjust working hours to suit family needs or personal preference.

formal kinship care Out-of-home placement with biological relatives of children who are in the custody of the state.

foster care Care provided to children by other than their parents as a result of state intervention.

Four Horsemen of the Apocalypse Contempt, criticism, defensiveness, and stonewalling—marital communication behaviors delineated by John Gottman that often indicate a couple's future divorce.

free-choice culture Culture or society in which individuals choose their own marriage partners, a choice usually based at least somewhat on romance.

friends with benefits Sexual activity between friends or acquaintances with no expectation of romance or emotional attachment; typically practiced by unattached people who want to have a sexual outlet without "complications."

gay A person whose sexual attraction is to people of the same sex. Used especially for males, but may include both sexes. This term is usually used rather than *homosexual.*

gender Attitudes and behavior associated with and expected of the two sexes. The term sex denotes biology, while *gender* refers to social role. See also **gendered.**

gender model of marriage [13]

gender role Prescription for masculine or feminine behavior. The masculine gender role demands instrumental character traits and behavior, whereas the feminine gender role specifies expressive character traits and behavior.

gender schema theory of gender socialization A framework of knowledge and beliefs about differences or similarities between males and females. Gender schema shape socialization into gender roles.

gender similarities hypothesis Assertion—backed by research—that there are few gender differences in characteristics and abilities.

gendered The way that aspects of people's lives and relationships are influenced by gender.

geographic availability Traditionally known in the marriage and family literature as propinquity or proximity and referring to the fact that people tend to meet potential mates who are present in their regional environment.

gerontologists Social scientists who study aging and the elderly.

GLBT An acronym for gay, lesbian, bisexual, or transgendered; a term commonly used when discussing sexual minorities.

globalization The interdependency of people, organizations, economies, or governments across national borders.

good provider role A specialized masculine role that emerged in this country around the 1830s and that emphasized the husband as the only or primary economic provider for his family. The good provider role had disappeared as an expected masculine role by the 1970s. See also **provider role.**

grandfamilies Families headed by grandparents.

grandparent families Families in which a grandparent acts as primary parent to grandchildren.

group home One type of foster-care setting in which several children are cared for around-the-clock by paid professionals who work in shifts and live elsewhere.

guaranteed child support Type of child support (provided in France and Sweden, for example) in which the government sends to the custodial parent the full amount of support awarded to the child and assumes responsibility for collecting what is owed by the noncustodial parent.

habituation The decreased interest in sex over time that results from the increased accessibility of a sexual partner and the predictability of sexual behavior with that partner.

habituation hypothesis Hypothesis that the decline in sexual frequency over a marriage results from habituation.

Healthy Marriage Initiative (HMI). Federal program initiated in 2004 and targeted to TANF ("welfare") recipients, consisting of workshops on listening, communication, and problem-solving skills, as well as presentations on the value of marriage.

hermaphrodite See **intersexual.**

heterogamy Marriage between partners who differ in race, age, education, religious background, or social class. Compare with **homogamy.**

heterosexism The taken-for-granted system of beliefs, values, and customs that places superior value on heterosexual behavior (as opposed to homosexual) and denies or stigmatizes nonheterosexual relations. This tendency also sees the heterosexual family as standard.

heterosexuals People who prefer sexual partners of the opposite sex.

hierarchical compensatory model of caregiving The idea that elderly people prefer their caregivers in ranked order as follows: an available spouse, adult children, siblings, grandchildren, nieces and nephews, friends, neighbors, and, finally, a formal service provider.

hierarchical parenting Concept used to describe a Hispanic parenting philosophy that blends warm emotional support for children with demand for significant respect for parents and other authority figures, including older extended-family members.

HIV/AIDS HIV is *human immunodeficiency virus,* the virus that causes AIDS, or *acquired immune deficiency syndrome.* AIDS is a sexually transmitted disease involving breakdown of the immune system defense against viruses, bacteria, fungi, and other diseases.

homogamy Marriage between partners of similar race, age, education, religious background, and social class. See also **heterogamy.**

homophobia Fear, dread, aversion to, and often hatred of homosexuals.

homosexuals People who are sexually attracted to people of the same sex. Preferred terms are *gay* or *gay man* for men and *lesbian* for women. See also **gay** and **lesbian.**

hooking up A sexual encounter between young people with the understanding that there is no obligation to see each other again or to endow the sexual activity with emotional meaning. Usually there is a group or network context for hooking up; that is, the individuals meet at a social event or have common acquaintances. On some college campuses and elsewhere, hooking up has replaced dating, which is courtship-oriented socializing and sexual activity.

hormonal processes Chemical processes within the body regulated by such hormones as testosterone (a "male" hormone) and estrogen (a "female" hormone). Hormonal processes are thought to shape behavior, as well as physical development and reproductive functions, although experts disagree as to their impact on behavior.

hormones Chemical substances secreted into the bloodstream by the endocrine glands.

household As a Census Bureau category, a household is any group of people residing together.

househusbands Men who take a full-time family care role, rather than being employed; male counterparts to housewives.

hyperparenting The situation in which parents are excessively involved in their children's lives.

impaired fertility Describes the situation of a woman who is not able to succeed in having a child due to a physical barrier or an inability to carry a pregnancy to full term.

incest Sexual relations between closely related individuals.

income effect Occurs when an increase in income contributes to the stability of a marriage by giving it a more adequate financial basis.

income-to-needs ratio An assessment of income as to the degree it meets the needs of the individual, family, or household.

incomplete institution Cherlin's description of a remarried family due to cultural ambiguity.

independence effect Occurs when an increase in income leads to marital dissolution because the partners are better able to afford to live separately.

individualism The cultural milieu that emerged in Europe with industrialization and that values personal self-actualization and happiness along with individual freedom.

individualistic society Society in which the main concern is with one's own interests (which may or may not include those of one's immediate family).

individualistic (self-fulfillment) values Values that encourage self-fulfillment, personal growth, autonomy, and independence over commitment to family or other communal needs.

individualized marriage Concept associated with the argument that contemporary marriage in the United States and other fully industrialized Western societies is no longer institutionalized. Four interrelated characteristics distinguish individualized marriage: (1) it is optional; (2) spouses' roles are flexible—negotiable and renegotiable; (3) its expected rewards involve love, communication, and emotional intimacy; and (4) it exists in conjunction with a vast diversity of family forms.

induced abortion The scientific term for what is commonly termed *abortion.* The removal of the fetus from the uterus is "induced"; that is, it requires a deliberate surgical or pharmaceutical act. What we commonly call "miscarriage" is termed *spontaneous abortion* in scientific language because it happens without any initiative on the part of individuals or medical personnel and, in fact, is usually not desired.

informal adoption Children are taken into a home and considered to be children of the parents, although the "adoption" is not legally formalized.

informal caregiving Unpaid caregiving, provided personally by a family member.

informational power One of the six power bases, or sources of power. This power is based on the persuasive content of what the dominant person tells another individual.

informed consent A requirement of research involving human subjects; before agreeing to participate, subjects are told the purpose of the research and the procedure and whether any risk is involved in participation. The process of obtaining informed consent is supervised by an *institutional review board*.

in-home caregiver A caregiver who provides child care in the child's home, either coming in by the day or as a live-in caregiver.

insecure/anxious attachment style One of three attachment styles in attachment theory, this style entails concern that the beloved will disappear, a situation often characterized as "fear of abandonment."

instability hypothesis The idea that the instability of postdivorce family structure(s) can be damaging to children. According to this hypothesis, a stable single-parent family might be better for a child than a single-parent family succeeded by a remarried family.

institution See **social institution.**

institutional marriage Marriage as a social institution based on dutiful adherence to the time-honored *marriage premise* (which see), particularly the norm of permanence. "Once ensconced in societal mandates for permanence and monogamous sexual exclusivity, the institutionalized marriage in the United States was centered on economic production, kinship network, community connections, the father's authority, and marriage as a functional partnership rather than a romantic relationship. . . . Family tradition, loyalty, and solidarity were more important than individual goals and romantic interest" (Doherty 1992, p. 33). Also referred to as *institutionalized marriage*.

institutional review board (IRB) A local body of experts and community representatives established by a university or research organization to scrutinize research proposals for adherence to professional ethical standards for the protection of human subjects.

instrumental traits See **agentic (instrumental) character traits.**

interaction–constructionist perspective Theoretical perspective that focuses on internal family dynamics; the ongoing action among and response to one another of family members.

interactionist perspective on human sexuality A perspective, derived from symbolic interaction theory, which holds that sexual activities and relationships are shaped by the sexual scripts available in a culture.

interethnic marriages Marriages between spouses who are not defined as of different races but do belong to different ethnic groups.

interference with visitation A legal term for actions of a custodial parent that hinder the noncustodial parent's scheduled visitation with a child. Such interference may consist of alleging (falsely) that the child is too ill to visit, has other plans, is not at home at the pickup time, and the like. Some states have legislated penalties for interference with visitation, whereas others make little effort to enforce the noncustodial parent's right to contact with her or his child. Often allegations of interference with visitation are difficult to evaluate.

intergenerational transmission of divorce risk The tendency for children of divorced parents to have a greater propensity to divorce than children from intact families.

internalize Make a part of oneself. Often refers to the socialization process by which children learn their parents' norms and values to the point that they become the child's own views.

interparental conflict perspective (on children's adjustment to divorce) One of the theoretical perspectives concerning the negative outcomes among children of divorced parents. From the interparental conflict perspective, the conflict between parents before, during, and after the divorce is responsible for the lowered well-being of the children of divorce.

interpersonal exchange model of sexual satisfaction A view of sexual relations, derived from exchange theory, that sees sexual satisfaction as shaped by the costs, rewards, and expectations of a relationship and the alternatives to it.

interracial marriages Marriages of a partner of one (socially defined) race to someone of a different race.

intersexual A person whose genitalia, secondary sex characteristics, hormones, or other physiological features are not unambiguously male or female.

intimacy (Sternberg's triangular theory of love) Committing oneself to a particular other and honoring that commitment in spite of some personal sacrifices while sharing one's inner self with the other. Intimacy requires interdependence.

intimate partner power Power in a relationship, whether of married or unmarried intimate partners.

intimate partner violence Violence against current or former spouses, cohabitants, or sexual or relationship partners.

intimate terrorism Abuse that is almost entirely male and that is oriented to controlling the partner through fear and intimidation.

intimate terrorism A man's systematic use of verbal or physical violence to gain or maintain control over his female partner.

involuntary infertility Situation of a couple or individual who would like to have a baby but cannot. Involuntary infertility is medically diagnosed when a woman has tried for twelve months to become pregnant without success.

job sharing Two people sharing one job.

joint custody A situation in which both divorced parents continue to take equal responsibility for important decisions regarding their child's general upbringing.

kin Parents and other relatives, such as in-laws, grandparents, aunts and uncles, and cousins. See also **extended family.**

labor force A social invention that arose with the industrialization of the nineteenth century, when people characteristically became wage earners, hiring out their labor to someone else.

laboratory observation Observation of behavior, including verbal behavior, in an environment controlled by the researcher. For example, a researcher may ask a father, a mother, and an adolescent to discuss an issue, solve a problem, or play a game and observe their responses and interactions, which may be audio- or video-recorded.

laissez-faire parenting style Overly permissive parenting. Children set their own standards for behavior, with little or no parental guidance or authority. Parents are indulgent but not necessarily involved in a supportive way with the child's everyday activities and problems.

latent kin matrix "A web of continually shifting linkages that provide the potential for activating and intensifying close kin relationships" (Riley 1983, p. 441).

legal divorce The dissolution of a marriage by the state through a court order terminating the marriage.

legitimate power One of the six power bases, or sources of power. Legitimate power stems from the more dominant individual's ability to claim authority, or the right to request compliance.

leisure gap Men enjoy more free time to spend engaged in activities of their own choosing than do women.

lesbian A woman who is sexually attracted to other women. This term is usually used rather than *homosexual.*

leveling Being transparent, authentic, and explicit about how one truly feels, especially concerning the more conflictive or hurtful aspects of an intimate relationship. Among other things, leveling between intimates implies self-disclosure and commitment (to intimacy).

Levinger's model of divorce decisions This model, derived from exchange theory, presents a decision to divorce as involving a calculus of the *barriers* to divorce (e.g., concerns about children and finances; religious prohibitions), the *rewards* of the marriage, and *alternatives* to the marriage (e.g., can the divorced person anticipate a new relationship, career development, or a single life that will be more rewarding and less stressful than the marriage?).

life chances The opportunities that exist for a social group or an individual to pursue education and economic advancement, to secure medical care and preserve health, to marry and have children, to have material goods and housing of desired quality, and so forth.

life stress perspective (on children's adjustment to divorce) One of the theoretical perspectives concerning the negative outcomes among children of divorced parents. From the life stress perspective, divorce involves the same stress for children as for adults. Divorce is not one single event but a process of stressful events—moving, changing schools, and so on.

limerence A psycho-emotional situation in which one obsesses about another person (the "limerent object") and yearns for reciprocation but has little, if any, concern for the other person's well-being. Not to be confused with love or the early, anxious stage of discovering love.

longitudinal study One technique of scientific investigation in which researchers study the same individuals or groups over an extended period, usually with periodic surveys.

looking-glass self The concept that people gradually come to accept and adopt as their own the evaluations, definitions, and judgments of themselves that they see reflected in the faces, words, and gestures of those around them.

love A deep and vital emotion resulting from significant need for satisfaction, coupled with a caring for and acceptance of the beloved, and resulting in an intimate relationship. Love may make the world go 'round, but it's a lot of work, too.

love style A distinctive characteristic or personality that loving or lovelike relationships can take. One social scientist has distinguished six: **agape, eros, ludus, mania, pragma,** and **storge.**

ludus The love style that focuses on love as play and on enjoying many sexual partners rather than searching for one serious relationship. This love style emphasizes the recreational aspect of sexuality. See also **agape, eros, mania, pragma,** and **storge.**

male dominance The cultural idea of masculine superiority; the idea that men should and do exercise the most control and influence over society's members.

mania The love style that combines strong sexual attraction and emotional intensity with extreme jealousy and moodiness, in which manic partners alternate between euphoria and depression. See also **agape, eros, ludus, pragma,** and **storge.**

manipulating Seeking to control the feelings, attitudes, and behavior of one's partner or partners in underhanded ways rather than by assertively stating one's case.

marital power Power exercised between spouses.

marital rape A husband's forcing a wife to submit to sexual contact that she does not want or that she finds offensive.

marital stability The quality or situation of remaining married.

market approach to child care Child-care arrangement of working parents; other people are hired to care for children while parents are at their jobs.

marriage market The sociological concept that potential mates take stock of their personal and social characteristics and then comparison shop or bargain for the best buy (mate) they can get.

marriage premise By getting married, partners accept the responsibility to keep each other primary in their lives and to work hard to ensure that their relationship continues.

martyring Doing all one can for others while ignoring one's own legitimate needs. Martyrs often punish the person to whom they are martyring by letting the person know "just how much I put up with."

masculinities Culturally defined ways of being a man. The plural conveys the idea that there are varied models of appropriate behavior.

mate selection risk The idea that children of divorce may be likely to select spouses who are unlikely to make good marriage partners.

Miller's typology of urban Native American families

> **bicultural:** Families that develop a successful blend of native beliefs and practices with those adaptive to living in urban settings.

> **marginal:** Urban families that have become alienated from both Indian and mainstream American cultures.

> **traditional:** Families that retain primarily Indian ways in their urban environment.

> **transitional:** Families that are tending to assimilate to the white working class.

minority stress The negative consequences of chronic stress that result from minority-group stigmatization such as discrimination and prejudice. Outcomes include suffering from psychological disorders and increased health risks. Manifestations include self-loathing, fear of attacks, substance abuse, and so on.

mixed message Two simultaneous messages that contradict each other; also called a *double message.* For example, society gives us mixed messages regarding family values and individualistic values and about premarital sex. People, too, can send mixed messages, as when a partner says, "Of course I always like to talk with you" while turning up the TV.

modern sexism Sexism that takes the form of (a) denial of the existence of discrimination against women, (b) resentment of complaints about discrimination, and (c) resentment of "special favors" for women.

motherhood penalty Negative lifetime impact on earnings for women who raise children.

mothering approach to child care A family's child care arrangement that gives preference to the mother's caregiving role. A couple balances nonemployment of the mother with extra jobs or hours for the father, or, if the mother must work to maintain the family economically, her employment role is minimized.

multipartnered fertility Having children in more than one marriage or relationship.

mutually economically dependent spouses (MEDS) Describes a dual-earner marriage in which each partner earns between 40 and 59 percent of the family income.

nadir of family disorganization Low point of family disorganization when a family is going through a family crisis.

nanny An in-home child care worker who cares for a family's children either on a live-in basis or by the day; may include traveling with the family.

naturalistic observation A technique of scientific investigation in which a researcher lives with a family or social group or spends extensive time with them, carefully recording their activities, conversations, gestures, and other aspects of everyday life.

near peer marriage (Schwartz's typology) Couples who believe in partner equality but fall short of a 60–40 division of household labor, usually because of the need for the husband's higher earnings.

negative affect Showing emotion(s) defined as negative, such as anger, sadness, whining, disgust, tension, fear, and/or belligerence.

neo-sexism Same as **modern sexism**.

neotraditional families Families that value traditional gender roles and organize their family life in these terms as far as practicable. Formal male dominance is softened by an egalitarian spirit.

no-fault divorce The legal situation in which a partner seeking a divorce no longer has to prove "fault" according to a state's legal definition but only needs to assert "irretrievable breakdown" or "irreconcilable differences." Sometimes termed **unilateral divorce.**

noncustodial grandparent A parent of a divorced, noncustodial parent.

no-power A situation in which partners are equally able to influence each other and, at the same time, are not concerned about their relative power vis-à-vis each other. No-power partners negotiate and compromise instead of trying to win.

normative order hypothesis The thesis that to proceed through the family life cycle "on time" provides the best chance for a good adjustment in each family stage. Applies as well to the sequencing of education, job, marriage, and parenthood.

nuclear family A family group comprising only the wife, the husband, and their children. See also **extended family.**

nuclear-family model monopoly The cultural assumption that the first-marriage family is the "real" model of family living, with all other family forms viewed as deficient.

occupational segregation The distribution of men and women into substantially different occupations. Women are overrepresented in clerical and service work, for example, whereas men dominate the higher professions and the upper levels of management.

occupations Employment—that is, work for pay—as contrasted with unpaid household work.

"on-time" transition Moving from one family life cycle stage to another according to the most common cultural pattern.

opportunity costs (of children) The economic opportunities for wage earning and investments that parents forgo when raising children.

opting out A woman's leaving the labor force, permanently or temporarily, in order to devote full time to child-raising.

para-parent An unrelated adult who informally plays a parentlike role for a child.

parental adjustment perspective (on children's adjustment to divorce) One of the theoretical perspectives concerning the negative outcomes among children of divorced parents. From the parental adjustment perspective, the parent's child-raising skills are impaired as a result of the divorce, with probable negative consequences for the children.

parental loss perspective (on children's adjustment to divorce) One of the theoretical perspectives concerning the negative outcomes among children of divorced parents. From the parental loss perspective, divorce involves the absence of a parent from the household, which deprives children of the optimal environment for their emotional, practical, and social support.

parenting alliance The degree to which partners agree with and support each other as parents.

parenting approach to child care In this approach, child care is shared by the parents on as equal a basis as possible. Working parents try to restructure their employment arrangements to make this possible.

parenting style A general manner of relating to and disciplining children.

passion (Sternberg's triangular theory of love) The drives that lead to romance, physical attraction, sexual consummation, and so on in a loving relationship; one dimension of the triangular theory of love.

passive-aggression Expressing anger at some person or situation indirectly, through nagging, nitpicking, or sarcasm, for example, rather than directly and openly. See also **displacement, sabotage.**

patriarchal norm (of marital power) The norm (cultural rule) that the man should be dominant in a marital relationship.

patriarchal sexuality The view of human sexuality in which men own everything in the society, including women and women's sexuality. Males' sexual needs are emphasized while females' needs are minimized.

patriarchy A social system in which males are dominant.

peer marriage (Schwartz's typology) Couples who have a close-to-equal split of household chores and money management and who consider themselves to have equal status in the marriage or cohabiting union.

period of family disorganization That period in a family crisis, after the stressor event has occurred, during which family morale and organization slump and habitual roles and routines become nebulous.

permissiveness with affection The standard that permits nonmarital sex for women and men equally, provided they have a fairly stable, affectionate relationship.

permissiveness without affection The standard that allows nonmarital sex for women and men regardless of how much stability or affection exists in their relationship. Also called the *recreational standard.*

permissive parenting style One of three parenting styles in this schema, permissive parenting gives children little parental guidance.

pileup (stressor overload) Concept from family stress and crisis theory that refers to the accumulation of family stressors and prior hardships.

pleasure bond The idea, from Masters and Johnson's book by the same name, that sexual expression between intimates is one way of expressing and strengthening the emotional bond between them.

pluralistic family Term used to designate the contemporary family, characterized by "tolerance and diversity, rather than a single family ideal" (Doherty 1992, p. 35). Taking many forms, the pluralistic family is also referred to as the *postmodern family.*

polyamory A marriage system in which one or both spouses retain the option to sexually love others in addition to their spouses.

polyandry A marriage system in which a woman has more than one spouse.

polygamy A marriage system in which a person takes more than one spouse.

polygyny A marriage system in which one man has multiple wives; a marriage of a woman with plural husbands is termed **polyandry**.

pool of eligibles A group of individuals who, by virtue of background or social status, are most likely to be considered eligible to make culturally compatible marriage partners.

positive affect The expression, either verbal or nonverbal, of one's feelings of affection toward another.

postmodern family Term used to describe the situation in which (1) families today exhibit multiple forms, and (2) new or altered family forms continue to emerge or develop.

postmodern theory Theoretical perspective that largely analyzes social interaction (discourse or narrative) in order to demonstrate that a phenomenon is socially constructed.

power The ability to exercise one's will. *Personal power,* or *autonomy,* is power exercised over oneself. *Social power* is the ability to exercise one's will over others.

power politics Power struggles between spouses in which each seeks to gain a power advantage over the other; the opposite of a **no-power** relationship.

pragma The love style that emphasizes the practical, or pragmatic, element in human relationships and involves the rational assessment of a potential (or actual) partner's assets and liabilities. See also **agape, eros, ludus, mania,** and **storge.**

primary group A group, usually relatively small, in which there are close, face-to-face relationships or equivalent ties that are technologically mediated. The family and a friendship group are primary groups. See also **secondary group.**

primary parent Parent who takes full responsibility for meeting the child's physical and emotional needs by providing the major part of the child's care directly or by managing the child's care by others or by doing both.

principle of least interest The postulate that the partner with the least interest in the relationship is the one who is more apt to control the relationship and to exploit the other.

private face of family The aspect of the family that provides individuals with intimacy, emotional support, and love.

private safety net Social support from family and friends, rather than from public sources.

pronatalist bias A cultural attitude that takes having children for granted.

provider role A term for the family role involving wage work to support the family. May be carried out by one spouse or partner only or by both.

psychic divorce Regaining psychological autonomy after divorce; emotionally separating oneself from the personality and influence of the former spouse.

psychic intimacy The sharing of people's minds and feelings. Psychic intimacy may or may not involve sexual intimacy.

psychological control Control over others by use of manipulative strategies, such as inducing guilt or withdrawing signs of affection.

psychological parent The parent, usually but not necessarily the mother, who assumes principal responsibility for raising the child.

public face of family The aspect of the family that produces public goods and services.

race A group or category thought of as representing a distinct biological heritage. In reality, there is only one human race. "Racial" categories are social constructs; the so-called races do not differ significantly in terms of basic biological makeup. But "racial" designations nevertheless have social and economic effects and cultural meanings.

race socialization The socialization process that involves developing a child's pride in his or her cultural heritage while warning and preparing him or her about the possibilities of encountering discrimination.

rape myths Beliefs about rape that function to blame the victim and exonerate the rapist.

rapport talk In Deborah Tannen's terms, this is conversation engaged in by women aimed primarily at gaining or reinforcing rapport or intimacy. See also **report talk.**

redivorce An emerging trend in U.S. society. Redivorces take place more rapidly than first divorces so that many who divorce (and their children) can expect several rapid and emotionally significant transitions in lifestyle and family unit.

referent power One of the six power bases, or sources of power. In a marriage or relationship, this form of power is based on one partner's emotional identification with the other and his or her willingness to agree to the other's decisions or preferences.

refined divorce rate Number of divorces per 1,000 married women over age fifteen. See also **crude divorce rate.**

relationship ideologies Expectations for closeness and/or distance as well as ideas about how partners should play their roles.

relatives of divorce Kinship ties established by marriage but retained after the marriage is dissolved—for example, the relationship of a former mother-in-law and daughter-in-law.

remarriages Marriages in which at least one partner has already been divorced or widowed. Remarriages are becoming increasingly common for Americans.

replacement level (of fertility) The average number of births per woman (a total fertility rate of 2.1) necessary to replace the population.

report talk In Deborah Tannen's terms, this is conversation engaged in by men aimed primarily at conveying information. See also **rapport talk.**

resilience The ability to recover from challenging situations.

resilient families Families that emphasize mutual acceptance, respect, and shared values; members rely on one another for emotional support.

resilient individuals Individuals with the capacity to recover from or rise above adverse situations and events.

resource hypothesis Hypothesis (originated by Robert Blood and Donald Wolfe) that the relative power between wives and husbands results from their relative resources as individuals.

resources In exchange theory or intimate partner power analysis, the assets an individual can bring to the relationship. Resources can be material (e.g., income, gifts) or nonmaterial (e.g., emotional support, practical assistance, personality qualities).

resources in cultural context The effect of resources on marital power depends on the cultural context. In a traditional society, norms of patriarchal authority may override personal resources. In a fully egalitarian society, a norm of intimate partner and marital equality may override personal resources. It is in a transitional society that the resource hypothesis is most likely to shape marital power relations.

reward power One of the six power bases, or sources of power. With regard to marriage or partner relationships, this power is based on an individual's ability to give material or nonmaterial gifts and favors to the partner.

rewards and costs In exchange theory or related theoretical analyses, the benefits and disadvantages of a relationship.

role The expectations associated with a particular position in society or in a family. The mother role, for example, calls for its occupant to provide physical care, emotional nurturance, social guidance, and the like to her children.

role ambiguity The situation in which there are few clear guidelines regarding what responsibilities, behaviors, and emotions family members are expected to exhibit.

role-making Improvising a course of action as a way of enacting a role. In role-making, we may use our acts to alter the traditional expectations and obligations associated with a role. This concept emphasizes the variability in the ways different individuals enact a particular role.

role-taking Role-taking has two meanings. It can mean playing a role associated with a status one occupies, such as taking the mother role when one has a child. It can also mean acting out a role that is not, or not yet, one's own, as when children play "mommy" or "daddy" or "police officer."

sabotage A passive-aggressive action in which a person tries to spoil or undermine some activity another has planned. Sabotage is not always consciously planned. See also **passive-aggression.**

sandwich generation Middle-aged (or older) individuals, usually women, who are sandwiched between the simultaneous responsibilities of caring for their dependent children (sometimes young adults) and aging parents.

science "A logical system that bases knowledge on . . . systematic observation,

empirical evidence, facts we verify with our senses" (Macionis 2006, p. 15).

scientific investigation In social science, the systematic gathering of information—using surveys, experiments, naturalistic observation, archival historical material, and case studies—from which it is often possible to generalize with a significant degree of predictability. Data collection and analysis are usually guided by theory or earlier scientific observations. They point to theory modification and a greater understanding of the phenomenon being studied.

second shift Sociologist Arlie Hochschild's term for the domestic work that employed women must perform after coming home from a day on the job.

secondary group A group, often large and geographically dispersed, characterized by distant, practical relationships. An impersonal society is characterized by secondary groups and relations. See also the opposite, **primary group.**

secure attachment style One of three attachment styles in attachment theory, this style involves trust that the relationship will provide necessary and ongoing emotional and social support.

segmented assimilation Assimilation may vary within an immigrant stream. Immigrants with professional education and skills or from favored national origin groups or both may do very well economically and socially and become culturally integrated. Other immigrants may not have the educational background or other human capital necessary to advance in the new environment and may even experience downward mobility.

selection hypothesis The idea that many of the changes found in a dependent variable, which might be assumed to be associated with the independent variable, are really due to sample selection. For instance, the selection hypothesis posits that many of the benefits associated with marriage—for example, higher income and wealth, along with better health—are not necessarily due to the fact of being married but, rather, to the personal characteristics of those who choose—or are selected into—marriage. Similarly, the selection hypothesis posits that many of the characteristics associated with cohabitation result not from the practice of cohabiting itself but from the personal characteristics of those who choose to cohabit. See also the antonym, **experience hypothesis.**

self-care An approach to child care for working parents in which the child is at

home or out without an adult caretaker. Parents may be in touch by phone.

self-concept The basic feelings people have about themselves, their characteristics and abilities, and their worth; how people think of or view themselves.

self-disclosure Letting others see one as one really is. Self-disclosure demands authenticity. Also see **self-revelation.**

self-identification theory A theory of gender socialization, developed by psychologist Lawrence Kohlberg, that begins with a child's categorization of self as male or female. The child goes on to identify sex-appropriate behaviors in the family, media, and elsewhere and to adopt those behaviors.

self-revelation Gradually sharing intimate information about oneself. Also see **self-disclosure**.

self-worth Part of a person's self-concept that involves feelings about one's own value; also called *self-esteem*.

seven-stage model of stepfamily development Model of stepfamily progression that proceeds through the following stages: fantasy, immersion, awareness, mobilization, action, contact, and resolution.

sex Refers to biological characteristics—that is, male or female anatomy or physiology. The term **gender** refers to the social roles, attitudes, and behavior associated with males or females.

sex ratio The number of men per 100 women in a society. If the sex ratio is above 100, there are more men than women; if it is below 100, there are more women than men.

sexting Using cell phones to send sexually explicit images or messages to others.

sexual abuse A form of child abuse that involves forced, tricked, or coerced sexual behavior—exposure, unwanted kissing, fondling of sexual organs, intercourse, rape, and incest—between a minor and an older person.

sexual intimacy A level of interpersonal interaction in which partners have a sexual relationship. Sexual intimacy may or may not involve psychic intimacy.

sexual orientation The attraction an individual has for a sexual partner of the same or opposite sex.

sexual responsibility The assumption by each partner of responsibility for his or her own sexual response.

sexual scripts *Scripts* are culturally written patterns or "plots" for human behavior.

Sexual scripts offer reasons for having sex and designate who should take the sexual initiative, how long an encounter should last, what positions are acceptable, and so forth.

shared parenting Mother and father, or two same-sex parents, who both take full responsibility as parents.

shift work As defined by the Bureau of Labor Statistics, any work schedule in which more than half of an employee's hours are before 8 a.m. or after 4 p.m.

sibling violence Family violence that takes place between siblings (brothers and sisters).

single mothers by choice Women who intentionally become mothers, although they are not married or with a partner. They are typically older, with economic and educational resources that enable them to be self-supporting.

single mothers by circumstance Women who become single mothers in ways other than by purposeful choice.

situational couple violence Mutual violence between partners that often occurs in conjunction with a specific argument. It involves fewer instances, is not likely to escalate, and tends to be less severe in terms of injuries.

social capital perspective (on parenthood) Motivation for parenthood in anticipation of the links parenthood provides to social networks and their resources.

social class Position in the social hierarchy, such as *upper class, middle class, working class,* or *lower class.* Can be viewed in terms of such indicators as education, occupation, and income or analyzed in terms of status, respect, and lifestyle.

social fathers Males who are not a biological father but are performing the role of father, such as a stepfather.

social institution A system of patterned and predictable ways of thinking and behaving—beliefs, values, attitudes, and norms—concerning important aspects of people's lives in society. Examples of major social institutions are the family, religion, government, the economy, and education.

social learning theory (of gender socialization) According to this theory, children learn gender roles as they are taught or modeled by parents, schools, and the media.

socialization The process by which society influences members to internalize attitudes, beliefs, values, and expectations.

socioeconomic status One's position in society, measured by educational achievement, occupation, and/or income.

spectatoring A term Masters and Johnson coined to describe the practice of emotionally removing oneself from a sexual encounter in order to watch oneself and see how one is doing.

spousal support Economic support of a separated spouse or ex-spouse by the other spouse ordered by a court following separation or divorce.

status exchange hypothesis Regarding interracial/interethnic marriage, the argument that an individual might trade his or her socially defined superior racial/ethnic status for the economically or educationally superior status of a partner in a less-privileged racial/ethnic group.

stepmother trap The conflict between two views: Society sentimentalizes the stepmother's role and expects her to be unnaturally loving toward her stepchildren but at the same time views her as a wicked witch.

stonewalling One of the **Four Horsemen of the Apocalypse** (which see) that involves refusing to listen to a partner's complaints.

storge An affectionate, companionate style of loving. See also **agape, eros, ludus, mania,** and **pragma.**

stress model of parental effectiveness The idea that stress experienced by parents causes parental frustration, anger, and depression, increasing the likelihood of household conflict and leading to poorer parenting practices.

stressors Precipitating events that cause a crisis; they are often situations for which the family has had little or no preparation. See also **ABC-X model.**

stressor overload A situation in which an unrelenting series of small crises adds up to a major crisis.

stress-related growth Personal growth and maturity attained in the context of a stressful life experience such as divorce.

structural antinatalism The structural, or societal, conditions in which bearing and raising children is discouraged either overtly or—as may be the case in the United States—covertly through inadequate support for parenting.

structural constraints Economic and social forces that limit options and, hence, personal choices.

structure–functional perspective Theoretical perspective that looks to the functions that institutions perform for society and the structural form of the institution.

survey A technique of scientific investigation using questionnaires or brief face-to-face interviews or both. An example is the U.S. census.

swinging A marriage agreement in which couples exchange partners to engage in purely recreational sex.

symbolic interaction theory (of gender socialization) Uses the concepts of Charles Cooley (primary group, looking-glass self) and George Herbert Mead ("me" and "I," "play, the game, and the generalized other") to explain how children are socialized into culturally defined gender roles.

system A combination of elements or components that are interrelated and organized as a whole. The human body is a system, as is a family.

Temporary Assistance for Needy Families (TANF) Federal legislation that replaces Aid to Families with Dependent Children and whereby government welfare assistance to poor parents is limited to five years for most families, with most adult recipients required to find work within two years.

theoretical perspective A way of viewing reality, or a lens through which analysts organize and interpret what they observe. Researchers on the family identify those aspects of families that are of interest to them, based on their own theoretical perspective.

theory of complementary needs Theory developed by social scientist Robert Winch suggesting that we are attracted to partners whose needs complement our own. In the positive view of this theory, we are attracted to others whose strengths are harmonious with our own so that we are more effective as a couple than either of us would be alone.

total fertility rate For a given year, the number of births that women would have over their reproductive lifetimes if all women at each age had babies at the rate for each age group that year; can be calculated for social or age categories as well as for nations as a whole.

traditional sexism Beliefs that men and women are essentially different and should occupy different social roles, that women are not as fit as men to perform certain tasks and occupations, and that differential treatment of men and women is acceptable.

traditionals (Schwartz's typology) Marriages or domestic partnerships in which the man dominates all areas of decision making except children. He is the primary breadwinner and she is the primary homemaker, even if employed. In Schwartz's typology, both spouses favor this arrangement.

trailing spouse The spouse of a relocated employee who moves with the other spouse.

transgendered A person who has adopted a gender identity that differs from sex/gender as recorded at birth; a person who declines to identify as either male or female.

transition to parenthood The circumstances involved in assuming the parent role.

transitional egalitarian situation (of marital power) Marriages or domestic partnerships in which neither patriarchal nor egalitarian norms prevail. The couple negotiate relationship power, with the relative resources of each individual playing an important role in the outcome.

transnational family A family of immigrants or immigrant stock that maintains close ties with the sending country. Identity and behavior connect the immigrant family to the new country and the old, and their social networks cross national boundaries.

transsexual An individual who has begun life identified as a member of one sex, but later comes to believe he or she belongs to the other sex. The person may undertake surgical reconstruction to attain a body type closer to that of the desired sex.

triangular theory of love Robert Sternberg's theory that consummate love involves three components: intimacy, passion, and commitment.

two-career marriage Marriage in which both partners have a strong commitment to the lifetime development of both careers. Also called *dual-career couple* or *dual-career family*.

two-earner marriages Marriages in which the wife as well as the husband is employed, but her work is not viewed as a lifetime career. His may be viewed as a "job" rather than a career, as well. Sometimes termed *dual-earner marriage* or *two-paycheck marriage*.

undocumented immigrant The preferred term for "illegal" immigrants, those who are present in a country but are not citizens or legal residents. The implication of the term *undocumented* (compared to *illegal*) is that immigrants may or should have legitimate claims to asylum or residence even if these have not been formally recognized.

unilateral divorce A divorce can be obtained under the no-fault system by one partner even if the other partner objects. The term *unilateral divorce* emphasizes this feature of current divorce law. See also **no-fault divorce.**

unpaid family work The necessary tasks of attending to both the emotional needs of all family members and the practical needs of dependent members, such as children or elderly parents, and maintaining the family domicile.

value of children perspective (on parenthood) Motivation for parenthood because of the rewards, including symbolic rewards, that children bring to parents.

voluntary childlessness The deliberate choice not to become a parent.

vulnerable families Families that have a low sense of common purpose, feel in little control over what happens to them, and tend to cope with problems by showing diminished respect and/or understanding for each other.

wage gap The persistent difference in earnings between men and women.

wheel of love An idea developed by Ira Reiss in which love is seen as developing through a four-stage, circular process, including rapport, self-revelation, mutual dependence, and personality need fulfillment.

References

AAUW Educational Foundation. 1999. *Gender Gaps: Where Schools Still Fail Our Children.* New York: Marlowe.

AAUW Educational Foundation. 2006. *Drawing the Line: Sexual Harassment on Campus Study.* Washington, DC: American Association of University Women. January 26. Retrieved February 5, 2007 (www.aauw.org).

Aasen, Eric. 2009. "More Men Becoming Stay-at-home Dads." *The Dallas Morning News,* July 29.

Abbey, L. 2009. "Elder Abuse and Neglect: When Home is Not Safe." *Clinics in Geriatric Medicine* 25(1):47–61.

Abboud, Soo Kim. 2006. *Top of the Class: How Asian Parents Raise High Achievers, and How You Can Too.* New York: Berkley Books.

Aber, J. Lawrence. 2007. *Child Development and Social Policy.* Washington, DC: American Psychological Association.

Abma, Joyce C. and Gladys M. Martinez. 2006. "Childlessness among Older Women in the United States: Trends and Profiles." *Journal of Marriage and Family* 68(4):1045–56.

Abma, Joyce C., Gladys M. Martinez, William D. Mosher, and D. S. Dawson. 2004. "Teenagers in the United States: Sexual Activity, Contraceptive Use, and Childbearing 2002." *Vital Health Statistics* 23(24). Hyattsville, MD: National Center for Health Statistics. December.

"The Aboriginal Aesthetic." 2009. *The Globe and Mail,* October 19. Retrieved December 27, 2009 (www.lexisnexis.com).

Abraham, Margaret. 1995. "Ethnicity, Gender, and Marital Violence: South Asian Women's Organizations in the U.S." *Gender and Society* 9:450–68.

———. 2000. *Speaking the Unspeakable: Marital Violence among South Asian Immigrants in the United States.* New Brunswick, NJ: Rutgers University Press.

Abrams, Michael. 2007. "Born Gay?" *Discover* 28:58–83.

"The Abstinence-Only Delusion" (editorial). 2007. *New York Times,* April 28.

Achen, Alexandra C. and Frank P. Stafford. 2005. *Data Quality of Housework Hours in the Panel Study of Income Dynamics: Who Really Does the Dishes?* Ann Arbor, MI: Institute for Social Research, University of Michigan.

Ackerman, Brian P., Kristen Schoff D'Eramo, Lina Umylny, David Schultz, and Carroll E. Izard. 2001. "Family Structure and the Externalizing Behavior of Children from Economically Disadvantaged Families." *Journal of Family Psychology* 15(2):288–301.

Adam, Barry D. 2007. "Relationship Innovation in Male Couples." Pp. 122–40 in *The Sexual Self: The Construction of Sexual Scripts,* edited by Michael S. Kimmel. Nashville: Vanderbilt University Press.

Adams, Lynn. 2010. *Parenting on the Autism Spectrum: A Survival Guide.* San Diego: Plural Publishers.

Adams, Michele and Scott Coltrane. 2004. "Boys and Men in Families." Pp. 189–98 in *Families and Society: Classic and Contemporary Readings,* edited by Scott Coltrane. Belmont, CA: Wadsworth.

Adamsons, Kari and Kay Pasley. 2006. "Coparenting Following Divorce and Relationship Dissolution." Pp. 241–61 in *Handbook of Divorce and Relationship Dissolution,* edited by Mark A. Fine and John H. Harvey. Mahwah, NJ: Erlbaum.

Adelman, Rebecca A. 2009. "Sold(i)ering Masculinity: Photographing the Coalition's Male Soldiers." *Men & Masculinities* 11(3):259–85.

Adimora, Adaora A., Victor J. Schoenbach, and Irene A. Doherty. 2007. "Concurrent Sexual Partnerships Among Men in the United States." *American Journal of Public Health* 97(12):2230–37.

Adkins, Sue. 2007. *Raising Happy Children for Dummies: Hands-on Parenting Skills for Happy Families.* Chichester, UK: Wiley.

Adler, Jerry. 1993. "Sex in the Snoring '90s." *Newsweek,* April 26, pp. 55–57.

Adler, Nancy E., Henry P. David, Brenda N. Major, Susan H. Roth, Nancy Felipe Russo, and Gail E. Wyatt. 1992. "Psychological Factors in Abortion: A Review." *American Psychologist* 47:1194–1204.

Adler-Baeder, Francesca, Christiana Russell, Jennifer Kerpelman, Joe Pittman, Scott Ketring, Thomas Smith, Mallory Lucier-Green, Angela Bradford, and Kate Stringer. 2010. "Thriving in Stepfamilies: Exploring Competence and Well-Being Among African American Youth." *Journal of Adolescent Health* 46:396–98.

African Wedding Guide. n.d. Retrieved October 15, 2006 (www.africanweddingguide.com).

Agee, Mark D., Scott E. Atkinson, and Thomas D. Crocker. 2008. "Multiple-Output Child Health Production Functions: The Impact of Time-Varying and Time-Invariant Inputs." *Southern Economic Journal* 75(2):410–28.

Ahmed, Ashraf Uddin. 1993. "Marriage and Its Transition in Bangladesh." Pp. 74–83 in *Next of Kin: An International Reader on Changing Families,* edited by Lorne Tepperman and Susannah J. Wilson. Englewood Cliffs, NJ: Prentice Hall.

Ahn, Annie, Bryan Kim, and Park Yong. 2008. "Asian Cultural Values Gap, Cognitive Flexibility, Coping Strategies, and Parent-Child Conflicts Among Korean Americans." *Cultural Diversity and Ethnic Minority Psychology* 14(4):353–63.

Ahrons, Constance. 1994. *The Good Divorce: Raising Your Family Together When Your Marriage Comes Apart.* New York: HarperCollins.

———. 2004. *We're Still Family: What Grown Children Have to Say about Their Parents' Divorce.* New York: HarperCollins.

———. 2007. "Family Ties After Divorce: Long-Term Implications for Children." Family Process 46(1):53–65.

Ahrons, Constance and Jennifer L. Tanner. 2003. "Adult Children and Their Fathers: Relationship Changes 20 Years after Parental Divorce." *Family Relations* 52:340–51.

Ainsworth, Mary D. S. 1967. *Infancy in Uganda: Infant Care and the Growth of Love.* Baltimore, MD: Johns Hopkins University Press.

Ainsworth, Mary D. S., M. C. Blehar, E. Waters, and S. Wall. 1978. *Patterns of Attachment: A Psychological Study of the Strange Situation.* Hillsdale, NJ: Erlbaum.

Alanen, Julia. 2007. "Remedies and Resources to Combat International Family Abduction." *American Journal of Family Law* 21:11–27.

Albrecht, Chris and Jay D. Teachman. 2003. "Childhood Living Arrangements and the Risk of Premarital Intercourse." *Journal of Family Issues* 24(7):867–94.

Aldous, Joan. 1978. *Family Careers: Developmental Change in Families.* New York: Wiley.

———. 1996. *Family Careers: Rethinking the Developmental Perspective.* Thousand Oaks, CA: Sage Publications.

Alexander, Brian. 2010. "Lovesick: Hooking Up Over a Shared Disease." *MSNBC,* February 12. www.msnbc.msn.com.

Ali, Lorraine and Julie Scelfo. 2002. "Choosing Virginity." *Newsweek,* December 9, pp. 61–71.

Allan, Graham, Graham Crow, and Sheila Hawker. 2008. *Stepfamilies: A Sociological Review.* New York: Pelgrave Macmillan.

Allen, Anne Wallace. 2009. "Recession Uproots Families, Takes Toll on Children." Associated Press, May 27.

Allen, C. M. and M. A. Straus. 1980. "Resources, Power, and Husband Wife Violence." Pp. 188–208 in *The Social Causes of Husband-Wife Violence,* edited by Murray A. Straus and Gerald T. Hotaling. Minneapolis, MN: University of Minnesota Press.

Allen, Katherine R. 1997. "Lesbian and Gay Families." Pp. 196–218 in *Contemporary Parenting: Challenges and Issues,* edited by Terry Arendell. Thousand Oaks, CA: Sage Publications.

———. 2007. "Ambiguous Loss After Lesbian Couples with Children Break Up: A Case for Same-Gender Divorce." *Family Relations* 56(April):175–83.

Allen, Katherine R., Rosemary Bleiszner, Karen A. Roberto, Elizabeth B. Farnsworth, and Karen L. Wilcox. 1999. "Older Adults and Their Children: Family Patterns of Structural Diversity." *Family Relations* 48(2):151–57.

Allen, Mike. 2002. "Law Extends Benefits to Same-Sex Couples." *Washington Post,* June 26.

Allen, Mike and Alan Cooperman. 2004. "Bush Plans to Back Marriage Amendment." *Washington Post.*

Allgeier, A. R. 1983. "Sexuality and Gender Roles in the Second Half of Life." Pp. 135–57 in *Changing Boundaries: Gender Roles and Sexual Behavior,* edited by Elizabeth Rice Allgeier and Naomi B. McCormick. Palo Alto, CA: Mayfield.

Almeling, Rene. 2007. "Selling Genes, Selling Gender: Egg Agencies, Sperm Banks, and the Medical Market in Genetic Material." *American Sociological Review* 72:319–40.

Altman, Irwin and Dalmas A. Taylor. 1973. *Social Penetration: The Development of Interpersonal Relations.* New York: Holt, Rinehart & Winston.

Altman, Lawrence K. 2004. "Study Finds That Teenage Virginity Pledges Are Rarely Kept." *New York Times,* March 10.

———. 2005. "More Living with HIV, But Concerns Remain." *New York Times,* June 14.

Alvarez, Lizette. 2003. "Helping Retrieve Muslim Children, Including Her Own." *New York Times,* June 16.

———. 2006a. "After Loss of a Parent to War, a Shared Grieving." *New York Times,* May 29.

———. 2006b. "Jane, We Hardly Knew Ye Died." *New York Times,* September 24.

———. 2009. "G.I. Jane Stealthily Breaks the Combat Barrier." *New York Times,* August 16.

Alvarez, Michelle. 2009. "Exclusive AARP Bulletin Poll Reveals New Trends in Multigenerational Housing." March 3. AARP Press Center. Retrieved November 21, 2009 (www.aarp.org).

Amato, Paul R. 1993. "Children's Adjustment to Divorce: Theories, Hypotheses, and Empirical Support." *Journal of Marriage and Family* 55(1):23–28.

———. 1996. "Explaining the Intergenerational Transmission of Divorce." *Journal of Marriage and Family* 58(3):628–40.

———. 1997. *A Generation at Risk: Growing Up in an Era of Family Upheaval.* Cambridge, MA: Harvard University Press.

———. 2000. "The Consequences of Divorce for Adults and Children." *Journal of Marriage and Family* 62:1269–87.

———. 2003. "Reconciling Divergent Perspectives: Judith Wallerstein, Quantitative Research, and Children of Divorce." *Family Relations* 52:332–30.

———. 2004. "Tension between Institutional and Individual Views of Marriage." *Journal of Marriage and Family* 66(4):959–65.

———. 2005. "The Impact of Family Formation Change on the Cognitive, Social and Emotional Well-Being of the Next Generation." *The Future of Children* 15(2):75–96.

———. 2007. "Transformative Processes in Marriage: Some Thoughts from a Sociologist." *Journal of Marriage and Family* 69(2):305–09.

———. 2010. "Research on Divorce: Continuing Trends and New Developments." *Journal of Marriage and Family* 72(3):650–66.

Amato, Paul R., Alan Booth, David R. Johnson, and S. J. Rogers. 2007. *Alone Together: How Marriage in America is Changing.* Cambridge, MA: Harvard University Press.

Amato, Paul R. and Jacob Cheadle. 2005. "The Long Reach of Divorce: Divorce and Child Well-Being across Three Generations." *Journal of Marriage and Family* 67(1):191–206.

Amato, Paul R. and Jacob E. Cheadle. 2008. "Parental Divorce, Marital Conflict and Children's Behavior Problems: A Comparison of Adopted and Biological Children." *Social Forces* 86(3):1139–61.

Amato, Paul R. and Danelle B. DeBoer. 2001. "The Transmission of Marital Instability across Generations: Relationship Skills or Commitment to Marriage?" *Journal of Marriage and Family* 63:1038–51.

Amato, Paul R. and Bryndl Hohmann-Marriott. 2007. "A Comparison of High- and Low-Distress Marriages That End in Divorce." *Journal of Marriage and Family* 69:621–38.

Amato, Paul R. and Shelley Irving. 2006. "Historical Trends in Divorce in the United States." Pp. 41–57 in *Handbook of Divorce and Relationship Dissolution,* edited by Mark A. Fine and John H. Harvey. Mahwah, NJ: Erlbaum.

Amato, Paul R., David R. Johnson, Alan Booth, and Stacy J. Rogers. 2003. "Continuity and Change in Marital Quality between 1980 and 2000." *Journal of Marriage and Family* 65(1):1–22.

Amato, Paul R., Nancy S. Landale, Tara C. Havasevich-Brooks, and Alan Booth. 2008. "Precursors of Young Women's Family Formation Pathways." *Journal of Marriage and Family* 70:1271–86.

Amato, Paul R., Catherine E. Meyers, and Robert E. Emery. 2009. "Changes in Nonresident Father-Child Contact from 1976 to 2002." *Family Relations* 58(1):41–53.

American Academy of Pediatrics. 1992. *Caring for Our Children: National Health and Safety Performance Standards as Guidelines for Out-of-Home Child Care Programs.* Elk Grove Village, IL: American Public Health Association and American Academy of Pediatrics.

———. 1998. "Guidance for Effective Discipline." *Pediatrics* 101:723–28.

American-Arab Anti-Discrimination Committee Research Institute. 2008. *Report on Hate Crimes and Discrimination Against Arab Americans 2003–2007.* Washington, DC: American-Arab Anti-Discrimination Committee Research Institute.

American Association of Retired Persons. 2004. *A Report of Multicultural Boomers Coping with Family and Aging Issues.* Washington, DC: American Association of Retired Persons.

American Bar Association. 2010. "The Elder Justice Act." Retrieved May 15, 2010 (www.abanet.org).

American Moslem Society. 2010. www.masjiddearborn.org.

American Sociological Association. 2002. "Statement of the American Sociological Association on the Importance of Collecting Data and Doing Social Science Research on Race." Washington, DC: American Sociological Association. Retrieved August 28, 2006 (www.asanet.org).

———. 2005a. "The Impact of Abortion on Women: What Does the Psychological Research Say?" APA Briefing Paper. Washington, DC: American Psychological Association. January 31. Retrieved November 10, 2006 (www.apa.org).

———. 2005b. "Statement on Parental Alienation Syndrome." Press Release, October 28. Washington, DC: American Psychological Association. Retrieved May 15, 2007 (www.apa.org).

———. 2007. "Answers to Your Questions about Sexual Orientation and Homosexuality." Washington, DC: American Psychological Association. Retrieved June 12, 2007 (www.apa.org).

American Psychological Association. 2009. "APS Survey Raises Concern about Parent Perceptions of Children's Stress." November 3. Retrieved January 21, 2010 (http://www.apa.org).

———. 2010. "Sexual Orientation and Homosexuality." APA Online (http://www.apa.org/helpcenter/sexual-orientation.aspx).

Ames, Barbara D., Whitney A. Brosi, and Karla M. Damiano-Teixeira. 2006. "'I'm Just Glad My Three Jobs Could Be during the Day': Women and Work in a Rural Community." *Family Relations* 55(1):119–31.

Ananat, Elizabeth O. and Guy Michaels. 2007. "The Effect of Marital Breakup on the Income Distribution of Women with Children." *Journal of Human Resources* 43(3):611–29.

Anderlini-D'Onofrio, Serina. 2004. "Introduction to Plural Loves: Bi and Poly Utopias for a New Millennium." *Journal of Bisexuality* 4(3/4):2–6.

Andersen, Jan. 2004. "Stepfamilies—How to Live in Harmony." SelfGrowth.com. Retrieved September 21, 2004 (www.selfgrowth.com).

Andersen, Julie Donner. 2002. "His Kids: Becoming a W.O.W. Stepmother." SelfGrowth.com (www.selfgrowth.com/articles/Andersen3.html).

Andersen, Margaret L. 1988. *Thinking about Women: Sociological Perspectives on Sex and Gender.* 2nd ed. New York: Macmillan.

———. 2005. "Thinking about Women: A Quarter Century's View." *Gender & Society* 19:437–55.

Andersen, Margaret L. and Patricia Hill Collins. 2007a. "Why Race, Class, and Gender Still Matter." Pp. 1–16 in *Race, Class and Gender: An Anthology,* 6th ed., edited by Margaret L. Andersen and Patricia Hill Collins. Belmont, CA: Wadsworth.

———. 2007b. "Why Race, Class, and Gender Still Matter." Pp. 1–16 in *Race, Class and Gender: An Anthology,* 6th ed., edited by Margaret L. Andersen and Patricia Hill Collins. Belmont, CA: Wadsworth.

Andersen, Margaret L., and Howard F. Taylor. 2002. *Sociology: Understanding A Diverse Society,* 2nd ed. Belmont, CA: Wadsworth.

Anderson, Jared R. and William J. Doherty. 2005. "Democratic Community Initiatives: The Case of Overscheduled Children." *Family Relations* 54(5):654–65.

Anderson, Kristin L. 1997. "Gender, Status, and Domestic Violence: An Integration of Feminist and Family Violence Approaches." *Journal of Marriage and Family* 59(3):655–69.

Anderson, Michael A., Paulette Marie Gillig, Marilyn Sitaker, Kathy McCloskey, Katherine Malloy, and Nancy Grigsby. 2003. "'Why Doesn't She Just Leave?' A Descriptive Study of Victim Reported Impediments to Her Safety." *Journal of Family Violence* 18:151–55.

Anetzberger, Georgia, Jill Korbin, and Craig Austin. 1994. "Alcoholism and Elder Abuse." *Journal of Interpersonal Violence* 9(2):184–93.

Angelo, Megan. 2010. "Fortune 500 Women CEOs." *Fortune Magazine*, April 23.

Ansay, Sylvia J., Daniel F. Perkins, and Colonel John Nelson. 2004. "Interpreting Outcomes: Using Focus Groups in Evaluation Research." *Family Relations* 53(3):310–16.

Antiretroviral Therapy Cohort Collaboration. 2008. "Life Expectancy of Individuals on Combination Antiretroviral Therapy in High-Income Countries: A Collaborative Analysis of 14 Cohort Studies." *The Lancet* 372(9635):293–99.

Anyiam, Thony. 2002. "Who Should Jump the Broom?" Retrieved October 2, 2006 (www.anyiams.com/jumping_the_broom).

Apple, Rima D. 2006. *Perfect Motherhood: Science and Childrearing in America.* New Brunswick, NJ: Rutgers University Press.

Apter, T. E. 2007. *The Confident Child: Raising Children to Believe in Themselves.* New York: W. W. Norton.

Aquilino, William S. 1994a. "Later Life Parental Divorce and Widowhood: Impact on Young Adults' Assessment of Parent–Child Relations." *Journal of Marriage and Family* 56(4):908–22.

Archer, John. 2000. "Sex Differences in Aggression between Heterosexual Partners: A Meta-Analytic Review." *Psychological Bulletin* 126:651–80.

Archibold, Randal C. 2009. "Octuplets, 6 Siblings, and Many Questions." *The New York Times*, February 3. Retrieved June 16, 2009 (www.nytimes.com).

Arditti, Joyce A. 2003. "Incarceration Is a Major Source of Family Stress." *Family Focus* (June):F15–F17. Minneapolis, MN: National Council on Family Relations.

Arditti, Joyce A. and Timothy Z. Keith. 1993. "Visitation Frequency, Child Support Payment, and the Father–Child Relationship Postdivorce." *Journal of Marriage and Family* 55(3):699–712.

Arditti, Joyce A., Jennifer Lambert-Shute, and Karen Joest. 2003. "Saturday Morning at the Jail: Implications of Incarceration for Families and Children." *Family Relations* 52(3):195–204.

Arendell, Terry. 1986. *Mothers and Divorce: Legal, Economic, and Social Dilemmas.* Berkeley, CA: University of California Press.

———. 1997. "Divorce and Remarriage." Pp. 154–95 in *Contemporary Parenting: Challenges and Issues*, edited by Terry Arendell. Thousand Oaks, CA: Sage Publications.

———. 2000. "Conceiving and Investigating Motherhood: The Decade's Scholarship." *Journal of Marriage and Family* 62(4):1192–1207.

Arenson, Karen W. 2005. "Little Advance Is Seen in Ivies' Hiring of Minorities and Women." *New York Times*, March 1.

Ariès, Phillipe. 1962. *Centuries of Childhood: A Social History of Family Life.* New York: Knopf.

———. 2006. "Cost-effective 'Homesourcing' Grows." *USA Today*, March 13.

———. 2007. "'Helicopter' Parents Hover When Kids Job Hunt." *USA Today*, April 23. Retrieved January 15, 2010 (www.usatoday.com).

———. 2009. "Love Isn't All That's Keeping Family Together Today." *USA Today*, February 2. Retrieved May 3, 2010 (www.usatoday.com).

Armstrong, Larry. 2003. "Your Mouse Knows Where Your Car Is." *Business Week* 16.

Arnall, Judy and Debbie Elicksen. 2007. *Discipline without Punishment: 135 Tools for Raising Caring, Responsible Children without Time-Out, Spanking, Punishment, or Bribery.* Calgary, Canada: Professional Parenting Canada.

Arnett, Jeffrey Jensen. 2000. "Emerging Adulthood: A Theory of Development from the Late Teens through the Twenties." *American Psychologist* 55(5):469–80. Retrieved March 28, 2003 (PsycARTICLES 0003-006X).

———. 2004. *Emerging Adulthood: The Winding Road from the Late Teens through the Twenties.* London: Oxford University Press.

Arnott, Teresa and Julie Matthaei. 2007. "Race, Class, and Gender and Women's Works." Pp. 283–92 in *Race, Class, and Gender*, 6th ed., edited by Margaret Andersen and Patricia Hill Collins. Belmont, CA: Wadsworth.

Aronson, Pamela. 2003. "Feminists or 'Postfeminists'? Young Women's Attitudes toward Feminism and Gender Relations." *Gender and Society* 17:903–22.

Artis, Julie E. 2004. "Judging the Best Interests of the Child: Judges' Accounts of the Tender Years Doctrine." *Law & Society Review* 38(4):769–806.

Artis, Julie E. and Eliza K. Pavalko. 2003. "Explaining the Decline in Women's Household Labor: Individual Change and Cohort Differences." *Journal of Marriage and Family* 65:746–61.

Aseltine, Robert H., Jr. and Ronald C. Kessler. 1993. "Marital Disruption and Depression in a Community Sample." *Journal of Health and Social Behavior* 34(September):237–51.

Ashe, Fidelma. 2004. "Deconstructing the Experiential Bar." *Men & Masculinities* 7(2):187–204.

"As Recession Erodes Family Inheritances, Brits Face Uncertain Future." 2009. *Finance.* May 5. Retrieved June 15, 2009 (www.easier.com).

Association for Psychological Science. 2008. "Is Empty Nest Best? Changes in Marital Satisfaction in Late Middle Age." Retrieved April 7, 2010 (www.physorg.com).

Athenstaedt, Ursula, Gerold Mikula, and Cornelia Bredt. 2009. "Gender Role Self-Concept and Leisure Activities of Adolescents." *Sex Roles* 60(5/6):399–409.

Austin, Algernon. 2009. "African Americans See Weekly Wage Decline." Washington, DC: Economic Policy Institute. Retrieved July 8, 2009 (www.epi.org).

Avellar, Sarah and Pamela J. Smock. 2003. "Has the Price of Motherhood Declined Over *Time*? A Cross-Cohort Comparison of the Motherhood Wage Penalty." *Journal of Marriage and Family* 65:597–607.

———. 2005. "The Economic Consequences of the Dissolution of Cohabiting Unions." *Journal of Marriage and Family* 67(2):315–27.

Aydt, Hilary and William A. Corsaro. 2003. "Differences in Children's Construction of Gender Across Culture." *American Behavioral Scientist* 46(10):1306–25.

Babbie, Earl. 2007. *The Practice of Social Research.* 11th ed. Belmont, CA: Wadsworth/Cengage.

———. 2009. *The Practice of Social Research.* 12th ed. Belmont, CA: Wadsworth/Cengage.

Babbitt, Charles E. and Harold J. Burbach. 1990. "A Comparison of Self-Orientation among College Students across the 1960s, 1970s and 1980s." *Youth & Society* 21(4):472–82.

Baca Zinn, Maxine, Pierette Hondagneu-Sotelo, and Michael A. Messner. 2004. "Gender through the Prism of Difference." Pp. 166–74 in *Race, Class, and Gender*, 5th ed., edited by Margaret L. Andersen and Patricia Hill Collins. Belmont, CA: Wadsworth.

———. 2007. "Sex and Gender through the Prism of Difference." Pp. 147–55 in *Race, Class, and Gender: An Anthology,* 6th ed, edited by Margaret L. Andersen and Patricia Hill Collins. Belmont, CA: Wadsworth.

Baca Zinn, Maxine and Angela Y. H. Pok. 2002. "Tradition and Transition in Mexican-Origin Families." Pp. 79–100 in *Minority Families in the United States,* 3rd ed., edited by Ronald L. Taylor. Upper Saddle River, NJ: Prentice Hall.

Baca Zinn, Maxine and Barbara Wells. 2007. "Diversity within Latino Families: New Lessons for Family Social Science." Pp. 422–47 in *Family in Transition,* 14th ed., edited by Arlene S. Skolnick and Jerome H. Skolnick. Boston, MA: Allyn and Bacon.

Bachrach, Christine, Patricia F. Adams, Soledad Sambrano, and Kathryn A. London. 1990. "Adoption in the 1980s." *Advance Data*, No. 181. Hyattsville, MD: U.S. National Center for Health Statistics, January 5.

Baer, Judith C. and Mark F. Schmitz. 2007. "Ethnic Differences in Trajectories of Family Cohesion for Mexican American and Non-Hispanic White Adolescents." *Journal of Youth and Adolescence* 36:583–92.

Bailey, Jo Daugherty and Dawn McCarty. 2009. "Assessing Empowerment in Divorce Mediation." *Negotiation Journal* 25(3):327–36.

Bainbridge, Daryl, Paul Krueger, Lynne Lohfeld, and Kevin Brazil. 2009. "Stress Processes in Caring for an End-of-Life Family

Member: Application of a Theoretical Model." *Aging and Mental Health* 13(4):537–45.

Baird, Julia. 2008. "From Seneca Falls to Sarah Palin?" *Newsweek,* September 22. Accessed January 17, 2010 (www.newsweek.com).

Bakalar, Nicholas. 2007. "Breast Cancer Not Linked to Abortion, Study Says." *New York Times,* April 24.

Baker, Carrie N. 2008. *The Women's Movement Against Sexual Harassment.* New York: Cambridge University Press.

Baker, Elizabeth H., Laura A. Sanchez, Steven L. Nock, and James D. Wright. 2009. "Covenant Marriage and the Sanctification of Gendered Marital Roles." *Journal of Family Issues* 30(2):147–78.

Baker, Katie. 1008. "Seeking a Samaritan." *Newsweek.* March 16.

"Balancing Work and Life: Family Friendly Workplace Policies." 2009. Washington, DC: American Association of University Women.

Ball, Derek and Peter Kivisto. 2006. "Couples Facing Divorce." Pp. 145–61 in *Couples, Kids, and Family Life,* edited by Jaber F. Gubrium and James A. Holstein. New York: Oxford University Press.

Ball, Victoria and Kristin A. Moore. 2008. "What Works For Adolescent Reproductive Health: Lessons from Experimental Evaluations of Programs and Interventions." *Fact Sheet* #2008-20. Washington, DC. Child Trends.

Balsam, Kimberly F. and Dawn M. Szymanski. 2005. "Relationship Quality and Domestic Violence in Women's Same-Sex Relationships: The Role of Minority Stress." *Psychology of Women Quarterly* 29:258–69.

Bandura, Albert and Richard H. Walters. 1963. *Social Learning and Personality Development.* New York: Holt, Rinehart & Winston.

Banerjee, Neela. 2006a. "Clergywomen Find Hard Path to Bigger Pulpit." *New York Times,* August 26.

———. 2006b. "A Woman Is Installed as Top Bishop of the Episcopal Church." *New York Times,* November 5.

Banks, Stephen P. 2009. "Intergenerational Ties Across Borders: Grandparenting narratives by Expatriate Retirees in Mexico." *Journal of Aging Studies* 23:178–87.

Banse, Rainer. 2004. "Adult Attachment and Marital Satisfaction: Evidence for Dyadic Configuration Effects." *Journal of Social and Personal Relationships* 21(2):273–82.

Banyard, Victoria L., Valerie J. Edwards, and Kathleen Kendall-Tackett, eds. 2009. *Trauma and Physical Health: Understanding the Effects of Extreme Stress and of Psychological Harm.* New York: Routledge.

Bar, Alan. 2009. *Survival Guide for the Stay at Home Dad.* Charleston, SC: Booksurge Publishing.

Barajas, Manuel and Elvia Ramirez. 2007. "Beyond Home-Host Dichotomies: A Comparative Examination of Gender Relations in a Transnational Mexican Community." *Sociological Perspectives* 50(3):367–92.

Barash, Susan Shapiro. 2000. *Second Wives: The Pitfalls and Rewards of Marrying Widowers and Divorced Men.* Far Hills, NJ: New Horizon.

Barber, Bonnie L. and David H. Demo. 2006. "The Kids Are Alright (at Least Most of Them): Links between Divorce and Dissolution and Child Well-Being." Pp. 289–311 in *Handbook of Divorce and Relationship Dissolution,* edited by Mark A. Fine and John H. Harvey. Mahwah, NJ: Erlbaum.

Barker, Susan E. n.d. "'Cuddle Hormone': Research Links Oxytocin and Sociosexual Behaviors." Retrieved August 30, 2006 (www.oxytocin.org/cuddle-hormone/).

Barlow, Daniel. 2005. "Vermont's — and Nation's — First Civil Union Breaking Up." *The Barre Montpelier Times Argus,* December 15. Retrieved from www.timesargus.com/apps/pbcs.dll/article?AID=/20051215/NEWS/512150369/1002.

Barnes, Norine R. 2006. "Children Need Guidance. . . . Developmentally Appropriate Interaction and Discipline." Retrieved December 30, 2009 (http://www.parentingweb.com).

Barnes, Richard E. 2009. *Estate Planning for Blended Families: Providing for Your Spouse and Children in a Second Marriage.* Berkeley, CA: Nolo Press.

Barnett, Melissa A. 2008. "Mother and Grandmother Parenting in Low-Income Three-Generation Rural Households." *Journal of Marriage and Family* 70(5):1241–57.

Barnett, Rosalind Chait, Karen C. Gareis, Laura Sabattini, and Nancy M. Carter. 2009. "Parental Concerns About After-School Time: Antecedents and Correlates Among Dual-Earner Parents." *Journal of Family Issues* 31(5):606–25.

Barrionuevo, Alexei. 2010. "Argentina Approves Gay Marriage, in a First for Region." *The New York Times,* July 15. Retrieved July 19, 2010 (www.nytimes.com).

Barron, David J. 2010. "Whether the Criminal Provisions of the Violence Against Women Act Apply To Otherwise Covered Conduct when the Offender and Victim are the Same Sex." *Memorandum Opinion for the Acting Deputy Attorney General.* Washington, DC: United States Department of Justice Office of Legal Counsel, April 27.

Barth, Richard P. and Marianne Berry. 1988. *Adoption and Disruption: Rates, Risks, and Responses.* New York: Aldine.

Bartkowski, John P. 2001. *Remaking the Godly Family: Gender Negotiation in Evangelical Families.* Piscataway, NJ: Rutgers University Press.

Bartkowski, John P. and Jen'nan Ghazal Read. 2003. "Veiled Submission: Gender, Power, and Identity among Evangelical and Muslim Women in the United States." *Qualitative Sociology* 26(1):71–92.

Bartlett, Thad Q. 2009. *The Gibbons of Khao Yai: Seasonal Variation In Behavior and Ecology.* Upper Saddle River, NJ: Pearson.

Basow, Susan H. 1992. *Gender: Stereotypes and Roles.* 3rd ed. Pacific Grove, CA: Brooks/Cole.

Bass, Brenda L., Adam B. Butler, Joseph G. Grzywacz, and Kirsten D. Linney. 2009. "Do Job Demands Undermine Parenting? A Daily Analysis of Spillover and Crossover Effects." *Family Relations* 58(April):201–15.

Bass, Linda A., Brent A. Taylor, Carmen Knudson-Martin, and Douglas Huenergardt. 2006. "Making Sense of Abuse: Case Studies in Sibling Incest." *Contemporary Family Therapy* 28(1):87–109.

Batson, Christie D., Zhenchao Qian, and Daniel T. Lichter. 2006. "Interracial and Intraracial Patterns of Mate Selection among America's Diverse Black Populations." *Journal of Marriage and Family* 68(3):658–72.

Battle, Danielle. 2009. *Characteristics of Public, Private, and Bureau of Indian Education Elementary and Secondary School Principals in the United States: Results From the 2007–08 Schools and Staffing Survey.* NCES 2009-323. Washington, DC: National Center for Education Statistics, Institute of Education Sciences, U.S. Department of Education.

Baucom, Donald H., Gordon, Kristina C., Snyder, Douglas K., Atkins, David C., and Christensen, Andrew 2006. "Treating Affair Couples: Clinical Considerations and Initial Findings." *Journal of Cognitive Psychotherapy* 20:375–92.

Baucom, Donald H., Douglas K. Snyder, and Kristina Coop Gordon. 2009. *Helping Couples Get Past the Affair: A Clinician's Guide.* New York: The Guilford Press.

Baum, Angela C., Sedahlia Jasper Crase, and Kirsten Lee Crase. 2001. "Influences on the Decision to Become or Not Become a Foster Parent." *Families in Society* 82(2):202–21.

Baumrind, Diana. 1978. "Parental Disciplinary Patterns and Social Competence in Children." *Youth and Society* 9:239–76.

Baumrind, Diana, Robert E. Larzelere, and Philip A. Cowan. 2002. "Ordinary Physical Punishment: Is it Harmful?" Comment on Gershoff." *Psychological Bulletin* 128(4):580–89.

Bausch, Robert S. 2006. "Predicting Willingness to Adopt a Child: A Consideration of Demographic and Attitudinal Factors." *Sociological Perspectives* 49(1):47–65.

Bauserman, Robert. 2002. "Child Adjustment in Joint-Custody versus Sole Custody Arrangements: A Meta-Analytic Review." *Journal of Family Psychology* 16:91–102.

Baxter, Christine C. 1989. "Investigating Stigma as Stress in Social Interactions of Parents." *Journal of Intellectual Disability Research* 33(6):455–66.

Baxter, Janeen. 2005. "To Marry or Not to Marry: Marital Status and the Household Division of Labor." *Journal of Family Issues* 26(3):300–321.

Baxter, Janeen, Belinda Hewitt, and Michele Haynes. 2008. "Life Course Transitions and Housework: Marriage, Parenthood, and Time on Housework." *Journal of Marriage and Family* 70(2):259–69.

Baxter, Leslie A., Dawn O. Braithwaite, and Leah E. Bryart. 2006. "Types of Communication Triads Perceived by young-Adult Stepchildren in Established Stepfamilies." *Communication Studies* 57(4):381–400.

Baxter, Leslie A., Dawn O. Braithwaite, Jody K. Kellas, Cassandra LeClaire-Underberg, Emily

Lamb Normand, Tracy Routsong, and Matthew Thatcher. 2009. "Empty Ritual: Young-adult Stepchildren's Perceptions of the Remarriage Ceremony." *Journal of Social and Personal Relationships* 26(4):467–87.

Bazelton, Emily. 2009. "Unwashed Coffee Mugs: How the Recession Is Affecting Family Relationships." *Slate*, February 20.

Beaman, Lori G. 2001. "Molly Mormons, Mormon Feminists and Moderates: Religious Diversity and the Latter Day Saints Church." *Sociology of Religion* 62:65–86.

Bean, Frank D., Jennifer Lee, Jeanne Batalova, and Mark Leach. 2004. *Immigration and Fading Color Lines in America.* New York: Russell Sage.

Bean, Roy A. and Jason C. Northrup. 2009. "Parental Psychological Control, Psychological Autonomy, and Acceptance as Predictors of Self-Esteem in Latino Adolescents." *Journal of Family Issues* 30(11):1486–1504.

Bearman, Peter. 2008. "Exploring Genetics and Social Structure." *American Journal of Sociology* 114(Supplement):v–x.

Bearman, Peter S. and Hannah Brückner. 2001. "Promising the Future: Virginity Pledges and First Intercourse." *American Journal of Sociology* 106:859–912.

Beck, Audrey N., Carey E. Cooper, Sara McLanahan, Jeanne Brooks-Gunn. 2010. "Partnership Transitions and Maternal Parenting." *Journal of Marriage and Family* 72(2):219–33.

Becker, Gary S. 1981/1991. *A Treatise on the Family.* 2nd ed. Cambridge, MA: Harvard University Press.

Becker, Gay. 1990. *Healing the Infertile Family.* New York: Bantam.

———. 2000. *The Elusive Embryo: How Women and Men Approach New Reproductive Technologies.* Berkeley and Los Angeles, CA: University of California Press.

Beer, William R. 1989. *Strangers in the House: The World of Stepsiblings and Half Siblings.* New Brunswick, NJ: Transaction Books.

———. 1992. *American Stepfamilies.* New Brunswick, NJ: Transaction Books.

———. 2007. "Just Say No to Bad Science." *Newsweek*, May 7, pp. 57–58.

———. 2009. "Pink Brain, Blue Brain." *Newsweek*, September 9, p. 28.

Behnke, Andrew O., Shelley M. MacDermid, Scott L. Coltrane, Ross D. Parke, Sharon Duffy, and Keith F. Widaman. 2008. "Family Cohesion in the Lives of Mexican American and European American Parents." *Journal of Marriage and Family* 70(4):1045–59.

Beins, Bernard. 2008. *Research Methods: A Tool for Life.* Boston, MA: Allyn and Bacon.

Beitin, Ben, Katherine Allen, and Maureen Bekheet. 2010. "A Critical Analysis of Western Perspectives on Families of Arab Descent." *Journal of Family Issues* 31(2):211–33.

Belch, Michael A., Kathleen A. Krentler, and Laura A. Willis-Flurry. 2005. "Teen Internet Mavens: Influence in Family Decision-Making." *Journal of Business Research* 58(5):569–75.

Belkin, Lisa. 2009. "Postponing a Baby in This Recession." *The New York Times*, February 9. Retrieved June 10, 2009 (parenting.blogs.nytimes.com).

———. 2010. "The Marrying Kind." *The New York Times*, March 22. Retrieved April 23, 2010 (www.nytimes.com).

Bell, Alan P., Martin S. Weinberg, and Sue Kiefer Hammersmith. 1981. *Sexual Preference: Its Development in Men and Women.* Bloomington, IN: University of Indiana Press.

Bell, Maya. 2003. "More Gays and Lesbians Than Ever Are Becoming Parents." Knight Ridder/Tribune News Service, October 1.

Bell, Richard Q. 1974. "Contributions of Human Infants to Caregiving and Social Interaction." Pp. 11–19 in *The Effect of the Infant on Its Care Giver: Origins of Behavior Series*, Vol. 1, edited by Michael Lewis and Leonard A. Rosenblum. New York: Wiley.

Bellafante, Ginia. 2004. "Two Fathers, with One Happy to Stay at Home." *New York Times*, January 12.

Bellah, Robert N., Richard Madsen, William M. Sullivan, Ann Swidler, and Steven M. Tipton. 1985. *Habits of the Heart: Individualism and Commitment in American Life.* Berkeley and Los Angeles, CA: University of California Press.

Belluck, Pam. 2000. "States Declare War on Divorce Rates Before Any 'I Dos.'" *New York Times*, April 21.

———. 2009. "In Turnabout, Children Take Caregiver Role." The New York Times, February 23. Retrieved March 16, 2010 (www.nytimes.com).

Belluck, Pam and Jim Yardley. 2006. "China Tightens Adoption Rules for Foreigners." *New York Times*, December 20.

Belsky, Jay. 1990. "Parental and Nonparental Child Care and Children's Socioemotional Development: A Decade in Review." *Journal of Marriage and Family* 52(4):885–903.

———. 1991. "Parental and Non-Parental Child Care and Children's Socioemotional Development." Pp. 122–40 in *Contemporary Families: Looking Forward, Looking Back*, edited by Alan Booth. Minneapolis, MN: National Council on Family Relations.

———. 2002. "Quantity Counts: Amount of Child Care and Children's Socioemotional Development." *Developmental and Behavioral Pediatrics* 23:167–70.

———. 2008. "Quality, Quantity and Type of Child Care: Effects on Child Development in the USA." In G. Bentley & R. Mace (Eds.), *Substitute Parenting: Alloparenting in Human Societies.* London, UK: Berghahn Books.

Belsky, Jay and K. H. Hsieh. 1998. "Patterns of Marital Change during the Early Childhood Years: Parent Personality, Coparenting, and Division-of-Labor Correlates." *Journal of Family Psychology* 12:511–26.

Belsky, Jay, Deborah Lowe Vandell, Margaret Burchinal, Alison Clarke-Stewart, Kathleen McCartney, Margaret Tresch Owen, and the NICHD Early Child Care Research Network. 2007. "Are There Long-Term Effects of Early Child Care?" *Child Development* 78(2):681–701.

Bem, Sandra Lipsitz. 1981. "Gender Schema Theory: A Cognitive Account of Sex Typing." *Psychological Review* 88:354–64.

Bendix, Jeffrey. 2009. "Elder Exploitation." *RN* 72(3):42–45.

Bengston, Vern L. 2001. "Beyond the Nuclear Family: The Increasing Importance of Multigenerational Bonds." *Journal of Marriage and Family* 63(1):1–16.

Bengston, Vern L., Timothy J. Biblarz, and Robert E. L. Roberts. 2002. *How Families Still Matter: A Longitudinal Study of Youth in Two Generations.* Cambridge, UK: Cambridge University Press.

———. 2007. "How Families Still Matter: A Longitudinal Study of Youth in Two Generations." Pp. 315–24 in *Family in Transition*, 14th ed., edited by Arlene S. Skolnick and Jerome H. Skolnick. Boston, MA: Allyn and Bacon.

Benkel, I., H. Wijk, and U. Molander. 2009. "Family and Friends Provide Most Social Support for the Bereaved." *Palliative Medicine* 23(2):141–9.

Benner, Aprile D. and Su Yeong Kim. 2009. "Intergenerational Experiences of Discrimination in Chinese American Families: Influences of Socialization and Stress." *Journal of Marriage and Family* 71(4):862–77.

———. 2010. "Understanding Chinese American Adolescents' Developmental Outcomes: Insights from the Family Stress Model." *Journal of Research on Adolescence* 20(1):1–12.

Bennett, Linda A., Steven J. Wolin, and David Reiss. 1988. "Deliberate *Family Process*: A Strategy for Protecting Children of Alcoholics." *British Journal of Addiction* 83:821–29.

Bennetts, Leslie. 2007. *The Feminine Mistake: Are We Giving Up Too Much?* New York: Voice.

Benoit, D. and K. Parker. 1994. "Stability and Transmission of Attachment across Three Generations." *Child Development* 65:1444–56.

Benson, Michael. L, John Wooldredge, Amy B. Thistlethwaite, and Greer Liton Fox. 2004. "The Correlation between Race and Domestic Violence is Confounded with Community Context." *Social Problems* 51(3):326–42.

Bentley, Evie. 2007. *Adulthood.* New York: Routledge.

Berg, Barbara. 1984. "Early Signs of Infertility." *Ms.*, May, pp. 68ff.

Berge, Jerica M. and Kristen E. Holm. 2007. "Boundary Ambiguity in Parents with Chronically Ill Children: Integrating Theory and Research." *Family Relations* 56(2):123–34.

Bergen, Raquel Kennedy. 2006. "Marital Rape: New Research and Directions." The National Online Resource Center on Violence Against Women. Retrieved June 9, 2010 (new.vawnet.org).

Berger, Lawrence M., Theresa Heintze, Wendy Naidich, and Marcia Meyers. 2008. "Subsidized Housing and Household Hardship among Low-Income Single-Mother Households." *Journal of Marriage and Family* 70(4):934–49.

Berger, Peter L. and Hansfried Kellner. 1970. "Marriage and the Construction of Reality." Pp. 49–72 in *Recent Sociology No. 2*, edited by Hans Peter Dreitzel. New York: Macmillan.

Berger, Peter L. and Thomas Luckman. 1966. *The Social Construction of Reality: A Treatise in the Sociology of Knowledge*. Garden City, NY: Anchor Books.

Berger, Roni. 1997. "Immigrant Stepfamilies." *Contemporary Family Therapy* 19(3):361–70.

———. 1998. *Stepfamilies: A Multi-dimensional Perspective*. New York: Haworth.

———. 2000. "Gay Stepfamilies: A Triple-Stigmatized Group." *Families in Society* 81(5):504–16.

Bergman, Mike. 2004. "Census Bureau Releases Information on Home Workers." Press Release CB04-183, October 20. Washington, DC: U.S. Census Bureau.

———. 2006a. "Americans Marrying Older, Living Alone More, See Households Shrinking, Census Bureau Reports." Press Release CB06-83, May 25. Washington, DC: U.S. Census Bureau.

———. 2006b. "Census Bureau Data Underscore Value of College Degree." Press Release CB06-159. Washington, DC: U.S. Census Bureau.

———. 2006c. "Dramatic Changes in U.S. Aging: Highlighted in New Census, NIH Report." Press Release CB06-36, March 9. Washington, DC: U.S. Census Bureau.

———. 2006d. "Growth of Hispanic-owned Businesses Triples the National Average." Press Release, March 21. Washington, DC: U.S. Census Bureau.

Berk, Richard A., Phyllis J. Newton, and Sarah Fenstermaker Berk. 1986. "What a Difference a Day Makes: An Empirical Study of the Impact of Shelters for Battered Women." *Journal of Marriage and Family* 48:481–90.

Berk, Sarah Fenstermaker. 1985. *The Gender Factory: The Apportionment of Work in American Households*. New York: Plenum.

Berke, Debra L. 2003. "Coming Home Again: The Challenges and Rewards of Home-Based Self-Employment." *Journal of Family Issues* 24:513–46.

Berkowitz, Dana and William Marsiglio. 2007. "Gay Men: Negotiating Procreative, Father, and Family Identities." Journal of Marriage & Family 69(2):366–81.

Berlin, Gordon. 2007. "Rewarding the Work of Individuals: A Counterintuitive Approach to Reducing Poverty and Strengthening Families." *The Future of Children* 17(2):17–42.

Berlin, Lisa, Jean Ispa, Mark Fine, Patrick Malone, Jeanne Brooks-Gunn, Christy Brady-Smith, Catherine Ayoub, and Yu Bai. 2009. "Correlates and Consequences of Spanking

and Verbal Punishment for Low-Income White, African American, and Mexican American Toddlers." *Child Development* 80(5):1403–20.

Bernard, Jessie. 1986. "The Good-Provider Role: Its Rise and Fall." Pp. 125–44 in *Family in Transition: Rethinking Marriage, Sexuality, Child Rearing, and Family Organization*, 5th ed., edited by Arlene S. Skolnick and Jerome H. Skolnick. Boston, MA: Little, Brown.

Bernstein, Aaron. 2004. "Shaking Up Trade Theory." *Business Week*, December 6, pp. 116–20.

Bernstein, Anne C. 1997. "Stepfamilies from Siblings' Perspectives." Pp. 153–75 in *Stepfamilies: History, Research, and Policy*, edited by Irene Levin and Marvin B. Sussman. New York: Haworth.

———. 1999. "Reconstructing the Brothers Grimm: New Tales for Stepfamily Life." *Family Process* 38(4):415–30.

———. 2007. "Re-visioning, Restructuring, and Reconciliation: Clinical Practice With Complex Postdivorce Families." Family Process 46(1):67–78.

Bernstein, Jeffrey and Susan Magee. 2004. *Why Can't You Read My Mind? Overcoming the 9 Toxic Thought Patterns That Get in the Way of a Loving Relationship*. New York: Marlowe.

Bernstein, Nina. 2004. "More Teenagers Are Striving for Restraint." *New York Times*, March 7.

———. 2007. "In Secret, Polygamy Follows Africans to N.Y." The New York Times. March 23. Retrieved November 6, 2009 (www.nytimes.com).

Berridge, K. and T. Robinson. 1998. "What Is the Role of Dopamine in Reward: Hedonic Impact, Reward Learning, or Incentive Salience?" *Brain Research Review* 28(3):309–69.

Berry, Brent. 2008. "Financial Transfers from Living Parents to Young Adult Children: Who Is Helped and Why?" *The American Journal of Economics and Sociology* 67(2):207–39.

Bersamin, Melina, Michael Todd, Deborah Fisher, Douglas Hill, Joel Grube, and Samantha Walker. 2008. "Parenting Practices and Adolescent Sexual Behavior: A Longitudinal Study." *Journal of Marriage and Family* 70(1):97–112.

Bertram, Rosalyn M. and Jennifer L. Dartt. 2009. "Post Traumatic Stress Disorder: A Diagnosis for Youth from Violent, Impoverished Communities." *Journal of Child and Family Studies*, 294–302.

Beutel, Ann M. and Margaret Mooney Marini. 1995. "Gender and Values." *American Sociological Review* 60(3):436–49.

Bhalla, Vibha. 2008. "Couch Potatoes and Super-Women: Gender, Migration, and the Emerging Discourse on Housework among Asian Indian Immigrants." *Journal of American Ethnic History* 27(4):71–99.

Bianchi, Suzanne M. 2000. "Maternal Employment and *Time* with Children: Dramatic Change or Surprising Continuity?" *Demography* 37:401–14.

Bianchi, Suzanne M., and Lynne M. Casper. 2000. "American Families." *Population Bulletin* 55(4). Washington, DC: Population Reference Bureau.

Bianchi, Suzanne M., John P. Robinson, and Melissa A. Milkie. 2006. *Changing Rhythms of American Family Life*. New York: Russell Sage.

Bianchi, Suzanne M., Lekha Subaiya, and Joan P. Kahn. 1999. "The Gender Gap in the Economic Well-Being of Nonresident Fathers and Custodial Mothers." *Demography* 36:185–203.

Bickman, Keonard and Debra J. Rog. 2009. *The SAGE Handbook of Applied Social Research Methods*. 2nd ed. Los Angeles: Sage.

Biddulph, Steve and Shaaron Biddulph. 2007. *Raising a Happy Child*. London: Doring Kindersley.

Bierman, Alex, Elena M. Fazio, and Melissa A. Milkie. 2006. "A Multifaceted Approach to the Mental Health Advantage of the Married." *Journal of Family Issues* 27(4):554–82.

"Big (Lack of) Men on Campus." 2005. *USA Today*, September 23.

Billingsley, Andrew. 1968. *Black Families in White America*. Englewood Cliffs, NJ: Prentice Hall.

Binkin, Martin, Mark J. Eitelberg, Alvin J. Schexnider, and Marvin M. Smith. 1982. *Blacks and the Military: Studies in Defense Policy*. Washington, DC: The Brookings Institution.

Binning, Kevin R., Miguel M. Unzueta, Yuen J. Huo, and Ludwin E. Molina. 2009. "The Interpretation of Multiracial Status and Its Relation to Social Engagement and Psychological Well-Being." *Journal of Social Issues* 65(1):35–49.

Binstock, Georgina and Arland Thornton. 2003. "Separations, Reconciliations, and Living Apart in Cohabiting and Marital Unions." *Journal of Marriage and Family* 65:432–43.

"Biological Fathers' Rights." 2007. *Adopting.org*, March 29. www.adopting.org.

"Biology of Social Bonds." 1999. *Science News*, August 7.

Bird, Chloe E. 1997. Gender Differences in the Social and Economic Burdens of Parenting and Psychological Distress." *Journal of Marriage and Family* 59(4):809–23.

Bird, Gloria W., Rick Peterson, and Stephanie Hotta Miller. 2002. "Factors Associated with Distress among Support-Seeking Adoptive Parents." *Family Relations* 51(3):215–20.

"Birds of a Feather—More Than a Place to Live, a Way to Live." 2010. Retrieved May 17, 2010 (flock2it.com).

Biskupic, Joan. 2003. "Same-Sex Couples Are Redefining Family Law in USA." *USA Today*, February 18.

Bjelde, Kristine E. and Gregory F. Sanders. 2009. "Snowbird Intergenerational Family Relationships." *Activities, Adaptation, and Aging* 33:81–95.

Black, Dan, Gary Gates, Seth Sanders, and Lowell Taylor. 2000. "Demographics of the Gay and Lesbian Population in the United States: Evidence from Available Systematic Data Sources." *Demography* 37:139–54.

Black, Michele C. and Matthew J. Breiding. 2008. "Adverse Health Conditions and Health Risk Behaviors Associated With Intimate Partner Violence—United States, 2005." *MMWR* 57(5):114-40. Also, see *JAMA* 300(6):646-49.

Blackman, Lorraine, Obie Clayton, Norval Glenn, Linda Malone-Colon, and Alex Roberts. 2006. *The Consequences of Marriage for African Americans: A Comprehensive Literature Review*. New York: Institute for American Values.

Blakely, Mary Kay. 1995. "An Outlaw Mom Tells All." *Ms.*, January/February, pp. 34–45.

Blakemore, Judith and Craig Hill. 2008. "The Child Gender Socialization Scale: A Measure to Compare Traditional and Feminist Parents." *Sex Roles* 58(3/4):192–207.

Blanchard, Victoria L., Alan J. Hawkins, Scott A. Baldwin, and Elizabeth B. Fawcett. 2009. "Investigating the Effects of Marriage and Relationship Education on Couples' Communication Skills: A Meta-analytic Study." *Journal of Family Psychology* 23(2):203–14.

Blanck, J. L. 2007. "Are helicopter Parents Landing in Graduate School?" *Journal of Career Planning and Employment* 68(2):35–39.

Blankenhorn, David. 1995. *Fatherless America: Confronting Our Most Urgent Social Problem*. New York: Basic Books.

———. 2007. *The Future of Marriage*. New York: Encounter Books.

Blau, Francine D., Mary C. Brinton, and David B. Grusky, eds. 2006. "The Declining Significance of Gender?" Pp. 3–34 in *The Declining Significance of Gender*, edited by Francine D. Blau, Mary C. Brinton, and David B. Grusky. New York: Russell Sage.

Blau, Francine D. and Lawrence M. Kahn. 2006. "The Gender Pay Gap: Going, Going, but Not Gone." Pp. 37–66 in *The Declining Significance of Gender*, edited by Francine Blau, Mary C. Brinton, and David B. Grusky. New York: Russell Sage.

Blau, Melinda. 1993. *Families Apart: Ten Keys to Successful Co-Parenting*. New York: Perigee.

Blazei, R. W., W. G. Iacono, and M. McGue. 2008. "Father-Child Transmission of Antisocial Behavior: The Moderating Role of Father's Presence in the Home." *Journal of the American Academy of Child and Adolescent Psychiatry* 47:406–15.

Blieszner, Rosemary, Karen A. Roberto, Karen L. Wilcox, Elizabeth J. Barham, and Brianne L. Winston. 2007. "Dimensions of Ambiguous Loss in Couples Coping with Mild Cognitive Impairment." *Family Relations* 56(2):196–209.

Blinn-Pike, Lynn. 1999. "Why Abstinent Adolescents Report They Have Not Had Sex: Understanding Sexually Resilient Youth." *Family Relations* 48:295–301.

Block, Joel D., and Susan S. Bartell 2001. *Stepliving for Teens: Getting along with Step-Parents, Parents, and Siblings*. New York: Penguin Young Readers Group.

Blood, Robert O., Jr. and Donald M. Wolfe. 1960. *Husbands and Wives: The Dynamics of Married Living*. New York: Free Press.

Blow, Adrian J. and Kelley Hartnett. 2005. "Infidelity in Committed Relationships II: A Substantive Review." *Journal of Marital and Family Therapy* 31(2):217–33.

Blum, Deborah. 1997. *Sex on the Brain: The Biological Differences between Men and Women*. New York: Penguin.

Blumberg, Rae Lesser and Marion Tolbert Coleman. 1989. "A Theoretical Look at the Gender Balance of Power in the American Couple." *Journal of Family Issues* 10:225–50.

Blumer, Herbert. 1969. *Social Interactionism: Perspective and Method*. Berkeley, CA: University of California Press.

Blumstein, Philip and Pepper Schwartz. 1983. *American Couples: Money, Work, Sex*. New York: Morrow.

Bly, Robert. 1990. *Iron John: A Book about Men*. Reading, MA: Addison-Wesley.

Bock, Jane D. 2000. "'Doing the Right Thing?' Single Mothers by Choice and the Struggle for Legitimacy." *Gender and Society* 14:62–86.

Bodenmann, Guy, Thomas Bradbury, and Sandrine Pihet. 2009. "Relative Contributions of Treatment-Related Changes in Communication Skills and Dyadic Coping Skills to the Longitudinal Course of Marriage in the Framework of Marital Distress Prevention." *Journal of Divorce* 50(1):1–21.

Bodenmann, Guy, Thomas Ledermann, and Thomas N. Bradbury. 2007. "Stress, Sex, and Satisfaction in Marriage." *Personal Relationships* 14(4):551–69.

Bodnar, Janet. 2007. "Advice for Parents of Boomerang Kids." *Kiplinger*, May 16. Retrieved December 21, 2009 (www.kiplinger.com).

"Body Image." n.d. Children, Youth, and Women's Health Service: Kids' Health. Retrieved May 6, 2008 (www.cyh.com).

Bogaert, Anthony F. 2004. "Asexuality: Prevalence and Associated Factors in a National Probability Sample." *Journal of Sex Research* 41(3):279–83.

Bogenschneider, Karen. 2006. *Family Policy Matters: How Policymaking Affects Families and What Professionals Can Do*. 2nd ed. Mahwah, NJ: Erlbaum.

Bogle, Kathleen A. 2004. "From Dating to Hooking Up: The Emergence of a New Sexual Script." Unpublished PhD dissertation, Department of Sociology, University of Delaware. Newark, DE.

———. 2008. *Hooking Up: Sex, Dating, and Relationships on Campus*. New York: New York University Press.

Bohannan, Paul. 1970. "The Six Stations of Divorce." Pp. 29–55 in *Divorce and After*, edited by Paul Bohannan. New York: Doubleday.

Bolzendahl, Catherine I. and Daniel J. Myers. 2004. "Feminist Attitudes and Support for Gender Equality: Opinion Change in Women and Men, 1974–1998." *Social Forces* 83(2):759–90.

Bonach, Kathryn. 2007. "Forgiveness Intervention Model: Application to Coparenting Post-Divorce." *Journal of Divorce and Remarriage* 48(1/2):105–23.

"Boomerang Kids Contract." n.d. Retrieved December 21, 2009 (boomerangkidshelp.com).

Boon, Susan D., Vicki L. Deveau, and Alishia M. Allbhal. 2009. "Payback: The Parameters of Revenge in Romantic Relationships." *Journal of Social and Personal Relationships* 26(6–7):747–68.

Boonstra, Heather D., Rachel Benson Gold, Cory L. Richards, and Lawrence B. Finer. 2006. *Abortion in Women's Lives*. New York: Guttmacher Institute.

Booth, Alan, Karen Carver, and Douglas A. Granger. 2000. "Biosocial Perspectives on the Family." *Journal of Marriage and Family* 62(4):1018–34.

Booth, Alan and Ann C. Crouter, eds. 2002. *Just Living Together: Implications of Cohabitation for Children, Families, and Social Policy*. Mahwah, NJ: Erlbaum.

Booth, Alan, Ann C. Crouter, and Mari Clements, eds. 2001. *Couples in Conflict*. Mahwah, NJ: Erlbaum.

Booth, Alan, Douglas Granger, Allan Mazur, and Katie Kivlighan. 2006. "Testosterone and Social Behavior." *Social Forces* 85(1):167–91.

Booth, Alan, David R. Johnson, and Douglas A. Granger. 2005. "Testosterone, Marital Quality, and Role Overload." *Journal of Marriage and Family* 67(2):483–98.

Booth, Alan, Elisa Rustenbach, and Susan McHale. 2008. "Early Family Transitions and Depressive Symptom Changes from Adolescence to Early Adulthood." *Journal of Marriage and Family* 70(1):3–14.

Booth, Alan, Mindy E. Scott, and Valarie King. 2010. "Father Residence and Adolescent Problem Behavior: Are Youth Always Better Off in Two-Parent Families?" *Journal of Family Issues* 31(5):585–605.

Booth, Cathy. 1977. "Wife-Beating Crosses Economic Boundaries." *Rocky Mountain News*, June 17.

Boraas, Stephanie and William R. Rodgers III. 2003. "How Does Gender Play a Role in the Earnings Gap? An Update." *Monthly Labor Review*, March, pp. 9–15.

Borbely, James Marschall. 2009. "U.S. Labor Market in 2008: Economy in Recession." *Monthly Labor Review* 132(3).

Bornstein, Danica R., Jake Fawcett, Marianne Sullivan, Kirsten D. Senturia, and Sharyne Shiu-Thornton. 2006. "Understanding the Experiences of Lesbian, Bisexual and Trans Survivors of Domestic Violence: A Qualitative Study." *Journal of Homosexuality* 51(1):159–81.

Bosman, Julie. 2006. "Hey, Just Because He's Divorced Doesn't Mean He Can't Sell Things." *New York Times*, August 17.

Boss, Pauline. 1980. "Normative Family Stress: Family Boundary Changes across the Lifespan." *Family Relations* 29:445–52.

———. 1987. "Family Stress." Pp. 695-723 in *Handbook of Marriage and Family*, edited by M. B. Sussman and Suzanne K. Steinmets. New York: Plenum.

———. 1997. "Ambiguity: A Factor in Family Stress Management." St. Paul, MN: University of Minnesota Extension Service (http://www.extension.umn.edu).

———. 2002. *Family Stress Management*. 2nd ed. Newbury Park, CA: Sage Publications.

———. 2004. "Ambiguous Loss Research, Theory, and Practice: Reflections after 9/11." *Journal of Marriage and Family* 66(3):551–66.

———. 2007. "Ambiguous Loss Theory: Challenges for Scholars and Practitioners." *Family Relations* 56(2):105–111.

Bosse, Irina. 1999. "Oxytocin: A Hormone for Love." *Futureframe: International Webzine for Science and Culture.* Retrieved August 30, 2006 (www.morgenwelt.de/futureframe/9908-oxytocin.htm).

Bost, Kelly K., Martha J. Cox, and Chris Payne. 2002. "Structural and Supportive Changes in Couples' Family and Friendship Networks across the Transition to Parenthood." *Journal of Marriage and Family* 64(2):517–31.

Boston Women's Health Book Collective. 1998. *Our Bodies, Ourselves for the New Century.* New York: Touchstone/Simon and Schuster.

Bottke, Allison. 2010. *Setting Boundaries with Your Aging Parents.* Eugene, OR: Harvest House Publishers.

Bouchard, Emily. n.d. "Navigating Parenting Differences." SelfGrowth.com. Retrieved September 21, 2004. (www.selfgrowth.com/articles/Bouchard2.html).

Bould, Sally. 2003. "Caring Neighborhoods: Bringing Up the Kids Together." *Journal of Family Issues* 24:427–47.

Bourdieu, Pierre and Jean-Claude Passeron. 1979. *The Inheritors: French Students and their Relations to Culture.* Chicago, IL: University of Chicago Press.

Boushey, Heather. 2005a. "Are Women Opting Out? Debunking the Myth." Briefing Paper. Washington, DC: Center for Economic and Policy Research. November. Retrieved March 28, 2007 (www.cepr.net).

———. 2005b. "'Baby Panic' Book Skews Data, Misses Actual Issue." *Viewpoints,* July 12. Washington, DC: Economic Policy Institute. Retrieved June 16, 2005 (www.epinet.org).

———. 2006. "Are Mothers Really Leaving the Workplace?" Issue Brief. Chicago, IL: Council on Contemporary Families. Retrieved March 28, 2007 (www.contemporaryfamiiles.org).

Bouton, Katherine. 1987. "Fertility and Family." *Ms.,* April, p. 92.

Bowen, Gary L., Jay A. Mancini, James A. Martin, William B. Ware, and John P. Nelson. 2003. "Promoting the Adaptation of Military Families: An Empirical Test of a Community Practice Model." *Family Relations* 52(1):33–44.

Bowen, Gary L., Roderick A. Rose, Joelle D. Powers, and Elizabeth J. Glennie. 2008. "The Joint Effects of Neighborhoods, Schools, Peers, and Families on Changes in the School Success of Middle School Children." *Family Relations* 57(4):504–516.

Bowers v. Hardwick. 1986. 478 U.S. 186, 92 L. Ed. 2d 140, 106 S. Ct. 2841.

Bowlby, John. 1982. *Attachment and Loss.* 2nd ed. New York: BasicBooks.

———. 1988. A Secure Base. London: Routledge.

Bowleg, Lisa, Jennifer Huang, Kelly Brooks, Amy Black, and Gary Burkholder. 2003. "Triple Jeopardy and Beyond: Multiple Minority Stress and Resilience Among Black Lesbians." *Journal of Lesbian Studies* 7(4):87–108.

Boylan, Jennifer Finney. 2009. "Is My Marriage Gay?" *The New York Times,* May 12. Retrieved October 25, 2009 (www.nytimes.com).

Boylorn, Robin M. 2008. "As Seen on TV: An Autoethnographic Reflection on Race and Reality Television." *Critical Studies in Media Communication* 25(4):413–33.

"Boys' Academic Slide Calls for Accelerated Attention." 2003. *USA Today,* December 22.

"Boys' Turn: Now *They* Need Help Getting into Colleges." 2006. *USA Today,* April 3.

Bracke, Piet, Wendy Christiaens, and Naomi Wauterickx. 2008. "The Pivotal Role of Women in Informal Care." *Journal of Family Issues* 29(10):1348–78.

Bradbury, Thomas N., Frank D. Fincham, and Steven R. H. Beach. 2000. "Research on the Nature and Determinants of Marital Satisfaction: A Decade in Review." *Journal of Marriage and Family* 62(4):964–80.

Bradbury, Thomas N. and Benjamin R. Karney. 2004. "Understanding and Altering the Longitudinal Course of Marriage." *Journal of Marriage and Family* 66(6):862–79.

Braithwaite, Dawn O. and Leslie A. Baxter. 2006. *Engaging Theories in Family Communication: Multiple Perspectives.* Thousand Oaks, California: Sage.

Bramlett, Matthew D. and William D. Mosher. 2001. "First Marriage Dissolution, Divorce, and Remarriage: United States." *Advance Data from Vital and Health Statistics,* No. 323. Hyattsville, MD: U.S. National Center for Health Statistics.

Brand, Bethany L. and Pamela C. Alexander. 2003. "Coping with Incest: The Relationship Between Recollections of Childhood Coping and Adult Functioning in Female Survivors of Incest." *Journal of Traumatic Stress* 16(3):285–92.

Branden, Nathaniel. 1988. "A Vision of Romantic Love." Pp. 218–31 in *The Psychology of Love,* edited by Robert J. Sternberg and Michael L. Barnes. New Haven, CT: Yale University Press.

Brandl, Bonnie. 2007. *Elder Abuse Detection and Intervention: A Collaborative Approach.* New York: Springer.

Brandon, Emily. 2008. "Baby Boomers Moving in with Adult Children." *U.S. News & World Report,* November 20. Retrieved February 6, 2009 (www.usnews.com).

Braschi v. Stahl Associates Company. 1989. 74 N.Y. 2d 201.

Bratter, Jenifer L. and Karl Eschbach. 2006. "'What About the Couple?' Interracial Marriage and Psychological Distress." *Social Science Research* 35(4):1025–47.

Bratter, Jenifer L. and Rosalind B. King. 2008. "But Will It Last?": Marital Instability Among Interracial and Same-Race Couples." *Family Relations* 57(April):160–71.

Braver, Sanford, Jennesa R. Shapiro, and Matthew R. Goodman. 2006. "Consequences of Divorce for Parents." Pp. 313–37 in *Handbook of Divorce and Relationship Dissolution,* edited by Mark A. Fine and John H. Harvey. Mahwah, NJ: Erlbaum.

Braver, Sanford L., Pamela J. Fitzpatrick, and R. Curtis Bay. 1991. "Noncustodial Parent's Report of Child Support Payments." *Family Relations* 40(2):180–85.

Bray, James H. 1999. "From Marriage to Remarriage and Beyond." Pp. 253–71 in *Coping with Divorce, Single Parenting and Remarriage,* edited by E. Mavis Hetherington. Mahwah, NJ: Erlbaum.

Bray, J. H., and I. Easling. 2005. Pp. 267–94 in *Family Psychology: The Art of the Science,* edited by W. M. Pinsof and J. L. Lebow. New York: Oxford Press.

Brazelton, T. Berry, and Stanley Greenspan. 2000. "Our Window to the Future." *Newsweek* Special Issue, Fall/Winter, pp. 34–36.

Brehm, Sharon S., Rowland S. Miller, Daniel Perlman, and Susan M. Campbell. 2002. *Intimate Relationships.* 3rd ed. New York: McGraw-Hill.

Breiding, Matthew J., Michele C. Black, and George W. Ryan. 2008. "Chronic Disease and Health Risk Behaviors Associated with Intimate Partner Violence—18 U.S. States/Territories, 2005." *Annals of Epidemiology* 18(7):538–44.

Breitenbecher, Kimberly Hanson. 2006. "The Relationships among Self-blame, Psychological Distress, and Sexual Victimization." *Journal of Interpersonal Violence* 21(5):597–611.

Brennan, Bridget. 2003. "No Time. No Sex. No Money." *First Years and Forever: A Monthly Online Newsletter for Marriages in the Early Years.* Chicago, IL: Archdiocese of Chicago, Family Ministries. Retrieved September 8, 2006 (www.familyministries.org).

Brennan, Robert T., Rosalind Chait Barnett, and Karen C. Gareis. 2001. "When He Earns More Than She Does: A Longitudinal Study of Dual Earner Couples." *Journal of Marriage and Family* 63:168–80.

Brenner, Aprile D. and Su Yeong Kim. 2009. "Intergenerational Experiences of Discrimination in Chinese American Families: Influences of Socialization and Stress." *Journal of Marriage and Family* 71(4):862–77.

Brenner, N. L. Kann, R. Lowry, H. Wechsler, and L. Romero. 2006. "Trends in HIV-Related Risk Behaviors among High School Students—United States, 1991–2005." *Morbidity and Mortality Weekly Review* 55(31):851–54.

Bretherton, I. 1992. "The Origins of Attachment Theory: John Bowlby and Mary Ainsworth." *Developmental Psychology* 28:759ff.

Brewster, Karin L. and Irene Padavic. 2002. "No More Kin Care? Change in Black Mothers' Reliance on Relatives for Child Care, 1977–94." *Gender and Society* 16:546–63.

Brickley, Margie, Aimee Gelnaw, Hilary Marsh, and Daniel Ryan. 1999. "Opening Doors: Lesbian and Gay Parents and Schools." Educational Advocacy Committee of the Family Pride Coalition (www.familypride.org).

Brimhall, Andrew S. and Mark H. Butler. 2007. "Intrinsic vs. Extrinsic Religious Motivation and the Marital Relationship." *American Journal of Family Therapy* 35:235–49.

Brimhall, Andrew, Karen Wampler, and Thomas Kimball. 2008. "Learning from the Past, Altering the Future: A Tentative Theory of the Effect of Past Relationships on Couples Who Remarry." *Family Process* 47(3):373–87.

Brines, Julie. 1994. "Economic Dependency, Gender, and the Division of Labor at Home." *American Journal of Sociology* 100(3):652–88.

Brinig, Margaret F. and Douglas W. Allen. 2000. "'These Boots Are Made for Walking': Why Most Divorce Filers Are Women." *American Law and Economics Review* 2(1):126–69.

Brink, Susan. 2008. "Modern Puberty." *Los Angeles Times*, January 21.

Britz, Jennifer Delahunty. 2006. "To All the Girls I've Rejected." *New York Times*, March 23.

Brock, Rebecca L. and Erika Lawrence. 2009. "Too Much of a Good Thing: Underprovision Versus Overprovision of Partner Support." *Journal of Family Psychology* 23(2):181–92.

Broderick, Carlfred B. 1979a. *Couples: How to Confront Problems and Maintain Loving Relationships*. New York: Simon and Schuster.

———. 1979b. *Marriage and the Family*. Englewood Cliffs, NJ: Prentice Hall.

Brodie, Deborah. 1999. *Untying the Knot: Ex-Husbands, Ex-Wives, and Other Experts on the Passage of Divorce*. New York: St. Martin's Griffin.

Brodkin, Adele M. 2006. *Raising Happy and Successful Kids: A Guide for Parents*. New York: Scholastic Books.

Brody, Gene, Yi Fu Chen, Steven Kogan, Velma McBride Murray, Patricia Logan, and Zupei Luo. 2008. "Linking Perceived Discrimination to Longitudinal Changes in African American Mothers' Parenting Practices." *Journal of Marriage and Family* 70(2):319–31.

Brody, Jane E. 2004. "Abstinence-Only: Does It Work?" *New York Times*, June 3.

———. 2008. "When Families Take Care of Their Own." *The New York Times*, November 11. Retrieved March 16, 2010 (www.nytimes.com).

Broman, Clifford L., Li Xin, and Mark Reckase. 2008. "Family Structure and Mediators of Adolescent Drug Use." *Journal of Family Issues* 29(12):1625–49.

Bronfenbrenner, Urie. 1979. *The Ecology of Human Development: Experiments by Nature and Design*. Cambridge, MA: Harvard University Press.

Bronte-Tinkew, Jacinta, Jennifer Carrano, Allison Horowitz, and Akemi Kinukawa. 2008. "Involvement among Resident Fathers and Links to Infant Cognitive Outcomes." *Journal of Family Issues* 29(9):1211–44.

Bronte-Tinkew, Jacinta and Allison Horowitz. 2010. "Factors Associated with Unmarried, Nonresident Fathers' Perceptions of Their Coparenting." *Journal of Family Issues* 31(1):31–65.

Bronte-Tinkew, Jacinta, Allison Horowitz, and Mindy E. Scott. 2009. "Fathering with Multiple Partners: Links to Children's Well-Being in Early Childhood." *Journal of Marriage and Family* 71(3):608–31.

Brooke, Jill. 2004. "Close Encounters with a Home Barely Known." *New York Times*, July 22.

———. 2006. "Home Alone Together." *New York Times*, May 4. Retrieved May 7, 2006 (www.nytimes.com).

Brooks, David. 2004. "The New Red-Diaper Babies." *New York Times*, December 7.

———. 2006. "Immigrants to Be Proud Of." *New York Times*, March 30.

Brooks, Robert and Sam Goldstein. 2001. *Raising Resilient Children: Fostering Strength, Hope, and Optimism in Your Child*. New York: Contemporary Books.

Broome, Claire V. 1995. "Change in the Marriage and Divorce Data Available from *the National* Center for Health Statistics." *Federal Register* 60, No. 241: 64437–38.

Brotherson, Sean. 2003. "Time, Sex, and Money: Challenges in Early Marriage." *The Meridian*. Retrieved September 8, 2006 (www.meridianmagazine.com).

Brougham, Ruby R. and David A. Walsh. 2009. "Early and Late Retirement Exits." *International Journal of Aging and Human Development* 69(4):267–86.

Brown, Carrie. 2008. "Gender-Role Implications on Same-Sex Intimate Partner Abuse." *Journal of Family Violence* 23:475–62.

Brown, Dave and Phil Waugh. 2004. *Covenant vs. Contract*. New York: Franklin, Son Publishers.

Brown, Edna, Terri L. Orbuch, and Artie Maharaj. 2010. "Social Networks and Marital Stability among Black American and White American Couples." Pp. 318–34 in *Support Processes in Intimate Relationships*, edited by Kieran T. Sullivan and Joanne Davila. New York: Oxford University Press.

Brown, Grant A. 2004. "Gender as a Factor in the Response of the Law-Enforcement System to Violence Against Partners." *Sexuality and Culture* 8(3–4):1–139.

Brown, Patricia Leigh. 2004. "For Children of Gays, Marriage Brings Joy." *New York Times*, March 19.

———. 2006. "Supporting Boys or Girls When the Line Isn't Clear." *New York Times*, December 4.

Brown, Susan L. 2004. "Family Structure and Child Well-Being: The Significance of Parental Cohabitation." *Journal of Marriage and Family* 66(2):351–67.

Brown, Susan L. and Jennifer Roebuck Bulanda. 2008. "Relationship Violence in Young Adulthood: A Comparison of Daters, Cohabitors, and Marrieds." *Social Science Research* 37(1):73–87.

Brown, Susan L. and Wendy D. Manning. 2009. "Family Boundary Ambiguity and the Measurement of Family Structure: The Significance of Cohabitation." *Demography* 46(1):85–101.

Brown, Tiffany, Miriam Linver, Melanie Evans, and Donna DeGennaro. 2009. "African American Parents' Racial and Ethnic Socialization and Adolescent Academic Grades: Teasing Out the Role of Gender." *Journal of Youth & Adolescence* 38(2):214–27.

Brown, Tony, Emily Tanner-Smith, Chase Lesane-Brown, and Michael Ezell. 2007. "Child, Parent, and Situational Correlates of Familiar Ethnic/Race Socialization." *Journal of Marriage and Family* 69(1):14–25.

Brownridge, Douglas A. 2006. "Partner Violence Against Women With Disabilities: Prevalence, Risk, and Explanations." *Violence Against Women* 12(9):805–22.

Brownridge, Douglas A. and Shiva Halli. 2002. "Understanding Male Partner Violence against Cohabiting and Married Women: An Empirical Investigation with a Synthesized Model." *Journal of Family Violence* 17(4):341–61.

Brubaker, Ellie, Mary Anne Gorman, and Michele Hiestand. 1990. "Stress Perceived by Elderly Recipients of Family Care." Pp. 267–81 in *Family Relationships in Later Life*, 2nd ed., edited by Timothy H. Brubaker. Newbury Park, CA: Sage Publications.

Brumbaugh, Stacey M., Laura A. Sanchez, Steven L. Nock, and James D. Wright. 2008. "Attitudes Toward Gay Marriage in States Undergoing Marriage Law Transformation." *Journal of Marriage and Family* 70(2):345–59.

Brush, Lisa D. 2008. "Book Review." *Gender and Society* 22(1):126–42.

Bryan, Willie V. 2010. *Sociopolitical Aspects of Disabilities: The Social Perspectives and Political History of Disabilities and Rehabilitation in the United States*. Springfield, IL: Charles C. Thomas.

Bubolz, Margaret M. and M. Suzanne Sontag. 1993. "Human Ecology Theory." Pp. 419–48 in *Sourcebook of Family Theories and Methods: A Contextual Approach*, edited by Pauline G. Boss, William J. Doherty, Ralph LaRossa, Walter R. Schumm, and Suzanne K. Steinmetz. New York: Plenum.

Buchanan, Christy M., Eleanor E. Maccoby, and Sanford M. Dornbusch. 1996. *Adolescents After Divorce*. Cambridge, MA: Harvard University Press.

———. 2006. "Poll Finds U.S. Warming to Gay Marriage." *San Francisco Chronicle*, March 23.

Buehler, Cheryl. 1995. "Divorce Law in the United States." Pp. 99–120 in *Families and Law*, edited by Lisa J. McIntyre and Marvin B. Sussman. New York: Haworth.

Buehler, Cheryl, Ambika Krishnakumar, Gaye Stone, Christine Anthony, Sharon Pemberton, Jean Gerard, and Brian K. Barber. 1998. "Interpersonal Conflict Styles and Youth Problem Behaviors." *Journal of Marriage and Family* 60(1):119–32.

Bukhari, Zahid Hussain. 2004. *Muslims' Place in the American Public Square: Hope, Fears, and Aspirations*. Walnut Creek, CA: AltaMira.

Bulcroft, Kris, Richard Bulcroft, Linda Smeins, and Helen Cranage. 1997. "The Social Construction of the North American Honeymoon, 1800–1995." *Journal of Family History* 22(4):462–91.

Bulcroft, Richard and Jay Teachman. 2004. "Ambiguous Constructions: Development of a Childless or Child-free Life Course." Pp. 116–35 in *Handbook of Contemporary Families: Considering the Past; Contemplating the Future,* edited by Marilyn Coleman and Lawrence H. Ganong. Thousand Oaks, CA: Sage Publications.

Bumpass, Larry L. and Hsien-Hen Lu. 2000. "Trends in Cohabitation and Implications for Children's Family Contexts in the United States." *Population Studies* 54:29–41.

Bumpass, Larry L., James A. Sweet, and Andrew Cherlin. 1991. "The Role of Cohabitation in Declining Rates of Marriage." *Journal of Marriage and Family* 53(4):913–27.

Bundy-Fazioli, Kimberly, Marc Winokur, and Tobi DeLong-Hamilton. 2009. "Placement Outcomes for Children Removed for Neglect." *Child Welfare* 88(3):85–102.

Burbach, Mary and Mary Ann Lamanna. 2000. "The Moral Mother: Motherhood Discourse in Biological Father and Third Party Cases." *Journal of Law and Family Studies* 2:153–97.

Burch, Rebecca and Gordon G. Gallup, Jr. 2004. "Pregnancy as a Stimulus for Domestic Violence." *Journal of Family Violence* 19:243–47.

Burchinal, Margaret, Nathan Vandergrift, Robert Pianta, and Andrew Mashburn. 2010. "Threshold Analysis of Association Between Child Care Quality and Child Outcomes for Low-Income Children in Pre-Kindergarten Programs." *Early Childhood Research Quarterly* 25(2):166–76.

Bures, Regina M. 2009. "Living Arrangements Over the Life Course." *Journal of Family Issues* 30(5):579–85.

Bures, Regina M., Tanya Koropeckyj-Cox, and Michael Loree. 2009. "Childlessness, Parenthood, and Depressive Symptoms among Middle-Aged and Older Adults." *Journal of Family Issues* 30(5):670–87.

Burgess, Ernest and Harvey Locke. 1953 [1945]. *The Family: From Institution to Companionship.* New York: American.

Burgess, Wes. 2008. *The Bipolar Handbook for Children, Teens, and Families: Real-Life Questions with Up-To-Date Answers.* New York: Avery.

Burke, Tod W. and Stephen S. Owen. 2006. "Same-sex Domestic Violence: Is Anyone Listening?" *Gay and Lesbian Review Worldwide* 13(1):6–7.

Burkett, Elinor. 2000. *The Baby Boon: How Family-Friendly America Cheats the Childless.* New York: Free Press.

Burns, A. and R. Homel. 1989. "Gender Division of Tasks by Parents and Their Children." *Psychology of Women Quarterly* 13:113–25.

Burns, George W. 2010. *Happiness, Healing, Enhancement: Your Casebook Collection for Applying Positive Psychology in Therapy.* Hoboken, NJ: Wiley.

Burpee, Leslie C. and Ellen J. Langer. 2005. "Mindfulness and Marital Satisfaction." *Journal of Adult Development* 12(1):1281–87.

Burr, Jeffrey A. and Jan E. Mutchler. 1999. "Race and Ethnic Variation in Norms of Filial Responsibility among Older Persons." *Journal of Marriage and Family* 61(3):674–87.

Burr, Wesley R., Shirley Klein, and Marilyn McCubbin. 1995. "Reexamining Family Stress: New Theory and Research." *Journal of Marriage and Family* 57(3):835–46.

Burrell, Ginger L. and Mark W. Roosa. 2009. "Mothers' Economic Hardship and Behavior Problems in Their Early Adolescents." *Journal of Family Issues* 30(4):511–31.

Burt, Sandra and Linda Perlis. 2006. *Raising a Successful Child: Discover and Nurture Your Child's Talents.* Berkeley, CA: Ulysses/Enfield Publishers Group.

Burton, Linda M., Andrew Cherlin, Donna-Marie Winn, Angela Estacion, and Clara Holder-Taylor. 2009. "The Role of Trust in Low-Income Mothers' Intimate Unions." *Journal of Marriage and Family* 71(December):1107–24.

Burton, Linda M. and M. Belinda Tucker. 2009. "Romantic Unions in an Era of Uncertainty: A Post-Moynihan Perspective on African American Women and Marriage." *The Annals of the American Academy of Political and Social Science* 621(1):132–48.

Busby, Dean M. and Thomas B. Holman. 2009. "Perceived Match or Mismatch on the Gottman Conflict Styles: Associations with Relationship Outcome Variables." *Family Process* 48(4):531–45.

Buss, David M. 2009. "Darwin and the Emergence of Evolutionary Psychology." *American Psychologist* 64(2):140–8.

Buss, David M. and Todd K. Shackelford. 2008. "Attractive Women Want It All: Good Genes, Economic Investment, Parenting Proclivities, and Emotional Commitment." *Evolutionary Psychology* 6:134–46.

Buss, D. M., Todd K. Shackelford, Lee A. Kirkpatrick, and Randy J. Larsen. 2001. "A Half Century of Mate Preferences: The Cultural Evolution of Values." *Journal of Marriage and Family* 63(2):491–503.

Bussey, K. and A. Bandura. 1999. "Social Cognitive Theory of Gender Development and Differentiation." *Psychological Review* 106:676–713.

Butler, Amy C. 2005. "Gender Differences in the Prevalence of Same-Sex Sexual Partnering: 1998–2002." *Social Forces* 84(1):421–39.

Butler, Judith. 1990. *Gender Trouble: Feminism and the Subversion of Identity.* New York: Routledge.

Butler, Katy. 2006a. "Beyond Rivalry: A Hidden World of Sibling Violence." *New York Times,* February 28.

———. 2006b. "Many Couples Must Negotiate Terms of 'Brokeback' Marriages." *New York Times,* March 7.

Butt, Riazat. 2009. "Church Throws Open Female Bishops Dispute. *The Guardian,* February 12. Retrieved January 14, 2010 (www.guardian.co.uk).

Button, Deeanna M. 2008. "Social Disadvantage and Family Violence: Neighborhood Effects on Attitudes About Intimate Partner Violence and Corporal Punishment." *American Journal of Criminal Justice* 33(September):130–47.

Button, Deeanna and Roberta Gealt. 2010. "High Risk Behaviors Among Victims of Sibling Violence." *Journal of Family Violence* 25(2):131–40.

Byers, E. Sandra. 2005. "Relationship Satisfaction and Sexual Satisfaction: A Longitudinal Study of Individuals in Long-Term Relationships." *Journal of Sex Research* 42(2):113–18.

Byrd, Stephanie Ellen. 2009. "The Social Construction of Marital Commitment." *Journal of Marriage and Family* 71(2):318–36.

Bystrom, Dianne. 2006. "Advertising, Web Sites and Media Coverage: Gender and Communication Along the Campaign Trail." Pp. 169–88 in *Gender and Elections: Shaping the Future of American Politics,* edited by Susan J. Carroll and Richard L. Fox. New York: Cambridge University Press.

Bzostek, Sharon H. 2008. "Social Fathers and Child Well-Being." *Journal of Marriage and Family* 70(4):950–61.

Bzostek, Sharon, M. J. Carlson, and Sara McLanahan. 2007. "Repartnering after a Nonmarital Birth: Does Mother Know Best?" Working paper #2006-27-FF, Center for Research on Child Wellbeing. Princeton, NJ: Princeton University.

Cabrera, Natasha J., Jay Fagan, and Danielle Farrie. 2008a. "Explaining the Long Reach of Fathers' Prenatal Involvement on Later Paternal Engagement" *Journal of Marriage and Family* 70(5):1094–1107.

Caldera, Y. M., A. C. Huston, and M. O'Brien. 1989. "Social Interactions and Play Patterns of Parents and Toddlers with Feminine, Masculine, and Neutral Toys." *Child Development* 60:70–76.

Caldwell, John. 1982. *Theory of Fertility Decline.* London, England: Academic Press.

Call, Vaughn, Susan Sprecher, and Pepper Schwartz. 1995. "The Incidence and Frequency of Marital Sex in a National Sample." *Journal of Marriage and Family* 57(3):639–52.

Callahan, Daniel. 1985. "What Do Children Owe Elderly Parents? Toward a Policy that Promotes, not Corrupts, Family Bonds." *Hastings Center Report* (April):32–37.

Calzo, Jerel P. and L. Monique Ward. 2009. "Contributions of Parents, Peers, and Media to Attitudes Toward Homosexuality: Investigating Sex and Ethnic Differences." *Journal of Homosexuality* 56(8):1101–16.

Campbell, Bernadette, E. Glenn Schellenberg, and Charlene Y. Senn. 1997. "Evaluating Measures of Contemporary Sexism." *Psychology of Women Quarterly* 1(1):89–102.

Campbell, Ross and Rob Suggs. 2006. *How to Really Parent Your Teenager: Raising Balanced Teens in an Unbalanced World.* Nashville, TN: W Publishing Group.

Campbell, Susan. 2004. *Truth in Dating: Finding Love by Getting Real.* Tiburon, CA: H. J. Kramer/New World Library.

Campbell, Susan Miller and Marcia L. Collaer. 2009. "Stereotype Threat and Gender Differences in Performance on a Novel Visuospatial Task." Psychology of Women Quarterly 33(4):437–44.

Canary, D. J. and K. Dindia. 1998. *Sex Differences and Similarities in Communication.* Mahwah, NJ: Erlbaum.

Cancian, Francesca M. 1985. "Gender Politics: Love and Power in the Private and Public Spheres." Pp. 253–64 in *Gender and the Life Course,* edited by Alice S. Rossi. New York: Aldine.

———. 1987. *Love in America: Gender and Self-Development.* New York: Cambridge University Press.

Cancian, Francesca M. and Stacey J. Oliker. 2000. *Caring and Gender.* Walnut Creek, CA: AltaMira.

Canedy, Dana. 2001. "Often Conflicted, Hispanic Girls Are Dropping Out at High Rates." *New York Times,* March 25.

Cantor, M. H. 1979. "Neighbors and Friends: An Overlooked Resource in the Informal Support System." *Research on Aging* 1:434–63.

Capizzano, Jeffrey, Gina Adams, and Jason Ost. 2006. *Caring for Children of Color: The Child Care Patterns of White, Black and Hispanic Children.* Washington, DC: The Urban Institute.

Capps, Randy, Rosa M. Castaneda, Ajay Chaudry, and Robert Santos. 2007. "Paying the Price: The Impact of Immigration Raids on America's Children." Washington, DC: Urban Institute For the National Council of La Raza.

Careerbuilder.com. 2007. "Thirty-Seven Percent of Working Dads Would Leave Their Jobs if Their Family Could Afford It." *Careerbuilder.com,* June 11.

Caregiver Support Blog. 2008 (October 13). Retrieved March 13, 2010, (caregiversupport. wordpress.com).

Carey v. Population Services International. 1977. 431 U.S. 678, 52 L. Ed. 2d 675, 97 S. Ct. 2010.

Carey, Benedict. 2004. "Long after Kinsey, Only the Brave Study Sex." *New York Times,* November 9.

———. 2007. "Study Finds Rise in Behavior Problems after Significant Time in Child Care." *New York Times,* March 26.

———. 2008. "U.S. Panel Calls for Brain Injury Screening for Troops." *New York Times,* December 5.

Carlin, Diana B. and Kelly L. Winfrey. 2009. "Have You Come a Long Way, Baby? Hillary Clinton, Sarah Palin, and Sexism in 2008 Campaign Coverage." *Communication Studies* 60(4):326–43.

Carlson, Bonnie E., Katherine Maciol, and Joanne Schneider. 2006. "Sibling Incest: Reports From Forty-One Survivors." *Journal of Child Sexual Abuse* 15(4):19–34.

Carlson, Elwood. 2009. "20th-Century U.S. Generations." *Population Bulletin* 64(1). Retrieved June 10, 2009 (www.prb.com).

Carlson, Marcia J. 2006. "Family Structure, Father Involvement, and Adolescent Behavior

Outcomes." *Journal of Marriage and Family* 68(1):137–54.

Carlson, Marcia J. and Frank F. Furstenberg, Jr. 2006. "The Prevalence and Correlates of Multipartnered Fertility among Urban U.S. Parents." *Journal of Marriage and Family* 68(3):718–32.

Carlton, Erik., Jason Whiting, Kay Bradford, Patricia Hyjer Dyk, and Ann Vail. 2009. "Defining Factors of Successful University-Community Collaborations: An Exploration of One Healthy Marriage Project." *Family Relations* 58(1):28–40.

Carney, Michelle, Fred Buttell, and Don Dutton. 2007. "Women Who Perpetrate Intimate Partner Violence: A Review of the Literature with Recommendations for Treatment." *Aggression and Violent Behavior* 12(1):108–15.

Carnoy, Martin and David Carnoy. 1995. *Fathers of a Certain Age: The Joys and Problems of Middle-Aged Fatherhood.* Minneapolis, MN: Fairview Press.

Carr, Anne and Mary Stewart Van Leeuwen, eds. 1996. *Religion, Feminism, and the Family.* Louisville, KY: Westminster John Knox Press.

Carr, Deborah. 2004. "The Desire to Date and Remarry among Older Widows and Widowers." *Journal of Marriage and Family* 66(4):1051–68.

Carré, Justin M. and Cheryl M. McCormick. 2008. "Aggressive Behavior and Change in Salivary Testosterone Concentrations Predict Willingness to Engage in a Competitive Task." *Hormones and Behavior* 54(3):403–09.

Carrigan, William D. and Clive Webb. 2009. "Repression and Resistance: The Lynching of Persons of Mexican Origin in the United States, 1848–1928." Pp. 69-86 in *How the United States Racializes Latinos: White Hegemony & Its Consequences,* edited by Jose A. Cobas, Jorge Duany, and Joe R. Feagin. Boulder, CO: Paradigm Publishers.

Carroll, Jason S., Sarah Badger, and Chongming Yang. 2006. "The Ability to Negotiate or the Ability to Love? Evaluating the Developmental Domains of Marital Competence." *Journal of Family Issues* 27(7):1001–32.

Carroll, Jason S., Chad D. Olson, and Nicolle Buckmiller. 2007. "Family Boundary Ambiguity: A 30-Year Review of Theory, Research, and Measurement." *Family Relations* 56(2):210–30.

Carroll, Joseph. 2005. "Society's Moral Boundaries Expand Somewhat This Year." *Public Opinion 2005.* May 16. Retrieved March 2, 2007 (www.gallup.com).

———. 2006. "One in Four Americans Think Most Mormons Endorse Polygamy." The Gallup Poll. September 7. Retrieved September 8, 2006 (www.galluppoll.com).

———. 2007a. "Most Americans Approve of Interracial Marriages." Gallup News Service, August 16. Retrieved February 19, 2009 (www .gallup.com/poll).

———. 2007b. "Public: 'Family Values' Important to Presidential Vote." December 26. Retrieved May 20, 2010 (www.gallup.com/poll).

———. 2007c. "Stress More Common Among Younger Americans, Parents, Workers." Gallup

News Service, January 24. Retrieved February 19, 2009 (www.gallup.com/poll).

———. 2007d. "Americans: 2.5 Children Is 'Ideal' Family Size." The Gallup Poll, June 26. Retrieved April 14, 2010 (http://www .galluppoll.com).

———. 2008. "Time Pressures, Stress Common for Americans." *Gallup Poll,* January 2. www .gallup.com.

Carter, Betty. 1991. "Children's TV, Where Boys Are King." *New York Times,* May 1.

Carter, Betty and Monica McGoldrick. 1988. *The Changing Family Life Cycle: A Framework for Family Therapy.* 2nd ed. New York: Gardner.

Carter, Gerard A. 2009. "Book Review: Unmarried Couples with Children." *Journal of Marriage and Family* 71(2):432–4.

Cartmell, Todd. 2006. *Respectful Kids: The Complete Guide to Bringing Out the Best in Your Child.* Colorado Springs, CO: NavPress.

Case, Anne, I-Fen Lin, and Sara McLanahan. 2000. "How Hungry Is the Selfish Gene?" *The Economic Journal* 110(October):781–804.

Casper, Lynne M. and Suzanne M. Bianchi. 2002. *Continuity and Change in the American Family.* Thousand Oaks, CA: Sage Publications.

Cass, Julia. 2007. *Katrina's Children: Still Waiting.* Washington, DC: Children's Defense Fund.

Castro Martin, Teresa. 2002. "Consensual Unions in Latin America: Persistence of a Dual Nuptiality System." *Journal of Comparative Family Studies* 33(1):35–56.

Catalano, Shannan. 2007. *Intimate Partner Violence in the United States.* NCJ 210675 Washington, DC: U.S. Bureau of Justice Statistics, December 19. Retrieved June 9, 2010. (www.ojp.usdog.gov/bjs).

Cataldi, Emily Forrest, Jennifer Laird, and Angelina KewalRamani. 2009. "High School Dropout and Completion Rates in the United States: 2007." NCES 2009-064. Washington, DC: National Center for Education Statistics, U.S. Department of Education.

Cavanagh, Shannon E. 2008. "Family Structure History and Adolescent Adjustment." *Journal of Family Issues* 29(7):944–80.

Ceballo, Rosario, Jennifer E. Lansford, Antonia Abbey, and Abigail J. Stewart. 2004. "Gaining a Child: Comparing the Experiences of Biological Parents, Adoptive Parents, and Stepparents." *Family Relations* 53(1):38–48.

Censky, Annalyn. 2010. "Women in Top-Paying Jobs Still Make Less Than Men." *CNN Money. com,* April 20. Atlanta, GA: Cable News Network. money.cnn.com.

Center for the Advancement of Women. 2003. "Progress and Perils: New Agenda for Women." (www.advancewomen.org).

Center for the Improvement of Child Caring. n.d. "Systematic Training for Effective Parenting Programs." Retrieved February 12, 2007 (www .ciccparenting.org).

Chabot, Jennifer M. and Barbara D. Ames. 2004. "'It Wasn't "Let's Get Pregnant and Go Do It'":

Decision Making in Lesbian Couples Planning Motherhood Via Donor Insemination." *Family Relations* 53(4):348–56.

Chacko, Anil, Brian Wymbs, Lizette Flammer-Rivera, William Pelham, Kathryn Walker, Fran Arnold, Hema Visweswaraiah, Michelle Swanger-Gagne, Erin Cirio, Lauma Pirvics, and Laura Herbst. 2008. "A Pilot Study of the Feasibility and Efficacy of the Strategies to Enhance Positive Parenting (STEPP) Program for Single Mothers of Children with ADHD." *Journal of Attention Disorders* 12(3):270–80.

Chadiha, Letha A., Jane Rafferty, and Joseph Pickard. 2003. "The Influence of Caregiving Stressors, Social Support, and Caregiving Appraisal on Marital Functioning among African American Wife Caregivers." *Journal of Marital and Family Therapy* 29(4):479–90.

Chafetz, Janet Saltzman. 1989. "Marital Intimacy and Conflict: The Irony of Spousal Equality." Pp. 149–56 in *Women: A Feminist Perspective,* 4th ed., edited by Jo Freeman. Mountain View, CA: Mayfield.

Champion, Jennifer E., Sarah S. Jaser, Dristen L. Reeslund, Lauren Simmons, Jennifer E. Potts, Angela R. Shears, and Bruce E. Compas. 2009. "Caretaking Behaviors by Adolescent Children of Mothers with and without a History of Depression." *Journal of Family Psychology* 23(2):156–66.

Chandra, Anjani, Gladys M. Martinez, William D. Mosher, Joyce C. Abma, and Jo Jones. 2005. "Fertility, Family Planning, and Reproductive Health of U.S. Women: Data from the 2002 National Survey of Family Growth." *Vital and Health Statistics* 23(25). Hyattsville, MD: U.S. National Center for Health Statistics. December.

Chandra, Anjani and Elizabeth Hervey Stephen. 1998. "Impaired Fecundity in the United States: 1982–1995." *Family Planning Perspectives* 30(1):35–42.

———. 2010. "Infertility Service Use Among U.S. Women: 1995 and 2002." *Fertility and Sterility* 93(3):725–36.

Chaney, Cassandra. 2009. "The Commitment Continuum: Cohabitation and Commitment among African America Couples." *Family Focus* (Summer):F4–F6. Minneapolis, MN: National Council on Family Relations.

Chao, R. K. 1994. "Beyond Parental Control and Authoritarian Parenting Style: Understanding Chinese Parenting through the Cultural Notion of Training." *Child Development* 65(4):1111–19.

Chaplin, Tara M., Pamela M. Cole, and Carolyn Zahn-Waxler. 2005. "Parental Socialization of Emotion Expression: Gender Differences and Relations to Child Adjustment." *Emotion* 5(1):80–8.

Chappell, Crystal Lee Hyun Joo. 1996. "Korean-American Adoptees Organize for Support." *Minneapolis Star Tribune,* December 29, p. E7.

Charles, Laurie L., Dina Thomas, and Matthew L. Thornton. 2005. "Overcoming Bias toward Same-Sex Couples." *Journal of Marital and Family Therapy* 31(3):239–49.

"Chart: State Marriage License and Blood Test Requirements." 2006. Nolo. Retrieved September 7, 2006 (www.nolo.com).

Chatzky, Jeann. 2006. "Just When You Thought It Was Safe to Retire. . . . " CNN Money.com, September 21. Retrieved October 29, 2006 (money.cnn.com).

Chaves, Mark, Shawna Anderson, and Jason Byassee. 2009. "American Congregations at the Beginning of the 21st Century." *National Congregations Study.*

Cheadle, Jacob E., Paul R. Amato, and Valerie King. 2010. "Patterns of Nonresident Father Contact." Demography 47(1):205–25.

Cheeseman, Jennifer Day and Barbara Downs. 2007. *Examining the Gender Earnings Gap: Occupational Differences and the Life Course.* Housing and Household Economic Statistics Division, U.S. Census Bureau, presented at Annual Meeting of the Population Association of America, New York, March 29–31.

Cheeseman, Jennifer Day and Jeffrey Rosenthal. 2008. *Detailed Occupations and Median Earnings: 2008.* Washington, DC: U.S. Census Bureau.

Chen, Gina. 2009. "Parents Struggle To Afford Child Care in Recession." The Post Standard, May 6.

Chen, Yingyu and Sarah E. Ullman. 2010. "Women's Reporting of Sexual and Physical Assaults to Police in the National Violence Against Women Survey." *Violence Against Women* 16(3):262–79.

Cherlin, Andrew J. 1978. "Remarriage as Incomplete Institution." *American Journal of Sociology* 84:634–50.

———. 1996. *Public and Private Families.* New York: McGraw-Hill.

———. 1999. "Going to Extremes: Family Structure, Children's Well-being, and Social Science." *Demography* 36:421–28.

———. 2000. "Generation Ex-." *The Nation,* December 11 (www.thenation.com).

———. 2003. "Should the Government Promote Marriage?" *Contexts* 2(4):22–29.

———. 2004. "The Deinstitutionalization of American Marriage." *Journal of Marriage and Family* 66(4):848–61.

———. 2005. "American Marriage in the Early Twenty-First Century." *The Future of Children* 15(2):33–55.

———. 2006. "On Single Mothers 'Doing' Family." *Journal of Marriage and Family* 68(4):800–3.

———. 2008. "Public Display: The Picture-Perfect American Family? These Days, It Doesn't Exist." *Washington Post,* September 7. Retrieved October 28, 2008 (www.washingtonpost.com).

———. 2009a. *The Marriage-Go-Round: The State of Marriage and the Family in America Today.* New York: Alfred A. Knopf.

———. 2009b. "Married with Bankruptcy." *The New York Times,* May 29. Retrieved June 8, 2009 (www.nytimes.com).

Cherlin, Andrew J. and Frank F. Furstenberg, Jr. 1986. *The New American Grandparent: A Place in the Family, a Life Apart.* New York: BasicBooks.

Chesler, Phyllis. 2005 [1972]. *Women and Madness.* New York: Palgrave/Macmillan.

Child Abuse Prevention Council of Sacramento. n.d. "About Child Abuse: Cultural Customs." Retrieved May 5, 2007 (www.capcsac.org).

"Child Support Collected: DHHS Press Release." 1995. Family Law List. lawlib.wuacc.edu).

Children's Defense Fund. 2008. *Annual Report* 2007. Washington, DC: Children's Defense Fund.

———. 2009. *Annual Report* 2008. Washington, DC: Children's Defense Fund.

Childress, Sarah. 2003. "9/11's Hidden Toll." *Newsweek,* August 4, p. 37.

———. 2006. "Fighting Over the Kids." *Newsweek,* September 25, p. 35.

Childs, Erica Chito. 2008. "Listening to the Interracial Canary: Contemporary Views on Interracial Relationships Among Blacks and Whites." *Fordham Law Review* 76(6):2772–86.

Chira, Susan. 1994. "Hispanic Families Avoid Using Day Care, Study Says." *New York Times,* April 6.

Choice, Pamela and Leanne K. Lamke. 1997. "A Conceptual Approach to Understanding Abused Women's Stay/Leave Decisions." *Journal of Family Issues* 18:290–314.

"Choosing Senior Housing." 2007. Retrieved May 15, 2010 (helpguide.org).

Chou, Rosalind R. and Joe R. Feagin. 2008. *The Myth of the Model Minority: Asian Americans Facing Racism.* Boulder, CO: Paradigm Publishers.

Christakis, Nicholas A. and Paul D. Allison. 2006. "Mortality After the Hospitalization of a Spouse." *New England Journal of Medicine* 354(7):719–30.

Christensen, Andrew and Neil Jacobson. 1999. *Reconcilable Differences.* London, England: Guilford.

Christiansen, Shawn L. and Rob Palkovitz. 2001. "Why the 'Good Provider' Role Still Matters: Providing as a Form of Paternal Involvement." *Journal of Family Issues* 22:84–106.

Christopher, F. Scott and Susan Sprecher. 2000. "Sexuality in Marriage, Dating, and Other Relationships." *Journal of Marriage and Family* 62:999–1017.

Christopherson, Brian. 2006. "Some Buck Trends, Marry Before Finishing College." *Lincoln Journal Star,* October 3. Retrieved October 4, 2006 (www.journalstar.com).

Chou, Rosalind R. and Joe R. Feagin. 2008. *The Myth of the Model Minority: Asian Americans Facing Racism.* Boulder, CO: Paradigm Publishers.

Christie-Mizell, C. Andre, Erin M. Pryor, and Elizabeth Grossman. 2008. "Child Depressive Symptoms, Spanking, and Emotional Support: Differences Between African American and European American Youth." *Family Relations* 57(June):335–50.

Chudacoff, Howard P. 2007. *Children at Play: An American History.* New York: NYU Press.

Chung, Juliet. 2006. "Hispanic Paradox: Income May Be Lower but Health Better Than Most." *Seattle Times*, August 29. Retrieved February 12, 2007 (www.seattletimes.com).

Ciabattari, Teresa. 2004. "Cohabitation and Housework: The Effects of Marital Intentions." *Journal of Marriage and Family* 66(1):118–125.

Ciaramigoli, Arthur P. and Katherine Ketcham. 2000. *The Power of Empathy: A Practical Guide to Creating Intimacy, Self-Understanding, and Lasting Love in Your Life*. New York: Dutton.

Cichocki, Mark. 2009. "HIV and Sperm Washing Sperm Washing: Hope For Serodiscordant Couples Wanting a Family." *About.com Health Disease and Condition*, December 18. aids.about.com.

Cicirelli, Victor G. 2000. "An Examination of the Trajectory of the Adult Child's Caregiving for an Elderly Parent." *Family Relations* 49(2):169–75.

Cieraad, Irene. 2006. At Home: *An Anthropology of Domestic Space*. Syracuse, NY: Syracuse University Press.

Clark, Lauren. 2009. "Scientific Inquiry: Focus Group Research with Children and Youth." *Journal for Specialists in Pediatric Nursing* 14(2):152–4.

Clark, Michele C. and Pamela M. Diamond. 2010. "Depression in Family Caregivers of Elders: A Theoretical Model of Caregiver Burden, Sociotropy, and Autonomy." *Research in Nursing and Health* 33:20–34.

Clark, Vicki, Catherine Huddleston-Casas, Susan Churchill, Denise O'Neil Green, and Amanda Garrett. 2008. "Mixed Methods Approaches in Family Science Research." *Journal of Family Issues* 29(11):1543–66.

Clarke, L. 2005. "Remarriage in Later Life: Older Women's Negotiation of Power, Resources, and Domestic Labor." *Journal of Women and Aging* 17(4):21–41.

"Class in America: The Unspoken Divide— Interview with Annette Lareau." 2003. WKSU (Public Radio) News. Retrieved December 21, 2010 (www.wksu.org/news/features).

Claxton-Oldfield, Stephen, Carla Goodyear, Tina Parsons, and Jane Claxton-Oldfield. 2002. "Some Possible Implications of Negative Stepfather Stereotypes." *Journal of Divorce and Remarriage* Spring-Summer: 77–89.

Clayton, Any and Maureen Perry-Jenkins. 2008. "No Fun Anymore: Leisure and Marital Quality Across the Transition to Parenthood." *Journal of Marriage and Family* 70(1):28–43.

Clayton, Obie and Joan Moore. 2003. "The Effects of Crime and Imprisonment on Family Formation." Pp. 84–102 in *Black Fathers in Contemporary American Society: Strengths, Weaknesses, and Strategies for Change*, edited by Obie Clayton, Ronald B. Mincy, and David Blankenhorn. New York: Russell Sage.

Clements, Mari L., Scott M. Stanley, and Howard J. Markman. 2004. "Before They Said 'I Do': Discriminating among Marital Outcomes over 13 Years." *Journal of Marriage and Family* 66(3):613–26.

Clemetson, Lynette. 2006a. "Adopted in China: Seeking Identity in America." *New York Times*, March 23.

———. 2006b. "Breaking the Biology Barrier." *New York Times*, August 30.

———. 2006c. "Weekends with Dad, Courtesy of D.S.L." *New York Times*, March 19.

———. 2007. "Working on Overhaul, Russia Halts Adoption Applications." *New York Times*, April 12.

Clemetson, Lynette and Ron Nixon. 2006. "Overcoming Adoption's Racial Barriers." *New York Times*, August 17.

Clifford, Denis, Frederick Hertz, and Emily Doskow. 2007. *A Legal Guide for Lesbian & Gay Couples*. Berkeley: Nolo Press.

Clinton, Hillary Rodham. 1990. "In France, Day Care Is Every Child's Right." *New York Times*, April 7.

Clinton, Timothy and Mark Laaser. 2010. *The Quick-Reference Guide to Sexuality and Relationship Counseling*. Grand Rapids, MI: Baker Books.

Clinton, Timothy E., and John Trent. 2009. *The Quick-Reference Guide to Marriage and Family Counseling*. Grand Rapids, Michigan: Baker Books.

Cloud, Henry and John Sims Townsend. 2005. *Rescue Your Love Life: Changing Those Dumb Attitudes and Behaviors That Will Sink Your Marriage*. Nashville, TN: Integrity Publishers.

Cloud, John. 2007. "Busy Is O.K." *Time*, January 29, p. 51.

Clunis, D. Merilee and G. Dorsey Green. 2005. *Lesbian Couples: A Guide to Creating Healthy Relationships*. Emeryville, CA: Seal Press.

Coalition for Marriage, Family, and Couples Education. 2009. "Smart Marriages." Retrieved October 11, 2009 (www.smartmarriages.com).

Coan, James A. and John M. Gottman. 2007. "Sampling, Experimental Control, and Generalizability in the Study of Marital Process Models." *Journal of Marriage and Family* 69(1):73–80.

Cobb, Nathan P., Jeffry H. Larson, and Wendy L. Watson. 2003. "Development of the Attitudes about Romance and Mate Selection Scale." *Family Relations* 52(3):222–31.

Cockerham, William C. 2007. "A Note on the Fate of Postmodern Theory and Its Failure to Meet the Basic Requirements for Success in Medical Sociology." *Social Theory & Health* 5(4):285–96.

Coffman, Ginger and Carol Markstrom-Adams. 1995. "A Model for Parent Education among Incarcerated Adults." Presented at the annual meeting of the *Nation*al Council on *Family Relations*, November 15–19, Portland, OR.

Cogan, Rosemary and Bud C. Ballinger III. 2006. "Alcohol Problems and the Differentiation of Partner, Stranger, and General Violence." *Journal of Interpersonal Violence* 21(7):924–35.

Cohen, Leonard. 2008. *Runaway Youth and Multisystemic Therapy (MST): A Program Model*. West Hartford, CT: University of Hartford.

Cohen, Patricia. 2007a. "As Ethics Panels Expand Grip, No Research Field Is Off Limits." *New York Times*, February 28.

———. 2007b. "Signs of Détente in the Battle between Venus and Mars." *New York Times*, May 31.

Cohen, Robin A. and Barbara Bloom. 2005. "Trends in Health Insurance and Access to Medical Care for Children under Age 19 Years: United States, 1998–2003." *Advance Data from Vital and Health Statistics*, No. 355. Hyattsville, MD: U.S. National Center for Health Statistics.

Cohen, Neil A., Thanh Tran, and Siyon Rhee. 2007. *Multicultural Approaches in Caring for Children, Youth, and Their Families*. Boston: Pearson, Allyn, and Bacon Publishers.

Cohn, Lisa, Debbie Glasser, and Steve Mark. 2008. *The Step-Tween Survival Guide: How to Deal with Life in a Stepfamily*. Minneapolis: Free Spirit Publishers.

"Cohousing in Today's Real Estate Market." 2006. *Cohousing Magazine*. The Cohousing Association of the United States. Retrieved October 3, 2006 (www.cohousing.org).

Cole, Harriette. 1993. *Jumping the Broom: The African-American Wedding Planner*. New York: Henry Holt.

Cole, Thomas. 1983. "The 'Enlightened' View of Aging." *Hastings Center Report* 13:34–40.

Coleman, Joshua. 2006. *The Lazy Husband: How to Get Men to Do More Parenting and Housework*. New York: St. Martin's Griffin.

Coleman, Marilyn and Lawrence H. Ganong. 1997. "Stepfamilies from the Stepfamily's Perspective." Pp. 107–22 in *Stepfamilies: History, Research, and Policy*, edited by Irene Levin and Marvin B. Sussman. New York: Haworth.

Coleman, Marilyn, Lawrence H. Ganong, and Mark Fine. 2000. "Reinvestigating Remarriage: Another Decade of Progress." *Journal of Marriage and Family* 62:1288–1307.

Coleman, Marilyn and Lynette Nickleberry. 2009. "An Evaluation of the Remarriage and Stepfamily Self-Help Literature." *Family Relations* 58(December):549–61.

Coleman, Priscilla and Anne Watson. 2000. "Infant Attachment as a Dynamic System." *Human Development* 43:295–313.

Coles, M. E., L. M. Cook, and T. R. Blake. 2007. "Assessing Obsessive Compulsive Symptoms and Cognitions on the Internet: Evidence for the Comparability of Paper and Internet Administration." *Behaviour Research and Therapy* 45:2232–40.

Coles, Roberta L. 2009. "Just Doing What They Gotta Do: Single Black Custodial Fathers Coping with the Stressors and Reaping the Rewards of Parenting." *Journal of Family Issues* 30(10):1311–38.

Collaborative Group on Hormonal Factors in Breast Cancer. 2004. "Breast Cancer and Abortion: Collaborative Reanalysis of Data from 53 Epidemiological Studies, Including 83,000 Women with Breast Cancer from 16 Countries." *Lancet* 363(9414):1007–16.

Collins, Chuck, Betsy Leondar-Wright, and Holly Sklar. 1999. *Shifting Fortunes: The Perils of the Growing American Wealth Gap*. Boston: United For A Fair Economy.

Collins, Patricia Hill. 1999. "Shifting the Center: Race, Class, and Feminist Theorizing about Motherhood." Pp. 197–217 in *American Families: A Multicultural Reader,* edited by Stephanie Coontz. New York: Routledge.

———. 2004. *Black Sexual Politics: African Americans, Gender, and the New Racism.* New York: Routledge.

Collins, Randall and Scott Coltrane. 1995. *Sociology of Marriage and the Family: Gender, Love, and Property.* 4th ed. Chicago, IL: Nelson-Hall.

Collymore, Yvette. 2002. "Risk of Homicide Is High for U.S. Infants." *Population Today,* May/June, p. 10.

Coltrane, Scott. 1990. "Birth Timing and the Division of Labor in Dual-Earner Families: Exploratory Findings and Suggestions for Further Research." *Journal of Family Issues* 11:157–81.

———. 1998. "Gender, Power, and Emotional Expression: Social and Historical Contexts for a Process Model of Men in Marriages and Families." Pp. 193–211 in *Men in Families: When Do They Get Involved? What Difference Does It Make?* edited by Alan Booth and Ann C. Crouter. Mahwah, NJ: Erlbaum.

———. 2000. "Research on Household Labor: Modeling and Measuring the Social Embeddedness of Routine Family Work." *Journal of Marriage and Family* 62:1208–33.

Coltrane, Scott and Masako Ishii-Kuntz. 1992. "Men's Housework: A Life Course Perspective." *Journal of Marriage and Family* 54(2):43–57.

Colvin, Jan, Lillian Chenoweth, Mary Bold, and Cheryl Harding. 2004. "Caregivers of Older Adults: Advantages and Disadvantages of Internet-Based Social Support." *Family Relations* 53(1):49–57.

Comerford, Lynn. 2006. "The Child Custody Mediation Policy Debate." *Family Focus* (March):F7, F12. National Council on Family Relations.

Comfort, Megan. 2008. *Doing Time Together: Love and Family in the Shadow of Prison.* Chicago: University of Chicago Press.

Concise Columbia Encyclopedia. 1994. New York: Columbia University Press.

"Conclusions Are Reported on Teaching of Abstinence." 2007. *New York Times,* April 15.

"Condoms and STDs: Fact Sheet for Public Health Personnel." 2010. Atlanta: U. S. Centers for Disease Control and Prevention, February 8. www.cdc.gov/condomeffectiveness/latex.htm.

Condon, Stephanie. 2010. "Reid: Unemployment Leads to Domestic Violence." *CBS News,* February 23. www.cbsnews.com.

Conger, Rand D., Katherine J. Conger, and Monica J. Martin. 2010. "Socioeconomic Status, Family Processes, and Individual Development." *Journal of Marriage and Family* 72(3):685–704.

Congregation for the Doctrine of the Faith. 1988. "Instruction on Respect for Human Life in Its Origin and on the Dignity of Procreation." Pp. 325–31 in *Moral Issues and Christian Response,*

4th ed., edited by Paul Jersild and Dale A. Johnson. New York: Holt, Rinehart & Winston.

Congress Must Act to Close the Wage Gap for Women. 2010. Washington, DC: National Women's Law Center.

Conlin, Michelle. 2003. "The New Gender Gap." *Business Week,* May 26, pp. 75–82.

Connell, R. W. 2005. *Masculinities.* 2nd ed. Berkeley: University of California Press.

Conner, Karen A. 2000. *Continuing to Care: Older Americans and Their Families.* New York: Falmer.

Connidis, Ingrid Arnet. 2007. "Negotiating Inequality Among Adult Siblings: Two Case Studies." *Journal of Marriage and Family* 69(2):482–99.

———. 2009. Family Ties and Aging. Thousand Oaks, CA: Pine Forge Press.

———. 2010. Family Ties and Aging, 2nd ed. Los Angeles: Pine Forge Press.

Connidis, Ingrid Arnet and Candace L. Kemp. 2008. "Negotiating Actual and Anticipated Parental Support: Multiple Sibling Voices in Three-Generation Families." *Journal of Aging Studies* 22:229–38.

Connor, James. 2007. *The Sociology of Loyalty.* New York: Springer-Verlag.

Conway, Tiffany and Rutledge Q. Hutson. 2007. "Is Kinship Care Good for Kids?" Center for Law and Social Policy, March 2. Retrieved September 14, 2009 (www.clasp.org).

Cook, Judith, Lynne Mock, Jessica Jonikas, Jane Burke-Miller, Tina Carter, Amanda Taylor, Carol Petersen, Dennis Grey, and David Gruenfelder. 2009. "Prevalence of Psychiatric and Substance Use Disorders among Single Mothers Nearing Lifetime Welfare Eligibility Limits. *Archives of General Psychiatry* 66(3):249–60.

Cook, Philip W. 2009. Abused Men: *The Hidden Side of Domestic Violence.* Santa Barbara, CA: Praeger.

Cooke, Lynn P. 2006. "'Doing' Gender in Context: Household Bargaining and Risk of Divorce in Germany and the United States." *American Journal of Sociology* 112(2):44–72.

Cooley, Charles Horton. 1902. *Human Nature and the Social Order.* New York: Scribner's.

———. 1909. *Social Organization.* New York: Scribner's.

Coombs-Orme, Terri and Daphne S. Cain. 2008. "Predictors of Mothers' Use of Spanking with Their Infants." *Child Abuse and Neglect* 32(6):649–57.

Cooney, Rosemary, Lloyd H. Rogler, Rose Marie Hurrel, and Vilma Ortiz. 1982. "Decision Making in Intergenerational Puerto Rican Families." *Journal of Marriage and Family* 44:621–31.

Coontz, Stephanie. 1992. *The Way We Never Were: American Families and the Nostalgia Trap.* New York: Basic Books.

———. 1997. "Divorcing Reality." *The Nation,* November 17, pp. 21–24.

———. 2005a. "The Heterosexual Revolution." *New York Times,* July 5.

———. 2005b. *Marriage, a History: From Obedience to Intimacy, or How Love Conquered Marriage.* New York: Viking.

———. 2005c. "The New Fragility of Marriage, for Better or for Worse." *Chronicle of Higher Education* 51(35):B7–10. Retrieved September 29, 2006 (chronicle.com).

———. 2007. "The Romantic Life of Brainiacs." *The Boston Globe,* February 18. www.boston.com.

———. 2009. "Till Children Do Us Part." *The New York Times,* February 5, A31.

———. 2010a. "Divorce, No-Fault Style." *The New York Times,* June 17, A29.

———. 2010b. "Why Gore Breakup Touched a Nerve." *CNN International Edition,* June 4. Retrieved June 20, 2010 (edition.cnn.com).

Cooper, Al. 2004. "Online Sexual Activity in the New Millennium." *Contemporary Sexuality* 38:i–vii.

Cooper, Carey E., Robert Crosnoe, Marie-Anne Suizzo, and Keenan A. Pituch. 2009. "Poverty, Race, and Parental Involvement During the Transition to Elementary School." *Journal of Family Issues* (October).

Cooper, Carey E., Sara McLanahan, Sarah Meadows, and Jeanne Brooks-Gunn. 2009. "Family Structure Transitions and Maternal Parenting Stress." *Journal of Marriage and Family* 71(3):558–74.

Cooper, Claudia, Amber Selwood, and Gill Livingston. 2008. "The Prevalence of Elder Abuse and Neglect: A Systematic Review." *Age and Ageing* 37(2):151–60.

Cooper, Frank Rudy. 2008. "Who's the Man?: Masculinities and Police Stops." *Suffolk University Law School* 8(23):1–50.

Coopersmith, Jared. 2009. *Characteristics of Public, Private, and Bureau of Indian Education Elementary and Secondary School Teachers in the United States: Results From the 2007–08 Schools and Staffing Survey.* NCES 2009-324. Washington, DC: National Center for Education Statistics, Institute of Education Sciences, U.S. Department of Education.

Coordinated Access for Child Care. 2001. "Choosing Quality Child Care" (www.cafcc. on.ca).

Cooter, Roger and Claudia Stein. 2010. "Positioning the Image of AIDS." *Endeavor* 34(1):12–15.

Corbett, Christianne, Catherine Hill, and Andresse St. Rose. 2008. *Where the Girls Are: The Facts About Gender Equity in Education.* Washington, DC: American Association of University Women.

Cordes, Henry J. 2009. "'He-Cession' Reshuffles Roles." *Omaha World-Herald,* November 29. www. omaha.com. Retrieved May 2, 2009.

Corliss, Richard and Sonja Steptoe. 2004. "The Marriage Savers." *Time,* January 19.

Cornwell, Erin York and Linda J. Waite. 2009. "Social Disconnectedness, Perceived Isolation, and Health among Older Adults." *Journal of Health and Social Behavior* 50(March):31–48.

Corra, Mamadi, Shannon K. Carter, J. Scott Carter, and David Knox. 2009. "Trends in Marital Happiness by Gender and Race, 1973 to 2006." *Journal of Family Issues* 30(10):1379–1404.

Cott, Nancy F. 2000. *Public Vows: A History of Marriage and the Nation.* Cambridge, MA: Harvard University Press.

Cotter, David A., Joan M. Hermsen, and Reeve Vanneman. 2004. *Gender Inequality at Work.* New York: Russell Sage.

Cotton, Sheila R., Russell Burton, and Beth Rushing. 2003. "The Mediating Effects of Attachment to Social Structure and Psychosocial Resources on the Relationship between Marital Quality and Psychological Distress." *Journal of Family Issues* 24(4):547–77.

Cottrell, Barbara and Peter Monk. 2004. "Adolescent-to-Parent Abuse: A Qualitative Overview of Common Themes." *Journal of Family Issues* 25(8):1072–95.

Coulter, Ann H. 2009. *Liberal "Victims" and Their Assault on America.* New York: Three Rivers Press.

"Couple Support." 2006. Covenant Marriage Movement. Retrieved October 16, 2006 (www.covenantmarriage.com).

"Court Treats Same-sex Breakup as Divorce." 2002. *Seattle Times,* November 3.

Covenant Marriage: A Fact Sheet. 2010. Fairfax, VA: National Healthy Marriage Resource Center.

"Covenant Marriages Ministry." 1998. www.covenantmarriages.com.

Covert, Juanita J. and Travis L. Dixon. 2008. "A Changing View: Representation and Effects of the Portrayal of Women of Color in Mainstream Women's Magazines." *Communication Research* 35(2):232–56.

Cowan, C. P. and P. A. Cowan. 1992. *When Partners Become Parents: The Big Life Change for Couples.* New York: Basic Books.

Cowan, Gloria. 2000. "Beliefs about the Causes of Four Types of Rape." *Sex Roles* 42(9/10):807–23.

Cowan, Philip and Carolyn Cowan. 2009. "News You Can Use: Are Babies Bad for Marriage?" *Press Release,* January 9. Chicago: Council on Contemporary Families. www.contemporaryfamilies.org.

Cowan, Philip A., Carolyn Pape Cowan, Marsha Kline Pruett, Kyle Pruett, and Jessie J. Wong. 2009. "Promoting Fathers' Engagement with Children: Preventive Interventions for Low-Income Families." *Journal of Marriage and Family* 71(4): 663–679.

Cowan, Ruth Schwartz. 1983. *More Work for Mother: The Ironies of Household Technology from the Open Hearth to the Microwave.* New York: Basic Books.

Cowdery, Randi S., Norma Scarborough, Carmen Knudson-Martin, Gita Seshadri, Monique E. Lewis, and Anne Rankin Mahoney. 2009. "Gendered Power in Cultural Contexts: Part II. Middle Class African American Heterosexual Couples with Young Children." *Family Process* 48(1): 25–39.

Cowley, Geoffrey. 2000. "For the Love of Language." *Newsweek* Special Issue, Fall/Winter, pp. 12–15.

Cox, Adam J. 2006. *Boys of Few Words: Raising Our Sons to Communicate and Connect.* New York: Guilford.

Cox, Erin. 2008. *Intimate Partner Violence among Pregnant and Parenting Women: Local Health Department Strategies for Assessment, Intervention, and Prevention.* Washington, DC: National Association of County & City Health Officials.

Coy, Peter, Michelle Conlin, and Moira Herbst. 2010. "The Disposable Worker." *Bloomberg Businessweek.* January 18:33–39.

Coyle, James, Thomas Nochajski, Eugene Maguin, Andrew Safyer, David DeWit, and Scott Macdonald. 2009. "An Exploratory Study of the Nature of Family Resilience in Families Affected by Parental Alcohol Abuse." *Journal of Family Issues* 30(12):1606–23.

Craig, Stephen. 1992. "The Effect of Television Day Part on Gender Portrayals in Television Commercials: A Content Analysis." *Sex Roles* 26(5/6):197–211.

Crary, David. 2007a. "More Couples Seeking Kinder, Gentler Divorces." *MSNBC,* December 18. Retrieved June 21, 2010 (www.msnbc.com).

———. 2007b. "U.S. Divorce Rate Lowest Since 1979." Associated Press, May 10. Retrieved May 10, 2007 (www.breitbart.com).

———. 2008. "Housework Gets You Laid." *The Huffington Post,* March 6. Retrieved May 20, 2010 (www.huffingtonpost.com).

Crawford, D., D. Feng, and J. Fischer. 2003. "The Influence of Love, Equity, and Alternatives on Commitment in Romantic Relationships." *Family and Consumer Sciences Research Journal* 31(3):253–71.

Crawford, Duane W., Renate M. Houts, Ted L. Huston, and Laura J. George. 2002. "Compatibility, Leisure, and Satisfaction in Marital Relationships." *Journal of Marriage and Family* 64(2):433–49.

Crawford, Lizabeth A. and Katherine B. Novak. 2008. "Parent-Child Relations and Peer Associations as Mediators of the Family Structure-Substance Use Relationship." *Journal of Family Issues* 29(2):155–84.

Cresswell, Mark. 2003. "Sex/Gender: Which Is Which? A Rejoinder to Mary Riege Laner." *Sociological Inquiry* 73:138–51.

Crimmins, Eileen M., Mark D. Hayward, Aaron Hagedorn, Yasuhiko Saito, and Nicolas Brouard. 2009. "Change in Disability-Free Life Expectancy for Americans 70 Years Old and Older." *Demography* 46(3):627–46.

Crittenden, Ann. 2001. *The Price of Motherhood: Why the Most Important Job in the World Is Still the Least Valued.* New York: Metropolitan.

Crook, Tylon, Chippewa M. Thomas, and Debra C. Cobia. 2009. "Masculinity and Sexuality: Impact on Intimate Relationships of African American Men." *Family Journal* 17(4):360–6.

Crooks, Robert and Karla Baur. 2005. *Our Sexuality.* 9th ed. Belmont, CA: Wadsworth.

Crosbie-Burnett, Margaret and Edith Lewis. 1999. "Use of African-American Family Structure and Functioning to Address the Challenges of European-American Post-Divorce Families." Pp. 455–68 in *American Families: A Multicultural Reader,* edited by Stephanie Coontz. New York: Routledge.

Crosbie-Burnett, Margaret, Edith A. Lewis, Summer Sullivan, Jessica Podolsky, Rosane Mantilla de Souza, and Victoria Mitrani. 2005. "Advancing Theory through Research: The Case of Extrusion in Stepfamilies." Pp. 213–30 in *Sourcebook of Family Theory and Research,* edited by Vern L. Bengston, Alan C. Acock, Katherine R. Allen, Peggye Dilworth-Anderson, and David M. Klein. Thousand Oaks, CA: Sage Publications.

Crosby, John F. 1991. *Illusion and Disillusion: The Self in Love and Marriage.* 4th ed. Belmont, CA: Wadsworth.

Crosse, Marcia. 2008. "Abstinence Education: Assessing the Accuracy and Effectiveness of Federally Funded Programs." *Testimony Before the Committee on Oversight and Government Reform, House of Representatives* April 23. Washington, DC: United States Government Accountability Office.

Crouter, Anne C., Megan E. Baril, and Kelly D. Davis, and Susan M. McHale. 2008. "Processes Linking Social Class and Racial Socialization in African American Dual-Earner Families." *Journal of Marriage and Family* 70(5):1311–25.

Crouter, Ann C. and Alan Booth, eds. 2003. *Children's Influence on Family Dynamics: The Neglected Side of Family Relationships.* Mahwah, NJ: Erlbaum.

Crouter, Ann C., S. D. Whiteman, S. M. McHale, and D. W. Osgood. 2007. "Development of Gender Attitude Traditionality Across Middle Childhood and Adolescence." *Child Development* 78:911–26.

Crowder, Kyle D. and Stewart E. Tolnay. 2000. "A New Marriage Squeeze for Black Women: The Role of Racial Intermarriage by Black Men." *Journal of Marriage and Family* 62(3):792–807.

Crowley, Martha, Daniel T. Lichter, and Zhenchao Qian. 2006. "Beyond Gateway Cities: Economic Restructuring and Poverty Among Mexican Immigrant Families and Children." *Family Relations* 55(3):345–60.

Cuber, John and Peggy Harroff. 1965. *The Significant Americans.* New York: Random House. (Published also as *Sex and the Significant Americans.* Baltimore, MD: Penguin, 1965.)

Cui, Ming and M. Brent Donnellan. 2009. "Trajectories of Conflict over Raising Adolescent Children and Marital Satisfaction." *Journal of Marriage and Family* 71(3):478–94.

Cui, Ming, Frederick O. Lorenz, Rand D. Conger, Janet N. Melby, and Chalandra M. Bryant. 2005. "Observer, Self-, and Partner Reports of Hostile Behaviors in Romantic Relationships." *Journal of Marriage and Family* 67(5):1169–81.

Cullen, Jennifer C., Leslie B. Hammer, Margaret B. Neal, and Robert R. Sinclair. 2009. "Development of a Typology of Dual-Earner Couples Caring for Children and Aging Parents." *Journal of Family Issues* 30(4):458–83.

Cullen, Lisa T. 2007. "Till Work Do Us Part." *Time Magazine*, September 27. Retrieved May 27, 2010 (www.time.com).

Cunningham, Mick. 2005. "Gender in Cohabitation and Marriage: The Influence of Gender Ideology on Housework Allocation over the Life Course." *Journal of Family Issues* 26(8):1037–61.

———. 2008. "Changing Attitudes toward the Male Breadwinner, Female Homemaker Family Model: Influences of Women's Employment and Education over the Lifecourse." *Social Forces* 87(1):299–323.

Cunningham-Burley, Sarah. 2001. "The Experience of Grandfatherhood." Pp. 92–96 in *Families in Later Life: Connections and Transitions*, edited by Alexis J. Walker, Margaret Manoogian-O'Dell, Lori A. McGraw, and Diana L. G. White. Thousand Oaks, CA: Pine Forge Press.

Currie, Janet, and Cathy Spatz Widom. 2010. "Long-Term Consequences of Child Abuse and Neglect on Adult Economic Well-Being." *Child Maltreatment* 15(2):111–20.

Curtis, Kristen Taylor and Christopher G. Ellison. 2002. "Religious Heterogamy and Marital Conflict." *Journal of Family Issues* 23(4):551–76.

Cyr, Mireille, Pierre McDuff, and John Wright. 2006. "Prevalence and Predictors of Dating Violence among Adolescent Female Victims of Child Sexual Abuse." *Journal of Interpersonal Violence* 21(8):1000–17.

Dahms, Alan M. 1976. "Intimacy Hierarchy." Pp. 85–104 in *Process in Relationship: Marriage and Family*, 2nd ed., edited by Edward A. Powers and Mary W. Lees. New York: West.

Dailard, Cynthia. 2002. "Abstinence Promotion and Teen Family Planning: The Misguided Drive for Equal Funding." *The Guttmacher Report on Public Policy* 5(1).

———. 2003. "Understanding 'Abstinence': Implications for Individuals, Programs and Policies." *The Guttmacher Report* 6(5). December (www.agi-usa.org).

Dailey, Rene M. 2009. "Confirmation from Family Members: Parent and Sibling Contributions to Adolescent Psychosocial Adjustment." *Western Journal of Communication* 73(3):273–99.

Dalla, Rochelle, Susan Jacobs-Hagen, Betsy Jareske, and Julie Sukup. 2009. "Examining the Lives of Navajo Native American Teenage Mothers in Context: A 12- to 15-Year Follow-Up." *Family Relations* 58(April):148–61.

Dalley, Timothy J. 2004. "Homosexual Parenting: Placing Children at Risk." Family Research Council, March 25 (www.frc.org).

Daly, Martin and Margo I. Wilson. 1994. "Some Differential Attributes of Lethal Assaults on Small Children by Stepfathers versus Genetic Fathers." *Ethology and Sociobiology* 15:207–17.

Dance, Theodore and Elizabeth Latrobe Place. 2006. *Raising Athletic Stars*. Erie, PA: First Books.

Dang, Alain and Samjen Frazer. 2004. *Black Same-Sex Households in the United States: A Report from the 2000 Census*. New York: National Gay and Lesbian Task Force Policy Institute and National Black Justice Coalition.

Daniel, Brigid, Julie Taylor, and Jane Scott. 2010. "Recognition of Neglect and Early Response: Overview of a Systematic Review of the Literature." *Child & Family Social Work* 15(2):248–57.

Danziger, Sheldon and Cecilia Elena Rouse. 2008. "The Price of Independence: The Economics of Early Adulthood." *Family Focus*. March: F8–F9.

D'Antonio, W. V., D. R. Hoge, K. Meyer, and J. D. Davidson. 1999. "American Catholics." *Catholic Reporter*, October 29, p. 20.

Dao, James. 2005. "Grandparents Given Rights by Ohio Court." *New York Times*, October 11.

Darwin, Charles. 1977 [1859]. *On the Origin of Species*. Fulcroft, PA: Fulcroft Library Editions.

Davey, Adam, Jyoti Savla, Megan Janke, and Shayne Anderson. 2009. "Grandparent-Grandchild Relationships: From Families in Context to Families As Contexts." *Aging and Human Development* 69(4):311–25.

Davey, Monica. 2006. "As Tribal Leaders, Women Still Fight Old Views." *New York Times*, February 4.

———. 2010. "Nebraska, Citing Pain, Sets Limits On Abortion." *New York Times*, April 14.

Davies, Lorraine, Marilyn Ford-Gilboe, Joanne Hammerton. 2009. "Gender Inequality and Patterns of Abuse Post Leaving." *Journal of Family Violence* 24(1):27–39.

Davis, Belinda Creel, and Valentina A. Bali. 2008. "Examining the Role of Race, NIMBY, and Local Politics in FEMA Trailer Park Placement." *Social Science Quarterly* 89(5):1181–94.

Davis, James. 1991. *Who Is Black? One Nation's Definition*. University Park, PA: Pennsylvania State University Press.

Davis, Kelly D., W. Benjamin Goodman, Amy E. Pirretti, and David M. Almeida. 2008. "Nonstandard Work Schedules, Perceived Family Well-Being, and Daily Stressors." *Journal of Marriage and Family* 70(4):991–1003.

Davis, Rebecca L. 2010. *More Perfect Unions: The American Search for Marital Bliss*. Boston: Harvard University Press.

Davis, Shannon, Theodore Greenstein, and Jennifer Gertelsen Marks. 2007. "Effects of Union Type on Division of Household Labor." *Journal of Family Issues* 28(5):1260–71.

Davis, Wendy. 2001. "Some Lawyers Are Growing Hostile to the 'Friendly Parent' Idea in Custody Fights." *American Bar Association Journal* 87(October):26ff.

Dawkins, Richard. 1976. *The Selfish Gene*. New York: Oxford University Press.

Dawn, Laura. 2006. *It Takes a Nation: How Strangers Became Family in the Wake of Hurricane Katrina*. San Rafael, CA: Earth Aware Editions.

"Day-Care Researchers in Retreat." 2001. *Omaha World-Herald*, April 26.

Deal, Ron L. 2010. "Marriage, Family, & Stepfamily Facts, Updated Jan. 2010." Retrieved April 17, 2010 (www.successfulstepfamilies.com).

Dee, Jonathan. 2005. "Their Unexpected Adolescence." *New York Times Magazine*. 35-40; 53.

DeFrain, John. 2002. *Creating a Strong Family: American Family Strengths Inventory*. Nebraska Cooperative Extension NF01-498 (ianrpubs.unl.edu/family/nf498.htm).

DeLamater, John and William N. Friedrich. 2002. "Human Sexual Development." *Journal of Sex Research* 39:10–14.

DeLamater, John D. and Morgan Sill. 2005. "Sexual Desire in Later Life." *Journal of Sex Research* 42(2):138–49.

DeLeire, Thomas and Ariel Kalil. 2005. "How Do Cohabiting Couples with Children Spend Their Money?" *Journal of Marriage and Family* 67(2):286–95.

della Cava, Marco R. 2009. "Women Step Up As Men Lose Jobs." *USA Today*, March 19, 1D.

Dellmann-Jenkins, Mary, Maureen Blankemeyer, and Odessa Pinkard. 2000. "Young Adult Children and Grandchildren in Primary Caregiver Roles to Older Relatives and Their Service Needs." *Family Relations* 49(2):177–86.

del Pinal, Jorge and Audrey Singer. 1997. "Generations of Diversity: Latinos in the United States." *Population Bulletin* 52(3). Washington, DC: Population Reference Bureau.

Del Vecchio, Tamara and Susan G. O'Leary. 2006. "Antecedents of Toddler Aggression: Dysfunctional Parenting in Mother-Toddler Dyads." *Journal of Clinical Child and Adolescent Psychology* 35(2):194–202.

DeMaria, Rita M. 2005. "Distressed Couples and Marriage Education." *Family Relations* 54(2):242–53.

DeMaris, Alfred. 2001. "The Influence of Intimate Violence on Transitions out of Cohabitation." *Journal of Marriage and Family* 63(1):235–46.

———. 2007. "The Role of Relationship Inequality in Marital Disruption." *Journal of Social and Personal Relationships* 24(2):177–95.

———. 2009. "Distal and Proximal Influences on the Risk of Extramarital Sex: A Prospective Study of Longer Duration Marriages." *Journal of Sex Research* 46(6):597–607.

DeMaris, Alfred and Steven Swinford. 1996. "Female Victims of Spousal Violence: Factors Influencing Their Level of Fearfulness." *Family Relations* 45(1):98–106.

D'Emilio, John and Estelle B. Freedman. 1988. *Intimate Matters: A History of Sexuality in America*. New York: HarperCollins.

Demo, David H. and Alan C. Acock. 1996. "Singlehood, Marriage, and Remarriage: The Effects of Family Structure and *Family Relations*hips on Mothers' Well-being." *Journal of Family Issues* 17(3):388–407.

Demo, David H., William S. Aquilino, and Mark A. Fine. 2005. "Family Composition and Family Transitions." Pp. 119–34 in *Sourcebook of Family Theory and Research*, edited by Vern L.

Bengston, Alan C. Acock, Katherine R. Allen, Peggye Dilworth-Anderson, and David M. Klein. Thousand Oaks, CA: Sage Publications.

Demo, David H. and Mark A. Fine. 2010. *Beyond the Average Divorce*. Thousand Oaks, CA: Sage Publications.

"The Demographics of Aging." n.d. Transgenerational Design Matters. Retrieved May 5, 2010 (www.transgenerational.org/aging/demographics.htm).

DeNavas-Walt, Carmen, Bernadette D. Proctor, and Cheryl Hill Lee. 2006. *Income, Poverty, and Health Insurance Coverage in the United States: 2005*. Current Population Reports P60-291. Washington, DC: U.S. Census Bureau.

DeNavas-Walt, Carmen, Bernadette D. Proctor, and Jessica C. Smith. 2009. *Income, Poverty, and Health Insurance Coverage in the United States: 2008*. Current Population Reports P60-236. Washington, DC: U.S. Census Bureau.

Denizet-Lewis, Benoit. 2003."Double Lives on the Down Low." *New York Times Magazine*, August 3.

———. 2004. "Friends, Friends with Benefits, and the Benefits of the Local Mall." *New York Times*, May 30.

DeNoon, Daniel J. 2006. "Many Straight Men Have Gay Sex." *WebMD Health News*, September 18. www.webmd.com.

DeParle, Jason. 2009. "The 'W' Word, Re-Engaged." *The New York Times*, February 8.

Depner, Charlene E. 1993. "Parental Role Reversal: Mothers as Nonresidential Parents." Pp. 37–57 in *Nonresidential Parenting: New Vistas in Family Living*, edited by Charlene E. Depner and James H. Bray. Newbury Park, CA: Sage Publications.

Dermott, Esther. 2008. *Intimate Fatherhood: A Sociological Analysis*. New York: Routledge.

Desmond-Harris, Jenee. 2010. "My Race-Based Valentine." *Time Magazine*. February 22:99.

Deveny, Kathleen. 2007. "Yummy vs. Slummy." *Newsweek*, August 13, pp. 44–45.

———. 2008. "Why Only-Children Rule." *Newsweek*, June 2. www.newsweek.com.

DeVisser, Richard and Dee McDonald. 2007. "Swings and Roundabouts: Management of Jealousy in Heterosexual 'Swinging' Couples." *British Journal of Social Psychology* 46(2): 459–76.

Devitt, Kerry and Debi Roker. 2009. "The Role of Mobile Phones in Family Communication." Children & Society 23:189–202.

DeVoe, Jill Fleury, and Kristen E. Darling-Churchill. 2008. *Status and Trends in the Education of American Indians and Alaska Natives: 2008*. NCES 2008-084. Washington, DC: National Center for Education Statistics, Institute of Education Sciences, U.S. Department of Education.

Dew, Jeffrey. 2008. "Debt Change and Marital Satisfaction Change in Recently Married Couples." *Family Relations* 57(1):60–71.

Dew, Jeffrey, and Jeremy Yorgason. 2010. "Economic Pressure and Marital Conflict in Retirement-Aged Couples." *Journal of Family Issues* 31(2):164–88.

Dewilde, Caroline, and Wilfred Uunk. 2008. "Remarriage As a Way to Overcome the Financial Consequences of Divorce—A Test of the Economic Need Hypothesis for European Women." *European Sociological Review* 24(3):393–407.

Dickinson, Amy. 2002. "An Extra-Special Relation." *Time*, November 18, pp. A1+.

Diduck, Alison and Katherine O'Donovan, eds. 2006. *Feminist Perspectives on Family Law*. London: Routledge-Cavendish.

"Digital Divide: What It Is and Why It Matters." n.d. Seattle, WA: Digital Divide.org. Retrieved June 25, 2009 (www.digitaldivide.org).

Dillon, Frank R., Hilda Pantin, Michael S. Robbins, and Jose Szapocznik. 2008. "Exploring the Role of Parental Monitoring of Peers on the Relationship Between Family Functioning and Delinquency in the Lives of African American and Hispanic Adolescents." *Crime & Delinquency* 54(1):65–94.

Dinkes, R., J. Kemp, and K. Baum. 2009. Indicators of School Crime and Safety:2009 (NCES 2010-012/NCJ 228478). Washington, DC: U.S. Department of Education, and U.S. Department of Justice.

Dinkmeyer, Don Jr. 2007. "A Systematic Approach to Marriage Education." *The Journal of Individual Psychology* 63(3):315–21.

Dinkmeyer, Don, Sr., Gary D. McKay, and Don Dinkmeyer, Jr. 1997. *The Parent's Handbook: Systematic Training for Effective Parenting*. Circle Pines, MN: American Guidance Service.

Dion, Karen K. 1995. "Delayed Parenthood and Women's Expectations about the Transition to Parenthood." *International Journal of Behavioral Development* 18(2):315–33.

Dion, Karen K. and Kenneth L. Dion. 1991. "Psychological Individualism and Romantic Love." *Journal of Social Behavior and Personality* 6:17–33.

Dion, M. Robin. 2005. "Healthy Marriage Programs: Learning What Works." *The Future of Children* 15(2):139–56.

DiStefano, Joseph. 2001. "Jumping the Broom." Retrieved October 2, 2006 (www.randomhouse.com).

Ditzen, Beate, Marcel Schaer, Barbara Gabriel, Guy Bodenmann, Ulrike Ehlert, and Markus Heinrichs. 2009. "Intranasal Oxytocin Increases Positive Communication and Reduces Cortisol Levels During Couple Conflict." *Biological Psychiatry* 65(9):728–32.

"The Divorce Dilemma." 1996. *U.S. News & World Report*, September 30, pp. 58–62.

"Divorce Rate Drops to Lowest Since 1970." 2007. *USA Today*, May 11.

Dixon, Nicholas. 2007. "Romantic Love, Appraisal, and Commitment." *The Philosophical Forum* 38(4):373–86.

Dixon, Patricia. 2009. "Marriage among African Americans: What Does the Research Reveal?" *Journal of African American Studies* 13(1):29–46.

Doan, Alesha E. and Jean Calterone Williams. 2008. *The Politics of Virginity: Abstinence in Sex Education*. Santa Barbara, CA: Praeger.

Doble, Richard deGaris. 2006. AbusiveLove.com. Retrieved August 16, 2006 (www.abusivelove.com).

Dodson, Jualynne E. 2007. "Conceptualization and Research of African American Family Life in the United States: Some Thoughts." Pp. 51–68 in *Black Families*, 4th ed., edited by Harriette Pipes McAdoo. Thousand Oaks, CA: Sage Publications.

Doherty, William J. 1992. "Private Lives, Public Values." *Psychology Today* 25(3):32–39.

———. 2008. "Public Policy and Couple Relationships: A Commentary on Cabrera et al. 2009." *Journal of Marriage and Family* 70(December):1114–17.

Doherty, William, Janet Jacob, and Beth Cutting. 2009. "Community Engaged Parent Education: Strengthening Civic Engagement among Parents and Parent Educators." *Family Relations* 58(3):303–15.

Doherty, William J., Susan Su, and Richard Needle. 1989. "Marital Disruption and Psychological Well-Being: A Panel Study." *Journal of Family Issues* 10:72–85.

Dolan, Elizabeth M., Bonnie Braun, and Jessica C. Murphy. 2003. "A Dollar Short: Financial Challenges of Working-poor Rural Families." *Family Focus* (June):F13–F15. National Council on Family Relations.

Dolan, Frances Elizabeth. 2008. *Marriage and Violence: The Early Modern Legacy*. Philadelphia, PA : University of Pennsylvania Press.

Dolbin-MacNab, Megan L. 2006. "Just Like Raising Your Own? Grandmothers' Perceptions of Parenting a Second Time Around." *Family Relations* 55(5):564–75.

Dolbin-MacNab, Megan I. and Margaret K. Keiley. 2009. "Navigating Interdependence: How Adolescents Raised Solely by Grandparents Experience Their Family Relationships." *Family Relations* 58(April):162–75.

Domestic Sex Trafficking of Minors. n.d. Washington, DC: U.S. Department of Justice Child Exploitation and Obscenity Section.

"Domestic Violence Decline Reported." 2006. *Omaha World-Herald/Los Angeles Times*, December 29.

Domitz, Michael J. 2003. *May I Kiss You? A Candid Look at Dating, Communication, Respect, and Sexual Assault Awareness*. Greenfield, WI: Awareness Publications.

Donahoe, Elizabeth. 2005. "Using Foster Grandparents as Mentors in Family Drug Court: A Case Study." *Family Focus on . . . Substance Abuse across the Life Span*, FF25:F17–F18. National Council on Family Relations.

Doohan, Eve-Anne M., Sybil Carrere, Chelsea Siler, and Cheryl Beardslee. 2009. "The Link Between the Marital Bond and Future Triadic Family Interactions." *Journal of Marriage and Family* 71(4):892–904.

Dore, Margaret K. 2004. "The 'Friendly Parent' Concept: A Flawed Factor for Child Custody." *Loyola Journal of Public Interest Law* 6:41–56.

Dorman, Clive. 2006. "The Social Toddler: Promoting Positive Behaviour." *Infant Observation* 9(1):95–97.

Dosani, Sabina and Peter Cross 2007. *Raising Young Children: 52 Brilliant Little Ideas for Parenting Under 5s.* Oxford, Canada: Infinite Ideas Press.

Doss, Brian D., Galena K. Rhoads, Scott M. Stanley, and Howard J. Markman. 2009. "The Effect of the Transition to Parenthood on Relationship Quality: An 8-Year Prospective Study." *Journal of Personality and Social Psychology* 96(3):601–19.

Dotinga, Randy. 2009. "'Macho' Men Visit Doctor Even Less." *USA Today*, August 12. Retrieved January 16, 2010 (www.usatoday.com).

Douglas, Edward and Sharon Douglas. 2000. *The Blended Family: Achieving Peace and Harmony in the Christian Home.* Franklin, TN: Providence House.

Douglas, Susan J. 2009. "Where Have You Gone, Roseanne Barr?" Pp. 281–389 in *The Shriver Report: A Woman's Nation Changes Everything*, edited by Heather Boushey and Ann O'Leary. Washington, DC: Maria Shriver and the Center for American Progress.

Dowd, Lynn S, Penny A, Leisring, and Alan Rosenbaum. 2005. "Partner Aggressive Women: Characteristics and Treatment Attrition." *Violence and Victims* 20(2):219–33.

Dowland, Seth. 2009. "'Family Values' and the Formation of a Christian Right Agenda." *Church History* 78(3):606–31.

Downey, Douglas B. 1995. "When Bigger Is Not Better: Family Size, Parental Resources, and Children's Educational Performance." *American Sociological Review* 60:746–61.

Downey, Douglas B. and Dennis J. Condron. 2004. "Playing Well with Others in Kindergarten: The Benefit of Siblings at Home." *Journal of Marriage and Family* 66(2):333–50.

Downey, Liam and Brian Hawkins. 2008. "Single-Mother Families and Air Pollution: A National Study." *Social Science Quarterly* 89(2):523–36.

Downs, Barbara. 2003. *Fertility of American Women: June 2002.* Current Population Reports P20-548. Washington, DC: U.S. Census Bureau. October.

Drasin, Harry, Kristin P. Beals, Marc N. Elliott, Janet Lever, David J. Klein and Mark A. Schuster. 2008. "Age Cohort Differences in the Developmental Milestones of Gay Men." *Journal of Homosexuality* 54(4):381–99.

Dreger, Alice D. and April Herndon. 2009. "Progress and Politics in the Intersex Rights Movement: Feminist Theory in Action. GLQ: *A Journal of Lesbian and Gay Studies* 15(2):199–224.

Dreifus, Claudia. 2006. "An Economist Examines the Business of Fertility." *New York Times*, February 26.

Driscoll, Anne K., Stephen R. Russell, and Lisa J. Crockett. 2008. "Parenting Styles and Youth Well-Being across Immigrant Generations." *Journal of Family Issues* 29(2):185–209.

Driver, Janice L. and John M. Gottman. 2004. "Daily Marital Interactions and Positive Affect During Marital Conflict among Newlywed Couples." *Family Process* 43(3):301–14.

Dronkers, Jaap and Juho Harkonen. 2008. "The Intergenerational Transmission of Divorce in Cross-National Perspective: Results from the Fertility and Family Surveys." *Population Studies* 62(3):273–88.

Drummet, Amy R., Marilyn Coleman, and Susan Cable. 2003. "Military Families Under Stress: Implications for Family Life Education." *Family Relations* 52(3):279–87.

Duba, Jill D. and Richard E. Watts. 2009. *Journal of Clinical Psychology* 65(2):210–23.

Duck, Steve W. and Julia T. Wood. 2006. "What Goes Up May Come Down: Sex and Gendered Patterns in Relationship Dissolution." Pp. 169–99 in *Handbook of Divorce and Relationship Dissolution*, edited by Mark A. Fine and John H. Harvey. Mahwah, NJ: Erlbaum.

Dudley, Kathryn M. 2007. "The Social Economy of Single Motherhood: Book Review." *Social Forces* 86(1):360–61.

Duenwald, Mary. 2005. "For Them, Just Saying No Is Easy." *New York Times*, June 9.

Dugan, Laura, Daniel S. Nagin, and Richard Rosenfeld. 2004. "Do Domestic Violence Services Save Lives?" *NIJ Journal* 250:20–5. Washington, DC: National Institute of Justice, U.S. Department of Justice.

Dugger, Celia W. 1998. "In India, an Arranged Marriage of Two Worlds." *New York Times*, July 20, pp. A1, A10.

Duhigg, Charles. 2007. "Shielding Money Clashes with Elders' Free Will." *The New York Times*, December 24. Retrieved December 25, 2007 (www.nytimes.com).

Duncan, Gabriel. 2005. "Don't Follow America: Tribes Should Lift Bans on Gay Marriage." Pacific News Service. Retrieved December 29, 2006 (www.imdiversity.com).

Duncan, Greg J. and P. Lindsay Chase-Lansdale, eds. 2004. *For Better and For Worse: Welfare Reform and the Well-Being of Children and Families.* New York: Russell Sage.

Duncan, Stephen F., Thomas B. Holman, and Chongming Yang. 2007. "Factors Associated with Involvement in Marriage Preparation Programs." *Family Relations* 56(3):270–78.

Dunifon, Rachel and Lori Kowaleski-Jones. 2007. "The Influence of Grandparents in Single-Mother Families." *Journal of Marriage and Family* 69(2):456–81.

Dunleavy, Victoria Orrego. 2004. "Examining Interracial Marriage Attitudes as Value Expressive Attitudes." *The Howard Journal of Communications* 15:21–38.

Dunn, Judith F. 2005. "State of the Art: Siblings." *Psychiatry* 13(5):244–49.

Dunne, John E., E. Wren Hudgins, and Julia Babcock. 2000. "Can Changing the Divorce Law Affect Post-Divorce Adjustment?" *Journal of Divorce & Remarriage* 33(3):35–55.

Dunnewind, Stephanie. 2003. "Book Helps Impart Coping Skills, Self-Esteem to Multiracial Children." Knight Ridder/Tribune News Service, August 5.

Dunphy v. Gregor. 1994. 136 N.J. 99.

Dupuis, Sara. 2007. "Examining Remarriage: A Look at Issues Affecting Remarried Couples and the Implications Towards Therapeutic Techniques." *Journal of Divorce and Remarriage* 48(1.2):91–104.

———. 2009. "An Ecological Examination of Older Remarried Couples." *Journal of Divorce & Remarriage* 50(6):369–87.

Duquaine-Watson, Jillian M. 2007. "'Pretty Darned Cold': Single Mother Students and the Community College Climate in Post-Welfare Reform America." *Equity and Excellence in Education* 40:229–40.

Durham, Ricky. 2010. Prescription 4 Love. http://www.prescription4love.com.

Durodoye, Beth A. and Angela D. Coker. 2008. "Crossing Culture in Marriage: Implications for Counseling African American/African Couples." *International Journal of Counseling* 30:25–37.

Durose, Matthew R., Caroline Wolf Harlow, Patrick A. Langan, Mark Motivans, Ramona R. Rantala, and Erica L. Smith. 2005. *Family Violence Statistics, Including Statistics on Strangers and Acquaintances.* NCJ 207846. June. Washington, DC: Bureau of Justice Statistics.

Durr, Marlese and Shirley Ann Hill. 2006. "The Family–Work Interface in African American Households." Pp. 73–85 in *Race, Work, and Family in the Lives of African Americans*, edited by Marlese Durr and Shirley Ann Hill. Lanham, MD: Rowman & Littlefield.

Dush, Claire M. Kamp, Catherine L. Cohan, and Paul R. Amato. 2003. "The Relationship between Cohabitation and Marital Quality and Stability: Change Across Cohorts?" *Journal of Marriage and Family* 65(3):539–49.

Dush, Claire M. Kamp, Miles G. Taylor, and Rhiannon A. Kroeger. 2008. "Marital Happiness and Psychological Well-Being across the Life Cycle." *Family Relations* 57(April):211–26.

Dutton, Donald G. 2006. *Rethinking Domestic Violence.* Vancouver: University of British Columbia Press.

Dutton, Donald G. and James J. Browning. 1988. "Concern for Power, Fear of Intimacy, and Aversive Stimuli for Wife Assault." Pp. 163–75 in *Family Abuse and Its Consequences: New Directions in Research*, edited by Gerald T. Hotaling, David Finkelhor, John T. Kirkpatrick, and Murray A. Straus. Newbury Park, CA: Sage Publications.

Dutton, Donald G. and Tanya L. Nicholls. 2005. "The Gender Paradigm in Domestic Violence Research and Theory: The Conflict of Theory and Data." *Aggression and Violent Behavior* 10:680–714.

Dworkin, Paul H. 2002. "Editor's Note." *Journal of Developmental and Behavioral Pediatrics* 23:167.

Dworkin, Shari L. and Michael A. Messner. 1999. "Just Do . . . What? Sports, Bodies, and Gender." Pp. 341–61 in *Revisioning Gender*, edited by Myra Marx Ferree, Judith Lorber, and Beth. B. Hess. Thousand Oaks, CA: Sage Publications.

Dye, Jane Lawler. 2005. "Fertility of American Women: June 2004." *Current Population Reports* 20-555. December. Washington, DC: U.S. Census Bureau.

————. 2008. *Fertility of American Women: 2006.* U.S. Census Bureau Current Population Reports P20-558.

Dye, Jane Lawler and Tallese Johnson. 2007. *A Child's Day: 2003.* Current Population Reports P70-109. Washington, DC: U.S. Census Bureau.

Dyer, Carmel Bitondo, Sabrina Pickens, and Jason Burnett. 2007. "Vulnerable Elders: When It Is No Longer Safe to Live Alone." *Journal of the American Medical Association (JAMA)* 298(12):1448–50.

Dyess, Drucilla. 2009. "Pregnancies & Sexually Transmitted Diseases on the Rise Among Teens." *HealthNews*, July 21. www.healthnews.com.

Dykstra, Pearl A. and Gunhild O. Hagestad. 2007. "Roads Less Taken: Developing a Nuanced View of Older Adults Without Children." *Journal of Family Issues* 28(10):1275–1310.

Dziech, Billie Wright. 2003. "Sexual Harassment on College Campuses." Pp. 147–71 in *Academic and Workplace Sexual Harassment: A Handbook of Cultural, Social Science, Management, and Legal Perspectives,* edited by Michele Paludi and Carmen A. Paludi, Jr. Westport, CT: Praeger.

Eagly, Alice, Wendy Wood, and Amanda Diekman. 2000. "Social Role Theory of Sex Differences and Similarities: A Current Appraisal." Pp. 123–74 in *The Developmental Social Psychology of Gender,* edited by Thomas Eckes and Hanns M. Trautner. Mahwah, NJ: Erlbaum.

Eaklor, Vicki L. 2008. *Queer America: A GLBT History of the 20th Century.* Westport, CT: Greenwood Press.

Early, Theresa J., Thomas K. Gregoire, and Thomas P. McDonald. 2002. "Child Functioning and Caregiver Well-being in Families of Children with Emotional Disorders." *Journal of Family Issues* 23(3):374–91.

East, Patricia L., Barbara T. Reyes, and Emily J. Horn. 2007. "Association between Adolescent Pregnancy and a Family History of Teenage Births." *Perspectives on Sexual and Reproductive Health* 39(2):108–15.

Easterlin, Richard. 1987. *Birth and Fortune: The Impact of Numbers on Personal Welfare.* 2nd rev. ed. Chicago, IL: University of Chicago Press.

Eaton, Danice K., Laura Kann, Steve Kinchen, Shari Shanklin, James Ross, Joseph Hawkins, William A. Harris, Richard Lowry, Tim McManus, David Chyen, Connie Lim, Nancy D. Brener, and Howell Wechsler. 2008. "Youth Risk Behavior Surveillance—United States, 2007." Surveillance Summaries. *Morbidity and Mortality Weekly Report* 57, No. SS-04, June 6.

Eaton, Leslie. 2004. "Divorced Parents Move, and Custody Gets Trickier." *New York Times,* August 8.

Ebaugh, Helen Rose and Mary Curry. 2000. "Fictive Kin as Social Capital in New Immigrant Communities." *Sociological Perspectives* 43(2):189–209.

Ebeling, Ashlea. 2007. "The Second Match." *Forbes,* November 12. Retrieved April 15, 2010 (www.forbes.com).

Eckel, Sara. 2010. "Role Reversal." *Working Mother* 33(2):26–33.

Eckholm, Eric. 2007. "Childhood Poverty Is Found to Portend High Adult Costs." *New York Times,* January 25.

————. 2008. "Special Session Called on Nebraska Safe-Haven Law." *The New York Times,* October 30. Retrieved December 12, 2008 (www.nytimes.com).

Edleson, Jeffrey L., A. L. Ellerton, E. A. Seagren, S. O. Schmidt, S. L. Kirchberg, and A. T. Ambrose. 2007. "Assessing Child Exposure to Adult Domestic Violence." *Children and Youth Services Review* 29:961–71.

Edin, Kathryn and Maria Kefalas. 2005. *Promises I Can Keep: Why Poor Women Put Motherhood before Marriage.* Berkeley and Los Angeles, CA: University of California Press.

————. 2007. "Unmarried with Children." Pp. 505–11 in *Family in Transition,* 14th ed., edited by Arlene S. Skolnick and Jerome H. Skolnick. Boston, MA: Allyn and Bacon.

Edin, Kathryn and Joanna M. Reed. 2005. "Why Don't They Just Get Married? Barriers to Marriage among the Disadvantaged." *The Future of Children* 15(2):117–38.

Edwards, Cody S. 2006. "Friends with Benefits." Presented at the annual meeting of the Midwest Sociological Society, March 31, Omaha, NE.

Edwards, Margie L. K. 2004. "We're Decent People: Constructing and Managing Family Identity in Rural Working-Class Communities." *Journal of Marriage and Family* 66(2):515–29.

Edwards, Tom. 2007. "Most People Make Only One Trip Down the Aisle, But First Marriages Shorter, Census Bureau Reports." *U.S. Census Bureau News,* September 19. Retrieved June 18, 2009 (www.census.gov/Press-Release/www/releases/).

————. 2009. "As Baby Boomers Age, Fewer Families Have Children Under 18 at Home." *U.S. Census Bureau News,* February 25. Retrieved June 20, 2009 (www.census.gov/Press-Release/www/releases/).

Egelko, Bob. 2008. "Churches on Both Sides of Marriage Law Debate." *San Francisco Chronicle,* February 18: A1, A10.

————. 2009. "The Role of Religion in Adolescence for Family Formation in Young Adulthood." *Journal of Marriage and Family* 71(1):108–21.

Eggebeen, David J., Jeffrey Dew, and Chris Knoester. 2009. "Fatherhood and Men's Lives at Middle Age." *Journal of Family Issues* 31(1):113–30.

Ehrenberg, Marion F., Margaret Gearing-Small, Michael A. Hunter, and Brent J. Small. 2001. "Childcare Task Division and Shared Parenting Attitudes in Dual-earner Families with Young Children." *Family Relations* 50(2):143–53.

Ehrenfeld, Temma. 2002. "Infertility: A Guy Thing." *Newsweek,* March 25, pp. 60–61.

Ehrenreich, Barbara. 2001. *Nickel and Dimed: On (Not) Getting By in America.* New York: Henry Holt.

————. 2004. "Owning Up to Abortion." *The New York Times,* July 22: A21.

Ehrenreich, Barbara and Arlie Russell Hochschild. 2002. *Global Woman: Nannies, Maids, and Sex Workers in the New Economy.* New York: Henry Hold and Company.

Ehrensaft, Diane. 2005. *Mommies, Daddies, Donors, Surrogates: Answering Tough Questions and Building Strong Families.* New York: Guilford.

Eicher-Catt, Deborah. 2004a. "Noncustodial Mothering: A Cultural Paradox of Competent Performance–Performative Competence." *Journal of Contemporary Ethnography* 33(1):72–108.

————.2004b. "Noncustodial Mothers and Mental Health: When Absence Makes the Heart Break." *Family Focus* (March):F7–F8. National Council on Family Relations.

Eikhof, Doris Ruth, Chris Warhurst, and Axel Haunschild. 2007. "Introduction: What Work? What Life? What Balance?: Critical Reflections on the Work-Life Balance Debate." *Employee Relations* 24(4):325–33.

Eisenstadt v. Baird. 1972. 405 U.S. 398.

Elder, Glen H., Jr. 1974. *Children of the Great Depression: Social Change in Life Experience.* Chicago, IL: University of Chicago Press.

"Eldercare Locator." n.d. Washington, DC. U.S. Administration on Aging. Retrieved May 4, 2010 (www.eldercare.gov).

Elfrink, Tim. 2006. "RX for Deployment Blues." *Omaha World-Herald,* December 4.

Elias, Marilyn. 2003. "Children on Heightened Alert." *USA Today,* March 24.

————. 2004. "Marriage Taken to Heart." *USA Today,* March 4.

————. 2009. "Mental Stress Spirals with Economy." *USA Today,* March 12. Retrieved March 8, 2010 (www.usatoday.com).

Elkind, David. 1988. *The Hurried Child: Growing Up Too Fast Too Soon.* Reading, MA: Addison-Wesley.

————. 2007a. *The Hurried Child: Growing Up Too Fast Too Soon.* 25th anniversary ed. Cambridge, MA: Da Capo Lifelong.

————. 2007b. *The Power of Play: How Spontaneous, Imaginative Activities Lead to Happier, Healthier Children.* Cambridge, MA: Da Capo Lifelong.

Ellin, Abby. 2009. "The Recession. Isn't It Romantic?" *The New York Times,* February 12. Retrieved June 8, 2009 (www.nytimes.com).

Elliott, Sinikka and Debra Umberson. 2008. "The Performance of Desire: Gender and Sexual Negotiation in Long-Term Marriages." *Journal of Marriage and Family* 70(2):391–406.

Ellis, D. 2006. "Male Abuse of a Married or Cohabiting Female Partner: The Application of Sociological Theory to Research Findings." *Violence and Victims* 4:235–55.

Ellison, Christopher G. and Matt Bradshaw. 2009. "Religious Beliefs, Sociopolitical Ideology, and Attitudes Toward Corporal Punishment." *Journal of Family Issues* 30(3):330–40.

Ellison, Christopher G., Jenny A. Trinitapoli, Kristin L. Anderson, and Byron R. Johnson. 2007. "Race/Ethnicity, Religious Involvement, and Domestic Violence." *Violence Against Women* 13(11):1094–112.

Ellison, Peter T. and Peter B. Gray, eds. 2009. *Endocrinology of Social Relationships* .Cambridge, MA: Harvard University Press.

El Nasser, Haya. 2001. "Minorities Make Choice to Live with Their Own." *USA Today,* July 9.

El Nasser, Haya and Lorrie Grant. 2005. "Diversity Tints New Kind of Generation Gap." *USA Today,* June 9.

Elrod, Linda D. and Robert G. Spector. 2002. "A Review of the Year in Family Law: State Courts React to *Troxel.*" *Family Law Quarterly* 35(4):577–617.

Elsby, Michael, Bart Hobijn, and Aysgül Şahin. 2010. "The Labor Market in the Great Recession." *Brookings Panel on Economic Activity,* March 18–19. Washington, DC: Brookings Institute.

Else-Quest, Nicole M., Janet Shibley Hyde, and Marcia C. Linn. 2010. "Cross-National Patterns of Gender Differences in Mathematics: A Meta-Analysis." *Psychological Bulletin* 136(1):103–27.

El-Sheikh, Mona and Elizabeth Flanagan. 2001. "Parental Problem Drinking and Children's Adjustment: Family Conflict and Parental Depression as Mediators and Moderators of Risk." *Journal of Abnormal Child Psychology* 29(5):417–35.

Emlen, Stephen T. 1995. "An Evolutionary Theory of the Family." *Proceedings of the National Academy of Sciences* 92:8092–99.

Employee Benefit Research Institute. 2009. "Income Statistics of the Population Aged 55 and Over." Retrieved May 8, 2010 (www.ebri.org).

Enander, Viveka and Carin Holmberg. 2008. "Why Does She Leave? The Leaving Process(es) of Battered Women." *Health Care for Women International* 29(3):200–26.

Enda, Jodi. 1998. "Women Have Made Gains, Seek More." Washington Bureau in Saint Paul, *Minneapolis Pioneer Press,* Early Edition, July 19.

Engel, Marjorie. 2000. "The Financial (In) Security of Women in Remarriages." *Research Findings.* Stepfamily Association of America (www.saafamilies.org).

———. 2003. "Stepfamily Resources." Retrieved October 5, 2004 (www.marriagepreparation.com/stepfamily_resources.html).

Engels, Friedrich. 1942 [1884]. *The Origin of the Family, Private Property, and the State.* New York: International.

England, Diane. 2009. *The Post Traumatic Stress Disorder Relationship: How To Support Your Partner and Keep Your Relationship Healthy.* Avon, Mass.: Adams Media.

England, Paula. 2006. "Toward Gender Equality: Progress and Bottlenecks." Pp. 245–64 in *The Declining Significance of Gender?,* edited by Francine D. Blau, Mary C. Brinton, and David B. Grusky. New York: Russell Sage.

England, Paula, Carmen Garcia-Beaulieu, and Mary Ross. 2004. "Women's Employment among Blacks, Whites, and Three Groups of Latinas: Do More Privileged Women Have Higher Employment?" *Gender and Society* 18(4):494–509.

England, Paula and Kathryn Edin (Eds.). 2007. *Unmarried Couples with Children.* New York: Russell Sage Foundation.

England, Paula, and Elizabeth A. McClintock. 2009. "The Gendered Double Standard of Aging in US Marriage Markets." *Population and Development Review* 35(4):797–816.

England, Paula and Reuben J. Thomas. 2007. "The Decline of the Date and the Rise of the College Hook Up." Pp. 151–71 in *Family in Transition,* 14th ed., edited by Arlene S. Skolnick and Jerome H. Skolnick. Boston, MA: Allyn and Bacon.

English, Bella. 2009. "Recession Spurs Egg and Sperm Donations." *The Boston Globe,* April 7. Retrieved June 8, 2009 (www.geneticsandsociety.org).

Enos, Sandra. 2001. *Mothering from the Inside: Parenting in a Women's Prison.* Albany: SUNY Press.

Ephron, Delia. 1988. *Funny Sauce: Us, the Ex, the Ex's New Mate, the New Mate's Ex, and the Kids.* New York: Penguin.

Epstein, Ann S. 2007. *The Intentional Teacher: Choosing the Best Strategies for Young Children's Learning.* Washington, DC: National Association for the Education of Young Children.

Epstein, Cynthia Fuchs. 1988. *Deceptive Distinctions: Sex, Gender, and the Social Order.* New Haven, CT: Yale University Press.

Epstein, Marina, Jerel P. Calzo, Andrew P. Smiler, and L. Monique Ward. 2009. "Anything From Making Out to Having Sex: Men's Negotiations of Hooking Up and Friends With Benefits Scripts." *Journal of Sex Research* 46(5)414–42.

Erera, Pauline and Karen Fredriksen. 1999. "Lesbian Stepfamilies: A Unique Family Structure." *Families in Society* 80(3):263–70.

Erickson, Martha F. and Enola G. Aird. 2005. *The Motherhood Study: Fresh Insights on Mothers' Attitudes and Concerns.* New York: Institute for American Values. www.motherhoodproject.org.

Erickson, Nancy S. 1991. "Battered Mothers of Battered Children: Using Our Knowledge of Battered Women to Defend Them against Charges of Failure to Act." *Current Perspectives in Psychological, Legal, and Ethical Issues,* Vol. 1A, *Children and Families: Abuse and Endangerment,* pp. 197–218.

Erickson, Rebecca J. 2005. "Why Emotion Work Matters: Sex, Gender, and the Division of Household Labor." *Journal of Marriage and Family* 67(2):337–51.

Eriksen, Shelley and Naomi Gerstel. 2002. "A Labor of Love or Labor Itself: Care Work among Brothers and Sisters." *Journal of Family Issues* 23(7):836–56.

Espelage Dorothy L. and Susan M. Swearer. 2004. *Bullying in American Schools: A Social-Ecological Perspective on Prevention and Intervention.* Mahwah, NJ: Erlbaum.

Essex, Elizabeth L. and Junkuk Hong. 2005. "Older Caregiving Parents: Division of Household Labor, Marital Satisfaction, and Caregiver Burden." *Family Relations* 54(3):448–60.

Estes, Richard and Neil Weiner. 2002. *The Commercial Sexual Exploitation of Children in the U.S., Canada, and Mexico.* Philadelphia, PA: University of Pennsylvania (www.ssw.upenn.edu/~restes/CSEC).htm.

Etcheverry, Paul E. and Benjamin Le. 2005. "Thinking about Commitment: Accessibility of Commitment and Prediction of Relationship Persistence, Accommodation, and Willingness to Sacrifice." *Personal Relationships* 23(1):103–23.

Etheridge, Tiara. 2007. "Displacement, Loss Still Blur American Indian Identities." *Oklahoma Daily,* April 25. Retrieved December 27, 2009 (www.lexisnexis.com).

Evans, Gary W. and Theodore D. Wachs. 2010. *Chaos and Its Influence on Children's Development: An Ecological Perspective.* Washington, DC: American Psychological Association.

Even, William E. and David A. Macpherson. 2004. "When Will the Gender Gap in Retirement Income Narrow?" *Southern Economic Journal* 71(1):182–201.

Evenson, Ranae J., and Robin W. Simon. 2005. "Clarifying the Relationship between Parenthood and Depression." *Journal of Health and Social Behavior* 46:341–58.

Faber, Adele, and Elaine Mazlish. 2006. *How To Talk So Kids Will Listen and Listen So Kids Will Talk.* New York: Collins.

"Fact Sheet: Elder Abuse Prevalence and Incidence." 2005. Washington, DC: National Center on Elder Abuse.

"Facts on American Teens' Sexual and Reproductive Health." 2010. In Brief, January. New York: The Guttmacher Institute.

Fadiman, Anne. 1998. *The Spirit Catches You and You Fall Down.* New York: Farrar, Straus and Giroux.

Fagan, Jay. 2008. "Randomized Study of a Prebirth Coparenting Intervention with Adolescent and Young Fathers." *Family Relations* 57(3):309–23.

———. 2009. "Relationship Quality and Changes in Depressive Symptoms Among Urban, Married African Americans, Hispanics, and Whites." *Family Relations* 58(July):259–74.

Fagan, Jay and Marina Barnett. 2003. "The Relationship between Maternal Gatekeeping, Paternal Competence, Mothers' Attitudes about the Father Role, and Father Involvement." *Journal of Family Issues* 24(8):1020–43.

Fairchild, Emily. 2006. "'I'm Excited to Be Married, But . . .': Romance and Realism in Marriage." Pp. 1–19 in *Couples, Kids, and Family Life,* edited by Jaber F. Gubrium and James A. Holstein. New York: Oxford University Press.

Falbo, T. 1976. "Does the Only Child Grow Up Miserable?" *Psychology Today* 9:60–65.

Falconer, Mary Kay, Mary E. Haskett, Linda McDaniels, Thelma Dirkes, and Edward C. Siegel. 2008. "Evaluation of Support Groups for Child Abuse Prevention: Outcomes of Four State Evaluations." *Social Work With Groups* 31(2):165–82.

Falicov, Celia J. 2007. Working With Transnational Immigrants: Expanding Meanings of Family, Community, and Culture. *Family Process* 46:157–71.

Families and Work Institute. 2004. *Gender and Generation in the Workplace*. Boston, MA: American Business Collaboration. October 5.

Families with Children from China. 1999. "Mongolian Spots." Retrieved April 26, 2006 (www.fwcc.org).

Family Support Act. 1988. Public Law 100-628. Washington, DC: U.S. Congress.

"Family Ties." 2008. *The Economist, U.S. Edition*, April 12.

Fanshel, David. 1972. *Far from the Reservation: The Transracial Adoption of American Indian Children*. Metuchen, NJ: Scarecrow.

Farkas, Steve, Jean Johnson, and Ann Duffett. 2002. "A Lot Easier Said Than Done: Parents Talk about Raising Children in Today's America." *Public Agenda:* A report prepared for State Farm Insurance Companies.

Farver, JoAnn M, Yiyuan Xu, Bakhtawar R. Bhadha, Sonia Narang, and Eli Lieber. 2007. "Ethnic Identity, Acculturation, Parenting Beliefs, and Adolescent Adjustment: A Comparison of Asian Indian and European American Families." *Merrill-Palmer Quarterly* 53(2):184–215.

Fass, Paula S. 2003. Review of *Anxious Parents* (www.amazon.com).

Fausto-Sterling, Anne. 2000. "The Five Sexes Revisited." *Sciences*, July/August, pp.19–23.

———. 2007. "Frameworks of Desire." *Daedalus* 136(2):47–57.

Fears, Darryl. 2004. "Black Baby Boomers' Income Gap Cited." *Washington Post*, December 17.

"Federal Court Dismisses Lawsuit from Lesbian Banned from Dying Partner's Bedside." 2009. *Southern Voice*, September 29. Retrieved November 21, 2009 (www.sovo.com).

Federal Interagency Forum on Aging-Related Statistics. 2008. *Older Americans 2008: Key Indicators of Well-Being*. Washington DC: U.S. Government Printing Office.

"Federal Marriage Benefits Denied to Same-Sex Couples." n.d. http://www.nolo.com.

Feigelman, Susan, Howard Dubowitz, Wendy Lane, Leslie Prescott, Walter Meyer, Kathleen Tracy, and Jeongeun Kim. 2009. "Screening for Harsh Punishment in a Pediatric Primary Care Clinic." *Journal of Child Abuse & Neglect* 33(5):269–77.

Feigelman, W. 2000. "Adjustments of Transracially and Inracially Adopted Young Adults." *Child and Adolescent Social Work Journal* 17:165–83.

Feijoo, Ammie N. 2004. "Trends in Sexual Risk Behavior among High School Students—United States, 1991 to 1997 and 1999 to 2003." Washington, DC: Advocates for Youth. September. Retrieved January 7, 2007 (www.advocatesforyouth.org).

Fein, Esther B. 1997. "Failing to Discuss Dying Adds to Pain of Patient and Family." *New York Times*, March 5, pp. A1, A14.

Feinberg, Mark E., Marni L. Kan, and E. Mavis Hetherington. 2007. "The Longitudinal Influence of Coparenting Conflict on Parental Negativity and Adolescent Maladjustment." *Journal of Marriage and Family* 69(3):687–702.

Feldman, Robert S. 2003. *Development across the Life Span*. 3rd ed. Upper Saddle River, NJ:Prentice Hall.

Felmlee, Diane H. 2001. "No Couple Is an Island: A Social Network Perspective on Dyadic Stability." *Social Forces* 79(4):1259–82.

Fergusson, David M., Joseph M. Boden, and L. John Horwood. 2007. "Abortion among Young Women and Subsequent Life Outcomes." *Perspectives on Sexual and Reproductive Health* 39(1):6–12.

Fergusson, David M., L. John Horwood, and Elizabeth M. Ridder. 2006. "Abortion in Young Women and Subsequent Mental Health." *Journal of Child Psychiatry and Psychology* 47(1):16–24.

Ferrante, Joan. 2000. *Sociology: The United States in a Global Community*. 4th ed. Belmont, CA: Wadsworth.

Ferrari, J. R. and R. A. Emmons. 1994. "Procrastination as Revenge: Do People Report Using Delays as a Strategy for Vengeance?" *Personality and Individual Differences* 17(4):539–42.

Ferrer-Chancy, Millie. 2009. "Stepping Stones for Stepfamilies: For Step-Grandparents." University of Florida Extension. Retrieved May 5, 2010 (www.edis.ifas.ufl.edu).

Ferro, Christine, Jill Cermele, and Ann Saltzman. 2008. "Current Perceptions of Marital Rape: Some Good and Not-So-Good News." *Journal of Interpersonal Violence* 23(6):764–79.

Festinger, Trudy B. 2005. "Adoption and Disruption." Pp. 452–68 in *Child Welfare for the 21st Century: A Handbook of Practices, Policies, and Programs*, edited by G. Mallon and P. Hess. New York: Columbia University Press.

Fetsch, Robert J., Raymond K. Yang, and Matthew J. Pettit. 2008. "The RETHINK Parenting and Anger Management Program." *Family Relations* 57(5):543–52.

Few, April L. and Karen H. Rosen. 2005. "Victims of Chronic Dating Violence: How Women's Vulnerabilities Link to Their Decisions to Stay." *Family Relations* 54(2):265–79.

Fidas, Deena and Samir Luther. 2010. *Corporate Equality Index 2010: A Report Card on Lesbian, Gay, Bisexual and Transgender Equality in Corporate America*. Washington, DC: Human Rights Campaign Foundation.

Fields, Jason. 2001. *Living Arrangements of Children: 1996*. Current Population Reports P70-74. Washington, DC: U.S. Census Bureau. April.

———. 2003. *Children's Living Arrangements and Characteristics: March 2003*. Current Population Reports P20-547. Washington, DC: U.S. Census Bureau.

———. 2004. *America's Families and Living Arrangements: 2003*. Current Population Reports

P20-553. Washington, DC: U.S. Census Bureau. November.

Fields, Julianna. 2010. *Families Living with Mental and Physical Challenges*. Broomall, PA: Mason Crest Publishers.

Fincham, Frank D. and Steven R. H. Beach. 2010. "Marriage in the New Millennium: A Decade in Review." *Journal of Marriage and Family* 72(3):630–49.

Fincham, Frank D., Julie Hall, and Steven R. H. Beach. 2006. "Forgiveness in Marriage: Current Status and Future Directions." *Family Relations* 55(4):415–27.

Fincham, Frank D., Scott M. Stanley, and Steven R. H. Beach. 2007. "Transformative Processes in Marriage: An Analysis of Emerging Trends." *Journal of Marriage and Family* 69(2):275–92.

Find Care. 2002. "Choosing Quality Child Care" (www.cafcc.on.ca).

Finer, Lawrence B. 2007. "Trends in Premarital Sex in the United States, 1954–2003." *Public Health Reports* 122(January/February):73–78. Retrieved December 20, 2006 (www.publichealthreports.org).

Finer, Lawrence B., Jacqueline E. Darroch, and Jennifer J. Frost. 2003. "Services for Men at Publicly Funded Family Planning Agencies, 1998–1999." *Perspectives on Sexual and Reproductive Health* 35:202–07.

Finer, Lawrence B., Lori Frohwirth, Lindsay A. Dauhiphinee, Sushella Singh, and Ann M. Moore. 2005. "Reasons U.S. Women Have Abortions: Quantitative and Qualitative Perspectives." *Perspectives on Sexual and Reproductive Health* 37(3):110–18.

Finer, Lawrence B. and Stanley K. Henshaw. 2006. "Disparities in Rates of Unintended Pregnancy in the United States, 1994 and 2001." *Perspectives on Sexual and Reproductive Health* 38(2):90–6.

Fingerman, Karen L., Elizabeth L. Hay, and Kira S. Birditt. 2004. "The Best of Ties, the Worst of Ties: Close, Problematic, and Ambivalent Social Relationships." *Journal of Marriage and Family* 66(3):792–808.

Fingerman, Karen, Laura Miller, Kira Birditt, and Steven Zarit. 2009. "Giving to the Good and the Needy: Parental Support of Grown Children." *Journal of Marriage and Family* 71(5):1220–33.

Finkelhor, David, Gerald Hotaling, and Andrea Sedlak. 1991. "Children Abducted by Family Members: A National Household Survey of Incidence and Episode Characteristics." *Journal of Marriage and Family* 53(3):805–17.

Finkelhor, David and Lisa M. Jones. 2004. "Explanations for the Decline in Child Sexual Abuse Cases." NCJ 199298, September. Washington, DC: Office of Juvenile Justice and Delinquency Prevention.

———. 2006. "Why Have Child Maltreatment and Child Victimization Declined?" *Journal of Social Issues* 62(4):685–716.

Finkelhor, David, Richard Ormrod, Heather Turner, and Sherry Hamby. 2005. "The Victimization of Children and Youth: A Comprehensive National Survey." *Child Maltreatment* 10:5–25.

Finley, Gordon E. 2000. "Adoptive Families: Dramatic Changes across Generations." *Family Focus* (June): F10, F12. National Council on Family Relations.

Firmin, Michael and Annie Phillips. 2009. "A Qualitative Study of Families and Children Possessing Diagnoses of ADHD." *Journal of Family Issues* 30(9):1155–74.

Fisher, Judith L. 1983. "Mothers Living Apart from Their Children." *Family Relations* 32:351–57.

Fisher, Linda L. 2010. *Sex, Romance, and Relationships: AARP Survey of Midlife and Older Adults.* Washington, DC: American Association of Retired Persons. Retrieved May 10, 2010 (www.aarp.org).

Fitzpatrick, Jacki, Elizabeth Sharp, and Alan Reifman. 2009. "Midlife Singles' Willingness to Date Partners with Heterogeneous Characteristics." *Family Relations* 58(1):121–33.

Fitzpatrick, Mary Anne. 1995. *Explaining Family Interactions.* Thousand Oaks, CA: Sage Publications.

Fivush, Robyn, Kelly Marin, Kelly McWilliams, and Jennifer Bohanek. 2009. "Family Reminiscing Style: Parent Gender and Emotional Focus in Relation to Child Well-Being." *Journal of Cognition and Development* 10(3):210–35.

Fleeson, W. and E. Noftle. (2008). "The End of the Person–Situation Debate: An Emerging Synthesis in the Answer to the Consistency Question." *Social and Personality Psychology Compass* 2:1667–84.

Fletcher, Garth. 2002. *The New Science of Intimate Relationships.* Malden, MA: Blackwell.

Flora, Carlin. 2007. "Can Grown-Up siblings Learn to Get Along?" *Psychology Today.* March/April:48–49.

Fomby, Paula and Andrew J. Cherlin. 2007. "Family Instability and Child Well-Being." *American Sociological Review* 72(2):181–204.

Fontaine, Romeo, Agnes Gramain, and Jerome Wittwer. 2009. "Providing Care for an Elderly Parent: Interactions among Siblings." *Health Economics* 18:1,011–29.

Foran, Heather M. and K. Daniel O'Leary. 2008. "Problem Drinking, Jealousy, and Anger Control: Variables Predicting Physical Aggression Against a Partner." *Journal of Family Violence* 23:141–48.

Ford, Melissa. 2009. *Navigating the Land of If: Understanding Infertility and Exploring Your Options.* Berkeley, CA: Seal Press.

Formichelli, Linda. 2001. "Baby Blues." *Psychology Today* (March/April):24.

Formoso, Diana, Nancy A. Gonzales, Manuel Barrera, Jr., and Larry E. Dumka. 2007. "Interparental Relations, Maternal Employment, and Fathering in Mexican American Families." *Journal of Marriage and Family* 69(1):26–39.

Forry, Nicole D., Leigh A. Leslie, and Bethany L. Letiecq. 2007. "Marital Quality in Interracial Relationships: The Role of Sex Role Ideology and Perceived Fairness." *Journal of Family Issues* 28(12):1538–52.

Fortuny, Karina, Randy Capps, Margaret Simms, and Ajay Chaudry. 2009. *Children of Immigrants: National and State Characteristics.* Washington, DC: The Urban Institute.

Foster, Brooke Lea. 2006. "The Way They Were." *AARP Magazine,* September. www.aarp.org.

"Foster Care." 2005. *Facts for Families.* American Academy of Child and Adolescent Psychiatry. Retrieved February 19, 2007 (www.aacap.org).

Foster, E. Michael, Damon Jones, and Saul D. Hoffman. 1998. "The Economic Impact of Nonmarital Childbearing: How Are Older, Single Mothers Faring?" *Journal of Marriage and Family* 60(1):163–74.

Foster, Holly and Jeanne Brooks-Gunn. 2009. "Toward a Stress Process Model of Children's Exposure to Physical Family and Community Violence." *Clinical Child and Family Psychology Review* 12:71–94.

Fountain, Kim, Maryse Mitchell-Brody, Stephanie A. Jones, and Kaitlin Nichols. 2009. *Lesbian, Gay, Bisexual, Transgender and Queer Domestic Violence in the United States In 2008.* New York: The National Coalition of Anti-Violence Programs. Available at www.avp.org/documents/2008NCAVPLGBTQDVReportFINAL.pdf.

Fox, Bonnie. 2009. *When Couples Become Parents: The Creation of Gender in the Transition to Parenthood.* Buffalo, NY: University of Toronto Press.

Fox, Greer Litton, Michael L. Benson, Alfred A. DeMaris, and Judy Van Wyk. 2002. "Economic Distress and Intimate Violence: Testing Family Stress and Resources Theory." *Journal of Marriage and Family* 64:793–807.

Fox, Greer Litton, Carol Bruce, and Terri Combs-Orme. 2000. "Parenting Expectations and Concerns of Fathers and Mothers of Newborn Infants." *Family Relations* 49(2):123–31.

Fox, Greer Litton and Robert F. Kelly. 1995. "Determinants of Child Custody Arrangements at Divorce." *Journal of Marriage and Family* 57(3):693–708.

Fox, James Alan and Marianne W. Zawitz. 2004. *Homicide Trends in the U.S.* Washington, DC: U.S. Bureau of Justice Statistics (www.ojp.usdoj.gov/bjs).

Fox, Maggie. 2010. "Study Shows Consistent Benefit of Early Daycare." *Reuters,* May 14. Chicago, IL: Thompson Reuters.

Fracher, Jeffrey and Michael S. Kimmel. 1992. "Hard Issues and Soft Spots: Counseling Men about Sexuality." Pp. 438–50 in *Men's Lives,* 2nd ed., edited by Michael S. Kimmel and Michael A. Messner. New York: Macmillan.

Francis, Meagan. 2007. "About LargerFamilies.com." Retrieved March 23, 2007 (www.largerfamilies.com).

Freeman, Marsha B. 2008. "Love Means Always Having To Say You're Sorry: Applying the Realities of Therapeutic Jurisprudence to Family Law." *UCLA Women's Law Journal* 17:215–41.

Freeman, Melissa and Sandra Mathison. 2009. *Researching Children's Experiences.* New York: Guilford Press.

French, J. R. P. and Bertram Raven. 1959. "The Basis of Power." In *Studies in Social Power,* edited by D. Cartwright. Ann Arbor: University of Michigan Press.

Frey, William H. 2002. "The New White Flight." *American Demographics* 24:20–23.

Friedan, Betty. 1963. *The Feminine Mystique.* New York: Dell.

Friedman, Jaclyn and Jessica Valenti. 2008. *Yes Means Yes! Visions of Female Sexual Power and a World Without Rape.* Berkeley, CA: Seal Press.

Friedman, Joel, Marcia M. Boumil, and Barbara Ewert Taylor. 1992. *Date Rape: What It Is, What It Isn't, What It Does to You, What You Can Do About It.* Deerfield Beach, FL: Health Communications.

Friedrich, William N., Jennifer Fisher, Daniel Broughton, Margaret Houston, and Constance R. Shafran. 1998. "Normative Sexual Behavior in Children: A Contemporary Sample." *Pediatrics* 104(April):E9 (www.pediatrics.org).

Frisco, Michelle L., Chandra Muller, and Kenneth Frank. 2007. "Parents' Union Dissolution and Adolescents' School Performance: Comparing Methodological Approaches." *Journal of Marriage and Family* 69(3):721–41.

Frisco, Michelle L. and Kristi Williams. 2003. "Perceived Housework Equity, Marital Happiness, and Divorce in Dual Earner Families." *Journal of Family Issues* 24:51–73.

Fritsch, Jane. 2001. "A Rise in Single Dads." *New York Times,* May 20.

Fromm, Erich. 1956. *The Art of Loving.* New York: Harper and Row.

Fruhauf, Christine A., Shannon E. Jarrott, and Katherine R. Allen. 2006. "Grandchildren's Perceptions of Caring for Grandparents." *Journal of Family Issues* 27(7):887–911.

Fry, Richard and D'Vera Cohn. 2010. *Women, Men and the New Economics of Marriage.* Washington, DC: Pew Research Center.

Fry, Richard and Jeffrey S. Passel. 2009. "Latino Children: A Majority Are U.S.-Born Offspring of Immigrants." Washington, DC: Pew Hispanic Center. Retrieved December 26, 2009 (pewhispanic.org).

Fryar, Cheryl D., Rosemarie Hirsch, Kathryn S. Porter, Benny Kottiri, Debra J. Brody, and Tatiana Louis. 2007. "Drug Use and Sexual Behaviors Reported by Adults: United States, 1999–2002." *Advance Data from Vital and Health Statistics* No. 384. Hyattsville, MD: National Center for Health Statistics.

Frye, Marilyn. 1992. "Lesbian 'Sex.'" Pp. 109–19 in *Essays in Feminism 1976–1992,* edited by Marilyn Frye. Freedom, CA: Crossing.

Fryer, Roland G. 2007. "Guess who's Been Coming To Dinner: Trends in Interracial Marriage over the Twentieth Century." *The Journal of Economic Perspectives* 21(2):71–90.

Fu, Xuanning and Tim B. Heaton. 2000. "Status Exchange in Intermarriage among Hawaiians, Japanese, Filipinos and Caucasians in Hawaii: 1983–1994." *Journal of Comparative Family Studies* 31(1):45–64.

———. 2008. "Racial and Educational Homogamy: 1980 to 2000." *Sociological Perspectives* 51(4):735–58.

Furdyna, Holly E., M. Belinda Tucker, and Angela D. James. 2008. "Relative Spousal Earnings and Marital Happiness Among African American and White Women." *Journal of Marriage and Family* 70(2):332–44.

Furstenberg, Frank F., Jr. 2000. "The Sociology of Adolescence and Youth in the 1990s: A Critical Commentary." *Journal of Marriage and Family* 62(4):896–910.

———. 2003. "The Future of Marriage." Pp. 171–77 in *Family in Transition*, 12th ed., edited by Arlene S. Skolnick and Jerome H. Skolnick. Boston, MA: Allyn and Bacon.

———. 2005. "Banking on Families: How Families Generate and Distribute Social Capital." *Journal of Marriage and Family* 67(4):809–21.

———. 2006. "Diverging Development: The Not-So-Invisible Hand of Social Class in the United States." Network on Transitions to Adulthood Research Network Working Paper. Presented at the biennial meeting of the Society for Research on Adolescence, March 23–26, 2006, San Francisco, CA.

———. 2008. "The Changing Landscape of Early Adulthood in the U.S." *Family Focus*, March:F2–F3, F18.

Furstenberg, Frank F., Jr., J. Brooks-Gunn, and S. Philip Morgan. 1987. *Adolescent Mothers in Later Life*. New York: Cambridge University Press.

Furstenberg, Frank F., Jr., Sheela Kennedy, Vonnie C. McLoyd, Rubén G. Rumbaut, and Richard A. Settersten, Jr. 2004. "Growing Up Is Harder to Do." *Contexts* 3(3):33–41.

Furstenberg, Frank F., Jr. and Kathleen E. Kiernan. 2001. "Delayed Parental Divorce: How Much Do Children Benefit?" *Journal of Marriage and Family* 63(2):446–57.

Gable, Shelly L., Harry T. Reis, Emily A. Impett, and Evan R. Asher. 2004. "Interpersonal Relations and Group Processes—What Do You Do When Things Go Right? The Intrapersonal and Interpersonal Benefits of Sharing Positive Events." *Journal of Personality and Social Psychology* 87(2):228–45.

Gaertner, Bridget M., Tracy I. Spinrad, Nancy Eisenberg, and Karissa A. Greving. 2007. "Parental Childrearing Attitudes as Correlates of Father Involvement During Infancy." *Journal of Marriage and Family* 69(4):962–76.

Gaffney, Dennis. 2006. "'American Indian' or 'Native American': Which Is Correct?" *Antiques Roadshow* (TV series). Washington, DC: Public Broadcasting System. Retrieved December 2, 2006 (www.pbs.org).

Gager, Constance T. and Laura Sanchez. 2003. "Two as One? Couples' Perceptions of *Time* Spent Together, Marital Quality, and the Risk of Divorce." *Journal of Family Issues* 24(1):21–50.

Gagnon, John H. and William Simon. 2005. *Sexual Conduct: The Social Sources of Human Sexuality (Social Problems and Social Issues)*. 2nd ed. New Brunswick, NJ: Transaction Books.

Galambos, Nancy I. and Harvey J. Krahn. 2008. "Depression and Anger Trajectories During the Transition to Adulthood." *Journal of Marriage and Family* 70(1):15–27.

Galinsky, Ellen. 2001. "Parent Tips." Families and Work Institute (www.familiesandwork.org).

Galinsky, Ellen, Kerstin Aumann, and James T. Bond. 2009. "Times are Changing: Gender and Generation at Work and at Home." *2008 National Study of the Changing Workforce*. New York: Families and Work Institute.

Gallagher, Charles A. 2006. "Interracial Dating and Marriage: Fact, Fantasy and the Problem of Survey Data." Pp. 141–53 in *African Americans and Whites: Changing Relationships on College Campuses*, edited by Robert M. Moore III. New York: University Press of America.

Gallagher, Sally K. 2004. "Where Are the Antifeminist Evangelicals? Evangelical Identity, Subcultural Location, and Attitudes toward Feminism." *Gender and Society* 18(4):451–72.

Gallagher, Sally K. and Naomi Gerstel. 2001. "Connections and Constraints: The Effects of Children on Caregiving." *Journal of Marriage and Family* 63(1):265–75.

Gallo, Eileen and Jon J. Gallo. 2005. *The Financially Intelligent Parent: 8 Steps to Raising Successful, Generous, Responsible Children*. New York: New American Library.

Gallup Poll News Service. 2007. "Gallup's Pulse of Democracy: Abortion." Washington, DC: Gallup Poll News Service. Retrieved March 21, 2007 (www.galluppoll.com).

———. 2009. "Abortion." Washington, DC: Gallup Poll News Service, July 17-19. www. galluppoll.com.

Gamache, Susan J. 1997. "Confronting Nuclear Family Bias in Stepfamily Research." *Marriage and Family Review* 26(1–2):41–50.

Games-Evans, Tina. 2009. "Finding Your Personal Identity as a Mom." Babyzone, June 1. www.babyzone.com.

Gandy, Kim. 2005. "The Patriarchy Isn't Falling." *USA Today*, September 22.

Gangstad, Jack. 2006. *From Diapers to Diplomas: A Common Sense Approach to Raising Well-Adjusted Children*. Bloomington, IN: AuthorHouse.

Ganong, Lawrence H. and Marilyn Coleman. 2000. "Remarried Families." Pp. 155–68 in *Close Relationships: A Sourcebook*, edited by Clyde Hendrick and Susan S. Hendrick. Thousand Oaks, CA: Sage Publications.

———.2004. *Stepfamily Relationships: Development, Dynamics, and Interventions*. New York: Kluwer Academic/Plenum.

———. 2005. "Leaving Whose Home? When Stepchildren Leave Is It Always Extrusion?" Pp. 233–36 in *Sourcebook of Family Theory and Research*, edited by Vern L. Bengston, Alan C. Acock, Katherine R. Allen, Peggye Dilworth-Anderson, and David M. Klein. Thousand Oaks, CA: Sage Publications.

Ganong, Lawrence H., Marilyn Coleman, and Tanja Rothrauff. 2009. "Patterns of Assistance between Adult Children and Their Older Parents: Resources, Responsibilities, and Remarriage." *Journal of Social and Personal Relationships* 26(2–3):161–78.

Gans, Daphna and Merril Silverstein. 2006. "Norms of Filial Responsibility for Aging Parents across Time and Generations." *Journal of Marriage and Family* 68(4):961–76.

Gans, Daphne, Merril Silverstein, and Anela Lowenstein. 2009. "Do Religious Children Care More and Provide More Care for Older Parents? A Study of Filial Norms and Behaviors across Five Nations." *Journal of Comparative Family Studies* 40(2):187–201.

Gans, Herbert J. 1982 [1962]. *The Urban Villagers: Group and Class in the Life of Italian-Americans*, updated and expanded edition. New York: Free Press.

Garasky, Steven, Elizabeth Peters, Laura Argys, Steven Cook, Lenna Nepomnyaschy, and Elaine Sorensen. 2007. "Measuring Support to Children by Nonresident Fathers." Pp. 399–428 in *Handbook of Measurement Issues in Family Research*, edited by Sandra L. Hofferth and Lynne M. Casper. Mahwah, NJ: Lawrence Erlbaum Associates.

Gardner, Jonathan and Andrew Oswald. 2006. "Do Divorcing Couples Become Happier by Breaking Up?" *Journal of the Royal Statistical Society*, Series A 169(2):319–36.

Garfield, Robert. 2010. "Male Emotional Intimacy: How Therapeutic Men's Groups Can Enhance Couples Therapy." *Family Process* 49(1):109–22.

Garfinkel, Irwin, Daniel R. Meyer, and Sara S. McLanahan. 1998. "A Brief History of Child Support Policies in the U.S." Pp. 14–30 in *Fathers Under Fire: The Revolution in Child Support Enforcement*, edited by Irwin Garfinkel, Sara S. McLanahan, Daniel R. Meyer, and Judith A. Seltzer. New York: Russell Sage.

Garner, Steve. 2007. *Whiteness: An Introduction*. New York: Routledge.

Garson, David G. n.d. "Economic Opportunity Act of 1964." Retrieved October 9, 2006 (wps. prenhall.com).

Gates, Gary J. 2006. *Same-Sex Couples and the Gay, Lesbian, and Bisexual Population: New Estimates from the American Community Survey*. Los Angeles, CA: The Williams Institute on Sexual Orientation, Law, and Public Policy, University of California at Los Angeles. October.

———. 2009a. "FAQ: Same-Sex Couples in the 2008 American Community Survey." UC Los Angeles: The Williams Institute. Retrieved November 20, 2009 (www.law.ucla.edu/williamsinstitute).

———. 2009b. *Same-Sex Spouses and Unmarried Partners in the American Community Survey, 2008*. Los Angeles, CA: The Williams Institute. Retrieved November 20, 2009 (www.law.ucla .edu/williamsinstitute).

Gaughan, Monica. 2002. "The Substitution Hypothesis: The Impact of Premarital Liaisons and Human Capital on Marital Timing." *Journal of Marriage and Family* 64(2):407–19.

Gaunt, Ruth. 2006. "Couple Similarity and Marital Satisfaction: Are Similar Spouses Happier?" *Journal of Personality* 74(5):1401–20.

Gavin, Lorrie, Andrea P. MacKay, Kathryn Brown, Sara Harrier, Stephanie J. Ventura, Laura Kann, Maria Rangel, Stuart Berman, Patricia Dittus, Nicole Liddon, Lauri Markowitz, Maya Sternberg, Hillard Weinstock, Corinne

David-Ferdon, George Ryan. 2009. "Sexual and Reproductive Health of Persons Aged 10–24 Years—United States, 2002–2007." *Morbidity and Mortality Weekly Report Surveillance Summaries* 58(SS06):1–58. Atlanta: Centers for Disease Control and Prevention.

"Gay and Lesbian Adolescents." 2006. *Facts for Families.* Academy of Child and Adolescent Psychiatry # 63. Retrieved July 1, 2010 (www.aacap.org.)

"Gay and Lesbian Rights." 2009. *Gallup Poll,* May 7–10. www.gallup.com.

Gayles, Jochebed G., J. Douglas Coatsworth, Hilda M. Pantin, and Jose Szapocznik. 2009. "Parenting and Neighborhood Predictors of Youth Problem Behaviors within Hispanic Families: The Moderating Role of Family Structure." *Hispanic Journal of Behavioral Sciences* 31(3):277–96.

"Gays Want the Right, but Not Necessarily the Marriage." 2004. *The Christian Science Monitor,* February 12.

Geary, David C. and Mark V. Flinn. 2001. "Evolution of Human Parental Behavior and the Human Family." *Parenting Science and Practice* 1:5–61.

Gecas, Viktor. 1982. "The Self-Concept." *Annual Review of Sociology* 8:1–33.

Gelles, Richard J. 1974. *The Violent Home: A Study of Physical Aggression between Husbands and Wives.* Beverly Hills, CA: Sage Publications.

———. 1996. *The Book of David: How Preserving Families Can Cost Children's Lives.* New York: Basic Books.

———. 1997. *Intimate Violence in Families.* 3rd ed. Thousand Oaks, CA: Sage Publications.

———. 2005. "Protecting Children Is More Important Than Preserving Families." Pp. 329–40 in *Current Controversies on Family Violence,* 2nd ed., edited by Donileen R. Loseke, Richard J. Gelles, and Mary M. Cavanaugh. Thousand Oaks, CA: Sage Publications.

Gelles, Richard J. and Mary M. Cavanaugh. 2005. "Violence, Abuse, and Neglect in Families and Intimate Relationships." Pp. 129–54 in *Families and Change: Coping with Stressful Events and Transitions,* 3rd ed., edited by Patrick C. McKenry and Sharon J. Price. Thousand Oaks, CA: Sage Publications.

Gelles, Richard J. and Jane B. Lancaster, eds. 1987. *Child Abuse and Neglect: Biosocial Dimensions.* Hawthorne, NY: Aldine.

Gelles, Richard J. and Murray A. Straus. 1988. *Intimate Violence: The Definitive Study of the Causes and Consequences of Abuse in the American Family.* New York: Simon and Schuster.

Gemelli, Marcella. 2008. "Understanding the Complexity of Attitudes of Low-Income Single Mothers Toward Work and Family in the Age of Welfare Reform." *Gender Issues* 25:101–13.

"Genetically, Race Doesn't Exist." 2003. *Washington University Magazine,* Fall, p. 4.

Gerbrandt, Roxanne. 2007. *Exposing the Unmentionable Class Barriers in Graduate Education.* Unpublished Doctoral Dissertation, University of Oregon.

Gerlach, Peter K. 2010. "What's *Normal* in a Multi-home Stepfamily? 60 Common Myths and Their (Typical) Realities." Retrieved April 17, 2010 (sfhelp.org/sf/basics/myths2.htm).

Geronimus, Arline T. 1991. "Teenage Childbearing and Social and Reproductive Disadvantage: The Evolution of Complex Questions and the Demise of Simple Answers." *Family Relations* 40(4):463–71.

Gerson, Kathleen. 1993. *No Man's Land: Men's Changing Commitments to Family and Work.* New York: HarperCollins, Basic Books.

———. 1997. "The Social Construction of Fatherhood." Pp. 119–53 in *Contemporary Parenting: Challenges and Issues,* edited by Terry Arendell. Thousand Oaks, CA: Sage Publications.

———. 2010. *The Unfinished Revolution: How a New Generation is Reshaping Family, Work, and Gender in America.* New York: Oxford University Press.

Gerstenfeld, Phyllis. 2010. "Hate Crimes." Pp. 257–75 in *Violent Crime: Clinical and Social Implications,* edited by Christopher J. Ferguson. Thousand Oaks: Sage Publications.

Gewertz, Catherine. 2009. "Report Probes Educational Challenges Facing Latinas." *Education Week* 29(2):12.

Ghosh, Shuvo. 2009. "Sexuality, Gender Identity." *eMedicine from WebMD.* Retrieved January 19, 2010 (emedicine.medscape.com).

Gibbs, Nancy and Michael Scherer. 2009. "Michelle Up Close." *Time,* June 1, pp. 16–25.

Gibson-Davis, Christina M. 2008. "Family Structure Effects on Maternal and Paternal Parenting in Low-Income Families." *Journal of Marriage and Family* 70(2):452–65.

———. 2009. "Money, Marriage, and Children: Testing the Financial Expectations and Family Formation Theory." *Journal of Marriage and Family* 71(1):146–60.

Giddens, Anthony. 2007. "The Global Revolution in Family and Personal Life." Pp. 26–31 in *Family in Transition,* edited by Arlene S. Skolnick and Jerome H. Skolnick. Boston, MA: Allyn and Bacon.

Giele, Janet Z. 2007. "Decline of the Family: Conservative, Liberal, and Feminist Views." Pp. 76–91 in *Family in Transition,* edited by Arlene S. Skolnick and Jerome H. Skolnick. Boston, MA: Allyn and Bacon.

Gilbert, Lucia Albino and Jill Rader. 2008. "Work, Family, and Dual-Earner Couples: Implications for Research and Practice." In *Handbook of Counseling Psychology,* 4th ed., edited by Steven D. Brown and Robert W. Lent. New York: John Wiley & Sons, Inc.

Gilbert, Neil. 2008. *A Mother's Work: How Feminism, the Market, and Policy Shape Family Life.* New Haven, CT: Yale University Press.

Gilbert, Susan. 1997. "Two Spanking Studies Indicate Parents Should Be Cautious." *New York Times,* August 20.

Gilkey, So'Nia L., JoAnne Carey, and Shari L. Wade. 2009. "Families in Crisis: Considerations for the Use of Web-Based Treatment Models in

Family Therapy." *Families in Society: The Journal of Contemporary Human Services* 90(1):37–46.

Gillespie, Dair. 1971. "Who Has the Power? The Marital Struggle." *Journal of Marriage and Family* 33:445–58.

Gilligan, Carol. 1982. *In a Different Voice: Psychological Theory and Women's Development.* Cambridge, MA: Harvard University Press.

Gilligan, Carol, Nona P. Lyons, and Trudy J. Hanmer, eds. 1990. *Making Connections: The Relational Worlds of Adolescent Girls at Emma Willard School.* Cambridge, MA: Harvard University Press.

Gillman, Todd J. 2008. "For Some Mixed-Race Couples, Barack Obama Is a Symbol of Acceptance." *The Dallas Morning News,* March 2.

Gillmore, Mary Rogers, Jungeun Lee, Diane M. Morrison, and Taryn Lindhorst. 2008. "Marriage Following Adolescent Parenthood: Relationship to Adult Well-Being." *Journal of Marriage and Family* 70(5):1136–44.

Gillum, Tameka L., Sullivan, Cris M., and Deborah I. Bybee. 2006. "The Importance of Spirituality in the Lives of Domestic Violence Survivors." *Violence Against Women* 12(3):240–50.

Gilmartin, B. 1977. "Swinging: Who Gets Involved and How." Pp. 161–85 in *Marriage and Alternatives,* edited by R. W. Libby and R. N. Whitehurst. Glenview, IL: Scott, Foresman.

Ginott, Haim G., Alice Ginott, and Wallace Goddard. 2003. *Between Parent and Child: The Bestselling Classic That Revolutionized Parent-Child Communication.* New York: Three Rivers Press.

Giordano, Peggy C., Monica A. Longmore, and Wendy D. Manning. 2006. "Gender and the Meanings of Adolescent Romantic Relationships: A Focus on Boys." *American Sociological Review* 71(2):260–87.

Glass, Jennifer and Leda E. Nath. 2006. "Religious Conservatism and Women's Market Behavior Following Marriage and Childbirth." *Journal of Marriage and Family* 68(3):611–29.

Glass, Shirley. 1998. "Shattered Vows." *Psychology Today* 31(4):34–52.

Glassner, Barry. 1999. *The Culture of Fear: Why Parents Are Afraid of the Wrong Things.* New York: Basic Books.

Glauber, Rebecca. 2007. "Marriage and the Motherhood Wage Penalty among African Americans, Hispanics, and Whites." *Journal of Marriage and Family* 69:951–61.

Glendon, Mary Ann. 1981. *The New Family and the New Property.* Toronto: Butterworths.

Glenn, Norval. 1990. "Quantitative Research on Marital Quality in the 1980s: A Critical Review." *Journal of Marriage and Family* 52(November):818–31.

———. 1997. "A Reconsideration of the Effect of No-Fault Divorce on Divorce Rates." *Journal of Marriage and Family* 59(4):1023–30.

———. 1998. "The Course of Marital Success and Failure in Five American 10-Year Marriage Cohorts." *Journal of Marriage and Family* 60(3):569–76.

Glenn, Norval and Elizabeth Marquardt. 2001. *Hooking Up, Hanging Out, and Hoping for Mr. Right: College Women on Dating and Mating Today.* New York: Institute for American Values.

Glenn, Norval D., Jeremy Uecker, and Robert Love. 2009. "Later First Marriage and Marital Success." Paper presented at the 2009 Annual Meetings of the American Sociological Association, San Francisco, August 10.

Glennon, Theresa. 2008. "Divided Parents, Shared Children Conflicting Approaches to Relocation Disputes in the USA." *Utrecht Law Review* 4(2):55–72.

Glick, Jennifer E., Frank D. Bean, and Jennifer Van Hook. 1997. "Immigration and Changing Patterns of Extended Family Household Structure in the United States: 1970–1990." *Journal of Marriage and Family* 59:177–91.

Glick, Jennifer E. and Jennifer Van Hook. 2002. "Parents' Coresidence with Adult Children: Can Immigration Explain Racial and Ethnic Variation?" *Journal of Marriage and Family* 64(1):240–53.

Glick, Paul C. and Sung-Ling Lin. 1986. "More Young Adults Are Living with Their Parents: Who Are They?" *Journal of Marriage and Family* 48:107–12.

Glick, Paul C. and Arthur J. Norton. 1979. "Marrying, Divorcing, and Living Together in the U.S. Today." *Population Bulletin* 32(5). Washington, DC: Population Reference Bureau.

"Glossary of Senior Housing Terms." 2010. Retrieved May 17, 2010 (http://senioroutlook.com/glossary.asp).

Goeke-Morey, Marcie, E. Mark Cummings, and Lauren M. Papp. 2007. "Children and Marital Conflict Resolution: Implications for Emotional Security and Adjustment." *Journal of Family Psychology* 21(4):744–53.

Goff, Philip Atiba, Claude M. Steele, and P. G. Davies. 2008. "The Space Between Us." *Journal of Personality and Social Psychology* 94(1):91–107.

Goffman, Erving. 1959. *The Presentation of Self in Everyday Life.* Garden City, NY: Doubleday.

———. 1961. *Asylums: Essays on the Social Situation of Mental Patients and Other Inmates.* Garden City, NY: Anchor.

———. 1963. *Stigma.* Englewood Cliffs, N.J.: Prentice-Hall.

———. 1967. *Interaction Ritual: Essays in Face-To-Face Behavior.* Chicago: Aldine Publishing Company.

Gold, Joshua M. 2009. "Negotiating the Financial Concerns of Stepfamilies: Directions for Family Counselors." *The Family Journal: Counseling and Therapy for Couples and Families* 17(2):185–88.

Gold, Steven J. 1993. "Migration and Family Adjustment: Continuity and Change among Vietnamese in the United States." Pp. 300–314 in *Family Ethnicity: Strength in Diversity,* edited by Harriette Pipes McAdoo. Newbury Park, CA: Sage Publications.

Gold, Steven J. and Mehdi Bozorgmehr. 2007. "Middle East and North Africa." Pp. 518–33 in

The New Americans: A Guide to Immigration Since 1965, edited by Mary C. Waters, Reed Ueda, and Helen B. Marrow. Cambridge, MA: Harvard University Press.

Goldberg, Abbie E. 2007. "Talking About Family." Journal of Family Issues 28(1):100–31.

———. 2010. *Lesbian and Gay Parents and Their Children: Research on the Family Life Cycle.* Washington DC: American Psychological Association.

Goldberg, Abbie E. and Aline Sayer. 2006. "Lesbian Couples' Relationship Quality across the Transition to Parenthood." *Journal of Marriage and Family* 68(1):87–100.

Goldberg, Abbie E. and JuliAnna Z. Smith. 2008. "Social Support and Psychological Well-Being in Lesbian and Heterosexual Preadoptive Couples." *Family Relations* 57(July):281–94.

Goldberg, Carey. 1998. "After Girls Get the Attention, Focus Shifts to Boys' Woes." *New York Times,* April 23.

Goldberg, Naomi G., M.V. Lee Badgett, and Peter J. Cooper. 2009. "Cost of Florida's Ban on Adoption by GLB Individuals and Same-Sex Couples." UCLA: The Williams Institute. Retrieved November 22, 2009 (www.law.ucla.edu/williamsinstitute).

Goldin, Claudia. 2006. "Working It Out." *New York Times,* March 15.

Goldman, Linda. 2006. *Raising Our children to Be Resilient: A Guide to Helping Children Cope with Trauma in Today's World.* New York: Brunner-Routledge.

Goldscheider, Francis, Sandra Hofferth, Carrie Spearin, and Sally Curtin. 2009. "Fatherhood Across Two Generations: Factors Affecting Early Family Roles." *Journal of Family Issues* 30(5):586–604.

Goldscheider, Frances, Gayle Kaufman, and Sharon Sassler. 2009. "Navigating the 'New' Marriage Market." *Journal of Family Issues* 30(6):719–37.

Goldscheider, Frances and Sharon Sassler. 2006. "Creating Stepfamilies: Integrating Children into the Study of Union Formation." *Journal of Marriage and Family* 68(2):275–91.

Goldstein, Arnold P., Harold Keller, and Diane Erne. 1985. *Changing the Abusive Parent.* Champaign, IL: Research Press.

Goldstein, Joshua R. 1999. "The Leveling of Divorce in the United States." *Demography* 36:409–14.

Goldstein, Joshua R. and Kristen Harknett. 2006. "Parenting across Racial and Class Lines: Assortative Mating Patterns of New Parents Who Are Married, Cohabiting, Dating, or No Longer Romantically Involved." *Social Forces* 85(1):121–43.

Goldstein, Joshua R., Tomáscaron Sobotka, and Aiva Jasilioniene. 2009. "The End of 'Lowest-Low' Fertility?" *Population and Development Review* 35(4):663–99.

Goldstein, Seth and R. P. Tyler. 1998. "Frustrations of Inquiry: Child Sexual Abuse

Allegations in Divorce and Custody Cases." *Law Enforcement Bulletin,* July, pp. 1–6.

Goleman, Daniel. 1985. "Patterns of Love Charted in Studies." *New York Times,* September 10.

———. 1992. "Family Rituals May Promote Better Emotional Adjustment." *New York Times,* March 11.

Gomes, Charlene. 2003. "Partners as Parents: Challenges Faced by Gays Denied Marriage." *The Humanist* 63(6):14–20.

Gomes, Peter J. 2004. "For Massachusetts, a Chance and a Choice." *Boston Globe,* February 8. Retrieved October 6, 2006 (www.boston.com/news/globe).

Gonzales, Felisa. 2008. *Hispanic Women in the United States, 2007.* Washington, DC: Pew Hispanic Center, May.

Gonzales, Jaymes. 2009. "Prefamily Counseling: Working with Blended Families." *Journal of Divorce & Remarriage* 50(2):148–57.

Gonzales, Nancy A., Julianna Deardorff, Diana Formoso, Alicia Barr, and Manuel Berrara, Jr. 2006. "Family Mediators of the Relation between Acculturation and Adolescent Mental Health." *Family Relations* 55(3):318–30.

Gonzales v. Carhart. 2007. U.S.S.C. 05-380. April 18.

Gonzalez, Cindy. 2006a. "Latino Leader Urges End to Concept of 'Minorities.'" *Omaha World-Herald,* November 17.

———. 2006b. "Short Supply of Visas Adds to Illegal Immigration." *Omaha World-Herald,* May 16.

Gonzalez, Cindy and Michael O'Connor. 2002. "Dialogue Key in Blending Cultures." *Omaha World-Herald,* May 4.

Good, Maria and Teena Willoughby. 2006. "The Role of Spirituality Versus Religiosity in Adolescent Psychosocial Adjustment." *Journal of Youth and Adolescence* 35:41–55.

Goode, Erica. 2002. "Deflating Self-Esteem's Role in Society's Ills." *New York Times,* October 1. Retrieved January 14, 2004 (web.lexis-nexis.com).

Goode, William J. 1971. "Force and Violence in the Family." *Journal of Marriage and Family* 33:624–36.

———. 1982. "Why Men Resist." Pp. 131–50 in *Rethinking the Family: Some Feminist Questions,* edited by Barrie Thorne and Marilyn Yalom. New York: Longman.

———. 1993. *World Changes in Divorce Patterns.* New Haven, CT: Yale University Press.

———. 2007 [1982]. "The Theoretical Importance of the Family." Pp. 14–25 in *Family in Transition,* 14th ed., edited by Arlene S. Skolnick and Jerome H. Skolnick. Boston, MA: Allyn and Bacon.

Goodman, W. Benjamin, Ann C. Crouter, Stephanie T. Lanza, and Martha J. Cox. 2008. "Paternal Work Characteristics and Father-Infant Interactions in Low-Income, Rural Families." *Journal of Marriage and Family* 70(3):640–53.

Goodman, Ellen. 2009. "This Isn't Exactly What I Had In Mind When I Wished For Women's Equality." *The Modesto Bee,* February 14. www.modbee.com.

Goodstein, Laurie. 2001. "New Christian Take on the Old Dating Ritual." *New York Times,* September 9.

Goodwin Paula, Brittany McGill, and Anjani Chandra. 2009. *Who Marries and When? Age at First Marriage in the United States, 2002.* NCHS data brief, no. 19. Hyattsville, MD: National Center for Health Statistics.

Goosby, Bridget J. 2007. "Poverty Duration, Maternal Psychological Resources, and Adolescent Socioemotional Outcomes." *Journal of Family Issues* 28(8):1113–34.

Gordon, Thomas. 2000. *Parent Effectiveness Training: The Proven Program for Raising Responsible Children.* New York: Three Rivers Press.

Gormley, Barbara and Frederick G. Lopez. 2010. "Psychological Abuse Perpetration in College Dating Relationships: Contributions of Gender, Stress, and Adult Attachment Orientations." *Journal of Interpersonal Violence* 25(2):204–18.

Gornick, Janet C., Harriet B. Presser and Caroline Batzdorf. 2009. "Outside the 9-to-5." *The American Prospect,* June 9. Retrieved May 26, 2010 (www.prospect.org).

Gott, Natalie. 2010. "Clergy Women Make Connections." *Faith & Leadership,* February 16.

Gottesman, Karen. 2006. *Raising Twins After the First Year.* New York: Marlowe.

Gottlieb, Alison Stokes. 1997. "Single Mothers of Children with Developmental Disabilities: The Impact of Multiple Roles." *Family Relations* 46(1):5–12.

Gottlieb, Laurie N., Ariella Lang, and Rhonda Amsel. 1996. "The Long-term Effects of Grief on Marital Intimacy Following an Infant's Death." *Omega* 33(1):1–9.

Gottlieb, Lori. 2006. "How Do I Love Thee?" *Atlantic Monthly,* March, pp. 58–70.

Gottman, John M. 1979. *Marital Interaction: Experimental Investigations.* New York: Academic.

———. 1994. *Why Marriages Succeed or Fail.* New York: Simon and Schuster.

———. 1996. *What Predicts Divorce? The Measures.* Hillsdale, NJ: Erlbaum.

———. 1998. "Toward a Process Model of Men in Marriages and Families." Pp. 149–92 in *Men in Families,* edited by Alan Booth and Ann C. Crouter. Mahwah, NJ: Erlbaum.

Gottman, John M., James Coan, Sybil Carrere, and Catherine Swanson. 1998. "Predicting Marital Happiness and Stability from Newlywed Interactions." *Journal of Marriage and Family* 60(1):5–22.

Gottman, John M. and Joan DeClaire. 2001. *The Relationship Cure: A Five-step Guide for Building Better Connections with Family, Friends, and Lovers.* New York: Crown.

Gottman, John M. and L. J. Krotkoff. 1989. "Marital Interaction and Satisfaction: A Longitudinal View." *Journal of Consulting and Clinical Psychology* 57:47–52.

Gottman, John M. and Robert W. Levenson. 2000. "The Timing of Divorce: Predicting When a Couple Will Divorce Over a 14-Year Period." *Journal of Marriage and Family* 62(3):737–45.

———. 2002. "A Two-factor Model for Predicting When a Couple Will Divorce: Exploratory Analyses Using 14-Year Longitudinal Data." *Family Process* 41(1):83–96.

Gottman, John M., Robert W. Levenson, James Gross, Barbara Frederickson, Leah Rosenthal, Anna Ruef, and Dan Yoshimoto. 2003. "Correlates of Gay and Lesbian Couples' Relationship Satisfaction and Relationship Dissolution." *Journal of Homosexuality* 45(1):23–45.

Gottman, John M. and Clifford I. Notarius. 2000. "Decade Review: Observing Marital Interaction." *Journal of Marriage and Family* 62(4):927–47.

———. 2003. "Marital Research in the 20th Century and a Research Agenda for the 21st Century." *Trends in Marriage, Family, and Society* 25(2):283–97.

Gottman, John M. and Nan Silver. 1999. *The Seven Principles for Making Marriage Work.* New York: Crown.

Gough, Brendan, Nicky Weyman, Julie Alderson, Gary Butler, and Mandy Stoner. 2008. "'They Did Not Have a Word': The Parental Quest to Locate a 'True Sex' For Their Intersex Children. Psychology & Health 23(4):493–507.

Gove, Walter R., Carolyn Briggs Style, and Michael Hughes. 1990. "The Effect of Marriage on the Well-Being of Adults." *Journal of Family Issues* 11(1):4–35.

Gowan, Annie. 2009. "Immigrants' Children Look Closer for Love: More Young Adults Are Seeking Partners of the Same Ethnicity." *Washington Post.* March 8: A01.

Grace, Gerald. 2002. *Catholic Schools: Mission, Markets and Morality.* London: Routledge Farmer.

Grady, Bill. 2009. "A Marriage Blending Family and Race." Pp. 90–93 in *Social Issues First Hand: Blended Families,* edited by Stefan Kiesbye. New York: Greenhaven Press/Cengage Learning.

Grady, Denise. 2007. "Girl or Boy? As Fertility Technology Advances, So Does an Ethical Debate." *New York Times,* February 6.

Graefe, Deborah R. and Daniel T. Lichter. 1999. "Life Course Transitions of American Children: Parental Cohabitation, Marriage, and Single Motherhood." *Demography* 36(2):205–17.

———. 2007. "When Unwed Mothers Marry." *Journal of Family Issues* 28(5):595–622.

Graham, Kathy T. 2008. "Same-Sex Couples: Their Rights as Parents, and Their Children's Rights as Children." *Santa Clara Law Review* 48:999–1037.

Grall, Timothy S. 2006. *Custodial Mothers and Fathers and Their Child Support: 2003.* Current Population Reports P60-230. Washington, DC: U.S. Census Bureau.

———. 2009. *Custodial Mothers and Fathers and Their Child Support: 2007.* Current Population Reports P60-237. Washington, DC: U.S. Census Bureau.

Grant, Jaime M. 2010. *Outing Age 2010.* Washington, DC: The National Gay and Lesbian Task Force Policy Institute. Retrieved May 15, 2010 (www.thetaskforce.org.).

Gray, Peter B. 2003. "Marriage, Parenting, and Testosterone Variation among Kenyan Swahili Men." *American Journal of Physical Anthropology* 122(3):279–86.

Gray, Peter B., Peter T. Ellison, and Benjamin C. Campbell. 2007. "Testosterone and Marriage among Ariaal Men of Northern Kenya." *Current Anthropology* 48(5):750–55.

Greeley, Andrew. 1991. *Faithful Attraction: Discovering Intimacy, Love, and Fidelity in American Marriage.* New York: Doherty.

Green, Adam Isaiah. 2006. "Until Death Do Us Part? The Impact of Differential Access to Marriage on a Sample of Urban Men." *Sociological Perspectives* 49(2):163–89.

Greenberg, Jerrold S., Clint E. Bruess, and Debra W. Haffner. 2002. *Exploring the Dimensions of Human Sexuality.* Sudbury, MA: Jones and Bartlett.

Greenberg, Susan H. and Anna Kuchment. 2006. "The 'Familymoon.'" *Newsweek,* January 9, pp. 44–45.

Greenblatt, Cathy Stein. 1983. "The Salience of Sexuality in the Early Years of Marriage." *Journal of Marriage and Family* 45:289–99.

Greenfield, Patricia M. and Lalita K. Suzuki. 2001. "Culture and Parenthood." Pp. 20–33 in *Parenthood in America,* edited by Jack C. Westman. Madison: University of Wisconsin Press.

Greenhouse, Steven. 2009. "Back to the Grind." *The New York Times,* September 19, B1.

Greenstein, Theodore N. 2009. "National Context, Family Satisfaction, and Fairness in the Division of Household Labor." *Journal of Marriage and Family* 71(4):1039–51.

Greenwald, John. 1999. "Elder Care: Making the Right Choice." *Time,* August 30, pp. 52–56.

Greif, Geoffrey and Mary S. Pabst. 1988. *Mothers Without Custody.* Lexington, MA: Heath.

Grekin, E. R., P. A. Brennan, and C. Hammen. 2005. "Parental Alcohol Use Disorders and Child Delinquency: The Mediating Effects of Executive Functioning and Chronic Family Stress." *Journal of Studies on Alcohol* 66(1):14–23.

Greksa, L. P . 2002. "Population Growth and Fertility Patterns in an Old Order Amish Settlement." *Annals of Human Biology* 29(2):192–201.

Grice, Helena. 2005. "Transracial Adoption Narratives: Prospects and Perspectives." *Meridians: Feminism, Race, Transnationalism* 5(2):124–48.

Grimsley, Kristen Downey. 2000. "Family a Priority for Young Workers." *Washington Post,* May 3.

Griswold v. Connecticut. 1965. 381 U.S. 479, 14 L. Ed.2d 510, 85 S. Ct. 1678.

Groat, Theodore, Peggy Giordano, Stephen Cernkovich, M. D. Puch, and Steven Swinford.

1997. "Attitudes Toward Childbearing Among Young Parents." *Journal of Marriage and Family* 59:568–81.

Grogan-Kaylor, Andrew and Melanie D. Otis. 2007. "The Predictors of Parental Use of Corporal Punishment." *Family Relations* 56(1):80–91.

Gross, Harriet Engel. 1980. "Dual-career Couples Who Live Apart: Two Types." *Journal of Marriage and Family* 42:567–76.

Gross, Jane. 2002. "U.S. Fund for Tower Victims Will Aid Some Gay Partners." *New York Times,* May 22.

———. 2004. "Alzheimer's in the Living Room: How One Family Rallies to Cope." *New York Times,* September 16.

———. 2006a. "As Parents Age, Baby Boomers and Business Struggle to Cope." *New York Times,* March 25.

———. 2006b. "Forensic Skills Seek to Uncover Hidden Patterns of Elder Abuse." *New York Times,* September 27.

———. 2006c. "Seeking Doctor's Advice in Adoptions from Afar." *New York Times,* January 3.

———. 2006d. "When the Beard Is Too Painful to Remove." *New York Times,* August 3.

———. 2007. "A Taste of Family Life in U.S., but Adoption Is in Limbo." *New York Times,* January 13.

Gross, Michael. 2006. "Bad Advice: How Not to Have Sex in an Epidemic." *American Journal of Public Health* 96(6):964–66.

Grossman, Lev. 2009. "Facebook Is for Old People." *Time: Mind and Body Special Issue,* February 23, p. 94.

Groves, Robert M. and Daniel L. Cork (Eds). 2008. *Surveying Victims: Options for Conducting the National Crime Victimization Survey.* Committee on National Statistics and Committee on Law and Justice, Division of Behavioral and Social Sciences and Education. Washington, DC: National Academies Press. Retrieved June 9, 2010 (www.nap.edu).

Groze, V. 1996. *Successful Adoptive Families: A Longitudinal Study.* Westport, CT: Praeger.

Guberman, Nancy, Eric Gagnon, Denyse Cote, Claude Gilbert, Nicole Thivierge, and Marielle Tremblay. 2005. "How the Trivialization of the Demands of High-tech Care in the Home Is Turning Family Members into Para-Medical Personnel." *Journal of Family Issues* 26(2):247–72.

Gubernskaya, Zoya. 2010. "Changing Attitudes Toward Marriage and Children in Six Countries." *Sociological Perspectives* 53(2):179–200.

Gubrium, Jaber F., and James A. Holstein. 1990. *What Is Family?* Mountain View, CA: Mayfield Press.

———. 2009. *Analyzing Narrative Reality.* Thousand Oaks: Sage.

Gudelunas, David. 2006. "Who's Hooking Up On-line?" *Gay and Lesbian Review Worldwide* 13(1):23–27.

Gudmunson, Clinton G., Sharon M. Danes, James D. Werbel, and Johnben Teik-Cheok

Loy. 2009. "Spousal Support and Work—Family Balance in Launching a Family Business." *Journal of Family Issues* 30(8):1098–1121.

Gueorguieva, Ralitza V., Randy L. Carter, Mario Ariet, Jeffrey Roth, Charles S. Mahan, and Michael B. Resneck. 2001. "Effect of Teenage Pregnancy on Educational Disabilities in Kindergarten." *American Journal of Epidemiology* 154:212–20.

Guilamo-Ramos, Vincent, James Jaccard, Patricia Dittus, and Alida M. Bouris. 2006. "Parental Experience, Trustworthiness, and Accessibility: Parent-Adolescent Communication and Adolescent Risk Behavior." *Journal of Marriage and Family* 68(5):1229–46.

Gullickson, Aaron. 2006. "Black/White Interracial Marriage Trends, 1850–2000." *Journal of Family History* 31(3):289–312.

Gupta, Sanjiv and Michael Ash. 2008. "Whose Money, Whose Time? A NonParametric Approach to Modeling Time Spent on Housework in the United States." *Feminist Economics* 14(1):93–120.

Gurel, Perin. 2009. "Transnational Feminism, Islam, and the Other Woman: How to Teach." *Radical Teacher* 86:66–70.

Gustafson, Teresa S. 2009. *Empowering Children to Lead Change.* Ft. Belvoir: Defense Technical Information Center.

Gute, Gary and Elaine Eshbaugh. 2008. "Personality As a Factor of Hooking Up Among College Students." *Journal of Community Health Nursing* 25:26–43.

Guttmacher Advisory. 2010. *Review of New Study on a Theory-Based Abstinence Program.* New York: Guttmacher Institute.

Guttmacher Institute. 2006. "Facts on Induced Abortion in the United States." May. New York: Guttmacher Institute.

Guzzo, Karen Benjamin. 2009a. "Marital Intentions and the Stability of First Cohabitations." *Journal of Family Issues* 30(2):179–205.

———. 2009b. "Maternal Relationships and Nonresidential Father visitation of Children Born Outside of Marriage." *Journal of Marriage and Family* 71(3):632–49.

Guzzo, Karen Benjamin and Frank F. Furstenberg, Jr. 2007. "Multipartnered Fertility among Young Women with a Nonmarital First Birth: Prevalence and Risk Factors." *Perspectives on Sexual and Reproductive Health* 39(1):29–38.

Guzzo, Karen Benjamin and Helen Lee. 2008. "Couple Relationship Status and Patterns in Early Parenting Practices." *Journal of Marriage and Family* 70(1):44–61.

Ha, Jung-Hwa. 2008. "Changes in Support from Confidants, Children, and Friends following Widowhood." *Journal of Marriage and Family* 70(2):306–18.

Ha, K. Oanh. 2006. "Filipinos Work Overseas, Help Kin." *San Jose Mercury News,* April 4.

Haas, Jane Glenn. 2009. "Hard Times Lead to Crowded Homes." *Orange County Register,* January 16, 2009. Retrieved February 3, 2009 (www.ocregister.com).

Haas, Stephen M., and Laura Stafford. 2005. "Maintenance Behaviors in Same-Sex and Marital Relationships: A Matched Sample Comparison." *The Journal of Family Communication* 5(1):43–60.

Haberman, Clyde. 2006. "A Battle for Freedom at the Mall." *New York Times,* November 24.

Hacker, Jacob S. 2006. *The Great Risk Shift: The Assault on American Jobs, Families, Health Care, and Retirement—And How You Can Fight Back.* New York: Oxford University Press.

Hackstaff, Karla B. 2007. "Divorce Culture: A Quest for Relational Equality in Marriage." Pp. 188–222 in *Family in Transition,* 4th ed., edited by Arlene S. Skolnick and Jerome H. Skolnick. Boston, MA: Pearson.

Haddad, Yvonne Y. and Jane I. Smith. 1996. "Islamic Values Among American Muslims." Pp. 19–40 in *Family and Gender among American Muslims: Issues Facing Middle Eastern Immigrants and Their Descendants,* edited by Barbara C. Aswad and Barbara Bilgé. Philadelphia, PA: Temple University Press.

Hagan, Frank E. 2010. *Crime Types and Criminals.* Thousand Oaks, CA: Sage.

Hagedoorn, Mariet, Nico W. Van Yperen, James C. Coyne, Cornelia van Jaarsveld, Adelita Ranchor, Eric van Sonderen, and Robbert Sanderman. 2006. "Does Marriage Protect Older People from Distress? The Role of Equity and Recency of Bereavement." *Psychology and Aging* 21(3):611–20.

Hagestad, G. 1996. "On-*Time,* Off-*Time,* Out of *Time?* Reflections on Continuity and Discontinuity from an Illness Process." Pp. 204–22 in *Adulthood and Aging,* edited by V. L. Bengston. New York: Springer.

Hagewen, Kellie J. and S. Philip Morgan. 2005. "Intended and Ideal Family Size in the United States, 1970–2002." *Population and Development Review* 31(3):507–27.

"The Hague Convention on Intercountry Adoption." 2010. *Child Welfare Information Gateway,* U.S. Department of Health and Human Services. www.childwelfare.gov.

Hakim, Danny. 2006. "Panel Asks New York to Join the Era of No-fault Divorce." *New York Times,* February 7.

Hale-Benson, J. E. 1986. *Black Children: Their Roots, Culture, and Learning Styles.* Provo, UT: Brigham Young University Press.

Halford, Kim, Jan Nicholson, and Matthew Sanders. 2007. "Couple Communication in Stepfamilies." *Family Process* 46(4):471–83.

Hall, Edie Jo and E. Mark Cummings. 1997. "The Effects of Marital and Parent–Child Conflicts on Other Family Members: Grandmothers and Grown Children." *Family Relations* 46(2):135–43.

Hall, Elaine J. and Marnie Salupo Rodriguez. 2003. "The Myth of Postfeminism." *Gender and Society* 17:878–902.

Hall, Scott S. and Shelley M. MacDermid. 2009. "A Typology of Dual Earner Marriages Based on Work and Family Arrangements." *Journal of Family and Economic Issues* 30(3):215–25.

Hall, Sharon K. 2008. *Raising Kids in the 21ˢᵗ Century.* West Sussex, UK: Wiley-Blackwell.

Halpern, Diane F., Joshua Aronson, Nona Reimer, Sandra Simpkins, Jon Star, and Kathryn Wentzel. 2007. *Encouraging Girls in Math and Science.* NCER 2007-2003. Washington, DC: National Center for Education Research, Institute of Education Sciences, U.S. Department of Education. Retrieved January 18, 2010 (http://ncer.ed.gov).

Halpern, Susan P. 2009. *Finding the Words: Candid Conversations with Loved Ones.* Berkeley, CA: North Atlantic Books.

Halpern-Felsher, Bonnie L., Jodi L. Cornell, Rhonda Y. Kropp, and Jeanne M. Tschann. 2005. "Oral versus Vaginal Sex among Adolescents: Perceptions, Attitudes, and Behavior." *Pediatrics* 115(4):845–51.

Halpern-Meekin, Sarah and Laura Tach. 2008. "Heterogeneity in Two-Parent Families and Adolescent Well-Being." *Journal of Marriage and Family* 70(May):435–51.

Halpin, John and Ruy Teixeira. 2009. "Battle of the Sexes Gives Way to Negotiations." In *The Shriver Report: A Woman's Nation Changes Everything,* edited by Heather Boushey and Ann O'Leary. Washington, DC: Maria Shriver and the Center for American Progress.

Halsall, Paul. 2001. *Internet Medieval Sourcebook.* Retrieved October 6, 2006 (www.fordham.edu).

Hamamci, Zeynep. 2005. "Dysfunction Relationship Beliefs in Marital Conflict." *Journal of Relational-Emotive and Cognitive-Behavior Therapy* 23(3):245–61.

Hamel, John C. and Tanya L. Nicholls. 2006. *Family Interventions in Domestic Violence A Handbook of Gender-Inclusive Theory and Treatment.* New York: Springer Publishing Company.

Hamer, Jennifer. 2001. *What It Means to Be Daddy: Fatherhood for Black Men Living Away from Their Children.* New York: Columbia University Press.

Hamilton, Brady E., Joyce A. Martin, and Stephanie J. Ventura. 2006. "Births: Preliminary Data for 2005." *National Vital Statistics Reports* 55(11). Hyattsville, MD: National Center for Health Statistics. Retrieved March 18, 2010 (www.cdc.gov/nchs/products).

———. 2007. *Births: Preliminary Data for 2005.* NCHS E-Stats. Hyattsville, MD: National Center for Health Statistics. Retrieved March 18, 2007 (www.cdc.gov/nchs).

———. 2009. *Births: Preliminary data for 2007.* National Vital Statistics Reports 57 (12). Hyattsville, MD: National Center for Health Statistics. March 18.

———. 2010. "Births: Preliminary Data for 2008." *National Vital Statistics Reports* 58(16). Hyattsville, MD: National Center for Health Statistics. April 20.

Hamilton, Brady E., Joyce A. Martin, Stephanie J. Ventura, Paul D. Sutton, and Fay Menacker. 2005. "Births: Preliminary Data for 2004." *National Vital Statistics Reports* 54(8). Hyattsville, MD: National Center for Health Statistics. December 29.

Hamilton, Jon. 2010. "Grandparents Often Help support Kids with Autism." National Public Radio, April 22. Retrieved May 16, 2010 (www.npr.org).

Hamilton, W. D. 1964. "The Genetical Evolution of Social Behavior: II." *Journal of Theoretical Biology* 7:17–52.

Hammen, Constance, Patricia A. Brennan, and Josephine H. Shih. 2004. "Family Discord and Stress Predictors of Depression and Other Disorders in Adolescent Children of Depressed and Nondepressed Women." *Journal of the American Academy of Child and Adolescent Psychiatry* 43(8):994–1003.

Hamon, Raeann R. and Bron B. Ingoldsby. 2003. *Mate Selection across Cultures.* Thousand Oaks, CA: Sage Publications.

Han, Chong-suk. 2008. "A Qualitative Exploration of the Relationship Between Racism and Unsafe Sex Among Asian Pacific Islander Gay Men." *Archives of Sexual Behavior* 37(5):827–37.

Han, Wen-Jui, Christopher J. Ruhm, Jane Waldfogel, and Elizabeth Washbrook. 2008. "The Timing of Mothers' Employment after Childbirth." *Monthly Labor Review* June. Washington, DC: U.S. Bureau of Labor Statistics.

Hanawalt, Barbara. 1986. *The Ties That Bound: Peasant Families in Medieval England.* New York: Oxford University Press.

Handler, Joel F. and Yeheskel Hasenfeld. 2006. *Blame Welfare, Ignore Poverty and Inequality.* New York: Cambridge University Press.

Hanna, Sharon L., Rose Suggett, and Doug Radtke. 2008. *Person to Person: Positive Relationships Don't Just Happen.* 5th ed. Upper Saddle River, NJ: Pearson/Prentice Hall.

Hans, Jason D. 2002. "Stepparenting after Divorce: Stepparents' Legal Position Regarding Custody, Access, and Support." *Family Relations* 51(4):301–07.

———. 2009. "Beliefs about Child Support Modification Following Remarriage and Subsequent Childbirth." *Family Relations* 58(February):65–78.

Hans, Jason D. and Marilyn Coleman. 2009. "The Experiences of Remarried Stepfathers Who Pay Child Support." *Personal Relationships* 16:597–618.

Hans, Jason D., Lawrence H. Ganong, and Marilyn Coleman. 2009. "Financial Responsibilities Toward Older Parents and Stepparents Following Divorce and Remarriage." *Journal of Family Economic Issues* 30:55–66.

Hansen, Donald A. and Reuben Hill. 1964. "Families Under Stress." Pp. 782–819 in *The Handbook of Marriage and the Family,* edited by Harold Christensen. Chicago, IL: Rand McNally.

Hansen, Matthew and Karyn Spencer. 2009. "Safe Haven Meant Kids Finally Got Right Help." *Omaha World Herald,* February 1. Retrieved February 3, 2009 (www.omaha.com).

Hansen, Thorn, Torbjorn Moum, and Adam Shapiro. 2007. "Relational and Individual Well-Being among Cohabitors and Married Individuals in Midlife." *Journal of Family Issues* 28(7):910–33.

Haraway, Donna. 1989. *Primate Visions: Gender, Race, and Nature in the World of Modern Science.* New York: Routledge, Chapman and Hall.

Hardesty, Jennifer L. and Grace H. Chung. 2006. "Intimate Partner Violence, Parental Divorce, and Child Custody: Directions for Intervention and Future Research." *Family Relations* 55:200–216.

Harding, Rosie. 2007. "Sir Mark Potter and the Protection of the Traditional Family: Why Same Sex Marriage Is (Still) a Feminist Issue." *Feminist Legal Studies* 15(2):223–34.

Hare, Jan and Denise Skinner. 2008. "'Whose Child Is This?' Determining Legal Status for Lesbian Parents Who Used Assisted Reproductive Technology." *Family Relations* 57(3):365–75.

Harknett, Kristen. 2006. "The Relationship between Private Safety Nets and Economic Outcomes among Single Mothers." *Journal of Marriage and Family* 69(1):172–91.

Harknett, Kristen and Jean Knab. 2007. "More Kin, Less Support: Multipartnered Fertility and Perceived Support among Mothers." *Journal of Marriage and Family* 69(1):237–53.

Harmanci, Reyhan. 2006. "The Neighbordaters." *San Francisco Chronicle Magazine,* February 12, pp. 13–14.

Harmon, Amy. 2005a. "Ask Them (All 8 of Them) about the Grandkids." *New York Times,* March 20.

———. 2005b. "Hello, I'm Your Sister, Our Father Is Donor 150." *New York Times,* November 20.

———. 2006. "Seeking Healthy Children, Couples Cull Embryos." *New York Times,* September 6.

———. 2007. "Sperm Donor Father Ends His Anonymity." *New York Times,* February 14.

Harpel, Tammy S. and Jodie Hertzog. 2010. "'I Thought My Heart Would Burst': The Role of Ultrasound Technology on Expectant Grandmotherhood." *Journal of Family Issues* 31(2):257–74.

Harper, Nevin J. 2009. "Family Crisis and the Enrollment of Children in Wilderness Treatment." *Journal of Experimental Education* 31(3):447–50.

Harris, Christine R. 2003a. "A Review of Sex Differences in Sexual Jealousy, Including Self-Report Data: Psychophysiological Responses, Interpersonal Violence, and Morbid Jealousy." *Personality and Social Psychology Review* 7(2):102–8.

Harris, Deborah and Domenico Parisi. 2008. "Looking for 'Mr. Right': The Viability of Marriage Initiatives for African American Women in Rural Settings." *Sociological Spectrum* 28(4):338–56.

Harris, Judith Rich. 1998. *The Nurture Assumption: Why Children Turn Out the Way They Do.* New York: Free Press.

Harris, Marian S. and Ada Skyles. 2008. "Kinship Care for African American Children: Disproportionate and Disadvantageous." *Journal of Family Issues* 29(8):1013–30.

Harris, Paul. 2009. "Revealed: The Shocking Rise of 'New Slavery' in US Midwest." *The Observer* November 22, p. 37.

Harris, Philip M. and Nicholas A. Jones. 2005. *We the People: Pacific Islanders in the United States.* CENSR-26. Washington, DC: U.S. Census Bureau. August. Retrieved December 15, 2006 (www.census.gov).

Harris, Scott R. 2008. "What Is Family Diversity? Objective and Interpretive Approaches." *Journal of Family Issues* 29(11):1407–25.

Harrist, Amanda W. and Ricardo C. Ainslie. 1998. "Marital Discord and Child Behavior Problems." *Journal of Family Issues* 19(2):140–63.

Harter, James and Raksha Arora. 2008. "Social Time Crucial to Daily Emotional Well-Being in U.S." June 5. Retrieved November 24, 2009 (www.gallup.com/poll).

Hartmann, Heidi, Ashley English, and Jeffrey Hayes. 2010. "Women and Men's Employment and Unemployment in the Great Recession." *Institute for Women's Research Briefing Paper C373.* Washington, DC: Institute for Women's Research.

Hartmann, Heidi, Stephen J. Rose, and Vicky Lovell. 2009. "How Much Progress in Closing the Long-term Earnings Gap?" Pp. 125–55 in *The Declining Significance of Gender?* edited by Francine D. Blau, Mary C. Brinton, and David B. Grusky. New York: Russell Sage.

Hartocollis, Anemona. 2006. "Meaning of 'Normal' Is at Heart of Gay Marriage Ruling." *New York Times,* July 8. Retrieved July 11, 2006 (www.nytimes.com).

———. 2007. "Married or Not, Gay Couple Are Ruled Legally Separated." *New York Times,* January 9.

Hartog, Henrik. 2000. *Man and Wife in America: A History.* Cambridge, MA: Harvard University Press.

Hartsoe, Steve. 2005. "Shacking Up: N.C. Anti-cohabitation Law Under Legal Attack." *Raleigh News and Observer,* May 9. Retrieved May 11, 2005 (www.newsobserver.com).

Harvard University. Mind/Brain/Behavior Initiative. 2005. "The Science of Gender and Science: Pinker vs. Spelke, a Debate." Cambridge, MA. May 16. Retrieved January 1, 2006 (www.edge.org).

Harvey, Elizabeth. 1999. "Short-term and Long-term Effects of Early Parental Employment on Children of the National Longitudinal Survey of Youth." *Developmental Psychology* 35:445–59.

Hass, Giselle Aguilar, Nawal Ammar, and Leslye Orloff. 2006. "Battered Immigrants and U.S. Citizen Spouses." *Legal Momentum,* April 24.

Hassrick, Elizabeth McGhee and Barbara Schneider. 2009. "Parent Surveillance in Schools: A Question of Social Class." *American Journal of Education* 115(February):195–211.

Hatch, Laurie R. and Kris Bulcroft. 2004. "Does Long-term Marriage Bring Less Frequent Disagreements?" *Journal of Family Issues* 25(4):465–95.

Haub, Carl. 2009. "Will the Economic Downturn Lower Birthrates?" Washington, DC: Population Reference Bureau. Retrieved June 14, 2009 (http://prbblog.org).

Hawke, Sharryl and David Knox. 1978. "The One-Child Family: A New Life-Style." *Family Coordinator* 27:215–19.

Hawkins, Alan J., Kimberly R. Lovejoy, Erin K. Holmes, Victoria L. Blanchard, and Elizabeth Fawcett. 2008. "Increasing Fathers' Involvement in child Care with a couple-Focused Intervention during the Transition to Parenthood." *Family Relations* 57(1):49–59.

Hawkins, Daniel N. and Alan Booth. 2005. "Unhappily Ever After: Effects of Long-Term, Low-Quality Marriages on Well-Being." *Social Forces* 84(1):451–71.

Hay, Elizabeth L., Karen L. Fingerman, and Eva S. Lefkowitz. 2007. "The Experience of Worry in Parent-Adult Child Relationships." *Personal Relationships* 14(4):605–22.

Hayden, Dolores. 1981. *The Grand Domestic Revolution: A History of Feminist Designs for American Homes, Neighborhoods, and Cities.* Cambridge, MA: MIT Press.

Hayghe, Howard. 1982. "Dual Earner Families: Their Economic and Demographic Characteristics." Pp. 27–40 in *Two Paychecks,* edited by Joan Aldous. Newbury Park, CA: Sage Publications.

Haynes, Faustina E. 2000. "Gender and Family Ideals: An Exploratory Study of Black Middle Class Americans." *Journal of Family Issues* 21:811–37.

Hazan, C. and P. R. Shaver. 1987. "Romantic Love Conceptualized as an Attachment Process." Journal of Personality and Social Psychology 52:511–24.

Hazen, C. and P. Shaver. 1994. "Attachment as an Organizing Framework for Research on Close Relationships." *Psychological Inquiry* 5:1–22.

Hazra, Rohan, George K. Siberry, and Lynne M. Mofenson. 2010. "Growing Up with HIV: Children, Adolescents, and Young Adults with Perinatally Acquired HIV Infection." *Annual Review of Medicine* 61:169–85.

Healthy Marriage Initiative. 2009. Retrieved October 9, 2009 (www.acf.hhs.gov/healthymarriage).

Heard, Holly E. 2007. "The Family Structure Trajectory and Adolescent School Performance: Differential Effects by Race and Ethnicity." *Journal of Family Issues* 28(3):319–54.

Heaton, Tim B. 2002. "Factors Contributing to Increasing Marital Stability in the United States." *Journal of Family Issues* 23(3):392–409.

Heaton, Tim B. and Edith L. Pratt. 1990. "The Effects of Religious Homogamy on Marital Satisfaction and Stability." *Journal of Family Issues* 11(2):191–207.

Hebert, Laura A. 2007. "Taking 'Difference' Seriously: Feminisms and the 'Man Question'." *Journal of Gender Studies* 16(1):31–45.

Heene, Els, Ann Buysse, and Paulette Van Oost. 2007. "An Interpersonal Perspective on Depression: The Role of Marital Adjustment, Conflict Communication, Attributions, and Attachment within a Clinical Sample." *Family Process* 46(4):499–514.

Heffner, Christopher L. 2003. "Counseling the Gay and Lesbian Client." *AllPsych Journal.* August 12. Retrieved march 3, 2010 (allpsych.com/journal).

Hefling, Kimberly. 2009. "Big Increase in Troops' Kids Seeking Mental Help." *Associated Press,* July 7.

Hegesisch, Ariane and Hannah Liepmann. 2010. "The Gender Wage Gap by Occupation." *Fact Sheet IWPR #C350a.* Washington, DC: Institute for Women's Policy Research.

Heim, Susan M. 2007. *It's Twins! Parent-to-Parent Advice from Infancy through Adolescence.* Charlottesville, VA: Hampton Roads.

Heldman, Caroline, Susan J. Carroll, and Stephanie Olson. 2005. "She Brought Only a Skirt": Print Media Coverage of Elizabeth Dole's Bid for the Republican Presidential Nomination." *Political Communication* 22(3):313–35.

Helfich, Christine A. and Emily K. Simpson. 2005. "Lesbian Survivors of Intimate Partner Violence: Provider Perspectives on Barriers to Accessing Service." *Journal of Gay and Lesbian Social Services* 18(2):39–59.

"Helicopter Parents Hover Over Kids' Lives: Experts Say They Shouldn't Inhibit Children's Decision Making." 2007. ABC News, October 7. Retrieved January 19, 2010 (www.abcnews.com).

Hench, David. 2004. "Is Anger Management a Remedy for Batterers?" *Portland Press Herald* (Maine), October 10.

Henderson, Craig E., Bert Hayslip Jr., Leah M. Sanders, and Linda Louden. 2009. "Grandmother-Grandchild Relationship Quality Predicts Psychological Adjustment among Youth from Divorced Families." *Journal of Family Issues* 30(9):1,245–1,264.

Henderson, Tammy L. 2005. "Grandparent Visitation Rights: Justices' Interpretation of the Best Interests of the Child Standard." *Journal of Family Issues* 26(5):638–64.

———. 2008. "Transforming the Discussion about Diversity, Policies, and Law." *Journal of Family Issues* 29(8):983–94.

Henderson, Tammy L. and Patricia B. Moran. 2001. "Grandparent Visitation Rights." *Journal of Family Issues* 22(5):619–38.

Hendrick, Clyde, Susan S. Hendrick, and Amy Dicke. 1998. "The Love Attitudes Scale: Short Form." *Journal of Social and Personal Relationships* 15(2):147–59.

Hendrick, Susan S. 2000. "Links between Sexuality and Love as Contributors to Relationship Satisfaction." Presented at the annual meeting of the National Council on Family Relations, November 10–13, Minneapolis, MN.

Henly, Julia R., Sandra K. Danziger, and Shira Offer. 2005. "The Contribution of Social Support to the Material Well-Being of Low-Income Families." *Journal of Marriage and Family* 67(1):122–40.

Henretta, J. C., M. S. Hill, L. Wei, B. J. Soldo, and D. A. Wolf. 1997. "Selection of Children to Provide Care: The Effect of Early Parental Transfers." *Journal of Gerontology* 52B:110–19.

Henry, Carolyn, Michael Merton, Scott Plunkett, and Tovah Sands. 2008. "Neighborhood, Parenting, and Adolescent Factors and Academic Achievement in Latino Adolescents from Immigrant Families." *Family Relations* 57(December):579–90.

Henry, Pamela J. and James McCue. 2009. "The Experience of Nonresidential Stepmothers." *Journal of Divorce and Remarriage* 50:185–205.

Henry, Ray. 2008. "Divorce Hard To Get For Some Gay Couples." *USA Today*, April 15. Retrieved June 19, 2010 (www.usatoday.com).

Henshaw, Stanley K. and Kathryn Kost. 2008. *Trends in the Characteristics of Women Obtaining Abortions, 1974 to 2004*. New York: Guttmacher Institute.

Hequembourg, Amy L. 2007. "Becoming Lesbian Mothers." *Journal of Homosexuality* 53(3):153–80.

Hequembourg, Amy L., and Sara Brallier. 2005. "Gendered Stories of Parental Caregiving among Siblings." *Journal of Aging Studies* 19:53–71.

Herd, Pamela. 2009. "Women, Public Pensions, and Poverty: What Can the United States Learn from Other Countries?" *Journal of Women, Politics, and Policy* 30(2–3):301–34.

Herek, Gregory M. 2009a. "Marriage Equality Attitudes: Simply Knowing Gay People Helps, But Isn't Enough." *Beyond Homophobia*, June 16. www.beyondhomophobia.com.

———. 2009b. "Sexual Stigma and Sexual Prejudice in the United States: A Conceptual Framework." Pp. 65–111 in *Contemporary Perspectives On Lesbian, Gay and Bisexual Identities: The 54th Nebraska Symposium on Motivation*, edited by D. A. Hope. New York: Springer.

Heriot, Gail. 2009. "A Professor Proposes to Examine Gender Bias in College Admissions." The Chronicle of Higher Education, October 31. (http://chronicle.com).

Hernandez, Raymond. 2001. "Children's Sexual Exploitation Underestimated, Study Finds." *New York Times*, September 10.

Heron, Melonie, Donna L. Hoyert, Sherry L. Murphy, Jiaquan Xu, Kenneth D. Kochanek, and Betzaida Tejada-Vera. 2009. "Deaths: Final Data for 2006." *National Vital Statistics Reports* 57(14):1–134.

Herring, Jeff. 2005. "Affairs Don't Have to Be Physical." Knight/Ridder Newspapers, November 6.

Hertz, Rosanna. 1986. *More Equal Than Others: Women and Men in Dual Career Marriages*. Berkeley: University of California Press.

———. 1997. "A Typology of Approaches to Child Care." *Journal of Family Issues* 18(4):355–85.

———. 2006a. *Single by Chance, Mothers by Choice: Women Are Choosing Parenthood without [Marriage] and Creating the New American Family.* [New York:] Oxford University Press.

———. 2006b. "Talking About 'Doing' Family." *Journal of Marriage and Family* 68(4):796–99.

Herzog, Dagmar. 2008. *Sex in Crisis: The New Sexual Revolution and the Future of American Politics*. New York: Basic Books.

Hesse-Biber, Sharlene Nagy. 2007. *Handbook of Feminist Research: Theory and Praxis*. Thousand Oaks, CA: Sage.

Hetherington, E. Mavis. 2003. "Intimate Pathways: Changing Patterns in Close Personal Relationships across *Time*." *Family Relations* 52(4):318–31.

———. 2005. "The Adjustment of Children in Divorced and Remarried Families." Pp. 137–39 in *Sourcebook of Family Theory and Research*, edited by Vern L. Bengston, Alan C. Acock, Katherine R. Allen, Peggye Dilworth-Anderson, and David M. Klein. Thousand Oaks, CA: Sage Publications.

Hetherington, E. Mavis and John Kelly. 2002. *For Better or for Worse: Divorce Reconsidered*. New York: Norton.

Hewitt, Sally. 2009. *My Stepfamily*. Mankato, MN: Smart Apple Media.

Hewlett, Sylvia Ann. 2002. *Creating a Life: Professional Women and the Quest for Children*. New York: Hyperion.

Heyman, Richard E. and Ashley N. Hunt. 2007. "Replication in Observational Couples Research: A Commentary." *Journal of Marriage and Family* 69(1):81–85.

Heyman, Richard, Ashley Hunt-Martorano, Jill Malik, and Amy M. Smith. 2009. "Desired Change in Couples: Gender Differences and Effects on Communication." *Journal of Family Psychology* 23(4):474–84.

Heyman, Richard A. and Amy M. Smith Slep. 2002. "Do Child Abuse and Interparental Violence Lead to Adulthood Family Violence?" *Journal of Marriage and Family* 64:864–70.

Heywood, Leslie and Jennifer Drake. 1997. "Introduction." Pp. 1–20 in *Third Wave Agenda: Being Feminist, Doing Feminism*, edited by Leslie Heywood and Jennifer Drake. Minneapolis, MN: University of Minnesota Press.

Higginbotham, Brian J., Julie J. Miller, and Sylvia Niehuis. 2009. "Remarriage Preparation: Usage, Perceived Helpfulness, and Dyadic Adjustment." *Family Relations* 58(July):316–29.

Hildreth, Carolyn J., Alison Burke, and Richard Glass. 2009. "Elder Abuse." *JAMA: The Journal of the American Medical Association* 302(5):588.

Hill, Jeffrey E., Alan J. Hawkins, and Brent C. Miller. 1996. "Work and Family in the Virtual Office: Perceived Influences of Mobile Telework." *Family Relations* 45(3):293–301.

Hill, Jennifer L., Jane Waldfogel, Jeanne Brooks-Gunn, and Wen-Juui Han. 2005. "Maternal Employment and Child Development: A Fresh Look Using Newer Methods." *Developmental Psychology* 41(6):833–50.

Hill, Nancy E., Cynthia Ramirez, and Larry E. Dumka. 2003. "Early Adolescents' Career Aspirations: A Qualitative Study of Perceived Barriers and Family Support among Low-income Ethnically Diverse Adolescents." *Journal of Family Issues* 24(7):934–59.

Hill, Reuben. 1958. "Generic Features of Families Under Stress." *Social Casework* 49:139–50.

Hill, Shirley A. 2004. *Black Intimacies: A Gender Perspective on Families and Relationships*. Lanham, MD: Rowman and Littlefield.

———. 2006. "Marriage among African American Women: A Gender Perspective." *Journal of Comparative Family Studies* 37(3):421–40.

Hillaker, Barbara, Holly Brophy-Herb, Francisco Villarruel, and Bruce Haas. 2008. "The Contributions of Parenting to Social Competencies and Positive Values in Middle School Youth: Positive Family Communication, Maintaining Standards, and Supportive Family Relationships." *Family Relations* 57(December):591–601.

Himes, Christine L. 2001. "Social *Demography* of Contemporary Families and Aging." Pp. 47–50 in *Families in Later Life: Connections and Transitions*, edited by Alexis J. Walker, Margaret Manoogian-O'Dell, Lori A. McGraw, and Diana L. G. White. Thousand Oaks, CA: Pine Forge.

Hines, Denise, Jan Brown, and Edward Dunning. 2007. "Characteristics of Callers to the Domestic Abuse Helpline for Men." *Journal of Family Violence* 22(2):63–72.

Hines, Denise A. and Emily M. Douglas. 2009. "Women's Use of Intimate Partner Violence against Men: Prevalence, Implications, and Consequences." *Journal of Aggression, Maltreatment & Trauma* 18(6)572–86.

Hines, Melissa, Susan Golombok, John Rust, Katie J. Johnston, Jean Golding, and the Avon Longitudinal Study of Parents and Children Study Team. 2002. "Testosterone during Pregnancy and Gender Role Behavior of Preschool Children: A Longitudinal Population Study." *Child Development* 73:1678–87.

Hirsch, Jennifer S. 2003. *A Courtship after Marriage: Sexuality and Love in Mexican Transnational Families*. Berkeley, CA: University of California Press.

———. 2008. "Catholics Using Contraceptives: Religion, Family Planning, and Interpretive Agency in Rural Mexico." Studies in Family Planning 39(2):93–104.

Hirsch, Jennifer S., Miguel Muñoz-Laboy, Christina M. Nyhus, Kathryn M. Yount, and José A. Bauermeister. 2009. "They Miss More Than Anything Their Normal Life Back Home": Masculinity and Extramarital Sex Among Mexican Migrants in Atlanta." *Perspectives on Sexual & Reproductive Health* 41(1):23–32.

Hirsch, Jennifer S., Holly Wardlow, Daniel Jordan Smith, Harriet Phinney, Shanti Parikh, and Constance A. Nathanson. 2010. *The Secret: Love, Marriage, and HIV*. Nashville: Vanderbilt University Press.

Hirschman, Linda. 2006. *Get to Work: A Manifesto for Women of the World*. New York: Viking.

Hitlin, Steven, J., Scott Brown, and Glen H. Elder. 2007. "Measuring Latinos: Racial vs. Ethnic Classification and Self-Understandings." Social Forces 86(2):587–611.

Hitt, Jack. 2005. "The Newest Indians." *New York Times,* August 21.

"HIV/AIDS Epidemic in the United States, The." 2009. *HIV/AIDS Policy Fact Sheet* September. Menlo Park, CA: Kaiser Family Foundation.

Hochschild, Arlie. 1983. *The Managed Heart: Commercialization of Human Feeling.* Berkeley: University of California Press.

———. 1989. *The Second Shift: Working Parents and the Revolution at Home.* New York: Viking/Penguin.

———. 1997. *Time Bind: When Work Becomes Home and Home Becomes Work.* New York: Henry Holt.

Hodapp, R.M., and R. C. Urbano. 2007. "Adult Siblings of Individuals with Down Syndrome Versus with Autism: Findings from a Large-Scale U.S. Survey." *Journal of Intellectual Disability Research* 51(12):1018–29.

Hoelter, Lynette F., William G. Axinn, and Dirgha J. Ghimire. 2004. "Social Change, Premarital Nonfamily Experiences, and Marital Dynamics." *Journal of Marriage and Family* 66(5):1131–51.

Hoffman, Jan. 2009. "Teenage Girls Stand by Their Man." *The New York Times.* March 19. Retrieved May 12, 2009 (www.nytimes.com).

Hoffman, Kristi L., Jill K. Kiecolt, and John N. Edwards. 2005. "Physical Violence Between Siblings: A Theoretical and Empirical Analysis." *Journal of Family Issues* 26(8):1103–30.

Hoffman, Kristin and John N. Edwards. 2004. "An Integrated Theoretical Model of Sibling Violence and Abuse." *Journal of Family Violence* 19(3):185–200.

Hoffman, Lois W. and Jean B. Manis. 1979. "The Value of Children in the United States: A New Approach to the Study of Fertility." *Journal of Marriage and Family* 41: 583–96.

Hogan, Dennis P. and Nan M. Astone. 1986. "The Transition to Adulthood." *Annual Review of Sociology* 12:109–30.

Hoge, Charles W. 2010. *Once a Warrior–Always a Warrior: Navigating the Transition from Combat to Home—Including Combat Stress, PTSD, and MTBI.* Guilford, CT: GPP Life.

Hohmann-Marriott, Bryndl. 2009. "Father Involvement Ideals and the Union Transitions of Unmarried Parents." *Journal of Family Issues* 30(7):898–920.

Hohmann-Marriott, Bryndl and Paul Amato. 2008. "Relationship Quality in Interethnic Marriages and Cohabitations." *Social Forces* 87(2):825–55.

Holland, Rochelle. 2009. "Perceptions of Mate Selection for Marriage among African American, College-Educated, Single Mothers." *Journal of Counseling & Development* 87(Spring):170–78.

Holloway, Angela Ann. 2008. "Pre-Marital Counseling." April 28. Retrieved October 12, 2009 (marriage.suite101.com).

Holmberg, Diane and Karen Blair. 2009. Sexual Desire, Communication, Satisfaction, and Preferences of Men and Women in Same-Sex versus Mixed-Sex Relationships. *Journal of Sex Research* 46:57–66.

Holmberg, Diane, Karen Blair, and Maggie Phillips. 2010. "Women's Sexual Satisfaction as a Predictor of Well-Being in Same-Sex Versus Mixed-Sex Relationships." *Journal of Sex Research* 47(1):1–11.

Holmes, Steven A. 1990. "Day Care Bill Marks a Turn toward Help for the Poor." *New York Times,* January 25.

———. 1998. "Children Study Longer and Play Less, a Report Says." *New York Times,* November 11.

Holstein, James A. and Jaber F. Gubrium. 2008. *Handbook of Constructionist Research.* New York: Guilford Press.

Holtzworth-Munroe, Amy, Amy G. Applegate, and Brian D'Onofrio. 2009. "Special Issue: For the Sake of the Children: Collaborations between Law and Social Science to Advance the Field of Family Dispute Resolution: Family Dispute Resolution: Charting a Course for the Future." *Family Court Review* 47(3):493–505.

"Homeless Families with Children." 2001. National Coalition for the Homeless, Fact Sheet #7 (www.nationalhomeless.org/families.html).

"Home School Laws." 2010. Home School Legal Defense Association. Retrieved January 26, 2010 (http://www.hslda.org).

Hondagneu-Sotelo, Pierrette. 1996. "Overcoming Patriarchal Constraints: The Reconstruction of Gender Relations among Mexican Immigrant Women and Men." Pp. 184–205 in *Race, Class, and Gender,* edited by Esther Ngan-Ling Chow, Doris Wilkinson, and Maxine Baca Zinn. Newbury Park, CA: Sage Publications.

Hondagneu-Sotelo, Pierrette and Michael A. Messner. 1994. "Gender Displays and Men's Power: The 'New Man' and the Mexican Immigrant." Pp. 200–18 in *Theorizing Masculinity,* edited by Harry Brod and Michael Kaufman. Newbury Park, CA: Sage Publications.

Hook, Jennifer and Satvika Chalasani. 2008. "Gendered Expectations? Reconsidering Single Fathers' Child-Care Time." *Journal of Marriage and Family* 70(November):978–90.

Hooven, Carole K., Christopher F. Chabris, Peter T. Ellison, and Stephen M. Kosslyn. 2004. "The Relationship of Male Testosterone to Components of Mental Rotation." *Neuropsychologia* 42:782–90.

Hopfensperger, Jean. 1990. "A Day for Courting, Hmong Style." *Minneapolis Star Tribune,* November 24, p. 1B.

Hopper, Joseph. 1993. "The Rhetoric of Motives in Divorce." *Journal of Marriage and Family* 55:801–13.

Hornik, Donna. 2001. "Can the Church Get in Step with Stepfamilies?" *U.S. Catholic* 66(7):30–41.

Horowitz, A. 1985. "Sons and Daughters as Caregivers to Older Parents: Differences in Role Performance and Consequences." *The Gerontologist* 25:612–17.

Horowitz, June Andrews. 1999. "Negotiating Couplehood: The Process of Resolving the December Dilemma among Interfaith Couples." *Family Process* 38:303–23.

House, Anthony. 2002. *A Problematic Solution: Responses to the Marriage Reform Act of 1753.* Retrieved October 5, 2006 (users.ox.ac.uk).

Houseknecht, Sharon K. 1987. "Voluntary Childlessness." Pp. 369–in *Handbook of Marriage and the Family,* edited by Marvin B. Sussman and Suzanne K. Steinmetz. New York: Plenum.

Houseknecht, Sharon K. and Susan K. Lewis. 2005. "Explaining Teen Childbearing and Cohabitation: Community Embeddedness and Primary Ties." *Family Relations* 54(5):607–20.

Houseknecht, Sharon K. and Jaya Sastry. 1996. "Family 'Decline' and Child Well-Being: A Comparative Assessment." *Journal of Marriage and Family* 58(3):726–39.

Houts, Carrie R. and Sharon G. Horne. 2008. "The Role of Relationship Attributions in Relationship Satisfaction among Cohabiting Gay Men." *The Family Journal: Counseling and Therapy for Couples and Families* 16(3):240–48.

Houts, Leslie A. 2005. "But Was It Wanted? Young Women's First Voluntary Sexual Intercourse." *Journal of Family Issues* 26(8):1082–1102.

Howard, J. A. 1988. "Gender Differences in Sexual Attitudes: Conservatism or Powerlessness?" *Gender and Society* 2:103–14.

Hoyt, Michael A. 2009. "Gender Role Conflict and Emotional Approach Coping in Men With Cancer." *Psychology & Health* 24(8):981–96.

"HPV Policy Becomes Political Issue." 2007. *Day to Day.* National Public Radio (NPR), February 27.

Hsia, Annie. 2002. "Considering Grandparents' Rights and Parents' Wishes: 'Special Circumstances' Litigation Alternatives." *The Legal Intelligencer* 227(83):7.

Huang, Chien-Chung, Ronald B. Mincy, and Irwin Garfinkel. 2005. "Child Support Obligations and Low-Income Fathers." *Journal of Marriage and Family* 67(5):1213–25.

Huber, Joan. 1980. "Will U.S. Fertility Decline toward Zero?" *Sociological Quarterly* 21:481–92.

Huebner, Angela J., Jay A. Mancini, Ryan M. Wilcox, Saralyn R. Grass, and Gabriel A. Grass. 2007. "Parental Deployment and Youth in Military Families: Exploring Uncertainty and Ambiguous Loss." *Family Relations* 56(2):112–22.

Hughes, Mary E. and Linda J. Waite. 2009. "Marital Biography and Health at Mid-Life." *Journal of Health and Social Behavior* 50:344–58.

Hughes, Michael and Walter R. Gove. 1989. "Explaining the Negative Relationship between Social Integration and Mental Health: The Case of Living Alone." Presented at the annual meeting of the American Sociological Association, August, San Francisco, CA.

Hughes, Patrick C. and Fran C. Dickson. 2005. "Communication, Marital Satisfaction, and Religious Orientation in Interfaith Marriages." *Journal of Family Communication* 5(1):25–41.

Hulbert, Ann. 2006. "Confidant Crisis." *New York Times,* July 16.

Hull, K.G. 2008. "Broken Trust: Pursuing Remedies for Victims of Elder Financial Abuse by Agents under Power-of-Attorney Agreements." *Clearinghouse Review* 42(5/6):223–31.

Human Rights Campaign. 2007. "Which States Permit Same-Sex Parents to Be Listed on a Birth Certificate?" Washington, DC: Human Rights Campaign. Retrieved March 30, 2007 (www.hrc .org).

———. 2009. "Human Rights Campaign Lauds Bicameral Introduction of the Domestic Partnership Benefits and Obligations Act." Washington, DC: Human Rights Campaign. Retrieved June 15, 2009 (www.hrc.org).

Humble, Aine M. 2009. "The Second Time 'Round: Gender Construction in Remarried Couples' Wedding Planning." Journal of Divorce and Remarriage 50:260–81.

Hunt, Gail, Carol Levine, and Linda Naiditch. 2005. *Young Caregivers in the U.S: Report of Findings September 2005*. National Alliance for Caregiving.

Hunt, Janet G. and Larry L. Hunt. 1977. "Dilemmas and Contradictions of Status: The Case of the Dual-career Family." *Social Problems* 24:407–16.

———. 1986. "The Dualities of Careers and Families: New Integrations or New Polarizations?" Pp. 275–89 in *Family in Transition: Rethinking Marriage, Sexuality, Child Rearing, and Family Organization*, 5th ed., edited by Arlene S. Skolnick and Jerome H. Skolnick. Boston, MA: Little, Brown.

Hunter, Andrea G. and Sherrill L. Sellers. 1998. "Feminist Attitudes among African American Women and Men." *Gender and Society* 12:81–99.

Hurd, James P . 2006. "The Shape of High Fertility in a Traditional Mennonite Population." *Annals of Human Biology* 33(5/6):557–69.

Hurley, Dan. 2005. "Divorce Rate: It's Not as High as You Think." *New York Times*, April 19.

Hurwitt, Sam. 2004. "Meet the Quirkyalones." *East Bay Express*, March 17.

Huston, Ted L. and Heidi Melz. 2004. "The Case for (Promoting) Marriage: The Devil Is in the Details." *Journal of Marriage and Family* 66(4):943–58.

Hutson, Matthew. 2009. "With God As My Wingman." *Psychology Today* 42(1):26.

Hwang, Sean-Shong and Rogelio Saenz. 1997. "Fertility of Chinese Immigrants in the U.S.: Testing a Fertility Emancipation Hypothesis." *Journal of Marriage and Family* 59:50–61.

Hyde, Janet Shibley. 2005. "The Gender Similarities Hypothesis." *American Psychologist* 60(6):581–92.

———. 2007. "New Directions in the Study of Gender Similarities and Differences." *Current Directions in Psychological Science* 16(5):259–63.

Hyde, Janet Shibley, Elizabeth Fennema, and Susan Lamon. 1990. "Gender Differences in Mathematics Performance: A Meta-Analysis." *Psychological Bulletin* 106:139–55.

Hymowitz, Kay. 2006. *Marriage and Caste in America: Separate and Unequal Families in a Post-Marital Age*. Chicago, IL: Ivan R. Dee.

Iacuone, David. 2005. "'Real Men Are Tough Guys': Hegemonic Masculinity and Safety in the Construction Industry." *Journal of Men's Studies* 13(2):247–67.

Iceland, John and Kyle Anne Nelson. 2008. "Hispanic Segregation in Metropolitan America: Exploring the Multiple Forms of Spatial Assimilation." *American Sociological Review* 73(5):741–65.

"If Your Child Is Gay or Lesbian." 2008. Retrieved March 13, 2010 (www.rollercoaster.ie.).

Ihinger-Tallman, Marilyn and Kay Pasley. 1997. "Stepfamilies in 1984 and Today—A Scholarly Perspective." *Marriage and Family Review* 26(1-2):19–41.

"Illegal Immigrant Population." 2001. *New York Times,* July 26.

Immigrant Battered Women Power and Control Wheel, produced and distributed by National Center on Domestic and Sexual Violence, Austin, TX. Available at www.endingviolence. org/files/uploads/ImmigrantWomenPCwheel. pdf and adapted from original wheel by Domestic Abuse Intervention Project, Duluth, MN.

"Improvements in Teen Sexual Risk Behavior Flatline." 2006. Press Release. Washington, DC: Advocates for Youth.

Ingersoll-Dayton, Berit, Margaret B. Neal, Jung-Hwa Ha, and Leslie B. Hammer. 2003. "Redressing Inequity in Parent Care among Siblings." *Journal of Marriage and Family* 65(1):201–12.

Ingoldsby, Bron B. 2006a. "Family Origin and Universality." Pp. 67–78 in *Families in Global and Multicultural Perspective,* 2nd ed., edited by Bron B. Ingoldsby and Suzanna D. Smith. New York: Guilford.

———. 2006b. "Mate Selection and Marriage." Pp. 133–46 in *Families in Global and Multicultural Perspective,* 2nd ed., edited by Bron B. Ingoldsby and Suzanna D. Smith. New York: Guilford.

Ingoldsby, Bron B. and Suzanna D. Smith, eds. 2005. *Families in Global and Multicultural Perspective.* 2nd ed. Thousand Oaks, CA: Sage Publications.

Ingram, Stephanie, Jay L. Ringle, Kristen Hallstrom, David E. Schill, Virginia M. Gohr, and Ronald W. Thompson. 2008. "Coping with Crisis across the Lifespan: The Role of a Telephone Hotline." Journal of Child and Family Studies 17:663–74.

Inman, Jessica and Larry J. Koenig. 2006. *Crash Course on Successful Parenting: 13 Dynamics of Raising Great Kids*. Nashville, TN: Countryman Press.

Institute for Social Research. University of Michigan. 2002. "U.S. Husbands Are Doing More Housework While Wives Are Doing Less." Press Release, March 12.

Institute for Women's Policy Research. 2009. *The Gender Wage Gap: 2008* (C350). Washington, DC: Institute for Women's Policy Research.

Interactive Autism Network. 2010. *IAN Research Report—April 2010: Grandparents of Children with ASD, Parts 1 and 2.* Retrieved May 15, 2010 (www.iancommunity.org).

"Iowa Supreme Court Rejects Antigay Activists' Claims in Lesbian Civil Union Dissolution Case; Lambda Legal Declares Victory." 2007. Press Release, February 26. New York: Lambda Legal. Retrieved May 13, 2007 (www.lambdalegal.org).

Isaacs, Marla Beth, Braulio Montalvo, and David Abelsohn. 1986. *The Difficult Divorce: Therapy for Children and Families.* New York: Basic Books.

Isaksen, Lise W. 2002. "Toward a Sociology of (Gendered) Disgust." *Journal of Family Issues* 23(7):791–811.

Ishii-Kuntz, Masako. 2000. "Diversity within Asian American Families." Pp. 274–92 in *Handbook of Family Diversity*, edited by David H. Demo, Katherine R. Allen, and Mark Fine. New York: Oxford University Press.

Ishii-Kuntz, Masako, Jessica N. Gomei, Barbara J. Tinsley, and Rose D. Parks. 2009. "Economic Hardship and Adaptation Among Asian American Families." *Journal of Family Issues* 31(3):407–20.

Island, David and Patrick Letellier. 1991. *Men Who Beat the Men Who Love Them: Battered Gay Men and Domestic Violence.* New York: Haworth.

Ispa, Jean M., Marjorie R. Sable, Noriko Porter, and Annamaria Csizmadia. 2007. "Pregnancy Acceptance, Parenting Stress, and Toddler Attachment in Low-Income Black Families." *Journal of Marriage and Family* 69(1):1–13.

Italie, Leanne. 2008. "Parents of Teens Watch Nebraska Safe Law." Associated Press, November 11. Retrieved December 12, 2008 (ap.google. com).

Jackson, Aurora P., Jeong-Kyun Choi, and Peter Bentler. 2009. "Parenting Efficacy and the Early School Adjustment of Poor and Near-Poor Black Children." *Journal of Family Issues* 30(10):1339–55.

Jackson, Aurora P., Jeong-Kyun Choi, and T. M. Franke. 2009. "Poor Single Mothers with Young Children: Mastery, Relations with Nonresident Fathers, and Child Outcomes." *Social Work Research* 33(2):95–106.

Jackson, Pamela Braboy. 2004. "Role Sequencing: Does Order Matter for Mental Health?" *Journal of Health and Social Behavior* 45:132–54.

Jackson, Robert Max. 2006. "Opposing Forces: How, Why, and When Will Gender Inequality Disappear?" Pp. 215–44 in *The Declining Significance of Gender?* edited by Francine D. Blau, Mary C. Brinton, and David B. Grusky. New York: Russell Sage.

Jackson, Robert M. 2007. "Destined for Equality." Pp. 109–16 in *Family in Transition*, 14th ed., edited by Arlene S. Skolnick and Jerome H. Skolnick. Boston, MA and New York: Pearson.

Jackson, Shelly, Lynette Feder, David R. Forde, Robert C. Davis, Christopher D. Maxwell, and Bruce G. Taylor. 2003. *Batterer Intervention Programs: Where Do We Go from Here?* Washington, DC: U.S. National Institute of Justice.

Jackson-Newsom, Julia, Christy M. Buchanan, and Richard M. McDonald. 2008. "Parenting and Perceived Warmth in European American and African American Adolescents." *Journal of Marriage and Family* 70(1):62–75.

Jacobs, Andres. 2006. "Extreme Makeover, Commune Edition." *New York Times,* June 11. Retrieved June 11, 2006 (www.nytimes.com).

Jacobs, Jerry A. and Kathleen Gerson. 2004. *The Time Divide: Work, Family, and Gender Inequality.* Cambridge, MA: Harvard University Press.

Jacobson, Cardell K. and Bryan R. Johnson. 2006. "Interracial Friendship and African American Attitudes about Interracial Marriage." *Journal of Black Studies* 36(4):570–84.

Jacoby, Susan. 1999. "Great Sex: What's Age Got to Do with It?" *Modern Maturity* (September–October), pp. 41–47.

———. 2005. "Sex in America." *AARP The Magazine,* July/August, pp. 57–62, 114.

Jaksch, Mary. 2002. *Learn to Love: A Practical Guide to Fulfilling Relationships.* San Francisco, CA: Chronicle Books.

Janofsky, Michael. 2001. "Conviction of a Polygamist Raises Fears Among Others." *New York Times,* May 24.

Janssen, Patricia A., Victoria L. Holt, Nancy K. Sugg, Irvin Emanuel, Cathy M. Critchlow, and Angela D Henderson. 2003. "Intimate Partner Violence and Adverse Pregnancy Outcomes: A Population-Based Study." *American Journal of Obstetrics & Gynecology* 188(5):1341–47.

Jarrell, Anne. 2007. "The Daddy Track." *The Boston Globe,* July 8. Retrieved January 18, 2009 (www.boston.com).

Jaschik, Scott. 2009. "Probe of Extra Help for Men." *Inside Higher Ed,* November 2. (www. insidehighered.com).

Jay, David. 2005. "Asexual: A Person Who Does Not Experience Sexual Attraction." Asexual Visibility and Education Network. Retrieved February 22, 2007 (www.asexuality.org).

Jayakody, R. and A. Kalil. 2002. "Social Fathering in Low-income, African American Families with Preschool Children." *Journal of Marriage and Family* 64(2):504–16.

Jayson, Sharon. 2005. "Yep, Life'll Burst That Self-esteem Bubble." *USA Today,* February 16.

———. 2006a. "Is 'Failure to Launch' Really a Failure?" *USA Today,* March 16.

———. 2006b. "Outside the Race Box." *USA Today,* February 8.

———. 2008. "Infidelity Today Has A New Face." *USA Today,* July 1.

———. 2009. "Gay Couples: A Close Look At This Modern Family, Parenting." USA Today, November 5. Retrieved November 20, 2009 (usatoday.com).

———. 2010. "Delaying Kids May Prevent Financial 'Motherhood Penalty'." *USA Today,* April 16.

Jemmott III, John B., Loretta S. Jemmott, and Geoffrey T. Fong. 2010. "Efficacy of a Theory-Based Abstinence-Only Intervention Over 24 Months: A Randomized Controlled Trial With Young Adolescents." *Archives of Pediatrics & Adolescent Medicine* 164(2):152–59.

Jenkins, Nate. 2008. "Neb. Parents Rush To Leave Kids Before Law Changes." Associated Press, November 13. Retrieved December 12, 2008 (news.yahoo.com).

Jenks, R. L. 1998. "Swinging: A Review of the Literature." *Archives of Sexual Behavior* 27:507–21.

Jeong, Jae Y. and Sharon G. Horne. 2009. "Relationship Characteristics of Women in Interracial Same-Sex Relationships." *Journal of Homosexuality* 56(4):443–56.

Jepsen, Lisa K. and Christopher A. Jepsen. 2002. "An Empirical Analysis of the Matching Patterns of Same-Sex and Opposite-Sex Couples?" *Demography* 39(3):435–53.

Jervey, Gay. 2004. "The Bad Mother." *Good Housekeeping,* August, pp. 132–82.

Jhally, Sut. 2007. *Dreamworlds 3: Desire, Sex & Power in Music Video.* Northampton, MA: Media Education Foundation.

———. 2009. *Codes of Gender.* Transcript. Northampton, MA: Media Education Foundation.

Jimenez, Thomas. 2008. "Changes in Attitudes Toward Race Change Slowly Despite Obama." San Jose Mercury News, December 2.

Jin, Lei and Nicholasa Chrisatakis. 2009. "Investigating the Mechanism of Marital Mortality Reduction: The Transition to Widowhood and Quality of Health Care." Demography 46(3):605–25.

Jio, Sarah. 2008. "Career Couples Fight Over Who's the 'Trailing Spouse'." *CNN,* June 26. Retrieved May 27, 2010 (www.cnn.com).

Jo, Moon H. 2002. "Coping with Gender Role Strains in Korean American Families." Pp. 78–83 in *Contemporary Ethnic Families in the United States,* edited by Nijole V. Benekraitis. Upper Saddle River, NJ: Prentice Hall.

John, Robert. 1998. "Native American Families." Pp. 382–421 in *Ethnic Families in America: Patterns and Variations,* edited by Charles H. Mindel, Robert W. Haberstein, and Roosevelt Wright, Jr. Upper Saddle River, NJ: Prentice Hall.

Johnson, Colleen L. 1988. *Ex Familia: Grandparents, Parents, and Children Adjust to Divorce.* New Brunswick, NJ: Rutgers University Press.

Johnson, Dirk. 1991. "Polygamists Emerge from Secrecy, Seeking Not Just Peace but Respect." *New York Times,* April 9.

Johnson, Elizabeth M. and Ted L. Huston. 1998. "The Perils of Love, or Why Wives Adapt to Husbands during the Transition to Parenthood." *Journal of Marriage and Family* 60(1):195–204.

Johnson, James. 2010. *Combat Trauma: A Personal Look at Long-Term Consequences.* Lanham, MD: Rowman & Littlefield Publishers.

Johnson, Jason B. 2000. "Something Akin to Family: Struggling Parents, Kids, Move in with Their Mentors." *San Francisco Chronicle,* November 10.

Johnson, Julia Overturf. 2005. *Who's Minding the Kids? Child Care Arrangements: Winter 2002.* Current Population Reports P70-101. Washington, DC: U.S. Census Bureau.

Johnson, Julia Overturf and Barbara Downs. 2005. *Maternity Leave and Employment Patterns of First-time Mothers 1961–2000.* Current Population Reports P70-103. Washington, DC: U.S. Census Bureau. October.

Johnson, Leanor Boulin and Robert Staples. 2005. *Black Families at the Crossroads: Challenges and Prospects, 2nd ed.* San Francisco, CA: Jossey-Bass.

Johnson, Matthew D., Joanne Davila, Ronald D. Rogge, Kieran T. Sullivan, Catherine L. Cohan, Erika Lawrence, Benjamin R. Karney, and Thomas N. Bradbury. 2005. "Problem-Solving Skills and Affective Expressions as Predictors of Change in Marital Satisfaction." *Journal of Consulting and Clinical Psychology* 73(1):15–27.

Johnson, Michael P. 1995. "Patriarchal Terrorism and Common Couple Violence: Two Forms of Violence Against Women." *Journal of Marriage and Family* 57(2):283–94.

———. 2008. *A Typology of Domestic Violence: Intimate Terrorism, Violent Resistance, and Situational Couple Violence.* Lebanon, NH: Northeastern University Press.

———. 2010. "Langhinrichsen-Rolling's Confirmation of the Feminist Analysis of Intimate Partner Violence: Comment on 'Controversies Involving Gender and Intimate Partner Violence in the United States.'" *Sex Roles* 32(3-4):212–19.

Johnson, Michael P. and Kathleen J. Ferraro. 2000. "Research on Domestic Violence in the 1990s: Making Distinctions." *Journal of Marriage and Family* 62(4):948–63.

Johnson, Michael P. and Janel M. Leone. 2005. "The Differential Effects of Intimate Terrorism and Situational Couple Violence: Findings From the National Violence Against Women Survey." *Journal of Family Issues, 26* (3):322–49.

Johnson, Phyllis J. 1998. "Performance of Household Tasks by Vietnamese and Laotian Refugees." *Journal of Family Issues* 10(3):245–73.

Johnson, Phyllis J. and Kathrin Stoll. 2008. "Remittance Patterns of Southern Sudanese Refugee Men: Enacting the Global Breadwinner Role." Family Relations 57(4):431–443.

Johnson, Richard W. and Joshua M. Wiener. 2006. "A Profile of Frail Older Americans and Their Caregivers." Urban Institute. March 1. Retrieved May 18, 2007 (www.urban.org/ publications).

Johnson, Suzanne M. and Elizabeth O'Connor. 2002. *The Gay Baby Boom: The Psychology of Gay Parenthood.* New York: New York University Press.

Johnson, Tallese D. 2007. *Maternity Leave and Employment Patterns: 2001–2003.* Current Population Report, P70-113. U.S. Census Bureau, Washington, DC.

Johnston, Alexandra. 2007. "Gatekeeping in the Family: How Family Members Position One Another as Decision Makers." Pp. 165–93 in *Family Talk: Discourse and Identity in Four American Families,* edited by Deborah Tannen, Shari Kendall, and Cynthia Gordon. New York: Oxford University Press.

Jones, A. J. and M. Galinsky. 2003. "Restructuring the Stepfamily: Old Myths, New Stories." *Social Work* 48(2):228–37.

Jones, A. R., R. M. Hyland, K. N. Parkinson, and A. J. Adamson. 2009. "Developing a Focus Group Approach for Exploring Parents'

Perspectives on Childhood Overweight." Nutrition Bulletin 34(2):214–9.

Jones, Anne C. 2004. "Transforming the Story: Narrative Applications to a Stepmother Support Group." *Families in Society: The Journal of Contemporary Human Services* 85(1):29–39.

Jones, B. J. 1995. "The Indian Child Welfare Act: The Need for a Separate Law." *GP Solo Magazine*, Fall. Retrieved December 28, 2006 (www.abanet.org/general practice).

Jones, Del 2006. "One of USA's Exports: Love, American Style." USA Today. February 14.

Jones, Jeffrey M. 2005. "Most Americans Approve of Interracial Dating." The Gallup Poll. Retrieved November 4, 2006 (http://www.galluppoll.com).

———. 2008. "Most Americans Not Willing to Forgive Unfaithful Spouse." Washington, DC: Gallup Poll News Service. Retrieved December 2, 2008 (www.gallup.com/poll).

Jones, Matthew D. 2006. *Raising Boys to Become Responsible Men.* New York: iUniverse.

Jones, Nicholas. 2005. *We the People of More Than One Race in the United States.* CENSR22. Washington, DC: U.S. Census Bureau. April. Retrieved April 30, 2005 (www.census.gov).

Jones, Nicholas A. and Amy Symens Smith. 2001. *The Two or More Races Population: 2000.* Census 2000 Brief C2KBR/01-6. Washington, DC: U.S. Census Bureau.

Jones, Rachel K., Jacqueline E. Darroch, and Stanley K. Henshaw. 2002. "Patterns in the Socioeconomic Characteristics of Women Obtaining Abortions in 2000–2001." *Perspectives on Sexual and Reproductive Health* 34:226–35.

Jong-Fast, Molly. 2003. "Out of Step and Having a Baby." *New York Times*, October 5.

Jonsson, Patrik. 2006. "Debate Grows on Out-of-Wedlock Laws." *Christian Science Monitor*, August 23. Retrieved November 21, 2009 (www.csmonitor.com).

Joseph, Elizabeth. 1991. "My Husband's Nine Wives." *New York Times*, May 23.

Joshi, Pamela and Karen Bogen. 2007. "Nonstandard Schedules and Young Children's Behavioral Outcomes Among Working Low-Income Families." *Journal of Marriage and Family* 60(1):139–56.

Joshi, Pamela, James M. Quane, and Andrew J. Cherlin. 2009. "Contemporary Work and Family Issues Affecting Marriage and Cohabitation Among Low-Income Single Mothers." *Family Relations* 58(5):647–61.

Josten, Edith. 2002. *The Effects of Extended Workdays.* The Netherlands: Van Gorcum Ltd.

Joyce, Amy. 2006. "Kid-Friendly Policies Don't Help Singles." *Washington Post*, September 16.

———. 2007a. "Caring for Dear Old Dad Becomes a Little Easier." *Washington Post*, March 4.

———. 2007b. "Developing Boomerang Mothers." *Washington Post*, March 11.

Joyner, Kara and Grace Kao. 2005. "Interracial Relationships and the Transition to Adulthood." *American Sociological Review* 70(4):563–81.

Juby, Heather, Jean-Michel Billette, Benoît Laplante, and Céline Le Bourdais. 2007. "Nonresident Fathers and Children: Parents' New Unions and Frequency of Contact." *Journal of Family Issues* 28(9):1220–45.

Juffer, Femmie and Marinus H. van Uzendoorn. 2005. "Behavior Problems and Mental Health Referrals of International Adoptees: A Meta-Analysis." *Journal of the American Medical Association* 293(20):2501–15.

Juffer, Jane. 2006. *Single Mother: The Emergence of the Domestic Intellectual.* New York; New York University Press.

Junn, Ellen Nan and Chris J. Boyatzis. 2005. *Child Growth and Development 05/06.* 12th ed. Guilford, CT: McGraw-Hill/Dushkin.

Kader, Samuel. 1999. *Openly Gay, Openly Christian: How the Bible Really Is Gay Friendly.* Leyland Publications.

Kagan, Marilyn and Neil Einbund. 2008. *Defenders of the Heart: Managing the Habits and Attitudes That Block You from a Richer, More-Satisfying Life.* Carlsbad, CA: Hay House.

Kallivayalil, Diya. 2004. "Gender and Cultural Socialization in Indian Immigrant Families in the United States." *Feminism and Psychology* 14(4):535–59.

Kalmijn, Matthijs. 1998. "Differentiation and Stratification—Intermarriage and Homogamy: Causes, Patterns, and Trends." *Annual Review of Sociology* 24:395–427.

Kamo, Yoshinori and Min Zhou. 1994. "Living Arrangements of Elderly Chinese and Japanese in the United States." *Journal of Marriage and Family* 56(3):544–58.

Kan, Marni L. and Alison C. Cares. 2006. "From 'Friends with Benefits' to 'Going Steady': New Directions in Understanding Romance and Sex in Adolescence and Emerging Adulthood." Pp. 241–58 in *Romance and Sex in Adolescence and Emerging Adulthood: Risks and Opportunities*, edited by Ann C. Crouter and Alan Booth. Mahwah, NJ: Erlbaum.

Kan, Marni L., Susan M. McHale, and Ann C. Crouter. 2008. "Interparental Incongruence in Differential Treatment of Adolescent Siblings: Links with Marital Quality." *Journal of Marriage and Family* 70(May):466–79.

Kane, Emily W. 2000. "Racial and Ethnic Variations in Gender-related Attitudes." *Annual Review of Sociology* 26:416–39.

Kane, John M. 2009. "Manliness and the Constitution." *Harvard Journal of Law & Public Policy* 32(1):261–332.

Kann, Mark E. 1986. "The Costs of Being on Top." *Journal of The National Association for Women Deans, Administrators, and Counselors* 49:29–37.

Kantor, Jodi. 2006. "Nanny Hunt Can Be a 'Slap in the Face' for Blacks." *New York Times*, December 26.

Kantrowitz, Barbara. 2000. "Busy around the Clock." *Newsweek*, July 17, pp. 49–50. Kantrowitz, Barbara and Karen Springen. 2005. "A Peaceful Adolescence." *Newsweek*, April 25, pp. 58–61.

Kantrowitz, Barbara and Peg Tyre. 2006. "The Fine Art of Letting Go." *Newsweek*, May 22, pp. 49–61.

Kaplan, Marion A., ed. 1985. *The Marriage Bargain: Women and Dowries in European History.* New York: Harrington Park.

Karasik, Rona J. and Raeann R. Hamon. 2007. "Cultural Diversity and Aging Families." Pp. 136–53 in *Cultural Diversity and Families*, edited by Bahira Sherif Trask and Raeann R. Hamon. Thousand Oaks, CA: Sage Publications.

Karim, Jamillah. 2009. *American Muslim Women Negotiating Race, Class, and Gender within the Ummah.* New York: NYU Press.

Karimzadeh, Mohammad Ali and Sedigheh Ghandi. 2008. "Early Marriage: a Policy for Infertility Prevention." *Journal of Family and Reproductive Health* 2(2):61–64.

Kates, Jen, Murray Penner, Dave Kern, Alicia Carbaugh, Britten Ginsburg, and Connie Jorstad. 2009. *The National HIV Prevention Inventory: The State of HIV Prevention Across the U.S.* A Report By NASTAD and the Kaiser Family Foundation.

Katz, Jackson. 2006. *The Macho Paradox: Why Some Men Hurt Women and and How All Men Can Help.* Naperville, IL: Sourcebooks, Inc.

Katz, Jonathan Ned. 2007. *The Invention of Heterosexuality.* Chicago: University of Chicago Press.

Kaufman, Gayle and Hiromi Taniguchi. 2006. "Gender and Marital Happiness in Later Life." *Journal of Family Issues* 27(6):735–57.

Kaufman, Gayle and Peter Uhlenberg. 1998. "Effects of Life Course Transitions on the Quality of Relationships between Adult Children and Their Parents." *Journal of Marriage and Family* 60(4):924–38.

———. 2000. "The Influence of Parenthood on the Work Effort of Married Men and Women." *Social Forces* 78:931–49.

Kaufman, Leslie. 2006. "Facing Hardest Choice in Child Safety, New York Tilts to Preserving Families." *New York Times*, February 4.

———. 2007. "In Custody Fights, a Hurdle for the Poor." *New York Times*, April 8.

Kaukinen, Catherine. 2004. "Status Compatibility, Physical Violence, and Emotional Abuse in Intimate Relationships." *Journal of Marriage and Family* 66:452–71.

Kaye, Sarah. 2005. "Substance Abuse Treatment and Child Welfare: Systematic Change Is Needed." *Family Focus on . . . Substance Abuse across the Life Span*: FF25: F15–F16. Minneapolis, MN: National Council of Family Relations.

Kaye, Kelleen, Suellentrop, Katherine, and Sloup, Corinna. 2009. *The Fog Zone: How Misperceptions, Magical Thinking, and Ambivalence Put Young Adults at Risk for Unplanned Pregnancy.* Washington, DC: The National Campaign to Prevent Teen and Unplanned Pregnancy.

Kazdin, Alan E. and Corina Benjet. 2003. "Spanking Children: Evidence and Issues." *Current Directions in Psychological Science* (March):99–103.

Kearney, Christopher A. 2010. *Casebook in Child Behavior Disorders.* 4th ed. Belmont, CA: Wadsworth/Cengage Learning.

Keaten, James and Lynne Kelly. 2008. "Emotional Intelligence as a Mediator of Family Communication Patterns and Reticence." *Communication Reports* 21(2):104–16.

Keefe, Janice and Pamela Fancey. 2000. "The Care Continues: Responsibility for Elderly Relatives Before and After Admission to a Long Term Care Facility." *Family Relations* 49(3):235–44.

Kellas, Jody Koenig, Cassandra LeClair-Underberg, and Emily Lamb Normand. 2008. "Stepfamily Address Terms: 'Sometimes They Mean Something and Sometimes They Don't.'" *Journal of Family Communication* 8:238–63.

Kelleher, Elizabeth. 2007. "In Dual-Earner Couples, Family Roles Are Changing in U.S." *USINFO* March 21. Washington, DC: Bureau of International Information Programs, U.S. Department of State. www.america.gov.

Kellogg, Nancy D. 2009. "Clinical Report: The Evaluation of Sexual Behaviors in Children." *Pediatrics* 124(3):992–98.

Kelly, Joan B. 2007. "Children's Living Arrangements Following Separation and Divorce: Insights From Empirical and Clinical Research." *Family Process* 46(1):35–52.

Kelly, Joan B. and Robert E. Emery. 2003. "Children's Adjustment Following Divorce: Risk and Resiliency Perspectives." *Family Relations* 52:352–62.

Kelly, Joan B. and Michael E. Lamb. 2003. "Developmental Issues in Relocation Cases Involving Young Children: When, Whether, and How?" *Journal of Family Psychology* 17:193–205.

Kelly, Maura. 2009. "Women's Voluntary Childlessness: A Radical Rejection of Motherhood?" *WSQ: Women's Studies Quarterly* 37(3/4):157–72.

Kemp, Candace L. 2007. "Grandparent-Grandchild Ties: Reflections on Continuity and Change across Three Generations." *Journal of Family Issues* 28(7):855–81.

Kempe, C. Henry, Frederic N. Silverman, Brandt F. Steele, William Droegemuller, and Henry K. Silver. 1962. "The Battered Child Syndrome." *Journal of the American Medical Association* 181:17–24.

Kendall, Shari. 2007. "Father as Breadwinner, Mother as Worker: Gendered Positions in Feminisht and Traditional Discourses of Work and Family." Pp. 123–63 in *Family Talk: Discourse and Identity in Four American Families*, edited by Deborah Tannen, Shari Kendall, and Cynthia Gordon. New York: Oxford University Press.

Kennedy, Denise E. and Laurie Kramer. 2008. "Improving Emotion Regulation and Sibling Relationship Quality: The More Fun with Sisters and Brothers Program." *Family Relations* 57(December):567–78.

Kenney, Catherine. 2004. "Cohabiting Couple, Filing Jointly? Resource Pooling and U.S. Poverty Policies." *Family Relations* 53(2):237–47.

———. 2006. "The Power of the Purse: Allocative Systems and Inequality in Couple Households." *Gender and Society* 20(3):354–81.

Kenney, Catherine and Sara S. McLanahan. 2006. "Why Are Cohabiting Relationships More Violent Than Marriages?" *Demography* 43(1):127–40.

Kent, Mary Mederios. 2007. "Immigration and America's Black Population." *Population Bulletin* 62(4). Washington, DC: Population Reference Bureau. Retrieved June 10, 2009 (www.prb .com).

Kent, Mary M., Kelvin M. Pollard, John Hagga, and Mark Mather. 2001. "First Glimpses from the 2000 U.S. Census." *Population Bulletin* 56(2). Washington, DC: Population Reference Bureau.

Kephart, William. 1971. "Oneida: An Early American Commune." Pp. 481–92 in *Family in Transition: Rethinking Marriage, Sexuality, Child Rearing, and Family Organization*, edited by Arlene S. Skolnick and Jerome H. Skolnick. Boston, MA: Little, Brown.

Kern, Louis J. 1981. *An Ordered Love: Sex Roles and Sexuality in Victorian Utopias—The Shakers, the Mormons, and the Oneida Community.* Chapel Hill: University of North Carolina Press.

Kershaw, Sarah. 2003. "Many Immigrants Decide to Embrace Homes for Elderly." *New York Times*, October 20, pp. A1, A10.

———. 2004. "For Native Alaskans, Tradition Is Yielding to Modern Customs." *New York Times*, August 21.

———. 2005. "Crisis of Indian Children Intensifies as Families Fail." *New York Times*, April 5.

Kershaw, Sarah. 2009a. "Mr. Moms (by Way of Fortune 500)." *The New York Times*, April 23. Retrieved December 23, 2009 (http://www.nytimes.com).

———. 2009b. "Rethinking the Older Woman-Younger Man Relationship." *The New York Times*, October 15.

Kheshgi-Genovese, Zareena and Thomas A. Genovese. 1997. "Developing the Spousal Relationship within Stepfamilies." *Families in Society* 78(3):255–64.

Kibria, Nazli. 2000. "Race, Ethnic Options, and Ethnic Binds: Identity Negotiations of Second-Generation Chinese and Korean Americans." *Sociological Perspectives* 43(1):77–95.

———. 2007. "Vietnamese Americans and the Rise of Women's Power." Pp. 220–27 in *Race, Class, and Gender: An Anthology*, 6th ed., edited by Margaret L. Andersen and Patricia Hill Collins. Belmont, CA: Wadsworth.

"KidsPoll." 2008. www.kidshealth.org.

Kiecolt, K. Jill. 2003. "Satisfaction with Work and Family Life: No Evidence of Cultural Reversal." *Journal of Marriage and Family* 65(1):23–35.

Kiernan, Kathleen. 2002. "Cohabitation in Western Europe: Trends, Issues, and Implications." Pp. 3–31 in *Just Living Together: Implication of Cohabitation on Families, Children, and Social Policy*, edited by Alan Booth and A. C. Crouter. Mahwah, NJ: Erlbaum.

Kiesbye, Stefan. 2009. *Blended Families.* Detroit: Greenhaven Press.

Kilbourne, Jean. 1994. "'Gender Bender' Ads: Same Old Sexism." *New York Times*, May 15, p. F13.

Killian, Timothy and Lawrence H. Ganong. 2002. "Ideology, Context, and Obligations to Assist Older Persons." *Journal of Marriage and Family* 64(4):1080–88.

Killoren, Sarah E., Shawna M. Thayer, and Kimberly A. Updegraff. 2008. "Conflict Resolution Between Mexican Origin Adolescent Siblings." *Journal of marriage and Family* 70(December):1200–12.

Kim, Haejeong and Sharon A. DeVaney. 2005. "The Selection of Partial or Full Retirement by Older Workers." *Journal of Family and Economic Issues* 26(3):371–94.

Kim, Hyoun K., Deborah M. Capaldi, and Lynn Crosby. 2007. "Generalizability of Gottman and Colleagues' Affective Process Models of Couples' Relationship Outcomes." *Journal of Marriage and Family* 69(1):55–72.

Kim, Hyoun K., and Patrick C. McKenry. 2002. "The Relationship between Marriage and Psychological Well-Being." *Journal of Family Issues* 23(8):885–911.

Kim, Ji-Yeon, Susan M. McHale, Ann C. Crouter, and D. Wayne Osgood. 2007. "Longitudinal Linkages Between Sibling Relationships and Adjustment From Middle Childhood through Adolescence." *Developmental Psychology* 43(4):960–73.

Kimbro, Rachel Tolbert. 2008. "Together Forever? Romantic Relationship Characteristics and Prenatal Health Behaviors." Journal of Marriage and Family 70(3):756–57.

Kimerling, Rachel, Jennifer Alvarez, Joanne Pavao, Katelyn P. Mack, Mark W. Smith, and Nikki Baumrind. 2009. "Unemployment among Women: Examining the Relationship of Physical and Psychological Intimate Partner Violence and Posttraumatic Stress Disorder." Journal of Interpersonal Violence 24(3):450–63.

Kimmel, Michael S. 1995. "Misogynists, Masculinist Mentors, and Male Supporters: Men's Responses to Feminism." Pp. 561–72 in *Women: A Feminist Perspective*, 5th ed., edited by Jo Freeman. Mountain View, CA: Mayfield.

———. 2000. *The Gendered Society.* New York: Oxford University Press.

———. 2001. "Manhood and Violence: The Deadliest Equation." *Newsday*, March 8.

———. 2002. "'Gender Symmetry' in Domestic Violence." *Violence Against Women* 8:1332–63.

Kimmel, Michael S. and Michael A. Messner. 1998. *Men's Lives.* 4th ed. Boston, MA: Allyn and Bacon.

Kimmel, Tim. 2006. *Raising Kids for True Greatness.* Nashville, TN: W Publishing Group.

Kindlon, Dan and Michael Thompson. 1999. *Raising Cain: Protecting the Emotional Life of Boys.* New York: Ballantine.

King, Anthony E. and Terrence T. Allen. 2009. "Personal Characteristics of the Ideal African American Marriage Partner." *Journal of Black Studies* 39(4):570–88.

King, Valarie. 1994. "Variation in the Consequences of Nonresident Father Involvement for Children's Well-Being." *Journal of Marriage and Family* 56(3):963–72.

———. 2009. "Stepfamily Formation: Implications for Adolescent Ties to Mothers, Nonresident Fathers, and Stepfathers." Journal of Marriage and Family 71(November):954–68.

King, Valarie and Mindy E. Scott. 2005. "A Comparison of Cohabiting Relationships among Older and Younger Adults." *Journal of Marriage and Family* 67(2):271–85.

King, Valarie, Katherine Stamps Mitchell, and Daniel N. Hawkins. 2010. "Adolescents with Two Nonresident Biological Parents: Living Arrangements, Parental Involvement, and Well-Being." *Journal of Family Issues* 31(1):3–30.

King, Valarie, Merril Silverstein, Glen H. Elder, Jr., Vern L. Bengston, and Rand D. Conger. 2003. "Relations with Grandparents: Rural Midwest versus Urban Southern California." *Journal of Family Issues* 24(8):1044–69.

King, Valarie and Juliana M. Sobolewski. 2006. "Nonresident Fathers' Contributions to Adolescent Well-Being." *Journal of Marriage and Family* 68(3):537–57.

Kingston, Anne. 2009. "No Kids, No Grief." *Maclean's* 122(29/30):38–41.

Kinsey, Alfred, Wardell B. Pomeroy, and Clyde E. Martin. 1948. *Sexual Behavior in the Human Male*. Philadelphia, PA: Saunders.

———. 1953. *Sexual Behavior in the Human Female*. Philadelphia, PA: Saunders.

Kirby, Carrie. 2005. "Picky Suitors Narrow Search to Niche Sites." *San Francisco Chronicle*, February 14.

Kirby, Douglas. 2001. *Emerging Answers*. Washington, DC: National Campaign to Prevent Teen Pregnancy.

Kirby, Douglas, B. A. Laris, and Lori Rolleri. 2006. *Sex and HIV Education Programs for Youth: Their Impact and Important Characteristics*. Research Triangle Park, NC: Family Health International. May 13.

Kirby, James B. 2006. "From Single-Parent Families to Stepfamilies: Is the Transition Associated with Adolescent Alcohol Initiation?" *Journal of Family Issues* 27(5):685–711.

Kiselica, Mark S. and Mandy Morrill-Richards. 2007. "Sibling Maltreatment: The Forgotten Abuse." *Journal of Counseling and Development* 85(2):148–61.

Kisler, Tiffani S. and F. Scott Christopher. 2008. "Sexual Exchanges and Relationship Satisfaction: Testing the Role of Sexual Satisfaction as a Mediator and Gender as a Moderator." *Journal of Social and Personal Relationships* 25(4):587–602.

Kitano, Harry and Roger Daniels. 1995. *Asian Americans: Emerging Minorities*, 2nd ed. Englewood Cliffs, NJ:Prentice Hall.

Kjobli, John and Kristine Amlund Hagen. 2009. "A Mediation Model of Interparental Collaboration, Parenting Practices, and Child Externalizing Behavior in a Clinical Sample." *Family Relations* 58(July):275–88.

Klaus, Daniela. 2009. "Why Do Adult Children Support Their Parents?" *Journal of Comparative Family Studies* 40(2):227–41.

Kleber, Rolf J., Charles R. Figley, P. R. Barthold, and John P. Wilson. 1997. "Beyond Trauma: Cultural and Societal Dynamics." *Contemporary Psychology* 42(6):516–27.

Klein, Barbara Schave. 2007. *Raising Gifted Kids: Everything You Need to Know to Help Your Exceptional Child Thrive*. New York: AMACOM (American Management Association).

Kleinfield, N. R. 2003. "Around Tree, Smiles Even for Wives No. 2 and 3." *New York Times*, December 24.

Kleinplatz, P. J., A. D. Me´nard, N. Paradis, M. Campbell, T. Dalgleish, A. Segovia, and K. Davis. 2009. "From Closet to Reality: Optimal Sexuality Among the Elderly." *The Irish Psychiatrist* 10:15–8.

Kliff, Sarah. 2010. "Outing Abortion, From Town Halls to Twitter." *Newsweek*, March 3. www.newsweek.com.

Klohnen, E. C. and S. J. Bera. 1998. "Behavioral and Experiential Patterns of Avoidantly and Securely Attached Women Across Adulthood: A 30-Year Longitudinal Perspective." *Journal of Personality and Social Psychology* 74:211–223.

Klohnen, Eva C. and Gerald A. Mendelsohn. 1998. "Partner Selection for Personality Characteristics: A Couple-Centered Approach." *Personality & Social Psychology Bulletin* 24(3):268–77.

Kluger, Jeffrey. 2004. "The Power of Love." *Time*, January 19.

———. 2006. "The New Science of Siblings." *Time*, July 10, pp. 47–56.

———. 2010. "Be Careful What You Wish For." *Time*, February 22, pp. 68–72.

Kluwer, Esther S., Jose A. M. Heesink, and Evert Van de Vliert. 2002. "The Division of Labor across the Transition to Parenthood: A Justice Perspective." *Journal of Marriage and Family* 64(4):930–43.

Knapp, Mark L. and Anita L. Vangelisti. 2009. *Interpersonal Communication and Human Relationships*. Boston: Pearson Allyn & Bacon.

Knobloch, Leanne and Lynne Knobloch-Fedders. 2010. "The Role of Relational Uncertainty in Depressive Symptoms and Relationship Quality: An Actor-Partner Interdependence Model." *Journal of Social and Personal Relationships* 27(1):137–59.

Knobloch, Leanne K., Denise H. Solomon, and Jennifer A. Theiss. 2006. "The Role of Intimacy in the Production and Perception of Relationship Talk within Courtship." *Communication Research* 33(4):211–41.

Knodel, John, Nathalie Williams, Sovan Kiry Kim, Sina Puch, and Chanpen Saengtienchai. 2010. "Community Reaction to Older Age Parental AIDS Caregivers and Their Families: Evidence From Cambodia." *Research on Aging* 32(1):122–51.

Knoester, Chris and Alan Booth. 2000. "Barriers to Divorce: When Are They Effective? When Are They Not?" *Journal of Family Issues* 21:78–99.

Knoester, Chris and Dana L. Haynie. 2005. "Community Context, Social Integration into Family, and Youth Violence." *Journal of Marriage and Family* 67:767–80.

Knoester, Chris, Dana L. Haynie, and Crystal M. Stephens. 2006. "Parenting Practices and Adolescents' Friendship Networks." *Journal of Marriage and Family* 68(5):1247–60.

Knoester, Chris, Richard J. Petts, and David J. Eggebeen. 2007. "Commitments to Fathering and the Well-Being and Social Participation of New, Disadvantaged Fathers." *Journal of Marriage and Family* 69(4):991–1004.

Knox, David and Marty E. Zusman. 2009. "Sexuality in Black and White: Data From 783 Undergraduates." *Electronic Journal of Human Sexuality* 12, June 26. www.ejhs.org.

Knox, Noelle. 2004. "Orphans Caught in the Middle." *USA Today*, May 18.

Knudson-Martin, Carmen and Martha J. Laughlin. 2005. "Gender and Sexual Orientation in Family Therapy: Toward a Postgender Approach." *Family Relations* 54(1):101–15.

Knudson-Martin, Carmen and Anne Rankin Mahoney. 1998. "Language and Processes in the Construction of Equality in New Marriages." *Family Relations* 47(1):81–91.

———. 2009. Couples, Gender, and Power: Creating Change in Intimate Relationships. New York: Springer Publishing Company.

Koch, Wendy. 2006. "Creative Efforts Ensure Parents Pay." USA Today, March 15.

———. 2009. "Savings Plan Benefits Teens Leaving Foster Care." USA Today, June 15. Retrieved February 10, 2010 (www.usatoday.com).

Koepke, Leslie. 2007. "A Call to Action: Five Policy Proposals on Behalf of All Families—President's Address to the 2007 Groves Conference on Marriage and Family." Journal of Feminist Family Therapy 19(3):1–12.

Koesten, Joy, Paul Schrodt, and Debra Ford. 2009. "Cognitive Flexibility as a Mediator of Family Communication Environments and Young Adults' Well-Being." Health Communication 24:82–94.

Kohlberg, Lawrence. 1966. "A Cognitive–Developmental Analysis of Children's Sex-role Concepts and Attitudes." Pp. 82–173 in *The Development of Sex Differences*, edited by Eleanor E. Maccoby. Palo Alto, CA: Stanford University Press.

Kolata, Gina. 2002. "Parenthood Help for Men with HIV." *New York Times*, April 30.

———. 2009. "Picture Emerging on Genetic Risks of IVF." New York Times, February 17, D1.

Komsi, Niina, Katri Raikkonen, Anu-Katriina Pesonen, Kati Heinonen, Pertti Keskivaara, Anna-liisa Japvenpaa, and Timo E. Strandberg. 2006. "Continuity of Temperament from Infancy to Middle Childhood." *Infant Behavior and Development* 29(4):494–508.

Komter, A. 1989. "Hidden Power in Marriage." *Gender and Society* 3:187–216.

Konner, Melvin. 1990. "Women and Sexuality." *New York Times Magazine*, April 29, pp. 24, 26.

Kools, S. 2008. "From Heritage to Postmodern Grounded Theorizing: Forty Years of Grounded Theory." Studies in Symbolic Interaction 32:73–86.

Koro-Ljungberg, Mirka, and Regina Bussing. 2009. "The Management of Courtesy Stigma in the Lives of Families with Teenagers with ADHD." Journal of Family Issues 30(9):1175–1200.

Koropeckyj-Cox, Tanya and Gretchen Pendell. 2007. "The Gender Gap in Attitudes About Childlessness in the United States." Journal of Marriage & Family 69(4):899–915.

Kosmin, Barry A., Egon Mayer, and Ariela Keysar. 2001. "American Religious Identification Survey, 2001." New York and Hartford, CT: Graduate Center of the City University of New York and the Institute for the Study of Secularism in Society. Retrieved July 5, 2007 (www.trincoll.edu/secularisminstitute).

Kossinets, Gueorgi and Duncan J. Watts. 2009. "Origins of Homophily in an Evolving Social Network." American Journal of Sociology 115(2):405–50.

Kost Kathryn, Stanley Henshaw, and Liz Carlin. 2010. U.S. Teenage Pregnancies, Births and Abortions: National and State Trends and Trends by Race and Ethnicity. New York: Guttmacher Institute. www.guttmacher.org.

Kouros, Chrystyna, Christine Merrilees, and E. Mark Cummings. 2008. "Marital Conflict and Children's Emotional Security in the Context of Parental Depression." Journal of Marriage and Family 70(3):684–97.

Krampe, Edythe M. 2009. "When Is the Father Really There? A Conceptual Reformulation of Father Presence." Journal of Family Issues 30(7):875–97.

Kravets, David. 2005. "Custody, Support Applied to Gays in Calif." USA Today, August 23.

Kreider, Rose M. 2003. Adopted Children and Stepchildren: 2000. Census 2000 Special Reports CENSR-6. August. Washington, DC: U.S. Census Bureau.

———. 2005. Number, Timing, and Duration of Marriages and Divorces: 2001. Current Population Reports P70-97. Washington, DC: U.S. Census Bureau.

Kreider, Rose M., and Dianna B. Elliott. 2009a. America's Families and Living Arrangements: 2007. Current Population Reports, P20-561. U.S. Census Bureau, Washington, DC.

———. 2009b. "The Complex Living Arrangements of Children and Their Unmarried Parents." Population Association of America 2009 Poster Presentation, Detroit, May 2.

Kreider, Rose M. and Jason M. Fields. 2005. Living Arrangements of Children: 2001. Current Population Reports P70-104. Washington, DC: U.S. Census Bureau.

Krementz, Jill. 1984. How It Feels When Parents Divorce. New York: Knopf.

Krivo, Lauren J. and Robert L. Kaufman. 2004. "Housing and Wealth Inequality: Racial-Ethnic Differences in Home Equity in the United States." Demography 41(3):585–605.

Kroneman, Leoniek, Rolf Loeber, Alison Hipwell, and Hans Koot. 2009. "Girls' Disruptive Behavior and its Relationship to Family Functioning: A Review." Journal of Child & Family Studies 18(3):259–73.

Kroska, Amy. 1997. "The Division of Labor in the Home: A Review and Reconceptualization." Social Psychology Quarterly 60(4):304–22.

Krueger, Alan. B. 2004. "Economic Scene." New York Times, April 29.

Krugman, Paul. 2006. "The Great Wealth Transfer." Rolling Stone, December 14, pp. 44–48.

Kuczynski, Alex. 2001. "Guess Who's Coming to Dinner Now?" New York Times, December 23.

Kulczycki, Andrzej and Arun Peter Lobo. 2002. "Patterns, Determinants, and Implications of Intermarriage among Arab Americans." Journal of Marriage and Family 64(1):202–10.

Kulkin, Heidi S., June Williams, Heath F. Borne, Dana de la Bretonne, Judy Laurendine. 2008. "A Review of Research on Violence in Same-Gender Couples." Journal of Homosexuality 53(4):71–87.

Kumpfer, Karol and Connie Tait. 2000. Family Skills Training for Parents and Children. Juvenile Justice Bulletin. April.

Kuperminc, Gabriel P., Gregory J. Jurkovic, and Sean Casey. 2009. "Relation of Filial Responsibility to the Personal and Social Adjustment of Latino Adolescents from Immigrant Families." Journal of Family Psychology 23(1):14–22.

Kurcinka, Mary S. 2006. Raising Your Spirited Child. New York: Harper.

Kurdek, Lawrence A. 1991. "The Relations between Reported Well-Being and Divorce History, Availability of a Proximate Adult, and Gender." Journal of Marriage and Family 53(1):71–78.

———. 1994. "Areas of Conflict for Gay, Lesbian, and Heterosexual Couples: What Couples Argue About Influences Relationship Satisfaction." Journal of Marriage and Family 56(4):923–34.

———. 2005. "Gender and Marital Satisfaction Early in Marriage: A Growth Curve Approach." Journal of Marriage and Family 67(1):68–84.

———. 2006. "Differences between Partners from Heterosexual, Gay, and Lesbian Cohabiting Couples." Journal of Marriage and Family 68(2):509–28.

———. 2007. "The Allocation of Household Labor by Partners in Gay and Lesbian Couples." Journal of Family Issues 28(1):132–48.

Kurdek, Lawrence A. and Mark A. Fine. 1991. "Cognitive Correlates of Satisfaction for Mothers and Stepfathers in Stepfather Families." Journal of Marriage and Family 53(3):565–72.

Kurtz, Stanley. 2003. "Beyond Gay Marriage: The Road to Polygamy." The Weekly Standard, August 11. Retrieved November 2, 2009 (www.weeklystandard.com).

———. 2006. "Big Love, from the Set." National Review, March 13.

Kurz, Demie. 1993. "Physical Assaults by Husbands: A Major Social Problem." Pp. 88–103

in Current Controversies on Family Violence, edited by Richard J. Gelles and Donileen R. Loseke. Newbury Park, CA: Sage Publications.

Kutner, Lawrence. 1988. "Parent and Child: Working at Home; or, The Midday Career Change." New York Times, December 8.

Labaton, Vivien and Dawn Lundy Martin. 2004. The Fire This Time: Young Activists and the New Feminism. New York: Random House.

Lacey, Mary. 2006. "With More Americans Adopting, Guatemala System Is Questioned." New York Times, November 5.

Lacey, Rachel Saul, Alan Reifman, Jean Pearson Scott, Steven M. Harris, and Jacki Fitzpatrick. 2004. "Sexual-Moral Attitudes, Love Styles, and Mate Selection." Journal of Sex Research 41(2):121–29.

Lacy, Karyn R. 2007. Blue-Chip Black: Race, Class, and Status in the New Black Middle Class. Berkeley, CA: University of California Press.

Lacy, Karyn and Angel L. Harris. 2008. "Breaking the Class Monolith: Understanding Class Differences in Black Adolescents' Attachment to Racial Identity." Pp. 152–78 in Social Class: How Does It Work?, edited by Annette Lareau and Dalton Conley. New York: The Russell Sage Foundation.

Laff, Michael. 2007. The Invisible Wall: Workplace Barriers and Self-Image Hinder Women's Leadership Advancement. Alexandria, VA: American Society for Training & Development.

Lally, Catherine F. and James W. Maddock. 1994. "Sexual Meaning Systems of Engaged Couples." Family Relations 43:53–60.

Lam, Brian Trung. 2005. "Self-Esteem Among Vietnamese American Adolescents: The Role of Self-Construal, Family Cohesion, and Social Support." Journal of Ethnic and Cultural Diversity in Social Work 14(3/4):21–34.

Lamb, Kathleen A., Gary R. Lee, and Alfred DeMaris. 2003. "Union Formation and Depression: Selection and Relationship Effects." Journal of Marriage and Family 65(4):953–62.

Lambert, Nathaniel M. and David C. Dollahite. 2006. "How Religiosity Helps Couples Prevent, Resolve, and Overcome Marital Conflict." Family Relations 55(4):439–49.

Lan, Pei-Chia. 2002. "Subcontracting Filial Piety: Elder Care in Ethnic Chinese Immigrant Families in California." Journal of Family Issues 23(7):812–35.

Landale, Nancy S., Robert Schoen, and Kimberly Daniels. 2009. "Early Family Formation Among White, Black, and Mexican American Women." Journal of Family Issues 31(4):445–74.

Landau, Iddo. 2008. "Problems with Feminist Standpoint Theory in Science Education." Science & Education 17(10):1081–8.

Landry-Meyer, Laura and Barbara M. Newman. 2004. "An Exploration of the Grandparent Caregiver Role." Journal of Family Issues 25(8):1005–25.

Lane, Wendy G., David M. Rubin, Ragin Monteith, and Cindy Christian. 2002. "Racial Differences in the Evaluation of Pediatric

Fractures for Physical Abuse." *Journal of the American Medical Association* 288(13). (www. jama.org).

Laner, Mary Riege. 2003. "A Rejoinder to Mark Cresswell." *Sociological Inquiry* 73:152–56.

Langston, Donna. 2007. "Tired of Playing Monopoly?" Pp. 118–27 in *Race, Class & Gender: An Anthology*, 6th ed., edited by Margaret L. Andersen and Patricia Hill Collins. Belmont, CA: Thomson Wadsworth.

Lannutti, Pamela J. 2007. "The Influence of Same-Sex Marriage on the Understanding of Same-Sex Relationships." *Journal of Homosexuality* 53(3):135–51.

Lansford, J. E. and K. A. Dodge. 2008. "Cultural Norms for Adult Corporal Punishment of Children and Societal Rates of Endorsement and Use of Violence." *Parenting Science & Practice* 8(3):257–70.

LaPierre, Tracey A. 2009. "Marital Status and Depressive Symptoms Over Time: Age and Gender Variations." *Family Relations* 58(4):404–16.

Lara, Teena and Jill Duba Onedera. 2008. "Inter-Religion Marriages." Pp. 213–28 in *The Role of Religion and Marriage and Family Counseling*, edited by Jill Duba Onedera. New York: Taylor and Francis.

Lareau, Annette. 2003a. "The Long-Lost Cousins of the Middle Class." *New York Times*, December 20.

———. 2003b. *Unequal Childhoods: Class, Race, and Family Life*. Berkeley, CA: University of California Press.

———. 2006. "Unequal Childhoods: Class, Race, and Family Life." Pp. 537–48 in *The Inequality Reader: Contemporary and Foundational Readings in Class, Race, and Gender*, edited by David B. Grusky and Szonja Szelenyi. Boulder, CO: Westview.

LaRossa, Ralph. 1979. *Conflict and Power in Marriage: Expecting the First Child*. Newbury Park, CA: Sage Publications.

———. 2009. "Single-Parent Family Discourse in Popular Magazines and Social Science Journals." *Journal of Marriage and Family* 71(2):235–39.

Larson, Jeffry H. and Rachel Hickman. 2004. "Are College Marriage Textbooks Teaching Students the Premarital Predictors of Marital Quality?" *Family Relations* 53(4):385–92.

Larzelere, Robert E. 2008. "Disciplinary Spanking: The Scientific Evidence." Journal of Developmental and Behavioral Pediatrics (JDBP) 29(4):334–35.

Lasch, Christopher. 1977. *Haven in a Heartless World: The Family Besieged*. New York: Basic Books.

———. 1980. *The Culture of Narcissism*. New York: Warner Books.

Latham, Melanie. 2008. "The Shape of Things To Come: Feminism, Regulation and Cosmetic Surgery." *Medical Law Review* 16:437–57.

Lau, Anna S., David T. Takeuchi, and Margarita Alegria. 2006. "Parent-to-Child Aggression

among Asian American Parents: Culture, Context, and Vulnerability." *Journal of Marriage and Family* 68(5):1261–75.

Laub, Gillian. 2005. "The Young and the Sexless." *Rolling Stone*, June 30–July 14, pp. 103–11.

Lauby, Mary R. and Sue Else. 2008. "Recession Can Be Deadly for Domestic Abuse Victims." *The Boston Globe*, December 25. Retrieved June 8, 2009 (bostonglobe.com).

Laughlin, Lynda and Joseph Rukus. 2009. "Who's Minding the Kids in the Summer? Child Care Arrangements for Summer 2006." Presented at the Annual Meeting of the *Population Association of America*, April 30–May 2.

Laumann, Edward, Aniruddha Das, and Linda Waite. 2008. "Sexual Dysfunction among Older Adults: Prevalence and Risk Factors from a Nationally Representative U.S. Probability Sample of Men and Women 57–85 Years of Age." *Journal of Sexual Medicine* 5(10):2,300–2,311.

Laumann, Edward, John H. Gagnon, Robert T. Michael, and Stuart Michaels. 1994. *The Social Organization of Sexuality: Sexual Practices in the United States*. Chicago, IL: University of Chicago Press.

Laumann, Edward, Sara Leitsch, and Linda Waite. 2008. "Elder Mistreatment in the United States: Prevalence Estimates from a Nationally Representative Study." *The Journals of Gerontology Series B: Social Sciences* 63(4):S248–54.

Lauro, Patricia Winters. 2000. "Advertising: The Subject of Divorce Is Becoming More Common as Another Backdrop in Campaigns." *New York Times*, October 12.Lavee, Yoav and Ruth Katz. 2002. "Division of Labor, Perceived Fairness, and Marital Quality: The Effect of Gender Ideology." *Journal of Marriage and Family* 64:27–39.

Lavietes, Stuart. 2003. "Richard Gardner, 72, Dies; Cast Doubt on Abuse Claims." *New York Times*, June 9.

Lavin, Judy. 2003. "Smoothing the Step-Parenting Transition." SelfGrowth.com. Retrieved April 26, 2007 (www.selfgrowth .com).

Lavy, Shiri, Mario Mikulincer, Phillip R. Shaver, and Omri Gillath. 2009. "Intrusiveness in Romantic Relationships: A Cross-cultural Perspective on Imbalances Between Proximity and Autonomy." *Journal of Social and Personal Relationships* 26(6–7):989–1008.

Lawrance, K. and E. S. Byers. 1995. "Sexual Satisfaction in Long-Term Heterosexual Relationships: The Interpersonal Exchange Model of Sexual Satisfaction." *Personal Relationships* 2:267–85.

Lawrence et al. v. Texas. 2003. 539 U.S. 558.

Lawson, Willow. 2004a. "Encouraging Signs: How Your Partner Responds to Your Good News Speaks Volumes." *Psychology Today* (January/February):22.

———. 2004b. "The Glee Club: Positive Psychologists Want to Teach You to Be Happier." *Psychology Today* (January/February):34–40.

LeBlanc, Allen J. and Richard G. Wright. 2000. "Reciprocity and Depression in AIDS Caregiving." *Sociological Perspectives* 43(4):631–49.

LeBlanc, Steve. 2006a. "Boston Catholic Charities Halts Adoptions." Associated Press. *San Francisco Chronicle*, March 10.

———. 2006b. "Massachusetts Lawmakers Delay Action on Gay Marriage Ban." Associated Press. Retrieved November 10, 2006 (http://www. boston.com/news).

Le, Thao N. 2005. "Narcissism and Immature Love As Mediators of Vertical Individualism and Ludic Love Style." *Journal of Social and Personal Relationships* 22(4):543–60.

Ledbetter, Andrew M. 2009. "Family Communication Patterns and Relational Maintenance Behavior: Direct and Mediated Associations with Friendship Closeness." *Human Communication Research* 36:130–47.

Ledermann, Thomas, Guy Bodenmann, Myriam Rudaz, and Thomas N. Bradbury. 2010. "Stress, Communication, and Marital Quality in Couples." *Family Relations* 59(2):195–206.

Lee, Arlene F., Philip M. Genty, and Mimi Laver. 2005. *The Impact of the Adoption and Safe Families Act on Children of Incarcerated Parents*. Washington, DC: Child Welfare League of America.

Lee, Cameron and Judith Iverson-Gilbert. 2003. "Demand, Support, and Perception in Family-related Stress among Protestant Clergy." *Family Relations* 52(3):249–57.

Lee, Carol E. 2006. "Sibling Seeks Same to Share Apartment." *New York Times*, January 29.

Lee, Chu-Yuan, Jared Anderson, Jason Horowitz, and Gerald August. 2009. "Family Income and Parenting: The Role of Parental Depression and Social Support." *Family Relations* 58(October):417–30.

Lee, Cynthia. 2008. "The Gay Panic Defense." *U.C. Davis Law Review* 42:471–566.

Lee, Eunju, Glenna Spitze, and John R. Logan. 2003. "Social Support to Parents-in-Law: The Interplay of Gender and Kin Hierarchies." *Journal of Marriage and Family* 65(2):396–403.

Lee, F. R. 2001. "Trying to Soothe the Fears Hiding Behind the Veil." *New York Times*, September 23.

Lee, Gary R., Julie K. Netzer, and Raymond T. Coward. 1994. "Filial Responsibility Expectations and Patterns of Intergenerational Assistance." *Journal of Marriage and Family* 56(3):559–65.

Lee, Gary R., Chuck W. Peek, and Raymond T. Coward. 1998. "Race Differences in Filial Responsibility Expectations among Older Parents." *Journal of Marriage and Family* 60(2):404–12.

Lee, Jacrim, Mary Jo Katras, and Jean W. Bauer. 2009. "Children's Birthday Celebrations from the Experiences of Low-Income Rural Mothers." *Journal of Family Issues* 30(4):532–53.

Lee, Jennifer. 2007. "The Incredible Flying Granny Nanny." *New York Times*, May 10. Retrieved May 11, 2007 (www.nytimes.com).

Lee, John Alan. 1973. *The Colours of Love.* Toronto: New Press.

———. 1981. "Forbidden Colors of Love: Patterns of Gay Love." Pp. 128–39 in *Single Life: Unmarried Adults in Social Context,* edited by Peter J. Stein. New York: St. Martin's.

Lee, Michael S. 2005. Healing the Nation: The Arab American Experience After September 11. Washington, DC: The Arab American Institute.

Lee, Mo-Yee. 2002. "A Model of Children's Postdivorce Behavioral Adjustment in Maternal- and Dual-residence Arrangements." *Journal of Family Issues* 23(5):672–97.

Lee, Sharon M. 1998. "Asian Americans: Diverse and Growing." *Population Bulletin* 53(2). Washington, DC: Population Reference Bureau.

Lee, Sharon M. and Barry Edmonston. 2005. "New Marriages, New Families: U.S. Racial and Hispanic Intermarriage." *Population Bulletin* 60(2). Washington, DC: Population Reference Bureau.

Lee, Shawna J., Jennifer L. Bellamy, and Neil B. Guterman. 2009. "Fathers, Physical Child Abuse, and Neglect." *Child Maltreatment* 14(3):227–31.

Lee, Wai-Yung, Man-Lun Ng, Ben Cheung, and Joyce Wayung. 2010. "Capturing Children's Response to Parental Conflict and Making Use of It." *Family Process* 49(1):43–58.

Lees, Janet and Jan Horwath. 2009. " 'Religious Parents . . . Just Want the Best for Their Kids': Young People's Perspectives on the Influence of Religious Beliefs on Parenting." *Children & Society* 23(3):162–75.

Leff, Lisa. 2006. "State Lawyer Faces Tough Questions Arguing Against Gay Marriage." Associated Press, July 11. Retrieved July 11, 2006 (www.mercurynews.com).

———. 2009. "Some Gays Seek Renewed Focus on Civil Unions." The Washington Post, November 28. Retrieved December 1, 2009. (www.washingtonpost.com).

"Legislation Introduced to Repeal Discriminatory Defense of Marriage Act." 2009. American Civil Liberties Union, September 15. Retrieved September 29, 2009 (www.aclu.org).

Lehmann-Haupt, Rachel. 2009. "Why I Froze My Eggs." Newsweek, May 18, pp. 50–52.

Leidy, Melinda, Ross D. Parke, Mina Cladis, Scott Coltrane, and Sharon Duffy. 2009. "Positive Marital Quality, Acculturative Stress, and Child Outcomes among Mexican Americans." *Journal of Marriage and Family* 71(4):833–47.

Leigh, Suzanne. 2004. "Fertility Patients Deserve to Know the Odds—and Risks." *USA Today,* July 7.

Leisenring, Amy. 2008. "Controversies Surrounding Mandatory Arrest Policies and the Police Response to Intimate Partner Violence." *Sociology Compass* 2(2):451–66.

Leisey, Monica, Paul Kupstas, and Aly Cooper. 2009. "Domestic Violence in the Second Half of Life." Journal of Elder Abuse and Neglect 21(2):141–55.

Leite, Randall. 2007. "An Exploration of Aspects of Boundary Ambiguity among Young, Unmarried Fathers during the Prenatal Period." *Family Relations* 56(2):162–74.

Leland, John. 2006. "A Spirit of Belonging, Inside and Out." *New York Times,* October 8.

———. 2010. "For a Graying Population, Care Increasingly Comes From a Graying Work Force." *The New York Times,* April 25, A14.

Leman, Kevin. 2006. *Single Parenting That Works: Six Keys to Raising Happy, Healthy Children in a Single-Parent Home.* Carol Stream, IL: Tyndale House Publishers.

Lento, Jennifer. 2006. "Relational and Physical Victimization by Peers and Romantic Partners in College Students." *Journal of Social and Personal Relationships* 23(3):331–48.

Leon, Kim. 2009. "Covenant Marriage: What Is It and Does It Work?" *Relationships.* Retrieved October 12, 2009 (missourifamilies.org).

Leone, Janel M., Michael P. Johnson, Catherine L. Cohan. 2007. "Victim Help Seeking: Differences between Intimate Terrorism and Situational Couple Violence." *Family Relations* 56(5):427–39.

Leong, Nancy. 2006. "Multiracial Identity and Affirmative Action." Paper 1126. Berkeley, CA: The Berkeley Electronic Press (bepress).

Leonhardt, David. 2003. "It's a Girl! (Will the Economy Suffer?)" *New York Times,* October 26.

———. 2006a "The New Inequality." *New York Times Magazine,* December 10.

———. 2006b. "Scant Progress in Closing Gap." *New York Times,* December 24.

Lepper, John M. 2009. *When Crisis Comes Home.* Macon, GA: Smyth & Helwys Publishers.

Lerner, Barron M. 2003. "If Biology Is Destiny, When Shouldn't It Be?" *New York Times,* May 27.

Lerner, Harriet. 2001. *The Dance of Connection: How to Talk to Someone When You're Mad, Hurt, Scared, Frustrated, Insulted, Betrayed, or Desperate.* New York: HarperCollins.

Lerner, Janet W. and Beverly Johns. 2009. *Learning Disabilities and Related Mild Disabilities: Characteristics, Teaching Strategies and New Directions.* Boston: Houghton Mifflin.

Lessane, Patricia Williams. 2007. "Women of Color Facing Feminism—Creating Our Space at Liberation's Table: A Report on the Chicago Foundation for Women's "F" Series." *The Journal of Pan African Studies* 1(7):3–10.

Letarte, Marie-Josée, Sylvie Normandeau, and Julie Allard. 2010. "Effectiveness of a Parent Training Program 'Incredible Years' in a Child Protection Service." *Child Abuse & Neglect* 34(4):253–61.

Letiecq, Bethany L., Sandra J. Bailey, and Marcia A. Kurtz. 2008. "Depression Among Rural Native American and European American Grandparents Rearing Their Grandchildren." Journal of Family Issues 29(3):334–56.

Letiecq, Bethany L., Sandra J. Bailey, and Fonda Porterfield. 2008. " 'We Have No Rights, We Get No Help': The Legal and Policy Dilemmas Facing Grandparent Caregivers." Journal of Family Issues 29(8):995–1012.

Levaro, Liz Bayler. 2009. "Living Together or Living Apart Together: New Choices for Old Lovers." Family Focus (Summer):F9–F10. Minneapolis, MN: National Council on Family Relations.

Levin, Irene. 1997. "Stepfamily as Project." Pp. 123–33 in *Stepfamilies: History, Research, and Policy,* edited by Irene Levin and Marvin B. Sussman. New York: Haworth.

Levine, Carol. 2008. "Family Caregiving." Pp. 63–68 in *From Birth to Death and Bench to Clinic: The Hastings Center Bioethics Briefing Book for Journalists, Policymakers, and Campaigns,* edited by Mary Crowley. Garrison, NY: The Hastings Center. Retrieved May 17, 2010 (www. thehastingscenter.org).

Levine, Judith A., Clifton R. Emery, and Harold Pollack. 2007. "The Well-Being of Children Born to Teen Mothers." *Journal of Marriage and Family* 69(1):105–22.

Levine, Kathryn. 2009. "Against All Odds: Resilience in Single Mothers of Children with Disabilities." *Social Work in Health Care* 48(4):402–19.

Levine, Madeline. 2006. *The Price of Privilege: How Parental Pressure and Material Advantage Are Creating a Generation of Disconnected and Unhappy Kids.* New York: HarperCollins.

Levine, Robert, Suguru Sato, Tsukasa Hashimoto, and Jyoti Verma. 1995. "Love and Marriage in Eleven Cultures." *Journal of Cross-Cultural Psychology* 26(5):554–71.

Levinger, George. 1965. "Marital Cohesiveness and Dissolution: An Integrative Review." *Journal of Marriage and Family* 27:19–28.

———. 1976. "A Social Psychological Perspective on Marital Discord." *Journal of Social Issues* 32:21–47.

Levy, Donald P. 2005. "Hegemonic Complicity, Friendship, and Comradeship: Validation and Causal Processes among White, Middle-class, Middle-aged Men." *Journal of Men's Studies* 13(2):199–225.

Lewin, Tamar. 2001a. "Report Looks at a Generation, and Caring for Young and Old." *New York Times,* July 11.

———. 2001b. "Study Says Little Has Changed." *New York Times,* September 10.

———. 2003. "For More People in Their 20s and 30s, Going Home Is Easier Because They Never Left." *New York Times,* December 22.

———. 2005. "A Marriage of Unequals: When Richer Weds Poorer, Money Isn't the Only Difference." *New York Times,* May 19.

———. 2006a. "At Colleges, Women Are Leaving Men in the Dust." *New York Times,* July 9.

———. 2006b. "Boys Are No Match for Girls in Completing High School." *New York Times,* April 19.

———. 2006c. "A More Nuanced Look at Men, Women, and College." *New York Times,* July 12.

———. 2006d. "Unwed Fathers Fight for Babies Placed for Adoption by Mothers." *New York Times,* March 19.

Lewin, Ellen. 2009. *Gay Fatherhood: Narratives of Family and Citizenship in America.* Chicago, IL: University of Chicago Press.

Lewis, Jane. 2008. "Childcare Policies and the Politics of Choice." *The Political Quarterly* 79(4):499–507.

Lewis, Robert. 2007. *Raising a Modern-Day Knight.* Carol Stream, IL: Tyndale House Publishers.

Lewis, Thomas, M.D., Fari Amini, M.D., and Richard Lannon, M.D. 2000. *A General Theory of Love.* New York: Random House.

Lewontin, Richard. 1993. *Biology as Ideology: The Doctrines of DNA.* New York: Harper Perennial.

———. 2004. "Dishonesty in Science." *New York Review of Books,* November 18, pp. 38–39.

Lewontin, Richard and Richard Levins. 2007. *Biology Under the Influence: Dialectical Essays on Ecology, Agriculture, and Health.* New York: Monthly Review Press.

Li, Jui-Chung Allen, and Lawrence L. Wu. 2008. "No Trend in the Intergenerational Transmission of Divorce." *Demography* 45(4):875–83.

Lichter, Daniel T., J. Brian Brown, Zhenchao Qian, and Julie H. Carmalt. 2007. "Marital Assimilation among Hispanics: Evidence of Declining Cultural and Economic Incorporation?" *Social Science Quarterly* 88(3):745–65.

Lichter, Daniel T. and Julie H. Carmalt. 2009. "Cohabitation and the Rise in Out-of-Wedlock Childbearing." *Family Focus* (Summer):F11–F13. Minneapolis, MN: National Council on Family Relations.

Lichter, Daniel T. and Zhenchao Qian. 2004. *Marriage and Family in a Multiracial Society.* New York: Russell Sage.

———. 2008. "Serial Cohabitation and the Marital Life Course." *Journal of Marriage and Family* 70(4):861–78.

Lichter, Daniel T., Zhenchao Qian, and Leanna M. Mellott. 2006. "Marriage or Dissolution? Union Transitions among Poor Cohabiting Women." *Demography* 43(2):223–41.

Liefbroer, Aart C. and Edith Dourleijn. 2006. "Unmarried Cohabitation and Union Stability: Testing the Role of Diffusion Using Data from 16 European Countries." *Demography* 43(2):203–21.

Lin, Chien and William T. Liu. 1993. "Relationships among Chinese Immigrant Families." Pp. 271–86 in *Family Ethnicity: Strength in Diversity,* edited by Harriette Pipes McAdoo. Newbury Park, CA: Sage Publications.

Lincoln, Karen, Robert Taylor, and James Jackson. 2008. "Romantic Relationships among Unmarried African Americans and Caribbean Blacks: Findings from the National Survey of American Life." *Family Relations* 57(April):254–66.

Lindau, Stacy Tessler and Natalia Gavrilova. 2010. "Sex, Health, and Years of Sexually Active Life Gained Due to Good Health: Evidence from Two U.S. Population Based Cross-Sectional Surveys of Ageing." *British Medical Journal* 340(7746):580.

Lindau, Stacy Tessler, L. Philip Schumm, Edward O. Laumann, Wendy Levinson, Colm A. O'Muircheartaigh, and Linda J. Waite. 2007. "A Study of Sexuality and Health among Older Adults in the United States." *New England Journal of Medicine* 357(8):762–74.

Lindberg, Laura Duberstein, John S. Santelli, and Susheela Singh. 2006. "Changes in Formal Sex Education: 1995–2002. *Perspectives on Sexual and Reproductive Health* 38(4):182–89.

Lindner Gunnoe, Marjorie, E. Mavis Hetherington, and David Reiss. 2006. *Journal of Family Psychology* 20(4):589–96.

Lindsey, Elizabeth W. 1998. "The Impact of Homelessness and Shelter Life on *Family Relation*ships." *Family Relations* 47(3):243–52.

Lindsey, Eric W., Yvonne M. Caldera, and Laura Tankersley. 2009. "Marital Conflict and the Quality of Young Children's Peer Play Behavior: The Mediating and moderating Role of Parent-Child Emotional Reciprocity and Attachment Security." *Journal of Family Psychology* 23(2):130–45.

Lindsey, Eric W., Jessico Campbell Chambers, James Frabutt, and Carol Mackinnon-Lewis. 2009. "Marital Conflict and Adolescents' Peer Aggression: The Mediating and Moderating Role of Mother-Child Emotional Reciprocity." *Family Relations* 58(December):593–606.

Lino, Mark. 2008. "Expenditures on Children by Families, 2007." Annual Report. Washington, DC: U.S. Department of Agriculture, Center for Nutrition Policy and Promotion. Miscellaneous Publication No. 1528-2007.

Lips, Hilary M. 2004. *Sex and Gender: An Introduction.* New York: McGraw-Hill.

Liptak, Adam. 2002. "Judging a Mother for Someone Else's Crime." *New York Times,* November 27.

Little, Betsi and Cheryl Terrance. 2010. "Perceptions of Domestic Violence in Lesbian Relationships: Stereotypes and Gender Role Expectations." *Journal of Homosexuality* 57(3):429–40.

Liu, Chien. 2000. "A Theory of Marital Sexual Life." *Journal of Marriage and Family* 62(2): 363-74.

Liu, Hesheng, Steven M. Stufflebeam, Jorge Sepulcre, Trey Hedden, and Randy L. Buckner. 2009. "Evidence From Intrinsic Activity That Asymmetry of the Human Brain Is Controlled By Multiple Factors." *Proceedings of the National Academy of Sciences of the United States of America* 106(48):20499–503.

Liu, Hui. 2009. "Till Death Do Us Part: Marital Status and U.S. Mortality Trends, 1986–2000." Journal of Marriage and Family 71(December):1158–73.

Lleras, Christy. 2008. "Employment, Work Conditions, and the Home Environment in Single-Mother Families." *Journal of Family Issues* 29(10):1268–97.

Lloyd, Kim M. 2006. "Latinas' Transition to First Marriage: An Examination of Four Theoretical Perspectives." *Journal of Marriage and Family* 68(4):993–1014.

Lloyd, Sally A., April L. Few, and Katherine R. Allen. 2007. "Feminist Theory, Methods, and Praxis in Family Studies: An Introduction to the Special Issue." *Journal of Family Issues* 28(4):447–51.

LoBiondo-Wood, Geri, Laurel Williams, and Charles McGhee. 2004. "Liver Transplantation in Children: Maternal and Family Stress, Coping, and Adaptation." *Journal of the Society of Pediatric Nurses* 9(2):59–67.

Lobo, Susan, ed. 2001. *American Indians and the Urban Experience.* Thousand Oaks, CA: AltaMira.

Lofas, Jeannette. n.d. "Ten Steps for Steps." The Stepfamily Foundation (www.stepfamily.org/ ten_steps_for_stepfamilies.html).

Loftin, Colin, David McDowall, and Matthew Fetzer. 2008. "The Accuracy of Supplementary Homicide Report Data for Large U.S. Cities." Paper presented at the Annual Meeting of the American Society of Criminology, November 12. St. Louis Adam's Mark, St. Louis, Missouri.

Loftus, Jeni. 2001. "America's Liberalization in Attitudes toward Homosexuality, 1973 to 1998." *American Sociological Review* 66:762–82.

Logan, Cassandra, Emily Holcombe, Jennifer Manlove, and Suzanne Ryan. 2007. *The Consequences of Unintended Childbearing: A White Paper.* Washington, DC: Child Trends.

Lohr, Steve. 2006. "Study Plays Down Export of Computer Jobs." *New York Times,* February 23.

Lohrer, Steven P., Ellen P. Lukens, and Helle Thorning. 2007. "Adult Siblings of Persons with Severe Mental Illness." *Community Mental Health Journal* 43(2):129–51.

Lois, Jennifer. 2010. "Gender and Emotion Management in the Stages of Edgework." Pp 333–44 in *The Kaleidoscope of Gender:* Prisms, Patterns, and Possibilities, 3rd ed., edited by Joan Z. Spade and Catherine G. Valentine. Newbury Park, CA: Pine Forge Press.

Long, George. 1875. "Patria Potestas." Pp. 873–75 in *A Dictionary of Greek and Roman Antiquities,* edited by Sir William Smith, William Wayte, and G. E. Marindin. London, England: J. Murray. Retrieved October 6, 2006 (penelope.uchicago. edu).

Longman, Phillip J. 1998. "The Cost of Children." *U.S. News & World Report,* March 30, pp. 51–58.

Longmore, Monica A., Abbey L. Eng, Peggy C. Giordano, and Wendy D. Manning. 2009. "Parenting and Adolescents' Sexual Initiation." *Journal of Marriage and Family* 71(4):969–82.

Lopez, Mark Hugo. 2009. *Latinos and Education: Explaining the Attainment Gap* Washington, DC: Pew Hispanic Center, October 7.

Lopez, Mark Hugo, Gretchen Livingston, and Rakesh Kochhar. 2009. *Hispanics and the Economic Downturn: Housing Woes and Remittance Cuts.* Washington, DC: Pew Hispanic Center, January 8.

Loseke, Donileen R., Richard J. Gelles, and Mary M. Cavanaugh, eds. 2005. *Current*

Controversies on Family Violence. 2nd ed. Thousand Oaks, CA: Sage Publications.

Loseke, Donileen R. and Demie Kurz. 2005. "Men's Violence toward Women Is the Serious Social Problem." Pp. 79–95 in *Current Controversies on Family Violence*, edited by Donileen R. Loseke, Richard J. Gelles, and Mary M. Cavanaugh. Thousand Oaks, CA: Sage Publications.

Loser, Rachel, Shirley Klein, E. Hill, and David Dollahite. 2008. "Religion and the Daily Lives of LSD Families: An Ecological Perspective." Family and Consumer Sciences Research Journal 37(1):52–70.

Losh-Hesselbart, Susan. 1987. "Development of Gender Roles." Pp. 535–64 in *Handbook of Marriage and the Family*, edited by Marvin B. Sussman and Suzanne K. Steinmetz. New York: Plenum.

Lott, Juanita Tamayo. 2004. "Asian-American Children Are Members of a Diverse and Urban Population." Washington, DC: Population Reference Bureau. January. Retrieved December 13, 2006 (www.prb.org).

Loukas, Alexandra, Hazel Prelow, Marie-Anne Suizzo, and Shane Allua. 2008. "Mothering and Peer Associations Mediate Risk Effects for Latino Youth." *Journal of Marriage and Family* 70(1):76–85.

Love, Patricia. 2001. *The Truth About Love.* New York: Simon and Schuster.

Love, Patricia, and Steven Stosny. 2007. *How To Improve Your Marriage Without Talking about It: Finding Love Beyond Words.* New York: Broadway Books.

Lovell, Vicky., Elizabeth O'Neill, and Skylar Olsen. 2007. "Maternity Leave in the United States." Washington, DC: Institute for Women's Policy Research.

Loving v. Virginia. 1967. 388 U.S. 1, 87 S. Ct. 1817, 18 L.Ed.2d 1010.

"Loving Your Partner as a Package Deal." 2000. *Newsweek*, March 20, p. 78.

Lowe, Chelsea and Bruce M. Cohen. 2010. *Living with Someone Who's Living with Bipolar Disorder: A Practical Guide for Family, Friends, and Coworkers.* San Francisco: Jossey-Bass.

Lubrano, Alfred. 2003. *Blue-Collar Roots, White-Collar Dreams.* New York: Wiley.

Lucas, Kristen. 2007. "Anticipatory Socialization in Blue-Collar Families: The Social Mobility-Reproduction Dialectic." Paper presented at the annual meeting of the International Communication Association, San Francisco CA, May 23. Retrieved January 19, 2010 (www .allacademic.com).

Luepnitz, Deborah Anne. 1982. *Child Custody: A Study of Families After Divorce.* Lexington, MA: Lexington Books.

Lugaila, Terry and Julia Overturf. 2004. *Children and the Households They Live In: 2000.* Census Special Report CENSR-14. Washington, DC: U.S. Census Bureau. February.

Lugo Steidel, Angel G. and Josefina M. Contreras. 2003. "A New Familism Scale for Use with Latino Populations." *Hispanic Journal of Behavioral Sciences* 25(3):312–30.

Luker, Kristin. 1984. *Abortion and the Politics of Motherhood.* Berkeley: University of California Press.

Lumpkin, James R. 2008. "Grandparents in a Parental or Near-Parental Role: Sources of Stress and Coping Mechanisms." *Journal of Family Issues* 29(3):357–72.

Lundberg, Michael. 2009. "Our Parents' Keepers: The Current Status of American Filial Responsibility Laws." *Utah Law Review* 11(2):534–82.

Lundquist, Jennifer Hickes. 2004. "When Race Makes No Difference: Marriage and the Military." *Social Forces* 83(2):731–57.

Lundquist, Jennifer Hickes, Michelle J. Budig, and Anna Curtis. 2009. "Race and Childlessness in America, 1988–2002." *Journal of Marriage & Family* 71(3):741–55.

Lundquist, Jennifer Hickes and Herbert L. Smith. 2005. "Family Formation among Women in the U.S. Military: Evidence from the NLSY." *Journal of Marriage and Family* 67:1–13.

Luscombe, Belinda. 2009. "Facebook and Divorce." *Time,* June 22.

Luster, Tom, Kelly Rhoades, and Bruce Haas. 1989. "The Relation between Parental Values and Parenting Behavior: A Test of the Kohn Hypothesis." *Journal of Marriage and Family* 51:139–47.

Luster, Tom and Stephen A. Small. 1997. "Sexual Abuse History and Problems in Adolescence: Exploring the Effects of Moderating Variables." *Journal of Marriage and Family* 59(1):131–42.

Lustig, Daniel C. 1999. "Family Caregiving of Adults with Mental Retardation: Key Issues for Rehabilitation Counselors." *Journal of Rehabilitation* 65(2):26–45.

Luthar, Suniya S. 2003. "The Culture of Affluence: Psychological Costs of Material Wealth." *Child Development* 74:1581–93.

Luthra, Rohini and Christine A. Gidycz. 2006. "Dating Violence among College Men and Women: Evolution of a Theoretical Model." *Journal of Interpersonal Violence* 21(6):717–31.

Lye, Diane N., Daniel H. Klepinger, Patricia Davis Hyle, and Anjanette Nelson. 1995. "Childhood Living Arrangements and Adult Children's Relations with Their Parents." *Demography* 32(2):261–80.

Lyman, Stanford M. and Marvin B. Scott. 1975. *The Drama of Social Reality.* New York: Oxford University Press.

Lynch, Jean M. and Kim Murray. 2000. "For the Love of the Children: The Coming Out Process for Lesbian and Gay Parents and Stepparents." *Journal of Homosexuality* 39(1):1–24.

Lyngstad, Torkild Hovde and Henriette Engelhardt. 2009. "The Influence of Offspring's Sex and Age at Parents' Divorce on the Intergenerational Transmission of Divorce, Norwegian First Marriages 1980–2003." *Population Studies* 63(2):173–85.

Lyons, Linda. 2003. "Who Can't Get No Satisfaction?" The Gallup Poll, February 18. Retrieved September 13, 2006 (www.galluppoll .com).

———. 2004. "How Many Teens Are Cool with Cohabitation?" The Gallup Poll, April 13. Retrieved September 13, 2006 (www.galluppoll. com).

MacCallum, Catriona and Emma Hill. 2006. "Being Positive About Selection." *PLoS Biology* 4(3):293–5.

Maccoby, Eleanor E. 1998. *The Two Sexes: Growing Up Apart; Coming Together.* Cambridge, MA: Belknap/Harvard University Press.

———. 2002. "Gender and Group Process: A Developmental Perspective." *Current Directions in Psychological Science* 11(2):54–8.

Maccoby, Eleanor E. and Carol Nagy Jacklin. 1974. *The Psychology of Sex Differences.* Stanford, CA: Stanford University Press.

Maccoby, Eleanor E. and Catherine C. Lewis. 2003. "Less Day Care or a Different Day Care?" *Child Development* 74:1069–75.

Maccoby, E. E., and J. A. Martin. 1983. "Socialization in the Context of the Family: Parent-Child Interaction." Pp. 1–101 in *Handbook of Child Psychology*, edited by P. Mussen. New York: Wiley.

Maccoby, Eleanor E. and Robert Mnookin. 1992. *Dividing the Child: Social and Legal Dilemmas of Custody.* Cambridge, MA: Harvard University Press.

MacDonald, William L. and Alfred DeMaris. 2002. "Stepfather-Stepchild Relationship Quality: The Stepfather's Demand for Conformity and the Biological Father's Involvement." *Journal of Family Issues* 23(1):121–37.

MacDorman, Marian F. and T. J. Mathews. 2008. Recent Trends in Infant Mortality in the United States. *NCHS Data Brief* 9. Hyattsville, MD: National Center for Health Statistics, October.

MacFarquhar, Neil. 2006. "It's Muslim Boy Meets Girl, but Don't Call It Dating." *New York Times,* September 19. Retrieved September 19, 2006 (www.nytimes.com).

Machir, John. 2003. "The Impact of Spousal Caregiving on the Quality of Marital Relationships in Later Life." *Family Focus* (September):F11–F13. National Council on Family Relations.

Maciel, Jose A., Zanetta Van Putten, and Carmen Knudson-Martin. 2009. "Gendered Power in Cultural Contexts: Part I. Immigrant Couples." *Family Process* 48(1):9–23.

Macionis, John J. 2006. *Society: The Basics.* 6th ed. Upper Saddle River, NJ: Prentice Hall.

MacInnes, Maryhelen D. 2008. "One's Enough for Now: Children, Disability, and the Subsequent Childbearing of Mothers." *Journal of Marriage and Family* 70(3):758–71.

Mackay, Judith. 2000. *The Penguin Atlas of Human Sexual Behavior.* New York: Penguin Putnam.

Mackey, Richard A., Matthew A. Diemer, and Bernard A. O'Brien. 2000. "Psychological Intimacy in the Lasting Relationships of Heterosexual and Same-gender Couples." *Sex Roles* (August):201–15.

Macklin, Eleanor D. 1987. "Nontraditional Family Forms." Pp. 317–53 in *Handbook of Marriage and the Family*, edited by Marvin B. Sussman and Suzanne K. Steinmetz. New York: Plenum.

MacNeil, Sheila and Sandra E. Byers. 2009. "Role of Sexual Self-Disclosure in the Sexual Satisfaction of Long-Term Heterosexual Couples." *Journal of Sex Research* 46(1):3–14.

MacQuarrie, Brian. 2006. "Guard Families Cope in Two Dimensions." *Boston Globe*, August 30.

Madathil, Jayamala and James M. Benshoff. 2008. "Importance of Marital Characteristics and Marital Satisfaction: A Comparison of Asian Indians in Arranged Marriages and Americans in Marriages of Choice." *The Family Journal: Counseling and Therapy for Couples and Families* 16(3):222–30.

Madden-Derdich, Debra A. and Joyce A. Arditti. 1999. "The Ties That Bind: Attachment between Former Spouses." *Family Relations* 48(3):243–48.

Madrigal, Luke. 2001. "Indian Child and Welfare Act: Partnership for Preservation." *American Behavioral Scientist* 44(9):1505–11.

Maestas, Nicole and Julie Zissimopoulos. 2009. "How Longer Work Lives Ease the Crunch of Population Aging." Working Paper WR-728, December. Rand Center for the Study of Aging.

Magdol, Lynn, Terrie E. Moffitt, Avshalom Caspi, and Phil A. Silva. 1998. "Hitting without a License: Testing Explanations for Differences in Partner Abuse between Young Adult Daters and Cohabitors." *Journal of Marriage and Family* 60(1):41–55.

Magnuson, Katherine and Lawrence Berger. 2009. "Family Structure States and Transitions: Associations with Children's Well-Being During Middle Childhood." *Journal of Marriage and Family* 71(3):575–91.

Mahay, Jenna and Alisa C. Lewin. 2007. "Age and the Desire to Marry." *Journal of Family Issues* 28(5):706–23.

Mahoney, Annette. 2005. "Religion and Conflict in Marital and Parent-Child Relationships." *Journal of Social Issues* 61(4):689–717.

Maier, Thomas. 1998. "Everybody's Grandfather." *U.S. News & World Report*, March 30, p. 59.

Maillard, Kevin Noble. 2008. "The Multiracial Epiphany of *Loving*." *Fordham Law Review* 76(6):2709–32.

Mainemer, Henry, Lorraine C. Gilman, and Elinor W. Ames. 1998. "Parenting Stress in Families Adopting Children from Romanian Orphanages." *Journal of Family Issues* 19(2):164–80.

Major, Brenda, Mark Appelbaum, Linda Beckman, Mary Ann Dutton, Nancy Felipe Russo, and Carolyn West. 2009. "Abortion and Mental Health." *American Psychologist* 64(9):863–90.

Majumdar, Debarun. 2004. "Choosing Childlessness: Intentions of Voluntary Childlessness in the United States." *Michigan Sociological Review* 18:108–35.

Malakh-Pines, Ayala. 2005. *Falling in Love: Why We Choose the Lovers We Choose.* New York: Routledge.

Malebranche, D. 2007. *Black Bisexual Men and HIV: Time to Think Deeper.* Paper presented at the Center for Sexual Health Promotion Sexual Health Seminar Series, Bloomington, Indiana.

Malia, Sarah E. C. 2005. "Balancing Family Members' Interests Regarding Stepparent Rights and Obligations: A Social Policy Challenge." *Family Relations* 54(2):298–319.

Manago, Adriana M., Christia Spears Brown, and Campbell Leaper. 2009. "Feminist Identity Among Latina Adolescents." *Journal of Adolescent Research* 24(6):750–76.

Mandara, Jelani, Jamie S. Johnston, Carolyn B. Murray, and Fatima Varner. 2008. "Marriage, Money, and African American Mothers' Self-Esteem." *Journal of Marriage and Family* 70(5):1188–99.

Mandara, Jelani and Crysta L. Pikes. 2008. "Guilt Trips and Love Withdrawal: Does Mothers' Use of Psychological Control Predict Depressive Symptoms among African American Adolescents?" *Family Relations* 57(December):602–12.

Manisses Communications Group. 2000. "When It Comes to Handling Your Hard-to-Handle Child, Are You an Authoritative, Authoritarian or Permissive Parent?" *The Brown University Child and Adolescent Behavior Letter* 16(3):S1–S2.

Manning, Wendy D. 2001. "Childbearing in Cohabiting Unions: Racial and Ethnic Differences." *Family Planning Perspectives* 33(5):217–34.

———. 2004. "Children and the Stability of Cohabiting Couples." *Journal of Marriage and Family* 66(3):674–89.

———. 2009. "Divorce-Proofing Marriage: Young Adults' Views on the Connection Between Cohabitation and Marital Longevity." *Family Focus* (Summer):F13–F15. Minneapolis, MN: National Council on Family Relations.

Manning, Wendy D. and Susan Brown. 2006. "Children's Economic Well-Being in Married and Cohabiting Parent Families." *Journal of Marriage and Family* 68(2):345–62.

Manning, Wendy D., Peggy C. Giordano, and Monica A. Longmore. 2006. "Hooking Up: The Relationship Contexts of 'Nonrelationship' Sex." *Journal of Adolescent Research* 21(5):459–83.

Manning, Wendy D. and Kathleen A. Lamb. 2003. "Adolescent Well-Being in Cohabiting, Married, and Single-Parent Families." *Journal of Marriage and Family* 65(4):876–93.

Manning, Wendy D. and Nancy S. Landale. 1996. "Racial and Ethnic Differences in the Role of Cohabitation in Premarital Childbearing." *Journal of Marriage and Family* 58(1):63–77.

Manning, Wendy D., Susan D. Stewart, and Pamela J. Smock. 2003. "The Complexity of Fathers' Parenting Responsibilities and Involvement with Nonresident Children." *Journal of Family Issues* 24(5):645–67.

Manning, Wendy D., Deanna Trella, Heidi Lyons, and Nola Cora DuToit. 2010. "Marriageable Women: A Focus on Participants in a Community Healthy Marriage Program." *Family Relations* 59(1):87–102.

———. 2002. "First Comes Cohabitation and Then Comes Marriage?" *Journal of Family Issues* 23(8):1065–87.

———. 2005. "Measuring and Modeling Cohabitation: New Perspectives from Qualitative Data." *Journal of Marriage and Family* 67(4):989–1002.

Mannis, Valerie S. 1999. "Single Mothers by Choice." *Family Relations* 48(2):121–28.

Mansson, Daniel H., Scott A. Myers, and Lynn H. Turner. 2010. "Relational Maintenance Behaviors in the Grandchild-Grandparent Relationship." *Communication Research Reports* 27(1):68–79.

Marano, Hara Estroff. 2000. "Divorced? (Remarriage in America)." *Psychology Today* 33(2):56–60.

———. 2010. "The Expectations Trap." *Psychology Today* 43(2):63–71.

Marchione, Marilynn and Lindsey Tanner. 2006. "Many U.S. Couples Seek Embryo Screening." Associated Press, September 20. Retrieved September 21, 2006 (www.apnews.myway.com).

Marech, Rona. 2004. "To Wed or Not to Wed." *San Francisco Chronicle Magazine*, January 18.

Marikar, Sheila. 2009. "Cher is Supporting Chasity's Sex Change, Though She Doesn't Understand It." *ABC News*, June 18. Retrieved January 12, 2010 (www.abcnews.go.com).

Markman, Howard, Scott Stanley, and Susan L. Blumberg. 2001. *Fighting for Your Marriage: Positive Steps for Preventing Divorce and Preserving a Lasting Love.* San Francisco, CA: Jossey-Bass.

Markon, Jerry. 2010. "The Baby He's Never Met; Va. Father Fights for Child His Girlfriend Sent to Utah for Adoption." *Washington Post*, April 14, p. A1.

Marks, Jaime, Chun Lam, and Susan McHale. 2009. *Sex Roles* 61(3/4):221–34.

Marks, Loren, Katrina Hopkins, Cassandra Chaney, Pamela Monroe, Olena Nesteruk, and Diane Sasser. 2008. "'Together We Are Strong': A Qualitative Study of Happy, Enduring, African American Marriages." *Family Relations* 57(2):172–85.

Marks, Nadine F., James D. Lambert, and Heejeong Choi. 2002. "Transitions to Caregiving, Gender, and Psychological Well-Being: A Prospective U.S. National Study." *Journal of Marriage and Family* 64(3):657–67.

Marks, Stephen R. 2001. "Teasing Out the Lessons of the 1960s: Family Diversity and Family Privilege." Pp. 66–79 in *Understanding Families in the New Millennium: A Decade in Review*, edited by Robert M. Milardo. Lawrence, KS: National Council on Family Relations.

Markway, Barbara and Gregory Markway. 2006. *Nurturing the Shy Child: Practical Help for Raising Confident and Socially Skilled Kids and Teens.* New York: Griffin.

Marquardt, Elizabeth. n.d. *The Revolution in Parenthood: The Emerging Global Clash between Adult Rights and Children's Needs.* Commission on Parenthood's Future. Retrieved October 4, 2006 (www.americanvalues.org).

"Marriage." 2008. The Gallup Poll. Retrieved October 9, 2009 (www.gallup.com).

"Marriage, Domestic Partnerships, and Civil Unions: An Overview of Relationship Recognition for Same-Sex Couples in the United States." 2009. National Center for Lesbian Rights. Retrieved December 20, 2009 (www.nclrights.org).

"Married Households Rise Again among Blacks, Census Finds." 2003. Associated Press, April 25.

Marschark, Marc. 2007. *Raising and Educating a Deaf Child.* New York: Oxford University Press.

Marsh, Kris, William A. Darity Jr., Philip N. Cohen, Lynne M. Casper, and Danielle Salters. 2007. "The Emerging Black Middle Class: Single and Living Alone." *Social Forces* 86(2):735–62.

Marshall, Barbara L. 2009. "Science, Medicine and Virility Surveillance: 'Sexy Seniors' in the Pharmaceutical Imagination." *Sociology of Health & Illness* 32(2):211–24.

Marshall, Catherine A. 2010. *Surviving Cancer as a Family and Helping Co-Survivors Thrive.* Santa Barbara, CA: Praeger.

Marshall, Nancy and Allison Tracy. 2009. "After the Baby: Work-Family Conflict and Working Mothers' Psychological Health." *Family Relations* 58(October):380–91.

Marsiglio, William. 2004. *Stepdads: Stories of Love, Hope, and Repair.* Boulder, CO: Rowman and Littlefield.

———. 2008. "Understanding Men's Prenatal Experience and the Father Involvement Connection: Assessing Baby Steps." Journal of Marriage and Family 20(December):1108–13.

Marsiglio, William, Paul Amato, Randal Day, and Michael E. Lamb. 2000. "Scholarship on Fatherhood in the 1990s and Beyond." *Journal of Marriage and Family* 62:1173–91.

Martin, Anne, Rebecca M. Ryan, and Jeanne Brooks-Gunn. 2007. "The Joint Influence of Mother and Father Parenting on Child Outcomes at Age 5." *Early Childhood Research Quarterly* 22:423–39.

Martin, Brandon E. and Frank Harris. 2006. "Examining Productive Conceptions of Masculinities: Lessons Learned From Academically Driven African American Male Student-Athletes." *Journal of Men's Studies* 14(3):359–78.

Martin, C. L. 1989. "Children's Use of Gender-Related Information in Making Social Judgments." *Developmental Psychology* 25:80–88.

Martin, Jacqueline L. and Hildy S. Ross. 2005. "Sibling Aggression: Sex Differences and Parents' Reactions." *International Journal of Behavioral Development* 29(2):129–38.

Martin, John Levi. 2005. "Is Power Sexy?" *American Journal of Sociology* 111(2):408–47.

Martin, Joyce A., Brady E. Hamilton, Paul D. Sutton, Stephanie J. Ventura, Fay Menacker,

and Sharon Kirmeyer. 2006. "Births: Final Data for 2004." *National Vital Statistics Report* 55(1). Hyattsville, MD: National Center for Health Statistics. September 29.

Martin, Joyce A., Brady E. Hamilton, Paul D. Sutton, Stephanie J. Ventura, Fay Menacker, Sharon Kirmeyer, and T. J. Mathews. 2009. "Births: Final Data for 2006." *National Vital Statistics Report* 57(7). Hyattsville, MD: National Center for Health Statistics. January 7.

Martin, Joyce A., Brady E. Hamilton, Paul D. Sutton, Stephanie J. Ventura, Fay Menacker, and Martha L. Munson. 2005. "Births: Final Data for 2003." *National Vital Statistics Report* 54(2). Hyattsville, MD: National Center for Health Statistics. September 8.

Martin, Philip and Elizabeth Midgley. 2003. "Immigration: Shaping and Reshaping America." *Population Bulletin* 58(2). Washington, DC: Population Reference Bureau. June.

———. 2006. "Immigration: Shaping and Reshaping America." *Population Bulletin* 61(4). Washington, DC: Population Reference Bureau.

Martin, Sandra L., April Harris-Britt, Yun Li, Kathryn E. Moracco, Lawrence L. Kupper, and Jacquelyn C. Campbell. 2004. "Change in Intimate Partner Violence during Pregnancy." *Journal of Family Violence* 19:243–47.

Martin, Steven P. 2006. "Trends in Marital Dissolution by Women's Education in the United States." *Demographic Research* 15(Article 20):537–60. Retrieved May 10, 2007 (www. demographic-research.org).

Martin, Steven P. and Sangeeta Parashar. 2006. "Women's Changing Attitudes toward Divorce, 1974–2002: Evidence for an Educational Crossover." *Journal of Marriage and Family* 68(1):29–40.

Martinez, Gladys M., Anjani Chandra, Joyce C. Abma, Jo Jones, and William D. Mosher. 2006. "Fertility, Contraception, and Fatherhood: Data on Men and Women from Cycle 6 (2002) of the National Survey of Family Growth." *Vital and Health Statistics* 23(26). May. Retrieved January 27, 2007 (www.nchs.gov).

Martino, Steven C., Rebecca L. Collins, and Phyllis L. Ellickson. 2004. "Substance Use and Early Marriage." *Journal of Marriage and Family* 66(1):244–57.

Martyn, Kristy, Carol Loveland-Cherry, Antonia Villarruel, Esther Gallegos Cabriales, Yan Zhou, David Ronis, and Brenda Eakin. 2009. "Mexican Adolescents' Alcohol Use, Family Intimacy, and Parent-Adolescent Communication." *Journal of Family Nursing* 15(2):152–70.

Masanori, Ishimori, Ikuo Daibo, and Yuji Kanemasa. 2004. "Love Styles and Romantic Love Experiences in Japan." *Social Behavior and Personality* 32(3):265–81.

Mason, Mary Ann, Mark A. Fine, and Sarah Carnochan. 2001. "Family Law in the New Millennium: For Whose Families?" *Journal of Family Issues* 22(7):859–81.

Mason, Mary Ann, Sydney Harrison-Jay, Gloria Messick Svare, and Nicholas H. Wolfinger. 2002. "Stepparents: De Facto Parents or Legal Strangers?" *Journal of Family Issues* 23(4):507–22.

Mason, Mary Ann and Ann Quirk. 1997. "Are Mothers Losing Custody? Read My Lips: Trends in Judicial Decision-making in Custody Disputes—1920, 1960, 1990, and 1995." *Family Law Quarterly* 31:215–36.

Masters, William H. and Virginia E. Johnson. 1966. *Human Sexual Response.* Boston, MA: Little, Brown.

———. 1976. *The Pleasure Bond: A New Look at Sexuality and Commitment.* New York: Bantam.

Masters, William H., Virginia E. Johnson, and Robert C. Kolodny. 1994. *Heterosexuality.* New York: HarperCollins.

Masuda, Masahiro. 2006. "Perspectives on Premarital Postdissolution Relationships: Account-Making of Friendships between Former Romantic Partners." Pp. 113–32 in *Handbook of Divorce and Relationship Dissolution,* edited by Mark A. Fine and John H. Harvey. Mahwah, NJ: Erlbaum.

Mather, Mark. 2009. "Children in Immigrant Families Chart New Path." Washington, DC: Population Reference Bureau.

Mather, Mark and Kelvin Pollard. 2009. "U.S. Hispanic and Asian Population Growth Levels Off." Washington, DC: Population Reference Bureau. Retrieved June 8, 2009 (www.prb.org).

Mather, Mark and Kerri L. Rivers. 2006. *"The Concentration of Negative Child Outcomes in Low-Income Neighborhoods."* Washington, DC: Annie E. Casey Foundation and Population Reference Bureau. Retrieved March 31, 2006 (www.prb .org).

Mathews, T. J. and Brady E. Hamilton. 2009. "Delayed Childbearing: More Women Are Having Their First Child Later in Life." *NCHS Data Brief* 21. Hyattsville, MD: National Center for Health Statistics.

Matiasko, Jennifer, Leslie Grunden, and Jody Ernst. 2007. "Structural and Dynamic Process Family Risk Factors: Consequences for Holistic Adolescent Functioning." *Journal of Marriage and Family* 69(3):654–74.

Matsunaga, Masaki, and Tadasu Todd Imahori. 2009. "Profiling Family Communication Standards." *Communication Research* 36(1):3–31.

Matta, William J. 2006. *Relationship Sabotage: Unconscious Factors That Destroy Couples, Marriages, and Family.* Westport, Conn.: Praeger Publishers.

Matthews, Ralph and Anne Martin Matthews. 1986. "Infertility and Involuntary Childlessness: The Transition to Nonparenthood." *Journal of Marriage and Family* 48:641–49.

Mattingly, Marybeth J. and Suzanne M. Bianchi. 2003. "Gender Differences in the Quantity and Quality of Free Time: The U.S. Experience." *Social Forces* 81(3):999–1030.

Mauer, Marc. 2009. "Racial Disparities in the Criminal Justice System." Testimony before the House Judiciary Subcommittee on Crime, Terrorism, and Homeland Security, October 29. Washington, DC: The Sentencing Project.

Mauldon, Jane. 2003. "Families Started by Teenagers." Pp. 40–65 in *All Our Families,* 2nd ed., edited by Mary Ann Mason, Arlene

Skolnick, and Stephen D. Sugarman. New York: Oxford University Press.

May, Rollo. 1969. *Love and Will.* New York: Norton.

———. 1975. "A Preface to Love." Pp. 114–19 in *The Practice of Love,* edited by Ashley Montagu. Englewood Cliffs, NJ: Prentice Hall.

Mayo Clinic Staff. 2005. "Marriage Counseling: Working Through Relationship Problems." Retrieved May 1, 2007 (www.mayoclinic.com/health/marriage-counseling/MH00104).

Mays, Vickie M. and Susan D. Cochran. 1999. "The Black Woman's Relationship Project: A National Survey of Black Lesbians." Pp. 59–66 in *The Black Family: Essays and Studies,* 6th ed., edited by Robert Staples. Belmont, CA: Wadsworth.

Mays, Vickie M., Susan D. Cochran, and Anthony Zamudio. 2004. "HIV Prevention Research: Are We Meeting the Needs of African American Men Who Have Sex With Men?" *Journal of Black Psychology* 30(1):78–105.

McAdoo, Harriette Pipes. 2007. *Black Families,* 4th edition. Thousand Oaks, CA: Sage Publications.

McBride-Chang, Catherine and Lei Chang. 1998. "Adolescent–Parent Relations in Hong Kong: Parenting Styles, Emotional Autonomy, and School Achievement." *Journal of Genetic Psychology* 159(4):421–35.

McCarthy, Kara. 2009. "Growth in Prison and Jail Populations Slowing." Bureau of Justice Statistics: Press Release. Retrieved March 13, 2010 (bjs.ojp.usdoj.gov).

McCone, David and Kathy O'Donnell. 2006. "Marriage and Divorce Trends for Graduates of the U.S. Air Force Academy." *Military Psychology* 18(1):61–75.

McCormick, Richard A., S.J. 1992. "Christian Approaches: Catholicism." Presented at the Surrogate Motherhood and Reproductive Technologies Symposium, January 13, Creighton University.

McCubbin, Hamilton I. and Marilyn A. McCubbin. 1991. "Family Stress Theory and Assessment: The Resiliency Model of Family Stress, Adjustment and Adaptation." Pp. 3–32 in *Family Assessment Inventories for Research and Practice,* 2nd ed., edited by Hamilton I. McCubbin and Anne I. Thompson. Madison: University of Wisconsin, School of Family Resources and Consumer Services.

———. 1994. "Families Coping with Illness: The Resiliency Model of Family Stress, Adjustment, and Adaptation." Chapter 2 in *Families, Health, and Illness.* St. Louis: Mosby.

McCubbin, Hamilton I. and Joan M. Patterson. 1983. "Family Stress and Adaptation to Crisis: A Double ABCX Model of Family Behavior." Pp. 87–106 in *Family Studies Review Yearbook,* Vol. 1, edited by David H. Olson and Brent C. Miller. Newbury Park, CA: Sage Publications.

McCubbin, Hamilton I., Anne I. Thompson, and Marilyn A. McCubbin, eds. 1996. *Family Assessment: Resiliency, Coping and Adaptation: Inventories for Research and Practice.* Madison: University of Wisconsin Press.

McCubbin, Hamilton I., Elizabeth Thompson, Anne Thompson, Jo A. Futrell, and Suniya Luthar. 2001. "The Dynamics of Resilient Families." *Contemporary Psychology* 48(2):154–56.

McCubbin, Marilyn A. 1995. "The Typology Model of Adjustment and Adaptation: A Family Stress Model." *Guidance and Counseling* 10(4):31–39.

McCubbin, Marilyn A. and Hamilton I. McCubbin. 1989. "Theoretical Orientations to Family Stress and Coping." Pp. 3–43 in *Treating Families Under Stress,* edited by Charles Figley. New York: Brunner/Mazel.

McCue, Margi Laird. 2008. *Domestic Violence: A Reference Handbook.* 2nd ed. Santa Barbara, CA: ABC-CLIO, Inc.

McDermott, Monica and Frank L. Samson. 2005. "White Racial and Ethnic Identity in the United States." *Annual Review of Sociology* 31:245–61.

McDonald, Merrilyn. 1998. "The Myth of Epidemic False Allegations of Sexual Abuse in Divorce Cases." *Court Review* (Spring). Retrieved January 14, 2002 (www.omsys.com).

McDonough, Tracy A. 2010. "A Policy Capturing Investigation of Battered Women's Decisions to Stay in Violent Relationships." *Violence and Victims* 25(2):165–84.

McGinn, Daniel. 2006a. "Getting Back on Track." *Newsweek,* September 26, pp. 62–64.

———. 2006b. "Marriage by the Numbers." *Newsweek,* June 5, pp. 40–47.

McGough, Lucy S. 2005. "Protecting Children in Divorce: Lessons from Caroline Norton." *Maine Law Review* 57:13–37.

McGraw, Lori A., Anisa M. Zvonkovic, and Alexis J. Walker. 2000. "Studying Postmodern Families: A Feminist Analysis of Ethical Tensions in Work and Family Research." *Journal of Marriage and the Family* 62(February):68–77.

McHale, Susan M., W. T. Bartko, Ann C. Crouter, and M. Perry-Jenkins. 1990. "Children's Housework and Psychological Functioning: The Mediating Effects of Parents' Sex-Role Behaviors and Attitudes." *Child Development* 61:1413–26.

McHale, Susan M. and Ann C. Crouter. 1992. "You Can't Always Get What You Want: Incongruence between Sex-role Attitudes and Family Work Roles and Its Implications for Marriage." *Journal of Marriage and Family* 54(3):537–47.

McHugh, Maureen F. 2007. "The Evil Stepmother." Second Wives Café: Online Support for Second Wives and Stepmothers. Retrieved May 2, 2007 (secondwivescafe.com).

McIntyre, Matthew H. and Carolyn Pope Edwards. 2009. "The Early Development of Gender Differences." *Annual Review of Anthropology* 38(1):83–97.

McIntyre, Matthew, Steven W. Gangestad, Peter B. Gray, Judith Flynn Chapman, Terence C. Burnham, Mary T. O'Rourke, and Randy Thornhill. 2006. Romantic Involvement Often Reduces Men's Testosterone Levels—But Not Always: The Moderating Role of Extrapair Sexual Interest." *Journal of Personality and Social Psychology* 91(4):642–51.

McKinley, Jesse, and John Schwartz. 2010. "Court Rejects Same-Sex Marriage Ban in California." *The New York Times.* August 4. Retrieved August 10, 2010 (www.nytimes.com).

McLanahan, Sara and Marcia J. Carlson. 2002. "Welfare Reform, Fertility, and Father Involvement." *The Future of Children* 12(1):147–65.

McLanahan, Sara, Elizabeth Donahue, and Ron Haskins. 2005. "Introducing the Issue." *The Future of Children* 15(2):3–12.

McLanahan, Sara, Irwin Garfinkel, Nana E. Reichman, and Julien O. Teitler. 2001. "Unwed Parents or Fragile Families? Implications for Welfare and Child Support Policy." Pp. 202–28 in *Out of Wedlock: Causes and Consequences of Nonmarital Fertility,* edited by Lawrence L. Wu and Barbara Wolfe. New York: Russell Sage.

McLoyd, Vonnie C., Ana Mari Cauce, David Takeuchi, and Leon Wilson. 2000. "Marital Processes and Parental Socialization in Families of Color: A Decade Review of Research." *Journal of Marriage and Family* 62:1070–93.

McMahon, Martha. 1995. *Engendering Motherhood.* New York: Guilford.

McMahon, Thomas, Justin Winkel, Nancy Suchman, and Bruce Rounsaville. 2007. "Drug-Abusing Fathers: Patterns of Pair Bonding, Reproduction, and Paternal Involvement." *Journal of substance Abuse Treatment* 33:295–302.

McMahon-Howard, Jennifer, Jody Clay-Warner, and Linda Renzulli. 2009. "Criminalizing Spousal Rape: The Diffusion of Legal Reforms." *Sociological Perspectives* 52(4):505–31.

McManus, Mike. 2006. "Inside/Out Dads: Helping Prisoners Reenter Society." Retrieved November 17, 2006 (smartmarriages.com).

McManus, Patricia A. and Thomas DiPrete. 2001. "Losers and Winners: The Financial Consequences of Separation and Divorce for Men." *American Sociological Review* 66:246–68.

McMenamin, Terence M. 2007. "A Time to Work: Recent Trends in Shift Work and Flexible Schedules." *Monthly Labor Review,* December. Washington, DC: U.S. Bureau of Labor Statistics.

McNamara, Robert Hartmann. 2008. *Homelessness in America.* Westport, CT: Praeger.

McPherson, Miller and Lynn Smith-Lovin. 2006. "Social Isolation in America: Changes in Core Discussion Networks over Two Decades." *American Sociological Review* 71(June):353–75.

McVeigh, Rory and Maria-Elena D. Diaz. 2009. "Voting to Ban Same-Sex Marriage: Interests, Values, and Communities." *American Sociological Review* 74(6):891–915.

McWey, Lenore M., Tammy L. Henderson, and Jenny Burroughs Alexander. 2008. "Parental Rights and the Foster Care System." *Journal of Family Issues* 29(8):1031–50.

Mead, George Herbert. 1934. *Mind, Self, and Society.* Chicago, IL: University of Chicago Press.

Meadows, Sarah O., Kenneth C. Land, and Vicki L. Lamb. 2005. "Assessing Gilligan vs. Sommers: Gender-Specific Trends in Child and Youth

Well-Being in the United States, 1985–2001." *Social Indicators Research* 70:1–52.

Meadows, Sarah O., Sara S. McLanahan, and Jeanne Brooks-Gunn. 2007. "Parental Depression and Anxiety and Early Childhood Behavior Problems Across Family Types." *Journal of Marriage and Family* 69(5):1162–77.

Meadows, Sarah O., Sara S. McLanahan, and Jean T. Knab. 2009. "Economic Trajectories in Non-Traditional Families with Children." *The Rand Corporation*, August 21, pp. 1–41.

Medeiros, Rose A. and Murray A. Straus. 2006. "Risk Factors for Physical Violence Between Dating Partners: Implications for Gender-Inclusive Prevention and Treatment of Family Violence." Pp. 59–87 in *Family Interventions in Domestic Violence A Handbook of Gender-Inclusive Theory and Treatment*, edited by John C. Hamel and Tanya L. Nicholls. New York: Springer Publishing Company.

Meezan, William, and Jonathan Rauch. 2005. "Gay Marriage, Same-Sex Parenting, and America's Children." *The Future of Children* 15(2):157–75.

Mehrota, Meela. 1999. "The Social Construction of Wife Abuse: Experiences of Asian Indian Women in the United States." *Gender and Society* 16:898–920.

Mehta, Pranjal H., Amanda C. Jones, and Robert A. Josephs. 2008. "The Social Endocrinology of Dominance: Basal Testosterone Predicts Cortisol Changes and Behavior Following Victory and Defeat." *Journal of Personality and Social Psychology* 94(6):1078–93.

Meier, Ann and Gina Allen. 2009. "Romantic Relationships from Adolescence to Young Adulthood." *The Sociological Quarterly* 50(2):308–35.

Meier, Ann, Kathleen E. Hull, and Timothy A. Ortyl. 2009. "Young Adult Relationship Values at the Intersection of Gender and Sexuality." *Journal of Marriage and Family* 71(3):510–25.

Meilander, Gilbert. 1992. "Christian Approaches: Protestantism." Presented at the Surrogate Motherhood and Reproductive Technologies Symposium, January 13, Creighton University.

Melby, Todd. 2010. "Sexuality for the Young and Old." *Contemporary Sexuality* 44(1):1, 4–6.

Memmott, Mark. 2001. "Sex Trade May Lure 325,000 U.S. Kids." *USA Today,* September 10.

Meneses, Paulina Angarita. 2008. "Hogares de Reinsertados Hay Violencia." *El Tiempo* (Colom.), March 31.

Menjívar, Cecilia and Olivia Salcido. 2002. "Immigrant Women and Domestic Violence: Common Experiences in Different Countries." *Gender and Society* 16:898–920.

Mental Health America. 2009. "Factsheet: What Every Child Needs for Good Mental Health." Retrieved January 2, 2010 (www.mentalhealthamerica.net).

Mercier, Laurie. 2010. "Historical Overview: Japanese Americans in the Columbia River Basin." Vancouver, WA: Columbia River Basin Ethnic History Archive, Washington State University. www.vancouver.wsu.edu/crbeha/ja/ja.htm#biblio.

Merkle, Erich R. and Rhonda A. Richardson. 2000. "Digital Dating and Virtual Relating: Conceptualizing Computer Mediated Romantic Relationships." *Family Relations* 49(2):187–92.

Merton, Robert K. 1973 [1942]. "The Normative Structure of Science." Chapter 3 in *The Sociology of Science: Theoretical and Empirical Investigations,* edited by Robert K. Merton. Chicago, IL: University of Chicago Press.

Merton, Robert K. 1968 [1949]. *Social Theory and Social Structure.* New York: The Free Press.

Messner, Michael A. 1997. *The Politics of Masculinity: Men in Movements.* Thousand Oaks, CA: Sage Publications.

Meston, Cindy M. and Tierney Ahrold. 2010. "Ethnic, Gender, and Acculturation Influences on Sexual Behaviors." *Archives of Sexual Behavior* 39:179–89.

Meteyer, Karen B. and Maureen Perry-Jenkins. 2009. "Dyadic Parenting and Children's Externalizing Symptoms." *Family Relations* 58(July):289–302.

Metropolitan Life Insurance Company. 1997. *The Metropolitan Life Survey of the American Teacher, 1997: Examining Gender Issues in the Public Schools.* New York: Metropolitan Life Insurance Company.

Meyer, Harris. 2009. "Getting Back in the Game." *The Oregonian,* November 4.

Meyer, Jennifer. 2007. "Making Good Decisions." *Omaha World Herald,* January 22.

Meyer, Madonna H. and Marcia L. Bellas. 2001. "U.S. Old-age Policy and the Family." Pp. 191–201 in *Families in Later Life: Connections and Transitions,* edited by Alexis J. Walker, Margaret Manoogian-O'Dell, Lori A. McGraw, and Diana L. G. White. Thousand Oaks, CA: Pine Forge.

Michaels, Marcia L. 2006. "Stepfamily enrichment Program: A Preventive Intervention for Remarried Couples." *Journal for Specialists in Group Work* 31(2):135–52.

Middlemiss, Wendy and William McGuigan. 2005. "Ethnicity and Adolescent Mothers' Benefit from Participation in Home-Visitation Services." *Family Relations* 54(2):212–24.

Mikelson, Kelly S. 2008. "He Said, She Said: Comparing Mother and Father Reports of Father Involvement." *Journal of Marriage and Family* 70(3):613–24.

Milardo, Robert M. 2005. "Generative Uncle and Nephew Relationships." *Journal of Marriage and Family* 67(5):1226–36.

Miles, Leonora. 2008. "The Hidden Toll—Financial Abuse Is one of the Most Common Forms of Elder Abuse." *Adults Learning* 19(9):28–30.

Milkie, Melissa A., Marybeth Mattingly, Kei M. Nomaguchi, Suzanne M. Bianchi, and John D. Robinson. 2004. "The *Time* Squeeze: Parental Statuses and Feelings about *Time* with Children." *Journal of Marriage and Family* 66:739–61.

Miller, Carol T. and Cheryl R. Kaiser. 2001. "A Theoretical Perspective on Coping with Stigma." *Journal of Social Issues* 57(1):73–92.

Miller, Courtney Waite and Michael E. Roloff. 2005. "Gender and Willingness to Confront Hurtful Messages from Romantic Partners." *Communication Quarterly* 53(3):323–38.

Miller, Dawn. 2004. "From the Author. TheStepfamilyLife: A Column from Life in the Blender" (www.thestepfamilylife.com).

———. n.d. "Don't Go Nuclear—Negotiate." SelfGrowth.com. Retrieved September 21, 2004 (www.selfgrowth.com/articles/Miller).

———. n.d. "Surviving the Blended Family Holiday: Five Tips for the Stressed Out." SelfGrowth.com. Retrieved September 21, 2004 (www.selfgrowth.com/articles/Miller).

Miller, Dorothy. 1979. "The Native American Family: The Urban Way." Pp. 441–84 in *Families Today: A Research Sampler on Families and Children,* edited by Eunice Corfman. Washington, DC: U.S. Government Printing Office.

Miller, Elizabeth. 2000. "Religion and Families over the Life Course." Pp. 173–86 in *Families Across Time: A Life Course Perspective,* edited by Sharon J. Price, Patrick C. McKenry, and Megan J. Murphy. Los Angeles, CA: Roxbury.

Miller, Laurie C. 2005a. *Handbook of International Adoptive Medicine.* New York: Oxford University Press.

———. 2005b. "International Adoption, Behavior, and Mental Health." *Journal of the American Medical Association* 293(20):2533–35.

Miller, J. Mitchell, Megan Kurlycheck, J. Andrew Hansen, and Kristine Wilson. 2008. "Examining Child Abduction by Offender Type Patterns." *Justice Quarterly* 25(3):523–43.

Mills, C. Wright. 1940. "Situated Actions and Vocabularies of Motive." *American Sociological Review* 5:904–13.

———. 2000 [1959]. *The Sociological Imagination.* 40th Anniversary Edition. New York: Oxford University Press.

Mills, Terry L., Melanie A. Wakeman, and Christopher B. Fea. 2001. "Adult Grandchildren's Perceptions of Emotional Closeness and Consensus with Their Maternal and Paternal Grandparents." *Journal of Family Issues* 22(4):427–55.

Mills, Terry L. and Janet M. Wilmoth. 2002. "Intergenerational Differences and Similarities in Life-Sustaining Treatment Attitudes and Decision Factors." *Family Relations* 51(1):46–54.

Milner, Joel S., Cynthia J. Thomsen, Julie L. Crouch, Mandy M. Rabenhorst, Patricia M. Martens, Christopher W. Dyslin, Jennifer M. Guimond, Valerie A. Stander, and Lex L. Merrill. 2010. "Do Trauma Symptoms Mediate the Relationship Between Childhood Physical Abuse and Adult Child Abuse Risk?" *Child Abuse & Neglect* 34(3):332–44.

Min, Pyong Gap. 2002. "Korean American Families." Pp. 193–211 in *Minority Families in the United States: A Multicultural Perspective,* 3rd ed., edited by Ronald L. Taylor. Upper Saddle River, NJ: Prentice Hall.

Mincy, Ronald B. 2006. *Black Males Left Behind.* Washington, DC: Urban Institute Press.

Mintz, Steven. 2004. *Huck's Raft: A History of American Childhood.* Cambridge, MA: Harvard University Press.

Mintz, Steven and Susan Kellogg. 1988. *Domestic Revolutions: A Social History of American Family Life.* New York: Free Press.

Mishel, Lawrence, Jared Bernstein, and Heidi Shierholz. 2009. *The State of Working America 2008/2009.* New York: Cornell University Press.

Mistry, Rashmita S., Edward D. Lowe, Aprile Benner, and Nina Chien. 2008. "Expanding the Family Economic Stress Model: Insights from a Mixed-Methods Approach." *Journal of Marriage and Family* 70(1):196–209.

Mitchell, Katherine Stamps, Alan Booth, and Valarie King. 2009. "Adolescents with Nonresident Fathers: Are Daughters More Disadvantaged Than Sons?" *Journal of Marriage and Family* 71:650–62.

Mitchell, Melanthia. 2007. "Cohousing for Lesbians Planned in Bremerton." The Associated Press, July 8. Retrieved May 17, 2010 (www.seattlepi.com).

"Mixed Race Americans Picture A 'Blended Nation'." 2009. National Public Radio, November 9. Retrieved February 12, 2010 (www.npr.org).

Mixon, Bobbie. 2008. "Chore Wars: Men, Women and Housework." *Discovery,* April 28. Arlington, VA: National Science Foundation. Retrieved June 7, 2010 (nsf.gov).

Moffett, Frances. 2007. "Why Adults in 30s and 40s Are Choosing to Abstain From Sex." *Jet* 112(6):46–9.

Mollborn, Stefanie. 2009. "Norms About Nonmarital Pregnancy and Willingness to Provide Resources to Unwed Parents." *Journal of Marriage and Family* 71(1):122–34.

Mollen, Debra. 2008. "Guidance and Guidelines for the Theory and Practice of Feminist Therapy." Sex Roles 59(11–12):900–2.

Molyneux, Guy. 1995. "Losing by the Rules." *Los Angeles Times,* September 3, pp. M1, M3.

"Mom Needs an A: Hovering, Hyper-Involved Parents." 2007. University of Texas at Austin Office of Public Affairs, April 2. Retrieved January 16, 2010 (www.utexas.edu/feaures).

Monahan Lang, Molly and Barbara J. Risman. 2007. "A 'Stalled' Revolution or a Still-unfolding One? The Continuing Convergence of Men's and Women's Roles." Discussion Paper, 10th anniversary conference of the Council on Contemporary Families, May 4–5, University of Chicago, Chicago, IL. Retrieved June 4, 2007 (www.contemporaryfamilies.org).

Monaghan, Robyn. 2010. "Though Illegal, Housing Discrimination Continues Against Families With Kids." *Chicago Parent* February 23. www.chicagoparent.com.

Monestero, Nancy. 1990. Personal communication.

"Monogamy: Is It for Us?" 1998. *The Advocate,* June 23, p. 29.

Monroe, Barbara, and Frances Kraus. 2010. *Brief Interventions with Bereaved Children.* 2nd ed. New York: Oxford University Press.

Monserud, Maria A. 2008. "Intergenerational Relationships and Affectual Solidarity Between Grandparents and Young Adults." *Journal of Marriage and Family* 71(1):182–95.

Montgomery, Marilyn J. and Gwendolyn T. Sorell. 1997. "Differences in Love Attitudes across Family Life Stages." *Family Relations* 46:55–61.

Montoro-Rodriguez, J. and D. Gallagher-Thompson. 2009. "The Role of Resources and Appraisals in Predicting Burden among Latina and Non-Hispanic White Female Caregivers." *Aging and Mental Health* 13(5):648–58.

Montoya, R. Matthew. 2008. "I'm Hot, So I'd Say You're Not: The Influence of Objective Physical Attractiveness on Mate Selection." *Personality and Social Psychology Bulletin* 34:1315–31.

Moody, Harry R. 2006. *Aging: Concepts and Controversies.* 5th ed. Thousand Oaks, CA: Pine Forge Press.

Moore, David W. 2003. "Family, Health Most Important Aspects of Life." The Gallup Poll. January 3. Retrieved September 13, 2006 (www.poll.gallup.com).

———.2004. "Modest Rebound in Public Acceptance of Homosexuals." The Gallup Poll, May 20 (www.gallup.com).

——— 2005. "Gender Stereotypes Prevail on Working Outside the Home." The Gallup Poll, August 17. Retrieved August 17, 2005 (www.gallup.com/poll).

Moore, Kathleen A., Marita P. McCabe, and Roger B. Brink. 2001. "Are Married Couples Happier in Their Relationships Than Cohabiting Couples? Intimacy and Relationship Factors." *Sexual and Relationship Therapy* 16(1):35–46.

Moore, Kristin Anderson, Zakia Redd, Mary Burkhauser, Kassim Mbwana, and Ashleigh Collins. 2009. "Research Brief: Children in Poverty: Trends, Consequences, and Policy Options." April. Washington DC: Child Trends. Retrieved December 12, 2009 (www.childtrends.org).

Moore, Mignon R. 2008. "Gendered Power Relations Among Women: A Study of Household Decision Making in Black, Lesbian Stepfamilies." *American Sociological Review* 73(2):335–56.

Moorman, Elizabeth and Eva Pomerantz. 2008. "The Role of Mothers' Control in Children's Mastery Orientation: A Time Frame Analysis." *Journal of Family Psychology* 22(5):734–41.

Moorman, Sara M., Alan Booth, and Karen L. Fingerman. 2006. "Women's Romantic Relationships after Widowhood." *Journal of Family Issues* 27(9):1281–1304.

Moran, Rachel F. 2001. *Interracial Intimacy: The Regulation of Race and Romance.* Chicago, IL: University of Chicago Press.

"More Binding Marriage Gets a Governor's Participation." 2004. *Omaha World-Herald,* November 14.

"More New Dads Seek Time Off." 2005. *Omaha World-Herald (Baltimore Sun),* May 9.

Morford, Mark. 2006. "My Baby Has Rainbow Hair, Gay Parents, Solo Moms, Sperm-Swappin' Friends." April 12. Retrieved April 12, 2006 (www.sfgate.com).

Morgan, S. Philip. 2003. "Is Low Fertility a 21st Century Demographic Crisis?" *Demography* 40:589–603.

Morgan, S. Philip and R. B. King. 2001. "Why Have Children in the 21st Century? Biological Predispositions, Social Coercion, Rational Choice." *European Journal of Population* 17:3–20.

Morgan, S. Philip and Heather Rackin. 2010. "The Correspondence Between Fertility Intentions and Behavior in the United States." *Population and Development Review* 36(1):91–118.

Morris, Frank. 2007. "War Strains Family Life for Military Couples." *Morning Edition.* National Public Radio, February 2.

Morrison, Donna R. and Amy Ritualo. 2000. "Routes to Children's Economic Recovery after Divorce: Are Cohabitation and Remarriage Equivalent?" *American Sociological Review* 65(4):560–80.

Morrow, Lance. 1992. "Family Values." *Time,* August 31, pp. 22–27.

Mosher, William D., Anjani Chandra, and Jo Jones. 2005. "Sexual Behavior and Selected Health Measures: Men and Women 15–44 Years of Age, United States, 2002." *Advance Data from Vital and Health Statistics,* No. 362. September 15. Hyattsville, MD: U.S. National Center for Health Statistics.

Mosher, W. D. and J. Jones. 2010. "Use of Contraception in the United States: 1982–2008." *Vital Health Statistics* 23(29). Hyattsville, MD: National Center for Health Statistics.

Mossaad, Nadwa. 2010. "The Impact of the Recession on Older Americans." Population Reference Bureau. March. Retrieved May 5, 2010 (www.prb.org).

Mueller, Karla A. and Janice D. Yoder. 1999. "Stigmatization of Non-Normative Family Size Status." *Sex Roles* 41(11/12):901–19.

Mueller, Margaret M. and Glen H. Elder, Jr. 2003. "Family Contingencies across the Generations: Grandparent-Grandchild Relationships in Holistic Perspective." *Journal of Marriage and Family* 65(2):404–17.

Muller, Martin and Richard Wrangham. 2009. *Sexual Coercion in Primates and Humans: An Evolutionary Perspective on Male Aggression Against Females.* Cambridge, MA: Harvard University Press.

Mulvaney, Matthew K. and Carolyn J. Mebert. 2007. "Parental Corporal Punishment Predicts Behavior Problems in early Childhood." *Journal of Family Psychology* 21(3):389–97.

Mumola, Christopher J. 2000. *Incarcerated Parents and Their Children.* Washington, DC: U.S. Bureau of Justice Statistics.

Mundy, Liza. 2007. *Everything Conceivable: How Assisted Reproduction Is Changing Men, Women, and the World.* New York: Knopf.

Munroe, Erin A. 2009. *The Everything Guide to Stepparenting.* Avon, MA: Aadams Media.

Munroe, Robert L. and A. Kimball Romney. 2006. "Gender and Age Differences in Same-Sex

Aggregation and Social Behavior." *Journal of Cross-Cultural Psychology* 37(1):3–19.

Murdock, George P. 1949. *Social Structure*. New York: Free Press.

Murphy, Dean E. and Carolyn Marshall. 2005. "Family Feuds over Soldier's Remains." *New York Times,* October 12.

Murphy, Tim and Loriann Hoff Oberlin. 2006. *Overcoming Passive-Aggression: How To Stop Hidden Anger from Spoiling Your Relationships, Career, and Happiness*. New York, NY: Marlowe & Company.

Murray, Bob and Alicia Fortinberry. 2006. *Raising an Optimistic Child: A Proven Plan for Depression-Proofing Young Children—for Life*. New York: McGraw-Hill.

Murray, Christine E. 2004. "The Relative Influence of Client Characteristics on the Process and Outcomes of Premarital Counseling." *Contemporary Family Therapy* 26(4):447–63.

———. 2006. "Professional Responses to Government-endorsed Premarital Counseling." *Marriage and Family Review* 40(1):53–67.

Murray, Christine E. and A. Keith Mobley. 2009. "Empirical Research About Same-Sex Intimate Partner Violence: A Methodological Review." *Journal of Homosexuality* 56(3):361–86.

Murry, Velma M., P. Adama Brown, Gene H. Brody, Carolyn E. Cutrona, and Ronald L. Simons. 2001. "Racial Discrimination as a Moderator of the Links among Stress, Maternal Psychological Functioning, and *Family Relation*ships." *Journal of Marriage and Family* 63(4):915–26.

Murstein, Bernard I. 1980. "Mate Selection in the 1970s." *Journal of Marriage and Family* 42:777–92.

Murtaugh, Paul A. and Michael G. Schlax. 2008. "Reproduction and the Carbon Legacies of Individuals." *Global Environmental Change* 19:14–20.

Musick, Kelly. 2002. "Planned and Unplanned Childbearing among Unmarried Women." *Journal of Marriage and Family* 64(4):915–29.

"Muslim Parents Seek Cooperation from Schools." 2005. CNN.com, September 5. Retrieved September 5, 2005 (www.cnn.com).

Mustafa, Nadia and Jeff Chu. 2006. "Between Two Worlds." *Time,* January 8. Retrieved January 14, 2006 (www.time.com).

Mustillo, Sarah, John Wilson, and Scott M. Lynch. 2004. "Legacy Volunteering: A Test of Two Theories of Intergenerational Transmission." *Journal of Marriage and Family* 66(2):530–41.

Myers, David G. and Letha Dawson Scanzoni. 2006. *What God Has Joined Together? A Christian Case for Gay Marriage*. San Francisco, CA: HarperSanFrancisco.

Myers, Jane E., Jayamala Madathil, and Lynne R. Tingle. 2005. "Marriage Satisfaction and Wellness in India and the United States: A Preliminary Comparison of Arranged Marriages and Marriages of Choice." *Journal of Counseling and Development* 83(2):183–90.

Myers, Scott M. 2006. "Religious Homogamy and Marital Quality: Historical and Generational Patterns, 1980–1997." *Journal of Marriage and Family* 68(2):292–304.

Myers-Walls, Judith A. 2002. "Talking to Children about Terrorism and Armed Conflict." *The Forum for Family and Consumer Issues* 7(1). Retrieved September 14, 2006 (www.ces.ncsu.edu/depts/fcs/pub/2002w/myers-wall.html).

Myrskylä, Mikko, Hans-Peter Kohler, and Francesco C. Billari. 2009. "Advances in Development Reverse Fertility Declines" *Nature* 460:741–43.

Nadir, Aneesah. 2009. "Preparing Muslims for Marriage." Retrieved October 11, 2009 (www.soundvision.com).

Nakonezny, Paul A. and Wayne H. Denton. 2008. "Marital Relationships: A Social Exchange Theory Perspective." *The American Journal of Family Therapy* 36:402–12.

Nakonezny, P. A., R. D. Shull, and J. L. Rodgers. 1995. "The Effect of No-fault Divorce Law on the Divorce Rate across the 50 States and Its Relation to Income, Education, and Religiosity." *Journal of Marriage and Family* 57(2):477–88.

Nanji, Azim A. 1993. "The Muslim Family in North America." Pp. 229–42 in *Family Ethnicity: Strength in Diversity,* edited by Harriette Pipes McAdoo. Newbury Park, CA: Sage Publications.

Narayan, Chetna. 2008. "Is There a Double Standard of Aging?" *Educational Gerontology* 34(9):782–87.

National Ag Safety Database. n.d. "From Family Stress to Family Strengths: Stress, Lesson 5" (www.cec.gov/niosh/nasd).

National Alliance for Caregiving. 2009. *Caregiving in the United States: A Focused Look at Those Caring for Someone Age 50 and Older, Executive Summary*. November. Retrieved May 15, 2010 (www.aarp.org).

National Association for Sick Child Day Care. 2010. Birmingham, AL: National Association for Sick Child Day Care. www.nascd.com.

National Clearinghouse on Marital and Date Rape. 2005. "State Law Chart." Retrieved May 3, 2007 (http://members.aol.com/ncmdr/state_law_chart).

National Coalition for the Homeless. 2009a. "How Many People Experience Homelessness?" Washington, DC. Retrieved January 20, 2010 (www.nationalhomeless.org).

———. 2009b. "Who Is Homeless?" July. Washington, D.C. Retrieved January 20, 2010 (www.nationalhomeless.org).

———. 2009c. "Why are People Homeless?" Washington, DC. Retrieved January 20, 2010 (www.nationalhomeless.org).

"National Compensation Survey: Employee Benefits in Private Industry in the United States, March 2007." 2007. Summary 07-05. Washington, DC: U.S. Bureau of Labor Statistics. www.bls.gov.

National Institute of Justice. 2007. *Commercial Sexual Exploitation of Children: What Do We Know and What Do We Do About It?* Washington, DC: U. S. Department of Justice.

National Latina Institute for Reproductive Health. 2009. "A White Paper on Supporting Healthy Pregnancies, Parenting, and Young Latinas' Sexual Health." New York: National Latina Institute for Reproductive Health, September. Retrieved October 31, 2009 (latinainstitute.org).

The National Marriage Project. 2009. "Figures Supplement to *The State of Our Unions: The Social Health of Marriage in America, 2008*. Rutgers University. Retrieved October 12, 2009 (http://marriage.rutgers.edu).

National Organization for Women. 1966. "Statement of Purpose." Washington, DC: N.O.W. Retrieved February 5, 2007 (www.now.org).

National Public Radio (NPR)/Kaiser Family Foundation/Kennedy School of Government. 2004. *Sex Education in America Survey*. Menlo Park, CA: Kaiser Family Foundation. January (www.kff.org).

National Resource Center on Children and Families of the Incarcerated. 2009a. "An Overview of Statistics" Retrieved March 13, 2010 (www.fcnetwork.org).

———. 2009b. "What Happens to Children?" Retrieved March 13, 2010 (www.fcnetwork.org).

National Right to Life. 2005. "Abortion's Physical Complications." Washington, DC: National Right to Life Educational Trust Fund. July.

———. 2006. "Abortion's Psycho-Social Consequences." Washington, DC: National Right to Life Educational Trust Fund, December 6. Retrieved March 29, 2007 (www.nrlc.org).

National Women's Law Center. 2007. *Congress Must Act to Close the Wage Gap for Women*. Washington, DC: National Women's Law Center.

———. 2010. *Congress Must Act to Close the Wage Gap for Women*. Washington, DC: National Women's Law Center.

Navarro v. LaMusga. 2004. (Cal. Lexis 6507).

Navarro, Mireya. 2004. "For Younger Latinas, a Shift to Smaller Families." *New York Times,* December 5.

Nazario, Sonia L. 1990. "Identity Crisis: When White Parents Adopt Black Babies, Race Often Divides." *Wall Street Journal,* September 20.

Neal, Margaret B. and Leslie B. Hammer. 2007. *Working Couples Caring for Children and Aging Parents: Effects on Work and Well-Being*. Mahwah, NJ: Erlbaum.

Neblett, Nicole Gardner. 2007. "Patterns of Single Mothers' Work and Welfare Use." *Journal of Family Issues* 28(9):1093–1112.

Neff, Kristen D. and Susan Harter. 2003. "Relationship Styles of Self-Focused Autonomy, Other-Focused Connectedness, and Mutuality Across Multiple Relationship Contexts." *Journal of Social and Personal Relationships* 20(1):81–99.

Negy, Charles and Douglas K. Snyder. 2000. "Relationship Satisfaction of Mexican American and Non-Hispanic White American Interethnic Couples: Issues of Acculturation and Clinical Intervention." *Journal of Marital and Family Therapy* 26(3):293–305.

Neimark, Jill. 2003. "All You Need Is Love: Why It's Crucial to Your Health—and How to Get More in Your Life." *Natural Health* 33(8):109–13.

Nelsen, Jane, Cheryl Erwin, and Roslyn Duffy. 2007. *Positive Discipline for Preschoolers: For Their Early Years, Raising Children Who Are Responsible, Respectful, and Resourceful.* New York: Three Rivers Press.

Nelson, Margaret K. 2006. "Families in Not-So-Free Fall: A Response to Comments." *Journal of Marriage and Family* 68(4):817–23.

———. 2008. "Watching Children." *Journal of Family Issues* 29(4):516–38.

Newman, Bernie Sue. 2007. "College Students' Attitudes About Lesbians: What Difference Does 16 Years Make? *Journal of Homosexuality* 52(3/4):249–65.

Newman, Louis. 1992. "Jewish Approaches." Presented at the Surrogate Motherhood and Reproductive Technologies Symposium, January 13, Creighton University.

Newport, Frank. 2008. "Wives Still Do Laundry, Men Do Yard Work." Gallup Poll, April 4. Retrieved December 2, 2009 (www.gallup.com/poll).

———. 2009. "Extramarital Affairs, Like Sanford's, Morally Taboo." Poll Analysis. Retrieved June 25 (www.gallup.com).

Newport, Frank and Brett Pelham. 2009. "Americans Least Happy in Their 50s and Late 80s." Gallup Poll, October 5. Retrieved May 6, 2010 (www.gallup.com/poll).

NICHD Early Child Care Research Network. 1999a. "Child Care and Mother–Child Interaction in the First Three Years of Life." *Developmental Psychology* 35:1399–1413.

———. 1999b. "Child Outcomes When Child Care Center Classes Meet Recommended Standards for Quality." *American Journal of Public Health* 89:1072–77.

———. 2000a. "Characteristics and Quality of Child Care for Toddlers and Preschoolers." *Applied Developmental Science* 4:116–35.

———. 2000b. "The Relation of Child Care to Cognitive and Language Development." *Child Development* 71:960–80.

———. 2003a. "Does the Amount of *Time* Spent in Child Care Predict Socioemotional Adjustment during the Transition to Kindergarten?" *Child Development* 74:976–1005.

———. 2003b. "Does Quality of Child Care Affect Child Outcomes at Age 4?" *Developmental Psychology* 39:451–69.

Nicholson v. Scopetta. 2004. N.Y. No. 113.

Nielsen, Rasmus. 2009. "Adaptionism—30 Years After Gould and Lewontin." *Evolution* 63(10):2487–90.

Nierenberg, Gerard and Henry H. Calero. 1973. *Meta-Talk: Guide to Hidden Meanings in Conversations.* New York: Trident Press.

Nock, Steven L. 1995. "A Comparison of Marriages and Cohabiting Relationships." *Journal of Family Issues* 16(1):53–76.

———.1998. *Marriage in Men's Lives.* New York: Oxford University Press.

———. 2001. "The Marriages of Equally Dependent Spouses." *Journal of Family Issues* 22:755–75.

———. 2005. "Marriage as a Public Issue." *The Future of Children* 15(2):13–32.

Nock, Steven L., Laura Sanchez, Julia C. Wilson, and James D. Wright. 2003. "Covenant Marriage Turns Five Years Old." *Michigan Journal of Gender and Law 10(1):169–88.*

Nock, Steven M., Laura A. Sanchez, and James D. Wright. 2008. *Covenant Marriage: The Movement to Reclaim Tradition in America.* New Brunswick, NJ: Rutgers University Press.

Noland, Virginia J., Karen D. Liller, Robert J. McDermott, Martha Coulter, and Anne E. Seraphine. 2004. "Is Adolescent Sibling Violence a Precursor to Dating Violence?" *American Journal of Health Behavior* 28(Supp. 1): 813–23.

Noller, Patricia and Mary Anne Fitzpatrick. 1991. "Marital Communication in the Eighties." Pp. 42–53 in *Contemporary Families: Looking Forward, Looking Back,* edited by Alan Booth. Minneapolis, MN: National Council on *Family Relations.*

Nomaguchi, Kei M. 2008. "Gender, Family Structure, and Adolescents' Primary Confidants." *Journal of Marriage and Family* 70(5):1213–27.

Nomaguchi, Kei M. and Melissa A. Milkie. 2003. "Costs and Rewards of Children: The Effects of Becoming a Parent on Adults' Lives." *Journal of Marriage and Family* 65:356–74.

Nordwall, Smita P. and Paul Leavitt. 2004. "Court Rules for Battered Women's Rights." *USA Today,* October 24.

Notter, Megan L., Katherine A. MacTavish, and Devora Shamah. 2008. "Pathways Toward Resilience among Women in Rural Trailer Parks." *Family Relations* 57(5):613–624.

Oakley, Ann. 1972. *Sex, Gender, and Society.* London, England: Maurice Temple Smith.

Obama, Barack. 2006. *The Audacity of Hope.* New York: Crown Publishers.

———. 2007 [1995]. *Dreams from My Father: A Story of Race and Inheritance.* New York: Crown Publishers.

Obejas, Achy. 1994. "Women Who Batter Women." *Ms.,* September/October, p. 53.

Oberlander, Sarah E., Avril Melissa Houston, Wendy R. Miller Agostini, and Maureen M. Black. 2010. "A Seven-Year Investigation of Marital Expectations and Marriage Among Urban, Low-Income, African American Adolescent Mothers." *Journal of Family Psychology* 24(1):31–40.

O'Brien, Karen M. and Kathy P. Zamostny. 2003. "Understanding Adoptive Families: An Integrative Review of Empirical Research and Future Directions for Counseling Psychology." *The Counseling Psychologist* 31(6):679–710.

O'Brien, Marion. 2007. "Ambiguous Loss in Families of Children with Autism Spectrum Disorders." *Family Relations* 56(2):135–46.

O'Connell, Martin and Daphne Lofquist. 2009. "Counting Same-Sex Couples: Official Estimates and Unofficial Guesses." Paper presented at Annual Meeting of the Population Association of America, Detroit, Michigan, April 30–May 2.

Odom, Samuel L. 2009. *Handbook of Developmental Disabilities.* New York: Guilford Press.

O'Donoghue, Margaret. 2005. "White Mothers Negotiating Race and Ethnicity in the Mothering of Biracial, Black-White Adolescents." *Journal of Ethnic and Cultural Diversity in Social Work* 14(3/4):125–56.

"The Office Closet Empties Out." 2006. *Business Week,* May 1, p. 16.

Offner, Paul. 2005. "Welfare Reform and Teenage Girls." *Social Science Quarterly* 86(2):306–22.

Ogunwole, Stella U. 2006. *We the People: American Indians and Alaska Natives in the United States.* CENSR 28. Washington, DC: U.S. Census Bureau. Retrieved February 28, 2006 (www.census.gov).

"Old Ways Bring Tears in a New World." 2003. *New York Times,* March 7.

Older Direct-Care Workers: Key Facts and Trends. 2010. Bronx, NY: Paraprofessional Healthcare Institute.

"Older Moms' Birth Risks Called Greater." 1999. *Omaha World-Herald,* January 2.

Oldham, J. Thomas. 2008a. "Changes in the Economic Consequences of Divorces, 1958–2008." *Family Law Quarterly* 42(3):419–47.

———. 2008b. "What if the Beckhams Move to L.A. and Divorce—Marital Property Rights of Mobile Spouses When They Divorce in the United States." *Family Law Quarterly* 42(2):263–93.

Olmsted, Maureen E., Judith A. Crowell, and Everett Waters. 2003. "Assortative Mating among Adult Children of Alcoholics and Alcoholics." *Family Relations* 52(1):64–71.

Olson, David H. and Dean M. Gorall. 2003. "Circumplex Model of Marital and Family Systems." Pp. 514–47 in *Normal Family Processes,* 3rd ed., edited by F. Walsh. New York: Guilford.

Olson, M., C. S. Russell, M. Higgins-Kessler, and R. B. Miller. 2002. "Emotional Processes Following Disclosure of an Extramarital Fidelity." *Journal of Marital and Family Therapy* 28:423–34.

Olvera, Mary M. 2009. *Fact Sheet: Mixed Race Reporting.* Chicago, IL: Cultural Marketing Public Relations. Retrieved January 1, 2010 (www.culturalmarketingpr.com).

Omaha Public Schools (OPS) Dual Language Research Group. 2006. *Examining the Impact of Parental Involvement in a Dual Language Program: Implications for Children and Schools.* Omaha, NE: Office of Latino and Latin American Studies, University of Nebraska at Omaha.

O'Malley, Jaclyn. 2002. "Abortion Grief Not Etched in Stone." *Omaha World-Herald,* October 16.

O'Neill, Natalie. 2009. "Mamas vs. Papas: Two Gay Couples Fight Over Custody of Child." *Miami New Times,* July 16. www.miaminewtimes.com.

O'Neill, Nena and George O'Neill. 1974. *Shifting Gears: Finding Security in a Changing World.* New York: M. Evans.

Ono, Hiromi. 2009. "Husbands' and Wives' Education and Divorce in the United States and Japan, 1946–2000." *Journal of Family History* 34(3):292–22.

Ontai, Lenna, Yoshie Sano, Holly Hatton, and Katherine Conger. 2008. "Low-Income Rural Mothers' Perceptions of Parent Confidence: The Role of Family Health Problems and Partner Status." *Family Relations* 57(July):324–34.

Ooms, Theodora. 2005. *The New Kid on the Block: What Is Marriage Education and Does It Work?* CLASP Policy Brief No. 7. Washington, DC: Center for Law and Social Policy. Retrieved September 4, 2005 (www.clasp.org).

Oppenheim, Keith. 2007. "Soldier Fathers Child Two Years after Dying in Iraq." CNN.com, March 20. Retrieved March 20, 2007 (http://cnn.usnews.com).

Oppenheimer, Valerie Kincade. 1997. "Women's Employment and the Gain to Marriage: The Specialization and Trading Model." *Annual Review of Sociology* 23:431–53.

Orchard, Ann L. and Kenneth B. Solberg. 2000. "Expectations of the Stepmother's Role." *Journal of Divorce and Remarriage* 31(1/2):107–24.

O'Reilly, Sally. 2009. "Sally O'Reilly Looks at the Increasing Evidence Linking the Economic Downturn to Workplace Stress and Mental Ill Health." *Personnel Today,* February 12. Retrieved June 15, 2009 (www.personneltoday.com).

Orenstein, Peggy. 1994. *School Girls: Young Women, Self-esteem, and the Confidence Gap.* New York: Doubleday.

———. 2007. "Your Gamete, Myself." New York Times Magazine, July 15.

Ornelas, India J., Krista M. Perreira, Linda Beeber, and Lauren Maxwell. 2009. "Challenges and Strategies to Maintaining Emotional Health: Qualitative Perspectives of Mexican Immigrant Mothers." *Journal of Family Issues* 30(11):1556–75.

Oropesa, R. S. 1996. "Normative Beliefs about Marriage and Cohabitation: A Comparison of Non-Latino Whites, Mexican Americans, and Puerto Ricans." *Journal of Marriage and Family* 58:49–62.

Oropesa, R. S. and Nancy S. Landale. 2004. "The Future of Marriage and Hispanics." *Journal of Marriage and Family* 66(4):901–20.

Orsmond, G. I. and M. M. Seltzer. 2007. "Siblings of Individuals with Autism or Down Syndrome: Effects on Adult Lives." *Journal of Intellectual Disability Research* 51(9):682–96.

Orthner, Dennis K., Hinckley Jones-Sanpei, and Sabrina Williamson. 2004. "The Resilience and Strengths of Low-Income Families." *Family Relations* 53(2):159–67.

Osborne, Cynthia and Lawrence M. Berger. 2009. "Parental Substance Abuse and Child Well-Being." *Journal of Family Issues* 30(3):341–70.

Osgood, D. Wayne, E. Michael Foster, Constance Flanagan, and Gretchen Ruth, eds. 2005. *On Your Own Without a Net: The Transition to Adulthood for Vulnerable Populations.* Chicago, IL: University of Chicago Press.

Osment, Steven. 2001. *Ancestors: The Loving Family of Old Europe.* Cambridge, MA: Harvard University Press.

Oswald, Ramona Faith and Katherine A. Kuvalanka. 2008. "Same-Sex Couples: Legal Complexities." *Journal of Family Issues* 29(8):1051–66.

Oswald, Ramona Faith and Brian P. Masciadrelli. 2008. "Generative Ritual Among Nonmetropolitan Lesbians and Gay Men: Promoting Social Inclusion." *Journal of Marriage and Family* 70:1060–73.

Otis, Melanie D., Sharon S. Rostosky, Ellen D. B. Riggle, and Rebecca Hamrin. 2006. "Stress and Relationship Quality in Same-sex Couples." *Journal of Social and Personal Relationships* 23(1):81–99.

Ott, Mary A., Susan G. Millstein, Susan Offner, and Bonnie L. Halpern-Felsher. 2006. "Greater Expectations for Adolescents' Positive Motivations for Sex." *Perspectives on Sexual and Reproductive Health* 38(2):84–89.

Overbeek, Geertjan, Thao Ha, Ron Scholte, Raymond DeKemp, and Rutger Engels. 2007. "Brief Report: Intimacy, Passion, and Commitment in Romantic Relationships—Validation of a 'Triangular Love Scale' for Adolescents." *Journal of Adolescence* 30(3):523–30.

"Oxytocin." 1997. *Britannica.* Retrieved August 30, 2006 (www.britannica.com).

Oyamot, Clifton M. Jr., Paul T. Fuglestad, and Mark Snyder. 2010. "Balance of Power and Influence in Relationships: The Role of Self-monitoring." *Journal of Social and Personal Relationships* 27(1):23–46.

Ozawa, Martha N. and Yongwoo Lee. 2006. "The Net Worth of Female-Headed Households: A Comparison to Other Types of Households." *Family Relations* 55(1):132–34.

Pacey, Susan. 2005. "Step Change: The Interplay of Sexual and Parenting Problems When Couples Form Stepfamilies." *Social and Relationship Therapy* 20(3):359–69.

Page, Susan and Richard Benedetto. 2004. "Bush Backs Gay-Marriage Ban." *USA Today,* February 24.

Pager, Devah, Bart Bonikowski, and Bruce Western. 2009. "Discrimination in a Low-Wage Labor Market: A Field Experiment." *American Sociological Association* 74(5): 777–99.

Pahl, Jan M. 1989. *Marriage and Money.* Basingstoke, UK: Macmillan.

Painter, Kim. 2006. "Male Life Span Increasing." *USA Today,* June 12.

Palmer, Kim. 2008. "Multigenerational Living." *Minneapolis Star Tribune,* November 20. Retrieved February 3, 2009 (www.startribune.com).

Palmer, Kimberly. 2007. "Accountability: His and Hers; Separate Finances Can Help, Especially When Boomers Remarry." *U. S. News and World Report,* October 8: 51ff.

Pan, En-ling and Michael P. Farrell. 2006. "Ethnic Differences in the Effects of Intergenerational Relations on Adolescent Problem Behavior in U.S. Single-Mother Families." *Journal of Family Issues* 27(8):1137–58.

Panayiotou, Georgia and Myroula Papageorgiou. 2007. "Depressed Mood: The

Role of Negative Thoughts, Self-Consciousness, and Sex Role Stereotypes." *International Journal of Psychology* 42(5):289–96.

Papernow, P. 1993. *Becoming a Stepfamily: Patterns of Development in Remarried Families.* San Francisco, CA: Jossey-Bass.

Papp, Lauren, E. Mark Cummings, and Marcie C. Goeke-Morey. 2009. "For Richer, for Poorer: Money as a Topic of Marital Conflict in the Home." *Family Relations* 55(February):91–103.

Parent, Claudine, Marie-Christine Saint-Jacques, Madeleine Beaudry, and Caroline Robitaille. 2007. "Stepfather Involvement in Social Interventions Made by Youth Protection Services in Stepfamilies." *Child and Family Social Work* 12:229–38.

"Parents and the High Price of Child Care: 2009 Update." 2009. Arlington, VA: National Association of Child Care Resource & Referral Agencies.

Paris, Ruth and Nicole Dubus. 2005. "Staying Connected while Nurturing an Infant: A Challenge of New Motherhood." *Family Relations* 54(1):72–83.

Park, Kristin. 2002. "Stigma Management among the Voluntarily Childless." *Sociological Perspectives* 45:21–45.

———. 2005. "Choosing Childlessness: Weber's Typology of Action and Motives of the Voluntarily Childless." *Sociological Inquiry* 75(3):372–402.

Park, Yong, Leyna Vo, and Yuying Tsong. 2009. "Family Affection As a Protective Factor Against Effects of Perceived Asian Values Gap on the Parent-Child Relationship for Asian American Male and Female College Students." *Cultural Diversity and Ethnic Minority Psychology* 51(1):18–26.

Parke, R. D. and R. Buriel. 2006. "Socialization in the Family: Ecological and Ethnic Perspectives." Pp. 429–504 in *Handbook of Child Psychology: Social, Emotional, and Personality Development,* 6th ed., edited by N. Eisenberg. New York: Wiley.

———. 2008. "Gay Unions Shed Light on Gender in Marriage." *The New York Times.* June 19. Retrieved September 9, 2008 (www.nytimes.com).

Parker-Pope, Tara. 2009. "Kept From a Dying Partner's Bedside." *The New York Times,* May 19. Retrieved November 21, 2009 (www.nytimes.com).

Parra-Cardona, Jose Ruben, Emily Meyer, Lawrence Schiamberg, and Lori Post. 2007. "Elder Abuse and Neglect in Latino Families: An Ecological and Culturally Relevant Theoretical Framework for Clinical Practice." *Family Process* 46(4):451–70.

Parsons, Talcott. 1943. "The Kinship System of the Contemporary United States." *American Anthropologist* 45:22–38.

Parsons, Talcott and Robert F. Bales. 1955. *Family, Socialization, and Interaction Process.* Glencoe, IL: Free Press.

Partners Task Force for Gay and Lesbian Couples. 2004. "Canada Offers Legal Marriage." Partners Task Force for Gay & Lesbian Couples. Retrieved April 3, 2003 (www.buddybuddy.com).

———. 2009. "Legal Marriage Report: Global Status of Legal Marriage." Retrieved November 23, 2009 (www.buddybuddy.com).

Paset, Pamela S. and Ronald D. Taylor. 1991. "Black and White Women's Attitudes toward Interracial Marriage." *Psychological Reports* 69:753–54.

Pasley, Kay. 1998a. "Contemplating Stepchild Adoption." *Research Findings*. Stepfamily Association of America (www.saafamilies.org).

———. 1998b. "Divorce and Remarriage in Later Adulthood." *Research Findings*. Stepfamily Association of America (www.saafamilies.org).

Pasley, Kay and Emily Lipe. 1998. "How Does Having a Mutual Child Affect Stepfamily Adjustment?" *Research Findings*. Stepfamily Association of America (www.saafamilies.org).

Passel, Jeffrey S. and D'Vera Cohn. 2009. *A Portrait of Unauthorized Immigrants in the United States*. Washington, DC: Pew Hispanic Center, April.

Passno, Diane. 2000. "The Feminist Mistake." *Focus on the Family*, September, pp. 12–13.

Pastore, Ann L. and Kathleen Maguire, eds. 2007. *Sourcebook of Criminal Justice Statistics*, Tables 3.131.2005 and Table 3.132.2005. Retrieved June 10, 2010 (www.albany.edu/sourcebook/pdf/t31322005.pdf).

Patterson, Charlotte. 2000. "*Family Relationships* of Lesbians and Gay Men." *Journal of Marriage and Family* 62(4):1052–69.

Patterson, J. M. "Families Experiencing Stress: The Family Adjustment and Adaptation Response Model." *Family Systems Medicine* 6(2):202–37.

———. 2002a. "Family Caregiving for Medically Fragile Children." *Family Focus* (December):F5–F7. National Council on Family Relations.

———. 2002b. "Integrating Family Resilience and Family Stress Theory." *Journal of Marriage and Family* 64(2):349–60.

Pattillo-McCoy, Mary. 1999. *Black Picket Fences: Privilege and Peril among the Black Middle Class*. Chicago, IL: University of Chicago Press.

Paul, Annie Murphy. 2006. "The Real Marriage Penalty." *New York Times*, November 19.

Paul, Pamela. 2008. *Parenting, Inc.: How We Are Sold on $800 Strollers, Fetal Education, Baby Sign Language, Sleeping Coaches, Toddler Culture, and Diaper Wipe Warmers—and What It Means for Our Children*. Henry Holt.

———. 2009. *Parenting, Inc.: How the Billion-Dollar Baby Business Has Changed the Way We Raise Our Children*. New York: Holt Paperbacks.

Pearce, Lisa D. 2002. "The Influence of Early Life Course Religious Exposure on Young Adults' Dispositions Toward Childbearing." *Journal for the Scientific Study of Religion* 41(2):325–40.

Pearson, Jessica and J. Anhalt. 1994. "The Enforcement of Visitation Rights: An Assessment of Five Exemplary Programs." *The Judges Journal* 33(2):2ff.

Peck, M. Scott. 1978. *The Road Less Traveled: A New Psychology of Love, Traditional Values and Spiritual Growth*. New York: Simon and Schuster.

Peck, M. Scott and Sharon Peck. 2006. *The Top 60 Love Skills You Were Never Taught*. Solana Beach, CA: Lifepath Publishers.

Peck, Peggy. 2001. "Cancer Hard on Marriages." WebMD Medical News Archive (my.webmd.com).

Peele, Thomas. 2006. "Court Hears State's Take on Gay Marriage Ban." *Contra Costa Times*, July 10. Retrieved July 11, 2006 (www.mercurynews.com).

Pelham, Brett. 2008. "Relationships, Financial Security Linked to Well-Being." November 17. Retrieved December 2, 2008 (www.gallup.com/poll).

Peltola, Pia, Melissa A. Milkie, and Stanley Presser. 2004. "The 'Feminist' Mystique: Feminist Identity in Three Generations of Women." *Gender and Society* 18(1):122–44.

Pence, E. and M. Paymar. 1993. *Education Groups for Men Who Batter: The Duluth Model*. New York: Springer.

Pendley, Elisabeth. 2006. *Marriage Works: Before You Say "I Do."* Bellevue, WA: Merril Press.

"*People v. Michael Fortin* [PAS Denied in New York]." 2000. *New York Law Journal*, March 27.

Peplau, Letitia Anne, and Adam Fingerhut. 2004. "The Paradox of the Lesbian Worker." *Journal of Social Issues* 60(4):719–35.

Peplau, Letitia Anne., A. Fingerhut, and K. P. Beals. 2004. "Sexuality in the Relations of Lesbians and Gay Men." Pp. 349–69 in *The Handbook of Sexuality in Close Relationships*, edited by J. H. Harvey, A. Wenzel, and S. Sprecher. Mahwah, NJ: Erlbaum.

Peralta, Robert L. and J. Michael Cruz. 2006. "Conferring Meaning onto Alcohol-Related Violence: An Analysis of Alcohol Use and Gender in a Sample of College Youth." *Journal of Men's Studies* 14(1):109–36.

Perez, Lisandro. 2002. "Cuban American Families." Pp. 114–33 in *Minority Families in the United States*, 3rd ed., edited by Ronald L. Taylor. Upper Saddle River, NJ: Prentice Hall.

Perper, Kate, Kristen Peterson, and Jennifer Manlove. 2010. "Diploma Attainment among Teen Mothers." *Child Trends Fact Sheet* 2010-01. Washington, DC: Child Trends.

Perrin, Ellen C. 2002. *Sexual Orientation in Child and Adolescent Health Care*. New York: Kluwer Academic/Plenum.

Perrine, Stephen. 2006. "Keeping Divorced Dads at a Distance." *New York Times*, June 18.

Perrow, Susan. 2009 (2003). "The Hurried Child Syndrome." *Kindred Natural Parenting Magazine*. Retrieved January 15, 2010 (www.kindredmedia.com).

Perry-Jenkins, Maureen, Abbie Goldberg, Courtney Pierce, and Aline Sayer. 2007. "Shift Work, Role Overload, and the Transition to Parenthood." *Journal of Marriage and Family* 69(1):123–38.

Peter, Tracey. 2009. "Exploring Taboos: Comparing Male- and Female-Perpetrated Child Sexual Abuse." *Journal of Interpersonal Violence* 24(7):1111–28.

Peters, Marie F. 2007. "Parenting of Young Children in Black Families." Pp. 203–18 in *Black Families*, 4th ed., edited by Harriette Pipes McAdoo. Thousand Oaks, CA: Sage Publications.

Petersen, Larry R. 1994. "Education, Homogamy, and Religious Commitment." *Journal for the Scientific Study of Religion* 33(2):122–28.

Peterson, Heather A. 2004. "The Daddy Track: Locating the Male Employee Within the Family and Medical Leave Act." *Journal of Law & Policy* 15:253–84.

Peterson, Iver. 2005a. "In Suburbs, New Housing Is Unfriendly to Children." *New York Times*, May 30.

———. 2005b. "Princeton Students Who Say 'No' and Mean 'Entirely No.'" *New York Times*, April 18.

Peterson, James L. and Christine Winquist Nord. 1990. "The Regular Receipt of Child Support: A Multistep Process." *Journal of Marriage and Family* 52(2):539–51.

Peterson, Karen S. 2001. "The Good in a Bad Marriage." *USA Today*, June 21.

———. 2002. "'Market Work, Yes; Housework, Hah'" *USA Today*, March 13.

Peterson, Richard R. 1996. "A Re-evaluation of the Economic Consequences of Divorce." *American Sociological Review* 61:528–36.

Pettit, Gregory S., Patrick S. Malone, Jennifer E. Lansford, Kenneth A. Dodge, and John E. Bates. 2010. "Domain Specificity in Relationship History, Social-Information Processing, and Violent Behavior in Early Adulthood." *Journal of Personality & Social Psychology* 98(2):190–200.

Pew Forum on Religion & Public Life. 2008. *U.S. Religious Landscape Survey: Religious Affiliation: Diverse and Dynamic*. Washington, DC: Pew Research Center.

———. 2009. *Mapping the Global Muslim Population: A Report on the Size and Distribution of the World's Muslim Population*. Washington, DC: Pew Research Center.

———. 2009a. *Eastern, New Age Beliefs Widespread: Many Americans Mix Multiple Faiths*. Washington, DC: Pew Research Center.

———. 2009b. *Faith in Flux: Changes in Religious Affiliation in the U.S.* Washington, DC: Pew Research Center.

Pew Hispanic Center. 2002. "Hispanic Health: Divergent and Changing." Washington, DC: Pew Hispanic Center. Retrieved February 12, 2007 (www.pewhispanic.org).

———. 2009. "Between Two Worlds: How Young Latinos Come of Age in America." Washington, DC: Pew Hispanic Center. Retrieved December 26, 2009 (www.pewresearch.org).

Pew Research Center. 2005. "Baby Boomers: From the Age of Aquarius to the Age of Responsibility." Washington, DC: Pew Research Center. Retrieved January 11, 2007 (www.pewresearch.org).

———. 2008. "More Americans Question Religion's Role in Politics." August 21. Retrieved November 23, 2009 (people-press.org).

———. 2009a. "Between Two Worlds: How Young Latinos Come of Age in America." Pew Hispanic Center. Retrieved December 23, 2009 (pewhispanic.org).

———. 2009b. "Inside the Middle Class: Bad Times Hit the Good Life." April 9. Retrieved December 23, 2009 (www.pewsocialtrends.org).

———. 2009c. "Majority Continues to Support Civil Unions; Most Still Oppose Same-Sex Marriage." October 6. Retrieved November 23, 2009 (people-press.org).

Pezzin, Liliana E., Robert A. Pollak, and Barbara Steinberg Schone. 2008. "Parental Marital Disruption, Family Type, and Transfers to Disabled Elderly Parents." *Journal of Gerontology: Social Sciences* 638(6):S349–S358.

Pezzin, Liliana E. and Barbara Steinberg Schone. 1999. "Parental Marital Disruption and Intergenerational Transfers: An Analysis of Lone Elderly Parents and Their Children." *Demography* 36(3):287–97.

Phelan, Amanda. 2009. "Practice Development—Preventing Neglect in Formal Care Settings: Elder Abuse and Neglect: The Nurse's Responsibility in Care of the Older Person." *International Journal of Older People Nursing* 4(2):115–19.

Phillips, Deborah and Gina Adams. 2001. "Child Care and Our Youngest Children." *The Future of Children* 11:35–51.

Phillips, Julie A. and Megan M. Sweeney. 2005. "Premarital Cohabitation and Marital Disruption among White, Black, and Mexican American Women." *Journal of Marriage and Family* 67(2):296–314.

Phillips, Kate. 2009. "Same-Sex Partner Benefits." *The New York Times,* June 17. Retrieved June 18, 2009 (www.nyt.com).

Phillips, Linda R. 1986. "Theoretical Explanations of Elder Abuse: Competing Hypotheses and Unresolved Issues." Pp. 197–217 in *Elder Abuse: Conflict in the Family,* edited by Karl A. Pillemer and Rosalie S. Wolf. Dover, MA: Auburn.

Phillips, Roderick. 1997. "Stepfamilies from a Historical Perspective." Pp. 5–18 in *Stepfamilies: History, Research, and Policy,* edited by Irene Levin and Marvin B. Sussman. New York: Haworth.

Picker, Lauren. 2005. "And Now, the Hard Part." *Newsweek,* April 25, pp. 46–51.

Pickhardt, Carl E. 2005. *The Everything Parent's Guide to the Strong-Willed Child: An Authoritative Guide to Raising a Respectful, Cooperative, and Positive Child.* Avon, MA: Adams Media.

Piercy, Kathleen W. 2007. "Characteristics of Strong Commitments to Intergenerational Family Care of Older Adults." *Journals of Gerontology: Series B* 62(6):S381–87.

———. 2010. *Working with Aging Families: Therapeutic Solutions for Caregivers, Spouses, and Adult Children.* New York: W.W. Norton.

Pietropinto, Anthony and Jacqueline Simenauer. 1977. *Beyond the Male Myth: What Women Want to Know about Men's Sexuality; A National Survey.* New York: *Times* Books.

Pillemer, Karl A. 1986. "Risk Factors in Elder Abuse: Results from a Case-control Study." Pp. 239–64 in *Elder Abuse: Conflict in the Family,* edited by Karl A. Pillemer and Rosalie S. Wolf. Dover, MA: Auburn.

Pinel, Philippe and Stanislas Dehaene. 2009. "Beyond Hemispheric Dominance: Brain Regions Underlying the Joint Lateralization of Language and Arithmetic to the Left Hemisphere." *Journal of Cognitive Neuroscience* 22(1):48–66.

Pines, Maya. 1981. "Only Isn't Lonely (or Spoiled or Selfish)." *Psychology Today* 15:15–19.

Pinker, Steven. 2005. Opening Remarks. "The Science of Gender and Science—Pinker vs. Spelke: A Debate." Cambridge, MA: Harvard University, Mind/Brain/Behavior Initiative. May 16. Retrieved January 1, 2006 (www.edge. org).

Pinto, Consuela A. 2009. "Eliminating Barriers to Women's Advancement: Focus on the Performance Evaluation Process." Chicago, IL: American Bar Association. Retrieved January 16, 2010 (www.abanet.org).

Pinto, Katy and Scott Coltrane. 2009. "Divisions of Labor in Mexican Origin and Anglo Families: Structure and Culture." *Sex Roles* 60(7/8):482–95.

Pipher, Mary. 1995. *Reviving Ophelia: Saving the Lives of Adolescent Girls.* New York: Ballantine.

Pirog-Good, Maureen A. and Lydia Amerson. 1997. "The Long Arm of Justice: The Potential for Seizing the Assets of Child Support Obligors." *Family Relations* 46(1):47–54.

Pisano, Marina. 2005. "Young Adults Live in Luxury at Family's Expense." *San Antonio Express.* Retrieved October 29, 2006 (www. mysanantonio.com).

Pittman, Joe F., Jennifer L. Kerpelman, and Jennifer M. McFadyen. 2004. "Internal and External Adaptation in Army Families: Lessons from Operations Desert Shield and Desert Storm." *Family Relations* 53(3):249–60.

Pitts, Leonard, Jr. 2008. "Overstressed Parents, Kids Need Help." *Miami Herald,* November 29. Retrieved December 8, 2008 (www.miamiherald. com).

Pitzer, Ronald L. 1997a. "Change, Crisis, and Loss in Our Lives." University of Minnesota Extension Service (www.extension.umn.edu).

———. 1997b. "Perception: A Key Variable in Family Stress Management." University of Minnesota Extension Service (www.extension. umn.edu).

Planned Parenthood of Central Missouri v. Danforth. 1976. 428 U.S. 52.

Pleck, Joseph H. 1992. "Prisoners of Manliness." Pp. 98–107 in *Men's Lives,* 2nd ed., edited by Michael S. Kimmel and Michael A. Messner. New York: Macmillan.

Ploeg, Jenny, Jana Fear, and Brian Hutchison. 2009. "A Systematic Review of Interventions for Elder Abuse." *Journal of Elder Abuse and Neglect* 21(3):187–210.

Plotnick, Robert D. 2007. "Adolescent Expectations and Desires About Marriage and Parenthood." *Journal of Adolescence* 30(6):943–63.

Poe, Marshall. 2004. "The Other Gender Gap." *Atlantic Monthly,* February, p. 137.

Poehlmann, Julie. 2005. "Children's Family Environments and Intellectual Outcomes during Maternal Incarceration." *Journal of Marriage and Family* 67(5):1275–85.

Polikoff, Nancy D. 2008. *Beyond (Straight and Gay) Marriage: Valuing All Families Under the Law.* Boston, MA: Beacon Press.

"Poll: Blacks Optimistic About Future." 2010. CNN.com, January 14, 2010. Retrieved January 13 (www.cnn.com).

Pollack, William. 1998. *Real Boys: Rescuing Our Sons from the Myths of Boyhood.* New York: Random House.

———. 2006. "The "War" For Boys: Hearing "Real Boys" Voices, Healing Their Pain." *Professional Psychology: Research and Practice* 37(2):190–95.

Pollet, Susan L. and Melissa Lombreglia. 2008. "A Nationwide Survey of Mandatory Parent Education." *Family Court Review* 46(2):375–94.

Pollitt, Katha. 2002. "Backlash Babies." *The Nation,* May 13, p. 10.

Polyamory Society. n.d. "Children Education Branch." Retrieved October 4, 2006 (www. polyamorysociety.org).

Pomerleau, A., D. Bolduc, G. Malcuit, and L. Cossetts. 1990. "Pink or Blue: Environmental Stereotypes in the First Two Years of Life." *Sex Roles* 22:359–67.

Pong, Suet-ling, Jamie Johnston, Vivien Chen. 2010. "Authoritarian Parenting and Asian Adolescent School Performance: Insights from the US and Taiwan." *International Journal of Behavioral Development* 34(1):62–72.

Poniewozik, James. 2002. "The Cost of Starting Families." *Time,* April 15, pp. 56–58.

Poortman, Anne-Rigt and Judith A. Seltzer. 2007. "Parents' Expectations about Children after Divorce: Does Anticipating Difficulty Deter Divorce?" *Journal of Marriage and Family* 69(1):254–69.

Popenoe, David. 1993. "American Family Decline, 1960–1990: A Review and Appraisal." *Journal of Marriage and Family* 55(3):527–55.

———. 1994. "The Evolution of Marriage and the Problem of Stepfamilies: A Biosocial Perspective." Pp. 3–28 in *Stepfamilies: Who Benefits? Who Does Not?,* edited by Alan Booth and Judy Dunn. Hillsdale, NJ: Erlbaum.

———. 2007. "Essay: The Future of Marriage in America." *The State of Our Unions 2007: The Social Health of Marriage in America.* Rutgers University. Retrieved October 12, 2009 (marriage.rutgers. edu).

———. 2008. "Cohabitation, Marriage, and Child Wellbeing: A Cross-National Perspective." The National Marriage Project at Rutgers University. Retrieved October 12, 2009 (marriage.rutgers.edu).

Popenoe, David and Barbara Dafoe Whitehead. 2000. *The State of Our Unions 2000: The Social Health of Marriage in America*. New Brunswick, NJ: Rutgers University, National Marriage Project.

———. 2005. "The State of Our Unions 2005: The Social Health of Marriage in America." New Brunswick, NJ: Rutgers University, National Marriage Project. Retrieved August 23, 2006 (http://marriage.rutgers.edu).

Popenoe, David and Barbara Dafoe Whitehead, eds. 2009. *The State of Our Unions 2009: Money & Marriage in America*. New Brunswick, NJ: Rutgers University, National Marriage Project. Retrieved December 21, 2009 (marriage. rutgers.edu).

Popkin, Michael H. and Elizabeth Einstein. 2006. *Active Parenting: A New Program to Help Create Successful Stepfamilies*. Kennesaw, GA: Active Parenting Publishers.

Population Reference Bureau. 2009. "Social Support, Networks, and Happiness." *Today's Research on Aging* 17: June. Retrieved May 8, 2010 (www.prb.org).

Porche, Michelle and Diane Purvin. 2008. "'Never in Our Lifetime': Legal Marriage for Same-Sex Couples in Long-Term Relationships." *Family Relations* 37(April):144–59.

Porter, Eduardo. 2006. "Stretched to Limit, Women Stall March to Work." *New York Times*, March 21.

Porter, Eduardo and Michelle O'Donnell. 2006. "Facing Middle Age with No Degree, and No Wife." *New York Times*, August 6.

Porterfield, Ernest. 1982. "Black-American Intermarriages in the United States." Pp. 17–34 in *Intermarriages in the United States*, edited by Gary Crester and Joseph J. Leon. New York: Haworth.

Porterfield, Shirley L. 2002. "Work Choices of Mothers in Families with Children with Disabilities." *Journal of Marriage and Family* 64(4):972–81.

Potter-Efron, Ronald T. and Patricia S. Potter-Efron. 2008. *The Emotional Affair: How to Recognize Emotional Infidelity and What to Do About It*. Oakland, CA: New Harbinger Publications, Inc.

"Pow Wow Culture." 2006. *Indian Country Diaries* (TV series). Washington, DC: Public Broadcasting System. Retrieved December 2, 2006 (http://www.pbs.org/indiancountry).

Powell, Brian and Douglas Downey. 1997. "Living in Single-parent Households: An Investigation of the Same-sex Hypothesis." *American Sociological Review* 62:521–39.

Powell, Jean W. 2004. *Older Lesbian Perspectives on Advance Care Planning*. University of Rhode Island.

Power, Kathryn. 2004. "Resilience and Recovery in Family Mental Health Care." *Family Focus*

(March):F1–F2. National Council on Family Relations.

Power, Paul W. 1979. "The Chronically Ill Husband and Father: His Role in the Family." *Family Coordinator* 28:616–21.

Prahlad, Anand. 2006. *The Greenwood Encyclopedia of African American Folklore*. Westport, CT: Greenwood Press.

"Pregnancy and Childbirth." 2007. Atlanta: U.S. Centers for Disease Control and Prevention. http://www.cdc.gov/hiv/topics/perinatal/index.htm.

"Preserving Families of African Ancestry." 2003. Washington, DC: *National Association of Black Social Workers*, January 10. www.nabsw.org.

Press, Julie E. 2004. "Cute Butts and Housework: A Gynocentric Theory of Assortive Mating." *Journal of Marriage and Family* 66(4):1029–33.

Presser, Harriet B. 2000. "Nonstandard Work Schedules and Marital Instability." *Journal of Marriage and Family* 62:93–110.

Preston, Julia. 2007. "As Pace of Deportation Rises, Illegal Families Are Digging In." *New York Times*, May 1.

———. 2009. "Bill Proposes Immigration Rights for Gay Couples." *The New York Times*, June 3. Retrieved October 25, 2009 (www.nytimes.com).

Preuschoff, Gisela. 2006. *Raising Girls*. Berkeley, CA: Celestial Press.

Preves, Sharon E. 2002. "Sexing the Intersexed: An Analysis of Sociocultural Responses to Intersexuality." *Signs* 27:523–56.

Previti, Denise and Paul R. Amato. 2003. "Why Stay Married? Rewards, Barriers, and Marital Stability." *Journal of Marriage and Family* 65:561–73.

Priem, Jennifer S., Rachel McLaren, and Denise Haunani Soloman. 2010. "Relational Messages, Perceptions of Hurt, and Biological Stress Reactions to a Disconfirming Interaction." *Communication Research* 37(1):48–72.

Priem, Jennifer S., Denise Haunani Solomon, and Keli Ryan Steuber. 2009. "Accuracy and Bias in Perceptions of Emotionally Supportive Communication in Marriage." *Personal Relationships* 16:531–51.

Proctor, Bernadette D. and Joseph Dalaker. 2003. *Poverty in the United States: 2002*. Current Population Reports P60-222. Washington, DC: U.S. Census Bureau.

"Professor Says Seminary Dismissed Her Over Gender." 2007. *New York Times*, January 27.

"Promoting Respectful, Nonviolent Intimate Partner Relationships through Individual, Community, and Societal Change." 2009. *Strategic Direction for Intimate Partner Violence Prevention*. Atlanta, GA: Centers for Disease Control and Prevention, National Center for Injury Prevention and Control. Retrieved June 10, 2010 (www.cdc.gov).

Prospero, Moises. 2006. "The Role of Perceptions in Dating Violence among Young Adolescents." *Journal of Interpersonal Violence* 21(4):470–84.

"The Psychological Impact of Infertility and Its Treatment." 2009. *Harvard Mental Health Letter* 24(11):1–3.

"PTSD and Relationships: A National Center for PTSD Fact Sheet." 2006. United States Department of Veterans Affairs National Center for PTSD. Retrieved August 16, 2006 (www.ncptsd.va.gov).

Pugh, Allison J. 2009. *Longing and Belonging: Parents, Children, and Consumer Culture*. Berkeley, CA: University of California Press.

Punyanunt-Carter, Narissra Maria. 2008. "Father-Daughter Relationships: Examining Family Communication Patterns and Interpersonal Communication Satisfaction." *Communication Research Reports* 25(1):23–33.

Purcell, Patrick. 2009. *Income and poverty Among Older Americans in 2008*. Congressional Research Service. October 2. Retrieved May 8, 2010 (www.crs.gov).

Purcell, Tom. 2007. "Baby Name Game Brings Shame." *Omaha World Herald*, August 6.

Purdy, Matthew. 1995. "A Sexual Revolution for the Elderly: At Nursing Homes, Intimacy Is Becoming a Matter of Policy." *New York Times*, November 6.

Purkayastha, Bandana. 2002. "Rules, Roles, and Realities: Indo-American Families in the United States." Pp. 212–26 in *Minority Families in the United States*, 3rd ed., edited by Ronald L. Taylor. Upper Saddle River, NJ: Prentice Hall.

Purnine, Daniel M. and Michael P. Carey. 1998. "Age and Gender Differences in Sexual Behavior Preferences: A Follow-Up Report." *Journal of Sex & Marital Therapy* 24:93–102.

Pyke, Karen D. 1999. "The Micropolitics of Care in Relationships between Aging Parents and Adult Children: Individualism, Collectivism, and Power." *Journal of Marriage and Family* 61(3):661–72.

———. 2007. " 'The Normal American Family' as an Interpretive Structure of Family Life among Grown Children and Vietnamese Immigrants." Pp. 469–90 in *Family in Transition*, 14th ed., edited by Arlene S. Skolnick and Jerome H. Skolnick. Boston, MA: Allyn and Bacon.

Pyke, Karen and Michelle Adams. 2010. "What's Age Got to Do With It? A Case Study Analysis of Power and Gender in Husband-Older Marriages." *Journal of Family Issues* 31(6):748–77.

Pyke, Karen D. and Vern L. Bengston. 1996. "Caring More or Less: Individualistic and Collectivist Systems of Family Eldercare." *Journal of Marriage and Family* 58(2):379–92.

Qian, Zhenchao and Daniel T. Lichter. 2007. "Social Boundaries and Marital Assimilation: Interpreting Trends in Racial and Ethnic Intermarriage." *American Sociological Review* 72(1):68–94.

Quinn, Christine C. 2009. "Letter to Hon. Gary Locke, Secretary, U.S. Department of Commerce." April 8. Retriever June 13, 2009 (council.nyc.gov).

Rabin, Roni. 2006. "Health Disparities Persist for Men, and Doctors Ask Why." *New York Times*, November 14.

———. 2007. "It Seems the Fertility Clock Ticks for Men, Too." *New York Times*, February 27.

Radey, Melissa and Karen A. Randolph. 2009. "Parenting Sources: How Do Parents Differ in Their Efforts to Learn About Parenting?" *Family Relations* (December):536–48.

Radina, M. Elise. 2007. "Mexican American Siblings Caring for Aging Parents: Processes of Caregiver Selection/Designation." *Journal of Comparative Family Studies* 38(1):143–68.

Raley, R. Kelly and Jennifer Bratter. 2004. "Not Even if You Were the Last Person on Earth! How Marital Search Constraints Affect the Likelihood of Marriage." *Journal of Family Issues* 25(2):167–81.

Raley, R. Kelly and Elizabeth Wildsmith. 2004. "Cohabitation and Children's Family Instability." *Journal of Marriage and Family* 66(1):210–19.

Ramirez, Robert R. and G. Patricia de la Cruz. 2003. *The Hispanic Population of the United States: March 2002*. Current Population Reports P-20-545. Washington, DC: U.S. Census Bureau. June.

Ramnarace, Cynthia. 2010. "Congress Passes Elder Justice Act." *AARP Bulletin Today*, March 25. Retrieved May 15, 2010 (bulletin.aarp.org).

Ramo, Joshua Cooper. 2009. "Unemployment Nation." *Time*, September 21, pp. 26–32.

Ramos-Sánchez, Lucila and Donald R. Atkinson. 2009. "The Relationships Between Mexican American Acculturation, Cultural Values, Gender, and Help-Seeking Intentions." *Journal of Counseling & Development* 87(1):62–71.

Rampell, Catherine. 2009a. "Money Fights Predict Divorce Rates." *The New York Times Economix Blog*. Retrieved January 11, 2010 (economix.blogs.nytimes.com).

———. 2009b. "SAT Scores and Family Income." *New York Times*, August 27.

Rampey, B. D., G. S. Dion, and P. L. Donahue. 2009. *NAEP 2008 Trends in Academic Progress* (NCES 2009–479). Washington, DC: National Center for Education Statistics, Institute of Education Sciences, U.S. Department of Education.

Rankin, Jane. 2005. *Parenting Experts: Their Advice, the Research and Getting It Right*. Westport, CT: Praeger.

Rao, Patricia A. and Deborah C. Beidel. 2009. "The Impact of Children with High-Functioning Autism on Parental Stress, Sibling Adjustment, and Family Functioning." *Behavior Modification* 33(4):437–51.

Rasberry, Catherine N. and Patricia Goodson. 2009. "Predictors of Secondary Abstinence in U.S. College Undergraduates." *Archives of Sexual Behavior* 38:74–86.

Raschick, Michael and Berit Ingersoll-Dayton. 2004. "Costs and Rewards of Caregiving among Aging Spouses and Adult Children." *Family Relations* 53(3):317–25.

Rasheed, Janice M., Mikal N. Rasheed, and James A. Marley. 2010. *Readings in Family Therapy: From Theory to Practice*. Los Angeles: Sage.

Rasmussen Reports. 2008. "71% Willing to Vote for Woman President, 73% for African-American." *Rasmussen Reports*, February

17. Retrieved November 27, 2009 (www.rasmussenreports.com).

———. 2009. "80% Say Parents Should Teach Their Children about Sex." *Rasmussen Reports*, January 14. Retrieved March, 2010 (www.rasmussenreports.com).

Rauch, Jonathan. 2004. "Considerations on Gay Marriage: Conservative Policy Dilemmas." *The Public Interest* 156:17–24.

Rauer, Amy J. and Brenda L. Volling. 2007. "Differential Parenting and Sibling Jealousy: Developmental Correlates of Young Adults' Romantic Relationships." *Personal Relationships* 14:495–511.

Raven, Bertram, Richard Centers, and Arnoldo Rodrigues. 1975. "The Bases of Conjugal Power." Pp. 217–32 in *Power in Families*, edited by Ronald E. Cromwell and David H. Olson. Beverly Hills, CA: Sage Publications.

Ravitz, Paula, Robert Maunder, and Carolina McBride. 2008. "Attachment, Contemporary Interpersonal Theory, and IPT: An Integration of Theoretical, Clinical, and Empirical Perspectives." *Journal of Contemporary Psychotherapy* 38(1):11–21.

Ray, S. Alan. 2006. "A Race or a Nation? Cherokee National Identity and the Status of Freedmen's Descendents." Legal Series August #1570. Berkeley, CA: Berkeley Electronic Press (bepress).

Raymond, Joan and Devin Gordon. 2008. "Won't You Be My Neighbor?" *Newsweek*, August 11, p. 66.

Reader, Steven K., Lindsay M. Stewart, and James H. Johnson. 2009. *Journal of Clinical Psychology in Medical Settings* 16:148–60.

Real, Terrence. 1997. *I Don't Want to Talk about It: Overcoming the Secret Legacy of Male Depression*. New York: Scribner's.

———. 2002. *How Can I Get Through to You? Reconnecting Men and Women*. New York: Scribner's.

Reardon-Anderson, Jane, Randy Capps, and Michael Fix. 2002. *The Health and Well-Being of Children in Immigrant Families*. Series B, No. B-52. Washington, DC: The Urban Institute. November.

"Recession Depression: Men Are Twice as Likely to Suffer Stress in Silence." 2009. *Daily Mail*, May 11. Retrieved June 15, 2009 (www.dailymail.co.uk).

Rector, Robert and Kirk A. Johnson. 2005. "Adolescent Virginity Pledges and Risky Sexual Behaviors." Presented at the Eighth Annual National Welfare Research and Evaluation Conference of the Administration for Children and Families, June 15, Washington, DC. Retrieved March 28, 2007 (www.heritage.org).

Rector, Robert E. and Melissa G. Pardue. 2004. *Understanding the President's Healthy Marriage Initiative*. Heritage Foundation (www.heritage.org/Research/Family/bg1741.cfm).

Reczek, Corinne, Sinikka Elliott, and Debra Umberson. 2009. "Commitment Without Marriage." *Journal of Family Issues* 30(6):738–56.

Reed, Acacia and Philippa Strum. 2008. *New Scholarship in Race and Ethnicity—The Long*

Shadow of the GI Bill: U.S. Social Policy and the Black-White Gap. Washington, DC: Woodrow Wilson International Center for Scholars.

Reeves, Terrance J. and Claudette E. Bennett. 2004. *We the People: Asians in the United States*. CDNSR-17. Washington, DC: U.S. Census Bureau. December.

"Registration of Holy Union for the Dignity USA National Registry." 2008. Retrieved November 22, 2009 (www.dignityusa.org).

Rehman, Uzma, Amy Holtzworth-Munroe, Katherine Herron, and Kahni Clements. 2009. "'My Way or No Way': Anarchic Power, Relationship Satisfaction, and Male Violence." *Personal Relationships* 16:475–88.

Reinert, Duane F. 2005. "Spirituality, Self-Representations, and Attachment to Parents: A Longitudinal Study of Roman Catholic College Seminarians." *Counseling and Values* 49(3):226–38.

Reiss, David, Sandra Gonzalez, and Norman Kramer. 1986. "Family Process, Chronic Illness, and Death: On the Weakness of Strong Bonds." *Archives of General Psychiatry* 43:795–804.

Reiss, Ira L. 1976. *Family Systems in America*. 2nd ed. Hinsdale, IL: Dryden.

———. 1986. *Journey into Sexuality: An Exploratory Voyage*. Englewood Cliffs, NJ: Prentice Hall.

"Religion's Generation Gap." 2007. *The Wall Street Journal*, March 2: W1, W12.

Renteln, Alison Dundes. 2004. *The Cultural Defense*. New York: Oxford University Press.

"Report: Gay Couples Similar to Straight Spouses in Age Income." 2009. Associated Press. Retrieved November 20, 2009 (www.usatoday).

Rest, Kathleen M. and Michael H. Halpern. 2007. "Politics and the Erosion of Federal Scientific Capacity: Restoring Scientific Integrity to Public Health Science." *American Journal of Public Health* 97(11):1939-44.

Retz, Wolfgang and Rachel G. Klein. 2010. *Attention-Deficit Hyperactivity Disorder (ADHD) in Adults*. New York: Karger.

Reynolds v. United States. 1878. 98 U.S. 145, 25 L. Ed. 244.

Rhoades, Galena Kline, Jocelyn N. Petrella, Scott M. Stanley, Howard J. Markman. 2006. "Premarital Cohabitation, Husbands' Commitment, and Wives' Satisfaction with the Division of Household Contributions." *Marriage & Family Review* 40(4):5–22.

Rhoades, Galena K., Scott M. Stanley, and Howard J. Markman. 2009. "Couples' Reasons for Cohabitation." *Journal of Family Issues* 30(2):233–58.

Rhodes, Angel R. 2002. "Long-Distance Relationships in Dual-Career Commuter Couples: Review of Counseling Issues." *The Family Journal* 10:398–404.

Rice, C. 2007. "Becoming 'the Fat Girl': Acquisition of an Unfit Identity." *Women's Studies International Forum* 30(2):158–74.

Richards, Marty. 2009. *Caresharing: A Reciprocal Approach to Caregiving and Care Receiving in the Complexities of Aging, Illness, or Disability.* Woodstock, VT: Skylight Paths Publishers.

Richardson, Joseph B., Jr. 2009. "Men Do Matter: Ethnographic Insights on the Socially Supportive Role of the African American Uncle in the Lives of Inner-City African American Male Youth." *Journal of Family Issues* 30(8):1041–69.

Richtel, Matt. 2005. "Past Divorce, Compassion at the End." *New York Times*, May 19.

Ridgeway, Cecilia. 2006. "Gender as an Organizing Force in Social Relations: Implications for the Future of Inequality." Pp. 245–64 in *The Declining Significance of Gender?*, edited by Francine D. Blau, Mary C. Brinton, and David B. Grusky. New York: Russell Sage.

Riedmann, Agnes and Lynn White. 1996. "Adult Sibling Relationships: Racial and Ethnic Comparisons." Pp. 105–26 in *Sibling Relationships: Their Causes and Consequences*, edited by Gene H. Brody. Norwood, NJ: Ablex.

Rigby, Jill M. 2006. *Raising Respectful Children in a Disrespectful World.* New York: Howard Books.

Riger, Stephanie, Susan L. Staggs, and Paul Schewe. 2004. "Intimate Partner Violence as an Obstacle to Employment among Mothers Affected by Welfare Reform." *Journal of Social Issues* 60(4):801–17.

Riley, Glenda. 1991. *Divorce: An American Tradition.* New York: Oxford University Press.

Riley, Matilda W. 1983. "The Family in an Aging Society: A Matrix of Latent Relationships." *Journal of Family Issues* (4):439–54.

Rinker, Austin G. 2009. "Recognition and Perception of Elder Abuse by Prehospital and Hospital-Based Care Providers." *Archives of Gerontology and Geriatrics* 48(1):110–16.

Ring, Wilson. 2009. "Vt. Judge: Birth Mom Must Give Child to Ex-Partner." Associated Press, December 29. Retrieved December 31, 2009 (www.comcast.net/articles).

Risch, Gail S., Lisa A. Riley, and Michael G. Lawler. 2004. "Problematic Issues in the Early Years of Marriage: Content for Premarital Education." *Journal of Psychology and Theology* 31(1):253–69.

Risman, Barbara J. and Pepper Schwartz. 2002. "After the Sexual Revolution: Gender Politics in Teen Dating." *Contexts* (Spring):16–23.

Roan, Shari. 2009a. "Recession May Be Behind Decline in Fertility Treatments." *Los Angeles Times*, May 28. Retrieved June 8, 2009 (www.newsday.com).

———. 2009b. "Unintended Pregnancies: A Sign of the Times." *Los Angeles Times*, May 7. Retrieved June 8, 2009 (www.newsday.com).

Robbers, Monica L. P. 2009. "Facilitating Fatherhood: A Longitudinal Examination of Father Involvement Among Young Minority Fathers." *Child & Adolescent Social Work Journal* 26(2):121–34.

Roberto, Anthony, Kellie Carlyle, and Catherine Goodall. 2007. "Communication and Corporal Punishment: The Relationship Between Self-Report Parent Verbal and Physical Aggression." *Communication Research Reports* 24(2):103–11.

Roberts, Alex. 2009. "Marriage & the Great Recession." Pp. 31–48 in *The State of Our Unions 2009: Money & Marriage in America*, edited by David Popenoe and Barbara Dafoe Whitehead. New Brunswick, NJ: Rutgers University, National Marriage Project. Retrieved December 21, 2009 (marriage.rutgers.edu).

Roberts, Alison. 2005. "Americans Are Alone Together." *The Sacramento Bee*, November 6, pp. H1, H5.

Roberts, James C., Loreen Wolfer, and Marie Mele. 2008. "Why Victims of Intimate Partner Violence Withdraw Protection Orders." *Journal of Family Violence* 23(5):369–75.

Roberts, Linda J. 2000. "Fire and Ice in Marital Communication: Hostile and Distancing Behaviors as Predictors of Marital Distress." *Journal of Marriage and Family* 62(3):693–707.

———. 2005. "Alcohol and the Marital Relationship." *Family Focus on . . . Substance Abuse across the Life Span*: FF25: F12–F13. Minneapolis, MN: National Council of Family Relations.

Roberts, Nicole A. and Robert W. Levenson. 2001. "The Remains of the Workday: Impact of Job Stress and Exhaustion on Marital Interaction in Police Couples." *Journal of Marriage and Family* 63(4):1052–67.

Roberts, Sam. 2007. "Children's Quality of Life Is on the Rise, Report Finds." *New York Times*, January 11.

Robinson, B. E. and R. L. Barret. 1986. *The Developing Father: Emerging Roles in Contemporary Society.* New York: Guilford.

Robinson, Robert Burton. 2009. "7 Steps to Adult Family Conflict Resolution." Retrieved February 21, 2010 (www.mindovermania.com).

Robinson, Russell K. 2008. "Structural Dimensions of Romantic Preferences." *Fordham Law Review* 76(6):2787–2819.

Robison, Jennifer. 2003. "Young Love, First Love, True Love?" The Gallup Poll, February 11. Retrieved August 28, 2006 (www.gallup.com/content).

Robitaille, Caroline and Marie-Christine Sant-Jacques. 2009. "Social Stigma and the Situation of Young People in Lesbian and Gay Stepfamilies." *Journal of Homosexuality* 56:421–42.

Rockquemore, Kerry Ann. 2002. "Negotiating the Color Line: The Gendered Process of Racial Identity Construction among Black/White Biracial Women." *Gender and Society* 16(4):485–503.

Rockquemore, Kerry Ann and Tracey A. Laszloffy. 2005. *Raising Biracial Children.* Lanham, MD: AltaMira.

Rodgers, Joseph Lee, Paul A. Nakonezny, and Robert D. Shull. 1997. "The Effect of No-fault Divorce Legislation on Divorce Rates: A Response to a Reconsideration." *Journal of Marriage and Family* 59:1020–30.

Rodman, Hyman. 1971. *Lower Class Families: The Culture of Poverty in Rural Trinidad.* New York: Oxford University Press.

Rodrigues, Amy E., Julie H. Hall, and Frank D. Fincham. 2006. "What Predicts Divorce and Relationship Dissolution?" Pp. 85–112 in *Handbook of Divorce and Relationship Dissolution*, edited by Mark A. Fine and John H. Harvey. Mahwah, NJ: Erlbaum.

Rodriguez, Gregory. 2003. "Mongrel America." *Atlantic Monthly*, January/February, pp. 95–97.

Roe v. Wade. 1973. 410 U.S. 113.

Roest, Annette M., Judith Semon Dubas, and Jan R. M. Gerris. 2009. "Value Transmissions Between Fathers, Mothers, and Adolescent and Emerging Adult Children: The Role of the Family Climate." *Journal of Family Psychology* 23(2):146–55.

Rogers, Lesley. 2001. *Sexing the Brain.* New York: Columbia University Press.

Rogers, Michelle L. and Dennis P. Hogan. 2003. "Family Life with Children with Disabilities: The Key Role of Rehabilitation." *Journal of Marriage and Family* 65(4):818–33.

Rogers, Stacy J. and Paul R. Amato. 2000. "Have Changes in Gender Relations Affected Marital Quality?" *Social Forces* 79:731–53.

Roisman, Glenn I. and R. Chris Fraley. 2006. "The Limits of Genetic Influence: A Behavior-Genetic Analysis of Infant-Caregiver Relationship Quality and Temperament." *Child Development* 77(6):1656–67.

Romero, Mary. 1992. *Maid in the U.S.A.* New York: Routledge.

Roper, Susanne Olsen and Jeffrey B. Jackson. 2007. "The Ambiguities of Out-of-Home Care: Children with Severe or Profound Disabilities." *Family Relations* 56(2):147–61.

Roper, Susanne Olsen, and Jeremy B. Yorgason. 2009. "Older Adults with Diabetes and Osteoarthritis and Their Spouses: Effects of Activity Limitations, Marital Happiness, and Social Contacts on Partners' Daily Mood." *Family Relations* 58(October):460–74.

Rosario, Margaret, Eric W. Schrimshaw, and Joyce Hunter. 2008. "Predicting Different Patterns of Sexual Identity Development Over Time Among Lesbian, Gay, and Bisexual Youths: A Cluster Analytic Approach." *American Journal of Community Psychology* 42(3/4):266–82.

Rosario, Margaret, Eric W. Schrimshaw, Joyce Hunter, and Lisa Braun. 2006. "Sexual Identity Development among Lesbian, Gay, and Bisexual Youths: Consistency and Change over Time." *Journal of Sex Research* 43(1):46–58.

Roscoe, Lori, Elizabeth Corsentino, Shirley Watkins, Marcia McCall, and Juan Sanchez-Ramos. 2009. "Well-Being of Family Caregivers of Persons with Late-Stage Huntington's Disease: Lessons in Stress and Coping." *Health Communication* 24(3):239–48.

Rose, A. J. and K. D. Rudolph. 2006. "A Review of Sex Differences in Peer Relationship Processes: Potential Trade-Offs for the Emotional and Behavioral Development of Girls and Boys." *Psychological Bulletin* 132:98–131.

Rosen, Ruth. 2007. "The Care Crisis." *The Nation*, March 12, pp. 11–16.

Rosenbaum, Janet Elise. 2009. "Patient Teenagers? A Comparison of the Sexual Behavior of Virginity Pledgers and Matched Nonpledgers." *Pediatrics* 123(1):e110–20.

Rosenblatt, Paul C. and Linda Hammer Burns. 1986. "Long-term Effects of Perinatal Loss." *Journal of Family Issues* 7:237–54.

Rosenbloom, Stephanie. 2005. "Supersize Strollers Ignite Sidewalk Drama." *New York Times,* September 22.

———. 2006. "Here Come the Great-Grandparents." *New York Times,* November 2. Retrieved November 3, 2006 (http://www.nytimes.com).

Rosenbluth, Susan C., Janice M. Steil, and Juliet H. Whitcomb. 1998. "Marital Equality: What Does It Mean?" *Journal of Family Issues* 19(3):227–44.

Roseneil, Sasha and Shelley Budgeon. 2004. "Cultures of Intimacy and Care Beyond 'the Family': Personal Life and Social Change in the Early 21st Century." *Current Sociology* 52(2):135–59.

Rosenfeld, Dana. 1999. "Identity Work among Lesbian and Gay Elderly." *Journal of Aging Studies* 13(2):121–44.

Rosenfeld, Michael J. 2007. *The Age of Independence: Interracial Unions, Same-Sex Unions, and the Changing American Family.* Cambridge, MA: Harvard University Press.

———. 2008. "Increasing Percentage of Marriages in the U.S. that are Interracial." Stanford, CA: Stanford University, Department of Sociology. Retrieved November 1, 2009 (www.stanford.edu).

Rosenfeld, Michael J. and Kim Byung-Soo. 2005. "The Independence of Young Adults and the Rise of Interracial and Same-Sex Unions." *American Sociological Review* 70(4):541–62.

Ross, Catherine E. 1995. "Reconceptualizing Marital Status as a Continuum of Social Attachment." *Journal of Marriage and Family* 57(1):129–40.

Ross, Mary Ellen Trail and Lu Ann Aday. 2006. "Stress and Coping in African American Grandparents Who Are Raising Their Grandchildren." *Journal of Family Issues* 27(7):912–32.

Rossi, Alice S. 1968. "Transition to Parenthood." *Journal of Marriage and Family* 30:26–39.

———. 1973. *The Feminist Papers.* New York: Bantam.

———. 1984. "Gender and Parenthood." *American Sociological Review* 49:1–19.

Roth, Benita. 2004. *Separate Roads to Feminism: Black, Chicana, and White Feminist Movements in America's Second Wave.* New York: Cambridge University Press.

Roth, Ilona and Chris Barson. 2010. *The Autism Spectrum in the 21st Century: Exploring Psychology, Biology, and Practice.* London: Jessica Kingsley Publishers.

Roth, Wendy. 2005. "The End of the One-Drop Rule? Labeling of Multiracial Children in Black Intermarriages." *Sociological Forum* 20(1):35–67.

Rothman, Barbara Katz. 1999. "Comment on Harrison: The Commodification of Motherhood." Pp. 435–38 in *American Families: A Multicultural Reader,* edited by Stephanie Coontz. New York: Routledge.

Roughgarden, Joan. 2009a. *Evolution's Rainbow: Diversity, Gender, and Sexuality in Nature and People.* Berkeley: University of California Press.

———. 2009b. *The Genial Gene: Deconstructing Darwinian Selfishness.* Berkeley: University of California Press.

Rovers, Martin W. 2006. *Healing the Wounds That Hurt Relationships.* Peabody, MA: Hendrickson.

Rowland, Donald T. 2007. "Historical Trends in Childlessness." *Journal of Family Issues* 28(10):1311–37.

Roy, Kevin M., Nicolle Buckmiller, and April McDowell. 2008. "Together but Not 'Together': Trajectories of Relationship Suspension for Low-Income Unmarried Parents." *Family Relations* 57(April):198–210.

Royster, Deirdre A. 2005. "Review of *White Out: The Continuing Significance of Racism,* edited by Ashley W. Doane and Eduardo Bonilla-Silva (Routledge, 2003)." *Contemporary Sociology* 35(3):255–56.

Ruane, Michael E. 2006. "Marriages Tested by Scars of War." *Washington Post,* May 27.

Rubin, Lillian B. 1976. *Worlds of Pain: Life in the Working-Class Family.* New York: Basic Books.

———. 2007. "The Approach-Avoidance Dance: Men, Women, and Intimacy." Pp. 319–24 in *Men's Lives,* 7th ed., edited by Michael S. Kimmel and Michael A. Messner. Boston, MA: Pearson.

Rubin, Lisa and Nancy Felipe Russo. 2004. "Abortion and Mental Health: What Therapists Need to Know." *Women and Therapy* 27(3/4):69–90.

Rubin, Roger H. 2001. "Alternative Lifestyles Revisited, or Whatever Happened to Swingers, Group Marriages, and Communes?" *Journal of Family Issues* 22(6):711–26.

Rudman, Laurie A. and Peter Glick. 2008. *The Social Psychology of Gender: How Power and Intimacy Shape Gender Relations.* New York: Guilford Press.

Rudolph, Bonnie, Cecily Cornelius-White, and Fernando Quintana. 2005. "Filial Responsibility among Mexican American College Students: A Pilot Investigation and Comparison." *Journal of Hispanic Higher Education* 4(1):64–78.

Ruefli, Terry, Olivia Yu, and Judy Barton. 1992. "Brief Report: Sexual Risk Taking in Smaller Cities." *Journal of Sex Research* 29(1):95–108.

Rueter, Martha A., and Ascan F. Koerner. 2008. "The Effect of Family Communication Patterns on Adopted Adolescent Adjustment." *Journal of Marriage and Family* 70(3):715–27.

Ruiz, Sarah A. and Merril Silverstein. 2007. "Relationships with Grandparents and the Emotional Well-Being of Late Adolescent and Young Adult Grandchildren." *Journal of Social Issues* 63(4):793–808.

Rumney, Avis. 2009. *Dying to Please: Anorexia, Treatment and Recovery.* Jefferson, NC: McFarland & Company.

Runner, Michael, Mieko Yoshihama, and Steve Novick. 2009. *Intimate Partner Violence in Immigrant and Refugee Communities: Challenges, Promising Practices and Recommendations.* Princeton, NJ: Family Violence Prevention Fund for the Robert Wood Johnson Foundation.

Russell, Stephen T., Thomas J. Clarke, and Justin Clary. 2009. "Are Teens 'Post-Gay'? Contemporary Adolescents' Sexual Identity Labels." *Journal of Youth and Adolescence* 38(7):884–90.

Russo, Nancy Felipe and Amy J. Dabul. 1997. "The Relationship of Abortion to Well-Being: Do Race and Religion Make a Difference?" *Professional Psychology: Research and Practice* 28:23–31.

Russo, Nancy Felipe and Kristin L. Zierk. 1992. "Abortion, Childbearing, and Women's Well-Being." *Professional Psychology: Research and Practice* 23:269–80.

Rutgers University School of Criminal Justice and the New Jersey Institute for Social Justice. *Bringing Families In: Recommendations of the Incarceration, Reentry and the Family Roundtables.* 2006. December. Retrieved March 11, 2010 (www.reentry.net/library/item.126583-Bringing_Families_In_Recommendations_of_the_Incarceration_Reentry_and_the_F).

Rutter, Michael. 2002. "Nature, Nurture, and Development: From Evangelism through Science toward Policy and Practice." *Child Development* 73(1):1–21.

Ryan, Joan. 2006. "War Without End." *San Francisco Chronicle,* March 26.

Ryan, Kathryn M. and Sharon Mohr. 2005. "Gender Differences in Playful Aggression during Courtship in College Students." *Sex Roles: A Journal of Research* 53(7-8):591–602.

Ryan, Kathryn M., Kim Weikel, and Gene Sprechini. 2008. "Gender Differences in Narcissism and Courtship Violence in Dating Couples." *Sex Roles* 58:802–13.

Ryan, Rebecca M., Ariel Kalil, and Lindsey Leininger. 2009. "Low-Income Mothers' Private Safety Nets and Children's Socioemotional Well-Being." *Journal of Marriage and Family* 71(2):278–97.

Ryan, Rebecca M., Ariel Kalil, and Kathleen M. Ziol-Guest. 2008. "Longitudinal Patterns of Nonresident Fathers' Involvement: The Role of Resources and Relations: Using Data from the Fragile Families and Child Wellbeing Study." *Journal of Marriage and Family* 70(November):962–977.

Ryan, Suzanne, Kerry Franzetta, Erin Schelar, and Jennifer Manlove. 2009. "Family Structure History" Links to Relationship Formation Behaviors in Young Adulthood." *Journal of marriage and Family* 71(November):935–53.

Ryan, Suzanne, Jennifer Manlove, and Kerry Franzetta. 2003. *The First Time: Characteristics of Teens' First Sexual Relationships.* Washington, DC: ChildTrends. August (www.childtrends.org).

Saad, Lydia. 2003. "*Roe v. Wade* Has Positive Public Image." Gallup News Service. January 20 (www.gallup.com/poll).

———. 2004. "No Time for R & R." Gallup Poll Tuesday Briefing. May 11 (www.gallup.com)

———. 2006a. "Americans at Odds over Gay Rights." The Gallup Poll, May 31. Retrieved July 24, 2006 (poll.gallup.com).

———. 2006b. "Americans Have Complex Relationship with Marriage." The Gallup Poll, May 31. Retrieved July 24, 2006 (poll.gallup.com).

———. 2006c. "Blacks Committed to the Idea of Marriage." The Gallup Poll, July 14. Retrieved October 10, 2006 (poll.gallup.com).

———. 2007. "Women Slightly More Likely to Prefer Working to Homemaking." *Gallup Poll,* August 31. www.gallup.com.

———. 2008a. "Americans Evenly Divided on Morality of Homosexuality." *Gallup Poll Briefing.* Retrieved November 19, 2009 (www.gallup .com).

———. 2008b. "By Age 24, Marriage Wins Out." *Gallup Poll Briefing.* Retrieved October 12, 2009 (www.gallup.com).

Saadeh, Wasim, Christopher P. Rizzo, and David G. Roberts. 2002. "This Month's Debate: Spanking." *Clinical Pediatrics* (March):87–91.

Sabatelli, Ronald M. and Constance L. Shehan. 1993. "Exchange and Resources Theories." Pp. 385–411 in *Sourcebook of Family Theories and Methods,* edited by Pauline Boss, William J. Doherty, Ralph La Rossa, Walter R. Schumm, and Suzanne K. Steinmetz. New York: Plenum.

Sabol, William J., and Heather C. West. 2009. *Prison and Jail Inmates at Midyear 2008.* Bureau of Justice Statistics Bulletin. Washington, DC. U.S. Department of Justice, March 31. (Retrieved March 13, 2010 (bjs.ojp.usdoj. gov).

Sachs, Andrea, Dorian Solot, and Marshall Miller. 2003. "Happily Unmarried." *Time,* March 3.

Sachs-Ericsson, Natalie, Mathew D. Gayman, Kathleen Kendall-Tackett, Donald A. Lloyd, Amanda Medley, Nicole Collins, Elizabeth Corsentino, and Kathryn Sawyer. 2010. "The Long-Term Impact of Childhood Abuse on Internalizing Disorders Among Older Adults: The Moderating Role of Self-Esteem." *Aging & Mental Health* 14(4):489–501.

Sadker, M. and D. Sadker. 1994. *Failing at Fairness: How America's Schools Cheat Girls.* New York: Scribner's.

Safilios-Rothschild, Constantina. 1967. "A Comparison of Power Structure and Marital Satisfaction in Urban Greek and French Families." *Journal of Marriage and Family* 29:345–52.

———. 1970. "The Study of Family Power Structure: A Review 1960–1969." *Journal of Marriage and Family* 32:539–43.

Sagarin, Brad J. 2005. "Reconsidering Evolved Sex Differences in Jealousy: Comment on Harris (2003)." *Personality and Social Psychology Review* 9(1):62–75.

Said, Edward W. 1978. *Orientalism.* New York: Vintage Books.

Salari, S. and W. Zhang. 2006. "Kin Keepers and Good Providers: Influence of Gender Socialization on Well-Being among USA Birth Cohorts." *Aging & Mental Health* 10(5):485–96.

Salisbury, Emily J., Kris Henning, and Robert Holdford. 2009. "Fathering by Partner-Abusive Men." *Child Maltreatment* 14(3):232–42.

Salmon, Catherine and Todd Shackelford. 2008. *Family Relationships: An Evolutionary Perspective.* New York: Oxford University Press.

Saluter, Arlene F. and Terry A. Lugaila. 1998. *Marital Status and Living Arrangements: March 1996.* Current Population Reports P20-496. Washington, DC: U.S. Bureau of the Census.

"Same-Sex Couples in the 2008 American Community Survey." 2009. UC Los Angeles: The Williams Institute. Retrieved November 20, 2009 (www.law.ucla.edu/williamsinstitute).

Samhan, Helen Hatab. 2005. "Who Are Arab Americans?" In *Healing the Nation: The Arab American Experience After September 11,* edited by Michael S. Lee. Washington, DC: The Arab American Institute.

Samuels, Gina Miranda. 2009. "'Being Raised by White People': Navigating Racial Difference Among Adopted Multiracial Adults." *Journal of Marriage and Family* 71(1):80–94.

Sanchez, Laura, Steven L. Nock, James D. Wright, and Constance T. Gager. 2002. "Setting the Clock Forward or Back? Covenant Marriage and the 'Divorce Revolution.'" *Journal of Family Issues* 23:91–120.

Sanders, Joshunda. 2004. "Breaking Free from the Tired Old Dating Game: Quirky Singles Forge New Attitudes on Relationships." *San Francisco Chronicle,* February 8.

Sanderson, Warren and Sergei Scherbov. 2008. "Rethinking Age and Aging." *Population Bulletin* 63(4): Population Reference Bureau.

Sandfort, Theo G. M. and Brian Dodge. 2008. ". . . And Then There was the Down Low: Introduction to Black and Latino Male Bisexualities." *Archives of Sexual Behavior* 37(5):675–82.

Sanford, Keith. 2006. "Communication during Marital Conflict: When Couples Alter Their Appraisal, They Change Their Behavior." *Journal of Family Psychology* 20(2):256–66.

Sanghavi, Darshak M. 2006. "Wanting Babies Like Themselves, Some Parents Choose Genetic Defects." *New York Times,* December 5.

Sano, Yoshie, Leslie N. Richards, and Anisa M. Zvonkovic. 2008. "Are Mothers Really 'Gatekeepers' of Children?: Rural Mothers' Perceptions of Nonresident Fathers' Involvement in Low-Income Families." *Journal of Family Issues* 29(12):1701–23.

Santelli, John S., Laura Duberstein Lindberg, Lawrence B. Finer, and Susheela Singh. 2007. "Explaining Recent Declines in Adolescent Pregnancy in the United States: The Contribution of Abstinence and Improved Contraceptive Use." *American Journal of Public Health* 97(1):150–57.

Santelli, John S., Roger Rochat, Kendra Hatfield-Timajchy, Brenda Colley Gilbert, Kathryn Curtis, Rebecca Cabral, Jennifer S. Hirsch, Laura Shieve, and Other Members of the Unintended Pregnancy Working Group. 2003. "The Measurement and Meaning of Unintentional Pregnancy." *Perspectives on Social and Reproductive Health* 35:94–101.

Sapone, Anna I. 2000. "Children as Pawns in Their Parents' Fight for Control: The Failure of the United States to Protect against International Child Abduction." *Women's Rights Law Reporter* 21(2):129–38.

Sard, Barbara. 2009. "Number of Homeless Families Climbing Due To Recession." Washington, DC: Center on Budget and Policy Priorities. Retrieved June 7, 2009 (www.cbpp.org).

Sarkar, N. N. 2008. "The Impact of Intimate Partner Violence on Women's Reproductive Health and Pregnancy Outcome." *Journal of Obstetrics and Gynecology* 28(3):266–71.

Sarkisian, Natalia, Mariana Gerena, and Naomi Gerstel. 2006. "Extended Family Ties among Mexicans, Puerto Ricans, and Whites: Superintegration or Disintegration?" *Family Relations* 55(3):331–44.

Sarkisian, Natalia and Naomi Gerstel. 2004. "Explaining the Gender Gap in Help to Parents: The Importance of Employment." *Journal of Marriage and Family* 66(2):431–51.

———. 2008. "Till Marriage Do Us Part: Adult Children's Relationships with Their Parents." *Journal of Marriage and Family* 70(May):360–76.

Sassler, Sharon. 2004. "The Process of Entering into Cohabiting Unions." *Journal of Marriage and Family* 66(2):491–505.

Sassler, Sharon and Anna Cunningham. 2008. "How Cohabitors View Childbearing." *Sociological Perspectives* 51(1):3–28.

Sassler, Sharon, Anna Cunningham, and Daniel T. Lichter. 2009. "Intergenerational Patterns of Union Formation and Relationship Quality." *Journal of Family Issues* 30(6):757–86.

Sassler, Sharon and Frances Goldscheider. 2004. "Revisiting Jane Austen's Theory of Marriage Timing: Changes in Union Formation among Men in the Late 20th Century." *Journal of Family Issues* 25(2):139–66.

Sassler, Sharon, Amanda Miller, and Sarah Favinger. 2009. "Planned Parenthood?" *Journal of Family Issues* 30(2):206–32.

Sattler, David N. 2006. "Family Resources, Family Strains, and Stress Following the Northridge Earthquake." *Stress, Trauma, and Crisis: An International Journal* 9(3-4):187–202.

Saulny, Susan. 2006. "In Baby Boomlet, Preschool Derby Is the Fiercest Yet." *New York Times,* March 3. Retrieved March 15, 2006 (www. nytimes.com).

Savage, Charlie. 2010. "Gay Couples Gain Under Violence Against Women Act." *The New York Times,* June 11, A18.

"Save the Date: Relationships Ward Off Disease and Stress." 2004. *Psychology Today* (January/ February):32.

Savin-Williams, Ritch. 2006. *The New Gay Teenager.* Cambridge, MA: Harvard University Press.

Sawhill, Isabel V. and John E. Morton. 2009. *Economic Mobility: Is the American Dream Alive and Well?* Washington, DC: Economic Mobility Project. Retrieved on January 11, 2010 (www.economicmobility.org).

Saxe, Rose. 2007. "Civil Rights Protections Needn't Threaten Religious Freedom." Pp. 58–59 in LGBT & AIDS Annual Update 2007. New York: American Civil Liberties Union Lesbian Gay Bisexual Transgender & AIDS Project.

Saxena, Divya and Gregory F. Sanders. 2009. "Quality of Grandparent-Grandchild Relationship in Asian-Indian Immigrant Families." International Journal of Aging and Human Development 68(4):321–37.

Sayeed, Almas. 2007. "Chappals and Gym Shorts: An Indian Muslim Woman in the Land of Oz." Pp. 358–364 in Race, Class & Gender: An Anthology, 6th ed., edited by Margaret L. Andersen and Patricial Hill Collins. Belmont, CA: Thomson Wadsworth.

Sayer, Liana. 2006. "Economic Aspects of Divorce and Relationship Dissolution." Pp. 385–406 in Handbook of Divorce and Relationship Dissolution, edited by Mark A. Fine and John H. Harvey. Mahwah, NJ: Erlbaum.

Sayer, Liana and Suzanne M. Bianchi. 2000. "Women's Economic Independence and the Probability of Divorce." Journal of Family Issues 21:906–42.

Sayer, Liana C., Philip N. Cohen, and Lynne M. Casper. 2004. Women, Men, and Work. New York: Russell Sage.

Sbarra, David A. and Robert E. Emery. 2006. "In the Presence of Grief: The Role of Cognitive-Emotional Adaptation in Contemporary Divorce Mediation." Pp. 553–73 in Handbook of Divorce and Relationship Dissolution, edited by Mark A. Fine and John H. Harvey. Mahwah, NJ: Erlbaum.

Scanzoni, John H. 1972. Sexual Bargaining: Power Politics in the American Marriage. Englewood Cliffs, NJ: Prentice Hall.

Scelfo, Julie. 2004. "Happy Divorce." Newsweek, December 6, pp. 42–43.

Schaie, K. Warner and Glen Elder, eds. 2005. Historical Influences on Lives and Aging. New York: Springer.

Schechtman, Morris R. and Arleah Schechtman. 2003. Love in the Present Tense: How to Have a High Intimacy, Low Maintenance Marriage. Boulder, CO: Bull Publishing.

Schieszer, John. 2008. "Male Birth Control Pill Soon A Reality." MSNBC, October 1. www.msnbc.com.

Schilling, E. A., D. H. Baucom, C. K. Burnett, E. S. Allen, and L. Ragland. 2003. "Altering the Course of Marriage: The Effect of PREP Communication Skills Acquisition on Couples' Risk of Becoming Martially Distressed." Journal of Family Psychology 17(1):41–53.

Schindler, Holly S. 2010. "The Importance of Parenting and Financial Contributions in Promoting Fathers' Psychological Health." Journal of Marriage and Family 72(2):318-32.

Schlesinger, Naomi J. (2006). "Treatment Implications of a Female Incest Survivor's Misplaced Guilt." Psychoanalytic Social Work 13(2):53–65.

Schlesinger, Robert. 2010. "Study Shows Abstinence Sex Education Works, But Not Well Enough." U.S. News and World Report, February 2. www.usnews.com.

Schmeeckle, Maria. 2007. "Gender Dynamics in Stepfamilies: Adult Stepchildren's Views." Journal of Marriage and Family 69(1):174–89.

Schmitt, Marina, Matthias Kliegel, and Adam Shapiro. 2007. "Marital Interaction in Middle and Old Age." International Journal of Aging and Human Development 65(4):283–300.

Schneider, Barbara, Sylvia Martinez, and Ann Owens. 2006. "Barriers to Educational Opportunities for Hispanics in the United States." Pp. 179ff. in Hispanics and the Future of America, edited by Marta Tienda and Faith Mitchel. Washington, DC: National Academy Press.

Schneider, Jodi. 2003. "Living Together." U.S. News & World Report, March 24.

Schnittker, Jason. 2007. "Working More and Feeling Better: Women's Health, Employment, and Family Life, 1974–2004." American Sociological Review 72(2):221–38.

Schnittker, Jason, Jeremy Freese, and Brian Powell. 2003. "Who Are the Feminists and What Do They Believe? The Role of Generations." American Sociological Review 68:607–22.

Schoen, Robert and Vladimir Canudas-Romo. 2006. "Timing Effects on Divorce: 20th Century Experience in the United States." Journal of Marriage and Family 68(3):749–58.

Schoen, Robert and Yen-Hsin Alice Cheng. 2006. "Partner Choice and the Differential Retreat from Marriage." Journal of Marriage and Family 68(1):1–10.

Schoen, Robert, Nancy S. Landale, Kimberly Daniels, and Yen-Hsin Alice Cheng. 2009. "Social Background Differences in Early Family Behavior." Journal of Marriage and Family 71(2):384–95.

Schoen, Robert, Stacy J. Rogers, and Paul R. Amato. 2006. "Wives' Employment and Spouses' Marital Happiness: Assessing the Direction of Influence Using Longitudinal Couple Data." Journal of Family Issues 27(4):506–28.

Schoen, Robert and Paula Tufis. 2003. "Precursors of Nonmarital Fertility in the United States." Journal of Marriage and Family 65:1030–40.

Schoppe-Sullivan, Sarah J., Sarah C. Mangeisdorf, Geoffrey L. Brown, and Margaret Sokolowski-Szewczyk. 2007. "Goodness-of-Fit in Family Context: Infant Temperament, Marital Quality, and Early Coparenting Behavior." Infant Behavior and Development 30(1):82–97.

Schoppe-Sullivan, Sarah J., Alice C. Schermerhorn, and E. Mark Cummings. 2007. "Marital Conflict and Children's Adjustment: Evaluation of the Parenting Process Model." Journal of Marriage and Family 69(5):1118–34.

Schor, Juliet B. 1991. The Overworked American: The Unexpected Decline of Leisure. New York: Basic Books.

Schott, Ben. 2007. "Who Do You Think We Are?" New York Times, February 27.

Schottenbauer, Michelle A., Stephanie M. Spernak, and Ingrid Hellstrom. 2007. "Relationship Between Family Religious Behaviors and Child Well-Being Among Third-Grade Children." Mental Health, Religion, & Culture 10(2):191–8.

Schramm, David G. and Brian J. Higginbotham. 2009. "A Revision of the Questionnaire for Couples in Stepfamilies." Journal of Divorce and Remarriage 50:341–55.

Schrodt, Paul. 2006. "Development and Validation of the Stepfamily Life Index." Journal of Social and Personal Relationships 23(3):427–44.

———. 2008. "Sex Differences in Stepchildren's Reports of Stepfamily Functioning." Communication Reports 21(1):46–58.

———. 2009. "Family Strength and Satisfaction as Functions of Family Communication Environments" Communication Quarterly 57(2):171–86.

Schrodt, Paul, Leslie A. Baxter, M. Chad McBride, Dawn Braithwaite, and Mark A. Fine. 2006. "The Divorce Decree, Communication, and the Structuration of Coparenting Relationships in Stepfamilies." Journal of Social and Personal Relationships 23(5):741–95.

Schrodt, Paul, Andrew M. Ledbetter, Kodiane A. Jernberg, Lara Larson, Nicole Brown, and Katie Glosnek. 2009. "Family Communication Patterns As Mediators of Communication Competence in the Parent-Child Relationship." Journal of Social and Personal Relationships 26(6–7):853–74.

Schrodt, Paul, Jordan Soliz, and Dawn O. Braithwaite. 2008. "A Social Relations Model of Everyday Talk and Relational Satisfaction in Stepfamilies." Communication Monographs 75(2):190–217.

Schrodt, Paul, Paul L. Witt, and Amber S. Messersmith. 2008. "A Meta-Analytical Review of Family Communication Patterns and Their Associations with Information Processing, Behavioral, and Psychosocial Outcomes." Communication Monographs 75(3):248–69.

Schultz, Keren Blankfeld. 2008. "Divorce During Recession." Forbes, July 7. Retrieved June 8, 2009 (www.forbes.com).

Schumacher, J. A., S. Feldbau-Kohn, A. Slep, and R. E. Heyman. 2001. "Risk Factors for Male-to-Female Partner Physical Abuse." Aggression and Violent Behavior 6:281–352.

Schuman, Michael. 2009. "Going Home: On the Road Again." Time, April 27, pp. 18–26.

Schwartz, Christine R. and Robert D. Mare. 2005. "Trends in Educational Assortative Marriage from 1940 to 2003." Demography 42(4):621–46.

Schwartz, John. 2010. "In Same-Sex Ruling, and Eye on the Supreme Court." The New York Times. August 4. Retrieved August 10, 2010 (www.nytimes.com).

Schwartz, Lita Linzer. 2003. "A Nightmare for King Solomon: The New Reproductive Technology." Journal of Family Psychology 17:229–37.

Schwartz, Pepper. 1994. Peer Marriage: How Love between Equals Really Works. New York: Free Press.

———. 2001. "Peer Marriage: What Does It Take to Create a Truly Egalitarian Relationship?" Pp. 182–89 in Families in Transition, 11th ed., edited by Arlene S. Skolnick and Jerome H. Skolnick. Boston, MA: Allyn and Bacon.

———. 2006. *Finding Your Perfect Match.* New York: Penguin Group.

———. 2007. "The Social Construction of Heterosexuality." Pp. 80–92 in *The Sexual Self: The Construction of Sexual Scripts*, edited by Michael Kimmel. Nashville: Vanderbilt University Press.

———. 2009. "Adultery: Everyone Thinks Its Wrong, But a Whole Lot of People Still Do It!" *Psychology Today*, June 30. http://www.psychologytoday.com.

———. 2010. "How Could They Break Up Now?" *AARP Magazine*, June. www.aarp.org.

Schwartz, Pepper and Virginia Rutter. 1998. *The Gender of Sexuality*. Thousand Oaks, CA: Pine Forge Press.

Schwartz, Seth J. and Gordon E. Finley. 2006. "Father Involvement, Nurturant Fathering, and Young Adult Psychosocial Functioning: Differences among Adoptive, Adoptive Stepfather, and Nonadoptive Stepfamilies." *Journal of Family Issues* 27(5):712–31.

Schwartz, Seth, Byron Zamboanga, Russell Ravert, Su Yeong Kim, Robert Weisskirch, Michelle Williams, Melina Bersamin, and Gordon Finley. 2009. "Perceived Parental Relationships and Health-Risk Behaviors in College-Attending Emerging Adults." *Journal of Marriage and Family* 71(3):727–40.

Schweiger, Wendi K. and Marion O'Brien. 2005. "Special Needs Adoption: An Ecological Systems Approach." *Family Relations* 54(4):512–22.

Scott, Janny. 2001. "Rethinking Segregation Beyond Black and White." *New York Times*, July 29.

Scott, Katreena L. 2004. "Predictors of Change Among Batterers: Application of Theories and Review of Empirical Findings. *Trauma, Violence and Abuse: A Review Journal* 5:184–260.

Scott, Lisa. 2006. *Little Did You Know—101 Truths about Raising Children from Happy to Sad to Inspirational!* Bloomington, IN: Authorhouse.

Scott, Mark. 2009. "Outsourcing: Thriving at Home and Abroad." Business Week Online, May 5. Retrieved July 8, 2009 (www.businessweek.com).

Scottham, Krista Maywalt and Ciara P. Smalls. 2009. "Unpacking Racial Socialization: Considering Female African American Primary Caregivers' Racial Identity." *Journal of Marriage and Family* 71(4):807–18.

Seager, Joni. 2009. "Murders of Women by Intimate Partners." *Environment and Planning* 41(10):2287.

Seal, David Wyatt and Anke A. Ehrhardt. 2003. "Masculinity and Urban Men: Perceived Scripts for Courtship, Romantic, and Sexual Interactions with Women." *Culture, Health, and Sexuality* 5:295–319.

Seal, David Wyatt, Lucia F. O'Sullivan, and Anke A. Ehrhardt. 2007. "Miscommunications and Misinterpretations: Men's Scripts about Sexual Communication and Unwanted Sex in Interactions with Women." Pp. 141–61 in *The Sexual Self: The Construction of Sexual Scripts*, edited by Michael Kimmel. Nashville: Vanderbilt University Press.

Sears, Heather A., E. Sandra Byers, John J. Whelan, and Marcelle Saint-Pierre. 2006. "'If It Hurts You, Then It Is Not a Joke.' Adolescents' Ideas about Girls' and Boys' Use and Experience of Abusive Behavior in Dating Relationships." *Journal of Interpersonal Violence* 21(9):1191–1207.

Seaton, Eleanor K. and Ronald D. Taylor. 2003. "Exploring Familial Processes in Urban, Low-income African American Families." *Journal of Family Issues* 24(5):627–44.

Seccombe, Karen. 2007. *Families in Poverty*. New York: Pearson Education.

Seery, Brenda L. and M. Sue Crowley. 2000. "Women's Emotion Work in the Family: Relationship Management and the Process of Building Father–Child Relationships." *Journal of Family Issues* 21(1):100–27.

Segal, David R. and Mady Wechsler Segal. 2004. "America's Military Population." *Population Bulletin* 59(4):1–40.

Segura, Denise A. and Beatriz M. Pesquera. 1995. "Chicana Feminisms: Their Political Context and Contemporary Expressions." Pp. 617–31 in *Women: A Feminist Perspective*, 5th ed., edited by Jo Freeman. Mountain View, CA: Mayfield.

Seideman, Ruth Young, Roma Williams, Paulette Burns, Sharon Jacobson, Francene Weatherby, and Martha Primeaux. 1994. "Cultural Sensitivity in Assessing Urban Native American Parenting." *Public Health Nursing* 11(2):98–103.

Seidman, Steven. 2003. *The Social Construction of Sexuality*. New York: W. W. Norton.

Self, Sharmistha and Richard Grabowski. 2009. "Modernization, Inter-caste Marriage, and Dowry." *Journal of Asian Economics* 20(1):69–77.

Seligman, Katherine. 2006. "I Married Myself." *San Francisco Chronicle Magazine*, February 12, pp. 9–10.

Seltzer, Judith A. 2000. "Families Formed Outside of Marriage." *Journal of Marriage and Family* 62(4):1247–68.

———. 2004. "Cohabitation in the United States and Britain: Demography, Kinship, and the Future." *Journal of Marriage and Family* 66(4):921–28.

Seltzer, Marsha and Tamar Heller, eds. 1997. "Family Caregiving for Persons with Disabilities." *Family Relations* Special Issue 46(4).

Şenyürekli, Ayşem R. and Daniel F. Detzner. 2009. "Communication Dynamics of the Transnational Family." *Marriage & Family Review* 45(6–8):807–24.

Sepler, Harvey J. 1981. "Measuring the Effects of No-Fault Divorce Laws Across Fifty States: Quantifying a Zeitgeist." *Family Law Quarterly* 15(1):65–103.

Serano, Julia. 2007. *Whipping Girl: A Transsexual Woman on Sexism and the Scapegoating of Femininity*. Berkeley: Seal Press.

Sevier, Mia, Kathleen Eldridge, Janice Jones, Brian Doss, and Andrew Christensen. 2008.

"Observed Communication and Associations with Satisfaction During Traditional and Integrative Behavioral Couple Therapy." *Behavior Therapy* 39(2):137–51.

Sex and Tech: Results from a Survey of Teens and Young Adults. The National Campaign to Prevent Teen and Unplanned Pregnancy and CosmoGirl.com (2009).

"Sexual Development and Behavior in Children." 2009. Los Angeles: National Center for Child Traumatic Stress. University of California. http://nctsn.org.

Shalit, Wendy. 1999. *A Return to Modesty: Discovering the Lost Virtue*. New York: Free Press.

Shanahan, Lilly, Susan M. McHale, Ann C. Crouter, and D. Wayne Osgood. 2008. "Linkages between Parents' Differential Treatment, Youth Depressive Symptoms, and Sibling Relationships." *Journal of Marriage and Family* 70(May):480–94.

Shanahan, Michael J. 2000. "Pathways to Adulthood in Changing Societies: Variabilities and Mechanisms in Life Course Perspective." *Annual Review of Sociology* 26:667–97.

Shannon, Joyce B. 2009. *Learning Disabilities Sourcebook*. 3rd ed. Detroit, MI: Omnigraphics.

Shapiro, Joseph P. 2001a. "The Assisted-living Dilemma." *U.S. News & World Report*, May 21, pp. 64–66.

———. 2001b. "Growing Old in a Good Home." *U.S. News & World Report*, May 21, pp. 57–61.

———. 2006a. "Caregiver Role Brings Purpose—and Risk—to Kids." National Public Radio. March 23. Retrieved March 23, 2006 (www.npr.org).

———. 2006b. "Family Ties Source of Strength for Elderly Caregivers." National Public Radio, March 23. Retrieved March 23, 2006 (www.npr.org).

———. 2009. "Donor-Conceived Kids Connect with Half Siblings." National Public Radio, March 2. Retrieved February 3, 2009 (www.npr.org).

Shauman, Kimberlee A. and Mary C. Noonan. 2007. "Family Migration and Labor Force Outcomes: Sex Differences in Occupational Context." *Social Forces* 85(4):1735–64.

Shellenbarger, Sue. 1991. "Companies Team Up to Improve Quality of Their Employees' Child-Care Choices." *Wall Street Journal*, October 17, pp. B1, B4.

———. 2007. "Men on the Daddy Track Find A Place of Their Own at Home." *Wall Street Journal*, November 8.

———. 2009. "Extreme Child-Care Maneuvers." *The Wall Street Journal*, May 20, D1.

Shelton, Beth Anne and Daphne John. 1993. "Ethnicity, Race, and Difference: A Comparison of White, Black, and Hispanic Men's Household Labor *Time*." Pp. 131–50 in *Men, Work, and Family*, edited by Jane C. Hood. Newbury Park, CA: Sage Publications.

———. 1996. "The Division of Household Labor." *Annual Review of Sociology* 22:299–322.

Sherif-Trask, Bahira. 2003. "Marriage from a Cross-Cultural Perspective." *National Council on Family Relations Report* 48(3):F13–F14.

Sherman, Carey Wexler. 2006. "Remarriage and Stepfamily in Later Life." *Family Focus on . . . Families and the Future* FF32:F8–F9. National Council on Family Relations.

Sherman, Carey Wexler and Pauline Boss. 2007. "Spousal Dementia Caregiving in the Context of Late-Life Remarriage." *Dementia* 6(2):245–70.

Sherman, Jake. 2009. "White House Looks to Include Same-Sex Unions in Census Count." *The Wall Street Journal*, June 19, A4.

Sherman, Lawrence W. and Richard A. Berk. 1984. "Deterrent Effects of Arrest for Domestic Assault." *American Sociological Review* 49:261–72.

Sherman, Paul J. and Janet T. Spence. 1997. "A Comparison of Two Cohorts of College Students in Responses to the Male–Female Relations Questionnaire." *Psychology of Women Quarterly* 21(2):265–78.

Shierholz, Heidi. 2009. "New 2008 Poverty, Income Data Reveal only Tip of the Recession Iceberg." Washington, DC: Economic Policy Institute. Retrieved September 10, 2009 (www.epi.org).

Shifren, Kim, ed. 2009. *How Caregiving Affects Development: Psychological Implications for Child, Adolescent, and Adult Caregivers.* Washington, DC: American Psychological Association.

Shinn, Lauren Keel and Marion O'Brien. 2008. "Parent-Child Conversational Styles in Middle Childhood: Gender and Social Class Differences." *Sex Roles* 59:61–67.

Shorter, Edward. 1975. *The Making of the Modern Family.* New York: Basic Books.

Showden, Carisa R. 2009. "What's Political about the New Feminisms?" *Frontiers: A Journal of Women Studies* 30(2):166–98.

Shriner, Michael. 2009. "Marital Quality in Remarriage A Review of Methods and Results." *Journal of Divorce and Remarriage* 50:81–99.

Shugrue, Noreen and Julie Robison. 2009. "Intensifying Individual, Family, and Caregiver Stress: Health and Social Effects of Economic Crisis." *Journal of American Society on Aging* (Fall):34–39.

Shulman, Seth. 2008. *Undermining Science: Suppression and Distortion in the Bush Administration.* Berkeley: University of California Press.

Sidelinger, Robert J. and Melanie Booth-Butterfield. 2007. "Mate Value Discrepancy as Predictor of Forgiveness and Jealousy in Romantic Relationships." *Communication Quarterly* 55(2):207–23.

Siebel, Cynthia C. 2006. "Fathers and Their Children: Legal and Psychological Issues of Joint Custody." *Family Law Quarterly* 40(2):213–36.

Siegel, Bernie S. 2006. *Love, Magic, and Mudpies: Raising Your Kids to Feel Loved, Be Kind, and Make a Difference.* New York: St. Martin's.

Siegel, Mark D. 2004. "To the Editor: Making Decisions about How to Die." *New York Times*, October 3.

Siegel, Michele, Judith Brisman, and Margot Weinshel. 2009. *Surviving An Eating Disorder: Strategies for Families and Friends.* 3rd ed. New York, NY: Collins Living.

Signorella, Margaret L. and Irene Hanson Frieze. 2008. "Interrelations of Gender Schemas in Children and Adolescents: Attitudes, Preferences, and Self-Perceptions." *Social Behavior and Personality* 36(7):941–54.

Siller, Sidney. 1984. "National Organization for Men." Privately printed pamphlet.

Silverstein, Merril. 2000. "The Impact of Acculturation on Intergenerational Relationships in Mexican American Families." *National Council on Family Relations Report* 45(2):F9.

Silverstein, Merril and Vern L. Bengston. 2001. "Intergenerational Solidarity and the Structure of Adult Child–Parent Relationships in American Families." Pp. 53–61 in *Families in Later Life: Connections and Transitions,* edited by Alexis J. Walker, Margaret Manoogian-O'Dell, Lori A. McGraw, and Diana L. G. White. Thousand Oaks, CA: Pine Forge Press.

Simmel, Cassandra, Richard P. Barth, and Devon Brooks. 2007. "Adopted Foster Youths' Psychosocial Functioning: A Longitudinal Perspective." *Child & Family Social Work* 12(4):336–48.

Simmons, Tavia and Martin O'Connell. 2003. *Married-Couple and Unmarried Partner Households: 2000.* Census Special Report CENSR-5 Washington, DC: U.S. Census Bureau.

Simmons, Tracy and Kevin Eckstrom. 2009. "Birth Control Puts Catholics in Tight Spot on Abortion." *Religion News Services,* July 21.

Simon, Rita J. 1990. "Transracial Adoptions Can Bring Joy: Letters to the Editor." *Wall Street Journal,* October 17.

Simon, Rita J. and Howard Altstein. 2002. *Adoption, Race, & Identity: From Infancy to Young Adulthood.* New Brunswick, NJ: Transaction Books.

Simoncelli, Tania with Jay Stanley. 2005. *Science under Siege: The Bush Administration's Assault on Academic Freedom and Scientific Inquiry.* New York: American Civil Liberties Union.

Simons, Leslie Gordon and Rand D. Conger. 2009. "Linking Mother-Father Differences in Parenting to a Typology of Family Parenting Styles and Adolescent Outcomes." *Journal of Family Issues* 28(2):212–41.

Simons, Leslie Gordon, Ronald L. Simons, Gene Brody, and Carolyn Cutrona. 2006. "Parenting Practices and Child Adjustment in Different Types of Households: A Study of African American Families." *Journal of Family Issues* 27(6):803–25.

Simpson, Emily K. and Christine A. Helfrich. 2005. "Lesbian Survivors of Intimate Partner Violence: Provider Perspectives on Barriers to Accessing Services." *Journal of Gay & Lesbian Social Services* 18(2):39–59.

Simpson, Ruth. 2005. "Men in Non-Traditional Occupations: Career Entry, Career Orientation, and Experience of Role Strain." *Gender, Work, & Organization* 12:363–80.

Sinclair, Stacey L. and Gerald Monk. 2004. "Couples—Moving Beyond the Blame Game: Toward a Discursive Approach to Negotiating Conflict within Couple Relationships." *Journal of Marital and Family Therapy* 30(3):335–49.

Singer v. Hara. 1974. 11 Wash. App. 247, 522 P. 2d 1187.

Sirjamaki, John. 1948. "Cultural Configurations in the American Family." *American Journal of Sociology* 53(6):464–70.

Sit, Dorothy, Anthony J. Rothschild, Mitchell D. Creinin, Barbara H. Hanusa, and Katherine L. Wisner. 2007. *Psychiatric Outcomes Following Medical and Surgical Abortions* 22:878–84.

Skarupski, Kimberly, Judy McCann, Julia Bienias, and Denis Evans. 2009. "Race Differences in Emotional Adaptation of Family Caregivers." *Aging and Mental Health* 13(5):715–24.

Skinner, Kevin B., Stephen J. Bahr, D. Russell Crane, and Vaughn A. Call. 2002. "Cohabitation, Marriage, and Remarriage: A Comparison of Relationship Quality over *Time*." *Journal of Family Issues* 23(1):74–90.

Skipp, Catharine and Dan Ephron. 2006. "Trouble at Home." *Newsweek,* October 23, pp. 48–49.

Skloot, Rebecca L. 2003. "The Other Baby Experiment." *New York Times,* February 22.

Skogrand, Linda, Abril Barrios-Bell, and Brian Higginbotham. 2009. "Stepfamily Education for Latino Families: Implications for Practice." *Journal of Couple and Relationship Therapy* 8(2):113–28.

Skolnick, Arlene S. 1978. *The Intimate Environment: Exploring Marriage and Family.* 2nd ed. Boston, MA: Little, Brown.

Sky, Natasha. 2009. "Leaving Behind Traditional Views of Belonging." Pp. 73–76 in *Social Issues First Hand: Blended Families,* edited by Stefan Kiesbye. New York: Greenhaven Press/Cengage Learning.

Slark, Samantha. 2004. "Are Anti-Polygamy Laws an Unconstitutional Infringement on the Liberty Interests of Consenting Adults?" *Journal of Law and Family Studies* 6(2):451–60.

Slater, Lauren. 2006. "Love: The Chemical Reaction." *National Geographic,* February, pp. 34–49.

Slota, Holly J. 2009. "Preventing and Responding to International Child Abduction." *American Journal of Family Law* 23(2):73–78.

Sluzki, C. E. 2008. "Migration and the Disruption of the Social Network." Pp. 39–47 in *Revision Family Therapy: Race, Culture, and Gender in Clinical Practice,* 2nd ed., edited by M. McGoldrick and K. V. Hardy. New York: Guilford Press.

Smalley, Gary. 2000. *Secrets to Lasting Love: Uncovering the Keys to Life-long Intimacy.* New York: Simon and Schuster.

Smerglia, Virginia, Nancy Miller, and Diane Sotnak. 2007. "Social Support and Adjustment to Caring for Elder Family Members: A Multi-Study Analysis." *Aging and Mental Health* 11(2):205–17.

Smiler, Andrew P. 2008. "'I Wanted To Get To Know Her Better': Adolescent Boys' Dating Motives, Masculinity Ideology, and Sexual Behavior." *Journal of Adolescence* 31(1):17–32.

Smith, Christian. 2003. "Religious Participation and Network Closure Among American Adolescents." *Journal for the Scientific Study of Religion* 42(2):259–67.

Smith, Jeremy Adam. 2009. *The Daddy Shift: How Stay-at-Home Dads, Breadwinning Moms, and Shared Parenting Are Transforming the American Family*. Boston: Beacon Press.

Smith, Kevin M., Patti A. Freeman, and Ramon B. Zabriskie. 2009. "An Examination of Family Communication within the Core and Balance model of Family Leisure Functioning." *Family Relations* 58(February):79–90.

Smith, Kristin. 2008. "Working Hard for the Money Trends in Women's Employment: 1970 to 2007." *A Carsey Institute Report on Rural America*, University of New Hampshire.

Smith, Maureen Margaret and Becky Beal. 2007. "So You Can See How the Other Half Lives: MTV Cribs' Use of 'the Other' in Framing Successful Athletic Masculinities." *Journal of Sport & Social Issues* 31(2):103–27.

Smith, Suzanna D. 2006. "Global Families." Pp. 3–24 in *Families in Global and Multicultural Perspective*, 2nd ed., edited by Bron B. Ingoldsby and Suzanna D. Smith. Thousand Oaks, CA: Sage Publications.

Smith, Tim. 2006. *The Danger of Raising Nice Kids: Preparing Our Children to Change Their World*. Downers Grove, IL: IVP Books.

Smith, Tom W. 1999. *The Emerging 21st Century American Family*. GSS Social Change Report No. 42. Chicago, IL: University of Chicago, National Opinion Research Center.

———. 2003. *Coming of Age in 21st Century America: Public Attitudes towards the Importance and Timing of Transitions to Adulthood*. GSS Topical Report No. 35. Chicago, IL: National Opinion Research Center.

Smithson, Michael and Cathy Baker. 2008. "Risk Orientation, Loving, and Liking in Long-Term Romantic Relationships." *Journal of Social and Personal Relationships* 25(1):87–103.

Smock, Pamela J. and Sanjiv Gupta. 2002. "Cohabitation in Contemporary North America." Pp. 53–84 in *Just Living Together: Implications of Cohabitation on Families*, edited by Alan Booth and Ann C. Crouter. Mahwah, NJ: Erlbaum.

Smokowski, Paul, Roderick Rose, and Martica Bacallao. 2008. "Acculturation and Latino Family Processes: How Cultural Involvement, Biculturalism, and Acculturation Gaps Influence Family Dynamics." *Family Relations* 57(July):295–308.

Smolowe, Jill. 1996a. "The Unmarrying Kind." *Time*, April 29, pp. 68–69.

Snider, Carolyn, Daniel Webster, Chris S. O'Sullivan, and Jacquelyn Campbell. 2009. "Intimate Partner Violence: Development of a Brief Risk Assessment for the Emergency Department." *Academic Emergency Medicine* 16(11):1208–16.

Sniezek, Tamara. 2002. "Getting Married: An Interactional Analysis of Weddings and Transitions into Marriage." Unpublished dissertation. University of California at Los Angeles.

———. 2005. "Is It Our Day or the Bride's Day? The Division of Wedding Labor and Its Meaning for Couples." *Qualitative Sociology* 28(3):215–34.

———. 2007. "When Does a Relationship Lead to Marriage?" Unpublished manuscript.

Snipp, C. Matthew. 2002. *American Indian and Alaska Native Children in the 2000 Census*. Washington, DC: Annie E. Casey Foundation, Population Reference Bureau.

———. 2005. *American Indian and Alaska Native Children: Results from the 2000 Census*. Washington, DC: Population Reference Bureau. August.

Snyder, Douglas K., Donald H. Baucom, and Kristina C. Gordon. 2008. "An Integrative Approach to Treating Infidelity." *The Family Journal: Counseling and Therapy for Couples and Families* 16(4):300–04.

Snyder, Karrie Ann. 2007. "A Vocabulary of Motives: Understanding How Parents Define Quality Time." *Journal of Marriage and Family* 69(2):320–40.

Snyder, Karrie Ann and Adam Isaiah Green. 2008. "Revisiting the Glass Escalator: The Case of Gender Segregation in a Female Dominated Occupation." *Social Problems* 55(2):271–99.

Snyder, Thomas D., Sally A. Dillow, and Charlene M Hoffman. 2009. *Digest of Education Statistics 2008* (NCES 2009-020). Washington, DC: National Center for Education Statistics, Institute of Education Sciences, U.S. Department of Education.

Soares, Rachel, Nancy M. Carter, and Jan Combopiano. 2009. *2009 Catalyst Census: Fortune 500 Women Board Directors*. New York: Catalyst. www.catalyst.org.

Sobolewski, Juliana and Paul R. Amato. 2005. "Economic Hardship in the Family of Origin and Children's Psychological Well-being in Adulthood." *Journal of Marriage and Family* 67(1):141–56.

Sobolewski, Juliana and Valarie King. 2005. "The Importance of the Coparental Relationship for Nonresident Fathers' Ties to Children." *Journal of Marriage and Family* 67:1196–1212.

Socha, Thomas J. and Glen H. Stamp. 2009. *Parents and Children Communicating with Society: Managing Relationships Outside of Home*. New York: Routledge.

Social Security Online. n.d. "Widows, Widowers, and Other Survivors." Retrieved May 12, 2010 (www.ssa.gov).

Soenens, Bart, Maarten Vansteenkiste, and Eline Sierens. 2009. "How Are Parental Psychological Control and Autonomy-Support Related? A Cluster-Analytic Approach." *Journal of Marriage and Family* 71(1):187–202.

Soliz, Jordan, Allison R. Thorson, and Christine E. Rittenour. 2009. "Communicative Correlates of Satisfaction, Family Identity, and Group Salience in Multiracial/Ethnic Families." *Journal of Marriage and Family* 71(4): 819–32.

Soll, Anna R., Susan M. McHale, and Mark E. Feinberg. 2009. "Risk and Protective Effects of Sibling Relationships Among African American Adolescents." *Family Relations* 58(December):578–92.

Sommers, Christina Hoff. 2000a. *The War Against Boys*. New York: Simon and Schuster.

———. 2000b. "The War Against Boys." *Atlantic Monthly*, May, pp. 59–75.

Sontag, Deborah. 2008. "An Iraq Veteran's Descent; a Prosecutor's Choice." *The New York Times*, January 20.

Sontag, Susan. 1976. "The Double Standard of Aging." Pp. 350–66 in *Sexuality Today and Tomorrow*, edited by Sol Gordon and Roger W. Libby. North Scituate, MA: Duxbury.

Soons, Judith and Aart Liefbroer. 2009. "The Long-Term Consequences of Relationship Formation for subjective Well-Being." *Journal of Marriage and Family* 71(December):1254–70.

Sorensen, Elaine. 2010. "Rethinking Public Policy Toward Low-Income Fathers in the Child Support Program." *Journal of Policy Analysis and Management* 29(3):604–10.

Sorenson, Susan B. and Kristie A. Thomas. 2009. "Views of Intimate Partner Violence in Same- and Opposite-Sex Relationships." *Journal of Marriage and Family* 71(2):337–52.

Soukhanov, Anne H. 1996. "Watch." *Atlantic Monthly*, August: 96ff.

Soukup, Elise. 2006. "Polygamists, Unite!" *Newsweek*, March 20, p. 52.

Sousa, Liliana and Elaine Sorensen. 2006. *The Economic Reality of Nonresident Mothers and Their Children*. Series B, No. B-69. Washington, DC: Urban Institute. May.

South, Scott J. and Kyle D. Crowder. 2010. "Neighborhood Poverty and Nonmarital Fertility: Spatial and Temporal Dimensions." *Journal of Marriage and Family* 72(1):89–104.

Souto, Claudia and Telma Garcia. 2002. "Construction and Validation of a Body-Image Rating Scale." *International Journal of Nursing Terminologies and Classifications* 13 (4): 117–26.

Spade, Joan Z. and Catherine G. Valentine. 2010. *The Kaleidoscope of Gender: Prisms, Patterns, and Possibilities*. 3rd ed. Newbury Park, CA: Pine Forge Press.

Span, Paula. 2009a. *When the Time Comes: Families with Aging Parents Share Their Struggles and Solutions*. New York: Springboard Press.

———. 2009b. "Years Later, Divorce Complicates Caregiving." *The New York Times*, New Old Age Blog: Caring and Coping. Retrieved May 5, 2010 (newoldage.blogs .nytimes.com).

Spar, Debora L. 2006. *The Baby Business: How Money, Science, and Politics Drive the Commerce of Conception*. Boston, MA: Harvard Business School Press.

Speer, Rebecca B. and April R. Trees. 2007. "The Push and Pull of Stepfamily Life: the Contribution of Stepchildren's Autonomy and Connection-Seeking Behaviors to Role

Development in Stepfamilies." *Communication Studies* 58(4):377–94.

Spelke, Elizabeth S. 2005. Opening Remarks. "The Science of Gender and Science—Pinker vs. Spelke: A Debate." Cambridge, MA: Harvard University, Mind/Brain/Behavior Initiative. May 16. Retrieved January 1, 2006 (www.edge.org)

Spencer, Liz and R. E. Pahl. 2006. *Rethinking Friendship: Hidden Solidarities Today.* Princeton, NJ: Princeton University Press.

Spiro, Melford. 1956. *Kibbutz: Venture in Utopia.* New York: Macmillan.

Spitze, Glenna, John R. Logan, Glenn Deane, and Suzanne Zerger. 1994. "Adult Children's Divorce and Intergenerational Relationships." *Journal of Marriage and Family* 56(2):279–93.

Spitze, Glenna and Katherine Trent. 2006. "Gender Differences in Adult Sibling Relations in Two-Child Families." *Journal of Marriage and Family* 68(4):977–92.

Spratling, Cassandra. 2009. "A Grand Living Arrangement." *Detroit Free Press*, January 18. Retrieved February 3, 2009 (www.modbee.com).

Sprecher, Susan. 2002. "Sexual Satisfaction in Premarital Relationships: Associations with Satisfaction, Love, Commitment, and Stability." *Journal of Sex Research* 39:190–96.

Sprecher, Susan, Maria Schmeeckle, and Diane Felmlee. 2006. "The Principle of Least Interest: Inequality in Emotional Involvement in Romantic Relationships." *Journal of Family Issues* 27(9):1255–80.

Sprecher, Susan and Pepper Schwartz. 1994. "Equity and Balance in the Exchange of Contributions in Close Relationships." Pp. 89–116 in *Entitlement and the Affectional Bond: Justice in Close Relationships,* edited by M. J. Lerner and G. Mikula. New York: Plenum.

Sprecher, Susan and Maura Toro-Morn. 2002. "A Study of Men and Women from Different Sides of Earth to Determine if Men Are from Mars and Women Are from Venus in Their Beliefs about Love and Romantic Relationships." *Sex Roles: A Journal of Research* (March):131–48.

Sprenkle, Douglas. 2003. "Effectiveness Research in Marriage and Family Therapy." *Journal of Marital and Family Therapy* 29(1):85–96.

Springer, Kristen W. 2010. "Economic Dependence in Marriage and Husbands' Midlife Health." *Gender & Society* 24(3):378–401.

Springer, Kristen W., Brenda K. Parker, and Catherine Leviten-Reid. 2009. "Making Space for Graduate Student Parents." *Journal of Family Issues* 30(4):435–57.

Springer, S. and G. Deutsch. 1994. *Left Brain/ Right Brain,* 4th edition. New York: Freeman.

Squires, Catherine. 2007. "Popular Sentiments and Black Womens Studies: The Scholarly and Experiential Divide." *Black Women, Gender, and Families* 1(1):74–93.

Srinivasan, Padma and Gary R. Lee. 2004. "The Dowry System in Northern India: Women's Attitudes and Social Change." *Journal of Marriage and Family* 66(5):1108–17.

St. George, Donna. 2006. "Home but Still Haunted." *Washington Post,* August 20.

———. 2010. "More Wives Are the Higher-Income Spouse, Pew Report Says." *The Washington Post,* January 9. Retrieved January 9, 2010 (www.washingtonpost.com).

Stacey, Judith. 1990. *Brave New Families: Stories of Domestic Upheaval in Late Twentieth Century America.* New York: Basic Books.

———. 1993. "Good Riddance to 'The Family': A Response to David Popenoe." *Journal of Marriage and Family* 55(3):545–47.

———. 1996. *In the Name of the Family: Rethinking Family Values in the Postmodern Age.* Boston, MA: Beacon Press.

———. 2006. "Feminism and Sociology in 2005: What Are We Missing?" *Social Problems* 53(4):479–82.

Stacey, Judith and Tey Meadow. 2009. "New Slants on the Slippery Slope: The Politics of Polygamy and Gay Family Rights in South Africa and the United States." *Politics & Society* 37(2):167–202).

Stack, Carol B. 1974. *All Our Kin: Strategies for Survival.* New York: Harper and Row.

Stanley, Scott M. 2009. "'Sliding vs. Deciding': Understanding a Mystery." Family Focus (Summer):F1–F3. Minneapolis, MN: National Council on Family Relations.

Stanley, Scott M., Galena K. Rhoades, and Howard J. Markman. 2006. "Sliding versus Deciding: Inertia and the Premarital Cohabitation Effect." *Family Relations* 55(4):499–509.

Stanton, Glenn T. 2004a. "Why Marriage Matters for Adults." June 21. *Focus on Social Issues: Marriage and Family.* Focus on the Family. Retrieved September 30, 2006 (www.family.org/ forum).

———. 2004b. "Why Marriage Matters for Children." June 21. *Focus on Social Issues: Marriage and Family.* Focus on the Family. Retrieved September 30, 2006 (www.family.org/forum).

Staples, Robert. 1994. *The Black Family: Essays and Studies.* 5th ed. Belmont, CA: Wadsworth.

———. 1999. *The Black Family: Essays and Studies.* 6th ed. Belmont, CA: Wadsworth.

Staples, Robert and Leanor Boulin Johnson. 1993. *Black Families at the Crossroads: Challenges and Prospects.* San Francisco, CA: Jossey-Bass.

"State Court Orders Lesbian Mother to Pay Support." 2005. *South Bend Tribune,* February 19.

"Statistics." 2009. Retrieved March 12, 2010 (www.stepfamily.org/statistics.html).

Stattin, H. and G. Klackenberg. 1992. "Discordant *Family Relations* in Intact Families: Developmental Tendencies over 18 Years." *Journal of Marriage and Family* 54:940–56.

Stearns, Peter N. 2003. *Anxious Parents: A History of Modern Childrearing in America.* New York: NYU Press.

Stein, Arlene. 1989. "Three Models of Sexuality: Drives, Identities, and Practices." *Sociological Theory* 7:1–13.

Stein, Rob. 2010. "Study Finds That Effects of Low-Quality Child Care Last Into Adolescence." *The Washington Post,* May 14. Retrieved May 31, 2010 (www.washingtonpost.com).

Steinberg, Julia Renee and Nancy F. Russo. 2008. "Abortion and Anxiety: What's the Relationship?" *Social Science and Medicine* 67:238–52.

Steingraber, Sandra. 2007. *The Falling Age of Puberty in U.S. Girls: What We Know, What We Need To Know.* San Francisco: The Breast Cancer Fund.

Steinhauer, Jennifer. 2007. "A Proposal to Ban Spanking Sparks Debate." *New York Times,* January 21.

Stempel, Jonathon. 2010. "Wal-Mart in $86 Million Settlement of Wage Lawsuit." *Reuters,* May 12. Chicago, IL: Thompson Reuters.

"Stepfamily Fact Sheet." 2010. National Stepfamily Resource Center. Retrieved April 17, 2010 (www.stepfamilies.info).

"Stepfamily Stages: A Thumbnail Sketch." n.d. Retrieved July 31, 2007 (www.stepfamilytips. com/advice.html).

Stephens, William N. 1963. *The Family in Cross-Cultural Perspective.* New York: Holt, Rinehart & Winston.

Stepp, Laura Sessions. 2007. *Unhooked: How Young Women Pursue Sex, Delay Love and Lose At Both.* New York: Riverhead Books.

Stern, Gabriella. 1991. "Young Women Insist on Career Equality, Forcing the Men in Their Lives to Adjust." *Wall Street Journal,* September 16, pp. B1, B3.

Sternberg, Robert J. 1988a. "Triangular Love." Pp. 119–38 in *The Psychology of Love,* edited by Robert J. Sternberg and Michael L. Barnes. New Haven, CT: Yale University Press.

———. 1988b. *The Triangle of Love: Intimacy, Passion, Commitment.* New York: Basic Books.

———. 2006. Robert J. Sternberg's home page. Retrieved August 16, 2006 (http://dis. ijs.si/mitjal/genre/online/data/file1353. htm).

Stets, Jan E. 1991. "Cohabiting and Marital Aggression: The Role of Social Isolation." *Journal of Marriage and Family* 53(3):669–80.

Stevenson, Betsey. 2010. "Al and Tipper: What Does It Mean?" *New York Times, Freakonomics,* June 4. Retrieved June 19, 2010 (freakonomics. blogs.nytimes.com).

Stevenson, Betsey and Justin Wolfers. 2004. *Bargaining in the Shadow of the Law: Divorce Laws and Family Distress.* National Bureau of Economic Research Working Paper 10175. Cambridge, MA: National Bureau of Economic Research. October 4.

———2006. "Bargaining in the Shadow of the Law: Divorce Laws and Family Distress." *Quarterly Journal of Economics* 121:267–88.

———. 2007. *Marriage and Divorce: Changes and Their Driving Forces.* National Bureau of Economic Research Working Paper 12944. Cambridge, MA: National Bureau of Economic Research. March. Retrieved April 27, 2007 (www.nber.org).

Stewart, Mary White. 1984. "The Surprising Transformation of Incest: From Sin to Sickness." Presented at the annual meeting of the Midwest Sociological Society, April 18, Chicago, IL.

Stewart. Susan D. 1999. "Disneyland Dads, Disneyland Moms?" *Journal of Family Issues* 20:539–56.

———. 2002. "The Effect of Stepchildren on Childbearing Intentions and Births." *Demography* 39:181–97.

———. 2005a. "Boundary Ambiguity in Stepfamilies." *Journal of Family Issues* 26(7):1002–29.

———. 2005b. "How the Birth of a Child Affects Involvement with Stepchildren." *Journal of Marriage and Family* 67(2):461–73.

———. 2007. *Brave New Stepfamilies: Diverse Paths toward Stepfamily Living*. Thousand Oaks, CA: Sage Publications.

Stinnett, Nick. 1985. *Secrets of Strong Families*. New York: Little, Brown.

———. 2008. *Fantastic Families: 6 Proven Steps to Building a Strong Family*. Howard Pub. Co.

Stinnett, Nick, Donnie Hilliard, and Nancy Stinnett. 2000. *Magnificent Marriage: 10 Beacons Show the Way to Marriage Happiness*. Montgomery, AL: Pillar Press.

Stith, Sandra M., Douglas B. Smith, Carrie E. Penn, David B. Ward, and Dari Tritt. 2004. "Intimate Partner Physical Abuse Perpetration and Victimization Risk Factors: A Meta-Analytic Review." *Aggression and Violent Behavior* 10(1):65–98.

Stoddard, Martha. 2006. "Grandparents Prevail in Visitation Challenge." *Omaha World-Herald*, June 6.

Stokes, Charles E. and Christopher G. Ellison. 2010. "Religion and Attitudes Toward Divorce Laws Among U.S. Adults." *Journal of Family Issues* XX(X):1–26. doi:10.1177/0192513X10363887.

Stokes-Brown, Atiya Kai. 2009. "The Hidden Politics of Identity: Racial Self-Identification and Latino Political Engagement." *Politics & Policy* 37(6):1281–305.

Stolberg, Sheryl Gay. 1997. "For the Infertile: A High Tech Treadmill." *New York Times*, December 14.

———. 2010. "Obama Widens Medical Rights for Same-Sex Partners." *The New York Times*, April 15. Retrieved April 16, 2010 (www.nytimes.com).

Stoll, Michael A. 2004. "African Americans and the Color Line." Washington, DC: Population Reference Bureau. Retrieved December 13, 2006 (www.prb.org).

Stone, Lawrence. 1980. *The Family, Sex, and Marriage in England, 1500–1800*. New York: Harper and Row.

Stone, Pamela. 2007. *Opting Out: Why Women Really Quit Careers and Head Home*. Berkeley, CA: University of California Press.

Story, T. Nathan, Cynthia Berg, Timothy Smith, Ryan Beveridge, Nancy Henry, and Gale Pearce.

2007. "Age, Marital Satisfaction, and Optimism as Predictors of Positive Sentiment Override in Middle-Aged and Older Married Couples." *Psychology and Aging* 22(4):719–27.

Stout, David. 2007. "Supreme Court Upholds Ban on Abortion Procedure." *New York Times*, April 18.

Stratton, Peter. 2003. "Causal Attributions during Therapy: Responsibility and Blame." *Journal of Family Therapy* 25(2):136–60.

Straus, Martha. 2009. "Bungee Families—While Some Warn That the Conveyor Belt That Once Transported Adolescents into Adulthood Has Broken Down, Others Insist the Increasing Numbers of Adult Children Living at Home is Less About Dysfunction Than the Changing Function of Family life Today." *Psychotherapy Networker* 33(5):30–38.

Straus, Murray A. 1993. "Physical Assaults by Wives: A Major Social Problem." Pp. 67–87 in *Current Controversies on Family Violence*, edited by Richard J. Gelles and Donileen R. Loseke. Newbury Park, CA: Sage Publications.

———. 1999a. "The Benefits of Avoiding Corporal Punishment: New and More Definitive Evidence." Paper No. CP40–59/CP41B.P. University of New Hampshire: Family Research Laboratory.

———. 1999b. "The Controversy over Domestic Violence by Women: A Methodological, Theoretical, and Sociology of Science Analysis." Pp. 17–44 in *Violence in Intimate Relationships*, edited by Ximena B. Arriaga and Stuart Oskamp. Thousand Oaks, CA: Sage Publications.

———. 2005. "Women's Violence toward Men Is a Serious Social Problem." Pp. 55–77 in *Current Controversies on Family Violence*, edited by Donileen Loseke, Richard J. Gelles, and Mary M. Cavanaugh. Thousand Oaks, CA: Sage Publications.

———. 2007. "Do We Need a Law to Prohibit Spanking?" *Family Focus on . . . Adolescence* FF34:F7. National Council on Family Relations.

———. 2008. "From Ideology to Inclusion." *National Family Violence Legislative Center Conference*. Sacramento, California, February.

Straus, Murray A. and Denise A. Donnelly. 2001. *Beating the Devil Out of Them: Corporal Punishment in American Families and Its Effect on Children*. 2nd ed. New Brunswick, NJ: Transaction Books.

Straus, Murray A. and Richard J. Gelles. 1986. "Societal Change and Change in Family Violence from 1975 to 1985 as Revealed by Two National Surveys." *Journal of Marriage and Family* 48:465–79.

———. 1988. "How Violent Are American Families? Estimates from *The National* Family Violence Resurvey and Other Studies." Pp. 14–36 in *Family Abuse and Its Consequences: New Directions in Research*, edited by Gerald T. Hotaling, David Finkelhor, John T. Kirkpatrick, and Murray A. Straus. Newbury Park, CA: Sage Publications.

———. 1995. *Physical Violence in American Families: Risk Factors and Adaptations to Violence in 8,145 Families*. New Brunswick, NJ: Transaction Books.

Straus, Murray A., Richard J. Gelles, and Suzanne K. Steinmetz. 1980. *Behind Closed Doors: Violence in the American Family*. New York: Doubleday.

Straus, Murray A., Sherry L. Hambey, Sue Boney-McCoy, and David B. Sugarman. 1996. "The Revised Conflict Tactics Scales (CTS2): Development and Preliminary Psychometric Data." *Journal of Family Issues* 17:283–316.

Straus, Murray A. and Ignacio Luis Ramirez. 2007. "Gender Symmetry in Prevalence, Severity, and Chronicity of Physical Aggression Against Dating Partners by University students in Mexico and USA." *Aggressive Behavior* 33:281–90.

Straus, Murray A. and Katreena Scott. Forthcoming. "Gender Symmetry in Partner Violence: The Evidence, the Denial, and the Implications for Primary Prevention and Treatment." In *Preventing Partner Violence: Foundations, Intervention, and Issues*, edited by Whitaker D. J. and J. R. Lutzker. American Psychological Association.

Strauss, Anselm and Barney Glaser. 1975. *Chronic Illness and the Quality of Life*. St. Louis: Mosby.

Strohschein, Lisa. 2005. "Household Income Histories and Child Mental Health Trajectories." *Journal of Health and Social Behavior* 46(4):359–75.

Strow, Claudia W. and Brian K. Strow. 2006. "A History of Divorce and Remarriage in the United States." *Humanomics* 22(4):239–51.

Stryker, Sheldon. 2003 [1980]. *Symbolic Interactionism: A Social Structural Version*. Caldwell, NJ: Blackburn.

"Study: Academic Gains for Women, Stagnation for Men." 2006. CNN. June 2. Retrieved June 4, 2006 (www.cnn.com).

"Study Says White Families' Wealth Advantage Has Grown." 2004. *New York Times*, October 18.

Sturge-Apple, Melissa, Patrick Davies, Dante Cicchetti, and E. Mark Cummings. 2009. "The Role of Mothers' and Fathers' Adrenocortical Reactivity in Spillover Between Interparental Conflict and Parenting Practices." *Journal of Family Psychology* 23(2):215–25.

Su, Eleanor Yang. 2007. "'Helicopter Parents' Still Hover Even As Grads Pound Pavement." *The San Diego Union-Tribune*, July 5. Retrieved January 20, 2010 (legacy.signonsandiego.com/uniontrib/20070705/news_1n5parents.html).

Suitor, J. Jill, Jori Sechrist, Mari Plikuhn, Seth T. Pardo, Megan Gilligan, and Karl Pillemer. 2009. "The Role of Perceived Maternal Favoritism in Sibling Relations in Midlife." *Journal of Marriage and Family* 71(November):1026–38.

Sullivan, Harmony B. and Maureen C. McHugh. 2009. "The Critical Eye: Whose Fantasy is This?" *Sex Roles* 60:745–47.

Sullivan, Jeremy R., Cynthia A Riccio, and Cecil R. Reynolds. 2008. "Variations in Students' School- and Teacher-Related Attitudes Across Gender, Ethnicity, and Age." *Journal of Instructional Psychology* 35(3):296–305.

Sullivan, Kieran T., Lauri A. Pasch, Matthew D. Johnson, Thomas N. Bradbury. 2010. "Social

Support, Problem Solving, and the Longitudinal Course of Newlywed Marriage." *Journal of Personality & Social Psychology* 98(4):631–44.

Sullivan, Oriel and Scott Coltrane. 2008. "Men's Changing Contribution to Housework and Childcare." Prepared for the *11th Annual Conference of the Council on Contemporary Families*, April 25–26.

Supple, Andres J. and Stephen A. Small. 2006. "The Influence of Parental Support, Knowledge, and Authoritative Parenting on Hmong and European American Adolescent Development." *Journal of Family Issues* 27(9):1214–32.

Sun, Tao and Christopher A. Walsh. 2006. "Molecular Approaches to Brain Asymmetry and Handedness." *Nature Reviews Neuroscience* 7(8):655–662.

Surdin, Ashley. 2009. "Benefits for Same-Sex Partners Are Expanding." *The Washington Post*, November 27. Retrieved December 1, 2009 (www.washingtonpost.com).

Suro, Roberto. 1992. "Generational Chasm Leads to Cultural Turmoil for Young Mexicans in U.S." *New York Times*, January 20.

Surra, Catherine A. 1990. "Research and Theory on Mate Selection and Premarital Relationships in the 1980s." *Journal of Marriage and Family* 52:844–65.

Surra, Catherine A. and Debra K. Hughes. 1997. "Commitment Processes in Accounts of the Development of Premarital Relationships." *Journal of Marriage and Family* 59:5–21.

Sussman, L. J. 2006. "A 'Delicate Balance': Interfaith Marriage, Rabbinic Officiation, and Reform Judaism in America, 1870–2005." *CCAR Journal: A Reform Jewish Quarterly* 53(2):38–67.

Sussman, Marvin B., Suzanne K. Steinmetz, and Gary W. Peterson. 1999. *Handbook of Marriage and the Family*. 2nd ed. New York: Plenum.

Suter, Elizabeth, Karla Mason Bergen, Karen Daas, and Wesley Durham. 2006. "Lesbian Couples' Management of Public-Private Dialectical Contradictions." *Journal of Social and Personal Relationships* 23(3):349–65.

Suter, Elizabeth A., Karen L. Daas, and Karla Mason Bergen. 2008. "Negotiating Lesbian Family Identity via Symbols and Rituals." *Journal of Family Issues* 29(1):26–47.

Sutphin, Suzanne. 2006. "The Division of Household Labor in Same Sex Couples." Paper presented at the annual meeting of the *American Sociological Association*, August 11.

Svensson, R. 2003. "Gender Differences in Adolescent Drug Use: The Impact of Parental Monitoring and Peer Deviance." *Youth & Society* 34:300–29.

Swager, Toby J. 2009. *Daddy, Do My Socks Match? Adventures of a Stay-at-Home Dad.* Tate Publishers and Enterprises.

Swain, Carol M. 2002. *The New White Nationalism: Its Challenge to Integration.* Cambridge, U.K.: Cambridge University Press.

Swanbrow, Diane. 2008. "Exactly How Much Housework Does A Husband Create?" *Michigan Today*, April 3. Retrieved May 26, 2010 (michigantoday.umich.edu).

Swartz, Susan. 2004. "Singular Lives: SSU Sociologist Kay Trimberger Is Documenting a New Trend of Unmarried Women Who Are Not Only Content, but Happy to Be Single." *Press Democrat*, February 29.

Swartz, Teresa. 2008. "Family Capital and the Invisible Transfer of Privilege: Intergenerational Support and Social Class in Early Adulthood." *New Directions for Child and Adolescent Development* 119:11–24.

Sweeney, Megan M. and Maria Cancian. 2004. "The Changing Importance of White Women's Economic Prospects for Assortative Mating." *Journal of Marriage and Family* 66(4):1015–28.

Sweeney, Megan M. and Julie A. Phillips. 2004. "Understanding Racial Differences in Marital Disruption: Recent Trends and Explanations." *Journal of Marriage and Family* 66:639–50.

Sweet, James, Larry Bumpass, and Vaughn Call. 1988. "The Design and Content of *The National Survey of Families and Households*." Working Paper NSFH–1, University of Wisconsin, Center for Demography and Ecology, Madison, WI.

Swim, Janet K., K. J. Aikin, W. S. Hall, and B. A. Hunter. 1995. "Sexism and Racism: Old-Fashioned and Modern Prejudices." *Journal of Personality and Psychology* 68(2):199–214.

Swiss, Liam and Céline Le Bourdais. 2009. "Father—Child Contact After Separation: The Influence of Living Arrangements." *Journal of Family Issues* 30(5):623–52.

Szinovacz, Maximiliane and Anne Schaffer. 2000. "Effects of Retirement on Marital Conflict Tactics." *Journal of Family Issues* 21(3):376–89.

Tach, Laura and Sarah Halpern-Meekin. 2009. "How Does Premarital Cohabitation Affect Trajectories of Marital Quality?" *Journal of Marriage and Family* 71(2):298–317.

Tach, Laura, Ronald Mincy, and Kathryn Edin. 2010. "Parenting As A 'Package Deal': Relationships, Fertility, and nonresident Father Involvement among Unmarried Parents." *Demography* 47(1):181–204.

Taffel, Selma. 1987. "Characteristics of American Indian and Alaska Native Births: United States, 1984." *Monthly Vital Statistics Report* 36(3), Supplement. Washington, DC: U.S. National Center for Health Statistics. June 19.

Tafoya, Sonya M., Hans Johnson, and Laura E. Hill. 2004. *Who Chooses to Chose Two? Multiracial Identification and Census 2000.* New York and Washington, DC: Russell Sage Foundation and Population Reference Bureau.

Taft, Casey, Candice Monson, Jeremiah Schumm, Laura Watkins, Jillian Panuzio, and Patricia Resick. 2009. "Posttraumatic Stress Disorder Symptoms, Relationship Adjustment, and Relationship Aggression in a Sample of Female Flood Victims." *Journal of Family Violence* 24:389–96.

Tak, Young Ran and Marilyn McCubbin. 2002. "Family Stress, Perceived Social Support, and Coping Following the Diagnosis of a Child's Congenital Heart Disease." *Journal of Advanced Nursing* 39(2):190–98.

Takagi, Dana Y. 2002. "Japanese American Families." Pp. 164–80 in *Multicultural Families*

in the United States, 3rd ed., edited by Ronald L. Taylor. Upper Saddle River, NJ: Prentice Hall.

Talan, Kenneth H. 2009. *Help Your Child or Teen Get Back on Track: What Parents and Professionals Can Do for Childhood Emotional and Behavioral Problems.* London & Philadelphia: Jessica Kingsley Publisher.

Talbot, Margaret. 2001. "Open Sperm Donation." *New York Times Magazine*, December 8, p. 88.

Tanne, Janice Hopkins. 2009. "Virginity Pledge Ineffective Against Teen Sex Despite Government Funding, US Study Finds." *British Medical Journal* 338(7686):69.

Tannen, Deborah. 1990. *You Just Don't Understand.* New York: Morrow.

———. 2006. *You're Wearing That? Understanding Mothers and Daughters in Conversation.* New York: Random House.

Tanner, Lindsey. 2005. "Study: Children Adopted from Foreign Countries Adjust Surprisingly Well." Associated Press, May 24. Retrieved July 22, 2007 (www.associatedpress.org).

Tarrance Group, The and Lake Research Partners. 2008. GWU Battleground 2008. Study 11633. Washington, DC: The Tarrance Group. Retrieved July 7, 2009 (www.lakeresearch.com/polls/pdf/BG080522/08BGfre1-Final.pdf).

Tashiro, Ty, Patricia Frazier, and Margit Berman. 2006. "Stress-Related Growth Following Divorce and Relationship Dissolution." Pp. 361–84 in *Handbook of Divorce and Relationship Dissolution*, edited by Mark A. Fine and John H. Howard. Mahwah, NJ: Erlbaum.

Taubman, Phoebe. 2009. "Free Riding on Families: Why the American Workplace Needs to Change and How to Do It." *Issue Brief* December. Washington, DC: America Constitution Society for Law and Policy.

Taylor, Paul. 2009. *America's Changing Workforce: Recession Turns a Graying Office Grayer.* September 3. Pew Research Center, Social and Demographic Trends Project. Retrieved May 8, 2010 (pewresearch.org).

Taylor, Paul, Richard Fry, D'Vera Cohn, Wendy Wang, Gabriel Velasco, and Daniel Dockterman. 2010. *Women, Men and the New Economics of Marriage.* Washington, DC: Pew Research Center.

Taylor, Paul, Cary Funk, and Peyton Craighill. 2006. "Are We Happy Yet?" Pew Research Center. Retrieved October 10, 2006 (pewresearch.org).

Taylor, Robert J., Karen D. Lincoln, and Linda M. Chatters. 2005. "Supportive Relationships with Church Members among African Americans." *Family Relations* 54(4):501–11.

Taylor, Ronald L. 2002a. "Black American Families." Pp. 19–47 in *Minority Families in the United States: A Multicultural Perspective*, 3rd ed., edited by Ronald L. Taylor. Upper Saddle River, NJ: Prentice Hall.

———. 2002b. "Minority Families and Social Change." Pp. 252–300 in *Minority Families in the United States: A Multicultural Perspective*, 3rd ed., edited by Ronald L. Taylor. Upper Saddle River, NJ: Prentice Hall.

———. 2002c. *Minority Families in the United States: A Multicultural Perspective.* 3rd ed. Upper Saddle River, NJ: Prentice Hall.

———. 2007. "Diversity within African American Families." Pp. 398–421 in *Family in Transition,* 14th ed., edited by Arlene S. Skolnick and Jerome H. Skolnick. Boston, MA: Allyn and Bacon.

Taylor, Ronald L., M. Belinda Tucker, and C. Mitchell-Kernan. 1999. "Ethnic Variations in Perceptions of Men's Provider Role." *Psychology of Women Quarterly* 23:741–61.

Taylor, Verta, Katrina Kimport, Nella Van Dyke, and Ellen Ann Andersen. 2009. "Culture and Mobilization: Factional Repertoires, Same-Sex Weddings, and the Impact on Gay Activism." *American Sociological Review* 74(6):865–90.

Teachman, Jay D. 2000. "Diversity of Family Structure: Economic and Social Influences." Pp. 32–58 in *Handbook of Family Diversity,* edited by David H. Demo, Katherine R. Allen, and Mark A. Fine. New York: Oxford University Press.

———. 2002a. "Childhood Living Arrangements and the Intergenerational Transmission of Divorce." *Journal of Marriage and Family* 64:717–29.

———. 2002b. "Stability across Cohorts in Divorce Risk Factors." *Demography* 39(2):331–51.

———. 2003. "Premarital Sex, Premarital Cohabitation, and the Risk of Subsequent Marital Dissolution among Women." *Journal of Marriage and Family* 65(2):444–55.

———. 2004. "The Childhood Living Arrangements of Children and the Characteristics of Their Marriages." *Journal of Family Issues* 25(1):86–111.

———. 2008a. "Complex Life Course Patterns and the Risk of Divorce in Second Marriages." *Journal of Marriage and Family* 70(2):294–305.

———. 2008b. "The Living Arrangements of Children and Their Educational Well-Being." *Journal of Family Issues* 29(6):734–61.

———. 2009. "Military Service, Race, and the Transition to Marriage and Cohabitation." *Journal of Family Issues* 30(10):1433–54.

———. 2010. "Wives' Economic Resources and Risk of Divorce." *Journal of Family Issues* 31(4):1–19.

Teachman, Jay D., Lucky M. Tedrow, and Kyle D. Crowder. 2000. "The Changing Demography of America's Families." *Journal of Marriage and Family* 62(4):1234–46.

Teachman, Jay D., Lucky M. Tedrow, and Matthew Hall. 2006. "The Demographic Future of Divorce and Dissolution." Pp. 59–82 in *Handbook of Divorce and Relationship Dissolution,* edited by Mark A. Fine and John H. Howard. Mahwah, NJ: Erlbaum.

Teitler, Julien and Nancy Reichman. 2008. "Mental Illness as a Barrier to Marriage among Unmarried Mothers." *Journal of Marriage and Family* 70(3):772–82.

Teitler, Julien, Nancy Reichman, Lenna Nepomnyaschy, and Irwin Garfinkel. 2009. "Effects of Welfare Participation on Marriage." *Journal of Marriage and Family* 71(3):878–91.

Tejada-Vera, B. and P. D. Sutton. 2009. "Births, Marriages, Divorces, and Deaths: Provisional Data for May 2008." *National Vital Statistics Reports* 57 (10). Hyattsville, MD: National Center for Health Statistics.

Tender, Lars. 2008. "Love, Technology, and Dating." *Theory & Event* 11 (3). Retrieved May 12, 2009 (http://muse.jhu.edu/).

Tennov, Dorothy. 1999 [1979]. *Love and Limerence: The Experience of Being in Love.* 2nd ed. New York: Scarborough House.

Tepperman, Lorne and Susannah J. Wilson, eds. 1993. *Next of Kin: An International Reader on Changing Families.* New York: Prentice Hall.

Tessina, Tina B. 2008. *The Commuter Marriage: Keep Your Relationship Close While You're Far Apart.* Avon, MA: Adams Media.

Theidon, Kimberly. 2009. "Reconstructing Masculinities: The Disarmament, Demobilization, and Reintegration of Former Combatants in Colombia." *Human Rights Quarterly* 31:1–34.

Therborn, Göran. 2004. *Between Sex and Power: Family in the World, 1900–2000.* London, England: Routledge.

Theran, Sally A. 2009. "Predictors of Level of Voice in Adolescent Girls: Ethnicity, Attachment, and Gender Role Socialization." *Journal of Youth & Adolescence* 38(8):1027–37.

"Third of New HIV Cases Acquired Heterosexually." 2004. *Omaha World-Herald,* February 20.

Thobaben, Marshelle. 2008. "Elder Abuse Prevention." *Home Health Care Management and Practice* 20(2):194–96.

Thoennes, Nancy and Patricia G. Tjaden. 1990. "The Extent, Nature, and Validity of Sexual Abuse Allegations in Custody/Divorce Disputes." *Child Abuse & Neglect* 14:151ff.

Thomas, Adam and Isabel Sawhill. 2005. "For Love *and* Money? The Impact of Family Structure on Family Income." *The Future of Children* 15(2):57–74.

Thomas, Alexander, Stella Chess, and Herbert G. Birch. 1968. *Temperament and Behavior Disorders in Children.* New York: New York University Press.

Thomas, Brett W. and Mark W. Roberts. 2009. "Sibling Conflict Resolution Skills: Assessment and Training." *Journal of Child and Family Studies* 18:447–53.

Thomas, J. 2009. "Virginity Pledgers Are Just as Likely as Matched Nonpledgers to Report Premarital Intercourse." *Perspectives on Sexual & Reproductive Health* 41(1):63.

Thomason, Deborah J. 2005. "Natural Disasters: Opportunity to Build and Reinforce Family Strengths." *Family Focus on . . . Family Strengths and Resilience* FF28:F11. Minneapolis, MN: National Council on Family Relations.

Thompson, Kacie M. 2009. "Sibling Incest: A Model for Group Practice with Adult Female Victims of Brother–Sister Incest." *Journal of Family Violence* 24(7):531–37.

Thompson, Steven S. 2006. "Was Ancient Rome a Dead Wives Society? What Did the Roman Paterfamilias Get Away With?" Journal of Family History 31(1):3–27.

Thomson, Elizabeth and Ugo Colella. 1992. "Cohabitation and Marital Stability: Quality or Commitment?" *Journal of Marriage and Family* 54(2):368–78.

Thomson, Elizabeth, Jane Mosley, Thomas L. Hanson, and Sara S. McLanahan. 2001. "Remarriage, Cohabitation, and Changes in Mothering Behavior." *Journal of Marriage and Family* 63(2):370–80.

Thorne, Barrie. 1992. "Girls and Boys Together . . . But Mostly Apart: Gender Arrangements in Elementary School." Pp. 108–23 in *Men's Lives,* 2nd ed., edited by Michael S. Kimmel and Michael A. Messner. New York: Macmillan.

Thornton, Arland. 2009. "Framework for Interpreting Long-Term Trends in Values and Beliefs Concerning Single-Parent Families." *Journal of Marriage and Family* 71(2):230–34.

Thornton, Arland, William G. Axinn, and Yu Xie. 2007. Marriage and Cohabitation. Chicago, IL: The University of Chicago Press.

Thornton, Arland and Deborah Freedman. 1983. "The Changing American Family." *Population Bulletin* 38. Washington, DC: Population Reference Bureau.

Thornton, Arland and Linda Young-DeMarco. 2001. "Four Decades of Trends in Attitudes toward Family Issues in the United States: The 1960s through the 1990s." *Journal of Marriage and Family* 63:1009–37.

Tichenor, Veronica Jaris. 1999. "Status and Income as Gendered Resources: The Case of Marital Power." *Journal of Marriage and Family* 61:638–50.

———. 2005. *Earning More and Getting Less: Why Successful Wives Can't Buy Equality.* New Brunswick, NJ: Rutgers University Press.

———. 2010. "Thinking About Gender and Power in Marriage." Pp. 415–25 in *The Kaleidoscope of Gender: Prisms, Patterns, and Possibilities,* 3rd ed., edited by Joan Z. Spade and Catherine G. Valentine. Newbury Park, CA: Pine Forge Press.

Tiefer, Leonore. 2008. "Female Genital Cosmetic Surgery: Freakish or Inevitable? Analysis from Medical Marketing, Bioethics, and Feminist Theory." *Feminism & Psychology* 18(4):466–79.

Tiger, Lionel. 1969. *Men in Groups.* New York: Vintage.

Tillman, Kathryn Harker. 2007. "Family Structure Pathways and Academic Disadvantage among Adolescents in Stepfamilies." *Sociological Inquiry* 77(3):383–424.

Time Opinion Poll on Gender. "Men and Women Have Similar Life Goals." *Time Magazine,* October 14. www.time.com.

Tjaden, Patricia and Nancy Thoennes. 2000. "Prevalence and Consequences of Male-to-Female and Female-to-Male Intimate Partner Violence as Measured by *The Nationa*l Violence Against Women Survey." *Violence Against Women* 6:142–61.

Tolan, Patrick H., Jose Szapocznik, and Soledad Sambrano. 2007. *Preventing Youth Substance Abuse*. Washington, DC: American Psychological Association.

Toledo, Sylvie de and Doborah Edler Brown. 1995. *Grandparents as Parents: A Survival Guide for Raising a Second Family*. New York: Guilford.

Tolin, David F., Randy O. Frost, Gail Steketee, and Kristin E. Fitch. 2008. "Family Burden of Compulsive Hoarding: Results of an Internet Survey." *Behaviour Research and Therapy* 46:334–344.

Tonelli, Bill. 2004. "Thriller Draws on Oppression of Italians in War*time* U.S." *New York Times*, August 2.

Toner, Robin. 2007. "Women Feeling Freer to Suggest 'Vote for Mom.'" *New York Times*, January 29.

Tong, Benson. 2004. *Asian American Children: An Historical Guide*. Westport, CT: Greenwood Press.

Topoleski, Julie. 2009. "The Long-Term Outlook for Medicare, Medicaid, and Total Health Care Spending." The Long-Term Budget Outlook. Washington, DC: Congressional Budget Office. (http://www.cbo.gov/ftpdocs/102xx/doc10297/06-25-LTBO.pdf).

Torres, Zenia. 1997. "Interracial Dating." Unpublished student paper.

Total Group Report: College-Bound Seniors 2009. 2009. *New York:* The College Board. www.collegeboard.com.

Toth, Joan. 2007. "NEW Women of Color Report Outlines Industry's Challenges, Opportunities." *Network of Executive Women*. Retrieved on August 11, 2010 (www.newnewsletter.org).

Totura, Christine M. Wienke, Carol MacKinnon-Lewis, Ellis L. Gesten, Ray Gadd, Katherine P. Divine, Sherri Dunham, and Dimitra Kamboukos. 2009. "Bullying and Victimization Among Boys and Girls in Middle School: The Influence of Perceived Family and School Contexts." *Journal of Early Adolescence* 29(4):571–609.

Tougas, F., R. Brown, A. M. Beaton, and S. Joly. 1995. "Neosexism: Plus ça change, plus c'est pareil." *Personality and Social Psychology Bulletin* 21:842–49.

Townsend, Aloen L. and Melissa M. Franks. 1997. "Quality of the Relationship between Elderly Spouses: Influence on Spouse Caregivers' Subjective Effectiveness." *Family Relations* 46(1):33–39.

Townsend, Nicholas W. 2002. *The Package Deal: Marriage, Work, and Fatherhood in Men's Lives*. Philadelphia, PA: Temple.

Tozzi, John. 2010. "Home-Based Businesses Increasing." *Bloomberg Businessweek*, January 25. Retrieved May 26, 2010 (www.businessweek.com).

Tracy, J. K. and J. Junginger. 2007. "Correlates of Lesbian Sexual Functioning." *Journal of Women's Health* 16:499–509.

Trask, Bahira Sherif, Jocelyn D. Taliaferro, Margaret Wilder, and Raheemah Jabbar-Bey. 2005. "Strengthening Low-Income Families through Community-Based Family Support Initiatives." . . .

Family Strengths and Resilience FF28. Minneapolis, MN: National Council on Family Relations.

Trask, Bahira Sherif, Bethany Willis Hepp, Barbara Settles, and Lilianah Shabo. 2009. "Culturally Diverse Elders and Their Families: Examining the Need for Culturally Competent Services." *Journal of Comparative Family Studies* 40(2):293–303.

Travis, Carol. 2006. "Letters to the Editor: The New Science of Love." *Atlantic Monthly*, May.

Treas, Judith. 1995. "Older Americans in the 1990s and Beyond." *Population Bulletin* 50(2). Washington, DC: Population Reference Bureau.

Treas, Judith and Deirdre Giesen. 2000. "Sexual Infidelity among Married and Cohabiting Americans." *Journal of Marriage and Family* 62(1):48–60.

Treas, Judith and Sonja Drobnič. 2010. *Dividing the Domestic: Men, Women, and Household Work in Cross-National Perspective*. Stanford, CA: Stanford University Press.

Trevathan, Melissa and Sissy Goff. 2007. *Raising Girls*. Grand Rapids, MI: Zondervan.

Trimberger, E. Kay. 2005. *The New Single Woman*. New York: Beacon Press.

Troilo, Jessica and Marilyn Coleman. 2008. "College Students' Perceptions of the Content of Father Stereotypes." *Journal of Marriage and Family* 70(1):218–27.

Troll, Lillian E. 1985. "The Contingencies of Grandparenting." Pp. 135–50 in *Grandparenthood*, edited by Vern L. Bengston and Joan F. Robertson. Newbury Park, CA: Sage Publications.

Troxel v. Granville. 2000. 530 U.S. 57.

Troy, Adam B., Jamie Lewis-Smith, and Jean-Phillippe Laurenceau. 2006. "Interracial and Intraracial Romantic Relationships: The Search for Differences in Satisfaction, Conflict, and Attachment Style." *Journal of Social and Personal Relationships* 23(1):65–80.

Trudeau, Michelle. 2006. "School, Study, SATs: No Wonder Teens Are Stressed." National Public Radio, October 9. Retrieved October 9, 2006 (www.npr.org).

Trumbull, Mark. 2009. "Boomerang Kids: Recession Sends More Young Adults Back Home." *The Christian Science Monitor*, November 24. Retrieved December 20, 2009 (http:///www.csmonitor.com).

Trussell, James and L. L. Wynn. 2008. "Reducing Unintended Pregnancy in the United States." *Contraception* 77:1–5.

Tsang, Laura Lo Wa, Carol D. H. Harvey, Karen A. Duncan, and Reena Sommer. 2003. "The Effects of Children, Dual Earner Status, Sex Role Traditionalism, and Marital Structure on Marital Happiness over *Time*." *Journal of Family and Economic Issues* 24:5–26.

Tshann, Jeane M., Lauri A. Pasch, Elena Flores, Barbara VanOss Marin, E. Marco Baisch, and Charles J. Wibbelsman. 2009. "Nonviolent Aspects of Interparental Conflict and Dating Violence Among Adolescents." *Journal of Family Issues* 30(3):295–319.

Tucker, Corinna J., Susan M. McHale, and Ann C. Crouter. 2003. "Conflict Resolution: Links with Adolescents' *Family Relation*ships and Individual Well-being." *Journal of Family Issues* 24(6):715–36.

Tucker, M. Belinda. 2000. "Marital Values and Expectations in Context: Results from a 21-City Survey." Pp. 166–87 in *The Ties That Bind: Perspectives on Cohabitation and Marriage*, edited by Linda J. Waite. New York: Aldine.

Tuhus-Dubrow, Rebecca. 2009. "The Female Advantage: A New Reason For Businesses to Promote Women: It's More Profitable." *The Boston Globe*, May 3 (www.boston.com).

Tuller, David. 2001. "Adoption Medicine Brings New Parents Answers and Advice." *New York Times*, September 4.

Turkat, Ira Daniel. 1997. "Management of Visitation Interference." *The Judges Journal* 36(Spring):17–47. Retrieved May 16, 2007 (www.fact.on.ca).

Turley, Ruth N. Lopez. 2003. "Are Children of Young Mothers Disadvantaged Because of Their Mother's Age or Family Background?" *Child Development* 74:465–74.

Turnbull, A. and H. Turnbull. 1997. *Families, Professionals, and Exceptionality: A Special Partnership*. 3rd ed. Upper Saddle River, NJ: Merrill.

Turner, Lynn H. and Richard L. West. 2006. *The Family Communication Sourcebook*. Thousand Oaks, California: Sage.

Turner, Ralph H. 1976. "The Real Self: From Institution to Impulse." *American Journal of Sociology* 81:989–1016.

Turow, Joseph. 2001. "Family Boundaries, Commercialism, and the Internet: A Framework for Research." *Journal of Applied Developmental Psychology* 22(1):73–86.

Turrell, Susan C. 2000. "A Descriptive Analysis of Same-Sex Relationship Violence for a Diverse Sample." *Journal of Family Violence* 15:281–93.

Twenge, Jean M. 1997a. "Attitudes toward Women, 1970–1995: A Meta-Analysis." *Psychology of Women Quarterly* 21(1):35–51.

———. 1997b. "'Mrs. His Name': Women's Preferences for Married Names." *Psychology of Women Quarterly* 21(3):417–30.

Twenge, Jean M., W. Keith Campbell, and Craig Foster. 2003. "Parenthood and Marital Satisfaction: A Meta-Analytic Review." *Journal of Marriage and Family* 65:574–83.

Tyler, Kimberly A. and Ana Mari Cauce. 2002. "Perpetrators of Early Physical and Sexual Abuse among Homeless and Runaway Adolescents." *Child Abuse & Neglect* 26(12):1261–74.

Tyre, Peg. 2004. "A New Generation Gap." *Newsweek*, January 19, pp. 68–71.

———. 2006a. "The New First Grade." *Newsweek*, September 11, pp. 34–43.

———. 2006b. "Smart Moms, Hard Choices." *Newsweek*, March 6, p. 55.

———. 2006c. "The Trouble With Boys." *Newsweek*, January 30.

———. 2009. "Daddy's Home, and a Bit Lost." *The New York Times,* January 11. Retrieved October 5, 2009 (www.nytimes.com).

Tyre, Peg and Daniel McGinn. 2003. "She Works, He Doesn't." *Newsweek,* May 12, pp. 45–52.

Udry, J. Richard. 1974. *The Social Context of Marriage.* 3rd ed. Philadelphia, PA: Lippincott.

———. 1994. "The Nature of Gender." *Demography* 31(4):561–73.

———. 2000. "The Biological Limits of Gender Construction." *American Sociological Review* 65:443–57.

Uecker, Jeremy E., Mark D. Regnerus, and Margaret L. Vaaler. 2007. "Losing My Religion: The Social Sources of Religious Decline in Early Adulthood." *Social Forces* 85(4):1667–92.

Uecker, Jeremy E. and Charles E. Stokes. 2008. "Early Marriage in the United States." *Journal of Marriage and Family* 70(November):835–46.

"UF Study: Sibling Violence Leads to Battering in College Dating." 2004. *UF News.* Gainesville, FL: University of Florida.

Uhlenberg, Peter. 1996. "Mortality Decline in the Twentieth Century and Supply of Kin over the Life Course." *The Gerontologist* 36:681–85.

Ulker, Aydogan. 2009. "Wealth Holdings and Portfolio Allocation of the Elderly: The Role of Marital History." *Journal of Family Economic Issues* 30:90–108.

Umaña-Taylor, Adriana J. and Edna C. Alfaro. 2006. "Divorce and Relationship Dissolution Among Latino Populations in the U.S." Pp. 515–30 in *Handbook of Divorce and Relationship Dissolution,* edited by Mark A. Fine and John H. Harvey. Mahwah, NJ: Erlbaum and Associates.

Umana-Taylor, Adriana J., Edna C. Alfaro, Mayra Y. Bamaca, and Amy B. Guimond. 2009. "The Central role of Familial Ethnic Socialization in Latino Adolescents' Cultural Orientation." *Journal of Marriage and Family* 71(1):46–60.

Umberson, Debra, Kristin Anderson, Jennifer Glick, and Adam Shapiro. 1998. "Domestic Violence, Personal Control, and Gender." *Journal of Marriage and Family* 60(2):442–52.

Umberson, Debra, Meichu D. Chen, James S. House, Kristine Hopkins, and Ellen Slaten. 1996. "The Effect of Social Relationships on Psychological Well-being: Are Men and Women Really So Different?" *American Sociological Review* 61:837–57.

UNICEF (United Nations Children's Fund). 2007. *Child Poverty in Perspective: An Overview of Child Well-being in Rich Countries.* Innocenti Research Center Report Card No. 7. Florence, Italy: UNICEF Innocenti Research Center.

"U.N.: U.S. Workers are World's Most Productive." 2007. *MSNBC,* September 3. Retrieved May 27, 2010 (www.msnbc.com).

Unitarian Universalist Association. n.d. *Premarital Counseling Guide for Same Gender Couples.* Boston, MA: Unitarian Universalist Association Office of Bisexual, Gay, Lesbian, and Transgender Concerns. Retrieved March 3, 2010 (www.uua.org/obgltc).

"United Nations Drops Gay Civil Rights." 2004. 365Gay.com Newscenter Staff, March 29.

Retrieved March 29, 2004 (www.sodomylaws. org/world/wonews023.htm).

"Urban Parents, Particularly Those Who Are Unmarried, Frequently Have Children by Multiple Partners." 2006. *Perspectives on Sexual and Reproductive Health* 38(4):225–26.

Uruk, Ayse, Thomas Sayger, and Pamela Cogdal. 2007. "Examining the Influence of Family Cohesion and Adaptability on Trauma Symptoms and Psychological Well-Being." *Journal of College Student Psychotherapy* 22(2):51–63.

U.S. Administration for Children and Families. African American Healthy Marriage Initiative. N.d. "A Targeted Strategy for Working Effectively with African American Communities." Washington, DC: Administration for Children and Families. Retrieved December 19, 2006 (www.acf.hhs.gov/health marriage).

U.S. Administration on Aging. 2010. *A Profile of Older Americans:2009.* Washington, DC: U.S. Department of Health and Human Services. Retrieved May 8, 2010 (www.aoa.gov).

U.S. Bureau of Labor Statistics 2006d. Women in the Labor Force: A Databook. Report 996. Washington, DC: U.S. Bureau of Labor Statistics. September. Retrieved January 18, 2007 (www.bls.gov).

U.S. Bureau of Labor Statisics. 2008. Married Parents' Use of Time 2003–06. Report 0219. Washington, DC: U.S. Bureau of Labor Statistics.

———. 2009a. "A Profile of the Working Poor, 2007." Report 1012. Washington, DC: U.S. Bureau of Labor Statistics.

———. 2009b. "Highlights of Women's Earnings in 2008." Report 1017. Washington, DC: U.S. Bureau of Labor Statistics.

———. 2009c. *Women in the Labor Force: A Databook.* Report 1018. Washington, DC: U.S. Bureau of Labor Statistics. www.bls.gov.

———. 2009d. *American Time Use Survey Summary.* Report 09-0704. Washington, DC: U.S. Bureau of Labor Statistics. www.bls.gov.

U.S. Census Bureau. 1989. *Statistical Abstract of the United States.* 109th ed. Washington, DC: U.S. Government Printing Office.

———. 1993. *We the Americans: Pacific Islanders.* WE-4. Washington, DC: U.S. Census Bureau. September.

———. 1998. *Statistical Abstract of the United States, 1998.* Washington, DC: U.S. Government Printing Office.

———. 2000. *Statistical Abstract of the United States.* 120th ed. Washington, DC: U.S. Government Printing Office (www.census.gov/ stat_abstract).

———. 2002. *Statistical Abstract of the United States: 2006.* Washington, DC: U.S. Census Bureau.

———. 2003a. *Statistical Abstract of the United States, 2003.* Washington, DC: U.S. Census Bureau.

———. 2003b. "U.S. Census Bureau Guidance on the Presentation and Comparison of Race

and Hispanic Origin Data." Washington, DC: U.S. Census Bureau.

———. 2005. "Survey of Income and Program Participation, 2004 Panel, Wave 4." Washington, DC: U.S. Census Bureau.

———. 2006a. "2005 American Community Survey." Washington, DC: U.S. Census Bureau. Retrieved August 15, 2006 (factfinder.census.gov).

———. 2006b. "America's Families and Living Arrangements: 2005." Washington, DC: U.S. Census Bureau. May 25. Retrieved June 20, 2006 (www.census.gov/population/www/socdemo/ hh-fam/cps2005.html).

———. 2006c. "Statistical Abstract of the United States: 2006." Washington, DC: U.S. Census Bureau.

———. 2007a. "The American Community— Pacific Islanders: 2004." Washington, DC: U.S. Census Bureau. Retrieved November 4, 2009 (factfinder.census.gov).

———. 2007b. "Number, Timing, and Duration of Marriages and Divorces: 2004." Washington, DC: U.S. Census Bureau. Retrieved June 18, 2009 (www.census.gov/population/www/ socdemo/marr-div/2004detailed_tables.html).

———. 2007c. "Statistical Abstract of the United States: 2007." Washington, DC: U.S. Census Bureau.

———. 2008a. "2007 American Community Survey." Washington, DC: U.S. Census Bureau. Retrieved June 17, 2009 (factfinder.census.gov).

———. 2008b. "Current Population Survey, 2007: Annual Social and Economic Supplement." Washington, DC: U.S. Census Bureau. Retrieved June 29, 2009 (www.census. gov/population/www/socdemo/hhfam/ cps2009.html).

———. 2008c. "2008 National Population Projections." Washington, DC: U.S. Census Bureau.

———. 2008d. *Annual State Population Estimates by Demographic Characteristics with 6 Race Groups.* Washington, DC: U.S. Census Bureau.

———. 2008e. *Income, Earnings, and Poverty Data From the 2007 American Community Survey.* American Community Survey Reports, ACS-09. Washington, DC: U.S. Census Bureau.

———. 2008f. *Population by Selected Ancestry Group and Region: 2005.* American Community Survey, B04006. Washington, DC: U.S. Census Bureau.

———. 2009a. "2006–2008 American Community Survey 3-Year Estimates." Washington, DC: U.S. Census Bureau. Retrieved November 6, 2009 (factfinder.census.gov/).

———. 2009b. "2008 American Community Survey 1-Year Estimates." Washington, DC: U.S. Census Bureau. Retrieved October 2, 2009 (factfinder.census.gov/).

———. 2009c. "America's Families and Living Arrangements: 2008." Washington, DC: U.S. Census Bureau. Retrieved June 15, 2009 (www. census.gov/population/www/socdemo/hh-fam/cps2008.html).

———. 2009d. "Black (African-American) History Month: February 2009." Washington,

DC: U.S. Census Bureau. Retrieved June 9, 2009 (www.census.gov/Press-Release).

———. 2009e. "Current Population Survey, 2008 Annual Social and Economic Supplement." Washington, DC: U.S. Census Bureau. Retrieved June 29, 2009 (www.census.gov/hhes/www/macro/032008/faminc/new01_001.htm).

———. 2009f. "Current Population Survey, March and Annual Social and Economic Supplements, 2008 and Earlier." Washington, DC: U.S. Census Bureau. Retrieved June 10, 2009 (www.census.gov/population/socdemo/).

———. 2009g. "Income, Poverty and Health Insurance Coverage in the United States: 2008." Washington, DC: U.S. Census Bureau. Retrieved September 11, 2009 (www.census.gov).

———. 2009h. "Selected Indicators of Child Well-Being—A Child's Day: 2006." Washington, DC: U.S. Census Bureau. Retrieved December 23, 2009 (www.census.gov).

———. 2010a. "Current Population Survey, March and Annual Social and Economic Supplements, 2009 and Earlier." Washington, DC: U.S. Census Bureau. Retrieved January 21, 2010 (www.census.gov/population).

———. 2010b. "Statistical Abstract of the United States." Washington, DC: U.S. Census Bureau.

———. 2010c. "America's Families and Living Arrangements: 2009." *Current Population Reports.* Washington, DC: U.S. Census Bureau. Retrieved on March 10, 2010 (www.census.gov/population/www/socdemo/hh-fam/cps2009.html).

U.S. Centers for Disease Control and Prevention. 2006a. *2004 Assisted Reproductive Technology Success Rates: National Summary and Fertility Clinic Report.* Atlanta, GA: U.S. Centers for Disease Control and Prevention. December.

———. 2006b. *HIV/AIDS Surveillance Report 2005.* Vol. 17. Atlanta, GA: U.S. Centers for Disease Control and Prevention.

———. 2007. "HIV/AIDS among Women." Fact Sheet. Atlanta, GA: U.S. Centers for Disease Control and Prevention. March.

———. 2008. "2006 Assisted Reproductive Technology Success Rates: National Summary and Fertility Clinic Reports." Atlanta, GA: U.S. Department of Health and Human Services, Centers for Disease Control and Prevention and American Society for Reproductive Medicine, Society for Assisted Reproductive Technology.

———. 2009. *HIV/AIDS Surveillance Report,* 2007 19. Atlanta: U.S. Department of Health and Human Services, Centers for Disease Control and Prevention. www.cdc.gov.

U.S. Department of Education. 2009. "Employees in Degree-Granting Institutions, by Employment Status, Sex, Control and Type of Institution, and Primary Occupation: Fall 2007." Washington, DC: National Center for Education Statistics.

U.S. Department of Health and Human Services. 2009. "Adoption and Foster Care Analysis Reporting system (AFCARS) FY 2008 Data." Retrieved February 10, 2010 (www.acf.hhs.gov).

———. 2010. *Child Maltreatment 2008.* Washington, DC: Administration for Children and Families, Administration on Children, Youth and Families, Children's Bureau. Available from www.acf.hhs.gov/programs/cb/stats_research/index.htm#can.

U.S. Department of Health, Education, and Welfare. 1975. *Child Abuse and Neglect. Vol. I, An Overview of the Problem.* Publication (OHD) 75–30073. Washington, DC: U.S. Government Printing Office.

U.S. Department of Health, Education, and Welfare. National Institute of Mental Health. 1978. *Yours, Mine, and Ours: Tips for Stepparents.* Washington, DC: U.S. Government Printing Office.

U.S. Department of Homeland Security. 2009. Yearbook of Immigration Statistics: 2008. Washington, DC: Office of Immigration Statistics.

U.S. Federal Interagency Forum on Child and Family Statistics. 2005. *"America's Children: Key National Indicators of Well-Being, 2005."* Washington, DC: U.S. Federal Interagency Forum on Child and Family Statistics.

———. 2006. *"America's Children in Brief: Key National Indicators of Well-Being, 2006."* Washington, DC: U.S. Federal Interagency Forum on Child and Family Statistics. Retrieved August 13, 2006 (www.childtrends.org).

———. 2009. *"America's Children: Key National Indicators of Well-Being, 2009."* Washington, DC: U.S. Federal Interagency Forum on Child and Family Statistics.

U.S. Food and Drug Administration. 2006. "Mifeprex (mifepristone) Information." Washington, DC: U.S. Food and Drug Administration. April 10. Retrieved March 29, 2007 (www.fda.gov).

U.S. General Accounting Office. 1997. Letter to The Honorable Henry J. Hyde. Office of the General Counsel, January 31 (www.frwebgate.access.gpo.gov/cgi-bin/useftp.cgi/).

U.S. National Cancer Institute. 2003. "Summary Report: Early Reproductive Events and Breast Cancer Workshop." Bethesda, MD: U.S. National Cancer Institute. March 23.

———. 2010. "Abortion, Miscarriage, and Breast Cancer Risk." Bethesda, MD: U.S. National Cancer Institute, January 12. www.cancer.gov.

U.S. National Center for Education Statistics. 2006. *Digest of Education Statistics, 2005.* Washington, DC: U.S. National Center for Education Statistics. June. Retrieved January 31, 2007 (www.nces.ed.gov).

U.S. National Center for Health Statistics. 1990a. "Advance Report of Final Divorce Statistics, 1987." *Monthly Vital Statistics Report* 38(12), Suppl. April 3.

———. 1990b. "Advance Report of Final Marriage Statistics, 1987." *Monthly Vital Statistics Report* 38(12), Suppl. April 3.

———. 1998. "Births, Marriages, Divorces, and Deaths for 1997." *Monthly Vital Statistics Report* 46(12). July 28.

———. 2006. "Births, Marriages, Divorces, and Deaths: Provisional Data for 2005." *National Vital Statistics Reports* 54(20). Hyattsville, MD: U.S. National Center for Health Statistics.

U.S. National Institute of Child Health and Human Development. 1999. "Only Small Link Found between Hours in Child Care and Mother–Child Interaction." News Release, November 7 ().

———. 2002. "The NICHD Study of Early Child Care." Bethesda, MD: U.S. Institute of Health and Human Development.

U.S. Office of Management and Budget. 1999. *Revisions to the Standards for Classification of Federal Data on Race and Ethnicity.* Washington, DC: U.S. Census Bureau.

"U.S. Recession Causing Increase in Child Abuse Reports." 2009. Red Orbit News, April 16. Retrieved June 8, 2009 (www.redorbit.com).

"U.S. Scraps Study of Teen-age Sex." 1991. *New York Times,* July 25.

U.S. Social Security Administration. 2009. "Fact Sheet: Social Security Is Important to Women." Retrieved May 8, 2010 (www.socialsecurity.gov/women).

Usdansky, Margaret L. 2009a. "Ambivalent Acceptance of Single-Parent Families: A Response to Comments." *Journal of Marriage and Family* 71(2):240–46.

———. 2009b. "A Weak embrace: Popular and Scholarly Depictions of Single-Parent Families, 1900–1998." *Journal of Marriage and Family* 71(2):209–25.

Uttal, Lynet. 1999. "Using Kin for Child Care." *Journal of Marriage and Family* 61:845–57.

———. 2004. "Racial Safety and Cultural Maintenance: The Child Care Concerns of Employed Mothers of Color." Pp. 295–304 in *Race, Class, and Gender,* 5th ed., edited by Margaret L. Andersen and Patricia Hill Collins. Belmont, CA: Wadsworth.

Vaaler, Margaret L., Christopher G. Ellison, and Daniel A. Powers. 2009. "Religious Influences on the Risk of Marital Dissolution." *Journal of Marriage and Family* 71(4):917–34.

Valentine, Kylie. 2008. "After Antagonism: Feminist Theory and Science." *Feminist Theory* 9(3):355–65.

Valiente, Carlos, Richard A. Fabes, Nancy Eisenberg, and Tracy L. Spinrad. 2004. "The Relations of Parental Expressivity and Support to Children's Coping with Daily Stress." *Journal of Family Psychology* 18(1):97–107.

van Anders, Sari M. and Neil V. Watson. 2006. "Relationship Status and Testosterone in North American Heterosexual and Nonheterosexual Men and Women: Cross-Sectional and Longitudinal Data." *Psychoneuroendocrinology* 31(6):715–23.

———. 2007. "Testosterone Levels in Women and Men Who Are Single, in Long-Distance Relationships, Or Same-City Relationships." *Hormones and Behaviors* 51(2):286–91.

Vandeleur, C. L., N. Jeanpretre, M. Perrez, and D. Schoebi. 2009. "Cohesion, Satisfaction with Family Bonds, and Emotional Well-Being in Families with Adolescents." *Journal of Marriage and Family* 71(5):1205–19.

Vandell, Deborah Lowe, Jay Belsky, Margaret Burchinal, Laurence Steinberg, and Nathan Vandergrift. 2010. "Do Effects of Early Child Care Extend to Age 15 Years? Results From the NICHD Study of Early Child Care and Youth Development." *Child Development* 81(3):737–56.

Vandell, Deborah L., Kathleen McCarthy, Margaret T. Owen, Cathryn Booth, and Alison Clarke-Stewart. 2003. "Variations in Child Care by Grandparents during the First Three Years." *Journal of Marriage and Family* 65(2):375–81.

Van den Haag, Ernest. 1974. "Love or Marriage." Pp. 134–42 in *The Family: Its Structures and Functions*, 2nd ed., edited by Rose Laub Coser. New York: St. Martin's.

Vanderber, Kathleen S., Rudolph Verderber, and Cynthia Berryman-Fink. 2010. *Inter-act: Interpersonal Communication Concepts, Skills, and Contexts*. New York: Oxford University Press.

VanderLaan, Doug P. and Paul L. Vasey. 2009. "Patterns of Sexual Coercion in Heterosexual and Non-Heterosexual Men and Women." *Archives of Sexual Behavior* 38(6):987–99.

van der Lippe, Tanja. 2010. "Women's Employment and Housework." In *Dividing the Domestic: Men, Women, and Household Work in Cross-National Perspective*, edited by Judith Treas and Sonja Drobnič. Stanford, CA: Stanford University Press.

Vander Ven, Thomas M., Francis T. Cullen, Mark A. Carrozza, and John Paul Wright. 2001. "Home Alone: The Impact of Maternal Employment on Delinquency." *Social Problems* 48: 236–57.

Vandewater, Elizabeth A. and Jennifer E. Lansford. 2005. "A Family Process Model of Problem Behaviors in Adolescents." *Journal of Marriage and Family* 67(1):100–109.

Vandivere, Sharon, Karin Malm, and Laura Radel. 2009. *Adoption USA: A Chartbook Based on the 2007 National Survey of Adoptive Parents*. Washington, DC: The U.S. Department of Health and Human Services, Office of the Assistant Secretary for Planning and Evaluation.

VanDorn, Richard A., Gary L. Bowen, and Judith R. Blau. 2006. "The Impact of Community Diversity and Consolidated Inequality on Dropping Out of High School." *Family Relations* 55(1):105–18.

van Eeden-Moorefield, Brad, Kari Henley, and Kay Pasley. 2005. "Identity Enactment and Verification in Gay and Lesbian Stepfamilies." Pp. 230–33 in *Sourcebook of Family Theory and Research*, edited by Vern L. Bengston, Alan C. Acock, Katherine R. Allen, Peggye Dilworth-Anderson, and David M. Klein. Thousand Oaks, CA Sage Publications.

van Honk, Jack, Dennis J. L. G. Schutter, Erno J. Hermans, and Peter Putman. 2004. "Testosterone, Cortisol, Dominance, and Submission: Biologically Prepared Motivation, No Psychological Mechanisms Involved." *Behavioral and Brain Sciences* 27(1):160–2.

VanNatta, Michelle. 2005. "Constructing the Battered Woman." *Feminist Studies* 31(2):416–29.

Van Pelt, Nancy L. 1985. *How to Turn Minuses into Pluses: Tips for Working Moms, Single Parents, and Stepparents*. Washington, DC: Review and Herald, Better Living Series.

Vaughan, Diane. 1986. *Uncoupling: Turning Points in Intimate Relationships*. New York: Oxford University Press.

Ventura, Stephanie J. 2009. "Changing Patterns of Nonmarital Childbearing in the United States." *NCHS Data Brief* 18. Hyattsville, MD: U.S. National Center for Health Statistics.

Ventura, Stephanie, Joyce C. Abma, William D. Mosher, and Stanley K. Henshaw. 2006. *Recent Trends in Teenage Pregnancy in the United States, 1990–2002*. Health E-Stats. Hyattsville, MD: U.S. National Center for Health Statistics. December 13. Retrieved January 20, 2007 (www.cdc.gov/nchs).

Ventura, Stephanie J., T. J. Mathews, and Brady E. Hamilton. 2001. "Births to Teenagers in the United States, 1940–2000." *National Vital Statistics Reports* 49(10). Hyattsville, MD: U.S. National Center for Health Statistics. September 25.

Verhofstadt, Lesley L., William Ickes, and Ann Buysse. 2010. "I Know What You Need Right Now:" Empathic Accuracy and Support Provision in Marriage." Pp. 71–88 in *Support Processes in Intimate Relationships*, edited by Kieran T. Sullivan and Joanne Davila. New York: Oxford University Press.

Vestal, Christine. 2009. "Gay Marriage Legal in Six States." Stateline.org, June 4. Retrieved November 20, 2009 (www.stateline.org).

Villarosa, Linda. 2002. "Once-Invisible Sperm Donors Get to Meet the Family." *New York Times*, May 21.

———. 2003. "Raising Awareness about AIDS and the Aging." *New York Times*, July 8.

Vinciguerra, Thomas. 2007. "He's Not My Grandpa, He's My Dad." *New York Times*, April 12.

Visher, Emily B. and John S. Visher. 1996. *Therapy with Stepfamilies*. New York: Brunner/Mazel.

Vlosky, Denese Ashbaugh and Pamela A. Monroe. 2002. "The Effective Dates of No-Fault Divorce Laws in the 50 States." *Family Relations* 51(4):317–24.

Vogel, Erin R., Livia Haag, Mitra-Setia Tatang, Carel P. van Schaik, and Nathaniel J. Dominy. 2009. "Foraging and Ranging Behavior During a Failback Episode: *Hylobates albibarbis and Pongo pygmaeus wurmbii* Compared." *American Journal of Physical Anthropology* 140(4):716–26.

Vogler, Carolyn. 2005. "Cohabiting Couples: Rethinking Money in the Household at the Beginning of the Twenty-First Century." *The Sociological Review* 53(1):1–29.

Vogler, Carolyn, Clare Lyonette, and Richard D. Wiggins. 2008. "Money, Power and Spending Decisions in Intimate Relationships." *Sociological Review* 56(1):117–43.

Von Drehle. 2010. "Why Crime Went Away." *Time*, February 22, pp. 32–36.

Von Rosenvinge, Kristina. n.d. "Seven Tips To Improve Couple Communication." Retrieved January 26, 2010 (ezinearticles.com).

Voorpostel, M. and Rosemary Blieszner. 2008. "Intergenerational Support and Solidarity between Adult Siblings." *Journal of Marriage and Family* 70(1):157–67.

Waddell, Lynn. 2005. "Gays in Florida Seek Adoption Alternatives." *New York Times*, January 21.

Wadler, Joyce. 2009. "Caught in the Safety Net." *New York Times*, May 14. Retrieved November 21, 2009 (www.nytimes.com).

Wadsworth, Martha and Lauren Berger. 2006. "Adolescents Coping with Poverty-Related Family Stress: Prospective Predictors of Coping and Psychological Symptoms." *Journal of Youth and Adolescence* 35(1):54–67.

Wagenaar, Deborah B. 2009. "Elder Abuse Education in Residency Programs: How Well Are We Doing?" *Academic Medicine: Journal of the Association of American Medical Colleges* 84(5):611–19.

Wagmiller, Robert L., Jr., Mary Clare Lennon, Li Kuang, Philip M. Alberti, and J. Lawrence Aber. 2006. "The Dynamics of Economic Advantage and Children's Life Chances." *American Sociological Review* 71(5):847–66.

Wagner-Raphael, Lynne I., David Wyatt Seal, and Anke A. Ehrhardt. 2001. "Close Emotional Relationships with Women versus Men." *Journal of Men's Studies* 9(2):243–56.

Waite, Linda J. 1995. "Does Marriage Matter?" *Demography* 32(4):483–507.

———. 2001. "The Family as Social Organization: Key Ideas for the Twenty-First Century." *Contemporary Sociology* 29:463–99.

Waite, Linda J., Don Browning, William J. Doherty, Maggie Gallagher, Ye Luo, and Scott M. Stanley. 2002. *Does Divorce Make People Happy? Findings from a Study of Unhappy Marriages*. New York: Institute for American Values.

Waite, Linda J. and Maggie Gallagher. 2000. *The Case for Marriage: Why Married People Are Happier, Healthier, and Better Off Financially*. New York: Doubleday.

Waite, Linda J. and Kara Joyner. 2001. "Emotional and Physical Satisfaction with Sex in Married, Cohabiting, and Dating Sexual Unions: Do Men and Women Differ?" Pp. 239–69 in *Sex, Love, and Health in America*, edited by Edward O. Laumann and Robert T. Michael. Chicago, IL: University of Chicago Press.

Waite, Linda J., Edward O. Laumann, Aniruddha Das, and Philip L. Schlimm. 2009. "Sexuality: Measures of Partnerships, Practices, Attitudes, and Problems in the National Social Life, Health, and Aging Study." *The Journals of Gerontology, Series B, Psychological Sciences and Social Sciences* 64(1):156.

Waite, Linda J., Ye Luo, and Alisa C. Lewin. 2009. "Marital Happiness and Marital Stability: Consequences for Psychological Well-Being." *Social Science Research* 38(1):201–12.

Waldfogel, Jane. 2006. "What Do Children Need?" *Public Policy Research* 13(1):26–34.

Walker, Alexis J. 1985. "Reconceptualizing Family Stress." *Journal of Marriage and Family* 47(4):827–37.

Walker, Alexis J., Margaret Manoogian-O'Dell, Lori A. McGraw, and Diana L. G. White, eds. 2001. *Families in Later Life: Connections and Transitions.* Thousand Oaks, CA: Pine Forge Press.

Walker, Lenore E. 2009. *The Battered Woman Syndrome.* 3rd ed. New York: Springer Publishing Company.

Walker, Samuel, Cassia Spohn, and Miriam DeLone. 2007. *The Color of Justice: Race, Ethnicity, and Crime in America.* 4th ed. Belmont, CA: Wadsworth.

Wallace, Harvey. 2008. *Family Violence: Legal, Medical, and Social Perspectives.* Boston: Pearson/Allyn and Bacon.

Wallace, Stephen G. 2007. "Hooking Up, Losing Out?" *Healthy Teens Camping Magazine* (March/April):26–30.

Waller, Maureen R. and H. Elizabeth Peters. 2008. "The Risk of Divorce As a Barrier to Marriage Among Parents of Young Children." *Social Science Research* 37(4):1188–99.

Waller, Willard. 1951. *The Family: A Dynamic Interpretation,* revised by Reuben Hill. New York: Dryden.

Wallerstein, Judith S. 2003. "Children of Divorce: A Society in Search of Policy." Pp. 66–96 in *All Our Families: New Policies for a New Century,* 2nd ed., edited by Mary Ann Mason, Arlene Skolnick, and Stephen D. Sugarman. New York: Oxford University Press.

Wallerstein, Judith. 2008. "The Transition to Adulthood: Children of Divorce Make Their Way." *Family Focus.* March: F11–F12.

Wallerstein, Judith S. and Sandra Blakeslee. 1989. *Second Chances: Men, Women, and Children a Decade After Divorce.* New York: Ticknor and Fields.

———. 1995. *The Good Marriage: How and Why Love Lasts.* Boston, MA: Houghton Mifflin.

———. 2003. *What About the Kids? Raising Your Children Before, During, and After Divorce.* New York: Hyperion.

Wallerstein, Judith S. and Joan Kelly. 1980. *Surviving the Break-Up: How Children Actually Cope with Divorce.* New York: Basic Books.

Wallerstein, Judith S. and Julia M. Lewis. 2007. "Disparate Parenting and Step-Parenting with Siblings in the Post-Divorce Family: Report From a 10-Year Longitudinal Study." *Journal of Family Studies* 13(2):224–35.

———. 2008. "Divorced Fathers and Their Adult Offspring: Report from a Twenty-Five-Year Longitudinal Study." *Family Law Quarterly* 42(4). Available through Academic Search Elite.

Walsh, David Allen. 2002. "Bouncing Forward: Resilience in the Aftermath of September 11." *Family Process* 41(1):34–36.

———. 2007. *No: Why Kids—of All Ages—Need to Hear It and Ways Parents Can Say It.* New York: Free Press.

Walter, Carolyn Ambler. 1986. *The Timing of Motherhood.* Lexington, MA: Heath.

Wang, Wendy and Rich Morin. 2009. "Home for the Holidays . . . and Every Other Day: Recession Brings Many Young Adults Back to the Nest." Washington, DC: Pew Research Center. Retrieved December 21, 2009 (pewresearch.org).

Ward, Margaret. 1997. "Family Paradigms and Older-child Adoption: A Proposal for Matching Parents' Strengths to Children's Needs." *Family Relations* 46(3):257–62.

Ward, Russell A., Glenna Spitze, and Glenn Deane. 2009. "The More the Merrier? Multiple Parent-Adult Child Relations." *Journal of Marriage and Family* 71(1):161–73.

Wardle, Francis. 2000. "Children of Mixed Race—No Longer Invisible—From Revising School Forms to Reviewing the Curriculum for Inclusiveness, How to Make School More Welcoming to Multicultural Children." *Educational Leadership: Journal of the Department of Supervision and Curriculum Development, N.E.A.* 57(4):68–73.

Wark, Linda and Shilpa Jobalia. 1998. "What Would It Take to Build a Bridge? An Intervention for Stepfamilies." *Journal of Family Psychotherapy* 9(3):69–77.

Warne, Garry L. and Vijayalakshmi Bhatia. 2006. "Intersex, East and West." Pp. 183–205 in *Ethics and Intersex,* edited by Sharon E. Sytsma. New York: Springer.

Warner, Judith. 2006. *Perfect Madness: Motherhood in the Age of Anxiety.* New York: Riverhead Books.

Warshak, Richard. 2000. "Remarriage as a Trigger of Parental Alienation Syndrome." *American Journal of Family Therapy* 28(3):229–41.

Wasserman, Jason Adam. 2009. "But Where Do We Go from Here: A Reply to Tomso on the State and Direction of Postmodern Theory." *Social Theory & Health* 7(1):78–80.

Waters, Mary C. 2007. "Optional Ethnicities: For Whites Only?" Pp. 198–207 in *Race, Class, and Gender: An Anthology,* edited by Margaret L. Andersen and Patricia Hill Collins. Belmont, CA: Wadsworth.

Watson, Russell. 1984. "Five Steps to Good Day Care." *Newsweek,* September 10, p. 21.

Weaver, David, Maxwell McCombs, and David L. Shaw. 2004. "Agenda-Setting Research: Issues, Attributes, and Influences." Pp. 257–282 in *Handbook of Political Communication Research,* edited by Linda L. Kaid. New York: Erlbaum.

Weaver, Shannon E., Marilyn Coleman, and Lawrence H. Ganong. 2003. "The Sibling Relationship in Young Adulthood." *Journal of Family Issues* 24(2):245–63.

Webster, Murray and Lisa Rashotte. "Fixed Roles and Situated Actions." *Sex Roles* 61(5/6):325–37.

Weeks, John R. 2002. *Population: An Introduction to Concepts and Issues.* 8th ed. Belmont, CA: Wadsworth.

———. 2007. *Population: An Introduction to Concepts and Issues.* 10th ed. Belmont, CA: Wadsworth.

Weger, H. 2005. "Disconfirming Communication and Self-Verification in Marriage: Associations among the Demand/ Withdraw Interaction Pattern, Feeling Understood, and Marital Satisfaction." *Journal of Social and Personal Relationships* 22(1):19–31.

Weibel-Orlando, J. 2001. "Grandparenting Styles: Native American Perspectives." Pp. 139–45 in *Families in Later Life: Connections and Transitions,* edited by Alexis J. Walker, Margaret Manoogian-O'Dell, Lori A. McGraw, and Diana L. G. White. Thousand Oaks, CA: Pine Forge Press.

Weigel, Daniel J. 2008. "The Concept of Family: An Analysis of Laypeople's Views of Family." *Journal of Family Issues* 29(11):1426–47.

Weigel, Daniel J., Kymberley K. Bennett, and Deborah S. Ballard-Reisch. 2006b. "Roles and Influence in Marriages: Both Spouses' Perceptions Contribute to Marital Commitment." *Family and Consumer Science Research Journal* 35(1):74–92.

Weil, Elizabeth. 2006. "What If It's (Sort of) a Boy and (Sort of) a Girl?" *New York Times Magazine,* September 24. Retrieved September 6, 2007 (www.nytimes.com).

Weinberg, Daniel H. 2004. *Evidence from Census 2000 about Earnings by Detailed Occupation for Men and Women.* Census 2000 Special Reports CENSR-15. May.

Weininger, Elliot B. and Annette Lareau. 2009. "Paradoxical Pathways: An Ethnographic Extension of Kohn's Findings on Class and Childrearing." *Journal of Marriage and Family* 71(3):680–95.

Weisman, Alan. 2008. *The World Without Us.* New York: Picador.

Weisman, Carol. 2006. *Raising Charitable Children.* St. Louis, MO: F. E. Robbins & Sons.

Weissman, M. M., J. C. Markowitz, and G. L. Klerman. 2007. *Clinician's Quick Guide to Interpersonal Psychotherapy.* New York: Oxford University Press.

Weitzman, Lenore J. 1985. *The Divorce Revolution: The Unexpected Social and Economic Consequences for Women and Children in America.* New York: Free Press.

Wejnert, Cyprian. 2008. "Strategies for Measuring and Promoting Mothers' Social Support Networks." *Marriage & Family Review* 44(2/3):380–8.

Welborn, Vickie. 2006. "Black Students Ordered to Give Up Seats to Whites." *Shreveport Times,* August 24. Retrieved October 2, 2006 (www.shreveporttimes.com).

Wells, Brooke E. and Jean M. Twenge. 2005. "Changes in Young People's Sexual Behavior and Attitudes, 1943–1999: A Cross-Temporal Meta-Analysis." *Review of General Psychology* 9(3):249–61.

Wells, Mary S., Mark A. Widmer, and J. Kelly McCoy. 2004. "Grubs and Grasshoppers: Challenge-based Recreation and the Collective Efficacy of Families with At-Risk Youth." *Family Relations* 53(3):326–33.

Wells, Robert V. 1985. *Uncle Sam's Family: Issues in and Perspectives on American Demographic*

History. Albany, NY: State University of New York Press.

Wen, Ming. 2008. "Family Structure and Children's Health and Behavior." *Journal of Family Issues* 29(11):1492–1519.

Wenger, G. Clare, Pearl Dykstra, Tuula Melkas, and Kees C.P.M. Knipscheer. 2007. "Social Embeddedness and Late-Life Parenthood." *Journal of Family Issues* 28(11):1419–56.

Wentzel, Jo Ann. 2001. "Foster Kids Really Are Ours." www.fosterparents.com.

Werner, Emmy E. and Ruth S. Smith. 2001. *Journeys from Childhood to Midlife: Risk, Resiliency, and Recovery.* Ithaca, NY: Cornell University Press.

Wessel, David. 2010. "Meet the Unemployable Man." *Wall Street Journal,* May 6. www.wsj.com.

West, Carolyn M. 2003. "'Feminism Is a Black Thing'? Feminist Contributions to Black Family Life." *State of Black America 2003.* Washington, DC: National Urban League.

West, Martha S. and John W. Curtis. 2006. *AAUP Faculty Gender Equity Indicators 2006.* Washington, DC: American Association of University Professors. Retrieved January 31, 2007. (www.aaup.org)

Wester, Stephen R., David L. Vogel, Meifen Wei, and Rodney McLain. 2006. "African American Men, Gender Role Conflict, and Psychological Distress: The Role of Racial Identity." Journal of Counseling & Development 84:419–29.

Western, Bruce, Deirdre Bloome, and Christine Percheski. 2008. "Inequality among American Families." American Sociological Review 73(6):903–20.

Western, Bruce and Sara McLanahan. 2000. "Fathers Behind Bars: The Impact of Incarceration on Family Formation." Pp. 309–24 in *Families, Crime, and Criminal Justice,* edited by Greer Litton Fox and Michael L. Benson. New York: Elsevier Science.

Wexler, Richard. 2005. "Family Preservation Is the Safest Way to Protect Most Children." Pp. 311–27 in *Current Controversies on Family Violence,* 2nd ed., edited by Donileen R. Loseke, Richard J. Gelles, and Mary M. Cavanaugh. Thousand Oaks, CA: Sage Publications.

"What Happened to the Wedding Bells? Cohabitation Is On the Rise, New Data from Census Reveals." 2003. *Forecast* 23(4):1–4.

"What Is Marriage and Family Therapy?" 2005. American Association for Marriage and Family Therapy. Retrieved March 3, 2010 (www.aamft.org/faqs/index_nm.asp).

"What's Eating Our Kids? Fears about 'Bad' Foods." 2009. *The New York Times,* February 26. Retrieved December 23, 2009 (www.nytimes.com).

Whealin, Julia and Ilona Pivar. 2006. "Coping when a Family Member Has Been Called to War: A National Center for PTSD Fact Sheet." U.S. Department of Veterans Affairs National Center for PTSD. Retrieved August 16, 2006 (www.ncptsd.va.gov).

Whitaker, Daniel J., Tadesse Haileyesus, Monica Swahn, and Linda S. Saltzman. 2007.

"Differences in Frequency of Violence and Reported Injury between Relationships With Reciprocal and Nonreciprocal Intimate Partner Violence." *American Journal of Public Health* 97(5):941–47.

White House Conference on Aging. 2005. *Report to the President and the Congress: The Booming Dynamics of Aging—from Awareness to Action.* Retrieved May 16, 2007 (www.whcoa.gov).

White, James M. and David M. Klein. 2002. *Family Theories: An Introduction.* 2nd ed. Thousand Oaks, CA: Sage Publications.

———. 2008. *Family Theories: An Introduction.* 3rd ed. Thousand Oaks, CA: Sage Publications.

White, Lynn K. 1994. "Growing Up with Single Parents and Stepparents: Long-term Effects on Family Solidarity." *Journal of Marriage and Family* 56(4):935–48.

———. 1998. "Who's Counting? Quasi-Facts and Stepfamilies in Reports of Number of Siblings." Journal of Marriage and Family 60(August):725–33.

———. 1999. "Contagion in Family Affection: Mothers, Fathers, and Young Adult Children." *Journal of Marriage and Family* 61(2):284–94.

White, Lynn K. and Alan Booth. 1985. "The Quality and Stability of Remarriages: The Role of Stepchildren." *American Sociological Review* 50:689–98.

———. 1991. "Divorce Over the Life Course: The Role of Marital Happiness." *Journal of Family Issues* 12:5–21.

White, Lynn K. and Joan G. Gilbreth. 2001. "When Children Have Two Fathers: Effects of Relationships with Stepfathers and Noncustodial Fathers on Adolescent Outcomes." *Journal of Marriage and Family* 63:155–67.

White, Lynn K. and Agnes Riedmann. 1992. "When the Brady Bunch Grows Up: Step/Half- and Full-sibling Relationships in Adulthood." *Journal of Marriage and Family* 54(1):197–208.

White, Rebecca M. B., Mark W. Roosa, Scott R. Weaver, and Rajni L. Nair. 2009. "Cultural and Contextual Influences on Parenting in Mexican American Families." *Journal of Marriage and Family* 71(1):61–79.

Whitehead, Barbara. 1997. *The Divorce Culture.* New York: Random House.

Whitehead, Barbara Dafoe and David Popenoe. 2001. "Who Wants to Marry a Soul Mate?" In *The State of Our Unions 2001: The Social Health of Marriage in America.* Piscataway, NJ: RutgersUniversity, National Marriage Project.

———. 2003. "Did a Family Turnaround Begin in the 1990s?" In *The State of Our Unions 2003: The Social Health of Marriage in America.* Piscataway, NJ: Rutgers University, National Marriage Project. Retrieved September 20, 2006 (marriage.rutgers.edu).

———. 2006. *The State of Our Unions 2006: The Social Health of Marriage in America* (includes essay "Life without Children"). Piscataway, NJ: Rutgers University, National Marriage Project. Retrieved July 18, 2006 (marriage.rutgers.edu).

———. 2008. "Life Without Children: The Social Retreat from Children and How It Is

Changing America." Piscataway, NJ: Rutgers University, National Marriage Project. Retrieved April 16, 2010 (marriage.rutgers.edu).

Whiteman, Shawn D., Susan M. McHale, and Ann C. Crouter. 2007. "Longitudinal Changes in Marital Relationships: The Role of Offspring's Pubertal Development." *Journal of Marriage and Family* 69(November):1005–20.

Whiting, Jason, Donna Smith, and Tammy Barnett. 2007. "Overcoming the Cinderella Myth: A Mixed Methods Study of Successful Stepmothers." *Journal of Divorce and Remarriage* 47(1/2):95–109.

Whitman, David. 1997. "Was It Good for Us?" *U.S. News & World Report,* May 19, pp. 56–64.

Whittaker, Terri. 1995. "Violence, Gender and Elder Abuse: Towards a Feminist Analysis and Practice." *Journal of Gender Studies* 4(1):35–45.

Whitty, Monica T. and Laura-Lee Quigley. 2008. "Emotional and Sexual Infidelity Offline and in Cyberspace." *Journal of Marital and Family Therapy* 34(4):461–68.

Who We Are. 2009. Louisville, CO: National Organization for Men Against Sexism.

"Why Interracial Marriages Are Increasing." 1996. *Jet,* June 3, pp. 12–15.

Whyte, Martin King. 1990. *Dating, Mating, and Marriage.* New York: Aldine.

Wickersham, Joan. 2008. *The Suicide Index: Putting My Father's Death in Order.* Orlando: Harcourt.

Widmer, Eric D., Francesco Giudici, Jean-Marie LeGoff, and Alexandre Pollien. 2009. "From Support to Control: A Configurational Perspective on Conjugal Quality." *Journal of Marriage and Family* 71(3):437–48.

Wienke, Chris and Gretchen J. Hill. 2009. "Does the 'Marriage Benefit' Extend to Partners in Gay and Lesbian Relationships?" *Journal of Family Issues* 30(2):259–89.

Wiersma, Jacquelyn D., H. Harrington Cleveland, Veronica Herrera, and Judith L. Fischer. 2010. "Intimate Partner Violence in Young Adult Dating, Cohabiting, and Married Drinking Partnerships." *Journal of Marriage & Family* 72(2):360–74.

Wiglesworth, Aileen, Bryan Kemp, and Laura Mosqueda. 2008. "Combating Elder and Dependent Adult Mistreatment: The Role of the Clinical Psychologist." *Journal of Elder Abuse and Neglect* 20(3):207–30.

Wight, Vanessa R. and Sara B. Raley. 2009. "When Home Becomes Work: Work and Family Time among Workers at Home." *Social Indicators Research* 93(1):197–202.

Wilcox, W. Bradford. 1998. "Conservative Protestant Childrearing: Authoritarian or Authoritative?" *American Sociological Review* 63:796–809.

———. 2002. "Religion, Convention, and Paternal Involvement." *American Sociological Review* 64:780–92.

———. 2004. *Soft Patriarchs, New Men: How Christianity Shapes Fathers and Husbands.* Chicago, IL: University of Chicago Press.

———. 2009. "The Evolution of Divorce." *National Affairs* 1(Fall):81–94.

Wilcox, W. Bradford and Elizabeth Marquardt. 2009. *The State of Our Unions 2009: Marriage in America: Money & Marriage.* Virginia: National Marriage Project and the Institute for American Values.

———. 2009. "The Great Recession's Silver Lining." Pp.15–22 in *The State of Our Unions 2009: Money & Marriage in America,* edited by David Popenoe and Barbara Dafoe Whitehead. New Brunswick, NJ: Rutgers University, National Marriage Project. Retrieved December 21, 2009 (marriage.rutgers.edu).

Wilcox, W. Bradford and Steven L. Nock. 2006. "What's Love Got to Do with It? Equality, Equity, Commitment and Women's Marital Quality." *Social Forces* 84(3):1321–45.

Wildeman, Christopher. 2009. "Parental Imprisonment, the Prison Boom, and the Concentration of Childhood Disadvantage." *Demography* 46(2):265–80.

Wildeman, Christopher and Christine Percheski. 2009. "Associations of Childhood Religious Attendance, Family Structure, and Nonmarital Fertility Across Cohorts." *Journal of Marriage and Family* 71(5):1294–308.

Wildman, Sarah. 2010. "Children Speak for Same-Sex Marriage." *The New York Times,* January 20. Retrieved January 22, 2010 (www.nytimes.com).

Wildsmith, Elizabeth and R. Kelley Raley. 2006. "Race-Ethnic Differences in Nonmarital Fertility: A Focus on Mexican American Women." *Journal of Marriage and Family* 68(2):491–508.

Wiley, Angela R., Henriette B. Warren, and Dale S. Montanelli. 2002. "Shelter in a *Time* of Storm: Parenting in Poor Rural African American Communities." *Family Relations* 51(3):265–73.

Wilkie, Jane Riblett. 1991. "The Decline in Men's Labor Force Participation and Income and the Changing Structure of Family Economic Support." *Journal of Marriage and Family* 53(1):111–22.

Wilkie, Jane Riblett, Myra Marx Ferree, and Kathryn Strother Ratcliff. 1998. "Gender and Fairness: Marital Satisfaction in Two-Earner Couples." *Journal of Marriage and Family* 60(3):577–94.

Wilkins, Amy C. 2009. "Masculinity Dilemmas: Sexuality and Intimacy Talk among Christians and Goths." *Signs: Journal of Women in Culture and Society* 34(2):343–68.

Wilkins, Victoria M., Martha L. Bruce, and Jo Anne Sirey. 2009. "Caregiving Tasks and Training Interest of Family Caregivers of Medically Ill Homebound Older Adults." *Journal of Aging and Health* 21(3):528–42.

Wilkinson, Deanna, Amanda Magora, Marie Garcia, and Atika Khurana. 2009. "Fathering at the Margins of Society: Reflections from Young, Minority, Crime-Involved Fathers." *Journal of Family Issues* 30(7):945–67.

Wilkinson, Doris. 1993. "Family Ethnicity in America." Pp. 15–59 in *Family Ethnicity: Strength in Diversity,* edited by Harriette Pipes McAdoo. Newbury Park, CA: Sage Publications.

———. 2000. "Rethinking the Concept of 'Minority': A Task for Social Scientists and Practitioners." *Journal of Sociology and Social Welfare* 27:115–32.

Willeto, Angela A. A. and Charlotte Goodluck. 2004. "Economic, Social and Demographic Losses and Gains among American Indians." Washington, DC: Population Reference Bureau (www.prb.org).

Willetts, Marion C. 2003. "An Exploratory Investigation of Heterosexual Licensed Domestic Partners." *Journal of Marriage and Family* 65(4):939–52.

———. 2006. "Union Quality Comparisons between Long-Term Heterosexual Cohabitation and Legal Marriage." *Journal of Family Issues* 27(1):110–27.

Williams, Joan C. and Heather Boushey. 2010. *The Three Faces of Work-Family Conflict: The Poor, the Professionals, and the Missing Middle.* Center for American Progress. Retrieved May 8, 2010 (www.americanprogress.org).

Williams, Kristi, Sharon Sassler, and Lisa M. Nicholson. 2008. "For Better or For Worse? The Consequences of Marriage and Cohabitation for Single Mothers." *Social Forces* 86(4):1481–1511.

Williams, Lee M. and Michael G. Lawler. 2003. "Marital Satisfaction and Religious Heterogamy." *Journal of Family Issues* 24(8):1070–92.

Williams, Wendy R. 2009. "Struggling with Poverty: Implications for Theory and Policy of Increasing Research on Social Class-Based Stigma." *Analyses of Social Issues and Public Policy* 9(1):37–56.

Willie, Charles Vert and Richard J. Reddick. 2003. *A New Look at Black Families.* 5th ed. Lanham, MD: Rowman and Littlefield.

Wilson, Brenda. 2009. "Sex Without Intimacy: No Dating, No Relationship." National Public Radio. May 18. Retrieved June 10, 2009 (www.npr.org).

Wilson, James Q. 2001. "Against Homosexual Marriage." Pp. 123–27 in *Debating Points: Marriage and Family Issues,* edited by Henry L. Tischler. Upper Saddle River, NJ: Prentice Hall.

———. 2002. "Why We Don't Marry." *City Journal,* Winter. Retrieved October 2, 2006 (www.city-journal.org).

Wilson, William Julius. 2009. *More Than Just Race: Being Black and Poor in the Inner City.* New York: W. W. Norton & Company.

Winch, Robert F. 1958. *Mate Selection: A Study of Complementary Needs.* New York: Harper and Row.

Wineberg, Howard. 1996. "The Resolutions of Separation: Are Marital Reconciliations Attempted?" *Population Research and Policy Review* 15:297–310.

Winerip, Michael. 2009. "Anything He Can Do, She Can Do." *The New York Times,* November 15.

Wingert, Pat and Barbara Kantrowitz. 2010. "The Rise of the 'Silver Divorce'." Newsweek, June 7. Retrieved June 19, 2010 (www.newsweek.com).

Wingett, Yvonne. 2007. "Foster Parents Needed to Help Hispanic Children." *Arizona Republic,* March 5.

Winner, Lauren F. 2006. *Real Sex: The Naked Truth about Chastity.* Grand Rapids, MI: Brazos Press.

Winter, Judy. 2006. *Breakthrough Parenting for Children with Special Needs: Raising the Bar of Expectations.* San Francisco, CA: Jossey-Bass.

Wiseman, Michael. 2008. "The Role of the Dentist in Recognizing Elder Abuse." *Journal of the Canadian Dental Association* 74(8):715–21.

"Wives Earning More than their Husbands, 1987–2006." 2009. The Editor's Desk, January 9. Washington, DC: U.S. Bureau of Labor Statistics. Retrieved April 23, 2010 (data.bls.gov).

Woldt, Veronica. 2010. "Elder Care Benefits: Retention and Recruitment Tools." *Corporate Wellness Magazine,* May 7. Retrieved May 30, 2010 (www.corporatewellnessmagazine.com).

Wolf, D. A., V. Freedman, and B. J. Soldo. 1997. "The Division of Family Labor: Care for Elderly Parents." *Journal of Gerontology* 52B:102–9.

Wolf, Marsha E., Uyen Ly, Margaret A. Hobart, and Mary A. Kernic. 2003. "Barriers to Seeking Police Help for Intimate Partner Violence." *Journal of Family Violence* 18:121–29.

Wolf, Naomi. 1991. *The Beauty Myth: How Images of Beauty Are Used Against Women.* New York: W. Morrow.

Wolf, Rosalie S. 1986. "Major Findings from Three Model Projects on Elderly Abuse." Pp. 218–38 in *Elder Abuse: Conflict in the Family,* edited by Karl A. Pillemer and Rosalie S. Wolf. Dover, MA: Auburn.

Wolfers, Justin. 2006. "Did Unilateral Divorce Raise Divorce Rates? A Reconciliation and New Results." *American Economic Review* 96(5):1802–20.

Wolfinger, Nicholas H. 1999. "Trends in the Intergenerational Transmission of Divorce." *Demography* 36:415–20.

———. 2005. *Understanding the Divorce Cycle: The Children of Divorce in Their Own Marriages.* New York: Cambridge University Press.

Wolfinger, Nicholas H. and Raymond E. Wolfinger. 2008. "Family Structure and Voter Turnout." *Social Forces* 86(4):1513–28.

"Women Catching Up to Men in College Degrees." 2008. Reuters, January 10.

"Women CEOs." 2010. *Fortune Magazine,* May 3. www.fortune.com.

"Women and Social Security." 2007. American Academy of Actuaries Issue Brief. June. Retrieved May 8, 2010 (www.actuary.org.)

"Women Still Lag White Males in Pay." 2004. CNNMoney, April 20 (cnnmoney.com).

Wong, Sabrina, Grace Yoo, and Anita Stewart. 2006. "The Changing Meaning of Family Support Among Older Chinese and Korean Immigrants." *Journal of Gerontology: Social Sciences* 61B(1):S4–9.

Wood, Wendy and Alice H. Eagly. 2002. "A Cross-cultural Analysis of Behavior of Women and Men: Implications for the Origins of Sex Differences." *Psychology Bulletin* 128:699–727.

Wooding, G. Scott. 2008. *Stepparenting and the Blended Family: Recognizing the Problems and Overcoming the Obstacles.* Markham, Ontario: Fitzhenry and Whiteside.

Woodward, Kenneth L. 2001. "A Mormon Moment." *Newsweek,* September 10, pp. 44–51.

Woolley, Michael E. and Andrew Grogan-Kaylor. 2006. "Protective Family Factors in the Context of Neighborhood: Promoting Positive School Outcomes." *Family Relations* 55(1):93–104.

Working Moms Refuge. 2001. "Factors to Consider When Selecting a Child Care Center." www.momsrefuge.com.

"World's 1st 'Test-Tube' Baby Gives Birth." 2007. CNN.com, January 15. Retrieved January 15, 2007 (http://cnn.health.com).

Wright, Eric and Brea Perry. 2006. "Sexual Identity Distress, Social Support, and the Health of Gay, Lesbian, and Bisexual Youth." *Journal of Homosexuality* 51(1):81–110.

Wright, H. Norman. 2006a. *Be a Great Parent: 12 Secrets to Raising Responsible Children.* Colorado Springs, CO: Life Journey Press.

————. 2006b. *How to Speak Your Spouse's Language: Ten Easy Steps to Great Communication from One of America's Foremost Counselors.* New York: Center Street.

Wright, Robert. 1994. *The Moral Animal.* New York: Pantheon.

Wright, Susan. 2005. "Autism: Willing the World to Listen." *Newsweek,* February 28. Retrieved March 13, 2009.

Wright, Suzanne. 2005. "Willing the World to Listen." *Newsweek,* February 28, p. 47.

Wrigley, Julia and Joanna Dreby. 2005. "Fatalities and the Organization of Child Care in the United States, 1985–2003." *American Sociological Review* 70:729–57.

Wu, Lawrence L. and Barbara Wolfe, eds. 2001. *Causes and Consequences of Nonmarital Fertility.* New York: Russell Sage.

Wu, Zheng and Feng Hou. 2008. "Family Structure and Children's Psychosocial Outcomes." *Journal of Family Issues* 29(12):1600–24.

Wurzel, Barbara J. n.d. "Extension Fact Sheet: Growing Up with Yours, Mine, and Ours in Stepfamilies." Ohio State University Family and Consumer Sciences. Retrieved April 26, 2007 (ohioline.osu.edu).

Wuthnow, Robert. 2002. *Loose Connections: Joining Together in America's Fragmented Communities.* Cambridge, MA: Harvard University Press.

Xie, Yu and Margaret Gough. 2009. *Ethnic Enclaves and the Earnings of Immigrants.* Research Report 09-685. Ann Arbor, MI: Populations Studies Center.

Xu, Xianohe, Clarke D. Hudspeth, and John P. Bartkowski. 2006. "The Role of Cohabitation in Remarriage." *Journal of Marriage and Family* 68(2):261–74.

Yabiku, Scott T. and Constance T. Gager. 2009. "Sexual Frequency and the Stability of Marital and Cohabiting Unions." *Journal of Marriage and Family* 71(November):983–1000.

Yakushko, Oksana and Oliva M. Espin. 2010. "The Experience of Immigrant and Refugee Women: Psychological Issues." Pp. 535–58 in *Handbook of Diversity in Feminist Psychology,* edited by Hope Landrine and Nancy Felipe Russo. New York: Springer Publishing Company.

Yalcin, Bektas Murat and Tevfik Fikret Karaban. 2007. "Effects of a Couple Communication Program on Marital Adjustment." *Journal of the American Board of Family Medicine* 20(1):36–44.

Yancey, G. 2007. "Experiencing Racism: Differences in the Experiences of Whites Married to Blacks and Non-Black Racial Minorities." *Journal of Comparative Family Studies* 38:197–213.

Yang, Chi-Fu Jeffrey, Carole K. Hooven, Matthew Boynes, Peter B. Gray, and Harrison G. Pope, Jr. 2007. "Testosterone Levels and Mental Rotation Performance in Chinese Men. *Hormones and Behavior* 51(3):373–8.

Yearbook of Immigration Statistics: 2008. 2009. Washington, DC: U.S. Department of Homeland Security, Office of Immigration Statistics, 2009.

Yellowbird, Michael and C. Matthew Snipp. 2002. "American Indian Families." Pp. 227–49 in *Multicultural Families in the United States,* 3rd ed., edited by Ronald L. Taylor. Upper Saddle River, NJ: Prentice Hall.

Yeung, King-To and John Levi Martin. 2003. "The Looking Glass Self: An Empirical Test and Elaboration." *Social Forces* 81(3):843–79.

Yin, Sandra. 2007. "New Restrictions Could Limit U.S. Adoptions from Top Two Countries of Origin: China and Guatemala." Washington, DC: Population Reference Bureau, March. Retrieved March 26, 2007 (www.prb.org).

————. 2008. *How Older Women Can Shield Themselves From Poverty.* Washington, DC: Population Reference Bureau. Retrieved June 21, 2010 (www.prb.org).

Yodanis, Carrie and Sean Lauer. 2007. "Managing Money in Marriage: Multilevel and Cross-National Effects of the Breadwinner Role." *Journal of Marriage and Family* 20(December):1307–25.

Yorburg, Betty. 2002. *Family Realities: A Global View.* Upper Saddle River, NJ: Prentice Hall.

Yoshihama, Mieko. 2008. "Literature on Intimate Partner Violence in Immigrant and Refugee Communities: Review and Recommendations" (paper prepared for RWJF, July 2008).

Yoshioka, Marianne R., Louisa Gilbert, Nabila El-Bassel, and Malahat Baig-Amin. 2003. "Social Support and Disclosure of Abuse: Comparing South Asian, African American, and Hispanic Battered Women." *Journal of Family Violence* 18:171–80.

"You Will Be a Parent to Your Parents." 2009. *Newsweek,* August 15. Retrieved May 8, 2010 (www.newsweek.com).

"Young Caregivers." 2009 (April 29). Retrieved March 13, 2010 (www.girlshealth.gov).

Young, Stacy L. 2010. "Positive Perceptions of Hurtful Communication: The Packaging Matters." *Communication Research Reports* 27(1):49–57.

Yuan, Anastasia S. Vogt, and Hayley A. Hamilton. 2006. "Stepfather Involvement and Adolescent Well-Being." *Journal of Family Issues* 27(9):1191–1213.

Yuval-Davis, Nira. 2006. "Intersectionality and Feminist Politics." *European Journal of Women's Studies* 13(3):1193–2009.

Zablocki v. Redhail. 1978. 434 U.S. 374, 54 L. Ed. 2d 618, 98 S. Ct. 673.

Zahn-Waxler, C. and N. Polanichka. 2004. "All Things Interpersonal: Socialization and Female Aggression." Pp. 48–68 in *Aggression, Antisocial Behavior, and Violence Among Girls: A Developmental Perspective,* edited by M. Putallaz and K. L. Bierman. New York: Guilford Press.

Zaman, Ahmed. 2008. "Gender Sensitive Teaching: A Reflective Approach for Early Childhood Education Teacher Training." *Education* 129(1):110–18.

Zang, Xiaowel. 2008. "Gender and Ethnic Variation in Arranged Marriages in a Chinese City." *Journal of Family Issues* 29(5):615–38.

Zeitz, Joshua M. 2003. "The Big Lie about the Little Pill." *New York Times,* December 27.

Zelizer, Viviana K. 1985. *Pricing the Priceless Child: The Changing Social Value of Children.* New York: Basic Books.

Zentgraf, Kristine A. 2002. "Immigration and Women's Empowerment: Salvadorans in Los Angeles." *Gender and Society* 16:625–46.

Zernike, Kate. 2006. "The Bell Tolls for the Future Merry Widow." *New York Times,* April 30. Retrieved April 30, 2007 (www.nytimes.com).

————. 2009. "And Baby Makes How Many?" *The New York Times,* February 9. Retrieved December 5, 2009 (www.nytimes.com).

Zezima, Katie. 2006. "When Soldiers Go to War, Flat Daddies Hold Their Place at Home." *New York Times,* September 30.

Zhang, Shuangyue. 2009. "Sender-Recipient Perspectives of Honest but Hurtful Evaluative Messages in Romantic Relationships." *Communication Reports* 22(2):89–101.

Zhang, Shuangyue and Susan L. Kline. 2009. "Can I Make My Own Decision? A Cross-Cultural Study of Perceived Social Network Influence in Mate Selection." *Journal of Cross-Cultural Psychology* 40(1):3–23.

Zhang, Yuanting and Jennifer Van Hook. 2009. "Marital Dissolution among Interracial Couples." *Journal of Marriage and Family* 71(February):95–107.

Zhao, Yilu. 2002. "Immersed in 2 Worlds, New and Old." *New York Times,* July 22.

Zielinski, David S. 2009. "Child Maltreatment and Adult Socioeconomic Well-Being." *Child Abuse & Neglect* 33(10):666–78.

Zimmerman, Jeffrey and Elizabeth Thayer. 2003. *Adult Children of Divorce: How to Overcome the Legacy of Your Parents' Breakup and Enjoy*

Love, Trust, and Intimacy. Oakland, CA: New Harbinger Publications.

Zink, D. W. 2008. "The Practice of Marriage and Family Counseling and Conservative Christianity." Pp. 55–72 in *The Role of Religion and Marriage and Family Counseling,* edited by Jill Duba Onedera. New York: Taylor and Francis.

Zoroya, Gregg. 2006. "Families Bear Catastrophic War Wounds." *USA Today,* September 25.

————. 2009. "Troops' Kids Feel War Toll." *USA Today,* June 25.

Zsembik, Barbara A. and Zobeida Bonilla. 2000. "Eldercare and the Changing Family in Puerto Rico." *Journal of Family Issues* 21(5):652–74.

Zuang, Yuanting. 2004. "Why Foreign Adoption?" Presented at the annual meeting of the American Sociological Association, August 15, San Francisco, CA.

Zurcher, Kristinia E. 2004. "'I Do' or 'I Don't'? Covenant Marriage after Six Years." *Notre Dame Journal of Law, Ethics, and Public Policy* 18:273–92.

Zwick, Rebecca. 2001. "Is the SAT a Wealth Test?" *Rethinking the SAT in University Admissions Conference.* Santa Barbara, CA: University of California, Santa Barbara.

Photo Credits

Chapter 1: pp. 2-3, Ariel Skelley/Getty Images; p. 4, ©Bill Aron/PhotoEdit; p. 4, ©Ron Chapple/Getty Images; p. 6, Image copyright Gorilla, 2010. Used under license from Shutterstock.com; p. 7, Blend Images/Getty Images; p. 12, Courtesy of Mitchell Gold and Trone Advertising; p. 12, Courtesy of MONTAUKSOFA; p. 15, Alex Wong/ Getty Images; p. 17, www.CartoonStock.com; p. 19, Mark Romaine/Stock Connection; p. 22, Ariel Skelley/Getty Images; p. 23, AP Images

Chapter 2: pp. 26-27, Ryan McVay/Getty Images; p. 28, ©Ken Benjamin; p. 32, ©Tony Freeman/Photo Edit; p. 34, ©Jeff Greenberg/The Image Works; p. 35, ©The New Yorker Collection 2000 Mick Stevens from cartoonbank.com. All rights reserved; p. 36, ©Lawrence Migdale; p. 38, Copyright ©David Young-Wolff/Photo Edit; p. 40, ©The New Yorker Collection 2003 Michael Shaw from; cartoonbank.com. All rights reserved"; p. 41, ©Martin Rodgers/Stock Boston, LLC; p. 47, ©Michael Newman/Photo Edit

Chapter 3: pp. 50-51, Kevin Dodge/Corbis; p. 55, ©Jeff Greenberg/The Image Works; p. 56, Linda Coan O'Kresik/ The New York Times/Redux; p. 60, ©The New Yorker Collection 2007 Glen Le Lievre from cartoonbank.com. All rights reserved.; p. 61, Charles Thatcher/Getty Images; p. 65, Copyright ©Mary Kate Denny/Photo Edit; p. 69, Copyright ©Kayte Deioma/Photo Edit; p. 70, Phil Schermeister/Corbis; p. 72, AP Photo/Susan Walsh; p. 73, Image copyright Rob Marmion, 2010. Used under license from Shutterstock.com; p. 75, AP Photo/East Valley Tribune, Heidi Huber

Chapter 4: pp. 78-79, Tom & Dee Ann McCarthy; p. 81, Image courtesy of The Advertising Archives; p. 83, Davis Factor/CORBIS; p. 85, "©The New Yorker Collection 2000 William Haefeli from cartoonbank.com.; All rights reserved."; p. 86, AP Photo/Paul Sakuma; p. 93, ©David Young-Wolff/PhotoEdit; p. 94, Courtesy Deb Glover and Celeste Wheeler; p. 96, Copyright ©Michael Newman/Photo Edit; p. 96, ©Michael Newman/Photo Edit; p. 98, ©The New Yorker Collection 1995 Robert Mankoff from cartoonbank.com. All rights; reserved"; p. 101, Dan Koeck/ The New York Times/Redux; p. 103, Reprinted by permission of Anne Gibbons

Chapter 5: pp. 106-107, Christian Michaels/Getty Images; p. 112, ©Paul Fusco/Magnum Photos; p. 115, Joseph Sohm/Visions of America/Corbis; p. 116, ©The New Yorker Collection 2009 William Haefeli from cartoonbank. com. All rights reserved; p. 120, ©The New Yorker Collection 2003 Michael Maslin from cartoonbank.com. All rights reserved.; p. 125, "©The New Yorker Collection 2005 Michael Maslin from cartoonbank.com.; All rights reserved"; p. 127, ©The Newark Museum/Art Resource, NY; p. 133, Eric K. K. Yu/CORBIS

Chapter 6: pp. 136-137, ©RubberBall/Alamy; p. 139, Jack Hollingsworth/Photolibrary; p. 146, Fredde Lieberman/ Photolibrary; p. 146, NGS Image Collection; p. 147, ©Edward Keating/The New York Times/Redux; p. 149, ©Joel McLeister/Star Tribune, Minneapolis-St. Paul; p. 152, Greg Vote/Getty Images; p. 155, ©Pi Kappa Phi Fraternity; p. 159, "©The New Yorker Collection 2003 Liza Donnelley from cartoonbank.com.; All rights reserved"; p. 161, Digital Vision/Getty Images

Chapter 7: pp. 164-165, Stockbyte/Jupiterimages; p. 167, Zigy Kaluzny-Charles Thatcher/Getty Images; p. 173, ©"The Dinner Quilt, 1986." Dyed painted story quilt, pieced fabric with beads, 45½" × 66", copyright ©by Faith; Ringgold. All rights reserved"; p. 173, New York Public Library Digital Image Collection; p. 176, ©Michael Newman/PhotoEdit; p. 180, Rhoberazzi/Getty Images; p. 182, ©Marmaduke St. John/Alamy; p. 185, Evelyn Hockstein/ MCT/Landov; p. 187, AP Photo/Erik S. Lesser; p. 190, ©Corbis/Jupiter Images

Chapter 8: pp. 192-193, ©Mode Images Limited/Alamy; p. 195, Image Source Pink; p. 198, Courtesy of Matt Kramer; p. 204, ©Royalty-Free/CORBIS; p. 209, Robert Mankoff/Bios/Cartoonbank; p. 210, San Francisco Examiner/Lucy Atkins; p. 214, Joel W. Rogers/Corbis; p. 215, ©Michael Ventura/Alamy; p. 216, Vicky Kasala; p. 220, ©Christie's Images/CORBIS

Chapter 9: pp. 218-219, "©IT Stock International/Creatas; p. 221, ©Michael Newman/PhotoEdit; p. 222, JED KIRSCHBAUM/Baltimore Sun; p. 225, ©Bill Bachmann/PhotoEdit; p. 227, ©Beth Huber; p. 229, ©Kirk Condyles/

Name Index

Subject Index